DIGITAL SIGNAL PROCESSING

A Computer-Based Approach

DIGITAL SIGNAL PROCESSING
A Computer Based Approach

Third Edition

Sanjit K. Mitra

Department of Electrical and Computer Engineering
University of California, Santa Barbara

Boston   Burr Ridge, IL   Dubuque, IA   Madison, WI   New York   San Francisco   St. Louis
Bangkok   Bogotá   Caracas   Kuala Lumpur   Lisbon   London   Madrid   Mexico City
Milan   Montreal   New Delhi   Santiago   Seoul   Singapore   Sydney   Taipei   Toronto

The McGraw-Hill Companies

DIGITAL SIGNAL PROCESSING: A COMPUTER-BASED APPROACH, THIRD EDITION
International Edition 2006

10 09 08 07 06 05 04 03 02 01
20 09 08 07 06
CTF SLP

Library of Congress Control Number: 2004016994

When ordering this title, use ISBN 007-125579-6

Printed in Singapore

www.mhhe.com

About the Author

Sanjit K. Mitra received the M.S. and Ph.D. degrees in Electrical Engineering from the University of California, Berkeley, in 1960 and 1962, respectively. After holding the position of assistant professor at Cornell University until 1965 and working at AT&T Bell Laboratories, Holmdel, New Jersey, until 1967, he joined the faculty of the University of California at Davis. Dr. Mitra then transferred to the Santa Barbara campus in 1977, where he served as department chairman from 1979 to 1982 and is now a Professor of Electrical and Computer Engineering. He has served IEEE in various capacities, including service as the President of the IEEE Circuits & Systems Society in 1986 and as a Member-at-Large of the Board of Governors of the IEEE Signal Processing Society from 1996 to 1999. He is currently a member of the editorial boards for four journals: *Signal Processing; Journal of the Franklin Institute; Automatika*; and *Ingeneiría*. He has published over 600 papers in signal and image processing, twelve books, and holds five patents. Dr. Mitra has received many distinguished industry and academic awards, including the 1973 F. E. Terman Award and the 1985 AT&T Foundation Award of the American Society of Engineering Education, the 1989 Education Award of the IEEE Circuits & Systems Society, the 1989 Distinguished Senior U.S. Scientist Award from the Alexander von Humboldt Foundation of Germany, the 1995 Technical Achievement Award and the 2001 Society Award of the IEEE Signal Processing Society, the 1999 Mac Van Valkenburg Society Award and the CAS Golden Jubilee Medal of the IEEE Circuits & Systems Society, the IEEE Millennium Medal in 2000, the 2002 Technical Achievement Award of the European Association for Signal Processing (EURASIP), and the 2005 SPIE Technical Achievement Award of the International Society for Optical Engineering. He is the co-recipient of the 2000 Blumlein-Browne-Willans Premium of the the Institution of Electrical Engineers (London) and the 2001 IEEE Transactions on Circuits & Systems for Video Technology Best Paper Award. He is a member of the U.S. National Academy of Engineering, an Academician of the Academy of Finland, a member of the Norwegian Academy of Technological Sciences, a foreign member of the Croatian Academy of Sciences and Arts, and a foreign member of the Academy of Engineering of Mexico. He has been awarded Honorary Doctorate degrees from the Tampere University of Technology, Finland and the "Politehnica" University of Bucharest, Romania, and the University Medal of the Slovak University of Technology, Bratislava, Slovakia. Dr. Mitra is a Fellow of the IEEE, AAAS, and SPIE, and a member of EURASIP.

To my students

Contents

Preface

The field of digital signal processing (DSP) has seen explosive growth during the past four decades, as phenomenal advances both in research and application have been made. Fueling this growth have been the advances in digital computer technology and software development. Almost every electrical and computer engineering department in this country and abroad now offers one or more courses in digital signal processing, with the first course usually being offered at the senior level. This book is intended for a two-semester course on digital signal processing for seniors or first-year graduate students. It is also written at a level suitable for self-study by the practicing engineer or scientist.

Even though the second edition of this book was published barely three years ago, based on the feedback received from professors who adopted this book for their courses and many readers, it became evident that a new edition was needed to incorporate the suggested changes to the contents. Three types of changes were made to the manuscript: inclusion of a number of new topics, elimination of some topics, and a major reorganization of the materials. We believe the materials in each chapter are now organized more logically. In addition more worked-out examples have been included to explain new and difficult concepts.

One major change occurring in the third edition is the splitting of the chapter on transform-domain representations of discrete-time signals and systems into three chapters: one chapter on the discrete-time Fourier transform (DTFT) representation, a second one on the discrete Fourier transform (DFT) representation, and the third one on the z-transform representation. The chapter on discrete-time Fourier transform representation also includes a brief review of the continuous-time Fourier transform (CTFT) representation of continuous-time signals and systems to point out the basic similarities and differences between the two transforms. The concept of the frequency response of a linear, time-invariant discrete-time system and its properties are discussed in this chapter. The chapter containing a discussion of the DFT includes a review of two other finite-length discrete transforms, namely, the discrete cosine transform (DCT) and the Haar transform. These two latter transforms are often used in signal compression. The concept of the transfer function of a linear, time-invariant discrete-time system is reviewed in the chapter on the z-transform representation.

The second major change implemented in this edition is to cover the design of infinite impulse response (IIR) and finite impulse response (FIR) digital filters in two separate chapters. The third major change involves the splitting of the chapter on multirate digital signal processing into two chapters. The first chapter covers a discussion on fundamental concepts of multirate digital signal processing, while the second chapter includes a discussion on filter banks and discrete wavelet transform. Finally, relevant materials on discrete-time random signals are included as an appendix.

The new topics included in the third edition are continuous-time Fourier transform (Section 3.1), unwrapped phase function (Section 3.7), phase and group delays (Section 3.9), Fourier-domain filtering (Section 5.8), discrete cosine transform (Section 5.11), Haar transform (Section 5.12), energy compaction properties of the finite-length discrete transforms (Section 5.13), transfer function classification based on magnitude and phase characteristics (Sections 7.1 and 7.2), minimum-phase FIR filter design (Section 10.4), spectral factorization (Section 10.4), computationally efficient FIR filter design (Section 10.6), fast discrete Fourier transform computation using index mapping (Section 11.4), sliding discrete Fourier transform (Section 11.6), discrete Fourier transform computation over a narrow frequency band (Section 11.7), spline interpolation (Section 13.5), discrete-wavelet transform (Section 14.6), digital music synthesis (Section 15.7), and wavelet-based signal compression (Section 15.9).

A key feature of this book is the extensive use of MATLAB® -based[1] examples that illustrate the program's powerful capability to solve signal processing problems. The book uses a three-stage pedagogical structure designed to take full advantage of MATLAB and to avoid the pitfalls of a "cookbook" approach to problem solving. First, each chapter begins by developing the essential theory and algorithms. Second, the material is illustrated with examples solved by hand calculation. And third, solutions are derived using MATLAB. From the beginning, MATLAB codes are provided with enough details to permit the students to repeat the examples on their computers. In addition to conventional theoretical problems requiring analytical solutions, each chapter also includes a large number of problems requiring solution via MATLAB. This book requires a minimal knowledge of MATLAB. We believe students learn the intricacies of problem solving with MATLAB faster by using tested, complete programs and then writing simple programs to solve specific problems that are included at the ends of Chapters 2 to 14.

Because computer verification enhances the understanding of the underlying theories, as in the first two editions, a large library of worked-out MATLAB programs are included in the third edition. The original MATLAB programs of the second edition have been updated to run on the newer versions of MATLAB and the *Signal Processing Toolbox*. In addition, new MATLAB programs and code fragments have been added in this edition. All MATLAB programs are included in the CD accompanying this text. The reader can run these programs to verify the results included in the book. All MATLAB programs and code fragments in the text have been tested under version 7 (Release 14) of MATLAB and version 6.2 of the *Signal Processing Toolbox*. Some of the programs listed in this book are not necessarily the fastest with regard to their execution speeds, nor are they the shortest. They have been written for maximum clarity without detailed explanations.

A second attractive feature of this book is the inclusion of extensive simple, but practical, examples that expose the reader to real-life signal processing problems, which has been made possible by the use of computers in solving practical design problems. This book also covers many topics of current interest not normally found in an upper-division text. Additional topics are also introduced to the reader through problems at the end of Chapters 2 through 14. Finally, the book concludes with a chapter that focuses on several important, practical applications of digital signal processing. These applications are easy to follow and do not require knowledge of other advanced-level courses.

The CD accompanying the book also contains several other useful materials, such as files of real signals, review materials, additional examples, frequently asked questions (FAQs), and a short tutorial on MATLAB. Where possible, pointers in the text with CD symbols have been used to direct the reader to relevant materials in the CD. From the feedback we hope to receive from the readers of the third edition, we hope to improve the contents on the CD for future editions.

The prerequisite for this book is a junior-level course in linear continuous-time and discrete-time

---

[1]**MATLAB** is a registered trademark of The MathWorks, Inc., 3 Apple Hill Dr., Natick, MA 01760, Phone: 508-647-7000, **http://www.mathworks.com**.

systems, which is usually required in most universities. A minimal review of linear systems and transforms is provided in the text, and basic materials from linear system theory are included, with important materials summarized in tables. This approach permits the inclusion of more advanced materials without significantly increasing the length of the book.

The book is divided into 15 chapters and an appendix. Chapter 1 presents an introduction to the field of signal processing and provides an overview of signals and signal processing methods.

Chapter 2 discusses the time-domain representations of discrete-time signals and discrete-time systems as sequences of numbers and describes classes of such signals and systems commonly encountered. Several basic discrete-time signals that play important roles in the time-domain characterization of arbitrary discrete-time signals and discrete-time systems are then introduced. Next, a number of basic operations to generate other sequences from one or more sequences are described. A combination of these operations is also used in developing a discrete-time system. The problem of representing a continuous-time signal by a discrete-time sequence is examined for a simple case.

Chapter 3 is devoted to the discrete-time Fourier transform (DTFT) representations of discrete-time sequences. It starts with a short review of the continuous-time Fourier transform (CTFT) representations of continuous-time signals and systems. The DTFT and its inverse are introduced, along with a discussion of the convergence of the DTFT. Properties of the DTFT are next reviewed, and the unwrapping of the phase function to remove certain discontinues in the DTFT is discussed. The concept of the frequency response of a linear, time-invariant (LTI) discrete-time system is then introduced, followed by a careful examination of the difference between phase and group delays associated with the frequency response.

Chapter 4 is concerned primarily with the discrete-time processing of continuous-time signals. The conditions for discrete-time representation of a band-limited continuous-time signal under ideal sampling and its exact recovery from the sampled version are first derived. Several interface circuits are used for the discrete-time processing of continuous-time signals. Two of these circuits are the anti-aliasing filter and the reconstruction filter, which are analog lowpass filters. As a result, a brief review of the basic theory behind some commonly used analog filter design methods is included, and their use is illustrated with MATLAB. Other interface circuits discussed in this chapter are the sample-and-hold circuit, the analog-to-digital converter, and the digital-to-analog converter.

The major part of Chapter 5 is concerned with the discrete Fourier transform (DFT), which plays an important role in some digital signal processing applications as it can be used to implement linear convolution efficiently using fast algorithm for its computation. The DFT and its inverse are introduced, along with a discussion of their properties. This chapter also includes a review of the discrete cosine transform (DCT) and the Haar transform. All three transforms discussed in this chapter are examples of orthogonal transforms of a finite-length sequence.

Chapter 6 is devoted to a discussion of z-transform. The transform and its inverse are introduced, along with a discussion of their properties. The convergence condition of the z-transform is examined in details. It also includes a discussion of the concept of the transfer function of a LTI discrete-time system and its relation to the frequency response of the system.

This book concentrates almost exclusively on the linear time-invariant discrete-time systems, and Chapter 7 discusses their transform-domain representations. Specific properties of such transform-domain representations are investigated, and several simple applications are considered.

A structural representation using interconnected basic building blocks is the first step in the hardware or software implementation of an LTI digital filter. The structural representation provides the relations between some pertinent internal variables with the input and the output, which, in turn, provides the keys to the implementation. There are various forms of the structural representation of a digital filter, and two such representations are reviewed in Chapter 8, followed by a discussion of some popular schemes for the realization of real causal IIR and FIR digital filters. In addition, it describes a method for the realization of IIR digital filter structures that can be used for the generation of a pair of orthogonal sinusoidal sequences.

Chapter 9 considers the IIR digital filter design problem. First, it discusses the issues associated with the filter design problem. Then, it describes the most popular approach to IIR filter design, based on the conversion of a prototype analog transfer function to a digital transfer function. The spectral transformation of one type of IIR transfer function into another type is discussed. The use of MATLAB in IIR digital filter design is illustrated.

Chapter 10 is concerned with the FIR digital filter design problem. A very simple approach to FIR filter design is described, followed by a discussion of a popular algorithm for the computer-aided design of equiripple linear-phase FIR digital filters. The use of MATLAB in FIR digital filter design is illustrated.

Chapter 11 is concerned with the implementation aspects of DSP algorithms. Two major issues concerning implementation are discussed first. The software implementations of digital filtering and DFT algorithms on a computer using MATLAB are reviewed to illustrate the main points. This is followed by a discussion of various schemes for the representation of number and signal variables on digital machines, which is basic to the development of methods for the analysis of finite wordlength effects considered in Chapter 12. Algorithms used to implement addition and multiplication, the two key arithmetic operations in digital signal processing, are reviewed next, along with operations developed to handle overflow. Finally, the chapter outlines two general methods for the design and implementation of tunable digital filters, followed by a discussion of algorithms for the approximation of certain special functions.

Chapter 12 is devoted to analysis of the effects of the various sources of quantization errors; it describes structures that are less sensitive to these effects. Included here are discussions on the effect of coefficient quantization.

Chapters 13 and 14 discuss multirate discrete-time systems with unequal sampling rates at various parts. The chapter includes a review of the basic concepts and properties of sampling rate alteration, design of decimation and interpolation digital filters, and multirate filter bank design.

The final chapter, Chapter 15, reviews a few simple practical applications of digital signal processing to provide a glimpse of its potential.

The materials in this book have been used in a two-quarter course sequence on digital signal processing at the University of California, Santa Barbara, and have been extensively tested in the classroom for over 12 years. Basically, Chapters 2 through 8 form the basis of an upper-division course, while Chapters 8 through 15 form the basis of a graduate-level course.

This text contains 352 examples, 163 MATLAB programs and code fragments, 783 problems, and 158 MATLAB exercises.

Every attempt has been made to ensure the accuracy of all materials in this book, including the MATLAB programs. I would, however, appreciate readers bringing to my attention any errors that may appear in the printed version for reasons beyond my control and that of the publisher. These errors and any other comments can be communicated to me by e-mail addressed to **mitra@ece.ucsb.edu**.

Finally, I have been particularly fortunate to have had the opportunity to work with outstanding students who were in my research group during my teaching career, which spans over 40 years. I have benefited immensely, and continue to do so, both professionally and personally, from my friendship and association with them, and to them I dedicate this book.

Sanjit K. Mitra

Acknowledgments

The preliminary versions of the complete manuscript for the first edition were reviewed by Dr. Hrvojc Babic of University of Zagreb, Croatia; Dr. James F. Kaiser of Duke University; Dr. Wolfgang F. G. Mecklenbräuker of Technical University of Vienna, Austria; and Dr. P. P. Vaidyanathan of California Institute of Technology. A later version was reviewed by Dr. Roberto H. Bambmerger of Microsoft; Dr. Charles Boumann of Purdue University; Dr. Kevin Buckley of University of Minnesota; Dr. John A. Flemming of Texas A & M University; Dr. Jerry D. Gibson of Southern Methodist University; Dr. John Gowdy of Clemson University; Drs. James Harris and Mahmood Nahvi of California Polytechnic University, San Louis Obispo; Dr. Yih-Chyun Jenq of Portland State University; Dr. Troung Q. Ngyuen of University of California, San Diego; and Dr. Andreas Spanias of Arizona State University. Various parts of the manuscript of the the first edition were reviewed by Dr. C. Sidney Burrus of Rice University; Dr. Richard V. Cox of AT&T Laboratories; Dr. Ian Galton of University of California, San Diego; Dr. Nikil S. Jayant of Georgia Institute of Technology; Dr. Tor Ramstad of Norwegian University of Science and Technology, Trondheim, Norway; Dr. B. Ananth Shenoi of Wright State University; Dr. Hans W. Schüssler of University of Erlangen-Nuremberg, Germany; Dr. Richard Schreier of Analog Devices and Dr. Gabor C. Temes of Oregon State University.

Reviews for the second edition were provided by Dr. Winser E. Alexander of North Carolina State University; Dr. Sohail A. Dianat of Rochester Institute of Technology; Dr. Suhash Dutta Roy of Indian Institute of Technology, New Delhi; Dr. David C. Farden of North Dakota State University; Dr. Abdulnasir Y. Hossein of Sultan Qaboos University, Sultanate of Omman; Dr. James F. Kaiser of Duke University; Dr. Ramakrishna Kakarala of Agilent Laboratories; Dr. Wolfgang F. G. Mecklenbräuker of Technical University of Vienna, Austria; Dr. Antonio Ortega of University of Southern California; Dr. Stanley J. Reeves of Auburn University; Dr. George Symos of University of Maryland, College Park; and Dr. Gregory A. Wornell of Massachusetts Institute of Technology. Various parts of the manuscript for the second edition were reviewed by Dr. Dimitris Anastassiou of Columbia University; Dr. Rajendra K. Arora of Florida State University; Dr. Ramdas Kumaresan of University of Rhode Island; Dr. Upamanyu Madhow of University of California, Santa Barbara; Dr. Urbashi Mitra of University of Southern California; Dr. Randolph Moses of Ohio State University; Dr. Ivan Selesnick of Polytechnic University, Brooklyn, New York; and Dr. Gabor C. Temes of Oregon State University.

Reviews for the third edition were provided by Dr. Donald G. Bailey of Massey University, New Zealand; Dr. Marco Carli of University of Rome 'TRE', Italy; Dr. Emad S. Ebbini of University of Minnesota; Dr. Chandrakanth H. Gowda of Tuskegee University; Dr. Robert W. Heath, Jr., of University of Texas at Austin; Dr. Hongbin Li of Stevens Institute of Technology; Dr. Ping Liang of University of California, Riverside; Dr. Luca Lucchese of Oregon State University; Dr. Kamal Premaratne of University of Miami; Dr. Lawrence R. Rabiner of Rutgers University; Dr. Ali M. Reza of University of Wisconsin-Milwaukee; Dr. Terry E. Riemer of University of New Orleans; Dr. Erchin Serepdin of Texas A&M University; and Dr. Okechukwu C. Ugweje of University of Akron. Various parts of the manuscript for the third edition were reviewed by Dr. Shivkumar Chandrasekaran of University of California, Santa Barbara; Dr. Charles D. Creusere of New Mexico State University at Las Cruces; Dr. Yong-Ching Lim of Nanyang Technological University, Singapore; Mr. Ricardo Losada of MathWorks, Inc.; Dr. Kai-Kuang Ma of Nanyang Technological University, Singapore; Dr. Julius O. Smith of Stanford University and Dr. Truong Ngyuen of University of California, San Diego.

I thank all of them for their valuable comments, which have improved the book tremendously.

Many of my former and present research students reviewed various portions of the manuscript of all editions and tested a number of the MATLAB programs. In particular, I would like to thank Drs. Charles D. Creusere, Rajeev Gandhi, Gabriel Gomes, Serkan Hatipoglu, Zhihai He, Michael Lightstone, Ing-Song Lin, Luca Lucchese, Michael Moore, Debargha Mukherjee, Norbert Strobel, and Stefan Thurnhofer, and Mylene Queiroz de Farias, and Messrs. Hsin-Han Ho and Eric Leipnik. I am also indebted to all former students in my ECE 158 and ECE 258A classes at the University of California, Santa Barbara, for their feedback over the years, which helped refine the book.

I thank Goutam K. Mitra and Alicia Rodriguez for the cover design of the book. Finally, I thank Patricia Monohon for her outstanding assistance in the preparation of the LaTeX files of the third edition.

Supplements

All MATLAB programs included in this book are in the CD accompanying this book and are also available via anonymous file transfer protocol (FTP) from the Internet site **iplserv.ece.ucsb.edu** in the directory **/pub/mitra/Book_3e**.

A solutions manual prepared by Chowdary Adsumilli, Chin-chaye Koh, Gabriel Gomes, Hsin-Han Ho, and Mylene Queiroz de Farias and containing the solutions to all problems and MATLAB exercises is available to instructors from the publisher. PowerPoint slides of most materials of this book are available to instructors from the author.

A companion book *Digital Signal Processing Laboratory Using MATLAB* by the author is also available from McGraw-Hill.

1 Signals and Signal Processing

Signals play an important role in our daily life. Examples of signals that we encounter frequently are speech, music, picture, and video signals. A signal is a function of independent variables such as time, distance, position, temperature, and pressure. For example, speech and music signals represent air pressure as a function of time at a point in space. A black-and-white picture is a representation of light intensity as a function of two spatial coordinates. The video signal in television consists of a sequence of images, called frames, and is a function of three variables: two spatial coordinates and time.

Most signals we encounter are generated by natural means. However, a signal can also be generated synthetically or by computer simulation. A signal carries information, and the objective of signal processing is to extract useful information carried by the signal. The method of information extraction depends on the type of signal and the nature of the information being carried by the signal. Thus, roughly speaking, signal processing is concerned with the mathematical representation of the signal and the algorithmic operation carried out on it to extract the information present. The representation of the signal can be in terms of basis functions in the domain of the original independent variable(s), or it can be in terms of basis functions in a transformed domain. Likewise, the information extraction process may be carried out in the original domain of the signal or in a transformed domain. This book is concerned with discrete-time representation of signals and their discrete-time processing.

This chapter provides an overview of signals and signal processing methods. The mathematical characterization of the signal is first discussed along with a classification of signals. Next, some typical signals are discussed in detail, and the type of information carried by them is described. Then, a review of some commonly used signal processing operations is provided and illustrated through examples. A brief review of some typical signal processing applications is discussed next. Finally, the advantages and disadvantages of digital processing of signals are discussed.

1.1 Characterization and Classification of Signals

Depending on the nature of the independent variables and the value of the function defining the signal, various types of signals can be defined. For example, independent variables can be continuous or discrete. Likewise, the signal can either be a continuous or a discrete function of the independent variables. Moreover, the signal can be either a real-valued function or a complex-valued function.

A signal can be generated by a single source or by multiple sources. In the former case, it is a scalar signal, and in the latter case, it is a vector signal, often called a multichannel signal.

A one-dimensional (1-D) signal is a function of a single independent variable. A two-dimensional (2-D) signal is a function of two independent variables. A multidimensional (M-D) signal is a function of more than one variable. The speech signal is an example of a 1-D signal where the independent variable is time. An image signal, such as a photograph, is an example of a 2-D signal where the two independent variables are the two spatial variables. Each frame of a black-and-white video signal is a 2-D image signal that is a function of two discrete spatial variables, with each frame occurring sequentially at discrete instants of time. Hence, the black-and-white video signal can be considered an example of a three-dimensional (3-D) signal where the three independent variables are the two spatial variables and time. A color video signal is a three-channel signal composed of three 3-D signals representing the three primary colors: red, green, and blue (RGB). For transmission purposes, the RGB television signal is transformed into another type of three-channel signal composed of a luminance component and two chrominance components.

The value of the signal at a specific value(s) of the independent variable(s) is called its *amplitude*. The variation of the amplitude as a function of the independent variable(s) is called its *waveform*.

For a 1-D signal, the independent variable is usually labeled as *time*. If the independent variable is continuous, the signal is called a *continuous-time signal*. If the independent variable is discrete, the signal is called a *discrete-time signal*. A continuous-time signal is defined at every instant of time. On the other hand, a discrete-time signal takes certain numerical values at specified discrete instants of time, and between these specified instants of time, the signal is not defined. Hence, a discrete-time signal is basically a sequence of numbers.

Speech Demo 1

A continuous-time signal with a continuous amplitude is usually called an *analog signal*. A speech signal is an example of an analog signal. Analog signals are commonly encountered in our daily life and are usually generated by natural means. A discrete-time signal with discrete-valued amplitudes represented by a finite number of digits is referred to as a *digital signal*. An example of a digital signal is the digitized music signal stored in a CD-ROM disk. A discrete-time signal with continuous-valued amplitudes is called a *sampled-data signal*. This last type of signal occurs in switched-capacitor (SC) circuits. A digital signal is thus a quantized sampled-data signal. Finally, a continuous-time signal with discrete-valued amplitudes has been referred to as a *quantized boxcar signal* [Ste93]. The latter type of signals occurs in digital electronic circuits where the signal is kept at fixed level (usually one of two values) between two instants of clocking. Figure 1.1 illustrates the four types of signals.

The functional dependence of a signal in its mathematical representation is often explicitly shown. For a continuous-time 1-D signal, the continuous independent variable is usually denoted by t, whereas for a discrete-time 1-D signal, the discrete independent variable is usually denoted by n. For example, $u(t)$ represents a continuous-time 1-D signal and $\{v[n]\}$ represents a discrete-time 1-D signal. Each member, $v[n]$, of a discrete-time signal is called a *sample*. In many applications, a discrete-time signal is generated from a parent continuous-time signal by sampling the latter at uniform intervals of time. If the discrete instants of time at which a discrete-time signal is defined are uniformly spaced, the independent discrete variable n can be normalized to assume integer values.

In the case of a continuous-time 2-D signal, the two independent variables are usually the spatial coordinates, which are usually denoted by x and y. For example, the intensity of a black-and-white image can be expressed as $u(x, y)$. A color image $u(x, y)$, is composed of three signals representing the three primary colors, red, green, and blue:

$$\mathbf{u}(x, y) = \begin{bmatrix} r(x, y) \\ g(x, y) \\ b(x, y) \end{bmatrix}.$$

Image Demo 1
Video Demo 1

On the other hand, a digitized image is a 2-D discrete signal, and its two independent variables are discretized spatial variables often denoted by m and n. Hence, a digitized image can be represented as $v[m, n]$. Likewise, a black-and-white video sequence is a 3-D signal and can be represented as $u(x, y, t)$, where x and y denote the two spatial variables and t denotes the temporal variable time. A color video signal is a vector signal composed of three video signals representing the three primary colors, red, green, and blue.

There is another classification of signals that depends on the certainty by which the signal can be uniquely described. A signal that can be uniquely determined by a well-defined process such as a mathematical expression or rule, or table look-up, is called a *deterministic signal*. A signal that is generated

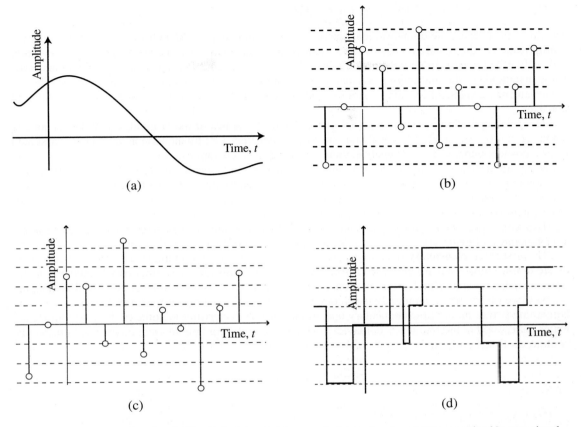

Figure 1.1: (a) An analog signal, (b) a digital signal, (c) a sampled-data signal, and (d) a quantized boxcar signal.

in a random fashion and cannot be predicted ahead of time is called a *random signal.* In th s text, we are primarily concerned with the processing of discrete-time deterministic signals. However, since practical discrete-time systems employ finite wordlengths for the storing of signals and the implementation of the signal processing algorithms, it is necessary to develop tools for the analysis of finite worclength effects on the performance of discrete-time systems. To this end, it has been found convenient to represent certain pertinent signals as random signals and employ statistical techniques for their analysis.

Some typical signal processing operations performed on analog signals are reviewed in the following section.

1.2 Typical Signal Processing Operations

Various types of signal processing operations are employed in practice. In the case of analog signals, most signal processing operations are usually carried out in the time-domain, whereas, in the case of discrete-time signals, both time-domain and frequency-domain operations are employed. In either case, the desired operations are implemented by a combination of some elementary operations. These operations are also usually implemented in real-time or near real-time, even though, in certain applicat ons, they may be implemented off-line.

1.2.1   Simple Time-Domain Operations

The three most basic time-domain signal operations are scaling, delay, and addition. *Scaling* is simply the multiplication of the signal by a positive or a negative constant. In the case of analog signals, this operation is usually called *amplification* if the magnitude of the multiplying constant, called *gain,* is greater than one. If the magnitude of the multiplying constant is less than one, the operation is called *attenuation.* Thus, if $x(t)$ is an analog signal, the scaling operation generates a signal $y(t) = \alpha x(t)$, where α is the multiplying constant.

The *delay* operation generates a signal that is a delayed replica of the original signal. For an analog signal $x(t)$, $y(t) = x(t - t_0)$ is the signal obtained by delaying $x(t)$ by the amount t_0, which is assumed to be a positive number. If t_0 is negative, then it is an *advance* operation.

Many applications require operations involving two or more signals to generate a new signal. For example, $y(t) = x_1(t) + x_2(t) - x_3(t)$ is the signal generated by the *addition* of the three analog signals $x_1(t)$, $x_2(t)$, and $x_3(t)$. Another elementary operation is the *product* of two signals. Thus, the product of two signals $x_1(t)$ and $x_2(t)$ generates a signal $y(t) = x_1(t)x_2(t)$.

Two other elementary operations are integration and differentiation. The *integration* of an analog signal $x(t)$ generates a signal $y(t) = \int_{-\infty}^{t} x(\tau)\, d\tau$, while its *differentiation* results in a signal $w(t) = dx(t)/dt$.

The first three elementary operations, namely, scaling, delay, and addition, are also carried out on discrete-time signals and are discussed further in later parts of this text. The other two elementary operations, namely, integration and differentiation, are implemented approximately in the discrete-time domain.

Next we review some commonly used complex signal processing operations that are implemented by combining two or more of the elementary operations. The characteristics of some of these operations are better understood in the frequency-domain by making use of the continuous-time Fourier transform. The continuous-time Fourier transform $X(j\Omega)$ of a continuous-time signal $x(t)$ is given by[1]

$$X(j\Omega) = \int_{-\infty}^{\infty} x(t)e^{-j\Omega t}\, dt. \tag{1.1}$$

$X(j\Omega)$ is called the spectrum of $x(t)$.

1.2.2   Filtering

One of the most widely used complex signal processing operations is *filtering,* whose main objective is to alter the spectrum according to some given specifications. The system implementing this operation is called a *filter.* For example, the filter may be designed to pass certain frequency components in a signal through the system and to block other frequency components. The range of frequencies of the signal components allowed to pass through the filter is called the *passband,* and the range of frequencies of the signal components blocked by the filter is called the *stopband.* Various types of filters can be defined, depending on the nature of the filtering operation. In most cases, the filtering operation for analog signals is performed by a linear, time-invariant filter. If the filter is characterized by an impulse response $h(t)$, then its output $y(t)$ to an input is given by the convolution integral

$$y(t) = \int_{-\infty}^{\infty} h(t - \tau)x(\tau)\, d\tau, \tag{1.2}$$

assuming the filter is relaxed with zero initial conditions at the time of application of the input signal. In the frequency-domain, the above equation can be expressed as

$$Y(j\Omega) = H(j\Omega)X(j\Omega), \tag{1.3}$$

---

[1] See Section 3.1 for a review of the continuous-time Fourier transform.

where $Y(j\Omega)$, $X(j\Omega)$, and $H(j\Omega)$, are, respectively, the continuous-time Fourier transforms of $y(t)$, $x(t)$, and $h(t)$.

A *lowpass filter* passes all low-frequency components below a certain specified frequency f_p, called the *passband edge frequency,* and blocks all high-frequency components above f_s, called the *stopband edge frequency.* A *highpass filter* passes all high-frequency components above a certain passband edge frequency f_p and blocks all low-frequency components below the stopband edge frequency f_s. A *bandpass filter* passes all frequency components between two passband edge frequencies f_{p1} and f_{p2} where $f_{p1} < f_{p2}$, and blocks all frequency components below the stopband edge frequency f_{s1} and above the stopband edge frequency f_{s2}. A *bandstop filter* blocks all frequency components between two stopband edge frequencies f_{s1} and f_{s2} and passes all frequency components below the passband edge frequency f_{p1} and above the passband edge frequency f_{p2}. Figure 1.2(a) shows a signal composed of three sinusoidal components of frequencies 50 Hz, 100 Hz, and 200 Hz, respectively. Figures 1.2(b) to (e) show the results of the above four types of filtering operations with appropriately chosen cutoff frequencies.

A bandstop filter designed to block a single frequency component is called a *notch filter.* A *multiband filter* has more than one passband and more than one stopband. A *comb filter* is designed to block frequencies that are integral multiples of a low frequency.

A signal may get corrupted unintentionally by an interfering signal called *interference* or *noise.* In many applications, the desired signal occupies a low-frequency band from dc to some frequency f_L Hz, and it is corrupted by a high-frequency noise with frequency components above f_H Hz with $f_H > f_L$. In such cases, the desired signal can be recovered from the noise-corrupted signal by passing the latter through a lowpass filter with a cutoff frequency f_c where $f_L < f_c < f_H$. In some applications, the noise corrupting the desired signal may be a single-frequency sinusoidal signal. For example, the noise generated by power lines radiating electric and magnetic fields appears as a 60-Hz sinusoidal signal. The desired signal can then be recovered by passing the corrupted signal through a notch filter with a notch frequency at 60 Hz.[2]

1.2.3 Generation of Complex-Valued Signals

As indicated earlier, a signal can be real-valued or complex-valued. For convenience, the former is usually called a *real signal,* while the latter is called a *complex signal.* All naturally generated signals are real-valued. In some applications, it is necessary to develop a complex signal from a real signal having more desirable properties. One approach to generating a complex signal from a real signal is by employing a *Hilbert transformer* that is characterized by an impulse response $h_{\mathrm{HT}}(t)$ given by [Fre94], [Opp83]

$$h_{\mathrm{HT}}(t) = \frac{1}{\pi t}, \tag{1.4}$$

with a continuous-time Fourier transform $H_{\mathrm{HT}}(j\Omega)$ given by

$$H_{\mathrm{HT}}(j\Omega) = \begin{cases} -j, & \Omega > 0, \\ j, & \Omega < 0. \end{cases} \tag{1.5}$$

Let $x(t)$ denote a real analog signal with a continuous-time Fourier transform $X(j\Omega)$. The magnitude spectrum of a real signal exhibits even symmetry, while the phase spectrum exhibits odd symmetry. Thus, the spectrum $X(j\Omega)$ of a real signal $x(t)$ contains both positive and negative frequencies and can therefore be expressed as

$$X(j\Omega) = X_p(j\Omega) + X_n(j\Omega), \tag{1.6}$$

---

[2]In many countries, power lines generate 50-Hz noise.

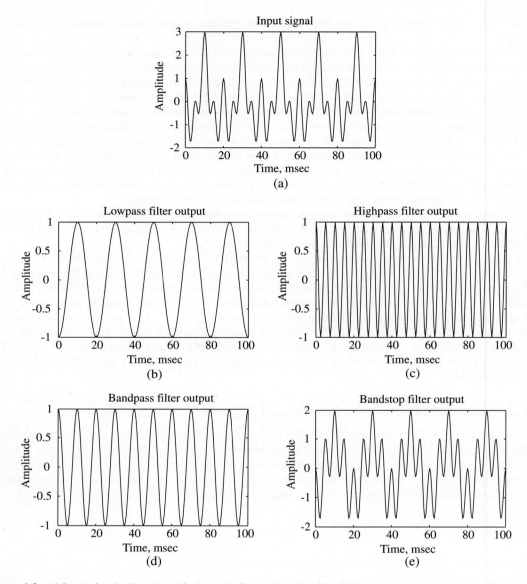

Figure 1.2: (a) Input signal, (b) output of a lowpass filter with a cutoff at 80 Hz, (c) output of a highpass filter with a cutoff at 150 Hz, (d) output of a bandpass filter with cutoffs at 80 Hz and 150 Hz, and (e) output of a bandstop filter with cutoffs at 80 Hz and 150 Hz.

where $X_p(j\Omega)$ is the portion of $X(j\Omega)$ occupying the positive frequency range and $X_n(j\Omega)$ is the portion of $X(j\Omega)$ occupying the negative frequency range. If $x(t)$ is passed through a Hilbert transformer, its output $\hat{x}(t)$ has a spectrum $\hat{X}(j\Omega)$ given by

$$\hat{X}(j\Omega) = H_{\text{HT}}(j\Omega)X(j\Omega) = -jX_p(j\Omega) + jX_n(j\Omega). \qquad (1.7)$$

It can be shown that $\hat{x}(t)$ is also a real signal. Consider the complex signal $y(t)$ formed by the sum of $x(t)$

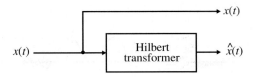

Figure 1.3: Generation of an analytic signal using a Hilbert transformer.

and $\hat{x}(t)$:

$$y(t) = x(t) + j\hat{x}(t). \tag{1.8}$$

The signals $x(t)$ and $\hat{x}(t)$ are called, respectively, the *in-phase* and *quadrature components* of $y(t)$. The continuous-time Fourier transform of $y(t)$ is given by (Problem 3.9)

$$Y(j\Omega) = X(j\Omega) + j\hat{X}(j\Omega) = 2X_p(j\Omega). \tag{1.9}$$

Thus, the complex signal $y(t)$, called an *analytic signal*, has only positive-frequency components.

A block diagram representation of the scheme for the analytic signal generation from a real signal is sketched in Figure 1.3. One application of the Hilbert transformer is in the implementation of a single-sideband modulation as indicated in Figure 1.8 and discussed in Section 1.2.4.

1.2.4 Amplitude Modulation

For transmission of signals over long distances, a transmission media such as cable, optical fiber, or the atmosphere is employed. Each such medium has a bandwidth that is more suitable for the efficient transmission of signals in the high-frequency range. As a result, for the transmission of a low-frequency signal over a channel, it is necessary to transform the signal to a high-frequency signal by means of a modulation operation. At the receiving end, the modulated high-frequency signal is demodulated, and the desired low-frequency signal is then extracted by further processing. There are four major types of modulation of analog signals: amplitude modulation, frequency modulation, phase modulation, and pulse amplitude modulation. Of these schemes, amplitude modulation is conceptually simple and is discussed here [Fre94], [Opp83].

In the *amplitude modulation* scheme, the amplitude of a high-frequency sinusoidal signal $A\cos(\Omega_o t)$, called the *carrier signal,* is varied by the low-frequency band-limited signal $x(t)$, called the *modulating signal*, generating a high-frequency signal, called the *modulated signal,* $y(t)$ according to

$$y(t) = Ax(t)\cos(\Omega_o t). \tag{1.10}$$

Thus, amplitude modulation can be implemented by forming the product of the modulating signal with the carrier signal. It can be shown that the spectrum $Y(j\Omega)$ of $y(t)$ is given by

$$Y(j\Omega) = \frac{A}{2}X(j(\Omega - \Omega_o)) + \frac{A}{2}X(j(\Omega + \Omega_o)), \tag{1.11}$$

where $X(j\Omega)$ is the spectrum of the modulating signal $x(t)$. Figure 1.4 shows the spectra of the modulating signal and that of the modulated signal under the assumption that the carrier frequency Ω_o is greater than Ω_m, the highest frequency contained in $x(t)$. As seen from this figure, $y(t)$ is now a band-limited high-frequency signal with a bandwidth $2\Omega_m$ centered at Ω_o.

The portion of the amplitude-modulated signal between Ω_o and $\Omega_o + \Omega_m$ is called the *upper sideband,* whereas the portion between Ω_o and $\Omega_o - \Omega_m$ is called the *lower sideband.* Because of the generation of two sidebands and the absence of a carrier component in the modulated signal, the process is called *double-sideband suppressed carrier* (DSB-SC) *modulation.*

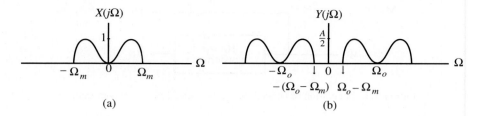

Figure 1.4: (a) Spectrum of the modulating signal $x(t)$ and (b) spectrum of the modulated signal $y(t)$. For convenience, both spectra are shown as real functions.

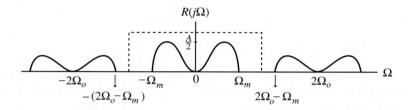

Figure 1.5: Spectrum of the product of the modulated signal and the carrier.

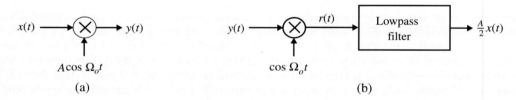

Figure 1.6: Schematic representations of the DSB-SC amplitude modulation and demodulation schemes: (a) modulator and (b) demodulator.

The *demodulation* of $y(t)$, assuming $\Omega_o > \Omega_m$, is carried out in two stages. First, the product of $y(t)$ with a sinusoidal signal of the same frequency as the carrier is formed. This results in

$$r(t) = y(t) \cos \Omega_o t = Ax(t) \cos^2 \Omega_o t, \tag{1.12}$$

which can be rewritten as

$$r(t) = y(t) \cos \Omega_o t = \tfrac{A}{2}x(t) + \tfrac{A}{2}x(t) \cos(2\Omega_o t). \tag{1.13}$$

This result indicates that the product signal is composed of the original modulating signal scaled by a factor 1/2 and an amplitude-modulated signal with a carrier frequency $2\Omega_o$. The spectrum $R(j\Omega)$ of $r(t)$ is as indicated in Figure 1.5. The original modulating signal can now be recovered from $r(t)$ by passing it through a lowpass filter with a cutoff frequency Ω_c satisfying the relation $\Omega_m < \Omega_c < 2\Omega_o - \Omega_m$. The output of the filter is then a scaled replica of the modulating signal.

Figure 1.6 shows the block diagram representations of the amplitude modulation and demodulation schemes. The underlying assumption in the demodulation process outlined above is that a sinusoidal signal identical to the carrier signal can be generated at the receiving end. In general, it is difficult to ensure

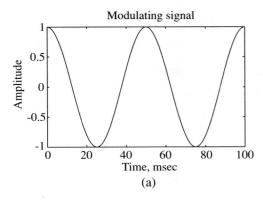 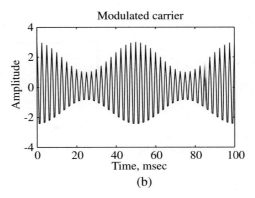

Figure 1.7: (a) A sinusoidal modulating signal of frequency 20 Hz and (b) modulated carrier with a carrier frequency of 400 Hz based on the DSB modulation.

that the demodulating sinusoidal signal has a frequency identical to that of the carrier all the time. To get around this problem, in the transmission of amplitude-modulated radio signals, the modulation process is modified so that the transmitted signal includes the carrier signal. This is achieved by redefining the amplitude modulation operation as follows:

$$y(t) = A[1 + mx(t)]\cos(\Omega_o t), \tag{1.14}$$

where m is a number chosen to ensure that $[1 + mx(t)]$ is positive for all t. Figure 1.7 shows the waveforms of a modulating sinusoidal signal of frequency 20 Hz and the amplitude-modulated carrier obtained according to Eq. (1.14) for a carrier frequency of 400 Hz and $m = 0.5$. Note that the envelope of the modulated carrier is essentially the waveform of the modulating signal. As here the carrier is also present in the modulated signal, the process is called simply *double-sideband* (DSB) *modulation.* At the receiving end, the carrier signal is separated first and then used for demodulation.

In the case of the conventional amplitude modulation, as can be seen from Figure 1.4, the modulated signal has a bandwidth of $2\Omega_m$, whereas the bandwidth of the modulating signal is Ω_m. To increase the capacity of the transmission medium, a modified form of the amplitude modulation is often employed in which either the upper sideband or the lower sideband of the modulated signal is transmitted. The corresponding procedure is called *single-sideband* (SSB) *modulation* to distinguish it from the double-sideband modulation scheme of Figure 1.6(a).

One way to implement the single-sideband amplitude modulation is indicated in Figure 1.8, where the Hilbert transformer is defined by Eq. (1.4). The spectra of pertinent signals in Figure 1.8 are shown in Figure 1.9.

1.2.5 Multiplexing and Demultiplexing

For an efficient utilization of a wideband transmission channel, many narrow-bandwidth low-frequency signals are combined to form a composite wideband signal that is transmitted as a single signal. The process of combining these signals is called *multiplexing,* which is implemented to ensure that a replica of the original narrow-bandwidth low-frequency signals can be recovered at the receiving end. The recovery process is called *demultiplexing.*

One widely used method of combining different voice signals in a telephone communication system is the *frequency-division multiplexing* (FDM) scheme [Cou83], [Opp83]. Here, each voice signal, typically band-limited to a low-frequency band of width $2\Omega_m$, is frequency-translated into a higher frequency band

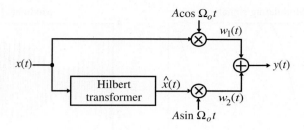

Figure 1.8: Single-sideband modulation scheme employing a Hilbert transformer.

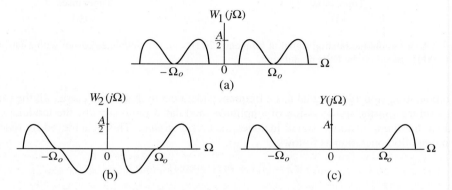

Figure 1.9: Spectra of pertinent signals in Figure 1.8.

using the amplitude modulation method of Eq. (1.10). The carrier frequency of adjacent amplitude-modulated signals is separated by Ω_o, with $\Omega_o > 2\Omega_m$ to ensure that there is no overlap in the spectra of the individual modulated signals after they are added to form a baseband composite signal. This signal is then modulated onto the main carrier developing the FDM signal and transmitted. Figure 1.10 illustrates the frequency-division multiplexing scheme.

At the receiving end, the composite baseband signal is first derived from the FDM signal by demodulation. Then each individual frequency-translated signal is first demultiplexed by passing the composite signal through a bandpass filter with a center frequency of identical value as that of the corresponding carrier frequency and a bandwidth slightly greater than $2\Omega_m$. The output of the bandpass filter is then demodulated using the method of Figure 1.6(b) to recover a scaled replica of the original voice signal.

1.2.6   Quadrature Amplitude Modulation

We observed earlier that DSB amplitude modulation is half as efficient as SSB amplitude modulation with regard to utilization of the spectrum. The *quadrature amplitude modulation* (QAM) method uses DSB modulation to modulate two different signals so that they both occupy the same bandwidth; thus, QAM takes up only as much bandwidth as the SSB modulation method. To understand the basic idea behind the QAM approach, let $x_1(t)$ and $x_2(t)$ be two band-limited low-frequency signals with a bandwidth of Ω_m as indicated in Figure 1.4(a). The two modulating signals are individually modulated by the two carrier signals $A\cos(\Omega_o t)$ and $A\sin(\Omega_o t)$, respectively, and are summed, resulting in a composite signal $y(t)$ given by

$$y(t) = Ax_1(t)\cos(\Omega_o t) + Ax_2(t)\sin(\Omega_o t). \tag{1.15}$$

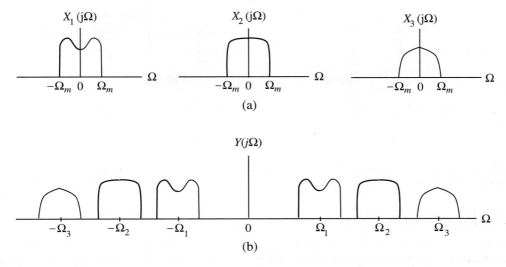

Figure 1.10: Illustration of the frequency-division multiplexing operation. (a) Spectra of three low-frequency signals and (b) spectra of the modulated composite signal.

Note that the two carrier signals have the same carrier frequency Ω_o but have a phase difference of 90°. In general, the carrier $A\cos(\Omega_o t)$ is called the *in-phase component,* and the carrier $A\sin(\Omega_o t)$ is called the *quadrature component.* The spectrum $Y(j\Omega)$ of the composite signal $y(t)$ is now given by

$$Y(j\Omega) = \tfrac{A}{2}\{X_1(j(\Omega - \Omega_o)) + X_1(j(\Omega + \Omega_o))\}$$

$$+\tfrac{A}{2j}\{X_2(j(\Omega - \Omega_o)) - X_2(j(\Omega + \Omega_o))\} \qquad (1.16)$$

and is seen to occupy the same bandwidth as the modulated signal obtained by a DSB modulation.

To recover the original modulating signals, the composite signal is multiplied by both the in-phase and the quadrature components of the carrier separately, resulting in two signals:

$$r_1(t) = y(t)\cos(\Omega_o t), \qquad r_2(t) = y(t)\sin(\Omega_o t). \qquad (1.17)$$

Substituting $y(t)$ from Eq. (1.15) in Eq. (1.17), we obtain, after some algebra,

$$r_1(t) = \tfrac{A}{2}x_1(t) + \tfrac{A}{2}x_1(t)\cos(2\Omega_o t) + \tfrac{A}{2}x_2(t)\sin(2\Omega_o t),$$

$$\qquad (1.18)$$

$$r_2(t) = \tfrac{A}{2}x_2(t) + \tfrac{A}{2}x_1(t)\sin(2\Omega_o t) - \tfrac{A}{2}x_2(t)\cos(2\Omega_o t).$$

Lowpass filtering of $r_1(t)$ and $r_2(t)$ by filters with a cutoff at Ω_m yields the two modulating signals. Figure 1.11 shows the block diagram representations of the quadrature amplitude modulation and demodulation schemes.

As in the case of the DSB suppressed carrier modulation method, the QAM method also requires at the receiver an exact replica of the carrier signal employed in the transmitting end for accurate demodulation. It is therefore not employed in the direct transmission of analog signals, but finds applications in the transmission of discrete-time data sequences and in the transmission of analog signals converted into discrete-time sequences by sampling and analog-to-digital conversion.

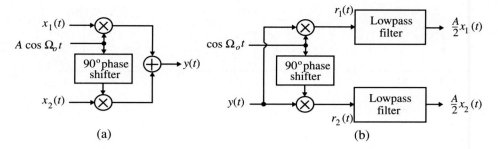

Figure 1.11: Schematic representations of the quadrature amplitude modulation and demodulation schemes: (a) modulator and (b) demodulator.

1.2.7 Signal Generation

An equally important part of signal processing is synthetic signal generation. One of the simplest such signal generators is a device generating a sinusoidal signal, called an *oscillator*. Such a device is an integral part of the amplitude-modulation and demodulation system described in the previous two sections. It also has various other signal processing applications.

There are applications that require the generation of other types of periodic signals such as square waves and triangular waves. Certain types of random signals with a spectrum of constant amplitude for all frequencies, called *white noise,* often find applications in practice. One such application is in the generation of discrete-time synthetic speech signals.

1.3 Examples of Typical Signals[3]

To better understand the breadth of the signal processing task, we now examine a number of examples of some typical signals and their subsequent processing in typical applications.

Electrocardiography (ECG) Signal

The electrical activity of the heart is represented by the ECG signal [Sha81]. A typical ECG signal trace is shown in Figure 1.12(a). The ECG trace is essentially a periodic waveform. One cycle of the blood transfer process from the heart to the arteries is represented by one period of the ECG waveform as shown in Figure 1.12(b). This part of the waveform is generated by an electrical impulse originating at the sinoatrial node in the right atrium of the heart. The impulse causes contraction of the atria, which forces the blood in each atrium to squeeze into its corresponding ventricle. The resulting signal is called the P-wave. The atrioventricular node delays the excitation impulse until the blood transfer from the atria to the ventricles is completed, resulting in the P–R interval of the ECG waveform. The excitation impulse then causes contraction of the ventricles, which squeezes the blood into the arteries. This generates the QRS part of the ECG waveform. During this phase, the atria are relaxed and filled with blood. The T-wave of the waveform represents the relaxation of the ventricles. The complete process is repeated periodically, generating the ECG trace.

Each portion of the ECG waveform carries various types of information for the physician analyzing a patient's heart condition [Sha81]. For example, the amplitude and timing of the P and QRS portions indicate

---

[3]This section has been adapted from *Handbook for Digital Signal Processing*, Sanjit K. Mitra and James F. Kaiser, Eds., ©1993, John Wiley & Sons. Adapted by permission of John Wiley & Sons.

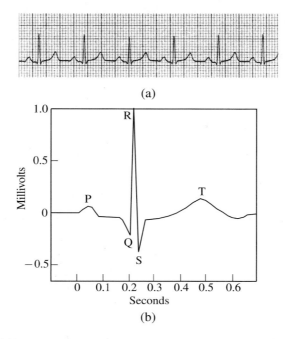

Figure 1.12: (a) A typical ECG trace and (b) one cycle of an ECG waveform.

the condition of the cardiac muscle mass. Loss of amplitude indicates muscle damage, whereas increased amplitude indicates abnormal heart rates. Too long a delay in the atrioventricular node is indicated by a very long P–R interval. Likewise, blockage of some or all of the contraction impulses is reflected by intermittent synchronization between the P- and QRS-waves. Most of these abnormalities can be treated with various drugs, and the effectiveness of the drugs can again be monitored by observing the new ECG waveforms taken after the drug treatment.

In practice, there are various types of externally produced artifacts that appear in the ECG signal [Tom81]. Unless these interferences are removed, it is difficult for a physician to make a correct diagnosis. A common source of noise is the 60-Hz power lines whose radiated electric and magnetic fields are coupled to the ECG instrument through capacitive coupling and/or magnetic induction. Other sources of interference are the electromyographic signals that are the potentials developed by contracting muscles. These and other interferences can be removed with careful shielding and signal processing techniques.

Electroencephalogram (EEG) Signal

The summation of the electrical activity caused by the random firing of billions of individual neurons in the brain is represented by the EEG signal [Coh86], [Tom81]. In multiple EEG recordings, electrodes are placed at various positions on the scalp with two common electrodes placed on the earlobes, and potential differences between the various electrodes are recorded. A typical bandwidth of this type of EEG ranges from 0.5 to about 100 Hz, with the amplitudes ranging from 2 to 100 mV. An example of multiple EEG traces is shown in Figure 1.13.

Both frequency-domain and time-domain analyses of the EEG signal have been used for the diagnosis of epilepsy, sleep disorders, psychiatric malfunctions, and so forth. To this end, the EEG spectrum is subdivided into the following five bands: (1) the *delta* range, occupying the band from 0.5 to 4 Hz; (2) the

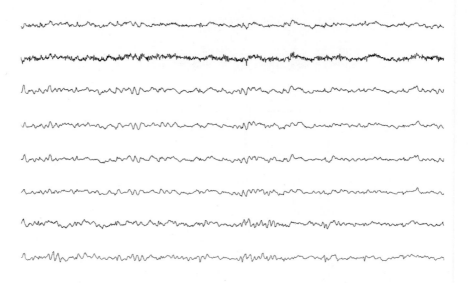

Figure 1.13: Multiple EEG signal traces.

theta range, occupying the band from 4 to 8 Hz; (3) the *alpha* range, occupying the band from 8 to 13 Hz; (4) the *beta* range, occupying the band from 13 to 22 Hz; and (5) the *gamma* range, occupying the band from 22 to 30 Hz.

The delta wave is normal in the EEG signals of children and sleeping adults. Since it is not common in alert adults, its presence indicates certain brain diseases. The theta wave is usually found in children even though it has been observed in alert adults. The alpha wave is common in all normal humans and is more pronounced in a relaxed and awake subject with closed eyes. Likewise, the beta activity is common in normal adults. The EEG exhibits rapid, low-voltage waves, called *rapid-eye-movement* (REM) waves, in a subject dreaming during sleep. Otherwise, in a sleeping subject, the EEG contains bursts of alpha-like waves, called *sleep spindles*. The EEG of an epileptic patient exhibits various types of abnormalities, depending on the type of epilepsy that is caused by uncontrolled neural discharges.

Seismic Signals

Seismic signals are caused by the movement of rocks resulting from an earthquake, a volcanic eruption, or an underground explosion [Bol93]. The ground movement generates elastic waves that propagate through the body of the earth in all directions from the source of movement. Three basic types of elastic waves are generated by the earth movement. Two of these waves propagate through the body of the earth, one moving faster with respect to the other. The faster moving wave is called the *primary* or *P-wave,* while the slower moving one is called the *secondary* or *S-wave.* The third type of wave is known as the *surface wave,* which moves along the ground surface. These seismic waves are converted into electrical signals by a seismograph and are recorded on a strip chart recorder or a magnetic tape.

Because of the three-dimensional nature of ground movement, a seismograph usually consists of three separate recording instruments that provide information about the movements in the two horizontal directions and one vertical direction and develops three records as indicated in Figure 1.14. Each such record is a one-dimensional signal. From the recorded signals, it is possible to determine the magnitude of the earthquake or nuclear explosion and the location of the source of the original earth movement.

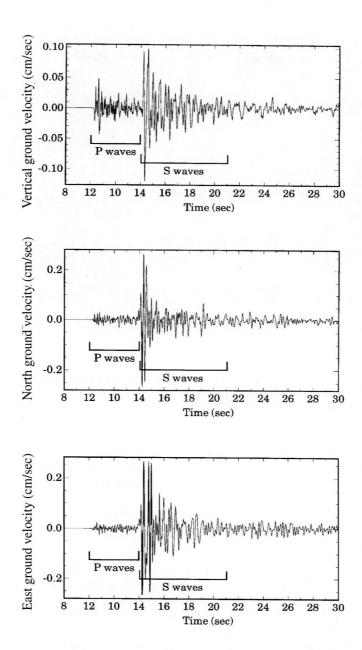

Figure 1.14: Seismograph record of the Northridge aftershock, January 29, 1994. Recorded at Stone Canyon Reservoir, Los Angeles, CA. (Courtesy of Institute for Crustal Research, University of California, Santa Barbara, CA.)

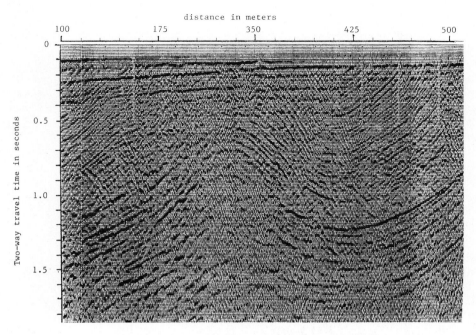

Figure 1.15: A typical seismic signal trace gather. (Courtesy of Institute for Crustal Research, University of California, Santa Barbara, CA.)

Seismic signals also play an important role in the geophysical exploration for oil and gas [Rob80]. In this type of application, linear arrays of seismic sources, such as high-energy explosives, are placed at regular intervals on the ground surface. The explosions cause seismic waves to propagate through the subsurface geological structures and reflect back to the surface from interfaces between geological strata. The reflected waves are converted into electrical signals by a composite array of geophones laid out in certain patterns and displayed as a two-dimensional signal that is a function of time and space, called a *trace gather,* as indicated in Figure 1.15. Before these signals are analyzed, some preliminary time and amplitude corrections are made on the data to compensate for different physical phenomena. From the corrected data, the time differences between reflected seismic signals are used to map structural deformations, whereas the amplitude changes usually indicate the presence of hydrocarbons.

Speech Signals

The acoustic theory of speech production has led to a range of mathematical models for the representation of speech signals. A speech signal is created by exciting the vocal tract using either quasi-periodic puffs of air or by creating turbulent air flow around a constriction in the vocal tract or by a mixture of these two sound sources [Del93], [Rab78]. So-called *voiced sounds* are generated when air is forced through the tensed glottis, causing it to vibrate in an oscillatory manner and generating pseudo-periodic pulses of air that excite the vocal tract. Included in the class of voiced sounds are vowels such as /I/ (as in 'big') or /ae/ (as in 'bad'); voiced consonants such as /b/, /d/, /g/, /m/, /n/ and so on; and so-called *liquids* and *glides* such as /w/, /l/, /r/, and /y/.[4] *Unvoiced sounds* are generated by forming a constriction at some point in the

---

[4]The sounds of a speech signal are usually represented pictorially using a "phonetic alphabet" by inserting the marker "/" on both sides of the letters representing the sound with uppercase letters representing voiced sounds that are stronger in amplitudes.

vocal tract, which causes the air flow to become turbulent (noise-like) and to act as the excitation source for sounds such as /f/, /s/, /sh/ and so forth. Finally, there is a class of sounds that utilizes both sources of excitation and hence has characteristics of both voiced sounds and unvoiced sounds. Among this class of sounds are the *voiced fricatives* such as /v/, /z/, and /zh/.

There are sounds, called *plosive sounds,* whose characteristics change dynamically during their production, such as /p/, /t/, and /k/. For these sounds, the vocal tract has a total constriction at some point along the tract (thereby totally blocking the flow of air and equivalently the production of sound in the vocal tract). Pressure builds up behind the constriction causing a sudden release of pressure as the constriction is removed, leading to a sound, followed by a gradual release of the built-up air flow.

Speech Demo 1

Figure 1.16(a) depicts the speech waveform of a male utterance "I like digital signal processing ." The total duration of the speech waveform shown is 3 s. Magnified versions of /I/ in the word "like" and the /S/ segment in the word "processing" are sketched in Figures 1.16(b) and (c), respectively. The slowly varying low-frequency voiced waveform of /I/ and the high-frequency unvoiced fricative waveform of /S/ are evident from the magnified waveforms. The voiced waveform in Figure 1.16(b) is seen to be quasi-periodic and can be modeled by a sum of a finite number of sinusoids. The lowest frequency of oscillation in this representation is called the *fundamental frequency* or *pitch frequency.* The unvoiced waveform in Figure 1.16(c) has no regular fine structure and is more noise-like.

One of the major applications of digital signal processing techniques is in the general area of speech processing. Problems in this area are usually divided into three groups: (1) speech analysis, (2) speech synthesis, and (3) speech analysis and synthesis [Opp78b]. Digital speech analysis methods are used in automatic speech recognition, speaker verification, and speaker identification. Applications of digital speech synthesis techniques include reading machines for the automatic conversion of written text into speech and retrieval of data from computers in speech form by remote access through terminals or telephones. One example belonging to the third group is voice scrambling for secure transmission. Speech data compression for an efficient use of the transmission medium is another example of the use of speech analysis followed by synthesis. A typical speech signal after conversion into a digital form contains about 64,000 bits per second (bps). Depending on the desired quality of the synthesized speech, the original data can be compressed considerably; for example, down to about 1000 bps.

Musical Sound Signal

The electronic synthesizer is an example of the use of modern signal processing techniques [Ler83], [Moo77]. The natural sound generated by most musical instruments is generally produced by mechanical vibrations caused by activating some form of oscillator that then causes other parts of the instrument to vibrate. All these vibrations together in a single instrument generate the musical sound. In a violin, the primary oscillator is a stretched piece of string (cat gut). Its movement is caused by drawing a bow across it; this sets the wooden body of the violin vibrating, which in turn sets up vibrations of the air inside as well as outside the instrument. In a piano, the primary oscillator is a stretched steel wire that is set into vibratory motion by the hitting of a hammer, which in turn causes vibrations in the wooden body (sounding board) of the piano. In wind or brass instruments, the vibration occurs in a column of air, and a mechanical change in the length of the air column by means of valves or keys regulates the rate of vibration.

The sound of orchestral instruments can be classified into two groups: *quasi-periodic* and *aperiodic.* Quasi-periodic sounds can be described by a sum of a finite number of sinusoids with independently varying amplitudes and frequencies. Figure 1.17(a) and (b) show the sound waveforms of two different instruments, the cello and the bass drum, respectively. In each figure, the top waveform is the plot of an entire isolated note, whereas the bottom plot shows an expanded version of a portion of the note: 10 ms for the cello and 80 ms for the bass drum. The waveform of the note from a cello is seen to be quasi-periodic. On the other hand, the bass drum waveform is clearly aperiodic. The tone of an orchestral instrument is commonly divided into three segments called the *attack* part, the *steady-state* part, and the *decay* part.

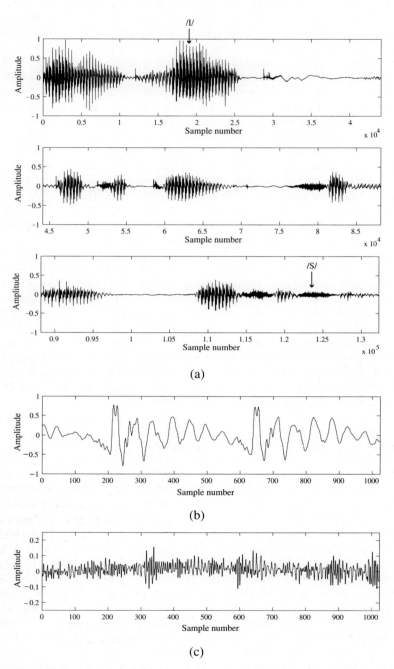

Figure 1.16: Speech waveform example: (a) sentence-length segment, (b) magnified version of the voiced segment (the letter /I/ in "like"), and (c) magnified version of the unvoiced segment (the letter /S/ in "processing").

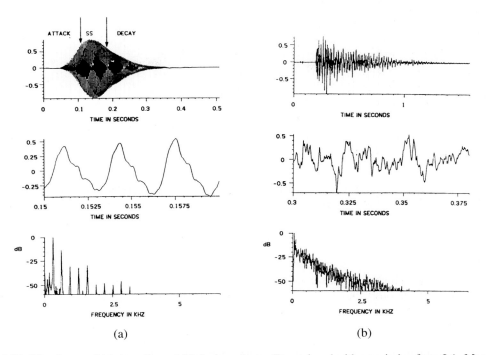

Figure 1.17: Waveforms of (a) the cello and (b) the bass drum. (Reproduced with permission from J A. Moorer, Signal processing aspects of computer music: A survey, *Proceedings of the IEEE*, vol. 65, August 1977, pp. 1108–1137 ©1977 IEEE.)

Figure 1.17 illustrates this division for the two tones. Note that the bass drum tone of Figure 1.17(b) shows no steady-state part. A reasonable approximation of many tones is obtained by splicing together these parts. However, high-fidelity reproduction requires a more complex model.

Time Series

The signals described thus far are continuous functions with time as the independent variable. In many cases, the signals of interest are naturally discrete functions of the independent variables. Often such signals are of finite duration. Examples of such signals are the yearly average number of sunspots, daily stock prices, the value of total monthly exports of a country, the yearly population of animal species in a certain geographical area, the annual yields per acre of crops in a country, and the monthly totals of international airline passengers over certain periods. This type of finite extent signal, usually called a *time series*, occurs in business, economics, physical sciences, social sciences, engineering, medicine, and many other fields. Plots of some typical time series are shown in Figures 1.18(a) and 1.18(b).

There are many reasons for analyzing a particular time series [Box70]. In some applications, there may be a need to develop a model to determine the nature of the dependence of the data on the independent variable and use it to forecast the future behavior of the series. As an example, in business planning, reasonably accurate sales forecasts are necessary. Some types of series possess seasonal or periodic components, and it is important to extract these components. The study of sunspot numbers is important for predicting climate variations. Invariably, the time series data are noisy, and their representations require models based on their statistical properties.

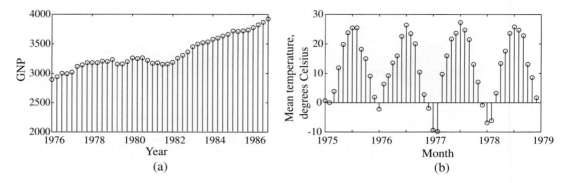

Figure 1.18: (a) Seasonally adjusted quarterly Gross National Product of the United States in 1982 dollars from 1976 to 1986 (adapted from [Lüt91]). (b) Monthly mean St. Louis, Missouri, temperature in degrees Celsius for the years 1975 to 1978 (adapted from [Mar87]).

Images

As indicated earlier, an image is a two-dimensional signal whose intensity at any point is a function of two spatial variables. Common examples are photographs, still video images, radar and sonar images, and chest and dental X-rays. An image sequence, such as that seen in a television, is essentially a three-dimensional signal for which the image intensity at any point is a function of three variables: two spatial variables and time. Figure 1.19(a) shows the photograph of a digital image.

The basic problems in image processing are image signal representation and modeling, enhancement, restoration, reconstruction from projections, analysis, and coding [Jai89].

Each picture element in a specific image represents a certain physical quantity; a characterization of the element is called the *image representation*. For example, a photograph represents the luminances of various objects as seen by the camera. An infrared image taken by a satellite or an airplane represents the temperature profile of a geographical area. Depending on the type of image and its applications, various types of image models are usually defined. Such models are also based on perception and on local or global characteristics. The nature and performance of the image processing algorithms depend on the image model being used.

Image enhancement algorithms are used to emphasize specific image features to improve the quality of the image for visual perception or to aid in the analysis of the image for feature extraction. These include methods for contrast enhancement, edge detection, sharpening, linear and nonlinear filtering, zooming, and noise removal. Figure 1.19(b) shows the contrast-enhanced version of the image of Figure 1.19(a), developed using a nonlinear filter [Thu2000].

The algorithms used for elimination or reduction of degradations in an image, such as blurring and geometric distortion caused by the imaging system and/or its surroundings, are known as *image restoration*. *Image reconstruction* from projections involves the development of a two-dimensional image slice of a three-dimensional object from a number of planar projections obtained from various angles. By creating a number of contiguous slices, a three-dimensional image giving an inside view of the object is developed.

Image analysis methods are employed to develop a quantitative description and classification of one or more desired objects in an image.

For digital processing, an image needs to be sampled and quantized using an analog-to-digital converter. A reasonable size digital image in its original form takes a considerable amount of memory space for storage. For example, an image of size 512×512 samples with 8-bit resolution per sample contains over 2 million bits. *Image coding* methods are used to reduce the total number of bits in an image without any

(a)   (b)

Figure 1.19: (a) A digital image and (b) its contrast-enhanced version. (Reproduced with permission from *Nonlinear Image Processing*, S.K. Mitra and G. Sicuranza, Eds., Academic Press, New York, ©2000.)

degradation in visual perception quality as in speech coding; for example, down to about 1 bit per sample on the average.

1.4 Typical Signal Processing Applications[5]

There are numerous applications of signal processing that we often encounter in our daily life without being aware of them. Originally the signal processing algorithms used in these applications were carried out in the continuous-time domain. However, they are now being increasingly implemented using discrete-signal processing algorithms. Due to space limitations, it is not possible to discuss all of these applications. However, an overview of selected applications is presented.

1.4.1 Sound Recording Applications

The recording of most musical programs nowadays is usually made in an acoustically inert studio. The sound from each instrument is picked up by its own microphone closely placed to the instrument and is recorded on a single track in a multitrack tape recorder containing as many as 48 tracks. The signals from individual tracks in the master recording are then edited and combined by the sound engineer in a *mix-down* system to develop a two-track stereo recording. There are a number of reasons for following this approach. First, the closeness of each individual microphone to its assigned instrument provides a high degree of separation between the instruments and minimizes the background noise in the recording. Second, the sound part of one instrument can be rerecorded later if necessary. Third, during the mix-down process, the sound engineer can manipulate individual signals by using a variety of signal processing devices to alter the musical balances between the sounds generated by the instruments, can change the timbre, and can add natural room acoustics effects and other special effects [Ble78], [Ear76].

Various types of signal processing techniques are utilized in the mix-down phase. Some are used to modify the spectral characteristics of the sound signal and to add special effects, whereas others are used to improve the quality of the transmission medium. The signal processing circuits most commonly used are (1) compressors and limiters, (2) expanders and noise gates, (3) equalizers and filters, (4) noise reduction

[5]This section has been adapted from *Handbook for Digital Signal Processing,* Sanjit K. Mitra and James F. Kaiser, Eds., ©1993, John Wiley & Sons. Adapted by permission of John Wiley & Sons.

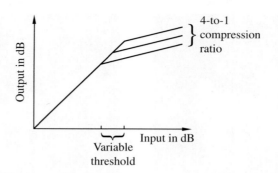

Figure 1.20: Transfer characteristic of a typical compressor.

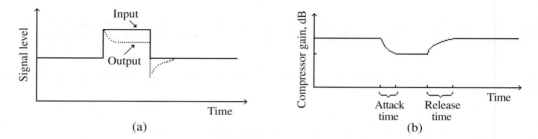

Figure 1.21: Parameters characterizing a typical compressor.

systems, (5) delay and reverberation systems, and (6) circuits for special effects [Ble78], [Ear76], [Hub89], [Wor89]. These operations are usually performed on the original analog audio signals and are implemented using analog circuit components. However, there is a growing trend toward all digital implementation and its use in the processing of the digitized versions of the analog audio signals [Ble78].

Compressors and limiters. These devices are used for the compression of the dynamic range of an audio signal. The compressor can be considered as an amplifier with two gain levels: the gain is unity for input signal levels below a certain threshold and less than unity for signals with levels above the threshold. The threshold level is adjustable over a wide range of the input signal. Figure 1.20 shows the transfer characteristic of a typical compressor.

The parameters characterizing a compressor are its compression ratio, threshold level, attack time, and release time, which are illustrated in Figure 1.21.

When the input signal level suddenly rises above a prescribed threshold, the time taken by the compressor to adjust its normal unity gain to the lower value is called the *attack time.* Because of this effect, the output signal exhibits a slight degree of overshoot before the desired output level is reached. A zero attack time is desirable to protect the system from sudden high-level transients. However, in this case, the impact of sharp musical attacks is eliminated, resulting in a dull "lifeless" sound [Wor89]. A longer attack time causes the output to sound more percussive than normal.

Similarly, the time taken by the compressor to reach its normal unity gain value when the input level suddenly drops below the threshold is called the *release time* or *recovery time.* If the input signal fluctuates rapidly around the threshold in a small region, the compressor gain also fluctuates up and down. In such a situation, the rise and fall of background noise results in an audible effect called *breathing* or *pumping,* which can be minimized with a longer release time for the compressor gain.

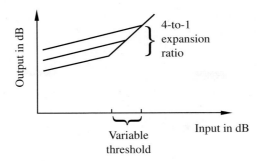

Figure 1.22: Transfer characteristic of a typical expander.

There are various applications of the compressor unit in musical recording [Ear76]. For example, it can be used to eliminate variations in the peaks of an electric bass output signal by clamping them to a constant level, thus providing an even and solid bass line. To maintain the original character of the instrument, it is necessary to use a compressor with a long recovery time compared to the natural decay rate of the electric bass. The device is also useful to compensate for the wide variations in the signal level produced by a singer who moves frequently, changing the distance from the microphone.

A compressor with a compression ratio of 10-to-1 or greater is called a *limiter* because its output levels are essentially clamped to the threshold level. The limiter is used to prevent overloading of amplifiers and other devices caused by signal peaks exceeding certain levels.

Expanders and noise gates.   The expander's function is opposite that of the compressor. It is also an amplifier with two gain levels: the gain is unity for input signal levels above a certain threshold and less than unity for signals with levels below the threshold. The threshold level is again adjustable over a wide range of the input signal. Figure 1.22 shows the transfer characteristic of a typical expander. The *expander* is used to expand the dynamic range of an audio signal by boosting the high-level signals and attenuating the low-level signals. The device can also be used to reduce noise below a threshold level.

The expander is characterized by its expansion ratio, threshold level, attack time, and release time. Here, the time taken by the device to reach the normal unity gain for a sudden change in the input signal to a level above the threshold is defined as the *attack time*. Likewise, the time required by the device to lower the gain from its normal value of one for a sudden decrease in the input signal level is called the *release time*.

The noise gate is a special type of expander that heavily attenuates signals with levels below the threshold. It is used, for example, to totally cut off a microphone during a musical pause so as not to pass the noise being picked up by the microphone.

Equalizers and filters.   Various types of filters are used to modify the frequency response of a recording or the monitoring channel. One such filter, called the *shelving filter,* provides boost (rise) or cut (drop) in the frequency response at either the low or at the high end of the audio frequency range while not affecting the frequency response in the remaining range of the audio spectrum, as shown in Figure 1.23. *Peaking filters* are used for midband equalization and are designed to have either a bandpass response to provide a boost or a bandstop response to provide a cut, as indicated in Figure 1.24.

The parameters characterizing a low-frequency shelving filter are the two frequencies f_{1L} and f_{2L}, where the magnitude response begins tapering up or down from a constant level and the low-frequency gain levels in dB. Likewise, the parameters characterizing a high-frequency shelving filter are the two frequencies f_{1H} and f_{2H}, where the magnitude response begins tapering up or down from a constant level

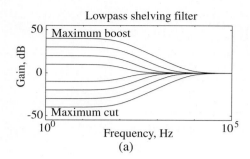
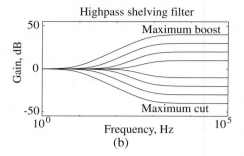

Figure 1.23: Frequency responses of (a) low-frequency shelving filter and (b) high-frequency shelving filter.

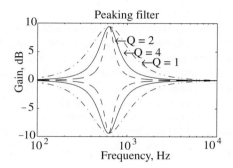

Figure 1.24: Peaking filter frequency response.

and the high-frequency gain levels in dB. In the case of a peaking filter, the parameters of interest are the center frequency f_0, the 3-dB bandwidth Δf of the bell-shaped curve, and the gain level at the center frequency. Most often, the quality factor $Q = f_0/\Delta f$ is used to characterize the shape of the frequency response instead of the bandwidth Δf.

A typical equalizer consists of a cascade of a low-frequency shelving filter, a high-frequency shelving filter, and three or more peaking filters with adjustable parameters to provide adjustment of the overall equalizer frequency response over a broad range of frequencies in the audio spectrum. In a *parametric equalizer*, each individual parameter of its constituent filter blocks can be varied independently without affecting the parameters of the other filters in the equalizer.

The *graphic equalizer* consists of a cascade of peaking filters with fixed center frequencies but adjustable gain levels that are controlled by vertical slides in the front panel. The physical position of the slides reasonably approximates the overall equalizer magnitude response, as shown schematically in Figure 1.25.

Other types of filters that also find applications in the musical recording and transfer processes are the *lowpass, highpass,* and *notch filters.* Their corresponding frequency responses are indicated in Figure 1.26. The notch filter is designed to attenuate a particular frequency component and has a narrow notch width so as not to affect the rest of the musical program.

Two major applications of equalizers and filters in recording are to correct certain types of problems that may have occurred during the recording or the transfer process and to alter the harmonic or timbral contents of a recorded sound purely for musical or creative purposes [Ear76]. For example, a direct transfer of a musical recording from old 78 rpm disks to a wideband playback system will be highly noisy due to the limited bandwidth of the old disks. To reduce this noise, a bandpass filter with a passband matching the bandwidth of the old records is utilized. Often, older recordings are made more pleasing by adding a

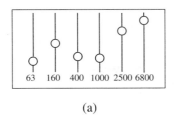
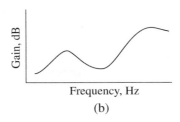

(a)   (b)

Figure 1.25: Graphic equalizer: (a) control panel settings and (b) corresponding frequency response. (Adapted from [Ear76].)

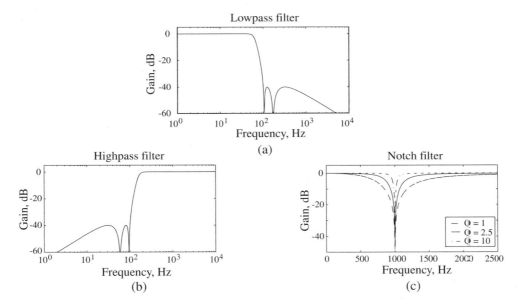

Figure 1.26: Frequency responses of other types of filters: (a) lowpass filter, (b) highpass filter, and (c) notch filter.

broad high-frequency peak in the 5- to 10-kHz range and by shelving out some of the lower frequencies. The notch filter is particularly useful in removing 60-Hz power supply hum.

In creating a program by mixing down a multichannel recording, the recording engineer usually employs equalization of individual tracks for creative reasons [Ear76]. For example, a "fullness" effect can be added to weak instruments such as the acoustical guitar by boosting frequency components in the range 100 to 300 Hz. Similarly, by boosting the 2- to 4-kHz range, the transients caused by the fingers against the string of an acoustical guitar can be made more pronounced. A high-frequency shelving boost above the 1- to 2-kHz range increases the "crispness" in percussion instruments such as the bongo or snare drums.

Noise reduction system.   The overall dynamic range of human hearing is over 120 dB. However, most recording and transmission mediums have a much smaller dynamic range. The music to be recorded must be above the sound background or noise. If the background noise is around 30 dB, the dynamic range available for the music is only 90 dB, requiring dynamic range compression for noise reduction.

A noise reduction system consists of two parts. The first part provides the compression during the recording mode, while the second part provides the complementary expansion during the playback mode.

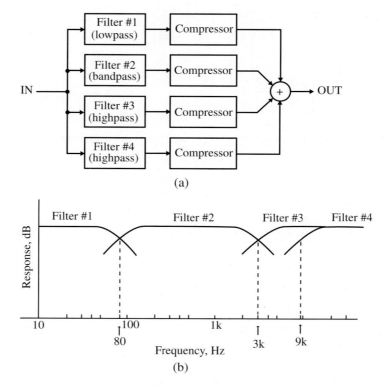

Figure 1.27: The Dolby A-type noise reduction scheme for the recording mode. (a) Block diagram and (b) frequency responses of the four filters with cutoff frequencies as shown.

To this end, the most popular methods in musical recording are the Dolby noise reduction schemes, of which there are several types [Ear76], [Hub89], [Wor89].

In the Dolby A-type method used in professional recording, for the recording mode, the audio signal is split into four frequency bands by a bank of four filters; separate compression is provided in each band and the outputs of the compressors are combined, as indicated in Figure 1.27. Moreover, the compression in each band is restricted to a 20-dB input range from -40 to -20 dB. Below the lower threshold (-40 dB), very low level signals are boosted by 10 dB, and above the upper threshold (-20 dB), the system has unity gain, passing the high-level signals unaffected. The transfer characteristic for the record mode is thus as shown in Figure 1.28.

In the playback mode, the scheme is essentially the same as that in the recording mode, except here the compressors are replaced by expanders with complementary transfer characteristics, as indicated in Figure 1.28. Here, the expansion is limited to a 10-dB input range from -30 to -20 dB. Above the upper threshold (-20 dB), very high level signals are cut by 10 dB, while below the lower threshold (-30 dB), the system has unity gain passing the low-level signals unaffected.

Note that for each band, a 2-to-1 compression is followed by a 1-to-2 complementary expansion such that the dynamic range of the signal at the input of the compressor is exactly equal to that at the expander output. This type of overall signal processing operation is often called *companding*. Moreover, the companding operation in one band has no effect on a signal in another band and may often be masked by other bands with no companding.

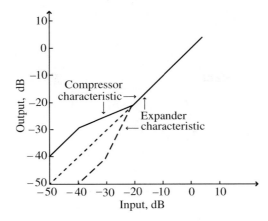

Figure 1.28: Compressor and expander transfer characteristic for the Dolby A-type noise reduction scheme.

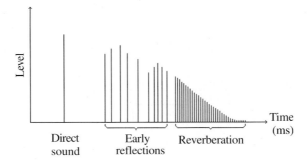

Figure 1.29: Various types of echoes generated by a single sound source in a room.

Delay and reverberation systems. Music generated in an inert studio does not sound natural compared to the music performed inside a room, such as a concert hall. In the latter case, the sound waves propagate in all directions and reach the listener from various directions and at various times, depending on the distance travelled by the sound waves from the source to the listener. The sound wave coming directly to the listener, called the *direct sound,* reaches first and determines the listener's perception of the location, size, and nature of the sound source. This is followed by a few closely spaced echoes, called *early reflections,* generated by reflections of sound waves from all sides of the room and reaching the listener at irregular times. These echoes provide the listener's subconscious cues as to the size of the room. After these early reflections, more and more densely packed echoes reach the listener due to multiple reflections. The latter group of echoes is referred to as the *reverberation.* The amplitude of the echoes decay exponentially with time as a result of attenuation at each reflection. Figure 1.29 illustrates this concept. The period of time in which the reverberation falls by 60 dB is called the *reverberation time.* Since the absorption characteristics of different materials are not the same at different frequencies, the reverberation time varies from frequency to frequency.

Delay systems with adjustable delay factors are employed to artificially create the early reflections. Electronically generated reverberation combined with artificial echo reflections are usually added to the recordings made in a studio. The block diagram representation of a typical delay-reverberation system in a monophonic system is depicted in Figure 1.30.

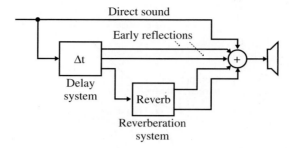

Figure 1.30: Block diagram of a complete delay-reverberation system in a monophonic system.

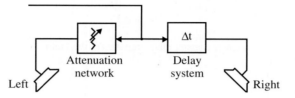

Figure 1.31: Localization of sound source using delay systems and attenuation network.

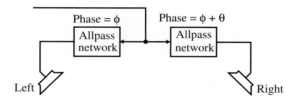

Figure 1.32: Sound broadening using allpass networks.

There are various other applications of electronic delay systems, some of which are described next [Ear76].

Special effects.   By feeding in the same sound signal through an adjustable delay and gain control, as indicated in Figure 1.31, it is possible to vary the localization of the sound source from the left speaker to the right for a listener located on the plane of symmetry. For example, in Figure 1.31, a 0-dB loss in the left channel and a few milliseconds delay in the right channel give the impression of a localization of the sound source at the left. However, lowering of the left-channel signal level by a few-dB loss results in a phantom image of the sound source moving toward the center. This scheme can be further extended to provide a degree of *sound broadening* by phase shifting one channel with respect to the other through allpass networks[6] as shown in Figure 1.32.

Another application of the delay-reverberation system is in the processing of a single track into a pseudo-stereo format while simulating a natural acoustical environment, as illustrated in Figure 1.33.

The delay system can also be used to generate a chorus effect from the sound of a soloist. The basic scheme used is illustrated in Figure 1.34. Each of the delay units has a variable delay controlled by a low-frequency, pseudo-random noise source to provide a random pitch variation [Ble78].

---

[6]An allpass network is characterized by a magnitude spectrum that is equal to one for all frequencies.

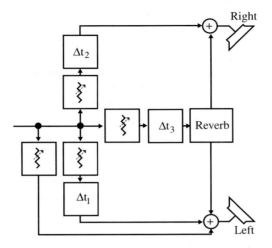

Figure 1.33: A possible application of delay systems and reverberation in a stereophonic system.

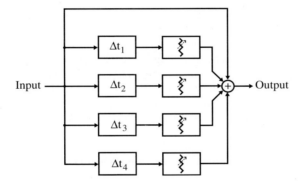

Figure 1.34: A scheme for implementing chorus effect.

It should be pointed out here that additional signal processing is employed to make the stereo submaster developed by the sound engineer more suitable for the record-cutting lathe or the cassette tape duplicator.

1.4.2 Telephone Dialing Applications

Signal processing plays a key role in the detection and generation of signaling tones for push-button telephone dialing [Dar76]. In telephones equipped with TOUCH-TONE® dialing, the pressing of each button generates a unique set of two-tone signals, called *dual-tone multifrequency (DTMF) signals*, that are processed at the telephone central office to identify the number pressed by determining the two associated tone frequencies. Seven frequencies are used to code the 10 decimal digits and the two special buttons marked " * " and " # ". The low-band frequencies are 697 Hz, 770 Hz, 852 Hz, and 941 Hz. The remaining three frequencies belonging to the highband are 1209 Hz, 1336 Hz, and 1477 Hz. The fourth high-band frequency of 1633 Hz is not presently in use and has been assigned for future applications to permit the use of four additional push-buttons for special services. The frequency assignments used in the TOUCH-TONE® dialing scheme are shown in Figure 1.35 [ITU84].

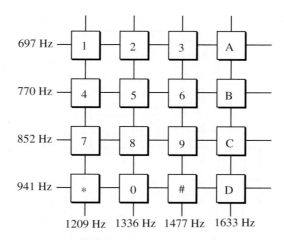

Figure 1.35: The tone frequency assignments for TOUCH-TONE® dialing.

The scheme used to identify the two frequencies associated with the button that has been pressed is shown in Figure 1.36. Here, the two tones are first separated by a lowpass and a highpass filter. The passband cutoff frequency of the lowpass filter is slightly above 1000 Hz, whereas that of the highpass filter is slightly below 1200 Hz. The output of each filter is next converted into a square wave by a limiter and then processed by a bank of bandpass filters with narrow passbands. The four bandpass filters in the low-frequency channel have center frequencies at 697 Hz, 770 Hz, 852 Hz, and 941 Hz. The four bandpass filters in the high-frequency channel have center frequencies at 1209 Hz, 1336 Hz, 1477 Hz, and 1633 Hz. The detector following each bandpass filter develops the necessary dc switching signal if its input voltage is above a certain threshold.

All the signal processing functions described above are usually implemented in practice in the analog domain. However, increasingly, these functions are being implemented using digital techniques.[7]

1.4.3 FM Stereo Applications

For wireless transmission of a signal occupying a low-frequency range, such as an audio signal, it is necessary to transform the signal to a high-frequency range by modulating it onto a high-frequency carrier. At the receiver, the modulated signal is demodulated to recover the low-frequency signal. The signal processing operations used for wireless transmission are modulation, demodulation, and filtering. Two commonly used modulation schemes for radio are amplitude modulation (AM) and frequency modulation (FM).

We next review the basic idea behind the FM stereo broadcasting and reception scheme as used in the United States [Cou83]. An important feature of this scheme is that at the receiving end, the signal can be heard over a standard monaural FM radio with a single speaker or over a stereo FM radio with two speakers. The system is based on the frequency-division multiplexing (FDM) method described earlier in Section 1.2.5.

The block diagram representations of the FM stereo transmitter and the receiver are shown in Figure 1.37(a) and (b), respectively. At the transmitting end, the sum and the difference of the left and right channel audio signals, $s_L(t)$ and $s_R(t)$, are first formed. Note that the summed signal $s_L(t) + s_R(t)$ is used in monaural FM radio. The difference signal $s_L(t) - s_R(t)$ is modulated using the double-sideband

---
[7]See Section 14.1.

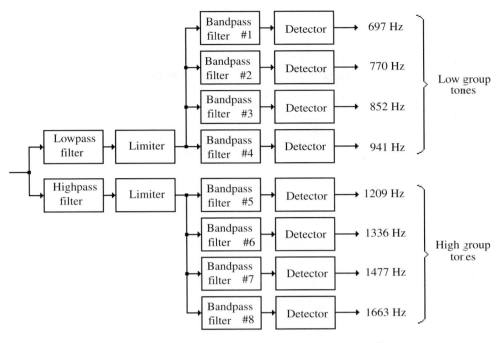

Figure 1.36: The tone detection scheme for TOUCH-TONE® dialing.

suppressed carrier (DSB-SC) scheme[8] using a subcarrier frequency f_{sc} of 38 kHz. The summed signal, the modulated difference signal, and a 19-kHz pilot tone signal are then added, developing the composite baseband signal $s_B(t)$. The spectrum of the composite signal is shown in Figure 1.37(c). The baseband signal is next modulated onto the main carrier frequency f_c using the frequency modulation method. At the receiving end, the FM signal is demodulated to derive the baseband signal $s_B(t)$, which is then separated into the low-frequency summed signal and the modulated difference signal, using a lowpass filter and a bandpass filter. The cutoff frequency of the lowpass filter is around 15 kHz, whereas the center frequency of the bandpass filter is at 38 kHz. The 19-kHz pilot tone is used in the receiver to develop the 38-kHz reference signal for a coherent subcarrier demodulation for recovering the audio difference signal. The sum and difference of the two audio signals create the desired left audio and right audio signals.

1.4.4 Musical Sound Synthesis

The generation of the sound of a musical instrument using electronic circuits or software algorithms is another example of the application of signal processing methods. There are basically four methods of musical sound synthesis: (1) *wavetable synthesis,* (2) *spectral modelling synthesis,* (3) *nonlinear synthesis,* and (4) *physical modelling synthesis* [Rab2001], [Smi91].

Wavetable Synthesis

In this widely used method, prerecorded database of musical sounds, called *wavetable,* stored in memory are played back when needed. The stored sounds contain both the attack and sustain portions. In the

Music Demo 1

---

[8]See Section 1.2.4.

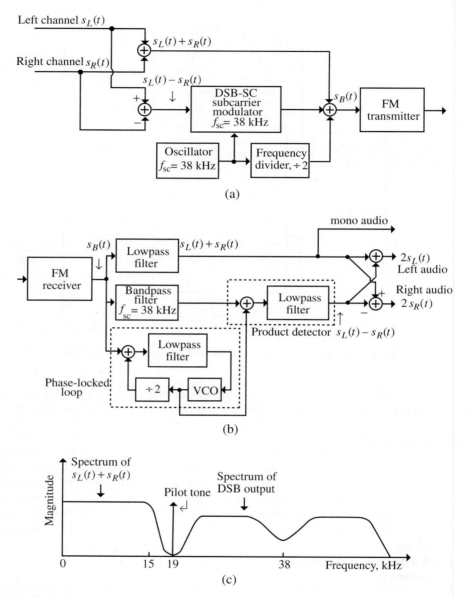

Figure 1.37: The FM stereo system: (a) transmitter, (b) receiver, and (c) spectrum of the composite baseband signal $s_B(t)$.

playback mode, a number of techniques are employed to produce variations in the sound reproduction. Typical techniques employed are pitch shifting, looping, enveloping, and filtering.

Pitch shifting is used to play a recorded tone at various pitches. To avoid excessive memory requirements, usually a musical note is recorded at a limited number of frequencies. A note with a frequency not recorded is then generated by varying the pitch of a stored note with the closest frequency.

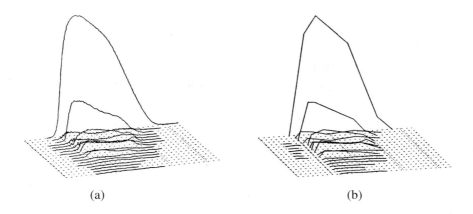

(a)   (b)

Figure 1.38: (a) Perspective plot of the amplitude functions $A_k(t)$ for an actual note from a clarinet. (b) A piecewise-linear approximation to the amplitude functions of Figure 1.38(a). (Reproduced with permission from J. A. Moorer, Signal processing aspects of computer music: A survey, *Proceedings of the IEEE*, vol. 65, August 1977, pp. 1108–1137 ©1977 IEEE.)

Looping is used to extend the playback time of a recorded note. It is implemented by reading out recursively the stored data.

As the attack, decay, sustain, and release portions of the envelope of a recorded note are often lost due to looping, they can be regenerated or changed by *enveloping*. It is implemented by changing the amplitude of the recorded note with a time-varying gain function.

Filtering is used to modify the spectral properties of a tone. It provides an improved time-dependent control of the recorded sound and is often used to amplitude change caused by enveloping.

Spectral Modelling Synthesis

The basis of this approach to music synthesis is the following representation of the sound signal $s(t)$ [Moo77], [All80]:

$$s(t) = \sum_{k=1}^{N} A_k(t) \sin\left(2\pi f_k(t)t\right), \tag{1.19}$$

Music Demo 2

where $A_k(t)$ is the the time-varying amplitude of the basis function $\sin\left(2\pi f_k(t)t\right)$ of the kth component of the signal with a frequency $f_k(t)$ varying with time. The frequency function $f_k(t)$ varies slowly with time. For an instrument playing an isolated tone, $f_k(t) = kf_o$; that is,

$$s(t) = A_k(t) \sin(2\pi k f_o t), \tag{1.20}$$

where f_o is called the *fundamental frequency*. In a musical sound with many tones, all other frequencies are usually integer multiples of the fundamental and are called *partial frequencies*, also called *harmonics*. Figure 1.38(a) shows, for example, the perspective plot of the amplitude functions as a function of time of 17 partial frequency components of an actual note from a clarinet. The aim of the synthesis is to produce electronically the $A_k(t)$ and $f_k(t)$ functions.

The partial frequency components are generated independently by oscillators with time-varying oscillation frequencies. The amplitudes of the required signals are then individually modified, approximating the actual variations obtained by analysis and combined (i.e., *added*) to produce the desired sound signal.

For example, a piecewise-linear approximation of the clarinet note of Figure 1.38(a) is sketched in Figure 1.38(b) and can be used to generate a reasonable replica of the note. Usually, some alterations to the amplitude and frequency functions may be needed before the music that is generated sounds as close as possible to that of the original instrument.

A recent variation to the additive synthesis is the *granular synthesis*. In this method, the basis functions of Eq. (1.19) are concentrated in frequency and time and are called *grains* or *atoms*. These basis functions are generated in one of several ways, such as windowed segments of sine waves, from wavetables, with wavelet expansions.[9]

Nonlinear Synthesis

Music Demo 3

A fairly simple nonlinear synthesis method is based on *frequency modulation* (FM). In its barest form, the sound signal $s(t)$ is generated by modulating the carrier signal $\sin(2\pi f_0 t)$ by the modulator $\phi(t)$, resulting in

$$s(t) = A(t)\sin(2\pi f_0 t + \phi(t)). \tag{1.21}$$

For a sinusoidal modulator

$$\phi(t) = \kappa \sin(2\pi f_m t), \tag{1.22}$$

the modulated carrier $s(t)$ contains sinusoidal components of frequencies $f_0 + n f_m$, $n = 0, 1, 2, \ldots$. The amplitudes of the sinusoidal components can be varied using the modulation index κ. A small value of the modulation frequency f_m combined with a high modulation index κ has been found to produce rich sounds. However, this method cannot be used to generate sounds of natural musical instruments due to the availability of a few adjustable parameters. In spite of this difficulty, the FM-based approach is an inexpensive method and often employed in sound cards used in personal computers and in synthesizers.

Physical Modelling Synthesis

Music Demo 4

In this method, the sound is generated by directly modelling the mechanism behind the sound production. It results in partial differential equations providing a physical description of the major vibrating structures of the instrument. Usually, the methods make use of the wave equation describing wave propagation in solids and in air. The partial differential equations can be solved using either finite difference methods or digital waveguides or using transfer function models.

1.4.5   Echo Cancellation in Telephone Networks

In a telephone network, the central offices perform the necessary switching to connect two subscribers [Dut80], [Fre78], [Mes84]. For economic reasons, a *two-wire* circuit is used to connect a subscriber to his/her central office, whereas the central offices are connected using *four-wire* circuits. A two-wire circuit is bidirectional and carries signals in both directions. A four-wire circuit uses two separate unidirectional paths for signal transmission in both directions. The latter is preferred for long-distance trunk connections since signals at intermediate points in the trunk can be equalized and amplified using repeaters and, if necessary, multiplexed easily. A hybrid coil in the central office provides the interface between a two-wire circuit and a four-wire circuit, as shown in Figure 1.39. The hybrid circuit ideally should provide a perfect impedance match to the two-wire circuit by impedance balancing so that the incoming four-wire receive signal is passed directly to the two-wire circuit connected to the hybrid with no portion appearing in the four-wire transmit path. However, to save cost, a hybrid coil is shared among several subscribers. Thus, it is not possible to provide a perfect impedance match in every case since the length of the subscriber lines

---

[9]See Section 13.1.3.

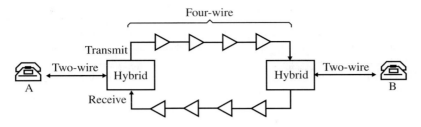

Figure 1.39: Basic 2/4-wire interconnection scheme.

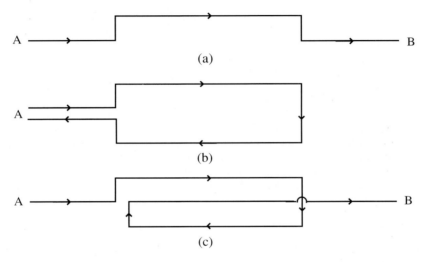

Figure 1.40: Various signal paths in a telephone network. (a) Transmission path from talker A to listener B, (b) echo path for talker A, and (c) echo path for listener B.

vary. The resulting imbalance causes a large portion of the incoming received signal from the distance talker to appear in the transmit path, and it is returned to the talker as an echo. Figure 1.40 illustrates the normal transmission between a talker and a listener, as well as two possible major echo paths.

The effect of the echo can be annoying to the talker, depending on the amplitude and delay of the echo, that is, on the length of the trunk circuit. The effect of the echo is worst for telephone networks involving geostationary satellite circuits, where the echo delay is about 540 ms.

Several methods are followed to reduce the effect of the echo. In trunk circuits up to 3000 km in length, adequate reduction of the echo is achieved by introducing additional signal loss in both directions of the four-wire circuit. In this scheme, an improvement in the signal-to-echo ratio is realized since the echo undergoes loss in both directions, while the signals are attenuated only once.

For distances greater than 3000 km, echoes are controlled by means of an echo suppressor inserted in the trunk circuit, as indicated in Figure 1.41(a). The device is essentially a voice-activated switch implementing two functions. It first detects the direction of the conversation and then blocks the opposite path in the four-wire circuit. Even though it introduces distortion when both subscribers are talking by clipping parts of the speech signal, the echo suppressor has provided a reasonably acceptable solution for terrestrial transmission.

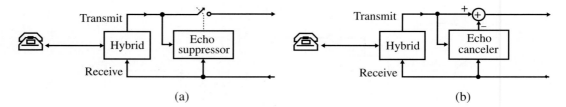

Figure 1.41: (a) Echo suppression scheme and (b) echo cancellation scheme.

Figure 1.42: Scheme for the digital processing of an analog signal.

For telephone conversation involving satellite circuits, an elegant solution is based on the use of an echo canceler. The circuit generates a replica of the echo using the signal in the receive path and subtracts it from the signal in the transmit path, as indicated in Figure 1.41(b). Basically, it is an adaptive filter structure whose parameters are adjusted using certain adaptation algorithms until the residual signal is satisfactorily minimized.[10] Typically, an echo reduction of about 40 dB is considered satisfactory in practice. To eliminate the problem generated when both subscribers are talking, the adaptation algorithm is disabled when the signal in the transmit path contains both the echo and the signal generated by the speaker closer to the hybrid coil.

1.5   Why Digital Signal Processing?[11]

In some sense, the origin of digital signal processing techniques can be traced back to the seventeenth century when finite difference methods, numerical integration methods, and numerical interpolation methods were developed to solve physical problems involving continuous variables and functions. The more recent interest in digital signal processing arose in the 1950s with the availability of large digital computers. Initial applications were primarily concerned with the simulation of analog signal processing methods. Around the beginning of the 1960s, researchers began to consider digital signal processing as a separate field by itself. Since then, there have been significant and myriad developments and breakthroughs in both theory and applications of digital signal processing.

Digital processing of an analog signal consists basically of three steps: conversion of the analog signal into a digital form; processing of the digital version; and finally, conversion of the processed digital signal back into an analog form. Figure 1.42 shows the overall scheme in a block diagram form.

Since the amplitude of the analog input signal varies with time, a *sample-and-hold* (S/H) circuit is used first to sample the analog input at periodic intervals and hold the sampled value constant at the input of the *analog-to-digital* (A/D) *converter* to permit accurate digital conversion. The input to the A/D converter is a staircase-type analog signal if the S/H circuit holds the sampled value until the next sampling instant. The output of the A/D converter is a binary data stream that is next processed by the digital processor implementing the desired signal processing algorithm. The output of the digital processor, another binary

---

[10]For a review of adaptive filtering methods, see [Cio93].

[11]This section has been adapted from *Handbook for Digital Signal Processing*, Sanjit K. Mitra and James F. Kaiser, Eds., ©1993, John Wiley & Sons. Adapted by permission of John Wiley & Sons.

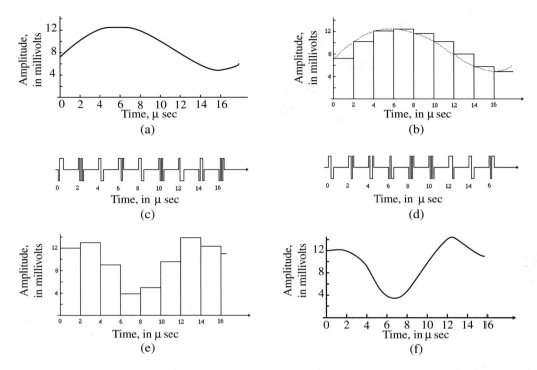

Figure 1.43: Typical waveforms of signals appearing at various stages in Figure 1.42. (a) Analog input signal, (b) output of the S/H circuit, (c) A/D converter output, (d) output of the digital processor, (e) D/A converter output, and (f) analog output signal. In (c) and (d), the digital HIGH and LOW levels are shown as positive and negative pulses for clarity.

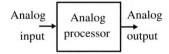

Figure 1.44: Analog processing of analog signals.

data stream, is then converted into a staircase-type analog signal by the *digital-to-analog* (D/A) *converter*. The lowpass filter at the output of the D/A converter then removes all undesired high-frequency components and delivers at its output the desired processed analog signal. Figure 1.43 illustrates the waveforms of the pertinent signals at various stages in the above process, where for clarity the two levels of the binary signals are shown as a positive and a negative pulse, respectively.

In contrast to the above, a direct analog processing of an analog signal is conceptually much simpler since it involves only a single processor, as illustrated in Figure 1.44. It is therefore natural to ask what the advantages are of digital processing of an analog signal.

There are of course many advantages in choosing digital signal processing. The most important ones are discussed next [Bel2000], [Pro96].

Unlike analog circuits, the operation of digital circuits does not depend on precise values of the digital signals. As a result, a digital circuit is less sensitive to tolerances of component values and is fairly

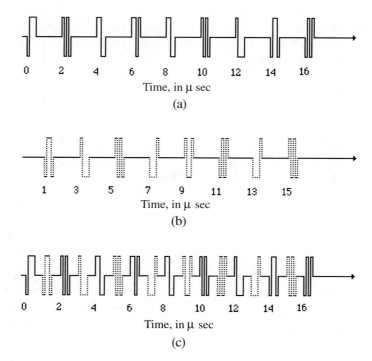

Figure 1.45: Illustration of the time-sharing concept. The signal shown in (c) has been obtained by time-multiplexing the signals shown in (a) and (b).

independent of temperature, aging, and most other external parameters. A digital circuit can be reproduced easily in volume quantities and does not require any adjustments either during construction or later while in use. Moreover, it is amenable to full integration, and with the recent advances in *very large scale integrated* (VLSI) circuits, it has been possible to integrate highly sophisticated and complex digital signal processing systems on a single chip.

In a digital processor, the signals and the coefficients describing the processing operation are represented as binary words. Thus, any desirable accuracy can be achieved by simply increasing the wordlength, subject to cost limitation. Moreover, the dynamic ranges for signals and coefficients can be increased still further by using floating-point arithmetic if necessary.

Digital processing allows the sharing of a given processor among a number of signals by timesharing, thus reducing the cost of processing per signal. Figure 1.45 illustrates the concept of timesharing, where two digital signals are combined into one by time-division multiplexing. The multiplexed signal can then be fed into a single processor. By switching the processor coefficients prior to the arrival of each signal at the input of the processor, the processor can be made to look like two different systems. Finally, by demultiplexing the output of the processor, the processed signals can be separated.

Digital implementation permits easy adjustment of processor characteristics during processing, such as that needed in implementing adaptive filters. Such adjustments can be simply carried out by periodically changing the coefficients of the algorithm representing the processor characteristics. Another application of the changing of coefficients is in the realization of systems with programmable characteristics, such as frequency selective filters with adjustable cutoff frequencies. Filter banks with guaranteed complementary frequency response characteristics are easily implemented in digital form.

Digital implementation allows the realization of certain characteristics not possible with analog implementation, such as exact linear phase and multirate processing. Digital circuits can be cascaded without any loading problems, unlike analog circuits. Digital signals can be stored almost indefinitely without any loss of information on various storage media, such as magnetic tapes and disks and optical disks. Such stored signals can later be processed off-line, such as in the compact disk player, the digital video disk player, the digital audio tape player, or simply by using a general purpose computer as in seismic data processing. On the other hand, stored analog signals deteriorate rapidly as time progresses and cannot be recovered in their original forms.

Another advantage is the applicability of digital processing to very low frequency signals, such as those occurring in seismic applications, where inductors and capacitors needed for analog processing would be physically very large in size.

Digital signal processing is also associated with some disadvantages. One obvious disadvantage is the increased system complexity in the digital processing of analog signals because of the need for additional pre- and postprocessing devices such as the A/D and D/A converters and their associated filters and complex digital circuitry.

A second disadvantage associated with digital signal processing is the limited range of frequencies available for processing. This property limits its application particularly in the digital processing of analog signals. As shown later, in general, an analog continuous-time signal must be sampled at a frequency that is at least twice the highest frequency component present in the signal. If this condition is not satisfied, then signal components with frequencies above half the sampling frequency appear as signal components below this particular frequency, totally distorting the input analog signal waveform. The available frequency range of operation of a digital signal processor is primarily determined by the S/H circuit and the A/D converter and, as a result, is limited by the state of the art of the technology. The highest sampling frequency reported in the literature presently is around 1 GHz [Pou87]. Such high sampling frequencies are not usually used in practice since the achievable resolution of the A/D converter, given by the wordlength of the digital equivalent of the analog sample, decreases with an increase in the speed of the converter. For example, the reported resolution of an A/D converter operating at 1 GHz is 6 bits [Pou87]. On the other hand, in most applications, the required resolution of an A/D converter is from 12 bits to around 16 bits. Consequently, a sampling frequency of at most 10 MHz is presently a practical upper limit. This upper limit, however, is getting larger and larger with advances in technology.

The third disadvantage stems from the fact that digital systems are constructed using active devices that consume electrical power. For example, the WE DSP32C Digital Signal Processor chip contains over 405,000 transistors and dissipates around 1 watt. On the other hand, a variety of analog processing algorithms can be implemented using passive circuits employing inductors, capacitors, and resistors that do not need power. Moreover, active devices are less reliable than passive components.

Another disadvantage of digital signal processing is due to the effects resulting from the algorithms implemented with finite precision arithmetic in hardware or software. By properly designing the algorithm and its implementation, the effects of the finite wordlength can be minimized in most cases.

However, the advantages far outweigh the disadvantages in various applications, and with the continuing decrease in the cost of digital processor hardware, applications of digital signal processing are increasing rapidly.

2 Discrete-Time Signals and Systems

As indicated in Figure 1.1(c), a discrete-time signal in its most basic form is defined at equally spaced discrete values of time, the independent variable, with the signal amplitude at these discrete times taking a continuous value. Consequently, a discrete-time signal can be represented as a sequence of numbers, with the independent time variable represented as an integer in the range from $-\infty$ to $+\infty$. Discrete-time signal processing then involves the processing of such a signal by a discrete-time system to develop another discrete-time signal with more desirable properties or to extract certain information about the original signal.

In many applications involving continuous-time signals, it is increasingly becoming more attractive to process the continuous-time signal by discrete-time signal processing methods. To this end, the continuous-time signal is first converted into an "equivalent" discrete-time signal by periodic sampling; the resulting signal is then processed by a discrete-time system to generate another discrete-time signal, and the latter is converted into an equivalent continuous-time signal, if necessary. As we shall show later in this book, under certain (ideal) conditions, the conversion of a continuous-time signal can be carried out such that the discrete-time equivalent has all the information contained in the original continuous-time signal and, if necessary, can be converted back into the original continuous-time signal without any distortion.

Thus, to understand the theory of digital signal processing and the design of discrete-time systems, we need to know the characterization of discrete-time signals and systems in the time-domain, a subject we discuss in this chapter. It turns out that it is often convenient to characterize the discrete-time signals and systems in a transformed domain. This alternative representation is considered in Chapters 3, 5, and 6.

In this chapter, we first discuss the time-domain representation of a discrete-time signal as a sequence of numbers and its various classifications. A number of basic operations that generate other sequences from one or more sequences are described next. As we show later, a discrete-time system is composed of a combination of these basic operations. We then describe several basic discrete-time signals or sequences that play important roles in the time-domain characterization of arbitrary discrete-time signals and discrete-time systems. The problem of representing a continuous-time signal by a discrete-time sequence is examined for a simple case. A more thorough mathematical treatment for the general case is deferred until Chapter 4 since it is based on a frequency-domain representation of the discrete-time signal discussed in Chapter 3.

In the latter half of this chapter, we introduce the general concept of the processing of a discrete-time signal by a discrete-time system and the classification of such systems. Of these systems, the class of linear, time-invariant type is almost of exclusive interest in this book, and we describe its time-domain characterization in several different forms. We also introduce the concept of cross-correlation between a pair of discrete-time sequences, which provides a measure of the degree of similarity between the pair. Finally, we describe several short-time characterizations of time-varying discrete-time signals.

Throughout this chapter and successive chapters, we make extensive use of MATLAB to illustrate through computer simulations the various concepts introduced.

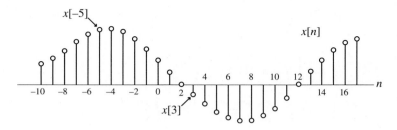

Figure 2.1: Graphical representation of a discrete-time sequence $\{x[n]\}$.

2.1   Discrete-Time Signals

We first consider here the time-domain representation of discrete-time signals and several concepts associated with such representations. We then describe several elementary operations on discrete-time signals and the devices used to implement these operations. These operations form the basic building blocks in developing discrete-time systems for the processing of discrete-time signals. We next discuss various classifications of the discrete-time signals that often can be used to simplify the signal processing algorithms.

2.1.1   Time-Domain Representation

As indicated earlier, in digital signal processing, signals are represented as sequences of numbers called *samples*. A sample value of a typical discrete-time signal or sequence is denoted as $x[n]$, with the argument n being an integer in the range $-\infty$ and ∞. It should be noted that $x[n]$ is defined only for integer values of n and is undefined for noninteger values of n. The discrete-time signal is represented by $\{x[n]\}$. If a discrete-time signal is written as a sequence of numbers inside braces, the location of the sample value associated with the time index $n = 0$ is indicated by an arrow $\uparrow$ under it. The sample values to its right are for positive values of n, and the sample values to its left are for negative values of n. An example of a discrete-time signal with real-valued samples is given by

$$\{x[n]\} = \{\ldots, 0.95, -0.2, 2.17, 1.1, 0.2, -3.67, 2.9, -0.8, 4.1, \ldots\}. \tag{2.1}$$
$$\uparrow$$

For the above signal, $x[-1] = -0.2$, $x[0] = 2.17$, $x[1] = 1.1$, and so on. The graphical representation of a sequence $\{x[n]\}$ with real-valued samples is illustrated in Figure 2.1.

In some applications a discrete-time sequence $\{x[n]\}$ is generated by periodically sampling a continuous-time signal $x_a(t)$ at uniform time intervals:

$$x[n] = x_a(t)|_{t=nT} = x_a(nT), \qquad n = \ldots, -2, -1, 0, 1, 2, \ldots \tag{2.2}$$

as illustrated in Figure 2.2. The spacing T between two consecutive samples in Eq. (2.2) is called the *sampling interval* or *sampling period*. The reciprocal of the sampling interval T, denoted as F_T, is called the *sampling frequency*

$$F_T = \frac{1}{T}. \tag{2.3}$$

The unit of sampling frequency here is cycles per second, or hertz (Hz), if the sampling period is in seconds (sec). In the case of an image, a spatial sampling is used in both horizontal and vertical directions. In this

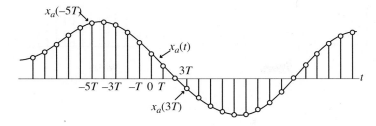

Figure 2.2: Sequence generated by sampling a continuous-time signal $x_a(t)$.

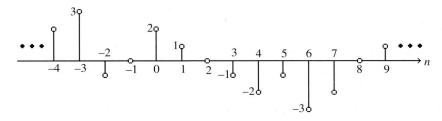

Figure 2.3: A digital signal.

case, the horizontal and vertical sampling frequencies are in cycles per unit of distance defining the two sampling periods.

It should be noted that, whether or not a sequence $\{x[n]\}$ has been obtained by sampling, the quantity $x[n]$ is called the *nth sample* of the sequence. For a sequence $\{x[n]\}$, the *n*th sample value $x[n]$ can, in general, take any real or complex value. If $x[n]$ is real for all values of n, then $\{x[n]\}$ is a *real sequence*. On the other hand, if the *n*th sample value is complex for one or more values of n, then it is a *complex sequence*. By separating the real and imaginary parts of $x[n]$, we can write a complex sequence $\{x[n]\}$ as

$$\{x[n]\} = \{x_{\mathrm{re}}[n]\} + j\,\{x_{\mathrm{im}}[n]\}, \tag{2.4}$$

where $x_{\mathrm{re}}[n]$ and $x_{\mathrm{im}}[n]$ are the real and the imaginary parts of $x[n]$, respectively, and are real sequences. For a real sequence, $x_{\mathrm{im}}[n] = 0$, and for a purely imaginary sequence, $x_{\mathrm{re}}[n] = 0$, for all values of n. The complex conjugate sequence of $\{x[n]\}$ is usually denoted by $\{x^*[n]\}$ and written as $\{x^*[n]\} = \{x_{\mathrm{re}}[n]\} - j\,\{x_{\mathrm{im}}[n]\}$.[1] Often the braces are ignored to denote a sequence if there is no ambiguity.

As defined in the previous chapter, there are basically two types of discrete-time signals: sampled-data signals, in which the samples are continuous-valued, and digital signals, in which the samples are discrete-valued. The pertinent signals in a practical digital signal processing system are digital signals obtained by quantizing the sample values either by *rounding* or by *truncation*. For example, the digital signal $\{\hat{x}[n]\}$ obtained by rounding the sample values of the discrete-time sequence $x[n]$ of Eq. (2.1) to the nearest integer values is given by

$$\{\hat{x}[n]\} = \{\ldots,\ 1,\ 0,\ 2,\ 1,\ 0,\ -4,\ 3,\ -1,\ 4,\ \ldots\}.$$
$$\uparrow$$

Figure 2.3 shows a digital signal with amplitudes taking discrete integer values in the range from -3 to 3.

For digital processing of a continuous-time signal, it is first converted into an equivalent digital signal by means of a sample-and-hold circuit, followed by an analog-to-digital converter. The processed digital signal

---

[1]The complex conjugation operation is denoted by the symbol *.

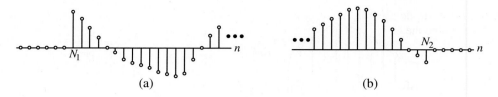

Figure 2.4: (a) A right-sided sequence and (b) a left-sided sequence.

is then converted back into an equivalent continuous-time signal by a digital-to-analog converter, followed by an analog reconstruction filter. Chapter 4 is concerned with the digital processing of continuous-time signals. It develops the mathematical foundation of the sampling process and describes the operations of various interface circuits between the continuous-time domain and the digital domain. Chapter 12 considers the effect of discretization of the amplitudes.

Length of a Discrete-Time Signal

The discrete-time signal may be a *finite-length* or an *infinite-length sequence*. A finite-length (also called *finite-duration* or *finite-extent*) sequence is defined only for a finite time interval:

$$N_1 \leq n \leq N_2, \tag{2.5}$$

where $-\infty < N_1$ and $N_2 < \infty$ with $N_2 \geq N_1$. The *length or duration N* of the above finite-length sequence is

$$N = N_2 - N_1 + 1. \tag{2.6}$$

A length-N discrete-time sequence consists of N samples and is often referred to as an N-point sequence. A finite-length sequence can also be considered as an infinite-length sequence by assigning zero values to samples whose arguments are outside the above range. The process of lengthening a sequence by adding zero-valued samples is called *appending with zeros* or *zero-padding*.

There are three types of infinite-length sequences. A *right-sided sequence $x[n]$* has zero-valued samples for $n < N_1$; that is,

$$x[n] = 0 \qquad \text{for } n < N_1, \tag{2.7}$$

where N_1 is a finite integer that can be positive or negative. If $N_1 \geq 0$, a right-sided sequence is usually called a *causal sequence.*[2] Likewise, a *left-sided sequence $x[n]$* has zero-valued samples for $n > N_2$; that is,

$$x[n] = 0 \qquad \text{for } n > N_2, \tag{2.8}$$

where N_2 is a finite integer that can be positive or negative. If $N_2 \leq 0$, a left-sided sequence is usually called an *anticausal sequence*. A general *two-sided sequence* is defined for both positive and negative values of n. Figure 2.4 illustrates the above two types of one-sided sequences.

For simplicity, for finite-length sequences defined for positive values of the time index n beginning at $n = 0$, the first sample in the sequence will always be assumed to be the one associated with the time index $n = 0$ without the arrow being explicitly shown under it.

We have so far considered the sequence of numbers representing a discrete-time signal in the time-domain. As we shall show in Chapter 5, a finite-length discrete-time signal can also be represented as a finite-length sequence in a transform-domain where the sample of the sequence is a function of discrete variable in the transform domain usually denoted by the integer variable k. It is a common practice to

---

[2]The term *causal sequence* has its origin in the causality condition of a discrete-time system as discussed in Section 2.5.4.

denote the transform-domain representation of a discrete-time signal $x[n]$ with capital letters; for example, $X[k]$. If we denote the vector of the time-domain samples of length N defined for $0 \leq n \leq N - 1$, as[3]

$$\mathbf{x} = [\, x[0] \quad x[1] \quad \cdots \quad x[N-1]\,]^t,$$

the vector of the transform-domain samples of length N defined for $0 \leq k \leq N - 1$,

$$\mathbf{X} = [\, X[0] \quad X[1] \quad \cdots \quad X[N-1]\,]^t,$$

is obtained by multiplying $\mathbf{x}$ by an $N \times N$ invertible matrix $\mathbf{D}$:

$$\mathbf{X} = \mathbf{D}\mathbf{x}.$$

The original time-domain sequence can be recovered from its transform-domain representation by applying an inverse transformation

$$\mathbf{x} = \mathbf{D}^{-1}\mathbf{X}.$$

In some cases, the transform-domain representation can be a complex-valued sequence for a real-valued time-domain sequence. The transform-domain sequence often is processed using some of the operations used in the processing of discrete-time signals in the time-domain. Some elementary operations on sequences are discussed in Section 2.1.2.

Size of a Discrete-Time Signal

The $\mathcal{L}_p$-*norm* of a sequence $\{x[n]\}$ is defined by

$$\| x \|_p = \left(\sum_{n=-\infty}^{\infty} |x[n]|^p \right)^{1/p}, \tag{2.9}$$

where p is a positive integer. The values of p used are typically 1 or 2 or ∞. It follows from the above definition that the $\mathcal{L}_2$-norm, $\| x \|_2$, is the root-mean-squared (rms) value of $\{x[n]\}$, and the $\mathcal{L}_1$-norm, $\| x \|_1$, is the mean absolute value of $\{x[n]\}$. The $\mathcal{L}_\infty$-norm given by

$$\| x \|_\infty = |x|_{\max}, \tag{2.10}$$

is the *peak absolute value* of $\{x[n]\}$. The norm provides an estimate of the size of a signal. For example, if the length-N sequence $y[n]$, $0 \leq n \leq N - 1$, is an approximation of the length-N sequence $x[n]$, $0 \leq n \leq N - 1$, then an estimate of the *relative error* is given by the ratio of the $\mathcal{L}_2$-norm of the difference signal and $\mathcal{L}_2$-norm of the original signal:

$$\mathrm{E_{rel}} = \left(\frac{\sum_{n=0}^{N-1} |y[n] - x[n]|^2}{\sum_{n=0}^{N-1} |x[n]|^2} \right)^{1/2}. \tag{2.11}$$

The norm of a finite-length sequence x can be computed using the M-file norm in MATLAB. The $\mathcal{L}_2$-*norm* of x is obtained using either norm(x) or norm(x,2). The $\mathcal{L}_1$-*norm* of x can be determined using norm(x,1). Finally, $\mathcal{L}_\infty$-*norm* of x is evaluated using norm(x,inf).

---

[3]The superscript "t" denotes matrix transposition.

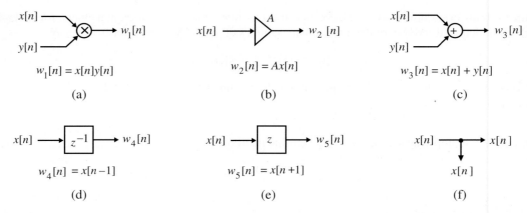

Figure 2.5: Schematic representations of basic operations on sequences: (a) modulator, (b) multiplier, (c) adder, (d) unit delay, (e) unit advance, and (f) pick-off node.

2.1.2   Operations on Sequences

A single-input, single-output discrete-time system operates on a sequence, called the *input sequence,* according to some prescribed rules and develops another sequence, called the *output sequence* usually with more desirable properties. For example, the input may be a signal corrupted by an additive noise, and the discrete-time system is designed to remove the noise component from the input. In some applications, the discrete-time system can have more than one input and more than one output. An M-input, N-output discrete-time system operates on M input signals generating N output signals. The FM stereo transmission system is a two-input, single-output system since here the left and right channel audio signals are combined into a high-frequency composite baseband signal. In most cases, the operation defining a particular discrete-time system is composed of some elementary operations that we describe next.

Elementary Operations

Let $x[n]$ and $y[n]$ be two known sequences. By forming the *product* of the sample values of these two sequences at each instant, we form a new sequence $w_1[n]$:

$$w_1[n] = x[n] \cdot y[n]. \tag{2.12}$$

In some applications, the product operation is also known as *modulation.* The device implementing the modulation operation is called a *modulator,* and its schematic representation is shown in Figure 2.5(a).

An application of the product operation is in forming a finite-length sequence from an infinite-length sequence by multiplying the latter with a finite-length sequence called a *window sequence.* This process of forming the finite-length sequence is usually called *windowing,* which plays an important role in the design of certain types of digital filters (Section 10.2). Another application of the product operation is illustrated in Example 2.10.

The second basic operation is the scalar *multiplication,* whereby a new sequence is generated by multiplying each sample of a sequence $x[n]$ by a scalar A:

$$w_2[n] = Ax[n]. \tag{2.13}$$

The device implementing the multiplication operation is called a *multiplier,* and its schematic representation is shown in Figure 2.5(b).

The third basic operation is the *addition* by which a new sequence $w_2[n]$ is obtained by adding the sample values of two sequences $x[n]$ and $y[n]$:

$$w_3[n] = x[n] + y[n]. \tag{2.14}$$

The device implementing the addition operation is called an *adder,* and its schematic representation is shown in Figure 2.5(c). By inverting the signs of all samples of the sequence $y[n]$, an adder can also be used to implement the *subtraction* operation.

A very simple application of the addition operation is in improving the quality of measured data that has been corrupted by an additive random noise. In many cases, the actual uncorrupted data vector **s** remains essentially the same from one measurement to the next, while the additive noise vector is random and not reproducible. Let $\mathbf{d}_i$ denote the noise vector corrupting the i-th measurement of the uncorrupted data vector **s**:

$$\mathbf{x}_i = \mathbf{s} + \mathbf{d}_i.$$

The average data vector, called the *ensemble average,* obtained after K measurements is then given by

$$\mathbf{x}_{\text{ave}} = \frac{1}{K} \sum_{i=1}^{K} (\mathbf{x}_i) = \frac{1}{K} \sum_{i=1}^{K} (\mathbf{s} + \mathbf{d}_i) = \mathbf{s} + \frac{1}{K} \left(\sum_{i=1}^{K} \mathbf{d}_i \right).$$

For a very large value of K, $\mathbf{x}_{\text{ave}}$ is usually a reasonable replica of the desired data vector **s**, as the samples of the average of the summed noise vector $\frac{1}{K}(\sum_{i=1}^{K} \mathbf{d}_i)$ become very small due to the randomness of the noise. Example 2.1 illustrates ensemble averaging.

EXAMPLE 2.1   Illustration of Ensemble Averaging

We assume for simplicity the original uncorrupted data is given by [Sch75]

$$s[n] = 2[n(0.9)^n]. \tag{2.15}$$

The MATLAB program to generate the above data $s[n]$, the noise $d[n]$, and the ensemble average is given in Program 2_1. The two sequences generated using the above program are shown in Figure 2.6. One sample of the noise-corrupted data $s[n] + d_i[n]$ and the ensemble average obtained after 50 measurements is shown in Figure 2.7. It can be seen that the ensemble average shown in Figure 2.7(b) is nearly the same as the original uncorrupted data shown in Figure 2.6(a).

Program 2_1.m

An application of ensemble averaging is in the power spectrum estimation of a random signal, discussed in Section 15.4.1.

The *time-shifting* operation illustrated below in Eq. (2.16) shows the relation between $x[n]$ and its time-shifted version $w_4[n]$:

$$w_4[n] = x[n - N], \tag{2.16}$$

where N is an integer. If $N > 0$, it is a *delaying* operation, and if $N < 0$, it is an *advancing* operation. For $N = 1$, we have the input–output relation

$$w_4[n] = x[n - 1],$$

and the device implementing the delay operation by one sample period is called a *unit delay* In terms of z-transform, introduced later in Chapter 6, the above relation can be rewritten as

$$W_4(z) = z^{-1} X(z),$$

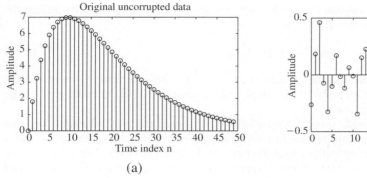
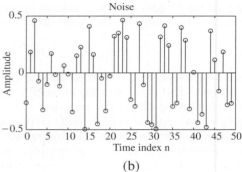

Figure 2.6: (a) The original uncorrupted sequence $s[n]$ and (b) the noise sequence $d_i[n]$.

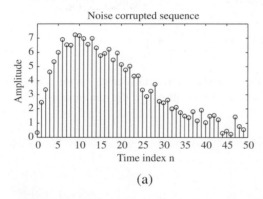
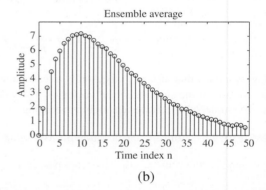

Figure 2.7: (a) A sample of the corrupted sequence and (b) the ensemble average after 50 measurements.

where $W_4(z)$ and $X(z)$ are, respectively, the z-transforms of the output sequence $w_4[n]$ and the input sequence $x[n]$. It is a usual practice to represent schematically the unit delay operation using the symbol z^{-1}, as shown in Figure 2.5(d).

The opposite of the unit delay operation is the unit advance operation defined by

$$w_5[n] = x[n+1],$$

which in terms of the z-transform is given by

$$W_5(z) = z\,X(z).$$

As a result, the *unit advance* operation is commonly represented schematically using the symbol z, as shown in Figure 2.5(e).

The *time-reversal* operation, also called the *folding operation*, is another useful scheme to develop a new sequence. An example is

$$w_6[n] = x[-n], \tag{2.17}$$

which is the time-reversed version of the sequence $x[n]$.

In Figure 2.5(f), we show a *pick-off node,* which is used to feed a sequence to different parts of a discrete-time system.

EXAMPLE 2.2   Basic Operations on Sequences of Equal Lengths

Consider the following two sequences of length 5 defined for $0 \leq n \leq 4$:

$$c[n] = \{3.2, \quad 41, \quad 36, \quad -9.5, \quad 0\},$$
$$d[n] = \{1.7, \quad -0.5, \quad 0, \quad 0.8, \quad 1\}.$$

Several new sequences of length 5 generated from the above sequences are given by

$$w_1[n] = c[n] \cdot d[n] = \{5.44, \quad -20.5, \quad 0, \quad -7.6, \quad 0\},$$
$$w_2[n] = c[n] + d[n] = \{4.9, \quad 40.5, \quad 36, \quad -8.7, \quad 1\},$$
$$w_3[n] = \tfrac{7}{2}c[n] = \{11.2, \quad 143.5, \quad 126, \quad -33.25, \quad 0\}.$$

As indicated by Example 2.2, operations on two or more sequences to generate a new sequence can be carried out if all sequences are of the same length and defined for the same range of the time index n. However, if the sequences are not of equal length, all sequences can be made to have the same range of the time index n by appending zero-valued samples to the sequence(s) of smaller lengths. This process is illustrated in Example 2.3.

EXAMPLE 2.3   Basic Operations on Sequences of Unequal Lengths

Consider a sequence $\{g[n]\}$ of length 3 defined for $0 \leq n \leq 2$ given by

$$\{g[n]\} = \{-21, \quad 1.5, \quad 3\}.$$

It is clear that we cannot develop another sequence by operating on this sequence and any one of the length-5 sequences of Example 2.2. However, it is possible to treat $\{g[n]\}$ as a sequence of length 5 and defined for $0 \leq n \leq 4$ by appending it with two zero-valued samples:

$$\{g_e[n]\} = \{-21, \quad 1.5, \quad 3, \quad 0, \quad 0\}.$$

Examples of new sequences generated from $\{g_e[n]\}$ and $c[n]$ of the previous example are indicated below:

$$\{w_4[n]\} = \{c[n] \cdot g_e[n]\} = \{-67.2, \quad 61.5, \quad 108, \quad 0, \quad 0\},$$
$$\{w_5[n]\} = \{c[n] + g_e[n]\} = \{-17.8, \quad 42.5, \quad 39, \quad -9.5, \quad 0\}.$$

Combination of Elementary Operations

In most applications, combinations of the above elementary operations are used. Illustrations of such combinations are given in Example 2.4.

EXAMPLE 2.4   Illustration of Combination of Basic Operations on Sequences

We next analyze the discrete-time system of Figure 2.8. Observe first that the two left-most delay blocks generate the sequences $x[n-1]$ and $x[n-2]$, whereas the two right-most delay blocks develop the sequences $y[n-1]$ and $y[n-2]$. These delayed sequences, along with $x[n]$, are then applied to the five multipliers, labeled b_0, b_1, b_2, a_1, and a_2, developing the sequences $b_0x[n]$, $b_1x[n-1]$, $b_2x[n-2]$, $a_1y[n-1]$, and $a_2y[n-2]$, which are then added to yield $y[n]$:

$$y[n] = b_0x[n] + b_1x[n-1] + b_2x[n-2] + a_1y[n-1] + a_2y[n-2]. \tag{2.18}$$

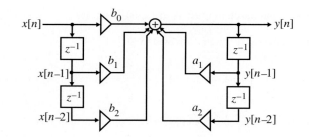

Figure 2.8: Discrete-time system of Example 2.4.

Figure 2.9: Representation of basic sampling rate alteration devices: (a) up-sampler and (b) down-sampler.

Sampling Rate Alteration

Another quite useful operation is the sampling rate alteration that is employed to generate a new sequence with a sampling rate higher or lower than that of a given sequence. Thus, if $x[n]$ is a sequence with a sampling rate of F_T Hz and it is used to generate another sequence $y[n]$ with a desired sampling rate of F_T' Hz, then the *sampling rate alteration ratio* is given by

$$\frac{F_T'}{F_T} = R. \tag{2.19}$$

If $R > 1$, the process is called *interpolation* and results in a sequence with a higher sampling rate. The discrete-time system implementing the interpolation process is called an *interpolator.* On the other hand, if $R < 1$, the sampling rate is decreased by a process called *decimation.* The discrete-time system implementing the decimation process is called a *decimator..*

The basic operations employed in the sampling rate alteration process are called *up-sampling* and *down-sampling.* These operations play important roles in multirate discrete-time systems and are considered in Chapters 13 and 14.

In up-sampling by an integer factor $L > 1$, $L - 1$ equidistant zero-valued samples are inserted by the up-sampler between each two consecutive samples of the input sequence $x[n]$ to develop an output sequence $x_u[n]$ according to the relation

$$x_u[n] = \begin{cases} x[n/L], & n = 0, \pm L, \pm 2L, \ldots, \\ 0, & \text{otherwise.} \end{cases} \tag{2.20}$$

The sampling rate of $x_u[n]$ is L times larger than that of the original sequence $x[n]$.

The block-diagram representation of the up-sampler, also called a *sampling rate expander*, is shown in Figure 2.9(a). Figure 2.10 illustrates the up-sampling operation for an up-sampling factor of $L = 3$. The interpolator consists of an up-sampler followed by a discrete-time system that replaces the inserted zero-valued samples of $x_u[n]$ with more appropriate values given by a linear combination of samples of $x[n]$. A simple example of an interpolator is given in Section 2.4.1.

Conversely, the down-sampling operation by an integer factor $M > 1$ on a sequence $x[n]$ consists of keeping every Mth sample of $x[n]$ and removing $M - 1$ between samples, generating an output sequence

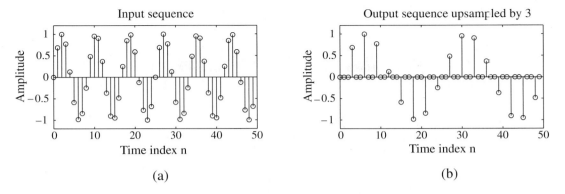

Figure 2.10: Illustration of the up-sampling process.

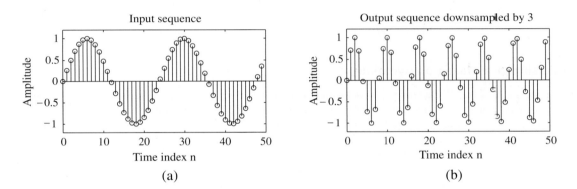

Figure 2.11: Illustration of the down-sampling process.

$y[n]$ according to the relation

$$y[n] = x[nM]. \tag{2.21}$$

This results in a sequence $y[n]$ whose sampling rate is $\frac{1}{M}$ th that of $x[n]$. Basically, all input samples with indices equal to an integer multiple of M are retained at the output, and all others are discarded.

The schematic representation of the *down-sampler* or *sampling rate compressor* is shown in Figure 2.9(b). Figure 2.11 illustrates the down-sampling operation for a down-sampling factor of $M = 3$. The decimator consists of a discrete-time system followed by a down-sampler. The discrete-time system preceding the down-sampler ensures that the input signal $x[n]$ is appropriately band-limited to prevent aliasing that is caused by the down-sampling operation.

2.1.3 Classification of Sequences

A discrete-time signal can be classified into several types based on its specific characteristics One classification discussed earlier is in terms of the number of samples defining the sequence. Another classification is with respect to the symmetry exhibited by the samples with respect to the time index $n = 0$. A discrete-time signal can also be classified in terms of its other properties such as periodicity, summability, energy, and power. These properties can often be exploited in simplifying some signal processing algorithms.

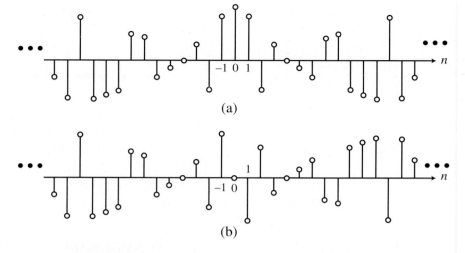

Figure 2.12: (a) An even sequence and (b) an odd sequence.

Classification Based on Symmetry

A sequence $x[n]$ is called a *conjugate-symmetric sequence* if $x[n] = x^*[-n]$. A real conjugate-symmetric sequence is called an *even sequence*. A sequence $x[n]$ is called a *conjugate-antisymmetric sequence* if $x[n] = -x^*[-n]$. A real conjugate-antisymmetric sequence is called an *odd sequence*. For a conjugate-antisymmetric sequence $x[n]$, the sample value at $n = 0$ must be purely imaginary. Consequently, for an odd sequence $x[0] = 0$. Examples of even and odd sequences are shown in Figure 2.12.

Any complex sequence $x[n]$ can be expressed as a sum of its conjugate-symmetric part $x_{cs}[n]$ and its conjugate-antisymmetric part $x_{ca}[n]$:

$$x[n] = x_{cs}[n] + x_{ca}[n], \tag{2.22}$$

where

$$x_{cs}[n] = \tfrac{1}{2}\left(x[n] + x^*[-n]\right), \tag{2.23a}$$

$$x_{ca}[n] = \tfrac{1}{2}\left(x[n] - x^*[-n]\right). \tag{2.23b}$$

As indicated by Eqs. (2.23a) and (2.23b), the computation of the conjugate-symmetric and conjugate-antisymmetric parts of a sequence involves conjugation, time-reversal, addition, and multiplication operations. Because of the time-reversal operation, the decomposition of a finite-length sequence into a sum of a conjugate-symmetric sequence and a conjugate-antisymmetric sequence is possible, if the parent sequence is of odd length defined for a symmetric interval, $-M \le 0 \le M$.

EXAMPLE 2.5   Generation of Symmetric Parts of a Complex Sequence

Consider the finite-length sequence of length 7 defined for $-3 \le n \le 3$:

$$\{g[n]\} = \{0, \ 1 + j4, \ -2 + j3, \ 4 - j2, \ -5 - j6, \ -j2, \ 3\}.$$
$$\uparrow$$

To determine its conjugate-symmetric part $g_{cs}[n]$ and its conjugate-antisymmetric part $g_{ca}[n]$, we form

$$\{g^*[n]\} = \{0, \ 1 - j4, \ -2 - j3, \ 4 + j2, \ -5 + j6, \ j2, \ 3\},$$
$$\uparrow$$

whose time-reversed version is then given by

$$\{g^*[-n]\} = \{3, \ j2, \ -5 + j6, \ 4 + j2, \ -2 - j3, \ 1 - j4, \ 0\}.$$
$$\uparrow$$

Using Eq. (2.23a), we thus arrive at

$$\{g_{cs}[n]\} = \{1.5, \ 0.5 + j3, \ -3.5 + j4.5, \ 4, \ -3.5 - j4.5, \ 0.5 - j3, \ 1.5\}.$$
$$\uparrow$$

Likewise, using Eq. (2.23b), we get

$$\{g_{ca}[n]\} = \{-1.5, \ 0.5 + j, \ 1.5 - j1.5, \ -j2, \ -1.5 - j1.5, \ -0.5 + j, \ 1.5\}.$$
$$\uparrow$$

It can be easily verified that $g_{cs}[n] = g_{cs}^*[-n]$ and $g_{ca}[n] = -g_{ca}^*[-n]$.

In a similar manner, any real sequence $x[n]$ can be expressed as a sum of its even part $x_{ev}[n]$ and its odd part $x_{od}[n]$:

$$x[n] = x_{ev}[n] + x_{od}[n], \tag{2.24}$$

where

$$x_{ev}[n] = \tfrac{1}{2} \left(x[n] + x[-n] \right), \tag{2.25a}$$

$$x_{od}[n] = \tfrac{1}{2} \left(x[n] - x[-n] \right). \tag{2.25b}$$

The symmetry properties of sequences often simplify their respective frequency-domain representations and can be exploited in signal analysis. Implications of the symmetry conditions are considered in the frequency-domain descriptions of sequences discussed in Chapter 3.

Periodic and Aperiodic Signals

A sequence $\tilde{x}[n]$ satisfying

$$\tilde{x}[n] = \tilde{x}[n + kN] \qquad \text{for all } n \tag{2.26}$$

is called a *periodic sequence* with a *period N*, where N is a positive integer and k is any integer. An example of a periodic sequence that has a period $N = 7$ samples is shown in Figure 2.13. A sequence is called an *aperiodic sequence* if it is not periodic. To distinguish a periodic sequence from an aperiodic sequence, we shall denote the former with a "~" on top. The *fundamental period* N_f of a periodic signal is the smallest value of N for which Eq. (2.26) holds.

Sum of two or more periodic sequences is also a periodic sequence. If $\tilde{x}_a[n]$ and $\tilde{x}_b[n]$ are two periodic sequences with fundamental periods N_a and N_b, respectively, then the sequence $\tilde{y}[n] = \tilde{x}_a[n] + \tilde{x}_b[n]$ is a periodic sequence with a fundamental period N given by

$$N = \frac{N_a N_b}{GCD(N_a, N_b)}, \tag{2.27}$$

where $GCD(N_a, N_b)$ is the greatest common divisor of N_a and N_b.

Likewise, the product of two or more periodic sequences is again a periodic sequence. Here, if $\tilde{x}_a[n]$ and $\tilde{x}_b[n]$ are two periodic sequences with fundamental periods N_a and N_b, respectively, then the sequence

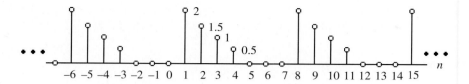

Figure 2.13: An example of a periodic sequence.

$\tilde{w}[n] = \tilde{x}_a[n]\tilde{x}_b[n]$ is a periodic sequence with a period given by Eq. (2.27). It should be noted that in some cases, the fundamental period can be smaller than that given by Eq. (2.27)

Energy and Power Signals

The total *energy* of a sequence $x[n]$ is defined by

$$\mathcal{E}_x = \sum_{n=-\infty}^{\infty} |x[n]|^2 . \tag{2.28}$$

An infinite-length sequence with finite sample values may or may not have finite energy, as illustrated in Example 2.6.

EXAMPLE 2.6   Examples of Finite Energy and Infinite Energy Signals

The infinite-length sequence $x_1[n]$ defined by

$$x_1[n] = \begin{cases} \frac{1}{n}, & n \geq 1, \\ 0, & n \leq 0, \end{cases} \tag{2.29}$$

has an energy equal to

$$\mathcal{E}_{x_1} = \sum_{n=1}^{\infty} \left(\frac{1}{n}\right)^2,$$

which converges to $\pi^2/6$, indicating that $x_1[n]$ has finite energy. However, the infinite-length sequence $x_2[n]$ defined by

$$x_2[n] = \begin{cases} \frac{1}{\sqrt{n}}, & n \geq 1, \\ 0, & n \leq 0, \end{cases} \tag{2.30}$$

has an energy equal to

$$\mathcal{E}_{x_2} = \sum_{n=1}^{\infty} \frac{1}{n},$$

which does not converge. As a result, the sequence $x_2[n]$ has infinite energy.

The *average power* of an aperiodic sequence $x[n]$ is defined by

$$\mathcal{P}_x = \lim_{K \to \infty} \frac{1}{2K+1} \sum_{n=-K}^{K} x[n]^2. \tag{2.31}$$

The average power of a sequence can be related to its energy by defining its energy over a finite interval $-K \leq n \leq K$ as

$$\mathcal{E}_{x,K} = \sum_{n=-K}^{K} |x[n]|^2 . \tag{2.32}$$

Then,

$$\mathcal{P}_x = \lim_{K \to \infty} \frac{1}{2K+1} \mathcal{E}_{x,K} . \tag{2.33}$$

The average power of a periodic sequence $\tilde{x}[n]$ with a period N is given by

$$\mathcal{P}_x = \frac{1}{N} \sum_{n=0}^{N-1} |\tilde{x}[n]|^2 . \tag{2.34}$$

The average power of an infinite-length sequence may be finite or infinite.

An infinite energy signal with finite average power is called a *power signal*. Likewise, a finite energy signal with zero average power is called an *energy signal*. An example of a power signal is a periodic sequence that has a finite average power but infinite energy. An example of an energy signal is a finite-length sequence which has finite energy but zero average power.

EXAMPLE 2.7   Example of a Power Signal

Consider the causal sequence defined by

$$x[n] = \begin{cases} 3(-1)^n, & n \geq 0, \\ 0, & n < 0. \end{cases}$$

It follows from Eq. (2.28) that $x[n]$ has infinite energy. On the other hand, from Eq. (2.31), its average power is given by

$$\mathcal{P}_x = \lim_{K \to \infty} \frac{1}{2K+1} \left(9 \sum_{n=0}^{K} 1 \right) = \lim_{K \to \infty} \frac{9(K+1)}{2K+1} = 4.5,$$

which is finite.

Other Types of Classification

A sequence $x[n]$ is said to be *bounded* if each of its samples is of magnitude less than or equal to a finite positive number B_x, that is,

$$|x[n]| \leq B_x < \infty. \tag{2.35}$$

The periodic sequence of Figure 2.13 is a bounded sequence with a bound $B_x = 2$.

A sequence $x[n]$ is said to be *absolutely summable* if

$$\sum_{n=-\infty}^{\infty} |x[n]| < \infty. \tag{2.36}$$

A sequence is said to be *square-summable* if

$$\sum_{n=-\infty}^{\infty} |x[n]|^2 < \infty. \tag{2.37}$$

A square-summable sequence therefore has finite energy and is an energy signal if it also has zero power. An example of a sequence that is square-summable but not absolutely-summable is

$$x_a[n] = \frac{\sin \omega_c n}{\pi n}, \qquad -\infty < n < \infty.$$

Examples of sequences that are neither absolutely-summable nor square-summable are

$$x_b[n] = \sin \omega_c n, \qquad -\infty < n < \infty,$$
$$x_c[n] = K, \qquad -\infty < n < \infty,$$

where K is a constant.

The sequence $\tilde{y}[n]$ obtained by adding an absolutely summable sequence $x[n]$ with its replicas shifted by integer multiples of N,

$$\tilde{y}[n] = \sum_{k=-\infty}^{\infty} x[n + kN], \qquad (2.38)$$

where N is a positive integer, is a periodic sequence with a period N. The periodic sequence $\tilde{y}[n]$ is called an *N-periodic extension* of $x[n]$.

2.2   Typical Sequences and Sequence Representation

We now consider several special sequences that play important roles in the analysis and design of discrete-time systems. For example, an arbitrary sequence can be expressed in terms of some of these basic sequences. Another fundamental application that is the key behind discrete-time signal processing is the representation of a class of discrete-time systems in terms of the response of the system to certain basic sequences. This representation permits the computation of the response of the discrete-time system to arbitrary discrete-time signals if the latter can be expressed in terms of these basic sequences.

2.2.1   Some Basic Sequences

The most common basic sequences are the unit sample sequence, the unit step sequence, the sinusoidal sequence, and the exponential sequence. These sequences are defined next.

Unit Sample Sequence

The simplest and one of the most useful sequences is the *unit sample sequence*, often called the *discrete-time impulse* or the *unit impulse,* as shown in Figure 2.14(a). It is denoted by $\delta[n]$ and defined by

$$\delta[n] = \begin{cases} 1, & n = 0, \\ 0, & n \neq 0. \end{cases} \qquad (2.39)$$

The unit sample sequence shifted by k samples is thus given by

$$\delta[n - k] = \begin{cases} 1, & n = k, \\ 0, & n \neq k. \end{cases}$$

Figure 2.14(b) shows $\delta[n - 2]$. We shall show later in this section that any arbitrary sequence can be represented as a sum of weighted time-shifted unit sample sequences. In Section 2.6.1, we demonstrate that a certain class of discrete-time systems is completely characterized in the time-domain by its output response to a unit impulse input. Furthermore, knowing this particular response of the system, we can compute its response to any arbitrary input sequence.

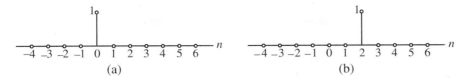

Figure 2.14: (a) The unit sample sequence $\{\delta[n]\}$ and (b) the shifted unit sample sequence $\{\delta[n - 2]\}$.

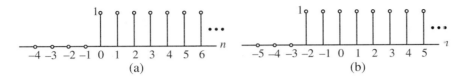

Figure 2.15: (a) The unit step sequence $\{\mu[n]\}$ and the shifted unit step sequence $\{\mu[n + 2]\}$.

Unit Step Sequence

A second basic sequence is the *unit step sequence,* shown in Figure 2.15(a). It is denoted by $\mu[n]$ and is defined by

$$\mu[n] = \begin{cases} 1, & n \geq 0, \\ 0, & n < 0. \end{cases} \tag{2.40}$$

The unit step sequence shifted by k samples is thus given by

$$\mu[n - k] = \begin{cases} 1, & n \geq k, \\ 0, & n < k. \end{cases}$$

Figure 2.15(b) shows $\mu[n + 2]$.

The unit sample and the unit step sequences are related as follows (Problem 2.3):

$$\mu[n] = \sum_{m=0}^{\infty} \delta[n - m] = \sum_{k=-\infty}^{n} \delta[k], \tag{2.41a}$$

$$\delta[n] = \mu[n] - \mu[n - 1]. \tag{2.41b}$$

Sinusoidal and Exponential Sequences

A commonly encountered sequence is the *real sinusoidal sequence* with constant amplitude of the form

$$x[n] = A \cos(\omega_o n + \phi), \qquad -\infty < n < \infty, \tag{2.42}$$

where A, ω_o, and ϕ are real numbers. The parameters A, ω_o, and ϕ are called, respectively, the *amplitude,* the *angular frequency,* and the *phase* of the sinusoidal sequence $x[n]$.

Figure 2.16 shows different types of sinusoidal sequences. The real sinusoidal sequence of Eq. (2.42) can be written alternatively as

$$x[n] = x_i[n] + x_q[n], \tag{2.43}$$

where $x_i[n]$ and $x_q[n]$ are, respectively, the *in-phase* and the *quadrature components* of $x[n]$, and are given by

$$x_i[n] = A \cos \phi \cos(\omega_o n), \qquad x_q[n] = -A \sin \phi \sin(\omega_o n). \tag{2.44}$$

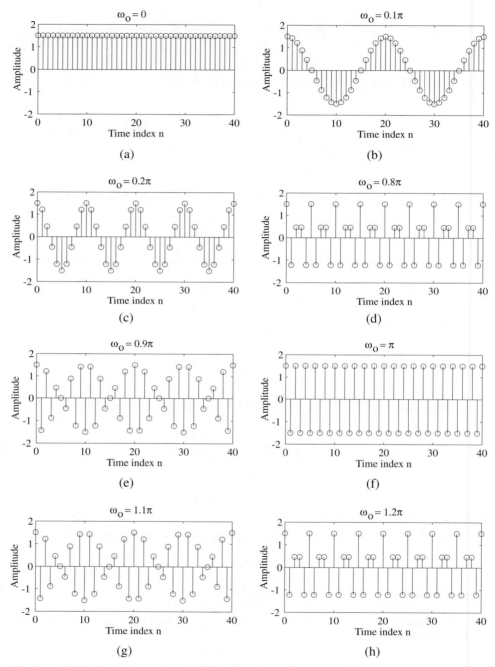

Figure 2.16: A family of sinusoidal sequences given by $x[n] = 1.5 \cos \omega_o n$: (a) $\omega_o = 0$, (b) $\omega_o = 0.1\pi$, (c) $\omega_o = 0.2\pi$, (d) $\omega_o = 0.8\pi$, (e) $\omega_o = 0.9\pi$, (f) $\omega_o = \pi$, (g) $\omega_o = 1.1\pi$, and (h) $\omega_o = 1.2\pi$.

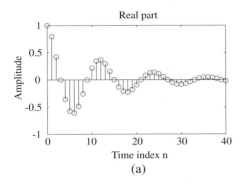
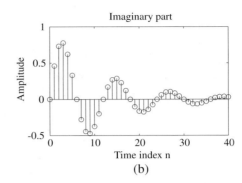

Figure 2.17: A complex exponential sequence $x[n] = e^{(-1/12 + j\pi/6)n}$. (a) Real part and (b) imaginary part.

Another set of basic sequences is formed by taking the nth sample value to be the nth power of a real or complex constant. Such sequences are termed *exponential sequences,* and their most general form is given by

$$x[n] = A\alpha^n, \qquad -\infty < n < \infty, \tag{2.45}$$

where A and α are real or complex numbers. By expressing

$$\alpha = e^{(\sigma_o + j\omega_o)}, \qquad A = |A|e^{j\phi},$$

we can rewrite Eq. (2.45) as

$$x[n] = Ae^{(\sigma_o + j\omega_o)n} = |A|e^{\sigma_o n}e^{j(\omega_o n + \phi)} \tag{2.46a}$$

$$= |A|e^{\sigma_o n}\cos(\omega_o n + \phi) + j|A|e^{\sigma_o n}\sin(\omega_o n + \phi), \tag{2.46b}$$

to arrive at an alternative general form of a *complex exponential sequence* where σ_o, ϕ, and ω_o are now real numbers. If we write $x[n] = x_{\text{re}}[n] + jx_{\text{im}}[n]$, then from Eq. (2.46b),

$$x_{\text{re}}[n] = |A|e^{\sigma_o n}\cos(\omega_o n + \phi),$$

$$x_{\text{im}}[n] = |A|e^{\sigma_o n}\sin(\omega_o n + \phi).$$

Thus, the real and imaginary parts of a complex exponential sequence are real sinusoidal sequences with constant ($\sigma_o = 0$), growing ($\sigma_o > 0$), or decaying ($\sigma_o < 0$) amplitudes for $n > 0$. Figure 2.17 depicts a complex exponential sequence with a decaying amplitude. Note that in the display of a complex exponential sequence, its real and imaginary parts are shown separately.

With both A and α real, the sequence of Eq. (2.45) reduces to a *real exponential sequence*. For $n \geq 0$, such a sequence with $|\alpha| < 1$ decays exponentially as n increases and with $|\alpha| > 1$ grows exponentially as n increases. Examples of real exponential sequences obtained for two values of α are shown in Figure 2.18.

We shall show later in Section 3.2 that a large class of sequences can be expressed in terms of complex exponential sequences of the form $e^{j\omega n}$.

Note that the sinusoidal sequence of Eq. (2.42) and the complex exponential sequence of Eq. (2.46a) with $\sigma_o = 0$ are periodic sequences of period N as long as $\omega_o N$ is an integer multiple of 2π; that is, $\omega_o N = 2\pi r$, where N and r are positive integers. The smallest possible N satisfying this condition is the *fundamental*

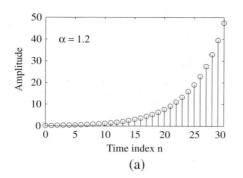
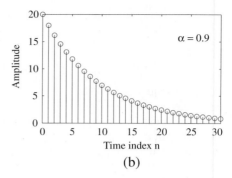

Figure 2.18: Examples of real exponential sequences: (a) $x[n] = 0.2(1.2)^n$, (b) $x[n] = 20(0.9)^n$.

period of the sequence. To verify this, consider the two sinusoidal sequences $x_1[n] = \cos(\omega_o n + \phi)$ and $x_2[n] = \cos(\omega_o(n + N) + \phi)$. Now

$$x_2[n] = \cos(\omega_o(n + N) + \phi)$$

$$= \cos(\omega_o n + \phi)\cos\omega_o N - \sin(\omega_o n + \phi)\sin\omega_o N,$$

which will be equal to $\cos(\omega_o n + \phi) = x_1[n]$ only if $\sin\omega_o N = 0$ and $\cos\omega_o N = 1$. These two conditions are satisfied if and only if $\omega_o N$ is an integer multiple of 2π; that is,

$$\omega_o N = 2\pi r \tag{2.47a}$$

or

$$\frac{2\pi}{\omega_o} = \frac{N}{r}. \tag{2.47b}$$

If $2\pi/\omega_o$ is a noninteger rational number, then the period will be a multiple of $2\pi/\omega_o$. If $2\pi/\omega_o$ is not a rational number, then the sequence is aperiodic even though it has a sinusoidal envelope. For example, $x[n] = \cos(\sqrt{3}n + \phi)$ is an aperiodic sequence.

EXAMPLE 2.8   Determination of the Period of a Sinusoidal Sequence

Let us determine the periods of the sinusoidal sequences of Figure 2.16. In Figure 2.16(a), $\omega_o = 0$, and Eq. (2.47a) is satisfied with $r = 0$ and any value of N. Here, the smallest value of N is equal to 1. Likewise, in Figure 2.16(b), $\omega_o = 0.1\pi$, and hence, applying Eq. (2.47a), we find that $N = 20$. Following similar lines, we arrive at $N = 10$ for Figure 2.16(c), $N = 5$ for Figure 2.16(d), $N = 20$ for Figure 2.16(e), $N = 2$ for Figure 2.16(f), $N = 20$ for Figure 2.16(g), and $N = 5$ for Figure 2.16(h). These numbers are also evident from the figures.

The parameter ω_o in the two sequences of Eqs. (2.42) and (2.47a) is called the *angular frequency*. Since the time instant n is dimensionless, the unit of angular frequency ω_o and *phase* ϕ is simply radians. If the unit of n is designated as samples, then the unit of ω_o and ϕ is radians per sample. Often in practice, the angular frequency ω is expressed as

$$\omega = 2\pi f, \tag{2.48}$$

where f is *frequency* in cycles per sample.

Two interesting properties of these sequences are discussed next. Consider two complex exponential sequences $x_1[n] = e^{j\omega_1 n}$ and $x_2[n] = e^{j\omega_2 n}$ with $0 \le \omega_1 < 2\pi$ and $2\pi k \le \omega_2 < 2\pi(k + 1)$, where k is any positive integer. If

$$\omega_2 = \omega_1 + 2\pi k, \qquad (2.49)$$

then

$$x_2[n] = e^{j\omega_2 n} = e^{j(\omega_1 + 2\pi k)n} = e^{j\omega_1 n} = x_1[n].$$

Thus, these two sequences are indistinguishable. Likewise, two sinusoidal sequences $x_1[n] = \cos(\omega_1 n + \phi)$ and $x_2[n] = \cos(\omega_2 n + \phi)$ with $0 \le \omega_1 < 2\pi$ and $2\pi k \le \omega_2 < 2\pi(k+1)$, where k is any positive integer, are indistinguishable from one another if $\omega_2 = 2\pi k + \omega_1$. In other words, any exponential or sinusoidal sequence of a frequency ω_2 with a value outside the frequency range $0 \le \omega < 2\pi$ is equivalent to an exponential or sinusoidal sequence of a frequency of value ω_2 modulo 2π.

The second interesting feature of the discrete-time sinusoidal signals follows from the property of the sine and cosine functions. To illustrate this feature, consider two sinusoidal sequences $x_1[n] = \cos(\omega_1 n)$ and $x_2[n] = \cos(\omega_2 n)$ with $0 \le \omega_1 < \pi$ and $\pi \le \omega_2 < 2\pi$. Let $\omega_1 = \pi - \omega_o$ and $\omega_2 = \pi + \omega_o$ with $0 \le \omega_o < \pi$. Then it follows that $x_1[n] = \cos(\pi - \omega_o n) = -\cos(\omega_o n)$ and $x_2[n] = \cos(\pi + \omega_o n) = -\cos(\omega_o n) = x_1[n]$. Thus, the two sinusoidal sequences $x_1[n]$ and $x_2[n]$ are indistinguishable from one another. Hence, a sinusoidal sequence with a frequency ω_2 in the range $\pi \le \omega_2 < 2\pi$ assumes the identity of a sinusoidal sequence with a frequency $\omega_o = \omega_2 - \pi$. The frequency π is thus called the *folding frequency*. An implication of this property of the sinusoidal sequences can be seen in Figure 2.16. The frequency of oscillation of the discrete-time sinusoidal sequence $x[n] = A\cos(\omega_o n)$ increases as ω_o increases from 0 to π, and then the frequency of oscillation decreases as ω_o increases from π to 2π. As a result of the first property, a frequency ω_o in the neighborhood of $\omega = 2\pi k$ is indistinguishable from a frequency $\omega_o - 2\pi k$ in the neighborhood of $\omega = 0$, and a frequency ω_o in the neighborhood of $\omega = \pi(2k+1)$ is indistinguishable from a frequency $\omega_o - \pi(2k+1)$ in the neighborhood of $\omega = \pi$ for any integer value of k. Therefore, frequencies in the neighborhood of $\omega_o = 2\pi k$ are usually called *low frequencies,* and frequencies in the neighborhood of $\omega_o = \pi(2k+1)$ are called *high frequencies.* For example, $v_1[n] = \cos(0.1\pi n) = \cos(1.9\pi n)$, shown in Figure 2.16(b), is a low-frequency signal, whereas, $v_2[n] = \cos(0.8\pi n) = \cos(1.2\pi n)$, shown in Figure 2.16(d) and (h), is a high-frequency signal.

The elementary operations described in Section 2.1.2 can be combined to generate sequences with more complex waveforms. We illustrate this application in Example 2.9.

EXAMPLE 2.9   Generation of a Square Wave Sequence

Consider the three sinusoidal sequences $x_1[n]$, $x_2[n]$, and $x_3[n]$ with normalized angular frequencies 0.5, 0.15, and 0.25, respectively, as given below:

$$x_1[n] = \sin(0.05\pi n)\mu[n],$$

$$x_2[n] = \sin(0.15\pi n)\mu[n],$$

$$x_3[n] = \sin(0.25\pi n)\mu[n].$$

A plot of these three sinusoidal sequences is given in Figure 2.19(a)–(c), respectively. A new sequence $y[n]$ is formed by combining the above three sequences as follows:

$$y[n] = x_1[n] + \tfrac{1}{3}x_2[n] + \tfrac{1}{5}x_3[n],$$

whose plot is indicated in Figure 2.19(d). It can be seen that this new sequence has almost a square waveform.

An application of the modulation operation discussed earlier in Section 2.1.2 is to transform a sequence with low-frequency sinusoidal components to a sequence with high-frequency components by modulating the former with a sinusoidal sequence of very high frequency, as illustrated in Example 2.10.

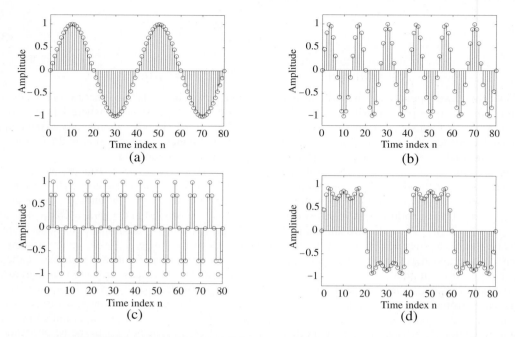

Figure 2.19: Sequences of Example 2.9: (a) $x_1[n]$, (b) $x_2[n]$, (c) $x_3[n]$, and (d) $y[n]$.

EXAMPLE 2.10   Illustration of the Modulation Operation

Let $x_1[n] = \cos(\omega_1 n)$ and $x_2[n] = \cos(\omega_2 n)$ with $\pi > \omega_2 >> \omega_1 > 0$. The product sequence $y[n] = x_1[n]x_2[n]$ is then given by

$$y[n] = \cos(\omega_1 n) \cdot \cos(\omega_2 n). \tag{2.50}$$

Using a trigonometric identity, we arrive at

$$y[n] = \tfrac{1}{2}\cos((\omega_1 + \omega_2)n) + \tfrac{1}{2}\cos((\omega_2 - \omega_1)n). \tag{2.51}$$

The new sequence $y[n]$ is thus composed of two sinusoidal sequences of frequencies $\omega_1 + \omega_2$ and $\omega_2 - \omega_1$. It should be noted that, because of the property given by Eq. (2.49), if $\omega_1 + \omega_2 > \pi$, the high-frequency sinusoidal sequence $\cos((\omega_1 + \omega_2)n)$ appears as a low-frequency sinusoidal sequence $\cos((2\pi - \omega_1 - \omega_2)n)$ with an angular frequency of value smaller than π.[4]

Figure 2.20 shows the plot of the sequence that is a product of two sinusoidal sequences of frequencies $\omega_1 = 0.1\pi$ and $\omega_2 = 0.01\pi$, respectively. As can be seen from this figure and also from Eq. (2.50), the product signal is a high-frequency sinusoidal sequence with an amplitude sinusoidally varying with time at a lower frequency.

Fundamental and Harmonic Components

It should be noted that the angular frequency of the sequence $x_2[n]$ in Example 2.9 is 3 times that of $x_1[n]$, and the angular frequency of the sequence $x_3[n]$ is 5 times that of $x_1[n]$. The sinusoidal sequences whose angular frequencies are integer multiples of a sinusoidal sequence of lower angular frequency are

---

[4]The appearance of a high-frequency signal $\cos((\omega_1 + \omega_2)n)$ as a low-frequency signal $\cos((2\pi - \omega_1 - \omega_2)n)$ is called aliasing (see Section 2.3).

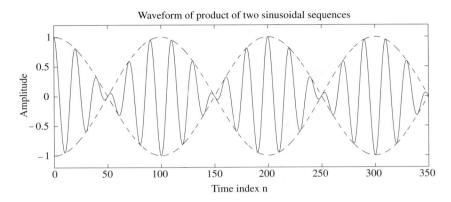

Figure 2.20: Product of two sinusoidal sequences with angular frequencies $\omega_1 = 0.1\pi$ and $\omega_2 = 0.01\pi$.

called the *harmonics,* whereas, the sinusoidal sequence with the lower frequency is called the *fundamental component,* with its frequency being called the *fundamental frequency.* The harmonic component with a frequency that is k times that of the fundamental frequency is called the *k-th harmonic.* In Example 2.9, the sequence $x_1[n]$ is the fundamental component with a fundamental angular frequency of 0.05 radian per sample. The sequences $x_2[n]$ and $x_3[n]$, are respectively, the third and fifth harmonic.

Any periodic sequence can be expressed in the form of linearly weighted combinations of a fundamental and a series of harmonic components called the *Fourier series* expansion. The weights associated with each component in the expansion are called the Fourier series coefficients. In Example 2.9, the Fourier series coefficients of $x_1[n]$, $x_2[n]$, and $x_3[n]$, are respectively, 1, $\frac{1}{3}$, and $\frac{1}{5}$. The development of the Fourier series expansion of a periodic sequence is considered in Problem 5.3.

Beat Notes

As shown in Example 2.10, the multiplication of two sinusoidal signals $x_1[n]$ and $x_2[n]$ with frequencies ω_1 and ω_2, respectively, results in a sum of two sinusoidal signals whose frequencies are $\omega_1 + \omega_2$ and $\omega_1 - \omega_2$, respectively, that is, the sum and differences of the frequencies of the original sinusoidal signals. If $\omega_2 \ll \omega_1$, then the two frequencies $\omega_1 + \omega_2$ and $\omega_1 - \omega_2$ are nearly equal to each other, and, as pointed out in Figure 2.20, the multiplication process generates a high-frequency signal, called a *beat note,* which fades in and out at a rate determined by the lower frequency. If the two frequencies are in the audible range, the beating effect can be heard. An application of the beat signal generation is in the tuning of two musical instruments. The beat note cannot be heard when both instruments have the same pitch.

2.2.2 Sequence Generation Using MATLAB

MATLAB includes a number of functions that can be used for signal generation. Some of these functions of interest are

```
exp,   sin,   cos,   square,   sawtooth
```

For example, the code fragments to generate a length-N complex exponential sequence with an exponent $a + jb$ and of the form shown in Figure 2.17 is given by

```
n = 1:N;
x = K*exp((a + b*i)*n);
```

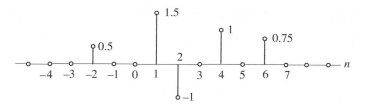

Figure 2.21: An arbitrary sequence $x[n]$.

Program 2_2.m
Program 2_3.m

The complete code is given in Program 2_2. Likewise, the code fragments to generate a length-$N + 1$ real exponential sequence with an exponent a and of the form shown in Figure 2.18 is given by

```
n = 0:N;
x = K*a.^n;
```

The complete code is given in Program 2_3. Another type of sequence generation using MATLAB can be found earlier in Example 2.1.

2.2.3   Representation of an Arbitrary Sequence

An arbitrary sequence can be represented in the time-domain as a weighted sum of a basic sequence and its delayed versions. A commonly used basic sequence in the representation of an arbitrary sequence is the unit sample sequence. For example, the sequence $x[n]$ of Figure 2.21 can be expressed as

$$x[n] = 0.5\,\delta[n + 2] + 1.5\,\delta[n - 1] - \delta[n - 2] + \delta[n - 4] + 0.75\,\delta[n - 6]. \tag{2.52}$$

An implication of this type of representation is considered later in Section 2.5.1, where we develop the general expression for calculating the output sequence of certain types of discrete-time systems for an arbitrary input sequence.

Since the unit step sequence and the unit sample sequence are simply related through Eq. (2.41a), it is also possible to represent an arbitrary sequence as a weighted combination of delayed unit step sequences (Problem 2.6).

2.3   The Sampling Process

We indicated earlier that often the discrete-time sequence is developed by uniformly sampling a continuous-time signal $x_a(t)$, as illustrated in Figure 2.2. The relation between the two signals is given by Eq. (2.2), where the time variable t of the continuous-time signal is related to the time variable n of the discrete-time signal only at discrete-time instants t_n given by

$$t_n = nT = \frac{n}{F_T} = \frac{2\pi n}{\Omega_T}, \tag{2.53}$$

with $F_T = 1/T$ denoting the sampling frequency and $\Omega_T = 2\pi F_T$ denoting the sampling angular frequency. For example, if the continuous-time signal is

$$x_a(t) = A\cos(2\pi f_o t + \phi) = A\cos(\Omega_o t + \phi), \tag{2.54}$$

the corresponding discrete-time signal is given by

$$x[n] = A \cos(\Omega_o nT + \phi)$$
$$= A \cos\left(\frac{2\pi \Omega_o}{\Omega_T} n + \phi\right) = A \cos(\omega_o n + \phi), \tag{2.55}$$

where

$$\omega_o = \frac{2\pi \Omega_o}{\Omega_T} = \Omega_o T \tag{2.56}$$

is the normalized digital angular frequency of the discrete-time signal $x[n]$. The unit of the normalized digital angular frequency ω_o is radians per sample, while the unit of the normalized analog angular frequency Ω_o is radians per second and the unit of the analog frequency f_o is hertz if the unit of the sampling period T is in seconds.

EXAMPLE 2.11   Ambiguity in the Discrete-Time Representation of Continuous-Time Signals

Consider the three sequences generated by uniformly sampling the three cosine functions of frequencies 3 Hz, 7 Hz, and 13 Hz, respectively: $g_1(t) = \cos(6\pi t)$, $g_2(t) = \cos(14\pi t)$, and $g_3(t) = \cos(26\pi t)$ with a sampling rate of 10 Hz; that is, with $T = 0.1$ sec. The derived sequences are therefore

$$g_1[n] = \cos(0.6\pi n), \qquad g_2[n] = \cos(1.4\pi n), \qquad g_3[n] = \cos(2.6\pi n).$$

Plots of these sequences (shown with circles) and their parent time functions are given in Figure 2.22 Note from these plots that each sequence has exactly the same sample value for any given n. This also can be verified by observing that

$$g_2[n] = \cos(1.4\pi n) = \cos((2\pi - 0.6\pi)n) = \cos(0.6\pi n)$$

and

$$g_3[n] = \cos(2.6\pi n) = \cos((2\pi + 0.6\pi)n) = \cos(0.6\pi n).$$

As a result, all three sequences above are identical, and it is difficult to associate a unique continuous-time function with any one of these sequences. In fact, all cosine waveforms of frequencies given by $(10k \pm 3)$ Hz, with k being any nonnegative integer, lead to the sequence $g_1[n] = \cos(0.6\pi n)$ when sampled at a 10-Hz rate.

In the general case, the family of continuous-time sinusoids

$$x_{a,k}(t) = A \cos(\pm(\Omega_o t + \phi) + k\Omega_T t), \qquad k = 0, \pm 1, \pm 2, \ldots \tag{2.57}$$

leads to identical sampled signals:

$$x_{a,k}(nT) = A \cos((\Omega_o + k\Omega_T)nT + \phi) = A \cos\left(\frac{2\pi(\Omega_o + k\Omega_T)n}{\Omega_T} + \phi\right)$$
$$= A \cos\left(\frac{2\pi \Omega_o n}{\Omega_T} + \phi\right) = A \cos(\omega_o n + \phi) = x[n]. \tag{2.58}$$

The above phenomenon of a continuous-time sinusoidal signal of higher frequency acquiring the identity of a sinusoidal sequence of lower frequency after sampling is called *aliasing*. Since there are an infinite number of continuous-time functions that can lead to a given sequence when sampled periodically, additional conditions need to be imposed so that the sequence $\{x[n]\} = \{x_a(nT)\}$ can uniquely represent the parent continuous-time function $x_a(t)$. In which case, $x_a(t)$ can be fully recovered from a knowledge of $\{x[n]\}$.

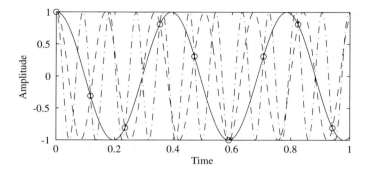

Figure 2.22: Ambiguity in the discrete-time representation of continuous-time signals. $g_1(t)$ is shown with the solid line, $g_2(t)$ is shown with the dashed line, $g_3(t)$ is shown with the dashed-dot line, and the sequence obtained by sampling is shown with circles.

EXAMPLE 2.12   Illustration of Aliasing

Determine the discrete-time signal $v[n]$ obtained by uniformly sampling a continuous-time signal $v_a(t)$ composed of a weighted sum of five sinusoidal signals of frequencies 30 Hz, 150 Hz, 170 Hz, 250 Hz, and 330 Hz, at a sampling rate of 200 Hz, as given below:

$$v_a(t) = 6\cos(60\pi t) + 3\sin(300\pi t) + 2\cos(340\pi t) + 4\cos(500\pi t) + 10\sin(660\pi t).$$

The sampling period $T = 1/200 = 0.005$ sec. Hence, the generated discrete-time signal $v[n]$ is given by

$$
\begin{aligned}
v[n] &= 6\cos(0.3\pi n) + 3\sin(1.5\pi n) + 2\cos(1.7\pi n) + 4\cos(2.5\pi n) \\
&\quad + 10\sin(3.3\pi n) \\
&= 6\cos(0.3\pi n) + 3\sin\left((2\pi - 0.5\pi)n\right) + 2\cos\left((2\pi - 0.3\pi)n\right) \\
&\quad + 4\cos\left((2\pi + 0.5\pi)n\right) + 10\sin\left((4\pi - 0.7\pi)n\right) \\
&= 6\cos(0.3\pi n) - 3\sin(0.5\pi n) + 2\cos(0.3\pi n) + 4\cos(0.5\pi n) \\
&\quad - 10\sin(0.7\pi n).
\end{aligned}
$$

As can be seen from the above, the components $3\sin(1.5\pi n)$, $2\cos(1.7\pi n)$, $4\cos(2.5\pi n)$, and $10\sin(3.3\pi n)$ have been aliased into the components $-3\sin(0.5\pi n)$, $2\cos(0.3\pi n)$, $4\cos(0.5\pi n)$, and $-10\sin(0.7\pi n)$, resulting in a discrete-time sequence

$$v[n] = 8\cos(0.3\pi n) + 5\cos(0.5\pi n + 0.6435) - 10\sin(0.7\pi n),$$

composed of only three sinusoidal sequences of normalized angular frequencies: 0.3π, 0.5π, and 0.7π.

It should be noted that an identical discrete-time signal is also generated by uniformly sampling at a 200-Hz sampling rate the following continuous-time signal composed of a weighted sum of three sinusoidal continuous-time signals of frequencies 30 Hz, 50 Hz, and 70 Hz:

$$w_a(t) = 8\cos(60\pi t) + 5\cos(100\pi t + 0.6435) - 10\sin(140\pi t).$$

Another example of a continuous-time signal generating the same discrete-time sequence is

$$u_a(t) = 2\cos(60\pi t) + 4\cos(100\pi t) + 10\sin(260\pi t) + 6\cos(460\pi t) + 3\sin(700\pi t),$$

which is composed of a weighted sum of five sinusoidal continuous-time signals of frequencies 30 Hz, 50 Hz, 130 Hz, 230 Hz, and 350 Hz, respectively.

Aliasing Demo

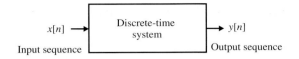

Figure 2.23: Schematic representation of a discrete-time system.

It follows from Eq. (2.56) that if $\Omega_T > 2\Omega_o$, then the corresponding normalized digital frequency ω_o of the discrete-time signal $x[n]$ obtained by sampling the parent continuous-time signal $x_a(t)$ will be in the range $-\pi < \omega < \pi$, implying no aliasing. On the other hand, if $\Omega_T < 2\Omega_o$, the normalized digital frequency will fold into a lower digital frequency ω_o in the range $-\pi < \omega < \pi$ because of aliasing. The value of ω_o is given by $2\pi\Omega_o/\Omega_T$ modulo 2π. Hence, to prevent aliasing, the sampling frequency Ω_T should be greater than 2 times the frequency Ω_o of the sinusoidal signal being sampled. Generalizing the above result, we observe that if we have an arbitrary continuous-time signal $x_a(t)$ that can be represented as a weighted sum of a number of sinusoidal signals, then $x_a(t)$ can also be represented uniquely by its sampled version $\{x[n]\}$ if the sampling frequency Ω_T is chosen to be greater than 2 times the highest frequency contained in $x_a(t)$. The condition to be satisfied by the sampling frequency to prevent aliasing is called the *sampling theorem,* which is formally derived later in Section 4.2.1.

The discrete-time signal obtained by sampling a continuous-time signal $x_a(t)$ may be represented as a sequence $\{x_a[nT]\}$. However, we shall use the more common notation $\{x[n]\}$ for simplicity (with T assumed to be normalized to 1 sec). It should be noted that when dealing with the sampled version of a continuous-time function, it is essential to know the numerical value of the sampling period T.

2.4 Discrete-Time Systems

The function of a discrete-time system is to process a given *input sequence* to generate an *output sequence*. In most applications, the discrete-time system used is a single-input, single-output system, as shown schematically in Figure 2.23. The output sequence is generated sequentially, beginning with a certain value of the time index n, and thereafter progressively increasing the value of n. If the beginning time index is n_o, the output $y[n_o]$ is first computed, then $y[n_o + 1]$ is computed, and so on. We restrict our attention in this text to this class of discrete-time systems with certain specific properties as described later in this section.

In a practical discrete-time system, all signals are digital signals, and operations on such signals also lead to digital signals. Such a discrete-time system is usually called a digital filter. However, if there is no ambiguity, we shall refer to a discrete-time system also as a digital filter, whether or not it has been implemented using finite precision arithmetic.

We consider here first several simple discrete-time systems that often find applications in practice. A study of these systems also will provide us with insights into the operation of more complex systems. We then describe several classifications of discrete-time systems. We next define two important characteristics of a discrete-time system.

2.4.1 Discrete-Time System Examples

The devices implementing the basic operations shown in Figures 2.5 and 2.9 can be considered as elementary discrete-time systems. The modulator and the adder are examples of two-input, single-output discrete-time systems. The remaining devices are examples of single-input, single-output discrete-time systems. More complex discrete-time systems are obtained by combining two or more of these elementary discrete-time systems, as illustrated in Figure 2.8. Some additional examples of discrete-time systems are given as follows.

Accumulator

A very simple example of a slightly more complex discrete-time system is the *accumulator,* defined by the input–output relation

$$y[n] = \sum_{\ell=-\infty}^{n} x[\ell]$$

$$= \sum_{\ell=-\infty}^{n-1} x[\ell] + x[n] = y[n-1] + x[n]. \tag{2.59}$$

The output $y[n]$ at time instant n is the sum of the input sample value $x[n]$ at time instant n and the previous output $y[n-1]$ at time instant $n-1$, which is the sum of all previous input sample values from $-\infty$ to the time instant $n-1$. The system therefore cumulatively adds; that is, it accumulates all input sample values from $-\infty$ to n. The accumulator can be considered as a discrete-time equivalent of a continuous-time integrator.

The above equation can also be written in the form

$$y[n] = \sum_{\ell=-\infty}^{-1} x[\ell] + \sum_{\ell=0}^{n} x[\ell] = y[-1] + \sum_{\ell=0}^{n} x[\ell], \qquad n \geq 0. \tag{2.60}$$

The second form of the accumulator is used for a causal input sequence, in which case $y[-1]$ is called the *initial condition.*

Moving-Average Filter

In Example 2.1, we pointed out that often data cannot be measured very accurately because of random variations in the measurements, and in the case of the data being corrupted by an additive noise, the n-th sample of the measured data $x[n]$ is modeled as $x[n] = s[n] + d[n]$, where $s[n]$ and $d[n]$ denote the n-th samples of the data and the noise. As demonstrated in this example, if multiple measurements of the same set of data samples are available, a reasonably good estimate of the uncorrupted data vector can be found by evaluating the ensemble average. In applications where data measurements cannot be repeated, a commonly used estimate of the data sample $s[n]$ at instant n from M measurements of the noise-corrupted data sample $x[\ell]$ available for the range $n - M + 1 \leq \ell \leq n$ is the *M-point average* or *mean* $y[n]$ given by

$$y[n] = \frac{1}{M} \sum_{\ell=0}^{M-1} x[n-\ell]. \tag{2.61}$$

An estimate of the spread of the mean value $y[n]$ from the actual value $s[n]$ is usually given by the *standard deviation* defined by [Tha98]

$$\sigma[n] = \sqrt{\frac{\sum_{\ell=0}^{M-1} (x[\ell] - y[n])^2}{M}}. \tag{2.62}$$

The discrete-time system implementing Eq. (2.61) is usually called the *M-point moving-average filter.* In most applications, the data $x[n]$ is a bounded sequence, and as a result, the M-point average $y[n]$ is also

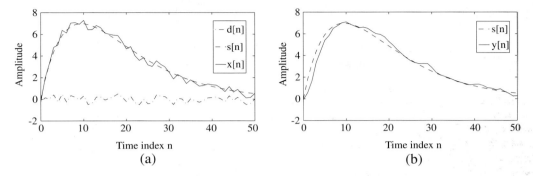

Figure 2.24: Pertinent signals of Example 2.13: $s[n]$ is the original uncorrupted sequence, $d[n]$ is the noise sequence, $x[n] = s[n] + d[n]$, and $y[n]$ is the output of the moving-average filter.

a bounded sequence. It follows from Eq. (2.62) that if there is no bias in the measurements, an improved estimate of the noisy data is obtained by simply increasing the number of measurements M.

A direct implementation of the M-point moving-average filter of Eq. (2.61) involves $M - 1$ additions, 1 division, and storage of $M - 1$ past input data samples. A more efficient implementation of the moving-average filter is developed next. From Eq. (2.61), it follows that we can write

$$y[n] = \frac{1}{M} \left(\sum_{\ell=0}^{M-1} x[n - \ell] + x[n - M] - x[n - M] \right)$$

$$= \frac{1}{M} \left(\sum_{\ell=1}^{M} x[n - \ell] + x[n] - x[n - M] \right)$$

$$= \frac{1}{M} \left(\sum_{\ell=0}^{M-1} x[n - 1 - \ell] + x[n] - x[n - M] \right),$$

which can be rewritten as

$$y[n] = y[n - 1] + \frac{1}{M} \left(x[n] - x[n - M] \right). \tag{2.63}$$

Computation of the M-point moving average $y[n]$ at time instant n of a sequence using the above recursive equation now requires 2 additions and 1 division, which is considerably smaller than that needed in the direct implementation of Eq. (2.61).

Using MATLAB, we illustrate the implementation of the moving-average filter of Eq. (2.61) and its application in removing the random variations in the measured data samples in Example 2.13

EXAMPLE 2.13   Moving-Average Filter

We assume for simplicity the original uncorrupted signal is given by Eq. (2.15). Program 2_4 has been used to generate the smoothed output $y[n]$ from the noise corrupted signal $x[n]$ using the moving-average system of Eq. (2.61). Figure 2.24 shows the plots of pertinent signals generated by Program 2_4 for $M = 3$. During execution, the program requests the input data, which is the desired number M of input samples to be added. It should be noted that to illustrate the effect of noise smoothing more clearly, the discrete-time signals have been plotted as continuous curves using the function `plot`.

Program 2_4.m

Figure 2.25: A factor-of-L interpolator.

We shall show in Example 3.18 that the moving-average filter of Eq. (2.61) acts like a lowpass filter and, as a result, smooths the input data by removing the high-frequency components. However, most random noise has frequency components throughout the frequency range $0 \leq \omega < \pi$, and hence, some low-frequency components of the noise will also be present in the output of the moving-average filter. An increase in the number of points M used for averaging decreases the passband width of the lowpass filter and may result in an overly smoothed output by removing some mid-frequency components of the original uncorrupted signal. The proper choice of M depends on the nature of the noise corrupting the original signal. In some applications, higher quality smoothed output may be obtained by using a cascade of identical moving average filters with a smaller value of M to process the noise-corrupted signal.

Note that in Figure 2.24(b), the output $y[n]$ of the 3-point moving-average filter is nearly equal to the desired uncorrupted input $s[n]$, except that it is delayed by one sample. We shall show later in Section 3.9 that a delay of $(M - 1)/2$ samples is inherent in an M-point moving-average filter.

Exponentially Weighted Running Average Filter

The average computed using Eq. (2.61) places equal emphasis on all M data samples. In some applications, it may be necessary to place more emphasis on data samples near the time instant n and less emphasis on data samples that are further away in determining the average. Such an average can be computed using the *exponentially weighted running average filter* given by [Tha98]:

$$y[n] = \alpha y[n - 1] + x[n], \qquad 0 < \alpha < 1. \tag{2.64}$$

Computation of the running average using the above equation requires only 2 additions and 1 multiplication. Moreover, it requires only the storage of the previous running average and does not require the storage of past input data samples.

For $0 < \alpha < 1$, the exponentially weighted average filter of Eq. (2.64) places more emphasis on current data samples and less emphasis on past data samples by exponentially weighting the data samples. To illustrate this property, we observe that $y[n]$ can be rewritten as

$$
\begin{aligned}
y[n] &= \alpha \left(\alpha y[n - 2] + x[n - 1] \right) + x[n] \\
&= \alpha^2 y[n - 2] + \alpha x[n - 1] + x[n] \\
&= \alpha^2 \left(\alpha y[n - 3] + x[n - 2] \right) + \alpha x[n - 1] + x[n] \\
&= \alpha^3 y[n - 3] + \alpha^2 x[n - 2] + \alpha x[n - 1] + x[n],
\end{aligned}
$$

by substituting the expression for $y[n - 1]$ in Eq. (2.64) and then substituting the expression for $y[n - 2]$ in the subsequent expression. Since $0 \leq \alpha < 1$, it can be seen from the last equation given above that the weights of past input data samples get progressively smaller at an exponential rate.

Linear Interpolator

Another example of a discrete-time system is the linear interpolator often employed to estimate sample values between pairs of adjacent sample values of a discrete-time sequence. The linear interpolation is

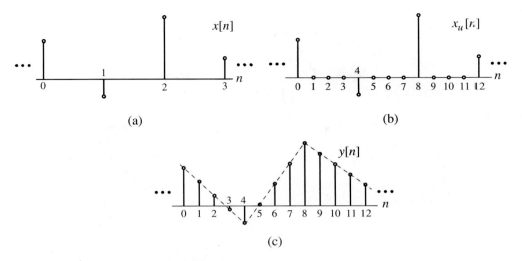

Figure 2.26: Illustration of the linear interpolation method.

implemented by first passing the input sequence $x[n]$ to be interpolated through an up-sampler whose output $x_u[n]$ is then passed through a second discrete-time system that "fills in" the zero-valued samples inserted by the up-sampler with values obtained by a linear interpolation of the pair of input samples surrounding the zero-valued samples, as indicated in Figure 2.25. The interpolated samples thus lie on a straight line joining the pair of input samples, as illustrated in Figure 2.26 for a factor-of-4 interpolation.

We develop the input–output relation of a linear factor-of-2 interpolator. Here, if $x_u[n]$ is a zero-valued sample inserted between a pair of input samples, it is replaced with the average of the two original input samples, $x_u[n-1]$ and $x_u[n+1]$:

$$y[n] = x_u[n] + \tfrac{1}{2}\left(x_u[n-1] + x_u[n+1]\right). \qquad (2.65)$$

On the other hand, if $x_u[n]$ is one of the original input samples, its neighbors, $x_u[n-1]$ and $x_u[n+1]$, are both equal to 0. In this case, it follows from Eq. (2.65) that $y[n] = x_u[n]$, or, in other words, the input sample is left unchanged. Thus, the cascade of the factor-of-2 up-sampler and the discrete-time system defined by Eq. (2.65) implements a factor-of-2 interpolator. Equation (2.65) is also known as the *bilinear interpolation.*

Likewise, it can be easily shown that for a factor-of-3 interpolator, the discrete-time system following the factor-of-3 up-sampler is characterized by the input–output relation given by

$$y[n] = x_u[n] + \tfrac{2}{3}\left(x_u[n-1] + x_u[n+1]\right) + \tfrac{1}{3}\left(x_u[n-2] + x_u[n+2]\right). \qquad (2.66)$$

The above type of linear interpolator is often used in some practical applications. One simple application is in the zooming of a digital image by integer factors both in the horizontal and vertical directions. For example, in the doubling of an image, zero-valued pixels are first inserted between each consecutive pixel in the horizontal direction, followed by an insertion of zero-valued pixels between each consecutive pixel in the vertical direction. Next, a bilinear interpolation is applied in the horizontal direction, followed by a bilinear interpolation in the vertical direction to replace the zero-valued pixels, with their interpolated values resulting in the zoomed image. An illustration of this approach to image zooming is shown in Figure 2.27.[5] Here, Figure 2.27(a) shows an original 512×512 size, gray-level image. The image of size

[5]Reproduced by permission from *http://vision.ai.uiuc.edu/ dugad/draft/dct.html.*

(a)   (b)

(c)

Figure 2.27: (a) Original 512×512 size gray-level image, (b) the down-sampled image of size 256×256, and (c) the zoomed version obtained using the bilinear interpolation.

256×256 obtained by down-sampling in both horizontal and vertical directions by a factor-of-2 is shown in Figure 2.27(b). The doubling of the down-sampled image obtained using the bilinear interpolation is indicated in Figure 2.27(c).

Another application of the linear interpolation is in the development of a full-color image from the data captured by a single CCD or CMOS imaging sensor array in inexpensive digital cameras.

The linear interpolator basically increases the sampling rate of the input signal by an integer factor. The theory behind interpolation is discussed in Section 13.2, where we also describe the interpolator design for increasing the sampling rate by an integer and by a ratio of integers. Arbitrary rate interpolation is discussed in Section 13.5.

Median Filter

The median of a set of $(2K+1)$ numbers is the number such that K numbers from the set have values greater than this number, while the other K numbers have values smaller. The median can be determined by rank-ordering the numbers in the set by their values and then choosing the number at the middle. For example,

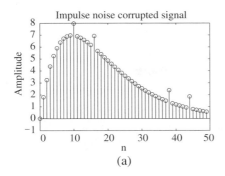
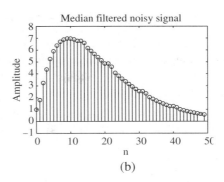

Figure 2.28: (a) Impulse-noise corrupted signal and (b) output of a length-3 median filter.

consider the set of numbers $\{2, -3, 10, 5, -1\}$. The rank-ordered set is given by $\{-3, -1, 2, 5, 10\}$. Hence, med$\{2, -3, 10, 5, -1\} = 2$.

The *median filter* is implemented by sliding a window of odd length over the input sequence $\{x[n]\}$ one sample at a time [Reg93],[Tuk74]. At any instant, the output of the filter is the median value of the input samples inside the window. More specifically, the output sample $y[n]$ at the nth instant of the median filter with a window of length-$(2K + 1)$ is given by

$$y[n] = \text{med}\{x[n - K], \ldots, x[n - 1], x[n], x[n + 1], \ldots, x[n + K]\}. \tag{2.67}$$

In practice, to process a finite-length sequence $\{x[n]\}$ of length N by a median filter with a window of length M, where $M < N$, $(M-1)/2$ zero-valued samples are appended to both sides of the input sequence $\{x[n]\}$ to create a new sequence $\{x_e[n]\}$ of length $N + M - 1$:

$$x_e[n] = \begin{cases} 0, & -\frac{(M-1)}{2} \leq n \leq -1, \\ x[n], & 0 \leq n \leq N - 1, \\ 0, & N \leq n \leq N - 1 + \frac{(M-1)}{2}. \end{cases}$$

The sequence $\{x_e[n]\}$ when processed by the median filter generates an output sequence $\{y[n]\}$ also of length N.

The median filter finds applications in removing additive random impulse noise, which show up as sudden large errors in the corrupted signal. In such cases, a linear lowpass filter, such as the moving-average filter or the exponentially weighted average filter, not only smooths out the sudden large-valued errors in the data but also distorts severely naturally occurring discontinuities in the original data. These types of discontinuities are often present in many practical situations, such as the sudden transitions between the voiced and unvoiced speech, and in naturally occurring edges in images and videos. The median filter is usually used for the smoothing of signals corrupted by impulse noise as illustrated in Example 2.14.

Program 2_5.m

EXAMPLE 2.14   Impulse Noise Removal Using Median Filter

The M-file `medfilt1` in MATLAB can be used to implement the median filter of Eq. (2.67) for the removal of the impulse noise in measured data samples. We assume for simplicity the original uncorrupted signal is given by Eq. (2.15). Figure 2.28 shows the plots of the pertinent signals generated by Program 2_5 for a median filter of length 3. It can be seen from Figure2.28(b) that the median filtered signal is nearly the same as the original uncorrupted signal as shown in Figure 2.6(a).

Median Filtering Demo

2.4.2   Classification of Discrete-Time Systems

There are various types of classification of discrete-time systems that are described next. These classifications are based on the input–output relation characterizing the system.

Linear System

The most widely used discrete-time system, and the one that we shall be concerned with throughout most of this text, is a *linear system* for which the superposition principle always holds. More precisely, for a linear discrete-time system, if $y_1[n]$ and $y_2[n]$ are the responses to the input sequences $x_1[n]$ and $x_2[n]$, respectively, then for an input

$$x[n] = \alpha x_1[n] + \beta x_2[n],$$

the response is given by

$$y[n] = \alpha y_1[n] + \beta y_2[n].$$

The superposition property must hold for any arbitrary constants, α and β, and for all possible inputs, $x_1[n]$ and $x_2[n]$. The above property makes it very easy to compute the response to a complicated sequence that can be decomposed as a weighted combination of some simple sequences, such as the unit sample sequences or the complex exponential sequences. In this case, the desired output is given by a similarly weighted combination of the outputs to the simple sequences.

    We examine the linearity property of the accumulator in Example 2.15.

EXAMPLE 2.15   Linearity Property of the Accumulator

From its input–output relation of the discrete-time accumulator given by Eq. (2.59), we observe that the outputs $y_1[n]$ and $y_2[n]$ for inputs $x_1[n]$ and $x_2[n]$ are given by

$$y_1[n] = \sum_{\ell=-\infty}^{n} x_1[\ell], \qquad\qquad y_2[n] = \sum_{\ell=-\infty}^{n} x_2[\ell].$$

The output $y[n]$ due to an input $\alpha x_1[n] + \beta x_2[n]$ is then given by

$$y[n] = \sum_{\ell=-\infty}^{n} (\alpha x_1[\ell] + \beta x_2[\ell])$$

$$= \alpha \sum_{\ell=-\infty}^{n} x_1[\ell] + \beta \sum_{\ell=-\infty}^{n} x_2[\ell] = \alpha y_1[n] + \beta y_2[n].$$

Hence, the discrete-time system of Eq. (2.59) is a linear system.

    Consider the modified form of the accumulator defined by Eq. (2.60) with a causal input applied at $n = 0$. Now, the outputs $y_1[n]$ and $y_2[n]$ for inputs $x_1[n]$ and $x_2[n]$ are given by

$$y_1[n] = y_1[-1] + \sum_{\ell=0}^{n} x_1[\ell], \qquad\qquad y_2[n] = y_2[-1] + \sum_{\ell=0}^{n} x_2[\ell].$$

Hence, the output $y[n]$ for an input $\alpha x_1[n] + \beta x_2[n]$ is of the form

$$y[n] = y[-1] + \sum_{\ell=0}^{n} (\alpha x_1[\ell] + \beta x_2[\ell]) = y[-1] + \alpha \sum_{\ell=0}^{n} x_1[\ell] + \beta \sum_{\ell=0}^{n} x_2[\ell]. \qquad (2.68)$$

On the other hand,

$$\alpha y_1[n] + \beta y_2[n] = \alpha y_1[-1] + \alpha \sum_{\ell=0}^{n} x_1[\ell] + \beta y_2[-1] + \beta \sum_{\ell=0}^{n} x_2[\ell], \tag{2.69}$$

which is equal to $y[n]$ of Eq. (2.68) if

$$y[-1] = \alpha y_1[-1] + \beta y_2[-1].$$

For the system of Eq. (2.60) to be linear, the above condition must hold for all initial conditions $y[-1]$, $y_1[-1]$, and $y_2[-1]$ and all constants α and β. This condition cannot be satisfied unless the system of Eq. (2.60) is initially at rest with a zero initial condition. For nonzero initial conditions, the discrete-time system of Eq. (2.60) is nonlinear.

It can be easily verified that the discrete-time systems of Eqs. (2.18), (2.20), (2.21), (2.61), (2.65), and (2.66) are linear systems (Problem 2.35). However, the linearity of the discrete-time system of Eq. (2.18) depends on the values of the initial conditions.

EXAMPLE 2.16   A Nonlinear Discrete-Time System

The median filter of Eq. (2.67) is a nonlinear discrete-time system. To show this, consider a median filter with a window of length 3. The output of the median filter for an input sequence $\{x_1[n]\} = \{3, 4, 5\}$, $0 \leq n \leq 2$, is a length-3 sequence $\{y_1[n]\} = \{3, 4, 4\}$, and that for an input sequence $x_2[n] = \{2, -1, -1\}$, $0 \leq n \leq 2$, is a length-3 sequence $\{y_2[n]\} = \{0, -1, -1\}$. On the other hand, the output for an input $\{x[n]\} = \{x_1[n] + x_2[n]\}$ is the sequence $\{y[n]\} = \{3, 4, 3\}$. It can be seen that $\{y_1[n] + y_2[n]\} = \{3, 3, 3\}$ is not equal to $\{y[n]\}$.

Shift-Invariant System

The *shift-invariance* property is the second condition imposed on most digital filters in practice. For a shift-invariant discrete-time system, if $y_1[n]$ is the response to an input $x_1[n]$, then the response to an input $x[n] = x_1[n - n_o]$ is simply $y[n] = y_1[n - n_o]$, where n_o is any positive or negative integer. This relation between the input and output must hold for any arbitrary input sequence and its corresponding output. In the case of sequences and systems with indices n related to discrete instants of time, the above restriction is more commonly called the *time-invariance* property. The time-invariance property ensures that for a specified input, the output of the system is independent of the time the input is being applied.

It can be shown that the median filter of Eq. (2.67) is a time-invariant system (Problem 2.37).

EXAMPLE 2.17   A Time-Varying Discrete-Time System

The up-sampler of Eq. (2.20) is a time-varying system. To show this, we observe from Eq. (2.20) that its output $x_{1,u}[n]$ for an input $x_1[n] = x[n - n_o]$ is given by

$$x_{1,u}[n] = \begin{cases} x_1[n/L], & n = 0, \pm L, \pm 2L, \ldots, \\ 0, & \text{otherwise}, \end{cases}$$

$$= \begin{cases} x[(n - Ln_o)/L], & n = 0, \pm L, \pm 2L, \ldots, \\ 0, & \text{otherwise}. \end{cases}$$

But from Eq. (2.20),

$$x_u[n - n_o] = \begin{cases} x[(n - n_o)/L], & n = n_o, n_o \pm L, n_o \pm 2L, \ldots \\ 0, & \text{otherwise}, \end{cases}$$

$$\neq x_{1,u}[n].$$

Likewise, it can be shown that the down-sampler of Eq. (2.21) is a time-varying system.

Causal System

In addition to the above two properties, we impose, for practicality, additional restrictions of causality and stability on the class of discrete-time systems we deal with in this text. In a *causal* discrete-time system, the n_oth output sample $y[n_o]$ depends only on input samples $x[n]$ for $n \leq n_o$ and does not depend on input samples for $n > n_o$. Thus, if $y_1[n]$ and $y_2[n]$ are the responses of a causal discrete-time system to the inputs $u_1[n]$ and $u_2[n]$, respectively, then

$$u_1[n] = u_2[n] \quad \text{for } n < N \qquad \text{implies also that} \qquad y_1[n] = y_2[n] \quad \text{for } n < N.$$

Simply speaking, for a causal system, changes in output samples do not precede changes in the input samples. It should be pointed out here that the definition of causality given above can be applied only to discrete-time systems with the same sampling rate for the input and the output.[6]

It can be easily shown that the discrete-time systems of Eqs. (2.18), (2.59), (2.60), and (2.61) are causal systems. However, the discrete-time systems defined by Eqs. (2.65) and (2.66) are noncausal systems. It should be noted that these two noncausal systems can be implemented as causal systems by simply delaying the output by one and two samples, respectively.

Stable System

There are various definitions of stability. We define a discrete-time system to be *stable* if and only if for every bounded input, the output is also bounded. This implies that, if the response to $x[n]$ is the sequence $y[n]$ and if

$$|x[n]| < B_x \qquad \text{for all values of } n, \text{ then,} \qquad |y[n]| < B_y$$

for all values of n where B_x and B_y are finite positive constants. This type of stability is usually referred to as *bounded-input, bounded-output* (BIBO) *stability*.

EXAMPLE 2.18   **Implication of the Stability Concept**

Consider the causal discrete-time system of Eq. (2.64). For a unit step sequence input, that is, for $x[n] = \mu[n], n \geq 0$, the output is given by

$$y[n] = \alpha^{n+1} y[-1] + \sum_{i=0}^{n} \alpha^i, \qquad n \geq 0.$$

Thus, if $0 < \alpha < 1$, $y[n]$ gets progressively smaller for increasing values of n, and $\lim_{n \to \infty} y[n] = 1/(1 - \alpha)$. On the other hand, if $\alpha > 1$, $y[n]$ becomes progressively larger for increasing values of n, and $\lim_{n \to \infty} y[n] = \infty$. In the latter case, for a bounded input, the output is unbounded, and hence the system is unstable.

EXAMPLE 2.19   **A Stable Discrete-Time System**

Consider the M-point moving-average filter of Example 2.13. It follows from Eq. (2.61) that for a bounded input $|x[n]| < B_x$, the magnitude of the nth sample of the output is given by

$$|y[n]| = \left| \frac{1}{M} \sum_{k=0}^{M-1} x[n-k] \right| \leq \frac{1}{M} \sum_{k=0}^{M-1} |x[n-k]| \leq \frac{1}{M} (M) B_x \leq B_x,$$

indicating that the system is BIBO stable.

---

[6]The definition of causality has to be modified if the input and output sampling rates are not the same.

Passive and Lossless Systems

A discrete-time system is said to be *passive* if, for every finite energy input sequence $x[n]$, the output sequence $y[n]$ has, at most, the same energy; that is,

$$\sum_{n=-\infty}^{\infty} |y[n]|^2 \leq \sum_{n=-\infty}^{\infty} |x[n]|^2 < \infty. \tag{2.70}$$

If the above inequality is satisfied with an equal sign for every input sequence, the discrete-time system is said to be *lossless*.

EXAMPLE 2.20   A Passive Discrete-Time System

Consider the discrete-time system defined by $y[n] = \alpha x[n - N]$, with N a positive integer. Its output energy is given by

$$\sum_{n=-\infty}^{\infty} |y[n]|^2 = |\alpha|^2 \sum_{n=-\infty}^{\infty} |x[n]|^2.$$

Hence, it is a passive system if $|\alpha| \leq 1$ and is a lossless system if $|\alpha| = 1$.

As we shall see later, in Section 12.9, the passivity and the losslessness properties are crucial to the design of discrete-time systems with very low sensitivity to changes in the filter coefficients.

2.4.3   Impulse and Step Responses

The response of a digital filter to a unit sample sequence $\{\delta[n]\}$ is called the *unit sample response,* or simply, the *impulse response,* and is denoted as $\{h[n]\}$. Correspondingly, the response of a discrete-time system to a unit step sequence $\{\mu[n]\}$, denoted as $\{s[n]\}$, is its *unit step response,* or simply, the *step response..* As we show in Section 2.5, a linear time-invariant digital filter is completely characterized in the time-domain by its impulse response or its step response.

EXAMPLE 2.21   Determination of the Impulse Response

Consider an LTI discrete-time system with an input-output relation

$$y[n] = \alpha_1 x[n] + \alpha_2 x[n - 1] + \alpha_3 x[n - 2] + \alpha_4 x[n - 3]. \tag{2.71}$$

Its impulse response $\{h[n]\}$ is obtained by setting $x[n] = \delta[n]$, resulting in

$$h[n] = \alpha_1 \delta[n] + \alpha_2 \delta[n - 1] + \alpha_3 \delta[n - 2] + \alpha_4 \delta[n - 3].$$

The impulse response is thus a finite-length sequence of length 4 given by

$$\{h[n]\} = \{\alpha_1, \quad \alpha_2, \quad \alpha_3, \quad \alpha_4\}, \quad 0 \leq n \leq 3.$$

EXAMPLE 2.22   Impulse Response of the Accumulator

The impulse response $\{h[n]\}$ of the discrete-time accumulator of Eq. (2.59) is obtained by setting $x[n] = \delta[n]$, resulting in

$$h[n] = \sum_{\ell=-\infty}^{n} \delta[\ell],$$

which from Eq. (2.41a) is precisely the unit step sequence $\mu[n]$.

EXAMPLE 2.23   **Impulse Response of the Linear Interpolator**

The impulse response $\{h[n]\}$ of the factor-of-2 interpolator of Eq. (2.65) is obtained by setting $x_u[n] = \delta[n]$ and is given by

$$h[n] = \delta[n] + \tfrac{1}{2}\left(\delta[n-1] + \delta[n+1]\right).$$

The impulse response is seen to be a finite-length sequence of length 3 and can be written alternately as

$$\{h[n]\} = \{1, 0.5, 1\}, \qquad -1 \le n \le 1.$$

2.5 Time-Domain Characterization of LTI Discrete-Time Systems

A *linear time-invariant* (LTI) discrete-time system satisfies both the linearity and the time-invariance properties. Such systems are mathematically easy to analyze and characterize and, as a consequence, easy to design. In addition, highly useful signal processing algorithms have been developed utilizing this class of systems over the last several decades. In this text, we consider almost entirely this type of discrete-time system.

In most cases, an LTI discrete-time system is designed as an interconnection of simple subsystems. Each subsystem, in turn, is implemented with the aid of the basic building blocks discussed earlier in Section 2.1.2. In order to be able to analyze such systems in the time domain, we need to develop the pertinent relationships between the input and the output of an LTI discrete-time system and the characterization of the interconnected system. We first show that the output sequence of a linear, time-invariant discrete-time system is given by the convolution sum of its impulse response sequence with the input sequence. We then outline a simple tabular method to compute the convolution sum of two finite-length sequences. From the convolution sum description of a linear, time-invariant discrete-time system, we next develop the stability condition and the causality condition in terms of its impulse response.

2.5.1 Input–Output Relationship

A consequence of the linear, time-invariance property is that an LTI discrete-time system is completely characterized by its impulse response; that is, knowing the impulse response, we can compute the output of the system to any arbitrary input. We develop this relationship now.

Let $h[n]$ denote the impulse response of the LTI discrete-time system of interest, that is, the response to an input $\delta[n]$. We first compute the response of this filter to the input $x[n]$ of Eq. (2.52). Since the discrete-time system is time-invariant, its response to $\delta[n-1]$ is $h[n-1]$. Likewise, the responses to $\delta[n+2]$, $\delta[n-4]$, and $\delta[n-6]$ are, respectively, $h[n+2]$, $h[n-4]$, and $h[n-6]$. Because of linearity, the response of the LTI discrete-time system to the input

$$x[n] = 0.5\delta[n+2] + 1.5\delta[n-1] - \delta[n-2] + \delta[n-4] + 0.75\delta[n-6]$$

will be simply

$$y[n] = 0.5h[n+2] + 1.5h[n-1] - h[n-2] + h[n-4] + 0.75h[n-6].$$

It follows from the above result that an arbitrary input sequence $x[n]$ can be expressed as a weighted linear combination of delayed and advanced unit sample sequences in the form

$$x[n] = \sum_{k=-\infty}^{\infty} x[k]\delta[n-k], \tag{2.72}$$

where the weight $x[k]$ on the right-hand side denotes specifically the kth sample value of the sequence $\{x[n]\}$. The response of the LTI discrete-time system to the sequence $x[k]\delta[n-k]$ is $x[k]h[n-k]$. As a result, the response $y[n]$ of the discrete-time system to $x[n]$ is given by

$$y[n] = \sum_{k=-\infty}^{\infty} x[k]h[n-k], \qquad (2.73a)$$

which can be alternately written as

$$y[n] = \sum_{k=-\infty}^{\infty} x[n-k]h[k] \qquad (2.73b)$$

by a simple change of variables. The above sum in Eqs. (2.73a) and (2.73b) is called the *convolution sum* of the sequences $x[n]$ and $h[n]$ and represented compactly as

$$y[n] = x[n] \circledast h[n], \qquad (2.74)$$

where the notation $\circledast$ denotes the convolution sum.[7]

EXAMPLE 2.24   Convolution of a Sequence with a Unit Sample Sequence

We evaluate

$$y[n] = x[n] \circledast \delta[n] = \sum_{k=-\infty}^{\infty} x[n-k]\delta[k].$$

To this end, we rewrite the above equation as

$$y[n] = \sum_{k=-\infty}^{-1} x[n-k]\delta[k] + x[n]\delta[0] + \sum_{k=1}^{\infty} x[n-k]\delta[k].$$

Since $\delta[k] = 0$ for $k \neq 0$ and $\delta[0] = 1$, the above equation reduces to $y[n] = x[n]$, or in other words,

$$x[n] \circledast \delta[n] = x[n].$$

EXAMPLE 2.25   Output Computation for a Delayed Impulse Response

An LTI discrete-time system with an impulse response $h[n]$ generates an output $y[n]$ for an input $x[n]$. We determine the output $y_1[n]$ of the LTI discrete-time system with an impulse response $h[n-m]$ for the input $x[n]$. From Eq. (2.73a), it follows that

$$y[n-m] = \sum_{k=-\infty}^{\infty} x[k]h[n-m-k] = x[n] \circledast h[n-m].$$

Hence,

$$y_1[n] = h[n-m].$$

---

[7] In the literature, the symbol for the convolution sum is $*$ without the circle. However, as the superscript $*$ is always used for denoting the complex conjugation operation, in this text we have adopted the symbol $\circledast$ to denote the convolution sum operation.

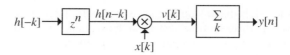

Figure 2.29: Schematic representation of the convolution sum operation.

The convolution sum operation satisfies several useful properties. First, the operation is *commutative;* that is,

$$x_1[n] \circledast x_2[n] = x_2[n] \circledast x_1[n]. \tag{2.75}$$

Second, the convolution operation, for stable and single-sided sequences, is *associative;* that is,

$$(x_1[n] \circledast x_2[n]) \circledast x_3[n] = x_1[n] \circledast (x_2[n] \circledast x_3[n]), \tag{2.76}$$

and last, the operation is *distributive;* that is,

$$x_1[n] \circledast (x_2[n] + x_3[n]) = x_1[n] \circledast x_2[n] + x_1[n] \circledast x_3[n]. \tag{2.77}$$

Proof of these properties is left as exercises (Problems 2.54 to 2.56).

The convolution sum operation of Eq. (2.73a) can be interpreted as follows. We first time-reverse the sequence $h[k]$, arriving at $h[-k]$. We then shift $h[-k]$ to the right by n sampling periods if $n > 0$, or to the left by n sampling periods if $n < 0$, to form the sequence $h[n - k]$. Next, we form the product sequence $v[k] = x[k]h[n - k]$. Summing all samples of $v[k]$ then yields the nth sample of $y[n]$ of the convolution sum. The process of generating $v[k]$ is illustrated in Figure 2.29. This process is implemented for each value of n in the range $-\infty < n < \infty$. The representation of the alternate form of the convolution sum operation given by Eq. (2.73b) is obtained by interchanging the sequences $x[k]$ and $h[k]$ in Figure 2.29.

It is clear from the above discussion that the impulse response $\{h[n]\}$ completely characterizes an LTI discrete-time system in the time domain because, knowing the impulse response, we can compute, in principle, the output sequence $y[n]$ for any given input sequence $x[n]$ using the convolution sum of Eq. (2.73a) or (2.73b). The computation of an output sample is simply a sum of products involving fairly simple arithmetic operations such as additions, multiplications, and delays. However, in practice, the convolution sum can be employed to compute the output sample at any instant only if either the impulse response sequence and/or the input sequence is of finite length, resulting in a finite sum of products. Note that if both the input and the impulse response sequences are of finite length, the output sequence is also of finite length. In the case of a discrete-time system with an infinite-length impulse response, it is obviously not possible to compute the output using the convolution sum if the input is also of infinite length. We shall therefore consider alternative time-domain descriptions of such systems that involve only finite sums of products.

EXAMPLE 2.26   Illustration of Output Computation Using the Convolution Operation

We systematically develop the sequence $y[n]$ generated by the convolution of the two finite-length sequences $x[n]$ and $h[n]$ shown in Figure 2.30(a). As can be seen from the plot of the shifted time-reversed version $\{h[n - k]\}$ for $n < 0$ sketched in Figure 2.30(b) for $n = -3$, for any value of the sample index k, the kth sample of either $\{x[k]\}$ or $\{h[n - k]\}$ is zero. As a result, the product of the kth samples is always zero for any k, and consequently, the convolution sum of Eq. (2.73a) leads to

$$y[n] = 0 \qquad \text{for } n < 0.$$

Consider now the calculation of $y[0]$. We form $\{h[-k]\}$ as shown in Figure 2.30(c). The product sequence $\{x[k]h[-k]\}$ is plotted in Figure 2.30(d), which has a single nonzero sample for $k = 0$, $x[0]h[0]$. Thus,

$$y[0] = x[0]h[0] = -2.$$

For the computation of $y[1]$, we shift $\{h[-k]\}$ to the right by one sample period to form $\{h[1-k]\}$, as sketched in Figure 2.30(e). The product sequence $\{x[k]h[1-k]\}$ shown in Figure 2.30(f) has one nonzero sample for $k = 0$, $x[0]h[1]$. As a result,

$$y[1] = x[0]h[1] + x[1]h[0] = -4 + 0 = -4.$$

To calculate $y[2]$, we form $\{h[2-k]\}$, as indicated in Figure 2.30(g). The resulting product sequence $\{x[k]h[2-k]\}$ is plotted in Figure 2.30(h), which leads to

$$y[2] = x[0]h[2] + x[1]h[1] + x[2]h[0] = 0 + 0 + 1 = 1.$$

This process is continued, resulting in

$$y[3] = x[0]h[3] + x[1]h[2] + x[2]h[1] + x[3]h[0] = 2 + 0 + 2 - 1 = 3,$$
$$y[4] = x[1]h[3] + x[2]h[2] + x[3]h[1] + x[4]h[0] = 0 + 0 - 2 + 3 = 1,$$
$$y[5] = x[2]h[3] + x[3]h[2] + x[4]h[1] = -1 + 0 + 6 = 5,$$
$$y[6] = x[3]h[3] + x[4]h[2] = 1 + 0 = 1,$$
$$y[7] = x[4]h[3] = -3.$$

From the plot of $\{h[n-k]\}$ for $n > 7$ as shown in Figure 2.30(i) and the plot of $\{x[k]\}$ in Figure 2.30(a), it can be seen that there is no overlap between these two sequences. Hence,

$$y[n] = 0 \qquad \text{for } n > 7.$$

The sequence $\{y[n]\}$ as obtained above is sketched in Figure 2.31.

It should be noted that the sum of the indices of each sample product inside the summation of either Eq. (2.73a) or Eq. (2.73b) is equal to the index of the sample being generated by the convolution sum operation. For example, the sum in the computation of $y[3]$ in the above example involves the products $x[0]h[3]$, $x[1]h[2]$, $x[2]h[1]$, and $x[3]h[0]$. The sum of the indices in each of these four products is equal to 3, which is the index of the sample $y[3]$.

As can be seen from Example 2.26, the convolution of two finite-length sequences results in a finite-length sequence. In this example, the convolution of a sequence $\{x[n]\}$ of length 5 with a sequence $\{h[n]\}$ of length 4 resulted in a sequence $\{y[n]\}$ of length 8. In general, if the lengths of the two sequences being convolved are M and N, then the resulting sequence after convolution is of length $M + N - 1$ (Problem 2.57).

EXAMPLE 2.27   Output Calculation Using the Convolution of the Input with the Impulse Response

Consider a causal LTI discrete-time system with an impulse response $h[n] = \beta^n \mu[n]$, where $|\beta| < 1$. We determine its output sequence $y[n]$ for a causal input $x[n] = \alpha^n \mu[n]$, with $|\alpha| < 1$.
The output $y[n]$ is given by

$$y[n] = h[n] \circledast x[n] = \beta^n \mu[n] \circledast \alpha^n \mu[n].$$

Since both $h[n]$ and $x[n]$ are causal sequences, $y[n]$ is also a causal sequence. For $n \geq 0$, we then have from Eq. (2.73a)

$$y[n] = \sum_{k=0}^{n} x[k]h[n-k] = \sum_{k=0}^{n} \alpha^k \beta^{n-k}.$$

In MATLAB, the M-file conv implements the convolution of two finite-length sequences. Its application is illustrated in Example 2.28.

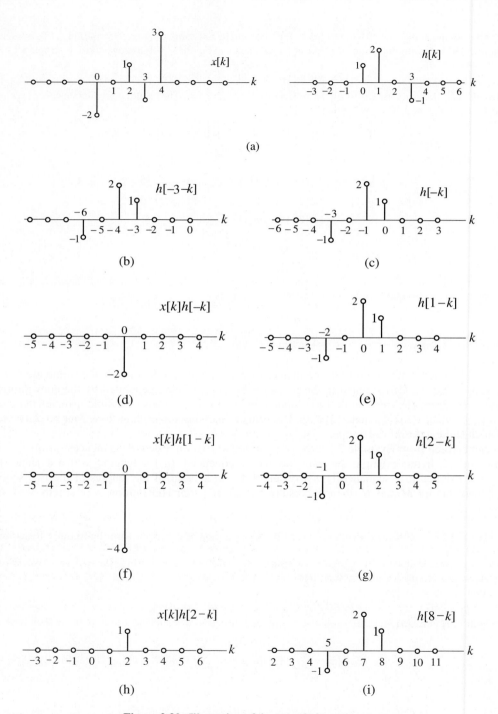

Figure 2.30: Illustration of the convolution process.

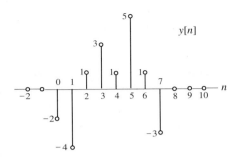

Figure 2.31: Sequence generated by the convolution.

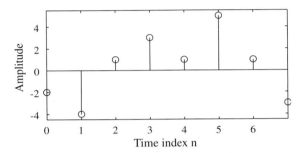

Figure 2.32: Sequence generated by convolution using MATLAB.

EXAMPLE 2.28   Convolution Computation Using MATLAB

Program 2_6 given in the Appendix can be used to compute the convolution of two finite-length sequences. The plot of the sequence generated by convolving the two sequences of Example 2.26 using Program 2_6 is indicated in Figure 2.32. As expected, the result is identical to that derived in Example 2.26.

Program 2_6.m

2.5.2 Tabular Method of Convolution Sum Computation

The calculation of the convolution sum of two finite-length sequences can be carried out simply either by using a tabular method similar to that used in conventional multiplication of two numbers [Pie96] or using a method based on the multiplication of two polynomials. These two method are simpler to carry out and easy to remember than the graphical method outlined in Section 2.5.1. We illustrate the first method in this section. The polynomial multiplication method of computing the convolution sum is described in Section 6.6.

Without any loss of generality we consider the evaluation of the convolution of the sequence $\{g[n]\}$, $0 \leq n \leq 3$, with the sequence $\{h[n]\}$, $0 \leq n \leq 2$ generating the sequence

$$y[n] = g[n] \circledast h[n], \qquad 0 \leq n \leq 5.$$

The samples of these two sequences are then multiplied using the conventional multiplication method but without any carry operations on the columns. First, each sample of $\{g[n]\}$ is multiplied with $h[0]$ and the samples of the product sequence are placed in a row beginning at time index $n = 0$. Next, each sample of $\{g[n]\}$ is multiplied with $h[1]$ and the samples of the product sequence are placed in a second row beginning at time index $n = 1$. Finally, each sample of $\{g[n]\}$ is multiplied with $h[2]$ and the samples of the product sequence are placed in a third row beginning at time index $n = 2$. The process is indicated below.

n :	0	1	2	3	4	5
$g[n]$:	$g[0]$	$g[1]$	$g[2]$	$g[3]$		
$h[n]$:	$h[0]$	$h[1]$	$h[2]$	-		
	$g[0]h[0]$	$g[1]h[0]$	$g[2]h[0]$	$g[3]h[0]$		
	-	$g[0]h[1]$	$g[1]h[1]$	$g[2]h[1]$	$g[3]h[1]$	
	-	-	$g[0]h[2]$	$g[1]h[2]$	$g[2]h[2]$	$g[3]h[2]$
$y[n]$:	$y[0]$	$y[1]$	$y[2]$	$y[3]$	$y[4]$	$y[5]$

It should be noted that each line in the above table corresponds to a delayed, weighted impulse response. The samples of the sequence $\{y[n]\}$ generated by the convolution sum are obtained by adding the three entries in the column above each sample and are given by

$$y[0] = g[0]h[0],$$

$$y[1] = g[1]h[0] + g[0]h[1],$$

$$y[2] = g[2]h[0] + g[1]h[1] + g[0]h[2],$$

$$y[3] = g[3]h[0] + g[2]h[1] + g[1]h[2],$$

$$y[4] = g[3]h[1] + g[2]h[2],$$

$$y[5] = g[3]h[2].$$

EXAMPLE 2.29   **Convolution of Two One-Sided Sequences Using the Tabular Method**

We develop the convolution sum of the two sequences $\{x[n]\}$ and $\{h[n]\}$ of Example 2.26 using the above method. The process is illustrated below.

n :	0	1	2	3	4	5	6	7
$x[n]$:	-2	0	1	-1	3			
$h[n]$:	1	2	0	-1	-			
	-2	0	1	-1	3			
	-	-4	0	2	-2	6		
	-	-	0	0	0	0	0	-
	-	-	-	2	0	-1	1	-3
$y[n]$:	-2	-4	1	3	1	5	1	-3

Thus, the convolution sum of the two sequences $x[n]$ and $h[n]$ yields

$$\{y[n]\} = \{-2 \quad -4 \quad 1 \quad 3 \quad 1 \quad 5 \quad 1 \quad -3\}, \qquad 0 \le n \le 7$$

which is seen to be identical to that derived in Example 2.26.

The tabular method can also be employed to evaluate the convolution sum of two finite-length two-sided sequences [Pie96]. In this case, a decimal point is first placed to the right of the sample with the time index $n = 0$. Next, the convolution is computed ignoring the locations of the decimal point. Finally, the decimal point is inserted according to the rules of conventional multiplication. The sample immediately to the left of the decimal point is then located at the time index $n = 0$.

EXAMPLE 2.30   Convolution of Two-Sided Sequences Using the Tabular Method

We determine the convolution sum $y[n]$ of the two sequences:

$$\{g[n]\} = \{3 \quad -2 \quad 4\}, \qquad \{h[n]\} = \{4 \quad 2 \quad -1\}.$$

The convolution sum of the above two sequences are next evaluated using the tabular method as shown below, where the samples at time index $n = 0$ are indicated by the solid circles placed immediately to their right.

			$g[n]$:		3	$-2_\bullet$	4	
			$h[n]$:		$4_\bullet$	2	-1	
					-3	2	-4	
				6	-4	8	-	
		12	-8	16	-	-		
		$y[n]$:	12	$-2_\bullet$	9	10	-4	

Hence, we have

$$\{y[n]\} = \{12 \quad -2 \quad 9 \quad 10 \quad -4\}.$$

2.5.3 Stability Condition in Terms of the Impulse Response

Recall from Section 2.4.2 that a discrete-time system is defined to be stable, or precisely, bounded-input, bounded-output (BIBO) stable, if the output sequence $\{y[n]\}$ of the system remains bounded for all bounded input sequences $\{x[n]\}$. We now develop the stability condition for an LTI discrete-time system. We show that an LTI digital filter is BIBO stable if and only if its impulse response sequence $\{h[n]\}$ is absolutely summable, that is,

$$S = \sum_{n=-\infty}^{\infty} |h[n]| < \infty. \tag{2.78}$$

We prove the above statement for a real impulse response $\{h[n]\}$. The extension of the proof for a complex impulse response sequence is left as an exercise (Problem 2.74). Now, if the input sequence $\{x[n]\}$ is bounded, that is, $|x[n]| \leq B_x < \infty$, then the output amplitude at time instant n, from Eq. (2.73b), is

$$|y[n]| = \left| \sum_{k=-\infty}^{\infty} h[k]x[n-k] \right| \leq \sum_{k=-\infty}^{\infty} |h[k]| \, |x[n-k]|$$

$$\leq B_x \sum_{k=-\infty}^{\infty} |h[k]| = B_x S < \infty. \tag{2.79}$$

Thus, $S < \infty$ implies $|y[n]| \leq B_y < \infty$, indicating that the sequence $\{y[n]\}$ is also bounded. To prove the converse, assume that the sequence $\{y[n]\}$ is bounded, that is, $|y[n]| \leq B_y$. Now, consider the input given by

$$x[n] = \text{sgn}\,(h[-n]), \tag{2.80}$$

where sgn(c) is the *signum function* defined by

$$\text{sgn}(c) = \begin{cases} +1, & \text{for } c \geq 0, \\ -1, & \text{for } c < 0. \end{cases}$$

Note that since $|x[n]| = 1$, $\{x[n]\}$ is obviously bounded. For this input, $y[n]$ at $n = 0$ is

$$y[0] = \sum_{k=-\infty}^{\infty} \text{sgn}\,(h[k])h[k] = \sum_{k=-\infty}^{\infty} |h[k]| = \mathcal{S}. \tag{2.81}$$

It follows from the above that if $\mathcal{S} = \infty$, then $\{y[n]\}$ is not a bounded sequence as one of its sample is equal to ∞.

We investigate the BIBO stability of some simple systems in Examples 2.31 through 2.33.

EXAMPLE 2.31   Stability Condition of a Causal First-Order LTI Discrete-Time System

Consider a causal LTI discrete-time system with an impulse response given by

$$h_1[n] = \alpha^n \mu[n].$$

For this system,

$$\mathcal{S} = \sum_{n=-\infty}^{\infty} \left| \alpha^n \mu[n] \right|$$

$$= \sum_{n=0}^{\infty} |\alpha|^n = \frac{1}{1 - |\alpha|}, \qquad \text{for } |\alpha| < 1.$$

Therefore, $\mathcal{S} < \infty$ if $|\alpha| < 1$ for which the above system is BIBO stable. On the other hand, if $|\alpha| \geq 1$, the infinite series $\sum_{n=0}^{\infty} |\alpha|^n$ does not converge, and as a result, the above causal system is not BIBO stable.

EXAMPLE 2.32   Stability Condition of an Anti-Causal First-Order LTI Discrete-Time System

We now examine the stability of an LTI discrete-time system with an anticausal impulse response

$$h_2[n] = -\beta^n \mu[-n-1].$$

Here,

$$\mathcal{S} = \sum_{n=-\infty}^{\infty} \left| \beta^n \mu[-n-1] \right| = \sum_{n=-\infty}^{-1} |\beta|^n$$

$$= \sum_{m=1}^{\infty} |\beta|^{-m} = |\beta|^{-1} \sum_{m=0}^{\infty} |\beta|^{-m} = \frac{|\beta|^{-1}}{1 - |\beta|^{-1}}, \qquad \text{for } |\beta| > 1.$$

Hence, in this case, $\mathcal{S} < \infty$ if $|\beta| > 1$ for which the above anticausal system is BIBO stable. However, for $|\beta| \leq 1$, the infinite series $\sum_{m=0}^{\infty} |\beta|^{-m}$ does not converge, in which case, the above anticausal system is not BIBO stable.

EXAMPLE 2.33 **Stability Condition of an LTI Discrete-Time System with a Finite Impulse Response**

Now an LTI discrete-time system with an impulse response given by

$$h[n] = \begin{cases} \alpha^n, & N_1 \le n \le N_2, \\ 0, & \text{otherwise,} \end{cases} \tag{2.82}$$

has only a finite number of nonzero impulse response samples for finite values of N_1 and N_2. Hence, the impulse response sequence is absolutely summable independent of the value of α as long as it is not infinite. Hence, the system of Eq. (2.82) is BIBO stable for finite values of α.

2.5.4   Causality Condition in Terms of the Impulse Response

We now develop the condition for an LTI discrete-time system to be causal. Let $x_1[n]$ and $x_2[n]$ be two input sequences with

$$x_1[n] = x_2[n] \qquad \text{for } n \le n_o. \tag{2.83}$$

From Eq. (2.73b), the corresponding output samples at $n = n_o$ of an LTI discrete-time system with an impulse response $\{h[n]\}$ are then given by

$$y_1[n_o] = \sum_{k=-\infty}^{\infty} h[k] x_1[n_o - k] = \sum_{k=0}^{\infty} h[k] x_1[n_o - k] + \sum_{k=-\infty}^{-1} h[k] x_1[n_o - k], \tag{2.84a}$$

$$y_2[n_o] = \sum_{k=-\infty}^{\infty} h[k] x_2[n_o - k] = \sum_{k=0}^{\infty} h[k] x_2[n_o - k] + \sum_{k=-\infty}^{-1} h[k] x_2[n_o - k]. \tag{2.84b}$$

If the LTI discrete-time system is also causal, then $y_1[n_o]$ must be equal to $y_2[n_o]$. Now, because of Eq. (2.83), the first sum on the right-hand side of Eq. (2.84a) is equal to the first sum on the right-hand side of Eq. (2.84b). This implies that the second sums on the right-hand side of the above two equations must be equal. As $x_1[n]$ may not be equal to $x_2[n]$ for $n > n_o$, the only way these two sums will be equal is if they are each equal to zero, which is satisfied if

$$h[k] = 0 \qquad \text{for } k < 0. \tag{2.85}$$

As a result, an LTI discrete-time system is *causal* if and only if its impulse response sequence $\{h[n]\}$ is a causal sequence satisfying the condition of Eq. (2.85) .

It follows from Example 2.21 that the discrete-time system of Eq. (2.71) is a causal system since its impulse response satisfies the causality condition of Eq. (2.85). Likewise, from Example 2.22, we observe that the discrete-time accumulator of Eq. (2.59) is also a causal system. On the other hand, from Example 2.23, it can be seen that the factor-of-2 linear interpolator defined by Eq. (2.65) is a noncausal system because its impulse response does not satisfy the causality condition of Eq. (2.85). However, a noncausal discrete-time system with a finite-length impulse response can often be realized as a causal system by inserting a delay of an appropriate amount. For example, a causal version of the discrete-time factor-of-2 linear interpolator is obtained by delaying the output by one sample period with an input–output relation given by

$$y[n] = x_u[n - 1] + \tfrac{1}{2} \left(x_u[n - 2] + x_u[n] \right) .$$

2.6   Simple Interconnection Schemes

Two widely used schemes for developing complex LTI discrete-time systems from simple LTI discrete-time system sections are described next.

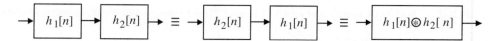

Figure 2.33: The cascade connection.

2.6.1 Cascade Connection

Figure 2.33 shows the *cascade connection* of two LTI discrete-time systems in which the output of one system is connected to the input of the second system. The overall impulse response $h[n]$ of the cascade of the two systems with impulse responses $h_1[n]$ and $h_2[n]$, respectively, is given by their linear convolution, that is,

$$h[n] = h_1[n] \circledast h_2[n]. \tag{2.86}$$

Hence, the cascaded system is equivalent to a system with an impulse response $h_1[n] \circledast h_2[n]$. If there are more than two LTI systems, the impulse response of the overall cascade is given by the linear convolution of the impulse responses of the individual systems. In general, the ordering of the filters in the cascade has no effect on the overall impulse response because of the commutative property of convolution.

It can be shown that the cascade connection of stable systems is stable. Likewise, the cascade connection of passive (lossless) systems is passive (lossless).

An application of the cascade connection scheme is in the development of an *inverse system*. If the two LTI systems in the cascade connection of Figure 2.33 are such that

$$h_1[n] \circledast h_2[n] = \delta[n], \tag{2.87}$$

then the LTI system $h_2[n]$ is said to be the inverse of the LTI system $h_1[n]$, and vice versa. As a result of the above relation, if the input to the cascaded system is $x[n]$, its output is also $x[n]$. An application of this concept is in the recovery of a signal from its distorted version appearing at the output of a transmission channel. This is accomplished by designing an inverse system if the impulse response of the channel is known.

Example 2.34 illustrates the development of an inverse system.

EXAMPLE 2.34   Inverse of a Discrete-Time Accumulator

From Example 2.22, the impulse response of the discrete-time accumulator is the unit step sequence $\mu[n]$. Therefore, from Eq. (2.87), the inverse system must satisfy the condition

$$\mu[n] \circledast h_2[n] = \delta[n]. \tag{2.88}$$

It follows from Eq. (2.88) that $h_2[n] = 0$ for $n < 0$ and

$$h_2[0] = 1, \quad \sum_{\ell=0}^{n} h_2[\ell] = 0, \qquad n \geq 1.$$

As a result,

$$h_2[1] = -1, \quad \text{and} \quad h_2[n] = 0, \quad \text{for} \quad n \geq 2.$$

Thus, the impulse response of the inverse system is given by

$$h_2[n] = \delta[n] - \delta[n-1],$$

which is called a *backward difference system*.

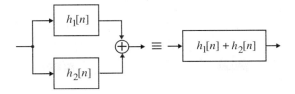

Figure 2.34: The parallel connection.

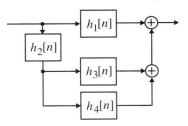

Figure 2.35: The discrete-time system of Example 2.35.

2.6.2 Parallel Connection

The connection scheme of Figure 2.34 is called the *parallel* connection, and here the outputs of the two
LTI discrete-time systems are added to form the new output while the same input is fed to both systems.
The impulse response of the overall system is thus given by

$$h[n] = h_1[n] + h_2[n]. \tag{2.89}$$

Likewise, if more than two LTI discrete-time systems are in parallel, the impulse response of the overall
system is given by the sum of the impulse responses of the individual systems.

It is a simple exercise to show that the parallel connection of stable systems is stable. However, the
parallel connection of passive (lossless) systems may or may not be passive (lossless).

**EXAMPLE 2.35   Analysis of a Discrete-Time System Composed of Cascade and Parallel Connections of
Several Systems**

Consider the discrete-time system of Figure 2.35 composed of an interconnection of four simple discrete-time
systems with impulse responses given by

$$h_1[n] = \delta[n] + \tfrac{1}{2}\delta[n-1], \qquad h_2[n] = \tfrac{1}{2}\delta[n] - \tfrac{1}{4}\delta[n-1],$$

$$h_3[n] = 2\delta[n], \qquad h_4[n] = -2\left(\tfrac{1}{2}\right)^n \mu[n].$$

The overall impulse response $h[n]$ is given by

$$h[n] = h_1[n] + h_2[n] \circledast (h_3[n] + h_4[n])$$

$$= h_1[n] + h_2[n] \circledast h_3[n] + h_2[n] \circledast h_4[n].$$

Now,

$$h_2[n] \circledast h_3[n] = (\tfrac{1}{2}\delta[n] - \tfrac{1}{4}\delta[n-1]) \circledast 2\delta[n] = \delta[n] - \tfrac{1}{2}\delta[n-1],$$

and

$$h_2[n] \circledast h_4[n] = \left(\tfrac{1}{2}\delta[n] - \tfrac{1}{4}\delta[n-1]\right) \circledast \left(-2\left(\tfrac{1}{2}\right)^n \mu[n]\right)$$

$$= -\left(\tfrac{1}{2}\right)^n \mu[n] + \tfrac{1}{2}\left(\tfrac{1}{2}\right)^{n-1} \mu[n-1] = -\left(\tfrac{1}{2}\right)^n \mu[n] + \left(\tfrac{1}{2}\right)^n \mu[n-1]$$

$$= -\left(\tfrac{1}{2}\right)^n \delta[n] = -\delta[n].$$

Therefore,

$$h[n] = \delta[n] + \tfrac{1}{2}\delta[n-1] + \delta[n] - \tfrac{1}{2}\delta[n-1] - \delta[n] = \delta[n].$$

2.7   Finite-Dimensional LTI Discrete-Time Systems

An important subclass of LTI discrete-time systems is characterized by a linear constant coefficient difference equation of the form

$$\sum_{k=0}^{N} d_k y[n-k] = \sum_{k=0}^{M} p_k x[n-k], \tag{2.90}$$

where $x[n]$ and $y[n]$ are, respectively, the input and the output of the system, and $\{d_k\}$ and $\{p_k\}$ are constants. The *order* of the discrete-time system is given by $\max(N, M)$, which is the order of the difference equation characterizing the system. It is possible to implement an LTI system characterized by Eq. (2.90) since the computation here involves two finite sums of products even though such a system, in general, has an impulse response of infinite length.

The output $y[n]$ can then be computed recursively from Eq. (2.90). If we assume the system to be causal, then we can rewrite Eq. (2.90) to express $y[n]$ explicitly as a function of $x[n]$:

$$y[n] = -\sum_{k=1}^{N} \frac{d_k}{d_0} y[n-k] + \sum_{k=0}^{M} \frac{p_k}{d_0} x[n-k], \tag{2.91}$$

provided $d_0 \neq 0$. The output $y[n]$ can be computed for all $n \geq n_o$, knowing $x[n]$ and the initial conditions $y[n_o - 1], y[n_o - 2], \ldots, y[n_o - N]$.

A simple finite-dimensional LTI system is considered in Example 2.36.

EXAMPLE 2.36   Monetary Dynamics of Hyperinflation

An example of a first-order LTI discrete-time system is the model for the portfolio equilibrium condition for the monetary dynamics of hyperinflation proposed by Cagan [Cag56]:

$$m[n] - p[n] = \alpha(p^e[n+1] - p[n]), \qquad \alpha < 0, \ n \geq 0, \tag{2.92}$$

where $m[n]$ is the log of the money supply, $p[n]$ is the log of the price level, and $p^e[n+1]$ is the log of the expected price level at the next period $n+1$ based on the data available at the current period n. Note that $p^e[n+1] - p[n]$ represents the expected inflation. If it is assumed that the expected inflation in the next period to be a scaled value of the current rate of inflation, $p[n] - p[n-1]$, that is,

$$p^e[n+1] - p[n] = \gamma(p[n] - p[n-1]), \tag{2.93}$$

then, Eq. (2.92) reduces to the first-order difference equation

$$m[n] - p[n] = \alpha\gamma p[n] - \alpha\gamma p[n-1],$$

or equivalently,

$$p[n] - \frac{\alpha\gamma}{1+\alpha\gamma} p[n-1] = \frac{1}{1+\alpha\gamma} m[n]. \tag{2.94}$$

It can be shown that the impulse response of the above system is given by

$$h[n] = \frac{1}{\alpha\gamma+1} \left(\frac{\alpha\gamma}{1+\alpha\gamma}\right)^n, \qquad n \geq 0. \tag{2.95}$$

From Example 2.31, it follows that the system of Eq. (2.36) is BIBO stable for finite α and γ if and only if

$$\left|\frac{\alpha\gamma}{1+\alpha\gamma}\right| < 1. \tag{2.96}$$

We first outline two different methods to compute the output of a causal LTI discrete-time system characterized by a linear-constant coefficient difference equation of the form of Eq. (2.91) for a prescribed input. We then describe the computation of the impulse response of the above type of systems. Computation of the output using MATLAB is next discussed. Stability condition for such systems is then developed.

2.7.1　Total Solution Calculation

The procedure for computing the solution of the constant coefficient difference equation of Eq. (2.90) is very similar to that employed in solving the constant coefficient differential equation in the case of an LTI continuous-time system. In the case of the discrete-time system of Eq. (2.90), the output response $y[n]$ also consists of two components that are computed independently and then added to yield the total solution:

$$y[n] = y_c[n] + y_p[n]. \tag{2.97}$$

In Eq. (2.97), the component $y_c[n]$ is the solution of Eq. (2.90) with the input $x[n] = 0$; that is, it is the solution of the homogeneous difference equation:

$$\sum_{k=0}^{N} d_k y[n-k] = 0, \tag{2.98}$$

and the component $y_p[n]$ is a solution of Eq. (2.90) with $x[n] \neq 0$. $y_c[n]$ is called the *complementary solution* or *homogeneous solution*, while $y_p[n]$ is called the *particular solution*, resulting from the specified input $x[n]$, often called the *forcing function*. The sum of the complementary and the particular solutions as given by Eq. (2.97) is called the *total solution*.

We first describe the method of computing the complementary solution $y_c[n]$. To this end, we assume that it is of the form

$$y_c[n] = \lambda^n. \tag{2.99}$$

Substituting the above in Eq. (2.98), we arrive at

$$\sum_{k=0}^{N} d_k y[n-k] = \sum_{k=0}^{N} d_k \lambda^{n-k}$$

$$= \lambda^{n-N}(d_0\lambda^N + d_1\lambda^{N-1} + \cdots + d_{N-1}\lambda + d_N) = 0. \tag{2.100}$$

The polynomial $\sum_{k=0}^{N} d_k \lambda^{N-k}$ is called the *characteristic polynomial* of the discrete-time system of Eq. (2.90). Let $\lambda_1, \lambda_2, \ldots, \lambda_N$ denote its N roots. If these roots are all distinct, then the general form of the complementary solution is given by

$$y_c[n] = \alpha_1 \lambda_1^n + \alpha_2 \lambda_2^n + \cdots + \alpha_N \lambda_N^n, \tag{2.101}$$

where $\alpha_1, \alpha_2, \ldots, \alpha_N$ are constants determined from the specified initial conditions of the discrete-time system. The complementary solution takes a different form in the case of multiple roots. For example, if λ_1 is of multiplicity L and the remaining $N - L$ roots, $\lambda_2, \lambda_3, \ldots, \lambda_{N-L}$, are distinct, then Eq. (2.101) takes the form

$$y_c[n] = \alpha_1 \lambda_1^n + \alpha_2 n \lambda_1^n + \alpha_3 n^2 \lambda_1^n + \cdots + \alpha_L n^{L-1} \lambda_1^n + \alpha_{L+1} \lambda_2^n + \cdots + \alpha_N \lambda_{N-L}^n. \tag{2.102}$$

Next, we consider the determination of the particular solution $y_p[n]$ of the difference equation of Eq. (2.90). Here the procedure is to assume that the particular solution is also of the same form as the specified input $x[n]$ if $x[n]$ has the form λ_0^n ($\lambda_0 \neq \lambda_i$, $i = 1, 2, \ldots, N$) for all n. Thus, if $x[n]$ is a constant, then $y_p[n]$ is also assumed to be constant. Likewise, if $x[n]$ is a sinusoidal sequence, then $y_p[n]$ is also assumed to be a sinusoidal sequence, and so on.

We illustrate the determination of the total solution in Example 2.37.

EXAMPLE 2.37   **Total Solution Computation of an LTI System for a Constant Input**

Let us determine the total solution for $n \geq 0$ of a discrete-time system characterized by the following difference equation:

$$y[n] + y[n-1] - 6y[n-2] = x[n], \tag{2.103}$$

for a step input $x[n] = 8\mu[n]$ and with initial conditions $y[-1] = 1$ and $y[-2] = -1$.

We first determine the form of the complementary solution. Setting $x[n] = 0$ and $y[n] = \lambda^n$ in Eq. (2.103), we arrive at

$$\lambda^n + \lambda^{n-1} - 6\lambda^{n-2} = \lambda^{n-2}(\lambda^2 + \lambda - 6)$$
$$= \lambda^{n-2}(\lambda + 3)(\lambda - 2) = 0,$$

and hence the roots of the characteristic polynomial $\lambda^2 + \lambda - 6$ are $\lambda_1 = -3, \lambda_2 = 2$. Therefore, the complementary solution is of the form

$$y_c[n] = \alpha_1(-3)^n + \alpha_2(2)^n. \tag{2.104}$$

For the particular solution, we assume

$$y_p[n] = \beta.$$

Substituting the above in Eq. (2.103), we get

$$\beta + \beta - 6\beta = 8\mu[n],$$

which for $n \geq 0$ yields $\beta = -2$.

The total solution is therefore of the form

$$y[n] = \alpha_1(-3)^n + \alpha_2(2)^n - 2, \qquad n \geq 0. \tag{2.105}$$

The constants α_1 and α_2 are chosen to satisfy the specified initial conditions. From Eqs. (2.103) and (2.105), we get

$$y[-2] = \alpha_1(-3)^{-2} + \alpha_2(2)^{-2} - 2 = -1,$$
$$y[-1] = \alpha_1(-3)^{-1} + \alpha_2(2)^{-1} - 2 = 1.$$

Solving these two equations, we arrive at

$$\alpha_1 = -1.8, \quad \alpha_2 = 4.8.$$

Thus, the total solution is given by

$$y[n] = -1.8(-3)^n + 4.8(2)^n - 2, \qquad n \geq 0. \qquad (2.106)$$

If the input excitation is of the same form as one of the terms in the complementary solution, then it is necessary to modify the form of the particular solution as illustrated in Example 2.38.

EXAMPLE 2.38   Total Solution Computation of an LTI System for an Exponential Input

We determine the total solution for $n \geq 0$ of the difference equation of Eq. (2.103) for an input $x[n] = 2^n \mu[n]$ with the same initial conditions as in Example 2.37.

As indicated in Example 2.37, the complementary solution contains a term $\alpha_2(2)^n$, which is of the same form as the specified input. Hence, we need to select a form for the particular solution that is distinct and does not contain any terms similar to those contained in the complementary solution. We assume

$$y_p[n] = \beta n(2)^n.$$

Substituting the above in Eq. (2.103), we get

$$\beta n(2)^n + \beta(n-1)(2)^{n-1} - 6\beta(n-2)(2)^{n-2} = (2)^n \mu[n].$$

For $n \geq 0$, we obtain from the above equation $\beta = 0.4$. The total solution is now of the form

$$y[n] = \alpha_1(-3)^n + \alpha_2(2)^n + 0.4n(2)^n, \qquad n \geq 0. \qquad (2.107)$$

To determine the values of α_1 and α_2, we make use of the specified initial conditions. From Eqs. (2.103) and (2.107), we arrive at

$$y[-2] = \alpha_1(-3)^{-2} + \alpha_2(2)^{-2} + 0.4(-2)(2)^{-2} = -1,$$
$$y[-1] = \alpha_1(-3)^{-1} + \alpha_2(2)^{-1} + 0.4(-1)(2)^{-1} = 1,$$

which when solved yields $\alpha_1 = -5.04, \alpha_2 = -0.96$. Therefore, the total solution is given by

$$y[n] = -5.04(-3)^n - 0.96(2)^n + 0.4n(2)^n, \qquad n \geq 0.$$

2.7.2   Zero-Input Response and Zero-State Response

An alternate approach to determining the total solution $y[n]$ of the difference equation of Eq. (2.90) is by computing its *zero-input response*, $y_{zi}[n]$, and *zero-state response*, $y_{zs}[n]$. The component $y_{zi}[n]$ is obtained by solving Eq. (2.90) by setting the input $x[n] = 0$, and the component $y_{zs}[n]$ is obtained by solving Eq. (2.90) by applying the specified input with all initial conditions set to zero. The total solution is then given by $y_{zi}[n] + y_{zs}[n]$.

This approach is illustrated in Example 2.39.

EXAMPLE 2.39   Total Solution Computation from Zero-Input and Zero-State Responses

We determine the total solution of the discrete-time system of Example 2.37 by computing the zero-input response and the zero-state response.

The zero-input response, $y_{zi}[n]$, of Eq. (2.103) is given by the complementary solution of Eq. (2.104), where the constants α_1 and α_2 are chosen to satisfy the specified initial conditions. Now, from Eq. (2.103), we get

$$y[0] = -y[-1] + 6y[-2] = -1 - 6 = -7, \qquad y[1] = -y[0] + 6y[-1] = 7 + 6 = 13.$$

Next, from Eq. (2.104), we get

$$y[0] = \alpha_1 + \alpha_2, \qquad y[1] = -3\alpha_1 + 2\alpha_2.$$

Solving these two sets of equations, we arrive at $\alpha_1 = -5.4$, $\alpha_2 = -1.6$. Therefore,

$$y_{zi}[n] = -5.4(-3)^n - 1.6(2)^n, \qquad n \geq 0.$$

The zero-state response is determined from Eq. (2.105) by evaluating the constants α_1 and α_2 to satisfy the zero initial conditions. From Eq. (2.103), we get

$$y[0] = x[0] = 8, \qquad y[1] = x[1] - y[0] = 0.$$

Next, from Eq. (2.105) and the above set of equations, we arrive at $\alpha_1 = 3.6$, $\alpha_2 = 6.4$. Thus, the zero-state response for $n \geq 0$ with initial conditions $y_{zs}[-2] = y_{zs}[-1] = 0$ is given by

$$y_{zs}[n] = 3.6(-3)^n + 6.4(2)^n - 2.$$

Hence, the total solution $y[n]$ is given by the sum $y_{zi}[n] + y_{zs}[n]$, resulting in

$$y[n] = -1.8(-3)^n + 4.8(2)^n - 2, \qquad n \geq 0,$$

which is identical to that derived in Example 2.37, as expected.

2.7.3 Impulse Response Calculation

The impulse response $h[n]$ of a causal LTI discrete-time system is the output observed with input $x[n] = \delta[n]$. Thus, it is simply the zero-state response with $x[n] = \delta[n]$. Now for such an input, $x[n] = 0$ for $n > 0$, and thus, the particular solution is zero, that is, $y_p[n] = 0$. Hence, the impulse response can be computed from the complementary solution of Eq. (2.101) in the case of simple roots of the characteristic equation by determining the constants α_i to satisfy the zero initial conditions. A similar procedure can be followed in the case of multiple roots of the characteristic equation. A system with all zero initial conditions is often called a *relaxed* system.

We illustrate the impulse response computation in Examples 2.40 and 2.41.

EXAMPLE 2.40   **Impulse Response Computation from Zero-State Response**

In this example, we determine the impulse response $h[n]$ of the causal discrete-time system of Example 2.37. From Eq. (2.104), we get

$$h[n] = \alpha_1(-3)^n + \alpha_2(2)^n, \qquad n \geq 0.$$

From the above, we arrive at

$$h[0] = \alpha_1 + \alpha_2, \qquad h[1] = -3\alpha_1 + 2\alpha_2.$$

Next, from Eq. (2.103) with $x[n] = \delta[n]$, we get

$$h[0] = 1, \qquad h[1] + h[0] = 0.$$

Solution of the above two sets of equations yields $\alpha_1 = 0.6$ and $\alpha_2 = 0.4$.

Thus, the impulse response is given by

$$h[n] = 0.6(-3)^n + 0.4(2)^n, \qquad n \geq 0.$$

EXAMPLE 2.41   Impulse Response Computation from Total Solution

A causal LTI discrete-time system with an impulse response $h[n]$ satisfies the following difference equation:

$$h[n] - ah[n-1] = \delta[n]. \tag{2.108}$$

We determine a closed-form expression for $h[n]$, and the input–output relation of the above system.

The total solution of the difference equation of Eq. (2.108) is given by

$$h[n] = h_c[n] + h_p[n], \tag{2.109}$$

where $h_c[n]$ and $h_p[n]$ are, respectively, the complementary and the particular solutions. To determine the complementary solution, we set the right-hand side of Eq. (2.108) equal to zero and set $h[n] = \lambda^n$, resulting in

$$\lambda^n - a\lambda^{n-1} = 0.$$

The nontrivial solution of the above equation is $\lambda = a$, and hence, $h_c[n] = a^n$. For the particular solution, we assume $h_p[n] = \beta$. Substituting the expressions for $h_c[n]$ and $h_p[n]$ in Eq. (2.109), we get

$$h[n] = a^n + \beta. \tag{2.110}$$

From Eqs. (2.108) and (2.110), we then have

$$h[0] = 1 = 1 + \beta,$$

implying $\beta = 0$. Hence, the total solution of the difference equation of Eq. (2.108) is given by

$$h[n] = \begin{cases} a^n, & n \geq 0, \\ 0, & n < 0. \end{cases} \tag{2.111}$$

It should be noted that the above result could also have been obtained by induction by evaluating Eq. (2.108) for $n = 0, 1, 2, \ldots$, and then solving for $h[0]$, $h[1]$, $h[2]$, and so on. (Problem 2.44).

To determine the general input–output relation of the above discrete-time system, we convolve both sides of Eq. (2.108) with $x[n]$ and make use of Eq. (2.74) to arrive at

$$y[n] - ay[n-1] = x[n]. \tag{2.112}$$

It follows from the form of the complementary solution given by Eq. (2.102) that the impulse response of a finite-dimensional LTI system characterized by a difference equation of the form of Eq. (2.90) is of infinite length. However, as illustrated in Example 2.42, there exist LTI discrete-time systems with an infinite impulse response that cannot be characterized by the difference equation form of Eq. (2.90).

EXAMPLE 2.42   A Causal Stable LTI Discrete-Time System with No Difference Equation Representation

The system defined by the impulse response

$$h[n] = \frac{1}{n^2} \mu[n-1]$$

does not have a representation in the form of a linear constant coefficient difference equation. It should be noted that the above system is causal and also BIBO stable.

Since the impulse response $h[n]$ of a causal discrete-time system is a causal sequence, Eq. (2.91) can also be used to calculate recursively the impulse response for $n \geq 0$ by setting initial conditions to zero values, that is, by setting $y[-1] = y[-2] = \cdots = y[-N] = 0$, and using a unit sample sequence $\delta[n]$ as the input $x[n]$. The step response of a causal LTI system can similarly be computed recursively by setting zero initial conditions and applying a unit step sequence as the input. It should be noted that the causal discrete-time system of Eq. (2.91) is linear only for zero initial conditions (Problem 2.67).

2.7.4 Output Computation Using MATLAB

The causal LTI system of the form of Eq. (2.91) can be simulated in MATLAB using the function `filter` already made use of in Program 2.4. The function implements Eq. (2.91) in the form of a set of equations as indicated below:

$$y[n] = \frac{p_0}{d_0}x[n] + s_1[n-1],$$

$$s_1[n] = \frac{p_1}{d_0}x[n] - \frac{d_1}{d_0}y[n] + s_2[n-1],$$

$$\vdots \qquad\qquad (2.113)$$

$$s_{N-1}[n] = \frac{p_{N-1}}{d_0}x[n] - \frac{d_{N-1}}{d_0}y[n] + s_{N-2}[n-1],$$

$$s_N[n] = \frac{p_N}{d_0}x[n] - \frac{d_N}{d_0}y[n],$$

where $s_i[n]$, $1 \leq i \leq N$, are N internal variables. By back substitution, it can be shown that the above set of equations indeed reduces to Eq. (2.91). The values of the internal variables $s_i[n]$ at the starting instant are called the *initial conditions*.

The basic forms of the function `filter` are as follows:

```
y = filter(p,d,x)
[y, sf] = filter(p,d,x, si]
```

In the first form, the input data vector x is processed by the system characterized by the coefficient vectors p and d to generate the output vector y, assuming zero initial conditions. The length of y is the same as the length of x. The second form permits the inclusion of nonzero initial conditions of the internal variables $s_i[n]$ in the vector si and provides an output that includes the vector sf as the final values of $s_i[n]$. Since the function implements Eq. (2.91), the coefficient d_0 must be nonzero.

Example 2.43 illustrates the use of the function `filter` in the computation of the total solution.

EXAMPLE 2.43   **Total Solution Computation Using MATLAB**

We verify the result of Example 2.37 using MATLAB. In this case, Eq. (2.103) is of the form:

$$y[n] = x[n] + s_1[n-1],$$

$$s_1[n] = s_2[n-1] - y[n],$$

$$s_2[n] = 6\,y[n].$$

We next determine the initial values of $s_1[n]$ and $s_2[n]$ from the specified initial conditions $y[-1]$ and $y[-2]$. From the above set of equations we arrive at

$$s_2[-1] = 6\,y[-1] = 6,$$

$$s_1[-1] = -y[-1] + s_2[-2] = -y[-1] + 6\,y[-2] = -7.$$

The code fragment used to determine the first eight samples of the output $y[n]$, $0 \leq n \leq 7$, is

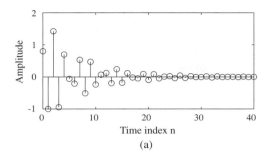
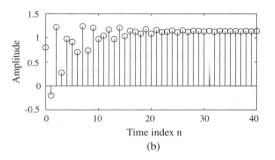

(a) (b)

Figure 2.36: (a) Impulse response and (b) step response of the system of Eq. (2.114).

```
[y,sf] = filter(1,[1 1 -6],8*ones(1,8), [-7 6]);
```

which resulted in the following values:

```
y =
Columns 1 through 8
1     13     1     85     -71     589     -1007     4549
```

The code fragments used to determine the first eight samples of the output $y[n]$, $0 \leq n \leq 7$, using Eq. (2.106) in MATLAB are

```
n = 0:7;
y = -1.8*(-3).^n + 4.8*(2).^n - 2;
```

The output samples generated using the above code are identical to those developed by the function `filter`.

2.7.5 Impulse and Step Response Computation Using MATLAB

The impulse and step responses of a causal LTI discrete-time system can be computed using the MATLAB M-files `impz` and `stepz`, respectively. Each function is available with several options. We illustrate the use of these two functions in Example 2.44.

EXAMPLE 2.44   Impulse and Step Response Computations Using MATLAB

Determine the first 41 samples of the impulse and response samples of the causal LTI system defined by

$$y[n] + 0.7y[n-1] - 0.45y[n-2] - 0.6y[n-3]$$
$$= 0.8x[n] - 0.44x[n-1] + 0.36x[n-2] + 0.02x[n-3]. \tag{2.114}$$

The code fragments that can be used to compute the impulse and step response samples are as follows:

```
p = [0.8 -0.44 0.36 0.02];
d = [1 0.7 -0.45 -0.6];
[h,m] = impz(p,d,41);
[s,m] = stepz(p,d,41);
```

The computed first 41 samples of the impulse and step response samples are indicated in Figures 2.36(a) and (b), respectively.

2.7.6   Location of Roots of Characteristic Equation for BIBO Stability

It should be noted that the impulse response samples of a stable LTI system decay to zero values as the time index n becomes very large. Likewise, the step response samples of a stable LTI system approach a constant value as n becomes very large. From the plots of Figure 2.36(a) and (b), we can conclude that most likely the LTI system of Eq. (2.114) is BIBO stable. However, it is impossible to check the stability of a system just by examining only a finite segment of its impulse or step response as in these figures.

The BIBO stability of a causal LTI system characterized by a constant coefficient difference equation of the form of Eq. (2.90) can be inferred from the values of the roots λ_i of its characteristic polynomial. To establish the stability conditions, recall that the form of the impulse response is the same as that of the complementary solution. From Eq. (2.101), assuming all the roots to be distinct, we have

$$h[n] = \sum_{i=1}^{N} \alpha_i \lambda_i^n \mu[n].$$

(2.115)

The constants α_i in the above expression are determined to satisfy zero initial conditions. From Eq. (2.115) we get

$$\sum_{n=0}^{\infty} |h[n]| = \sum_{n=0}^{\infty} \left| \sum_{i=1}^{N} \alpha_i (\lambda_i)^n \right| \leq \sum_{i=1}^{N} |\alpha_i| \sum_{n=0}^{\infty} |\lambda_i|^n.$$

(2.116)

It follows from the above equation that if $|\lambda_i| < 1$ for all values of i, then $\sum_{n=0}^{\infty} |\lambda_i|^n < \infty$, and as a result, $\sum_{n=0}^{\infty} |h[n]| < \infty$; that is, the impulse response is absolutely summable, implying BIBO stability of the causal LTI discrete-time system. However, the impulse response sequence is not absolutely summable if one or more of the roots λ_i has a magnitude greater than or equal to one. It should be noted that the discrete-time system of Example 2.37 described in Eq. (2.103) is clearly an unstable system as both roots of the characteristic equation have magnitudes greater than one.

In the case of multiple roots of the characteristic equation, the impulse response will contain terms of the form $n^K \lambda_i^n$. As a result, the expression for $\sum_{n=0}^{\infty} |h[n]|$ will contain the term

$$\sum_{n=0}^{\infty} |n^K (\lambda_i)^n|,$$

which converges if $|\lambda_i| < 1$ (Problem 2.89), and as a result, here also the impulse response is absolutely summable.

Summarizing, a causal LTI system characterized by a linear constant coefficient difference equation of the form of Eq. (2.90) is BIBO stable if the magnitude of each of the roots of its characteristic equation is less than one. This condition is both necessary and sufficient.

2.8   Classification of LTI Discrete-Time Systems

Linear time-invariant (LTI) discrete-time systems are usually classified either according to the length of their impulse response sequences or according to the method of calculation employed to determine the output samples.

2.8.1   Classification Based on Impulse Response Length

If $h[n]$ is of finite length, that is,

$$h[n] = 0 \qquad \text{for} \quad n < N_1 \quad \text{and} \quad n > N_2 \quad \text{with} \quad N_1 < N_2,$$

(2.117)

then it is known as a *finite impulse response* (FIR) discrete-time system, for which the convolution sum reduces to

$$y[n] = \sum_{k=N_1}^{N_2} h[k]x[n-k]. \tag{2.118}$$

Note that the above convolution sum, being a finite sum, can be used to calculate $y[n]$ directly. The basic operations involved are simply multiplication and addition. Note that the calculation of the present value of the output sequence involves the value of the input sample at $n = N_1$ and $N_2 - N_1$ previous values of the input sequence along with the $N_2 - N_1 + 1$ impulse response samples describing the FIR discrete-time system.

Examples of FIR discrete-time systems are the moving-average system of Eq. (2.61) and the linear interpolators of Eqs. (2.65) and (2.66).

If $h[n]$ is of infinite length, then it is known as an *infinite impulse response* (IIR) discrete-time system. For a causal IIR discrete-time system with a causal input $x[n]$, the convolution sum can be expressed in the form

$$y[n] = \sum_{k=0}^{n} x[k]h[n-k],$$

which can be used to compute the output samples. However, for increasing n, the computational complexity to compute the output sample increases as the number of products to be summed also increases.

The class of IIR filters we are concerned with in this text is the causal system characterized by the linear constant coefficient difference equation of Eq. (2.91). Note that here also the basic operations needed in the output calculations are multiplication and addition and involve a finite sum of terms for all values of n. An example of such an IIR system is the accumulator of Eqs. (2.59) and (2.60). Another example is the exponentially weighted running average filter of Eq. (2.64). A third example is described in Example 2.45.

EXAMPLE 2.45   Trapezoidal Integration as an IIR Discrete-Time System

The familiar numerical integration formulas that are used to numerically solve integrals of the form

$$y(t) = \int_0^t x(\tau)\,d\tau$$

can be shown to be characterized by linear constant coefficient difference equations and, hence, are examples of IIR systems. If we divide the interval of integration into n equal parts of length T, then the above integral can be rewritten as

$$y(nT) = y((n-1)T) + \int_{(n-1)T}^{nT} x(\tau)\,d\tau, \tag{2.119}$$

where we have set $t = nT$ and used the notation

$$y(nT) = \int_0^{nT} x(\tau)\,d\tau.$$

Using the trapezoidal method, we integrate the integral on the right-hand side of Eq. (2.119) and arrive at (with $y(0) = 0$)

$$y(nT) = y((n-1)T) + \frac{T}{2}[x((n-1)T) + x(nT)].$$

Denoting $y(nT) = y[n]$ and $x(nT) = x[n]$, we rewrite the above as

$$y[n] = y[n-1] + \frac{T}{2}(x[n] + x[n-1]), \tag{2.120}$$

which is recognized as the difference equation representation of a first-order IIR discrete-time system.

2.8.2   Classification Based on the Output Calculation Process

If the output sample can be calculated sequentially, knowing only the present and past input samples, the filter is said to be a *nonrecursive* discrete-time system. If, on the other hand, the computation of the output involves past output samples in addition to the present and past input samples, it is known as a *recursive* discrete-time system. An example of a nonrecursive system is the FIR discrete-time system implemented using Eq. (2.118). The IIR discrete-time system implemented using the difference equation of Eq. (2.91) is an example of a recursive system. This equation permits the recursive computation of the output response beginning at some instant $n = n_o$ and for progressively higher values of n provided the initial conditions $y[n_o - 1]$ through $y[n_o - N]$ are known. However, it is possible to implement an FIR system using a recursive computational scheme and an IIR system using a nonrecursive computational scheme [Gol68]. For example, a recursive implementation of the FIR discrete-time system of Eq. (2.61) is given by Eq. (2.63).

A different terminology is used to classify a causal finite-dimensional LTI discrete-time system in some applications, such as model-based spectral analysis (see Section 14.4.2). The classes assigned here are based on the form of the linear constant coefficient difference equation modeling the system. The simplest model is described by the input–output relation:

$$y[n] = \sum_{k=0}^{M} p_k x[n - k].$$

(2.121)

It is called a *moving average* (MA) model and is an FIR discrete-time system. It can be considered as a generalization of the M-point moving average filter of Eq. (2.61) with different weights assigned to input samples. The other two models are IIR discrete-time systems. The simplest IIR system, called an *autoregressive* (AR) model, is characterized by the input–output relation

$$y[n] = x[n] - \sum_{k=1}^{N} d_k y[n - k].$$

(2.122)

The second type of IIR system, called an *autoregressive moving average* (ARMA) model, is described by the input–output relation

$$y[n] = \sum_{k=0}^{M} p_k x[n - k] - \sum_{k=1}^{N} d_k y[n - k].$$

(2.123)

2.8.3   Classification Based on the Impulse Response Coefficients

A third classification scheme is based on the real or complex nature of the impulse response sequence. Thus, a discrete-time system with a real-valued impulse response is defined as a *real discrete-time system*. Likewise, for a *complex discrete-time system,* the impulse response is a complex-valued sequence.

2.9   Correlation of Signals

There are applications where it is necessary to compare one reference signal with one or more signals to determine the similarity between the pair and to determine additional information based on the similarity. A measure of the similarity between signals is given by the correlation sequence. For example, in digital communications, a set of data symbols are represented by a set of unique discrete-time sequences. If one of these sequences is transmitted, the receiver has to determine which particular sequence has been received by comparing the received signal with every member of possible sequences from the set. Similarly, in radar and sonar applications, the received signal reflected from the target is the delayed version of the transmitted signal, and by measuring the delay, one can determine the location of the target. The detection problem gets more complicated in practice, as often the received signal is corrupted by additive random noise. In this section, we define the correlation sequence and study its properties.

2.9.1 Definitions

A measure of similarity between a pair of energy signals, $x[n]$ and $y[n]$, is given by the *cross-correlation sequence* $r_{xy}[\ell]$ defined by

$$r_{xy}[\ell] = \sum_{n=-\infty}^{\infty} x[n]y[n - \ell], \quad \ell = 0, \pm 1, \pm 2, \ldots \tag{2.124}$$

The parameter ℓ called *lag*, indicates the time-shift between the pair. The time sequence $y[n]$ is said to be shifted by ℓ samples with respect to the reference sequence $x[n]$ to the right for positive values of ℓ and shifted by ℓ samples to the left for negative values of ℓ.

The ordering of the subscripts xy in Eq. (2.124) specifies that $x[n]$ is the reference sequence that remains fixed in time, whereas the sequence $y[n]$ is being shifted with respect to $x[n]$. If we wish to make $y[n]$ the reference sequence and shift the sequence $x[n]$ with respect to $y[n]$, then the corresponding cross-correlation sequence is given by

$$r_{yx}[\ell] = \sum_{n=-\infty}^{\infty} y[n]x[n - \ell]$$

$$= \sum_{m=-\infty}^{\infty} y[m + \ell]x[m] = r_{xy}[-\ell]. \tag{2.125}$$

Thus, $r_{yx}[\ell]$ is obtained by time-reversing the sequence $r_{xy}[\ell]$.

The *autocorrelation* sequence of $x[n]$ is given by

$$r_{xx}[\ell] = \sum_{n=-\infty}^{\infty} x[n]x[n - \ell], \tag{2.126}$$

obtained by setting $y[n] = x[n]$ in Eq. (2.124). Note from Eq. (2.126) that $r_{xx}[0] = \sum_{n=-\infty}^{\infty} x^2[n] = \mathcal{E}_x$, the energy of the signal $x[n]$. From Eq. (2.125), it follows that $r_{xx}[\ell] = r_{xx}[-\ell]$, implying that $r_{xx}[\ell]$ is an even function for real $x[n]$.

An examination of Eq. (2.125) reveals that the expression for the cross-correlation looks quite similar to that of the convolution given by Eq. (2.73a). This similarity is much clearer if we rewrite Eq. (2.124) as

$$r_{xy}[\ell] = \sum_{n=-\infty}^{\infty} x[n]y[-(\ell - n)] = x[\ell] \circledast y[-\ell]. \tag{2.127}$$

It thus follows then that the cross-correlation of the sequence $x[n]$ with the reference sequence $y[n]$ can be computed by processing $x[n]$ with an LTI discrete-time system of impulse response $y[-n]$. Likewise, the autocorrelation of $x[n]$ can be determined by passing it through an LTI discrete-time system of impulse response $x[-n]$.

2.9.2 Properties of Autocorrelation and Cross-correlation Sequences

We next derive some basic properties of the autocorrelation and cross-correlation sequences [Pro92]. Consider two finite-energy sequences $x[n]$ and $y[n]$. Now, the energy of the combined sequence $ax[n] + y[n - \ell]$ is also finite and nonnegative. That is,

$$\sum_{n=-\infty}^{\infty} (ax[n] + y[n - \ell])^2 = a^2 \sum_{n=-\infty}^{\infty} x^2[n] + 2a \sum_{n=-\infty}^{\infty} x[n]y[n - \ell] + \sum_{n=-\infty}^{\infty} y^2[n - \ell]$$

$$= a^2 r_{xx}[0] + 2a r_{xy}[\ell] + r_{yy}[0] \geq 0, \tag{2.128}$$

where $r_{xx}[0] = \mathcal{E}_x > 0$ and $r_{yy}[0] = \mathcal{E}_y > 0$ are energies of the sequences $x[n]$ and $y[n]$, respectively. We can rewrite Eq. (2.128) as

$$[a \quad 1] \begin{bmatrix} r_{xx}[0] & r_{xy}[\ell] \\ r_{xy}[\ell] & r_{yy}[0] \end{bmatrix} \begin{bmatrix} a \\ 1 \end{bmatrix} \geq 0$$

for any finite value of a. Or, in other words, the matrix

$$\begin{bmatrix} r_{xx}[0] & r_{xy}[\ell] \\ r_{xy}[\ell] & r_{yy}[0] \end{bmatrix}$$

is positive semi-definite. This implies

$$r_{xx}[0]r_{yy}[0] - r_{xy}^2[\ell] \geq 0$$

or equivalently,

$$|r_{xy}[\ell]| \leq \sqrt{r_{xx}[0]r_{yy}[0]} = \sqrt{\mathcal{E}_x \mathcal{E}_y}. \tag{2.129}$$

The above inequality provides an upper bound for the cross-correlation sequence samples. If we set $y[n] = x[n]$, the above reduces to

$$|r_{xx}[\ell]| \leq r_{xx}[0] = \mathcal{E}_x. \tag{2.130}$$

This is a significant result as it states that at zero lag ($\ell = 0$), the sample value of the autocorrelation sequence has its maximum value.

To derive an additional property of the cross-correlation sequence consider the case

$$y[n] = \pm b\, x[n - N],$$

where N is an integer and $b > 0$ is an arbitrary number. In this case $\mathcal{E}_y = b^2 \mathcal{E}_x$, and therefore,

$$\sqrt{\mathcal{E}_x \mathcal{E}_y} = \sqrt{b^2 \mathcal{E}_x^2} = b\mathcal{E}_x.$$

Using the above result in Eq. (2.129) we get

$$-b\, r_{xx}[0] \leq r_{xy}[\ell] \leq b\, r_{xx}[0].$$

2.9.3 Correlation Computation Using MATLAB

The cross-correlation and the autocorrelation sequences can be easily computed using MATLAB, as illustrated in Examples 2.46 and 2.47.

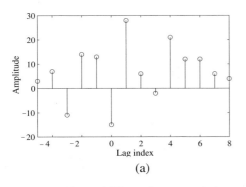 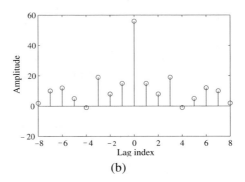

Figure 2.37: (a) Cross-correlation sequence and (b) autocorrelation sequence.

EXAMPLE 2.46   Cross-Correlation Computation Using MATLAB

Consider the two finite-length sequences

$$x[n] = [1 \quad 3 \quad -2 \quad 1 \quad 2 \quad -1 \quad 4 \quad 4 \quad 2],$$
$$y[n] = [2 \quad -1 \quad 4 \quad 1 \quad -2 \quad 3].$$

Using MATLAB, we determine and plot the cross-correlation sequence $r_{xy}[\ell]$.

As indicated by Eq. (2.127), the cross-correlation sequence $r_{xy}[\ell]$ of two sequences $x[n]$ and $y[n]$ is given by the linear convolution of $x[n]$ and the time-reversed sequence $y[-n]$. Hence, the cross-correlation of two finite-length sequences can be computed using Program 2_7, which makes use of the M-file conv. The pertinent code fragment is

```
r = conv(x,fliplr(y));
```

The cross-correlation sequence of the two finite-length sequences given above obtained using Program 2_7 is shown in Figure 2.37(a).

EXAMPLE 2.47   Autocorrelation Computation Using MATLAB   Program 2_7.m

In this example, we evaluate and plot the autocorrelation of the sequence $x[n]$ of Example 2.46. Next, by adding a random noise $d[n]$ to $x[n]$, we compute and plot the autocorrelation of the noise-corrupted sequence.

Program 2_7 can also be used to compute the autocorrelation sequence of a finite-length sequence. The plot of the autocorrelation sequence $r_{xx}[\ell]$ generated by this program for the sequence $x[n]$ of Example 2.46 is shown in Figure 2.37(b). As expected at zero lag, $r_{xx}[0]$ is the maximum.

Program 2_7 is run again by computing the cross-correlation of $x[n]$ and $y[n] = x[n - N]$ for $N = 4$. As can be seen from the plot in Figure 2.38(a), the peak of the cross-correlation is precisely the value of N used, thus demonstrating the fact that the cross-correlation can be employed to compute the exact value of the delay N.

Next, Program 2_7 is modified to generate a sequence formed by adding a random noise computed using the function rand to $x[n]$, and the autocorrelation of the noise-corrupted sequence is plotted in Figure 2.38(b). As expected, the autocorrelation still exhibits a pronounced peak at zero lag.

It should be noted that the autocorrelation and cross-correlation sequences can also be computed using the MATLAB function xcorr. However, the correlation sequences generated using this function are the time-reversed version of those generated using Programs 2_7 and 2_8. The cross-correlation $r_{xy}[\ell]$ of two sequences $x[n]$ and $y[n]$ can be computed using the statement r = xcorr(x,y), while the autocorrelation $r_{xx}[\ell]$ of the sequence $x[n]$ is determined using the statement r = xcorr(x).

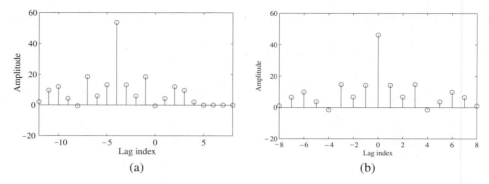

Figure 2.38: (a) Delay estimation from cross-correlation sequence and (b) autocorrelation sequence of a noise-corrupted aperiodic sequence.

2.9.4   Normalized Forms of Correlation

For convenience in comparing and displaying, normalized forms of autocorrelation and cross-correlation given by

$$\rho_{xx}[\ell] = \frac{r_{xx}[\ell]}{r_{xx}[0]},\tag{2.131}$$

$$\rho_{xy}[\ell] = \frac{r_{xy}[\ell]}{\sqrt{r_{xx}[0]\, r_{yy}[0]}},\tag{2.132}$$

are often used. It follows from Eqs. (2.129) and (2.130) that $|\rho_{xx}[\ell]| \leq 1$ and $|\rho_{xy}[\ell]| \leq 1$, independent of the range of values of $x[n]$ and $y[n]$.

2.9.5   Correlation Computation for Power and Periodic Signals

In the case of power and periodic signals, the autocorrelation and cross-correlation sequences are defined slightly differently.

For a pair of power signals, $x[n]$ and $y[n]$, the cross-correlation sequence is defined as

$$r_{xy}[\ell] = \lim_{K\to\infty} \frac{1}{2K+1} \sum_{n=-K}^{K} x[n]y[n-\ell],\tag{2.133}$$

and the autocorrelation sequence of $x[n]$ is given by

$$r_{xx}[\ell] = \lim_{K\to\infty} \frac{1}{2K+1} \sum_{n=-K}^{K} x[n]x[n-\ell].\tag{2.134}$$

Likewise, if $\tilde{x}[n]$ and $\tilde{y}[n]$ are two periodic signals with period N, then their cross-correlation sequence is given by

$$r_{\tilde{x}\tilde{y}}[\ell] = \frac{1}{N} \sum_{n=0}^{N-1} \tilde{x}[n]\, \tilde{y}[n-\ell],\tag{2.135}$$

and the autocorrelation sequence of $x[n]$ is given by

$$r_{\tilde{x}\tilde{x}}[\ell] = \frac{1}{N} \sum_{n=0}^{N-1} \tilde{x}[n]\,\tilde{x}[n-\ell]. \tag{2.136}$$

It follows from the above definitions that both $r_{\tilde{x}\tilde{y}}[\ell]$ and $r_{\tilde{x}\tilde{x}}[\ell]$ are also periodic sequences with a period N.

The periodicity properties of the autocorrelation sequence can be exploited to determine the period N of a periodic signal that may have been corrupted by an additive random disturbance. Let $\tilde{x}[n]$ be a positive periodic signal corrupted by the random noise $d[n]$, resulting in the signal

$$w[n] = \tilde{x}[n] + d[n],$$

which is observed for $0 \leq n \leq M - 1$ where $M \gg N$. The autocorrelation of $w[n]$ is given by

$$
\begin{aligned}
r_{ww}[\ell] &= \frac{1}{M} \sum_{n=0}^{M-1} w[n]\,w[n-\ell] \\
&= \frac{1}{M} \sum_{n=0}^{M-1} (\tilde{x}[n] + d[n])\,(\tilde{x}[n-\ell] + d[n-\ell]) \\
&= \frac{1}{M} \sum_{n=0}^{M-1} \tilde{x}[n]\,\tilde{x}[n-\ell] + \frac{1}{M} \sum_{n=0}^{M-1} d[n]\,d[n-\ell] \\
&\quad + \frac{1}{M} \sum_{n=0}^{M-1} \tilde{x}[n]\,d[n-\ell] + \frac{1}{M} \sum_{n=0}^{M-1} d[n]\,\tilde{x}[n-\ell] \\
&= r_{\tilde{x}\tilde{x}}[\ell] + r_{dd}[\ell] + r_{\tilde{x}d}[\ell] + r_{d\tilde{x}}[\ell]. \tag{2.137}
\end{aligned}
$$

Now in the above equation, $r_{\tilde{x}\tilde{x}}[\ell]$ is a periodic sequence with a period N, and hence, it will have peaks at $\ell = 0, N, 2N, \ldots$, with the same amplitudes as ℓ approaches M. As $\tilde{x}[n]$ and $d[n]$ are not correlated, samples of cross-correlation sequences $r_{\tilde{x}d}[\ell]$ and $r_{d\tilde{x}}[\ell]$ are likely to be very small relative to the amplitudes of $r_{xx}[\ell]$. The autocorrelation of the disturbance signal $d[n]$ shows a peak at $\ell = 0$, with other samples having rapidly decreasing amplitudes with increasing values of $|\ell|$. Hence, the peaks of $r_{ww}[\ell]$ for $\ell > 0$ are essentially due to the peaks of $r_{\tilde{x}\tilde{x}}[\ell]$ and can be used to determine whether $\tilde{x}[n]$ is a periodic sequence and its period if the peaks occur at periodic intervals.

2.9.6 Correlation Computation of a Periodic Sequence Using MATLAB

Using MATLAB we determine the period of a noise-corrupted periodic sequence in Example 2.48.

Program 2_8.m

EXAMPLE 2.48   **Determination of the Period of a Noise-Corrupted Periodic Sequence**

We determine the period of the sinusoidal sequence $x[n] = \cos(0.25\pi n)$, $0 \leq n \leq 95$, corrupted by an additive uniformly distributed random noise of amplitude in the range $[-0.5, 0.5]$.

To this end, we use Program 2_8 to compute the autocorrelation sequence of the noise-corrupted sinusoidal sequence. The plot generated by running this program is shown in Figure 2.39(a). As can be seen from this plot,

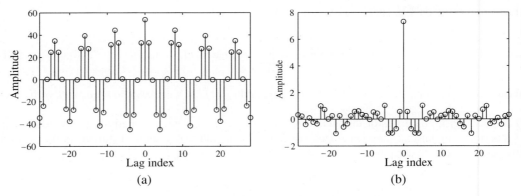

Figure 2.39: (a) Autocorrelation sequence of the noise-corrupted sinusoid and (b) autocorrelation sequence of the noise.

there is a strong peak at zero lag. However, there are distinct peaks at lags that are multiples of 8, indicating the period of the sinusoidal sequence to be 8, as expected.[8] Figure 2.39(b) shows the plot of the autocorrelation sequence $r_{dd}[n]$ of the noise component. As can be seen from this plot, $r_{dd}[n]$ shows a very strong peak only at zero lag. The amplitudes are considerably smaller at other values of the lag, as sample values of the noise sequence are uncorrelated with each other.

2.10   Random Signals

Most of the book deals with the processing of signals that are deterministic in nature. The underlying assumption of this type of discrete-time signals is that they can be uniquely determined by well-defined processes such as a mathematical expression or a rule or a lookup table. Such a signal is usually called a *deterministic signal* since all sample values of the sequence are well defined for all values of the time index. For example, the sinusoidal sequence of Eq. (2.42) and the exponential sequence of Eq. (2.45) are deterministic sequences.

Signals for which each sample value is generated in a random fashion and cannot be predicted ahead of time comprise another class of signals. Such a signal, called a *random signal* or a *stochastic signal,* cannot be reproduced at will, not even using the process generating the signal, and therefore needs to be modeled using statistical information about the signal. Some common examples of random signals are speech, music, and seismic signals. The error signal generated by forming the difference between the ideal sampled version of a continuous-time signal and its quantized version generated by a practical analog-to-digital converter is usually modeled as a random signal for analysis purposes.[9] The noise sequence $d[n]$ of Figure 2.6(b) generated using the `rand` function of MATLAB is also an example of a random signal.

The discrete-time *random signal* or process consists of a typically infinite collection or *ensemble* of discrete-time sequences $\{X[n]\}$. One particular sequence in this collection $\{x[n]\}$ is called a *realization* of the random process. At a given time index n, the observed sample value $x[n]$ is the value taken by the *random variable* $X[n]$. Thus, a random process is a family of random variables $\{X[n]\}$. In general, the range of sample values is a continuum. We review in the Appendix the important statistical properties of the random variable and the random process.

---

[8]The decaying amplitudes of the peaks for increasing values of the lag index ℓ are due to the finite lengths of the periodic sequences, which causes a reduction in the number of nonzero products in the computation of the convolution sum.

[9]See Section 12.5.1.

2.11 Summary

In this chapter, we introduced some important and fundamental concepts regarding the characterization of discrete-time signals and systems in the time domain. Certain basic discrete-time signals that play important roles in discrete-time signal processing have been defined, along with basic mathematical operations used for generating more complex signals and systems. The relation between a continuous-time signal and the discrete-time signal generated by sampling the former at uniform time intervals has been examined.

This text deals almost exclusively with linear, time-invariant (LTI) discrete-time systems that find numerous applications in practice. These systems are defined, and their convolution sum representation in the time domain is derived. The concepts of causality and stability of LTI systems are introduced. Also discussed is an important class of LTI systems characterized by an input–output relation composed of a linear constant coefficient difference equation and the procedure for computing its output for a given input and initial conditions. Several types of classification of LTI discrete-time systems are then considered, of which the more common one is in terms of the length of the impulse response. Finally, the concepts of the autocorrelation of a sequence and the cross-correlation between a pair of sequences are introduced.

Further insights can often be obtained by considering the frequency-domain representations of discrete-time signals and LTI discrete-time systems. These are discussed in Chapters 3, 5, 6, and 7.

2.12 Problems

2.1 Determine the $\mathcal{L}_1$-, $\mathcal{L}_2$-, and $\mathcal{L}_\infty$-norms of the following finite-length sequences:

(a) $\{x_1[n]\} = \{4.50 \quad -2.68 \quad -0.14 \quad 3.91 \quad 2.62 \quad -0.43 \quad -4.81 \quad 3.21 \quad -0.55\}$,
(b) $\{x_2[n]\} = \{0.92 \quad 2.34 \quad 3.37 \quad 1.90 \quad -2.59 \quad -0.75 \quad 3.48 \quad 3.33\}$.

2.2 Express the sequence $x[n] = 1$, $-\infty < n < \infty$, in terms of the unit step sequence $\mu[n]$.

2.3 Verify the relation between the unit sample sequence $\delta[n]$ and the unit step sequence $\mu[n]$ given in Eq. (2.31).

2.4 Express the length-4 sequence $x[n] = \{1 \quad 3 \quad -2 \quad 4\}$, $0 \le n \le 3$, in terms of the unit step sequence $\mu[n]$.

2.5 Consider the following sequences:

$$x[n] = \{-4 \quad 5 \quad 1 \quad -2 \quad -3 \quad 0 \quad 2\}, \quad -3 \le n \le 3$$

$$y[n] = \{6 \quad -3 \quad -1 \quad 0 \quad 8 \quad 7 \quad -2\}, \quad -1 \le n \le 5$$

$$w[n] = \{3 \quad 2 \quad 2 \quad -1 \quad 0 \quad -2 \quad 5\}, \quad 2 \le n \le 8.$$

The sample values of each of the above sequences outside the ranges specified are all zeros. Generate the following sequences: (a) $c[n] = x[-n + 2]$, (b) $d[n] = y[-n - 3]$, (c) $e[n] = w[-n]$, (d) $u[n] = x[n] + y[n - 2]$, (e) $v[n] = x[n] \cdot w[n + 4]$, (f) $s[n] = y[n] - w[n + 4]$, and (g) $r[n] = 3.5y[n]$.

2.6 (a) Express the sequences $x[n]$, $y[n]$, and $w[n]$ of Problem 2.5 as a linear combination of delayed unit sample sequences.

(b) Express the sequences $x[n]$, $y[n]$, and $w[n]$ of Problem 2.5 as a linear combination of delayed unit step sequences.

2.7 Analyze the block diagrams of Figure P2.1 and develop the relation between $y[n]$ and $x[n]$.

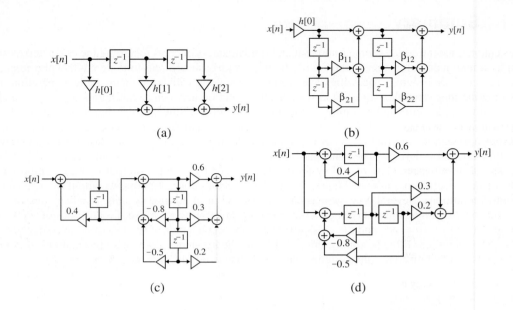

Figure P2.1

2.8 Determine the conjugate symmetric and conjugate antisymmetric parts of the following sequences:
(a) $x_1[n] = \{1 + j4 \quad -2 + j5 \quad 3 - j2 \quad -7 + j3 \quad -1 + j\}, -2 \le n \le 2$, (b) $x_2[n] = e^{j\pi n/3}$,
(c) $x_3[n] = j e^{-j\pi n/5}$.

2.9 Determine the even and odd parts of the sequences $x[n]$, $y[n]$, and $w[n]$ of Problem 2.5.

2.10 Determine the even and odd parts of the following real sequences:
(a) $x_1[n] = \mu[n + 2]$, (b) $x_2[n] = \alpha^n \mu[n - 3]$, (c) $x_3[n] = n\alpha^n \mu[n]$, (d) $x_4[n] = \alpha^{|n|}$.

2.11 Show that the even and odd parts of a real sequence are, respectively, even and odd sequences.

2.12 Let $x_{ev}[n]$ and $x_{od}[n]$ be even and odd real sequences, respectively. Which one of the following sequences is an even sequence, and which one is an odd sequence?
(a) $g[n] = x_{ev}[n]x_{ev}[n]$, (b) $u[n] = x_{ev}[n]x_{od}[n]$, (c) $v[n] = x_{od}[n]x_{od}[n]$.

2.13 (a) Show that a causal real sequence $x[n]$ can be fully recovered from its even part $x_{ev}[n]$ for all $n \ge 0$, whereas it can be recovered from its odd part $x_{od}[n]$ only for all $n > 0$.
(b) Is it possible to fully recover a causal complex sequence $y[n]$ from its conjugate antisymmetric part $y_{ca}[n]$? Can $y[n]$ be fully recovered from its conjugate symmetric part $y_{cs}[n]$? Justify your answers.

2.14 Determine the causal sequence $x[n]$ whose even part is given by $x_{ev}[n] = \cos(\omega_o n)$.

2.15 Which ones of the following sequences are bounded sequences?
(a) $x[n] = A\alpha^n$, where A and α are complex numbers, and $|\alpha| < 1$,
(b) $y[n] = A\alpha^n \mu[n]$, where A and α are complex numbers, and $|\alpha| < 1$,
(c) $h[n] = C\beta^n \mu[n]$, where C and β are complex numbers, and $|\beta| > 1$,
(d) $g[n] = 4\cos(\omega_a n)$,   (e) $v[n] = \left(1 - \frac{1}{n^2}\right)\mu[n - 1]$.

2.16 Show that the sequence $x[n] = \frac{(-1)^{n+1}}{n}\mu[n-1]$ is not absolutely summable even though $\sum_{n=-\infty}^{\infty} x[n] = \ln 2$.

2.17 Show that the following sequences are absolutely summable: (a) $x_1[n] = \alpha^n\mu[n-1]$, (b) $x_2[n] = n\alpha^n\mu[n-1]$, (c) $x_3[n] = n^2\alpha^n\mu[n-1]$, where $|\alpha| < 1$.

2.18 Show that the following sequences are absolutely summable. (a) $x_a[n] = \frac{1}{2^n}\mu[n]$, (b) $x_b[n] = \frac{1}{(n+1)(n+2)}\mu[n]$.

2.19 Show that an absolutely summable sequence has finite energy, but a finite energy sequence may not be absolutely summable.

2.20 Show that the square-summable sequence $x_1[n] = \frac{1}{n}$ of Eq. (2.29) is not absolutely summable.

2.21 Show that the sequence $x_2[n] = \frac{\cos \omega_c n}{\pi n}\mu[n-1]$ is square-summable but not absolutely summable.

2.22 The odd part of a real-valued sequence $x[n]$ is given by $x_{od} = \left(\frac{1}{3}\right)^3\mu[n]$. If the average power of $x[n]$ is $\mathcal{P}_x = 10$, determine the average power of its even part x_{ev}.

2.23 Compute the energy of the length-N sequence $x[n] = \sin(2\pi kn/N)$, $0 \le n \le N-1$.

2.24 Compute the energy of the following sequences:
(a) $x_a[n] = A\alpha^n\mu[n]$, $|\alpha| < 1$, (b) $x_b[n] = \frac{1}{n^2}\mu[n-1]$.

2.25 Determine the average power and the energy of the following sequences:
(a) $x_1[n] = (-1)^n$, (b) $x_2[n] = \mu[n]$, (c) $x_3[n] = n\mu[n]$, (d) $x_4[n] = A_o e^{j\omega_o n}$, (e) $x_5[n] = A\cos\left(\frac{2\pi n}{M} + \phi\right)$.

2.26 Determine the period and the average power of the following periodic sequences:
(a) $\tilde{x}_1[n] = 4\cos(2\pi n/5)$, (b) $\tilde{x}_2[n] = 3\cos(3\pi n/5)$, (c) $\tilde{x}_3[n] = 2\cos(3\pi n/7)$, (d) $\tilde{x}_4[n] = 4\cos(5\pi n/3)$,
(e) $\tilde{x}_5[n] = 4\cos(2\pi n/5) + 3\cos(3\pi n/5)$, (f) $\tilde{x}_6[n] = 4\cos(5\pi n/3) + 3\cos(3\pi n/5)$.

2.27 Let $x[n]$ be an absolutely summable sequence. Show that the sequence $\tilde{y}[n]$ formed by an N-periodic extension according to Eq. (2.38) is a periodic sequence with a period N.

2.28 Determine the samples of one period of the periodic sequences obtained by an N-periodic extension of the sequences of Problem 2.5 for the following values of N: (a) $N = 5$, and $N = 7$.

2.29 The following sequences represent one period of a sinusoidal sequence of the form $\tilde{x}[n] = A\cos(\omega_o n + \phi)$:
(a) {1 -1 -1 1   1 -1 -1 1}, (b) {0 $-\sqrt{3}$ 0 $\sqrt{3}$ $0-\sqrt{3}$ 0 $\sqrt{3}$},
(c) {1 -0.366 -1.366 -1 0.366   1.366}, (d) {2   0 -2 0   2 $0-2$ 0}.
Determine the values of the parameters A, ω_o, and ϕ for each case.

2.30 Determine the fundamental period of the following periodic sequences:
(a) $\tilde{x}_a[n] = e^{j0.5\pi n}$, (b) $\tilde{x}_b[n] = \sin(0.8\pi n + 0.8\pi)$, (c) $\tilde{x}_c[n] = \text{Re}\left(e^{j\pi n/5}\right) + \text{Im}\left(e^{j\pi n/10}\right)$,
(d) $\tilde{x}_4[n] = 3\cos(1.3\pi n) - 4\sin(0.5\pi n + 0.5\pi)$, (e) $\tilde{x}_5[n] = 5\cos(1.5\pi n + 0.75\pi) + 4\cos(0.6\pi n) - \sin(0.5\pi n)$.

2.31 Determine the fundamental period of the sinusoidal sequence $x[n] = A\sin(\omega_o n)$ for the following values of the angular frequency ω_o:
(a) 0.6π, (b) 0.28π, (c) 0.45π, (d) 0.55π, (e) 0.65π.

2.32 Determine the period of the sinusoidal sequence $x_1[n] = \sin(0.08\pi n)$. Determine at least two other distinct sinusoidal sequences having the same period as $x_1[n]$.

2.33 Show that the continuous-time signal $x_a(t) = A\cos(\Omega_o t + \phi)$ can be uniquely recovered from its sampled version $x[n] = x_a(nT)$, $-\infty < n < \infty$, if the sampling frequency $\Omega_T = 2\pi/T > 2\Omega_o$.

2.34 A continuous-time sinusoidal signal $x_a(t) = \cos\Omega_o t$ is sampled at $t = nT$, $-\infty < n < \infty$, generating the discrete-time sequence $x[n] = x_a(nT) = \cos(\Omega_o nT)$. For what values of T is $x[n]$ a periodic sequence? What is the fundamental period of $x[n]$ if $\Omega_o = 20$ radians and $T = \pi/8$ seconds?

2.35 Show that the discrete-time systems described by the following equations are linear systems:
  (a) Eq. (2.18),   (b) Eq. (2.20),   (c) Eq. (2.21),   (d) Eq. (2.61),   (e) Eq. (2.65),   and   (f) Eq. (2.66).

2.36 For each of the following discrete-time systems, where $y[n]$ and $x[n]$ are, respectively, the output and the input sequences, determine whether or not the system is (1) linear, (2) causal, (3) stable, and (4) shift-invariant:
  (a) $y[n] = n^3 x[n]$;   (b) $y[n] = (x[n])^5$;   (c) $y[n] = \beta + \sum_{\ell=0}^{3} x[n-\ell]$ β is a nonzero constant;
  (d) $y[n] = \ln(2 + |x[n]|)$;   (e) $y[n] = \alpha x[-n+2]$; α is a nonzero constant;   (f) $y[n] = x[n-4]$.

2.37 Show that the median filter defined by Eq. (2.67) is a time-invariant system.

2.38 The second derivative $y[n]$ of a sequence $x[n]$ at time instant n is usually approximated by

$$y[n] = x[n+1] - 2x[n] + x[n-1].$$

If $y[n]$ and $x[n]$ denote the output and input of a discrete-time system, is the system linear? Is it time-invariant? Is it causal?

2.39 A discrete-time system is characterized by [Kai80]

$$y[n] = x^2[n] - x[n-1]x[n+1],$$

where $y[n]$ and $x[n]$ are the output and input sequences. Is the above system linear? Is it time-invariant? Is it causal?

2.40 Consider the discrete-time system characterized by the input–output relation [Cad87]

$$y[n] = \frac{1}{2}\left(y[n-1] + \frac{x[n]}{y[n-1]}\right), \tag{2.138}$$

where $x[n]$ and $y[n]$ are, respectively, the input and output sequences. Show that the output $y[n]$ of the above system for an input $x[n] = \alpha\mu[n]$ with $y[-1] = 1$ converges to $\sqrt{\alpha}$ as $n \to \infty$ when α is a positive number. Is the above system linear or nonlinear? Is it time-invariant? Justify your answer.

2.41 An algorithm for the calculation of the square root of a number α is given by [Mik92]

$$y[n] = x[n] - y^2[n-1] + y[n-1], \tag{2.139}$$

where $x[n] = \alpha\mu[n]$ with $0 < \alpha < 1$. If $x[n]$ and $y[n]$ are considered as the input and output of a discrete-time system, is the system linear or nonlinear? Is it time-invariant? As $n \to \infty$, show that $y[n] \to \sqrt{\alpha}$. Note that $y[-1]$ is a suitable initial approximation to $\sqrt{\alpha}$.

2.42 Determine the expression for the impulse response of the factor-of-3 linear interpolator of Eq. (2.66).

2.43 Determine the expression for the impulse response of the factor-of-L linear interpolator.

2.44 Derive Eq. (2.111) by induction by first evaluating Eq. (2.108) for $n = 0, 1, 2, \ldots$, and then solving for $h[0], h[1], h[2]$, etc.

2.45 Let $y[n] = h[n] \circledast x[n]$, where $h[n]$ and $x[n]$ are right-sided sequences. Show that $\sum y[n] = \left(\sum h[n] \right) \left(\sum x[n] \right)$.

2.46 Develop closed-form expressions for the following convolution sums: (a) $\alpha^n \mu[n] \circledast \mu[n]$, (b) $nc^n \mu[n] \circledast \mu[n]$.

2.47 Develop a general expression for the output $y[n]$ of an LTI discrete-time system in terms of its input $x[n]$ and the unit step response $s[n]$ of the system.

2.48 A periodic sequence $\tilde{x}[n]$ with a period N is applied as an input to an LTI discrete-time system characterized by an impulse response $h[n]$ generating an output $y[n]$. Is $y[n]$ a periodic sequence? If it is, what is its period?

2.49 Consider the following sequences:
(i) $x_1[n] = 3\delta[n-2] - 2\delta[n+1]$, (ii) $x_2[n] = 5\delta[n-3] + 2\delta[n+1]$, (iii) $h_1[n] = -\delta[n+2] + 4\delta[n] - 2\delta[n-1]$, (iv) $h_2[n] = 3\delta[n-4] + 1.5\delta[n-2] - \delta[n+1]$.
Determine the following sequences obtained by a linear convolution of a pair of the above sequences:
(a) $y_1[n] = x_1[n] \circledast h_1[n]$, (b) $y_2[n] = x_2[n] \circledast h_2[n]$, (c) $y_3[n] = x_1[n] \circledast h_2[n]$, (d) $y_4[n] = x_2[n] \circledast h_1[n]$.

2.50 Determine the following sequences obtained by a linear convolution of the sequences given in Problem 2.5:
(a) $u[n] = x[n] \circledast y[n]$, (b) $v[n] = x[n] \circledast w[n]$, (c) $g[n] = w[n] \circledast y[n]$.

2.51 Let $g[n]$ be a finite-length sequence defined for $-3 \le n \le 4$, and let $h[n]$ be a finite-length sequence defined for $2 \le n \le 6$. Define $y[n] = g[n] \circledast h[n]$. (a) What is the length of $y[n]$? (b) What is the range of the index n for which $y[n]$ is defined?

2.52 Let $y[n] = x_1[n] \circledast x_2[n]$ and $v[n] = x_1[n - N_1] \circledast x_2[n - N_2]$. Express $v[n]$ in terms of $y[n]$.

2.53 Let $g[n] = x_1[n] \circledast x_2[n] \circledast x_3[n]$ and $h[n] = x_1[n - N_1] \circledast x_2[n - N_2] \circledast x_3[n - N_3]$. Express $h[n]$ in terms of $g[n]$.

2.54 Prove that the convolution sum operation is commutative and distributive.

2.55 Consider the following three sequences:

$$x_1[n] = A \text{ (a constant)}, \qquad x_2[n] = \mu[n], \qquad x_3[n] = \begin{cases} 1, & \text{for } n = 0, \\ -1, & \text{for } n = 1, \\ 0, & \text{otherwise.} \end{cases}$$

Show that $x_3[n] \circledast x_2[n] \circledast x_1[n] \ne x_2[n] \circledast x_3[n] \circledast x_1[n]$.

2.56 Prove that the convolution operation is associative for stable and single-sided sequences.

2.57 Show that the convolution of a length-M sequence with a length-N sequence leads to a sequence of length $(M + N - 1)$.

2.58 Let $\{x[n]\}$ be a length-N sequence defined for $0 \le n \le N - 1$, with its n-th sample given by a_n. If all samples of $x[n]$ are non-negative, determine the location and the value of the largest positive sample of $y[n] = x[n] \circledast x[n]$ without performing the convolution operation.

2.59 Let $\{x[n]\}$ and $\{h[n]\}$ be two length-N sequences defined for $0 \le n \le N - 1$, with their n-th samples given by a_n and b_n, respectively. If all samples of $x[n]$ and $h[n]$ are non-negative, determine the location and the value of the largest positive sample of $y[n] = x[n] \circledast h[n]$ without performing the convolution operation.

2.60 Consider two real sequences $h[n]$ and $g[n]$ expressed as a sum of their respective even and odd parts, i.e., $h[n] = h_{\text{ev}}[n] + h_{\text{od}}[n]$, and $g[n] = g_{\text{ev}}[n] + g_{\text{od}}[n]$. For each of the following sequences, determine if it is even or odd.

    (a) $h_{\text{ev}}[n] \circledast g_{\text{ev}}[n]$,   (b) $h_{\text{od}}[n] \circledast g_{\text{ev}}[n]$,   (c) $h_{\text{od}}[n] \circledast g_{\text{od}}[n]$.

2.61 Consider a cascade of two causal stable LTI systems characterized by impulse responses $\alpha^n \mu[n]$ and $\beta^n \mu[n]$, where $0 < \alpha < 1$ and $0 < \beta < 1$. Determine the expression for the impulse response $h[n]$ of the cascade.

2.62 Determine the impulse response $g[n]$ of the inverse system of the LTI discrete-time system of Example 2.31.

2.63 Determine the impulse response $g[n]$ characterizing the inverse system of the cascaded LTI discrete-time system of Problem 2.61.

2.64 Determine the expression for the impulse response of each of the LTI systems shown in Figure P2.2.

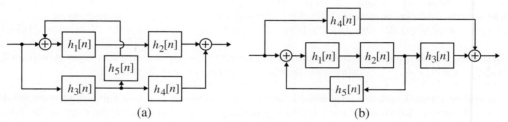

(a)   (b)

Figure P2.2

2.65 Determine the overall impulse response of the system of Figure P2.3, where the impulse responses of the component systems are: $h_1[n] = 2\delta[n - 2] - 3\delta[n + 1]$, $h_2[n] = \delta[n - 1] + 2\delta[n + 2]$, and $h_3[n] = 5\delta[n - 5] + 7\delta[n - 3] + 2\delta[n - 1] - \delta[n] + 3\delta[n + 1]$.

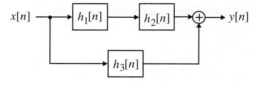

Figure P2.3

2.66 Let $y[n]$ be the sequence obtained by a linear convolution of two causal finite-length sequences $h[n]$ and $x[n]$. For each pair of $y[n]$ and $h[n]$ listed below, determine $x[n]$. The first sample in each sequence is at time index $n = 0$.

    (a) $\{y[n]\} = \{6,\ 11,\ -13,\ 16,\ 1,\ 9,\ 2,\ 8\}$, $\{h[n]\} = \{2,\ 5,\ -1,\ 4\}$,

    (b) $\{y[n]\} = \{2,\ 1,\ 2,\ 6,\ 0\ 17\ 12\}$, $\{h[n]\} = \{2,\ -3,\ 2,\ 3\}$,

    (c) $\{y[n]\} = \{-4,\ 10,\ -15,\ 8,\ 0,\ 1,\ -3,\ 1\}$, $\{h[n]\} = \{-4,\ 2,\ 1,\ 0,\ -1\}$.

2.67 Consider a causal discrete-time system characterized by a first-order linear, constant-coefficient difference equation given by

$$y[n] = ay[n - 1] + bx[n], \quad n \geq 0,$$

where $y[n]$ and $x[n]$ are, respectively, the output and input sequences. Compute the expression for the output sample $y[n]$ in terms of the initial condition $y[-1]$ and the input samples.

(a) Is the system time-invariant if $y[-1] = 1$? Is the system linear if $y[-1] = 1$?

(b) Repeat part (a) if $y[-1] = 0$.

(c) Generalize the results of parts (a) and (b) to the case of an Nth-order causal discrete-time system given by Eq. (2.82).

2.68 Consider the causal LTI system described by the difference equation

$$y[n] = p_0 x[n] + p_1 x[n-1] - d_1 y[n-1],$$

where $x[n]$ and $y[n]$ denote, respectively, its input and output. Determine the difference equation representation of its inverse system.

2.69 A causal LTI discrete-time system is said to have an *overshoot* in its step response if the response exhibits an oscillatory behavior with decaying amplitudes around a final constant value. Show that the system has no overshoot in its step response if the impulse response $h[n]$ of the system is nonnegative for all $n \geq 0$.

2.70 The sequence of Fibonacci numbers $f[n]$ is a causal sequence defined by

$$f[n] = f[n-1] + f[n-2], \quad n \geq 2,$$

with $f[0] = 0$ and $f[1] = 1$.

(a) Develop an exact formula to calculate $f[n]$ directly for any n.

(b) Show that $f[n]$ is the impulse response of a causal LTI system described by the difference equation [Joh89]

$$y[n] = y[n-1] + y[n-2] + x[n-1].$$

2.71 Consider a first-order complex digital filter characterized by a difference equation

$$y[n] = \alpha y[n-1] + x[n],$$

where $x[n]$ is the real input sequence, $y[n] = y_{\text{re}}[n] + j y_{\text{im}}[n]$ is the complex output sequence with $y_{\text{re}}[n]$ and $y_{\text{im}}[n]$ denoting its real and imaginary parts, and $\alpha = a + jb$ is a complex constant. Develop an equivalent two-output, single-input real difference equation representation of the above complex digital filter. Show that the single-input, single-output digital filter relating $y_{\text{re}}[n]$ to $x[n]$ is described by a second-order difference equation.

2.72 Let $h[0]$, $h[1]$, and $h[2]$ denote the first three impulse response samples of the first-order causal LTI system of Problem 2.68. Show that the coefficients of the difference equation characterizing this system can be uniquely determined from these impulse response samples.

2.73 Let a causal IIR digital filter be described by the difference equation

$$\sum_{k=0}^{N} d_k y[n-k] = \sum_{k=0}^{M} p_k x[n-k], \tag{2.140}$$

where $y[n]$ and $x[n]$ denote the output and the input sequences, respectively. If $h[n]$ denotes its impulse response, show that

$$p_k = \sum_{n=0}^{k} h[n] d_{k-n}, \quad k = 0, 1, \ldots, M.$$

From the above result, show that $p_n = h[n] \circledast d_n$.

2.74 Prove that the BIBO stability condition of Eq. (2.74) also holds for an LTI digital filter with a complex impulse response.

2.75 Is the cascade connection of two stable LTI systems also stable? Justify your answer.

2.76 Is the parallel connection of two stable LTI systems also stable? Justify your answer.

2.77 Prove that the cascade connection of two passive (lossless) LTI systems is also passive (lossless).

2.78 Is the parallel connection of two passive (lossless) LTI systems also passive (lossless)? Justify your answer.

2.79 Consider a causal FIR filter of length $L+1$ with an impulse response given by $\{g[n]\}, n = 0, 1, \ldots, L$. Develop the difference equation representation of the form of Eq. (2.90) where $M + N = L$ of a causal finite-dimensional IIR digital filter with an impulse response $\{h[n]\}$ such that $h[n] = g[n]$ for $n = 0, 1, \ldots, L$.

2.80 Compute the output of the accumulator of Eq. (2.60) for an input $x[n] = n\mu[n]$ and the following initial conditions: (a) $y[-1] = 0$, and (b) $y[-1] = -2$.

2.81 In the rectangular method of numerical integration, the integral on the right-hand side of Eq. (2.85) is expressed as

$$\int_{(n-1)T}^{nT} x(\tau)d\tau = T \cdot x\left((n-1)T\right). \tag{2.141}$$

Develop the difference equation representation of the rectangular method of numerical integration.

2.82 Develop a recursive implementation of the time-varying linear discrete-time system characterized by

$$y[n] = \begin{cases} \frac{1}{n} \sum_{\ell=1}^{n} x[\ell], & n > 0, \\ 0, & n \leq 0. \end{cases}$$

2.83 Determine the total solution for $n \geq 0$ of the difference equation

$$y[n] - 0.35y[n-1] = 2.4\mu[n],$$

with the initial condition $y[-1] = 3$.

2.84 Determine the total solution for $n \geq 0$ of the difference equation

$$y[n] - 0.3y[n-1] - 0.04y[n-2] = 3^n \mu[n],$$

with the initial condition $y[-1] = 2$, and $y[-2] = 1$.

2.85 Determine the total solution for $n \geq 0$ of the difference equation

$$y[n] - 0.3y[n-1] - 0.04y[n-2] = x[n] + 2x[n-1],$$

with the initial condition $y[-1] = 2$, and $y[-2] = 1$, when the forcing function is $x[n] = 3^n \mu[n]$.

2.86 Determine the impulse response $h[n]$ of the LTI system described by the difference equation

$$y[n] - 0.35y[n-1] = x[n].$$

2.87 Determine the impulse response $h[n]$ of the LTI system described by the difference equation

$$y[n] - 0.3y[n-1] - 0.04y[n-2] = x[n] + 2x[n-1].$$

2.88 Determine the step response of an LTI discrete-time system characterized by an impulse response $h[n] = (-\alpha)^n \mu[n]$, $0 < \alpha < 1$.

2.89 Show that the sum $\sum_{n=0}^{\infty} |n^K (\lambda_i)^n|$ converges if $|\lambda_i| < 1$.

2.90 (a) Evaluate the autocorrelation sequence of each of the sequences of Problem 2.5.

(b) Evaluate the cross-correlation sequence $r_{xy}[\ell]$ between the sequences $x[n]$ and $y[n]$, and the cross-correlation sequence $r_{xw}[\ell]$ between the sequences $x[n]$ and $w[n]$ of Problem 2.5.

2.91 Determine the autocorrelation sequence of each of the following sequences, and show that it is an even sequence in each case. What is the location of the maximum value of the autocorrelation sequence in each case?

(a) $x_1[n] = \alpha^n \mu[n]$, (b) $x_2[n] = \begin{cases} 1, & 0 \le n \le N - 1, \\ 0, & \text{otherwise.} \end{cases}$

2.92 Determine the autocorrelation sequence and its period of each of the following periodic sequences.

(a) $\tilde{x}_1[n] = \cos(\pi n/M)$, where M is a positive integer, (b) $\tilde{x}_2[n] = n$ modulo 6, (c) $\tilde{x}_3[n] = (-1)^n$.

2.13 MATLAB Exercises

M 2.1 (a) Using Program 2_2, generate the sequences shown in Figures 2.17 and 2.18.

(b) Generate and plot the complex exponential sequence $-3.6e^{(-0.5+j\pi/4)n}$ for $0 \le n \le 82$ using Program 2_2.

M 2.2 Generate the sequences of Problem 2.30(b) to 2.30(e) using MATLAB.

M 2.3 (a) Write a MATLAB program to generate a sinusoidal sequence $x[n] = A \sin(\omega_o n + \phi)$, and plot the sequence using the stem function. The input data specified by the user are the desired length L, amplitude A, the angular frequency ω_o, and the phase ϕ where $0 < \omega_o < \pi$ and $0 \le \phi \le 2\pi$. Using this program, generate the sinusoidal sequences shown in Figure 2.16.

(b) Generate sinusoidal sequences with the angular frequencies given in Problem 2.31. Determine the period of each sequence from the plot, and verify the result theoretically.

M 2.4 Write a MATLAB program to plot a continuous-time sinusoidal signal and its sampled version, and verify Figure 2.22. You need to use the hold function to keep both plots.

M 2.5 Using the program developed in the previous problem, verify experimentally that the family of continuous-time sinusoids given by Eq. (2.57) lead to identical sampled signals.

M 2.6 Using Program 2_4, investigate the effect of signal smoothing by a moving-average filter of lengths 5, 7, and 9. Does the signal smoothing improve with an increase in the length? What is the effect of the length on the delay between the smoothed output and the noisy input?

M 2.7 Write a MATLAB program implementing the discrete-time system of Eq. (2.138) in Problem 2.40, and show that the output $y[n]$ of this system for an input $x[n] = \alpha\mu[n]$ with $y[-1] = 1$ converges to $\sqrt{\alpha}$ as $n \to \infty$.

M 2.8 Write a MATLAB program to compute the square root using the algorithm of Eq. (2.139) in Problem 2.41, and show that the output $y[n]$ of this system for an input $x[n] = \alpha\mu[n]$ with $y[-1] = 1$ converges to $\sqrt{\alpha}$ as $n \to \infty$. Plot the error as a function of n for several different values of α. How would you compute the square-root of a number α with a value greater than one?

M 2.9 Using Program 2_7, determine the autocorrelation and the cross-correlation sequences of Problem 2.90. Are your results the same as those determined in Problem 2.90?

M 2.10 Modify Program 2_7 to determine the autocorrelation sequence of a sequence corrupted with a uniformly distributed random signal generated using the M-function rand. Using the modified program, demonstrate that the autocorrelation sequence of a noise-corrupted signal exhibits a peak at zero lag.

3 Discrete-Time Fourier Transform

In Section 2.2.3, we pointed out that any arbitrary sequence can be represented in the time domain as a weighted linear combination of delayed unit sample sequences $\{\delta[n-k]\}$. An important consequence of this representation, derived in Section 2.5.1, is the input–output characterization of a linear, time-invariant (LTI) digital filter in the time domain by means of the convolution sum describing the output sequence in terms of a weighted linear combination of its delayed impulse responses. In many applications, it is convenient to consider an alternate description of a sequence in terms of complex exponential sequences of the form $\{e^{-j\omega n}\}$, where ω is the normalized frequency variable in radians. This leads to a particularly useful representation of discrete-time sequences and LTI discrete-time systems in the frequency domain.[1]

The frequency-domain representation of a discrete-time sequence discussed in this chapter is the discrete-time Fourier transform by which a time-domain sequence is mapped into a continuous function of the frequency variable ω. Because of the periodicity of the discrete-time Fourier transform, the corresponding discrete-time sequence can be simply obtained by computing its Fourier series representation. Since the representation is in terms of an infinite series, existence of the discrete-time Fourier transform is examined along with its properties.

The frequency-domain representation of a linear, time-invariant (LTI) discrete-time system is its frequency response given by the discrete-time Fourier transform of its impulse response. The frequency response is often more convenient to use in the analysis and design of discrete-time systems. It is shown that the product of the frequency response and the discrete-time Fourier transform of the input sequence is precisely the discrete-time Fourier transform of the output sequence, which is the frequency-domain equivalent of the convolution sum description in the time domain. In most applications, these transforms are rational functions and are thus more convenient to deal with than the infinite-length sequences in the time domain. We discuss the properties of the frequency response and its application in the frequency selective processing of an input signal.

In this chapter also, we make extensive use of MATLAB to illustrate through computer simulations the various concepts introduced.

3.1   The Continuous-Time Fourier Transform

We begin with a brief review of the continuous-time Fourier transform, a frequency-domain representation of a continuous-time signal, and its properties, as it will provide a better understanding of the frequency-domain representation of the discrete-time signals and systems, in addition to pointing out the major differences between these two transforms.

---

[1] Periodic sequences can be represented in the frequency domain by means of the discrete Fourier series (see Problem 5.3).

3.1.1 The Definition

The frequency-domain representation of a continuous-time signal $x_a(t)$ is given by the *continuous-time Fourier transform* (CTFT):

$$X_a(j\Omega) = \int_{-\infty}^{\infty} x_a(t)e^{-j\Omega t}\,dt. \tag{3.1}$$

The CTFT often is referred to as the *Fourier spectrum,* or simply, the spectrum of the continuous-time signal. The continuous-time signal $x_a(t)$ can be recovered from its CTFT $X_a(j\Omega)$ via the *Fourier integral*

$$x_a(t) = \frac{1}{2\pi}\int_{-\infty}^{\infty} X_a(j\Omega)e^{j\Omega t}\,d\Omega. \tag{3.2}$$

We denote the CTFT pair of Eqs. (3.1) and (3.2) as

$$x_a(t) \overset{\text{CTFT}}{\longleftrightarrow} X_a(j\Omega).$$

In Eqs. (3.1) and (3.2), Ω is real and denotes the continuous-time angular frequency variable in radians. The inverse Fourier transform given by Eq. (3.2) can be interpreted as a linear combination of infinitesimally small complex exponential signals of the form $\frac{1}{2\pi}e^{j\Omega t}\,d\Omega$, weighted by the complex constant $X_a(j\Omega)$ over the angular frequency range from $-\infty$ to ∞. It can be seen from the definition given by Eq. (3.1) that, in general, the CTFT is a complex function of Ω in the range $-\infty < \Omega < \infty$. It can be expressed in polar form as

$$X_a(j\Omega) = |X_a(j\Omega)|e^{j\theta_a(\Omega)},$$

where

$$\theta_a(\Omega) = \arg\{X_a(j\Omega)\}.$$

The quantity $|X_a(j\Omega)|$ is called the *magnitude spectrum,* and the quantity $\theta_a(\Omega)$ is called the *phase spectrum,* with both spectrums being real functions of Ω.

In general, the CTFT $X_a(j\Omega)$ defined by Eq. (3.1) exists if the continuous-time signal $x_a(t)$ satisfies the *Dirichlet conditions:*

(a) The signal has a finite number of finite discontinuities and a finite number of maxima and minima in any finite interval.

(b) The signal is absolutely integrable; that is,

$$\int_{-\infty}^{\infty} |x_a(t)|\,dt < \infty. \tag{3.3}$$

If the Dirichlet conditions are satisfied, the integral on the right-hand side of Eq. (3.2) converges to $x_a(t)$ at all values of t except at values of t where $x_a(t)$ has discontinuities.

It can be easily shown that if $x_a(t)$ is absolutely integrable, then $|X_a(j\Omega)| < \infty$, proving the existence of the CTFT (Problem 3.1). We illustrate the CTFT computation in Example 3.1.

EXAMPLE 3.1   An Absolutely Integrable Continuous-Time Signal

Consider the real signal

$$x_a(t) = \begin{cases} e^{-\alpha t}, & t \geq 0, \\ 0, & t < 0, \end{cases} \tag{3.4}$$

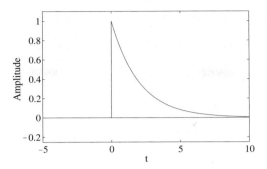

Figure 3.1: Plot of the continuous-time function of Eq. (3.4) for $\alpha = 0.5$.

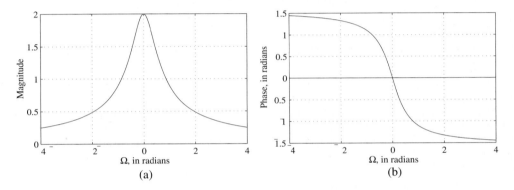

Figure 3.2: (a) Magnitude and (b) phase of $X_a(j\Omega) = 1/(0.5 + j\Omega)$.

where $\alpha > 0$. A plot of the above signal for $\alpha = 0.5$ is shown in Figure 3.1. This function is absolutely integrable as

$$\int_{-\infty}^{\infty} |x_a(t)| dt = \int_{0}^{\infty} e^{-\alpha t} dt = -\frac{e^{-\alpha t}}{\alpha}\Big|_{0}^{\infty} = \frac{1}{\alpha} < \infty.$$

Its CTFT obtained using Eq. (3.1) is given by

$$X_a(j\Omega) = \int_{0}^{\infty} e^{-\alpha t} e^{-j\Omega t} dt = \int_{0}^{\infty} e^{-(\alpha + j\Omega)t} dt$$

$$= -\frac{1}{\alpha + j\Omega} e^{-(\alpha + j\Omega)t}\Big|_{0}^{\infty} = \frac{1}{\alpha + j\Omega}. \tag{3.5}$$

We can express the above CTFT as

$$X_a(j\Omega) = \frac{1}{\sqrt{\alpha^2 + \Omega^2}} e^{-j \tan^{-1}(\Omega/\alpha)},$$

where $1/(\sqrt{\alpha^2 + \Omega^2})$ is the magnitude spectrum $|X_a(j\Omega)|$ and $-\tan^{-1}(\Omega/\alpha)$ is the phase spectrum $\theta_a(\Omega)$. A plot of these two functions is shown in Figure 3.2.

It can be shown that for $\alpha < 0$, $x_a(t)$ is not absolutely integrable, and hence, the CTFT $X_a(j\Omega)$ does not exist in this case.

EXAMPLE 3.2   Continuous-Time Fourier Transform of an Impulse Function

Determine the CTFT $\mathbf{\Delta}(j\Omega)$ of an ideal impulse $\delta(t)$. Applying Eq. (3.1), we get

$$\mathbf{\Delta}(j\Omega) = \int_{-\infty}^{\infty} \delta(t)\, e^{-j\Omega t}\, dt = 1,$$

using the sampling property of the delta function.

EXAMPLE 3.3   Continuous-Time Fourier Transform of a Shifted Impulse Function

Consider the shifted impulse function $x_a(t) = \delta(t - t_o)$. Its CTFT is thus given by

$$X_a(j\Omega) = \int_{-\infty}^{\infty} \delta(t - t_o)\, e^{-j\Omega t}\, dt = e^{-j\Omega t_o}.$$

It should be noted that an absolutely integrable continuous-time signal $x_a(t)$ always has finite energy; that is,

$$\int_{-\infty}^{\infty} |x_a(t)|^2\, dt < \infty. \tag{3.6}$$

However, the CTFT may exist for a finite-energy continuous-time signal that is not absolutely integrable (Problem 3.5). Hence, Eq. (3.6) is a milder condition than that given by Eq. (3.3).

The CTFT can also be defined using ideal impulses for some functions that do not satisfy either Eq. (3.3) or Eq. (3.6).

3.1.2   Energy Density Spectrum

The total energy $\mathcal{E}_x$ of a finite-energy continuous-time complex signal $x_a(t)$ is given by

$$\mathcal{E}_x = \int_{-\infty}^{\infty} |x_a(t)|^2\, dt = \int_{-\infty}^{\infty} x_a(t) x_a^*(t)\, dt. \tag{3.7}$$

The energy can also be expressed in terms of the CTFT $X_a(j\Omega)$. To this end, we first replace $x_a^*(t)$ in the above equation by its inverse CTFT expression as shown below:

$$\mathcal{E}_x = \int_{-\infty}^{\infty} x_a(t) \left[\frac{1}{2\pi} \int_{-\infty}^{\infty} X_a^*(j\Omega) e^{-j\Omega t}\, d\Omega \right] dt.$$

Interchanging the order of the integrations in the above, we get

$$\mathcal{E}_x = \frac{1}{2\pi} \int_{-\infty}^{\infty} X_a^*(j\Omega) \left[\int_{-\infty}^{\infty} x_a(t) e^{-j\Omega t}\, dt \right] d\Omega$$

$$= \frac{1}{2\pi} \int_{-\infty}^{\infty} X_a^*(j\Omega) X_a(j\Omega)\, d\Omega = \frac{1}{2\pi} \int_{-\infty}^{\infty} |X_a(j\Omega)|^2\, d\Omega. \tag{3.8}$$

Combining Eqs. (3.7) and (3.8), we arrive at

$$\int_{-\infty}^{\infty} |x_a(t)|^2\, dt = \frac{1}{2\pi} \int_{-\infty}^{\infty} |X_a(j\Omega)|^2\, d\Omega, \tag{3.9}$$

which is more commonly known as the *Parseval's relation* for finite-energy continuous-time signals.

EXAMPLE 3.4   Energy of a Continuous-Time Signal

We determine the total energy $\mathcal{E}_x$ of the continuous-time signal of Eq. (3.4). Using Eq. (3.9), we have

$$\mathcal{E}_x = \frac{1}{2\pi} \int_{-\infty}^{\infty} \frac{1}{\alpha^2 + \Omega^2} \, d\Omega = \frac{1}{2\pi} \cdot \frac{\pi}{\alpha} = \frac{1}{2\alpha}.$$

For $\alpha = 0.5$, the total energy is $\mathcal{E}_x = 1$.

The integrand $|X_a(j\Omega)|^2$ on the right-hand side of Eq. (3.8) is called the *energy density spectrum* of the continuous-time signal $x_a(t)$ and is usually denoted by the symbol $S_{xx}(\Omega)$; that is,

$$S_{xx}(\Omega) = |X_a(j\Omega)|^2.$$

The energy $\mathcal{E}_{x,r}$ over a specified range of frequencies $\Omega_a \leq \Omega \leq \Omega_b$ of the signal $x_a(t)$ can be computed by integrating $S_{xx}(\Omega)$ over this range:

$$\mathcal{E}_{x,r} = \frac{1}{2\pi} \int_{\Omega_a}^{\Omega_b} S_{xx}(\Omega) \, d\Omega.$$

3.1.3 Band-limited Continuous-Time Signals

A full-band, finite-energy, continuous-time signal has a spectrum occupying the whole frequency range $-\infty < \Omega < \infty$, whereas a *band-limited* continuous-time signal has a spectrum that is limited to a portion of the above frequency range. An ideal band-limited signal has a spectrum that is zero outside a finite frequency range $\Omega_a \leq |\Omega| \leq \Omega_b$; that is,

$$X_a(j\Omega) = \begin{cases} 0, & 0 \leq |\Omega| < \Omega_a, \\ 0, & \Omega_b < |\Omega| < \infty. \end{cases}$$

However, an ideal band-limited signal cannot be generated in practice, and, for practical purposes, it is sufficient to ensure that for a band-limited signal outside the specified frequency range, the signal energy is arbitrarily small.

Band-limited signals are classified according to the frequency range where most of the signal's energy is concentrated. A *lowpass* continuous-time signal has a spectrum occupying the frequency range $0 \leq |\Omega| \leq \Omega_p < \infty$, where Ω_p is called the *bandwidth* of the signal. Likewise, a *highpass* continuous-time signal has a spectrum occupying the frequency range $0 < \Omega_p \leq |\Omega| < \infty$, where the *bandwidth* of the signal is from Ω_p to ∞. Finally, a *bandpass* continuous-time signal has a spectrum occupying the frequency range $0 < \Omega_L \leq |\Omega| \leq \Omega_H < \infty$, where $\Omega_H - \Omega_L$ is its *bandwidth*. A precise definition of the bandwidth depends on applications. As can be seen from Figure 3.2(a), the continuous-time signal of Eq. (3.4) is a lowpass signal. It can be shown that 80% of the energy of this signal is contained in the frequency range $0 \leq |\Omega| \leq 0.4898\pi$, and hence, we can define the 80% bandwidth of the signal to be 0.4898π radians (Problem 3.10).

3.1.4 The Frequency Response of an LTI Continuous-Time System

As indicated by Eq. (1.2), the output response $y_a(t)$ of an initially relaxed linear, time-invariant continuous-time system characterized by an impulse response $h_a(t)$ for an input signal $x_a(t)$ is given by the convolution integral

$$y_a(t) = \int_{-\infty}^{\infty} h_a(t - \tau) x_a(\tau) d\tau. \tag{3.10}$$

Applying the CTFT to both sides of Eq. (3.10), we have

$$Y_a(j\Omega) = H_a(j\Omega)X_a(j\Omega), \tag{3.11}$$

where $Y_a(j\Omega)$, $X_a(j\Omega)$ and $H_a(j\Omega)$ are, respectively, the CTFTs of $y_a(t)$, $x_a(t)$, and $h_a(t)$. The function $H_a(j\Omega)$ is called the *frequency response* of the LTI continuous-time system.

3.2   The Discrete-Time Fourier Transform

The *discrete-time Fourier transform* (DTFT) of a discrete-time sequence $x[n]$ is a representation of the sequence in terms of the complex exponential sequence $\{e^{j\omega n}\}$, where ω is the real frequency variable. If there is no ambiguity, for brevity often the discrete-time Fourier transform is called simply the *Fourier transform* (FT). The Fourier transform representation of a sequence, if it exists, is unique, and the original sequence can be computed from its transform representation by an inverse transform operation. We first define the forward transform and derive its inverse transform. We then describe the condition for its existence and review its important properties.

3.2.1   Definition

The *discrete-time Fourier transform* $X(e^{j\omega})$ of a sequence $x[n]$ is defined by

$$X(e^{j\omega}) = \sum_{n=-\infty}^{\infty} x[n]e^{-j\omega n}. \tag{3.12}$$

We illustrate the Fourier transform computation in Examples 3.5 and 3.6.

EXAMPLE 3.5   **Discrete-Time Fourier Transform of the Unit Sample Sequence**

Consider the unit sample sequence $\delta[n]$ defined in Eq. (2.39). Its Fourier transform $\Delta(e^{j\omega})$ obtained using Eq. (3.12) is given by

$$\Delta(e^{j\omega}) = \sum_{n=-\infty}^{\infty} \delta[n]e^{-j\omega n} = 1, \tag{3.13}$$

where we have used the sampling property of the unit sample sequence; that is, $\delta[0] = 1$, and $\delta[n] = 0$ for $n \neq 0$.

EXAMPLE 3.6   **Discrete-Time Fourier Transform of an Exponential Sequence**

Consider the causal sequence

$$x[n] = \alpha^n \mu[n], \quad |\alpha| < 1, \tag{3.14}$$

shown in Figure 3.3 for $\alpha = 0.5$. Its Fourier transform $X(e^{j\omega})$ obtained using Eq. (3.12) is given by

$$X(e^{j\omega}) = \sum_{n=-\infty}^{\infty} \alpha^n \mu[n]e^{-j\omega n} = \sum_{n=0}^{\infty} \alpha^n e^{-j\omega n}$$

$$= \sum_{n=0}^{\infty} (\alpha e^{-j\omega})^n = \frac{1}{1 - \alpha e^{-j\omega}}, \tag{3.15}$$

as $|\alpha e^{-j\omega}| = |\alpha| < 1$.

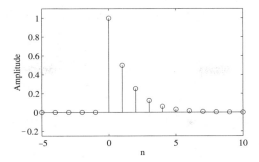

Figure 3.3: Plot of the discrete-time sequence of Eq. (3.14) for $\alpha = 0.5$.

It should be noted here that the Fourier transforms of most practical discrete-time sequences can be expressed in terms of a sum of a convergent geometric series, which can be summed in a simple closed form as illustrated by Example 3.6. We take up the issue of the convergence of a general Fourier transform in Section 3.2.4.

As can be seen from the definition, the discrete-time Fourier transform $X(e^{j\omega})$ of a sequence $x[n]$ is a continuous function of ω. However, unlike the continuous-time Fourier transform, it is a periodic function in ω with a period 2π. To verify this latter property, observe that for any integer k,

$$X(e^{j(\omega+2\pi k)}) = \sum_{n=-\infty}^{\infty} x[n]e^{-j(\omega+2\pi k)n} = \sum_{n=-\infty}^{\infty} x[n]e^{-j\omega n}e^{-j2\pi kn}$$

$$= \sum_{n=-\infty}^{\infty} x[n]e^{-j\omega n} = X(e^{j\omega}), \qquad \text{for all values of } k,$$

where we have used the fact $e^{-j2\pi kn} = 1$. It therefore follows that Eq. (3.12) represents the Fourier series expansion of the periodic function $X(e^{j\omega})$. As a result, the Fourier coefficients $x[n]$ can be computed from $X(e^{j\omega})$ using the Fourier integral given by

$$x[n] = \frac{1}{2\pi} \int_{-\pi}^{\pi} X(e^{j\omega})e^{j\omega n} \, d\omega, \qquad (3.16)$$

called the *inverse discrete-time Fourier transform.* It should be noted that even though the integration in Eq. (3.16) can be carried out over any interval of duration 2π, it is a common practice to choose the interval $[-\pi, \pi]$. The inverse discrete-time Fourier transform given by Eq. (3.16) can be interpreted as a linear combination of infinitesimally small complex exponential signals of the form $\frac{1}{2\pi}e^{j\omega n}d\omega$, weighted by the complex constant $X(e^{j\omega})$ over the angular frequency range from $-\pi$ to π.

Equations (3.12) and (3.16) constitute a discrete-time Fourier transform pair for the sequence $x[n]$. Equation (3.12) usually is referred to as the *analysis equation,* because it analyzes how much of each complex exponential signal is present in the original signal. On the other hand, Eq. (3.16) is referred to as the *synthesis equation,* because it synthesizes an arbitrary signal from its complex exponential components. For notational convenience, we shall use the operator symbol

$$\mathcal{F}\{x[n]\}$$

to denote the $X(e^{j\omega})$ of the sequence $x[n]$. Likewise, we shall use the operator symbol

$$\mathcal{F}^{-1}\{X(e^{j\omega})\}$$

to denote the inverse Fourier transform $x[n]$ of the transform $X(e^{j\omega})$. A discrete-time Fourier transform pair will be denoted as

$$x[n] \xleftrightarrow{\mathcal{F}} X(e^{j\omega}). \tag{3.17}$$

To verify that the integral on the right-hand side of Eq. (3.16) indeed results in the inverse FT $x[n]$, we substitute the expression for $X(e^{j\omega})$ from Eq. (3.12) in Eq. (3.16), arriving at

$$x[n] = \frac{1}{2\pi} \int_{-\pi}^{\pi} \left(\sum_{\ell=-\infty}^{\infty} x[\ell] e^{-j\omega\ell} \right) e^{j\omega n} d\omega.$$

The order of integration and the summation on the right-hand side of the above equation can be interchanged if the summation inside the brackets converges uniformly, that is, if $X(e^{j\omega})$ exists. Under this condition, we get from the above

$$\sum_{\ell=-\infty}^{\infty} x[\ell] \left(\frac{1}{2\pi} \int_{-\pi}^{\pi} e^{j\omega(n-\ell)} d\omega \right) = \sum_{\ell=-\infty}^{\infty} x[\ell] \frac{\sin \pi(n-\ell)}{\pi(n-\ell)}$$

$$= \sum_{\ell=-\infty}^{\infty} x[\ell] \operatorname{sinc}(n-\ell).$$

For $n \neq \ell$, $\sin \pi(n-\ell) = 0$, and as a result, $\operatorname{sinc}(n-\ell) = 0$. For $n = \ell$, $\operatorname{sinc}(n-\ell) = 0/0$. In this case, using L'Hospital rule, we get

$$\lim_{n \to \ell} \frac{\sin \pi(n-\ell)}{\pi(n-\ell)} = \frac{\pi \cos(\pi n)}{\pi} = 1.$$

Thus,

$$\operatorname{sinc}(n-\ell) = \begin{cases} 1, & n = \ell, \\ 0, & n \neq \ell, \end{cases}$$
$$= \delta[n-\ell]. \tag{3.18}$$

Hence,

$$\sum_{\ell=-\infty}^{\infty} x[\ell] \operatorname{sinc}(n-\ell) = \sum_{\ell=-\infty}^{\infty} x[\ell] \delta[n-\ell] = x[n],$$

using the sampling property of the unit sample sequence.

3.2.2 Basic Properties

We have already demonstrated one basic property of the Fourier transform in Section 3.2.1, namely, the periodicity property of the transform. We examine here a few additional basic properties of the Fourier transform of a complex sequence.

In general, the Fourier transform $X(e^{j\omega})$ is a complex function of the real variable ω and can be written in rectangular form as

$$X(e^{j\omega}) = X_{\text{re}}(e^{j\omega}) + j X_{\text{im}}(e^{j\omega}), \tag{3.19}$$

where $X_{\text{re}}(e^{j\omega})$ and $X_{\text{im}}(e^{j\omega})$ are, respectively, the real and imaginary parts of $X(e^{j\omega})$, and are real functions of ω. From Eq. (3.19), it follows that

$$X_{\text{re}}(e^{j\omega}) = \tfrac{1}{2}\{X(e^{j\omega}) + X^*(e^{j\omega})\}, \tag{3.20a}$$

$$X_{\text{im}}(e^{j\omega}) = \tfrac{1}{2j}\{X(e^{j\omega}) - X^*(e^{j\omega})\}, \tag{3.20b}$$

where $X^*(e^{j\omega})$ denotes the complex conjugate of $X(e^{j\omega})$.

The Fourier transform $X(e^{j\omega})$ can alternately be expressed in the polar form as

$$X(e^{j\omega}) = |X(e^{j\omega})|e^{j\theta(\omega)}, \tag{3.21}$$

where

$$\theta(\omega) = \arg\{X(e^{j\omega})\}. \tag{3.22}$$

The quantity $|X(e^{j\omega})|$ is called the *magnitude function,* and the quantity $\theta(\omega)$ is called the *phase function,* with both functions again being real functions of ω. In many applications, the Fourier transform is called the *Fourier spectrum* and likewise, $|X(e^{j\omega})|$ and $\theta(\omega)$ are referred to as the *magnitude spectrum* and *phase spectrum,* respectively.

The relations between the rectangular and polar forms of $X(e^{j\omega})$ follow from Eqs. (3.19) and (3.21) and are given by

$$X_{\text{re}}(e^{j\omega}) = |X(e^{j\omega})| \cos \theta(\omega), \tag{3.23a}$$

$$X_{\text{im}}(e^{j\omega}) = |X(e^{j\omega})| \sin \theta(\omega), \tag{3.23b}$$

$$|X(e^{j\omega})|^2 = X(e^{j\omega})X^*(e^{j\omega}) = X_{\text{re}}^2(e^{j\omega}) + X_{\text{im}}^2(e^{j\omega}), \tag{3.23c}$$

$$\tan \theta(\omega) = \frac{X_{\text{im}}(e^{j\omega})}{X_{\text{re}}(e^{j\omega})}. \tag{3.23d}$$

As in the case of the continuous-time Fourier transform, the phase function is also not uniquely specified for the discrete-time Fourier transform. Note from Eq. (3.21) that if we replace $\theta(\omega)$ with $\theta(\omega) + 2\pi k$, where k is any integer, we get

$$X(e^{j\omega}) = |X(e^{j\omega})|e^{j[\theta(\omega)+2\pi k]} = |X(e^{j\omega})|e^{j\theta(\omega)},$$

indicating that the Fourier transform $X(e^{j\omega})$ remains unchanged. Consequently, the phase function $\theta(\omega)$ cannot be uniquely specified for any discrete-time Fourier transform for all values of ω. Unless otherwise stated, we will assume that the phase function $\theta(\omega)$ is restricted to the following range of values,

$$-\pi \leq \theta(\omega) < \pi,$$

called the *principal value.*

3.2.3 Symmetry Relations

We describe here some additional properties of the Fourier transform that are based on the symmetry relations. These properties can simplify the computational complexity and are often useful in digital signal processing applications.

Complex Sequences

For a given sequence $x[n]$ with a Fourier transform $X(e^{j\omega})$, it is easy to determine the Fourier transforms of its time-reversed sequence $x[-n]$ and the complex conjugate sequence $x^*[n]$. It follows from Eq. (3.12) that

$$\mathcal{F}\{x[-n]\} = \sum_{n=-\infty}^{\infty} x[-n]e^{-j\omega n} = \sum_{m=\infty}^{-\infty} x[m]e^{j\omega m} = X(e^{-j\omega});$$

that is,

$$x[-n] \overset{\mathcal{F}}{\longleftrightarrow} X(e^{-j\omega}). \tag{3.24}$$

Likewise, from Eq. (3.12), we get

$$\mathcal{F}\{x^*[n]\} = \sum_{n=-\infty}^{\infty} x^*[n]e^{-j\omega n} = \left(\sum_{n=-\infty}^{\infty} x[n]e^{j\omega n} \right)^* = X^*(e^{-j\omega});$$

that is,

$$x^*[n] \overset{\mathcal{F}}{\longleftrightarrow} X^*(e^{-j\omega}). \tag{3.25}$$

In a similar manner or by combining Eqs. (3.24) and (3.25), we can show

$$x^*[-n] \overset{\mathcal{F}}{\longleftrightarrow} X^*(e^{j\omega}). \tag{3.26}$$

The real and imaginary parts of $X(e^{j\omega})$ can be expressed in terms of the real and imaginary parts of $x[n]$. Substituting

$$x[n] = x_{\text{re}}[n] + jx_{\text{im}}[n] \tag{3.27}$$

in Eq. (3.12), we get

$$X(e^{j\omega}) = \sum_{n=-\infty}^{\infty} (x_{\text{re}}[n] + jx_{\text{im}}[n]) (\cos \omega n + j \sin \omega n)$$

$$= \sum_{n=-\infty}^{\infty} (x_{\text{re}}[n] \cos \omega n - x_{\text{im}}[n] \sin \omega n) + j$$

$$\sum_{n=-\infty}^{\infty} (x_{\text{im}}[n] \cos \omega n + x_{\text{re}}[n] \sin \omega n). \tag{3.28}$$

From the above equation, we then have

$$X_{\text{re}}(e^{j\omega}) = \sum_{n=-\infty}^{\infty} (x_{\text{re}}[n] \cos \omega n - x_{\text{im}}[n] \sin \omega n), \tag{3.29a}$$

$$X_{\text{im}}(e^{j\omega}) = \sum_{n=-\infty}^{\infty} (x_{\text{im}}[n] \cos \omega n + x_{\text{re}}[n] \sin \omega n). \tag{3.29b}$$

A Fourier transform $X(e^{j\omega})$ is defined to be a *conjugate-symmetric* function of ω if

$$X(e^{j\omega}) = X^*(e^{-j\omega}),$$

that is, $X_{\text{re}}(e^{j\omega}) = X_{\text{re}}(e^{-j\omega})$ and $X_{\text{im}}(e^{j\omega}) = -X_{\text{im}}(e^{-j\omega})$. Hence, the real and imaginary parts of a conjugate-symmetric Fourier transform are, respectively, even and odd functions of ω. Likewise, the Fourier transform $X(e^{-j\omega})$ is a *conjugate-antisymmetric* function of ω if

$$X(e^{j\omega}) = -X^*(e^{-j\omega}),$$

that is, $X_{re}(e^{j\omega}) = -X_{re}(e^{-j\omega})$ and $X_{im}(e^{j\omega}) = X_{im}(e^{-j\omega})$. Therefore, the real and imaginary parts of a conjugate-antisymmetric Fourier transform are, respectively, odd and even functions of ω.

A complex-valued Fourier transform $X(e^{j\omega})$, in general, can be expressed as a sum of a conjugate-symmetric part $X_{cs}(e^{j\omega})$ and a conjugate-antisymmetric part $X_{ca}(e^{j\omega})$:

$$X(e^{j\omega}) = X_{cs}(e^{j\omega}) + X_{ca}(e^{j\omega}), \tag{3.30}$$

where

$$X_{cs}(e^{j\omega}) = \tfrac{1}{2}\left[X(e^{j\omega}) + X^*(e^{-j\omega})\right], \tag{3.31a}$$

$$X_{ca}(e^{j\omega}) = \tfrac{1}{2}\left[X(e^{j\omega}) - X^*(e^{-j\omega})\right]. \tag{3.31b}$$

We next derive the Fourier transforms of $x_{re}[n]$ and $x_{im}[n]$, the real and imaginary parts of the sequence $x[n]$. We rewrite Eq. (3.12) as

$$X(e^{j\omega}) = \sum_{n=-\infty}^{\infty} (x_{re}[n] + jx_{im}[n])e^{-j\omega n}.$$

Substituting the above in Eq. (3.31a) and making use of Eq. (3.25), we get

$$X_{cs}(e^{j\omega}) = \frac{1}{2}\left(\sum_{n=-\infty}^{\infty} x_{re}[n]e^{-j\omega n} + j\sum_{n=-\infty}^{\infty} x_{im}[n]e^{-j\omega n}\right.$$
$$\left. + \sum_{n=-\infty}^{\infty} x_{re}[n]e^{-j\omega n} - j\sum_{n=-\infty}^{\infty} x_{im}[n]e^{-j\omega n}\right)$$
$$= \sum_{n=-\infty}^{\infty} x_{re}[n]e^{-j\omega n} = \mathcal{F}\{x_{re}[n]\};$$

that is,

$$x_{re}[n] \overset{\mathcal{F}}{\longleftrightarrow} X_{cs}(e^{j\omega}). \tag{3.32}$$

In a similar manner, we can show

$$jx_{im}[n] \overset{\mathcal{F}}{\longleftrightarrow} X_{ca}(e^{j\omega}). \tag{3.33}$$

We can also derive the Fourier transforms of the conjugate-symmetric and conjugate-antisymmetric parts of a sequence $x[n]$. From Eqs. (2.23a) and (3.26), we get

$$\mathcal{F}\{x_{cs}[n]\} = \tfrac{1}{2}\left(\mathcal{F}\{x[n]\} + \mathcal{F}\{x^*[-n]\}\right)$$
$$= \tfrac{1}{2}\{X(e^{j\omega}) + X^*(e^{j\omega})\} = X_{re}(e^{j\omega}),$$

that is,

$$x_{cs}[n] \overset{\mathcal{F}}{\longleftrightarrow} X_{re}(e^{j\omega}). \tag{3.34}$$

Similarly, from Eqs. (2.23b) and (3.25), we arrive at

$$x_{ca}[n] \overset{\mathcal{F}}{\longleftrightarrow} jX_{im}(e^{j\omega}). \tag{3.35}$$

For future reference, we summarize in Table 3.1 the symmetry relations of the Fourier transform of a complex sequence.

Table 3.1: Symmetry relations of the discrete-time Fourier transform of a complex sequence.

Sequence	Discrete-Time Fourier Transform
$x[n]$	$X(e^{j\omega})$
$x[-n]$	$X(e^{-j\omega})$
$x^*[-n]$	$X^*(e^{j\omega})$
$\text{Re}\{x[n]\}$	$X_{cs}(e^{j\omega}) = \frac{1}{2}\{X(e^{j\omega}) + X^*(e^{-j\omega})\}$
$j\text{Im}\{x[n]\}$	$X_{ca}(e^{j\omega}) = \frac{1}{2}\{X(e^{j\omega}) - X^*(e^{-j\omega})\}$
$x_{cs}[n]$	$X_{re}(e^{j\omega})$
$x_{ca}[n]$	$jX_{im}(e^{j\omega})$

Note: $X_{cs}(e^{j\omega})$ and $X_{ca}(e^{j\omega})$ are the conjugate-symmetric and conjugate-antisymmetric parts of $X(e^{j\omega})$, respectively. Likewise, $x_{cs}[n]$ and $x_{ca}[n]$ are the conjugate-symmetric and conjugate-antisymmetric parts of $x[n]$, respectively.

Real and Purely Imaginary Sequences

For real sequence, $x_{im}[n] = 0$. Hence, from Eq. (3.33), $X_{ca}(e^{j\omega}) = 0$, and thus, from Eq. (3.31b), we have $X(e^{j\omega}) = X^*(e^{j\omega})$, implying that $X(e^{j\omega})$ is a conjugate-symmetric function. As a result, the real part $X_{re}(e^{j\omega})$ and imaginary part $X_{im}(e^{j\omega})$ of the Fourier transform of a real sequence are, respectively, even and odd functions of ω.

Next, we observe from Eq. (3.23c),

$$|X(e^{j\omega})| = \sqrt{X_{re}^2(e^{j\omega}) + X_{im}^2(e^{j\omega})}. \tag{3.36}$$

Thus, for a real signal, it follows from the above that

$$|X(e^{-j\omega})| = \sqrt{X_{re}^2(e^{-j\omega}) + X_{im}^2(e^{-j\omega})}$$

$$= \sqrt{X_{re}^2(e^{j\omega}) + X_{im}^2(e^{j\omega})} = |X(e^{j\omega})|, \tag{3.37}$$

indicating that $|X(e^{j\omega})|$ is an even function of ω.

Likewise, for a real signal, we note from Eq. (3.23d) that

$$\tan(-\theta) = \frac{X_{im}(e^{-j\omega})}{X_{re}(e^{-j\omega})}$$

$$= -\frac{X_{im}(e^{j\omega})}{X_{re}(e^{j\omega})} = -\tan(\theta), \tag{3.38}$$

implying that $\tan\theta(\omega)$, and hence, $\theta(\omega)$, is an odd function of ω.

Since, for a real signal $X^*(e^{j\omega}) = X(e^{-j\omega})$, it follows from Eq. (3.23c) that, in this case, the magnitude function can be easily computed using

$$|X(e^{j\omega})|^2 = X(e^{j\omega})X(e^{-j\omega}). \tag{3.39}$$

Example 3.7 illustrates the symmetry properties of the Fourier transform of a real sequence.

EXAMPLE 3.7 **Real and Imaginary Parts and Magnitude and Phase Functions of a Discrete-Time Fourier Transform**

The Fourier transform given by Eq. (3.15) of the real sequence of Eq. (3.14) can be rewritten as

$$X(e^{j\omega}) = \frac{1}{1 - \alpha e^{-j\omega}}$$

$$= \frac{1}{1 - \alpha e^{-j\omega}} \cdot \frac{1 - \alpha e^{j\omega}}{1 - \alpha e^{j\omega}}$$

$$= \frac{1 - \alpha \cos\omega - j\alpha \sin\omega}{1 - 2\alpha \cos\omega + \alpha^2}. \tag{3.40}$$

Therefore, the real and imaginary parts of $X(e^{j\omega})$ are given by

$$X_{\text{re}}(e^{j\omega}) = \frac{1 - \alpha \cos\omega}{1 - 2\alpha \cos\omega + \alpha^2},$$

$$X_{\text{im}}(e^{j\omega}) = -\frac{\alpha \sin\omega}{1 - 2\alpha \cos\omega + \alpha^2}.$$

The above real and imaginary parts have been plotted in Figures 3.4(a) and (b), respectively, for $\alpha = 0.5$. Since $\cos\omega$ and $\sin\omega$ are periodic functions of ω with a period 2π, it follows from the above and also from the plots that $X_{\text{re}}(e^{j\omega})$ and $X_{\text{im}}(e^{j\omega})$ are both periodic functions of ω with a period 2π. In addition, $\cos\omega$ and $\sin\omega$ are, respectively, even and odd functions of ω. Hence, $X_{\text{re}}(e^{j\omega})$ and $X_{\text{im}}(e^{j\omega})$ are also, respectively, even and odd functions of ω. Now,

$$|X(e^{j\omega})|^2 = X(e^{j\omega}) \cdot X^*(e^{j\omega})$$

$$= \frac{1}{1 - \alpha e^{-j\omega}} \cdot \frac{1}{1 - \alpha e^{j\omega}}$$

$$= \frac{1}{1 - 2\alpha \cos\omega + \alpha^2}.$$

Hence,

$$|X(e^{j\omega})| = \frac{1}{\sqrt{1 - 2\alpha \cos\omega + \alpha^2}}.$$

Likewise,

$$\tan(\theta) = \frac{X_{\text{im}}(e^{j\omega})}{X_{\text{re}}(e^{j\omega})} = -\frac{\alpha \sin\omega}{1 - \alpha \cos\omega}.$$

Therefore,

$$\theta(\omega) = \tan^{-1}\left(-\frac{\alpha \sin\omega}{1 - \alpha \cos\omega}\right).$$

The magnitude and the phase of the Fourier transform of Eq. (3.15) have been plotted in Figures 3.4(c) and (d), respectively, for $\alpha = 0.5$. It can be seen from these plots and the expressions given above that both $|X(e^{j\omega})|$ and $\theta(\omega)$ are periodic functions of ω with a period 2π. In addition, $|X(e^{j\omega})|$ and $\theta(\omega)$ are, respectively, even and odd functions of ω.

For a purely imaginary sequence, $x_{\text{re}}[n] = 0$. As a result, here Eqs. (3.29a) and (3.29b) reduce to

$$X_{\text{re}}(e^{j\omega}) = -\sum_{n=-\infty}^{\infty} x_{\text{im}}[n] \sin\omega n, \tag{3.41a}$$

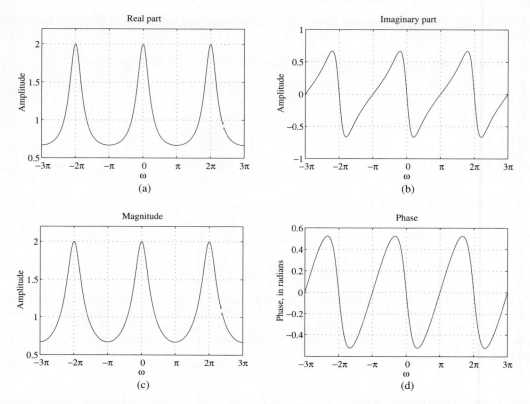

Figure 3.4: (a) Real and (b) imaginary parts and (c) magnitude and (d) phase of $X(e^{j\omega}) = 1/(1 - 0.5e^{-j\omega})$.

$$X_{im}(e^{j\omega}) = \sum_{n=-\infty}^{\infty} x_{im}[n] \cos \omega n. \tag{3.41b}$$

It follows from Eqs. (3.41a) and (3.41b) that for a purely imaginary sequence, $X_{re}(e^{j\omega})$ and $X_{im}(e^{j\omega})$ are, respectively, odd and even functions of ω. It can also be shown that $|X(e^{j\omega})|$ and $\theta(\omega)$ are, respectively, even and odd functions of ω.

As the conjugate-symmetric and the conjugate-antisymmetric parts of a real sequence are, respectively, even and odd sequences, it follows from Eqs. (3.34) and (3.35) that

$$x_{ev}[n] \overset{\mathcal{F}}{\longleftrightarrow} X_{re}(e^{j\omega}) \qquad \text{and} \qquad x_{od}[n] \overset{\mathcal{F}}{\longleftrightarrow} jX_{im}(e^{j\omega}).$$

For future reference, we summarize in Table 3.2 the symmetry relations of the Fourier transform of a real sequence.

3.2.4 Convergence Condition

Now, an infinite series of the form of Eq. (3.12) may or may not converge. The Fourier transform $X(e^{j\omega})$ of $x[n]$ is said to exist if the series in Eq. (3.12) converges in some sense. Let

Table 3.2: Symmetry relations of the discrete-time Fourier transform of a real sequence.

Sequence	Discrete-Time Fourier Transform
$x[n]$	$X(e^{j\omega}) = X_{\text{re}}(e^{j\omega}) + jX_{\text{im}}(e^{j\omega})$
$x_{\text{ev}}[n]$	$X_{\text{re}}(e^{j\omega})$
$x_{\text{od}}[n]$	$jX_{\text{im}}(e^{j\omega})$
	$X(e^{j\omega}) = X^*(e^{-j\omega})$
	$X_{\text{re}}(e^{j\omega}) = X_{\text{re}}(e^{-j\omega})$
Symmetry relations	$X_{\text{im}}(e^{j\omega}) = -X_{\text{im}}(e^{-j\omega})$
	$|X(e^{j\omega})| = |X(e^{-j\omega})|$
	$\arg\{X(e^{j\omega})\} = -\arg\{X(e^{-j\omega})\}$

Note: $x_{\text{ev}}[n]$ and $x_{\text{od}}[n]$ denote the even and odd parts of $x[n]$, respectively.

$$X_K(e^{j\omega}) = \sum_{n=-K}^{K} x[n]e^{-j\omega n}, \qquad (3.42)$$

denote the partial sum of the weighted complex exponentials in Eq. (3.12). Then for *uniform convergence* of $X(e^{j\omega})$,

$$\lim_{K \to \infty} X_K(e^{j\omega}) = X(e^{j\omega}).$$

Now, if $x[n]$ is an *absolutely summable sequence*, that is, if

$$\sum_{n=-\infty}^{\infty} |x[n]| < \infty, \qquad (3.43)$$

then

$$\left| X(e^{j\omega}) \right| = \left| \sum_{n=-\infty}^{\infty} x[n]\,e^{-j\omega n} \right| \le \sum_{n=-\infty}^{\infty} |x[n]|\,\left| e^{-j\omega n} \right| \le \sum_{n=-\infty}^{\infty} |x[n]| < \infty,$$

for all values of ω guaranteeing the existence of $X(e^{j\omega})$. Thus, Eq. (3.43) is a sufficient condition for the existence of the Fourier transform $X(e^{j\omega})$ of the sequence $x[n]$. Moreover, it can be shown that for an absolutely summable sequence, the infinite series of Eq. (3.12) defining the Fourier transform converges uniformly for all values of ω.

A large class of sequences encountered in practice are of finite length with finite sample values. These sequences are absolutely summable, and hence, their Fourier transforms converge uniformly. On the other hand, infinite length sequences may or may not converge uniformly. The sequence $x[n] = \alpha^n \mu[n]$, $|\alpha| < 1$, of Example 3.5 is absolutely summable as

$$\sum_{n=-\infty}^{\infty} |\alpha^n| \mu[n] = \sum_{n=0}^{\infty} |\alpha|^n = \frac{1}{1 - |\alpha|} < \infty,$$

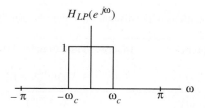

Figure 3.5: Plot of the Fourier transform of Eq. (3.45).

and its discrete-time Fourier transform $X(e^{j\omega})$ therefore converges to $1/(1 - \alpha e^{-j\omega})$ uniformly. However, the sequence $x[n] = \alpha^n \mu[n]$, $|\alpha| > 1$, is not absolutely summable, and its Fourier transform does not exist.

Since

$$\sum_{n=-\infty}^{\infty} |x[n]|^2 \leq \left(\sum_{n=-\infty}^{\infty} |x[n]| \right)^2,$$

an absolutely summable sequence always has a finite energy. However, a finite-energy sequence is not necessarily absolutely summable. For example, the sequence $x_1[n]$ of Example 2.6 is such a sequence. To represent such sequences by a discrete-time Fourier transform, it is necessary to consider a *mean-square convergence* of $X(e^{j\omega})$, in which case the total energy of the error $\mathcal{E}(\omega)$ must approach zero at each value of ω as K goes to ∞; that is,

$$\lim_{K \to \infty} \int_{-\pi}^{\pi} \left| X(e^{j\omega}) - X_K(e^{j\omega}) \right|^2 d\omega = 0. \tag{3.44}$$

In such a case, $X_K(e^{j\omega})$ is not a limit of $X_K(e^{j\omega})$ as K goes to ∞, and the Fourier transform is no longer bounded. Example 3.8 considers such a sequence.

EXAMPLE 3.8   Illustration of Mean-Square Convergence of a Discrete-Time Fourier Transform

Consider the Fourier transform

$$H_{LP}(e^{j\omega}) = \begin{cases} 1, & 0 \leq |\omega| \leq \omega_c, \\ 0, & \omega_c < |\omega| \leq \pi, \end{cases} \tag{3.45}$$

shown in Figure 3.5. This Fourier transform finds application in digital filtering, and we shall consider it again in Section 10.2.3. The inverse DTFT of $H_{LP}(e^{j\omega})$ is given by

$$h_{LP}[n] = \frac{1}{2\pi} \int_{-\pi}^{\pi} H_{LP}(e^{j\omega}) e^{j\omega n} d\omega = \frac{1}{2\pi} \int_{-\omega_c}^{\omega_c} e^{j\omega n} d\omega$$

$$= \frac{1}{2\pi} \left(\frac{e^{j\omega_c n}}{jn} - \frac{e^{-j\omega_c n}}{jn} \right) = \frac{\sin \omega_c n}{\pi n}, \qquad -\infty < n < \infty, n \neq 0. \tag{3.46}$$

For $n = 0$, the inverse Fourier transform expression reduces to

$$h_{LP}[0] = \frac{1}{2\pi} \int_{-\pi}^{\pi} H_{LP}(e^{j\omega}) d\omega = \frac{1}{2\pi} \int_{-\omega_c}^{\omega_c} d\omega = \frac{\omega_c}{\pi}. \tag{3.47}$$

Combining Eqs. (3.46) and (3.47), we can write

$$h_{LP}[n] = \begin{cases} \frac{\omega_c}{\pi}, & n = 0, \\ \frac{\sin \omega_c n}{\pi n}, & n \neq 0. \end{cases} \tag{3.48}$$

It should be noted that often the above sequence is expressed in a compact form as

$$h_{LP}[n] = \frac{\sin \omega_c n}{\pi n}, \qquad -\infty < n < \infty, \tag{3.49}$$

with the tacit assumption that at $n = 0$, $h_{LP}[n] = \frac{\omega_c}{\pi}$. We shall show later in Example 3.14 that the energy of the above sequence is given by $\frac{\omega_c}{\pi}$, and therefore, $h_{LP}[n]$ is a finite-energy sequence. However, it is not absolutely summable. As a result,

$$\sum_{n=-\infty}^{\infty} h_{LP}[n] e^{-j\omega n} = \sum_{n=-\infty}^{\infty} \frac{\sin \omega_c n}{\pi n} e^{-j\omega n}$$

does not uniformly converge to $H_{LP}(e^{j\omega})$ of Eq. (3.45) for all values of ω, but converges to $H_{LP}(e^{j\omega})$ in the mean-square sense.

The mean-square convergence property of the sequence $h_{LP}[n]$ discussed in Example 3.8 can be further illustrated by examining the plot of the function

$$H_{LP,K}(e^{j\omega}) = \sum_{n=-K}^{K} \frac{\sin \omega_c n}{\pi n} e^{-j\omega n}, \tag{3.50}$$

for various values of K, as shown in Figure 3.6. It can be seen from this figure that, independent of the number of terms in the above sum, there are ripples in the plot of $H_{LP}(e^{j\omega})$ around both sides of the point $\omega = \omega_c$. The number of ripples increases as K increases, with the height of the largest ripple remaining the same for all values of K. As K goes to infinity, the condition of Eq. (3.44) holds, indicating the convergence of $H_{LP,K}(e^{j\omega})$ to $H_{LP}(e^{j\omega})$. The oscillatory behavior in the plot of $H_{LP,K}(e^{j\omega})$ approximating a Fourier transform $H_{LP}(e^{j\omega})$ in the mean-square sense at a point of discontinuity, as indicated in Figure 3.6, is commonly known as the *Gibbs phenomenon*. We shall return to this phenomenon in the design of FIR filters based on the windowed Fourier series discussed in Section 10.2.3.

The Fourier transform can also be defined for a certain class of sequences that are neither absolutely summable nor square-summable. Examples of such sequences are the unit step sequence of Eq. (2.40), the sinusoidal sequence of Eq. (2.42), and the complex exponential sequence of Eq. (2.45), which are neither absolutely summable nor square-summable. For this type of sequence, a Fourier transform representation is possible by using Dirac delta functions. A *Dirac delta function*, also called an *ideal impulse function*, $\delta(\omega)$ is a function of ω with infinite height, zero width, and unit area. It is commonly defined by the equation

$$\int_{-\infty}^{\infty} \delta(\omega) \, d\omega = 1, \qquad \delta(\omega) = 0, \ \omega \neq 0. \tag{3.51}$$

It is also defined as the limiting form of a unit area pulse function $p_\Delta(\omega)$, shown in Figure 3.7 as Δ goes to 0:

$$\delta(\omega) = \lim_{\Delta \to 0} p_\Delta(\omega),$$

where

$$\int_{-\infty}^{\infty} p_\Delta(\omega) d\omega = 1, \qquad p_\Delta(\omega) = 0, \ \omega \neq 0.$$

The sampling property of the delta function is given by

$$\int_{-\infty}^{\infty} D(\omega) \delta(\omega - \omega_o) d\omega = D(\omega_o),$$

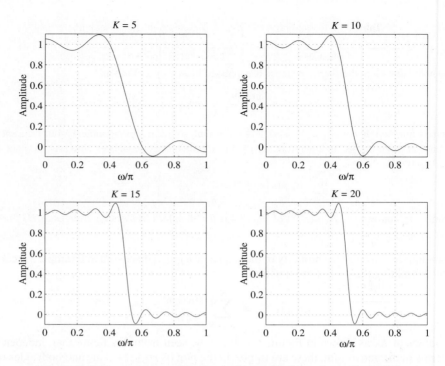

Figure 3.6: Frequency response plots of Eq. (3.51) for various values of K.

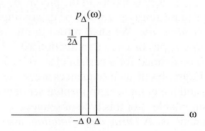

Figure 3.7: Unit area pulse function.

where $D(\omega)$ is an arbitrary function of ω that is continuous at ω_o.[2]

The Fourier transforms resulting from the use of Dirac delta functions are not continuous functions of ω.

EXAMPLE 3.9   Fourier Transform of a Complex Exponential Sequence

Consider the complex exponential sequence

$$x[n] = e^{j\omega_o n},$$

with ω_o real. Its Fourier transform is given by

---

[2]A detailed discussion of the Dirac delta function can be found in [Pap62].

Table 3.3: Commonly used discrete-time Fourier transform pairs.

Sequence	Discrete-Time Fourier Transform
$\delta[n]$	1
$1, \quad (-\infty < n < \infty)$	$\displaystyle\sum_{k=-\infty}^{\infty} 2\pi\delta(\omega + 2\pi k)$
$\mu[n]$	$\displaystyle\frac{1}{1 - e^{-j\omega}} + \sum_{k=-\infty}^{\infty} \pi\delta(\omega + 2\pi k)$
$e^{j\omega_o n}$	$\displaystyle\sum_{k=-\infty}^{\infty} 2\pi\delta(\omega - \omega_o + 2\pi k)$
$\alpha^n \mu[n], \quad (|\alpha| < 1)$	$\displaystyle\frac{1}{1 - \alpha e^{-j\omega}}$
$(n+1)\alpha^n \mu[n], \quad (|\alpha| < 1)$	$\displaystyle\frac{1}{(1 - \alpha e^{-j\omega})^2}$
$h_{LP}[n] = \dfrac{\sin \omega_c n}{\pi n}, \quad (-\infty < n < \infty)$	$H_{LP}(e^{j\omega}) = \begin{cases} 1, & 0 \le |\omega| \le \omega_c, \\ 0, & \omega_c < |\omega| \le \pi \end{cases}$

$$X(e^{j\omega}) = \sum_{k=-\infty}^{\infty} 2\pi\delta(\omega - \omega_o + 2\pi k), \tag{3.52}$$

where $\delta(\omega)$ is an impulse function of ω and $-\pi \le \omega_o < \pi$. The function on the right-hand side of the above equation is a periodic function of ω with a period 2π and is called a *periodic impulse train*.

To show the above result, we compute the inverse Fourier transform of $X(e^{j\omega})$ of Eq. (3.52):

$$x[n] = \frac{1}{2\pi} \int_{-\pi}^{\pi} \sum_{k=-\infty}^{\infty} 2\pi\delta(\omega - \omega_o + 2\pi k)\, e^{j\omega n}\, d\omega$$

$$= \int_{-\pi}^{\pi} \delta(\omega - \omega_o)\, e^{j\omega n}\, d\omega = e^{j\omega_o n},$$

where we have used the sampling property of the impulse function $\delta(\omega)$.

Table 3.3 lists the discrete-time Fourier transforms of some commonly encountered sequences.

3.2.5 Norm of a Discrete-Time Fourier Transform

A measure of the strength of a Fourier transform $X(e^{j\omega})$ is given by its norm. The $\mathcal{L}_p$-norm of $X(e^{j\omega})$ is defined by

$$\|X\|_p \triangleq \left(\frac{1}{2\pi} \int_{-\pi}^{\pi} \left| X(e^{j\omega}) \right|^p d\omega \right)^{1/p}. \tag{3.53}$$

From Eq. (3.53), it follows that the $\mathcal{L}_2$-norm, $\|X\|_2$, is the root-mean-squared (rms) value of $X(e^{j\omega})$, and the $\mathcal{L}_1$-norm, $\|X\|_1$, is the mean absolute value of $F(e^{j\omega})$ over $[-\pi, \pi]$. Moreover, $\lim_{p \to \infty} \|X\|_p$ exists for a continuous $X(e^{j\omega})$ and is given by the peak absolute value:

$$\|X\|_\infty = \max_{-\pi < \omega \leq \pi} \left| X(e^{j\omega}) \right|. \tag{3.54}$$

In this book, many of the Fourier transforms we shall encounter are rational functions in $e^{-j\omega}$, that is, ratios of polynomials in $e^{-j\omega}$, and are of the form

$$X(e^{j\omega}) = \frac{P(e^{j\omega})}{D(e^{j\omega})} = \frac{p_0 + p_1 e^{-j\omega} + \cdots + p_M e^{-j\omega M}}{d_0 + d_1 e^{-j\omega} + \cdots + d_N e^{-j\omega N}}. \tag{3.55}$$

The $\mathcal{L}_2$-norm or the $\mathcal{L}_\infty$-norm of such Fourier transforms can be determined using the M-file `filternorm` of MATLAB. This function is available with three options.

3.3   Discrete-Time Fourier Transform Theorems

There are a number of important theorems of the discrete-time Fourier transform that are useful in digital signal processing applications. These theorems can be used to determine the Fourier transforms of sequences obtained by combining sequences with known transforms. We review these theorems in this section. For compactness, the theorems are stated using the operator notation introduced in Eq. (3.17), and we shall make use of the following Fourier transform pairs:

$$g[n] \overset{\mathcal{F}}{\longleftrightarrow} G(e^{j\omega}), \tag{3.56a}$$

$$h[n] \overset{\mathcal{F}}{\longleftrightarrow} H(e^{j\omega}). \tag{3.56b}$$

The proofs of most of the theorems given here are quite straightforward and are left as exercises.

Linearity Theorem

Consider a sequence $x[n] = \alpha g[n] + \beta h[n]$ obtained by a linear combination of $g[n]$ and $h[n]$, where α and β are arbitrary constants. The Fourier transform $X(e^{j\omega})$ of $x[n]$ is then given by $\alpha G(e^{j\omega}) + \beta H(e^{j\omega})$; that is,

$$\alpha\, g[n] + \beta\, h[n] \overset{\mathcal{F}}{\longleftrightarrow} \alpha\, G(e^{j\omega}) + \beta H(e^{j\omega}). \tag{3.57}$$

It should be noted that the Fourier transforms of the conjugate-symmetric and conjugate-antisymmetric parts of a sequence derived in Eqs. (3.34) and (3.35) made use of the linearity theorem.

Time-Reversal Theorem

The Fourier transform of the time-reversed sequence $g[-n]$ is given by $G(e^{-j\omega})$; that is,

$$g[-n] \overset{\mathcal{F}}{\longleftrightarrow} G(e^{-j\omega}). \tag{3.58}$$

Time-Shifting Theorem

The Fourier transform of the delayed sequence $x[n] = g[n - n_o]$, with n_o an integer, is given by $X(e^{j\omega}) = e^{-j\omega n_o} G(e^{j\omega})$; that is,

$$g[n - n_o] \overset{\mathcal{F}}{\longleftrightarrow} e^{-j\omega n_o} G(e^{j\omega}). \tag{3.59}$$

It follows from Equation 3.59 that since $|e^{-j\omega n_o}| = 1$, $|G(e^{j\omega})| = |X(e^{j\omega})|$; that is, the magnitude spectrum is unchanged by shifting a signal in time.

EXAMPLE 3.10   Fourier Transform of a Finite-Length Exponential Sequence

Determine the Fourier transform $Y(e^{j\omega})$ of the sequence

$$y[n] = \begin{cases} \alpha^n, & 0 \leq n \leq M-1, \\ 0, & \text{otherwise,} \end{cases} \qquad |\alpha| < 1. \tag{3.60}$$

We first rewrite $y[n]$ as

$$y[n] = \alpha^n \mu[n] - \alpha^n \mu[n-M] = \alpha^n \mu[n] - \alpha^M \alpha^{n-M} \mu[n-M].$$

The Fourier transform of $x[n] = \alpha^n \mu[n]$ was computed in Example 3.6 and is given in Eq. (3.15). From the time-shifting theorem, the Fourier transform of $\alpha^{n-M} \mu[n-M]$ is thus given by $e^{-j\omega M}/(1-\alpha e^{-j\omega})$. Therefore, using the linearity theorem, we arrive at the Fourier transform of $y[n]$:

$$Y(e^{j\omega}) = \frac{1}{1-\alpha e^{-j\omega}} - \alpha^M \cdot \frac{e^{-j\omega M}}{1-\alpha e^{-j\omega}} = \frac{1-\alpha^M e^{-j\omega M}}{1-\alpha e^{-j\omega}}. \tag{3.61}$$

EXAMPLE 3.11   Fourier Transform of a Sequence Defined by a Difference Equation

Determine the Fourier transform $V(e^{j\omega})$ of the sequence $v[n]$ given by

$$d_0 v[n] + d_1 v[n-1] = p_0 \delta[n] + p_1 \delta[n-1], \qquad |d_1/d_0| < 1. \tag{3.62}$$

From Example 3.5 and also Table 3.3, the Fourier transform of $\delta[n]$ is simply 1. Next, from Table 3.4, using the time-shifting property of the Fourier transform, we observe that the Fourier transform of $\delta[n-1]$ is $e^{-j\omega}$ and the Fourier transform of $v[n-1]$ is $e^{-j\omega}V(e^{j\omega})$. Using the linearity property of Table 3.4, we then obtain from Eq. (3.62) the following equation

$$d_0 V(e^{j\omega}) + d_1 e^{-j\omega} V(e^{j\omega}) = p_0 + p_1 e^{-j\omega}.$$

Solving the above equation, we arrive at

$$V(e^{j\omega}) = \frac{p_0 + p_1 e^{-j\omega}}{d_0 + d_1 e^{-j\omega}}.$$

Frequency-Shifting Theorem

The Fourier transform of a sequence $x[n] = e^{j\omega_o n} g[n]$ is given by $X(e^{j\omega}) = G(e^{j(\omega-\omega_o)})$; that is,

$$e^{j\omega_o n} g[n] \xleftrightarrow{\mathcal{F}} G(e^{j(\omega-\omega_o)}). \tag{3.63}$$

EXAMPLE 3.12   Fourier Transform of an Exponential Sequence with Alternating Signs

Consider the sequence
$$y[n] = (-1)^n \alpha^n \mu[n], \qquad |\alpha| < 1.$$

The sequence $y[n]$ can be expressed as $y[n] = e^{j\pi n} x[n]$, where $x[n]$ is the complex exponential sequence of Example 3.5 whose Fourier transform $X(e^{j\omega})$ is given by Eq. (3.15). Hence, the Fourier transform of $y[n]$ is given by

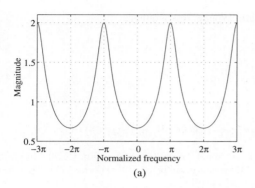
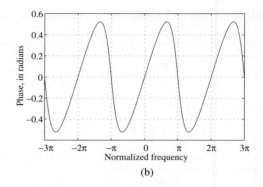

$$\text{(a)} \qquad\qquad\qquad\qquad\qquad\qquad \text{(b)}$$

Figure 3.8: (a) Magnitude and (b) phase of $X(e^{j\omega}) = 1/(1 + 0.5e^{-j\omega})$.

Table 3.4: Discrete-time Fourier transform theorems.

Theorem	Sequence	Discrete-Time Fourier Transform
	$g[n]$	$G(e^{j\omega})$
	$h[n]$	$H(e^{j\omega})$
Linearity	$\alpha g[n] + \beta h[n]$	$\alpha G(e^{j\omega}) + \beta H(e^{j\omega})$
Time-reversal	$g[-n]$	$G(e^{-j\omega})$
Time-shifting	$g[n - n_o]$	$e^{-j\omega n_o} G(e^{j\omega})$
Frequency-shifting	$e^{j\omega_o n} g[n]$	$G\left(e^{j(\omega-\omega_o)}\right)$
Differentiation-in-frequency	$n g[n]$	$j\dfrac{dG(e^{j\omega})}{d\omega}$
Convolution	$g[n] \circledast h[n]$	$G(e^{j\omega}) H(e^{j\omega})$
Modulation	$g[n] h[n]$	$\frac{1}{2\pi} \int_{-\pi}^{\pi} G(e^{j\theta}) H(e^{j(\omega-\theta)}) \, d\theta$
Parseval's Relation	$\displaystyle\sum_{n=-\infty}^{\infty} g[n] h^*[n] =$	$\dfrac{1}{2\pi} \displaystyle\int_{-\pi}^{\pi} G(e^{j\omega}) H^*(e^{j\omega}) \, d\omega$

$$Y(e^{j\omega}) = X(e^{j(\omega-\pi)}) = \frac{1}{1 - \alpha e^{-j(\omega-\pi)}} = \frac{1}{1 + \alpha e^{-j\omega}}.$$

A plot of the magnitude and phase of the above Fourier transform is shown in Figure 3.8 for $\alpha = 0.5$. Note that the spectrum of the sequence $(-1)^n \alpha^n \mu[n]$ is identical to the spectrum of $\alpha^n \mu[n]$ shown in Figure 3.4(c) and (d), except shifted by π radians.

An application of the frequency-shifting theorem is in the design of filter banks considered in Section 14.1.2.

Differentiation-in-Frequency Theorem

The Fourier transform of a sequence $x[n] = n\,g[n]$ is given by $X(e^{j\omega}) = j\frac{dG(e^{j\omega})}{d\omega}$; that is,

$$n\,g[n] \stackrel{\mathcal{F}}{\longleftrightarrow} j\frac{dG(e^{j\omega})}{d\omega}. \qquad (3.64)$$

EXAMPLE 3.13   **Fourier Transform Computation Using the Differentiation-in-Frequency Theorem**

Determine the Fourier transform of the sequence

$$y[n] = (n+1)\alpha^n \mu[n], \qquad |\alpha| < 1.$$

Let $x[n] = \alpha^n \mu[n]$, $|\alpha| < 1$. We can therefore write

$$y[n] = n\,x[n] + x[n].$$

From Table 3.3, the Fourier transform of $x[n]$ is given by

$$X(e^{j\omega}) = \frac{1}{1 - \alpha e^{-j\omega}}.$$

Using the differentiation property of the Fourier transform given in Table 3.4, we observe that the Fourier transform of $n\,x[n]$ is given by

$$j\frac{dX(e^{j\omega})}{d\omega} = j\frac{d}{d\omega}\left(\frac{1}{1 - \alpha e^{-j\omega}}\right) = \frac{\alpha e^{-j\omega}}{(1 - \alpha e^{-j\omega})^2}.$$

Next, using the linearity property of the Fourier transform given in Table 3.4, we arrive at the Fourier transform of $y[n]$ as

$$Y(e^{j\omega}) = \frac{\alpha e^{-j\omega}}{(1 - \alpha e^{-j\omega})^2} + \frac{1}{1 - \alpha e^{-j\omega}} = \frac{1}{(1 - \alpha e^{-j\omega})^2}.$$

Convolution Theorem

The Fourier transform $Y(e^{j\omega})$ of the convolution sum of two sequences, $y[n] = g[n] \circledast h[n]$, is given by the product of their Fourier transforms $G(e^{j\omega})H(e^{j\omega})$; that is,

$$g[n] \circledast h[n] \stackrel{\mathcal{F}}{\longleftrightarrow} G(e^{j\omega})H(e^{j\omega}). \qquad (3.65)$$

Proof. From Eq. (2.73a), we have

$$y[n] = \sum_{k=-\infty}^{\infty} g[k]h[n-k].$$

Applying Eq. (3.12) to the above equation, we get

$$Y(e^{j\omega}) = \sum_{n=-\infty}^{\infty}\left(\sum_{k=-\infty}^{\infty} g[k]h[n-k]\right) e^{-j\omega n}.$$

Substituting $m = n - k$ in the above and rearranging, we arrive at

$$Y(e^{j\omega}) = \sum_{m=-\infty}^{\infty}\sum_{k=-\infty}^{\infty} g[k]h[m]e^{-j\omega(m+k)}$$

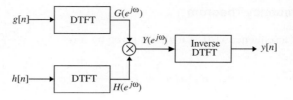

Figure 3.9: Linear convolution using the discrete-time Fourier transform.

$$= \sum_{k=-\infty}^{\infty} g[k] \left(\sum_{m=-\infty}^{\infty} h[m]e^{-j\omega m} \right) e^{-j\omega k}$$

$$= \sum_{k=-\infty}^{\infty} g[k]H(e^{j\omega})e^{-j\omega k} = G(e^{j\omega})H(e^{j\omega}).$$

An implication of the convolution theorem in Eq. (3.65) is that the linear convolution $y[n]$ of two sequences, $g[n]$ and $h[n]$, can be implemented by computing first their Fourier transforms, $G(e^{j\omega})$ and $H(e^{j\omega})$, forming the product $Y(e^{j\omega}) = G(e^{j\omega})H(e^{j\omega})$, and then computing the inverse Fourier transform of the product. This process is shown in Figure 3.9 in a block-diagram form. In some applications, particularly in the case of infinite-length sequences, this Fourier transform-based approach may be more convenient to carry out than the direct convolution.

The convolution theorem plays an important role in developing the input–output characterization of an LTI discrete-time system in the frequency domain and is discussed in Section 3.8.2.

Modulation Theorem

The Fourier transform $Y(e^{j\omega})$ of the product of two sequences given by $y[n] = g[n]h[n]$ is given by the convolution integral of their Fourier transforms $\frac{1}{2\pi} \int_{-\pi}^{\pi} G(e^{j\theta})H(e^{j(\omega-\theta)}) \, d\theta$; that is,

$$g[n]h[n] \overset{\mathcal{F}}{\longleftrightarrow} \frac{1}{2\pi} \int_{-\pi}^{\pi} G(e^{j\theta})H(e^{j(\omega-\theta)}) \, d\theta. \qquad (3.66)$$

Proof. Applying Eq. (3.12), we observe that the Fourier transform of $y[n] = g[n]h[n]$ is given by

$$Y(e^{j\omega}) = \sum_{n=-\infty}^{\infty} g[n]h[n]e^{-j\omega n}.$$

Using the inverse Fourier transform of Eq. (3.16) to express $g[n]$, we can rewrite the above equation as

$$Y(e^{j\omega}) = \frac{1}{2\pi} \sum_{n=-\infty}^{\infty} \int_{-\pi}^{\pi} h[n]e^{-j\omega n} G(e^{j\theta})e^{j\theta n} d\theta$$

$$= \frac{1}{2\pi} \int_{-\pi}^{\pi} G(e^{j\theta}) \left(\sum_{n=-\infty}^{\infty} h[n]e^{-j(\omega-\theta)n} \right) d\theta$$

$$= \frac{1}{2\pi} \int_{-\pi}^{\pi} G(e^{j\theta})H(e^{j(\omega-\theta)}) d\theta.$$

The modulation theorem plays a key role in the amplitude modulation scheme used in digital communications. The modulation theorem is also known as the *windowing theorem*. One application of this theorem, to be considered later in Section 10.2, is in the design of linear-phase FIR filter based on the windowing of the impulse response of an ideal linear-phase filter with a doubly-infinite impulse response.

Parseval's Relation

This theorem expresses the sum of sample-by-sample product of two complex sequences in terms of an integral of the product of their Fourier transforms. Specifically, the most general form of this theorem is given by

$$\sum_{n=-\infty}^{\infty} g[n]h^*[n] = \frac{1}{2\pi} \int_{-\pi}^{\pi} G(e^{j\omega})H^*(e^{j\omega})d\omega. \tag{3.67}$$

Proof. To prove this theorem, we express first one of the sequences in terms of its inverse Fourier transform form, interchange the summation and the integral, and after a rearrangement, express the remaining sum as a Fourier transform of the other sequence as indicated below:

$$\sum_{n=-\infty}^{\infty} g[n]h^*[n] = \sum_{n=-\infty}^{\infty} g[n] \left(\frac{1}{2\pi} \int_{-\pi}^{\pi} H^*(e^{j\omega})e^{-j\omega n}d\omega \right)$$

$$= \frac{1}{2\pi} \int_{-\pi}^{\pi} H^*(e^{j\omega}) \left(\sum_{n=-\infty}^{\infty} g[n]e^{-j\omega n} \right) d\omega$$

$$= \frac{1}{2\pi} \int_{-\pi}^{\pi} H^*(e^{j\omega})G(e^{j\omega})d\omega.$$

One important application of Parseval's relation discussed in Section 3.4 is in the computation of the energy of a finite-energy sequence.

3.4 Energy Density Spectrum of a Discrete-Time Sequence

Recall from Eq. (2.28) that the total energy of a finite-energy sequence $g[n]$ is given by

$$\mathcal{E}_g = \sum_{n=-\infty}^{\infty} |g[n]|^2.$$

If $h[n] = g[n]$, then from Parseval's theorem, we observe

$$\mathcal{E}_g = \sum_{n=-\infty}^{\infty} |g[n]|^2 = \frac{1}{2\pi} \int_{-\pi}^{\pi} |G(e^{j\omega})|^2 \, d\omega. \tag{3.68}$$

Thus, the energy of the sequence $g[n]$ can be computed by evaluating the integral on the right. The quantity

$$S_{gg}(e^{j\omega}) = |G(e^{j\omega})|^2 \tag{3.69}$$

is called the *energy density spectrum* of the sequence $g[n]$. The area under this curve in the range $-\pi \leq \omega < \pi$ divided by 2π is the energy of the sequence.

We illustrate the application of Eq. (3.68) in Example 3.14.

EXAMPLE 3.14   Energy of the Lowpass Discrete-Time Signal

We compute the energy of the sequence $h_{LP}[n]$ of Eq. (3.48). From Eq. (3.68) and Example 3.8, we get

$$\sum_{n=-\infty}^{\infty} |h_{LP}[n]|^2 = \frac{1}{2\pi} \int_{-\pi}^{\pi} |H_{LP}(e^{j\omega})|^2 \, d\omega = \frac{1}{2\pi} \int_{-\omega_c}^{\omega_c} d\omega = \frac{\omega_c}{\pi} < \infty.$$

Hence, $h_{LP}[n]$ is a finite-energy sequence.

EXAMPLE 3.15   Energy of the Exponential Discrete-Time Signal

We compute the energy of the exponential sequence of Eq. (3.14) whose Fourier transform is given by Eq. (3.15). Applying Eq. (3.68), we then get

$$\mathcal{E}_x = \frac{1}{2\pi} \int_{-\pi}^{\pi} \left| \frac{1}{1 - \alpha e^{j\omega}} \right|^2 d\omega = \frac{1}{2\pi} \int_{-\pi}^{\pi} \frac{1}{1 + \alpha^2 - 2\alpha \cos \omega} \, d\omega.$$

For $\alpha = 0.5$, we have

$$\mathcal{E}_x = 0.8488 \tan^{-1} \left(3 \tan(\pi/2) \right) = 1.3333.$$

For a discrete-time sequence $x[n]$ with exponentially decaying sample values for increasing $|n|$, the total energy can be computed using the approximation

$$\mathcal{E}_x \cong \mathcal{E}_{x,M} = \sum_{n=-M}^{M} |x[n]|^2,$$

with M chosen such that $|x[M + 1]|^2$ is less than some predetermined very small value such as 10^{-6}. It can be shown that for the above example, $\sum_0^{10} |x[n]|^2 = 1.3333$.

Recall from Eq. (2.126) that the autocorrelation sequence $r_{gg}[\ell]$ of $g[n]$ can be expressed as

$$r_{gg}[\ell] = \sum_{n=-\infty}^{\infty} g[n]g[-(\ell - n)] = g[\ell] \circledast g[-\ell]. \tag{3.70}$$

Now from Table 3.4, the Fourier transform of $g[-\ell]$ is $G(e^{-j\omega})$. Therefore, using the convolution property of the Fourier transform given in Table 3.4, we observe that the Fourier transform of $g[\ell] \circledast g[-\ell]$ is given by $G(e^{j\omega})G(e^{-j\omega}) = |G(e^{j\omega})|^2$, where we have used the fact that for a real sequence $g[n]$, $G(e^{-j\omega}) = G^*(e^{j\omega})$. As a result, the energy density spectrum $S_{gg}(e^{j\omega})$ of a real sequence $g[n]$ can be computed by taking the Fourier transform of its autocorrelation sequence $r_{gg}[\ell]$; that is,

$$S_{gg}(e^{j\omega}) = \sum_{\ell=-\infty}^{\infty} r_{gg}[\ell]e^{-j\omega\ell}, \tag{3.71}$$

or, equivalently,

$$r_{gg}[n] \overset{\mathcal{F}}{\longleftrightarrow} S_{gg}(e^{j\omega}). \tag{3.72}$$

Equation (3.72) is often referred to as the *Wiener-Khintchine theorem*.

Analogously, the Fourier transform $S_{gh}(e^{j\omega})$ of the cross-correlation sequence $r_{gh}[\ell]$ of two sequences $g[n]$ and $h[n]$ is called the *cross-energy density spectrum*:

$$S_{gh}(e^{j\omega}) = \sum_{\ell=-\infty}^{\infty} r_{gh}[\ell]e^{-j\omega\ell}. \tag{3.73}$$

3.5 Band-Limited Discrete-Time Signals

Since the spectrum of a discrete-time signal is a periodic function of ω with a period 2π, a full-band discrete-time signal has a spectrum occupying the whole frequency range $-\pi \leq \omega < \pi$. A *band-limited* discrete-time signal has a spectrum that is limited to a portion of the above frequency range. An ideal band-limited signal has a spectrum that is zero outside a finite frequency range $0 \leq \omega_a \leq |\omega| \leq \omega_b < \pi$; that is,

$$X(e^{j\omega}) = \begin{cases} 0, & 0 \leq |\omega| < \omega_a, \\ 0, & \omega_b < |\omega| < \pi. \end{cases}$$

However, as in the case of the continuous-time signal, an ideal band-limited signal cannot be generated in practice, and for practical purposes, it is sufficient to ensure that for a band-limited signal, outside the specified frequency range, the signal energy is very small.

A classification of the band-limited discrete-time signal is based on the frequency range where most of the signal's energy is concentrated. A *lowpass* discrete-time real signal has a spectrum occupying the frequency range $-\pi < -\omega_p \leq \omega \leq \omega_p < \pi$. However, because of symmetry, half of the energy of the signal is in the positive frequency range $0 \leq \omega_p < \pi$, where ω_p is called the *bandwidth* of the signal. Likewise, a *highpass* discrete-time real signal has a spectrum occupying the frequency ranges $\omega_p \leq |\omega| < \pi$. Again the energy of the signal is split evenly between the positive and the negative frequency ranges, where $\pi - \omega_p$ is the *bandwidth* of the signal. Similarly, a *bandpass* discrete-time real signal has a spectrum occupying the frequency ranges $0 < \omega_L \leq |\omega| \leq \omega_H < \pi$, with $\omega_H - \omega_L$ representing its *bandwidth*. A bandpass signal with a bandwidth that is much smaller than the average value $(\omega_H + \omega_L)/2$ of the bandedges is referred to as a *narrow-band* signal [Pro96].

As in the case of continuous-time signals, here also a precise definition of the bandwidth depends on applications. An example of an ideal band-limited lowpass discrete-time signal is the finite-energy sequence $h_{LP}[n]$ of Eq. (3.49), whose spectrum $H_{LP}(e^{j\omega})$ given by Eq. (3.45) and shown in Figure 3.5 is zero in the range $\omega_c \leq |\omega| < \pi$. The 100% bandwidth of $h_{LP}[n]$ is ω_c. Since $h_{LP}[n]$ is doubly-infinite in length, it can never be realized in practice. As demonstrated in Figure 3.6, a truncation of $h_{LP}[n]$ to a finite-length sequence results in a practical lowpass signal with most of its energy contained in the frequency range $0 \leq |\omega| < \omega_c$.

Another example of a lowpass discrete-time signal is the sequence given in Eq. (3.14), with a spectrum shown in Figure 3.4(c). It can be shown that 80% of the energy of this signal is contained in the frequency range $0 \leq |\omega| \leq 0.5081\pi$, and hence, we can define the 80% bandwith of the signal to be 0.5081π radians (Problem 3.43). An example of a highpass signal is the sequence $(-1)^n(0.5)^n\mu[n]$, whose spectrum is shown in Figure 3.8.

3.6 DTFT Computation Using MATLAB

The *Signal Processing Toolbox* in MATLAB includes a number of M-files to aid in the DTFT-based analysis of discrete-time signals. Specifically, the functions that can be used are `freqz`, `abs`, `angle`, and `unwrap`. In addition, the built-in MATLAB functions `real` and `imag` are also useful in some applications.

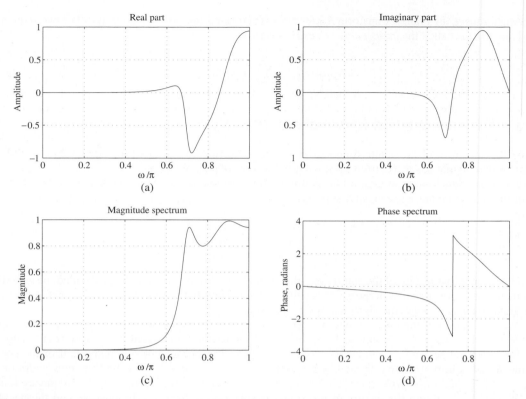

Figure 3.10: Plots of the (a) real and (b) imaginary parts and the (c) magnitude and (d) phase spectrums of the Fourier transform of Eq. (3.74).

The function `freqz` can be used to compute the values of the Fourier transform of a sequence, described as a rational function in $e^{j\omega}$ in the form of Eq. (3.55) at a prescribed set of discrete frequency points $\omega = \omega_\ell$. For a reasonably accurate plot, a fairly large number of frequency points should be selected. There are various forms of this function. We illustrate its use in Fourier transform computation in Example 3.16.

EXAMPLE 3.16   Illustration of Fourier Transform Computation Using MATLAB

Program 3_1.m

Program 3_1 can be employed to determine the values of the Fourier transform of a real sequence described as a rational function in $e^{-j\omega}$. The program computes the Fourier transform values at the prescribed frequency points and plots the real and imaginary parts and the magnitude and phase spectrums. It should be noted that because of the symmetry relations of the Fourier transform of a real sequence as indicated in Table 3.3, the Fourier transform is evaluated only at specified equally spaced values of ω between 0 and π.

We consider the evaluation of the following Fourier transform:

$$X(e^{j\omega}) = \frac{0.008 - 0.033e^{-j\omega} + 0.05e^{-j2\omega} - 0.033e^{-j3\omega} + 0.008e^{-j4\omega}}{1 + 2.37e^{-j\omega} + 2.7e^{-j2\omega} + 1.6e^{-j3\omega} + 0.41e^{-j4\omega}}. \tag{3.74}$$

The plots generated using Program 3_1 are shown in Figure 3.10.

3.7   The Unwrapped Phase Function

In numerical computation, when the computed phase function is outside the range $[-\pi, \pi]$, the phase is computed modulo 2π, to bring the computed value to this range. As a result, the phase functions of some sequences exhibit discontinuities of 2π radians in the plot. The occurrence of such discontinuities can be seen in Figure 3.10(d), in which the phase spectrum displays a discontinuity of 2π at around $\omega = 0.72$. In such cases, it is often useful to consider an alternate type of phase function that is a continuous function of ω derived from the original phase function by removing the discontinuities of 2π. The process of removing the discontinuities is called "*unwrapping the phase*," and the new phase function will be denoted as $\theta_c(\omega)$, with the subscript "*c*" indicating that it is a continuous function of ω.[3] We now outline the conditions under which the phase function will be a continuous function of ω [Tri77].

From Eq. (3.21), it follows that the natural logarithm of the Fourier transform $X(e^{j\omega})$ of the sequence $x[n]$ can be expressed as

$$\ln X(e^{j\omega}) = |X(e^{j\omega})| + j\theta(\omega), \tag{3.75}$$

where $\theta(\omega) = \arg\{X(e^{j\omega})\}$. If $\ln X(e^{j\omega})$ exists, then its derivative with respect to ω also exists and is given by

$$\frac{d \ln X(e^{j\omega})}{d\omega} = \frac{1}{X(e^{j\omega})}\left[\frac{dX(e^{j\omega})}{d\omega}\right] = \frac{1}{X(e^{j\omega})}\left[\frac{dX_{\mathrm{re}}(e^{j\omega})}{d\omega} + j\frac{dX_{\mathrm{im}}(e^{j\omega})}{d\omega}\right]. \tag{3.76}$$

From Eq. (3.75), the derivative of $\ln X(e^{j\omega})$ with respect to ω is also given by

$$\frac{d \ln X(e^{j\omega})}{d\omega} = \frac{d|X(e^{j\omega})|}{d\omega} + j\frac{d\theta(\omega)}{d\omega}. \tag{3.77}$$

Therefore, the derivative of $\theta(\omega)$ with respect to ω is given by the imaginary part of the right-hand side of Eq. (3.76); that is,

$$\frac{d\theta(\omega)}{d\omega} = \frac{1}{|X(e^{j\omega})|^2}\left[X_{\mathrm{re}}(e^{j\omega})\frac{dX_{\mathrm{im}}(e^{j\omega})}{d\omega} - X_{\mathrm{im}}(e^{j\omega})\frac{dX_{\mathrm{re}}(e^{j\omega})}{d\omega}\right]. \tag{3.78}$$

The phase function $\theta(\omega)$ can thus be defined unequivocally by its derivative $d\theta(\omega)/d\omega$:

$$\theta(\omega) = \int_0^\omega \left[\frac{d\theta(\eta)}{d\eta}\right] d\eta, \tag{3.79}$$

with the constraint

$$\theta(0) = 0. \tag{3.80}$$

The phase function as defined by Eq. (3.79) is called the *unwrapped phase function* of $X(e^{j\omega})$. It follows from Eqs. (3.79) and (3.80) that the unwrapped phase is a continuous function of ω. As a result, $\ln X(e^{j\omega})$ given in Eq. (3.75) exists. Moreover, the phase function will be an odd function of ω if

$$\frac{1}{\pi}\int_0^{2\pi}\left[\frac{d\theta(\omega)}{d\eta}\right] d\eta = 0.$$

If the above constraint is not satisfied, then the computed phase function will exhibit absolute jumps greater than π [Tri77]. For unwrapping the phase, these jumps should be replaced with their 2π complements. In MATLAB, this can be done using the M-file unwrap. The unwrapped phase spectrum of the Fourier transform of Eq. (3.74) obtained using this function is shown in Figure 3.11.

---

[3]In some cases, discontinuities of π may still be present in the phase response after phase unwrapping (see Table 3.3 for an example).

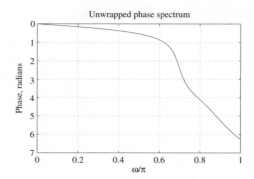

Figure 3.11: Unwrapped phase spectrum of the Fourier transform of Example 3.16.

Figure 3.12: An LTI discrete-time system.

3.8   The Frequency Response of an LTI Discrete-Time System

Most discrete-time signals encountered in practice can be represented as a linear combination of a very large, maybe infinite, number of sinusoidal discrete-time signals of different angular frequencies. Thus, knowing the response of the LTI system to a single sinusoidal signal, we can determine its response to more complicated signals by making use of the superposition property of the system. Since a sinusoidal signal can be expressed in terms of an exponential signal, the response of the LTI system to an exponential input is of practical interest. This leads to the concept of frequency response, a transform-domain representation of the LTI discrete-time system. We first define the frequency response, investigate its properties, and describe some of its applications. The computation of the time-domain representation of the LTI system from its frequency response is outlined.

3.8.1   Definition

An important property of an LTI system is that for certain types of input signals, called *eigenfunctions,* the output signal is the input signal multiplied by a complex constant. We consider here one such eigenfunction as the input. Recall from Section 2.5.1, the input–output relationship of an LTI discrete-time system as shown in Figure 3.12, with an impulse response $h[n]$, is given by the convolution sum of Eq. (2.73b) and is of the form

$$y[n] = \sum_{k=-\infty}^{\infty} h[k]x[n-k], \tag{3.81}$$

where $y[n]$ and $x[n]$ are, respectively, the output and the input sequences. Now, if the input $x[n]$ is a complex exponential sequence of the form

$$x[n] = e^{j\omega n}, \qquad -\infty < n < \infty, \tag{3.82}$$

then, from Eq. (3.81), the output is given by

$$y[n] = \sum_{k=-\infty}^{\infty} h[k]e^{j\omega(n-k)} = \left(\sum_{k=-\infty}^{\infty} h[k]e^{-j\omega k} \right) e^{j\omega n}, \qquad (3.83)$$

which can be rewritten as

$$y[n] = H(e^{j\omega})e^{j\omega n}, \qquad (3.84)$$

where we have used the notation

$$H(e^{j\omega}) = \sum_{n=-\infty}^{\infty} h[n]e^{-j\omega n}. \qquad (3.85)$$

It can be seen from Eq. (3.85) that for a complex exponential input signal $e^{j\omega n}$, the output of an LTI discrete-time system is also a complex exponential signal of the same frequency multiplied by a complex constant $H(e^{j\omega})$. Thus, $e^{j\omega n}$ is an eigenfunction of the system. Another example of an eigenfunction is given in Problem 3.45.

The quantity $H(e^{j\omega})$ defined above is called the *frequency response* of the LTI discrete-time system, and it provides a frequency-domain description of the system. Note from Eq. (3.85) that $H(e^{j\omega})$ is precisely the Fourier transform of the impulse response $h[n]$ of the system.

Equation (3.84) implies that for a complex sinusoidal input sequence $x[n]$ of angular frequency ω as in Eq. (3.82), the output $y[n]$ is also a complex sinusoidal sequence of the same angular frequency but weighted by a complex amplitude $H(e^{j\omega})$ that is a function of the input frequency ω and the system's impulse response coefficients $h[n]$. We shall show in the following section that $H(e^{j\omega})$ completely characterizes the LTI discrete-time system in the frequency domain.

Just like any other discrete-time Fourier transform, in general, $H(e^{j\omega})$ is also a complex function of ω with a period 2π and can be expressed in terms of its real and imaginary parts or its magnitude and phase. Thus,

$$H(e^{j\omega}) = H_{\text{re}}(e^{j\omega}) + j H_{\text{im}}(e^{j\omega})$$

$$= |H(e^{j\omega})|e^{j\theta(\omega)},$$

where $H_{\text{re}}(e^{j\omega})$ and $H_{\text{im}}(e^{j\omega})$ are, respectively, the real and imaginary parts of $H(e^{j\omega})$, and

$$\theta(\omega) = \arg\{H(e^{j\omega})\}. \qquad (3.86)$$

The quantity $|H(e^{j\omega})|$ is called the *magnitude response,* and the quantity $\theta(\omega)$ is called the *phase response* of the LTI discrete-time system. Design specifications for the discrete-time systems, in many applications, are given in terms of the magnitude response or the phase response or both. In some cases, the magnitude function is specified in decibels as defined below:

$$\mathcal{G}(\omega) = 20 \log_{10} |H(e^{j\omega})| \quad \text{dB}, \qquad (3.87)$$

where $\mathcal{G}(\omega)$ is called the *gain function* The negative of the gain function, $\mathcal{A}(\omega) = -\mathcal{G}(\omega)$, is called the *attenuation* or *loss function.*

It should be noted that the magnitude and phase functions are real functions of ω, whereas the frequency response is a complex function of ω. For a discrete-time system characterized by a real impulse response $h[n]$, it follows from Table 3.2 that the magnitude function is an even function of ω, that is, $|H(e^{j\omega})| = |H(e^{-j\omega})|$, and the phase function is an odd function of ω, that is, $\theta(\omega) = -\theta(-\omega)$. Likewise, $H_{\text{re}}(e^{j\omega})$ is even, and $H_{\text{im}}(e^{j\omega})$ is odd.

3.8.2 Frequency-Domain Characterization of the LTI Discrete-Time System

We now derive the frequency-domain representation of an LTI discrete-time system. If $Y(e^{j\omega})$ and $X(e^{j\omega})$ denote the Fourier transforms of the output and input sequences, $y[n]$ and $x[n]$, respectively, then applying the convolution theorem of Table 3.4 to Eq. (3.81), we arrive at

$$Y(e^{j\omega}) = H(e^{j\omega})X(e^{j\omega}), \tag{3.88}$$

where $H(e^{j\omega})$ is the frequency response of the LTI system as defined in Eq. (3.85). Equation (3.88) thus relates the input and the output of an LTI system in the frequency domain.

From Eq. (3.88), we obtain

$$H(e^{j\omega}) = \frac{Y(e^{j\omega})}{X(e^{j\omega})}. \tag{3.89}$$

Thus, the frequency response of an LTI discrete-time system is given by the ratio of the Fourier transform $Y(e^{j\omega})$ of the output sequence $y[n]$ to the Fourier transform $X(e^{j\omega})$ of the input sequence $x[n]$.

It follows from the input–output relation of Eq. (3.88) of an LTI discrete-time system in the frequency domain that the output cannot contain sinusoidal components of frequencies that are not present in the input and the system. As a result, if the output of a system has new frequency components, then the system is either nonlinear or time-varying or both (Problem 3.50).

EXAMPLE 3.17   Convolution Sum Computation Using Fourier Transform

We consider here the implementation of the convolution of the two sequences of Example 2.27 via DTFT-based approach. We observe from Table 3.3 that the Fourier transform of the input sequence $x[n] = \alpha^n \mu[n]$, with $|\alpha| < 1$, and the frequency response of the causal LTI system $h[n] = \beta^n \mu[n]$, with $|\beta| < 1$, are, respectively, given by $X(e^{j\omega}) = 1/(1 - \alpha e^{-j\omega})$ and $H(e^{j\omega}) = 1/(1 - \beta e^{-j\omega})$. Hence, using the convolution theorem, we note that the Fourier transform $Y(e^{j\omega})$ is given by

$$Y(e^{j\omega}) = H(e^{j\omega})X(e^{j\omega}) = \frac{1}{(1 - \alpha e^{-j\omega})(1 - \beta e^{-j\omega})}. \tag{3.90}$$

The above expression for $Y(e^{j\omega})$ can be rewritten as

$$Y(e^{j\omega}) = \frac{A}{1 - \alpha e^{-j\omega}} + \frac{B}{1 - \beta e^{-j\omega}} = \frac{(A+B) - (A\beta + B\alpha)e^{-j\omega}}{(1 - \alpha e^{-j\omega})(1 - \beta e^{-j\omega})}. \tag{3.91}$$

Comparing the numerators in the right-hand side of the above two equations we get

$$A + B = 1, \qquad A\beta + B\alpha = 0,$$

which when solved yields

$$A = \frac{\alpha}{\alpha - \beta}, \qquad B = -\frac{\beta}{\alpha - \beta}.$$

Substituting the above values of A and B in Eq. (3.91), we arrive at

$$Y(e^{j\omega}) = \frac{\frac{\alpha}{\alpha - \beta}}{1 - \alpha e^{-j\omega}} - \frac{\frac{\beta}{\alpha - \beta}}{1 - \beta e^{-j\omega}}.$$

An inverse Fourier transform of the above equation obtained using Table 3.3 leads to

$$y[n] = \frac{\alpha}{\alpha - \beta}\alpha^n \mu[n] - \frac{\beta}{\alpha - \beta}\beta^n \mu[n] = \frac{\alpha^{n+1} - \beta^{n+1}}{\alpha - \beta}\mu[n] = \left(\sum_{k=0}^{n} \alpha^k \beta^{n-k}\right)\mu[n],$$

which is the same as that derived by the direct evaluation of the convolution sum in Example 2.27.

3.8.3 Frequency Responses of LTI Discrete-Time Systems

We derive in this section the expressions for the frequency responses of LTI FIR and IIR discrete-time systems.

Frequency Response of LTI FIR Discrete-Time Systems

The LTI FIR discrete-time systems are characterized by an input–output relation of the form of Eq. (2.118) and repeated below for convenience:

$$y[n] = \sum_{k=N_1}^{N_2} h[k]x[n - k], \qquad N_1 < N_2.$$

Applying the discrete-time Fourier transform (DTFT) to the above equation and making use of the linearity and the time-shifting properties of Table 3.4, we arrive at the input–output relation of the LTI system in the transform-domain given by

$$Y(e^{j\omega}) = \sum_{k=N_1}^{N_2} h[k]e^{-j\omega k} X(e^{j\omega}), \qquad (3.92)$$

where $Y(e^{j\omega})$ and $X(e^{j\omega})$ are the Fourier transforms of the sequences $y[n]$ and $x[n]$, respectively. In developing Eq. (3.92), it has been tacitly assumed that $Y(e^{j\omega})$ and $X(e^{j\omega})$ exist. From the above equation, we arrive at the expression for its frequency response $H(e^{j\omega})$ as given below:

$$H(e^{j\omega}) = \sum_{k=N_1}^{N_2} h[k]e^{-j\omega k}, \qquad (3.93)$$

which is seen to be a polynomial in $e^{j\omega}$.

Frequency Response of LTI IIR Discrete-Time Systems

The LTI IIR discrete-time systems we shall be concerned with in this book are characterized by linear constant coefficient difference equations of the form of Eq. (2.90) and repeated below for convenience:

$$\sum_{k=0}^{N} d_k y[n - k] = \sum_{k=0}^{M} p_k x[n - k].$$

Applying the discrete-time Fourier transform (DTFT) to the above equation and making use of the linearity and the time-shifting properties of Table 3.4, we arrive at the input–output relation of the LTI system in the transform-domain given by

$$\sum_{k=0}^{N} d_k e^{-j\omega k} Y(e^{j\omega}) = \sum_{k=0}^{M} p_k e^{-j\omega k} X(e^{j\omega}). \qquad (3.94)$$

The above equation can be alternately written as

$$\left(\sum_{k=0}^{N} d_k e^{-j\omega k}\right) Y(e^{j\omega}) = \left(\sum_{k=0}^{M} p_k e^{-j\omega k}\right) X(e^{j\omega}).$$

(3.95)

Thus, from Eq. (3.95), the expression for its frequency response $H(e^{j\omega})$ is given by

$$H(e^{j\omega}) = \frac{Y(e^{j\omega})}{X(e^{j\omega})} = \frac{\sum_{k=0}^{M} p_k e^{-j\omega k}}{\sum_{k=0}^{N} d_k e^{-j\omega k}},$$

(3.96)

which is a rational function in $e^{j\omega}$.

3.8.4   Frequency Response Computation Using MATLAB

The M-file function `freqz(h,w)` in MATLAB can be used to determine the values of the frequency response of a prescribed impulse response vector h at a set of given frequency points w. From these frequency response values, one can then compute the real and imaginary parts using the functions `real` and `imag`, and the magnitude and phase using the functions `abs` and `angle`, as illustrated in Example 3.18.

EXAMPLE 3.18   Frequency Response of the Moving-Average Filter

Consider the moving-average filter of Eq. (2.61). Comparing Eq. (2.61) with Eq. (2.118), we note that the impulse response of the moving-average filter is given by

$$h[n] = \begin{cases} \frac{1}{M}, & 0 \le n \le M-1, \\ 0, & \text{otherwise.} \end{cases}$$

(3.97)

From Eq. (3.85), its frequency response is thus given by

$$H(e^{j\omega}) = \frac{1}{M} \sum_{n=0}^{M-1} e^{-j\omega n} = \frac{1}{M} \left(\sum_{n=0}^{\infty} e^{-j\omega n} - \sum_{n=M}^{\infty} e^{-j\omega n} \right)$$

$$= \frac{1}{M} \left(\sum_{n=0}^{\infty} e^{-j\omega n} \right) \left(1 - e^{-jM\omega} \right) = \frac{1}{M} \frac{1 - e^{-jM\omega}}{1 - e^{-j\omega}}$$

$$= \frac{1}{M} \frac{\sin(M\omega/2)}{\sin(\omega/2)} e^{-j(M-1)\omega/2}.$$

(3.98)

From the above, the magnitude and phase responses of the moving-average filter of Eq. (3.97) are obtained as

$$|H(e^{j\omega})| = \left| \frac{1}{M} \frac{\sin(M\omega/2)}{\sin(\omega/2)} \right|,$$

(3.99)

and

$$\theta(\omega) = -\frac{(M-1)\omega}{2} + \pi \sum_{k=1}^{\lfloor M/2 \rfloor} \mu\left(\omega - \frac{2\pi k}{M}\right),$$

(3.100)

where $\mu(\omega)$ is a step function in ω defined by

$$\mu(\omega) = \begin{cases} 1, & \omega \ge 0, \\ 0, & \omega < 0. \end{cases}$$

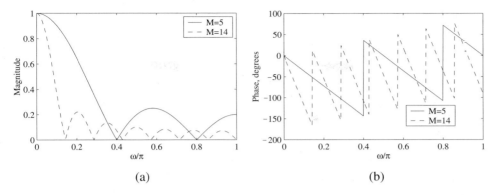

Figure 3.13: (a) Magnitude and (b) phase responses of the moving-average filters of length 5 and 14.

Figure 3.13 shows the magnitude and phase responses of the moving-average filters for $M = 5$ and $M = 14$. These plots have been obtained using Program 3.2. It can be seen from Figure 3.13(a) that in the range from $\omega = 0$ to $\omega = \pi$, the magnitude has a maximum value of unity at $\omega = 0$, and has zeros at $\omega = 2\pi k/M$ with $k = 1, 2, \ldots, \lfloor M/2 \rfloor$.[4] The phase function exhibits discontinuities of π at each zero of $H(e^{j\omega})$ and is linear elsewhere with a slope of $-(M-1)/2$. It should be noted that both the magnitude and phase functions are periodic in ω with a period 2π.

Program 3_2.m

The phase responses of discrete-time systems when determined by a computer may also exhibit jumps by an amount of 2π, caused by the way the arctangent function is computed, for example, in the function `angle` in MATLAB. The phase response can be made a continuous function of ω by unwrapping the phase response across the jumps by adding multiples of $\pm 2\pi$. The MATLAB function `unwrap` can be employed to this end, provided the computed phase is in radians.[5] The application of `unwrap` is illustrated later in Section 7.1.3 in Figure 7.5. These jumps should not be confused with the jumps by π caused by the zeros of the frequency response as shown, for example, in Figure 3.13(b).

3.8.5 Steady-State and Transient Responses

We indicated in Section 2.7.1 that the output $y[n]$ of an LTI discrete-time system characterized by a constant coefficient difference equation is a sum of two parts: the complementary solution $y_c[n]$ resulting with the input $x[n] = 0$ and a particular solution $y_p[n]$ resulting from the specified input $x[n]$. For a causal LTI system, the complementary solution $y_c[n]$ is of the form of Eq. (2.101), repeated below for convenience:

$$y_c[n] = \alpha_1 \lambda_1^n + \alpha_2 \lambda_2^n + \cdots + \alpha_N \lambda_N^n,$$

where $\alpha_1, \alpha_2, \ldots, \alpha_N$ are constants determined from the specified initial conditions of the discrete-time system and λ_i are the roots of the characteristic polynomial. Moreover, as pointed out in Section 2.7.6, for a stable system, $|\lambda_i| < 1$, and as a result, the complementary solution $y_c[n]$ decays to zero for large values of n. The complementary solution part $y_c[n]$ of the output is also referred to as the *transient response*. We now examine the behavior of the particular solution $y_p[n]$ for two specific types of input.

Consider first an input sequence with a constant amplitude starting at some time instant, say n_o, and then continuing forever afterwards. The output of a causal stable LTI discrete-time system will then be

---

[4] $\lfloor x \rfloor$ denotes the integer part of x.

[5] Care should be taken in using `unwrap` as it can give wrong answers sometimes if the computed phase response is sparse with rapidly changing values.

composed of a *steady-state response* (the particular solution), which is also a constant amplitude sequence, and a transient response with zero-valued samples after some time instant $n_1 > n_o$, resulting in a constant amplitude output after the time instant n_1. Example 3.19 illustrates this behavior of the LTI system.

EXAMPLE 3.19   Output of an FIR Filter for a Constant Input

A causal FIR discrete-time system is characterized by an impulse response $\{h[n]\} = \{4 \quad -5 \quad 6 \quad -3\}, 0 \le n \le 3$. Its output samples for an input $x[n]$ are then computed using

$$y[n] = 4x[n] - 5x[n-1] + 6x[n-2] - 3x[n-3].$$

For a unit step sequence input, the first 6 output samples are thus given by

$$y[0] = 4x[0] = 4,$$

$$y[1] = 4x[1] - 5x[0] = -1,$$

$$y[2] = 4x[2] - 5x[1] + 6x[0] = 5,$$

$$y[3] = 4x[3] - 5x[2] + 6x[1] - 3x[0] = 2,$$

$$y[4] = 4x[4] - 5x[3] + 6x[2] - 3x[1] = 2,$$

$$y[5] = 4x[5] - 5x[4] + 6x[3] - 3x[2] = 2.$$

It follows from the above that $y[n] = 2$ for $n \ge 3$, or the output has reached the steady-state at $n = 3$. The output samples for $n = 0, 1,$ and 2 are then composed of the samples of the transient and the steady-state responses.

Similarly, the output of a causal stable LTI system with a causal input that is a constant amplitude sinusoidal sequence will have after some time instant a steady-state output that is also a constant amplitude sinusoidal sequence of the same angular frequency as that of the input. In this case also, before the steady-state output is reached, the initial samples of the output will be composed of samples of the transient response and the steady-state response.

It is straightforward to develop the expression for the steady-state response of a LTI discrete-time system with a real impulse response $h[n]$ for a sinusoidal input in terms of its frequency response function $H(e^{j\omega})$. Let the input to the LTI system be

$$x[n] = A\cos(\omega_o n + \phi), \qquad -\infty < n < \infty, \qquad (3.101)$$

where A is real.

Now using trigonometric identities, we can express the input $x[n]$ as a sum of two complex exponential sequences:

$$x[n] = g[n] + g^*[n],$$

where $g[n] = \frac{1}{2}Ae^{j\phi}e^{j\omega_o n}$. From Eq. (3.84), the output of the LTI discrete-time system for the input $e^{j\omega_o n}$ is simply $H(e^{j\omega_o})e^{j\omega_o n}$. Because of linearity, the response $v[n]$ to the input $g[n]$ is then given by

$$v[n] = \frac{1}{2}Ae^{j\phi}H(e^{j\omega_o})e^{j\omega_o n}.$$

Likewise, the output $v^*[n]$ to the input $g^*[n]$ is the complex conjugate of $v[n]$; that is,

$$v^*[n] = \frac{1}{2}Ae^{-j\phi}H(e^{-j\omega_o})e^{-j\omega_o n}.$$

Combining these two expressions, we arrive at the expression for the desired output $y[n]$ as

$$y[n] = v[n] + v^*[n]$$
$$= \tfrac{1}{2}Ae^{j\phi}H(e^{j\omega_o})e^{j\omega_o n} + \tfrac{1}{2}Ae^{-j\phi}H(e^{-j\omega_o})e^{-j\omega_o n}$$
$$= \tfrac{1}{2}Ae^{j\phi}|H(e^{j\omega_o})|e^{j\theta(\omega_o)}e^{j\omega_o n} + \tfrac{1}{2}Ae^{-j\phi}|H(e^{-j\omega_o})|e^{j\theta(-\omega_o)}e^{-j\omega_o n}$$
$$= \tfrac{1}{2}Ae^{j\phi}|H(e^{j\omega_o})|e^{j\theta(\omega_o)}e^{j\omega_o n} + \tfrac{1}{2}Ae^{-j\phi}|H(e^{j\omega_o})|e^{-j\theta(\omega_o)}e^{-j\omega_o n},$$

where we have used the fact that the magnitude function of the LTI system with a real impulse response is an even function of ω, that is, $|H(e^{j\omega})| = |H(e^{-j\omega})|$, and the phase function is an odd function of ω, that is, $\theta(\omega) = -\theta(-\omega)$. The output of the LTI system thus can be expressed as

$$y[n] = \tfrac{1}{2}A|H(e^{j\omega_o})| \left\{ e^{j\theta(\omega_o)}e^{j\phi}e^{j\omega_o n} + e^{-j\theta(\omega_o)}e^{-j\phi}e^{-j\omega_o n} \right\}$$
$$= A|H(e^{j\omega_o})| \cos(\omega_o n + \theta(\omega_o) + \phi). \tag{3.102}$$

Thus, the output signal $y[n]$ has the same sinusoidal waveform as the input signal $x[n]$ of Eq. (3.101), with two differences: (1) the amplitude is multiplied by $|H(e^{j\omega_o})|$, the value of the magnitude function of the discrete-time system at $\omega = \omega_o$; and (2) the output signal has a phase *lag* relative to the input by an amount $\theta(\omega_o)$, the value of the phase of the discrete-time system at $\omega = \omega_o$.

For $\omega_o = 0$ and $\phi = 0$, the input $x[n]$ reduces to a constant amplitude sequence: $x[n] = A$. Hence, from Eq. (3.102), the steady-state output is of the form

$$y[n] = A|H(e^{j0})|. \tag{3.103}$$

In the case of Example 3.19, the frequency response of the FIR filter is $H(e^{j\omega}) = 4 - 5e^{j\omega} + 6e^{j2\omega} - 3e^{j3\omega}$. Hence, the steady-state output is $y[n] = |H(e^{j0})| = 4 - 5 + 6 - 3 = 2$, verifying the result obtained in this example.

3.8.6   Response to a Causal Exponential Sequence

In developing Eq. (3.102) for an input of the form of Eq. (3.101), it has been tacitly assumed that the input has been present for all time prior to time instant n, and the system therefore is in steady-state However, in practice, the excitation to an LTI discrete-time system is usually a causal sequence applied at some finite sample index $n = n_o$. Hence, the output for such an input when observed at sample instants beginning at $n = n_o$ will consist of a transient part along with a steady-state component and thus will be different from the one shown in Eq. (3.102). Without any loss of generality, we develop next the expression for the output response $y[n]$ of a causal LTI discrete-time system when the input is a causal exponential sequence applied at $n = 0$; that is,

$$x[n] = e^{j\omega n}\mu[n],$$

where $\mu[n]$ is the unit step sequence defined in Eq. (2.40). Since $x[n] = 0$, for $n < 0$, we have $y[n] = 0$, for $n < 0$. For $n \geq 0$, from Eq. (3.81), the output response is given by

$$y[n] = \sum_{k=0}^{\infty} h[k]e^{j\omega(n-k)}\mu[n-k] = \left(\sum_{k=0}^{n} h[k]e^{-j\omega k} \right) e^{j\omega n},$$

as $\mu[n-k] = 0$ for $k > n$. Rewriting the last expression of the above equation, we get

$$y[n] = \left(\sum_{k=0}^{\infty} h[k]e^{-j\omega k} \right) e^{j\omega n} - \left(\sum_{k=n+1}^{\infty} h[k]e^{-j\omega k} \right) e^{j\omega n}$$
$$= H(e^{j\omega})e^{j\omega n} - \left(\sum_{k=n+1}^{\infty} h[k]e^{-j\omega k} \right) e^{j\omega n}, \qquad n \geq 0. \tag{3.104}$$

The first term in the last expression of Eq. (3.104) is the same as that given by Eq. (3.84) and is the *steady-state response:*

$$y_{sr}[n] = H(e^{j\omega})e^{j\omega n}.$$

The second term in Eq. (3.104) is the *transient response:*

$$y_{tr}[n] = -\left(\sum_{k=n+1}^{\infty} h[k]e^{-j\omega k}\right)e^{j\omega n}.$$

To determine the effect of the second term on the output response, we observe that

$$|y_{tr}[n]| = \left|\sum_{k=n+1}^{\infty} h[k]e^{-j\omega(k-n)}\right| \leq \sum_{k=n+1}^{\infty} |h[k]| \leq \sum_{k=0}^{\infty} |h[k]|. \qquad (3.105)$$

Now, for a causal and stable IIR LTI discrete-time system, the impulse response is absolutely summable, and as a result, the transient response $y_{tr}[n]$ is a bounded sequence. Moreover, as $n \to \infty$, $\sum_{k=n+1}^{\infty} |h[k]| \to 0$, and hence, the transient response decays to zero as n gets very large. In most practical cases, the transient response becomes negligibly small after some finite amount of time, and the system can be assumed to be in a steady-state. On the other hand, for a causal FIR LTI discrete-time system with an impulse response of length $N + 1$, $h[n] = 0$ for $n > N$. Hence, $y_{tr}[n] = 0$ for $n > N - 1$. Thus, here the output $y[n]$ is in the steady-state at $n = N$, with output sample values $y_{sr}[n] = H(e^{j\omega})e^{j\omega n}$ for $n \geq N$. It can be seen that for the FIR filter of Example 3.19, $N = 3$, and hence, the steady-state is reached at $n = 3$ as demonstrated in the example.

It should be noted that transients will occur whenever an input signal is applied or changed.

3.8.7 The Concept of Filtering

One application of an LTI discrete-time system is to pass certain frequency components in an input sequence without any distortion (if possible) and to block other frequency components. Such systems are called *digital filters* and are one of the main subjects of discussion in this text. The key to the filtering process is the inverse discrete-time Fourier transform given in Eq. (3.16) which expresses an arbitrary input sequence as a linear weighted sum of an infinite number of exponential sequences, or equivalently, as a linear weighted sum of sinusoidal sequences. As a result, by appropriately choosing the values of magnitude function of the LTI digital filter at frequencies corresponding to the frequencies of the sinusoidal components of the input, some of these sinusoidal sequences can be selectively heavily attenuated or filtered with respect to the others.

We now explain the concept of filtering and then define the most commonly desired filter characteristics. To understand the mechanism behind the design of such a system, consider a real coefficient LTI discrete-time system characterized by a magnitude function

$$|H(e^{j\omega})| \cong \begin{cases} 1, & 0 \leq |\omega| \leq \omega_c, \\ 0, & \omega_c < |\omega| < \pi. \end{cases} \qquad (3.106)$$

We apply an input $x[n] = A\cos\omega_1 n + B\cos\omega_2 n$ to this system, where $0 < \omega_1 < \omega_c < \omega_2 < \pi$. Because of linearity, it follows from Eq. (3.102) that the output $y[n]$ of this system is of the form

$$y[n] = A|H(e^{j\omega_1})|\cos(\omega_1 n + \theta(\omega_1)) + B|H(e^{j\omega_2})|\cos(\omega_2 n + \theta(\omega_2)). \qquad (3.107)$$

Making use of Eq. (3.106) in Eq. (3.107), we get

$$y[n] \cong A|H(e^{j\omega_1})|\cos(\omega_1 n + \theta(\omega_1)),$$

indicating the LTI discrete-time system acts like a lowpass filter.

EXAMPLE 3.20   Design of a Simple Digital Filter

In this example, we consider the design of a very simple digital filter [Ham89]. The input signal consists of a sum of two cosine sequences of angular frequencies 0.1 rad/samples and 0.4 rad/samples, respectively. We need to design a highpass filter that will pass the high-frequency component of the input but block the low-frequency part.

For simplicity, we assume the filter to be an FIR filter of length 3 with an impulse response

$$h[0] = h[2] = \alpha_0, \quad h[1] = \alpha_1.$$

Hence, from Eq. (3.81), the digital filtering is performed using the difference equation

$$y[n] = h[0]x[n] + h[1]x[n-1] + h[2]x[n-2]$$
$$= \alpha_0 x[n] + \alpha_1 x[n-1] + \alpha_0 x[n-2], \tag{3.108}$$

where $y[n]$ and $x[n]$ represent, respectively, the output and the input sequences. Thus, our design objective is to choose suitable values for the filter parameters, α_0 and α_1, so that the output of the filter is a cosine sequence with a frequency 0.4 rad/samples.

Now, from Eq. (3.93), the frequency response of the above FIR filter is given by

$$H(e^{j\omega}) = h[0] + h[1]e^{-j\omega} + h[2]e^{-j2\omega}$$
$$= \alpha_0(1 + e^{-j2\omega}) + \alpha_1 e^{-j\omega} = 2\alpha_0 \left(\frac{e^{j\omega} + e^{-j\omega}}{2} \right) e^{-j\omega} + \alpha_1 e^{-j\omega}$$
$$= (2\alpha_0 \cos \omega + \alpha_1)e^{-j\omega}. \tag{3.109}$$

The magnitude and phase functions of this filter are

$$|H(e^{j\omega})| = |2\alpha_0 \cos \omega + \alpha_1|, \tag{3.110}$$
$$\theta(\omega) = -\omega + \beta, \tag{3.111}$$

where $\beta = 0$ when $2\alpha_0 \cos \omega + \alpha_1 > 0$, and $\beta = \pi$ when $2\alpha_0 \cos \omega + \alpha_1 < 0$.

In order to stop the low-frequency component from appearing at the output of the filter, the magnitude function at $\omega = 0.1$ should be equal to zero. Similarly, to pass the high-frequency component without any attenuation, we need to ensure that the magnitude function at $\omega = 0.4$ is equal to 1. Thus, the two conditions that must be satisfied are

$$2\alpha_0 \cos(0.1) + \alpha_1 = 0,$$
$$2\alpha_0 \cos(0.4) + \alpha_1 = 1.$$

Solving the above two equations, we arrive at

$$\alpha_0 = -6.76195, \qquad \alpha_1 = 13.456335. \tag{3.112}$$

Substituting Eq. (3.112) in Eq. (3.109), we obtain the input–output relation of the desired FIR filter as

$$y[n] = -6.76195(x[n] + x[n-2]) + 13.456335x[n-1], \tag{3.113}$$

with

$$x[n] = \{\cos(0.1n) + \cos(0.4n)\}\mu[n]. \tag{3.114}$$

To verify the filtering action, we implement the filter of Eq. (3.113) on MATLAB and calculate the first 100 output samples beginning from $n = 0$. Note that the input has been assumed to be a causal sequence with the first nonzero sample occurring at $n = 0$, and for calculating $y[0]$ and $y[1]$, we set $x[-1] = x[-2] = 0$. The MATLAB program Program 3_3 can be used to calculate the output of the above filter.

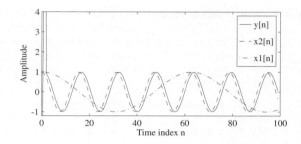

Figure 3.14: Output $y[n]$ (solid line), low-frequency input $x_1[n]$ (dash-dotted line), and high-frequency input $x_2[n]$ (dashed line) signals of the FIR filter of Eq. (3.105).

Table 3.5: Input and output sequences of the filter of Example 3.20.

n	$\cos(0.1n)$	$\cos(0.4n)$	$x[n]$	$y[n]$
0	1.0	1.0	2.0	-13.52390
1	0.9950041	0.9210609	1.9160652	13.956333
2	0.9800665	0.6967067	1.6767733	0.9210616
3	0.9553364	0.3623577	1.3176942	0.6967064
4	0.9210609	-0.0291995	0.8918614	0.3623572
5	0.8775825	-0.4161468	0.4614357	-0.0292002
6	0.8253356	-0.7373937	0.0879419	-0.4161467

Program 3_3.m

Figure 3.14 shows the plots of the output $y[n]$ generated using Program 3_3 and the two sinusoids of the input $x[n]$. In this figure, the output values have been clipped to the range $(-1.2, 4)$ to show the steady-state output more clearly. The first seven samples of the output are shown in Table 3.5, along with that of the two sinusoids. From this table and also from Figure 3.14, we observe that, neglecting the least significant digit,

$$y[n] = \cos(0.4(n-1)), \quad \text{for } n = 2, 3, 4, 5 \text{ and } 6.$$

Several comments are in order here. First, computation of the present value of output requires the knowledge of the present and two previous input samples. Hence, the first two output samples are the result of the assumed zero input sample values at $n = -1$ and $n = -2$ and, therefore, include the transient part of the output in addition to the steady-state part. Since the impulse response is of length $N + 1 = 3$, the steady-state is reached at $n = N = 2$. Second, the output is a delayed version of the high-frequency component $\cos(0.4n)$ of the input, and the delay is one sample period.

3.9   Phase and Group Delays

We conclude this chapter with a discussion on two important additional parameters that characterize the form of the output response $y[n]$ of an LTI discrete-time system excited by an input signal $x[n]$ composed of a weighted linear combination of sinusoidal sequences. These two parameters are associated with the frequency response $H(e^{j\omega})$ of the system. As pointed out in Eq. (3.102), the steady-state response of a stable LTI system for a sinusoidal input has the same form as the input except it suffers a change in its magnitude determined by the value of the magnitude function $|H(e^{j\omega_o})|$ of the LTI system at the frequency ω_o of the input sinusoidal signal and a phase difference relative to the input signal phase by an amount given by the value of the phase function $\theta(\omega_o) = \arg\{H(e^{j\omega_o})\}$ of the LTI system at ω_o. For a narrow-band input signal, we can assume that the magnitude function is essentially constant at all frequencies of the constituent sinusoidal signals comprising the input, and only the phase of each term in the output response relative to the phase of its corresponding components in the input affects the behavior of the output signal.

3.9.1 Definition

If the input is a sinusoidal signal of frequency ω_o as given by Eq. (3.101), the output is also a sinusoidal signal of the same frequency ω_o but lagging in phase by $\theta(\omega_o)$ radians, as demonstrated in Eq. (3.102). We can rewrite Eq. (3.102) as

$$
\begin{aligned}
y[n] &= A|H(e^{j\omega_o})|\cos\left(\omega_o\left(n + \frac{\theta(\omega_o)}{\omega_o}\right) + \phi\right), \\
&= A|H(e^{j\omega_o})|\cos\left(\omega_o\left(n - \tau_p(\omega_o)\right) + \phi\right),
\end{aligned}
\tag{3.115}
$$

where[6]

$$
\tau_p(\omega_o) = -\frac{\theta(\omega_o)}{\omega_o}
\tag{3.116}
$$

is called the *phase delay*. As Eq. (3.115) points out, the output $y[n]$ is a time-delayed version of the input $x[n]$. However, in general, the output $y[n]$ will not be a delayed replica of the input $x[n]$ unless the phase delay is an integer. Therefore, the phase delay has a physical meaning only with respect to the underlying continuous-time functions associated with the input and output sequences.

When the input signal contains many sinusoidal components with different frequencies that are not harmonically related, each component will go through different phase delays when processed by a frequency-selective LTI discrete-time system, and the signal delay is determined using a different parameter called the *group delay*, as defined below:

$$
\tau_g(\omega) = -\frac{d\theta(\omega)}{d\omega}.
\tag{3.117}
$$

It has been tacitly assumed here that the phase function is unwrapped so that its derivative exists. As in the case of the phase delay, the group delay also has a physical meaning only with respect to the underlying continuous-time functions associated with the input and output sequences.

A graphical comparison of the two types of delays are indicated in Figure 3.15. As can be seen from this figure, the group delay $\tau_g(\omega_o)$ is the negative of the slope of the phase function $\theta(\omega)$ at a frequency ω_o, whereas the phase delay $\tau_p(\omega_o)$ is the negative of the slope of the straight line from the origin to the point $[\omega_o, \theta(\omega_o)]$ on the phase function plot.

The physical significance of the above two delays are better understood by examining the continuous-time case [Pap62]. We consider an LTI continuous-time system with a frequency response

$$
H_a(j\Omega) = |H_a(j\Omega)|e^{j\theta_a(\Omega)}
$$

and excited by a narrow-band amplitude modulated continuous-time signal given by

$$
x_a(t) = a(t)\cos(\Omega_c t),
\tag{3.118}
$$

---

[6]The minus sign indicates phase lag.

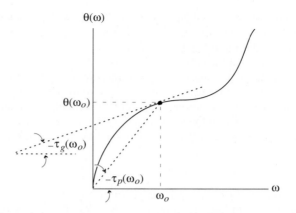

Figure 3.15: Evaluation of the phase delay and the group delay. Adapted from [Pap62].

where $a(t)$ is a lowpass modulating signal with a band-limited continuous-time Fourier transform given by

$$|A(j\Omega)| = 0, \qquad |\Omega| > \Omega_o, \tag{3.119}$$

and $\cos(\Omega_c t)$ is the high-frequency carrier signal.

We assume that in the frequency range $\Omega_c - \Omega_o < |\Omega| < \Omega_c + \Omega_o$, the frequency response of the continuous-time system has a constant magnitude and a linear phase; that is,

$$|H_a(j\Omega)| = |H_a(j\Omega_c)|,$$

$$\theta_a(\Omega) = \theta_a(\Omega_c) - (\Omega - \Omega_c) \left.\frac{d\theta_a(\Omega)}{d\Omega}\right|_{\Omega=\Omega_c}$$

$$= -\Omega_c \tau_p(\Omega_c) + (\Omega - \Omega_c)\tau_g(\Omega_c). \tag{3.120}$$

Now, the continuous-time Fourier transform of the input signal $x_a(t)$ is of the form

$$X_a(j\Omega) = \tfrac{1}{2}\left(A(j[\Omega + \Omega_c]) + A(j[\Omega - \Omega_c])\right),$$

obtained using the frequency-shifting property. Moreover, because of the constraint imposed by Eq. (3.119), $X_a(j\Omega) = 0$ outside the frequency range $\Omega_c - \Omega_o < |\Omega| < \Omega_c + \Omega_o$. As a result, the output response $y_a(t)$ of the LTI continuous-time system is given by

$$y_a(t) = a\left(t - \tau_g(\Omega_c)\right)\cos\Omega_c\left(t - \tau_p(\Omega_c)\right). \tag{3.121}$$

In deriving the above equation, it has been assumed that $|H_a(j\Omega_c)| = 1$. As can be seen from the above equation, the group delay $\tau_g(\Omega_c)$ is precisely the delay of the envelope $a(t)$ of the input signal $x_a(t)$, whereas the phase delay $\tau_p(\Omega_c)$ is the delay of the carrier signal. Figure 3.16 illustrates the effect of the two delays on an amplitude-modulated sinusoidal signal.

It should be noted that the above derivation for the output response holds provided the frequency response of the lTI system satisfies Eq. (3.120). In the case of LTI systems with a wide-band frequency response, the two delays do not have any physical meanings.

The waveform of the underlying continuous-time output shows distortion when the group delay of the LTI system is not constant over the bandwidth of the modulated signal. If the distortion is unacceptable, a delay equalizer is usually cascaded with the LTI system so that the overall group delay of the cascade

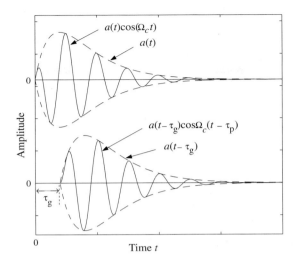

Figure 3.16: Illustration of the physical meanings of the phase and group delays. Adapted from [Pap62].

is approximately linear over the band of interest. However, to keep the magnitude response of the parent LTI system unchanged, the equalizer must have a constant magnitude response at all frequencies.[7]

For the filter of Example 3.20, the phase function $\theta(\omega) = -\omega$. Hence, the group delay is given by $\tau_g(\omega) = 1$, which is also evident from Figure 3.14, and pointed out earlier. Likewise, for the moving-average filter of Eq. (3.97), the group delay is given by

$$\tau_g(\omega) = \frac{M-1}{2}, \tag{3.122}$$

or, in other words, the moving-average filter exhibits a constant group delay for all frequencies.

3.9.2 Phase and Group Delay Computation Using MATLAB

The phase delay can be determined in MATLAB using the M-file `phasedelay`. Various options are available with this function. Its use is illustrated in Example 3.21.

EXAMPLE 3.21   Phase Delay Computation Using MATLAB

Determine the phase delay of the digital filter characterized by a frequency response

$$H(e^{j\omega}) = \frac{0.136728736(1 - e^{-j2\omega})}{1 - 0.53353098e^{-j\omega} + 0.726542528e^{-j2\omega}}.$$

The MATLAB code fragments used to compute the phase delay vector are

```
num = 0.136728736*[1 0 -1];
den = [1 -0.53353098 0.726542528];
[phi,w] = phasedelay(num,den,1024);
```

The plot of the phase delay evaluated using MATLAB is shown in Figure 3.17.

---

[7]See Section 7.1.3.

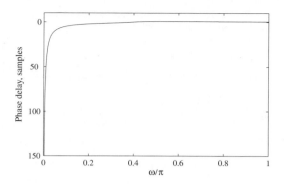

Figure 3.17: Phase delay of the frequency response function of Example 3.21.

Likewise, the group delay can be determined in MATLAB using the M-file `grpdelay`. Several options are also available with this function. We illustrate its use in Example 3.22.

EXAMPLE 3.22   Group Delay Computation Using MATLAB

Using MATLAB, we evaluate the group delay of the frequency response function of Example 3.21. The MATLAB code fragments used to compute the phase delay vector are

```
num = 0.136728736*[1 0 -1];
den = [1 -0.53353098 0.726542528];
[gd,w] = grpdelay(num,den,1024);
```

The plot of the group delay evaluated using MATLAB is shown in Figure 3.18.

3.10 Summary

This chapter provided a short review of the continuous-time Fourier transform (CTFT) representations of continuous-time signals and systems. The discrete-time Fourier transform (DTFT) and its inverse are introduced next along with a discussion of the convergence of the DTFT. Properties of the DTFT are reviewed and the unwrapping of the phase function to remove certain discontinuities in the DTFT is discussed. The concept of the frequency response of a linear, time-invariant (LTI) discrete-time system is then introduced followed by a careful examination of the difference between phase and group delays associated with the frequency response.

3.11 Problems

3.1 Show that the absolute value of the CTFT $X_a(j\Omega)$ defined in Eq. (3.1) is finite if $x_a(t)$ is absolutely integrable.

3.2 Determine the CTFT of the following continuous-time functions defined for $-\infty < t < \infty$:
   (a) $y_a(t) = \cos(\Omega_o t)$,   (b) $u_a(t) = e^{-\alpha|t|}$,   (c) $v_a(t) = e^{j\Omega_o t}$,   (d) $p_a(t) = \sum_{\ell=-\infty}^{\infty} \delta(t - \ell T)$.

3.3 Determine the CTFT of the following continuous-time functions defined for $-\infty < t < \infty$:

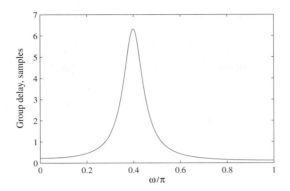

Figure 3.18: Group delay of the frequency response function of Example 3.21.

(a) $v_a(t) = 1$, (b) $\mu(t) = \begin{cases} 1, & t \geq 0, \\ 0, & t < 0, \end{cases}$ (c) $x_a(t) = \begin{cases} 1, & |t| < \frac{1}{2}, \\ \frac{1}{2}, & |t| = \frac{1}{2}, \\ 0, & |t| > \frac{1}{2}, \end{cases}$ (d) $y_a(t) = \begin{cases} 1 - 2|t|, & |t| < \frac{1}{2}, \\ 0, & |t| \geq \frac{1}{2}, \end{cases}$

3.4 The Gaussian density function, defined in Eq. (A.6), repeated below for convenience

$$h(t) = \frac{1}{\sigma\sqrt{(2\pi)}} e^{-(t-\mu)^2/2\sigma^2},$$ (3.123)

where σ and μ are, respectively, the variance and the mean of the density function. A continuous-time filter with an impulse response as given by Eq. (3.123) with zero mean is called a *Gaussian filter.* Show that the CTFT of $h(t)$ is also a Gaussian function of Ω.

3.5 The finite-energy function $x_a(t) = \sin(t)/\pi t$ is not absolutely summable. Show that its CTFT is given by

$$X_a(j\Omega) = \begin{cases} 1, & |\Omega| \leq 1, \\ 0, & |\Omega| > 1. \end{cases}$$

3.6 Consider the CTFT pair

$$x_a(t) \overset{\text{CTFT}}{\longleftrightarrow} X_a(j\Omega).$$

Prove the following properties of the CTFT.

(a) **Time-shifting Property:** $x_a(t - t_o) \overset{\text{CTFT}}{\longleftrightarrow} X_a(j\Omega)e^{-j\Omega t_o}$,

(b) **Frequency-shifting Property:** $x_a(t)e^{j\Omega_o t} \overset{\text{CTFT}}{\longleftrightarrow} X_a(j(\Omega - \Omega_o))$,

(c) **Symmetry Property:** $X_a(t) \overset{\text{CTFT}}{\longleftrightarrow} 2\pi x_a(-j\Omega)$,

(d) **The Scaling Property:** $x_a(at) \overset{\text{CTFT}}{\longleftrightarrow} \frac{1}{|a|} X_a\left(j\frac{\Omega}{a}\right)$,

(e) **Time Differentiation Property:** $\frac{dx_a(t)}{dt} \overset{\text{CTFT}}{\longleftrightarrow} j\Omega X_a(j\Omega)$.

3.7 Let $X_a(j\Omega)$ denote the CTFT of a real-valued continuous-time function $x_a(t)$. Show that the magnitude spectrum $|X_a(j\Omega)|$ is an even function of Ω and the phase spectrum $\theta(\Omega) = \arg\{X_a(j\Omega)\}$ is an odd function of Ω.

3.8 Show that the CTFT of the Hilbert transformer defined by Eq. (1.4) is

$$H_{\text{HT}}(j\Omega) = \begin{cases} -j, & \Omega > 0, \\ j, & \Omega < 0. \end{cases}$$

3.9 Let $x(t)$ be a real-valued input signal with a CTFT $X(j\Omega) = X_p(j\Omega) + X_n(j\Omega)$, where $X_p(j\Omega)$ is the portion of $X(j\Omega)$ occupying the positive frequency range and $X_n(j\Omega)$ is the portion of $X(j\Omega)$ occupying the negative frequency range. Let $\hat{x}(t)$ denote the output of an Hilbert transformer with an input $x(t)$. Show that the CTFT $Y(j\Omega)$ of the complex-valued signal $y(t) = x(t) + j\hat{x}(t)$ is given by $Y(j\Omega) = 2X_p(j\Omega)$. Thus, the spectrum of $y(t)$ occupies only the positive frequency range.

3.10 Compute the total energy of the continuous-time signal of Eq. (3.4) with $\alpha = 0.5$ and determine its 80% bandwidth.

3.11 Show that the DTFT of $\mu[n]$ is given by $\frac{1}{1-e^{-j\omega}} + \sum_{k=-\infty}^{\infty} \pi\delta(\omega + 2\pi k)$.

3.12 Show that the DTFT of the sequence $x[n] = 1, \ -\infty < n < \infty$, is given by $X(e^{j\omega}) = \sum_{k=-\infty}^{\infty} 2\pi\delta(\omega + 2\pi k)$.

3.13 Determine the DTFT of the two-sided sequence $y[n] = \alpha^{|n|}, |\alpha| < 1$.

3.14 In Example 3.8, we showed that the inverse DTFT $h_{LP}[n]$ of the DTFT $H_{LP}(e^{j\omega})$ shown in Figure 3.5 is given by Eq. (3.49). Determine and plot the DTFT of $g[n] = \delta[n] - \frac{\sin\omega_c n}{\pi n}, \ -\infty < n < \infty$.

3.15 Let $X(e^{j\omega})$ denote the DTFT of a real sequence $x[n]$.

   (a) Show that if $x[n]$ is even, then it can be computed from $X(e^{j\omega})$ using $x[n] = \frac{1}{\pi}\int_0^{\pi} X(e^{j\omega})\cos\omega n \, d\omega$.

   (b) Show that if $x[n]$ is odd, then it can be computed from $X(e^{j\omega})$ using $x[n] = \frac{j}{\pi}\int_0^{\pi} X(e^{j\omega})\sin\omega n \, d\omega$.

3.16 Determine the DTFT of the causal sequence $x[n] = A\alpha^n \sin(\omega_0 n + \phi)\mu[n]$, where A, α, ω_0, and ϕ are real, and $|\alpha| < 1$.

3.17 Determine the DTFT of each of the following sequences:

   (a) $x_1[n] = \alpha^n \mu[n-1], \quad |\alpha| < 1$,   (b) $x_2[n] = n\alpha^n \mu[n], \quad |\alpha| < 1$,   (c) $x_3[n] = \alpha^n \mu[n+1], \quad |\alpha| < 1$,

   (d) $x_4[n] = n\alpha^n \mu[n+2], \quad |\alpha| < 1$,   (e) $x_5[n] = \alpha^n \mu[-n-1], \quad |\alpha| > 1$,   (f) $x_6[n] = \begin{cases} \alpha^{|n|}, & |n| \le M, \\ 0, & \text{otherwise.} \end{cases}$

3.18 Determine the DTFT of each of the following sequences:

   (a) $x_a[n] = \mu[n] - \mu[n-5]$,   (b) $x_b[n] = \alpha^n\left(\mu[n] - \mu[n-8]\right), |\alpha| < 1$,   (c) $x_c[n] = (n+1)\alpha^n \mu[n], |\alpha| < 1$.

3.19 Determine the DTFT of each of the following finite-length sequences:

   (a) $y_1[n] = \begin{cases} 1, & -N \le n \le N, \\ 0, & \text{otherwise,} \end{cases}$ (b) $y_2[n] = \begin{cases} 1, & 0 \le n \le N, \\ 0, & \text{otherwise,} \end{cases}$ (c) $y_3[n] = \begin{cases} 1 - \frac{|n|}{N}, & -N \le n \le N, \\ 0, & \text{otherwise,} \end{cases}$

   (d) $y_4[n] = \begin{cases} N + 1 - |n|, & -N \le n \le N, \\ 0, & \text{otherwise,} \end{cases}$ (e) $y_f[n] = \begin{cases} \cos(\pi n/2N), & -N \le n \le N, \\ 0, & \text{otherwise.} \end{cases}$

3.20 Show that the inverse DTFT of

$$X(e^{j\omega}) = \frac{1}{(1 - \alpha e^{-j\omega})^m}, \quad |\alpha| < 1,$$

is given by

$$x[n] = \frac{(n + m - 1)!}{n!(m - 1)!}\alpha^n \mu[n].$$

3.21 Evaluate the inverse DTFT of each of the following DTFTs:

(a) $X_a(e^{j\omega}) = \sum_{k=-\infty}^{\infty} \delta(\omega + 2\pi k)$, (b) $X_b(e^{j\omega}) = \dfrac{e^{j\omega}(1 - e^{j\omega N})}{1 - e^{j\omega}}$,

(c) $X_c(e^{j\omega}) = 1 + 2\sum_{\ell=0}^{N} \cos \omega \ell$, (d) $X_d(e^{j\omega}) = \dfrac{-\alpha e^{-j\omega}}{(1 - \alpha e^{-j\omega})^2}$, $|\alpha| < 1$.

3.22 Determine the inverse DTFT of each of the following DTFTs:

(a) $H_a(e^{j\omega}) = \sin(4\omega)$, (b) $H_b(e^{j\omega}) = \cos(4\omega)$, (c) $H_c(e^{j\omega}) = \sin(5\omega)$, (d) $H_d(e^{j\omega}) = \cos(5\omega)$.

3.23 Determine the inverse DTFT of each of the following DTFTs:

(a) $H_1(e^{j\omega}) = 1 + 2\cos \omega + 3\cos 2\omega$, (b) $H_2(e^{j\omega}) = (3 + 2\cos \omega + 4\cos 2\omega)\cos(\omega/2)e^{-j\omega/2}$,
(c) $H_3(e^{j\omega}) = j(3 + 4\cos \omega + 2\cos 2\omega)\sin \omega$, (d) $H_4(e^{j\omega}) = j(4 + 2\cos \omega + 3\cos 2\omega)\sin(\omega/2)e^{j\omega/2}$.

3.24 Prove the following theorems of the discrete-time Fourier transform: (a) Linearity theorem, (b) Time-reversal theorem, (c) Time-shifting theorem, and (d) Frequency-shifting theorem.

3.25 Determine and plot the DTFT of the cascade of the LTI discrete-time systems with two-sided impulse responses given by $h_1[n] = \delta[n] - \dfrac{\sin \omega_1 n}{\pi n}$, and $h_2[n] = \dfrac{\sin \omega_2 n}{\pi n}$, respectively, where $0 < \omega_1 < \omega_2 < \pi$.

3.26 Let $X(e^{j\omega})$ denote the DTFT of a real sequence $x[n]$. Express the inverse DTFT $y[n]$ of $Y(e^{j\omega}) = X(e^{j4\omega})$ in terms of $x[n]$.

3.27 Let $X(e^{j\omega})$ denote the DTFT of a real sequence $x[n]$. Define $Y(e^{j\omega}) = \frac{1}{2}\left\{X(e^{j\omega/2}) + X(-e^{j\omega/2})\right\}$. Determine the inverse DTFT $y[n]$ of $Y(e^{j\omega})$.

3.28 Prove Eq. (3.26).

3.29 Prove Eq. (3.33).

3.30 The magnitude function $|X(e^{j\omega})|$ of a discrete-time sequence $x[n]$ is shown in Figure P3.1 for a portion of the angular frequency axis. Sketch the magnitude function for the frequency range $-\pi \leq \omega < \pi$. What type of sequence is $x[n]$?

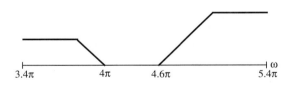

Figure P3.1

3.31 Without computing the DTFT, determine which of the following sequences have real-valued DTFTs and which have imaginary-valued DTFTs:

(a) $x_1[n] = \begin{cases} n, & -N \leq n \leq N, \\ 0, & \text{otherwise,} \end{cases}$ (b) $x_2[n] = \begin{cases} n^2, & -N \leq n \leq N, \\ 0, & \text{otherwise,} \end{cases}$ (c) $x_3[n] = \dfrac{\sin \omega_c n}{\pi n}$,

(d) $x_4[n] = \begin{cases} 0, & \text{for } n \text{ even,} \\ \frac{2}{\pi n}, & \text{for } n \text{ odd,} \end{cases}$ (e) $x_5[n] = \begin{cases} 0, & n = 0, \\ \frac{\cos \pi n}{n}, & |n| > 0. \end{cases}$

3.32 Without computing the inverse DTFT, determine which of the following DTFTs have an inverse that is an even sequence and which have an inverse that is an odd sequence:

(a) $Y_1(e^{j\omega}) = \begin{cases} |\omega|, & 0 \leq |\omega| \leq \omega_c, \\ 0, & \omega_c < |\omega| \leq \pi, \end{cases}$ (b) $Y_2(e^{j\omega}) = j\omega$, $0 \leq |\omega| \leq \pi$,   (c) $Y_3(e^{j\omega}) = \begin{cases} j, & -\pi < \omega < 0, \\ -j & 0 < \omega < \pi. \end{cases}$

3.33 Without computing the inverse DTFT, determine which of the DTFTs of Figure P3.2 has an inverse that is an even sequence and which has an inverse that is an odd sequence.

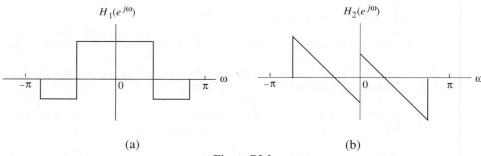

(a)   (b)

Figure P3.2

3.34 Let $X(e^{j\omega})$ denote the DTFT of a complex sequence $x[n]$. Determine the DTFT $Y(e^{j\omega})$ of the sequence $y[n] = x[n] \circledast x^*[-n]$ in terms of $X(e^{j\omega})$, and show that it is a real-valued function of ω.

3.35 A sequence $x[n]$ has a zero-phase DTFT $X(e^{j\omega})$ as sketched in Figure P3.3. Sketch the DTFT of the sequence $x[n]e^{-j\pi n/3}$.

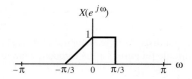

Figure P3.3

3.36 Using Parseval's relation, evaluate the following integrals: (a) $\int_0^\pi \frac{4}{5+4\cos\omega} d\omega$, (b) $\int_0^\pi \frac{1}{3.25-3\cos\omega} d\omega$, and (c) $\int_0^\pi \frac{4}{(5-4\cos\omega)^2} d\omega$.

3.37 Let $X(e^{j\omega})$ denote the DTFT of the sequence:

$$x[n] = \{1.2 \quad 2.9 \quad -4.2 \quad 2.4 \quad -3.2 \quad -0.9 \quad 4.4 \quad 4.2 \quad -0.8 \quad 3.9\}, -3 \leq n \leq 6.$$

Evaluate the integral

$$\int_{-\pi}^{\pi} \left| \frac{dX(e^{j\omega})}{d\omega} \right|^2 d\omega$$

without computing $X(e^{j\omega})$.

3.38 Let $X(e^{j\omega})$ denote the DTFT of a length-9 sequence $x[n]$ given by

$$x[n] = \{2 \quad 3 \quad -1 \quad 0 \quad -4 \quad 3 \quad 1 \quad 2 \quad 4\}, \quad -2 \le n \le 6.$$

Evaluate the following functions of $X(e^{j\omega})$ without computing the transform itself:

(a) $X(e^{j0})$, (b) $X(e^{j\pi})$, (c) $\int_{-\pi}^{\pi} X(e^{j\omega}) \, d\omega$, (d) $\int_{-\pi}^{\pi} \left| X(e^{j\omega}) \right|^2 d\omega$, (e) $\int_{-\pi}^{\pi} \left| \frac{dX(e^{j\omega})}{d\omega} \right|^2 d\omega$.

3.39 Repeat Problem 3.38 for the length-9 sequence

$$x[n] = \{4 \quad 2 \quad -1 \quad 5 \quad -3 \quad 1 \quad -2 \quad 4 \quad 2\}, \quad -6 \le n \le 2.$$

3.40 (a) A measure of the *time delay* of a sequence $x[n]$ is usually given by its *center of gravity*, defined by

$$C_g = \frac{\sum_{n=-\infty}^{\infty} n x[n]}{\sum_{n=-\infty}^{\infty} x[n]}.$$

Express C_g in terms of the DTFT $X(e^{j\omega})$ of $x[n]$.
(b) Determine the center of gravity of the sequence $x[n] = \alpha^n \mu[n]$.

3.41 Let $G_1(e^{j\omega})$ denote the discrete-time Fourier transform of the sequence $g_1[n]$ shown in Figure P3.4(a). Express the DTFTs of the remaining sequences in Figure P3.4 in terms of $G_1(e^{j\omega})$. Do not evaluate $G_1(e^{j\omega})$.

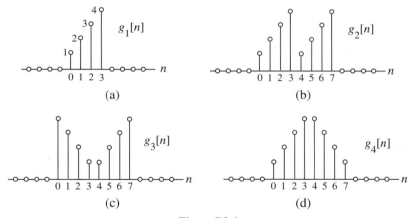

(a)   (b)

(c)   (d)

Figure P3.4

3.42 Let $y[n]$ denote the linear convolution of two sequences, $h[n]$ and $x[n]$; that is, $y[n] = h[n] \circledast x[n]$. Show that

(a) $\sum_{n=-\infty}^{\infty} y[n] = \left(\sum_{n=-\infty}^{\infty} h[n] \right) \left(\sum_{n=-\infty}^{\infty} x[n] \right)$, (b) $\sum_{n=-\infty}^{\infty} x[n] = \left(\sum_{n=-\infty}^{\infty} y[n] \right) \Big/ \left(\sum_{n=-\infty}^{\infty} h[n] \right)$,

(c) $\sum_{n=-\infty}^{\infty} (-1)^n y[n] = \left(\sum_{n=-\infty}^{\infty} (-1)^n h[n] \right) \left(\sum_{n=-\infty}^{\infty} (-1)^n x[n] \right)$.

3.43 Show that the 80% bandwidth of the discrete-time signal of Eq. (3.14) for $\alpha = 0.5$ is 0.5081π radians.

3.44 Let $x[n]$ be a causal and absolutely summable real sequence with a DTFT $X(e^{j\omega})$. If $X_{re}(e^{j\omega})$ and $X_{im}(e^{j\omega})$ denote the real and imaginary parts of $X(e^{j\omega})$, show that they are related as

$$X_{im}(e^{j\omega}) = -\frac{1}{2\pi} \int_{-\pi}^{\pi} X_{re}(e^{jv}) \cot\left(\frac{\omega - v}{2}\right) dv, \tag{3.124a}$$

$$X_{re}(e^{j\omega}) = \frac{1}{2\pi} \int_{-\pi}^{\pi} X_{im}(e^{jv}) \cot\left(\frac{\omega - v}{2}\right) dv + x[0]. \tag{3.124b}$$

The above equations are called the *discrete Hilbert transform relations*.

3.45 Show that the function $u[n] = z^n$, where z is a complex constant, is an eigenfunction of an LTI discrete-time system. Is $v[n] = z^n \mu[n]$ with z a complex constant also an eigenfunction of an LTI discrete-time system?

3.46 Determine the expression for the frequency response of the LTI discrete-time system of Figure 2.35 in terms of the frequency responses $H_i(e^{j\omega})$, $1 \le i \le 4$, of the individual blocks.

3.47 Determine the expression for the frequency response of each of the LTI discrete-time systems of Figure P2.2 (see Problem 2.64), in terms of the frequency responses $H_i(e^{j\omega})$, $1 \le i \le 5$, of the individual blocks.

3.48 Determine the expression for the frequency response of the LTI discrete-time system of Figure P2.3 (see Problem 2.65).

3.49 (a) Consider an LTI discrete-time system with a real and causal impulse response $h[n]$ and a frequency response $H(e^{j\omega})$. Show that $h[n]$ can be determined uniquely from the real part $H_{re}(e^{j\omega})$ of $H(e^{j\omega})$.
   (b) The real part of the frequency response of a real and causal LTI discrete-time system is given by $H_{re}(e^{j\omega}) = 1 + 2\cos\omega + 3\cos 2\omega + 4\cos 3\omega$. Determine the impulse response $h[n]$ of the system.

3.50 If the input to each of the following discrete-time systems is $x[n] = \cos(\omega_o n)$, determine the frequencies present in their outputs:
   (a) $y_a[n] = \sin(\pi n/3) x[n]$,   (b) $y_b[n] = x^3[n]$,   (c) $y_c[n] = x[3n]$.

3.51 Determine a closed-form expression for the frequency response $H(e^{j\omega})$ of the LTI discrete-time system characterized by an impulse response

$$h[n] = \delta[n] - \alpha\delta[n - R], \tag{3.125}$$

where $|\alpha| < 1$. What are the maximum and the minimum values of its magnitude response? How many peaks and dips of the magnitude response occur in the range $0 \le \omega < 2\pi$? What are the locations of the peaks and the dips? Sketch the magnitude and the phase responses for $R = 6$.

3.52 Determine a closed-form expression for the frequency response $G(e^{j\omega})$ of an LTI discrete-time system with an impulse response given by

$$g[n] = \begin{cases} \alpha^n, & 0 \le n \le M - 1, \\ 0 & \text{otherwise,} \end{cases}$$

where $|\alpha| < 1$. What is the relation of $G(e^{j\omega})$ to $H(e^{j\omega})$ of Eq. (3.98)? Scale the impulse response by multiplying it with a suitable constant so that the dc value of the magnitude response is unity.

3.53 A noncausal LTI FIR discrete-time system is characterized by an impulse response $h[n] = a_1\delta[n - 2] + a_2\delta[n - 1] + a_3\delta[n] + a_4\delta[n + 1]$. For what values of the impulse response samples will its frequency response $H(e^{j\omega})$ have a zero phase?

3.54 A causal LTI FIR discrete-time system is characterized by an impulse response $h[n] = a_1\delta[n] + a_2\delta[n - 1] + a_3\delta[n - 2] + a_4\delta[n - 3] + a_5\delta[n - 4]$. For what values of the impulse response samples will its frequency response $H(e^{j\omega})$ have a linear phase?

3.55 An FIR LTI discrete-time system is described by the difference equation

$$y[n] = a_1 x[n + k] + a_2 x[n + k - 1] + a_2 x[n + k - 2] + a_1 x[n + k - 3],$$

where $y[n]$ and $x[n]$ denote, respectively, the output and the input sequences. Determine the expression for its frequency response $H(e^{j\omega})$. For what values of the constant k will the system have a frequency response $H(e^{j\omega})$ that is a real function of ω?

3.56 Consider the cascade of three causal LTI systems: $h_1[n] = a\delta[n] + b\delta[n-1] + \delta[n-2]$; $h_2[n] = c^n \mu[n]$, $|c| < 1$; and $h_3[n] = d^n \mu[n]$, $|d| < 1$. Determine the frequency response $H(e^{j\omega})$ of the overall system. For what values of the constants $a, b, c,$ and d will $|H(e^{j\omega})| = 1$?

3.57 The input–output relation of a nonlinear discrete-time system in the frequency domain is given by

$$Y(e^{j\omega}) = |X(e^{j\omega})|^{\alpha} e^{j\arg X(e^{j\omega})}, \tag{3.126}$$

where $0 < \alpha \le 1$, and $X(e^{j\omega})$ and $Y(e^{j\omega})$ denote the DTFTs of the input and output sequences. Determine the expression for its frequency response $H(e^{j\omega}) = Y(e^{j\omega})/X(e^{j\omega})$, and show that it has zero phase. The nonlinear algorithm described by Eq. (3.126) is known as the *alpha-rooting method* and has been used in image enhancement [Jai89].

3.58 Determine the expression for the frequency response $H(e^{j\omega})$ of a causal IIR LTI discrete-time system characterized by the input–output relation

$$y[n] = x[n] - \alpha y[n - R], \qquad |\alpha| < 1,$$

where $y[n]$ and $x[n]$ denote, respectively, the output and the input sequences. Determine the maximum and the minimum values of its magnitude response. How many peaks and dips of the magnitude response occur in the range $0 \le \omega < 2\pi$? What are the locations of the peaks and the dips? Sketch the magnitude and the phase responses for $R = 5$.

3.59 An IIR LTI discrete-time system with input $x[n]$ and output $y[n]$ is described by the difference equation

$$y[n] + a_1 y[n - 1] = b_0 x[n] + b_1 x[n - 1],$$

where the constants $a_1, b_0,$ and b_1 are real. Determine the expression for its frequency response. For what values of the constants b_i will the magnitude response be a constant for all values of ω?

3.60 Repeat Problem 3.59 when the constants $a_1, b_0,$ and b_1 are complex numbers.

3.61 Determine the difference equation representation of each of the LTI discrete-time with frequency responses as given below:
    (a) $H_a(e^{j\omega}) = \mathrm{cosec}(\omega)$,   (b) $H_b(e^{j\omega}) = \sec(\omega)$,   (c) $H_c(e^{j\omega}) = \cot(\omega)$,   (d) $H_d(e^{j\omega}) = \tan(\omega/2)$.

3.62 Determine the input–output relation of a factor-of-L up-sampler in the frequency domain.

3.63 Determine the inverse DTFT of $G(e^{j\omega}) = 1/(1 - \alpha e^{-jL\omega})$, $|\alpha| < 1$, where L is a positive integer.

3.64 Consider an LTI discrete-time system with an impulse response $h[n] = (0.5)^n \mu[n]$. Determine the frequency response $H(e^{j\omega})$ of the system and evaluate its value at $\omega = \pm\pi/5$. What is the steady-state output $y[n]$ of the system for an input $x[n] = \sin(\pi n/3)\mu[n]$?

3.65 An FIR filter of length 3 is defined by a symmetric impulse response; that is, $h[0] = h[2]$. Let the input to this filter be a sum of two cosine sequences of angular frequencies 0.3 rad/samples and 0.6 rad/samples, respectively. Determine the impulse response coefficients so that the filter passes only the low-frequency component of the input.

3.66 An FIR filter of length 3 is defined by a symmetric impulse response; that is, $h[0] = h[2]$. Let the input to this filter be a sum of two cosine sequences of angular frequencies 0.3 rad/samples and 0.6 rad/samples, respectively. Determine the impulse response coefficients so that the filter passes only the high-frequency component of the input.

3.67 Determine the output response $y[n]$ of an LTI discrete-time system with an impulse response

$$h[n] = \frac{\sin\left((n-2)\frac{\pi}{3}\right)}{(n-2)\pi},$$

for an input

$$x[n] = 3\sin\left(\frac{\pi n}{3}\right) + 5\cos\left(\frac{2\pi n}{5}\right).$$

3.68 An FIR filter of length 5 is defined by a symmetric impulse response; that is, $h[n] = h[4-n], 0 \le n \le 4$. Let the input to this filter be a sum of three cosine sequences of angular frequencies: 0.2 rad/samples, 0.5 rad/samples, and 0.8 rad/samples, respectively. Determine the impulse response coefficients so that the filter passes only the midfrequency component of the input.

3.69 The frequency response $H(e^{j\omega})$ of a length-4 FIR filter with real impulse response has the following specific values: $H(e^{j0}) = 2$, $H(e^{j\pi/2}) = 7 - j3$, and $H(e^{j\pi}) = 0$. Determine its impulse response $h[n]$.

3.70 The frequency response $H(e^{j\omega})$ of a length-4 FIR filter with a real and antisymmetric impulse response has the following specific values: $H(e^{j\pi}) = 8$, and $H(e^{j\pi/2}) = -2 + j2$. Determine its impulse response $h[n]$.

3.71 (a) Design a length-3 FIR notch filter with a symmetric impulse response $h[n]$; that is, $h[n] = h[2-n], 0 \le n \le 2$, and with a notch frequency at 0.4π and a 0 dB dc gain.
(b) Determine the exact expression for the frequency response of the filter designed, and plot its magnitude and phase responses.

3.72 (a) Design a length-4 FIR lowpass filter with a symmetric impulse response $h[n]$; that is, $h[n] = h[3-n], 0 \le n \le 3$, satisfying the following magnitude response values: $|H(e^{j0.2\pi})| = 0.8$ and $|H(e^{j0.5\pi})| = 0.5$.
(b) Determine the exact expression for the frequency response of the filter designed, and plot its magnitude and phase responses.

3.73 (a) Design a length-4 FIR highpass filter with an antisymmetric impulse response $h[n]$; that is, $h[n] = -h[3-n], 0 \le n \le 3$, satisfying the following magnitude response values: $|H(e^{j0.5\pi})| = 0.2$ and $|H(e^{j0.8\pi})| = 0.7$.
(b) Determine the exact expression for the frequency response of the filter designed, and plot its magnitude and phase responses.

3.74 (a) Design a length-5 FIR bandpass filter with an antisymmetric impulse response $h[n]$; that is, $h[n] = -h[4-n], 0 \le n \le 4$, satisfying the following magnitude response values: $|H(e^{j0.4\pi})| = 0.8$ and $|H(e^{j0.8\pi})| = 0.2$.
(b) Determine the exact expression for the frequency response of the filter designed, and plot its magnitude and phase responses.

3.75 Consider the two LTI causal digital filters with impulse responses given by

$$h_A[n] = 0.3\delta[n] - \delta[n-1] + 0.3\delta[n-2],$$
$$h_B[n] = 0.3\delta[n] + \delta[n-1] + 0.3\delta[n-2].$$

(a) Sketch the magnitude responses of the two filters and compare their characteristics.

(b) Let $h_A[n]$ be the impulse response of a causal digital filter with a frequency response $H_A(e^{j\omega})$. Define another digital filter whose impulse response $h_C[n]$ is given by

$$h_C[n] = (-1)^n h_A[n], \qquad \text{for all } n.$$

What is the relation between the frequency response $H_C(e^{j\omega})$ of this new filter and the frequency response $H_A(e^{j\omega})$ of the parent filter?

3.76 As indicated in Example 2.45, the trapezoidal integration formula can be represented as an IIR digital filter represented by a difference equation given by

$$y[n] = y[n-1] + \tfrac{1}{2}\{x[n] + x[n-1]\},$$

with $y[-1] = 0$. Determine the frequency response of the above filter.

3.77 A recursive difference equation representation of the Simpson's numerical integration formula is given by [Ham89]

$$y[n] = y[n-2] + \tfrac{1}{3}\{x[n] + 4x[n-1] + x[n-2]\}.$$

Evaluate the frequency response of the above filter and compare it with that of the trapezoidal method of Problem 6.52.

3.78 The frequency response of an LTI FIR discrete-time system is given by $G(e^{j\omega}) = g_0 + g_1 e^{-j\omega} + g_2 e^{-j2\omega} + g_3 e^{-j3\omega}$. For what relations between the coefficients g_0, g_1, g_2, and g_3 will $G(e^{j\omega})$ have a constant group delay?

3.79 Determine the expressions for the group delay of each of the LTI systems whose frequency responses are given below.

(a) $H_a(e^{j\omega}) = a + be^{-j\omega}$, (b) $H_b(e^{j\omega}) = \frac{1}{1+ce^{-j\omega}}$, (c) $H_c(e^{j\omega}) = \frac{a+be^{-j\omega}}{1+ce^{-j\omega}}$, $|c| < 1$,

(d) $H_b(e^{j\omega}) = \frac{1}{(1+ce^{-j\omega})(1+de^{-j\omega})}$, $|c| < 1$, $|d| < 1$.

3.80 Show that the group delay $\tau_g(\omega)$ of an LTI discrete-time system characterized by a frequency response $H(e^{j\omega})$ can be expressed as

$$\tau_g(\omega) = \text{Re}\left\{\frac{j\frac{dH(e^{j\omega})}{d\omega}}{H(e^{j\omega})}\right\}. \tag{3.127}$$

3.81 Let $H(e^{j\omega})$ denote the frequency response of an LTI discrete-time system with an impulse response $h[n]$ and let $G(e^{j\omega})$ denote the Fourier transform of the sequence $nh[n]$. Show that the group delay of the LTI system can be computed using

$$\tau_g(\omega) = \frac{H_{\text{re}}(e^{j\omega})G_{\text{re}}(e^{j\omega}) + H_{\text{im}}(e^{j\omega})G_{\text{im}}(e^{j\omega})}{|H(e^{j\omega})|^2}, \tag{3.128}$$

where $H_{\text{re}}(e^{j\omega})$ and $H_{\text{im}}(e^{j\omega})$ denote the real and imaginary parts of $H(e^{j\omega})$, respectively, and $G_{\text{re}}(e^{j\omega})$ and $G_{\text{im}}(e^{j\omega})$ denote the real and imaginary parts of $G(e^{j\omega})$, respectively.

3.82 Using Eq. (3.128) determine the group delays of the LTI discrete-time systems with frequency responses as given below:

(a) $H_a(e^{j\omega}) = 1 + 0.4\,e^{-j\omega}$, (b) $H_b(e^{j\omega}) = \frac{1}{1+0.6\,e^{-j\omega}}$, (c) $H_c(e^{j\omega}) = \frac{1-0.5\,e^{-j\omega}}{1+0.3\,e^{-j\omega}}$,

(d) $H_d(e^{j\omega}) = \frac{1}{(1-0.3\,e^{-j\omega})(1+0.5\,e^{-j\omega})}$.

3.83 Consider an LTI discrete-time system with an impulse response $h[n] = (-0.5)^n \mu[n]$. Determine the frequency response $H(e^{j\omega})$ of the system, and evaluate its value at $\omega = \pm\pi/5$. What is the steady-state output $y[n]$ of the system for an input $x[n] = \sin(\pi n/5)\mu[n]$?

3.84 An FIR filter of length 3 is defined by a symmetric impulse response, that is, $h[0] = h[2]$. Let the input to this filter be a sum of two cosine sequences of angular frequencies 0.3 rad/samples and 0.7 rad/samples, respectively. Determine the impulse response coefficients so that the filter passes only the low-frequency component of the input.

3.12 MATLAB Exercises

M 3.1 Using Program 3_1, determine and plot the real and imaginary parts and the magnitude and phase spectra of the following DTFT for various values of r and θ:

$$G(e^{j\omega}) = \frac{1}{1 - 2r(\cos\theta)e^{-j\omega} + r^2 e^{-j2\omega}}, \qquad 0 < r < 1.$$

M 3.2 Using Program 3_1, determine and plot the real and imaginary parts and the magnitude and phase spectra of the DTFTs of the sequences of Problem 3.19 for $N = 10$.

M 3.3 Using Program 3_1, determine and plot the real and imaginary parts, and the magnitude and phase spectra of the following DTFTs:

$$\text{(a) } X(e^{j\omega}) = \frac{0.2418(1 + 0.139e^{-j\omega} - 0.3519e^{-j2\omega} + 0.139e^{-j3\omega} + e^{-j4\omega})}{1 + 0.2386e^{-j\omega} + 0.8258e^{-j2\omega} + 0.1393e^{-j3\omega} + 0.4153e^{-j4\omega}},$$

$$\text{(b) } X(e^{j\omega}) = \frac{0.1397(1 - 0.0911e^{-j\omega} + 0.0911e^{-j2\omega} - e^{-j3\omega})}{1 + 1.1454e^{-j\omega} + 0.7275e^{-j2\omega} + 0.1205e^{-j3\omega}}.$$

M 3.4 Using MATLAB, verify the symmetry relations of the DTFT of a complex sequence as listed in Table 3.1.

M 3.5 Using MATLAB, verify the symmetry relations of the DTFT of a real sequence as listed in Table 3.2.

M 3.6 Using MATLAB, verify the following general properties of the DTFT as listed in Table 3.4: (a) linearity, (b) time-shifting, (c) frequency-shifting, (d) differentiation-in-frequency, (e) convolution, (f) modulation, and (g) Parseval's relation. Since all data in MATLAB have to be finite-length vectors, the sequences used to verify the properties are thus restricted to be of finite length.

M 3.7 Write a MATLAB program to compute the group delay using the expression of Problem 3.80 at a prescribed set of discrete frequencies.

M 3.8 Write a MATLAB program to simulate the filter designed in Problem 3.65, and verify its filtering operation.

M 3.9 Write a MATLAB program to simulate the filter designed in Problem 3.68, and verify its filtering operation.

4 Digital Processing of Continuous-Time Signals

4.1 Introduction

Even though this book is concerned primarily with the processing of discrete-time signals, most signals we encounter in the real world are continuous in time, such as speech, music, and images. Increasingly, discrete-time signal processing algorithms are being used to process such signals and are implemented employing discrete-time analog or digital systems. For processing by digital systems, the discrete-time signals are represented in digital form with each discrete-time sample as a binary word. Therefore, we need the analog-to-digital and digital-to-analog interface circuits to convert the continuous-time signals into discrete-time digital form, and vice versa. As a result, it is necessary to develop the relations between the continuous-time signal and its discrete-time equivalent in the time-domain and also in the frequency-domain.[1] The latter relations are important in determining conditions under which the discrete-time processing of continuous-time signals can be done free of error under ideal situations.

The interface circuit performing the conversion of a continuous-time signal into a digital form is called the *analog-to-digital* (A/D) *converter.* Likewise, the reverse operation of converting a digital signal into a continuous-time signal is implemented by the interface circuit called the *digital-to-analog* (D/A) *converter.* In addition to these two devices, we need several additional circuits. Since the analog-to-digital conversion usually takes a finite amount of time, it is often necessary to ensure that the analog signal at the input of the A/D converter remains constant in amplitude until the conversion is complete to minimize the error in its representation. This is accomplished by a device called the *sample-and-hold* (S/H) circuit, which has dual purposes. It not only samples the input continuous-time signal at periodic intervals but also holds the analog sampled value constant at its output for sufficient time to permit accurate conversion by the A/D converter. In addition, the output of the D/A converter is a staircase-like waveform. It is therefore necessary to smooth the D/A converter output by means of an analog *reconstruction (smoothing) filter.* Finally, in most applications, the continuous-time signal to be processed usually has a larger bandwidth than the bandwidth of the available discrete-time processors. To prevent a detrimental effect called *aliasing,* an analog *anti-aliasing filter* is often placed before the S/H circuit. The complete block diagram illustrating the functional requirements for the discrete-time digital processing of a continuous-time signal is thus as indicated in Figure 4.1.

To understand the conditions under which the above system can work properly, we need to examine each of the interface circuits indicated in Figure 4.1. First, we assume a simpler mathematical equivalent of Figure 4.1, which enables us to derive the most fundamental condition that permits the discrete-time processing of continuous-time signals. To this end, we assume that the A/D and D/A converters have infinite precision wordlengths, resulting in the simplified representation of Figure 4.2.[2] In this representation, the S/H circuit in cascade with an infinite precision A/D converter has been replaced with the ideal continuous-

---

[1]These relations also apply to discrete-time analog signal processing systems, such as the switched-capacitor networks.

[2]Effects of finite wordlength are considered in Chapter 11.

Figure 4.1: Block diagram representation of the discrete-time digital processing of a continuous-time signal.

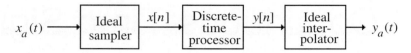

Figure 4.2: A simplified representation of Figure 4.1.

time–to–discrete-time (CT-DT) converter (i.e., an ideal sampler), which develops a discrete-time equivalent $x[n]$ of the continuous-time signal $x_a(t)$. Likewise, the infinite-precision D/A converter in cascade with the ideal reconstruction filter has been replaced with the ideal discrete-time–to–continuous-time (DT-CT) converter (i.e., an ideal interpolator), which develops a continuous-time equivalent $y_a(t)$ of the processed discrete-time signal $y[n]$.

In this chapter, we first derive the conditions for discrete-time representation of a band-limited continuous-time signal under ideal sampling and its exact recovery from the sampled version. If these conditions are not met, it is not possible to recover the original continuous-time signal from its sampled discrete-time equivalent, resulting in an undesirable distorted representation caused by an effect called aliasing. Since both the anti-aliasing filter and the reconstruction filter in Figure 4.1 are analog lowpass filters, we briefly review next the basic theory behind some commonly used analog filter design methods and illustrate them using MATLAB. It should be noted also that the most widely used digital filter design methods are based on the conversion of an analog prototype, and a knowledge of analog filter design is thus useful in digital signal processing. We then examine the basic properties of the various interface circuits depicted in Figure 4.1.

4.2   Sampling of Continuous-Time Signals

As indicated above, discrete-time signals in many applications are generated by sampling continuous-time signals. We also observed in Example 2.12 that identical discrete-time sequences may result from the sampling of more than one distinct continuous-time function. In fact, in general, there exists an infinite number of continuous-time signals, which when sampled at the same sampling rate lead to the same discrete-time signal. However, under certain conditions, it is possible to relate a unique continuous-time signal to a given discrete-time sequence, and it is possible to recover the original continuous-time signal from its sampled values. We develop this correspondence and the associated conditions next by considering the relation between the spectra of the original continuous-time signal and the discrete-time signal obtained by sampling.

4.2.1   Effect of Sampling in the Frequency-Domain

Let $g_a(t)$ be a continuous-time signal that is sampled uniformly at $t = nT$, generating the sequence $g[n]$ where

$$g[n] = g_a(nT), \qquad -\infty < n < \infty, \tag{4.1}$$

with T being the *sampling period.* The reciprocal of T is called the *sampling frequency F_T*; that is, $F_T = 1/T$. Now, the frequency-domain representation of $g_a(t)$ is given by its continuous-time Fourier

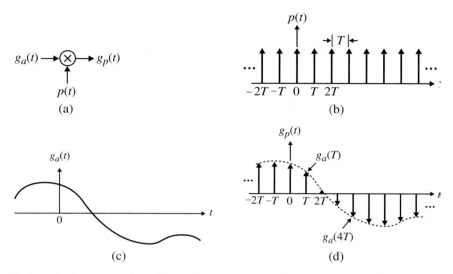

Figure 4.3: Mathematical representation of the uniform sampling process: (a) ideal sampling model, (b) impulse train, (c) continuous-time signal, and (d) its sampled version.

transform (CTFT) defined by Eq. (3.1):

$$G_a(j\Omega) = \int_{-\infty}^{\infty} g_a(t)e^{-j\Omega t} \, dt, \tag{4.2}$$

whereas the frequency-domain representation of $g[n]$ is given by its discrete-time Fourier transform defined by Eq. (3.12):

$$G(e^{j\omega}) = \sum_{n=-\infty}^{\infty} g[n]e^{-j\omega n}. \tag{4.3}$$

To establish the relations between these two different types of Fourier spectra, $G_a(j\Omega)$ and $G(e^{j\omega})$, we treat the sampling operation mathematically as a multiplication of the continuous-time signal $g_a(t)$ by a periodic impulse train $p(t)$:

$$p(t) = \sum_{n=-\infty}^{\infty} \delta(t - nT), \tag{4.4}$$

consisting of a train of ideal impulse functions[3] with a period T, as indicated in Figure 4.3. The multiplication operation yields an impulse train $g_p(t)$:

$$g_p(t) = g_a(t)p(t) = \sum_{n=-\infty}^{\infty} g_a(nT)\delta(t - nT), \tag{4.5}$$

which is seen to be a continuous-time signal consisting of a train of uniformly spaced impulses with the impulse at $t = nT$ weighted by the sampled value $g_a(nT)$ of $g_a(t)$ at that instant.

There are two different forms of the continuous-time Fourier transform $G_p(j\Omega)$ of $g_p(t)$. One form is obtained by taking the CTFT of Eq. (4.4), which results in a weighted sum of the continuous-time Fourier

---

[3] See Section 3.2.4 for a definition of the ideal impulse function.

transforms of the shifted impulse functions $\delta(t - nT)$. From Example 3.3 using the time-shifting property, we note that the CTFT of $\delta(t - nT)$ is given by $e^{-j\Omega nT}$. Hence, from Eq. (4.5), $G_p(j\Omega)$ is given by

$$G_p(j\Omega) = \sum_{n=-\infty}^{\infty} g_a(nT) e^{-j\Omega nT}. \tag{4.6}$$

To derive the second form, we make use of the *Poisson's sum formula,* given by [Pap62]

$$\sum_{n=-\infty}^{\infty} \phi(t + nT) = \frac{1}{T} \sum_{k=-\infty}^{\infty} \Phi(jk\Omega_T) e^{jk\Omega_T t}, \tag{4.7}$$

where $\Omega_T = 2\pi/T$ denotes the angular sampling frequency, and $\Phi(j\Omega)$ is the CTFT of the continuous-time function $\phi(t)$. A proof of the above identity is left as an exercise (Problem 4.1). For $t = 0$, Eq. (4.7) reduces to a simpler form

$$\sum_{n=-\infty}^{\infty} \phi(nT) = \frac{1}{T} \sum_{k=-\infty}^{\infty} \Phi(jk\Omega_T), \tag{4.8}$$

which is used to derive the second form of $G_p(j\Omega)$. Now from the frequency-shifting property of the CTFT, the CTFT of $g_a(t) e^{-j\Psi t}$ is given by $G_a(j(\Omega + \Psi))$. Substituting $\phi(t) = g_a(t) e^{-j\Psi t}$ in Eq. (4.8), we arrive at

$$\sum_{n=-\infty}^{\infty} g_a(nT) e^{-j\Psi nT} = \frac{1}{T} \sum_{k=-\infty}^{\infty} G_a(j(k\Omega_T + \Psi)). \tag{4.9}$$

By replacing Ψ with Ω in the above equation, we arrive at the alternative form of the continuous-time Fourier transform of $g_p(t)$ given by

$$G_p(j\Omega) = \frac{1}{T} \sum_{k=-\infty}^{\infty} G_a(j(\Omega + k\Omega_T)). \tag{4.10}$$

As can be seen from the above equation, $G_p(j\Omega)$ is a periodic function of frequency Ω consisting of a sum of shifted and scaled replicas of $G_a(j\Omega)$, shifted by integer multiples of Ω_T and scaled by $1/T$. The term on the right-hand side of Eq. (4.10) for $k = 0$ is the *baseband* portion of $G_p(j\Omega)$, and each of the remaining terms are the *frequency-translated* portions of $G_p(j\Omega)$. The frequency range $-\Omega_T/2 \leq \Omega < \Omega_T/2$ is called the *baseband* or *Nyquist band.*

Figure 4.4 illustrates the frequency-domain effects of time-domain sampling. Assume $g_a(t)$ is a band-limited signal with a frequency spectrum $G_a(j\Omega)$, as sketched in Figure 4.4(a), where Ω_m is the highest frequency contained in $g_a(t)$.[4] The spectrum $P(j\Omega)$ of the periodic impulse train $p(t)$ with a sampling period $T = 2\pi/\Omega_T$ is indicated in Figures 4.4(b) and (d). Two possible spectra of $G_p(j\Omega)$ are shown in Figures 4.4(c) and (e). It is evident from Figure 4.4(c) that if $\Omega_T \geq 2\Omega_m$, there is no overlap between the shifted replicas of $G_a(j\Omega)$ generating $G_p(j\Omega)$. On the other hand, as indicated in Figure 4.4(e), if $\Omega_T < 2\Omega_m$, there is an overlap of the spectra of the shifted replicas of $G_a(j\Omega)$ generating $G_p(j\Omega)$. Consequently, if $\Omega_T \geq 2\Omega_m$, $g_a(t)$ can be recovered exactly from $g_p(t)$ by passing it through an ideal lowpass filter $H_r(j\Omega)$ with a gain T and a cutoff frequency Ω_c greater than Ω_m and less than $\Omega_T - \Omega_m$, as illustrated in Figure 4.5. However, if $\Omega_T < 2\Omega_m$, due to the overlap of the shifted replicas of $G_a(j\Omega)$, the spectrum $G_p(j\Omega)$ cannot be separated by filtering to recover $G_a(j\Omega)$ because of the distortion caused by a part of the replicas immediately outside the baseband being folded back, that is, *aliased* into the baseband. The frequency $\Omega_T/2$ is often referred to as the *folding frequency* or *Nyquist frequency.* It should be noted, however, that if the spectrum $G_a(j\Omega)$ of $g_a(t)$ (band-limited to Ω_m) also contained an impulse at Ω_m, the sampling rate Ω_T must be greater than $2\Omega_m$ to recover fully $g_a(t)$ from the sampled version (Problem 4.2).

---

[4]The asymmetric CTFT has been purposely chosen to illustrate more clearly the effect of sampling.

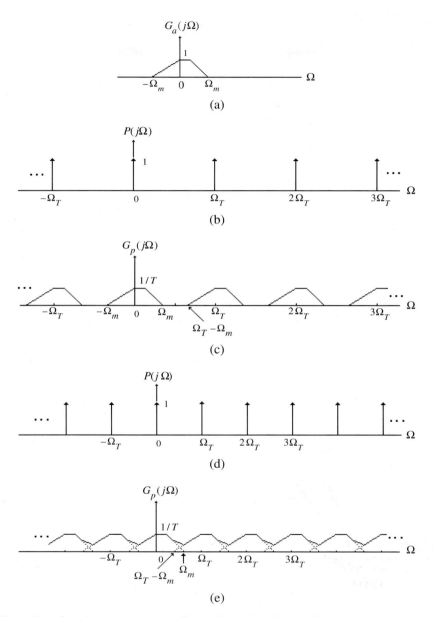

Figure 4.4: Illustration of the frequency-domain effects of time-domain sampling. (a) Spectrum of original continuous-time signal $g_a(t)$, (b) spectrum of the periodic impulse train $p(t)$, (c) spectrum of the sampled signal $g_p(t)$ with $\Omega_T > 2\omega_m$, (d) spectrum of the periodic impulse train $p(t)$ with a sampling period longer than that shown in (b), and (e) spectrum of the sampled signal $g_p(t)$ with $\Omega_T < 2\Omega_m$. [Note: The spectrum of $g_a(t)$ is shown as not being an even function to emphasize the effect of sampling.]

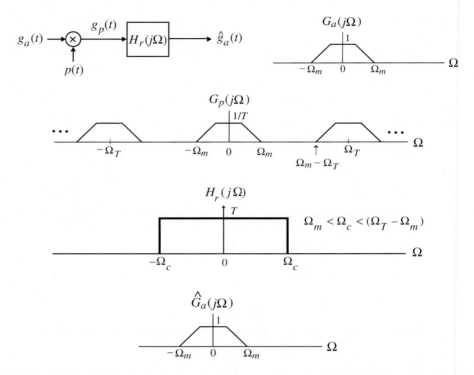

Figure 4.5: Reconstruction of the original continuous-time signal from its sampled version obtained by ideal sampling.

The above result is more commonly known as the *sampling theorem,*[5] which can be summarized as follows.

Sampling Theorem.   Let $g_a(t)$ be a band-limited signal with $G_a(j\Omega) = 0$ for $|\Omega| > \Omega_m$. Then $g_a(t)$ is uniquely determined by its samples $g_a(nT)$, $-\infty < n < \infty$, if

$$\Omega_T \geq 2\Omega_m, \tag{4.11}$$

where

$$\Omega_T = 2\pi/T. \tag{4.12}$$

Equation (4.11) is often referred to as the *Nyquist condition*. Given $\{g_a(nT)\}$, we can recover exactly $g_a(t)$ by generating an impulse train $g_p(t)$ of the form of Eq. (4.5) and then passing $g_p(t)$ through an ideal lowpass filter $H_r(j\Omega)$ with a gain T and a cutoff frequency Ω_c greater than Ω_m and less than $\Omega_T - \Omega_m$; that is,

$$\Omega_m < \Omega_c < (\Omega_T - \Omega_m). \tag{4.13}$$

The highest frequency Ω_m contained in $g_a(t)$ is usually called the *Nyquist frequency* since it determines the minimum sampling frequency $\Omega_T = 2\Omega_m$ that must be used to fully recover $g_a(t)$ from its sampled version. The frequency $2\Omega_m$ is called the *Nyquist rate*.

---

[5]Also called either the *Nyquist sampling theorem* or *Shannon sampling theorem*.

It should be pointed out here that to recover $g_a(t)$ from its samples $g_a(nT)$, $-\infty < n < \infty$, obtained by sampling at the Nyquist rate, the spectrum $G_a(j\Omega)$ of $g_a(t)$ should not contain an impulse function at the highest frequency Ω_m [Lat98]. If it does, then the sampling rate should be greater than $2\Omega_m$ to recover $g_a(t)$. For example, the signal $g_a(t) = \sin(\Omega_m t)$ is band-limited to Ω_m. However, when sampled at the Nyquist rate, $g_a(nT) = 0$ for $-\infty < n < \infty$, and as a result, $g_a(t)$ cannot be recovered from its samples.

If the sampling frequency is higher than the Nyquist rate, the sampling operation is referred to as *oversampling*. On the other hand, if the sampling frequency is lower than the Nyquist rate, it is called *undersampling*. Finally, if the sampling frequency is exactly equal to the Nyquist rate, it is called *critical sampling*.

Typical sampling rates used in practice are, for example, 8 kHz in digital telephony and 44.1 kHz in compact disk (CD) music systems. In the former case, a 3.4-kHz signal bandwidth is adequate for telephone conversation. Hence, a sampling rate of 8 kHz, which is greater than 2 times that of the acceptable signal bandwidth of 3.4 kHz, is used. In high-quality analog music signal processing, on the other hand, a bandwidth of about 20 kHz is required to preserve the fidelity of the most critical sections of the music. As a result, the analog music signal is sampled at a rate of 44.1 kHz, that is slightly higher than twice the highest frequency, to ensure negligible aliasing distortion.

EXAMPLE 4.1   Effect of Sampling of Analog Sinusoidal Signals of Different Frequencies

Consider the three pure sinusoidal signals of Example 2.11: $g_1(t) = \cos(6\pi t)$, $g_2(t) = \cos(14\pi t)$, and $g_3(t) = \cos(26\pi t)$. The corresponding continuous-time Fourier transforms consist of two impulses each and are given by

$$G_1(j\Omega) = \pi[\delta(\Omega - 6\pi) + \delta(\Omega + 6\pi)],$$
$$G_2(j\Omega) = \pi[\delta(\Omega - 14\pi) + \delta(\Omega + 14\pi)],$$
$$G_3(j\Omega) = \pi[\delta(\Omega - 26\pi) + \delta(\Omega + 26\pi)].$$

These transforms are plotted in Figures 4.6(a) to (c), respectively.

The above signals are sampled at a rate of $T = 0.1$ sec, that is, with a sampling frequency $\Omega_T = 20\pi$ rad/sec, generating the continuous-time impulse trains $g_{1p}(t)$, $g_{2p}(t)$, and $g_{3p}(t)$. From Eq. (4.10), their corresponding continuous-time Fourier transforms are given by

$$G_{\ell p}(j\Omega) = \frac{1}{T} \sum_{k=-\infty}^{\infty} G_\ell(j\Omega - jk\Omega_T), \qquad \ell = 1, 2, 3,$$

which have been plotted in Figures 4.6(d) to (f). These figures also indicate by dotted lines the frequency response of an ideal lowpass filter with a cutoff at $\Omega_c = \Omega_T/2 = 10\pi$ and a gain of $T = 0.1$ (not shown to scale). The continuous-time Fourier transforms of the output of the lowpass filter in each case are also indicated in Figures 4.6(d) to (f). Note that in the case of $g_1(t)$, the sampling rate satisfies the Nyquist condition and there is no aliasing. The continuous-time reconstructed output is precisely the original continuous-time signal $g_1(t)$. However, in the other two cases, the sampling rate does not satisfy the Nyquist condition, resulting in aliasing, and the outputs here are all equal to the aliased signal $\cos(6\pi t)$. Note that in Figure 4.6(e), the impulse appearing at $\Omega = 6\pi$ in the positive frequency passband of the lowpass filter results from the aliasing of the impulse in $G_2(j\Omega)$ at $\Omega = -14\pi$. On the other hand, the impulse appearing at $\Omega = 6\pi$ in Figure 4.6(f) has resulted from the aliasing of the impulse in $G_3(j\Omega)$ at $\Omega = 26\pi$.

We now establish the relation between the discrete-time Fourier transform $G(e^{j\omega})$ of the sequence $g[n]$ and the continuous-time Fourier transform $G_a(j\Omega)$ of the analog signal $g_a(t)$. If we compare Eq. (4.3) with Eq. (4.6) and make use of Eq. (4.1), we observe that

$$G(e^{j\omega}) = G_p(j\Omega)\big|_{\Omega=\omega/T}, \tag{4.14a}$$

or equivalently,

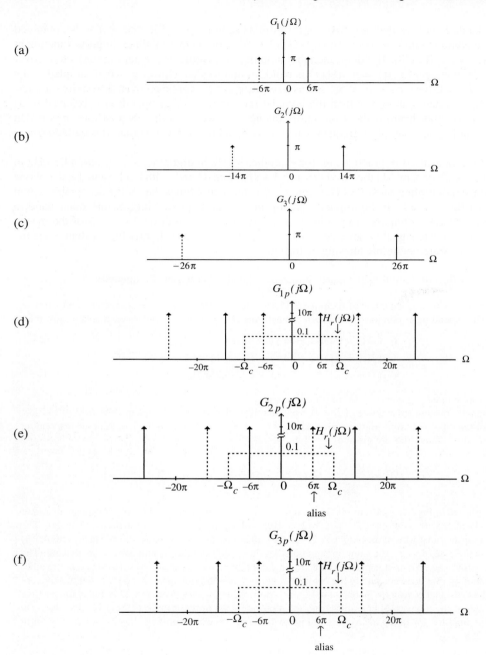

Figure 4.6: Effect of sampling on a pure cosine signal: (a) spectrum of $\cos(6\pi t)$, (b) spectrum of $\cos(14\pi t)$, (c) spectrum of $\cos(26\pi t)$, (d) spectrum of sampled version of $\cos(6\pi t)$ with $\Omega_T = 20\pi > 2\Omega_m = 12\pi$, (e) spectrum of sampled version of $\cos(14\pi t)$ with $\Omega_T = 20\pi < 2\Omega_m = 28\pi$, and (f) spectrum of sampled version of $\cos(26\pi t)$ with $\Omega_T = 20\pi < 2\Omega_m = 52\pi$.

$$G_p(j\Omega) = G(e^{j\omega})\Big|_{\omega=\Omega T}. \tag{4.14b}$$

Therefore, from the above and Eq. (4.10), we arrive at the desired result given by

$$G(e^{j\omega}) = \frac{1}{T} \sum_{k=-\infty}^{\infty} G_a(j\Omega - jk\Omega_T)\Big|_{\Omega=\omega/T}$$

$$= \frac{1}{T} \sum_{k=-\infty}^{\infty} G_a\left(j\frac{\omega}{T} - jk\Omega_T\right)$$

$$= \frac{1}{T} \sum_{k=-\infty}^{\infty} G_a\left(j\frac{\omega}{T} - j\frac{2\pi k}{T}\right), \tag{4.15a}$$

which can be alternatively expressed as

$$G(e^{j\Omega T}) = \frac{1}{T} \sum_{k=-\infty}^{\infty} G_a(j\Omega - jk\Omega_T). \tag{4.15b}$$

As can be seen from Eq. (4.14a) or (4.14b), $G(e^{j\omega})$ is obtained from $G_p(j\Omega)$ simply by scaling the frequency axis Ω according to the relation

$$\Omega = \frac{\omega}{T}. \tag{4.16}$$

Now, the continuous-time Fourier transform $G_p(j\Omega)$ is a periodic function of Ω with a period $\Omega_T = 2\pi/T$. Because of the above normalization, the discrete-time Fourier transform $G(e^{j\omega})$ is a periodic function of ω with a period 2π.

The effect of sampling in the frequency domain can be investigated using MATLAB. To this end, we use an exponentially decaying continuous-time signal with a continuous-time Fourier transform that is approximately band-limited as illustrated in Example 4.2.

EXAMPLE 4.2   Sampling of a Continuous-Time Signal at Two Different Rates

We consider the sampling of the continuous-time signal

$$x_a(t) = \begin{cases} 2t\,e^{-t}, & t \geq 0, \\ 0, & t < 0, \end{cases} \tag{4.17}$$

for two different values of the sampling period T to generate the discrete-time signal $x[n]$ and compare its continuous-time Fourier transform $X_a(j\Omega)$ with the discrete-time Fourier transform $TX(e^{j\omega})$ of $x[n]$. Plots of $x_a(t)$ and its continuous-time Fourier transform $X_a(j\Omega)$ are shown in Figure 4.7. The signal $x_a(t)$ is seen to be reasonably band-limited to about 1 Hz.

The continuous-time signal of Eq. (4.17) is then sampled at a rate of 2 Hz, which is twice the approximate bandwidth of $x_a(t)$ generating the discrete-time signal $x[n]$ depicted in Figure 4.8, along with its discrete-time Fourier transform $X(e^{j\omega})$ scaled by T. As expected, $T\,X(e^{j\omega})$ in the baseband is a reasonably good replica of $X_a(j\Omega)$, indicating that the effect of aliasing in the frequency-domain is minimal.

Next, we sample the continuous-time signal of Eq. (4.17) at a rate of 2/3 Hz, resulting in the discrete-time signal $x[n]$ shown in Figure 4.9, along with its discrete-time Fourier transform scaled by T. In this case, since $x_a(t)$ is being sampled at less than twice the approximate bandwidth of 1 Hz, there is a considerable amount of aliasing, and $TX(e^{j\omega})$ is no longer a replica of $X_a(j\Omega)$.

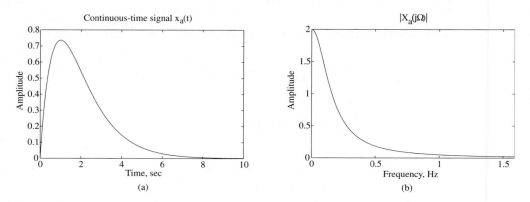

Figure 4.7: (a) The continuous-time signal of Eq. (4.17) and (b) its continuous-time Fourier transform.

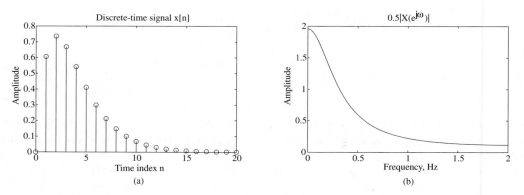

Figure 4.8: (a) The discrete-time signal obtained by sampling the signal of Eq. (4.17) at 2 kHz rate and (b) its discrete-time Fourier transform.

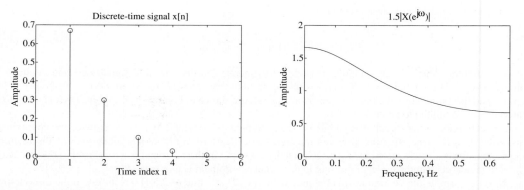

Figure 4.9: (a) The discrete-time signal obtained by sampling the signal of Eq. (4.17) at 2/3 kHz rate and (b) its discrete-time Fourier transform.

4.2.2 Recovery of the Analog Signal

We indicated earlier that if the discrete-time sequence $g[n]$ has been obtained by uniformly sampling a band-limited continuous-time signal $g_a(t)$ with a highest frequency Ω_m at a rate $\Omega_T = 2\pi/T$ satisfying the condition of Eq. (4.11), then the original continuous-time signal $g_a(t)$ can be fully recovered by passing the equivalent impulse train $g_p(t)$ through an ideal lowpass filter $H_r(j\Omega)$ with a cutoff at Ω_c, satisfying Eq. (4.13) and with a gain of T. We next derive the expression for the output $\hat{g}_a(t)$ of the ideal lowpass filter as a function of the samples $g[n]$.

Now, the impulse response $h_r(t)$ of the above ideal lowpass filter is obtained simply by taking the inverse continuous-time Fourier transform of its frequency response $H_r(j\Omega)$:

$$H_r(j\Omega) = \begin{cases} T, & |\Omega| \leq \Omega_c, \\ 0, & |\Omega| > \Omega_c, \end{cases} \tag{4.18}$$

and is given by

$$h_r(t) = \frac{1}{2\pi} \int_{-\infty}^{\infty} H_r(j\Omega)e^{j\Omega t}\, d\Omega = \frac{T}{2\pi} \int_{-\Omega_c}^{\Omega_c} e^{j\Omega t}\, d\Omega$$

$$= \frac{\sin(\Omega_c t)}{\Omega_T t/2}, \quad -\infty < t < \infty. \tag{4.19}$$

Observe that the impulse train $g_p(t)$ is given by

$$g_p(t) = \sum_{n=-\infty}^{\infty} g[n]\delta(t - nT). \tag{4.20}$$

Therefore, the output $\hat{g}_a(t)$ of the ideal lowpass filter is given by the convolution of $g_p(t)$ with the impulse response $h_r(t)$ of the analog reconstruction filter:

$$\hat{g}_a(t) = \sum_{n=-\infty}^{\infty} g[n]h_r(t - nT). \tag{4.21}$$

Substituting $h_r(t)$ from Eq. (4.19) in Eq. (4.21) and assuming for simplicity $\Omega_c = \Omega_T/2 = \pi/T$, we arrive at

$$\hat{g}_a(t) = \sum_{n=-\infty}^{\infty} g[n]\frac{\sin[\pi(t - nT)/T]}{\pi(t - nT)/T}. \tag{4.22}$$

The above expression indicates that the reconstructed continuous-time signal $\hat{g}_a(t)$ is obtained by shifting in time the impulse response of the lowpass filter $h_r(t)$ by an amount nT and scaling it in amplitude by the factor $g[n]$ for all integer values of n in the range $-\infty < n < \infty$ and then summing up all shifted versions. The ideal band-limited interpolation process is illustrated in Figure 4.10.

Now, it can be shown that when $\Omega_c = \Omega_T/2$ in Eq. (4.19), $h_r(0) = 1$ and $h_r(nT) = 0$ for $n \neq 0$ (Problem 4.10). As a result, from Eq. (4.22), $\hat{g}_a(rT) = g[r] = g_a(rT)$ for all integer values of r in the range $-\infty < r < \infty$, whether or not the condition of the sampling theorem of Eq. (4.11) has been satisfied. However, $\hat{g}_a(t) = g_a(t)$ for all values of t only if the sampling frequency Ω_T satisfies the condition of the sampling theorem.

It should be noted that the ideal analog lowpass filter of Eq. (4.19) has a doubly infinite length impulse response and, thus, is unstable and noncausal and does not have a rational transfer function, making it unrealizable. An analog lowpass filter is also needed to band-limit the continuous-time signal before it

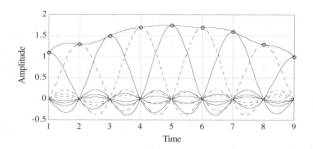

Figure 4.10: The ideal band-limited reconstruction by interpolation.

is sampled to ensure that the conditions of the sampling theorem are satisfied. The magnitude response specification for both the anti-aliasing lowpass filter and the analog reconstruction filter therefore must be modified to make them realizable. We review in Section 4.4 some commonly used methods for the determination of the transfer function of realizable and stable analog lowpass filters approximating the ideal magnitude characteristic of Eq. (4.18). In Sections 4.6 and 4.10, we consider specifically the issues concerning the design of an anti-aliasing filter and the reconstruction filter, respectively.

4.2.3   Implications of the Sampling Process

Consider again the sampling of the three continuous-time signals of Example 4.1. From Figure 4.6(d), it should be apparent that from the sampled version $g_{1p}(t)$ of the continuous-time signal $g_1(t) = \cos(6\pi t)$, we can also recover any of its frequency-translated versions $\cos[(20k \pm 6)\pi t]$ outside the baseband by passing $g_{1p}(t)$ through an ideal analog bandpass filter with a passband centered at $\Omega = (20k \pm 6)\pi$. For example, to recover the signal $\cos(34\pi t)$, it will be necessary to employ a bandpass filter with a frequency response

$$H_r(j\Omega) = \begin{cases} 0.1, & (34 - \Delta)\pi \leq |\Omega| \leq (34 + \Delta)\pi, \\ 0, & \text{otherwise,} \end{cases} \tag{4.23}$$

where Δ is an incrementally small number. Likewise, we can recover the aliased baseband component $\cos(6\pi t)$ from the sampled version of either $g_{2p}(t)$ or $g_{3p}(t)$ by passing it through an ideal lowpass filter:

$$H_r(j\Omega) = \begin{cases} 0.1, & (6 - \Delta)\pi \leq |\Omega| \leq (6 + \Delta)\pi, \\ 0, & \text{otherwise.} \end{cases} \tag{4.24}$$

There is no *aliasing distortion* unless the original continuous-time signal also contains the component $\cos(6\pi t)$. Similarly, from the sampled versions of either $g_{2p}(t)$ or $g_{3p}(t)$, we can recover any one of the frequency-translated versions, including the original continuous-time signal, $\cos(14\pi t)$ or $\cos(26\pi t)$ as the case may be, by employing suitable ideal bandpass filters.

The frequency-translation effect of sampling is further illustrated in Examples 4.3 and 4.4.

EXAMPLE 4.3   Illustration of Frequency-Translation Effect of Sampling

Figure 4.11(a) shows the continuous-time sinusoidal signal $g_1(t) = \cos(6\pi t)$ of frequency 3 Hz. The impulse train $g_{1p}(t)$ generated from $g_1(t)$ by uniformly sampling it at a rate of 10 Hz is depicted in Figure 4.11(b). The continuous-time output $y(t)$ by passing $g_{1p}(t)$ through an analog bandpass reconstruction filter with a passband from 5 Hz to 10 Hz is shown in Figure 4.11(c), which is seen to be a continuous-time sinusoidal signal $\cos(14\pi t)$ of frequency 7 Hz. In general, the continuous-time output of an analog bandpass reconstruction filter with a passband from $5k$ Hz to $5k + 5$ Hz, with k a positive integer, will be a sinusoidal signal of frequency of $5k + 3$ for k even and $5(k + 1) - 3$ for k odd.

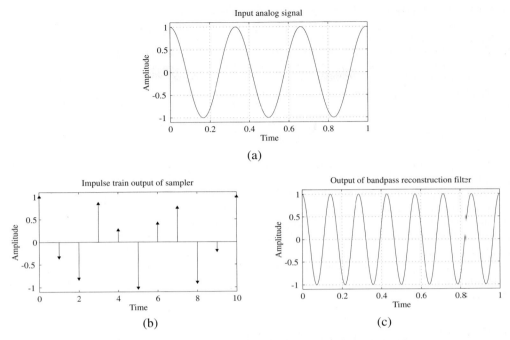

Figure 4.11: Illustration of frequency-translation effect of sampling.

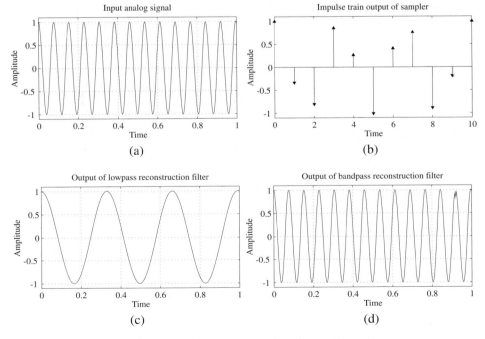

Figure 4.12: Illustration of the effect of undersampling.

EXAMPLE 4.4   Illustration of the Effect of Undersampling

The continuous-time sinusoidal signal $g_3(t) = \cos(26\pi t)$ of frequency 13 Hz shown in Figure 4.12(a) is sampled uniformly at a 10-Hz rate, resulting in an impulse train $g_{3p}(t)$ sketched in Figure 4.12(b). The output generated from $g_{3p}(t)$ by passing it through an analog lowpass reconstruction filter with a passband edge at 5 Hz is a continuous-time sinusoidal signal of frequency 3 Hz, as shown in Figure 4.12(c). Note that the original high-frequency sinusoidal signal can be fully recovered from $g_{3p}(t)$ by passing it instead through a bandpass filter with a passband from 10 Hz to 15 Hz, as indicated in Figure 4.12(d).

Generalizing the above discussions, consider the continuous-time signals $g_1(t)$, $g_2(t)$, and $g_3(t)$ with band-limited frequency spectrums $G_1(j\Omega)$, $G_2(j\Omega)$, and $G_3(j\Omega)$, as shown in Figures 4.13(a) to (c), respectively. Each of these continuous-time signals, when sampled at a sampling frequency of Ω_T, develops a continuous-time signal $g_{\ell p}(t)$, $\ell = 1, 2, 3$, with an identical periodic frequency spectrum, as indicated in Figure 4.13(d). Therefore, by passing the sampled signal through an appropriate analog lowpass or bandpass filter of bandwidth greater than $\Omega_2 - \Omega_1$ but less than or equal to $\Omega_T/2$, we can recover either the original continuous-time signal or any one of its frequency-translated versions. Note that as long as the spectrum of the continuous-time signal being sampled at a sampling rate of Ω_T is band-limited to the frequency range $k\Omega_T/2 \leq |\Omega| \leq (k+1)\Omega_T/2$, there is no aliasing distortion due to sampling, and its frequency-translated version can always be recovered from the sampled signal by appropriate filtering. There will be aliasing distortion only if there are frequency components in a wider frequency range than that indicated.

4.3   Sampling of Bandpass Signals

The conditions developed in Section 4.2.1 for the unique representation of a continuous-time signal by the discrete-time signal obtained by uniform sampling assumed that the spectrum of the continuous-time signal is band-limited in the frequency range from dc to some frequency Ω_m. Such continuous-time signals are commonly referred to as *lowpass* signals. There are applications where the continuous-time signal is band-limited to a higher range $\Omega_L \leq |\Omega| \leq \Omega_H$, where $\Omega_L > 0$. Such a signal is usually referred to as a *bandpass* signal and is often obtained by modulating a lowpass signal. We can, of course, sample such a bandpass continuous-time signal with a sampling rate greater than twice the highest frequency, that is, by ensuring

$$\Omega_T \geq 2\Omega_H,$$

to prevent aliasing. However, in this case, due to the bandpass spectrum of the continuous-time signal, the spectrum of the discrete-time signal obtained by sampling will have spectral gaps with no signal components present in these gaps. Moreover, if Ω_H is very large, the sampling rate also has to be very large, which may not be practical in some situations.

We outline next a more practical and efficient approach [Por97]. Let $\Delta\Omega = \Omega_H - \Omega_L$ define the *bandwidth* of the bandpass signal. Assume first that the highest frequency Ω_H contained in the signal is an integer multiple of the bandwidth; that is,

$$\Omega_H = M(\Delta\Omega).$$

We choose the sampling frequency Ω_T to satisfy the condition

$$\Omega_T = 2(\Delta\Omega) = \frac{2\Omega_H}{M}, \tag{4.25}$$

which is smaller than $2\Omega_H$, the Nyquist rate. Substituting Eq. (4.25) in Eq. (4.9) we arrive at the expression for the Fourier transform $G_p(j\Omega)$ of the impulse-sampled signal $g_p(t)$:

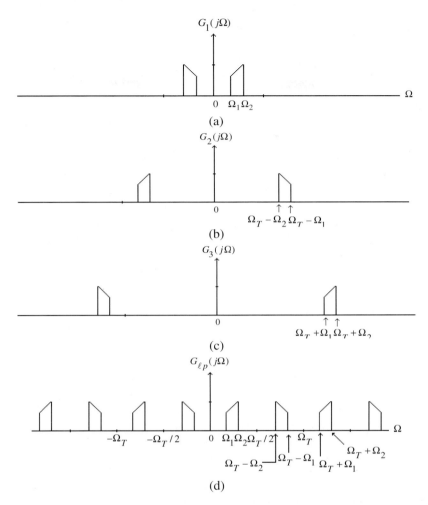

Figure 4.13: Further illustration of the effect of sampling.

$$G_p(j\Omega) = \frac{1}{T} \sum_{k=-\infty}^{\infty} G_a\left(j(\Omega - 2k(\Delta\Omega))\right). \tag{4.26}$$

As before, $G_p(j\Omega)$ consists of a sum of the original Fourier transform $G_a(j\Omega)$ and replicas of $G_a(j\Omega)$ shifted by integer multiples of twice the bandwidth $\Delta\Omega$, and then scaled by $1/T$. The amount of the shift for each value of k ensures that there will be no overlap between all shifted replicas and, hence, no aliasing. Figure 4.14 shows the spectrum of the original continuous-time signal $g_a(t)$ and that of the sampled version $g_p(t)$, sampled at the rate given by Eq. (4.25) for $M = 4$. As can be seen from this figure, $g_a(t)$ can be recovered from $g_p(t)$ by passing the latter through an ideal bandpass filter with a passband given by $\Omega_L \le |\Omega| \le \Omega_H$ and a gain of T.

Note that any of the replicas in the lower frequency bands can be retained by passing $g_p(t)$ through bandpass filters with passbands $\Omega_L - k(\Delta\Omega) \le |\Omega| \le \Omega_H - k(\Delta\Omega)$, $1 \le k \le M - 1$, providing a

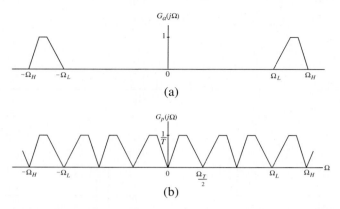

Figure 4.14: Illustration of the effect in the frequency-domain of sampling below the Nyquist rate. (a) Spectrum of a bandpass signal with highest frequency that is an integer multiple of its bandwidth and (b) spectrum of the bandpass signal sampled at a rate that is twice the bandwidth of the original signal.

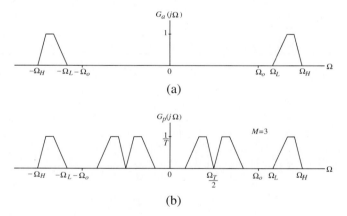

Figure 4.15: Illustration of the effect in the frequency-domain of sampling below the Nyquist rate. (a) Spectrum of a bandpass signal with highest frequency that is not an integer multiple of its bandwidth and (b) spectrum of the bandpass signal sampled at a rate that is slightly higher than twice the bandwidth of the original signal.

translation of the original bandpass signal to lower frequency ranges. If the bandpass signal has been obtained by modulating a lowpass signal, then the latter can be recovered by passing $g_p(t)$ through a lowpass filter with passband $0 \leq |\Omega| \leq \Delta\Omega$, which retains the replica in the baseband. This approach is often employed in digital radio receivers.

If Ω_H is not an integer multiple of the bandwidth $\Omega_H - \Omega_L$, we can artificially extend the bandwidth either to the right or to the left so that the highest frequency contained in the bandpass signal is an integer multiple of the extended bandwidth. For example, if we extend the bandwidth to the left by assuming the lowest frequency contained in the bandpass signal to be Ω_o, then Ω_o is chosen such that the extended bandwidth $\Omega_H - \Omega_o$ is an integer multiple of Ω_H. In both cases, the spectrum of the sampled signal obtained by sampling $g_a(t)$ will have small spectral gaps between the replicas. This is illustrated in Figure 4.15 when the bandwidth is extended to the left and M is chosen as 3.

As in the previous case, any of the replicas in the lower frequency bands can be retained by passing $g_p(t)$ through appropriate bandpass filters.

4.4   Analog Lowpass Filter Design

There are a number of established approximation techniques for the design of analog lowpass filters [Vla69], [Tem73], [Tem77]. We describe four widely used design techniques here without their detailed derivations. Further details of these methods can be found in texts on analog filter design. Extensive tables for the design of analog lowpass filters are also available [Chr66], [Skw65], [Zve67]. As indicated earlier, a commonly used technique for the design of IIR digital filters is based on the conversion of a prototype analog transfer function that has been designed employing one of the methods discussed here. The IIR digital filter design techniques are the subject of discussion in Chapter 9.

4.4.1   Filter Specifications

Both the anti-aliasing filter and the reconstruction filter of Figure 4.1 are of the lowpass type, and ideally they should have a magnitude response of the form shown in Figure 4.5. In practice, the magnitude response characteristics in the passband and in the stopband cannot be constant and are therefore specified with some acceptable tolerances. Moreover, a transition band is specified between the passband and the stopband to permit the magnitude to drop off smoothly. For example, the magnitude $|H_a(j\Omega)|$ of a lowpass filter may be given as shown in Figure 4.16. As indicated in the figure, in the *passband* defined by $0 \leq \Omega \leq \Omega_p$, we require

$$1 - \delta_p \leq |H_a(j\Omega)| \leq 1 + \delta_p, \qquad \text{for } |\Omega| \leq \Omega_p, \tag{4.27}$$

or in other words, the magnitude approximates unity within an error of $\pm\delta_p$. In the *stopband*, defined by $\Omega_s \leq \Omega < \infty$, we require

$$|H_a(j\Omega)| \leq \delta_s, \qquad \text{for } \Omega_s \leq |\Omega| < \infty, \tag{4.28}$$

implying that the magnitude approximates zero within an error of δ_s. The frequencies Ω_p and Ω_s are, respectively, called the *passband edge frequency* and the *stopband edge frequency*.

The limits of the tolerances in the passband and stopband, δ_p and δ_s, are called *ripples*. Usually, these ripples are specified in dB in terms of the *peak passband ripple* α_p and the *minimum stopband attenuation* α_s, defined by

$$\alpha_p = -20 \log_{10}(1 - \delta_p) \, \text{dB}, \tag{4.29}$$

$$\alpha_s = -20 \log_{10}(\delta_s) \, \text{dB}. \tag{4.30}$$

Often, the filter specifications are given in terms of the *loss function* or *attenuation function* $a(\Omega)$ in dB, which is defined as the negative of the gain in dB; that is, $-20 \log_{10} |H_a(j\Omega)|$.

EXAMPLE 4.5   Passband and Stopband Ripple Computation

Let the desired peak passband ripple of a lowpass filter be 0.01 dB, and the minimum attenuation in the stopband be 70 dB. Determine δ_p and δ_s. Substituting $\alpha_p = 0.01$ in Eq. (4.29) and solving, we get

$$\delta_p = 1 - 10^{-\alpha_p/20} = 0.00115.$$

Likewise, substituting $\alpha_s = 70$ in Eq. (4.30), we obtain $\delta_s = 10^{-\alpha_s/20} = 0.0003162.$

The magnitude response specifications for an analog lowpass filter, in some applications, are given in a normalized form, as indicated in Figure 4.17. Here the maximum value of the magnitude in the passband is assumed to be unity, and the passband ripple, denoted as $1/\sqrt{1 + \varepsilon^2}$, is given by the minimum value of the magnitude in the passband. The maximum passband gain or the minimum passband loss is therefore

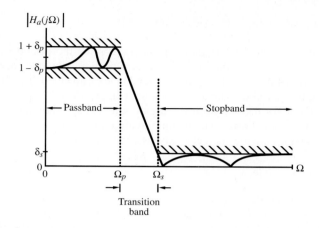

Figure 4.16: Typical magnitude specifications for an analog lowpass filter.

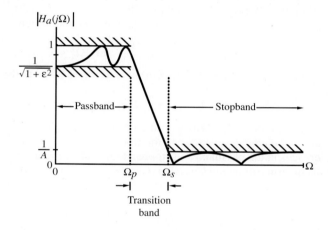

Figure 4.17: Normalized magnitude specifications for an analog lowpass filter.

0 dB. The maximum stopband ripple is denoted by $1/A$, and the minimum stopband attenuation is therefore given by $-20 \log_{10}(1/A)$.

In analog filter theory, two additional parameters are defined. The first one, called the *transition ratio* or *selectivity parameter,* is defined by the ratio of the passband edge frequency Ω_p and the stopband edge frequency Ω_s and is usually denoted by k; that is,

$$k = \frac{\Omega_p}{\Omega_s}. \tag{4.31}$$

Note that for a lowpass filter, $k < 1$. The second one, called the *discrimination parameter* and denoted by k_1, is defined as

$$k_1 = \frac{\varepsilon}{\sqrt{A^2 - 1}}. \tag{4.32}$$

Usually, $k_1 \ll 1$.

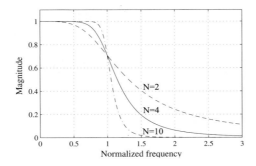

Figure 4.18: Typical Butterworth lowpass filter responses.

4.4.2 Butterworth Approximation

The magnitude-squared response of an analog lowpass Butterworth filter $H_a(s)$ of Nth order is given by

$$|H_a(j\Omega)|^2 = \frac{1}{1 + (\Omega/\Omega_c)^{2N}}. \tag{4.33}$$

It can be easily shown that the first $2N - 1$ derivatives of $|H_a(j\Omega)|^2$ at $\Omega = 0$ are equal to zero, and as a result, the Butterworth lowpass filter is said to have a *maximally flat magnitude* at $\Omega = 0$. The gain of the Butterworth filter in dB is given by

$$\mathcal{G}(\Omega) = 10\log_{10}|H_a(j\Omega)|^2 \text{ dB.}$$

At dc, that is, at $\Omega = 0$, the gain in dB is equal to zero, and at $\Omega = \Omega_c$, the gain is

$$\mathcal{G}(\Omega_c) = 10\log_{10}\left(\tfrac{1}{2}\right) = -3.0103 \cong -3\text{ dB,}$$

and, therefore, Ω_c is often called the *3-dB cutoff frequency*. Since the derivative of the squared-magnitude response or, equivalently, of the magnitude response is always negative for positive values of Ω, the magnitude response is monotonically decreasing with increasing Ω. For $\Omega \gg \Omega_c$, the squared-magnitude function can be approximated by

$$|H_a(j\Omega)|^2 \approx \frac{1}{(\Omega/\Omega_c)^{2N}}.$$

The gain $\mathcal{G}(\Omega_2)$ in dB at $\Omega_2 = 2\Omega_1$ with $\Omega_1 \gg \Omega_c$ is given by

$$\mathcal{G}(\Omega_2) = -10\log_{10}\left(\frac{\Omega_2}{\Omega_c}\right)^{2N} = -10\log_{10}\left(\frac{2\Omega_1}{\Omega_c}\right)^{2N} = \mathcal{G}(\Omega_1) - 6N \text{ dB,}$$

where $\mathcal{G}(\Omega_1)$ is the gain in dB at Ω_1. As a result, the gain in the stopband decreases by 6 dB per octave or, equivalently, by 20 dB per decade for an increase of the filter order by one. In other words, the passband and the stopband behaviors of the magnitude response improve with a corresponding decrease in the transition band as the filter order N increases. A plot of the magnitude response of the normalized Butterworth lowpass filter with $\Omega_c = 1$ for some typical values of N is shown in Figure 4.18.

The two parameters completely characterizing a Butterworth filter are therefore the 3-dB cutoff frequency Ω_c and the order N. These are determined from the specified passband edge Ω_p, the minimum passband magnitude $1/\sqrt{1+\varepsilon^2}$, the stopband edge Ω_s, and the maximum stopband ripple $1/A$. From Eq. (4.33), we get

$$|H_a(j\Omega_p)|^2 = \frac{1}{1 + (\Omega_p/\Omega_c)^{2N}} = \frac{1}{1+\varepsilon^2}, \tag{4.34a}$$

$$|H_a(j\Omega_s)|^2 = \frac{1}{1 + (\Omega_s/\Omega_c)^{2N}} = \frac{1}{A^2}. \tag{4.34b}$$

Solving the above, we arrive at the expression for the order N as

$$N = \frac{1}{2} \frac{\log_{10}\left[\left(A^2 - 1\right)/\varepsilon^2\right]}{\log_{10}\left(\Omega_s/\Omega_p\right)} = \frac{\log_{10}(1/k_1)}{\log_{10}(1/k)}. \tag{4.35}$$

Since the order N of the filter must be an integer, the value of N computed using the above expression is rounded up to the next higher integer. This value of N can be used next in either Eq. (4.34a) or (4.34b) to solve for the 3-dB cutoff frequency Ω_c. If Eq. (4.34a) is used to solve for Ω_c, then the passband specification at Ω_p is met exactly, while the stopband specification at Ω_s is exceeded. On the other hand, if Ω_c is determined using Eq. (4.34b), then the stopband specification at Ω_s is satisfied exactly, while the passband specification is exceeded, providing a safety margin at Ω_p [Tem77]. However, it is preferable to use the latter since this ensures the smallest ripple in the passband or, in other words, the smallest amplitude distortion to the signal being filtered in its band of interest.

The expression for the transfer function of the Butterworth lowpass filter is given by

$$H_a(s) = \frac{C}{D_N(s)} = \frac{\Omega_c^N}{s^N + \sum_{\ell=0}^{N-1} d_\ell s^\ell} = \frac{\Omega_c^N}{\prod_{\ell=1}^{N}(s - p_\ell)}, \tag{4.36}$$

where

$$p_\ell = \Omega_c e^{j[\pi(N+2\ell-1)/2N]}, \qquad \ell = 1, 2, \ldots, N. \tag{4.37}$$

The denominator $D_N(s)$ of Eq. (4.36) is known as the *Butterworth polynomial* of order N and is easy to compute. These polynomials have been tabulated for easy design reference [Chr66], [Skw65], [Zve67]. The analog lowpass Butterworth filters can be readily designed using MATLAB (see Section 4.4.6).

EXAMPLE 4.6   Order of an Analog Maximally Flat Lowpass Filter

Determine the lowest order of a transfer function $H_a(s)$ having a maximally flat lowpass characteristic with a 1-dB cutoff frequency at 1 kHz and a minimum attenuation of 40 dB at 5 kHz.

We first determine ε and A. From Eq. (4.34a),

$$10 \log_{10}\left(\frac{1}{1+\varepsilon^2}\right) = -1,$$

which yields $\varepsilon^2 = 0.25895$. Likewise, from Eq. (4.34b),

$$10 \log_{10}\left(\frac{1}{A^2}\right) = -40,$$

which leads to $A^2 = 10,000$. Substituting the values in Eq. (4.32), we get

$$1/k_1 = 196.51334.$$

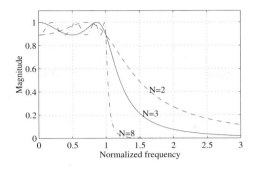

Figure 4.19: Typical Type 1 Chebyshev lowpass filter responses with 1-dB passband ripple.

The inverse transition ratio here is $1/k = 5000/1000 = 5$. Substituting these values in Eq. (4.35), we get

$$N = \frac{\log_{10}(1/k_1)}{\log_{10}(1/k)} = \frac{\log_{10}(196.51334)}{\log_{10}(5)} = 3.2811022. \tag{4.38}$$

Since the order of the transfer function must be an integer, we round the above up to the nearest integer $N = 4$.

4.4.3 Chebyshev Approximation

In this case, the approximation error, defined as the difference between the ideal brickwall characteristic and the actual response, is minimized over a prescribed band of frequencies. In fact, the magnitude error is equiripple in the band. There are two types of *Chebyshev transfer functions*. In the *Type 1* approximation, the magnitude characteristic is equiripple in the passband and monotonic in the stopband, whereas in the *Type 2* approximation, the magnitude response is monotonic in the passband and equiripple in the stopband.

Type 1 Chebyshev Approximation

The magnitude-squared response of the analog lowpass Type 1 Chebyshev filter $H_a(s)$ of Nth order is given by

$$|H_a(j\Omega)|^2 = \frac{1}{1 + \varepsilon^2 T_N^2(\Omega/\Omega_p)}, \tag{4.39}$$

where $T_N(\Omega)$ is the Chebyshev polynomial of order N:

$$T_N(\Omega) = \begin{cases} \cos(N\cos^{-1}\Omega), & |\Omega| \leq 1, \\ \cosh(N\cosh^{-1}\Omega), & |\Omega| > 1. \end{cases} \tag{4.40}$$

The above polynomial can also be derived via a recurrence relation given by

$$T_r(\Omega) = 2\Omega T_{r-1}(\Omega) - T_{r-2}(\Omega), \qquad r \geq 2, \tag{4.41}$$

with $T_0(\Omega) = 1$ and $T_1(\Omega) = \Omega$.

Typical plots of the magnitude responses of the Type 1 Chebyshev lowpass filter are shown in Figure 4.19 for three different values of filter order N with the same passband ripple ε. From these plots, it is seen that the squared-magnitude response is equiripple between $\Omega = 0$ and $\Omega = 1$, and it decreases monotonically for all $\Omega > 1$.

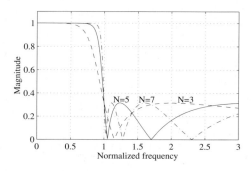

Figure 4.20: Typical Type 2 Chebyshev lowpass filter responses with 10-dB minimum stopband attenuation.

The order N of the transfer function is determined from the attenuation specification in the stopband at a particular frequency. For example, if at $\Omega = \Omega_s$, the magnitude is equal to $1/A$, then from Eqs. (4.39) and (4.40),

$$|H_a(j\Omega_s)|^2 = \frac{1}{1 + \varepsilon^2 T_N^2(\Omega_s/\Omega_p)}$$

$$= \frac{1}{1 + \varepsilon^2 \left\{ \cosh[N \cosh^{-1}(\Omega_s/\Omega_p)] \right\}^2} = \frac{1}{A^2}. \tag{4.42}$$

Solving the above, we get

$$N = \frac{\cosh^{-1}(\sqrt{A^2 - 1}/\varepsilon)}{\cosh^{-1}(\Omega_s/\Omega_p)} = \frac{\cosh^{-1}(1/k_1)}{\cosh^{-1}(1/k)}. \tag{4.43}$$

In computing N using the above expression, it is usually convenient to use the identity $\cosh^{-1}(y) = \ln\left(y + \sqrt{y^2 - 1}\right)$. As in the case of the Butterworth filter, the order N of the filter is chosen as the nearest integer greater than or equal to the number given by Eq. (4.43).

The transfer function $H_a(s)$ is again of the form of Eq. (4.36), with

$$p_\ell = \sigma_\ell + j\Omega_\ell, \qquad \ell = 1, 2, \dots, N, \tag{4.44}$$

where

$$\sigma_\ell = -\Omega_p \xi \sin\left[\frac{(2\ell - 1)\pi}{2N}\right], \qquad \Omega_\ell = \Omega_p \zeta \cos\left[\frac{(2\ell - 1)\pi}{2N}\right], \tag{4.45a}$$

$$\xi = \frac{\gamma^2 - 1}{2\gamma}, \qquad \zeta = \frac{\gamma^2 + 1}{2\gamma}, \qquad \gamma = \left(\frac{1 + \sqrt{1 + \varepsilon^2}}{\varepsilon}\right)^{1/N}. \tag{4.45b}$$

Type 2 Chebyshev Approximation

The magnitude-squared response of the the analog lowpass Type 2 Chebyshev filter, also known as the *inverse Chebyshev approximation,* exhibits a monotonic behavior in the passband with a maximally flat response at $\Omega = 0$ and an equiripple behavior in the stopband. The squared-magnitude response expression here is given by

$$|H_a(j\Omega)|^2 = \frac{1}{1 + \varepsilon^2 \left[\frac{T_N(\Omega_s/\Omega_p)}{T_N(\Omega_s/\Omega)}\right]^2}. \tag{4.46}$$

Typical responses are as shown in Figure 4.20. The transfer function of a Type 2 Chebyshev lowpass filter is no longer an all-pole function and has both poles and zeros. If we write

$$H_a(s) = C_0 \frac{\prod_{\ell=1}^{N}(s - z_\ell)}{\prod_{\ell=1}^{N}(s - p_\ell)}, \tag{4.47}$$

the zeros are on the $j\Omega$-axis and are given by

$$z_\ell = j\frac{\Omega_s}{\cos\left[\frac{(2\ell-1)\pi}{2N}\right]}, \qquad \ell = 1, 2, \ldots, N. \tag{4.48}$$

If N is odd, then for $\ell = (N+1)/2$, the zero is at $s = \infty$. The poles are located at

$$p_\ell = \sigma_\ell + j\Omega_\ell, \quad \ell = 1, 2, \ldots, N, \tag{4.49}$$

where

$$\sigma_\ell = \frac{\Omega_s \alpha_\ell}{\alpha_\ell^2 + \beta_\ell^2}, \qquad \Omega_\ell = -\frac{\Omega_s \beta_\ell}{\alpha_\ell^2 + \beta_\ell^2}, \tag{4.50a}$$

$$\alpha_\ell = -\Omega_p \zeta \sin\left[\frac{(2\ell-1)\pi}{2N}\right], \qquad \beta_\ell = \Omega_p \xi \cos\left[\frac{(2\ell-1)\pi}{2N}\right], \tag{4.50b}$$

$$\zeta = \frac{\gamma^2 - 1}{2\gamma}, \qquad \xi = \frac{\gamma^2 + 1}{2\gamma}, \qquad \gamma = \left(A + \sqrt{A^2 - 1}\right)^{1/N}. \tag{4.50c}$$

The order N of the Type 2 Chebyshev lowpass filter is determined from given ε, Ω_s, and A, using Eq. (4.43).

EXAMPLE 4.7 **Order of an Analog Chebyshev Lowpass Filter**

We wish to determine the minimum order N required to design a lowpass filter with a Chebyshev or an inverse Chebyshev response with the specifications given in Example 4.6.

From Example 4.6, we have the following parameters: $1/k_1 = 196.51334$ and $1/k = 5$. Substituting these values in Eq. (4.43), we arrive at

$$N = \frac{\cosh^{-1}(1/k_1)}{\cosh^{-1}(1/k)} = \frac{\cosh^{-1}(196.51334)}{\cosh^{-1}(5)} \tag{4.51}$$

$$= 2.60591. \tag{4.52}$$

Since the order of the filter must be an integer, we choose the next highest integer value 3 for N. Note that the order of the Chebyshev lowpass filter is lower than that of a Butterworth lowpass filter meeting the same specifications as given by Eq. (4.38). This is usually the case for $N \geq 2$.

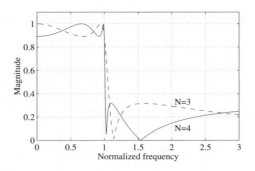

Figure 4.21: Typical elliptic lowpass filter responses with 1-dB passband ripple and 10-dB minimum stopband attenuation.

4.4.4 Elliptic Approximation

An *elliptic filter,* also known as a *Cauer filter,* has an equiripple passband and an equiripple stopband magnitude response, as indicated in Figure 4.21 for typical elliptic lowpass filters. The transfer function of an elliptic filter meets a given set of filter specifications, passband edge frequency Ω_p, stopband edge frequency Ω_s, passband ripple ε, and minimum stopband attenuation A, with the lowest filter order N. The theory of elliptic filter approximation is mathematically quite involved, and a detailed treatment of this topic is beyond the scope of this text. Interested readers are referred to the books by Antoniou [Ant93], Parks and Burrus [Par87], and Temes and LaPatra [Tem77].

The square-magnitude response of an elliptic lowpass filter is given by

$$|H_a(j\Omega)|^2 = \frac{1}{1 + \varepsilon^2 R_N^2(\Omega/\Omega_p)}, \tag{4.53}$$

where $R_N(\Omega)$ is a rational function of order N satisfying the property $R_N(1/\Omega) = 1/R_N(\Omega)$, with the roots of its numerator lying within the interval $0 < \Omega < 1$ and the roots of its denominator lying in the interval $1 < \Omega < \infty$. For most applications, the filter order meeting a given set of specifications of passband edge frequency Ω_p, passband ripple ε, stopband edge frequency Ω_s, and the minimum stopband ripple A can be estimated using the approximate formula [Ant93]

$$N \cong \frac{2 \log_{10}(4/k_1)}{\log_{10}(1/\rho)}, \tag{4.54}$$

where k_1 is the discrimination parameter defined in Eq. (4.32), and ρ is computed as follows:

$$k' = \sqrt{1 - k^2}, \tag{4.55a}$$

$$\rho_0 = \frac{1 - \sqrt{k'}}{2(1 + \sqrt{k'})}, \tag{4.55b}$$

$$\rho = \rho_0 + 2(\rho_0)^5 + 15(\rho_0)^9 + 150(\rho_0)^{13}. \tag{4.55c}$$

In Eq. (4.55a), k is the selectivity parameter defined in Eq. (4.31).

EXAMPLE 4.8   Order of an Analog Lowpass Elliptic Filter

We wish to determine the minimum order N required to design a lowpass elliptic filter with the specifications given in Example 4.6.

From Example 4.6, we note that $k = 1/5 = 0.2$ and $1/k_1 = 196.51334$. Substituting these values in Eqs. (4.55a) to (4.55c), we arrive at

$$k' = 0.979796, \quad \rho_0 = 0.00255135, \quad \rho = 0.0025513525.$$

Using the above values in Eq. (4.54), we get

$$N = 2.23308. \tag{4.56}$$

Rounding the above value to the nearest higher integer, we obtain $N = 3$ as the appropriate order for the elliptic lowpass filter.

4.4.5 Linear-Phase Approximation

The previous three approximation techniques are for developing analog lowpass transfer functions meeting specified magnitude or gain response specifications without any concern for their phase responses. In a number of applications, it is desirable that the analog lowpass filter being designed has a linear-phase characteristic in the passband, in addition to approximating the magnitude specifications. One way to achieve this goal is to cascade an analog allpass filter with the filter designed to meet the magnitude specifications so that the phase response of the overall cascade realization approximates linear-phase response in the passband. This approach increases the overall hardware complexity of the analog filter and may not be desirable for designing an analog anti-aliasing filter in some A/D conversion or designing an analog reconstruction filter in D/A conversion applications.

It is possible to design a lowpass filter that approximates a linear-phase characteristic in the passband. Such a filter has an all-pole transfer function of the form

$$H_a(s) = \frac{d_0}{B_N(s)} = \frac{d_0}{s^N + d_{N-1}s^{N-1} + \cdots + d_1 s + d_0} \tag{4.57}$$

and provides a maximally flat approximation to the linear-phase characteristic at $\Omega = 0$, that is, has a maximally flat constant group delay at dc ($\Omega = 0$). For a normalized group delay of unity at dc, the denominator polynomial $B_N(s)$ of the transfer function, called the *Bessel polynomial* can be derived via the recursion relation

$$B_N(s) = (2N - 1)B_{N-1}(s) + s^2 B_{N-2}(s), \tag{4.58}$$

starting with $B_1(s) = s + 1$ and $B_2(s) = s^2 + 3s + 3$. Alternatively, the coefficients of the Bessel polynomial $B_N(s)$ can be found from

$$d_\ell = \frac{(2N - \ell)!}{2^{N-\ell}\ell!(N - \ell)!}, \qquad \ell = 0, 1, \ldots, N - 1. \tag{4.59}$$

These filters are often referred to as *Bessel filters*. Figure 4.22 shows the phase responses of some typical Bessel filters. It should be noted that the Bessel filter has a poorer magnitude response than that of the lowpass filter of the same order designed using any one of the previous three techniques.

4.4.6 Analog Filter Design Using MATLAB

The *Signal Processing Toolbox* of MATLAB includes a number of M-files to directly develop analog transfer functions for each one of the above approximation techniques. We next review these functions.

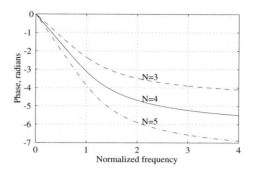

Figure 4.22: The phase responses of typical normalized Bessel lowpass filters.

Butterworth filter

The M-files for the design of analog Butterworth filters are

```
[z,p,k]    = buttap(N),
[num,den]  = butter(N,Wn,'s')
[num,den]  = butter(N,Wn,'type','s')
[N, Wn]    = buttord(Wp,Ws,Rp,Rs,'s')
```

The M-file `buttap(N)` computes the zeros, poles, and gain factor of the normalized analog Butterworth lowpass filter transfer function of order N with a 3-dB cutoff frequency of 1. The output files are the length N column vector p providing the locations of the poles, a null vector z for the zero locations, and the gain factor k. The form of the transfer function obtained is given by

$$H_a(s) = \frac{P_a(s)}{D_a(s)} = \frac{k}{(s - p(1))\,(s - p(2)) \cdots (s - p(N))}. \tag{4.60}$$

To determine the numerator and denominator coefficients of the transfer function from the computed zeros and poles, we need to use the M-file `zp2tf(z,p,k)`.[6]

Alternatively, we can use the M-file `butter(N,Wn,'s')` to design an order-N lowpass transfer function with a prescribed 3-dB cutoff frequency at Wn rad/sec, a nonzero number. The output data of this M-file are the numerator and the denominator coefficient vectors, num and den, respectively, in descending powers of s. If Wn is a two-element vector [W1,W2] with W1 < W2, the M-file generates an order-2N bandpass transfer function with 3-dB bandedge frequencies at W1 and W2, with both being nonzero numbers. To design an order-N highpass or an order-2N bandstop filter, the M-file `butter(N,Wn,'type','s')` is employed where type = high for a highpass filter with a 3-dB cutoff frequency at Wn or type = stop for a bandstop filter with 3-dB stopband edges given by a two-element vector of nonzero numbers Wn = [W1,W2] with W1 < W2.

The M-file `buttord(Wp,Ws,Rp,Rs,'s')` computes the lowest order N of a Butterworth analog transfer function meeting the specifications given by the filter parameters, Wp, Ws, Rp, and Rs, where Wp is the passband edge angular frequency in rad/sec, Ws is the stopband edge angular frequency in rad/sec, Rp is the maximum passband attenuation in dB, and Rs is the minimum stopband attenuation in dB. The output data are the filter order N and the 3-dB cutoff angular frequency Wn in rad/sec. This M-file can also be used to calculate the order of any one of the four basic types of analog Butterworth filters. For lowpass design, Wp < Ws, whereas for highpass design, Wp > Ws. For the other two types, Wp and Ws are two-element vectors specifying the passband and stopband edge frequencies.

---

[6]See Section 6.3.

Type 1 Chebyshev Filter

The M-files for the design of analog Type 1 Chebyshev filters are as follows:

```
[z,p,k]     = cheb1ap(N,Rp)
[num,den]   = cheby1(N,Rp,Wn,'s')
[num,den]   = cheby1(N,Rp,Wn,'type','s')
[N,Wn]      = cheb1ord(Wp,Ws,Rp,Rs,'s')
```

The M-file `cheb1ap(N,Rp)` computes the zeros, poles, and gain factor of the normalized analog Type 1 Chebyshev lowpass filter transfer function of order N with a passband ripple in dB given by Rp. The normalized passband edge frequency is set to 1. The output files are the column vector p providing the locations of the poles, a null vector z for the zero locations, and the gain factor k. The form of the transfer function is as in Eq. (4.60).

As in the previous case, the numerator and denominator coefficients of the transfer function can be determined using the M-file `zp2tf(z,p,k)`. The rational form of the Type 1 Chebyshev lowpass filter transfer function can also be determined directly using the M-file `cheby1(N,Rp,Wn,'s')`, where Wn is the passband edge angular frequency in rad/sec and Rp is the passband ripple in dB. The output data are the vectors, num and den, containing the numerator and the denominator coefficients of the transfer function in descending powers of s. If Wn is a two-element vector [W1,W2] with W1 < W2 the M-file computes the transfer function of an order-2N bandpass filter with passband edge angular frequencies in rad/sec given by W1 and W2. The M-file `cheby1(N,Rp,Wn,'type','s')` is employed for the other two types of filter designs, where the type = high for the highpass case and type = stop for the bandstop case. Wn is a scalar representing the passband edge frequency for the highpass filter design and is a two-element vector defining the stopband edge frequencies for a bandstop filter design.

The M-file `cheb1ord(Wp,Ws,Rp,Rs,'s')` determines the lowest order N of a Type 1 Chebyshev analog transfer function meeting the specifications given by the filter parameters, Wp, Ws, Rp, and Rs, where Wp is the passband edge angular frequency, Ws is the stopband edge angular frequency, Rp is the passband ripple in dB, and Rs is the minimum stopband attenuation in dB. The output data are the filter order N and the cutoff angular frequency Wn. This M-file can also be used to calculate the order of any one of the four basic types of analog Type 1 Chebyshev filters. For the lowpass design, Wp < Ws, whereas for the highpass design, Wp > Ws. For the other two types, Wp and Ws are two-element vectors specifying the passband and stopband edge frequencies. All bandedge frequencies are specified in rad/sec.

Type 2 Chebyshev Filter

The M-files for the design of analog Type 2 Chebyshev filters are

```
[z,p,k]     = cheb2ap(N,Rs)
[num,den]   = cheby2(N,Rs,Wn,'s')
[num,den]   = cheby2(N,Rs,Wn,'type','s')
[N,Wn]      = cheb2ord(Wp,Ws,Rp,Rs,'s')
```

The M-file `cheb2ap(N,Rs)` returns the zeros, poles, and gain factor of a normalized analog Type 2 Chebyshev lowpass filter of order N with a minimum stopband attenuation of Rs in dB. The normalized stopband edge angular frequency is set to 1. The output data are the length-N column vectors z and p, providing the locations of the zeros and the poles, respectively, and the gain factor k. If N is odd, z is of length N-1. The form of the transfer function obtained is given by

$$H_a(s) = \frac{P_a(s)}{D_a(s)} = k\frac{(s - z(1))\,(s - z(2))\cdots(s - z(N))}{(s - p(1))\,(s - p(2))\cdots(s - p(N))}. \tag{4.61}$$

The M-file cheby2(N,Rs,Wn,'s') can be employed to determine the transfer function of a Type 2 Chebyshev lowpass filter when Wn is a scalar defining the stopband edge angular frequency in rad/sec or a bandpass filter when Wn is a two-element vector defining the stopband edge angular frequencies in rad/sec. The M-file cheby2(N,Rs, Wn,'type','s') provides the transfer function of a Type 2 Chebyshev highpass filter when type = high or a bandstop filter when type = stop. In all cases, the specified minimum stopband attenuation is Rs in dB. The output data are the vectors, num and den, containing the numerator and denominator coefficients in descending powers of s.

The M-file cheb2ord(Wp,Ws,Rp,Rs,'s') determines the lowest order N of a Type 2 Chebyshev analog transfer function meeting the specifications given by the filter parameters, Wp, Ws, Rp, and Rs, as defined for the Type 1 Chebyshev filter.

Elliptic (Cauer) Filter

The M-files for the design of analog elliptic filters are

```
[z,p,k]     = ellipap(N,Rp,Rs)
[num,den]   = ellip(N,Rp,Rs,Wn,'s')
[num,den]   = ellip(N,Rp,Rs,Wn,'type','s')
[N,Wn]      = ellipord(Wp,Ws,Rp,Rs,'s')
```

The M-file ellipap(N,Rp,Rs) determines the zeros, poles, and gain factor of a normalized analog elliptic lowpass filter of order N with a passband ripple of Rp dB and a minimum stopband attenuation of Rs dB. The normalized passband edge angular frequency is set to 1. The output files are the length-N column vectors z and p, providing the locations of the zeros and the poles, respectively, and the gain factor k. If N is odd, z is of length N-1. The form of the transfer function obtained is as given in Eq. (4.61).

The M-file ellip(N,Rp,Rs,Wn,'s') returns the transfer function of an elliptic analog low-pass filter when Wn is a scalar defining the passband edge angular frequency in rad/sec or a bandpass filter when Wn is a two-element vector defining the passband edge frequencies in rad/sec. The M-file ellip(N,Rp,Rs,Wn,'type','s') is used to determine the transfer function of an elliptic highpass when type = high, and Wn is a scalar defining the stopband edge angular frequency in rad/sec, or a bandstop filter when type = stop, and Wn is a two-element vector defining the stopband edge angular frequencies in rad/sec. In all cases, the specified passband ripple is Rp dB, and the minimum stopband attenuation is Rs in dB. The output files are the vectors, num and den, containing the numerator and denominator coefficients in descending powers of s.

The M-file ellipord(Wp,Ws,Rp,Rs,'s') determines the lowest order N of an elliptic analog transfer function meeting the specifications given by the filter parameters, Wp, Ws, Rp, and Rs, as defined for the Type 1 Chebyshev filter.

Bessel Filter

For the design of a Bessel filter, the available M-files are

```
[z,p,k]     = besselap(N)
[num,den]   = besself(N,Wn)
[num,den]   = besself(N,Wn,'type')
```

The M-file besselap(N) is employed to compute the zeros, poles, and gain factor of an order-N Bessel lowpass filter prototype. The output data of these M-files are the length-N column vector p, providing the locations of the poles, and the gain factor k. Since there are no zeros, the output vector z is a null vector. The form of the transfer function is as in Eq. (4.60).

The M-file besself(N,Wn) is used to design an Nth-order analog lowpass Bessel filter with a 3-dB cutoff angular frequency given by the scalar Wn in rad/sec. It generates the length-(N+1) vectors, num and den, containing the numerator and the denominator coefficients in descending powers of s. If Wn is a two-element vector, then it returns the transfer function of an order-2N analog bandpass Bessel filter. For designing the other two types of Bessel filters, the function besself(N,Wn,'type') is used. Here the type = high with Wn representing the 3-dB stopband edge frequency in rad/sec for the highpass case, or type = stop with Wn a two-element vector defining the 3-dB stopband edge frequencies in rad/sec for the bandstop case.

Limitations

The zero-pole-gain form is more accurate than the transfer function form for the design of Butterworth, Type 2 Chebyshev, elliptic, or Bessel filter design. It is recommended that the filter design function in these cases be used only for filter orders less than 15, since numerical problems may arise for filter orders equal to or greater than 15.

Analog Lowpass Filter Design Examples

We provide below several examples illustrating the use of some of the above functions in the design of analog filters. In Examples 4.9 through 4.11, we repeat Examples 4.6 through 4.8 to determine the order of the transfer function using the respective M-files. In Examples 4.12 through 4.15, we determine the corresponding transfer functions and then compute the frequency response using the M-file freqs(num,den,w), where num and den are the vectors of the numerator and denominator coefficients in descending powers of s, and w is a set of specified discrete angular frequencies. This function generates a complex vector of frequency response samples from which magnitude and/or phase response samples can be readily computed.

EXAMPLE 4.9   Order of an Analog Butterworth Lowpass Filter

To determine the order of the analog Butterworth lowpass filter meeting the frequency response specifications given in Example 4.6, we use the command [N,Wn] = buttord (Wp,Ws,Rp,Rs,'s'), where Wp = $2\pi(1000)$, Ws = $2\pi(5000)$, Rp = 1, and Rs = 40. The outputs generated are given by N = 4 and Wn = 9934.7, where Wn is the 3-dB cutoff angular frequency. The computed order is the same as that determined in Example 4.6 using the formula of Eq. (4.35).

EXAMPLE 4.10   Order of an Analog Chebyshev Lowpass Filter

We next determine the order of analog Type 1 and Type 2 Chebyshev lowpass filters meeting the same specifications as above. To this end, we employ the commands [N,Wn] = cheb1ord (Wp,Ws,Rp,Rs,'s') and [N,Wn] = cheb2ord(Wp,Ws,Rp, Rs,'s'), respectively. For the Type 1 Chebyshev filter, the computed output data are N = 3 and Wn = 6283.18, and for the Type 2 Chebyshev filter, the computed output data are N = 3 and Wn = 23440.97. The computed Wn in the former case is the 1-dB passband edge angular frequency in rad/sec, whereas in the second case, it is the 40-dB stopband edge angular frequency in rad/sec. The order determined is identical to that derived in Example 4.7 using the formula of Eq. (4.43).

EXAMPLE 4.11   Order of an Analog Elliptic Lowpass Filter

To determine the order of an analog elliptic lowpass filter meeting the above specifications, we use the command [N,Wn] = ellipord(Wp,Ws,Rp,Rs,'s'), resulting in the output data N = 3 and Wn = 6283.185 and verifying the order obtained in Example 4.8. Here Wn is the 1-dB passband edge angular frequency.

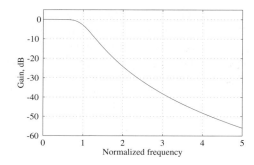

Figure 4.23: The gain response of the normalized fourth-order Butterworth lowpass filter of Example 4.12.

Program 4_1.m

EXAMPLE 4.12   Transfer Function of an Analog Fourth-Order Maximally Flat Lowpass Filter

We consider the design of an analog fourth-order maximally flat lowpass filter with a 3-dB cutoff frequency at $\Omega = 1$, using Program 4_1. The output data generated by running this program are as follows:

```
Poles are at
 -0.38268343236509 + 0.92387953251129i
 -0.92387953251129 + 0.38268343236509i
 -0.92387953251129 - 0.38268343236509i
 -0.38268343236509 - 0.92387953251129i

Numerator polynomial
       0      0      0      0      1

Denominator polynomial
Columns 1 through 4
1.0    2.6131259297    3.4142135624      2.6131259297

Column 5
1.0
```

The plot of the gain response generated by this program is sketched in Figure 4.23.

Program 4_2.m

EXAMPLE 4.13   Design of a Butterworth Lowpass Filter Using MATLAB

We now use MATLAB to complete the design of the Butterworth lowpass filter of Example 4.6. To this end, we modify Program 4_1 used in Example 4.12 as indicated in Program 4_2. During execution, the program asks for the order of the filter and the 3-dB cutoff angular frequency (determined in Example 4.9 to be equal to 4 and 9934.7, respectively) to be typed in. It then generates the gain response plot, as shown in Figure 4.24. As can be seen from this plot, the passband ripple at 1 kHz is less than 1 dB, whereas at 5 kHz, the stopband attenuation is 40 dB, as desired.

Program 4_3.m

EXAMPLE 4.14   Design of an Analog Type 1 Chebyshev Lowpass Filter Using MATLAB

We next illustrate the design of a Type 1 Chebyshev lowpass filter meeting the specifications of Example 4.6. To this end, we use Program 4_3. As the program is run, it asks for the order of the filter, the passband edge angular frequency (determined in Example 4.10 to be equal to 3 and 6283.18, respectively), and the passband ripple (1 dB) to be typed in. It then generates the gain response plot, as shown in Figure 4.25.

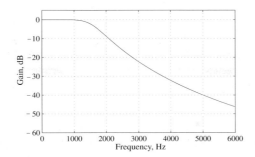

Figure 4.24: The gain response of the Butterworth lowpass filter of Example 4.13.

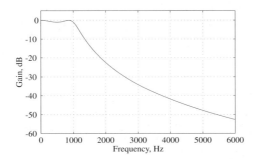

Figure 4.25: The gain response of the Type 1 Chebyshev lowpass filter of Example 4.14.

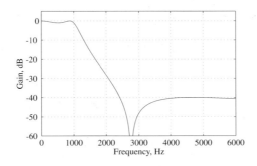

Figure 4.26: The gain response of the elliptic lowpass filter of Example 4.15.

EXAMPLE 4.15   Design of an Analog Elliptic Lowpass Filter Using MATLAB

The design of an analog elliptic lowpass filter meeting the specifications of Example 4.6 follows a similar format. The pertinent MATLAB program is Program 4_4. As the program is run, it asks for the order of the filter (determined in Example 4.11 to be equal to 3), the passband edge angular frequency ($2000\pi = 6283.185$), the passband ripple (1 dB), and the minimum stopband attenuation (40 dB) to be typed in. It then generates the gain response plot, as shown in Figure 4.26.

Program 4_4.m

4.4.7   A Comparison of the Filter Types

In the previous four sections, we have described four types of analog lowpass filter approximations, three of which have been developed primarily to meet the magnitude response specifications, while the fourth has been developed primarily to provide a near linear-phase approximation. In order to determine which filter type to choose to meet a given magnitude response specification, we need to compare the performances of the four types of approximations. To this end, we compare here the frequency responses of the normalized Butterworth, Chebyshev, and elliptic analog lowpass filters of the same order. The passband ripple of the Type 1 Chebyshev and the equiripple filters are assumed to be the same, while the minimum stopband attenuation of the Type 2 Chebyshev and the equiripple filters are assumed to be the same. The filter specifications used for comparison are as follows: filter order of 6, passband edge at $\Omega = 1$, maximum passband deviation of 1 dB, and minimum stopband attenuation of 40 dB. The frequency responses computed using MATLAB are plotted in Figure 4.27.

As can be seen from Figure 4.27, the Butterworth filter has the widest transition band, with a monotonically decreasing gain response. Both types of Chebyshev filters have a transition band of equal width that is smaller than that of the Butterworth filter but greater than that of the elliptic filter. The Type 1 Chebyshev filter provides a slightly faster roll-off in the transition band than the Type 2 Chebyshev filter. The magnitude response of the Type 2 Chebyshev filter in the passband is nearly identical to that of the Butterworth filter. The elliptic filter has the narrowest transition band, with an equiripple passband and an equiripple stopband response.

The Butterworth and Chebyshev filters have a nearly linear phase response over about three-fourths of the passband, whereas the elliptic filter has a nearly linear phase response over about one-half of the passband. On the other hand, the Bessel filter may be more attractive if the linearity of the phase response over a larger portion of the passband is desired at the expense of a poorer gain response. Figure 4.28 shows the gain and phase responses of a sixth-order Bessel filter frequency scaled to have a passband edge at $\Omega = 1$ with a maximum passband deviation of 1 dB. However, the Bessel filter provides a minimum of 40 dB attenuation at approximately $\Omega = 6.4$ and, as a result, has the largest transition band compared to the other three types.

Another way of comparing the performances of the Butterworth, Chebyshev, and elliptic filters would be to compare the order of these filters required to meet the same filter specifications. For example, consider the specifications of a lowpass filter: passband edge at $\Omega = 1$, maximum passband deviation of 1 dB, stopband edge at $\Omega = 1.2$, and minimum stopband attenuation of 40 dB. These specifications are met by a Butterworth filter of order 29, a Chebyshev Type 1 or 2 filter of order 10, and an elliptic filter of order 6.

4.5   Design of Analog Highpass, Bandpass, and Bandstop Filters

All of the four types of approximations discussed in Section 4.4 dealt with the design of analog lowpass filters meeting the prescribed specifications. Design of the other three classes of analog filters, namely, the highpass, bandpass, and bandstop filters, can be carried out by simple spectral transformations of the frequency variables [Tem77]. The design process involves the development of the specifications of a prototype analog lowpass filter from the specifications of the desired analog filter using a frequency transformation, design of the analog prototype lowpass filter, and then determination of the transfer function of the desired analog filter by applying the inverse of the frequency transformation used to determine the specifications of the prototype lowpass filter.

To eliminate the confusion between the Laplace transform variable of the prototype analog lowpass transfer function $H_{LP}(s)$ and that of the desired analog transfer function $H_D(s)$, we shall use different symbols. Thus, we shall use s to denote the Laplace transform variable of the prototype analog lowpass

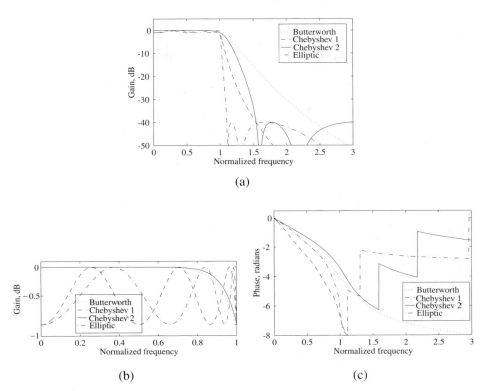

Figure 4.27: A comparison of the frequency responses of the four types of analog lowpass filters: (a) gain responses, (b) passband details, and (c) phase responses.

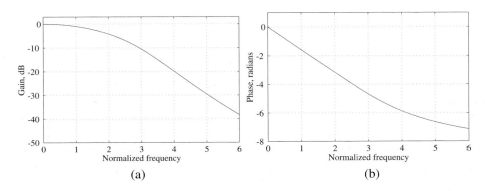

Figure 4.28: The frequency responses of a sixth-order analog Bessel filter: (a) gain response and (b) unwrapped phase response.

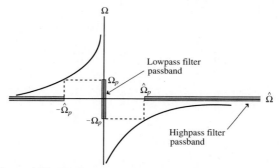

Figure 4.29: The lowpass-to-highpass mapping of Eq. (4.63).

filter $H_{LP}(s)$ and $\hat{s}$ to denote the Laplace transform variable of the desired analog filter $H_D(\hat{s})$. The angular frequency variables in the s- and $\hat{s}$-domains are given by Ω and $\hat{\Omega}$, respectively.

The mapping from s-domain to $\hat{s}$-domain is given by the invertible transformation

$$s = F(\hat{s}).$$

The transfer functions $H_{LP}(s)$ and $H_D(\hat{s})$ are related through

$$H_D(\hat{s}) = H_{LP}(s)\big|_{s=F(\hat{s})}\,,$$

$$H_{LP}(s) = H_D(\hat{s})\big|_{\hat{s}=F^{-1}(s)}\,.$$

4.5.1   Analog Highpass Filter Design

A prototype analog lowpass transfer function $H_{LP}(s)$ with a passband edge frequency Ω_p can be transformed into an analog highpass transfer function $H_{HP}(\hat{s})$ with a passband edge frequency $\hat{\Omega}_p$ using the spectral transformation

$$s = \frac{\Omega_p \hat{\Omega}_p}{\hat{s}}. \tag{4.62}$$

On the imaginary axis, the above transformation reduces to

$$\Omega = -\frac{\Omega_p \hat{\Omega}_p}{\hat{\Omega}}. \tag{4.63}$$

The above mapping implies that the passband of the lowpass filter in the positive frequency range $0 \le \Omega \le \Omega_p$ is mapped into the passband of the highpass filter in the negative frequency range $-\infty < \hat{\Omega} \le -\hat{\Omega}_p$, and the passband of the lowpass filter in the negative frequency range $-\Omega_p \le \Omega \le 0$ is mapped into the passband of the highpass filter in the positive frequency range $\hat{\Omega}_p \le \hat{\Omega} < \infty$. Likewise, the stopband of the lowpass filter in the positive frequency range $\Omega_s \le \Omega < \infty$ is mapped into the stopband of the highpass filter in the negative frequency range $\hat{\Omega}_s \le \hat{\Omega} \le 0$, and the stopband of the lowpass filter in the negative frequency range $-\infty < \Omega \le -\Omega_s$ is mapped into the stopband of the highpass filter in the positive frequency range $0 \le \hat{\Omega} \le \hat{\Omega}_s$. The lowpass-to-highpass mapping of Eq. (4.63) is illustrated in Figure 4.29.

The above mapping ensures that the gain value of $|H_{LP}(j\Omega)|$ of the prototype lowpass filter in its passband will appear in the passband $|\hat{\Omega}| \ge \hat{\Omega}_p$ of the desired highpass filter. Likewise, the gain value of the prototype lowpass filter in its stopband $|\Omega| \geq \Omega_s$ will appear in the stopband $0 \leq |\hat{\Omega}| \leq \hat{\Omega}_s$ of the desired highpass filter.

EXAMPLE 4.16   **Design of an Analog Butterworth Highpass Filter Using MATLAB**

Design an analog Butterworth highpass filter with the following specifications: passband edge at 4 kHz, stopband edge at 1 kHz, passband ripple of 0.1 dB, and minimum stopband attenuation of 40 dB.

For the prototype analog lowpass filter, we choose the normalized passband edge to be $\Omega_p = 1$. Hence, from Eq. (4.63), the normalized stopband edge of the lowpass filter is given by

$$\Omega_s = \frac{2\pi \times 4000}{2\pi \times 1000} = 4.$$

The specifications of the lowpass filter are therefore as follows: passband edge at $\Omega_p = 1$ rad/sec, stopband edge $\Omega_s = 4$ rad/sec, passband ripple of 0.1 dB, and minimum stopband attenuation of 40 dB.

We first use the function buttord to determine the order N and the 3-dB cutoff frequency Wn of the lowpass filter. Next, we use the function butter to determine the transfer function $H_{LP}(s)$ of the prototype lowpass filter. The lowpass filter is then transformed into the desired highpass filter $H_{HP}(s)$ using the function lp2hp. The code fragments used to design $H_{HP}(s)$ are given below.

```
[N,Wn]    = buttord(1,4,0.1,40,'s');
[B,A]     = butter(N,Wn,'s');
[num,den] = lp2hp(B,A,2*pi*4000);
```

The transfer function $H_{LP}(s)$ of the analog lowpass filter can be obtained by displaying the numerator coefficient vector B and denominator coefficient vector A and is given by

$$H_{LP}(s) = \frac{10.2405}{s^5 + 5.1533s^4 + 13.278s^3 + 21.1445s^2 + 20.8101s + 10.2405}.$$

The gain response of the above lowpass filter is shown in Figure 4.30(a). The transfer function $H_{HP}(s)$ of the desired analog highpass filter can be obtained by displaying the numerator coefficient vector num and denominator coefficient vector den. The gain response of the designed highpass filter is shown in Figure 4.30(b).

It should be noted that the desired highpass filter could have been designed directly using the statements

```
[N,Wn]    = buttord(8000*pi, 2000*pi, 0.1, 40, 's');
[num,den] = butter(N,Wn,'high','s');
```

4.5.2 Analog Bandpass Filter Design

A prototype analog lowpass transfer function $H_{LP}(s)$ with a passband edge frequency Ω_p can be transformed into an analog bandpass transfer function $H_{BP}(\hat{s})$ with a lower passband edge frequency $\hat{\Omega}_{p1}$ and an upper passband edge frequency $\hat{\Omega}_{p2}$ using the spectral transformation

$$s = \Omega_p \frac{\hat{s}^2 + \hat{\Omega}_o^2}{\hat{s}(\hat{\Omega}_{p2} - \hat{\Omega}_{p1})}. \tag{4.64}$$

On the imaginary axis, the above transformation reduces to

$$\Omega = -\Omega_p \frac{\hat{\Omega}_o^2 - \hat{\Omega}^2}{\hat{\Omega} \, B_w}, \tag{4.65}$$

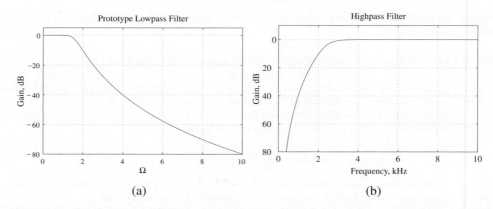

(a)   (b)

Figure 4.30: (a) Gain response of the prototype analog lowpass filter and (b) gain response of the desired analog highpass filter of Example 4.16.

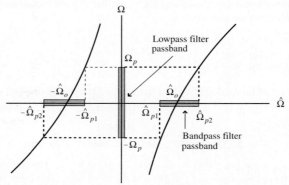

Figure 4.31: The lowpass-to-bandpass mapping of Eq. (4.65).

where $B_w = (\hat{\Omega}_{p2} - \hat{\Omega}_{p1})$ denotes the width of the passband of the bandpass filter. It follows from the above equation that the frequency $\Omega = 0$ is mapped onto the frequency $\hat{\Omega}_o$ which is called the *passband center frequency* of the bandpass filter. It also follows from Eq. (4.65) that the passband edge frequency Ω_p of the lowpass filter maps into the passband edge frequencies $\hat{\Omega}_{p2}$ and $-\hat{\Omega}_{p1}$ of the bandpass filter. Moreover, the frequency range $-\Omega_p \leq \Omega \leq \Omega_p$ of the lowpass filter is mapped into the frequency ranges $-\hat{\Omega}_{p2} \leq \hat{\Omega} \leq -\hat{\Omega}_{p1}$ and $\hat{\Omega}_{p1} \leq \hat{\Omega} \leq \hat{\Omega}_{p2}$ of the bandpass filter. Likewise, the stopband edge frequency Ω_s of the lowpass filter maps into the stopband edge frequencies $\hat{\Omega}_{s2}$ and $-\hat{\Omega}_{s1}$ of the bandpass filter. Similarly, the stopband edge frequency $-\Omega_s$ of the lowpass filter maps into the stopband edge frequencies $\hat{\Omega}_{s1}$ and $-\hat{\Omega}_{s2}$ of the bandpass filter. The lowpass-to-bandpass mapping of Eq. (4.65) is illustrated in Figure 4.31.

It can be shown that

$$\hat{\Omega}_{p1}\hat{\Omega}_{p2} = \hat{\Omega}_{s1}\hat{\Omega}_{s2} = \hat{\Omega}_o^2. \tag{4.66}$$

Thus, the two passband edge frequencies exhibit geometric symmetry with respect to the center frequency $\hat{\Omega}_o$. Likewise, the stopband edge frequencies exhibit geometric symmetry with respect to the center frequency. If the bandedge frequencies do not satisfy the condition of Eq. (4.66), one of them needs to be changed to a new value so that it is satisfied while introducing some safety margin [Tem77]. For example,

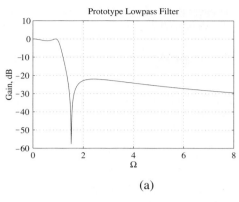

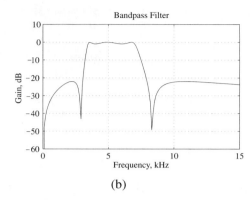

Figure 4.32: (a) Gain response of the prototype analog lowpass filter and (b) gain response of the desired analog bandpass filter of Example 4.17.

if

$$\hat{\Omega}_{p1}\hat{\Omega}_{p2} > \hat{\Omega}_{s1}\hat{\Omega}_{s2},$$

then either $\hat{\Omega}_{p1}$ can be decreased to $\hat{\Omega}_{s1}\hat{\Omega}_{s2}/\hat{\Omega}_{p2}$ or $\hat{\Omega}_{s1}$ can be increased to $\hat{\Omega}_{p1}\hat{\Omega}_{p2}/\hat{\Omega}_{s2}$ to satisfy the condition of Eq. (4.66). In the former case, the new passband will be larger than the original desired passband, whereas in the latter case, the left transition band will be smaller than the original value. On the other hand, if

$$\hat{\Omega}_{p1}\hat{\Omega}_{p2} < \hat{\Omega}_{s1}\hat{\Omega}_{s2},$$

then either $\hat{\Omega}_{p2}$ can be increased or $\hat{\Omega}_{s2}$ can be decreased to satisfy the condition of Eq. (4.66). Moreover, if the gain value of the lowpass filter at a frequency is α dB, the same gain value is obtained for the bandpass filter at the positive frequencies $\hat{\Omega}_a$ and $\hat{\Omega}_b$, with the latter frequencies exhibiting geometric symmetry with respect to $\hat{\Omega}_o$.

EXAMPLE 4.17   Design of an Analog Elliptic Bandpass Filter Using MATLAB

Design an analog elliptic bandpass filter with the following specifications: passband edges at 4 kHz and 7 kHz, stopband edges at 3 kHz and 8 kHz, passband ripple of 1 dB, and a minimum stopband attenuation of 22 dB.

The product of the passband edge frequencies is 28×10^6, whereas the product of the stopband edge frequencies is 24×10^6. Since the former product is greater than the latter, we decrease the first passband edge to $24/7 = 3.42857$ kHz. The passband center frequency is thus at $\sqrt{24} = 4.8989795$ kHz. The width of the passband then increases from 3 kHz to $25/7 = 3.571428$ kHz.

For the prototype analog lowpass filter, we choose the passband edge frequency $\Omega_p = 1$. From Eq. (4.65), we get the stopband edge frequency of the lowpass filter to be

$$\Omega_s = \frac{24 - 9}{(25/7) \times 3} = 1.4.$$

The specifications of the elliptic lowpass filter are thus given by normalized passband edge at 1 rad/sec, normalized stopband edge at 1.4 rad/sec, passband ripple of 1 dB, and minimum stopband attenuation of 22 dB.

We first use the function `ellipord` to determine the filter order N and the passband edge angular frequency Wn of the prototype lowpass filter $H_{LP}(s)$. We then use the function `ellip` to design the desired lowpass filter

$H_{LP}(s)$. Next, using the function lp2bp, we transform the above lowpass transfer function to the transfer function of the desired bandpass filter $H_{BP}(s)$. The code fragments used to design $H_{BP}(s)$ are given below.

```
[N,Wn] = ellipord(1,1.4,1,22,'s');
[B,A] = ellip(N,1,22,Wn,'s');
[num,den] = lp2bp(B,A,2*pi*4.8989795,2*pi*25/7);
```

The transfer function of the prototype analog lowpass filter $H_{LP}(s)$ can be obtained by printing the numerator coefficient vector B and the denominator coefficient vector A and is given by

$$H_{LP}(s) = \frac{0.275032211648s^2 + 0.638449761}{s^3 + 0.965577206s^2 + 1.243426s + 0.63844976}.$$

The gain response of the lowpass filter is plotted in Figure 4.32(a). The transfer function of the desired analog bandpass filter $H_{BP}(s)$ can be obtained by printing the numerator coefficient vector num and the denominator coefficient vector den. The gain response of $H_{BP}(s)$ is plotted in Figure 4.32(b).

As in the case of Example 4.16, the desired bandpass filter could have also been designed directly using the statements

```
Wp = [3.42857 7]*2*pi; Ws = [3 8]*2*pi;
[N,Wn] = ellipord(Wp,Ws,1,22,'s');
[num,den] = ellip(N,1,22,Wn,'s');
```

4.5.3   Analog Bandstop Filter Design

An analog prototype lowpass transfer function $H_{LP}(s)$ with a passband edge frequency Ω_p can be transformed into an analog bandstop transfer function $H_{BP}(\hat{s})$ with a lower stopband edge frequency $\hat{\Omega}_{s1}$ and an upper stopband edge frequency $\hat{\Omega}_{s2}$ using the spectral transformation

$$s = \Omega_s \frac{\hat{s}(\hat{\Omega}_{s2} - \hat{\Omega}_{s1})}{\hat{s}^2 + \hat{\Omega}_o^2}. \tag{4.67}$$

On the imaginary axis, the above transformation reduces to

$$\Omega = \Omega_s \frac{\hat{\Omega} B_w}{\hat{\Omega}_o^2 - \hat{\Omega}^2}, \tag{4.68}$$

where $B_w = (\hat{\Omega}_{s2} - \hat{\Omega}_{s1})$ is the width of the stopband of the bandpass filter. Here the frequency $\Omega = \pm\infty$ is mapped onto the frequency $\pm\hat{\Omega}_o$, which is called the *stopband center frequency* of the bandpass filter. It follows from Eq. (4.68) that the frequency range $-\Omega_p \leq \Omega \leq \Omega_p$ of the lowpass filter is mapped into the frequency ranges $-\hat{\Omega}_{p1} \leq \hat{\Omega} \leq \hat{\Omega}_{p1}$, $-\infty < \hat{\Omega} \leq -\hat{\Omega}_{p2}$, and $\hat{\Omega}_{p2} \leq \hat{\Omega} < \infty$ of the bandstop filter. The stopband edge frequency Ω_s of the lowpass filter is mapped onto the stopband edge frequencies $\hat{\Omega}_{s1}$ and $-\hat{\Omega}_{s2}$, and stopband edge frequency $-\Omega_s$ of the lowpass filter is mapped onto the stopband edge frequencies $-\hat{\Omega}_{s1}$ and $\hat{\Omega}_{s2}$. Moreover, as in the case of the analog bandpass filter, the bandedge frequencies here also exhibit geometric symmetry with respect to the center frequency; that is,

$$\hat{\Omega}_{s1}\hat{\Omega}_{s2} = \hat{\Omega}_o^2, \qquad \hat{\Omega}_{p1}\hat{\Omega}_{p2} = \hat{\Omega}_o^2.$$

The mapping of Eq. (4.68) is illustrated in Figure 4.33.

Since the design of the analog bandstop filter is very similar to that of the analog bandpass filter, we leave it as an exercise for the reader (Problem 4.33 and Exercise M4.7).

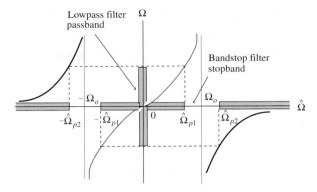

Figure 4.33: The lowpass-to-bandstop mapping of Eq. (4.68).

4.6   Anti-Aliasing Filter Design

According to the sampling theorem of Section 4.2.1, a band-limited continuous-time signal $g_a(t)$ can be fully recovered from its uniformly sampled version if the condition of Eq. (4.11) is satisfied; that is, if $g_a(t)$ is sampled at a sampling frequency Ω_T that is equal or higher than twice the highest frequency Ω_m contained in $g_a(t)$. If this condition is not satisfied, the original continuous-time signal $g_a(t)$ cannot be recovered from its sampled version because of distortion caused by aliasing. In practice, $g_a(t)$ is passed through an analog anti-aliasing lowpass filter prior to sampling to enforce the condition of Eq. (4.11). This analog filter is the first circuit in the interface between the continuous-time and the discrete-time domains and is studied in this section.

Ideally, the anti-aliasing filter $H_a(s)$ should have a lowpass frequency response $H_a(j\Omega)$ given by

$$H_a(j\Omega) = \begin{cases} 1, & |\Omega| < \Omega_T/2, \\ 0, & |\Omega| \geq \Omega_T/2. \end{cases} \tag{4.69}$$

Such a "brickwall" type frequency response cannot be realized using practical analog circuit components and, hence, must be approximated. A practical anti-aliasing filter therefore should have a magnitude response approximating unity in the passband with an acceptable tolerance, a stopband magnitude response exceeding a minimum attenuation level, and an acceptable transition band separating the passband and the stopband, with a transmission zero at infinity. In addition, in many applications, it is also desirable to have a linear-phase response in the passband. The passband edge frequency Ω_p, the stopband edge frequency Ω_s, and the sampling frequency Ω_T must satisfy the relation

$$\Omega_p < \Omega_s \leq \frac{\Omega_T}{2}. \tag{4.70}$$

The passband edge frequency Ω_p is determined by the highest frequency in the continuous-time signal $g_a(t)$ that must be faithfully preserved in the sampled version. Since signal components with frequencies greater than $\Omega_T/2$ appear as frequencies less than $\Omega_T/2$ due to aliasing, the attenuation level of the anti-aliasing filter at frequencies greater than $\Omega_T/2$ is determined by the amount of aliasing that can be tolerated in the passband. The maximum aliasing distortion comes from the signal components in the replicas of the input spectrum adjacent to the baseband.[7] It follows from Figure 4.34 that the frequency $\Omega_o = \Omega_T - \Omega_p$ is aliased into Ω_p, and if the acceptable amount of aliased spectrum at Ω_p is $\alpha_p = -20\log_{10}(1/A)$, then the minimum attenuation of the anti-aliasing filter at Ω_0 must also be α_p [Jac96].

---

[7]It is tacitly assumed here that the magnitude response of the anti-aliasing filter is monotonically decreasing in the stopband.

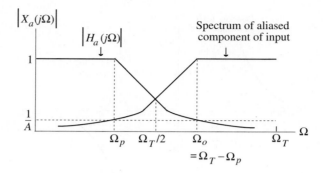

Figure 4.34: Anti-aliasing filter magnitude response and its effect in the signal band of interest.

Table 4.1: Approximate minimum stopband attenuation of a Butterworth lowpass filter.

Ω_0	$2\Omega_p$	$3\Omega_p$	$4\Omega_p$
Attenuation (dB)	$6.02N$	$9.54N$	$12.04N$
Ω_T	$3\Omega_p$	$4\Omega_p$	$5\Omega_p$

EXAMPLE 4.18   **Determination of the Order of the Anti-Aliasing Filter**

Consider the design of an anti-aliasing filter with a Butterworth lowpass magnitude response. From Eq. (4.33), the difference in dB in the attenuation levels at Ω_p and Ω_0 is given by

$$10\log_{10}\left[\frac{1+(\Omega_o/\Omega_c)^{2N}}{1+(\Omega_p/\Omega_c)^{2N}}\right] \cong 10\log_{10}\left(\frac{\Omega_o}{\Omega_p}\right)^{2N} \tag{4.71}$$

This difference has been tabulated in Table 4.1 for several values of the ratio Ω_o/Ω_p as a function of the filter order N.

This table can be used to estimate the minimum sampling frequency Ω_T as a function of the passband edge Ω_p for a specified filter order and a specified minimum stopband attenuation level at a frequency $\Omega_o = \Omega_T - \Omega_p$. For example, if the minimum stopband attenuation at $\Omega_T - \Omega_p$ is 60 dB, then from Table 4.1, we observe that for

a sampling frequency $\Omega_T = 3\Omega_p$, the filter order should be $N = \lceil 60/6.02 \rceil = \lceil 9.967 \rceil = 10$.[8] If we increase the sampling frequency to $4\Omega_p$, then the filter order reduces to $N = \lceil 60/9.54 \rceil = \lceil 6.29 \rceil = 7$. For a still higher sampling frequency of $5\Omega_p$, the filter order becomes $N = \lceil 60/12.04 \rceil = \lceil 4.98 \rceil = 5$.

In practice, the sampling frequency chosen depends on the specific application. In applications requiring minimal aliasing, the sampling rate is typically chosen to be 3 to 4 times the passband edge Ω_p of the anti-aliasing analog filter. In noncritical applications, a sampling rate of twice the passband edge Ω_p of the anti-aliasing analog filter is more than adequate. For example, in pulse-code modulation (PCM) telephone systems, the voice signal is first band-limited to 4 kHz by an anti-aliasing analog filter with a passband edge at 3.6 kHz and a stopband edge at 4 kHz. These specifications are typically met by a third-order elliptic lowpass filter. The output of the filter is then sampled at 8 kHz.

Requirements for the analog anti-aliasing filter can be relaxed by oversampling the analog signal and then decimating the high-sampling-rate digital signal to the desired low-rate digital signal. The decimation process can be implemented completely in the digital domain by first passing the high-rate digital signal

---

[8] $\lceil x \rceil$ denotes the smallest integer greater than or equal to x.

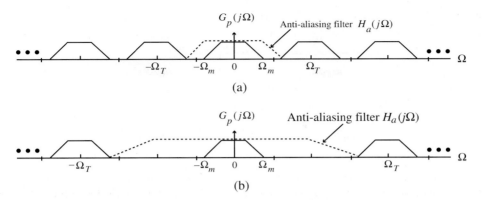

Figure 4.35: Analog anti-aliasing filter requirements for two different oversampling rates

through a digital anti-aliasing filter and then downsampling its output. To understand the advantages of the oversampling approach, consider the sampling of an analog signal band-limited to a frequency Ω_m. Figure 4.35 shows the spectra of the sampled version of this signal sampled at two different rates, Ω_T and $\Omega'_T = 2\Omega_T$, where Ω_T is slightly higher than the Nyquist rate of $2\Omega_m$. These figures also show the desired frequency response of the analog anti-aliasing filter in both cases. Note that the transition band of the analog anti-aliasing filter in the latter case is considerably more than 3 times that needed in the former situation. As a result, the filter specifications are met more easily with a much lower order analog filter.

For the design of the anti-aliasing filter, any one of the four approximations described in Section 4.4 can be employed. Of the four types, the Butterworth approximation provides a reasonable compromise between the desired magnitude response and a linear-phase response in the passband for a given filter order. For an improved phase response at the expense of a poorer magnitude response, the Bessel approximation can be used. On the other hand, for an improved magnitude response with a poorer phase response, either the Chebyshev or the elliptic approximation can be used, with the latter providing the smallest aliasing error for a given filter order. However, in the latter cases, it is necessary to ensure that the transfer function has a zero at infinity. Otherwise, the tails of all the shifted spectra will add up to infinity.

Once the transfer function $H_a(s)$ for the anti-aliasing filter meeting the requirements has been determined, it can be implemented in a number of ways, such as a passive RLC filter, an active-RC filter, or a switched-capacitor filter [Lar93]. A detailed discussion of these implementations is beyond the scope of this book, and we refer the reader to the texts listed at the end of this book [Dar76], [Tem73], [Tem77].

4.7 Sample-and-Hold Circuit[9]

As indicated earlier, the band-limited output of the analog anti-aliasing filter is fed into a sample-and-hold (S/H) circuit, which is the second circuit in the interface between the continuous-time and discrete-time domains. It samples the analog signal at uniform intervals and holds the sampled value after each sampling operation for sufficient time for accurate conversion by the A/D converter.

The basic S/H circuit, shown in Figure 4.36, operates as follows: During the sampling phase, the periodically operated analog switch S remains closed, allowing the capacitor C to track the input analog signal $x_a(t)$ and to charge up to a voltage equal to that of the input. During the hold period the switch remains open, permitting the charged capacitor to hold the voltage across it until the next sampling phase begins. The operation of the switch is controlled by a digital clock signal. The voltage follower at the output of the S/H circuit acts as a buffer between the capacitor C and the input stage of the A/D converter.

---

[9]This section has been adapted from [Mit80] by permission of the author and the publisher.

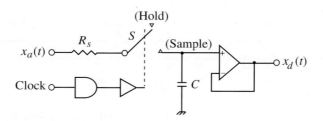

Figure 4.36: The basic sample-and-hold circuit.

Typical input–output waveforms of an S/H circuit are as shown in Figure 4.37, where the dotted line represents the input continuous-time signal $x_a(t)$ and the solid line represents the output $x_d(t)$, assuming instantaneous change from the sample mode to the hold mode, and vice versa.

Practical S/H circuits are often much more complex than the basic circuit of Figure 4.36. They typically include an additional operational amplifier at the input to provide better isolation between the source and the capacitor and better tracking of the input signal. They may contain additional circuit components to minimize the effect of the hold voltage decay, which otherwise may occur due to the leakage through the input resistance of the output buffer amplifier and through the finite OFF resistance of the switch. The major parameters characterizing the performance of a practical S/H circuit are acquisition time, aperture time, and droop. The total time needed to switch from the hold mode to the sample mode and acquire the input analog signal within a specified accuracy is defined as its *acquisition time.* This parameter depends on the switching delay time, the time constant of the RC circuit, and the dynamic performance of the output operational amplifier determined primarily by its slew time and settling time. The time taken by the switch to change from the sample mode to the hold mode is defined as the *aperture time.* The hold voltage drift per second due to leakage out of the holding capacitor is called the *droop.* In addition to these parameters, nonideal properties of the operational amplifiers used in the design of the S/H circuit should also be taken into account in determining the overall performance of the circuit.

4.8   Analog-to-Digital Converter

The next step in the digital processing of an analog signal is the conversion of the output of the S/H circuit in its hold mode to a digital form by means of an analog-to-digital (A/D) converter. For digital signal processing applications, the output of the A/D converter is usually in binary code. The output is a sequence of *words,* with each word representing a sample of the sequence. The *wordlength* of the A/D converter output, given by the number of bits, limits the achievable dynamic range of the converter and its accuracy in representing the input analog signal. The accuracy of conversion of an ideal A/D converter is expressed in terms of its *resolution,* which is determined by the number of discrete levels that can be assumed by the A/D converter output. For an output coded in natural binary form with an N-bit wordlength, the number of available discrete levels is 2^N, and as a result, the resolution or accuracy is 1 part in 2^N or $100/2^N$ percent.

There is a variety of A/D converters that are used in signal processing applications [Lar93]. In all of these converters, the analog comparator is an important circuit component. The analog comparator is a device that compares two analog voltages at its input and develops a binary output indicating which input voltage level is larger. With respect to its circuit symbol shown in Figure 4.38, the input–output relation of an analog comparator is as follows:

$$V_0 = \begin{cases} V^+, & \text{if } V_1 > V_2, \\ V^-, & \text{if } V_1 < V_2, \end{cases} \qquad (4.72)$$

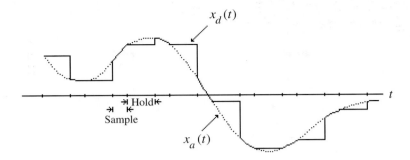

Figure 4.37: Input–output waveforms of a sample-and-hold circuit.

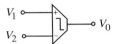

Figure 4.38: Block diagram representation of an analog comparator.

where $V^+ > V^-$. Usually these circuits are designed such that their output voltage levels are compatible with the logic levels of conventional digital circuits.

We briefly review next the operations of the following types of A/D converters: (1) flash A/D converter, (2) serial-parallel A/D converter, (3) successive-approximation A/D converter, (4) oversampling A/D converter. A detailed discussion of their design and implementation is beyond the scope of this book.

4.8.1   Flash A/D Converters

In an N-bit flash converter, shown in block diagram form in Figure 4.39, the input analog voltage V_A is compared simultaneously with a set of $2^N - 1$ uniformly separated reference voltage levels by means of a set of $2^N - 1$ analog comparators, and the location of the adjacent comparators for which the outputs changes from V^- to V^+ indicates the range of the input voltage. A logic encoder circuit is then used to convert the location information into an N-bit binary code. The set of reference voltages is usually derived by a potential-divider resistor string. In a flash A/D converter, all output bits are developed simultaneously, and as a result, it is the fastest converter with a conversion time given by the comparator switching time plus the propagation delay of the encoder circuit. However, the hardware requirements of this type of converter increase very rapidly (exponentially) with an increase in the resolution. As a result, flash converters are employed for low-resolution (typically 8-bit or less) and high-speed conversion applications.

4.8.2   Serial-Parallel A/D Converter

Two $N/2$-bit flash converters in a serial-parallel configuration can be employed to reduce the hardware complexity of an N-bit flash converter at a slight increase in the conversion time [Lar93]. One such scheme, called a *subranging A/D converter,* is shown in Figure 4.40. Here, a coarse approximation of the input analog voltage V_A composed of the $N/2$ most significant bits (MSBs) is first generated by means of one of the $N/2$-bit flash converters. These MSBs are then fed into a D/A converter whose analog output is subtracted from the V_A. The difference voltage is scaled by an amplifier of gain $2^{N/2}$ and converted into digital form by the second (fine) $N/2$-bit flash converter, which provides the least significant bits (LSBs).

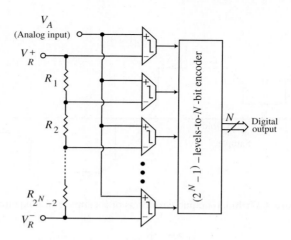

Figure 4.39: Block diagram representation of a flash unipolar A/D converter.

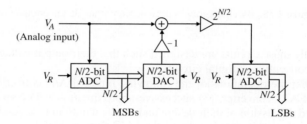

Figure 4.40: Block diagram representation of a subranging A/D converter.

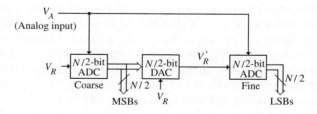

Figure 4.41: Block diagram representation of a ripple A/D converter.

The second scheme utilizing also two $N/2$-bit flash converters is called the *ripple A/D converter*, as shown in Figure 4.41. In this scheme, the coarse $N/2$-bit A/D converter has two functions. It generates the $N/2$ MSBs, and it controls a reference voltage generator that develops a reference voltage V_R' for the fine $N/2$-bit A/D converter generating the $N/2$ LSBs.

In both of the two serial-parallel A/D converters, while one of the $N/2$-bit A/D converters is operating, the other is idle, permitting the use of a single $N/2$-bit A/D converter twice in one conversion period. A generalization of the above two schemes employing N 1-bit A/D converters is called the *pipelined A/D converter* [Lar93].

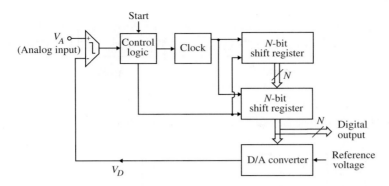

Figure 4.42: Block diagram representation of a successive-approximation A/D converter. (Adapted from [Mit80] by permission of the author and the publisher.)

4.8.3 Successive-Approximation A/D Converter

In this type of converter, essentially a trial-and-error approach is successively used to obtain the digital word representing the input analog voltage V_A [Mit80]. The basic idea behind the operation of this converter can be explained with the aid of its block diagram representation given in Figure 4.42. The conversion procedure is an iterative process. At the kth step of the iteration, the digital approximation stored in the shift register is converted into an analog voltage V_D by the D/A converter in the system. If $V_D < V_A$, then the digital number is increased by setting to ONE the $(k + 1)$th bit to the right of the kth bit, which is assumed to be a ONE. If, on the other hand, $V_D > V_A$, then the digital number is decreased by setting the kth bit to a ZERO and the $(k + 1)$th bit to a ONE. The above process is followed for all k from $k = 1, 2, \ldots, N$. After the Nth bit has been examined, the conversion process is terminated, with the contents of the shift register representing the digital equivalent of the input analog voltage. In practice, to round off the approximation to the nearest discrete level, an analog equivalent to one-half of the value of the LSB is added to the analog input signal before conversion begins.

The successive-approximation A/D converter can be designed with high resolution and reasonably high speed at a moderate cost and is therefore widely used in digital signal processing applications. The performance of this type of A/D converter depends primarily on the performance of its constituent D/A converter and the analog comparator.

4.8.4 Oversampling Sigma-Delta A/D Converter

As the name implies, in this type of converter, the analog signal is sampled at a rate much higher than the Nyquist rate, resulting in very closely spaced samples. As a consequence, the difference between the amplitudes of two consecutive samples is very small, permitting it to be represented in digital form using very few bits, usually by one bit. The sampling rate is then decreased by passing the digital signal first through a factor-of-M decimator to lower the sampling rate from $M F_T$ to F_T. The decimator is designed by cascading an anti-aliasing lowpass Mth band digital filter to reduce its bandwidth to π/M and a factor-of-M down-sampler.[10] The wordlength of the down-sampler output determines the resolution of the oversampling A/D converter, and it is much higher than that of the high-rate digital signal due to the effect of digital filtering. The basic block diagram representation of the oversampling A/D converter is shown in Figure 4.43. This type of converter, often called a *sigma-delta A/D converter*, is discussed in more detail in Section 15.11.

---

[10]The design of a decimator is considered in Section 13.2.

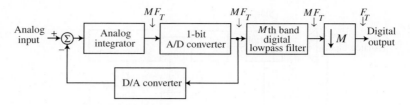

Figure 4.43: Block diagram representation of an oversampling sigma-delta A/D converter.

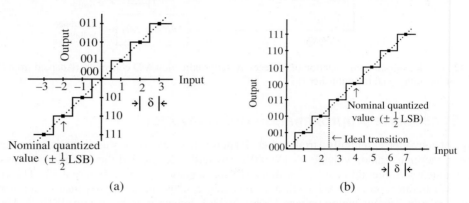

Figure 4.44: Input–output characteristics of an ideal 3-bit A/D converter: (a) bipolar converter and (b) unipolar converter.

4.8.5　Characteristics of a Practical A/D Converter [11]

A practical A/D converter is a nonideal device and exhibits a variety of errors that are usually determined in terms of the input analog values at which the transition in the digital output takes place, since these transitions can be measured accurately. In order to understand the effects of these errors on the performance of an A/D converter, consider first its operation as an ideal device. The input–output relation of an ideal 3-bit A/D converter is shown in Figure 4.44. The error introduced by an ideal A/D converter is simply the difference between the value of the analog input and the analog equivalent of the digital representation; this difference is called the *quantization error.* It follows from Figure 4.44 that the quantization error $e[n]$ for an ideal converter satisfies

$$-\frac{\delta}{2} < e[n] \le \frac{\delta}{2},\tag{4.73}$$

where δ is called the quantization step and is given by

$$\delta = 2^{-N}\tag{4.74}$$

for an N-bit wordlength. Note that δ is precisely the value of the LSB.

　　A practical A/D converter exhibits *linearity error* if the difference between two consecutive transition values of the input is not equal for the complete range of the input, as illustrated in Figure 4.45. The maximum value of this difference value over the full range is called the *differential nonlinearity (DNL) error.* Note from Figure 4.45(b) that in some cases severe nonlinearity in the input–output relation may result in missing codes at the output.

---

[11]This section has been adapted from [Mit80] by permission of the author and the publisher.

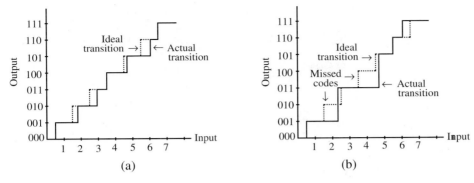

Figure 4.45: Linearity errors in a practical A/D converter.

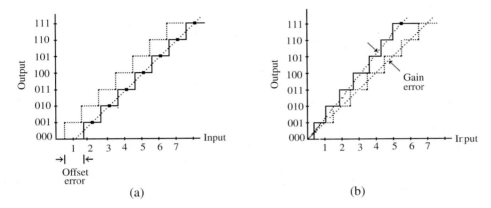

Figure 4.46: (a) Offset and (b) gain errors in a practical ideal A/D converter.

The A/D converter exhibits *gain* or *scale-factor error* if the difference between the last and first transitions is not equal to the full-scale value minus 1/2 LSB. An *offset error* occurs if all transitions are shifted by an equal amount from the ideal transition locations. These errors are illustrated in Figure 4.46.

The time needed by the A/D converter to generate the full digital equivalent of the input analog signal is called its *word-conversion time,* whereas the time required to generate a single bit is called the *bit-conversion time.* In the flash A/D converter, since all bits of the output digital word are generated at the same time, the word-conversion time is essentially equal to the bit-conversion time. On the other hand, in the successive approximation A/D converter, the word-conversion time is equal to the bit-conversion time multiplied by the wordlength.

It should be noted that an *overflow error* occurs if the input analog voltage V_A exceeds the dynamic range of the A/D converter. It is therefore important to ensure that V_A is properly scaled before it is fed into the A/D converter.

4.9   Digital-to-Analog Converter

The final step in the digital processing of analog signals is the conversion of the digital filter output into an analog form, which is accomplished by a digital-to-analog (D/A) converter followed by a reconstruction filter. The basic idea behind the most commonly used D/A converter can be explained by means of the

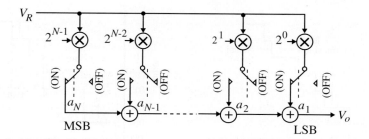

Figure 4.47: Block diagram representation of an N-bit D/A converter.

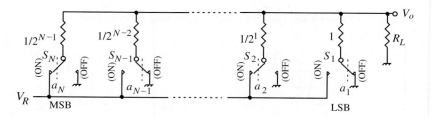

Figure 4.48: Schematic representation of an N-bit weighted resistor unipolar D/A converter.

simplified block diagram representation shown in Figure 4.47, where we have assumed, without any loss of generality, that the digital sample is positive and represented in a natural binary code. Here the ℓth switch S_ℓ is in its ON position if the ℓth binary bit $a_\ell = 1$, and it is in the OFF position if $a_\ell = 0$. The output V_o of the D/A converter is then given by

$$V_o = \sum_{\ell=1}^{N} 2^{\ell-1} a_\ell V_R. \tag{4.75}$$

There are a variety of D/A converters that are used in signal processing applications [Lar93]. We discuss below only the following types: (1) weighted-resistor D/A converter, (2) resistor-ladder D/A converter, and (3) oversampling D/A converter.

4.9.1   Weighted-Resistor D/A Converter

The schematic of an N-bit weighted-resistor D/A converter is shown in Figure 4.48. It can be shown that the output V_o of the D/A converter is given simply by

$$V_o = \sum_{\ell=1}^{N} 2^{\ell-1} a_\ell \left(\frac{R_L}{(2^N - 1)R_L + 1} \right) V_R. \tag{4.76}$$

The full-scale output voltage $V_{o,FS}$ is obtained when all a_ℓ's are ONEs. Then from Eq. (4.76),

$$V_{o,FS} = \left(\frac{(2^N - 1)R_L}{(2^N - 1)R_L + 1} \right) V_R.$$

In practice, usually $(2^N - 1)R_L \gg 1$, and, as a result, $V_{o,FS} \cong V_R$.

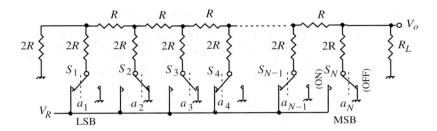

Figure 4.49: Schematic representation of an N-bit resistor-ladder ($R - 2R$ ladder) unipolar D/A converter.

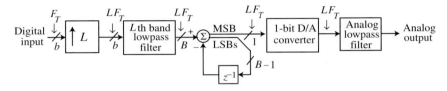

Figure 4.50: Block diagram representation of an oversampling sigma-delta D/A converter.

Usually a buffer amplifier is placed at the output to provide gain and prevent loading. For a D/A converter with a moderate to high resolution, the spread of the resistor values becomes very large, making this type of converter unsuitable for many applications.

Based on the same principle as discussed above, a weighted-capacitor D/A converter can be designed. Such circuits are more popular in IC design.

4.9.2 Resistor-Ladder D/A Converter

This type of converter is probably the most widely used in practice. From its schematic representation shown in Figure 4.49, it can be shown that the D/A converter output V_o is given by

$$V_o = \sum_{\ell=1}^{N} 2^{\ell-1} a_\ell \left(\frac{R_L}{2(R_L + R)} \right) \frac{V_R}{2^{N-1}}. \tag{4.77}$$

Because of the resistor values used and the ladder-like circuit connection, the structure is often referred to as the R–$2R$ ladder D/A converter. In practice, often $2R_L \gg R$ and, hence, the full-scale output voltage $V_{o,FS}$ is given by

$$V_{o,FS} \cong \left(\frac{2^N - 1}{2^N} \right) V_R.$$

As in the previous case, a buffer amplifier is also placed at the output to provide gain and prevent loading.

4.9.3 Oversampling Sigma-Delta D/A Converter

The basic scheme employed in an oversampling sigma-delta D/A converter is shown in Figure 4.50. Here, the sampling rate of the input b-bit digital signal is first increased from F_T to LF_T by a factor-of-L interpolator implemented by an up-sampler followed by an Lth band digital lowpass filter.[12] The output

---

[12]The design of an interpolator is treated in Section 13.2.

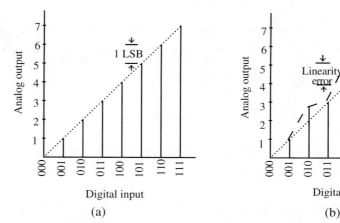

Figure 4.51: Input–output relation of a unipolar D/A converter: (a) ideal converter and (b) linearity errors in a practical D/A converter.

of the interpolator is fed into a digital sigma-delta quantizer, which creates a single bit output by extracting only the MSB. The MSB is then converted to analog form by a 1-bit D/A converter followed by an analog lowpass filter. The LSBs form the error signal that is subtracted from the interpolator output in the summer of the sigma-delta quantizer. A detailed analysis of the operation of the oversampling sigma-delta D/A converter is provided in Section 15.12.

4.9.4   Characteristics of a Practical D/A Converter[13]

A practical D/A converter is characterized by a number of parameters. The effects of these parameters on the performance of a D/A converter are best understood by first examining the input–output relation of an idealized device. Figure 4.51(a) shows the input–output relation of a 3-bit unipolar D/A converter. Here the analog outputs for all possible digital inputs are shown as vertical "bars."

The *resolution* of a D/A converter is defined in a manner identical to that of an A/D converter. For an N-bit D/A converter with the input digital word coded in natural binary form, the resolution is therefore 1 part in 2^{N-1}.

In an ideal D/A converter, the analog outputs as a function of the input discrete levels will be on a straight line going through the origin, as shown in Figure 4.51(a), with the difference between the outputs for two consecutive input digital signals being 1 LSB. In a practical D/A converter, the actual outputs evaluated for each possible input digital signal may be unevenly distributed instead of being on a straight line, as indicated in Figure 4.51(b). The *integral linearity* (INL) *error* is defined as the maximum deviation from the straight line. A measure of the variation in the difference of the analog outputs corresponding to two consecutive digital inputs is called the *differential nonlinearity*. If the analog outputs are on a straight line that is uniformly shifted by an equal amount for each input discrete value, as shown in Figure 4.52(a), the D/A converter is said to have an *offset error*. If the differences between the actual analog outputs and their corresponding ideal analog outputs increase linearly for increases in the digital inputs, as indicated in Figure 4.52(b), the D/A converter is said to exhibit a *gain error*.

The *accuracy* of a practical D/A converter is defined by the maximum deviation of its measured output from that of an ideal D/A converter. The *word-conversion time* of a D/A converter is given by the time taken to decode a digital word.

---

[13]This section has been adapted from [Mit80] by permission of the author and publisher.

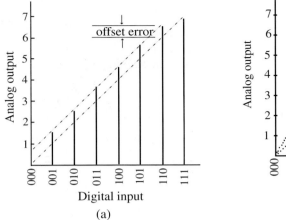
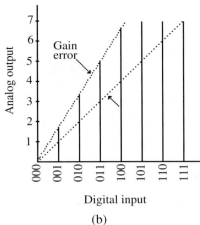

Figure 4.52: Input–output relation of a practical D/A converter: (a) offset error and (b) gain error.

The finite turn-on and turn-off times of analog switches in the D/A converter, in general, are not equal, and as a result, they give rise to a dynamic error called a *glitch*. Consider the situation when at time $t = t_n$, the N-bit word is given as the MSB being ONE and the remaining bits as ZEROs. Assume that at time $t = t_{n+1}$, the MSB changes to a ZERO with all other bits becoming ONEs. If the turn-on time of the analog switch is greater than its turn-off time, then the Nth switch is turned off first and momentarily all switches in the D/A converter are in their OFF positions, resulting in a false analog output equal to zero before the switches settle to their correct positions. The temporary state of all switches being in the OFF positions causes a narrow pulse or spike of half the height of full scale to appear at the converter output. These pulses appearing during the transition periods are called *glitches*. If such glitches are undesirable, they can be eliminated by placing an S/H circuit at the output of the D/A converter, which holds the previous D/A converter output until the glitches disappear and then acquires and holds the new output.

4.10 Reconstruction Filter Design

The output of the D/A converter is finally passed through an analog reconstruction or smoothing filter to eliminate all the replicas of the spectrum outside the baseband. As indicated earlier in Section 4.2.2, this filter ideally should have a frequency response such as given by Eq. (4.18). If the cutoff frequency Ω_c of the reconstruction filter is chosen as $\Omega_T/2$, where Ω_T is the sampling angular frequency, the corresponding frequency response is given by

$$H_r(j\Omega) = \begin{cases} T, & |\Omega| \le \Omega_T/2, \\ 0, & |\Omega| > \Omega_T/2. \end{cases} \tag{4.78}$$

If we denote the input to the D/A converter as $y[n]$, then from Eq. (4.22), the reconstructed analog equivalent $y_a(t)$ is given by

$$y_a(t) = \sum_{n=-\infty}^{\infty} y[n] \frac{\sin[\pi(t - nT)/T]}{\pi(t - nT)/T}. \tag{4.79}$$

Since the ideal reconstruction filter of Eq. (4.78) has a doubly infinite impulse response, it is noncausal and thus unrealizable. In practice, it is necessary to use filters that approximate the ideal lowpass frequency response.

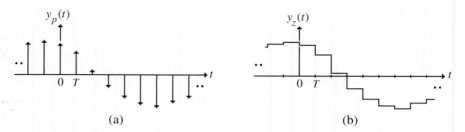

Figure 4.53: Typical output waveforms: (a) ideal D/A converter and (b) a practical D/A converter.

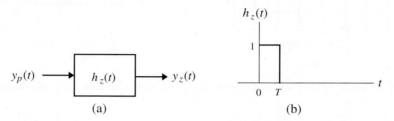

Figure 4.54: (a) Modeling of the zero-order hold operation and (b) the impulse response of the zero-order hold circuit.

The output of an ideal D/A converter is an impulse train as indicated in Figure 4.53(a). However, almost always, a practical D/A converter unit contains a *zero-order hold circuit* at its output, producing a staircase-like analog waveform $y_z(t)$, as shown in Figure 4.53(b). It is therefore important to analyze the effect of the zero-order hold circuit in order to determine the specifications for the smoothing lowpass filter that should follow the overall D/A converter structure.

The zero-order hold operation can be modeled by an ideal impulse-train D/A output $y_p(t)$, followed by a linear, time-invariant analog circuit with an impulse response $h_z(t)$ that is a rectangular pulse of width T and unity height, as indicated in Figure 4.54. It follows from this figure that if $Y_p(j\Omega)$ denotes the continuous-time Fourier transform of $y_p(t)$, the output of the ideal D/A converter, then the continuous-time Fourier transform $Y_z(j\Omega)$ of $y_z(t)$, the output of the zero-order hold circuit, is simply given by

$$Y_z(j\Omega) = H_z(j\Omega)Y_p(j\Omega), \tag{4.80}$$

where

$$
\begin{aligned}
H_z(j\Omega) &= \int_0^T e^{-j\Omega t}\, dt = -\left.\frac{e^{-j\Omega t}}{j\Omega}\right|_0^T \\
&= \frac{1 - e^{-j\Omega T}}{j\Omega} = e^{-j\frac{\Omega T}{2}}\left[\frac{\sin(\Omega T/2)}{\Omega/2}\right].
\end{aligned}
\tag{4.81}
$$

The magnitude response of the zero-order hold circuit, as indicated in Figure 4.55(a), has a lowpass characteristic with zeros at $\pm\Omega_T$, $\pm2\Omega_T$, $\ldots$, where $\Omega_T = 2\pi/T$ is the sampling angular frequency. Figure 4.55(b) shows the magnitude response $|Y_p(j\Omega)|$, which is a periodic function of Ω with a period Ω_T. Since the frequency response $Y_z(j\Omega)$ of the analog output $y_z(t)$ of the zero-order hold circuit is a product of $Y_p(j\Omega)$ and $H_z(j\Omega)$, the zero-order hold circuit somewhat attenuates the unwanted replicas centered at multiples of the sampling frequency Ω_T, as sketched in Figure 4.55(c). An *analog reconstruction filter*, also called a *smoothing filter*, $H_r(j\Omega)$ thus follows a practical D/A converter unit and is designed to further

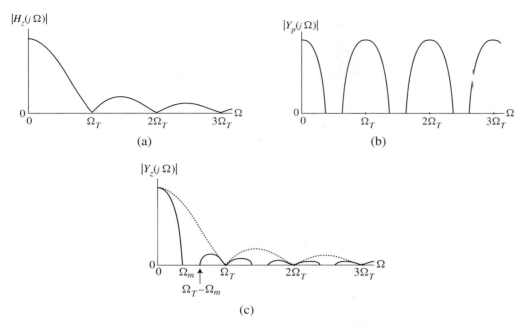

Figure 4.55: Magnitude responses of (a) the zero-order hold circuit, (b) the output of the ideal D/A converter, and (c) the output of the practical D/A converter.

attenuate the residual portions of the signal spectrum centered at multiples of the sampling frequency Ω_T. Moreover, it should also compensate for the amplitude distortion, more commonly called *droop,* caused by the zero-order hold circuit in the band from dc to $\Omega_T/2$.

The general specifications for the analog reconstruction filter $H_r(j\Omega)$ can be easily determined if the effect of the droop is neglected. If Ω_c denotes the highest frequency of the signal $y_p(t)$ that should be preserved at the output of the reconstruction filter, then the lowest-frequency component present in the residual images in the output of the zero-order hold circuit is of frequency $\Omega_o = \Omega_T - \Omega_c$. The zero-order hold circuit has a gain at frequency Ω_o, given by

$$20 \log_{10} |H_z(j\Omega_o)| = 20 \log_{10} \left[\frac{\sin(\Omega_o T/2)}{\Omega_o/2} \right]. \tag{4.82}$$

Therefore, if the system specification calls for a minimum attenuation of A_s dB of all frequency components in the residual images, then the reconstruction filter should provide at least an attenuation of $A_s + 20 \log_{10} |H_z(j\Omega_o)|$ dB at Ω_o. For example, if the normalized value of Ω_c is 0.7π, then the gain of the zero-order hold circuit at 0.7π is -7.2 dB. Now, the lowest normalized frequency of the residual images is given by 1.3π. For a minimum attenuation of 50 dB of all signal components in the residual images at the output of the zero-order hold, the reconstruction filter must therefore provide at least an attenuation of 42.8 dB at frequency 1.3π.

The droop caused by the zero-order hold circuit can be compensated either before the D/A converter by means of a digital filter or after the zero-order hold circuit by the analog reconstruction filter. For the latter approach, we observe that the cascade of the zero-order hold circuit and the analog reconstruction filter must have a frequency response of an ideal reconstruction filter following an ideal D/A converter. If we denote the frequency response of the ideal reconstruction filter as $H_r(j\Omega)$ and that of the actual

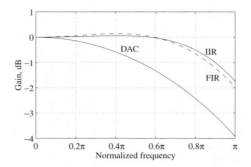

Figure 4.56: Gain responses of the uncompensated and compensated DAC in the baseband.

reconstruction filter as $\hat{H}_r(j\Omega)$, then we require

$$H_r(j\Omega) = H_z(j\Omega)\hat{H}_r(j\Omega), \tag{4.83}$$

where $H_r(j\Omega)$ is as given by Eq. (4.18). Therefore, from Eq. (4.74), the desired frequency response of the actual reconstruction filter is given by

$$\hat{H}_r(j\Omega) = \begin{cases} \frac{\Omega T/2}{\sin(\Omega T/2)}, & |\Omega| \le \Omega_c, \\ 0, & |\Omega| > \Omega_c, \end{cases} \tag{4.84}$$

to ensure a faithful reconstruction of the original signal $g_a(t)$. The modified reconstruction filter has also a noncausal impulse response defined for $-\infty < t < \infty$ and is therefore unrealizable. As a result, an analog filter approximating the magnitude response of $\hat{H}_r(j\Omega)$ must therefore be designed.

Alternatively, the effect of the droop can be compensated by including a digital compensation filter $G(z)$ prior to the D/A converter circuit with a modest increase in the digital hardware requirements. The digital compensation filter can be either an FIR or an IIR type. The frequency response of the digital compensation filter is given by

$$G(e^{j\omega}) = \frac{\omega/2}{\sin(\omega/2)}, \qquad 0 \le |\omega| \le \pi. \tag{4.85}$$

Two very low order digital compensation filters are as follows [Jac96]:

$$G_{\text{FIR}}(z) = -\frac{1}{16} + \frac{9}{8}z^{-1} - \frac{1}{16}z^{-2}, \tag{4.86a}$$

$$G_{\text{IIR}}(z) = \frac{9}{8 + z^{-1}}. \tag{4.86b}$$

Figure 4.56 shows the gain responses of the uncompensated and the droop-compensated D/A converters in the baseband. Since the above digital compensation filters have a periodic frequency response of period Ω_T, the replicas of the baseband magnitude response outside the baseband need to be suppressed sufficiently to ensure minimal effect from aliasing. Even though the zero-order hold circuit in the D/A converter provides some attenuation of these unwanted replicas [see Figure 4.55(c)], it may be necessary for the analog reconstruction filter following the D/A converter to provide additional attenuation.

4.11   Effect of Sample-and-Hold Operation

The frequency-domain analysis of the sampling of a continuous-time signal discussed in Section 4.2.1 assumed ideal sampling generating an impulse train representation of the sampled signal. As indicated

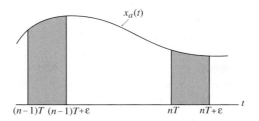

Figure 4.57: Illustration of the averaging operation caused by the S/H circuit.

in Figure 4.1, in most applications, the sampling operation is provided by an S/H circuit. In principle, the S/H circuit samples the analog signal at each sampling instant and holds a constant value equal to the sampled value for a finite and short period of time to permit the A/D converter to convert it into its digital form. However, in practice, as indicated in Figure 4.37, the S/H circuit tracks the analog signal $x_a(t)$ over a small interval ε. The overall effect, as illustrated in Figure 4.57, is to develop an average of the analog signal over this interval, which is held constant at the input of the A/D converter. We now analyze in the frequency domain the effect of the averaging operation of the S/H circuit [Por97].

From Figure 4.57, it follows that the nth sample value $x[n]$ of the impulse train output $x_p(t)$ of a practical S/H circuit is given by

$$x[n] = \frac{1}{\varepsilon} \int_{nT}^{nT+\varepsilon} x_a(t)dt. \tag{4.87}$$

To understand the effect of the above averaging operation, denote

$$g_a(t) = \int_{-\infty}^{t} x_a(\tau)\, d\tau + K, \tag{4.88}$$

where K is a constant of arbitrary value. Then, from Eq. (4.87), we get

$$x[n] = \frac{1}{\varepsilon} \{g_a(nT + \varepsilon) - g_a(nT)\}. \tag{4.89}$$

The impulse train $g_p(t)$ with sample value $g_a(nT)$ is obtained by an ideal sampling of the analog signal $g_a(t)$, and it follows from Eq. (4.88) and the differentiation property of the continuous-time Fourier transform (CTFT), that the CTFT of $g_p(t)$ is simply $\frac{1}{j\Omega} X_a(j\Omega)$, where $X_a(j\Omega)$ is the CTFT of $x_a(t)$. Using the time-shifting property, the CTFT of the impulse train with sample value $g_a(nT + \varepsilon)$ is therefore given by $\frac{e^{-j\Omega\varepsilon}}{j\Omega} X_a(j\Omega)$. Hence, from Eqs. (4.15a) and (4.89), the discrete-time Fourier transform (DTFT) of the discrete-time signal $x[n]$ appearing as the input to the A/D converter can be expressed as

$$X(e^{j\omega}) = \frac{1}{T} \sum_{k=-\infty}^{\infty} \bar{X}_a\left(j\frac{\omega - 2\pi k}{T}\right), \tag{4.90}$$

where

$$\bar{X}_a(j\Omega) = \frac{1 - e^{-j\Omega\varepsilon}}{j\Omega\varepsilon} X_a(j\Omega) = e^{-j\Omega\varepsilon/2}\left(\frac{\sin(\Omega\varepsilon/2)}{\Omega\varepsilon/2}\right) X_a(j\Omega). \tag{4.91}$$

It follows from Eq. (4.91) that the averaging operation performed by a practical S/H circuit is equivalent to passing the continuous-time signal $x_a(t)$ through an LTI discrete-time system with a frequency response $e^{-j\Omega\varepsilon/2}(\sin(\Omega\varepsilon/2))/(\Omega\varepsilon/2)$, followed by an ideal impulse train sampling, as indicated in Figure 4.58.

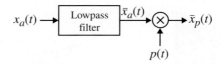

Figure 4.58: Equivalent representation of a practical S/H circuit.

Note that the frequency response of the discrete-time system is similar in form to that of the zero-order hold circuit as given in Eq. (4.81) and shown in Figure 4.55(a). Thus, the discrete-time system of Figure 4.58 acts like a narrowband lowpass filter that performs the averaging operation. If the tracking period ε is much smaller compared to the sampling period T, as is usually the case, the effect of the lowpass filter can be neglected, and the practical S/H circuit can be considered as an ideal sampler.

4.12 Summary

Various issues concerned with the digital processing of continuous-time signals are studied in this chapter. A discrete-time signal is obtained by uniformly sampling a continuous-time signal. The discrete-time representation is unique if the sampling frequency is greater than twice the highest frequency contained in the continuous-time signal, and the latter can be fully recovered from its discrete-time equivalent by passing it through an ideal analog lowpass reconstruction filter with a cutoff frequency that is half the sampling frequency. If the sampling frequency is lower than twice the highest frequency contained in the continuous-time signal, in general, the latter cannot be recovered from its discrete-time version due to aliasing. In practice, the continuous-time signal is first passed through an analog lowpass anti-aliasing filter, with the cutoff frequency chosen as half of the sampling frequency whose output is sampled to prevent aliasing. It is also shown that a bandpass continuous-time signal can be recovered from its discrete-time equivalent by undersampling, provided the highest frequency is an integer multiple of the bandwidth of the continuous-time signal and the sampling frequency is greater than twice the bandwidth.

A brief review of the theory behind some popular analog lowpass filter design techniques is included, and their design using MATLAB is illustrated. Also discussed are the procedures for designing analog highpass, bandpass, and bandstop filters and their implementations using MATLAB. The specifications of the analog filters are usually given in terms of the locations of the passband and stopband edge frequencies and the passband and stopband ripples. Effects of these parameters on the performances of the anti-aliasing and reconstruction filters are examined.

Other interface devices involved in the digital processing of continuous-time signals are the sample-and-hold circuit, comparator, analog-to-digital converter, and digital-to-analog converter. A brief introduction to these devices is included for completeness.

4.13 Problems

4.1 Prove the Poisson's sum formula of Eq. (4.7).

4.2 Show that if the spectrum $G_a(j\Omega)$ of $g_a(t)$ (band-limited to Ω_m) also contained an impulse at Ω_m, the sampling rate Ω_T must be greater than $2\Omega_m$ to recover fully $g_a(t)$ from the sampled version.

4.3 The Nyquist frequency of a continuous-time signal $g_a(t)$ is Ω_m. Determine the Nyquist frequency of each of the following continuous-time signals derived from $g_a(t)$:
    (a) $y_1(t) = g_a(t)g_a(t)$,   (b) $y_2(t) = g_a(t/3)$,   (c) $y_3(t) = g_a(3t)$,   (d) $y_4(t) = \int_{-\infty}^{\infty} g_a(t-\tau)g_a(\tau)d\tau$,

(e) $y_5(t) = \frac{dg_a(t)}{dt}$.

4.4 A finite-energy continuous-time signal $g_a(t)$ is sampled at a rate satisfying the Nyquist condition of Eq. (4.11), generating a discrete-time sequence $g[n]$. Develop the relation between the total energy $\mathcal{E}_{g_a(t)}$ of the continuous-time signal $g_a(t)$ and the total energy $\mathcal{E}_{g[n]}$ of the discrete-time signal $g[n]$.

4.5 A 2.5 s long segment of a continuous-time signal is uniformly sampled without aliasing and generating a finite-length sequence containing 5001 samples. What is the highest frequency component that could be present in the continuous-time signal?

4.6 A continuous-time signal $x_a(t)$ is composed of a linear combination of sinusoidal signals of frequencies 300 Hz, 500 Hz, 1.2 kHz, 2.15 kHz, and 3.5 kHz. The signal $x_a(t)$ is sampled at a 2.0-kHz rate, and the sampled sequence is passed through an ideal lowpass filter with a cutoff frequency of 900 Hz, generating a continuous-time signal $y_a(t)$. What are the frequency components present in the reconstructed signal $y_a(t)$?

4.7 A continuous-time signal $x_a(t)$ is composed of a linear combination of sinusoidal signals of frequencies F_1 Hz, F_2 Hz, F_3 Hz, and F_4 Hz. The signal $x_a(t)$ is sampled at an 10-kHz rate, and the sampled sequence is then passed through an ideal lowpass filter with a cutoff frequency of 4 kHz, generating a continuous-time signal $y_a(t)$ composed of three sinusoidal signals of frequencies 350 Hz, 575 Hz, and 815 Hz, respectively. What are the possible values of F_1, F_2, F_3, and F_4? Is your answer unique? If not, indicate another set of possible values of these frequencies.

4.8 The continuous-time signal $x_a(t) = 4\sin(20\pi t) - 5\cos(24\pi t) + 3\sin(120\pi t) + 2\cos(176\pi t)$ is sampled at a 50 Hz rate, generating the sequence $x[n]$. Determine the exact expression of $x[n]$.

4.9 The left and right channels of an analog stereo audio signal are sampled at a 45-kHz rate, with each channel then being converted into a digital bit stream using a 12-bit A/D converter. Determine the combined bit rate of the two channels after sampling and digitization.

4.10 Show that the impulse response $h_r(t)$ of an ideal lowpass filter as derived in Eq. (4.19) takes the value $h_r(nT) = \delta[n]$ for all n if the cutoff frequency $\Omega_c = \Omega_T/2$, where Ω_T is the sampling frequency.

4.11 Consider the system of Figure 4.2, where the input continuous-time signal $x_a(t)$ has a band-limited spectrum $X_a(j\Omega)$, as sketched in Figure P4.1(a), and is being sampled at the Nyquist rate. The discrete-time processor is an ideal lowpass filter with a frequency response $H(e^{j\omega})$, as shown in Figure P4.1(b), and has a cutoff frequency $\omega_c = \Omega_m T/3$, where T is the sampling period. Sketch as accurately as possible the spectrum $Y_a(j\Omega)$ of the output continuous-time signal $y_a(t)$.

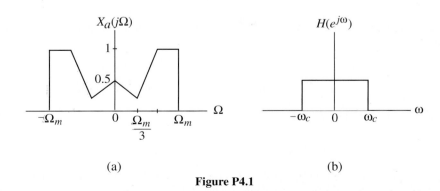

(a)   (b)

Figure P4.1

4.12 A continuous-time signal $x_a(t)$ has a band-limited spectrum $X_a(j\Omega)$, as indicated in Figure P4.2. Determine the smallest sampling frequency F_T that can be employed to sample $x_a(t)$ so that it can be fully recovered from its sampled version $x[n]$ for each of the following sets of values of the bandedges Ω_1 and Ω_2. Sketch the Fourier transform of the sampled version $x[n]$ obtained by sampling $x_a(t)$ at the smallest sampling rate F_T and the frequency response of the ideal reconstruction filter needed to fully recover $x_a(t)$ for each case.

    (a) $\Omega_1 = 100\pi$, $\Omega_2 = 150\pi$;   (b) $\Omega_1 = 160\pi$, $\Omega_2 = 250\pi$;   (c) $\Omega_1 = 110\pi$, $\Omega_2 = 180\pi$.

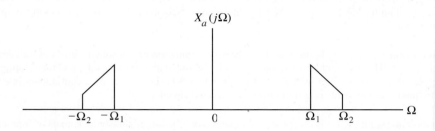

Figure P4.2

4.13 For each set of desired peak passband deviation α_p and the minimum stopband attenuation α_s of an analog lowpass filter given below, determine the corresponding passband and stopband ripples, δ_p and δ_s :

    (a) $\alpha_p = 0.21$ dB, $\alpha_s = 52$ dB;   (b) $\alpha_p = 0.03$ dB, $\alpha_s = 69$ dB;   (c) $\alpha_p = 0.33$ dB, $\alpha_s = 57$ dB.

4.14 Show that the analog transfer function

$$H_a(s) = \frac{a}{s+a}, \qquad a > 0, \tag{4.92}$$

has a lowpass magnitude response with a monotonically decreasing magnitude response with $|H_a(j0)| = 1$ and $|H_a(j\infty)| = 0$. Determine the 3-dB cutoff frequency Ω_c at which the gain response is 3 dB below the maximum value of 0 dB at $\Omega = 0$.

4.15 Show that the analog transfer function

$$G_a(s) = \frac{s}{s+a}, \qquad a > 0, \tag{4.93}$$

has a highpass magnitude response with a monotonically increasing magnitude response with $|G_a(j0)| = 0$ and $|G_a(j\infty)| = 1$. Determine the 3-dB cutoff frequency Ω_c at which the gain response is 3 dB below the maximum value of 0 dB at $\Omega = \infty$.

4.16 The lowpass transfer function $H_a(s)$ of Eq. (4.92) and the highpass transfer function $G_a(s)$ of Eq. (4.93) can be expressed in the form

$$H_a(s) = \tfrac{1}{2}\{A_0(s) - A_1(s)\}, \qquad G_a(s) = \tfrac{1}{2}\{A_0(s) + A_1(s)\},$$

where $A_0(s)$ and $A_1(s)$ are stable analog allpass transfer functions. Determine $A_0(s)$ and $A_1(s)$.

4.17 Show that the analog transfer function

$$H_a(s) = \frac{bs}{s^2 + bs + \Omega_o^2}, \qquad b > 0, \tag{4.94}$$

has a bandpass magnitude response with $|H_a(j0)| = |H_a(j\infty)| = 0$ and $|H_a(j\Omega_o)| = 1$. Determine the frequencies Ω_1 and Ω_2 at which the gain is 3 dB below the maximum value of 0 dB at Ω_o. Show that $\Omega_1\Omega_2 = \Omega_o^2$. The difference $\Omega_2 - \Omega_1$ is called the 3-dB bandwidth of the bandpass transfer function. Show that $b = \Omega_2 - \Omega_1$.

4.18 Show that the analog transfer function

$$G_a(s) = \frac{s^2 + \Omega_o^2}{s^2 + bs + \Omega_o^2}, \qquad b > 0, \tag{4.95}$$

has a bandstop magnitude response with $|G_a(j0)| = |G_a(j\infty)| = 1$ and $|G_a(j\Omega_o)| = 0$. Since the magnitude is exactly zero at Ω_o, it is called the notch frequency, and $G_a(s)$ is often called the notch transfer function. Determine the frequencies Ω_1 and Ω_2 at which the gain is 3 dB below the maximum value of 0 dB at $\Omega = 0$ and $\Omega = \infty$. Show that $\Omega_1\Omega_2 = \Omega_o^2$. The difference $\Omega_2 - \Omega_1$ is called the 3-dB notch bandwidth of the bandpass transfer function. Show that $b = \Omega_2 - \Omega_1$.

4.19 The bandpass transfer function $H_a(s)$ of Eq. (4.94) and the bandstop transfer function $G_a(s)$ of Eq. (4.95) can be expressed in the form

$$H_a(s) = \tfrac{1}{2}\{A_0(s) - A_1(s)\}, \qquad G_a(s) = \tfrac{1}{2}\{A_0(s) + A_1(s)\},$$

where $A_0(s)$ and $A_1(s)$ are stable analog allpass transfer functions. Determine $A_0(s)$ and $A_1(s)$.

4.20 An analog real-coefficient allpass transfer function $A(s)$ is defined by $|A(j\Omega)|^2 = 1$, where $A(j\Omega)$ is the magnitude function of the transfer function.

(a) Show that an analog real-coefficient causal and stable allpass transfer function $A(s)$ is given by

$$A(s) = \prod_{i=1}^{N} \left(\frac{s + \lambda_i^*}{s - \lambda_i} \right), \tag{4.96}$$

where $\mathrm{Re}\{\lambda_i\} < 0$.

(b) Show that an analog real-coefficient causal and stable allpass transfer function $A(s)$ satisfies the following property:

$$|A(s)| \begin{cases} < 1 & \text{for } \mathrm{Re}(s) > 0, \\ = 1 & \text{for } \mathrm{Re}(s) = 0, \\ > 1 & \text{for } \mathrm{Re}(s) < 0. \end{cases}$$

4.21 Show that the first $2N - 1$ derivatives of the squared-magnitude response $|H_a(j\Omega)|^2$ of a Butterworth filter of order N as given by Eq. (4.33) are equal to zero at $\Omega = 0$.

4.22 Using Eq. (4.35), determine the lowest order of a lowpass Butterworth filter with a 0.25-dB cutoff frequency at 1.5 kHz and a minimum attenuation of 25 dB at 6 kHz. Verify your result using `buttord`.

4.23 Using Eq. (4.37), determine the pole locations and the coefficients of a sixth-order Butterworth polynomial with unity 3-dB cutoff frequency.

4.24 Show that the Chebyshev polynomial $T_N(\Omega)$ defined in Eq. (4.40) satisfies the recurrence relation given in Eq. (4.41) with $T_0(\Omega) = 1$, and $T_1(\Omega) = \Omega$.

4.25 Using Eq. (4.43), determine the lowest order of a lowpass Type 1 Chebyshev filter with a 0.25-dB cutoff frequency at 1.5 kHz and a minimum attenuation of 25 dB at 6 kHz. Verify your result using `cheb1ord`.

4.26 Using Eq. (4.54) determine the lowest order of a lowpass elliptic filter with a 0.25-dB cutoff frequency at 1.5 kHz and a minimum attenuation of 25 dB at 6 kHz. Verify your result using `ellipord`.

4.27 Determine the Bessel polynomials $B_N(s)$ for the following values of N: (a) $N = 4$ and (b) $N = 5$.

4.28 The transfer function of a third-order analog Butterworth lowpass filter with a passband edge at 0.24 Hz and a passband ripple of 0.5 dB is given by

$$H_{LP}(s) = \frac{10}{s^3 + 4.309s^2 + 9.2835s + 10}.$$

Determine the transfer function $H_{HP}(s)$ of an analog highpass filter with a passband edge at 3 Hz and a passband ripple of 0.5 dB by applying the spectral transformation of Eq. (4.62).

4.29 The transfer function of a third-order analog Butterworth highpass filter with a passband edge at 0.9 Hz and a passband ripple of 1 dB is given by

$$H_{HP}(s) = \frac{s^3}{s^3 + 9.283s^2 + 40.087s + 100}.$$

Determine the transfer function $H_{LP}(s)$ of an analog lowpass filter with a passband edge at 3 Hz and a passband ripple of 1 dB by applying the spectral transformation of Eq. (4.62).

4.30 The transfer function of a second-order analog elliptic lowpass filter with a passband edge at 0.25 Hz and a passband ripple of 0.5 dB is given by

$$H_{LP}(s) = \frac{0.01(s^2 + 367.93)}{s^2 + 2.269s + 3.895}.$$

Determine the transfer function $H_{BP}(s)$ of an analog bandpass filter with a center frequency at 3 Hz and a bandwidth of 0.5 Hz by applying the spectral transformation of Eq. (4.64).

4.31 A Butterworth analog highpass filter is to be designed with the following specifications: $F_p = 6.5$ kHz, $F_s = 1.5$ kHz, $\alpha_p = 0.5$ dB, and $\alpha_s = 40$ dB. What are the bandedges and the order of the corresponding analog lowpass filter? What is the order of the highpass filter? Verify your results using the function `buttord`.

4.32 An elliptic analog bandpass filter is to be designed with the following specifications: passband edges at 20 kHz and 45 kHz, stopband edges at 15 kHz and 50 kHz, peak passband ripple of 0.25 dB, and minimum stopband attenuation of 50 dB. What are the bandedges and the order of the corresponding analog lowpass filter? What is the order of the bandpass filter? Verify your results using the function `ellipord`.

4.33 A Type 1 Chebyshev analog bandstop filter is to be designed with the following specifications: passband edges at 10 MHz and 70 MHz, stopband edges at 20 MHz and 45 MHz, peak passband ripple of 0.5 dB, and minimum stopband attenuation of 30 dB. What are the bandedges and the order of the corresponding analog lowpass filter? What is the order of the bandstop filter? Verify your results using the function `cheb1ord`.

4.34 Verify Table 4.1.

4.35 Derive Eq. (4.76).

4.36 Derive Eq. (4.77).

4.37 An alternative to the zero-order hold circuit of Figure 4.54 used for signal reconstruction at the output of a D/A converter is the *first-order hold circuit,* which approximates $y_a(t)$ according to the following relation:

$$y_f(t) = y_p(nT) + \frac{y_p(nT) - y_p(nT - T)}{T}(t - nT), \quad nT \leq t < (n+1)T.$$

As indicated by the above equation, the first-order hold circuit approximates $y_a(t)$ by straight-line segments. The slope of the segment between $t = nT$ and $t = (n+1)T$ is determined from the sample values $y_p(nT)$ and $y_p(nT - T)$. Determine the impulse response $h_f(t)$ and the frequency response $H_f(j\Omega)$ of the first-order hold circuit, and compare its performance with that of the zero-order hold circuit.

4.38 A more improved signal reconstruction at the output of a D/A converter is provided by a linear interpolation circuit, which approximates $y_a(t)$ by connecting successive sample points of $y_p(t)$ with straight-line segments. The input–output relation of this circuit is given by

$$y_f(t) = y_p(nT - T) + \frac{y_p(nT) - y_p(nT - T)}{T}(t - nT), \quad nT \le t < (n+1)T.$$

Determine the impulse response $h_f(t)$ and the frequency response $H_f(j\Omega)$ of the linear interpolation circuit, and compare its performance with that of the first-order hold circuit.

4.14 MATLAB Exercises

M 4.1 Determine the transfer function of a lowpass Butterworth analog filter with specifications as given in Problem 4.22, using Program 4_2. Plot the gain response and verify that the filter designed meets the given specifications. Show all steps.

M 4.2 Determine the transfer function of a lowpass Type 1 Chebyshev analog filter with specifications as given in Problem 4.25, using Program 4_3. Plot the gain response and verify that the filter designed meets the given specifications. Show all steps.

M 4.3 Modify Program 4_3 to design lowpass Type 2 Chebyshev analog filters. Using this program, determine the transfer function of a lowpass Type 2 Chebyshev analog filter with specifications as given in Problem 4.25. Plot the gain response and verify that the filter designed meets the given specifications. Show all steps.

M 4.4 Determine the transfer function of a lowpass elliptic analog filter with specifications as given in Problem 4.26, using Program 4_4. Plot the gain response and verify that the filter designed meets the given specifications. Show all steps.

M 4.5 Design, using MATLAB, a Butterworth analog highpass filter with specifications given in Problem 4.31. Show the transfer functions of the prototype analog lowpass and the highpass filters. Plot their gain responses and verify that both filters meet their respective specifications. Show all steps.

M 4.6 Design an elliptic analog bandpass filter with specifications given in Problem 4.32. Show the transfer functions of the prototype analog lowpass and the bandpass filters. Plot their gain responses and verify that both filters meet their respective specifications. Show all steps.

M 4.7 Design a Type 1 analog bandstop filter with specifications given in Problem 4.33. Show the transfer functions of the prototype analog lowpass and the bandstop filters. Plot their gain responses and verify that both filters meet their respective specifications. Show all steps.

M 4.8 Write a MATLAB program to verify the plots of Figure 4.56.

5 Finite-Length Discrete Transforms

Often, in practice, it is convenient to map a finite-length sequence from the time domain into a finite-length sequence of the same length in a different domain, and vice-versa. Such transformations are usually collectively called *finite-length transforms* and are the subject of this chapter. In the forward transform, the samples of the transform are unique and represented as a linear combination of the samples of the time-domain sequence. The original time-domain sequence can be obtained by applying an inverse transform in which the time-domain samples are expressed as a linear combination of the samples of its transform-domain representation.

In some applications, a very long length time-domain sequence is broken up into a set of short-length time-domain sequences and a finite-length transform is applied to each short-length sequence. The transformed sequences are next processed in the transform domain, and their time-domain equivalents are generated by applying the inverse transforms. The processed short-length sequences are then grouped together appropriately to develop the final long-length sequence.

A variety of finite-dimensional transforms have been advanced, and a discussion on each of these transforms is beyond the scope of this book. We restrict our attention in this chapter to the so-called class of *orthogonal transforms*. In particular, we discuss here three such transforms, namely, the *discrete Fourier transform*, the *discrete cosine transform*, and the *Haar transform*. The former transform is widely used in a number of digital signal processing applications, whereas the latter two find applications primarily in signal compression.

5.1 Orthogonal Transforms

Let $x[n]$ denote a length-N time-domain sequence with $\mathcal{X}[k]$ denoting the coefficients of its N-point orthogonal transform. Then, a general form of the orthogonal transform pair is of the form

$$\mathcal{X}[k] = \sum_{n=0}^{N-1} x[n]\psi^*[k, n], \qquad 0 \le k \le N - 1, \tag{5.1}$$

$$x[n] = \frac{1}{N} \sum_{k=0}^{N-1} \mathcal{X}[k]\psi[k, n], \qquad 0 \le n \le N - 1. \tag{5.2}$$

Equation (5.1) usually is referred to as the *analysis equation,* whereas Eq. (5.2) is referred to as the *synthesis equation.* In these equations, $\psi[k, n]$, called the *basis sequences,* are also length-N sequences in both domains. In the class of finite-dimensional transforms we shall deal with in this chapter, the basis sequences satisfy the condition

$$\frac{1}{N} \sum_{n=0}^{N-1} \psi[k, n]\psi^*[\ell, n] = \begin{cases} 1, & \ell = k, \\ 0, & \ell \ne k. \end{cases} \tag{5.3}$$

Basis sequences $\psi[k, n]$ satisfying the above condition are said to be *orthogonal* to each other.

To verify the inverse transform expression of Eq. (5.2), we substitute it into Eq. (5.1), yielding

$$\sum_{n=0}^{N-1} x[n]\psi^*[\ell, n] = \sum_{n=0}^{N-1} \left(\frac{1}{N} \sum_{k=0}^{N-1} \mathcal{X}[k]\psi[k, n] \right) \psi^*[\ell, n]. \tag{5.4}$$

An interchange of the order of summation on the right-hand side of the above equation yields

$$\sum_{n=0}^{N-1} x[n]\psi^*[\ell, n] = \sum_{k=0}^{N-1} \mathcal{X}[k] \left(\frac{1}{N} \sum_{n=0}^{N-1} \psi[k, n]\psi^*[\ell, n] \right) = \mathcal{X}[\ell], \tag{5.5}$$

by virtue of the orthogonality condition of Eq. (5.3) satisfied by the basis sequences $\psi[k, n]$, thus verifying that Eq. (5.2) indeed is the inverse transform formula.

An important consequence of the orthogonality of the basis sequence is the energy preservation property of the transform that allows one to compute the energy $\sum_{n=0}^{N-1} |x[n]|^2$ of the time-domain sequence $x[n]$ in the transform domain. To demonstrate this property, we observe

$$\sum_{n=0}^{N-1} |x[n]|^2 = \sum_{n=0}^{N-1} x[n]x^*[n].$$

In the last term, we can express $x[n]$ in terms of its transform-domain representation given by Eq. (5.2) and obtain

$$\sum_{n=0}^{N-1} x[n]x^*[n] = \sum_{n=0}^{N-1} \left(\frac{1}{N} \sum_{k=0}^{N-1} \mathcal{X}[k]\psi[k, n] \right) x^*[n].$$

Interchanging the summations in the last term of the above equation and making use of Eq. (5.1), we arrive at

$$\frac{1}{N} \sum_{k=0}^{N-1} \mathcal{X}[k] \left(\sum_{n=0}^{N-1} x^*[n]\psi[k, n] \right) = \frac{1}{N} \sum_{k=0}^{N-1} \mathcal{X}[k]\mathcal{X}^*[k].$$

Hence, we have

$$\sum_{n=0}^{N-1} |x[n]|^2 = \frac{1}{N} \sum_{k=0}^{N-1} |\mathcal{X}[k]|^2, \tag{5.6}$$

which is more commonly known as the *Parseval's relation*.

In some applications, it is also desirable to use transforms that decorrelate the transform coefficients $\mathcal{X}[k]$, $0 \leq k \leq N - 1$. Energy compaction is another attractive property that is taken advantage of in signal compression applications. In transforms with good energy compaction properties, most of the signal energy is concentrated in a subset of the transform coefficients, allowing the remaining coefficients with very low energy to be set to zero values. This process leads to an efficient approximation of the time-domain signal in the transform domain and is the basis of many signal compression methods.

5.2 The Discrete Fourier Transform

In this section, we define the discrete Fourier transform, usually known as the DFT, and develop the inverse transformation, often abbreviated as IDFT. We relate the DFT to the Fourier transform of the time-domain sequence, review its major properties, and study especially two of its unique properties in the following section. Several important applications of the DFT, such as the numerical computation of the Fourier transform and implementation of linear convolution, are discussed in a later section.

5.2.1   Definition

The *discrete Fourier transform* (DFT) of the length-N time-domain sequence $x[n]$ is defined by

$$X[k] = \sum_{n=0}^{N-1} x[n] e^{-j 2\pi kn/N}, \qquad 0 \le k \le N - 1, \tag{5.7}$$

obtained by setting the basis sequences in Eq. (5.1) as

$$\psi[k, n] = e^{j 2\pi kn/N}, \tag{5.8}$$

which are complex exponential sequences. As a result, the DFT coefficients, $X[k]$, in general are complex numbers even when $x[n]$ are real. It can be easily shown that the basis sequences $e^{j 2\pi kn/N}$ are orthogonal: that is, they satisfy the condition of Eq. (5.3). To this end, we observe that

$$\frac{1}{N} \sum_{n=0}^{N-1} e^{j 2\pi kn/N} e^{-j 2\pi \ell n/N} = \frac{1}{N} \sum_{n=0}^{N-1} e^{j 2\pi (k-\ell)n/N}.$$

By setting $u = e^{j 2\pi (k-\ell)/N}$, the right-hand side of the above equation reduces to a finite sum of the form

$$S_{N-1} = \sum_{n=0}^{N-1} u^n.$$

It follows from the above that

$$\sum_{n=0}^{N} u^n = 1 + u \sum_{n=0}^{N-1} u^n = 1 + u S_{N-1} = S_{N-1} + u^N,$$

which, when solved, yields

$$S_{N-1} = \frac{1 - u^N}{1 - u}. \tag{5.9}$$

Substituting $u = e^{j 2\pi (k-\ell)/N}$ in the above equation we get

$$\sum_{n=0}^{N-1} e^{j 2\pi (k-\ell)n/N} = \frac{1 - e^{j 2\pi (k-\ell)}}{1 - e^{j 2\pi (k-\ell)/N}}. \tag{5.10}$$

For $k \ne \ell$, the numerator of the right-hand side of the above equation is equal to zero. For $k = \ell + rN$, the right-hand side of the above equation is of the form $0/0$. However, it can be seen from the left-hand side that for $k = \ell$, the sum is equal to N. Hence,

$$\frac{1}{N} \sum_{n=0}^{N-1} e^{j 2\pi (k-\ell)n/N} = \begin{cases} 1, & \text{for } k = \ell + rN, \\ 0, & \text{for } k \ne \ell, \end{cases} \tag{5.11}$$

verifying the orthogonality property of the basis sequences $e^{j 2\pi kn/N}$. It follows from Eq. (5.7) that the DFT $X[k]$ is a length-N sequence in the transform domain. Often, the length-N DFT sequence is referred to as the *N-point DFT*. Applying the commonly used notation

$$W_N = e^{-j 2\pi/N}, \tag{5.12}$$

we can rewrite Eq. (5.7) as

$$X[k] = \sum_{n=0}^{N-1} x[n]W_N^{kn}, \qquad 0 \le k \le N - 1. \tag{5.13}$$

The *inverse discrete Fourier transform* (IDFT) is given by

$$x[n] = \frac{1}{N} \sum_{k=0}^{N-1} X[k]W_N^{-kn}, \qquad 0 \le n \le N - 1, \tag{5.14}$$

obtained using the basis sequences of Eq. (5.8) in Eq. (5.2). As can be seen from the above expression, the inverse DFT $x[n]$ can be a complex sequence even when the DFT $X[k]$ is a real sequence.

Equations (5.13) and (5.14) constitute a discrete Fourier transform pair for the sequence $x[n]$. A discrete Fourier transform pair will often be denoted as

$$x[n] \xrightarrow{\text{DFT}} X[k]. \tag{5.15}$$

The DFT computation is illustrated in Examples (5.1) and (5.2).

EXAMPLE 5.1   DFT Computation of a Finite-Length Sequence with a Single Non-Zero Sample

Consider the length-N sequence defined for $0 \le n \le N - 1$:

$$x[n] = \begin{cases} 1, & n = 0, \\ 0, & 1 \le n \le N - 1. \end{cases} \tag{5.16}$$

Its N-point DFT is obtained by applying Eq. (5.7), resulting in

$$X[k] = 1. \tag{5.17}$$

Now consider the length-N sequence defined for $n = 0, 1, \ldots, N - 1$

$$y[n] = \begin{cases} 1, & n = m, \ \ 0 \le m \le N - 1, \\ 0, & \text{otherwise.} \end{cases} \tag{5.18}$$

Its DFT is given by

$$Y[k] = W_N^{km}. \tag{5.19}$$

EXAMPLE 5.2   DFT Computation of a Finite-Length Sinusoidal Sequence

Compute the N-point DFT of the length-N sequence

$$x[n] = \cos(2\pi r n / N), \qquad 0 \le n \le N - 1, \tag{5.20}$$

where r is an integer in the range $0 < r \le N - 1$. Note that using a trigonometric identity and the notation of Eq. (5.12), we can rewrite $x[n]$,

$$x[n] = \tfrac{1}{2}\left(e^{j2\pi r n/N} + e^{-j2\pi r n/N} \right) = \tfrac{1}{2}\left(W_N^{-rn} + W_N^{rn} \right). \tag{5.21}$$

Substituting the above in Eq. (5.13), we arrive at its DFT,

$$X[k] = \frac{1}{2}\left[\sum_{n=0}^{N-1} W_N^{-(r-k)n} + \sum_{n=0}^{N-1} W_N^{(r+k)n} \right]. \tag{5.22}$$

Making use of the identity

$$\sum_{n=0}^{N-1} W_N^{-(k-\ell)n} = \begin{cases} N, & \text{for } k - \ell = rN, r \text{ an integer,} \\ 0, & \text{otherwise,} \end{cases} \tag{5.23}$$

in Eq. (5.22), we obtain the DFT of the length-N sequence $x[n]$ of Eq. (5.20):

$$X[k] = \begin{cases} N/2, & \text{for } k = r, \\ N/2, & \text{for } k = N - r, \\ 0, & \text{otherwise.} \end{cases} \tag{5.24}$$

As can be seen from Eqs. (5.13) and (5.14), the computation of the DFT and the IDFT requires, respectively, approximately N^2 complex multiplications and $N(N - 1)$ complex additions. However, elegant methods have been developed to reduce the computational complexity to about $N(\log_2 N)$ operations. These techniques are usually called fast Fourier transform (FFT) algorithms and are discussed in Sections 11.3.2 and 11.4. As a result of the availability of these fast algorithms, the DFT and the IDFT, and their variations, are often used in digital signal processing applications for various purposes.

5.2.2 Matrix Relations

The DFT samples defined in Eq. (5.13) can be expressed in matrix form as

$$\mathbf{X} = \mathbf{D}_N \mathbf{x}, \tag{5.25}$$

where $\mathbf{X}$ is the vector composed of the N DFT samples,

$$\mathbf{X} = [\, X[0] \quad X[1] \quad \cdots \quad X[N - 1] \,]^T, \tag{5.26}$$

$\mathbf{x}$ is the vector of N input samples,

$$\mathbf{x} = [\, x[0] \quad x[1] \quad \cdots \quad x[N - 1] \,]^T, \tag{5.27}$$

and $\mathbf{D}_N$ is the $N \times N$ DFT matrix given by

$$\mathbf{D}_N = \begin{bmatrix} 1 & 1 & 1 & \cdots & 1 \\ 1 & W_N^1 & W_N^2 & \cdots & W_N^{N-1} \\ 1 & W_N^2 & W_N^4 & \cdots & W_N^{2(N-1)} \\ \vdots & \vdots & \vdots & \ddots & \vdots \\ 1 & W_N^{N-1} & W_N^{2(N-1)} & \cdots & W_N^{(N-1)(N-1)} \end{bmatrix}. \tag{5.28}$$

Likewise, the IDFT relations can be expressed in matrix form as

$$\begin{bmatrix} x[0] \\ x[1] \\ \vdots \\ x[N-1] \end{bmatrix} = \mathbf{D}_N^{-1} \begin{bmatrix} X[0] \\ X[1] \\ \vdots \\ X[N-1] \end{bmatrix}, \tag{5.29}$$

where $\mathbf{D}_N^{-1}$ is the $N \times N$ *IDFT matrix* given by

$$
\mathbf{D}_N^{-1} = \frac{1}{N}
\begin{bmatrix}
1 & 1 & 1 & \cdots & 1 \\
1 & W_N^{-1} & W_N^{-2} & \cdots & W_N^{-(N-1)} \\
1 & W_N^{-2} & W_N^{-4} & \cdots & W_N^{-2(N-1)} \\
\vdots & \vdots & \vdots & \ddots & \vdots \\
1 & W_N^{-(N-1)} & W_N^{-2(N-1)} & \cdots & W_N^{-(N-1)(N-1)}
\end{bmatrix}.
\tag{5.30}
$$

It follows from Eqs. (5.28) and (5.30) that

$$
\mathbf{D}_N^{-1} = \frac{1}{N} \mathbf{D}_N^*.
\tag{5.31}
$$

5.2.3 DFT Computation Using MATLAB

There are four built-in functions in MATLAB for the computation of the DFT and the IDFT:

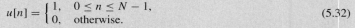

All of these functions make use of FFT algorithms, which are computationally highly efficient compared to the direct computation of DFT and the inverse DFT. In addition, the function `dftmtx(N)` in the *Signal Processing Toolbox* of MATLAB can be used to compute the $N \times N$ DFT matrix $\mathbf{D}_N$, defined in Eq. (5.28). To compute the inverse of the $N \times N$ DFT matrix, one can use the function `conj(dftmtx(N))/N`.

We illustrate the application of the above M-files in Examples 5.3 and 5.4.

EXAMPLE 5.3   DFT Computation Using MATLAB

Using MATLAB, we determine the M-point DFT $U[k]$ of the following N-point sequence:

$$
u[n] = \begin{cases} 1, & 0 \le n \le N-1, \\ 0, & \text{otherwise.} \end{cases}
\tag{5.32}
$$

To this end, we can use Program 5_1. During execution, the program requests the input data N and M. To ensure correct DFT values, M must be greater than or equal to N. After they are entered, it computes the M-point DFT and plots the original N-point sequence and the magnitude and phase of the DFT sequence, as indicated in Figure 5.1 for $N = 8$ and $M = 16$.

EXAMPLE 5.4   IDFT Computation Using MATLAB

We next illustrate the computation of the IDFT using MATLAB. The K-point DFT sequence is given by

$$
V[k] = \begin{cases} k/K, & 0 \le k \le K-1, \\ 0, & \text{otherwise.} \end{cases}
\tag{5.33}
$$

We make use of Program 5_2 to determine its IDFT sequence $v[n]$.

As the program is run, it calls for the input data consisting of the length of the DFT and the length of the IDFT. It then computes the IDFT of the ramp DFT sequence of Eq. (5.23) and plots the original DFT sequence and its IDFT, as indicated in Figure 5.2. Note that even though the DFT sequence is real, its IDFT is a complex time-domain sequence as expected.

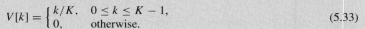

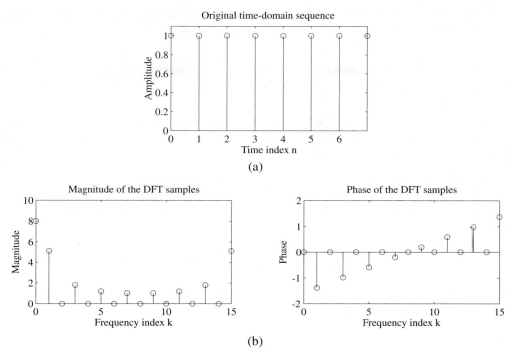

Figure 5.1: (a) Original length-N sequence of Eq. (5.32) and (b) its M-point DFT for $N = 8$ and $M = 16$.

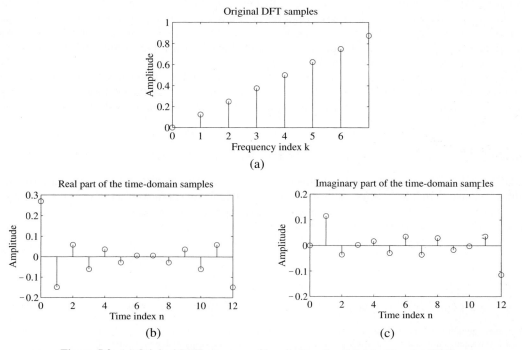

Figure 5.2: (a) Original DFT sequence of length $K = 8$ and (b) its 13-point IDFT.

The M-file `fftshift` shifts the zero-frequency sample at the frequency index $k = 0$ to the center of the spectrum and is often useful in visualizing the spectrum of a sequence.

5.3   Relation Between the Fourier Transform and the DFT and Their Inverses

We now examine the explicit relation between the Fourier transform and the N-point DFT of a length-N sequence and the relation between the Fourier transform of a length-M sequence and the N-point DFT obtained by sampling the Fourier transform.

5.3.1   Relation with Discrete-Time Fourier Transform

From Eq. (3.12), the Fourier transform $X(e^{j\omega})$ of the length-N sequence $x[n]$, defined for $0 \leq n \leq N-1$, is given by

$$X(e^{j\omega}) = \sum_{n=-\infty}^{\infty} x[n]e^{-j\omega n} = \sum_{n=0}^{N-1} x[n]e^{-j\omega n}. \tag{5.34}$$

By uniformly sampling $X(e^{j\omega})$ at N equally spaced frequencies $\omega_k = 2\pi k/N, 0 \leq k \leq N-1$, on the ω-axis between $0 \leq \omega < 2\pi$, we get from the above equation

$$X(e^{j\omega})\big|_{\omega=2\pi k/N} = \sum_{n=0}^{N-1} x[n]e^{-j2\pi k/N}, \qquad 0 \leq k \leq N-1. \tag{5.35}$$

Comparing Eq. (5.35) with Eq. (5.7), we observe that the N-point DFT sequence $X[k]$ is precisely the set of frequency samples of the Fourier transform $X(e^{j\omega})$ of the length-N sequence $x[n]$ at N equally spaced frequencies $\omega_k = 2\pi k/N, 0 \leq k \leq N-1$. Hence, Eq. (5.7) represents a frequency-domain representation of the sequence $x[n]$.[1] Since the computation of the DFT samples involve a finite sum, for time-domain sequences with finite sample values, the DFT always exists.

Because of the explicit relation between the DFT samples and the frequency samples of the Fourier transform, the normalized angular frequency associated with the index k of the DFT sample $X[k]$ is $2\pi k/N$ radians. For example, for $N = 32$, the sample index 11 represents the normalized angular frequency $\omega = 11\pi/16$.

5.3.2   Numerical Computation of the Fourier Transform Using the DFT

The DFT provides a practical approach to the numerical computation of the Fourier transform of a finite-length sequence, particularly if fast algorithms are available for the computation of the DFT. Let $X(e^{j\omega})$ be the Fourier transform of a length-N sequence $x[n]$. We wish to evaluate $X(e^{j\omega})$ at a dense grid of frequencies $\omega_k = 2\pi k/M, 0 \leq k \leq M-1$, where $M \gg N$:

$$X(e^{j\omega_k}) = \sum_{n=0}^{N-1} x[n]e^{-j\omega_k n} = \sum_{n=0}^{N-1} x[n]e^{-j2\pi kn/M}. \tag{5.36}$$

Define a new sequence $x_e[n]$ obtained from $x[n]$ by augmenting with $M - N$ zero-valued samples:

---

[1] A generalization of the DFT concept is the *nonuniform discrete Fourier transform* (NDFT), obtained by sampling the Fourier transform at nonuniformly spaced frequency points [Bag98]. The NDFT is investigated in Problem 6.28.

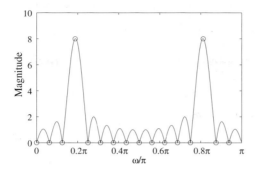

Figure 5.3: The magnitudes of the Fourier transform $X(e^{j\omega})$ and the DFT $X[k]$ of the sequence $x[n]$ of Eq. (5.20), with $r = 3$ and $N = 16$. The Fourier transform is plotted as a solid line, and the DFT samples are shown by circles.

$$x_e[n] = \begin{cases} x[n], & 0 \le n \le N - 1, \\ 0, & N \le n \le M - 1. \end{cases} \tag{5.37}$$

Making use of $x_e[n]$ in Eq. (5.36), we arrive at

$$X(e^{j\omega_k}) = \sum_{n=0}^{M-1} x_e[n]e^{-j2\pi kn/M}, \tag{5.38}$$

which is seen to be an M-point DFT $X_e[k]$ of the length-M sequence $x_e[n]$. The DFT $X_e[k]$ can be computed very efficiently using the FFT algorithm if M is an integer power of 2.

The MATLAB function `freqz`, described in Section 3.6, employs the above approach to evaluate the frequency response of a rational Fourier transform expressed as a rational function in $e^{-j\omega}$ at a prescribed set of discrete frequencies. It computes the DFTs of the numerator and the denominator separately by considering each as finite-length sequences and then expresses the ratio of the DFT samples at each frequency point to evaluate the Fourier transform.

EXAMPLE 5.5   DTFT Computation Using MATLAB

Let $r = 3$ and $N = 16$ for the finite-length sequence $x[n]$ of Eq. (5.20). From Eq. (5.24), its 16-point DFT is therefore given by

$$X[k] = \begin{cases} 8, & \text{for } k = 3, \\ 8, & \text{for } k = 13, \\ 0, & \text{otherwise.} \end{cases}$$

Program 5_3.m

Since the Fourier transform $X(e^{j\omega})$ is a continuous function of ω, we can plot it as a function of ω by computing the DFT of the sequence $x[n]$ using MATLAB at a dense grid of frequencies. Program 5_3 computes first a 512-point DFT, determines next the 16-point DFT of $x[n]$ given by Eq. (5.20), and then places the DFT samples on top of the Fourier transform $X(e^{j\omega})$ verifying that the 16-point DFT of the length-N sequence $x[n]$ are simply the 16 samples of the Fourier transform evaluated at equally spaced samples starting at $\omega = 0$. Figure 5.3 shows the plot of Fourier transform $X(e^{j\omega})$ along with the DFT samples $X[k]$ obtained using this program. As indicated in this figure, the DFT values $X[k]$, indicated by circles, are precisely the frequency samples of the Fourier transform $X(e^{j\omega})$ at $\omega = \pi k/8$, $0 \le k \le 15$.

5.3.3 Fourier Transform from DFT by Interpolation

Given the N-point DFT $X[k]$ of a length-N sequence $x[n]$, it is also possible to determine its Fourier transform $X(e^{j\omega})$ uniquely by interpolating $X[k]$ at all values of ω. To develop the desired relation, we first observe that the Fourier transform $X(e^{j\omega})$ of $x[n]$ is given by Eq. (5.34). Using the IDFT expression for $x[n]$ given by Eq. (5.14) in Eq. (5.34), we get

$$X(e^{j\omega}) = \frac{1}{N} \sum_{n=0}^{N-1} \left[\sum_{k=0}^{N-1} X[k] W_N^{-kn} \right] e^{-j\omega n}$$

$$= \frac{1}{N} \sum_{k=0}^{N-1} X[k] \sum_{n=0}^{N-1} e^{j2\pi kn/N} e^{-j\omega n}, \tag{5.39}$$

using the notation of Eq. (5.12). Now the right-hand summation in the above expression can be rewritten as

$$\sum_{n=0}^{N-1} e^{-j(\omega - 2\pi k/N)n} = \frac{1 - e^{-j(\omega N - 2\pi k)}}{1 - e^{-j[\omega - (2\pi k/N)]}}$$

$$= \frac{e^{-j[(\omega N - 2\pi k)/2]}}{e^{-j[(\omega N - 2\pi k)/2N]}} \cdot \frac{\sin\left(\frac{\omega N - 2\pi k}{2}\right)}{\sin\left(\frac{\omega N - 2\pi k}{2N}\right)},$$

$$= \frac{\sin\left(\frac{\omega N - 2\pi k}{2}\right)}{\sin\left(\frac{\omega N - 2\pi k}{2N}\right)} \cdot e^{-j[\omega - (2\pi k/N)][(N-1)/2]}, \tag{5.40}$$

by making use of the identity of Eq. (5.9). Substituting Eq. (5.40) in Eq. (5.39), we arrive at

$$X(e^{j\omega}) = \frac{1}{N} \sum_{k=0}^{N-1} X[k] \frac{\sin\left(\frac{\omega N - 2\pi k}{2}\right)}{\sin\left(\frac{\omega N - 2\pi k}{2N}\right)} \cdot e^{-j[\omega - (2\pi k/N)][(N-1)/2]}. \tag{5.41}$$

The above equation can be rewritten as

$$X(e^{j\omega}) = \sum_{k=0}^{N-1} X[k] \Phi\left(\omega - \frac{2\pi k}{N}\right), \tag{5.42}$$

where

$$\Phi(\omega) = \frac{\sin\left(\frac{\omega N}{2}\right)}{N \sin\left(\frac{\omega}{2}\right)} \cdot e^{-j\omega[(N-1)/2]}. \tag{5.43}$$

Equation (5.42) is the desired relation expressing the Fourier transform $X(e^{j\omega})$ of a length-N sequence in terms of its N-point DFT $X[k]$. It can be shown that the interpolating polynomial $\Phi(\omega)$ satisfies the condition

$$\Phi(\omega)\big|_{\omega=2\pi\ell/N} = \begin{cases} 1, & \ell = 0, \\ 0, & 1 \le \ell \le N - 1. \end{cases} \tag{5.44}$$

As a result,

$$X(e^{j\omega})\big|_{\omega=2\pi\ell/N} = X[\ell], \qquad 0 \le \ell \le N - 1. \tag{5.45}$$

Thus, the samples of the Fourier transform $X(e^{j\omega})$ obtained by interpolation at the discrete frequency points $\omega = 2\pi\ell/N, 0 \le \ell \le N - 1$, are indeed the DFT samples $X[\ell]$. Values of $X(e^{j\omega})$ at other values of ω are obtained by interpolating the DFT samples $X[\ell]$.

5.3.4 Sampling the Fourier Transform

Consider a sequence $\{x[n]\}$ with a discrete-time Fourier transform $X(e^{j\omega})$. We sample $X(e^{j\omega})$ at N equally spaced points $\omega_k = 2\pi k/N$, $0 \le k \le N-1$, developing the N *frequency samples* $\{X(e^{j\omega_k})\}$. These N frequency samples can be considered as an N-point DFT $Y[k]$ whose N-point inverse DFT is a length-N sequence $\{y[n]\}$, $0 \le n \le N-1$.

Now, $X(e^{j\omega})$ is a periodic function of ω with a Fourier series representation given by Eq. (3.12). Its Fourier coefficients $x[n]$ are given by Eq. (3.16). It is instructive to develop the relation between $x[n]$ and $y[n]$.

From Eq. (3.12),

$$Y[k] = X(e^{j\omega_k}) = X(e^{j(2\pi k/N)}) = \sum_{\ell=-\infty}^{\infty} x[\ell]W_N^{k\ell}, \qquad 0 \le k \le N-1. \tag{5.46}$$

An N-point inverse DFT of $Y[k]$ yields

$$y[n] = \frac{1}{N}\sum_{k=0}^{N-1} Y[k]W_N^{-kn}, \qquad 0 \le n \le N-1. \tag{5.47}$$

Substituting Eq. (5.46) in Eq. (5.47), we get

$$y[n] = \frac{1}{N}\sum_{k=0}^{N-1}\sum_{\ell=-\infty}^{\infty} x[\ell]W_N^{k\ell}W_N^{-kn}$$

$$= \sum_{\ell=-\infty}^{\infty} x[\ell]\left[\frac{1}{N}\sum_{k=0}^{N-1} W_N^{-k(n-\ell)}\right], \qquad 0 \le n \le N-1. \tag{5.48}$$

Recall from Eq. (5.23) that

$$\frac{1}{N}\sum_{k=0}^{N-1} W_N^{-k(n-\ell)} = \begin{cases} 1, & \text{for } \ell = n + mN, \\ 0, & \text{otherwise.} \end{cases}$$

Making use of the above identity in Eq. (5.48), we finally arrive at the desired relation

$$y[n] = \sum_{m=-\infty}^{\infty} x[n+mN], \qquad 0 \le n \le N-1. \tag{5.49}$$

The above relation indicates that $y[n]$ is obtained from $x[n]$ by adding an infinite number of shifted replicas of $x[n]$, with each replica shifted by an integer multiple of N sampling instants, and observing the sum only for the interval $0 \le n \le N-1$. To apply Eq. (5.49) to finite-length sequences, we assume the samples outside the specified range are zeros. Thus, if $x[n]$ is a finite-length sequence of length M defined for $0 \le n \le M-1$, then it is assumed that $x[n] = 0$ for $n < 0$ and $n \ge M$. It follows from Eq. (5.49) that if M is less than or equal to N, then $y[n] = x[n]$ for $0 \le n \le N-1$ and $x[n]$ can be recovered from $y[n]$ by extracting N samples of $y[n]$ for $0 \le n \le N-1$. If $M > N$, there is a time-domain aliasing of samples of $x[n]$ in generating $y[n]$, and as a result, $x[n]$ cannot be recovered from $y[n]$, as illustrated in Example 5.6.

EXAMPLE 5.6   **Illustration of the Effect of Sampling the Fourier Transform**

Let $x[n]$ be a length-6 sequence defined for $0 \leq n \leq 5$ and given by

$$\{x[n]\} = \{0 \quad 1 \quad 2 \quad 3 \quad 4 \quad 5\}.$$

We first consider sampling the discrete-time Fourier transform $X(e^{j\omega})$ of $\{x[n]\}$ at 8 equally spaced points given by $\omega_k = 2\pi k/8, 0 \leq k \leq 7$. Applying a 8-point inverse-DFT to these DFT samples, we arrive at a sequence $y[n]$, which from Eq. (5.49) is given by

$$y[n] = x[n] + x[n+8] + x[n-8], \quad 0 \leq n \leq 7;$$

that is,

$$\{y[n]\} = \{0 \quad 1 \quad 2 \quad 3 \quad 4 \quad 5 \quad 0 \quad 0\}.$$

It can be seen from the above that the original sequence $x[n]$ can be obtained from $y[n]$ by extracting the first 6 samples.

On the other hand, sampling the discrete-time Fourier transform $X(e^{j\omega})$ of $\{x[n]\}$ at 4 equally spaced points given by $\omega_k = 2\pi k/4, 0 \leq k \leq 3$, and then applying a 4-point inverse-DFT to these samples, we arrive at a sequence $y[n]$, which from Eq. (5.49) is given by

$$y[n] = x[n] + x[n+4] + x[n-4], \quad 0 \leq n \leq 3;$$

that is,

$$\{y[n]\} = \{4 \quad 6 \quad 2 \quad 3\}.$$

The first 2 samples of the above sequence are not the same as the first 2 samples of the original sequence $x[n]$ because of the aliasing caused by the overlap of the last 2 samples of the sequence $x[n+4]$. As a result, the original sequence $x[n]$ cannot be recovered from the above $y[n]$.

The above results can easily be verified using MATLAB.

5.4   Operations on Finite-Length Sequences

Like the Fourier transform, the DFT also satisfies a number of properties that are useful in signal processing applications. Some of these properties are essentially identical to those of the Fourier transform, while some others are somewhat different. We first point out the differences between two important operations on sequences and then review the properties of the DFT in the following section. One of the operations is concerned with time shifting and the other one with convolution.

5.4.1   Circular Shift of a Sequence

As several DFT properties and theorems involve shifting in the time domain and the frequency domain, it is important first to develop the correct procedure for such shifting operations. In the time domain, the operation is referred to as the *circular time-shifting* operation, whereas in the frequency domain, it is referred to as the *circular frequency-shifting* operation. Without any loss of generality, we restrict our attention to the time domain as the operation is identical in the frequency domain. Consider length-N sequences defined for $0 \leq n \leq N - 1$. Such sequences have sample values equal to zero for $n < 0$ and $n \geq N$. If $x[n]$ is such a sequence, then, for any arbitrary integer n_o, the shifted sequence $x_1[n] = x[n-n_o]$ is no longer defined for the range $0 \leq n \leq N - 1$. We therefore need to define another type of a shift that will always keep the shifted sequence in the range $0 \leq n \leq N - 1$. This is achieved by using the *modulo operation*.

Let $0, 1, \ldots, N$ be a set of N positive integers, and let m be any integer. The integer r obtained by evaluating m modulo N is called the *residue* and is an integer with a value between 0 and $N - 1$. The

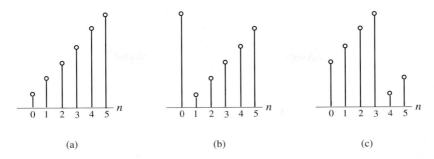

Figure 5.4: Illustration of a circular shift of a finite-length sequence. (a) $x[n]$, (b) $x[\langle n - 1 \rangle_6] = x[\langle n + 5 \rangle_6]$, and (c) $x[\langle n - 4 \rangle_6] = x[\langle n + 2 \rangle_6]$.

modulo operation is denoted by the notation

$$\langle m \rangle_N = m \text{ modulo } N.$$

If we let $r = \langle m \rangle_N$, then

$$r = m + \ell N,$$

where ℓ is an integer chosen to make $m + \ell N$ a number between 0 and $N - 1$. For example, for $N = 7$, $\langle 25 \rangle_7 = 4$ and $\langle -16 \rangle_7 = 5$. The modulo operation can be carried out in MATLAB using the M-file mod.

Using the modulo operation, we define the circular shift of a length-N sequence $x[n]$ by the equation

$$x_c[n] = x\left[\langle n - n_o \rangle_N\right], \tag{5.50}$$

where $x_c[n]$ is also a length-N sequence. If $n_o > 0$, it is a right circular shift, and if $n_o < 0$, it is a left circular shift. For $n_o > 0$, the above equation implies

$$x_c[n] = \begin{cases} x[n - n_o], & \text{for } n_o \leq n \leq N - 1, \\ x[N - n_o + n], & \text{for } 0 \leq n < n_o. \end{cases} \tag{5.51}$$

The concept of a circular shift of a finite-length sequence is illustrated in Figure 5.4. Figure 5.4(a) shows a length-6 sequence $x[n]$. Figure 5.4(b) shows its circularly shifted version shifted to the right by 1 sample period or, equivalently, shifted to the left by 5 sample periods. Likewise, Figure 5.4(c) depicts its circularly shifted version shifted to the right by 4 sample periods or, equivalently, shifted to the left by 2 sample periods.

As can be seen from Figures 5.4(b) and (c), a right circular shift by n_o is equivalent to a left circular shift by $N - n_o$ sample periods. It should be noted that a circular shift by an integer number n_o greater than N is equivalent to a circular shift by $\langle n_o \rangle_N$. For finite-length sequences defined for a specific time interval, the *circular time-reversal* operation, a time-reversal like operation, is defined using the modulo arithmetic. Specifically, for a length-N sequence $x[n]$, $0 \leq n < N - 1$, the circularly time-reversed sequence is also of length N and given by

$$x[\langle -n \rangle_N] = x[\langle N - n \rangle_N].$$

Since $\langle -n \rangle_N = N - n$ for $1 \leq n < N - 1$, the circular time-reversal operation can be implemented using

$$x[\langle -n \rangle_N] = \begin{cases} x[N - n], & \text{for } 1 \leq n \leq N - 1, \\ x[n], & \text{for } n = 0. \end{cases}$$

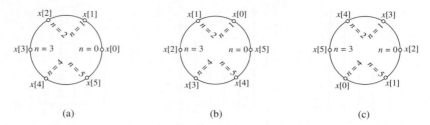

(a)   (b)   (c)

Figure 5.5: Alternate illustration of a circular shift of a finite-length sequence. (a) $x[n]$, (b) $x[\langle n-1\rangle_6] = x[\langle n+5\rangle_6]$, and (c) $x[\langle n-4\rangle_6] = x[\langle n+2\rangle_6]$.

If we view the length-N sequence displayed on a circle at N equally spaced points, then the circular shift operation can be considered as a clockwise or anticlockwise rotation of the sequence by n_o sample spacings on the circle. This is illustrated in Figure 5.5.

In the frequency domain, the circular shifting operation by k_o samples on the length-N DFT sequence $X[k]$ is defined by

$$X_c[k] = X\left[\langle k - k_o\rangle_N\right], \tag{5.52}$$

where $X_c[k]$ is also a length-N DFT.

The circular time-shifting operation can be implemented in MATLAB using the M-file `circshift1`.

5.4.2 Circular Convolution

This operation is analogous to the linear convolution of Eq. (2.73a) but with a subtle difference. Consider two length-N sequences, $g[n]$ and $h[n]$, respectively. Their linear convolution results in a length-$(2N-1)$ sequence $y_L[n]$ given by

$$y_L[n] = \sum_{m=0}^{N-1} g[m]h[n-m], \qquad 0 \le n \le 2N-2, \tag{5.53}$$

where we have assumed that both length-N sequences have been zero-padded to extend their lengths to $2N - 1$.[2] The longer length of $y_L[n]$ results from the time-reversal of the sequence $h[n]$ and its linear shifting to the right. The first nonzero value of $y_L[n]$ is $y_L[0] = g[0]h[0]$, and the last nonzero value of $y_L[n]$ is $y_L[2N-2] = g[N-1]h[N-1]$.

Definition

To develop a convolution-like operation that results in a length-N sequence $y_C[n]$, we first apply the circular time-reversal operation and then apply a circular time-shift. The resulting operation, called a *circular convolution*, is defined below:[3]

$$y_C[n] = \sum_{m=0}^{N-1} g[m]h\left[\langle n-m\rangle_N\right]. \tag{5.54}$$

---

[2]As indicated in Section 2.5.1, the sum of the indices of each sample product inside the summation is equal to the index of the sample being generated by the linear convolution operation.

[3]Note that here the sum of the indices of each sample product inside the summation modulo N is equal to the index of the sample being generated by the circular convolution operation.

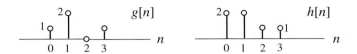

Figure 5.6: Two length-4 sequences.

The above operation is often referred to as an *N-point circular convolution* and is denoted as

$$y_C[n] = g[n] \circledN h[n]. \tag{5.55}$$

Like the linear convolution, the circular convolution is commutative (Problem 5.46); that is,

$$g[n] \circledN h[n] = h[n] \circledN g[n]. \tag{5.56}$$

The N-point circular convolution operation of Eq. (5.55) can be written in matrix form as

$$\begin{bmatrix} y_C[0] \\ y_C[1] \\ y_C[2] \\ \vdots \\ y_C[N-1] \end{bmatrix} = \begin{bmatrix} h[0] & h[N-1] & h[N-2] & \cdots & h[1] \\ h[1] & h[0] & h[N-1] & \cdots & h[2] \\ h[2] & h[1] & h[0] & \cdots & h[3] \\ \vdots & \vdots & \vdots & \ddots & \vdots \\ h[N-1] & h[N-2] & h[N-3] & \cdots & h[0] \end{bmatrix} \begin{bmatrix} g[0] \\ g[1] \\ g[2] \\ \vdots \\ g[N-1] \end{bmatrix}. \tag{5.57}$$

The elements in each row of the matrix of Eq. (5.57) are obtained by circularly rotating the elements of the previous row to the right by 1 position. Such a matrix is called a *circulant matrix*.

Example 5.7 illustrates the computation of the circular convolution.

EXAMPLE 5.7   Circular Convolution of Two Finite-Length Sequences

Determine the 4-point circular convolution of the two length-4 sequences $g[n]$ and $h[n]$ given by

$$g[n] = \{1 \quad 2 \quad 0 \quad 1\}, \qquad h[n] = \{2 \quad 2 \quad 1 \quad 1\}, \qquad 0 \le n \le 3, \tag{5.58}$$

and sketched in Figure 5.6. The result is a length-4 sequence $y_C[n]$ given by

$$y_C[n] = g[n] \circledfour h[n] = \sum_{m=0}^{3} g[m]h[\langle n-m\rangle_4], \qquad 0 \le n \le 3. \tag{5.59}$$

From Eq. (5.59), $y_C[0]$ is given by a

$$y_C[0] = \sum_{m=0}^{3} g[m]h[\langle -m\rangle_4].$$

The circularly time-reversed sequence $h[\langle -m\rangle_4]$ is given by

$$h[\langle -m\rangle_4] = [h[0] \quad h[3] \quad h[2] \quad h[1]] = [2 \quad 1 \quad 1 \quad 2]$$

and is shown in Figure 5.7(a). By forming the products of $g[m]$ with that of $h[\langle -m\rangle_4]$ for each value of m in the range $0 \le m < 3$ and summing the products, we arrive at

$$y_C[0] = g[0]h[0] + g[1]h[3] + g[2]h[2] + g[3]h[1]$$
$$= (1 \times 2) + (2 \times 1) + (0 \times 1) + (1 \times 2) = 6. \tag{5.60}$$

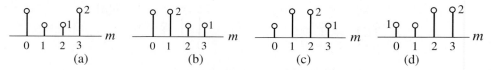

Figure 5.7: The circularly time-reversed sequence and its circularly shifted versions: (a) $h[\langle -m \rangle_4]$, (b) $h[\langle 1 - m \rangle_4]$, (c) $h[\langle 2 - m \rangle_4]$, and (d) $h[\langle 3 - m \rangle_4]$.

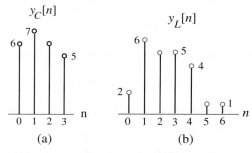

Figure 5.8: Results of convolution of the two sequences of Figure 5.7. (a) Circular convolution and (b) linear convolution.

Next, from Eq. (5.59), we compute $y_C[1]$ as

$$y_C[1] = \sum_{m=0}^{3} g[m]h[\langle 1 - m \rangle_4].$$

The sequence $h[\langle 1 - m \rangle_4]$ obtained by circularly time-shifting $h[\langle -m \rangle_4]$ to the right by 1 sample period is given by

$$h[\langle 1 - m \rangle_4] = [h[1] \quad h[0] \quad h[3] \quad h[2]] = [2 \quad 2 \quad 1 \quad 1].$$

It is also shown in Figure 5.7(b). Summing the products $g[m]h[\langle 1 - m \rangle_4]$ for each value of m, we arrive at

$$y_C[1] = g[0]h[1] + g[1]h[0] + g[2]h[3] + g[3]h[2]$$
$$= (1 \times 2) + (2 \times 2) + (0 \times 1) + (1 \times 1) = 7. \tag{5.61}$$

Continuing this process, we determine the remaining two samples of $y_C[n]$ as

$$y_C[2] = g[0]h[2] + g[1]h[1] + g[2]h[0] + g[3]h[3]$$
$$= (1 \times 1) + (2 \times 2) + (0 \times 2) + (1 \times 1) = 6. \tag{5.62}$$
$$y_C[3] = g[0]h[3] + g[1]h[2] + g[2]h[1] + g[3]h[0]$$
$$= (1 \times 1) + (2 \times 1) + (0 \times 2) + (1 \times 2) = 5. \tag{5.63}$$

The length-4 sequence $y_C[n]$ obtained by a 4-point circular convolution of the two length-4 sequences $g[n]$ and $h[n]$ is sketched in Figure 5.8(a).

A graphical interpretation of the circular convolution operation can be developed by viewing the length-N sequence $x[m]$ as samples placed at N equally spaced points on a circle and the length-N circularly time-reversed and shifted sequence $h[n - m]$ as samples placed at N equally spaced points on a concentric circle. By forming the sum of the products of adjacent samples, we then arrive at $y_C[n]$. We illustrate this process in Figure 5.9 for the computation of a 4-point circular convolution.

circconv.m

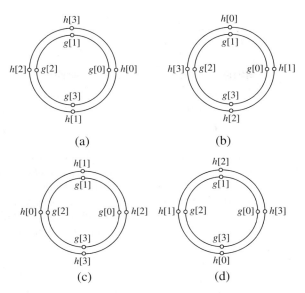

Figure 5.9: Graphical interpretation of circular convolution computation: (a) $y_C[0]$, (b) $y_C[1]$, (c) $y_C[2]$, and (d) $y_C[3]$. Note that the positions of the samples of $g[n]$ remain fixed on the inner circle, whereas the samples of $h[-m]$ located on the outer circle rotate anti-clockwise by 1 sample distance for each increasing value of n.

The circular convolution can be implemented in MATLAB using the M-file `circonv`. We illustrate the use of this function in Example 5.8.

EXAMPLE 5.8   Implementation of Circular Convolution Using MATLAB

We determine the 4-point circular convolution of the two sequences of the previous example, namely, `g = [1 2 0 1]` and `h = [2 2 1 1]` using MATLAB. The code fragment `y = circonv(g,h)` yields the output

```
y =
    6      7      6      5
```

which is seen to be identical to that computed in Example 5.7 and given by Eq. (5.60) and Eqs. (5.61)–(5.63).

Tabular Method

The tabular method for the computation of the linear convolution, as illustrated in Section 2.5.2, can be modified to implement the circular convolution [Pie96]. To this end, it is necessary to apply a circular shift to the partial products from the second and higher rows before they are added. We illustrate the method below.

Without any loss of generality, we consider the evaluation of the 4-point circular convolution of the sequences $\{g[n]\}$ and $\{h[n]\}$, $0 \leq n \leq 3$:

$$y[n] = g[n] \textcircled{4} h[n], \qquad 0 \leq n \leq 3.$$

The samples of these two sequences are then multiplied using the conventional multiplication method, but the partial products generated in the second, the third, and the fourth rows are circularly shifted to the left. The process is indicated below.

n:	0	1	2	3	$\langle 4 \rangle_4$	$\langle 5 \rangle_4$	$\langle 6 \rangle_4$
$g[n]$:	$g[0]$	$g[1]$	$g[2]$	$g[3]$			
$h[n]$:	$h[0]$	$h[1]$	$h[2]$	$h[3]$			
	$g[0]h[0]$	$g[1]h[0]$	$g[2]h[0]$	$g[3]h[0]$			
	-	$g[0]h[1]$	$g[1]h[1]$	$g[2]h[1]$	$g[3]h[1]$		
	-	-	$g[0]h[2]$	$g[1]h[2]$	$g[2]h[2]$	$g[3]h[2]$	
	-	-	-	$g[0]h[3]$	$g[1]h[3]$	$g[2]h[3]$	$g[3]h[3]$

After a circular shift to the left, the partial product $g[3]h[1]$ in the second row is moved to the position to the left in the column belonging to $n = 0$. Likewise, the partial products $g[3]h[2]$ and $g[2]h[2]$ in the third row are circularly shifted to the left and moved to the positions on the left in the columns belonging to $n = 1$ and $n = 0$, respectively. Finally, the partial products $g[3]h[3]$, $g[2]h[3]$, and $g[1]h[3]$ in the fourth row are circularly shifted to the left and placed in the columns corresponding to $n = 2, n = 1$, and $n = 0$, respectively. The result after all circular shifts of partial products is shown below.

n:	0	1	2	3
$g[n]$:	$g[0]$	$g[1]$	$g[2]$	$g[3]$
$h[n]$:	$h[0]$	$h[1]$	$h[2]$	$h[3]$
	$g[0]h[0]$	$g[1]h[0]$	$g[2]h[0]$	$g[3]h[0]$
	$g[3]h[1]$	$g[0]h[1]$	$g[1]h[1]$	$g[2]h[1]$
	$g[2]h[2]$	$g[3]h[2]$	$g[0]h[2]$	$g[1]h[2]$
	$g[1]h[3]$	$g[2]h[3]$	$g[3]h[3]$	$g[0]h[3]$
$y_C[n]$:	$y_C[0]$	$y_C[1]$	$y_C[2]$	$y_C[3]$

The samples of the sequence $\{y_C[n]\}$ generated by the convolution sum are obtained by adding the four partial products in the column above of each sample and are given by

$$y_C[0] = g[0]h[0] + g[3]h[1] + g[2]h[2] + g[1]h[3],$$
$$y_C[1] = g[1]h[0] + g[0]h[1] + g[3]h[2] + g[2]h[3],$$
$$y_C[2] = g[2]h[0] + g[1]h[1] + g[0]h[2] + g[3]h[3],$$
$$y_C[3] = g[3]h[0] + g[2]h[1] + g[1]h[2] + g[0]h[3].$$

EXAMPLE 5.9   Implementation of Circular Convolution Using the Tabular Method

We determine the 4-point circular convolution of the two sequences of Example 5.7, namely, $\{g[n]\} = \{1, 2, 0, 1\}$ and $\{h[n]\} = \{2, 2, 1, 1\}$, $0 \le n \le 3$, using the above approach. The process is illustrated below.

n:	0	1	2	3
$g[n]$:	1	2	0	1
$h[n]$:	2	2	1	1
	2	4	0	2
	2	2	4	0
	0	1	1	2
	2	0	1	1
$y_C[n]$:	6	7	6	5

From the last row of the above table, we observe $\{y_C[n]\} = \{6, 7, 6, 5\}$, $0 \leq n \leq 3$, which is the same as that derived in Example 5.7.

5.5 Classifications of Finite-Length Sequences

For a length-N sequence defined for $0 \leq n \leq N - 1$, the definitions of symmetry discussed in Section 2.1.3 are not applicable. The definitions of symmetry in the case of finite-length sequences are given such that the symmetric and antisymmetric parts of a length-N sequence are also of length N and defined for the same range of values of the time index n.

As in the case of infinite-length sequences, the symmetry properties of finite-length sequences often simplify their respective frequency-domain representations and can be exploited in signal analysis. Implications of the symmetry conditions of a sequence on its DFT are considered in the next section.

5.5.1 Classification Based on Conjugate Symmetry

One symmetry definition is given using a modulo operation. Here, a length-N sequence $x[n]$ is expressed as

$$x[n] = x_{\mathrm{cs}}[n] + x_{\mathrm{ca}}[n], \quad 0 \leq n \leq N - 1, \tag{5.64}$$

where $x_{\mathrm{cs}}[n]$ and $x_{\mathrm{ca}}[n]$ denote, respectively, the *circular conjugate-symmetric part* and the *circular conjugate-antisymmetric part,* defined by

$$x_{\mathrm{cs}}[n] = \tfrac{1}{2}\left(x[n] + x^*[\langle -n \rangle_N]\right), \quad 0 \leq n \leq N - 1, \tag{5.65a}$$

$$x_{\mathrm{ca}}[n] = \tfrac{1}{2}\left(x[n] - x^*[\langle -n \rangle_N]\right), \quad 0 \leq n \leq N - 1. \tag{5.65b}$$

For a real sequence $x[n]$, the conjugate-symmetric part is a real sequence, called the *circular even part,* and is denoted by $x_{\mathrm{ev}}[n]$. Likewise, for a real sequence $x[n]$, , the conjugate-antisymmetric part is also a real sequence, called the *circular odd part,* and is denoted by $x_{\mathrm{od}}[n]$. A length-N complex sequence $x[n]$, $0 \leq n \leq N-1$, is said to be *circular conjugate-symmetric* if $x[n] = x^*[\langle -n \rangle_N] = x^*[\langle N-n \rangle_N]$ and is said to be *circular conjugate-antisymmetric* if $x[n] = -x^*[\langle -n \rangle_N] = -x^*[\langle N - n \rangle_N]$. Likewise a length-$N$ real sequence $x[n]$, $0 \leq n \leq N - 1$, is called a *circular even sequence* if $x[n] = x[\langle -n \rangle_N] = x[\langle N - n \rangle_N]$ and is called a *circular odd sequence* if $x[n] = -x[\langle -n \rangle_N] = -x[\langle N - n \rangle_N]$.

EXAMPLE 5.10   Conjugate-Symmetric and Conjugate-Antisymmetric Parts of a Complex Sequence

Consider the finite-length sequence of length 4 defined for $0 \leq n \leq 3$:

$$\{u[n]\} = \{1 + j4, \ -2 + j3, \ 4 - j2, \ -5 - j6\}.$$

To determine its circular conjugate-symmetric part $u_{\mathrm{cs}}[n]$ and its circular conjugate-antisymmetric part $u_{\mathrm{ca}}[n]$, we form

$$\{u^*[n]\} = \{1 - j4, \ -2 - j3, \ 4 + j2, \ -5 + j6\}.$$

To compute the modulo-4 circularly time-reversed version $\{u^*[\langle -n \rangle_4]\}$, we observe that

$$u^*[\langle -0 \rangle_4] = u^*[0] = 1 - j4,$$
$$u^*[\langle -1 \rangle_4] = u^*[3] = -5 + j6,$$
$$u^*[\langle -2 \rangle_4] = u^*[2] = 4 + j2,$$
$$u^*[\langle -3 \rangle_4] = u^*[1] = -2 - j3.$$

Hence,

$$\{u^*[\langle -n \rangle_4]\} = \{1 - j4, \ -5 + j6, \ 4 + j2, \ -2 - j3\}.$$

Using Eq. (5.65a), we thus arrive at

$$\{u_{cs}[n]\} = \{1, \ -3.5 + j4.5, \ 4, \ -3.5 - j4.5\}.$$

Likewise, using Eq. (5.65b), we get

$$\{u_{ca}[n]\} = \{j4, \ 1.5 - j1.5, \ -j2, \ -1.5 - j1.5\}.$$

It can be easily verified that $u_{cs}[n] = u_{cs}^*[\langle -n \rangle_4]$ and $u_{ca}[n] = -u_{ca}^*[\langle -n \rangle_4]$, and $u[n] = u_{cs}[n] + u_{ca}[n]$.

Since the DFT is a finite-length sequence in the frequency domain, we can also classify it according to its symmetry properties. An N-point DFT $X[k]$ is defined to be a *circular conjugate-symmetric* sequence if

$$X[k] = X^*[\langle -k \rangle_N] = X^*[\langle N - k \rangle_N].$$

Likewise, a DFT $X[k]$ is defined to be a *circular conjugate-antisymmetric* sequence if

$$X[k] = -X^*[\langle -k \rangle_N] = -X^*[\langle N - k \rangle_N].$$

A complex DFT $X[k]$ can also be expressed as a sum of a circular conjugate-symmetric part $X_{cs}[k]$ and a circular conjugate-antisymmetric part $X_{ca}[k]$ as

$$X[k] = X_{cs}[k] + X_{ca}[k], \qquad 0 \le k \le N - 1, \tag{5.66}$$

where

$$X_{cs}[k] = \tfrac{1}{2} \left(X[k] + X^*[\langle -k \rangle_N] \right), \qquad 0 \le k \le N - 1, \tag{5.67a}$$

$$X_{ca}[k] = \tfrac{1}{2} \left(X[k] - X^*[\langle -k \rangle_N] \right), \qquad 0 \le k \le N - 1. \tag{5.67b}$$

5.5.2 Classification Based on Geometric Symmetry

Finite-length sequences exhibiting geometric symmetry play an important role in digital signal processing. Two types of geometric symmetries are usually defined: (1) symmetric and (2) antisymmetric. A length-N *symmetric sequence* $x[n]$ satisfies the condition

$$x[n] = x[N - 1 - n], \tag{5.68}$$

whereas a length-N *antisymmetric sequence* $x[n]$ satisfies the condition

$$x[n] = -x[N - 1 - n]. \tag{5.69}$$

Since the length N of a sequence can be either even or odd, four types of geometric symmetry are thus defined, as demonstrated in Figure 5.10. Note from this figure that the odd-length sequences are symmetric (or antisymmetric) with respect to the time index $n = (N-1)/2$, whereas the even-length sequences exhibit symmetry with respect to the half-sample point at $n = (N - 1)/2$.

We next develop the expressions for the Fourier transforms of these four types of sequences exhibiting geometric symmetry.

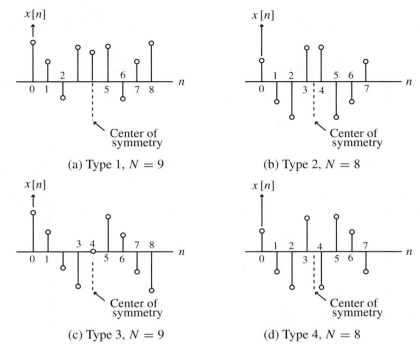

Figure 5.10: Illustration of the four types of geometric symmetry of a sequence.

Type 1: Symmetric Sequence with Odd Length. For simplicity, assume first, $N = 9$. The Fourier transform of the length-9 sequence $x[n]$ is then given by

$$X(e^{j\omega}) = x[0] + x[1]\,e^{-j\omega} + x[2]\,e^{-j2\omega} + x[3]\,e^{-j3\omega} + x[4]\,e^{-j4\omega}$$
$$+x[5]\,e^{-j5\omega} + x[6]\,e^{-j6\omega} + x[7]\,e^{-j7\omega} + x[8]\,e^{-j8\omega}. \tag{5.70}$$

But, from Eq. (5.68) for $N = 9$, $x[0] = x[8]$, $x[1] = x[7]$, $x[2] = x[6]$, and $x[3] = x[5]$. Ther., Eq. (5.70) reduces to

$$X(e^{j\omega}) = x[0]\left(1 + e^{-j8\omega}\right) + x[1]\left(e^{-j\omega} + e^{-j7\omega}\right) + x[2]\left(e^{-j2\omega} + e^{-j6\omega}\right)$$
$$+x[3]\left(e^{-j3\omega} + e^{-j5\omega}\right) + x[4]\,e^{-j4\omega}$$
$$= x[0]\,e^{-j4\omega}\left(e^{j4\omega} + e^{-j4\omega}\right) + x[1]\,e^{-j4\omega}\left(e^{j3\omega} + e^{-j3\omega}\right) + x[2]\,e^{-j4\omega}\left(e^{j2\omega} + e^{-j2\omega}\right)$$
$$+x[3]\,e^{-j4\omega}\left(e^{j\omega} + e^{-j\omega}\right) + x[4]\,e^{-j4\omega}. \tag{5.71}$$

Factoring out $e^{-j4\omega}$ in the right-hand side of Eq. (5.71), we get

$$X(e^{j\omega}) = e^{-j4\omega}\{x[0](e^{j4\omega} + e^{-j4\omega}) + x[1](e^{j3\omega} + e^{-j3\omega}) + x[2](e^{j2\omega} + e^{-j2\omega})$$
$$+x[3](e^{j\omega} + e^{-j\omega}) + x[4]\}$$
$$= e^{-j4\omega}\{2x[0]\cos(4\omega) + 2x[1]\cos(3\omega) + 2x[2]\cos(2\omega) + 2x[3]\cos(\omega) + x[4]\}. \tag{5.72}$$

Note that the quantity inside the braces in the above expression is a real function of ω and can assume positive or negative values in the range $0 \leq \omega \leq \pi$. Therefore, from Eq. (5.72), it follows that the phase of the sequence is given by

$$\theta(\omega) = -4\omega + \beta,$$

where β is either 0 or π, and hence, the phase is a linear function of ω.

In the general case for a Type 1 linear-phase sequence of length N, the Fourier transform of the sequence is of the form

$$X(e^{j\omega}) = e^{-j(N-1)\omega/2} \left\{ x\left[\frac{N-1}{2}\right] + 2 \sum_{n=1}^{(N-1)/2} x\left[\frac{N-1}{2} - n\right] \cos(\omega n) \right\}. \qquad (5.73)$$

The expression for the N-point DFT of the length-N Type 1 linear-phase sequences can be easily derived by uniformly sampling its Fourier transform given above at $\omega = 2\pi k/N, 0 \leq k \leq N - 1$:

$$X[k] = e^{-j(N-1)\pi k/N} \left\{ x\left[\frac{N-1}{2}\right] + 2 \sum_{n=1}^{(N-1)/2} x\left[\frac{N-1}{2} - n\right] \cos\left(\frac{2\pi kn}{N}\right) \right\}. \qquad (5.74)$$

Type 2: Symmetric Sequence with Even Length.   Let $N = 8$. By making use of the symmetry of the sequence coefficients given by Eq. (5.68), the Fourier transform of the sequence $x[n]$ can be written as

$$X(e^{j\omega}) = e^{-j7\omega/2}\{2x[0]\cos(\tfrac{7\omega}{2}) + 2x[1]\cos(\tfrac{5\omega}{2}) + 2x[2]\cos(\tfrac{3\omega}{2}) + 2x[3]\cos(\tfrac{\omega}{2})\}. \qquad (5.75)$$

As before, here, the quantity inside the braces in the above expression is a real function of ω, and as in the previous case, it can assume positive or negative values in the range $0 \leq |\omega| \leq \pi$. Here, the phase is given by

$$\theta(\omega) = -\tfrac{7}{2}\omega + \beta,$$

where β is again either 0 or π. As a result, the phase is also a linear function of ω.

For a Type 2 length-N linear-phase sequence, the expression for the Fourier transform in the general case is of the form

$$X(e^{j\omega}) = e^{-j(N-1)\omega/2} \left\{ 2\sum_{n=1}^{N/2} x\left[\frac{N}{2} - n\right] \cos\left(\omega(n - \tfrac{1}{2})\right) \right\}. \qquad (5.76)$$

The expression for the N-point DFT of the length-N Type 2 linear-phase sequences is obtained by uniformly sampling the above Fourier transform at $\omega = 2\pi k/N, 0 \leq k \leq N - 1$:

$$X[k] = e^{-j(N-1)\pi k/N} \left\{ 2\sum_{n=1}^{N/2} x\left[\frac{N}{2} - n\right] \cos\left(\frac{\pi k(2n-1)}{N}\right) \right\}. \qquad (5.77)$$

Type 3: Antisymmetric Sequence with Odd Length.   Consider $N = 9$. Then, applying the symmetry condition of Eq. (5.69) on the Fourier transform of the sequence, we arrive at

$$X(e^{j\omega}) = e^{-j4\omega}e^{j\pi/2}\{2x[0]\sin 4\omega + 2x[1]\sin 3\omega + 2h[2]\sin 2\omega = +2x[3]\sin \omega\}. \qquad (5.78)$$

It also exhibits a linear phase given by

$$\theta(\omega) = -4\omega + \tfrac{\pi}{2} + \beta,$$

where β is either 0 or π.

The expression for the Fourier transform in the general case of Type 3 length-N linear-phase sequence is given by

$$X(e^{j\omega}) = je^{-j(N-1)\omega/2} \left\{ 2 \sum_{n=1}^{(N-1)/2} x\left[\tfrac{N-1}{2} - n\right] \sin(\omega n) \right\}. \tag{5.79}$$

By uniformly sampling the above Fourier transform at $\omega = 2\pi k/N, 0 \le k \le N - 1$, we arrive at the expression for the N-point DFT of a length-N Type 3 linear-phase sequence as indicated below:

$$X[k] = je^{-j(N-1)\pi k/N} \left\{ 2 \sum_{n=1}^{(N-1)/2} x\left[\tfrac{N-1}{2} - n\right] \sin\left(\tfrac{2\pi kn}{N}\right) \right\}. \tag{5.80}$$

Type 4: Antisymmetric Sequence with Even Length.   For $N = 8$, the Fourier transform can be expressed as

$$X(e^{j\omega}) = e^{-j7\omega/2} e^{j\pi/2} \{ 2x[0] \sin(7\omega/2) + 2x[1] \sin(5\omega/2)$$
$$+ 2x[2] \sin(3\omega/2) + 2x[3] \sin(\omega/2) \}. \tag{5.81}$$

It has a linear phase given by

$$\theta(\omega) = -\tfrac{7}{2}\omega + \tfrac{\pi}{2} + \beta,$$

where β is either 0 or π.

The Fourier transform in the general case for Type 4 length-N linear-phase sequence is given by

$$X(e^{j\omega}) = je^{-j(N-1)\omega/2} \left\{ 2 \sum_{n=1}^{N/2} x\left[\tfrac{N}{2} - n\right] \sin\left(\omega(n - \tfrac{1}{2})\right) \right\}. \tag{5.82}$$

The N-point DFT of a length-N Type 4 linear-phase sequence is thus given by

$$X[k] = je^{-j(N-1)\pi k/N} \left\{ 2 \sum_{n=1}^{N/2} x\left[\tfrac{N}{2} - n\right] \sin\left(\tfrac{\pi k(2n-1)}{N}\right) \right\}. \tag{5.83}$$

It should be noted that the quantities inside the braces in Eqs. (5.75), (5.77), (5.80), and (5.83) are real numbers. Since the phase for each DFT sample is known for a given length N, the expressions inside the braces uniquely characterize the corresponding DFTs.

An alternate designation given to the above four types of sequences is based on the location of the point of symmetry [Mar94]. From Figure 5.10, it can be seen that in the case of Type 1 and 3 sequences, the point of symmetry is coincident with one of the samples of the sequences, leading to the designation as *whole-sample symmetry*. On the other hand, in the case of Type 2 and 4 sequences, the point of symmetry is halfway between two center samples of the sequences, leading to the designation as *half-sample symmetry*. Thus, the Type 1, 2, 3, and 4 sequences, respectively, have been referred to as a *whole-sample symmetric sequence*, a *half-sample symmetric sequence*, a *whole-sample antisymmetric sequence*, and a *half-sample antisymmetric sequence*.

5.6 DFT Symmetry Relations

As indicated earlier, in general, the DFT $X[k]$ of a sequence $x[n]$ is a sequence of complex numbers and can be expressed as

$$X[k] = X_{re}[k] + jX_{im}[k], \tag{5.84}$$

where $X_{re}[k]$ and $X_{im}[k]$ are, respectively, the real and imaginary parts of $X[k]$. The sequences $X_{re}[k]$ and $X_{im}[k]$ can be derived from $X[k]$ using

$$X_{re}[k] = \tfrac{1}{2}\left(X[k] + X^*[k]\right), \tag{5.85a}$$

$$X_{im}[k] = \tfrac{1}{2}\left(X[k] - X^*[k]\right). \tag{5.85b}$$

The real and imaginary parts of $X[k]$ can be expressed in terms of the real and imaginary parts of $x[n]$. Substituting

$$x[n] = x_{re}[n] + jx_{im}[n]$$

in Eq. (5.7), we arrive at

$$X[k] = \sum_{n=0}^{N-1} (x_{re}[n] + jx_{im}[n]) \left(\cos(\tfrac{2\pi k}{N}) - j\sin(\tfrac{2\pi k}{N})\right)$$

$$= \sum_{n=0}^{N-1} \left(x_{re}[n]\cos(\tfrac{2\pi k}{N}) + x_{im}[n]\sin(\tfrac{2\pi k}{N})\right) + j\sum_{n=0}^{N-1} \left(x_{im}[n]\cos(\tfrac{2\pi k}{N}) - x_{re}[n]\sin(\tfrac{2\pi k}{N})\right). \tag{5.86}$$

From the above equation, we thus have

$$X_{re}[k] = \sum_{n=0}^{N-1} \left(x_{re}[n]\cos(\tfrac{2\pi k}{N}) + x_{im}[n]\sin(\tfrac{2\pi k}{N})\right) = \sum_{n=0}^{N-1} x_{cs}[n]\,e^{-j\omega kn} \tag{5.87a}$$

$$X_{im}[k] = \sum_{n=0}^{N-1} \left(x_{im}[n]\cos(\tfrac{2\pi k}{N}) - x_{re}[n]\sin(\tfrac{2\pi k}{N})\right) = \sum_{n=0}^{N-1} x_{ca}[n]\,e^{-j\omega kn}. \tag{5.87b}$$

Finite-Length Complex Sequences

For a given length-N sequence with an N-point DFT $X[k]$, it is quite straightforward to derive the DFTs of its circularly time-reversed length-N sequence $x[\langle -n\rangle_N]$ and the length-N complex conjugate sequence $x^*[n]$. From Eq. (5.7), it follows that

$$X^*[k] = \sum_{n=0}^{N-1} x^*[n]e^{j2\pi kn/N}.$$

From the above, we get

$$X^*[\langle -k\rangle_N] = \sum_{n=0}^{N-1} x^*[n]e^{j2\pi(\langle -k\rangle_N)n/N}.$$

For $k = 0$, we have

$$X^*[\langle -k\rangle_N] = X^*[k] = \sum_{n=0}^{N-1} x^*[n].$$

For $1 \le k \le N - 1$, we have

$$X^*[\langle -k \rangle_N] = X^*[N - k] = \sum_{n=0}^{N-1} x^*[n] e^{j2\pi(N-k)n/N}$$

$$= \sum_{n=0}^{N-1} x^*[n] e^{j2\pi n} e^{-j2\pi kn/N} = \sum_{n=0}^{N-1} x^*[n] e^{-j2\pi kn/N},$$

as $e^{j2\pi} = 1$. Combining these two results, we thus get

$$X^*[\langle -k \rangle_N] = \sum_{n=0}^{N-1} x^*[n] e^{-j2\pi kn/N},$$

which leads to the DFT pair

$$x^*[n] \xleftrightarrow{\text{DFT}} X^*[\langle -k \rangle_N]. \qquad (5.88)$$

In a similar manner, we can derive the DFT pairs (Problem 5.26)

$$x^*[\langle -n \rangle_N] \xleftrightarrow{\text{DFT}} X^*[k], \qquad (5.89a)$$

$$x_{\text{re}}[n] \xleftrightarrow{\text{DFT}} X_{\text{cs}}[k], \qquad (5.89b)$$

$$j x_{\text{im}}[n] \xleftrightarrow{\text{DFT}} X_{\text{ca}}[k]. \qquad (5.89c)$$

Determination of the DFTs of the circular conjugate-symmetric part $x_{\text{cs}}[n]$ and the circular conjugate-antisymmetric part $x_{\text{ca}}[n]$ of the length-N sequence is also straightforward. For example, by applying Eq. (5.7) to Eq. (5.65a), we arrive at

$$\tfrac{1}{2} (X[k] + X^*[k]) = X_{\text{re}}[k],$$

resulting in the DFT pair

$$x_{\text{cs}}[n] \xleftrightarrow{\text{DFT}} X_{\text{re}}[k]. \qquad (5.90)$$

Likewise, applying Eq. (5.7) to Eq. (5.65b), we get the DFT pair

$$x_{\text{ca}}[n] \xleftrightarrow{\text{DFT}} j X_{\text{im}}[k]. \qquad (5.91)$$

For future reference, we summarize in Table 5.1 the symmetry properties of the DFT of a finite-length complex sequence.

It should be noted that often we encounter real sequences, and the DFTs of two real sequences can be computed efficiently using a single DFT of a complex sequence generated from the two real sequences by exploiting the symmetry properties of a DFT. This scheme is discussed in Section 5.9.1.

Finite-Length Real Sequences

For a real sequence, $x_{\text{im}}[n] = 0$, which when substituted in Eq. (5.86) yields

$$X[k] = \sum_{n=0}^{N-1} x_{\text{re}}[n] \cos\left(\frac{2\pi k}{N}\right) - j \sum_{n=0}^{N-1} x_{\text{re}}[n] \sin\left(\frac{2\pi k}{N}\right). \qquad (5.92)$$

Table 5.1: Symmetry properties of the DFT of a complex sequence.

Length-N Sequence	N-point DFT
$x[n] = x_{\text{re}}[n] + jx_{\text{im}}[n]$	$X[k] = X_{\text{re}}[k] + jX_{\text{im}}[k]$

$x^*[n]$	$X^*[\langle -k \rangle_N]$
$x^*[\langle -n \rangle_N]$	$X^*[k]$
$x_{\text{re}}[n]$	$X_{\text{cs}}[k] = \frac{1}{2}\{X[k] + X^*[\langle -k \rangle_N]\}$
jx_{im}	$X_{\text{ca}}[k] = \frac{1}{2}\{X[k] - X^*[\langle -k \rangle_N]\}$
$x_{\text{cs}}[n]$	$X_{\text{re}}[k]$
$x_{\text{ca}}[n]$	$jX_{\text{im}}[k]$

Note: $x_{\text{cs}}[n]$ and $x_{\text{ca}}[n]$ are the circular conjugate-symmetric and circular conjugate-antisymmetric parts of $x[n]$, respectively. Likewise, $X_{\text{cs}}[k]$ and $X_{\text{ca}}[k]$ are the circular conjugate-symmetric and circular conjugate-antisymmetric parts of $X[k]$, respectively.

From the above, we thus have

$$X_{\text{re}}[k] = \sum_{n=0}^{N-1} x_{\text{re}}[n] \cos\left(\frac{2\pi k}{N}\right), \tag{5.93a}$$

$$X_{\text{im}}[k] = -\sum_{n=0}^{N-1} x_{\text{re}}[n] \sin\left(\frac{2\pi k}{N}\right). \tag{5.93b}$$

Recall from Section 5.5.1 that the circular conjugate-symmetric part of a real sequence $x[n]$ is called the circular even part and is denoted by $x_{\text{ev}}[n]$. In this case, the DFT pair given by Eq. (5.90) reduces to

$$x_{\text{ev}}[n] \overset{\text{DFT}}{\longleftrightarrow} X_{\text{re}}[k]. \tag{5.94}$$

Likewise, the circular conjugate-antisymmetric part of a real sequence $x[n]$ is called the circular odd part and is denoted by $x_{\text{od}}[n]$. Here the DFT pair Eq. (5.91) reduces to

$$x_{\text{od}}[n] \overset{\text{DFT}}{\longleftrightarrow} jX_{\text{im}}[k]. \tag{5.95}$$

Next, we develop the symmetry relations for the DFT of a real sequence. From Eq. (5.92), we have

$$X[\langle -k \rangle_N] = \sum_{n=0}^{N-1} x_{\text{re}}[n] \cos\left(\frac{2\pi(\langle -k \rangle_N)}{N}\right) - j \sum_{n=0}^{N-1} x_{\text{re}}[n] \sin\left(\frac{2\pi(\langle -k \rangle_N)}{N}\right)$$

$$= \sum_{n=0}^{N-1} x_{\text{re}}[n] \cos\left(\frac{2\pi(N-k)}{N}\right) - j \sum_{n=0}^{N-1} x_{\text{re}}[n] \sin\left(\frac{2\pi(N-k)}{N}\right)$$

Table 5.2: Symmetry properties of the DFT of a length-N real sequence.

Length-N Sequence	N-point DFT
$x[n] = x_{ev}[n] + x_{od}[n]$	$X[k] = X_{re}[k] + jX_{im}[k]$
$x_{ev}[n]$ $x_{od}[n]$	$X_{re}[k]$ $jX_{im}[k]$
Symmetry relations	$X[k] = X^*[\langle -k\rangle_N]$ $X_{re}[k] = X_{re}[\langle -k\rangle_N]$ $X_{im}[k] = -X_{im}[\langle -k\rangle_N]$ $\|X[k]\| = \|X[\langle -k\rangle_N]\|$ $\arg X[k] = -\arg X[\langle -k\rangle_N]$

Note: $x_{ev}[n]$ and $x_{od}[n]$ are the circular even and circular odd parts of $x[n]$, respectively.

$$= \sum_{n=0}^{N-1} x_{re}[n]\cos\left(\frac{2\pi k}{N}\right) + j\sum_{n=0}^{N-1} x_{re}[n]\sin\left(\frac{2\pi k}{N}\right) = X^*[k].$$

This leads to the symmetry relation

$$X[k] = X^*[\langle -k\rangle_N]. \tag{5.96}$$

From the above equation, we arrive at the following symmetry relations:

$$X_{re}[k] = X_{re}[\langle -k\rangle_N], \tag{5.97a}$$
$$X_{im}[k] = -X_{im}[\langle -k\rangle_N], \tag{5.97b}$$
$$\|X[k]\| = \|X[\langle -k\rangle_N]\|, \tag{5.97c}$$
$$\arg X[k] = \arg X[\langle -k\rangle_N]. \tag{5.97d}$$

A proof of the above symmetry relations is left as an exercise (Problem 5.27). For future reference, we summarize in Table 5.2 the symmetry relations of a length-N real sequence.

5.7   Discrete Fourier Transform Theorems

As in the case of the Fourier transform, there are a number of important theorems of the DFT that are also useful in digital signal processing applications. These theorems can be used to determine the DFTs of sequences obtained by combining sequences with known transforms. We review these theorems in this section. All time-domain sequences are assumed to be of length-N with N-point DFTs. For compactness, the theorems are stated using the notation introduced in Eq. (5.15) and shall make use of the following

DFT pairs:

$$g[n] \xleftrightarrow{\text{DFT}} G[k],$$

$$h[n] \xleftrightarrow{\text{DFT}} H[k].$$

The proofs of many of these theorems are quite straightforward and are left as exercises.

Linearity Theorem

Consider a sequence $x[n] = \alpha g[n] + \beta h[n]$ obtained by a linear combination of $g[n]$ and $h[n]$, where α and β are arbitrary constants, then the DFT $X[k]$ of $x[n]$ is given by $\alpha G[k] + \beta H[k]$; that is,

$$\alpha g[n] + \beta h[n] \xleftrightarrow{\text{DFT}} \alpha G[k] + \beta H[k]. \tag{5.98}$$

It should be noted that the Fourier transforms of the periodic conjugate-symmetric and periodic conjugate-antisymmetric parts of a sequence derived in Eqs. (5.90) and (5.91) made use of the linearity theorem.

Circular Time-Shifting Theorem

The DFT of the circularly time-shifted sequence $x[n] = g[\langle n - n_o \rangle_N]$, with n_o an integer, is given by $X[k] = W_N^{kn_o} G[k]$; that is,

$$g[\langle n - n_o \rangle_N] \xleftrightarrow{\text{DFT}} W_N^{kn_o} G[k]. \tag{5.99}$$

Circular Frequency-Shifting Theorem

The inverse DFT of the circularly frequency-shifted DFT $X[k] = G[\langle k - k_o \rangle_N]$, with k_o an integer, is given by $x[n] = W_N^{-k_o n} g[n]$; that is,

$$W_N^{-k_o n} g[n] \xleftrightarrow{\text{DFT}} G[\langle k - k_o \rangle_N]. \tag{5.100}$$

Duality Theorem

If the N-point DFT of the length-N sequence $g[n]$ is $G[k]$, then the N-point DFT of the length-N sequence $G[n]$ is given by $Ng[\langle -k \rangle_N]$; that is,

$$G[n] \xleftrightarrow{\text{DFT}} Ng[\langle -k \rangle_N]. \tag{5.101}$$

Circular Convolution Theorem

The N-point DFT $Y[k]$ of the length-N sequence $y[n] = g[n] \circledN h[n]$ is given by $G[k]H[k]$; that is,

$$g[n] \circledN h[n] \xleftrightarrow{\text{DFT}} G[k]H[k]. \tag{5.102}$$

Proof. Applying Eq. (5.7) to the sequence $y[n] = g[n] \circledN h[n]$, we obtain

$$Y[k] = \sum_{n=0}^{N-1} y[n] W_N^{kn} = \sum_{n=0}^{N-1} \left(\sum_{m=0}^{N-1} g[m] h[\langle n - m \rangle_N] \right) W_N^{kn}.$$

Interchanging the order of the summations and substituting $n - m = \ell + Nr$, where ℓ and r are integers with $0 \le \ell \le N - 1$, in the above we get

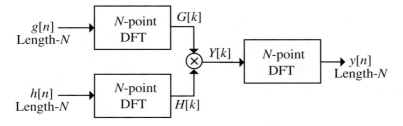

Figure 5.11: Circular convolution using the DFT.

$$Y[k] = \sum_{m=0}^{N-1} g[m] \left(\sum_{n=0}^{N-1} h[\langle n-m\rangle_N] W_N^{kn} \right) = \sum_{m=0}^{N-1} g[m] \left(\sum_{\ell=0}^{N-1} h[\ell] W_N^{k(\ell+m+Nr)} \right)$$

$$= \sum_{m=0}^{N-1} g[m] \left(\sum_{\ell=0}^{N-1} h[\ell] W_N^{k\ell} \right) W_N^{km} = \left(\sum_{m=0}^{N-1} g[m] W_N^{km} \right) H[k] = G[k]H[k].$$

Thus, an N-point circular convolution of two length-N sequences can be implemented by forming the product of their N-point DFTs and then taking the inverse DFT of the product, as shown in Figure 5.11. The process is illustrated in the Example 5.11.

EXAMPLE 5.11   Circular Convolution Computation Using DFT

We compute the circular convolution of the two length-4 sequences of Eq. (5.58) via a DFT-based approach. Now the 4-point DFT $G[k]$ of the length-4 sequence $g[n]$ of Eq. (5.58) is given by

$$G[k] = g[0] + g[1]e^{-j2\pi k/4} + g[2]e^{j4\pi k/4} + g[3]e^{-j6\pi k/4}$$

$$= 1 + 2e^{-j\pi k/2} + e^{-j3\pi k/2}, \qquad k = 0, 1, 2, 3. \tag{5.103}$$

Therefore,

$$G[0] = 1 + 2 + 1 = 4,$$

$$G[1] = 1 - j2 + j = 1 - j,$$

$$G[2] = 1 - 2 - 1 = -2,$$

$$G[3] = 1 + j2 - j = 1 + j. \tag{5.104}$$

The above DFT samples can also be computed using the matrix relation of Eq. (5.28):

$$\begin{bmatrix} G[0] \\ G[1] \\ G[2] \\ G[3] \end{bmatrix} = \mathbf{D}_4 \begin{bmatrix} g[0] \\ g[1] \\ g[2] \\ g[3] \end{bmatrix} = \begin{bmatrix} 1 & 1 & 1 & 1 \\ 1 & -j & -1 & j \\ 1 & -1 & 1 & -1 \\ 1 & j & -1 & -j \end{bmatrix} \begin{bmatrix} 1 \\ 2 \\ 0 \\ 1 \end{bmatrix} = \begin{bmatrix} 4 \\ 1-j \\ -2 \\ 1+j \end{bmatrix}, \tag{5.105}$$

where $\mathbf{D}_4$ is the 4×4 DFT matrix. Likewise, the 4-point DFT of the sequence $h[n]$ of Eq. (5.58) is given by

$$\begin{bmatrix} H[0] \\ H[1] \\ H[2] \\ H[3] \end{bmatrix} = \mathbf{D}_4 \begin{bmatrix} h[0] \\ h[1] \\ h[2] \\ h[3] \end{bmatrix} = \begin{bmatrix} 1 & 1 & 1 & 1 \\ 1 & -j & -1 & j \\ 1 & -1 & 1 & -1 \\ 1 & j & -1 & -j \end{bmatrix} \begin{bmatrix} 2 \\ 2 \\ 1 \\ 1 \end{bmatrix} = \begin{bmatrix} 6 \\ 1-j \\ 0 \\ 1+j \end{bmatrix}. \tag{5.106}$$

Applying the 4-point IDFT to the product $Y_C[k] = G[k]H[k]$,

$$\begin{bmatrix} Y_C[0] \\ Y_C[1] \\ Y_C[2] \\ Y_C[3] \end{bmatrix} = \begin{bmatrix} G[0]H[0] \\ G[1]H[1] \\ G[2]H[2] \\ G[3]H[3] \end{bmatrix} = \begin{bmatrix} 24 \\ -j2 \\ 0 \\ j2 \end{bmatrix}, \tag{5.107}$$

we arrive at the desired circular convolution result:

$$\begin{bmatrix} y_C[0] \\ y_C[1] \\ y_C[2] \\ y_C[3] \end{bmatrix} = \frac{1}{4}\mathbf{D}_4^* \begin{bmatrix} 24 \\ -j2 \\ 0 \\ j2 \end{bmatrix} = \frac{1}{4} \begin{bmatrix} 1 & 1 & 1 & 1 \\ 1 & j & -1 & -j \\ 1 & -1 & 1 & -1 \\ 1 & -j & -1 & j \end{bmatrix} \begin{bmatrix} 24 \\ -j2 \\ 0 \\ j2 \end{bmatrix} = \begin{bmatrix} 6 \\ 7 \\ 6 \\ 5 \end{bmatrix}, \tag{5.108}$$

which is identical to the results obtained in Example 5.7 by direct evaluation.

The circular convolution can also be implemented easily in MATLAB via the DFT-based approach and is left as an exercise (Exercise M5.2).

We show in Example 5.12 that a linear convolution can also be implemented using the circular convolution. This is an attractive option in many applications, as the DFT and the IDFT can be computed efficiently using fast algorithms (see Section 11.3.2).

EXAMPLE 5.12   Implementation of the Linear Convolution of Two Finite-Length Sequences Using the Circular Convolution

Now let us extend the two length-4 sequences of Eq. (5.58) to length 7 by appending each with three zero-valued samples, that is, by forming

$$g_e[n] = \begin{cases} g[n], & 0 \le n \le 3, \\ 0, & 4 \le n \le 6, \end{cases} \tag{5.109}$$

$$h_e[n] = \begin{cases} h[n], & 0 \le n \le 3, \\ 0, & 4 \le n \le 6. \end{cases} \tag{5.110}$$

We next determine the 7-point circular convolution of $g_e[n]$ and $h_e[n]$:

$$y[n] = \sum_{m=0}^{6} g_e[m]h_e[\langle n-m\rangle_7]. \tag{5.111}$$

From the above,

$$y[0] = g_e[0]h_e[0] + g_e[1]h_e[6] + g_e[2]h_e[5] + g_e[3]h_e[4]$$

$$+ g_e[4]h_e[3] + g_e[5]h_e[2] + g_e[6]h_e[1]. \tag{5.112}$$

It can be seen in the above equation, because of zero-padding, non-zero samples of $g_e[n]$ for $1 \le n \le 3$ are multiplied by zero-valued samples of $h_e[n]$ for $4 \le n \le 6$, and non-zero samples of $h_e[n]$ for $1 \le n \le 3$ are multiplied by zero-valued samples of $g_e[n]$ for $4 \le n \le 6$. Hence

$$y[0] = g[0]h[0] = (1 \times 2) = 2.$$

Next we compute $y[1]$. From Eq. (5.111),

$$y[1] = g_e[0]h_e[1] + g_e[1]h_e[0] + g_e[2]h_e[6] + g_e[3]h_e[5] + g_e[4]h_e[4] + g_e[5]h_e[3] + g_e[6]h_e[2]. \quad (5.113)$$

Here, non-zero samples of $g_e[n]$ for $2 \le n \le 3$ are multiplied by zero-valued samples of $h_e[n]$ for $5 \le n \le 6$, and non-zero samples of $h_e[n]$ for $2 \le n \le 3$ are multiplied by zero-valued samples of $g_e[n]$ for $5 \le n \le 6$. In the remaining product term $g_e[4]h_e[4]$, both samples are zeros. As a result, we have

$$y[1] = g[0]h[1] + g[1]h[0] = (1 \times 2) + (2 \times 2) = 6.$$

Continuing the above procedure, we arrive at the remaining samples of $y[n]$:

$$y[2] = g[0]h[2] + g[1]h[1] + g[2]h[0] = (1 \times 1) + (2 \times 2) + (0 \times 2) = 5,$$
$$y[3] = g[0]h[3] + g[1]h[2] + g[2]h[1] + g[3]h[0]$$
$$= (1 \times 1) + (2 \times 1) + (0 \times 2) + (1 \times 2) = 5,$$
$$y[4] = g[1]h[3] + g[2]h[2] + g[3]h[0] = (2 \times 1) + (0 \times 1) + (1 \times 2) = 4,$$
$$y[5] = g[2]h[3] + g[3]h[2] = (0 \times 1) + (1 \times 1) = 1,$$
$$y[6] = g[3]h[3] = (1 \times 1) = 1.$$

It is evident from the above that $y[n]$ is precisely the result $y_L[n]$, shown in Figure 5.8(b), obtained by a linear convolution of $g[n]$ and $h[n]$.

Modulation Theorem

The N-point DFT $Y[k]$ of the length-N product sequence $y[n] = g[n]h[n]$ is given by a circular convolution of their DFTs, $G[k]$ and $H[k]$, divided by N; that is,

$$g[n]h[n] \stackrel{\text{DFT}}{\longleftrightarrow} \frac{1}{N} \sum_{\ell=0}^{N-1} G[\ell]H[\langle k - \ell \rangle_N]. \quad (5.114)$$

We leave the proof of the above theorem as an exercise (Problem 5.30). It should be noted that the above theorem is the dual of the convolution theorem.

Parseval's Relation

The total energy of a length-N sequence $g[n]$ can be computed by summing the square of the absolute values of the DFT samples $G[k]$ divided by N; that is,

$$\sum_{n=0}^{N-1} |g[n]|^2 = \frac{1}{N} \sum_{k=0}^{N-1} |G[k]|^2, \quad (5.115)$$

The proof of this relation follows from the Parseval's relation of Eq. (5.6) derived for general orthogonal transforms. Equation (5.115) can be used to compute the total energy of a finite-length sequence in the frequency domain. A more general version of the above theorem is given by

$$\sum_{n=0}^{N-1} g[n]h^*[n] = \frac{1}{N} \sum_{k=0}^{N-1} G[k]H^*[k], \quad (5.116)$$

whose proof is left as an exercise (Problem 5.31).

For future reference, we summarize the DFT theorems in Table 5.3.

Table 5.3: DFT Theorems.

Theorem	Length-N Sequence	N-point DFT
	$g[n]$	$G[k]$
	$h[n]$	$H[k]$
Linearity	$\alpha g[n] + \beta h[n]$	$\alpha G[k] + \beta H[k]$
Circular time-shifting	$g[\langle n - n_o \rangle_N]$	$W_N^{-kn_o} G[k]$
Circular frequency-shifting	$W_N^{-k_o} g[n]$	$G[\langle k - k_o \rangle_N]$
Duality	$G[n]$	$Ng[\langle -k \rangle_N]$
N-point circular convolution	$\displaystyle\sum_{m=0}^{N-1} g[m]h[\langle n - m \rangle_N]$	$G[k]H[k]$
Modulation	$g[n]h[n]$	$\displaystyle\frac{1}{N}\sum_{m=0}^{N-1} G[m]H[\langle k - m \rangle_N]$
Parseval's relation	$\displaystyle\sum_{n=0}^{N-1} |g[n]|^2 = \frac{1}{N}\sum_{k=0}^{N-1} |G[k]|^2$	

5.8   Fourier-Domain Filtering

Often one is interested in removing the components of a finite-length discrete-time signal in one or more frequency bands. If the location of these bands are known a priori, then a straightforward approach to remove the unwanted components from a signal is to implement the filtering in the Fourier domain. Let $X(e^{j\omega})$ and $H(e^{j\omega})$ denote, respectively, the Fourier transforms of the signal $x[n]$ and the impulse response $h[n]$ of the filter. We then form the product $H(e^{j\omega})X(e^{j\omega})$, which is the Fourier transform $Y(e^{j\omega})$ of the filtered output $y[n]$. An inverse Fourier transform of $Y(e^{j\omega})$ results in the desired filtered response $y[n]$. In general, since the Fourier transform $X(e^{j\omega})$ is a complex function of ω, the Fourier transform $Y(e^{j\omega})$ is also a complex function of ω. In the processing of a real-valued signal $x[n]$, if the filter selected has also real-valued impulse response $h[n]$, the imaginary parts of the samples of the inverse Fourier transform of the product of their Fourier transforms will be theoretically all zeros. However, in practice, the imaginary parts are very small numbers due to computational errors. Hence, the real part of the inverse Fourier transform $y[n]$ of the product $H(e^{j\omega})X(e^{j\omega})$ is kept as the filtered response.

A simple approach to design the filter is to set the Fourier transform $H(e^{j\omega})$ to zero in the band containing the components of the signal $x[n]$ that need to be suppressed, and to set $H(e^{j\omega})$ equal to 1 in the band where the components of the signal $x[n]$ are to be preserved. Since the Fourier transform $H(e^{j\omega})$ has zero-phase, multiplying $X(e^{j\omega})$ with $H(e^{j\omega})$ does not change the phase of $X(e^{j\omega})$ in the band to be preserved.

The Fourier-domain filtering is usually implemented using the DFT. In the time domain, this approach is equivalent to the circular convolution of the finite-length signal $x[n]$ and the finite-length ideal filter. However, since the ideal filter with a Fourier transform $H(e^{j\omega})$ that is 1 in the passband and zero in the

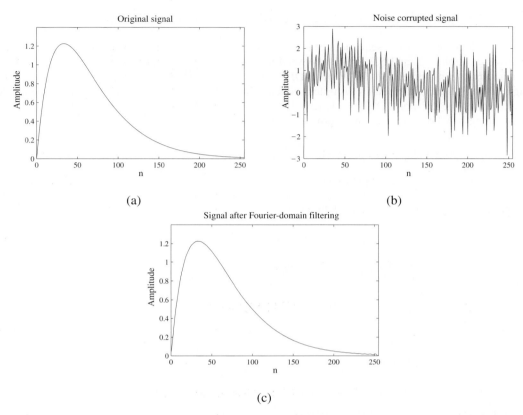

Figure 5.12: (a) Original signal, (b) the noise-corrupted signal, and (c) the signal obtained after a Fourier-domain filtering.

stopband has an inverse Fourier transform that is of infinite length, sampling the Fourier transform to create the DFT samples leads to the time-domain aliasing as mentioned in Section 5.3.4. As a result, the DFT-based filtering will always lead to some small ripples in the filtered response.

Usually such an approach to filtering a signal is employed in removing the high-frequency noise samples from a noise-corrupted low-frequency signal. We illustrate this approach in Example 5.13.

EXAMPLE 5.13   Illustration of Fourier-Domain Filtering

Consider the narrow-band lowpass signal

$$x[n] = 0.1n\, e^{-0.03n}, \qquad 0 \le n \le 255,$$

whose plot is shown in Figure 5.12(a). The signal $x[n]$ corrupted with a high-frequency random noise is indicated in Figure 5.12(b). The signal obtained by taking the 256-point DFT of the noise-corrupted signal and setting all DFT samples in the range $50 \le k \le 206$ to zero values and then forming a 256-point IDFT of the result is indicated in Figure 5.12(c). As can be seen from this plot, the filtered response is reasonably close to the original uncorrupted signal. However, some amount of ripples is present at the tail end of the response due to the time-domain aliasing.

The Fourier-domain filtering sometimes finds applications in image processing [Gon2002].

5.9 Computation of the DFT of Real Sequences

In most practical applications, sequences of interest are real. In such cases, the symmetry properties of the DFT given in Tables 5.1 and 5.2 can be exploited to make the DFT computations more efficient. As we shall demonstrate in this section, the N-point DFTs of two length-N real sequences can be computed from a single N-point DFT of a length-N complex sequence formed from the two length-N real sequences. In a similar manner, the $2N$-point DFT of a length-$2N$ real sequence can be determined from a single N-point DFT of a length-N complex sequence formed from the length-$2N$ real sequence.

5.9.1 N-Point DFTs of Two Real Sequences Using a Single N-Point DFT

Let $g[n]$ and $h[n]$ be two real sequences of length N each, with $G[k]$ and $H[k]$ denoting their respective N-point DFTs. These two N-point DFTs can be computed efficiently using a single N-point DFT $X[k]$ of a complex length-N sequence $x[n]$ defined by

$$x[n] = g[n] + jh[n]. \tag{5.117}$$

From the above, $g[n] = \text{Re}\{x[n]\}$ and $h[n] = \text{Im}\{x[n]\}$.

From Table 5.1, we arrive at

$$G[k] = \tfrac{1}{2}\left\{X[k] + X^*[\langle -k\rangle_N]\right\}, \tag{5.118}$$

$$H[k] = \tfrac{1}{2j}\left\{X[k] - X^*[\langle -k\rangle_N]\right\}. \tag{5.119}$$

Note that $X^*[\langle -k\rangle_N] = X^*[\langle N - k\rangle_N]$.

EXAMPLE 5.14   **Computation of the DFTs of Two Real Sequences Using a Single DFT**

In this example, we illustrate the computation of the 4-point DFTs of the two real sequences of Example 5.7, using a single 4-point DFT.

From Eq. (5.58), the complex sequence $\{x[n]\} = \{g[n] + jh[n]\}$ is given by

$$\{x[n]\} = \{1 + j2 \quad 2 + j2 \quad j \quad 1 + j\}.$$

Its DFT is then

$$
\begin{bmatrix} X[0] \\ X[1] \\ X[2] \\ X[3] \end{bmatrix}
=
\begin{bmatrix} 1 & 1 & 1 & 1 \\ 1 & -j & -1 & j \\ 1 & -1 & 1 & -1 \\ 1 & j & -1 & -j \end{bmatrix}
\begin{bmatrix} 1 + j2 \\ 2 + j2 \\ j \\ 1 + j \end{bmatrix}
=
\begin{bmatrix} 4 + j6 \\ 2 \\ -2 \\ j2 \end{bmatrix}. \tag{5.120}
$$

From the above,

$$\{X^*[k]\} = \{4 - j6 \quad 2 \quad -2 \quad -j2\}.$$

Therefore,

$$\{X^*[\langle 4 - k\rangle_4]\} = \{4 - j6 \quad -j2 \quad -2 \quad 2\}. \tag{5.121}$$

Substituting Eqs. (5.120) and (5.121) in Eqs. (5.118) and (5.119), we get

$$\{G[k]\} = \{4 \quad 1 - j \quad -2 \quad 1 + j\},$$

$$\{H[k]\} = \{6 \quad 1 - j \quad 0 \quad -1 + j\}, \tag{5.122}$$

verifying the results obtained in Example 5.8.

5.9.2 2N-Point DFT of a Real Sequence Using a Single N-Point DFT

Let $v[n]$ be a real sequence of length $2N$ with $V[k]$ denoting its $2N$-point DFT. Define two real sequences $g[n]$ and $h[n]$ of length N each as

$$g[n] = v[2n], \qquad h[n] = v[2n+1], \qquad 0 \le n < N, \tag{5.123}$$

with $G[k]$ and $H[k]$ denoting their N-point DFTs. Now define a complex length-N sequence $x[n]$ according to Eq. (5.117). The DFTs $G[k]$ and $H[k]$ can be computed from the N-point DFT $X[k]$ of the sequence $x[n]$ by means of Eqs. (5.118) and (5.119).

Now,

$$V[k] = \sum_{n=0}^{2N-1} v[n]W_{2N}^{nk} = \sum_{n=0}^{N-1} v[2n]W_{2N}^{2nk} + \sum_{n=0}^{N-1} v[2n+1]W_{2N}^{(2n+1)k}$$

$$= \sum_{n=0}^{N-1} g[n]W_N^{nk} + \sum_{n=0}^{N-1} h[n]W_N^{nk} W_{2N}^{k}$$

$$= \sum_{n=0}^{N-1} g[n]W_N^{nk} + W_{2N}^{k} \sum_{n=0}^{N-1} h[n]W_N^{nk}.$$

Note that the first sum on the last expression is simply an N-point DFT $G[k]$ of the length-N sequence $g[n]$, whereas the second sum is an N-point DFT $H[k]$ of the length-N sequence $h[n]$. Therefore, we can express the $2N$-point DFT $V[k]$ as

$$V[k] = G[\langle k\rangle_N] + W_{2N}^{k} H[\langle k\rangle_N], \qquad 0 \le k \le 2N - 1, \tag{5.124}$$

where we have used the identity $W_{2N}^{2k} = W_N^{k}$ and used the modulo sign for the argument of the two N-point DFTs on the right-hand side since k here is in the range $0 \le k \le 2N - 1$.

EXAMPLE 5.15   **Computation of the DFT of a Real Sequence Using a Single DFT of a Shorter-Length Complex Sequence**

Let us determine the 8-point DFT $V[k]$ of the length-8 real sequence $v[n]$ given below:

$$v[n] = \{1 \quad 2 \quad 2 \quad 2 \quad 0 \quad 1 \quad 1 \quad 1\}.$$

We form two length-4 real sequences

$$g[n] = v[2n] = \{1 \quad 2 \quad 0 \quad 1\}, \quad h[n] = v[2n+1] = \{2 \quad 2 \quad 1 \quad 1\}, \qquad 0 \le n \le 3.$$

From Eq. (5.124),

$$V[k] = G[\langle k\rangle_4] + W_8^{k} H[\langle k\rangle_4], \qquad 0 \le k \le 7.$$

The 4-point DFTs, $G[k]$ and $H[k]$ of the two length-4 sequences $g[n]$ and $h[n]$, were computed in Example 5.14 and are given in Eq. (5.122). Substituting these DFT values in the above equation, we finally arrive at

$$V[0] = G[0] + H[0] = 4 + 6 = 10,$$
$$V[1] = G[1] + W_8^1 H[1] = (1 - j) + (1 - j) \cdot e^{-j\pi/4} = 1.0 - j2.4142,$$
$$V[2] = G[2] + W_8^2 H[2] = -2 + 0 \cdot e^{-j\pi/2} = -2,$$

$$V[3] = G[3] + W_8^3 H[3] = (1+j) + (1+j) \cdot e^{-j3\pi/4} = 1.0 - j0.4142,$$

$$V[4] = G[0] + W_8^4 H[0] = 4 + 4 \cdot e^{-j\pi} = -2,$$

$$V[5] = G[1] + W_8^5 H[1] = (1-j) + (1-j) \cdot e^{-j5\pi/4} = 1.0 + j0.4142,$$

$$V[6] = G[2] + W_8^6 H[2] = -2 + 0 \cdot e^{-j3\pi/2} = -2,$$

$$V[7] = G[3] + W_8^7 H[3] = (1+j) + (1+j) \cdot e^{-j7\pi/4} = 1.0 + j2.4142.$$

The above result can be verified by computing the 8-point DFT of the sequence $v[n]$ using MATLAB.

5.10 Linear Convolution Using the DFT

Linear convolution is a key operation in most signal processing applications. Since an N-point DFT can be implemented very efficiently using approximately $N(\log_2 N)$ arithmetic operations, it is of interest to investigate methods for the implementation of the linear convolution using the DFT. Earlier, in Example 5.12, we have already illustrated for a very specific case how to implement a linear convolution using a circular convolution. We first generalize this example for the linear convolution of two finite-length sequences of unequal lengths. Then we consider the implementation of the linear convolution of a finite-length sequence with an infinite-length sequence.

5.10.1 Linear Convolution of Two Finite-Length Sequences

Recall from the definition of the circular convolution given in Eq. (5.54) that the process involves the sum of products of the samples of two sequences of same length with the second sequence being time-reversed and circularly shifted with respect to the first sequence. The circular shifting operation causes the wraparound of the second sequence to bring all sample indices to remain in the range of that of the first sequence. Hence, the number of sample products is always equal to the length of the sequences. On the other hand, in the case of linear convolution, the process involves the sum of products of the samples of two sequences with the second sequence being time-reversed and linearly shifted with respect to the first sequence. As a result, the number of sample products increases linearly starting with one and then decreases again linearly ending with one. It therefore follows that to implement a linear convolution using the circular convolution operation, we need to extend both sequences by zero-padding with sufficient number of zero-valued samples to avoid the wraparound of the original samples of the second sequence.

Let $g[n]$ and $h[n]$ be finite-length sequences of lengths N and M, respectively. Our objective is to implement their linear convolution

$$y_L[n] = g[n] \circledast h[n], \tag{5.125}$$

using a circular convolution. Denote the length of the sequence $y_L[n]$ where $L = M + N - 1$. Define two length-L sequences,

$$g_e[n] = \begin{cases} g[n], & 0 \le n \le N-1, \\ 0, & N \le n \le L-1, \end{cases} \tag{5.126}$$

$$h_e[n] = \begin{cases} h[n], & 0 \le n \le M-1, \\ 0, & M \le n \le L-1, \end{cases} \tag{5.127}$$

obtained by appending $g[n]$ and $h[n]$ with zero-valued samples. Then,

$$y_L[n] = y_C[n] = g_e[n] \copyright h_e[n]. \tag{5.128}$$

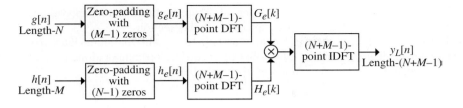

Figure 5.13: DFT-based implementation of the linear convolution of two finite-length sequences.

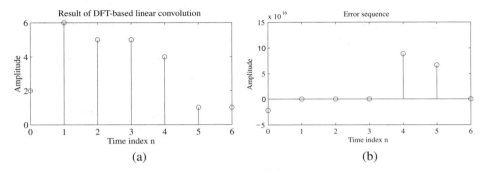

Figure 5.14: Plots of the output and the error sequences.

In other words, to implement Eq. (5.125) using the DFT-based approach, we first zero-pad $g[n]$ with $(M-1)$ zeros to obtain a length-L sequence $g_e[n]$, and zero-pad $h[n]$ with $(N-1)$ zeros to obtain a length-L sequence $h_e[n]$. Then we compute the $L = (N + M - 1)$-point DFTs of $g_e[n]$ and $h_e[n]$, respectively, resulting in $G_e[k]$ and $H_e[k]$. An L-point IDFT of the product $G_e[k]H_e[k]$ results in $y_L[n]$. The process involved is sketched in Figure 5.13.

It should be noted that even though we have used $L = M + N - 1$ here, any value $L \geq M + N - 1$ can be used. In practice, because the DFT is usually computed using fast algorithms, which are computationally very efficient for $L = 2^\nu$, L is often rounded up to the next highest power of 2.

Example 5.16 uses MATLAB to illustrate the above approach.

EXAMPLE 5.16   Linear Convolution Computation of Two Finite-Length Sequences Using the DFT

Program 5_4 can be used to determine the linear convolution of two finite-length sequences via the DFT-based approach and to compare the result using a direct linear convolution. The input data to be entered in vector format inside square brackets are the two sequences to be convolved. The program plots the sequence obtained using the DFT-based approach and the sequence obtained using the M-file `conv` and their difference. Using this program, we verify the result of Example 5.12 as demonstrated in Figure 5.14. It should be noted that the two non-zero samples of very small amplitudes are a result of finite-precision arithmetic employed in the computation of the DFT in MATLAB.

Program 5_4.m

5.10.2 Linear Convolution of a Finite-Length Sequence with an Infinite-Length Sequence

There are applications where we need to perform a linear convolution of a finite-length sequence with a sequence that is of infinite length or of length that is much greater than that of the first sequence. For example, the application may involve the processing of a speech signal by an FIR filter. In such cases, the method described in the previous section is impractical. There are two different approaches to solving this problem [Sto66]. We describe both of these approaches in this section.

Let $h[n]$ be a finite-length sequence of length M and $x[n]$ a sequence of infinite length (or a finite-length sequence of length much greater than M). Our objective is to develop computationally efficient DFT-based methods to implement the linear convolution of $h[n]$ and $x[n]$:

$$y[n] = \sum_{\ell=0}^{M-1} h[\ell]x[n-\ell] = h[n] \circledast x[n]. \tag{5.129}$$

Overlap-Add Method

In this method, we first segment $x[n]$, assumed to be a causal sequence here without any loss of generality, into a set of contiguous finite-length subsequences $x_m[n]$ of length N each:

$$x[n] = \sum_{m=0}^{\infty} x_m[n-mN], \tag{5.130}$$

where

$$x_m[n] = \begin{cases} x[n+mN], & 0 \le n \le N-1, \\ 0, & \text{otherwise.} \end{cases} \tag{5.131}$$

Substituting Eq. (5.130) in Eq. (5.129), we get

$$\begin{aligned} y[n] &= \sum_{\ell=0}^{M-1} h[\ell] \left(\sum_{m=0}^{\infty} x_m[n-\ell-mN] \right) \\ &= \sum_{m=0}^{\infty} \left(\sum_{\ell=0}^{M-1} h[\ell]x_m[n-\ell-mN] \right) \\ &= \sum_{m=0}^{\infty} y_m[n-mN], \end{aligned} \tag{5.132}$$

where

$$y_m[n] = h[n] \circledast x_m[n].$$

Since $h[n]$ is of length M and $x_m[n]$ is of length N, the linear convolution $h[n] \circledast x_m[n]$ is of length $(N+M-1)$. As a result, the desired linear convolution of Eq. (5.129) has been broken up into a sum of a large (possibly infinite) number of short-length linear convolutions of length $(N+M-1)$ each. Each of these short convolutions can be implemented using the method outlined in Figure 5.13, where now the DFTs (and the IDFT) are computed on the basis of $(N+M-1)$ points. There is one more subtlety to take care of before we can implement Eq. (5.132) using the DFT-based method.

Now the first short convolution in Eq. (5.132), given by $h[n] \circledast x_0[n]$, which is of length $(N+M-1)$, is defined for $0 \le n \le N+M-2$. The second short convolution in Eq. (5.132), given by $h[n] \circledast x_1[n]$, is also of length $(N+M-1)$ but is defined for $N \le n \le 2N+M-2$. This implies that there is an overlap of $M-1$ samples between these two short linear convolutions in the range $N \le n \le N+M-2$. Likewise, the third convolution in Eq. (5.132), given by $h[n] \circledast x_2[n]$, is defined for $2N \le n \le 3N+M-2$, causing an overlap between the samples of $h[n] \circledast x_1[n]$ and $h[n] \circledast x_2[n]$ for $2N \le n \le 2N+M-2$. In general, there will be an overlap of $M-1$ samples between the samples of the short convolutions $h[n] \circledast x_{r-1}[n]$ and $h[n] \circledast x_r[n]$ for $rN \le n \le rN+M-2$.

This process is illustrated in Figure 5.15. Figure 5.15(b) shows the first three length-7 ($N = 7$) segments $x_m[n]$ of the sequence $x[n]$ of Figure 5.15(a). Each of these segments is convolved with a length-5 ($M = 5$) sequence $h[n]$, resulting in length-11 ($N + M - 1 = 11$) short linear convolutions $y_m[n]$, shown in Figure 5.15(c). As can be seen from Figure 5.15(c), the last $M - 1 = 4$ samples of $y_0[n]$ overlap with the first 4 samples of $y_1[n]$. Likewise, the last $M - 1 = 4$ samples of $y_1[n]$ overlap with the first 4 samples of $y_2[n]$, and so on. Therefore, the desired sequence $y[n]$ obtained by a linear convolution of $x[n]$ and $h[n]$ is given by

$$y[n] = y_0[n], \qquad\qquad\qquad\qquad 0 \le n \le 6,$$
$$y[n] = y_0[n] + y_1[n - 7], \qquad\quad 7 \le n \le 10,$$
$$y[n] = y_1[n - 7], \qquad\qquad\qquad 11 \le n \le 13,$$
$$y[n] = y_1[n - 7] + y_2[n - 14], \quad 14 \le n \le 17,$$
$$y[n] = y_2[n - 14], \qquad\qquad\qquad 18 \le n \le 20,$$
$$\vdots$$

The above procedure is called the *overlap-add method* since the results of the short linear convolutions overlap and the overlapped portions are added to get the correct final result.

The M-file `fftfilt` can be used to implement the above method. We illustrate its use in Example 5.17.

EXAMPLE 5.17   Illustration of the Overlap-Add Method Using MATLAB

We consider the filtering of the noise-corrupted signal of Example 2.13 using a length-3 moving average filter. To this end, we use Program 5_5. The computer-generated results are indicated in Figure 5.16.

Program 5_5.m

Overlap-Save Method

In implementing the previous method using the DFT, we need to compute two $(N + M - 1)$-point DFTs and one $(N + M - 1)$-point IDFT since the overall linear convolution of Eq. (5.129) was expressed as a sum of short-length linear convolutions of length $(N + M - 1)$ each. It is possible to implement the linear convolution of Eq. (5.129) by performing, instead, circular convolutions of length shorter than $(N + M - 1)$. To this end, it is necessary to segment $x[n]$ into overlapping blocks $x_m[n]$, keep the terms of the circular convolution of $h[n]$ with $x_m[n]$ that corresponds to the terms obtained by a linear convolution of $h[n]$ and $x_m[n]$, and throw away the other parts of the circular convolution. To understand the correspondence between the linear and circular convolutions, consider a length-4 sequence $x[n]$ and a length-3 sequence $h[n]$. Let $y_L[n]$ denote the result of a linear convolution of $x[n]$ with $h[n]$. The six samples of $y_L[n]$ are given by

$$y_L[0] = h[0]x[0],$$
$$y_L[1] = h[0]x[1] + h[1]x[0],$$
$$y_L[2] = h[0]x[2] + h[1]x[1] + h[2]x[0],$$
$$y_L[3] = h[0]x[3] + h[1]x[2] + h[2]x[1],$$
$$y_L[4] = h[1]x[3] + h[2]x[2],$$
$$y_L[5] = h[2]x[3]. \tag{5.133}$$

If we append $h[n]$ with a single zero sample and convert it into a length-4 sequence $h_e[n]$, the 4-point circular convolution $y_C[n]$ of $h_e[n]$ and $x[n]$ is given by

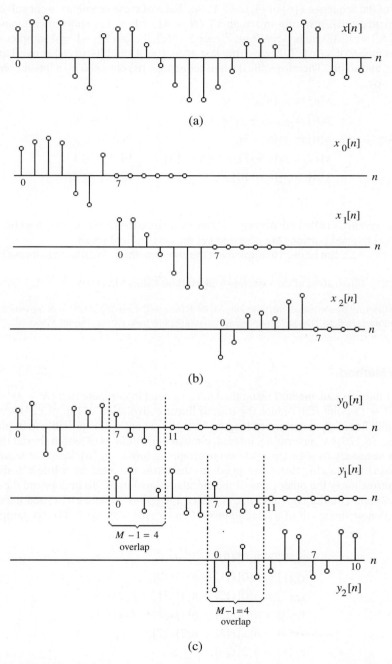

Figure 5.15: (a) Original $x[n]$, (b) segments $x_m[n]$ of $x[n]$, and (c) linear convolution of $x_m[n]$ with $h[n]$.

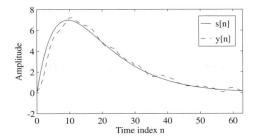

Figure 5.16: Uncorrupted input signal $s[n]$ (shown with solid line) and the filtered noisy signal $y[n]$ (shown with dashed line).

$$y_C[0] = h[0]x[0] + h[1]x[3] + h[2]x[2],$$
$$y_C[1] = h[0]x[1] + h[1]x[0] + h[2]x[3],$$
$$y_C[2] = h[0]x[2] + h[1]x[1] + h[2]x[0],$$
$$y_C[3] = h[0]x[3] + h[1]x[2] + h[2]x[1]. \tag{5.134}$$

Comparing Eqs. (5.133) and (5.134), we observe that the first two terms of the circular convolution do not correspond to the first two terms of the linear convolution, whereas the last two terms of the circular convolution are precisely the same as the third and fourth terms of the linear convolution; that is,

$$y_L[0] \neq y_C[0], \qquad y_L[1] \neq y_C[1],$$
$$y_L[2] = y_C[2], \qquad y_L[3] = y_C[3].$$

In the general case of an N-point circular convolution of a length-M sequence $h[n]$ with a length-N sequence $x[n]$ with $N > M$, the first $M - 1$ samples of the circular convolution are incorrect and are rejected, while the remaining $N - M + 1$ samples correspond to the correct samples of the linear convolution of $h[n]$ and $x[n]$.

Now consider an infinitely long or a very long sequence $x[n]$. We break it up as a collection of smaller length (length-4) sequences $x_m[n]$, as indicated below:

$$x_m[n] = x[n + 2m], \qquad 0 \leq n \leq 3, \quad 0 \leq m \leq \infty. \tag{5.135}$$

Next, we form

$$w_m[n] = h[n] \circledast_4 x_m[n],$$

or equivalently,

$$w_m[0] = h[0]x_m[0] + h[1]x_m[3] + h[2]x_m[2],$$
$$w_m[1] = h[0]x_m[1] + h[1]x_m[0] + h[2]x_m[3],$$
$$w_m[2] = h[0]x_m[2] + h[1]x_m[1] + h[2]x_m[0],$$
$$w_m[3] = h[0]x_m[3] + h[1]x_m[2] + h[2]x_m[1]. \tag{5.136}$$

Computing the above for $m = 0, 1, 2, 3, \ldots$, and substituting the values of $x_m[n]$ from Eq. (5.135), we arrive at

$$w_0[0] = h[0]x[0] + h[1]x[3] + h[2]x[2], \qquad \leftarrow \quad \textit{Reject}$$
$$w_0[1] = h[0]x[1] + h[1]x[0] + h[2]x[3], \qquad \leftarrow \quad \textit{Reject}$$
$$w_0[2] = h[0]x[2] + h[1]x[1] + h[2]x[0] = y[2], \qquad \leftarrow \quad \textit{Save}$$
$$w_0[3] = h[0]x[3] + h[1]x[2] + h[2]x[1] = y[3], \qquad \leftarrow \quad \textit{Save}$$

$$w_1[0] = h[0]x[2] + h[1]x[5] + h[2]x[4], \qquad \leftarrow \quad \textit{Reject}$$
$$w_1[1] = h[0]x[3] + h[1]x[2] + h[2]x[5], \qquad \leftarrow \quad \textit{Reject}$$
$$w_1[2] = h[0]x[4] + h[1]x[3] + h[2]x[2] = y[4], \qquad \leftarrow \quad \textit{Save}$$
$$w_1[3] = h[0]x[5] + h[1]x[4] + h[2]x[3] = y[5], \qquad \leftarrow \quad \textit{Save}$$

$$w_2[0] = h[0]x[4] + h[1]x[5] + h[2]x[6], \qquad \leftarrow \quad \textit{Reject}$$
$$w_2[1] = h[0]x[5] + h[1]x[4] + h[2]x[7], \qquad \leftarrow \quad \textit{Reject}$$
$$w_2[2] = h[0]x[6] + h[1]x[5] + h[2]x[4] = y[6], \qquad \leftarrow \quad \textit{Save}$$
$$w_2[3] = h[0]x[7] + h[1]x[6] + h[2]x[5] = y[7], \qquad \leftarrow \quad \textit{Save}$$

It should be noted that to determine $y[0]$ and $y[1]$, we need to form $x_{-1}[n]$:

$$x_{-1}[0] = 0, \quad x_{-1}[1] = 0, \quad x_{-1}[2] = x[0], \quad x_{-1}[3] = x[1],$$

and compute $w_{-1}[n] = h[n] \textcircled{4} x_{-1}[n]$ for $0 \leq n \leq 3$, reject $w_{-1}[0]$ and $w_{-1}[1]$, and save $w_{-1}[2] = y[0]$ and $w_{-1}[3] = y[1]$.

Generalizing the above, let $h[n]$ be a sequence of length M, and $x_m[n]$, the mth section of an infinitely long sequence $x[n]$ defined by

$$x_m[n] = x[n + m(N - M + 1)], \qquad 0 \leq n \leq N - 1, \tag{5.137}$$

be of length N, with $M \leq N$. If $w_m[n]$ denotes the N-point circular convolution of $h[n]$ and $x_m[n]$, that is, $w_m[n] = h[n] \textcircled{N} x_m[n]$, then we reject the first $M - 1$ samples of $w_m[n]$ and "abut" the remaining $N - M + 1$ saved samples of $w_m[n]$ to form $y_L[n]$, the linear convolution of $h[n]$ and $x[n]$. If we denote the saved portion of $w_m[n]$ as $y_m[n]$, that is,

$$y_m[n] = \begin{cases} 0, & 0 \leq n \leq M - 2, \\ w_m[n], & M - 1 \leq n \leq N - 1, \end{cases} \tag{5.138}$$

then,

$$y_L[n + m(N - M + 1)] = y_m[n], \qquad M - 1 \leq n \leq N - 1. \tag{5.139}$$

The above process is illustrated in Figure 5.17. The approach is called the *overlap-save method* since the input is segmented into overlapping sections and part of the results of the circular convolution are saved and abutted to determine the linear convolution result.

5.11 Discrete Cosine Transform

As pointed out earlier, in general, the N-point DFT $X[k]$ of a length-N real sequence is a complex sequence satisfying the symmetry condition $X[k] = X^*[\langle -k \rangle_N]$. For N even, the DFT samples $X[0]$ and $X[(N - 2)/2]$ are real and distinct. The remaining $N - 2$ DFT samples are complex, and only half of these samples are distinct as the rest are complex conjugates of these samples. For N odd, the DFT sample $X[0]$ is real, and the remaining $N - 1$ DFT samples are complex, of which only half of these samples

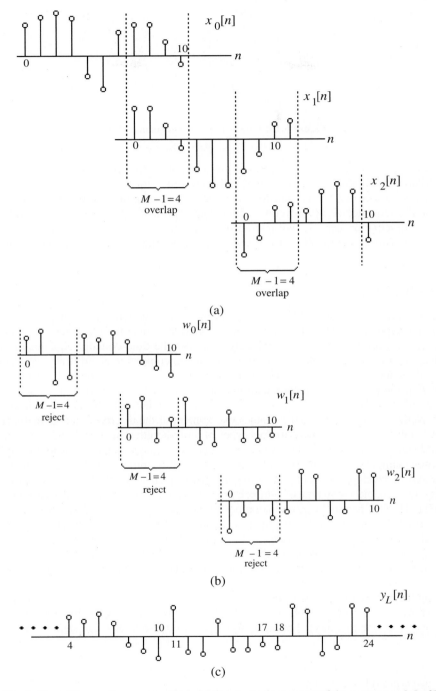

Figure 5.17: Illustration of the overlap-save method. (a) Overlapped segments of the sequence $x[n]$ of Figure 5.15(a), (b) sequences generated by an 11-point circular convolution, and (c) sequence obtained by rejecting the first four samples of $w_i[n]$ and abutting the remaining samples.

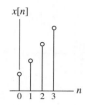

Figure 5.18: Original length-4 sequence $x[n]$.

are distinct as the rest are complex conjugates of these samples. As a result, there is a redundancy in the DFT-based frequency-domain representation of a discrete-time sequence.

It is of interest to have orthogonal transforms that represent a real time-domain sequence $x[n]$ by a real transform-domain sequence $\mathcal{X}[k]$. Several such orthogonal transforms have been developed, with each having certain attractive properties. We restrict our attention here to two such transforms, namely, the *discrete cosine transform* and the *Haar transform*. We discuss the former in this section and the latter in Section 5.12.

5.11.1   Definition

Recall from Section 5.5.2 that the DFT of a real symmetric or antisymmetric finite-length sequence is a product of a linear-phase term and a real amplitude function. Since the phase term is known for given length sequence, the amplitude function uniquely describes the time-domain sequence in the transform domain. One class of real orthogonal transforms is based on converting an arbitrary sequence into either a symmetric or an antisymmetric sequence and then extracting the real orthogonal transform coefficients from the DFT of the generated sequence with geometric symmetry. The transforms developed via this approach are called the *discrete cosine transform,* often abbreviated as DCT, and *discrete sine transform,* often abbreviated as DST. We consider here the development of the discrete cosine transform.

As indicated in Section 5.5.2, there are four types of finite-length sequences exhibiting geometric symmetry, depending on the location of the point of symmetry or antisymmetry: (1) whole-sample symmetry (WS), (2) whole-sample antisymmetry (WA), (3) half-sample symmetry (HS), and (4) half-sample antisymmetry (HA). To develop a symmetric or antisymmetric sequence by periodic extension from a specified finite-length sequence, the above type of symmetry or antisymmetry can be applied at each end of the given sequence [Mar94]. This results in 16 distinct types of periodic extensions, of which 8 extensions will be symmetric-periodic, leading to 8 different types of DCTs, and 8 extensions will be antisymmetric-periodic, leading to 8 different types of DSTs. In this section, we consider only the symmetric-periodic extensions.

To illustrate the periodic extension process, consider a length-4 sequence $x[n]$, shown in Figure 5.18. The eight symmetric periodic extensions generated from $\{x[n]\}$ are shown in Figures 5.19 and 5.20, respectively.

To develop the expression for the discrete-cosine transform for each of the above symmetric periodic sequences, we first extract one period from each. For example, to develop the *Type 1 discrete cosine transform,* we extract one period $y[n]$ of the symmetric periodic sequence $\tilde{x}_{\text{WSWS}}[n]$. Thus, if

$$\{x[n]\} = \{a \quad b \quad c \quad d\},$$

then the extracted period is given by

$$\{y[n]\} = \{a \quad b \quad c \quad d \quad c \quad b\}.$$

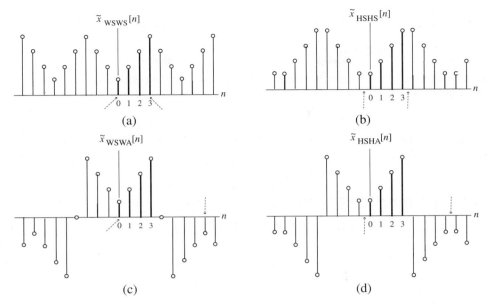

Figure 5.19: Four symmetric periodic extensions of $x[n]$ of Figure 5.18: (a) $\tilde{x}_{WSWS}[n]$ (for DCT-1), (b) $\tilde{x}_{HSHS}[n]$ (for DCT-2), (c) $\tilde{x}_{WSWA}[n]$ (for DCT-3), and (d) $\tilde{x}_{HSHA}[n]$ (for DCT-4). The dashed arrows indicate the points of symmetry.

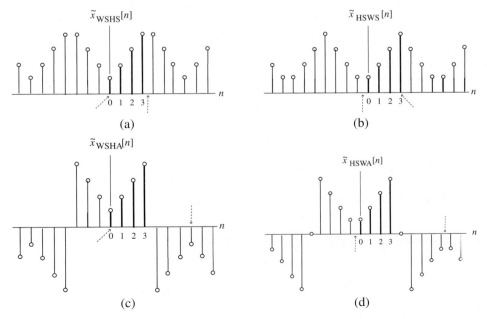

Figure 5.20: Four additional symmetric periodic extensions of $x[n]$ of Figure 5.18: (a) $\tilde{x}_{WSHS}[n]$ (for DCT-5), (b) $\tilde{x}_{HSWS}[n]$ (for DCT-6), (c) $\tilde{x}_{WSHA}[n]$ (for DCT-7), and (d) $\tilde{x}_{HSWA}[n]$ (for DCT-8). The dashed arrows indicate the points of symmetry.

Likewise, to develop the *Type 2 discrete cosine transform,* we extract one period $y[n]$ of the symmetric periodic sequence $\tilde{x}_{\text{HSHS}}[n]$. For the length-4 sequence $x[n]$ given above, the extracted period is given by

$$\{y[n]\} = \{a \quad b \quad c \quad d \quad d \quad c \quad b \quad a\}.$$

The discrete cosine transform is then determined from the discrete Fourier transform of $y[n]$.

As the Type 2 DCT has been employed in a number of international standards for image and video compression, such as JPEG, MPEG, and H.261, because of its better energy compaction property, we consider here only this type of DCT, also called the *even symmetrical discrete cosine transform* [Ahm74].

Let $x[n]$ be a length-N sequence defined for $0 \leq n \leq N - 1$. First, $x[n]$ is extended to a length-$2N$ sequence by zero-padding:

$$x_e[n] = \begin{cases} x[n], & 0 \leq n \leq N - 1, \\ 0, & N \leq n \leq 2N - 1. \end{cases}$$

Next, a Type 2 symmetric sequence $y[n]$ of length $2N$ is formed from $x_e[n]$ according to

$$y[n] = x_e[n] + x_e[2N - 1 - n], \qquad 0 \leq n \leq 2N - 1$$

$$= \begin{cases} x[n], & 0 \leq n \leq N - 1, \\ x[2N - 1 - n], & N \leq n \leq 2N - 1. \end{cases} \tag{5.140}$$

Note from Eq. (5.140) that the generated sequence $y[n]$ satisfies the symmetry property

$$y[n] = y[2N - 1 - n].$$

From Eq. (5.13), the $2N$-point DFT $Y[k]$ of the length-$2N$ sequence $y[n]$ is thus

$$Y[k] = \sum_{n=0}^{2N-1} y[n] W_{2N}^{kn}, \qquad 0 \leq k \leq 2N - 1. \tag{5.141}$$

We can rewrite Eq. (5.141) as

$$Y[k] = \sum_{n=0}^{N-1} y[n] W_{2N}^{kn} + \sum_{n=N}^{2N-1} y[n] W_{2N}^{kn}$$

$$= \sum_{n=0}^{N-1} x[n] W_{2N}^{kn} + \sum_{n=N}^{2N-1} x[2N - 1 - n] W_{2N}^{kn}, \qquad 0 \leq k \leq 2N - 1. \tag{5.142}$$

By making a change of variables, the DFT $Y[k]$ can be expressed as

$$Y[k] = \sum_{n=0}^{N-1} x[n] W_{2N}^{kn} + \sum_{n=0}^{N-1} x[n] W_{2N}^{-kn} W_{2N}^{k(2N-1)}$$

$$= W_{2N}^{-k/2} \sum_{n=0}^{N-1} x[n] \left(W_{2N}^{kn} W_{2N}^{k/2} + W_{2N}^{-kn} W_{2N}^{-k/2} \right)$$

$$= W_{2N}^{-k/2} \sum_{n=0}^{N-1} 2 x[n] \cos\left(\frac{\pi k (2n + 1)}{2N} \right), \qquad 0 \leq k \leq 2N - 1. \tag{5.143}$$

The Type 2 N-point *discrete cosine transform* (DCT) $X_{DCT}[k]$ of the length-N sequence $x[n]$ is obtained from Eq. (5.143) by extracting the first N samples of $Y[k]$ and multiplying them with $W_{2N}^{k/2}$:

$$X_{DCT}[k] = \sum_{n=0}^{N-1} 2 x[n] \cos\left(\frac{\pi k(2n+1)}{2N}\right), \qquad 0 \le k \le N-1. \tag{5.144}$$

Note that the samples of $X_{DCT}[k]$ are real for a real sequence $x[n]$.

The inverse discrete cosine transform (IDCT) of an N-point DCT $X_{DCT}[k]$ is given by

$$x[n] = \frac{1}{N} \sum_{k=0}^{N-1} \alpha[k] X_{DCT}[k] \cos\left(\frac{\pi k(2n+1)}{2N}\right), \qquad 0 \le n \le N-1, \tag{5.145}$$

where

$$\alpha[k] = \begin{cases} 1/2, & k = 0, \\ 1, & 1 \le k \le N-1. \end{cases} \tag{5.146}$$

Equations (5.144) and (5.145) constitute a discrete cosine transform pair for the sequence $x[n]$, where Eq. (5.144) is the analysis equation and Eq. (5.145) is the synthesis equation. A discrete cosine transform pair may often be denoted as

$$x[n] \overset{DCT}{\longleftrightarrow} X_{DCT}[k]. \tag{5.147}$$

It can be shown that

$$\frac{1}{N} \sum_{n=0}^{N-1} \cos\left(\frac{\pi k(2n+1)}{2N}\right) \cos\left(\frac{\pi m(2n+1)}{2N}\right) = \begin{cases} 1, & k = m = 0, \\ 1/2, & k = m \ne 0, \\ 0, & k \ne m, \end{cases} \tag{5.148}$$

or, in other words, the basis sequences $\cos\left(\frac{\pi k(2n+1)}{2N}\right)$ are orthogonal to each other.

To verify that $x[n]$ as given in Eq. (5.145) is indeed the IDCT of $X_{DCT}[k]$, we substitute Eq. (5.145) in Eq. (5.144). This leads to

$$X_{DCT}[k] = \frac{2}{N} \sum_{n=0}^{N-1} \left(\sum_{\ell=0}^{N-1} \alpha[\ell] X_{DCT}[\ell] \cos\left(\frac{\pi \ell(2n+1)}{2N}\right) \right) \cos\left(\frac{\pi k(2n+1)}{2N}\right)$$

$$= 2 \sum_{\ell=0}^{N-1} \alpha[\ell] X_{DCT}[\ell] \left(\frac{1}{N} \sum_{n=0}^{N-1} \cos\left(\frac{\pi \ell(2n+1)}{2N}\right) \cos\left(\frac{\pi k(2n+1)}{2N}\right) \right),$$

$$0 \le k \le N-1. \tag{5.149}$$

Making use of the identity of Eq. (5.148) on the right-hand side of the above equation, we get

$$X_{DCT}[k] = \begin{cases} 2\alpha[k] X_{DCT}[k], & k = 0, \\ \alpha[k] X_{DCT}[k], & 1 \le k \le N-1, \end{cases}$$

$$= X_{DCT}[k], \qquad 0 \le k \le N-1. \tag{5.150}$$

The inverse discrete cosine transform (IDCT) of an N-point DCT $X_{DCT}[k]$ can also be computed using the DFT. To this end, we first form a $2N$-point DFT $Y[k]$ according to

$$Y[k] = \begin{cases} W_{2N}^{-k/2} X_{DCT}[k], & 0 \le k \le N-1, \\ 0, & k = N, \\ -W_{2N}^{-k/2} X_{DCT}[2N-k], & N+1 \le k \le 2N-1, \end{cases} \tag{5.151}$$

compute next its length-$2N$ IDFT $y[n]$, and then extract the first N samples.

From Eq. (5.14), the $2N$-point IDFT $y[n]$ of $Y[k]$ is given by

$$y[n] = \frac{1}{2N} \sum_{k=0}^{2N-1} Y[k] W_{2N}^{-kn}, \qquad 0 \le n \le 2N - 1. \tag{5.152}$$

Substituting Eq. (5.151) in Eq. (5.152), we get

$$y[n] = \frac{1}{2N} \sum_{k=0}^{N-1} X_{\mathrm{DCT}}[k] W_{2N}^{-(n+\frac{1}{2})k} - \frac{1}{2N} \sum_{k=N+1}^{2N-1} X_{\mathrm{DCT}}[2N - k] W_{2N}^{-(n+\frac{1}{2})k}$$

$$= \frac{1}{2N} \sum_{k=0}^{N-1} X_{\mathrm{DCT}}[k] W_{2N}^{-(n+\frac{1}{2})k} - \frac{1}{2N} \sum_{k=1}^{N-1} X_{\mathrm{DCT}}[k] W_{2N}^{-(n+\frac{1}{2})(2N-k)}$$

$$= \frac{1}{2N} \sum_{k=0}^{N-1} X_{\mathrm{DCT}}[k] W_{2N}^{-(n+\frac{1}{2})k} + \frac{1}{2N} \sum_{k=1}^{N-1} X_{\mathrm{DCT}}[k] W_{2N}^{(n+\frac{1}{2})k}$$

$$= \frac{X_{\mathrm{DCT}}[0]}{2N} + \frac{1}{N} \sum_{k=1}^{N-1} X_{\mathrm{DCT}}[k] \cos\left(\frac{\pi k(2n + 1)}{2N}\right), \qquad 0 \le n \le 2N - 1. \tag{5.153}$$

The length-N IDCT $x[n]$ of the N-point DCT $X_{\mathrm{DCT}}[k]$ is given by the first N samples of $y[n]$:

$$x[n] = y[n], \qquad 0 \le n \le N - 1.$$

5.11.2   DCT Properties

The DCT satisfies a number of properties that are useful in certain applications. Because of the close relation between the DCT and the DFT, some of these properties can be derived easily by making use of the DFT properties. We review some of the important properties in this section. We assume all time-domain sequences to be of length N with an N-point DCT. In addition, for compactness, these properties are stated using the notation introduced in Eq. (5.147) and make use of the following DCT pairs:

$$g[n] \overset{\mathrm{DCT}}{\longleftrightarrow} G_{\mathrm{DCT}}[k],$$

$$h[n] \overset{\mathrm{DCT}}{\longleftrightarrow} H_{\mathrm{DCT}}[k].$$

Linearity Property

The DCT $X_{\mathrm{DCT}}[k]$ of a sequence $x[n] = \alpha g[n] + \beta h[n]$, obtained by a linear combination of two sequences, $g[n]$ and $h[n]$, where α and β are arbitrary constants, is given by the same linear combination of their individual DCTs, $G_{\mathrm{DCT}}[k]$ and $H_{\mathrm{DCT}}[k]$; that is,

$$\alpha g[n] + \beta h[n] \overset{\mathrm{DCT}}{\longleftrightarrow} \alpha G_{\mathrm{DCT}}[k] + \beta H_{\mathrm{DCT}}[k]. \tag{5.154}$$

The proof of the above property is straightforward and is left as an exercise (Problem 5.61).

Symmetry Property

The DCT of the conjugate sequence $g^*[n]$ is given by $G^*_{\mathrm{DCT}}[k]$; that is,

$$g^*[n] \xrightarrow{\text{DCT}} G^*_{\mathrm{DCT}}[k]. \tag{5.155}$$

The proof of the above property is also quite straightforward and is left as an exercise (Problem 5.61).

Energy Preservation Property

The energy preservation property of the DCT is similar to the Parseval's relation for the DFT. In the case of the Type 2 DCT, it is given by

$$\sum_{n=0}^{N-1} |g[n]|^2 = \frac{1}{2N} \sum_{k=0}^{N-1} \alpha[k]|G_{\mathrm{DCT}}[k]|^2. \tag{5.156}$$

The above result can be verified by making use of the inverse DCT expression of Eq. (5.145) and the symmetry property of Eq. (5.155) and is left as an exercise (Problem 5.61).

5.11.3 DCT Computation Using MATLAB

Recall the computation of the N-point DCT involves the computation of a $2N$-point DFT $Y[k]$ of a length-$2N$ sequence $y[n]$ exhibiting even symmetry. However, an alternate approach, as followed in MATLAB, is described next. A similar approach can be employed in computing the IDCT.

We first observe that the basis functions $\psi[k, n]$ employed in defining the DCT can be expressed as

$$\psi[k, n] = \cos\left(\frac{\pi k(2n + 1)}{2N}\right) = \mathrm{Re}\left\{W_{2N}^{k(n + \frac{1}{2})}\right\}.$$

Hence, Eq. (5.144) can be rewritten as

$$X_{\mathrm{DCT}}[k] = 2\,\mathrm{Re}\left(W_{2N}^{k/2} \sum_{n=0}^{2N-1} x[n]\, W_{2N}^{nk}\right), \qquad 0 \le k \le N - 1. \tag{5.157}$$

The above equation suggests that the N-point Type 2 DCT of a length-N sequence $x[n]$ can be computed as follows:

(a) Extend $x[n]$ to a length-$2N$ sequence $x_e[n]$ by zero-padding and compute its $2N$-point DFT $X_e[k]$.

(b) Extract the first N samples of $X_e[k]$ and multiply each sample with $W_{2N}^{k/2}$.

(c) Determine the real part of each of the above samples and multiply each with 2.

The DCT and IDCT definitions used in MATLAB are in a normalized form as shown below:

$$X_{\mathrm{DCT}}^{(n)}[k] = \sqrt{\frac{2}{N}}\beta[k] \sum_{n=0}^{N-1} x[n] \cos\left(\frac{\pi k(2n + 1)}{2N}\right), \qquad 0 \le k \le N - 1 \tag{5.158}$$

$$x[n] = \sqrt{\frac{2}{N}} \sum_{k=0}^{N-1} \beta[k] X_{\mathrm{DCT}}^{(n)}[k] \cos\left(\frac{\pi k(2n + 1)}{2N}\right), \qquad 0 \le n \le N - 1, \tag{5.159}$$

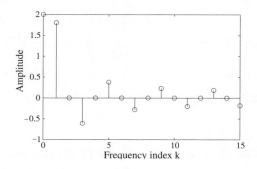

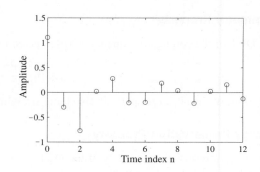

Figure 5.21: DCT samples.   **Figure 5.22:** IDCT samples.

where

$$\beta[k] = \begin{cases} 1/\sqrt{2}, & k = 0, \\ 1, & 1 \le k \le N - 1. \end{cases} \tag{5.160}$$

The normalization factors used in Eqs. (5.158) and (5.159) make the basis functions orthonormal and change the energy preservation property of Eq. (5.156) to the form indicated below:

$$\sum_{n=0}^{N-1} |x[n]|^2 = \sum_{k=0}^{N-1} |X_{\mathrm{DCT}}^{(n)}[k]|^2. \tag{5.161}$$

The M-files `dct` and `idct` can be used to compute the Type 2 DCT and its inverse as illustrated in Examples 5.18 and 5.19.

Program 5_2.m

EXAMPLE 5.18   DCT Computation Using MATLAB

Program 5_2 can be modified to evaluate the M-point DCT of the length-N sequence of Eq. (5.32). The key code fragment in the modified program is

```
U = dct(u,M);
```

A plot of the 16-point DCT generated by the modified program for $N = 8$ is shown in Figure 5.21.

Program 5_3.m

EXAMPLE 5.19   IDCT Computation Using MATLAB

Likewise, Program 5_3 can be modified to evaluate the N-point IDCT of the K-point DCT of Eq. (5.33). Here, the key code fragment in the modified program is

```
u = idct(U,N);
```

A plot of the 13-point IDCT generated by the modified program for $K = 8$ is shown in Figure 5.22.

The unnormalized form of the DCT coefficients can be obtained easily from the normalized form of the DCT coefficients. To this end, the following MATLAB code fragments can be used:

```
Cun(1)   = 2*sqrt(N)*Cn(1);
Cun(2:N) = sqrt(2*N)*Cn(2:N);
```

where `Cun` and `Cn` denote the unnormalized and the normalized DCT coefficient vectors, respectively.

5.12 The Haar Transform

The discrete-time Haar transform is derived by sampling the continuous-time *Haar functions* originally introduced to represent a continuous-time function $x_a(t)$ in the range $0 \leq t \leq 1$ [Haa10]. We first define the continuous-time Haar function, then derive the discrete-time Haar transform, and review some of the properties of the latter.

The set of Haar functions $h_\ell(t)$ contains N members, with N a power-of-2 positive integer; that is, $N = 2^{\nu+1}$, where $\nu \geq 0$. In defining the Haar function, the integer subscript ℓ is uniquely represented as a function of two nonnegative integer variables, r and s :

$$\ell = 2^r + s - 1.$$

It can be shown that the variables r and s have the following ranges for a given N: $0 \leq r \leq \nu$; $1 \leq s \leq 2^r$ for $r \neq 0$ and $s = 0$ or 1 for $r = 0$. The explicit relation between ℓ, r, and s is illustrated in Example 5.20.

EXAMPLE 5.20   Illustration of the Relation Between the Parameters Defining the Haar Functions

Consider $N = 8$, implying $\nu = 2$. In this case, r is in the range $0 \leq r \leq 2$. For $r = 1$, the range of s is $1 \leq s \leq 2$, and for $r = 2$, the range is $1 \leq s \leq 4$. Thus, the explicit relations between ℓ, r, and s are as shown below:

ℓ	0	1	2	3	4	5	6	7
r	0	0	1	1	2	2	2	2
s	0	1	1	2	1	2	3	4

The Haar functions are thus defined using the representation of ℓ as a function of r and s:

$$h_0(t) = h_{0,0}(t) = 1, \qquad 0 \leq t \leq 1, \tag{5.162}$$

$$h_\ell(t) = h_{r,s}(t) = \begin{cases} 2^{r/2}, & \frac{s-1}{2^r} \leq t < \frac{s-0.5}{2^r}, \\ -2^{r/2}, & \frac{s-0.5}{2^r} \leq t < \frac{s}{2^r}, \\ 0, & \text{otherwise for } 0 \leq t \leq 1. \end{cases} \tag{5.163}$$

5.12.1   Definition

The $N \times N$ discrete-time *Haar transform matrix* $\mathbf{H}_N$ is obtained by discretizing the Haar functions at discrete values of t, given by $t = n/N, 0 \leq n \leq N - 1$. We first derive the 2×2 Haar transform and then show how to derive a higher-order Haar transform from a lower-order Haar transform using a recursion.

EXAMPLE 5.21   Derivation of the 2 × 2 Haar Transform Matrix

Derive the 2×2 Haar transform matrix $\mathbf{H}_2$.

Here, $\nu = 0$ and $0 \leq \ell \leq 1$. Therefore, $r = 0$ and $0 \leq s \leq 1$. The relations between ℓ, r, and s are then as follows:

ℓ	0	1
r	0	0
s	0	1

The continuous-time Haar functions are now given by

$$\hbar_0(t) = \hbar_{0,0}(t) = 1, \qquad 0 \le t \le 1,$$

$$\hbar_1(t) = \hbar_{0,s}(t) = \begin{cases} 1, & s-1 \le t < s-0.5, \\ -1, & s-0.5 \le t < s, \\ 0, & \text{otherwise for } 0 \le t \le 1. \end{cases}$$

The 2×2 Haar transform matrix is obtained by sampling the above Haar functions at $t = n/2, n = 0, 1$, that is, at $t = 0$ and $t = 0.5$. This leads to

$$\hbar_{0,0}(0) = 1, \qquad \hbar_{0,0}(0.5) = 1,$$
$$\hbar_{0,1}(0) = 1, \qquad \hbar_{0,1}(0.5) = -1.$$

Hence,

$$\mathbf{H}_2 = \begin{bmatrix} 1 & 1 \\ 1 & -1 \end{bmatrix}. \tag{5.164}$$

Higher-order Haar transform matrices can be derived in an analogous manner. A more convenient approach is to derive the higher-order Haar transform matrix using the recursion

$$\mathbf{H}_{2^{\nu+1}} = \begin{bmatrix} \mathbf{H}_{2^\nu} \otimes [1 & 1] \\ 2^{\nu/2}\mathbf{I}_{2^\nu} \otimes [1 & -1] \end{bmatrix}, \quad \nu \ge 1, \tag{5.165}$$

where $\otimes$ denotes the *Kronecker product* and $\mathbf{I}_K$ is the $K \times K$ identity matrix.[4] Example 5.22 illustrates the above approach.

EXAMPLE 5.22   Derivation of the 4×4 Haar Transform Matrix Using Recursion

Derive the 4×4 Haar transform matrix using the recursion of Eq. (5.165). Here $\nu = 1$, which leads to

$$\mathbf{H}_4 = \begin{bmatrix} \mathbf{H}_2 \otimes [1 & 1] \\ \sqrt{2}\mathbf{I}_2 \otimes [1 & -1] \end{bmatrix}$$

$$= \begin{bmatrix} 1 & 1 & 1 & 1 \\ 1 & 1 & -1 & -1 \\ \sqrt{2} & -\sqrt{2} & 0 & 0 \\ 0 & 0 & \sqrt{2} & -\sqrt{2} \end{bmatrix}. \tag{5.166}$$

The N-point transform $\mathbf{X}_{\text{Haar}}$ of a length-N sequence $x[n]$ is thus given by

$$\mathbf{X}_{\text{Haar}} = \mathbf{H}_N \cdot \mathbf{x}, \tag{5.167}$$

where

$$\mathbf{X}_{\text{Haar}} = [X_{\text{Haar}}[0] \ X_{\text{Haar}}[1] \ \cdots X_{\text{Haar}}[N-1]]^T$$

is the vector of Haar transform coefficients and

$$\mathbf{x} = [x[0] \ x[1] \ \cdots x[N-1]]^T$$

is the vector of time-domain samples. The basis sequences of the Haar transform are thus given by the rows of the Haar transform matrix $\mathbf{H}_N$.

---

[4]If $\mathbf{A}$ and $\mathbf{B}$ are two arbitrary matrices, with the (r, s)-th element of $\mathbf{A}$ given by $[a_{rs}]$, then $\mathbf{A} \otimes \mathbf{B} = [a_{rs}\mathbf{B}]$.

In the case of the DFT and DCT matrices, the basis vectors are associated with specific frequencies, which can be defined as a function of the number of zero crossings of the basis vectors. A concept similar to frequency, called *sequency*, is usually associated with the basis vectors of the Haar transform matrix. It is defined by the number of zero crossings of the row vector. For example, in the case of the 4×4 Haar transform matrix $\mathbf{H}_4$ given by Eq. (5.166), the sequency of the first row is 0, that of the second row is 1, and that of the remaining two rows is 2.

The one-dimensional Haar transform can be computed using the M-file `haar_1D` which implements Haar matrix $\mathbf{H}_N$ using Eq. (5.165).

haar_1D.m

From Eq. (5.167), it follows that the inverse Haar transform is given by

$$\mathbf{x} = \mathbf{H}_N^{-1}\mathbf{X}_{\text{Haar}}. \tag{5.168}$$

Often, in practice, a normalized form of the Haar transform is used. Here, a 2×2 normalized Haar transform matrix is given by

$$\mathbf{H}_{2,n} = \begin{bmatrix} \frac{1}{\sqrt{2}} & \frac{1}{\sqrt{2}} \\ \frac{1}{\sqrt{2}} & -\frac{1}{\sqrt{2}} \end{bmatrix}. \tag{5.169}$$

The higher-order normalized Haar transform matrices are determined using the recursion

$$\mathbf{H}_{2^{v+1},n} = \begin{bmatrix} \mathbf{H}_{2^v,n} \otimes \begin{bmatrix} \frac{1}{\sqrt{2}} & \frac{1}{\sqrt{2}} \end{bmatrix} \\ \mathbf{I}_{2^v} \otimes \begin{bmatrix} \frac{1}{\sqrt{2}} & -\frac{1}{\sqrt{2}} \end{bmatrix} \end{bmatrix}, \quad v \geq 0. \tag{5.170}$$

5.12.2 Haar Transform Properties

Like the DFT and the DCT, the Haar transform also satisfies some useful properties. We review two of these properties next.

Orthogonality Property

The Haar transform matrix is orthogonal, and hence,

$$\mathbf{H}_N^{-1} = \frac{1}{N}\mathbf{H}_N^t. \tag{5.171}$$

The above result can be easily verified for $N = 2$ and $N = 4$ by computing the inverse of $\mathbf{H}_2$ from Eq. (5.164) and the inverse of $\mathbf{H}_4$ from Eq. (5.166). The proof of Eq. (5.171) for the general case is left as an exercise (Problem 5.63). As a result of the orthogonality property, the inverse Haar transform expression of Eq. (5.15) reduces to

$$\mathbf{x} = \frac{1}{N}\mathbf{H}_N^t \, \mathbf{X}_{\text{Haar}}. \tag{5.172}$$

If we denote the (k, ℓ)-th element of $\mathbf{H}_N$ as $h_N(k, \ell)$, then Eq. (5.172) can be rewritten as

$$x[n] = \frac{1}{N} \sum_{k=0}^{N-1} h_N(k, \ell)X_{\text{Haar}}[k], \qquad 0 \leq n \leq N - 1. \tag{5.173}$$

Energy Conservation Property

An expression similar to the Parseval's relation for the DFT also exists for the Haar transform, which can be employed to compute the total energy in the transform domain. It is given by

$$\sum_{n=0}^{N-1} |x[n]|^2 = \frac{1}{N} \sum_{k=0}^{N-1} |X_{\text{Haar}}[k]|^2 . \tag{5.174}$$

The proof of the above property is left as an exercise (Problem 5.63).

5.13 Energy Compaction Properties

In signal compression applications, it is desirable to use orthogonal transforms with high-energy compaction property. In such transforms, the dominant samples of the transform with high values are usually in the low-frequency range and contain most of the energy. In addition, transform samples with very small values tend to be in the high-frequency range and can be set to zero values. An inverse transform of the modified transform samples with zero-valued high-frequency samples is thus an approximate representation in the time domain of the original time-domain sequence.

If L samples of the transform with indices in the range $\mathcal{R}$ are set to zero with $L << N$ and if $x^{(m)}[n]$ denotes the inverse of the modified transform, then a measure of the energy compaction property caused by the removal of samples is given by the mean-square approximation error:

$$\mathcal{E}(L) = \frac{1}{N} \sum_{n=0}^{N-1} \left| x[n] - x^{(m)}[n] \right|^2 .$$

We now examine the energy compaction property of the three transforms considered in this chapter, namely, the DFT, the DCT, and the Haar transform.

The Discrete Fourier Transform

In the case of the N-point discrete Fourier transform, the high-frequency samples have indices around $N/2$. Here, the modified DFT is given by

$$X_{\text{DFT}}^{(m)}[k] = \begin{cases} X_{\text{DFT}}[k], & 0 \le k \le \frac{N-1-L}{2}, \\ 0, & \frac{N+1-L}{2} \le k \le \frac{N-1+L}{2}, \\ X_{\text{DFT}}[k], & \frac{N+1+L}{2} \le k \le N-1, \end{cases} \quad 0 \le k \le N-1. \tag{5.175}$$

For N even, the DFT index $N/2$ corresponds to the angular frequency $\omega = \pi$. Moreover, in the case of the N-point DFT of a real sequence $x[n]$, except for the samples $X[0]$ and $X[N/2]$, all other DFT samples are complex valued and satisfy the symmetry relation $X[k] = X^*[N-k]$. Hence, the $(\frac{N}{2}+1)$ DFT samples for $0 \le k \le \frac{N}{2}$ are sufficient to represent uniquely the length-N time domain sequence $x[n]$ in the transform domain. Since the DFT samples $X[k]$, $1 \le k \le \frac{N}{2} - 1$, are complex-valued, the real and imaginary parts of these DFT samples contain all together $N - 2$ real numbers. These samples, along with $X[0]$ and $X[N/2]$, thus constitute N real numbers and fully represent the length-N real time-domain sequence $x[n]$. An approximate representation $x_{\text{DFT}}^{(m)}[n]$ of the original time-domain sequence $x[n]$ is obtained by computing the IDFT of $X_{\text{DFT}}^{(m)}[k]$:

$$x_{\text{DFT}}^{(m)}[n] = \frac{1}{N} \sum_{k=0}^{N-1} X_{\text{DFT}}^{(m)}[k] W^{-kn}, \quad 0 \le k \le N-1. \tag{5.176}$$

The corresponding approximation error is given by

$$\mathcal{E}_{\text{DFT}}(L) = \frac{1}{N} \sum_{n=0}^{N-1} \left| x[n] - x_{\text{DFT}}^{(m)}[n] \right|^2. \tag{5.177}$$

The Discrete Cosine Transform

In this case, the high-frequency samples have high indices, and thus the modified DCT obtained by setting to zero the L samples with high indices is given by

$$X_{\text{DCT}}^{(m)}[k] = \begin{cases} X_{\text{DCT}}[k], & 0 \le k \le N - 1 - L, \\ 0, & N - L \le k \le N - 1. \end{cases} \tag{5.178}$$

The corresponding approximate representation $x_{\text{DCT}}^{(m)}[n]$ of the time-domain sequence $x[n]$ is given by an inverse DCT of $X_{\text{DCT}}^{(m)}[k]$:

$$x_{\text{DCT}}^{(m)}[n] = \frac{1}{N} \sum_{k=0}^{N-1-L} \alpha[k] X_{\text{DCT}}^{(m)}[k] \cos\left(\frac{\pi k(2n+1)}{2N}\right), \qquad 0 \le n \le N - 1. \tag{5.179}$$

Here the approximation error is given by

$$\mathcal{E}_{\text{DCT}}(L) = \frac{1}{N} \sum_{n=0}^{N-1} \left| x[n] - x_{\text{DCT}}^{(m)}[n] \right|^2. \tag{5.180}$$

The Discrete Haar Transform

Here the transform samples with high indices are associated with high sequency numbers. Hence, the modified Haar transform obtained by setting to zero the L samples with high indices is given by

$$X_{\text{Haar}}^{(m)}[k] = \begin{cases} X_{\text{Haar}}[k], & 0 \le k \le N - 1 - L, \\ 0, & N - L \le k \le N - 1. \end{cases} \tag{5.181}$$

From Eq. (5.173), the corresponding approximate representation $x_{\text{Haar}}^{(m)}[n]$ of the time-domain sequence $x[n]$ is given by

$$x_{\text{Haar}}^{(m)}[n] = \frac{1}{N} \sum_{k=0}^{N-1-L} h(k, n) X_{\text{Haar}}[k], \qquad 0 \le n \le N - 1, \tag{5.182}$$

resulting in the approximation error

$$\mathcal{E}_{\text{Haar}}(L) = \frac{1}{N} \sum_{n=0}^{N-1} \left| x[n] - x_{\text{Haar}}^{(m)}[n] \right|^2. \tag{5.183}$$

5.13.1 A Comparison

We compare the energy compaction properties of the discrete Fourier transform, the discrete cosine transform, and the Haar transform by computing the mean-square approximation error as a function of the number L of high-frequency transform coefficients set to zero for each transform for a typical signal, as illustrated in Example 5.23.

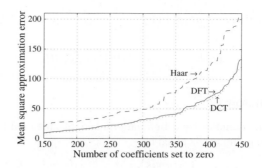

Figure 5.23: Energy compaction efficiency comparison.

EXAMPLE 5.23   A Comparison of the Energy Compaction Properties of the DFT, DCT, and the Haar Transform

The one-dimensional signal used for the comparison is the row 200 of the 512×512 image called "Goldhill," shown in Figure 1.19(a). The mean-square errors were computed by evaluating Eqs. (5.177), (5.180), and (5.183) using MATLAB. The resulting plot is shown in Figure 5.23. As can be seen from this plot, both the DCT and the Haar transform provide considerably more energy compaction than the DFT. Almost all signal compression methods used in practice are based on either the quantization of the DCT or the Haar transform coefficients.

5.14 Summary

Three different finite-length orthogonal transforms of a finite-length discrete-time sequence have been introduced and their properties reviewed. In each case, the length of the transform coefficients is the same as the length of the discrete-time sequence. Each of these representations consists of a pair of expressions: the analysis equation and the synthesis equation. The analysis equation is used to convert from the time-domain representation to the transform-domain representation, while the synthesis equation is used for the reverse process.

The first orthogonal transform discussed here is the discrete-Fourier transform (DFT), which is simply the samples of the Fourier transform of the discrete-time sequence evaluated at equally spaced frequency points at $\omega = 2\pi k/N$, where N is the length of the sequence and $0 \leq k \leq N - 1$. In general, the DFT samples are complex numbers whether or not the time-domain sequence is real or complex. The DFT is widely used in a number of digital signal processing applications, some of which will be discussed later in this book.

The two other finite-length orthogonal transforms discussed in this chapter are the discrete cosine transform (DCT) and the Haar transform. For a real discrete-time sequence, the samples of these two transforms are real numbers. These two transforms are often used in signal compression applications.

In some applications, a very long length time-domain sequence is broken up into a set of short-length time-domain sequences, and a finite-length transform is applied to each short-length sequence. The transformed sequences are next processed in the transform domain and their time-domain equivalents are generated by applying the inverse transforms. The processed short-length sequences are then grouped together appropriately to develop the final long-length sequence.

5.15 Problems

5.1 The *periodic convolution* of two periodic sequences, $\tilde{x}[n]$ and $\tilde{h}[n]$, of period N each, is defined by

$$\tilde{y}[n] = \sum_{r=0}^{N-1} \tilde{x}[r]\tilde{h}[n-r]. \tag{5.184}$$

Show that $\tilde{y}[n]$ is also a periodic sequence of period N.

5.2 Determine the periodic sequence $\tilde{y}[n]$ obtained by a periodic convolution of each pair of periodic sequences of period 5 given below:

(a) $\tilde{x}[n] = \{1 \quad 2 \quad -2 \quad -1 \quad 3\}, \quad \tilde{h}[n] = \{2 \quad 0 \quad 1 \quad 3 \quad -4\}, \ 0 \le n \le 4,$
(b) $\tilde{x}[n] = \{-1 \quad 5 \quad 3 \quad 0 \quad 3\}, \quad \tilde{h}[n] = \{-2 \quad 0 \quad 5 \quad 3 \quad -2\}, \ 0 \le n \le 4.$

5.3 Let $\tilde{x}[n]$ be a periodic sequence with period N, i.e., $\tilde{x}[n] = \tilde{x}[n + \ell N]$, where ℓ is any integer. The sequence $\tilde{x}[n]$ can be represented by a Fourier series given by a weighted sum of periodic complex exponential sequences $\tilde{\Psi}_k[n] = e^{j2\pi kn/N}$. Show that, unlike the Fourier series representation of a periodic continuous-time signal, the Fourier series representation of a periodic discrete-time sequence requires only N of the periodic complex exponential sequences $\tilde{\Psi}_k[n]$, $k = 0, 1, \ldots, N-1$, and is of the form

$$\tilde{x}[n] = \frac{1}{N} \sum_{k=0}^{N-1} \tilde{X}[k]e^{j2\pi kn/N}, \tag{5.185a}$$

where the Fourier coefficients $\tilde{X}[k]$ are given by

$$\tilde{X}[k] = \sum_{n=0}^{N-1} \tilde{x}[n]e^{-j2\pi kn/N}. \tag{5.185b}$$

Show that $\tilde{X}[k]$ is also a periodic sequence in k with a period N. The set of equations in Eqs. (5.185a) and (5.185b) represent the *discrete Fourier series* pair.

5.4 Determine the discrete Fourier series coefficients, defined in Eq. (5.185b), of the following periodic sequences:
(a) $\tilde{x}_1[n] = \cos(\pi n/4)$, (b) $\tilde{x}_2[n] = \sin(\pi n/3) + 3\cos(\pi n/4)$.

5.5 Show, using Eqs. (5.185a) and (5.185b), that the periodic impulse train $\tilde{p}[n] = \sum_{r=-\infty}^{\infty} \delta[n + rN]$ can be expressed in the form $\tilde{p}[n] = \frac{1}{N} \sum_{\ell=0}^{N-1} e^{j2\pi \ell n/N}$.

5.6 Let $x[n]$ be an aperiodic sequence with a DTFT $X(e^{j\omega})$. Define

$$\tilde{X}[k] = X(e^{j\omega})\Big|_{\omega=2\pi k/N} = X(e^{j2\pi k/N}), \qquad -\infty < k < \infty.$$

Show that $\tilde{X}[k]$ is a periodic sequence in k with a period N. Let $\tilde{X}[k]$ be the discrete Fourier series coefficients, defined in Eq. (5.185b), of the periodic sequence $\tilde{x}[n]$. Show, using Eqs. (5.185a) and (5.185b), that

$$\tilde{x}[n] = \sum_{r=-\infty}^{\infty} x[n + rN].$$

5.7 Let $\tilde{x}[n]$ and $\tilde{y}[n]$ be two periodic sequences with period N. Denote their discrete Fourier series coefficients, defined in Eq. (5.185b), as $\tilde{X}[k]$ and $\tilde{Y}[k]$, respectively.

(a) Let $\tilde{g}[n] = \tilde{x}[n]\tilde{y}[n]$ with $\tilde{G}[k]$ denoting its discrete Fourier series coefficients. Show, using Eqs. (5.185a) and (5.185b), that $\tilde{G}[k]$ can be expressed in terms of $\tilde{X}[k]$ and $\tilde{Y}[k]$ as

$$\tilde{G}[k] = \frac{1}{N} \sum_{\ell=0}^{N-1} \tilde{X}[\ell]\tilde{Y}[k - \ell]. \tag{5.186}$$

(b) Let $\tilde{H}[k] = \tilde{X}[k]\tilde{Y}[k]$ denote the discrete Fourier series coefficients of a periodic sequence $\tilde{h}[n]$. Show, using Eqs. (5.185a) and (5.185b), that $\tilde{h}[n]$ can be expressed in terms of $\tilde{x}[n]$ and $\tilde{y}[n]$ as

$$\tilde{h}[n] = \sum_{r=0}^{N-1} \tilde{x}[r]\tilde{y}[n - r]. \tag{5.187}$$

5.8 Determine the N-point DFTs of the following length-N sequences defined for $0 \leq n \leq N - 1$:
(a) $x_a[n] = \sin(2\pi n/N)$, (b) $x_b[n] = \cos^2(2\pi n/N)$, (c) $x_c[n] = \cos^3(2\pi n/N)$.

5.9 Determine the N-point DFTs of the following length-N sequences defined for $0 \leq n \leq N - 1$:
(a) $y_a[n] = \alpha^n$, (b) $y_b[n] = \begin{cases} 2, & \text{for } n \text{ even,} \\ -3, & \text{for } n \text{ odd.} \end{cases}$

5.10 Determine the N-point DFT $X[k]$ of the N-point sequence $x[n] = \cos(\omega_o n)$, $0 \leq n \leq N - 1$, for $\omega_o \neq 2\pi r/N$, $0 < r < N - 1$.

5.11 Consider a length-N sequence $x[n]$, $0 \leq n \leq N - 1$, with N even. Define 2 subsequences of length-$\frac{N}{2}$ each: $x_0[n] = x[2n]$ and $x_1[n] = x[2n + 1]$, $0 \leq n \leq \frac{N}{2}$. Let $X[k]$, $0 \leq k \leq N - 1$, denote the N-point DFT of $x[n]$, and $X_0[k]$ and $X_1[k]$, $0 \leq k \leq \frac{N}{2} - 1$, denote the $\frac{N}{2}$-point DFTs of $x_0[n]$ and $x_1[n]$, respectively. Express $X[k]$ as a function of $X_0[k]$ and $X_1[k]$.

5.12 Consider a length-N sequence $x[n]$, $0 \leq n \leq N - 1$, with N even. Define two length-$\frac{N}{2}$ sequences given by

$$x_0[n] = \left(x[n] + x[\tfrac{N}{2} + n]\right), \qquad x_1[n] = \left(x[n] - x[\tfrac{N}{2} + n]\right) W_N^n, \qquad 0 \leq n \leq \tfrac{N}{2} - 1.$$

If $X_0[k]$ and $X_1[k]$, $0 \leq k \leq \frac{N}{2} - 1$, denote the $\frac{N}{2}$-point DFTs of $x_0[n]$ and $x_1[n]$, respectively, determine the N-point DFT $X[k]$, $0 \leq k \leq N - 1$, of $x[n]$ from these two $\frac{N}{2}$-point DFTs.

5.13 Let $X[k]$ denote the N-point DFT of a length-N sequence $x[n]$, with N even. Define two length-$\frac{N}{2}$ sequences given by
$$g[n] = \tfrac{1}{2}(x[2n] + x[2n + 1]), \quad h[n] = \tfrac{1}{2}(x[2n] - x[2n + 1]), \quad 0 \leq n \leq \tfrac{N}{2} - 1.$$
If $G[k]$ and $H[k]$, $0 \leq k \leq \frac{N}{2} - 1$, denote $\frac{N}{2}$-point DFTs of $g[n]$ and $h[n]$, respectively, determine the N-point DFT $X[k]$ from these two $\frac{N}{2}$-point DFTs.

5.14 Let $X[k]$, $0 \leq k \leq N - 1$, denote the N-point DFT of a length-N sequence $x[n]$, with N even. Define two length-$\frac{N}{2}$ sequences given by

$$g[n] = a_1 x[2n] + a_2 x[2n + 1], \quad h[n] = a_3 x[2n] + a_4 x[2n + 1], \quad 0 \leq n \leq \tfrac{N}{2} - 1,$$

where $a_1 a_4 \neq a_2 a_3$. If $G[k]$ and $H[k]$, $0 \leq k \leq \frac{N}{2} - 1$, denote $\frac{N}{2}$-point DFTs of $g[n]$ and $h[n]$, respectively, determine the N-point DFT $X[k]$ from these two $\frac{N}{2}$-point DFTs.

5.15 Let $x[n], 0 \le n \le N-1$, be a length-N sequence with an N-point DFT given by $X[k], 0 \le k \le N-1$. Determine the $2N$-point DFT of each of the following length-$2N$ sequences:

(a) $g[n] = \begin{cases} x[n], & 0 \le n \le N-1, \\ 0, & N \le n \le 2N-1, \end{cases}$ (b) $h[n] = \begin{cases} 0, & 0 \le n \le N-1, \\ x[n-N], & N \le n \le 2N-1. \end{cases}$

5.16 Let $G[k]$ and $H[k], 0 \le k \le 2N-1$, denote, respectively, the $2N$-point DFTs of the length-$2N$ sequences $g[n]$ and $h[n]$ of Problem 5.15. Define a new length-$2N$ sequence given by $y[n] = g[n] + h[n]$, with a $2N$-point DFT $Y[k], 0 \le k \le 2N-1$. Develop the relation between $Y[k], G[k], H[k]$, and $X[k]$.

5.17 Let $Y[k]$ denote the MN-point DFT of a length-N sequence $x[n]$ appended with $(M-1)N$ zeros. Show that the N-point DFT $X[k]$ can be simply obtained from $Y[k]$ as follows:

$$X[k] = Y[kM], \qquad 0 \le k \le N-1.$$

5.18 Let $x[n], 0 \le n \le N-1$, be a length-N sequence with an N-point DFT given by $X[k], 0 \le k \le N-1$. Assume N is odd. Let $R = LN$, where L is a positive integer. Define an R-point DFT $Y[k], 0 \le k \le R-1$, given by

$$Y[k] = \begin{cases} LX[k], & 0 \le k \le \frac{N-1}{2}, \\ LX[k-R+N], & R - \frac{N-1}{2} \le k \le R-1, \\ 0, & \text{otherwise.} \end{cases}$$

Determine the length-R IDFT $y[n], 0 \le n \le R-1$, of $Y[k]$ as a function of $x[n]$.

5.19 Let $x[n], 0 \le n \le N-1$, be a length-N sequence with an N-point DFT $X[k], 0 \le k \le N-1$.

(a) If $x[n]$ is a symmetric sequence satisfying the condition $x[n] = x[\langle N-1-n \rangle_N]$, show that $X[N/2] = 0$ for N even.

(b) If $x[n]$ is a antisymmetric sequence satisfying the condition $x[n] = -x[\langle N-1-n \rangle_N]$, show that $X[0] = 0$.

(c) If $x[n]$ is a sequence satisfying the condition $x[n] = -x[\langle n+M \rangle_N]$ with $N = 2M$, show that $X[2\ell] = 0$ for $\ell = 0, 1, \ldots, M-1$.

5.20 Let $x[n], 0 \le n \le N-1$, be an even-length sequence with an N-point DFT $X[k], 0 \le k \le N-1$. If $X[2m] = 0$ for $0 \le m \le \frac{N}{2}-1$, show that $x[n] = -x[\langle n+\frac{N}{2} \rangle_N]$.

5.21 Let $x[n], 0 \le n \le N-1$, be a length-N sequence with an N-point DFT $X[k], 0 \le k \le N-1$. Determine the N-point DFTs of the following length-N sequences in terms of $X[k]$:

(a) $w[n] = \alpha x[\langle n-m_1 \rangle_N] + \beta x[\langle n-m_2 \rangle_N]$, where m_1 and m_2 are positive integers less than N,

(b) $g[n] = \begin{cases} x[n], & \text{for } n \text{ even,} \\ 0, & \text{for } n \text{ odd,} \end{cases}$

(c) $y[n] = x[n] \circledN x[n]$.

5.22 Let $x[n], 0 \le n \le N-1$, be an even-length sequence with an N-point DFT $X[k], 0 \le k \le N-1$. Determine the N-point DFTs of the following length-N sequences in terms of $X[k]$:

(a) $u[n] = x[n] - x[\langle n-\frac{N}{2} \rangle_N]$,   (b) $v[n] = x[n] + x[\langle n-\frac{N}{2} \rangle_N]$,   (c) $y[n] = (-1)^n x[n]$.

5.23 Let $x[n], 0 \le n \le N-1$, be a length-N sequence with an N-point DFT $X[k], 0 \le k \le N-1$. Determine the N-point inverse DFTs of the following length-N DFTs in terms of $x[n]$:

(a) $W[k] = \alpha X[\langle k-m_1 \rangle_N] + \beta X[\langle k-m_2 \rangle_N]$, where m_1 and m_2 are positive integers less than N,

(b) $G[k] = \begin{cases} X[k], & \text{for } k \text{ even,} \\ 0, & \text{for } k \text{ odd,} \end{cases}$

(c) $Y[k] = X[k] \circledN X[k]$.

5.24 Let $x[n]$, $0 \leq n \leq N - 1$, be a length-N sequence with an N-point DFT $X[k]$, $0 \leq k \leq N - 1$.

(a) Show that if N is even and if $x[n] = -x[\langle n + \frac{N}{2} \rangle_N]$ for all n, then $X[k] = 0$ for k even.

(b) Show that if N is an integer multiple of 4 and if $x[n] = -x[\langle n + \frac{N}{4} \rangle_N]$ for all n, then $X[k] = 0$ for $k = 4\ell$, $0 \leq \ell \leq \frac{N}{4} - 1$.

5.25 Let $x[n]$, $0 \leq n \leq N - 1$, be a length-N real sequence with an N-point DFT $X[k]$, $0 \leq k \leq N - 1$.

(a) Show that $X[\langle N - k \rangle_N] = X^*[k]$.

(b) Show that $X[0]$ is real.

(c) If N is even, show that $X[N/2]$ is real.

5.26 Let $x[n]$ be a length-N complex sequence with an N-point DFT $X[k]$. Determine the N-point DFTs of the following length-N sequences in terms of $X[k]$:

(a) $x^*[\langle -n \rangle_N]$,   (b) $x_{\mathrm{re}}[n]$,   (c) $jx_{\mathrm{im}}[n]$,   (d) $x_{\mathrm{cs}}[n]$,   (e) $x_{\mathrm{ca}}[n]$.

5.27 Let $x[n]$ be a length-N real sequence with an N-point DFT $X[k]$. Prove the following symmetry properties of $X[k]$:

(a) $X[k] = X^*[\langle -k \rangle_N]$,   (b) $X_{\mathrm{re}}[k] = X_{\mathrm{re}}[\langle -k \rangle_N]$,   (c) $X_{\mathrm{im}}[k] = -X_{\mathrm{im}}[\langle -k \rangle_N]$,   (d) $|X[k]| = |X[\langle -k \rangle_N]|$,
(e) $\arg X[k] = -\arg X[\langle -k \rangle_N]$.

5.28 Without computing the DFT, determine which one of the following length-9 sequences defined for $0 \leq n \leq 8$ has a real-valued 9-point DFT and which one has an imaginary-valued 9-point DFT.

(a) $\{x_1[n]\} = \{4 \quad\quad 3 \quad\quad -5 \quad\quad 1 \quad\quad -2 \quad\quad -2 \quad\quad 1 \quad\quad -5 \quad\quad 3\}$,

(b) $\{x_2[n]\} = \{0 \quad\quad 5 \quad\quad 1 \quad\quad 4 \quad\quad -3 \quad\quad 3 \quad\quad -4 \quad\quad -1 \quad\quad -5\}$,

(c) $\{x_3[n]\} = \{0 \quad\quad -5 \quad\quad 2 \quad\quad 4 \quad\quad -3 \quad\quad 3 \quad\quad -4 \quad\quad -1 \quad\quad -5\}$,

(d) $\{x_4[n]\} = \{-5 \quad\quad 5 \quad\quad -2 \quad\quad 2 \quad\quad 4 \quad\quad 4 \quad\quad 2 \quad\quad -2 \quad\quad 5\}$.

5.29 Let $G[k]$ and $H[k]$, $0 \leq k \leq 7$, denote the 8-point DFTs of two length-8 sequences, $g[n]$ and $h[n]$, $0 \leq n \leq 7$, respectively.

(a) If $G[k] = \{2.6 + j4.1 \quad 3 - j2.7 \quad -4.2 + j1.4 \quad 3.5 - j2.6 \quad 0.5 \quad 1.3 + j4.4 \quad 2.4 - j1.6 \quad -3 + j1.6\}$ and $h[n] = g[\langle n - 5 \rangle_8]$, determine $H[k]$ without forming $h[n]$ and then computing its DFT.

(b) If $g[n] = \{-0.1 - j0.7 \quad 1.3 + j \quad 2 + j0.7 \quad 1.1 + j2.2 \quad -0.8 + j0.2 \quad 3.4 - j0.1 \quad -1.2 + j3.1 \quad j1.5\}$ and $H[k] = G[\langle k + 3 \rangle_8]$, determine $h[n]$ without computing the DFT $G[k]$, forming $H[k]$, and then finding its inverse DFT.

5.30 Prove the following general properties of the DFT listed in Table 5.3: (a) linearity, (b) circular time-shifting, (c) circular frequency-shifting, (d) duality, and (e) N-point circular convolution.

5.31 Prove Eq. (5.116).

5.32 Consider two length-N real-valued sequences $x[n]$ and $y[n]$ defined for $0 \leq n \leq N - 1$ with N-point DFTs $X[k]$ and $Y[k]$, $0 \leq k \leq N - 1$, respectively. The *circular correlation* of $x[n]$ and $y[n]$ is given by

$$r_{xy}[\ell] = \sum_{n=0}^{N-1} x[n]y[\langle \ell + n \rangle_N], \quad 0 \leq \ell \leq N - 1. \tag{5.188}$$

Express the DFT of $r_{xy}[\ell]$ in terms $X[k]$ and $Y[k]$.

5.33 Let $x[n]$, $0 \leq n \leq N - 1$, be a length-N sequence with an MN-point DFT $X[k]$, $0 \leq k \leq MN - 1$. Define

$$y[n] = x[\langle n \rangle_N], \qquad 0 \leq n \leq MN - 1.$$

How would you compute the MN-point DFT $Y[k]$ of $y[n]$ knowing only $X[k]$?

5.34 Consider the length-10 sequence, defined for $0 \leq n \leq 9$,

$$\{x[n]\} = \{-3 \quad 5 \quad 45 \quad -15 \quad -9 \quad -19 \quad -8 \quad 21 \quad -10 \quad 23\},$$

with a 10-point DFT given by $X[k]$, $0 \leq k \leq 9$. Evaluate the following functions of $X[k]$ without computing the DFT:

(a) $X[0]$,   (b) $X[5]$,   (c) $\displaystyle\sum_{k=0}^{9} X[k]$,   (d) $\displaystyle\sum_{k=0}^{9} e^{-j2\pi k/5} X[k]$,   (e) $\displaystyle\sum_{k=0}^{9} |X[k]|^2$.

5.35 Let $X[k]$, $0 \leq k \leq 11$, be a 12-point DFT of a length-12 real sequence $x[n]$ with first 7 samples of $X[k]$ given by

$$X[k] = \{11 \quad 8 - j2 \quad 1 - j12 \quad 6 + j3 \quad -3 + j2 \quad 2 + j \quad 15\}, \ 0 \leq k \leq 6.$$

Determine the remaining samples of $X[k]$. Evaluate the following functions of $x[n]$ without computing the IDFT of $X[k]$:

(a) $x[0]$,   (b) $x[6]$,   (c) $\displaystyle\sum_{n=0}^{11} x[n]$,   (d) $\displaystyle\sum_{n=0}^{11} e^{j2\pi n/3} x[n]$,   (e) $\displaystyle\sum_{n=0}^{11} |x[n]|^2$.

5.36 Let $g[n]$ and $h[n]$ be two finite-length sequences of length 7 each. If $y_L[n]$ and $y_C[n]$ denote the linear and 7-point circular convolutions of $g[n]$ and $h[n]$, respectively, express $y_C[n]$ in terms of $y_L[n]$.

5.37 The even samples of the 9-point DFT of a length-9 real sequence are given by $X[0] = -5.7$, $X[2] = 1.2 - j4.1$, $X[4] = -3.5 + j5.3$, $X[6] = 8.6 - j9.6$, and $X[8] = -7.7 - j3.2$. Determine the missing odd samples of the DFT.

5.38 The following 5 samples of the 9-point DFT $X[k]$ of a real length-9 sequence are given: $X[0] = 11$, $X[2] = 1.2 - j2.3$, $X[3] = -7.2 - j4.1$, $X[5] = -3.1 + j8.2$, and $X[8] = 4.5 + j1.6$. Determine the remaining 4 samples.

5.39 The following 7 samples of a length-12 real sequence $x[n]$ with a real-valued 12-point DFT $X[k]$ are given by $x[0] = 3.8, x[2] = 0.7, x[3] = -3.25, x[5] = 4.1, x[6] = 2.87, x[8] = 9.3$, and $x[11] = -2$. Find the remaining 5 samples of $x[n]$.

5.40 The first 7 samples of a length-12 real sequence $x[n]$ with an imaginary-valued 12-point DFT $X[k]$ are given by $x[0] = 0, x[1] = 0.7, x[2] = -3.25, x[3] = 4.1, x[4] = 2.87, x[5] = -9.3$, and $x[6] = 0$. Find the remaining 5 samples of $x[n]$.

5.41 A 174-point DFT $X[k]$ of a real-valued sequence $x[n]$ has the following DFT samples: $X[0] = 1, X[9] = -3.4 + j5.9, \ X[k_1] = 7.1 + j2.4, \ X[51] = 5 - j1.6, \ X[k_2] = 8.7 + j4.9, \ X[87] = 4.5, \ X[113] = 8.7 - j4.9, \ X[k_3] = 5 + j1.6, \ X[162] = 7.1 - j2.4$, and $X[k_4] = -3.4 - j5.9$. Remaining DFT samples are assumed to be of zero value.

(a) Determine the values of the indices k_1, k_2, k_3, and k_4.
(b) What is the dc value of $\{x[n]\}$?
(c) Determine the expression for $\{x[n]\}$ without computing the IDFT.
(d) What is the energy of $\{x[n]\}$?

5.42 A 126-point DFT $X[k]$ of a real-valued sequence $x[n]$ has the following DFT samples: $X[0] = 12.8 + j\alpha$, $X[13] = -3.7 + j2.2$, $X[k_1] = 9.1 - j5.4$, $X[k_2] = 6.3 + j2.3$, $X[51] = -j1.7$, $X[63] = 13 + j\beta$, $X[k_3] = \gamma + j1.7$, $X[79] = 6.3 + j\delta$, $X[108] = \epsilon + j5.4$, and $X[k_4] = -3.7 - j2.2$. Remaining DFT samples are assumed to be of zero value.

  (a) Determine the values of the indices k_1, k_2, k_3, and k_4.
  (b) Determine the values of α, β, δ, and ϵ.
  (c) What is the dc value of $\{x[n]\}$?
  (d) Determine the expression for $\{x[n]\}$ without computing the IDFT.
  (e) What is the energy of $\{x[n]\}$?

5.43 A length-9 sequence is given by $\{x[n]\} = \{3, 5, 1, 4, -3, 5, -2, -2, 4\}$, $0 \le n \le 8$, with an 9-point DFT given by $X[k]$, $0 \le k \le 8$. Without computing the IDFT, determine the sequence $y[n]$ whose 9-point DFT is given by $Y[k] = W_3^{-2k} X[k]$.

5.44 The first 5 samples of the 9-point DFT $H[k]$, $0 \le k \le 8$, of a length-9 real sequence $h[n]$, $0 \le n \le 8$, are given by

$$H[k] = \{15 \qquad 6.8414 - j6.0572 \qquad 6.0346 - j1.957 \qquad j8.6603 \qquad -6.876 - j11.4883\}.$$

Determine the 9-point DFT $G[k]$ of the length-9 sequence $e^{j2\pi n/3}h[n]$ without computing $h[n]$, forming the sequence $g[n]$, and then taking its DFT.

5.45 Consider the two finite-length sequences $g[n] = \{2 \quad -1 \quad 3\}$, $0 \le n \le 2$ and $h[n] = \{-2 \ 4 \ 2 \ -1\}$, $0 \le n \le 3$.
  (a) Determine $y_L[n] = g[n] \circledast h[n]$.
  (b) Extend $g[n]$ to a length-4 sequence $g_e[n]$ by zero-padding and compute $y_C[n] = g_e[n] \textcircled{4} h[n]$.
  (c) Determine $y_C[n]$ using the DFT-based approach.
  (d) Extend $g[n]$ and $h[n]$ to length-6 sequences by zero-padding and compute the 6-point circular convolution $y[n]$ of the extended sequences. Is $y[n]$ the same as $y_L[n]$ determined in Part (a)?

5.46 Show that the circular convolution is commutative.

5.47 Consider two length-N sequences $x_1[n]$ and $x_2[n]$ defined for $0 \le n \le N-1$. Let $y[n] = x_1[n] \text{Ⓝ} x_2[n]$. Prove the following equalities:

(a) $\displaystyle\sum_{n=0}^{N-1} y[n] = \left(\sum_{n=0}^{N-1} x_1[n]\right)\left(\sum_{n=0}^{N-1} x_2[n]\right)$,

(b) $\displaystyle\sum_{n=0}^{N-1} (-1)^n y[n] = \left(\sum_{n=0}^{N-1} (-1)^n x_1[n]\right)\left(\sum_{n=0}^{N-1} (-1)^n x_2[n]\right)$ for N even.

5.48 Let $x[n]$ be a length-N sequence with an N-point DFT given by $X[k]$. Assume N is divisible by 3. Define a sequence

$$y[n] = x[3n], \qquad 0 \le n \le \tfrac{N}{3} - 1.$$

Express the $\frac{N}{3}$-point DFT $Y[k]$ of $y[n]$ in terms of $X[k]$.

5.49 The 8-point DFT of a length-8 complex sequence $v[n] = x[n] + jy[n]$ is given by

$$V[0] = 3 + j7, \quad V[1] = -2 + j6, \quad V[2] = 1 - j5, \quad V[3] = 4 - j9,$$

$$V[4] = 5 + j2, \quad V[5] = 3 - j2, \quad V[6] = j4, \quad V[7] = -3 - j8,$$

where $x[n]$ and $y[n]$ are, respectively, the real and imaginary parts of $v[n]$. Without computing the IDFT of $V[k]$, determine the 8-point DFTs $X[k]$ and $Y[k]$ of the real sequences $x[n]$ and $y[n]$, respectively. Verify your result by computing the IDFT of $V[k]$ using MATLAB.

5.50 Determine the 4-point DFTs of each sequence of the following pairs of length-4 sequences defined for $0 \leq n \leq 3$ by computing a single DFT:
(a) $g[n] = \{2 \quad -1 \quad 3 \quad 0\}$, $\quad h[n] = \{-2 \quad 4 \quad 2 \quad -1\}$,
(b) $x[n] = \{-3 \quad -2 \quad 2 \quad 4\}$, $\quad y[n] = \{1 \quad 2 \quad 3 \quad 4\}$.

5.51 Consider a rational discrete-time Fourier transform $X(e^{j\omega})$ with real coefficients of the form of

$$X(e^{j\omega}) = \frac{P(e^{j\omega})}{D(e^{j\omega})} = \frac{p_0 + p_1 e^{-j\omega} + \cdots + p_{M-1} e^{-j\omega(M-1)}}{d_0 + d_1 e^{-j\omega} + \cdots + d_{N-1} e^{-j\omega(N-1)}}.$$

Let $P[k]$ denote the M-point DFT of the numerator coefficients $\{p_i\}$ and $D[k]$ denote the N-point DFT of the denominator coefficients $\{d_i\}$. Determine the exact expressions of the DTFT $X(e^{j\omega})$ for $M = N = 4$ if the 4-point DFTs of its numerator and denominator coefficients are given by

$$P[k] = \{-5, \; -2 + j5, \; 4, \; -2 - j5\}, \qquad D[k] = \{3, \; 4 + j, \; -7, \; 4 - j\}.$$

Verify your result using MATLAB.

5.52 Repeat Problem 5.51 for the 4-point DFTs of the numerator and denominator coefficients given by

$$P[k] = \{8, \; -5 - j6, \; -3, \; -5 + j6\}, \qquad D[k] = \{-6, \; 6 + j2, \; 5, \; 6 - j2\}.$$

5.53 Consider a length-N sequence $x[n]$ with a DTFT $X(e^{j\omega})$. Define an M-point DFT $\hat{X}[k] = X(e^{j\omega_k})$, where $\omega_k = 2\pi k/M, k = 0, 1, \ldots, M - 1$. Denote the inverse DFT of $\hat{X}[k]$ as $\hat{x}[n]$, which is a length-M sequence. Express $x[n]$ in terms of $\hat{x}[n]$ and show that $x[n]$ can be fully recovered from $\hat{x}[n]$ only if $M \geq N$.

5.54 Let $X(e^{j\omega})$ denote the DTFT of the length-9 sequence $x[n] = \{1 \quad -2 \quad 3 \quad -4 \quad 5 \quad -4 \quad 3 \quad -2 \quad 1\}$.
(a) For the DFT sequence $X_1[k]$, obtained by sampling $X(e^{j\omega})$ at uniform intervals of $\pi/6$ starting from $\omega = 0$, determine the IDFT $x_1[n]$ of $X_1[k]$ without computing $X(e^{j\omega})$ and $X_1[k]$. Can you recover $x[n]$ from $x_1[n]$?
(b) For the DFT sequence $X_2[k]$, obtained by sampling $X(e^{j\omega})$ at uniform intervals of $\pi/4$ starting from $\omega = 0$, determine the IDFT $x_2[n]$ of $X_2[k]$ without computing $X(e^{j\omega})$ and $X_2[k]$. Can you recover $x[n]$ from $x_2[n]$?

5.55 Let $x[n]$ be a length-N sequence with $X[k]$ denoting its N-point DFT. We represent the DFT operation as $X[k] = \mathcal{F}\{x[n]\}$. Determine the sequence $y[n]$ obtained by applying the DFT operation 4 times to $x[n]$, i.e.,

$$y[n] = \mathcal{F}\{\mathcal{F}\{\mathcal{F}\{\mathcal{F}\{x[n]\}\}\}\}.$$

5.56 Let $x[n]$ and $h[n]$ be two length-51 sequences defined for $0 \leq n \leq 50$. It is known that $h[n] = 0$ for $0 \leq n \leq 16$ and $37 \leq n \leq 50$. Denote the 51-point circular convolution of these two sequences as $u[n]$ and their linear convolution as $y[n]$. Determine the range of n for which $y[n] = u[n]$.

5.57 The linear convolution of a length-110 sequence with a length-1300 sequence is to be computed using 128-point DFTs and IDFTs.
(a) Determine the smallest number of DFTs and IDFTs needed to compute the above linear convolution using the overlap-add approach.
(b) Determine the smallest number of DFTs and IDFTs needed to compute the above linear convolution using the overlap-save approach.

5.58 (a) Consider a length-N sequence $x[n]$, $0 \leq n \leq N - 1$, with an N-point DFT $X[k]$, $0 \leq k \leq N - 1$. Define a sequence $y[n]$ of length $LN, 0 \leq n \leq NL - 1$, given by

$$y[n] = \begin{cases} x[n/L], & n = 0, L, 2L, \ldots, (N-1)L, \\ 0, & \text{otherwise,} \end{cases} \qquad (5.189)$$

where L is a positive integer. Express the NL-point DFT $Y[k]$ of $y[n]$ in terms of $X[k]$.
(b) The 5-point DFT $X[k]$ of a length-5 sequence $x[n]$ is shown in Figure P5.1. Sketch the 21-point DFT $Y[k]$ of a length-21 sequence $y[n]$ generated using Eq. (5.189).

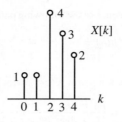

Figure P5.1

5.59 Consider two real, symmetric length-N sequences, $x[n]$ and $y[n]$, $0 \le n \le N - 1$ with N even. Define the length-$\frac{N}{2}$ sequences:

$$x_0[n] = x[2n + 1] + x[2n], \quad x_1[n] = x[2n + 1] - x[2n],$$
$$y_0[n] = y[2n + 1] + y[2n], \quad y_1[n] = y[2n + 1] - y[2n],$$

where $0 \le n \le \frac{N}{2} - 1$. It can be easily shown that $x_0[n]$ and $y_0[n]$ are real, symmetric sequences of length-$\frac{N}{2}$ each. Likewise, the sequences $x_1[n]$ and $y_1[n]$ are real and antisymmetric sequences. Denote the $\frac{N}{2}$-point DFTs of $x_0[n]$, $x_1[n]$, $y_0[n]$, and $y_1[n]$ by $X_0[k]$, $X_1[k]$, $Y_0[k]$, and $Y_1[k]$, respectively. Define a length-$\frac{N}{2}$ sequence $u[n]$:

$$u[n] = x_0[n] + y_1[n] + j(x_1[n] + y_0[n]).$$

Determine $X_0[k]$, $X_1[k]$, $Y_0[k]$, and $Y_1[k]$ in terms of the $\frac{N}{2}$-point DFT $U[k]$ of $u[n]$.

5.60 The *generalized discrete Fourier transform* (GDFT) is a generalization of the conventional DFT to allow shifts in either or both indices of the transform kernel [Bon76]. The N-point generalized discrete Fourier transform $X_{\mathrm{GDFT}}[k, a, b]$ of a length-N sequence $x[n]$ is defined by

$$X_{\mathrm{GDFT}}[k, a, b] = \sum_{n=0}^{N-1} x[n] \exp\left(-j\frac{2\pi(n + a)(k + b)}{N}\right). \tag{5.190}$$

Show that the inverse GDFT is given by

$$x[n] = \frac{1}{N} \sum_{k=0}^{N-1} X_{\mathrm{GDFT}}[k, a, b] \exp\left(j\frac{2\pi(n + a)(k + b)}{N}\right). \tag{5.191}$$

5.61 Prove the following properties of the DCT: (a) linearity property of the DCT, given by Eq. (5.154), (b) symmetry property of the DCT, given by Eq. (5.155), and (c) energy preservation property, given by Eq. (5.156).

5.62 The N coefficients of the normalized DCT $X_{\mathrm{DCT}}^{(n)}[k]$, $0 \le k \le N - 1$, given by Eq. (5.158), can be written in a matrix form $\mathbf{X}_{\mathrm{DCT}} = \mathbf{C}_N \mathbf{x}$, where

$$\mathbf{X}_{\mathrm{DCT}} = \left[\, X_{\mathrm{DCT}}^{(n)}[0] \quad X_{\mathrm{DCT}}^{(n)}[1] \quad \cdots \quad X_{\mathrm{DCT}}^{(n)}[N - 1] \,\right]^t, \quad \mathbf{x} = [\, x[0] \quad x[1] \quad \cdots \quad x[N - 1] \,]^t,$$

and $\mathbf{C}_N$ is the $N \times N$ DCT matrix whose (k, n)th element is given by

$$X_{\mathrm{DCT}}^{(n)}[k] = \sqrt{\frac{2}{N}} \beta[k] \sum_{n=0}^{N-1} \cos\left(\frac{\pi k(2n + 1)}{2N}\right),$$

with $\beta[k]$ given by Eq. (5.160). Even though the DCT matrix $\mathbf{C}_N$ is orthogonal, i.e., $\mathbf{x} = \mathbf{C}_N^{-1} \mathbf{X}_{\mathrm{DCT}} = \mathbf{C}_N^t \mathbf{X}_{\mathrm{DCT}}$, its elements are irrational numbers and do not produce the original input vector $\mathbf{x}$ by applying the inverse DCT transformation to $\mathbf{X}_{\mathrm{DCT}}$ when implemented with finite-precision arithmetic. It is thus desirable in practice to make use of integer-valued orthogonal transform matrix with a uniform frequency decomposition similar to that of the DCT.

(a) One such transform proposed for the H.26L video compression standard is the 4×4 matrix [Bjø98]:

$$\mathbf{H}_N = \begin{bmatrix} 13 & 13 & 13 & 13 \\ 17 & 7 & -7 & -17 \\ 13 & -13 & -13 & 13 \\ 7 & -17 & 17 & -7 \end{bmatrix}.$$

Show that the above transform matrix is orthogonal and all its rows have the same $\mathcal{L}_2$ norm.

(b) A recently proposed simpler 4×4 transform matrix [Mal2002]:

$$\mathbf{G}_N = \begin{bmatrix} 1 & 1 & 1 & 1 \\ 2 & 1 & -1 & -2 \\ 1 & -1 & -1 & 1 \\ 1 & -2 & 2 & -1 \end{bmatrix},$$

has a much smaller dynamic range than $\mathbf{H}_N$. Show that the rows of the above transform matrix are orthogonal but do not have the same $\mathcal{L}_2$ norm.

5.63 Prove the following properties of the Haar transform: (a) orthogonality property, given by Eq. (5.171) and (b) energy conservation property, given by Eq. (5.174).

5.64 The N-point *discrete Hartley transform* (DHT) $X_{\mathrm{DHT}}[k]$ of a length-N sequence $x[n]$ is defined by [Bra83]

$$X_{\mathrm{DHT}}[k] = \sum_{n=0}^{N-1} x[n] \left(\cos \left(\frac{2\pi nk}{N} \right) + \sin \left(\frac{2\pi nk}{N} \right) \right), \quad k = 0, 1, \ldots, N-1. \tag{5.192}$$

As can be seen from the above, the DHT of a real sequence is also a real sequence. Show that the inverse discrete Hartley transform (DHT) is given by

$$x[n] = \frac{1}{N} \sum_{k=0}^{N-1} X_{\mathrm{DHT}}[k] \left(\cos \left(\frac{2\pi nk}{N} \right) + \sin \left(\frac{2\pi nk}{N} \right) \right), \quad n = 0, 1, \ldots, N-1. \tag{5.193}$$

5.65 Let $X_{\mathrm{DHT}}[k]$ denote the N-point DHT of a length-N sequence $x[n]$.

(a) Show that the DHT of $x[\langle n - n_0 \rangle_N]$ is given by

$$X_{\mathrm{DHT}}[k] \cos \left(\frac{2\pi n_0 k}{N} \right) + X_{\mathrm{DHT}}[-k] \sin \left(\frac{2\pi n_0 k}{N} \right).$$

(b) Determine the N-point DHT of $x[\langle -n \rangle_N]$.

(c) Prove the Parseval's relation:

$$\sum_{n=0}^{N-1} x^2[n] = \frac{1}{N} \sum_{k=0}^{N-1} X_{\mathrm{DHT}}^2[k]. \tag{5.194}$$

5.66 Develop the relation between the N-point DHT $X_{\mathrm{DHT}}[k]$ and the N-point DFT $X[k]$ of a length-N sequence $x[n]$.

5.67 Let the N-point DHTs of the three length-N sequences $x[n]$, $g[n]$, and $y[n]$ be denoted by $X_{\mathrm{DHT}}[k]$, $G_{\mathrm{DHT}}[k]$, and $Y_{\mathrm{DHT}}[k]$, respectively. If $y[n] = x[n] \circledN g[n]$, show that

$$Y_{\mathrm{DHT}}[k] = \tfrac{1}{2} X_{\mathrm{DHT}}[k](G_{\mathrm{DHT}}[k] + G_{\mathrm{DHT}}[\langle -k \rangle_N])$$
$$+ \tfrac{1}{2} X_{\mathrm{DHT}}[\langle -k \rangle_N](G_{\mathrm{DHT}}[k] - G_{\mathrm{DHT}}[\langle -k \rangle_N]). \tag{5.195}$$

5.68 The *discrete combined Fourier transform* (DCFT) of a length-N sequence $x[n]$, $0 \leq n \leq N - 1$, is defined as a linear combination of its N-point DFT and the N-point IDFT given by [Ans85]

$$X_{\text{DCFT}}[k] = \sum_{n=0}^{N-1} \left(\alpha_1 W_N^{nk} + \alpha_2 W_N^{-nk} \right) x[n], \quad 0 \leq k \leq N - 1, \tag{5.196}$$

where at least one of the constants α_1 and α_2 is nonzero.

(a) Consider the sequence

$$y[n] = \sum_{n=0}^{N-1} \left(\beta_1 W_N^{-nk} + \beta_2 W_N^{nk} \right) X_{\text{DCFT}}[k], \quad 0 \leq n \leq N - 1. \tag{5.197}$$

Show that $y[n] = x[n]$, the inverse DCFT of $X_{\text{DCFT}}[k]$ if the following two conditions are satisfied:

$$\alpha_2 \beta_1 + \alpha_1 \beta_2 = 0,$$
$$N(\alpha_1 \beta_1 + \alpha_2 \beta_2) = 1.$$

(b) If $\alpha_2^2 \neq \alpha_2^2$, then show that the inverse DCFT of $X_{\text{DCFT}}[k]$ can be expressed as

$$x[n] = \frac{1}{N\left(\alpha_1^2 - \alpha_2^2\right)} \sum_{k=0}^{N-1} \left(\alpha_1 W_N^{-nk} - \alpha_2 W_N^{nk} \right) X_{\text{DCFT}}[k], \quad 0 \leq n \leq N - 1. \tag{5.198}$$

(c) Show that $X_{\text{DCFT}}[k]$ is a real sequence if $\alpha_1 = \alpha_2^\star = \alpha_{\text{re}} + j\alpha_{\text{im}}$, provided $\alpha_{\text{re}} \neq 0$, and $\alpha_{\text{im}} \neq 0$.

(d) Show that the discrete Hartley transform is a special case of the real-valued DCFT.

5.69 The Hadamard transform $X_{\text{HT}}[k]$ of a length-N sequence $x[n]$, $n = 0, 1, \ldots, N - 1$, is given by [Gon2002]

$$X_{\text{HT}}[k] = \frac{1}{N} \sum_{n=0}^{N-1} x[n](-1)^{\sum_{i=0}^{\ell-1} b_i(n)b_i(k)}, \quad k = 0, 1, \ldots, N - 1, \tag{5.199}$$

where $b_i(r)$ is the ith bit in the binary representation of r, and $N = 2^\ell$. In matrix form, the Hadamard transform can be represented as

$$\mathbf{X}_{\text{HT}} = \mathbf{H}_N \mathbf{x},$$

where

$$\mathbf{X}_{\text{HT}} = [X_{\text{HT}}[0] \quad X_{\text{HT}}[1] \quad \cdots \quad X_{\text{HT}}[N - 1]]^t,$$
$$\mathbf{x} = [x[0] \quad x[1] \quad \cdots \quad x[N - 1]]^t.$$

(a) Determine the form of the Hadamard matrix $\mathbf{H}_N$ for $N = 2, 4$, and 8.

(b) Show that

$$\mathbf{H}_4 = \begin{bmatrix} \mathbf{H}_2 & \mathbf{H}_2 \\ \mathbf{H}_2 & -\mathbf{H}_2 \end{bmatrix}, \qquad \mathbf{H}_8 = \begin{bmatrix} \mathbf{H}_4 & \mathbf{H}_4 \\ \mathbf{H}_4 & -\mathbf{H}_4 \end{bmatrix}.$$

(c) Determine the expression for the inverse Hadamard transform.

5.16 MATLAB Exercises

M 5.1 Using MATLAB, compute the N-point DFTs of the length-N sequences of Problem 3.19 for $N = 4, 6, 8$, and 10. Compare your results with that obtained by evaluating the DTFTs computed in Problem 3.19 at $\omega = 2\pi k/N$, $k = 0, 1, \ldots, N - 1$.

M 5.2 Write a MATLAB program to compute the circular convolution of two length-N sequences via the DFT-based approach. Using this program, determine the circular convolution of the following pairs of sequences:

(a) $g[n] = \{5, -2, 2, 0, 4, 3\}$, $h[n] = \{3, 1, -2, 2, -4, 4\}$,

(b) $x[n] = \{2 - j, -1 - j3, 4 - j3, 1 + j2, 3 + j2\}$, $v[n] = \{-3, 2 + j4, -1 + j4, 4 + j2, -3 + j\}$,

(c) $x[n] = \cos(\pi n/2)$, $y[n] = 3^n$, $0 \leq n \leq 4$.

Verify your result using the function `circonv`.

circonv.m

M 5.3 Using MATLAB, verify the symmetry relations of the DFT of a complex sequence as listed in Table 5.1.

M 5.4 Using MATLAB, verify the symmetry relations of the DFT of a real sequence as listed in Table 5.2.

M 5.5 Using MATLAB, prove the following general properties of the DFT listed in Table 5.3: (a) linearity, (b) circular time-shifting, (c) circular frequency-shifting, (d) duality, (e) N-point circular convolution, (f) modulation, and (g) Parseval's relation.

M 5.6 Write a MATLAB program to compute the DFTs of two real sequences of equal lengths based on the method outlined in Section 5.9.1.

M 5.7 Verify the results of Problem 5.34 by computing the DFT $X[k]$ of the sequence $x[n]$ given using MATLAB, and then evaluate the functions of $X[k]$ listed.

M 5.8 Verify the results of Problem 5.35 by computing the IDFT $x[n]$ of the DFT $X[k]$ given using MATLAB, and then evaluate the functions of $x[n]$ listed.

M 5.9 Write a MATLAB program to implement the Fourier-domain filtering illustrated in Example 5.13. Using this program, verify the results of this example.

M 5.10 Write a MATLAB function to implement the overlap-save method. Using this function, demonstrate the filtering of the noise-corrupted signal of Example 2.13 using a length-3 moving average filter by modifying Program 3.6.

Program 3_6

6 z-Transform

The discrete-time Fourier transform defined in Eq. (3.12) is a complex function of the angular frequency variable ω. It provides a frequency-domain representation of discrete-time signals and LTI systems. Moreover, because of the convergence condition, in many cases, the discrete-time Fourier transform of a sequence may not exist, and as a result, it is not possible to make use of such frequency-domain characterization in these cases.

A generalization of the discrete-time Fourier transform leads to the z-transform, which is a function of the complex variable z. It can exist for many sequences for which the discrete-time Fourier transform does not exist. In addition, for real-valued sequences with which we shall be concerned with in this text, the z-transform, in general, is a real-rational function of the complex variable z. Also, the use of z-transform techniques permits simple algebraic manipulations. Consequently, the z-transform has become an important tool in the analysis and design of digital filters. In this chapter, we discuss this alternate transform-domain representation of sequences and its properties. The representation of an LTI discrete-time system in the z-domain is given by its transfer function which is the z-transform of the impulse response of the system. We study the properties of the transfer function and relate it to the frequency response of the system, which is the discrete-time Fourier transform of the impulse response. Finally, the BIBO stability condition of an LTI system is derived in terms of its transfer function.

As in the case of the DTFT and the DFT, MATLAB has been used extensively to illustrate various concepts and implement a number of useful algorithms.

6.1 Definition and Properties

We first define the z-transform of a sequence and study its properties by treating it as a generalization of the discrete-time Fourier transform. This leads to the concept of the region of convergence of a z-transform that is investigated in detail. We then describe the inverse transform operation and point out two straightforward approaches for the computation of the inverse of a real rational z-transform. Next, the properties of the z-transform are reviewed.

For a given sequence $g[n]$, its z-transform $G(z)$ is defined as

$$G(z) = \sum_{n=-\infty}^{\infty} g[n]z^{-n}, \tag{6.1}$$

where $z = \text{Re}(z) + j\text{Im}(z)$ is a continuous complex variable. Often, it is convenient to express the z-transform as an operator as indicated below:

$$\mathcal{Z}\{g[n]\} = G(z) = \sum_{n=-\infty}^{\infty} g[n]z^{-n}.$$

301

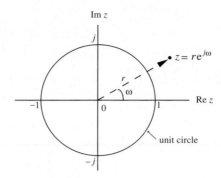

Figure 6.1: The point $z = re^{j\omega}$ in the complex z-plane.

Thus, the operator $\mathcal{Z}\{\cdot\}$ transforms a discrete-time sequence $g[n]$ into a function $G(z)$ of the complex variable z.

The relation between the sequence $g[n]$ and its z-transform $G(z)$ can be expressed in a compact form as

$$g[n] \xleftrightarrow{\ \mathcal{Z}\ } G(z). \tag{6.2}$$

We shall develop the expression for the computation of the inverse transform $g[n]$ from its z-transform $G(z)$ in Section 6.4.

If we express the complex variable z in polar form, $z = re^{j\omega}$, then Eq. (6.1) reduces to

$$G(re^{j\omega}) = \sum_{n=-\infty}^{\infty} g[n]r^{-n}e^{-j\omega n}. \tag{6.3}$$

Comparing the above with the discrete-time Fourier transform $G(e^{j\omega})$ of the sequence $g[n]$,

$$G(e^{j\omega}) = \sum_{n=-\infty}^{\infty} g[n]\,e^{-j\omega n},$$

we observe that Eq. (6.3) can be interpreted as the discrete-time Fourier transform of the modified sequence $\{g[n]r^{-n}\}$.

A geometrical interpretation of the z-transform can be given by considering the location of the point z in the complex z-plane. For fixed r and ω, the point $z = re^{j\omega}$ in the complex z-plane is at the tip of a vector of length r originating at the point $z = 0$ and subtending an angle ω with respect to the real axis, as shown in Figure 6.1. The contour $|z| = 1$ is a circle in the z-plane of unity radius and is called the *unit circle*. For $r = 1$ (i.e., $|z| = 1$), the z-transform $G(z)$ of $g[n]$ reduces to its Fourier transform $G(e^{j\omega})$, provided the latter exists. At $z = 1$, the value of $G(z)$ is $G(e^{j0})$, that is, the value of $G(e^{j\omega})$ at $\omega = 0$; at $z = j$, the value of $G(z)$ is $G(e^{j\pi/2})$, that is, the value of $G(e^{j\omega})$ at $\omega = \pi/2$, and so on. Hence, if we evaluate $G(z)$ on the unit circle counterclockwise at all values of ω beginning at $z = 1$, and terminating back at $z = 1$, effectively we have computed $G(e^{j\omega})$ in the frequency range $0 \leq \omega < 2\pi$. If, on the other hand, by going clockwise on the unit circle, we have determined $G(e^{j\omega})$ in the frequency range $-2\pi \leq \omega < 0$. Thus, it follows that by traversing the unit circle either clockwise or counterclockwise, we can evaluate the Fourier transform $G(e^{j\omega})$ at all values of the frequency $-\infty < \omega < \infty$ with the Fourier transform exhibiting a periodic response with a period 2π.

Like the discrete-time Fourier transform, there are conditions on the convergence of the infinite series of Eq. (6.1). For a given sequence, the set $\mathcal{R}$ of values of z for which its z-transform converges is called the *region of convergence* (ROC). It follows from our earlier discussion on the uniform convergence of the discrete-time Fourier transform in Section 3.2.4 and the interpretation of the z-transform $G(z)$ as the discrete-time Fourier transform of the sequence $g[n]r^{-n}$ that the series of Eq. (6.1) converges if $g[n]r^{-n}$ is absolutely summable, that is, if

$$\sum_{n=-\infty}^{\infty} \left| g[n]\, r^{-n} \right| < \infty. \tag{6.4}$$

It follows from Eq. (6.4) that the sequence $g[n]r^{-n}$ can be made absolutely summable by choosing the value of r properly even though the parent sequence $g[n]$ is not absolutely summable. Thus, a sequence $g[n]$ for which the discrete-time Fourier transform does not converge uniformly can have a z-transform with an ROC $\mathcal{R}_g$.

It also follows from Eq. (6.4) that if there is a specific value of $z = r\,e^{j\omega}$ for which the z-transform $G(z)$ exists, then it also exists everywhere on the circle of radius r in the z-plane. In general, the region of convergence $\mathcal{R}$ of a z-transform of a sequence $g[n]$ is an annular region of the z-plane:

$$R_{g-} < |z| < R_{g+}, \tag{6.5}$$

where $0 \leq R_{g-} < R_{g+} < \infty$. We shall show later that more than one sequence can have the same z-transform. Hence, it is important also to specify the associated ROC $\mathcal{R}_g$ for a sequence $g[n]$ and its z-transform $G(z)$ as indicated below:

$$g[n] \overset{\mathcal{Z}}{\longleftrightarrow} G(z), \qquad \text{ROC} = \mathcal{R}_g. \tag{6.6}$$

It should be noted that the z-transform as defined by Eq. (6.1) is a form of a Laurent series and is an analytic function at every point in its ROC [Chu90]. This in turn implies that the z-transform and all its derivatives are continuous functions of the complex variable z in the ROC.

EXAMPLE 6.1 z-Transform of a Causal Exponential Sequence

Let us determine the z-transform $X(z)$ of the causal sequence $x[n] = \alpha^n \mu[n]$ and its region of convergence. Applying the definition of Eq. (6.1), we obtain

$$X(z) = \sum_{n=-\infty}^{\infty} \alpha^n \mu[n] z^{-n} = \sum_{n=0}^{\infty} \alpha^n z^{-n}. \tag{6.7}$$

The above power series converges to

$$X(z) = \frac{1}{1 - \alpha z^{-1}}, \qquad \text{for } |\alpha z^{-1}| < 1, \tag{6.8}$$

indicating that the ROC is the annular region $|z| > |\alpha|$.

The z-transform $\mu(z)$ of the unit step sequence $\mu[n]$ can be obtained from Eq. (6.8) by setting $\alpha = 1$:

$$\mu(z) = \frac{1}{1 - z^{-1}}, \qquad \text{for } |z^{-1}| < 1. \tag{6.9}$$

The ROC of $\mu(z)$ is thus the annular region $1 < |z| < \infty$. Note that the unit step sequence is not absolutely summable, and, as a result, its Fourier transform does not converge uniformly.

EXAMPLE 6.2 z-Transform of an Anticausal Exponential Sequence

Consider the anticausal sequence $x[n] = -\alpha^n \mu[-n-1]$. Using Eq. (6.1), we arrive at the expression for its z-transform:

$$X(z) = -\sum_{n=-\infty}^{-1} \alpha^n z^{-n} = -\sum_{m=1}^{\infty} \alpha^{-m} z^m = -\alpha^{-1} z \sum_{n=0}^{\infty} \alpha^{-m} z^m$$

$$= -\frac{\alpha^{-1} z}{1 - \alpha^{-1} z} = \frac{1}{1 - \alpha z^{-1}}, \qquad \text{for } |\alpha^{-1} z| < 1, \tag{6.10}$$

where now the ROC is the annular region $|z| < |\alpha|$.

In Examples 6.1 and 6.2, the z-transforms are identical even though their parent sequences are different. The only way a unique sequence can be associated with a z-transform is by specifying its ROC. We shall discuss further the importance of the ROC in Section 6.3.

EXAMPLE 6.3 z-Transform of a Finite-Length Sequence

Determine the z-transform of the finite-length sequence

$$x[n] = \begin{cases} \alpha^n, & M \leq n \leq N-1, \\ 0, & \text{otherwise.} \end{cases} \tag{6.11}$$

Here we have

$$X(z) = \sum_{n=M}^{N-1} \alpha^n z^{-n} = z^{-M} \sum_{n=0}^{N-M-1} (\alpha z^{-1})^n$$

$$= z^{-M} \left(\frac{1 - \alpha^{N-M} z^{-(N-M)}}{1 - \alpha z^{-1}} \right)$$

$$= \frac{z^{-M} - \alpha^{N-M} z^{-N}}{1 - \alpha z^{-1}}, \tag{6.12}$$

using the identity of Eq. (5.9).

To determine the ROC of the above $X(z)$, we need to examine the sum $\sum_{n=M}^{N-1} |\alpha z^{-1}|^n$ and determine the values of z for which this sum is finite. Since the sum involves a finite number of terms, the sum is finite everywhere in the z-plane except possibly $z = 0$ and $z = \infty$, provided $|\alpha|$ is finite. For $N > M \geq 0$, the ROC is the entire z-plane excluding the origin $z = 0$. For $M < 0$ and $N > 0$, the ROC is the entire z-plane excluding $z = 0$ and $z = \infty$. For $M < N < 0$, the ROC is the entire z-plane excluding $z = \infty$.

It follows from Examples 6.1 through 6.3 that the Fourier transform $G(e^{j\omega})$ of a sequence $g[n]$ converges uniformly if and only if the ROC of the z-transform $G(z)$ of the sequence includes the unit circle. On the other hand, the existence of the Fourier transform does not always imply the existence of the z-transform. For example, the finite-energy sequence $h_{LP}[n]$ of Eq. (3.49) has a Fourier transform $H_{LP}(e^{j\omega})$, given by Eq. (3.45), which converges in the mean-square sense. However, this sequence does not have a z-transform because $h_{LP}[n] r^{-n}$ is not absolutely summable for any value of r.

Some commonly used z-transform pairs are listed in Table 6.1.

Table 6.1: Some commonly used z-transform pairs.

Sequence	z-Transform	ROC
$\delta[n]$	1	All values of z
$\mu[n]$	$\dfrac{1}{1 - z^{-1}}$	$|z| > 1$
$\alpha^n \mu[n]$	$\dfrac{1}{1 - \alpha z^{-1}}$	$|z| > |\alpha|$
$n\, \alpha^n \mu[n]$	$\dfrac{\alpha z^{-1}}{(1 - \alpha z^{-1})^2}$	$|z| > |\alpha|$
$(n + 1)\, \alpha^n \mu[n]$	$\dfrac{1}{(1 - \alpha z^{-1})^2}$	$|z| > |\alpha|$
$(r^n \cos \omega_o n)\mu[n]$	$\dfrac{1 - (r \cos \omega_o)z^{-1}}{1 - (2r \cos \omega_o)z^{-1} + r^2 z^{-2}}$	$|z| > |r|$
$(r^n \sin \omega_o n)\mu[n]$	$\dfrac{(r \sin \omega_o)z^{-1}}{1 - (2r \cos \omega_o)z^{-1} + r^2 z^{-2}}$	$|z| > |r|$

6.2 Rational z-Transforms

In the case of LTI discrete-time systems that we are concerned with in this text, all pertinent z-transforms are rational functions of z^{-1}, that is, ratios of two polynomials $P(z)$ and $D(z)$ in z^{-1}:

$$H(z) = \frac{P(z)}{D(z)} = \frac{p_0 + p_1 z^{-1} + \cdots + p_{M-1} z^{-(M-1)} + p_M z^{-M}}{d_0 + d_1 z^{-1} + \cdots + d_{N-1} z^{-(N-1)} + d_N z^{-N}}, \tag{6.13}$$

where the *degree* of the numerator polynomial $P(z)$ is M and that of the denominator polynomial $D(z)$ is N. An alternate representation of a rational z-transform is as a ratio of two polynomials in z:

$$H(z) = z^{(N-M)} \frac{p_0 z^M + p_1 z^{M-1} + \cdots + p_{M-1} z + p_M}{d_0 z^N + d_1 z^{N-1} + \cdots + d_{N-1} z + d_N}. \tag{6.14}$$

Equation (6.14) can also be written in factored form as

$$H(z) = \frac{p_0}{d_0} \frac{\prod_{\ell=1}^{M}(1 - \xi_\ell z^{-1})}{\prod_{\ell=1}^{N}(1 - \lambda_\ell z^{-1})} = z^{(N-M)} \frac{p_0}{d_0} \frac{\prod_{\ell=1}^{M}(z - \xi_\ell)}{\prod_{\ell=1}^{N}(z - \lambda_\ell)}. \tag{6.15}$$

At a root $z = \xi_\ell$ of the numerator polynomial, $H(\xi_\ell) = 0$, and as a result, these values of z are known as the *zeros* of $H(z)$. Likewise, at a root $z = \lambda_\ell$ of the denominator polynomial, $H(\lambda_\ell) \to \infty$, and these points in the z-plane are called the *poles* of $H(z)$. Observe from the expression in Eq. (6.15) that there are M finite zeros and N finite poles of $H(z)$.[1] It also follows from the above expression that there are

---

[1] In some cases, there may be common factors between the numerator $P(z)$ and the denominator $D(z)$ of the z-transform $H(z)$, resulting in less than M finite zeros and less than N finite poles caused by pole-zero cancellations.

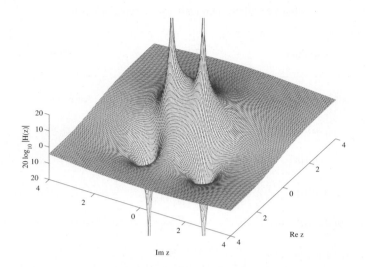

Figure 6.2: The 3-D plot of $20\log_{10}|H(z)|$ as a function of $\text{Re}(z)$ and $\text{Im}(z)$.

additional $(N - M)$ zeros at $z = 0$ (the origin in the z-plane) if $N > M$ or additional $(M - N)$ poles at $z = 0$ if $N < M$. For example, the z-transform $\mu(z)$ of Eq. (6.9) can be rewritten as

$$\mu(z) = \frac{z}{z - 1}, \qquad \text{for } |z| > 1, \tag{6.16}$$

which has a zero at $z = 0$ and a pole at $z = 1$.

It should be noted that a rational z-transform can be represented completely by the locations of its poles $\{\lambda_\ell\}$ and zeros $\{\xi_\ell\}$ and the gain constant p_0/d_0.

A physical interpretation of the concepts of poles and zeros can be given by plotting the log-magnitude $20\log_{10}|H(z)|$, which is a two-dimensional function of $\text{Re}(z)$ and $\text{Im}(z)$. Hence, its plot describes a surface in the complex z-plane, as illustrated in Figure 6.2 for the rational z-transform

$$H(z) = \frac{1 - 2.4\,z^{-1} + 2.88\,z^{-2}}{1 - 0.8\,z^{-1} + 0.64\,z^{-2}}.$$

It can be seen from this figure that the magnitude plot exhibits very large peaks around the points $z = 0.4 \pm j0.6928$, which are the poles of $H(z)$, and very narrow and deep wells around the location of the zeros at $z = 1.2 \pm j1.2$.

In most practical cases, the rational z-transforms are ratios of polynomials with real coefficients. The complex poles and zeros of such transforms occur as complex conjugate pairs. To verify this fact, let $z = a_i \pm jb_i$ be a pair of complex conjugate poles of the rational z-transform $H(z)$, with a_i and b_i being real numbers. Then the denominator of $H(z)$ contains the two factors $(z - a_i - jb_i)$ and $(z - a_i + jb_i)$. The product of these two first-order factors is a second-order factor given by

$$(z - a_i - jb_i)(z - a_i + jb_i) = (z - a_i)^2 + b_i^2 = z^2 - 2a_iz + (a_i^2 + b_i^2),$$

which has all real coefficients.

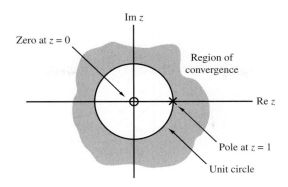

Figure 6.3: The pole-zero plot and the region of convergence of $\mathcal{Z}\{\mu[n]\}$.

6.3   Region of Convergence of a Rational z-Transform

The ROC of a z-transform is an important concept for a variety of reasons. As demonstrated earlier in Examples 6.1 and 6.2, without the knowledge of the ROC, there is no unique relationship between a sequence and its z-transform. Hence, the z-transform must always be specified with its ROC. Moreover, the Fourier transform of the sequence can be obtained simply by evaluating the z-transform of the sequence on the unit circle provided its ROC includes the unit circle. Later in this chapter, we shall point out the relationship between the ROC of the z-transform of the impulse response of a causal LTI system and its BIBO stability. It is thus of interest to investigate the ROC more thoroughly.

Now, the ROC of a rational z-transform is bounded by the locations of its poles. To understand this relationship between the poles and the ROC, it is instructive to examine the plot of the poles and the zeros of a z-transform. Figure 6.3 shows the pole-zero plot of the z-transform $\mu(z)$ of Eq. (6.16), where the location of the pole is indicated by a cross "×" and the location of the zero is indicated by a circle "o." In this figure, the ROC, shown as the shaded area, is the region of the z-plane just outside the circle centered at the origin and going through the pole at $z = 1$ and extending all the way to $|z| = \infty$.

EXAMPLE 6.4   ROC of the z-Transform of an Exponential Sequence

Determine the ROC of the z-transform $H(z)$ of the sequence $h[n] = (-0.6)^n \mu[n]$. From Eq. (6.8), we arrive at

$$H(z) = \frac{1}{1 + 0.6 z^{-1}} = \frac{z}{z + 0.6}, \qquad |z| > 0.6. \tag{6.17}$$

This implies that the ROC is just outside the circle going through the point $z = -0.6$ and extending all the way to $z = \infty$, as shown in Figure 6.4. Note that $H(z)$ has a zero at $z = 0$ and a pole at $z = -0.6$.

In general, the ROC depends on the type of the sequence of interest as defined earlier in Section 2.1.1. We examine in Examples 6.5 through 6.8 the ROCs of the z-transforms of several different types of sequences.

EXAMPLE 6.5   ROC of the z-Transform of a Two-Sided Finite-Length Sequence

Consider a finite-length sequence $g[n]$ defined for $-M \le n \le N$, where M and N are nonnegative integers and $|g[n]| < \infty$. Its z-transform is given by

$$G(z) = \sum_{n=-M}^{N} g[n] z^{-n} = \frac{\sum_{n=0}^{N+M} g[n-M] z^{N+M-n}}{z^N}. \tag{6.18}$$

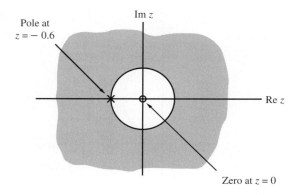

Figure 6.4: Pole-zero plot of $\mathcal{Z}\{(-0.6)^n \mu[n]\}$.

Note from Eq. (6.18) that $G(z)$ has M poles at $z = \infty$ and N poles at $z = 0$. As can be seen from the above, in general, the z-transform of a finite-length bounded sequence converges everywhere in the z-plane except possibly at $z = 0$ and/or at $z = \infty$.

EXAMPLE 6.6   ROC of the z-Transform of a Right-Sided Infinite-Length Sequence

A right-sided sequence $u_1[n]$ with nonzero sample values only for $n \geq 0$ is called a *causal sequence*. Its z-transform is given by

$$U_1(z) = \sum_{n=0}^{\infty} u_1[n]z^{-n}. \tag{6.19}$$

It can be shown that $U_1(z)$ converges exterior to a circle $|z| = \mathcal{R}_1$, including the point $z = \infty$. On the other hand, a right-sided sequence $u_2[n]$ with nonzero sample values only for $n \geq -M$ with M nonnegative has a z-transform $U_2(z)$ with M poles at $z = \infty$, and therefore, its ROC is exterior to a circle $|z| = \mathcal{R}_2$, excluding the point $z = \infty$.

EXAMPLE 6.7   ROC of the z-Transform of a Left-Sided Infinite-Length Sequence

A left-sided sequence $v_1[n]$ with nonzero sample values only for $n \leq 0$, called an *anticausal sequence*, has a z-transform given by

$$V_1(z) = \sum_{n=-\infty}^{0} v_1[n]z^{-n}, \tag{6.20}$$

and converges interior to a circle $|z| = \mathcal{R}_3$, including the point $z = 0$. However, a left-sided sequence $v_2[n]$ with nonzero sample values only for $n \leq N$ with N nonnegative has a z-transform $V_2(z)$ with N poles at $z = 0$. As a result, the ROC of $V_2(z)$ is interior to a circle $|z| = \mathcal{R}_4$, excluding the point $z = 0$.

The z-transform of a two-sided sequence $w[n]$ can be expressed as

$$W(z) = \sum_{n=-\infty}^{\infty} w[n]z^{-n} = \sum_{n=0}^{\infty} w[n]z^{-n} + \sum_{n=-\infty}^{-1} w[n]z^{-n}. \tag{6.21}$$

The first term on the right-hand side of Eq. (6.21) can be interpreted as the z-transform of a right-sided sequence, and it converges exterior to the circle $|z| = \mathcal{R}_5$. The second term, on the other hand, is the z-transform of a left-sided sequence, and it converges interior to the circle $|z| = \mathcal{R}_6$. As a result, if $\mathcal{R}_5 < \mathcal{R}_6$, there is an overlapping ROC given by $\mathcal{R}_5 < |z| < \mathcal{R}_6$. If, however, $\mathcal{R}_6 < \mathcal{R}_5$, the z-transform does not exist, as illustrated by Example 6.8.

EXAMPLE 6.8   **Nonexistence of the z-Transform of a Two-Sided Infinite-Length Sequence**

The two-sided sequence defined by

$$u[n] = \alpha^n,$$

where α can be a real or complex number, does not have a z-transform, regardless of the absolute value $|\alpha|$. This follows by noting that the z-transform expression can be rewritten as

$$U(z) = \sum_{n=0}^{\infty} \alpha^n z^{-n} + \sum_{n=-\infty}^{-1} \alpha^n z^{-n}. \qquad (6.22)$$

The first term on the right-hand side of Eq. (6.22) converges for $|z| > |\alpha|$, whereas the second term converges for $|z| < |\alpha|$, and hence, there is no overlap of the two ROCs.

A rational z-transform is infinite at a pole and, hence, does not converge by definition. As a result, the ROC of a rational z-transform cannot contain any poles. Moreover, the ROC of a rational z-transform is bounded by the poles. These two properties of such z-transforms can be seen from the z-transform of a unit step sequence given by Eq. (6.16) and illustrated by the pole-zero plot of Figure 6.3. Another example is the sequence defined in Example 6.4 and its z-transform given by Eq. (6.17), with the corresponding pole-zero plot given in Figure 6.4.

To show that the ROC is bounded by the poles, assume that the z-transform $X(z)$ has simple poles at α and β, with $|\alpha| < |\beta|$. If the sequence is also assumed to be a right-sided sequence, then it is of the form

$$x[n] = \left(r_1(\alpha)^n + r_2(\beta)^n\right) \mu[n - N_o], \qquad (6.23)$$

where N_o is a positive or negative integer. Now the z-transform of a right-sided sequence $(\gamma)^r \mu[n - N_o]$ exists if

$$\sum_{n=N_o}^{\infty} \left|(\gamma)^n z^{-n}\right| < \infty$$

for some z. It can be seen that the above holds for $|z| > |\gamma|$ but not for $|z| \leq |\gamma|$. The right-sided sequence of Eq. (6.23) has thus an ROC defined by $|\beta| < |z| < \infty$. A similar argument shows that if $X(z)$ is the z-transform of a left-sided sequence of the form

$$x[n] = \left(r_1(\alpha)^n + r_2(\beta)^n\right) \mu[-n - N_o], \qquad (6.24)$$

then the ROC of its z-transform $X(z)$ is defined by $0 \leq |z| < |\alpha|$.

For a two-sided sequence, some of the poles contribute to terms for $n < 0$ and the others to terms for $n \geq 0$. The ROC is thus bounded on the outside by the pole with the smallest magnitude that contributes for $n < 0$ and on the inside by the pole with the largest magnitude that contributes for $n \geq 0$.

Figure 6.5 shows the three possible ROCs of a rational z-transform with poles at $z = \alpha$ and $z = \beta$, with each ROC being associated with a unique sequence. In general, if the rational z-transform has N poles with R distinct magnitudes, then it has $R + 1$ ROCs and, as a result, $R + 1$ distinct sequences having the same rational z-transform. Consequently, a rational z-transform with a specified ROC has a unique sequence as its inverse z-transform. A rational z-transform without a specified ROC is thus not meaningful

Summarizing the above discussion, we thus have the following observations with regard to the ROC of a rational z-transform:

(a) The ROC of the z-transform of a finite-length sequence defined for $M \leq n \leq N$ is the entire z-plane except possibly $z = 0$ and/or $z = \infty$.

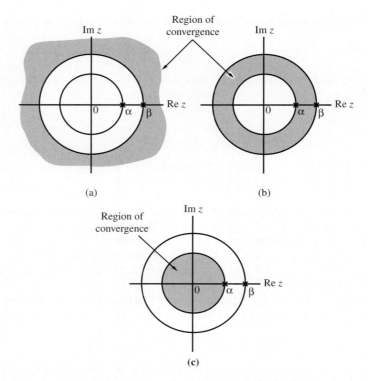

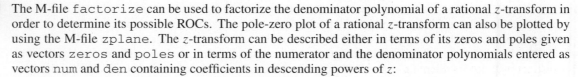

Figure 6.5: The pole-zero plot of a rational z-transform with three possible ROCs corresponding to three different sequences: (a) right-sided sequence, (b) two-sided sequence, and (c) left-sided sequence.

(b) The ROC of the z-transform of a right-sided sequence defined for $M \leq n < \infty$ is exterior to a circle in the z-plane passing through the pole furthest from the origin $z = 0$.

(c) The ROC of the z-transform of a left-sided sequence defined for $-\infty < n \leq N$ is interior to a circle in the z-plane passing through the pole nearest to the origin $z = 0$.

(d) The ROC of the z-transform of a two-sided sequence of infinite length is a ring bounded by two circles in the z-plane passing through two poles with no poles inside the ring.

factorize.m

The M-file `factorize` can be used to factorize the denominator polynomial of a rational z-transform in order to determine its possible ROCs. The pole-zero plot of a rational z-transform can also be plotted by using the M-file `zplane`. The z-transform can be described either in terms of its zeros and poles given as vectors `zeros` and `poles` or in terms of the numerator and the denominator polynomials entered as vectors `num` and `den` containing coefficients in descending powers of z:

$$\texttt{zplane(zeros,poles),} \qquad \texttt{zplane(num,den)}$$

It should be noted that the argument `zeros` and `poles` must be entered as column vectors, whereas the arguments `num` and `den` need to be entered as row vectors. The function `zplane` plots the zeros and poles in the current figure window, with the zero indicated by the symbol "o" and the pole indicated by the symbol "×." The unit circle is also included in the plot for reference. The automatic scaling included in the function can be overwritten if necessary.

Example 6.9 illustrates the application of the above functions.

EXAMPLE 6.9   Determination of Possible ROCs of a Rational z-Transform

Express the following z-transform in factored form, plot its poles and zeros, and then determine its ROCs.

Program 6_1.m

$$G(z) = \frac{2z^4 + 16z^3 + 44z^2 + 56z + 32}{3z^4 + 3z^3 - 15z^2 + 18z - 12}. \tag{6.25}$$

To carry out the factorization, we use the Program 6_1. The coefficients of the first- and second-order factors of the numerator and denominator polynomials generated by this program are as shown below:

```
Numerator factors
1.0     4.0     0
1.0     2.0     0
1.0     2.0     2.0

Denominator factors
1.0     3.236   0
1.0    -1.236   0
1.0    -1.0     1.0

Gain constant
0.6667
```

Therefore, the factored form of the z-transform of Eq. (6.25) is given by

$$G(z) = 0.6667 \frac{(1 + 4z^{-1})(1 + 2z^{-1})(1 + 2z^{-1} + 2z^{-2})}{(1 + 3.236z^{-1})(1 - 1.236z^{-1})(1 - z^{-1} + z^{-2})}. \tag{6.26}$$

The pole-zero plot developed by the program is shown in Figure 6.6. From Eq. (6.26), the four regions of convergence are thus seen to be

$$\mathcal{R}_1 : \infty \geq |z| > 3.2361,$$
$$\mathcal{R}_2 : 3.236 > |z| > 1.236,$$
$$\mathcal{R}_3 : 1.236 > |z| > 1,$$
$$\mathcal{R}_4 : 1 > |z| \geq 0.$$

The statement `[num,den] = zp2tf(z,p,k)` can be employed to determine the rational form of a z-transform from its factored form. The input arguments are the column vectors `z` and `p` containing the zeros $\{\xi_\ell\}$ and poles $\{\lambda_\ell\}$ of the z-transform, and the gain constant $k = p_0/d_0$ (see Eq. (6.15)). The output data are the row vectors `num` and `den` containing the coefficients of the numerator and the denominator polynomials in descending powers of z. Program 6_2 makes use of the function `zp2tf` and has been applied in Example 6.10.

EXAMPLE 6.10   The Rational Form of a z-Transform from Its Zero and Pole Locations

We now consider the determination of the rational z-transform from its zero and pole locations. The zeros are at $\xi_1 = 0.21$, $\xi_2 = 3.14$, $\xi_3 = -0.3 + j0.5$, $\xi_4 = -0.3 - j0.5$; the poles are at $\lambda_1 = -0.45$, $\lambda_2 = 0.67$, $\lambda_3 = 0.81 + j0.72$, $\lambda_4 = 0.81 - j0.72$; and the gain constant k is 2.2.

Program 6_2 during execution asks for the input data, the zero and pole locations, and the gain constant. The program then displays the coefficients of the numerator and the denominator polynomials in descending powers of z, as indicated in the following:

Program 6_2.m

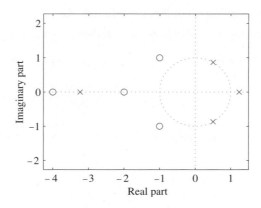

Figure 6.6: Pole-zero plot of the z-transform of Eq. (6.25).

```
Numerator polynomial coefficients
   Columns 1 through 5
    2.2   -6.05   -2.22332   -1.63592     -0.4932312


Denominator polynomial coefficients
   Columns 1 through 5
    1.0   -1.840   1.22940    0.23004    -0.35411175
```

From the above, we thus arrive at the desired expression:

$$G(z) = \frac{2.2z^4 - 6.05z^3 - 2.22332z^2 - 1.63592z + 0.4932312}{z^4 - 1.84z^3 + 1.2294z^2 + 0.23004z - 0.35411175}.$$

6.4 The Inverse z-Transform

We shall show later in Section 6.5 that the z-transform of a sequence obtained by a convolution of two sequences is given by the product of the z-transforms of these two sequences. As a result, an alternate approach to the implementation of the convolution sum is to form the product of the z-transforms of the individual sequences being convolved and then evaluating the inverse z-transform of the product. In many applications, this approach is more convenient as it leads to a closed-form answer. It is thus important to develop methods to compute the inverse of a specified z-transform. We now derive the general expression for the inverse z-transform and outline two methods for its computation.

6.4.1 General Expression

Multiplying both sides of Eq. (6.1) by z^{n-1} and integrating both sides of the result counterclockwise over a closed contour C inside the ROC of the z-transform $G(z)$ and encircling the point $z = 0$, we get

$$\oint_C G(z)z^{n-1}\,dz = \oint_C \sum_{\ell=-\infty}^{\infty} g[\ell]\,z^{-\ell}z^{n-1}\,dz. \tag{6.27}$$

We can interchange the order of integration and summation on the right-hand side of Eq. (6.27) as $G(z)$, and hence, $\sum_{\ell=-\infty}^{\infty} g[\ell]\,z^{-\ell}z^{n-1}\,dz$, converges on the contour C. As a result, Eq. (6.27) can be rewritten as, after multiplying both sides with $\frac{1}{2\pi j}$,

$$\frac{1}{2\pi j} \oint_C G(z)z^{n-1}\, dz = \frac{1}{2\pi j} \sum_{\ell=-\infty}^{\infty} g[\ell] \oint_C z^{n-1-\ell}\, dz. \tag{6.28}$$

The right-hand side of Eq. (6.28) reduces to $g[n]$ as by Cauchy's integral theorem [Chu90]

$$\frac{1}{2\pi j} \oint_C z^{n-1-\ell}\, dz = \delta(n - \ell).$$

Hence, the general expression for the inverse-z transform is given by

$$g[n] = \frac{1}{2\pi j} \oint_C G(z)z^{n-1}\, dz. \tag{6.29}$$

In the case of rational z-transforms, the contour integral in Eq. (6.29) can be evaluated using the Cauchy's residue theorem [Chu90], resulting in

$$g[n] = \sum \left[\text{residues of } G(z)z^{n-1} \text{ at the poles inside } C \right]. \tag{6.30}$$

If the pole at $z = \lambda_o$ of $G(z)z^{n-1}$ is of multiplicity m, we can express the function as

$$G(z)z^{n-1} = \frac{\Gamma(z)}{(z - \lambda_o)^m}.$$

Then, the residue of $G(z)z^{n-1}$ at the pole at $z = \lambda_o$ is given by

$$\text{Residue}[G(z)z^{n-1} \text{ at } z = \lambda_o] = \frac{1}{(m - 1)!} \left[\frac{d^{m-1}(z - \lambda_o)^m G(z)z^{n-1}}{d\, z^{m-1}} \right]_{z=\lambda_o}$$

$$= \frac{1}{(m - 1)!} \left[\frac{d^{m-1}\Gamma(z)}{d\, z^{m-1}} \right]_{z=\lambda_o}. \tag{6.31}$$

We illustrate the inverse-z transform computation using Eq. (6.30) in Example 6.11.

EXAMPLE 6.11   Inverse z-Transform Computation Using Cauchy's Residue Theorem

Consider

$$X(z) = \frac{z}{(z - 1)^2}, \qquad |z| > 1. \tag{6.32}$$

The location and the number of poles of $X(z)z^{n-1} = z^n/(z - 1)^2$ depend on the values of n. For $n \geq 0$, there are only 2 poles at $z = 1$, whereas for $n < 0$, the function has n poles at $z = 0$ and 2 poles at $z = 1$. In both cases, the poles are in the region that is exterior to the ROC of $X(z)$.

We first evaluate the residue ρ_1 of $X(z)z^{n-1}$ at the poles at $z = 1$.

$$\rho_1 = \frac{d}{dz} \left[(z - 1)^2 X(z)z^{n-1} \right]_{z=1}$$

$$= \frac{d\, z^n}{dz} \Big|_{z=1} = nz^{n-1} \Big|_{z=1} = n, \qquad -\infty < n < \infty. \tag{6.33}$$

We next compute the residue ρ_0 of $X(z)z^{n-1}$ at the pole at $z = 0$ for $n < 0$. To this end, we let $r = -n$ and evaluate the residue ρ_0 of $X(z)z^{-r-1}$ for $r > 0$.

$$\rho_0 = \frac{1}{(r-1)!}\left[\frac{d^{r-1}}{dz^{r-1}}\left(z^r X(z)z^{-r-1}\right)\right]_{z=0}$$

$$= \frac{1}{(r-1)!}\left[\frac{d^{r-1}}{dz^{r-1}}\left(\frac{1}{(z-1)^2}\right)\right]_{z=0} = \frac{r!}{(r-1)!} = r, \qquad r > 0.$$

Hence,

$$\rho_0 = -n, \qquad n < 0. \tag{6.34}$$

Combining Eqs. (6.33) and (6.34), we get

$$x[n] = \begin{cases} n, & n \geq 0, \\ 0, & n < 0. \end{cases} \tag{6.35}$$

Since Eq. (6.30) needs to be evaluated at all values of n, it is difficult to arrive at a closed-form answer in most cases, and, as a result, it is not pursued here further. In many practical situations, one of the following simple methods can be used to evaluate the inverse transform.

6.4.2 Inverse z-Transform by Table Look-Up Method

Often, the inverse z-transform can be obtained by inspection of z-transform tables that are included in many books [Jur73]. We illustrate this direct inspection method in Example 6.12.

EXAMPLE 6.12   **Inverse z-Transform Determination Using the Table Look-Up Method**

We determine the inverse transform $h[n]$ of

$$H(z) = \frac{0.5\,z}{z^2 - z + 0.25}, \qquad |z| > 0.5.$$

To this end, we rewrite $H(z)$ as

$$H(z) = \frac{0.5\,z}{(z-0.5)^2} = \frac{0.5\,z^{-1}}{(1 - 0.5\,z^{-1})^2}. \tag{6.36}$$

Now, from Table 6.1, we observe that

$$n\,\alpha^n\mu[n] \overset{\mathcal{Z}}{\longleftrightarrow} \frac{\alpha\,z^{-1}}{(1 - \alpha\,z^{-1})^2}, \qquad |z| > |\alpha|.$$

Comparing Eq. (6.36) with the above, we conclude that

$$h[n] = n\,(0.5)^n\mu[n].$$

6.4.3 Inverse z-Transform by Partial-Fraction Expansion

The expression of Eq. (6.29) can be computed in a number of ways. A rational z-transform $G(z)$ with a causal inverse transform $g[n]$ has an ROC that is exterior to a circle. In this case, it is more convenient to express $G(z)$ in a partial-fraction expansion form and then determine $g[n]$ by summing the inverse transforms of the individual simpler terms in the expansion. A rational $G(z)$ can be expressed as

$$G(z) = \frac{P(z)}{D(z)}, \tag{6.37}$$

where $P(z)$ and $D(z)$ are polynomials in z^{-1}, as indicated in Eq. (6.13). If the degree M of the numerator polynomial $P(z)$ is greater than or equal to the degree N of the denominator polynomial $D(z)$, the rational function $P(z)/D(z)$ is called an *improper fraction*. In this case, we can divide $P(z)$ by $D(z)$ and re-express $G(z)$ as

$$G(z) = \sum_{\ell=0}^{M-N} \eta_\ell z^{-\ell} + \frac{P_1(z)}{D(z)}, \tag{6.38}$$

where the degree of the polynomial $P_1(z)$ is less than that of $D(z)$. The rational function $P_1(z)/D(z)$ is called a *proper fraction*.

EXAMPLE 6.13   Determination of the Proper Fraction of a Rational z-Transform

Consider the rational z-transform

$$\frac{2 + 0.8z^{-1} + 0.5z^{-2} + 0.3z^{-3}}{1 + 0.8z^{-1} + 0.2z^{-2}}.$$

It can be seen that the degree of its numerator polynomial is 3, whereas the degree of its denominator polynomial is 2. Since the numerator degree is greater than the denominator degree, the above rational z-transform is not a proper fraction. We can express it as a sum of a polynomial in z^{-1} and a proper fraction in the form of Eq. (6.38) by dividing the numerator by the denominator so that the numerator of the remainder is a first-degree polynomial. To this end, we apply the long division with both polynomials expressed in a reverse order, which results in

$$-3.5 + 1.5z^{-1} + \frac{5.5 + 2.1z^{-1}}{1 + 0.8z^{-1} + 0.2z^{-2}}.$$

The rational function $\frac{5.5+2.1z^{-1}}{1+0.8z^{-1}+0.2z^{-2}}$ is a proper fraction.

Simple Poles

In most practical cases, $G(z)$ is a proper fraction with simple poles. Let the poles of $G(z)$ be at $z = \lambda_k$, $1 \le k \le N$, where λ_k are distinct. A partial-fraction expansion of $G(z)$ then is of the form

$$G(z) = \sum_{\ell=1}^{N} \frac{\rho_\ell}{1 - \lambda_\ell z^{-1}}, \tag{6.39}$$

where the constants ρ_ℓ in the above expression, called the *residues*, are given by

$$\rho_\ell = (1 - \lambda_\ell z^{-1})G(z)\Big|_{z=\lambda_\ell}. \tag{6.40}$$

Each term of the sum on the right-hand side of Eq. (6.39) has an ROC given by $|z| > |\lambda_\ell|$ and, thus, an inverse transform of the form $\rho_\ell(\lambda_\ell)^n \mu[n]$. Therefore, the inverse transform $g[n]$ of $G(z)$ is given by

$$g[n] = \sum_{\ell=1}^{N} \rho_\ell(\lambda_\ell)^n \mu[n]. \tag{6.41}$$

Note that the above approach, with slight modifications, can also be used to determine the inverse z-transform of a noncausal sequence with a rational z-transform.

The inverse transform computation via partial-fraction expansion of a rational z-transform with simple poles is considered in Example 6.14.

EXAMPLE 6.14   Inverse z-Transform Computation via Partial-Fraction Expansion

Let the z-transform of a causal sequence $h[n]$ be given by

$$H(z) = \frac{z(z + 2.0)}{(z - 0.2)(z + 0.6)} = \frac{1 + 2.0z^{-1}}{(1 - 0.2z^{-1})(1 + 0.6z^{-1})}. \tag{6.42}$$

A partial-fraction expansion of $H(z)$ is then of the form

$$H(z) = \frac{\rho_1}{1 - 0.2z^{-1}} + \frac{\rho_2}{1 + 0.6z^{-1}}. \tag{6.43}$$

Using Eq. (6.40), we obtain

$$\rho_1 = (1 - 0.2z^{-1})H(z)\Big|_{z=0.2} = \frac{1 + 2.0z^{-1}}{1 + 0.6z^{-1}}\Big|_{z=0.2} = 2.75,$$

and

$$\rho_2 = (1 + 0.6z^{-1})H(z)\Big|_{z=-0.6} = \frac{1 + 2.0z^{-1}}{1 - 0.2z^{-1}}\Big|_{z=-0.6} = -1.75.$$

Substituting the above in Eq. (6.43), we arrive at

$$H(z) = \frac{2.75}{1 - 0.2z^{-1}} - \frac{1.75}{1 + 0.6z^{-1}}.$$

The inverse transform of the above is given by

$$h[n] = 2.75(0.2)^n \mu[n] - 1.75(-0.6)^n \mu[n]. \tag{6.44}$$

An alternate approach to the determination of the partial-fraction expansion is to derive the rational form of a z-transform from its partial-fraction expansion form given by Eq. (6.39) by cross-multiplying terms, equating the expressions for the coefficients of the numerator with the specified numbers, and then solving for the residues. The method is illustrated in Example 6.15.

EXAMPLE 6.15   Residue Computation Using Coefficient Matching Approach

We repeat here Example 6.14. From Eq. (6.43), we arrive at

$$H(z) = \frac{(\rho_1 + \rho_2) + (0.6\rho_1 - 0.2\rho_2)z^{-1}}{(1 - 0.2z^{-1})(1 + 0.6z^{-1})}.$$

Comparing the coefficients of the numerator of the above equation with the corresponding coefficients in Eq. (6.42), we arrive at the following two equations:

$$\rho_1 + \rho_2 = 1, \qquad 0.6\rho_1 - 0.2\rho_2 = 2,$$

whose solutions yields

$$\rho_1 = 2.75, \qquad \rho_2 = -1.75.$$

The above values of the residues are seen to be the same as those derived in Example 6.14.

In the case of a real rational z-transform, a complex pole appears in complex-conjugate pairs. Here, if the residue of a pole at $z = \lambda_\ell$ is ρ_ℓ, then the residue of the pole at $z = \lambda_\ell^*$ is ρ_ℓ^*. Hence, the terms corresponding to a complex-conjugate pole-pair in the partial-fraction expansion can be combined to form a second-order factor in the expansion with real coefficients.

Multiple Poles

If $G(z)$ is a proper fraction with multiple poles, the partial-fraction expansion is of slightly different form. For example, if the pole at $z = \nu$ is of multiplicity L and the remaining $N - L$ poles are simple and at $z = \lambda_\ell$, $1 \leq \ell \leq N - L$, then the partial-fraction expansion of $G(z)$ is of the form

$$G(z) = \sum_{\ell=1}^{N-L} \frac{\rho_\ell}{1 - \lambda_\ell z^{-1}} + \sum_{i=1}^{L} \frac{\gamma_i}{(1 - \nu z^{-1})^i}, \tag{6.45}$$

where the constants γ_i (no longer called the residues for $i \neq 1$) are computed using the formula

$$\gamma_i = \frac{1}{(L-i)!(-\nu)^{L-i}} \frac{d^{L-i}}{d(z^{-1})^{L-i}} \left[(1 - \nu z^{-1})^L G(z) \right] \Big|_{z=\nu}, \qquad 1 \leq i \leq L, \tag{6.46}$$

and the residues ρ_ℓ are calculated using Eq. (6.40). Techniques for determining the inverse z-transform of terms like $\gamma_i / (1 - \nu z^{-1})^i$ are described later in Example 6.27.

6.4.4   Partial-Fraction Expansion Using MATLAB

The M-file `residuez` can be used to develop the partial-fraction expansion of a rational z-transform and to convert a z-transform expressed in a partial-fraction form to its rational form. For the former case, the statement is `[r,p,k] = residuez(num,den)`, where the input data are the vectors `num` and `den` containing the coefficients of the numerator and the denominator polynomials, respectively, expressed in descending powers of z, and the output files are the vector `r` of the residues and the numerator constants, the vector `p` of corresponding poles, and the vector `k` containing the constants η_ℓ. The statement `[num,den] = residuez(r,p,k)` is employed to carry out the reverse operation. The applications of these functions are considered in Examples 6.16 and 6.17.

EXAMPLE 6.16   Partial-Fraction Expansion Using MATLAB

Using MATLAB, we determine the partial-fraction expansion of the z-transform $G(z)$ given by

$$G(z) = \frac{18z^3}{18z^3 + 3z^2 - 4z - 1}. \tag{6.47}$$

Program 6_3.m

To this end we use Program 6_3. During execution, the program calls for the input data that are the vectors `num` and `den` of the coefficients of the numerator and the denominator polynomials, respectively, in descending powers of z. These data are entered inside square brackets as follows:

```
num = [18]
den = [18   3   -4   -1]
```

The output data are the residues and the constants, the poles of the expansion of $G(z)$ in the form of Eq. (6.45). For our above example, these are as given below.

```
Residues
    0.3600      0.2400      0.4000

Poles
    0.5000     -0.3333     -0.3333

Constants
```

Note also that the z-transform of Eq. (6.47) has double poles at $z = -1/3 = -0.3333$. The first entry in both the residues and poles given above corresponds to the simple pole factor $(1 - 0.5z^{-1})$, the second entry corresponds to the simple pole factor $(1 + 0.3333z^{-1})$, and the third entry corresponds to the factor $(1 + 0.3333z^{-1})^2$ in the partial-fraction expansion. Thus, the desired expansion is given by

$$G(z) = \frac{0.36}{1 - 0.5z^{-1}} + \frac{0.24}{1 + 0.3333z^{-1}} + \frac{0.4}{(1 + 0.3333z^{-1})^2}. \tag{6.48}$$

Program 6_4.m

EXAMPLE 6.17   Rational Form of a z-Transform from Its Partial-Fraction Expansion

We now consider the determination of the rational form of a z-transform from its partial-fraction expansion representation as given by Eq. (6.48). Program 6_4 can be used to find the rational form. During execution, the program requests the input data vector r of residues, the vector p of pole locations, and the vector k of constants, with each entered using square brackets as follows:

```
r = [0.36 0.24 0.4]
p = [0.5  -0.3333  -0.3333]
k = [0]
```

It then prints the coefficients of the numerator and the denominator polynomials, respectively, in descending powers of z, as indicated below. It can be seen that the coefficients are exactly the same as in Eq. (6.47) if we multiply each coefficient by 18.

```
Numerator polynomial coefficients
    1.0000     -0.0000          0          0

Denominator polynomial coefficients
    1.0000      0.1667     -0.2222     -0.0556
```

6.4.5   Inverse z-Transform via Long Division

For causal sequences, the z-transform $G(z)$ can be expanded into a power series in z^{-1}. In the series expansion, the coefficient multiplying the term z^{-n} is then the nth sample $g[n]$. We illustrate this approach through Examples 6.18 and 6.19.

EXAMPLE 6.18   Inverse z-Transform via Power Series Expansion

We consider the evaluation of the inverse-z transform $x[n]$ of the z-transform $X(z)$ of Example 6.11 using the power series expansion method. Since the ROC of $X(z)$ is given by $|z| > 1$, $x[n]$ is a causal sequence, and $X(z)$ can be expanded into a power series in z^{-1}.

Now, the specified z-transform can be expressed as

$$X(z) = \frac{z^{-1}}{(1 - z^{-1})^2}.$$

A binomial series expansion of $1/(1 - z^{-1})^2$ is given by

$$\frac{1}{(1 - z^{-1})^2} = 1 + 2z^{-1} + 3z^{-2} + 4z^{-3} + \cdots.$$

Thus,

$$\frac{z^{-1}}{(1 - z^{-1})^2} = z^{-1} + 2z^{-2} + 3z^{-3} + 4z^{-4} + \cdots.$$

As a result,

$$\{x[n]\} = \{0 \quad 1 \quad 2 \quad 3 \quad 4 \ldots\}, \qquad \text{for } n \geq 0.$$
$$\uparrow$$

It follows therefore that

$$x[n] = \begin{cases} n, & n \geq 0, \\ 0, & n < 0. \end{cases}$$

For a rational $G(z)$, a convenient way to determine the power series is to express the numerator and the denominator as polynomials in z^{-1} and then obtain the power series expansion by long division.

EXAMPLE 6.19   Inverse z-Transform Computation Using Long Division

Consider the causal $H(z)$ of Example 6.14 as given in Eq. (6.42). Multiplying the denominator factors, we get

$$H(z) = \frac{1 + 2.0\,z^{-1}}{1 + 0.4\,z^{-1} - 0.12\,z^{-2}}. \tag{6.49}$$

Long division of the numerator by the denominator yields

$$
\begin{array}{r}
1 + 1.6z^{-1} - 0.52z^{-2} \quad + 0.4z^{-3} - 0.2224z^{-4} + \cdots \\
\hline
1 + 0.4z^{-1} - 0.12z^{-2} \,\big)\, 1 + 2.0z^{-1} \\
1 + 0.4z^{-1} - 0.12z^{-2} \\
\hline
1.6z^{-1} + 0.12z^{-2} \\
1.6z^{-1} + 0.64z^{-2} - 0.192z^{-3} \\
\hline
-0.52z^{-2} + 0.192z^{-3} \\
-0.52z^{-2} - 0.208z^{-3} + 0.0624z^{-4} \\
\hline
0.400z^{-3} - 0.0624z^{-4} \\
0.400z^{-3} + 0.1600z^{-4} - 0.0480z^{-5} \\
\hline
-0.2224z^{-4} + 0.0480z^{-5} \\
\cdots \cdots \\
\cdots \cdots
\end{array}
$$

As a result, $H(z)$ can be expressed as

$$H(z) = 1.0 + 1.6z^{-1} - 0.52z^{-2} + 0.4z^{-3} - 0.2224z^{-4} + \cdots,$$

which leads to

$$\{h[n]\} = \{\underset{\uparrow}{1.0}, \quad 1.6, \quad -0.52, \quad 0.4, \quad -0.2224, \ldots\}, \qquad \text{for } n \geq 0.$$

By computing the values of $h[n]$ from Eq. (6.44) for $n = 0, 1, 2, 3,$ and 4, the correctness of the sample values obtained via the long-division method can be easily verified.

6.4.6   Inverse z-Transform Using MATLAB

The inverse of a rational z-transform of causal sequences can also be readily calculated using MATLAB. To this end, both the M-file `impz` and the M-file `filter` can be used. We present Examples 6.20 and 6.21 to illustrate the use of both functions.

Program 6_5.m

EXAMPLE 6.20   Inverse z-Transform Using MATLAB—First Method

We first illustrate the use of the function `impz` to determine the first 11 coefficients of the inverse z-transform of Eq. (6.49). To this end, we use Program 6_5. The input data called by the program are the desired number of coefficients and the vectors of the numerator and denominator coefficients of the rational z-transform, with these last entered inside square brackets. The output data are the desired inverse z-transform coefficients, as indicated below:

```
Coefficients of the power series expansion

    Columns 1 through 7
      1.0000    1.6000    -0.5200    0.4000    -0.2224    0.1370    -0.0815

    Columns 8 through 11
      0.0490    -0.0294    0.0176    -0.0106
```

whose first five coefficients are seen to be identical to that derived in Example 6.19 using the long-division approach.

Program 6_6.m

EXAMPLE 6.21   Inverse z-Transform Using MATLAB—Second Method

Program 6_6 can be used to determine the inverse z-transform of a rational z-transform. As the program is run, it asks for the desired length of the output vector y, followed by the vectors containing the numerator and denominator coefficients entered inside square brackets and in descending powers of z. After computing the output vector, it prints its elements that are the coefficients of the power series expansion of the rational z-transform.

An application of Program 6_6 to the z-transform of Eq. (6.49) yields the same output as that computed in Example 6.20 using the function `impz`.

6.5 z-Transform Properties

As in the case of the discrete-time Fourier transform, the z-transform also satisfies some important properties. We summarize these properties in Table 6.2. For compactness, the properties are stated using the operator notation introduced in Eq. (6.6) and make use of the following z-transform pairs:

$$g[n] \overset{\mathcal{Z}}{\longleftrightarrow} G(z), \qquad \text{ROC} = \mathcal{R}_g, \tag{6.50}$$

Table 6.2: Some useful properties of the z-transform.

Property	Sequence	z-Transform	ROC
	$g[n]$	$G(z)$	$\mathcal{R}_g$
	$h[n]$	$H(z)$	$\mathcal{R}_h$
Conjugation	$g^*[n]$	$G^*(z^*)$	$\mathcal{R}_g$
Time-reversal	$g[-n]$	$G(1/z)$	$1/\mathcal{R}_g$
Linearity	$\alpha g[n] + \beta h[n]$	$\alpha G(z) + \beta H(z)$	Includes $\mathcal{R}_g \cap \mathcal{R}_h$
Time-shifting	$g[n - n_o]$	$z^{-n_o} G(z)$	$\mathcal{R}_g$, except possibly the point $z = 0$ or ∞
Multiplication by an exponential sequence	$\alpha^n g[n]$	$G(z/\alpha)$	$|\alpha|\mathcal{R}_g$
Differentiation of $G(z)$	$ng[n]$	$-z\dfrac{dG(z)}{dz}$	$\mathcal{R}_g$, except possibly the point $z = 0$ or ∞
Convolution	$g[n] \circledast h[n]$	$G(z)H(z)$	Includes $\mathcal{R}_g \cap \mathcal{R}_h$
Modulation	$g[n]h[n]$	$\frac{1}{2\pi j} \oint_C G(v)H(z/v)v^{-1}\, dv$	Includes $\mathcal{R}_g \mathcal{R}_h$
Parseval's relation	$\displaystyle\sum_{n=-\infty}^{\infty} g[n]h^*[n] = \frac{1}{2\pi j} \oint_C G(v)H^*(1/v^*)v^{-1}\, dv$		

Note: If $\mathcal{R}_g$ denotes the region $R_{g-} < |z| < R_{g+}$ and $\mathcal{R}_h$ denotes the region $R_{h-} < |z| < R_{h+}$, then $1/\mathcal{R}_g$ denotes the region $1/R_{g+} < |z| < 1/R_{g-}$ and $\mathcal{R}_g\mathcal{R}_h$ denotes the region $R_{g-}R_{h-} < |z| < R_{g+}R_{h+}$.

$$h[n] \overset{\mathcal{Z}}{\longleftrightarrow} H(z), \qquad \text{ROC} = \mathcal{R}_h, \tag{6.51}$$

where $\mathcal{R}_g$ is given by $R_{g-} < |z| < R_{g+}$ and $\mathcal{R}_h$ is given by $R_{h-} < |z| < R_{h+}$. An understanding of these properties makes the application of the z-transform techniques to the analysis and design of digital filters often easier and less tedious. We prove some of the properties here. The reader is encouraged to verify the remaining properties.

Conjugation Property

The z-transform of the sequence $g^*[n]$ obtained by conjugating the sequence $g[n]$ is given by $G^*(z^*)$ with an ROC $\mathcal{R}_g$; that is,

$$g^*[n] \overset{\mathcal{Z}}{\longleftrightarrow} G^*(z^*), \qquad \text{ROC} = \mathcal{R}_g. \tag{6.52}$$

The proof of the conjugation property is straightforward and is left as an exercise (Problem 6.27).

Time-Reversal Property

The z-transform of the time-reversed sequence $g[-n]$ is given by $G(1/z)$ with an ROC $1/\mathcal{R}_g$; that is,

$$g[-n] \overset{\mathcal{Z}}{\longleftrightarrow} G(1/z), \qquad \text{ROC} = 1/\mathcal{R}_g. \tag{6.53}$$

Note that the notation $1/\mathcal{R}_g$ for the ROC implies the region $1/R_{g+} < |z| < 1/R_{g-}$. The proof of the time-reversal property is also straightforward and is left as an exercise (Problem 6.27).

Linearity Property

The z-transform of the sequence $\alpha g[n] + \beta h[n]$ obtained by a linear combination of the two sequences $g[n]$ and $h[n]$ is given by $\alpha G(z) + \beta H(z)$ with an associated ROC that includes an intersection of the ROCs $\mathcal{R}_g$ and $\mathcal{R}_h$; that is,

$$\alpha g[n] + \beta h[n] \xleftrightarrow{\ \mathcal{Z}\ } \alpha G(z) + \beta H(z), \qquad \text{ROC includes } \mathcal{R}_g \cap \mathcal{R}_h. \tag{6.54}$$

The proof of the linearity theorem is simple and is left as an exercise (Problem 6.27).

EXAMPLE 6.22 z-**Transform of a Causal Sinusoidal Sequence**

Verify the z-transform $X(z)$ of the causal real sequence $x[n] = r^n \cos(\omega_o n)\mu[n]$ and its ROC given in Table 6.1. Now we can express $x[n]$ as a sum of two causal exponential sequences:

$$x[n] = \tfrac{1}{2} r^n e^{j\omega_o n} \mu[n] + \tfrac{1}{2} r^n e^{-j\omega_o n} \mu[n].$$

We can write the above expression as $x[n] = v[n] + v^*[n]$, where

$$v[n] = \tfrac{1}{2}\alpha^n \mu[n],$$

with $\alpha = r e^{j\omega_o}$. The z-transform $V(z)$ of $v[n]$ from Table 6.1 is given by

$$V(z) = \frac{1}{2} \cdot \frac{1}{1 - \alpha z^{-1}} = \frac{1}{2} \cdot \frac{1}{1 - r e^{j\omega_o} z^{-1}}, \qquad |z| > |\alpha| = r.$$

Using the conjugation property, we obtain the z-transform of $v^*[n]$ as $V^*(z^*)$:

$$V^*(z^*) = \frac{1}{2} \cdot \frac{1}{1 - \alpha^* z^{-1}} = \frac{1}{2} \cdot \frac{1}{1 - r e^{-j\omega_o} z^{-1}}, \qquad |z| > |\alpha| = r.$$

Therefore, by the linearity property of the z-transform, we obtain

$$X(z) = V(z) + V^*(z^*)$$

$$= \frac{1}{2}\left(\frac{1}{1 - r e^{j\omega_o} z^{-1}} + \frac{1}{1 - r e^{-j\omega_o} z^{-1}} \right)$$

$$= \frac{1 - (r \cos \omega_o)\, z^{-1}}{1 - (2r \cos \omega_o)\, z^{-1} + r^2 z^{-2}}, \qquad |z| > r.$$

EXAMPLE 6.23 z-**Transform of a Sum of Two Causal Exponential Sequences**

We determine the z-transform of the right-sided sequence

$$w[n] = \left((-0.5)^{n-2} + (0.2)^{n-1} \right) \mu[n]. \tag{6.55}$$

We can rewrite $w[n]$ as

$$w[n] = 4(-0.5)^n \mu[n] + 2(0.2)^n \mu[n].$$

Using Table 6.1, we have

$$(-0.5)^n \mu[n] \overset{\mathcal{Z}}{\longleftrightarrow} \frac{1}{1+0.5\,z^{-1}}, \qquad |z| > 0.5,$$

$$(0.2)^n \mu[n] \overset{\mathcal{Z}}{\longleftrightarrow} \frac{1}{1-0.2\,z^{-1}}, \qquad |z| > 0.2.$$

The ROC of a linear combination of the sequences $(-0.5)^n \mu[n]$ and $(0.2)^n \mu[n]$ is given by at least the overlap of the two ROCs given by $|z| > 0.5$ and $|z| > 0.2$, which is simply the region $|z| > 0.5$. Using the linearity property, we get the *z*-transform of $w[n]$ as

$$4 \left(\frac{1}{1+0.5\,z^{-1}} \right) + 2 \left(\frac{1}{1-0.2\,z^{-1}} \right) = \frac{6+0.2\,z^{-1}}{(1+0.5\,z^{-1})(1-0.2\,z^{-1})}.$$

Thus, we have

$$\left((-0.5)^{n-2} + (0.2)^{n-1} \right) \mu[n] \overset{\mathcal{Z}}{\longleftrightarrow} \frac{6+0.2\,z^{-1}}{(1+0.5\,z^{-1})(1-0.2\,z^{-1})}, \qquad |z| > 0.5. \qquad (6.56)$$

In Example 6.23, since there was no pole-zero cancellation after the linear combination, the ROC of the *z*-transform of the linearly combined sequence $w[n]$ is given by the smaller ROC. However, in certain cases, the ROC of the *z*-transform of the linearly combined sequence may be larger because of pole-zero cancellation.

EXAMPLE 6.24   Sum of Left-Sided and Right-Sided Sequences with Nonoverlapping ROCs

Consider the two-sided sequence
$$v[n] = \alpha^n \mu[n] - \beta^n \mu[-n-1].$$
Denote $x_1[n] = \alpha^n \mu[n]$ and $x_2[n] = -\beta^n \mu[-n-1]$. Therefore, due to the linearity property of the *z*-transform, the *z*-transform $V(z)$ of the sequence $v[n]$ is given by the sum $X_1(z) + X_2(z)$, where $X_1(z)$ and $X_2(z)$ are the *z*-transforms of $x_1[n]$ and $x_2[n]$, respectively.

Now, from Eq. (6.8),

$$X_1(z) = \frac{1}{1-\alpha z^{-1}}, \qquad \text{for } |z| > |\alpha|,$$

and from Eq. (6.10),

$$X_2(z) = \frac{1}{1-\beta z^{-1}}, \qquad \text{for } |z| < |\beta|.$$

Therefore, using the linearity property, we arrive at

$$V(z) = \frac{1}{1-\alpha z^{-1}} + \frac{1}{1-\beta z^{-1}}.$$

The ROC of $V(z)$ is given by the overlap regions of the ROCs of $X_1(z)$ and $X_2(z)$, respectively. If $|\beta| > |\alpha|$, then there is an overlap between the ROCs of $X_1(z)$ and $X_2(z)$, and the ROC of $V(z)$ is an annular region $|\alpha| < |z| < |\beta|$. However, if $|\beta| < |\alpha|$, then there is no overlap, and as a result, $V(z)$ does not exist.

Time-Shifting Property

The *z*-transform of the delayed sequence $x[n] = g[n - n_o]$, where n_o is an integer, is given by $X(z) = z^{-n_o} G(z)$ with an ROC given by $\mathcal{R}_g$, excluding possibly the point $z = 0$ or $z = \infty$; that is,

$$g[n - n_o] \overset{\mathcal{Z}}{\longleftrightarrow} z^{-n_o} G(z), \qquad \text{ROC} = \mathcal{R}_g, \text{ excluding possibly the point } z = 0 \text{ or } z = \infty. \qquad (6.57)$$

It should be noted that if $n_o > 0$, the factor z^{-n_o} adds n_o poles at $z = 0$ and, as a result, the function $z^{-n_o} G(z)$ may have one or more poles at $z = 0$ excluding this point from the ROC $\mathcal{R}_g$. Likewise, when $n_o < 0$, the factor z^{-n_o} adds n_o poles at $z = \infty$, causing the function $z^{-n_o} G(z)$ to have one or more poles at $z = \infty$, which then excludes this point from the ROC $\mathcal{R}_g$. A proof of the time-shifting property follows from a direct computation of the z-transform of $x[n] = g[n - n_o]$ using Eq. (6.1) and is left as an exercise (Problem 6.27).

EXAMPLE 6.25   z-Transform of a First-Order Linear Difference Equation

Determine the z-transform $V(z)$ of the sequence $v[n]$ given by Eq. (3.62) and repeated below for convenience:

$$d_0 v[n] + d_1 v[n - 1] = p_0 \delta[n] + p_1 \delta[n - 1], \qquad |d_1/d_0| < 1. \qquad (6.58)$$

From Table 6.1, the z-transform of $\delta[n]$ is simply 1. Because of the time-shifting property of the z-transform, the z-transform of $\delta[n - 1]$ is z^{-1}, and the z-transform of $v[n - 1]$ is $z^{-1} V(z)$. Applying z-transforms to both sides of Eq. (6.58) and invoking the linearity property, we therefore arrive at

$$d_0 V(z) + d_1 z^{-1} V(z) = p_0 + p_1 z^{-1},$$

which yields

$$V(z) = \frac{p_0 + p_1 z^{-1}}{d_0 + d_1 z^{-1}}.$$

Note that $V(z)$ has a pole at $z = -d_1/d_0$. Hence, if $v[n]$ is a right-sided sequence, the ROC of $V(z)$ is exterior to the circle going through the point $z = -d_1/d_0$. On the other hand, if $v[n]$ is a left-sided sequence, the ROC of $V(z)$ is interior to the circle going through the point $z = -d_1/d_0$.

Multiplication by an Exponential Sequence

The z-transform of the scaled sequence $x[n] = \alpha^n g[n]$, where α is a real or a complex number, is given by $X(z) = G(z/\alpha)$ with an ROC given by $|\alpha| \mathcal{R}_g$; that is,

$$\alpha^n g[n] \overset{\mathcal{Z}}{\longleftrightarrow} G(z/\alpha), \qquad \text{ROC} = |\alpha| \mathcal{R}_g. \qquad (6.59)$$

It should be noted that if the ROC $\mathcal{R}_g$ is the region $\mathcal{R}_{g-} < |z| < \mathcal{R}_{g+}$, then the ROC $|\alpha| \mathcal{R}_g$ denotes the scaled region $|\alpha| \mathcal{R}_{g-} < |z| < |\alpha| \mathcal{R}_{g+}$. Eq. (6.59) can be proved easily by evaluating the z-transform of $\alpha^n g[n]$ using Eq. (6.1) (Problem 6.27).

Differentiation Property

The z-transform of the sequence $x[n] = n \, g[n]$ is given by $X(z) = -z \frac{dG(z)}{dz}$ with an ROC given by $\mathcal{R}_g$, except possibly the point $z = 0$ or ∞; that is,

$$n \, g[n] \overset{\mathcal{Z}}{\longleftrightarrow} -z \frac{dG(z)}{dz}, \qquad \text{ROC} = \mathcal{R}_g, \text{ excluding possibly the point } z = 0 \text{ or } z = \infty. \qquad (6.60)$$

The above property can be proved by determining the z-transform of $n \, g[n]$ and comparing it to the derivative of $G(z)$ with respect to z (Problem 6.27).

EXAMPLE 6.26   z-Transform Determination Using the Differentiation Property

Verify the z-transform $Y(z)$ and the ROC of the sequence $y[n] = (n+1)\alpha^n \mu[n]$ given in Table 6.1. Let $x[n] = \alpha^n \mu[n]$. Then we can write $y[n] = n\,x[n] + x[n]$. From Eq. (6.8), the z-transform of $x[n]$ is given by

$$X(z) = \frac{1}{1 - \alpha z^{-1}}, \qquad |z| > |\alpha|.$$

Next, using the differentiation property, we arrive at the z-transform of $n\,x[n]$ as

$$-z\frac{dX(z)}{dz} = \frac{\alpha z^{-1}}{(1 - \alpha z^{-1})^2}, \qquad |z| > |\alpha|.$$

Using the linearity property of the z-transform, we obtain

$$Y(z) = \frac{1}{1 - \alpha z^{-1}} + \frac{\alpha z^{-1}}{(1 - \alpha z^{-1})^2} = \frac{1}{(1 - \alpha z^{-1})^2}, \qquad |z| > |\alpha|.$$

EXAMPLE 6.27   Inverse z-Transform of a Rational Transform with Multiple Poles

Determine the inverse z-transform $g[n]$ of

$$G(z) = \frac{z^3}{\left(z - \frac{1}{2}\right)\left(z + \frac{1}{3}\right)^2}, \qquad |z| > \frac{1}{2}. \tag{6.61}$$

As the ROC is exterior to a circle of radius $1/2$, the inverse transform is a right-sided sequence. The partial-fraction expansion of $G(z)$ is given by Eq. (6.48) as determined in Example 6.15 and repeated below for convenience:

$$G(z) = \frac{0.36}{1 - \frac{1}{2}z^{-1}} + \frac{0.24}{1 + \frac{1}{3}z^{-1}} + \frac{0.4}{(1 + \frac{1}{3}z^{-1})^2}.$$

From Table 6.5, the inverse z-transform of $1/(1 - \frac{1}{2}z^{-1})$ is $(\frac{1}{2})^n \mu[n]$, and the inverse z-transform of $1/(1 + \frac{1}{3}z^{-1})$ is $(-\frac{1}{3})^n \mu[n]$.

From Example 6.26 we note that the inverse z-transform of $1/(1 + \frac{1}{3}z^{-1})^2$ is given by $(n+1)(-\frac{1}{3})^n \mu[n]$. Therefore, the inverse z-transform of $G(z)$ of Eq. (6.61) is

$$g[n] = \left[0.24\left(-\tfrac{1}{3}\right)^n + 0.4(n+1)\left(-\tfrac{1}{3}\right)^n + 0.36\left(\tfrac{1}{2}\right)^n\right]\mu[n].$$

Using MATLAB, we compute the first 10 samples of the above sequence resulting in

```
Columns 1 through 7
1.0000    -0.1667     0.2500    -0.0231     0.0502     0.0004     0.0098

Columns 8 through 10
0.0012     0.0020     0.0005
```

The above result can be verified by computing the first 10 samples of the inverse z-transform of $G(z)$ using the M-file impz.

Convolution Theorem

The z-transform of the convolution sum of the sequences $g[n]$ and $h[n]$ is given by $G(z)H(z)$ with an ROC that includes an intersection of the ROCs $\mathcal{R}_g$ and $\mathcal{R}_h$; that is,

$$g[n] \circledast h[n] \overset{\mathcal{Z}}{\longleftrightarrow} G(z)H(z), \qquad \text{ROC includes } \mathcal{R}_g \cap \mathcal{R}_h. \qquad (6.62)$$

Proof. Recall that the convolution sum of the sequences $g[n]$ and $h[n]$ is given by

$$x[n] = g[n] \circledast h[n] = \sum_{k=-\infty}^{\infty} g[k]h[n-k].$$

Hence, applying Eq. (6.1), we get

$$X(z) = \sum_{n=-\infty}^{\infty} x[n]z^{-n} = \sum_{n=-\infty}^{\infty} \left(\sum_{k=-\infty}^{\infty} g[k]h[n-k] \right) z^{-n}.$$

Interchanging the order of the summation on the right-hand side of the above equation we have

$$X(z) = \sum_{k=-\infty}^{\infty} g[k] \left(\sum_{n=-\infty}^{\infty} h[n-k]z^{-n} \right).$$

Substituting $\ell = n - k$ in the above equation and rearranging, we arrive at

$$X(z) = \sum_{k=-\infty}^{\infty} g[k] \left(\sum_{\ell=-\infty}^{\infty} h[\ell]z^{-k-\ell} \right) = \sum_{k=-\infty}^{\infty} g[k] \left(\sum_{\ell=-\infty}^{\infty} h[\ell]z^{-\ell} \right) z^{-k}$$

$$= \left(\sum_{k=-\infty}^{\infty} g[k]z^{-k} \right) H(z) = G(z)H(z).$$

The ROC of $X(z)$ includes an intersection of the ROCs of $G(z)$ and $H(z)$. However, in some cases, the ROC could be larger than $\mathcal{R}_g \cap \mathcal{R}_h$, caused by the cancellation of poles and zeros of $G(z)H(z)$ as illustrated in Example 6.28.

EXAMPLE 6.28   Enlargement of ROC Caused by Pole-Zero Cancellation

Consider two causal sequences, $g[n]$ and $h[n]$, with z-transforms $G(z)$ and $H(z)$, respectively, as given below:

$$G(z) = \frac{2 + 1.2\,z^{-1}}{1 - 0.2\,z^{-1}}, \qquad |z| > 0.2,$$

$$H(z) = \frac{3}{1 + 0.6\,z^{-1}} \qquad |z| > 0.6.$$

The intersection of the two ROCs is $|z| > 0.6$. The product of the above two z-transforms is

$$G(z)H(z) = \left(\frac{2 + 1.2\,z^{-1}}{1 - 0.2\,z^{-1}} \right) \left(\frac{3}{1 + 0.6\,z^{-1}} \right) = \frac{6}{1 - 0.2\,z^{-1}},$$

whose ROC is given by $|z| > 0.2$, which is larger than the region $|z| > 0.6$.

The convolution theorem can be employed to compute the time-domain convolution sum by forming the product of the z-transforms of the sequences to be convolved and then evaluating the inverse z-transform of the product. In many applications involving sequences with rational z-transforms, this approach is simpler and yields a closed-form answer. An important signal processing application of the convolution theorem to be discussed later in Section 6.7.1 is in the characterization of the input–output relation of an LTI discrete-time system in the z-domain. Another application of the convolution theorem is in the development of the expression for the z-transform of the cross-correlation sequence $r_{gh}[\ell]$ of two sequences $g[n]$ and $h[n]$ in terms of their z-transforms. Now, recall from Eq. (2.127) that the cross-correlation sequence $r_{gh}[\ell]$ can be expressed in terms of convolution as

$$r_{gh}[\ell] = g[\ell] \circledast h[-\ell].$$

Using the time-reversal property, we observe that the z-transform of $h[-\ell]$ is simply $H(z^{-1})$. Therefore, using the convolution theorem, we obtain from the above equation that

$$\mathcal{Z}\{r_{gh}[\ell]\} = G(z)H(z^{-1}), \tag{6.63}$$

with the ROC given by at least $\mathcal{R}_g \cap (1/\mathcal{R}_h)$.

Modulation Theorem

The z-transform of the product sequence $g[n]h^*[n]$ is given by $\frac{1}{2\pi j}\oint_C G(v)H^*(z^*/v^*)v^{-1}\,dv$, with an associated ROC that includes $\mathcal{R}_g\mathcal{R}_h$; that is,

$$g[n]h^*[n] \overset{\mathcal{Z}}{\longleftrightarrow} \frac{1}{2\pi j}\oint_C G(v)H^*(z^*/v^*)v^{-1}\,dv, \qquad \text{ROC includes } \mathcal{R}_g\mathcal{R}_h, \tag{6.64}$$

where C is a closed anticlockwise contour encircling the origin in the common regions of the ROCs $\mathcal{R}_g$ and $\mathcal{R}_h$. Note that if $\mathcal{R}_g$ denotes the region $\mathcal{R}_{g-} < |z| < \mathcal{R}_{g+}$ and $\mathcal{R}_h$ denotes the region $\mathcal{R}_{h-} < |z| < \mathcal{R}_{h+}$, then $\mathcal{R}_g\mathcal{R}_h$ represents the region $\mathcal{R}_{g-}\mathcal{R}_{h-} < |z| < \mathcal{R}_{g+}\mathcal{R}_{h+}$. The modulation theorem is also called the *complex convolution theorem*.

Proof. From Eq. (6.1), the z-transform of $x[n] = g[n]h^*[n]$ is given by

$$X(z) = \sum_{n=-\infty}^{\infty} g[n]h^*[n]z^{-n}. \tag{6.65}$$

Substituting the expression for $g[n]$ from Eq. (6.29) in Eq. (6.65), we get

$$X(z) = \frac{1}{2\pi j}\sum_{n=-\infty}^{\infty}\left(\oint_C G(v)v^{n-1}dv\right)h^*[n]z^{-n}.$$

Interchanging the order of summation and integration and making use of the conjugation property in the above, we arrive at

$$X(z) = \frac{1}{2\pi j}\oint_C G(v)\left(\sum_{n=-\infty}^{\infty}h^*[n]\left(\frac{z^*}{v^*}\right)^{-n}\right)v^{-1}dv$$

$$= \frac{1}{2\pi j}\oint_C G(v)H^*\left(\frac{z^*}{v^*}\right)v^{-1}dv.$$

Parseval's Relation

If $g[n]$ and $h[n]$ are two complex-valued sequences, then the sum of the samples of the product sequence $g[n]h^*[n]$ can be computed via a contour integration as follows:

$$\sum_{n=-\infty}^{\infty} g[n]h^*[n] = \frac{1}{2\pi j} \oint_C G(v)H^* \left(\frac{1}{v^*} \right) v^{-1} dv, \tag{6.66}$$

where C is a counterclockwise closed contour encircling the origin in the common region of the ROCs of $G(v)$ and $H^*(1/v^*)$ in the complex v-plane.

  Proof. From Eq. (6.64), we have

$$\sum_{n=-\infty}^{\infty} g[n]h^*[n]z^{-n} = \frac{1}{2\pi j} \oint_C G(v)H^*(z^*/v^*)v^{-1} dv.$$

By setting $z = 1$ in the above, we arrive at

$$\sum_{n=-\infty}^{\infty} g[n]h^*[n] = \frac{1}{2\pi j} \oint_C G(v)H^*(1/v^*)v^{-1} dv.$$

If the ROCs of G(z) and H(z) include the unit circle, we can let $v = e^{j\omega}$ in Eq. (6.66), yielding

$$\sum_{n=-\infty}^{\infty} g[n]h^*[n] = \frac{1}{2\pi} \int_{-\pi}^{\pi} G(e^{j\omega})H^*(e^{j\omega})d\omega,$$

which is seen to be identical to the Parseval's relation in the frequency domain, as given by Eq. (3.67).

  As in the case of the discrete-time Fourier transform, the Parseval's relation can also be used to compute the energy of a sequence in the z-domain. To establish the required formula, we let $g[n] = h[n]$ in Eq. (6.66), which leads to

$$\mathcal{E}_g = \sum_{n=-\infty}^{\infty} |g[n]|^2 = \frac{1}{2\pi j} \oint_C G(z)G^*(1/z^*)z^{-1} dz, \tag{6.67}$$

where C is a counterclockwise closed contour encircling the origin in the ROC of $G(z)G^*(1/z^*)$. Note that if the ROC of $G(z)$ includes the unit circle, then that of $G(1/z^*)$ will also include the unit circle. In fact, for an absolutely summable sequence $g[n]$, the ROC of its z-transform $G(z)$ includes the unit circle. In this case, we can let $z = e^{j\omega}$ in Eq. (6.67), resulting in

$$\sum_{n=-\infty}^{\infty} |g[n]|^2 = \frac{1}{2\pi} \int_{-\pi}^{\pi} G(e^{j\omega})G^*(e^{j\omega})d\omega.$$

For a real-valued sequence $g[n]$, the Parseval's relation reduces to

$$\sum_{n=-\infty}^{\infty} |g[n]|^2 = \frac{1}{2\pi j} \oint_C G(z)G(z^{-1})z^{-1} dz = \frac{1}{2\pi} \int_{-\pi}^{\pi} |G(e^{j\omega})|^2 d\omega. \tag{6.68}$$

EXAMPLE 6.29   Computation of the Energy of a Sequence Using Parseval's Relation

We determine the energy of the sequence $x[n] = \alpha^n \mu[n], 0 < \alpha < 1$, using the Parseval's relation. Using Eq. (6.68), we have

$$\mathcal{E}_x = \frac{1}{2\pi j} \oint_C X(z) X(z^{-1}) z^{-1} \, dz,$$

where $X(z)$ is the z-transform of $x[n]$. The contour integral on the right-hand side of the above equation can be evaluated using Cauchy's residue theorem:

$$\frac{1}{2\pi j} \oint_C X(z) X(z^{-1}) z^{-1} \, dz = \Big[\text{sum of residues of } X(z) X(z^{-1}) z^{-1} \text{ inside} C\Big].$$

Now, from Eq. (6.8), we observe

$$X(z) = \frac{1}{1 - \alpha z^{-1}}, \ |z| > \alpha.$$

Since $\alpha < 1$, the unit circle is in the ROCs of both $X(z)$ and $X(z^{-1})$ and can be chosen as the contour C. The function

$$X(z) X(z^{-1}) z^{-1} = \frac{z^{-1}}{(1 - \alpha z^{-1})(1 - \alpha z)} = \frac{1}{(z - \alpha)(1 - \alpha z)}$$

has one pole at $z = \alpha$ inside the unit circle, with a residue given by

$$\text{residue} = \frac{1}{1 - \alpha z}\Big|_{z=\alpha} = \frac{1}{1 - \alpha^2}.$$

Hence,

$$\mathcal{E}_x = \frac{1}{1 - \alpha^2}.$$

6.6   Computation of the Convolution Sum of Finite-Length Sequences

We outlined in Sections 2.5.2 and 5.4.2, respectively, tabular methods for the computations of the linear convolution of two finite-length sequences and the circular convolution of two finite-length sequences of same lengths. In this section, we describe alternate methods for the computation of the linear and circular convolutions that are based on the multiplication of two polynomials.

6.6.1   Linear Convolution

Let $x[n]$ and $h[n]$ be two causal sequences of lengths $L + 1$ and $M + 1$, respectively. Their z-transforms, $X(z)$ and $H(z)$, are polynomials in z^{-1} of degree L and M, respectively; that is,

$$\begin{align} X(z) &= x[0] + x[1] z^{-1} + x[2] z^{-2} + \ldots + x[L] z^{-L}, \\ H(z) &= h[0] + h[1] z^{-1} + h[2] z^{-2} + \ldots + h[M] z^{-M}. \end{align} \tag{6.69}$$

Now, from the convolution theorem of Eq. (6.62), the z-transform $Y(z)$ of the sequence $y[n]$ obtained by the convolution sum of $x[n]$ and $h[n]$ is given by the product of the z-transforms of $x[n]$ and $h[n]$, that is, $Y(z) = H(z)X(z)$, which is a polynomial in z^{-1} of degree $L + M$,

$$Y(z) = y[0] + y[1] z^{-1} + y[2] z^{-2} + \ldots + y[L + M] z^{-(L+M)}, \tag{6.70}$$

whose nth coefficient of $Y(z)$ is precisely

$$y[n] = \sum_{k=0}^{L+M} x[k]h[n-k], \qquad 0 \le n \le L+M.$$

In carrying out the above sum, it is assumed that $x[n] = 0$, $n > L$, and $h[n] = 0$, $n > M$. We illustrate this approach in Example 6.30.

EXAMPLE 6.30   Convolution of One-Sided Sequences Using the Polynomial Multiplication Method

We develop the convolution sum of the two sequences considered in Example 2.29 using the polynomial multiplication method. Here,

$$X(z) = -2 + z^{-2} - z^{-3} + 3z^{-4}, \qquad H(z) = 1 + 2z^{-1} - z^3.$$

Their product results in

$$\begin{aligned} Y(z) = X(z)H(z) &= (-2 + z^{-2} - z^{-3} + 3z^{-4})(1 + 2z^{-1} - z^{-3}) \\ &= -2 - 4z^{-1} + z^{-2} + 3z^{-3} + z^{-4} + 5z^{-5} + z^{-6} - 3z^{-7}. \end{aligned}$$

Note again that the coefficients of the polynomial $Y(z)$ are precisely the samples of the sequence $y[n]$ derived in Example 2.29 using the tabular method.

The polynomial multiplication based method can also be applied to determine the convolution sum of two noncausal finite-length sequences as illustrated in Example 6.31.

EXAMPLE 6.31   Convolution of Two-Sided Sequences Using the Polynomial Multiplication Method

We consider the computation of the convolution sum of the two two-sided sequences of Example 2.30. Here we now have the polynomials

$$X(z) = 3z - 2 + 4z^{-1}, \qquad H(z) = 4 + 2z^{-1} - z^{-2}.$$

Therefore,

$$\begin{aligned} Y(z) = X(z)H(z) &= (3z - 2 + 4z^{-1})(4 + 2z^{-1} - z^{-2}) \\ &= 12z - 2 + 9z^{-1} + 10z^{-2} - 4z^{-3}. \end{aligned}$$

The coefficients of $Y(z)$ are again seen to be the samples of the sequence $\{y[n]\}$ determined in Example 2.30.

6.6.2 Circular Convolution

As in the case of the linear convolution, the circular convolution can also be related to polynomial multiplication but requires a modulo operation after the multiplication. Consider the two polynomials $X(z)$ and $H(z)$ of degree $N - 1$ in z^{-1}; that is,

$$\begin{aligned} X(z) &= x[0] + x[1]z^{-1} + x[2]z^{-2} + \ldots + x[N-1]z^{-(N-1)}, \\ H(z) &= h[0] + h[1]z^{-1} + h[2]z^{-2} + \ldots + h[N-1]z^{-(N-1)}. \end{aligned} \tag{6.71}$$

Then their product $Y_L(z) = H(z)X(z)$ is a polynomial in z^{-1} of degree $2N - 1$,

$$Y_L(z) = y_L[0] + y_L[1]z^{-1} + y_L[2]z^{-2} + \ldots + y_L[2N-1]z^{-(2N-1)}. \tag{6.72}$$

As pointed out in Section 2.5.1, the coefficients $y_L[n]$ are given by the convolution of $x[n]$ and $h[n]$; that is, $v[n] = x[n] \circledast h[n]$. Let $y_C(z)$ denote the polynomial of degree $N - 1$ whose coefficients $y_C[n]$ are given by Eq. (5.55). It can be shown that (Problem 6.17)

$$Y_C(z) = \langle Y_L(z) \rangle_{(z^{-N}-1)}. \tag{6.73}$$

The modulo operation with respect to $z^{-N} - 1$ is taken by setting $z^{-N} = 1$ in Eq. (6.72). We illustrate the process in Example 6.32.

EXAMPLE 6.32   Circular Convolution of Causal Sequences Using the Polynomial Multiplication Method

Consider the two length-4 sequences $g[n]$ and $h[n]$ defined for $0 \le n \le 3$. We define

$$G(z) = g[0] + g[1]z^{-1} + g[2]z^{-2} + g[3]z^{-3},$$
$$H(z) = h[0] + h[1]z^{-1} + h[2]z^{-2} + h[3]z^{-3}.$$

Their product $Y_L(z) = G(z)H(z)$ is given by

$$Y_L(z) = y_L[0] + y_L[1]z^{-1} + y_L[2]z^{-2} + y_L[3]z^{-3} + y_L[4]z^{-4} + y_L[5]z^{-5} + y_L[6]z^{-6},$$

where

$$y_L[0] = g[0]h[0],$$
$$y_L[1] = g[0]h[1] + g[1]h[0],$$
$$y_L[2] = g[0]h[2] + g[1]h[1] + g[2]h[0],$$
$$y_L[3] = g[0]h[3] + g[1]h[2] + g[2]h[1] + g[3]h[0],$$
$$y_L[4] = g[1]h[3] + g[2]h[2] + g[3]h[1],$$
$$y_L[5] = g[2]h[3] + g[3]h[2],$$
$$y_L[6] = g[3]h[3].$$

We next form

$$Y_C(z) = \langle Y_L(z) \rangle_{(z^{-4}-1)}$$
$$= y_L[0] + y_L[1]z^{-1} + y_L[2]z^{-2} + y_L[3]z^{-3} + y_L[4] + y_L[5]z^{-1} + y_L[6]z^{-2}$$
$$= \left(y_L[0] + y_L[4]\right) + \left(y_L[1] + y_L[5]\right)z^{-1} + \left(y_L[2] + y_L[6]\right)z^{-2} + y_L[3]z^{-3}$$
$$= y_C[0] + y_C[1]z^{-1} + y_C[2]z^{-2} + y_C[3]z^{-3},$$

where

$$y_C[0] = y_L[0] + y_L[4] = g[0]h[0] + g[1]h[3] + g[2]h[2] + g[3]h[1],$$
$$y_C[1] = y_L[1] + y_L[5] = g[0]h[1] + g[1]h[0] + g[2]h[3] + g[3]h[2],$$
$$y_C[2] = y_L[2] + y_L[6] = g[0]h[2] + g[1]h[1] + g[2]h[0] + g[3]h[3],$$
$$y_C[3] = y_L[3] = g[0]h[3] + g[1]h[2] + g[2]h[1] + g[3]h[0].$$

Note from the above that the coefficients $y_C[n]$, $0 \le n \le 3$, are the same as those derived in Example 5.7.

Figure 6.7: An LTI discrete-time system.

6.7 The Transfer Function

The frequency response function $H(e^{j\omega})$ of an LTI discrete-time system introduced in Section 3.8 is given by the discrete-time Fourier transform of the impulse response of the system. As we have seen, the frequency response function does provide valuable information on the behavior of an LTI digital filter in the frequency domain. However, being a complex function of the frequency variable ω, it is difficult to manipulate it for the realization of a digital filter. On the other hand, the z-transform of the impulse response of an LTI system, called the transfer function, is a polynomial in z^{-1}, and for a system with a real impulse response, it is a polynomial with real coefficients. Moreover, in most practical cases, the LTI digital filter of interest is characterized by a linear difference equation with constant and real coefficients. The transfer function of such a filter is a real rational function of the variable z^{-1}, that is, a ratio of two polynomials in z^{-1} with real coefficients, and is thus more amenable for the synthesis.

We first develop the input–output relation of an LTI system in the z-domain from its various time-domain descriptions and arrive at different forms of the transfer function representation of the system. We then study its properties and in particular develop the conditions for the BIBO stability of a causal LTI system.

6.7.1 Definition

Consider the LTI digital discrete-time system of Figure 6.7 with an impulse response $h[n]$. The input–output relation of this system is given by

$$y[n] = \sum_{k=-\infty}^{\infty} h[k]x[n-k], \tag{6.74}$$

where $x[n]$ and $y[n]$ are, respectively, the input and the output of the system. Let $Y(z)$, $X(z)$, and $H(z)$ denote the z-transforms of $y[n]$, $x[n]$, and $h[n]$, respectively. Then applying the convolution theorem of Table 6.2, we arrive at the input–output relation of the LTI discrete-time system in the z-domain as

$$Y(z) = H(z)X(z), \tag{6.75}$$

where

$$H(z) = \sum_{n=-\infty}^{\infty} h[n]\, z^{-n}. \tag{6.76}$$

The quantity $H(z)$ is more commonly called the *transfer function* or the *system function*. From Eq. (6.75), we have

$$H(z) = \frac{Y(z)}{X(z)}. \tag{6.77}$$

Program 6_3.m
Program 6_5.m
Program 6_6.m

Thus, the transfer function $H(z)$ of an LTI discrete-time system is given by the ratio of the z-transform $Y(z)$ of the output sequence $y[n]$ to the z-transform $X(z)$ of the input sequence $x[n]$.

The inverse z-transform of the transfer function $H(z)$ yields the impulse response $h[n]$. For a causal rational transfer function, the methods outlined in Section 6.4 can be used to compute its impulse response. For example, an analytical form of the impulse response can be determined via a partial-fraction expansion using Program 6_3. On the other hand, a fixed number of impulse response samples starting at $n = 0$ can be computed using Programs 6_5 or 6_6.

6.7.2 Transfer Function Expression

We now develop the expressions for the transfer functions of linear, time-invariant FIR and IIR digital filters.

FIR Digital Filter

In the case of an LTI FIR digital filter, the input–output relation in the time domain is given by Eq. (2.118). Its impulse response $h[n]$ is defined for $N_1 \leq n \leq N_2$, and thus, $h[n] = 0$ for $n < N_1$ and $n > N_2$. Therefore, the transfer function is given by

$$H(z) = \sum_{n=N_1}^{N_2} h[n]z^{-n}. \tag{6.78}$$

For a causal FIR filter, $0 \leq N_1 \leq N_2$. Note that all poles of $H(z)$ of a causal FIR filter are at the origin in the z-plane, and as a result, the ROC of $H(z)$ is the entire z-plane, excluding the point $z = 0$.

EXAMPLE 6.33   Transfer Function of the Moving-Average Filter

Consider the moving-average filter of Example 3.18 with an impulse response $h[n]$ given by Eq. (3.97). The transfer function of the moving-average FIR filter is then given by

$$H(z) = \frac{1}{M} \sum_{n=0}^{M-1} z^{-n} \tag{6.79}$$

$$= \frac{1 - z^{-M}}{M(1 - z^{-1})} = \frac{z^M - 1}{M[z^{M-1}(z - 1)]}. \tag{6.80}$$

From Eq. (6.80), it is seen that the transfer function has M zeros on the unit circle at $z = e^{j2\pi k/M}$, $k = 0, 1, 2, \ldots, M - 1$. There is an $(M - 1)$th-order pole at the origin $(z = 0)$ and a single pole at $z = 1$. But the pole at $z = 1$ exactly cancels a zero at the same place, resulting in a transfer function with all poles at the origin. This is always the characteristic of a causal FIR filter for which the ROC is the entire z-plane except the origin.

Finite-Dimensional Linear Time-Invariant IIR Discrete-Time System

For an LTI IIR filter, characterized by the difference equation of Eq. (2.90), the expression for the transfer function can be easily derived. Here, the input–output relation of the LTI system in the z-domain is obtained by applying the z-transform to both sides of Eq. (2.90) and making use of the linearity and time-shifting properties of the z-transform, resulting in

$$\sum_{k=0}^{N} d_k z^{-k} Y(z) = \sum_{k=0}^{M} p_k z^{-k} X(z), \tag{6.81}$$

where $Y(z)$ and $X(z)$ denote the z-transforms of $y[n]$ and $x[n]$ with associated ROCs, respectively. A more convenient form of Eq. (6.81) is given by

$$\left(\sum_{k=0}^{N} d_k z^{-k}\right) Y(z) = \left(\sum_{k=0}^{M} p_k z^{-k}\right) X(z). \tag{6.82}$$

From Eq. (6.82), we arrive at the expression for the transfer function $H(z) = Y(z)/X(z)$ as given by Eq. (6.13), which is a rational function in z^{-1}; that is, it is a ratio of two polynomials in z^{-1}. By multiplying the numerator and the denominator of the right-hand side by z^M and z^N, respectively, the transfer function can be expressed as a rational function in z as given in Eq. (6.14). An alternate way to express the transfer function $H(z)$ of Eq. (6.13) is to factor out the numerator and denominator polynomials, as indicated in Eq. (6.15), where $\xi_1, \xi_2, \ldots, \xi_M$ are the finite zeros and $\lambda_1, \lambda_2, \ldots, \lambda_N$ are the finite poles of $H(z)$. For a causal IIR filter, the impulse response is a causal sequence. The ROC of the causal IIR transfer function $H(z)$ is thus exterior to the circle going through the pole furthest from the origin; that is, the ROC is given by

$$|z| > \max_k |\lambda_k|.$$

We illustrate the determination of the transfer function of a third-order causal LTI discrete-time system described by a constant coefficient difference equation in Example 6.34.

Program 6_1.m

EXAMPLE 6.34   Transfer Function of a Finite-Dimensional LTI IIR Digital Filter

A causal finite-dimensional linear, time-invariant IIR digital filter is characterized by a constant coefficient difference equation given by

$$y[n] = x[n-1] - 1.2x[n-2] + x[n-3] + 1.3y[n-1] - 1.04y[n-2] + 0.222y[n-3]. \tag{6.83}$$

From Eqs. (6.82) and (6.13), we arrive directly at the expression for the transfer function $H(z)$ as a rational function in z^{-1}:

$$H(z) = \frac{Y(z)}{X(z)} = \frac{z^{-1} - 1.2z^{-2} + z^{-3}}{1 - 1.3z^{-1} + 1.04z^{-2} - 0.222z^{-3}}. \tag{6.84}$$

Multiplying the numerator and the denominator of the right-hand-side term in the above equation by z^3, we obtain the expression for the transfer function as a rational function in z:

$$H(z) = \frac{z^2 - 1.2z + 1}{z^3 - 1.3z^2 + 1.04z - 0.222}. \tag{6.85}$$

The above transfer function can also be expressed in a factored form. To this end, we make use of Program 6_1 of Example 6.9, with the numerator and the denominator coefficients entered in vector form. The output data generated by the program are:

```
 Numerator factors
1.00000000000000   -1.20000000000000   1.00000000000000


 Denominator factors
1.00000000000000   -1.00000000000000   0.74000000000000
1.00000000000000   -0.30000000000000                  0


 Gain constant
  1
```

Thus, the factored form of the transfer function is given by

$$H(z) = \frac{(z^2 - 1.2\,z + 1))}{(z - 0.3)(z^2 - z + 0.74)}. \tag{6.86}$$

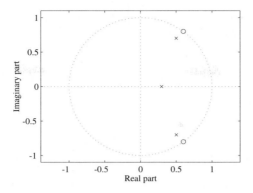

Figure 6.8: Pole-zero plot of the IIR transfer function of Eq. (6.85).

The zeros, poles, and the gain constant of the transfer function of Eq. (6.85) can be obtained using the code fragments

```
[z,p,k] = tf2zp(num,den)
```

which results in

```
z =
0.6000 + 0.8000i
0.6000 - 0.8000i

p =
0.5000 + 0.7000i
0.5000 - 0.7000i
0.3000

k =
1
```

A pole-zero plot of the above transfer function obtained using the function `zplane` is sketched in Figure 6.8. Note from Eq. (6.86) and also from Figure 6.8 that the two poles furthest from the origin are of magnitude $\sqrt{0.74}$. Hence, the ROC of the transfer function of Eq. (6.85) is given by $|z| > \sqrt{0.74}$.

6.7.3 Frequency Response from Transfer Function

The transfer function $H(z)$ of an LTI digital filter can be written in rectangular form as

$$H(z) = H_{\text{re}}(z) + j H_{\text{im}}(z), \qquad (6.87)$$

or, alternately, in polar form,

$$H(z) = |H(z)| e^{j \arg H(z)}, \qquad (6.88)$$

where

$$\arg H(z) = \tan^{-1}\left[\frac{H_{\text{im}}(z)}{H_{\text{re}}(z)}\right]. \qquad (6.89)$$

If the ROC of $H(z)$ includes the unit circle, then the frequency response $H(e^{j\omega})$ of the LTI digital filter can be obtained from its transfer function $H(z)$ by simply evaluating it on the unit circle; that is,

$$H(e^{j\omega}) = H(z)|_{z=e^{j\omega}}. \tag{6.90}$$

The z-transform $H(z)$ of a sequence $h[n]$ can be determined from its Fourier transform $H(e^{j\omega})$ by analytic continuation [Chu90]:

$$H(z) = H(e^{j\omega})\Big|_{\omega=\frac{1}{j}\ln z}. \tag{6.91}$$

As indicated in Eq. (3.86), the frequency response $H(e^{j\omega})$ can be written in terms of its real and imaginary parts, $H_{\text{re}}(e^{j\omega})$ and $H_{\text{im}}(e^{j\omega})$, or in terms of its magnitude and phase functions, $|H(e^{j\omega})|$ and $\arg[H(e^{j\omega})]$, respectively.

For a stable rational transfer function $H(z)$ in the form of Eq. (6.13), the factored form of the frequency response $H(e^{j\omega})$ is obtained by substituting $z = e^{j\omega}$ in Eq. (6.15), resulting in

$$H(e^{j\omega}) = \frac{p_0}{d_0} e^{j\omega(N-M)} \frac{\prod_{k=1}^{M}(e^{j\omega} - \xi_k)}{\prod_{k=1}^{N}(e^{j\omega} - \lambda_k)}. \tag{6.92}$$

The above form is convenient to visualize the contributions of the zero factor $(z - \xi_k)$ and the pole factor $(z - \lambda_k)$ of the transfer function $H(z)$ to the overall frequency response. >From Eq. (6.92), the expression for the magnitude function is thus given by

$$|H(e^{j\omega})| = \left|\frac{p_0}{d_0}\right| \left|e^{j\omega(N-M)}\right| \frac{\prod_{k=1}^{M}|e^{j\omega} - \xi_k|}{\prod_{k=1}^{N}|e^{j\omega} - \lambda_k|}$$

$$= \left|\frac{p_0}{d_0}\right| \frac{\prod_{k=1}^{M}|e^{j\omega} - \xi_k|}{\prod_{k=1}^{N}|e^{j\omega} - \lambda_k|}. \tag{6.93}$$

Likewise, from Eq. (6.92), the phase response of a rational transfer function is of the form

$$\arg H(e^{j\omega}) = \arg(p_0/d_0) + \omega(N - M) + \sum_{k=1}^{M}\arg(e^{j\omega} - \xi_k) - \sum_{k=1}^{N}\arg(e^{j\omega} - \lambda_k). \tag{6.94}$$

For a real-coefficient transfer function $H(z)$, it can be shown that

$$|H(e^{j\omega})|^2 = H(e^{j\omega})H^*(e^{j\omega}) = H(e^{j\omega})H(e^{-j\omega}) = H(z)H(z^{-1})|_{z=e^{j\omega}}. \tag{6.95}$$

The magnitude-squared function for a real-coefficient rational transfer function can be computed using Eq. (6.93), which leads to

$$|H(e^{j\omega})|^2 = H(e^{j\omega})H(e^{-j\omega}) = \left|\frac{p_0}{d_0}\right|^2 \frac{\prod_{k=1}^{M}(e^{j\omega} - \xi_k)(e^{-j\omega} - \xi_k^*)}{\prod_{k=1}^{N}(e^{j\omega} - \lambda_k)(e^{-j\omega} - \lambda_k^*)}. \tag{6.96}$$

The phase response $\theta(\omega)$ of an LTI discrete-time system with a real-coefficient transfer function is given by

$$\theta(\omega) = \tan^{-1}\left[\frac{H_{\text{im}}(z)}{H_{\text{re}}(z)}\right]_{z=e^{j\omega}} = \frac{1}{2j}\ln\left[\frac{H(z)}{H(z^{-1})}\right]_{z=e^{j\omega}}. \tag{6.97}$$

The group delay $\tau_g(\omega)$ of of an LTI discrete-time system with a real-coefficient transfer function is given by

$$\tau_g(\omega) = -\frac{d\theta(\omega)}{d\omega} = -\text{Re}\left(z\frac{d[\ln H(z)]}{dz}\right)\Big|_{z=e^{j\omega}}. \tag{6.98}$$

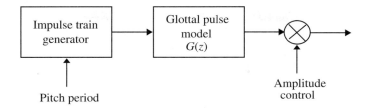

Figure 6.9: Schematic representation for the generation of the glottal wave for voiced speech.

EXAMPLE 6.35   Glottal Pulse Model

A schematic representation of the generation of the excitation signal for voiced speech is shown in Figure 6.9 [Rab78]. Here, the impulse train source generates a periodic sequence of unit impulses whose period is determined by the fundamental period of the desired voiced sound. These impulses then excite the LTI discrete-time system $G(z)$ with an impulse response $g[n]$ that simulates the desired glottal wave shape. A reasonably good approximation of the glottal pulse model is given by the causal second-order IIR transfer function [Ros71]

$$G(z) = \frac{\alpha z^{-1}}{(1 - \alpha z^{-1})^2}, \qquad 0 < \alpha < 1. \tag{6.99}$$

From Table 6.1, the inverse-z transform $g[n]$ of the above transfer function is given by

$$g[n] = n\alpha^n \mu[n], \tag{6.100}$$

which is glottal pulse model impulse response. A plot of the above impulse response for $\alpha = 0.9$ is shown in Figure 6.10(a). The corresponding magnitude response of the glottal model is sketched in Figure 6.10(b). To include the effect of lip radiation, a single zero at $z = 1$ is added to the numerator to the transfer function of Eq. (6.99), resulting in a modified transfer function given by [Rab78]

$$G(z) = \frac{\alpha z^{-1}(1 - z^{-1})}{(1 - \alpha z^{-1})^2}, \qquad 0 < \alpha < 1. \tag{6.101}$$

The magnitude response of the above transfer function is shown in Figure 6.10(c).

6.7.4   Geometric Interpretation of Frequency Response Computation

For an LTI digital filter with a rational transfer function $H(z)$, the factored form of the frequency response expression given by Eq. (6.92) is convenient to develop a geometric interpretation of the frequency response computation from the pole-zero plot of the transfer function, as ω is varied from 0 to 2π on the unit circle in the z-plane. The geometric interpretation can be used to obtain a sketch of the response as a function of the frequency.

If we examine the expression for the frequency response given in Eq. (6.92), we observe that a typical factor is of the form

$$(e^{j\omega} - \rho e^{j\phi}),$$

where $\rho e^{j\phi}$ is a zero if the factor is from the numerator, that is, a zero factor, or is a pole if it is from the denominator, that is, a pole factor. In the z-plane, the factor $(e^{j\omega} - \rho e^{j\phi})$ represents a vector starting from the point $z = \rho e^{j\phi}$ and ending on the unit circle at $z = e^{j\omega}$, as shown in Figure 6.11. As ω is varied from 0 to 2π, the tip of the vector moves counterclockwise from the point $z = 1$, tracing the unit circle, and back to the point $z = 1$.

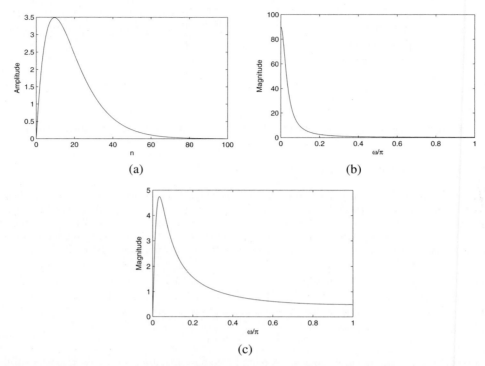

Figure 6.10: (a) The glottal pulse model impulse response for $\alpha = 0.9$, (b) corresponding magnitude response, and (c) magnitude response of the glottal pulse model including the lip radiation.

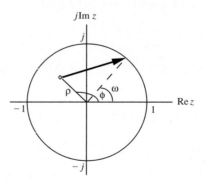

Figure 6.11: Geometric interpretation of frequency response computation of a rational transfer function.

As indicated by Eq. (6.93), the magnitude response $|H(e^{j\omega})|$ at a specific value of ω is given by the product of the magnitudes of all zero vectors divided by the product of the magnitudes of all pole vectors. Likewise, from Eq. (6.94), we observe that the phase response $\arg H(e^{j\omega})$ at a specific value of ω is obtained by adding the phase of the term p_0/d_0 and the linear-phase term $\omega(N - M)$ to the sum of the angles of all zero vectors minus the sum of the angles of all pole vectors. Thus, an approximate plot of the magnitude and phase responses of the transfer function of an LTI digital filter can be developed by examining its pole and zero locations.

Now, a zero (pole) vector has the smallest magnitude when $\omega = \phi$. If the digital filter is to be designed to highly attenuate signal components in a specified range of frequencies, we need to place zeros of the transfer function very close to or on the unit circle in this range. Similarly, to highly emphasize signal components in a specified range of frequencies, we need to place poles of the transfer function very close to the unit circle in this range.

6.7.5 Stability Condition in Terms of Pole Locations

We describe in this text several methods for determining the transfer function of a causal FIR or IIR filter meeting the prescribed frequency-domain specifications. However, before the digital filter is implemented, we need to ensure that the transfer function derived will lead to a stable structure.

We have shown earlier in Section 2.5.3 that an LTI digital filter is BIBO stable if and only if its impulse response sequence $h[n]$ is absolutely summable; that is,

$$\sum_{n=-\infty}^{\infty} |h[n]| < \infty. \tag{6.102}$$

The above stability condition is difficult to test for a system with an impulse response of infinite length. In this section, we develop a stability condition in terms of the pole locations of the transfer function $H(z)$, which is much easier to test if the pole locations are known.

Examples 6.36 and 6.37 consider, respectively, the stability of a first-order and a second-order LTI IIR digital filter.

EXAMPLE 6.36   Stability Condition for a First-Order LTI IIR Digital Filter

We first examine the stability of a first-order LTI IIR digital filter. We showed in Example 2.31 that an LTI IIR digital filter with a causal impulse response $h_1[n] = \alpha^n \mu[n]$ is BIBO stable for $|\alpha| < 1$, as $h_1[n]$ satisfies the stability condition of Eq. (6.102). The transfer function of this causal system is $H_1(z) = 1/(1 - \alpha z^{-1})$ with an ROC given by $|z| > |\alpha|$. The transfer function $H_1(z)$ has a pole at $z = \alpha$, which is inside the unit circle if $|\alpha| < 1$, and as a result, the unit circle is contained in the ROC of the transfer function.

In Example 2.32, we showed that the first-order LTI digital filter with an anticausal impulse response $h_2[n] = -\beta^n \mu[-n - 1]$ is BIBO stable for $|\beta| > 1$, as here $h_2[n]$ also satisfies the stability condition of Eq. (6.102). The transfer function of this anticausal system is $H_2(z) = 1/(1 - \beta z^{-1})$ with an ROC given by $|z| < |\beta|$. The transfer function $H_2(z)$ has a pole at $z = \beta$, which is outside the unit circle if $|\beta| > 1$, and as a result, the unit circle is also contained in the ROC of the transfer function.

Since both of the above two first-order LTI systems have absolutely summable impulse responses, the Fourier transforms of both impulse responses exist.

EXAMPLE 6.37   Stability Condition for a Second-Order LTI IIR Digital Filter

Consider the transfer function of a second-order LTI IIR digital filter given by

$$H(z) = \frac{(\alpha - \beta)z^{-1}}{(1 - \alpha z^{-1})(1 - \beta z^{-1})}, \qquad |\alpha| < 1 < |\beta|.$$

The transfer function has a pole inside the unit circle at $z = \alpha$ and a pole outside the unit circle at $z = \beta$. As indicated in Figure 6.5, the above transfer function can have three possible ROCs. We examine the impulse responses associated with each ROC and investigate the stability of the above transfer function for each case.

To determine the impulse response $h[n]$, we first rewrite $H(z)$ as

$$H(z) = \frac{1}{1 - \alpha z^{-1}} - \frac{1}{1 - \beta z^{-1}},$$

and then determine the inverse-z transform of each first-order term on the right-hand side for a given ROC. For the ROC defined by $|z| > |\beta|$, the impulse response is a right-sided sequence given by the sum of two causal sequences according to

$$h[n] = \alpha^n \mu[n] - \beta^n \mu[n].$$

Since $|\beta| > 1$, the second sequence on the right-hand side in the above equation is not an absolutely summable sequence, and hence, the LTI system is unstable.

For the ROC defined by $|\alpha| < |z| < |\beta|$, the impulse response is a two-sided sequence given by the sum of a causal sequence and an anticausal sequence as

$$h[n] = \alpha^n \mu[n] + \beta^n \mu[-n - 1].$$

As demonstrated in Examples 2.32 and 2.33, both sequences in the above equation are absolutely summable, and hence, the LTI system is stable. Note that the given ROC includes the unit circle as $|\beta| > 1$.

Finally, for the ROC defined by $|z| < |\alpha|$, the impulse response is a left-sided sequence given by the sum of two anticausal sequences according to

$$h[n] = -\alpha^n \mu[-n - 1] + \beta^n \mu[-n - 1].$$

The first sequence on the right-hand side of the above equation is not absolutely summable, and hence, the LTI system is unstable.

We shall now demonstrate that the absolute summability of an impulse response $h[n]$ guarantees the existence of its Fourier transform $H(e^{j\omega})$. From Eq. (6.76), we get

$$|H(z)| = \left| \sum_{n=-\infty}^{\infty} h[n] z^{-n} \right| \leq \sum_{n=-\infty}^{\infty} |h[n] z^{-n}| = \sum_{n=-\infty}^{\infty} |h[n]| \, |z^{-n}|.$$

Hence, on the unit circle, the above reduces to

$$|H(z)|_{z=e^{j\omega}} = |H(e^{j\omega})| \leq \sum_{n=-\infty}^{\infty} |h[n]|.$$

If $h[n]$ satisfies the BIBO stability condition of Eq. (6.102), then $\sum_{n=-\infty}^{\infty} |h[n]| < \infty$, and as a result, for a BIBO stable LTI digital filter, $|H(e^{j\omega})| < \infty$, indicating the existence of the Fourier transform of the impulse response. Therefore, the ROC of the transfer function $H(z)$ of a BIBO stable LTI digital filter contains the unit circle. Conversely, if the ROC of the transfer function of an LTI digital filter includes the unit circle, then the filter is BIBO stable.

As indicated earlier, a causal LTI FIR digital filter with bounded impulse response coefficients is always stable as all its poles are at the origin in the z-plane. On the other hand, a causal LTI IIR filter may be unstable if not designed properly. In addition, an originally stable IIR filter characterized by infinite precision coefficients and with all poles inside the unit circle may also become unstable after implementation due to the unavoidable quantization of all coefficients causing one or more poles to move to the unit circle or even outside the unit circle, as illustrated by Example 6.38.

EXAMPLE 6.38   Effect of Coefficient Quantization on the Stability of an LTI IIR Digital Filter

Consider the causal second-order IIR transfer function given by

$$H(z) = \frac{1}{1 - 1.845z^{-1} + 0.850586z^{-2}}. \tag{6.103}$$

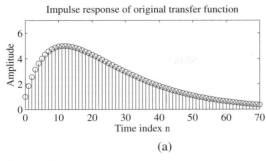

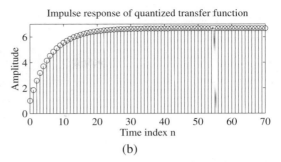

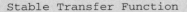

(a)

(b)

Figure 6.12: (a) Impulse response of original transfer function of Eq. (6.103) and (b) impulse response of transfer function of Eq. (6.104).

The plot of the impulse response coefficients of the above transfer function using a slightly modified Program 6_5 is shown in Figure 6.12(a). As seen from this figure, the impulse response coefficient $h[n]$ decays rapidly to zero value as n increases, indicating that the absolute summability condition of Eq. (6.102) is satisfied. Hence, $H(z)$ is a stable transfer function.

Now consider the case when the transfer function coefficients are rounded to values with two digits after the decimal point. Then the transfer function of Eq. (6.103) takes the form

$$\hat{H}(z) = \frac{1}{1 - 1.85z^{-1} + 0.85z^{-2}}. \tag{6.104}$$

A plot of its impulse response coefficients is shown in Figure 6.12(b). In this case, the impulse response coefficient $h[n]$ increases rapidly to a constant value as n increases. As a result, the absolute summability condition of Eq. (6.102) is violated, implying that $\hat{H}(z)$ is an unstable transfer function.

The stability testing of an IIR transfer function is therefore an important problem. However, it is difficult to compute the sum $\mathcal{S}$ of Eq. (6.102) analytically in most cases. For a causal IIR transfer function, it can be computed approximately by replacing the right-hand side of Eq. (6.102) with the following finite sum

$$\mathcal{S}_K = \sum_{n=0}^{K-1} |h[n]|, \tag{6.105}$$

and iteratively computing Eq. (6.105) until the difference between a series of consecutive values of $\mathcal{S}_K$ is smaller than some arbitrarily chosen small number, which is typically 10^{-6}.

The numerical evaluation of Eq. (6.105) to check for stability of a rational transfer function can be carried out using Program 6_7.

EXAMPLE 6.39   Stability Testing Using MATLAB

We test the stability of the transfer function of Eq. (6.103) using Program 6_7. The output data generated by running this program is

Program 6_7.m

```
Stable Transfer Function
```

indicating that the transfer function is stable. On the other hand, by running the program with the coefficients of the transfer function of Eq. (6.104), we get the printout

```
Unstable Transfer Function
```

verifying that this transfer function is unstable.

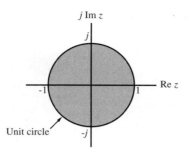

Figure 6.13: Stability region (shown shaded) in the z-plane for pole locations of a stable casual transfer function.

It should be noted that, in some cases, the above approach may not guarantee the absolute summability of the impulse response and, hence, the stability of the LTI discrete-time system, as illustrated in Example 6.40.

EXAMPLE 6.40   An Unstable Causal LTI IIR System with a Decaying Impulse Response

Consider the causal LTI discrete-time system with an impulse response given by

$$h[n] = \begin{cases} \frac{1}{n}, & n \geq 1, \\ 0, & n \leq 0. \end{cases} \tag{6.106}$$

Even though the impulse response decays with increasing n, it is not absolutely summable, as $\sum_{n=1}^{\infty} 1/|n|$ does not converge.

We now develop an alternate stability condition based on the location of the poles of the transfer function. We first consider causal transfer functions. Consider the IIR digital filter described by the rational transfer function $H(z)$ of Eq. (6.13). If the digital filter is assumed to be causal, that is, the impulse response sequence $\{h[n]\}$ is a right-sided sequence, the ROC of $H(z)$ is exterior to the circle going through the pole that is farthest from the origin. But stability requires that $\{h[n]\}$ be absolutely summable, which, in turn, implies that the discrete-time Fourier transform $H(e^{j\omega})$ of $\{h[n]\}$ exists. Now, the z-transform $H(z)$ of a sequence $\{h[n]\}$ reduces to the Fourier transform $H(e^{j\omega})$ by letting $z = e^{j\omega}$ if the unit circle lies within the ROC of the transfer function. Therefore, we conclude that all poles of a stable causal transfer function $H(z)$ must be strictly inside the unit circle, as indicated in Figure 6.13.

Example 6.41 considers the stability testing of a causal second-order transfer function from the pole locations.

EXAMPLE 6.41   Stability Determination from Pole Locations

The factored form of the transfer function $H(z)$ of the causal LTI digital filter of Eq. (6.103) is given by

$$H(z) = \frac{1}{(1 - 0.902z^{-1})(1 - 0.943z^{-1})},$$

which has a real pole at $z = 0.902$ and a real pole at $z = 0.943$. Since the magnitude of each pole of $H(z)$ is less than 1, all poles are inside the unit circle in the z-plane. Hence, the transfer function is stable.

On the other hand, the factored form of the transfer function $\hat{H}(z)$ of Eq. (6.104) is given by

$$\hat{H}(z) = \frac{1}{(1 - z^{-1})(1 - 0.85z^{-1})},$$

which now has one pole inside the unit circle at $z = 0.85$ and the other pole on the unit circle at $z = 1$, indicating that it is unstable.

Example 6.42 reexamines the stability of the causal LTI system given by Eq. (6.106) of Example 6.40 from the pole locations of the transfer function of the system.

EXAMPLE 6.42   Causal LTI IIR System with Infinite Number of Unit Circle Poles

The transfer function $H(z)$ of the causal LTI discrete-time system of Example 6.40 is given by the z-transform of the impulse response of Eq. (6.106). It is given by

$$H(z) = \sum_{n=1}^{\infty} \frac{z^{-n}}{n} = \log_e \left(\frac{1}{1 - z^{-1}} \right), \qquad |z| > 1, \tag{6.107}$$

which has infinite number of poles on the unit circle at $z = 1$, and hence, it is unstable.

On the other hand, an anticausal digital filter has a left-sided impulse response $\{h[n]\}$ and as a result, the ROC of its transfer function $H(z)$ is interior to the circle going through the pole that is closest to the origin. However, for BIBO stability, the Fourier transform $H(e^{j\omega})$ of $\{h[n]\}$ must exist, implying that the unit circle of $H(z)$ lies in its ROC. Hence, in this case, all poles of a stable anticausal transfer function $H(z)$ must be strictly outside the unit circle. This type of filter will thus have a stable response by running time backwards. In practice, such an anticausal transfer function can be implemented by storing a finite length of the output data in a buffer and reading it in a reverse order.

6.8 Summary

The z-transform of an aperiodic sequence has been introduced and its properties reviewed. As in the case of the discrete-time Fourier transform discussed in Chapter 3, and the discrete Fourier transform, the discrete cosine transform, and the Haar transform discussed in Chapter 5, this alternate representation reviewed in this chapter also consists of a pair of expressions: the analysis equation and the synthesis equation. The analysis equation is used to convert from the time-domain representation to the transform-domain representation, while the synthesis equation is used for the reverse process.

An important and useful characterization of an LTI discrete-time system is its transfer function given by the z-transform of its impulse response. The properties of the transfer function have been studied, and the stability condition of an LTI system in terms of the pole locations of its transfer function has been derived.

6.9 Problems

6.1 Show that for a causal sequence $x[n]$ defined for $n \geq 0$ and with a z-transform $X(z)$,

$$x[0] = \lim_{z \to \infty} X(z).$$

The above result is known as the *initial value theorem*.

6.2 Derive the z-transforms and the ROCs of the following sequences given in Table 6.1:
(a) $\delta[n]$, (b) $n\alpha^n \mu[n]$, and (c) $(r^n \sin \omega_0 n)\mu[n]$.

6.3 Determine the z-transforms of the following sequences and their respective ROCs:
(a) $x_1[n] = \alpha^n \mu[n-2]$, (b) $x_2[n] = -\alpha^n \mu[-n-3]$,   (c) $x_3[n] = \alpha^n \mu[n+4]$, (d) $x_4[n] = \alpha^n \mu[-n]$.

6.4 Determine which one of the following four sequences has the same z-transform:
(a) $x_1[n] = (0.4)^n \mu[n] + (-0.6)^n \mu[n]$,   (b) $x_2[n] = (0.4)^n \mu[n] - (-0.6)^n \mu[-n-1]$,
(c) $x_3[n] = -(0.4)^n \mu[-n-1] - (-0.6)^n \mu[-n-1]$,   (d) $x_4[n] = -(0.4)^n \mu[-n-1] + (-0.6)^n \mu[n]$.

6.5 Consider the following sequences:
(i) $x_1[n] = (0.3)^n \mu[n+1]$,   (ii) $x_2[n] = (0.7)^n \mu[n-1]$,   (iii) $x_3[n] = (0.4)^n \mu[n-5]$,
(iv) $x_4[n] = (-0.4)^n \mu[-n-2]$.
(a) Determine the ROCs of the z-transform of each of the above sequences.
(b) From the ROCs determined in Part (a), determine the ROCs of the following sequences:
(i) $y_1[n] = x_1[n] + x_2[n]$,   (ii) $y_2[n] = x_1[n] + x_3[n]$,   (iii) $y_3[n] = x_1[n] + x_4[n]$,
(iv) $y_4[n] = x_2[n] + x_3[n]$,   (v) $y_5[n] = x_2[n] + x_4[n]$,   (vi) $y_6[n] = x_3[n] + x_4[n]$.

6.6 Determine the z-transform of the two-sided sequence $v[n] = \alpha^{|n|}$, $|\alpha| < 1$. What is its ROC?

6.7 Determine the z-transform of each of the following sequences and their respective ROCs. Assume $|\beta| > |\alpha| > 0$. Show their pole-zero plots and indicate clearly the ROC in these plots.
(a) $x_1[n] = (\alpha^n + \beta^n)\mu[n+2]$,   (b) $x_2[n] = \alpha^n \mu[-n-2] + \beta^n \mu[n-1]$,
(c) $x_3[n] = \alpha^n \mu[n+1] + \beta^n \mu[-n-2]$.

6.8 Consider the z-transform

$$G(z) = \frac{(z^2 + 0.2z + 0.1)(z^2 - z + 0.5)}{(z^2 + 0.3z - 0.18)(z^2 - 2z + 4)}. \tag{6.108}$$

There are four possible nonoverlapping regions of convergence (ROCs) of this z-transform. Discuss the type of inverse z-transform (left-sided, right-sided, or two-sided sequences) associated with each of the four ROCs. It is not necessary to compute the exact inverse transform.

6.9 Let the z-transform of a sequence $x[n]$ be $X(z)$, with $\mathcal{R}_x$ denoting its ROC. Express the z-transforms of the real and imaginary parts of $x[n]$ in terms of $X(z)$. Show also their respective ROCs.

6.10 The z-transform $X(z)$ of the length-9 sequence of Problem 3.38 is sampled at seven points $\omega_k = 2\pi k/7$, $0 \le k \le 6$, on the unit circle yielding the frequency samples

$$\tilde{X}[k] = X(z)|_{z=e^{j2\pi k/7}}, \qquad 0 \le k \le 6.$$

Determine, without evaluating $\tilde{X}[k]$, the periodic sequence $\tilde{x}[n]$ whose discrete Fourier series coefficients are given by $\tilde{X}[k]$. What is the period of $\tilde{x}[n]$?

6.11 Repeat Problem 6.10 for the length-9 sequence of Problem 3.39.

6.12 Let $X(z)$ denote the z-transform of the length-12 sequence $x[n]$ of Problem 5.34. Let $X_0[k]$ represent the samples of $X(z)$ evaluated on the unit circle at nine equally spaced points given by $z = e^{j(2\pi k/9)}$, $0 \le k \le 8$, i.e.,

$$X_0[k] = X(z)|_{z=e^{j(2\pi k/9)}}, \qquad 0 \le k \le 8.$$

Determine the 9-point IDFT $x_0[n]$ of $X_0[k]$ without computing the latter function.

6.13 Consider the causal sequence $x[n] = (-0.5)^n \mu[n]$, with a z-transform given by $X(z)$.
(a) Determine the inverse z-transform of $X(z^3)$ without computing $X(z)$.
(b) Determine the inverse z-transform of $(1 + z^{-2})X(z^3)$ without computing $X(z)$.

6.14 Determine the z-transforms of the sequences of Problem 3.18 and their ROCs. Show that the ROC includes the unit circle for each z-transform. Evaluate the z-transform evaluated on the unit circle for each sequence and show that it is precisely the DTFT of the respective sequence computed in Problem 3.18.

6.15 Determine the z-transforms of the sequences of Problem 3.19 and their ROCs. Show that the ROC includes the unit circle for each z-transform. Evaluate the z-transform evaluated on the unit circle for each sequence and show that it is precisely the DTFT of the respective sequence computed in Problem 3.19.

6.16 Evaluate the linear convolutions of Problem 2.50 using the polynomial multiplication method.

6.17 Prove Eq. (6.73).

6.18 Evaluate the linear and circular convolutions of the sequences $g[n]$ and $h[n]$ of Problem 5.45 using the polynomial multiplication method. Verify your results in MATLAB using the functions `conv` and `circonv`.

circonv.m

6.19 Consider a rational z-transform $G(z) = P(z)/D(z)$, where $P(z)$ and $D(z)$ are polynomials in z^{-1}. Let ρ_ℓ denote the residue of $G(z)$ at a simple pole at $z = \lambda_\ell$. Show that

$$\rho_\ell = -\lambda_\ell \left. \frac{P(z)}{D'(z)} \right|_{z=\lambda_\ell},$$

where $D'(z) = \frac{dD(z)}{dz^{-1}}$.

6.20 Each one of following z-transforms

$$X_a(z) = \frac{3z}{z^2 + 0.3z - 0.18}, \qquad X_b(z) = \frac{3z^2 + 0.1z + 0.87}{(z+0.6)(z-0.3)^2}$$

has three ROCs. Evaluate their respective inverse z-transforms corresponding to each ROC.

6.21 Consider the z-transform $G(z)$ of Eq. (6.13), with $M < N$. If $G(z)$ has only simple poles, show that p_0/d_0 is equal to the sum of the residues in the partial-fraction expansion of $G(z)$ [Mit98].

6.22 Show that the inverse z-transform $h[n]$ of the following rational z-transform

$$H(z) = \frac{1}{1 - 2r(\cos\theta)z^{-1} + r^2 z^{-2}}, \qquad |z| > r > 0$$

is given by

$$h[n] = \frac{r^n \sin(n+1)\theta}{\sin\theta} \cdot \mu[n].$$

6.23 Determine the z-transform of each of the following left-sided sequences:
(a) $x[n] = \alpha^n \mu[-n-1]$, (b) $y[n] = (n+1)\alpha^n \mu[-n-1]$.

6.24 Determine the inverse z-transforms, $x_1[n]$ and $x_2[n]$, of the following rational z-transforms
(a) $X_1(z) = \dfrac{1}{1 - z^{-3}}$, $|z| > 1$, (b) $X_2(z) = \dfrac{1}{1 - z^{-4}}$, $|z| > 1$,
by expanding each in a power series and computing the inverse z-transform of the individual terms in the power series. Compare the results with that obtained using a partial-fraction approach.

6.25 Determine the inverse z-transforms of the following z-transforms:
(a) $X_1(z) = \log(1 - \alpha z^{-1})$, $\quad |z| > |\alpha|$, (b) $X_2(z) = \log\left(\frac{\alpha - z^{-1}}{\alpha}\right)$, $\quad |z| > 1/|\alpha|$,
(c) $X_3(z) = \log\left(\frac{1}{1 - \alpha z^{-1}}\right)$, $\quad |z| > |\alpha|$, (d) $X_4(z) = \log\left(\frac{\alpha}{\alpha - z^{-1}}\right)$, $\quad |z| > 1/|\alpha|$.

6.26 The z-transform of a right-sided sequence $h[n]$ is given by

$$H(z) = \frac{z + 1.7}{(z - 0.3)(z + 0.5)}.$$

Find its inverse z-transform $h[n]$ via the partial-fraction approach. Verify the partial fraction expansion using MATLAB.

6.27 Prove the following properties of the z-transform listed in Table 6.5: (a) conjugation, (b) time-reversal, (c) linearity, (d) time-shifting, (e) multiplication by an exponential sequence, and (f) differentiation.

6.28 A generalization of the DFT concept leads to the *nonuniform discrete Fourier transform* (NDFT) $X_{\text{NDFT}}[k]$, defined by [Bag98]

$$X_{\text{NDFT}}[k] = X(z_k) = \sum_{n=0}^{N-1} x[n] z_k^{-n}, \qquad 0 \le k \le N - 1, \tag{6.109}$$

where $z_0, z_1, \ldots, z_{N-1}$, are N distinct points located arbitrarily in the z-plane. The NDFT has been applied to the efficient design of digital filters, antenna array design, and dual-tone multifrequency detection [Bag98]. The NDFT can be expressed in a matrix form as

$$\begin{bmatrix} X_{\text{NDFT}}[0] \\ X_{\text{NDFT}}[1] \\ \vdots \\ X_{\text{NDFT}}[N-1] \end{bmatrix} = \mathbf{D}_N \begin{bmatrix} x[0] \\ x[1] \\ \vdots \\ x[N-1] \end{bmatrix}, \tag{6.110}$$

where

$$\mathbf{D}_N = \begin{bmatrix} 1 & z_0^{-1} & z_0^{-2} & \cdots & z_0^{-(N-1)} \\ 1 & z_1^{-1} & z_1^{-2} & \cdots & z_1^{-(N-1)} \\ 1 & z_2^{-1} & z_2^{-2} & \cdots & z_2^{-(N-1)} \\ \vdots & \vdots & \vdots & \ddots & \vdots \\ 1 & z_{N-1}^{-1} & z_{N-1}^{-2} & \cdots & z_{N-1}^{-(N-1)} \end{bmatrix} \tag{6.111}$$

is the $N \times N$ NDFT matrix. The matrix $\mathbf{D}_N$ is known as the *Vandermonde matrix*. Show that it is nonsingular provided the N sampling points z_k are distinct. In which case, the inverse NDFT is given by

$$\begin{bmatrix} x[0] \\ x[1] \\ \vdots \\ x[N-1] \end{bmatrix} = \mathbf{D}_N^{-1} \begin{bmatrix} X_{\text{NDFT}}[0] \\ X_{\text{NDFT}}[1] \\ \vdots \\ X_{\text{NDFT}}[N-1] \end{bmatrix}. \tag{6.112}$$

6.29 In general, for large N, the Vandermonde matrix is usually ill-conditioned (except for the case when the NDFT reduces to the conventional DFT), and a direct inverse computation is not advisable. A more efficient way is to directly determine the z-transform $X(z)$,

$$X(z) = \sum_{n=0}^{N-1} x[n] z^{-n}, \tag{6.113}$$

and hence, the sequence $x[n]$, from the given N-point NDFT $X_{\text{NDFT}}[k]$ by using some type of polynomial interpolation method [Bag98]. One popular method is the Lagrange interpolation formula, which expresses $X(z)$ as

$$X(z) = \sum_{k=0}^{N-1} \frac{I_k(z)}{I_k(z_k)} X_{\text{NDFT}}[k], \tag{6.114}$$

where

$$I_k(z) = \prod_{\substack{i=0 \\ i \neq k}}^{N-1} (1 - z_i z^{-1}). \tag{6.115}$$

Consider the z-transform $X(z) = 1 - 2z^{-1} + 3z^{-2} - 4z^{-3}$ of a length-4 sequence $x[n]$. By evaluating $X(z)$ at $z_0 = -1/2$, $z_1 = 1$, $z_2 = 1/2$, and $z_3 = 1/3$, determine the 4-point NDFT of $x[n]$ and then use the Lagrange interpolation method to show that $X(z)$ can be uniquely determined from these NDFT samples.

6.30 Consider a sequence $x[n]$ with a z-transform $X(z)$. Define a new z-transform $\hat{X}(z)$ given by the complex natural logarithm of $X(z)$; that is, $\hat{X}(z) = \ln X(z)$. The inverse z-transform of $\hat{X}(z)$ to be denoted by $\hat{x}[n]$ is called the *complex cepstrum* of $x[n]$ [Tri79]. Assume that the ROCs of both $X(z)$ and $\hat{X}(z)$ include the unit circle.
  (a) Relate the DTFT $X(e^{j\omega})$ of $x[n]$ to the DTFT $\hat{X}(e^{j\omega})$ of its complex cepstrum $\hat{x}[n]$.
  (b) Show that the complex cepstrum of a real sequence is a real-valued sequence.
  (c) Let $\hat{x}_{ev}[n]$ and $\hat{x}_{od}[n]$ denote, respectively, the even and odd parts of a real-valued complex cepstrum $\hat{x}[n]$. Express $\hat{x}_{ev}[n]$ and $\hat{x}_{od}[n]$ in terms of $X(e^{j\omega})$, the DTFT of $x[n]$.

6.31 Determine the complex cepstrum $\hat{x}[n]$ of a sequence $x[n] = a\delta[n] + b\delta[n-1]$, where $|b/a| < 1$. Comment on your results.

6.32 Let $x[n]$ be a sequence with a rational z-transform $X(z)$ given by

$$X(z) = K \frac{\prod_{k=1}^{N_\alpha}(1 - \alpha_k z^{-1}) \prod_{k=1}^{N_\gamma}(1 - \gamma_k z)}{\prod_{k=1}^{N_\beta}(1 - \beta_k z^{-1}) \prod_{k=1}^{N_\delta}(1 - \delta_k z)},$$

where α_k and β_k are the zeros and poles of $X(z)$ that are strictly inside the unit circle and $1/\gamma_k$ and $1/\delta_k$ are the zeros and poles of $X(z)$ that are strictly outside the unit circle [Rab78].
  (a) Determine the exact expression for the complex cepstrum $\hat{x}[n]$ of $x[n]$.
  (b) Show that $\hat{x}[n]$ is a decaying bounded sequence as $|n| \to \infty$.
  (c) If $\alpha_k = \beta_k = 0$, show that $\hat{x}[n]$ is an anticausal sequence.
  (d) If $\gamma_k = \delta_k = 0$, show that $\hat{x}[n]$ is a causal sequence.

6.33 Let $x[n]$ be a sequence with a rational z-transform $X(z)$ with poles and zeros strictly inside the unit circle. Show that the complex cepstrum $\hat{x}[n]$ of $x[n]$ can be computed using the recursion relation [Rab78]:

$$\hat{x}[n] = \begin{cases} 0, & n < 0, \\ \log(x[0]), & n = 0, \\ \frac{x[n]}{x[0]} - \sum_{k=0}^{n-1} \frac{k}{n} \cdot \hat{x}[k] \cdot \frac{x[n-k]}{x[0]}, & n > 0. \end{cases}$$

6.34 The magnitude response of a digital filter with a real-coefficient transfer function $H(z)$ is shown in Figure P6.1. Plot the magnitude response of the filter $H(z^5)$.

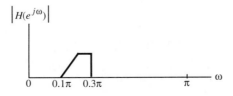

Figure P6.1

6.35 In this problem, we consider the determination of a real rational, causal, stable discrete-time transfer function

$$H(z) = \frac{P(z)}{D(z)} = \frac{\sum_{i=0}^{N} p_i z^{-i}}{\sum_{i=0}^{N} d_i z^{-i}}$$

from the specified real part of its frequency response [Dut83]:

$$H_{re}(e^{j\omega}) = \frac{\sum_{i=0}^{N} a_i \cos(i\omega)}{\sum_{i=0}^{N} b_i \cos(i\omega)} = \frac{A(e^{j\omega})}{B(e^{j\omega})}.$$ (6.116)

(a) Show that

$$H_{re}(e^{j\omega}) = \frac{1}{2}\left[H(z) + H(z^{-1})\right]\Big|_{z=e^{j\omega}} = \frac{1}{2}\left[\frac{P(z)D(z^{-1}) + P(z^{-1})D(z)}{D(z)D(z^{-1})}\right]\Big|_{z=e^{j\omega}}.$$ (6.117)

(b) Comparing Eqs. (6.116) and (6.117), we get

$$B(e^{j\omega}) = D(z)D(z^{-1})\Big|_{z=e^{j\omega}}$$

$$A(e^{j\omega}) = \frac{1}{2}\left[P(z)D(z^1) + P(z^{-1})D(z)\right]\Big|_{z=e^{j\omega}}.$$ (6.118)

The spectral factor $D(z)$ can be determined, except for the scale factor K, from the roots of $B(z) = B(e^{j\omega})\big|_{z=e^{j\omega}}$ inside the unit circle. Show that

$$K = \sqrt{B(1)}\Big/ \prod_{i=1}^{N}(1 - z_i).$$

(c) To determine $P(z)$, Eq. (6.118) can be rewritten through analytic continuation as

$$A(z) = \frac{1}{2}\left[P(z)D(z^{-1}) + P(z^{-1})D(z)\right].$$

Substituting the polynomial forms of $P(z)$ and $D(z)$ and equating coefficients of $(z^i + z^{-i})/2$ on both sides of the above equation, we arrive at a set of $N + 1$ equations that can be solved for the numerator coefficients $\{p_i\}$. Using the above approach, determine $H(z)$ for which

$$H_{re}(e^{j\omega}) = \frac{1 + \cos\omega + \cos 2\omega}{17 - 8\cos 2\omega}.$$

6.36 Let $H(z)$ be the transfer function of a causal stable LTI discrete-time system. Let $G(z)$ be the transfer function obtained by replacing z^{-1} in $H(z)$ with a stable allpass function $A(z)$; that is, $G(z) = H(z)|_{z^{-1}=A(z)}$. Show that $G(1) = H(1)$ and $G(-1) = H(-1)$.

6.37 Consider the digital filter structure of Figure P6.2, where

$$H_1(z) = 1.2 + 3.3z^{-1} + 0.7z^{-2}, \quad H_2(z) = -4.1 - 2.5z^{-1} + 0.9z^{-2}, \quad H_3(z) = 2.3 + 4.3z^{-1} + 0.8z^{-2}.$$

Determine the transfer function $H(z)$ of the composite filter.

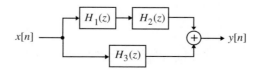

Figure P6.2

6.38 Determine the transfer function of each of the following causal LTI discrete-time systems described by the difference equations. Express each transfer function in a factored form and sketch its pole-zero plot. Is the corresponding system BIBO stable?

(a) $y[n] = 5x[n] + 9.5x[n-1] + 1.4x[n-2] - 24x[n-3] + 0.1y[n-1] - 0.14y[n-2] - 0.49y[n-3]$.

(b) $y[n] = 5x[n] + 16.5x[n-1] + 14.7x[n-2] - 22.04x[n-3] - 33.6x[n-4] + 0.5y[n-1] - 0.1y[n-2] -0.3y[n-3] + 0.0936y[n-4]$.

6.39 Determine the expression for the impulse response $\{h[n]\}$ of the following causal IIR transfer function:

$$H(z) = \frac{4.5 - 1.3z^{-1} + 1.12z^{-2}}{(1 + 0.5z^{-1} + 0.3z^{-2})(1 - 0.4z^{-1})}.$$

6.40 The transfer function of a causal LTI discrete-time system is given by

$$H(z) = \frac{1 - 3.3z^{-1} + 0.36z^{-2}}{1 + 0.3z^{-1} - 0.18z^{-2}}.$$

(a) Determine the impulse response $h[n]$ of the above system.

(b) Determine the output $y[n]$ of the above system for all values of n for an input

$$x[n] = 2.1(0.4)^n \mu[n] + 0.3(-0.3)^n \mu[n].$$

6.41 Using z-transform methods, determine the explicit expression for the output $y[n]$ of each of the following causal LTI discrete-time systems, with impulse responses and inputs as indicated:

(a) $h[n] = (-0.4)^n \mu[n]$, $x[n] = (0.2)^n \mu[n]$, (b) $h[n] = (-0.2)^n \mu[n]$, $x[n] = (-0.2)^n \mu[n]$.

6.42 Using z-transform methods, determine the explicit expression for the impulse response $h[n]$ of a causal LTI discrete-time system that develops an output $y[n] = 2(-0.3)^n \mu[n]$ for an input $x[n] = 4(0.6)^n \mu[n]$.

6.43 A causal LTI discrete-time system is described by the difference equation

$$y[n] = 0.2y[n-1] + 0.08y[n-2] + 2x[n],$$

where $x[n]$ and $y[n]$ are, respectively, the input and the output sequences of the system.

(a) Determine the transfer function $H(z)$ of the system.

(b) Determine the impulse response $h[n]$ of the system.

(c) Determine the step response $s[n]$ of the system.

6.44 Determine the frequency response $H(e^{j\omega})$ of the transfer function

$$H(z) = \frac{1 - z^{-2}}{1 - (1 + \alpha)\cos(\omega_c)z^{-1} + \alpha z^{-2}}.$$

Show that the magnitude response $|H(e^{j\omega})|$ assumes its maximum value of $2/(1-\alpha)$ at $\omega = \omega_c$.

6.45 Determine a closed-form expression for the frequency response $H(e^{j\omega})$ of the LTI discrete-time system characterized by an impulse response

$$h[n] = \delta[n] - \alpha\delta[n - R], \tag{6.119}$$

where $|\alpha| < 1$. What are the maximum and the minimum values of its magnitude response? How many peaks and dips of the magnitude response occur in the range $0 \le \omega < 2\pi$? What are the locations of the peaks and the dips? Sketch the magnitude and the phase responses for $R = 6$.

6.46 Determine a closed-form expression for the frequency response $G(e^{j\omega})$ of the LTI discrete-time system characterized by an impulse response

$$g[n] = h[n] \circledast h[n] \circledast h[n], \tag{6.120}$$

where $h[n]$ is given by Eq. (6.119).

6.47 Determine a closed-form expression for the frequency response $G(e^{j\omega})$ of an LTI discrete-time system with an impulse response given by

$$g[n] = \begin{cases} \alpha^n, & 0 \le n \le M - 1, \\ 0 & \text{otherwise}, \end{cases}$$

where $|\alpha| < 1$. What is the relation of $G(e^{j\omega})$ to $H(e^{j\omega})$ of Eq. (3.98)? Scale the impulse response by multiplying it with a suitable constant so that the dc value of the magnitude response is unity.

6.48 Determine the expression for the frequency response $H(e^{j\omega})$ of a causal IIR LTI discrete-time system characterized by the input–output relation

$$y[n] = x[n] + \alpha y[n - R], \qquad |\alpha| < 1,$$

where $y[n]$ and $x[n]$ denote, respectively, the output and the input sequences. Determine the maximum and the minimum values of its magnitude response. How many peaks and dips of the magnitude response occur in the range $0 \le \omega < 2\pi$? What are the locations of the peaks and the dips? Sketch the magnitude and the phase responses for $R = 6$.

6.49 An IIR LTI discrete-time system is described by the difference equation

$$y[n] + a_1 y[n - 1] + a_2 y[n - 2] = b_0 x[n] + b_1 x[n - 1] + b_2 x[n - 2],$$

where $y[n]$ and $x[n]$ denote, respectively, the output and the input sequences. Determine the expression for its frequency response. For what values of the constants b_i will the magnitude response be a constant for all values of ω?

6.50 Determine the input–output relation of a factor-of-2 up-sampler in the frequency domain. The time-domain input–output relation of the factor-of-L up-sampler is given by Eq. (2.20).

6.51 Consider an LTI discrete-time system with an impulse response $h[n] = (0.5)^n \mu[n]$. Determine the frequency response $H(e^{j\omega})$ of the system and evaluate its value at $\omega = \pm\pi/5$. What is the steady-state output $y[n]$ of the system for an input $x[n] = \cos(\pi n/5)\mu[n]$?

6.52 Let $H(z)$ be the transfer function of a causal, stable LTI discrete-time system. Consider the transfer function

$$G(z) = H(z)|_{z=F(z)}.$$

What conditions need to be satisfied by $F(z)$ so that $G(z)$ remain stable?

6.53 Determine the z-transform $F(z)$ of the Fibonacci sequence $\{f[n]\}$ of Problem 2.70. Evaluate the inverse z-transform $f[n]$ of $F(z)$.

6.54 The *time constant* K of an LTI stable causal discrete-time system with an impulse response $h[n]$ is given by the value of the total time interval n at which the partial energy of the impulse response is within 95% of the total energy; that is,

$$\sum_{n=0}^{K} |h[n]|^2 = 0.95 \sum_{n=0}^{\infty} |h[n]|^2.$$

Determine the time constant K of the first-order causal transfer function $H(z) = 1/(1 - \beta z^{-1})$, $|\beta| < 1$.

6.55 Figures P6.3(a) and P6.3(b) show, respectively, the DPCM (*differential pulse-code modulation*) coder and decoder often employed for the compression of digital signals [Jay84]. The linear predictor $P(z)$ in the encoder develops a prediction $\hat{x}[n]$ of the input signal $x[n]$, and the difference signal $d[n] = x[n] - \hat{x}[n]$ is quantized by the quantizer Q developing the quantized output $u[n]$, which is represented with fewer bits than that of $x[n]$. The output of the encoder is transmitted over a channel to the decoder. In the absence of any errors due to transmission and quantization, the input $v[n]$ to the decoder is equal to $u[n]$, and the decoder generates the output $y[n]$, which is equal to the input $x[n]$. Determine the transfer function $H(z) = U(z)/X(z)$ of the encoder in the absence of any quantization and the transfer function $G(z) = Y(z)/V(z)$ of the decoder for the case of each of the following predictors, and show that $G(z)$ is the inverse of $H(z)$ in each case.

(a) $P(z) = h_1 z^{-1}$, and (b) $P(z) = h_1 z^{-1} + h_2 z^{-2}$.

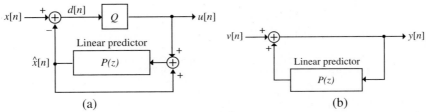

(a)   (b)

Figure P6.3

6.56 Consider the discrete-time system of Figure P6.4. For $H_0(z) = 1 + \alpha z^{-1}$, find a suitable $F_0(z)$ so that the output $y[n]$ is a delayed and scaled replica of the input.

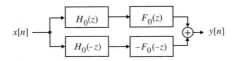

Figure P6.4

6.57 A causal stable LTI discrete-time system is characterized by an impulse response $h_1[n] = 1.2\delta[n] + 0.5(-0.5)^n \mu[n] + -0.6(0.2)^n \mu[n]$. Determine the impulse response $h_2[n]$ of its inverse system, which is causal and stable.

6.58 Show that the group delay $\tau_g(\omega)$ of an LTI transfer function $H(z)$ can be expressed as [Fot2001]

$$\tau_g(\omega) = \left. \frac{T(z) + T(z^{-1})}{2} \right|_{z=e^{j\omega}}, \qquad (6.121)$$

where $T(z) = z\dfrac{dH(z)/dz}{H(z)}$.

6.10 MATLAB Exercises

M 6.1 Using Program 6_1, determine the factored form of the following z-transforms:

(a) $G_1(z) = \dfrac{2z^4 - 5z^3 + 13.48z^2 - 7.78z + 9}{4z^4 + 7.2z^3 + 20z^2 - 0.8z + 8}$,

Program 6_1.m

(b) $G_2(z) = \dfrac{5z^4 + 3.5z^3 + 21.5z^2 - 4.6z + 18}{5z^4 + 15.5z^3 + 31.7z^2 + 22.52z + 4.8}$,

and show their pole-zero plots. Determine all possible ROCs of each of the above z-transforms, and describe the type of their inverse z-transforms (left-sided, right-sided, two-sided sequences) associated with each of the ROCs.

Program 6_3.m

M 6.2 Using Program 6_3, determine the partial-fraction expansions of the z-transforms listed in Problem 6.20, and then determine their inverse z-transforms.

Program 6_4.m

M 6.3 Using Program 6_4, determine the z-transform as a ratio of two polynomials in z^{-1} from each of the partial-fraction expansions listed below:

(a) $X_1(z) = 3 - \dfrac{4}{5 + z^{-1}} - \dfrac{7}{6 - z^{-1}}$, $|z| > 0.2$,

(b) $X_2(z) = -2.5 + \dfrac{3}{1 + 0.4z^{-1}} - \dfrac{1.4 + z^{-1}}{1 + 0.6z^{-2}}$, $|z| > 0.25$,

(c) $X_3(z) = \dfrac{-4}{(4 + 2z^{-1})^2} + \dfrac{6}{4 + 2z^{-1}} + \dfrac{5}{1 + 0.64z^{-2}}$, $|z| > 0.8$,

(d) $X_4(z) = -5 + \dfrac{2}{4 + 3z^{-1}} + \dfrac{z^{-1}}{4 + 3z^{-1} + 0.9z^{-2}}$, $|z| > 0.75$.

Program 6_5

M 6.4 Using Program 6_5, determine the first 30 samples of the inverse z-transforms of the rational z-transforms determined in Problem M6.3. Show that these samples are identical to those obtained by explicitly evaluating the exact inverse z-transforms.

M 6.5 Repeat Problem 6.57 using MATLAB.

M 6.6 Write a MATLAB program to compute the NDFT and the inverse NDFT using the Lagrange interpolation method. Verify your program by computing the NDFT of a length-20 sequence and reconstructing the sequence from its computed NDFT.

7 LTI Discrete-Time Systems in the Transform Domain

The time-domain classification of a digital transfer function based on the length of its impulse response sequence leads to the *finite impulse response* (FIR) and the *infinite impulse response* (IIR) transfer functions. We describe here several other types of classifications based on the behavior of the magnitude and phase responses of the transfer function.

There are several types of classifications of transfer functions based on their magnitude characteristics. In the case of digital transfer functions with frequency-selective frequency responses, four types of ideal filters are usually defined based on the shape of the magnitude function $|H(e^{j\omega})|$. These ideal filters have doubly infinite impulse responses and are unrealizable as they are not causal. In practice, realizable approximations to the ideal filters can be made. Another classification is based on a unity maximum value of the magnitude function.

Likewise, there are several types of classifications of transfer functions based on their phase characteristics. An important class is based on the linearity of the phase function. As there are usually several transfer functions with the same magnitude function, these are classified according to their relative phase responses.

We next describe very simple realizable FIR and IIR digital filter approximations. In a number of applications, these simple filters are quite adequate and provide satisfactory performances. The importance of transfer functions with linear phase is pointed out and the possible realizations of these transfer functions with FIR filters are discussed. Transfer functions with complementary characteristics are then considered.

The chapter introduces digital two-pairs, which are two-input, two-output structures that are characterized by a set of four transfer functions. A number of filter realizations discussed in Chapter 8 are based on the extraction of two-pairs.

Finally, the chapter concludes with the development of an algebraic stability test for causal IIR transfer functions.

7.1 Transfer Function Classification Based on Magnitude Characteristics

Transfer functions of LTI discrete-time systems are usually classified according to the characteristics of their phase responses or their magnitude responses. In this section, we consider several types of transfer functions that are classified according to their magnitude response characteristics. Later in this chapter, we consider transfer functions that are classified according to their phase response characteristics.

7.1.1 Digital Filters with Ideal Magnitude Responses

One commonly used classification of transfer functions is based on an ideal magnitude response. Even though such transfer functions are not realizable, they can be approximated in practice with some accept-

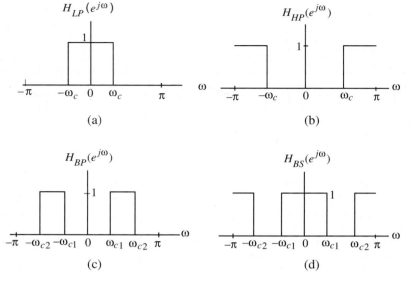

Figure 7.1: Four types of ideal filters: (a) ideal lowpass filter, (b) ideal highpass filter, (c) ideal bandpass filter, and (d) ideal bandstop filter.

able tolerances. As pointed out in Example 3.20, a digital filter designed to pass signal components of certain frequencies without any distortion should have a frequency response of value equal to one at these frequencies and should have a frequency response of value equal to zero at all other frequencies to totally block signal components with those frequencies. The range of frequencies where the frequency response takes the value of one is called the *passband,* and the range of frequencies where the frequency response is equal to zero is called the *stopband* of the filter.

The frequency responses of the four popular types of ideal digital filters with real impulse response coefficients are shown in Figure 7.1. For the *lowpass filter* of Figure 7.1(a), the passband and the stopband are given by $0 \leq \omega \leq \omega_c$ and $\omega_c < \omega \leq \pi$, respectively. For the *highpass filter* of Figure 7.1(b), the stopband is given by $0 \leq \omega < \omega_c$, while the passband is given by $\omega_c \leq \omega \leq \pi$. The passband region of the *bandpass filter* of Figure 7.1(c) is $\omega_{c1} \leq \omega \leq \omega_{c2}$, and the stopband regions are given by $0 \leq \omega < \omega_{c1}$ and $\omega_{c2} < \omega < \pi$. Finally, for the *bandstop filter* of Figure 7.1(d), the passband regions are $0 \leq \omega \leq \omega_{c1}$ and $\omega_{c2} \leq \omega \leq \pi$, while the stopband is from $\omega_{c1} < \omega < \omega_{c2}$. The frequencies ω_c, ω_{c1}, and ω_{c2} are called the *cutoff frequencies* of their respective filters. Note from this figure that an ideal filter thus has a magnitude response equal to unity in the passband and zero in the stopband and has a zero phase everywhere.

We have already encountered the frequency response $H_{LP}(e^{j\omega})$ of the ideal lowpass filter of Figure 7.1(a) in Example 3.8 where we computed its impulse response given in Eq. (3.49). We repeat it here for convenience:

$$h_{LP}[n] = \frac{\sin \omega_c n}{\pi n}, \qquad -\infty < n < \infty. \tag{7.1}$$

In this example, we have shown that the above impulse response is not absolutely summable, and hence, the corresponding transfer function is not BIBO stable. Note also that the above impulse response is not causal and is of doubly infinite length. The remaining three frequency responses of Figure 7.1 also are characterized by doubly infinite, noncausal impulse responses and are not absolutely summable. As a result, the ideal filters with the ideal "brick wall" characteristics of Figure 7.1 cannot be realized by an LTI filter with a transfer function of finite order.

In order to develop stable and realizable transfer functions, the ideal frequency response specifications of Figure 7.1 are relaxed by including a *transition band* between the passband and the stopband to permit the magnitude response to decay gradually from its maximum value in the passband to the zero value in the stopband. Moreover, the magnitude response is allowed to vary by a specified amount both in the passband and the stopband. Typical magnitude specifications used for the design of a lowpass filter are shown in Figure 9.1. Chapters 9 and 10 are devoted to a discussion of various filter design methods that lead to stable and realizable transfer functions meeting such relaxed specifications. In Section 7.4, we describe several very simple low-order FIR and IIR digital filters that exhibit selective frequency response characteristics providing a first-order approximation to the ideal characteristics of Figure 7.1. Frequency responses with sharper characteristics can often be obtained by cascading one or more of these simple filters, which, in many applications, is quite satisfactory.

7.1.2 Bounded Real Transfer Functions

A causal stable real-coefficient transfer function $H(z)$ is defined as a *bounded real* (BR) *transfer function* [Vai84] if

$$|H(e^{j\omega})| \leq 1, \qquad \text{for all values of } \omega . \tag{7.2}$$

Note that any stable real-coefficient transfer function can be made into a BR function by appropriate scaling.

EXAMPLE 7.1   Construction of a Bounded Real Function

Consider the causal stable transfer function

$$H(z) = \frac{K}{1 - \alpha z^{-1}}, \qquad 0 < |\alpha| < 1, \tag{7.3}$$

where K is a real constant. Its square-magnitude function is given by

$$|H(e^{j\omega})|^2 = H(z)H(z^{-1})\Big|_{z=e^{j\omega}} = \frac{K^2}{(1+\alpha^2) - 2\alpha \cos \omega}. \tag{7.4}$$

The maximum value of $|H(e^{j\omega})|^2$ is obtained when $2\alpha \cos \omega$ in the denominator of the above equation is a maximum, and the minimum value is obtained when $2\alpha \cos \omega$ is a minimum. For $\alpha > 0$, the maximum value of $2\alpha \cos \omega$ is equal to 2α at $\omega = 0$, and the minimum value is equal to -2α at $\omega = \pi$. As a result, the maximum value of $|H(e^{j\omega})|^2$ is equal to $K^2/(1 - \alpha)^2$ at $\omega = 0$, and the minimum value is equal to $K^2/(1 + \alpha)^2$ at $\omega = \pi$. On the other hand, for $\alpha < 0$, the maximum value of $2\alpha \cos \omega$ is equal to -2α at $\omega = \pi$, and the minimum value is equal to 2α at $\omega = 0$. As a result in this case, the maximum value of $|H(e^{j\omega})|^2$ is equal to $K^2/(1 - \alpha)^2$ at $\omega = \pi$, and the minimum value is equal to $K^2/(1 + \alpha)^2$ at $\omega = 0$.

The maximum value can be made equal to 1 by choosing $K = \pm(1 - \alpha)$, in which case the minimum value becomes $(1 - \alpha)^2/(1 + \alpha)^2$. Hence, the transfer function $H(z)$ of Eq. (7.3) is a BR function for $K \leq \pm(1 - \alpha)$. Plots of the magnitude functions for $\alpha = 0.5$ and $\alpha = -0.5$ with the values of K chosen to yield a maximum value of 1 are shown in Figure 7.2.

It can be seen from Figure 3.13 that the transfer function of a moving average lowpass filter of Eq. (3.97) is a BR function for $M = 5$ and $M = 14$. In fact, it can be shown that the BR property of the transfer function of a moving-average filter holds for all values of M (Problem 7.1).

If the input and output of a digital filter characterized by a BR transfer function $H(z)$ are given by $x[n]$ and $y[n]$, respectively, with $X(e^{j\omega})$ and $Y(e^{j\omega})$ denoting their respective discrete-time Fourier transforms, then Eq. (7.2) implies that

$$|Y(e^{j\omega})|^2 \leq |X(e^{j\omega})|^2. \tag{7.5}$$

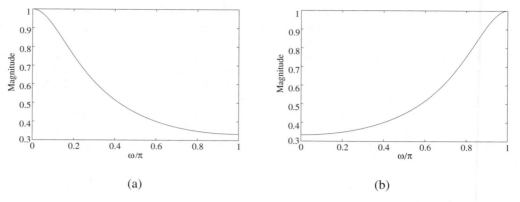

(a)   (b)

Figure 7.2: Magnitude responses of the transfer function of Eq. (7.3) for (a) $\alpha = 0.5$ and (b) $\alpha = -0.5$.

Integrating Eq. (7.5) from $-\pi$ to π and applying Parseval's relation (see Table 3.4), we arrive at

$$\sum_{n=-\infty}^{\infty} y^2[n] \le \sum_{n=-\infty}^{\infty} x^2[n]. \tag{7.6}$$

In other words, for all finite-energy inputs, the output energy is less than or equal to the input energy, implying that a digital filter characterized by a BR transfer function can be viewed as a *passive* structure.

If Eq. (7.2) is satisfied with an equal sign, then from Eq. (7.6), the output energy is equal to the input energy, and such a digital filter is therefore a *lossless* system. A causal stable real-coefficient transfer function $H(z)$ with a frequency response $H(e^{j\omega})$ of unity magnitude is thus called a *lossless bounded real* (LBR) *transfer function* [Vai84].

The BR and LBR transfer functions are the keys to the realization of digital filters with low coefficient sensitivity (see Section 12.9).

7.1.3   Allpass Transfer Function

A very special type of IIR transfer function is characterized by unity magnitude for all frequencies. Such a transfer function, called an *allpass transfer function,* has many useful applications in digital signal processing [Reg88]. We define the allpass transfer function, examine some of its key properties, and outline one of its common applications. Later in this chapter and elsewhere in the text, we discuss various other applications. One important application considered later in this chapter is the development of an algebraic test for the BIBO stability of a causal IIR transfer function.

Definition

An IIR transfer function $A(z)$ with unity magnitude response for all frequencies, that is,

$$|A(e^{j\omega})|^2 = 1, \qquad \text{for all } \omega \tag{7.7}$$

is called an *allpass transfer function.* Now, an Mth-order causal real-coefficient allpass transfer function is of the form

$$A_M(z) = \pm \frac{d_M + d_{M-1}z^{-1} + \cdots + d_1 z^{-M+1} + z^{-M}}{1 + d_1 z^{-1} + \cdots + d_{M-1}z^{-M+1} + d_M z^{-M}}. \tag{7.8}$$

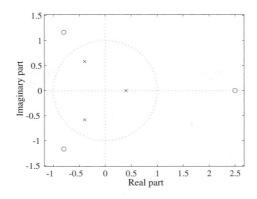

Figure 7.3: Pole-zero plot of the real coefficient allpass transfer function of Eq. (7.11).

If we denote the denominator polynomial of the allpass function $A_M(z)$ as $D_M(z)$,

$$D_M(z) = 1 + d_1 z^{-1} + \cdots + d_{M-1} z^{-M+1} + d_M z^{-M}, \tag{7.9}$$

then it follows that $A_M(z)$ can be written as

$$A_M(z) = \pm \frac{z^{-M} D_M(z^{-1})}{D_M(z)}. \tag{7.10}$$

Note from above that if $z = re^{j\phi}$ is a pole of a real-coefficient allpass transfer function, then it has a zero at $z = \frac{1}{r} e^{-j\phi}$. The numerator of an allpass transfer function is said to be the *mirror-image polynomial* of the denominator, and vice versa. We shall use the notation $\tilde{D}_M(z)$ to denote the mirror-image polynomial of a degree-M polynomial; that is, $\tilde{D}_M(z) = z^{-M} D_M(z^{-1})$. Equation (7.10) implies that the poles and the zeros of a real-coefficient allpass function exhibit *mirror-image symmetry* in the z-plane, as shown in Figure 7.3 for the following third-order allpass function:

$$A_3(z) = \frac{-0.2 + 0.18z^{-1} + 0.4z^{-2} + z^{-3}}{1 + 0.4z^{-1} + 0.18z^{-2} - 0.2z^{-3}}. \tag{7.11}$$

An M-th order causal real coefficient allpass transfer function can also be expressed in a factored form as

$$A_M(z) = \pm \prod_{i=1}^{M} \left(\frac{-\lambda_i^* + z^{-1}}{1 - \lambda_i z^{-1}} \right), \tag{7.12}$$

where λ_i is the pole of $A_M(z)$. As shown earlier in Section 6.7.5, the poles of a causal stable transfer function must lie inside the unit circle. Hence, for stability of $A_M(z)$, $|\lambda_i| < 1$, $1 \le i \le M$. This implies then that all zeros of a causal stable allpass transfer function lie outside the unit circle, while its poles are situated inside the unit circle with a mirror-image symmetry with respect to the zeros.

To show that the magnitude of $A_M(e^{j\omega})$ is indeed equal to one for all ω, it follows from Eq. (7.10) that

$$A_M(z^{-1}) = \pm \frac{z^M D_M(z)}{D_M(z^{-1})}. $$

Therefore,

$$A_M(z) A_M(z^{-1}) = \frac{z^{-M} D_M(z^{-1})}{D_M(z)} \frac{z^M D_M(z)}{D_M(z^{-1})} = 1. $$

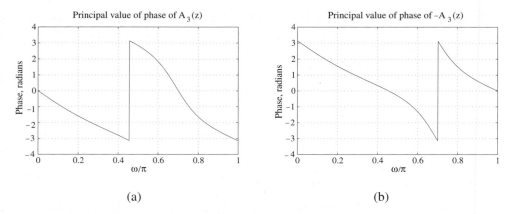

Figure 7.4: The principal value of the phase function: (a) $A_3(z)$, and (b) $-A_3(z)$.

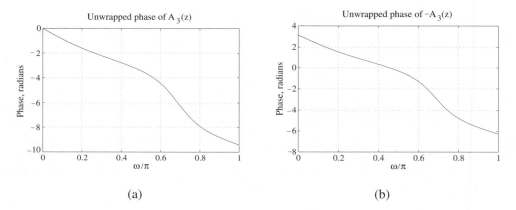

Figure 7.5: The unwrapped phase function of (a) $A_3(z)$ and (b) $-A_3(z)$.

Hence,

$$|A_M(e^{j\omega})|^2 = A_M(z)A_M(z^{-1})|_{z=e^{j\omega}} = 1. \tag{7.13}$$

Properties

It is interesting to examine the behavior of the phase function of an allpass transfer function. Figure 7.4 shows the principal value of the phase functions of the allpass transfer function $A_3(z)$ of Eq. (7.11) and the negative of this allpass function. Note that the phase of $-A_3(z)$ is obtained by adding π radians to the phase of $A_3(z)$ caused by the minus sign. Both allpass functions have a discontinuity by the amount of 2π in their respective phase functions $\theta(\omega)$. If we unwrap the phase by removing the discontinuity, we arrive at the unwrapped phase functions $\theta_c(\omega)$ indicated in Figure 7.5. As can be seen from this figure, the unwrapped phase function in both cases is a monotonically decreasing function of ω in the range $0 \leq \omega \leq \pi$. To remove the ambiguity caused by the minus sign, we restrict our discussion here to the allpass function of the form of Eq. (7.8) with a positive sign, in which case the unwrapped phase function is seen to be a nonpositive function of ω.

The monotonically decreasing behavior with respect to the angular frequency of the unwrapped phase function holds for any arbitrary causal stable allpass function. To show this, consider a first-order causal stable real coefficient allpass transfer function

$$A(z) = \frac{-\lambda^* + z^{-1}}{1 - \lambda z^{-1}}.$$ (7.14)

Its frequency response is then given by

$$A(e^{j\omega}) = \frac{-\lambda^* + e^{-j\omega}}{1 - \lambda e^{-j\omega}},$$ (7.15)

which can be rewritten as

$$A(e^{j\omega}) = \frac{-re^{-j\phi} + e^{-j\omega}}{1 - re^{j\phi}e^{-j\omega}} = e^{-j\omega} \left[\frac{1 - re^{j(\omega - \phi)}}{1 - re^{-j(\omega - \phi)}} \right]$$

by setting $\lambda = re^{j\phi}$. From Eq. (7.15), we arrive at the expression for the phase function as

$$\theta_c(\omega) = -\omega - 2 \tan^{-1} \left[\frac{r \sin(\omega - \phi)}{1 - r \cos(\omega - \phi)} \right].$$ (7.16)

It can be shown after some algebra that

$$\frac{d\theta_c(\omega)}{d\omega} = -\frac{(1 - r^2)}{|1 + re^{-j(\omega - \phi)}|^2},$$ (7.17)

and hence, the group delay of the first-order allpass function of Eq. (7.14) is given by

$$\tau_g(\omega) = -\frac{d\theta_c(\omega)}{d\omega} = \frac{(1 - r^2)}{|1 + re^{-j(\omega - \phi)}|^2}.$$ (7.18)

Since for stability, $|r| < 1$, it follows from the above that

$$\frac{d\theta_c(\omega)}{d\omega} < 0, \qquad 0 \le \omega \le \pi,$$

implying that the phase function $\theta_c(\omega)$ is a monotonically decreasing function of ω. Since, a higher-order real-coefficient causal stable allpass transfer function can be expressed as a product of first-order causal stable allpass sections, the (unwrapped) phase function of the overall allpass section is a sum of the phase functions of individual sections and is, therefore, a monotonically decreasing function of ω. Correspondingly, the group delay function of a real-coefficient causal stable allpass transfer function is always positive. Figure 7.6 shows the group delay of the allpass transfer function $A_3(z)$ of Eq. (7.11).

We next show that the (unwrapped) phase function of the first-order allpass function of Eq. (7.14) is a nonpositive function. To this end, we integrate the group delay function of Eq. (7.18) to arrive at

$$\theta_c(\omega) = -\int_0^\omega \tau_g(\varphi) d\varphi + \theta_c(0).$$ (7.19)

From Eq. (7.14), $A(e^{j0}) = 1$, and hence, $\theta_c(0) = 0$. Since $\tau_g(\omega) > 0$ for all values of ω, it follows then from Eq. (7.19) that $\theta_c(\omega) < 0$ for all values of ω. As a result, the (unwrapped) phase function of the Mth

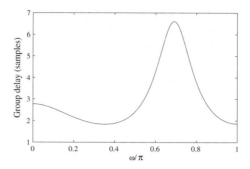

Figure 7.6: The group delay function of the allpass transfer function of Eq. (7.11).

order stable real-coefficient allpass function of Eq. (7.12) with a plus sign in front is a negative function of ω.

We now state three very useful and important properties of a causal stable allpass function without proof [Reg88].

Property 1. It follows from our discussion in Section 7.1.2 that a causal stable real coefficient allpass transfer function is a lossless bounded real (LBR) transfer function or, equivalently, a causal stable allpass filter is a lossless structure.

Property 2. The second property is concerned with the magnitude of a stable allpass function $A_M(z)$. It can be shown very simply that (Problem 7.2)

$$|A_M(z)| \begin{cases} < 1 & \text{for } |z| > 1, \\ = 1 & \text{for } |z| = 1, \\ > 1 & \text{for } |z| < 1. \end{cases} \qquad (7.20)$$

Property 3. The last property of interest is with regard to the change in phase for a real stable allpass function over the frequency range $\omega = 0$ to $\omega = \pi$. We have demonstrated earlier that the group delay function $\tau_g(\omega)$ of a causal stable real-coefficient allpass function $A_M(z)$ is everywhere positive in the range $0 \leq \omega \leq \pi$. It can be shown that an Mth-order stable real allpass transfer function satisfies the property (Problem 7.3)

$$\int_0^\pi \tau_g(\omega) \, d\omega = M\pi. \qquad (7.21)$$

In other words, the change in the phase of an Mth-order allpass function is $M\pi$ radians as ω goes from 0 to π.

A Simple Application

A simple but often used application of an allpass filter is as a *delay equalizer*. Let $G(z)$ be the transfer function of a digital filter that has been designed to meet a prescribed magnitude response. The nonlinear phase response of this filter can be corrected by cascading it with an allpass filter section $A(z)$ so that the overall cascade with a transfer function $G(z)A(z)$ has a constant group delay over the frequency domain of interest (see Figure 7.7). Since the allpass filter has a unity magnitude response, the magnitude response of the cascade is still equal to $|G(e^{j\omega})|$, while the overall delay is given by the sum of the group delays of $G(z)$ and $A(z)$. The allpass is designed so that the overall group delay is approximately a constant in the frequency region of interest.

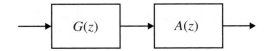

Figure 7.7: Use of an allpass filter as a delay equalizer.

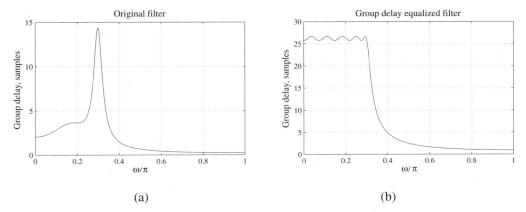

Figure 7.8: (a) The group delay of the elliptic filter and (b) the group delay of the cascade of the original filter and the allpass section.

Figure 7.8(a) shows the group delay of a 4th order elliptic filter with a passband edge at 0.3π, a passband ripple of 1 dB, and a minimum stopband attenuation of 35 dB. Figure 7.8(b) shows the group delay of the original filter cascaded with an 8th order allpass section designed to equalize the group delay in the passband. The allpass filter has been designed using the M-file `iirgrpdelay` (See Example 9.17).

Various other applications of the allpass filter are described in the latter parts of this book.

7.2 Transfer Function Classification Based on Phase Characteristics

We now consider classification of transfer functions based on the characteristics of their phase responses. Three such classes are discussed below.

7.2.1 Zero-Phase Transfer Function

In many applications, it is necessary to ensure that the designed digital filter does not distort the phase of the input signal components for frequencies in the passband. One way to avoid any phase distortions is to make the frequency response of the filter real and nonnegative, that is, to design the filter with a *zero phase* characteristic. Let $H(z)$ be a real-coefficient rational z-transform with no poles on the unit circle. Then the function

$$F(z) = H(z)H(z^{-1}) \tag{7.22}$$

has a zero phase on the unit circle, as here we have

$$\begin{aligned} F(e^{j\omega}) = F(z)|_{z=e^{j\omega}} &= H(z)H(z^{-1})|_{z=e^{j\omega}} \\ &= H(e^{j\omega})H(e^{-j\omega}) = |H(e^{j\omega})|^2. \end{aligned} \tag{7.23}$$

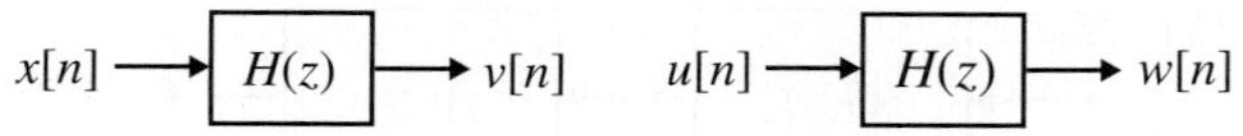

$$x[n] \longrightarrow \boxed{H(z)} \longrightarrow v[n] \qquad u[n] \longrightarrow \boxed{H(z)} \longrightarrow w[n]$$

$$u[n] = v[-n], \quad y[n] = w[-n]$$

Figure 7.9: Implementation of a zero-phase filtering scheme.

Note that if $z = \nu$ is a pole (zero) of $F(z)$, it also has a pole (zero) at $z = 1/\nu$. Thus, the zeros and poles of the function $F(z)$ are located in the z-plane with a mirror-image symmetry. A polynomial $A(z) = \sum_{i=0}^{L} a_i z^{-i}$ with roots located in the z-plane with a mirror-image symmetry is called a *symmetric polynomial,* as its coefficients satisfy the property

$$a_i = a_{L-i}.$$

A *zero-phase polynomial* $B(z)$ therefore must be of the form

$$B(z) = \sum_{\ell=0}^{N} b_\ell (z^\ell + z^{-\ell}).$$

It is impossible to implement a causal digital filter with a zero phase. For non-real-time processing of real-valued input signals of finite length, zero-phase filtering can be very simply implemented if the causality requirement is relaxed. To this end, one of two feasible schemes can be followed. In one scheme, the finite-length input data is processed through a causal real-coefficient filter $H(z)$ whose output is then time-reversed and processed by the same filter, as indicated in Figure 7.9.

To verify the above scheme, let $v[-n] = u[n]$. Also, let $X(e^{j\omega})$, $V(e^{j\omega})$, $U(e^{j\omega})$, $W(e^{j\omega})$, and $Y(e^{j\omega})$ denote the discrete-time Fourier transforms of $x[n]$, $v[n]$, $u[n]$, $w[n]$, and $y[n]$, respectively. Now, from Figure 7.9 and making use of the symmetry relations given in Tables 3.1 and 3.2, we arrive at the relations between the various Fourier transforms as

$$V(e^{j\omega}) = H(e^{j\omega})X(e^{j\omega}), \quad W(e^{j\omega}) = H(e^{j\omega})U(e^{j\omega}),$$
$$U(e^{j\omega}) = V^*(e^{j\omega}), \qquad Y(e^{j\omega}) = W^*(e^{j\omega}).$$

Combining the above equations, we obtain

$$Y(e^{j\omega}) = W^*(e^{j\omega}) = H^*(e^{j\omega})U^*(e^{j\omega}) = H^*(e^{j\omega})V(e^{j\omega})$$
$$= H^*(e^{j\omega})H(e^{j\omega})X(e^{j\omega}) = |H(e^{j\omega})|^2 X(e^{j\omega}).$$

Therefore, the overall arrangement of Figure 7.9 implements a *zero-phase filter* with a frequency response $|H(e^{j\omega})|^2$.

The function `filtfilt` implements the above scheme. A second method to achieve zero-phase filtering is outlined in Problem 7.5.

In a number of digital filter design methods, the square-magnitude function $|H(e^{j\omega})|^2$ in the form of a ratio of polynomials in $\cos\omega$ is first developed from the given filter specifications. The zero-phase function $H(z)H(z^{-1})$ is then derived from $|H(e^{j\omega})|^2$ by replacing ω with $\frac{1}{j}\ln z$, that is, by replacing $\cos\omega$ with $(z + z^{-1})/2$:

$$H(z)H(z^{-1}) = |H(e^{j\omega})|^2 \Big|_{\omega = \frac{1}{j}\ln z}.$$

The above equality holds everywhere in the z-plane where the z-transform $H(z)$ is an analytic function; that is, it has no poles.[1] The transfer function $H(z)$ is next determined by assigning half of the poles

---

[1] The conversion process is called *analytic continuation* in complex variable theory [Chu90].

and zeros of $H(z)H(z^{-1})$ to $H(z)$, with the other half of the poles and zeros located at the mirror-image locations belonging to $H(z^{-1})$. In the case of an IIR transfer function $H(z)$, for stability reasons, only the poles of $H(z)H(z^{-1})$ that are inside the unit circle are assigned to $H(z)$. However, there is no such restriction on the assignment of the zeros in most situations, as illustrated in Example 7.2.

EXAMPLE 7.2   **Determination of the Transfer Function from Specified Square-Magnitude Function**

Consider the square-magnitude function

$$|H(e^{j\omega})|^2 = \frac{4(1.09 + 0.6\cos\omega)(1.16 - 0.8\cos\omega)}{(1.04 - 0.2\cos\omega)(1.25 + \cos\omega)}. \tag{7.24}$$

By replacing $\cos\omega$ with $(z + z^{-1})/2$ on the right-hand side of the above equation, we get

$$\begin{aligned}
H(z)H(z^{-1}) &= \frac{4[1.09 + 0.3(z + z^{-1})][1.16 - 0.4(z + z^{-1})]}{[1.04 - 0.4(z + z^{-1})][1.25 + 0.5(z + z^{-1})]} \\
&= \frac{4(1 + 0.3z^{-1})(1 + 0.3z)(1 - 0.4z^{-1})(1 - 0.4z)}{(1 - 0.2z^{-1})(1 - 0.2z)(1 + 0.5z^{-1})(1 + 0.5z)} \\
&= \frac{4(1 + 0.3z^{-1})(0.3 + z^{-1})(1 - 0.4z^{-1})(0.4 - z^{-1})}{(1 - 0.2z^{-1})(0.2 - z^{-1})(1 + 0.5z^{-1})(0.5 + z^{-1})}.
\end{aligned}$$

As can be seen, there are four possible choices for a stable transfer function $H(z)$, all with the same square-magnitude function as given in Eq. (7.24):

$$H(z) = \frac{2(1 + 0.3z^{-1})(1 - 0.4z^{-1})}{(1 - 0.2z^{-1})(1 + 0.5z^{-1})}, \tag{7.25a}$$

or

$$H(z) = \frac{2(1 + 0.3z^{-1})(0.4 - z^{-1})}{(1 - 0.2z^{-1})(1 + 0.5z^{-1})}, \tag{7.25b}$$

or

$$H(z) = \frac{2(0.3 + z^{-1})(1 - 0.4z^{-1})}{(1 - 0.2z^{-1})(1 + 0.5z^{-1})}, \tag{7.25c}$$

or

$$H(z) = \frac{2(0.3 + z^{-1})((0.4 - z^{-1})}{(1 - 0.2z^{-1})(1 + 0.5z^{-1})}. \tag{7.25d}$$

7.2.2 Linear-Phase Transfer Function

In the case of a causal LTI system with a nonzero phase response, the phase distortion can be avoided by allowing the output to be a delayed version of the input:

$$y[n] = x[n - D].$$

By taking the Fourier transform of both sides of the above equation and making use of the time-shifting property,[2] we get

$$Y(e^{j\omega}) = e^{-j\omega D} X(e^{j\omega}).$$

Hence, from Eq. (3.88), the frequency response of the LTI system is given by

$$H(e^{j\omega}) = \frac{Y(e^{j\omega})}{X(e^{j\omega})} = e^{-j\omega D}. \tag{7.26}$$

---

[2]See Table 3.4.

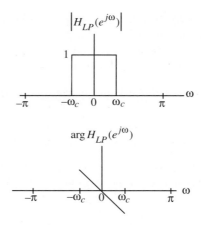

Figure 7.10: Frequency response of an ideal lowpass filter with a linear-phase response in the passband.

Note that the frequency response given by Eq. (7.26) has a unity magnitude response and a linear phase with a group delay of D samples at all frequencies; that is,

$$|H(e^{j\omega})| = 1, \qquad \tau(\omega) = D. \tag{7.27}$$

The output of this filter to an input $x[n] = Ae^{j\omega n}$ is then given by

$$y[n] = Ae^{-j\omega D}e^{j\omega n} = Ae^{j\omega(n-D)}.$$

If $x_a(t)$ and $y_a(t)$ represent the continuous-time signals whose sampled versions, sampled at $t = nT$, are $x[n]$ and $y[n]$ given above, then the delay between $x_a(t)$ and $y_a(t)$ is precisely the group delay D. Note that if D is an integer, then the output sequence $y[n]$ is identical to the input sequence $x[n]$, but delayed by D samples. If D is not an integer, $y[n]$, being delayed by a fractional part, is not identical to $x[n]$. But in this latter case, the waveform of the underlying continuous-time output is identical to the waveform of the underlying continuous-time input and delayed by D units of time.

If we desire to pass input signal components in a certain frequency range undistorted in both magnitude and phase, then the transfer function should exhibit a unity magnitude response and linear-phase response in the band of interest. Figure 7.10 shows the frequency response of a lowpass transfer function with a linear-phase characteristic in the passband. Since the signal components in the stopband are blocked, the phase response in the stopband can be of any shape.

EXAMPLE 7.3   **Impulse Response of an Ideal Linear-Phase Lowpass Filter**

Determine the impulse response of an ideal lowpass filter with a linear-phase response. The desired frequency response is now given by

$$H_{LP}(e^{j\omega}) = \begin{cases} e^{-j\omega n_o}, & 0 < |\omega| < \omega_c, \\ 0 & \omega_c \leq |\omega| \leq \pi. \end{cases} \tag{7.28}$$

Applying the frequency-shifting property of the Fourier transform given in Table 3.4 to Eq. (3.49), we arrive at the impulse response of the linear-phase lowpass filter of Eq. (7.28) as

$$h_{LP}[n] = \frac{\sin \omega_c(n - n_o)}{\pi(n - n_o)}, \qquad -\infty < n < \infty. \tag{7.29}$$

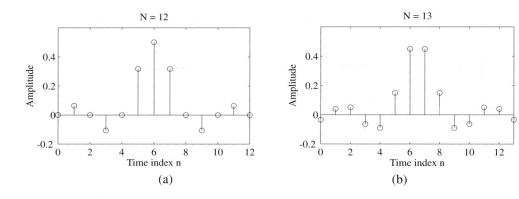

Figure 7.11: FIR approximation to the ideal linear-phase lowpass filter.

As before, the above filter is noncausal and of doubly infinite length and, hence, is unrealizable. However, by truncating the impulse response to a finite number of terms, we can develop a realizable FIR filter approximation to the ideal linear-phase lowpass filter. If we choose $n_o = N/2$, with N a positive integer, the truncated and shifted approximation

$$\hat{h}_{LP}[n] = \frac{\sin \omega_c(n - N/2)}{\pi(n - N/2)}, \qquad 0 \le n \le N, \tag{7.30}$$

will be a length $N + 1$ causal linear-phase FIR filter. Figure 7.11 shows the filter coefficients obtained using the M-file sinc for two different values of N. Because of the symmetry of the impulse response coefficients as indicated in this figure, the frequency response of the truncated approximation can be expressed as

$$\hat{H}_{LP}(e^{j\omega}) = \sum_{n=0}^{N} \hat{h}_{LP}[n]e^{-j\omega n} = e^{-j\omega N/2}\breve{H}_{LP}(\omega), \tag{7.31}$$

where $\breve{H}_{LP}(\omega)$, called the *zero-phase response* or *amplitude response*, is a real function of ω.

7.2.3 Minimum-Phase and Maximum-Phase Transfer Functions

Another useful classification of a transfer function is based on the location of its zeros, which, in turn, influences its phase response.

Definitions

Consider the two first-order transfer functions $H_1(z)$ and $H_2(z)$:

$$H_1(z) = \frac{z + b}{z + a}, \qquad 0 < |a| < 1, \tag{7.32a}$$

$$H_2(z) = \frac{bz + 1}{z + a}, \qquad 0 < |b| < 1. \tag{7.32b}$$

As can be seen from the pole-zero plots given in Figure 7.12, both transfer functions have a pole inside the unit circle at the same location at $z = -a$ and are therefore stable. On the other hand, the zero of $H_1(z)$ is inside the unit circle at $z = -b$, whereas the zero of $H_2(z)$ is outside the unit circle at $z = -1/b$ situated

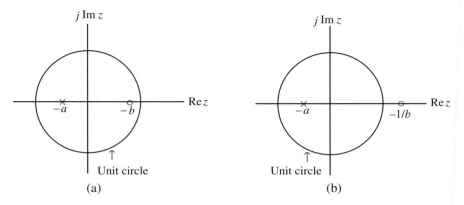

Figure 7.12: Pole-zero plots of the transfer functions of Eqs. (7.32a) and (7.32b): (a) $H_1(z)$ and (b) $H_2(z)$.

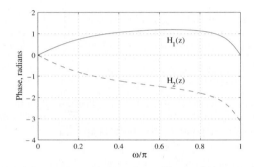

Figure 7.13: Unwrapped phase responses of the transfer functions of Eqs. (7.32a) and (7.32b) for $a = 0.5, b = -0.8$.

in a mirror-image symmetry with respect to the zero of $H_1(z)$. However, the two transfer functions have an identical magnitude function, since $H_1(z)H_1(z^{-1}) = H_2(z)H_2(z^{-1})$.

From Eqs. (7.32a) and (7.32b),

$$\arg[H_1(e^{j\omega})] = \theta_1(\omega) = \tan^{-1}\frac{\sin\omega}{b + \cos\omega} - \tan^{-1}\frac{\sin\omega}{a + \cos\omega}, \tag{7.33a}$$

$$\arg[H_2(e^{j\omega})] = \theta_2(\omega) = \tan^{-1}\frac{b\sin\omega}{1 + b\cos\omega} - \tan^{-1}\frac{\sin\omega}{a + \cos\omega}. \tag{7.33b}$$

Figure 7.13 shows the unwrapped phase responses of the two transfer functions. From this figure, it can be seen that $H_2(z)$ has an excess phase lag with respect to $H_1(z)$.

The excess phase lag property of $H_2(z)$ with respect to $H_1(z)$ can also be explained by observing that we can write the former transfer function in terms of the latter as

$$H_2(z) = \frac{bz + 1}{z + a} = \left(\frac{z + b}{z + a}\right)\left(\frac{bz + 1}{z + b}\right)$$

$$= H_1(z)A(z),$$

where

$$A(z) = \frac{bz + 1}{z + b},$$

is a stable allpass function. Hence, the relation between the phase functions of $H_1(z)$ and $H_2(z)$ are related through

$$\arg[H_2(e^{j\omega})] = \arg[H_1(e^{j\omega})] + \arg[A(e^{j\omega})]. \tag{7.34}$$

We have shown in Section 7.1.3 that the (unwrapped) phase function of a stable first-order allpass function is a negative function of ω. Hence, it follows from Eq. (7.34) that $H_2(z)$ has indeed an excess phase lag with respect to $H_1(z)$.

Generalizing the above result, let $H_m(z)$ be a causal stable transfer function with all zeros inside the unit circle and $H(z)$ be another causal stable transfer function with the same magnitude function as that of $H_m(z)$; that is, $|H(e^{j\omega})| = |H_m(e^{j\omega})|$. These two transfer functions then can be expressed as

$$H(z) = H_m(z)A(z), \tag{7.35}$$

where $A(z)$ is a stable allpass transfer function. The unwrapped phase functions of these transfer functions are therefore related as

$$\arg[H(e^{j\omega})] = \arg[H_m(e^{j\omega})] + \arg[A(e^{j\omega})].$$

It follows that $H(z)$ has an excess phase lag with respect to $H_m(z)$. As a result, a causal stable transfer function with all zeros inside the unit circle is called a *minimum-phase transfer function*, whereas a causal stable transfer function with all zeros outside the unit circle is called a *maximum-phase transfer function*. For example, the transfer function of Eq. (7.25a) is a minimum-phase function, whereas the transfer function of Eq. (7.25d) is a maximum-phase function. A transfer function with zeros inside and outside the unit circle is called a *mixed-phase transfer function*. The two transfer functions of Eq. (7.25b) and (7.25c) are mixed-phase functions.

EXAMPLE 7.4   Relation Between Mixed-Phase, Minimum-Phase and Maximum-Phase Transfer Functions

Consider the mixed-phase transfer function of Eq. (7.25b) in Example 7.2. We can rewrite it as

$$H(z) = \frac{2(1 + 0.3z^{-1})(0.4 - z^{-1})}{(1 - 0.2z^{-1})(1 + 0.5z^{-1})}$$

$$= \frac{2(1 + 0.3z^{-1})(1 - 0.4z^{-1})(0.4 - z^{-1})}{(1 - 0.2z^{-1})(1 + 0.5z^{-1})(1 - 0.4z^{-1})}$$

$$= \left[\frac{2(1 + 0.3z^{-1})(1 - 0.4z^{-1})}{(1 - 0.2z^{-1})(1 + 0.5z^{-1})} \right] \left[\frac{0.4 - z^{-1}}{1 - 0.4z^{-1}} \right],$$

which can be seen as a product of the minimum-phase transfer function of Eq. (7.25a) and a stable allpass transfer function

$$A_1(z) = \frac{0.4 - z^{-1}}{1 - 0.4z^{-1}}.$$

Likewise, the maximum-phase transfer function of Eq. (7.25d) can be expressed as a product of the minimum-phase transfer function of Eq. (7.25a) and a stable allpass transfer function

$$A_2(z) = \frac{(0.3 + z^{-1})(0.4 - z^{-1})}{(1 + 0.3z^{-1})(1 - 0.4z^{-1})}.$$

Properties

A minimum-phase causal stable transfer function $H_m(z)$ also has a group delay $\tau_g^{(H_m)}(\omega)$ that is smaller than the group delay $\tau_g^{(H)}(\omega)$ of a non-minimum-phase causal transfer function $H(z)$ having the same magnitude response function as that of $H_m(z)$. To demonstrate this property, we note from Eq. (7.35) that

$$\tau_g^{(H)}(\omega) = \tau_g^{(H_m)}(\omega) + \tau_g^{(A)}(\omega),$$

where $\tau_g^{(A)}(\omega)$ is the group delay of the allpass function $A(z)$. Since the group delay functions are always nonnegative functions of ω, it follows that $\tau_g^{(H)}(\omega) > \tau_g^{(H_m)}(\omega)$.

Another interesting property of a causal minimum-phase transfer function $H_m(z)$ is with regard to its impulse response magnitudes relative to that of a causal non-minimum-phase transfer function $H(z)$ with an identical magnitude response function; that is, $|H(e^{j\omega})| = |H_m(e^{j\omega})|$. Let $h_m[n]$ and $h[n]$ denote the impulse responses of $H_m(z)$ and $H(z)$, respectively. Then, it can be shown that

$$|h_m[0]| \geq |h[0]|, \tag{7.36}$$

and

$$\sum_{\ell=0}^{n} |h_m[n]|^2 \geq \sum_{\ell=0}^{n} |h[n]|^2. \tag{7.37}$$

The quantity $\sum_{\ell=0}^{n} |h[n]|^2$ is usually referred to as the *partial energy* of $H(z)$. It should be noted, however, that the total energy of $h[n]$ is the same as that of $h_m[n]$; that is,

$$\sum_{\ell=0}^{\infty} |h_m[n]|^2 = \sum_{\ell=0}^{\infty} |h[n]|^2. \tag{7.38}$$

The above result follows from Parseval's theorem.

We shall show later in Section 7.6 that a stable inverse system of a causal discrete-time system can be designed only if the latter has a minimum-phase transfer function. Hence, if a causal discrete-time system has a non-minimum-phase transfer function, it must be first cascaded with an appropriate allpass section so that the transfer function of the overall system has a minimum-phase transfer function and then used for the design of a causal stable inverse system.

## 7.3	Types of Linear-Phase FIR Transfer Functions

In the previous section, we pointed out why it is important to have a transfer function with a linear-phase property. It turns out it is always possible to design an FIR transfer function with an exact linear-phase response, whereas it is not possible to design an IIR transfer function with exact linear-phase. In this section, we develop the forms of the linear-phase FIR transfer function $H(z)$ with real impulse response $h[n]$ [Cav2000]:[3]

$$H(z) = \sum_{n=0}^{N} h[n] z^{-n}.$$

If $H(z)$ is required to have a linear-phase, its frequency response must be of the form

$$H(e^{j\omega}) = e^{j(c\omega+\beta)} \breve{H}(\omega), \tag{7.39}$$

---

[3]It should be noted that in Chapter 5, N denoted the length of a finite-length sequence, whereas in this and subsequent chapters, following common practice, N is being used to denote the order of an FIR transfer function.

where c and β are constants, and $\breve{H}(\omega)$, called the *amplitude response,* also called the *zero-phase response,* is a real function of ω. The zero-phase response at specified frequency points can be computed in MATLAB using the M-file zerophase, which is available with several options.

For a real impulse response, the magnitude response $|H(e^{j\omega})|$ is an even function of ω; that is, $|H(e^{j\omega})| = |H(e^{-j\omega})|$. Since $|H(e^{j\omega})| = |\breve{H}(\omega)|$, the amplitude response is then either an even or an odd function of ω; that is, $\breve{H}(-\omega) = \pm\breve{H}(\omega)$, and the frequency response satisfies the relation

$$H(e^{j\omega}) = H^*(e^{-j\omega}),$$

or, equivalently, the relation

$$e^{j(c\omega+\beta)}\breve{H}(\omega) = e^{-j(c(-\omega)+\beta)}\breve{H}(-\omega). \tag{7.40}$$

If $\breve{H}(\omega)$ is an even function, then Eq. (7.40) leads to

$$e^{j\beta} = e^{-j\beta},$$

implying that either $\beta = 0$ or $\beta = \pi$. Therefore, from Eq. (7.40), we get

$$\breve{H}(\omega) = \pm e^{-jc\omega}H(e^{j\omega}) = \pm\sum_{n=0}^{N} h[n]\,e^{-j\omega(c+n)}. \tag{7.41}$$

Since $\breve{H}(-\omega) = \breve{H}(\omega)$, we also have

$$\breve{H}(-\omega) = \pm\sum_{\ell=0}^{N} h[\ell]\,e^{j\omega(c+\ell)}. \tag{7.42}$$

Making a change of variable $\ell = N - n$, we rewrite the above equation as

$$\breve{H}(-\omega) = \pm\sum_{n=0}^{N} h[N - n]\,e^{j\omega(c+N-n)}. \tag{7.43}$$

Equating Eqs. (7.41) and (7.43), we arrive at the condition

$$h[n] = h[N - n], \qquad 0 \le n \le N, \tag{7.44}$$

with

$$c = -\frac{N}{2}.$$

Therefore, the FIR filter with an even amplitude response will have linear-phase response if it has a symmetric impulse response.

If, on the other hand, $\breve{H}(\omega)$ is an odd function, then from Eq. (7.40), we get

$$e^{j\beta} = -e^{-j\beta},$$

which is satisfied if either $\beta = \pi/2$ or $\beta = -\pi/2$. Then Eq. (7.39) reduces to

$$H(e^{j\omega}) = je^{-jc\omega}\breve{H}(\omega),$$

from which we arrive at

$$\breve{H}(\omega) = -je^{-jc\omega}H(e^{j\omega}) = -j\sum_{n=0}^{N}h[n]\,e^{-j\omega(c+n)}. \tag{7.45}$$

As here $\breve{H}(-\omega) = -\breve{H}(\omega)$, we also have

$$\breve{H}(-\omega) = j\sum_{\ell=0}^{N}h[\ell]\,e^{j\omega(c+\ell)}. \tag{7.46}$$

Making a change of variable $\ell = N - n$, we rewrite the above equation as

$$\breve{H}(-\omega) = j\sum_{n=0}^{N}h[N-n]\,e^{j\omega(c+N-n)}. \tag{7.47}$$

Finally, equating Eqs. (7.45) and (7.47), we arrive at the condition

$$h[n] = -h[N-n], \qquad 0 \le n \le N, \tag{7.48}$$

with

$$c = -\frac{N}{2}.$$

Therefore, the FIR filter with an odd amplitude response will have linear-phase response if it has an antisymmetric impulse response.

Since the filter length can be either even or odd and there are two forms of impulse response symmetries, there are four types of linear-phase FIR filters as described next.

Type 1 FIR Transfer Function

The impulse response of a Type 1 FIR transfer function is of odd length, that is, the degree N is even and satisfies the symmetry condition of Eq. (7.44). Its amplitude response $\breve{H}(\omega)$ is given by

$$\breve{H}(\omega) = h\left[\tfrac{N}{2}\right] + 2\sum_{n=1}^{N/2}h\left[\tfrac{N}{2}-n\right]\cos(\omega n). \tag{7.49}$$

Note that the FIR transfer function of Example 3.20 has a symmetric impulse response of the form $h[0] = h[2]$ and has a linear-phase response as indicated in Eq. (3.111). This filter is an example of a Type 1 FIR filter.

Useful Properties. Three types of Type 1 FIR transfer functions with different frequency responses can be generated from a given Type 1 FIR transfer function $H(z)$, as indicated below:

$$E(z) = z^{-N/2} - H(z), \tag{7.50a}$$

$$F(z) = (-1)^{N/2}H(-z), \tag{7.50b}$$

$$G(z) = z^{-N/2} - (-1)^{N/2}H(-z). \tag{7.50c}$$

It follows from Eq. (7.50a) that

$$H(z) + E(z) = z^{-N/2}.$$

The filters $H(z)$ and $E(z)$ are thus called *delay-complementary filters*.[4]

The amplitude responses of the transfer functions $E(z)$ and $G(z)$ are related to that of $H(z)$ according to

$$\breve{E}(\omega) = \breve{H}(\pi - \omega), \tag{7.51a}$$

$$\breve{F}(\omega) = 1 - \breve{H}(\omega), \tag{7.51b}$$

$$\breve{G}(\omega) = 1 - \breve{H}(\pi - \omega). \tag{7.51c}$$

The transformation given by Eq. (7.50a) is applicable to any Type 1 FIR transfer function. Moreover, the transfer function $E(z)$ defined in Eq. (7.50a) is a highpass transfer function if $H(z)$ is a lowpass transfer function, and vice versa. Similarly, $E(z)$ is a bandstop transfer function if $H(z)$ is a bandpass transfer function, and vice versa. The widths of the passband and the stopband $E(z)$ are the same as those of $H(z)$. However, the transformations given by Eqs. (7.50b) and (7.50c) are applicable only to bounded-real Type 1 FIR transfer functions. In this case, the transfer function $F(z)$ is of the same type as $H(z)$, except with passbands of different widths. For example, if $H(z)$ is a lowpass transfer function with a narrow passband, then $F(z)$ is also a lowpass transfer function but with a wider passband.

The transfer functions $E(z)$ and $G(z)$ can be generated from a specified Type 1 FIR transfer function $H(z)$ using the M-file `firlp2hp`, whereas the transfer function $F(z)$ can be generated using the M-file `firlp2lp`. Their use is illustrated in Example 7.5.

EXAMPLE 7.5 **Illustration of the Properties of the Type 1 FIR Transfer Function**

Consider the length-9 Type 1 FIR lowpass transfer function[5]

$$H(z) = -0.0637 - 0.0691\, z^{-1} + 0.101\, z^{-2} + 0.2857\, z^{-3} + 0.4107\, z^{-4} + 0.2857\, z^{-5}$$
$$+0.101\, z^{-6} - 0.0691\, z^{-7} - 0.0637\, z^{-8}.$$

The highpass transfer function $E(z)$ can be generated from the above $H(z)$ using the function `firlp2hp` and is given by

$$E(z) = 0.0637 - 0.0691\, z^{-1} - 0.1010\, z^{-2} + 0.2857\, z^{-3} + 0.5893\, z^{-4} + 0.2857\, z^{-5}$$
$$-0.101\, z^{-6} - 0.0691\, z^{-7} + 0.0637 z^{-8}.$$

Plots of the magnitude responses of $H(z)$ and $E(z)$ are shown in Figure 7.14(a).

On the other hand, the transfer function $F(z)$ defined in Eq. (7.50b) is a wideband lowpass transfer function. It can be generated using the function `firlp2lp` and is given below:

$$F(z) = -0.0637 + 0.0691\, z^{-1} + 0.101\, z^{-2} - 0.2857\, z^{-3} + 0.4107\, z^{-4} - 0.2857\, z^{-5}$$
$$+0.101\, z^{-6} + 0.0691\, z^{-7} - 0.0637\, z^{-8}.$$

Plots of the magnitude responses of $H(z)$ and $F(z)$ are shown in Figure 7.14(b).

Type 2 FIR Transfer Function

The impulse response of a Type 2 FIR filter is of even length; that is, the degree N is odd and satisfies the symmetric condition of Eq. (7.44). Its amplitude response $\breve{H}(\omega)$ is given by

---

[4]See Section 7.5.1.

[5]Designed using the M-file `remez`. See Section 10.5.2.

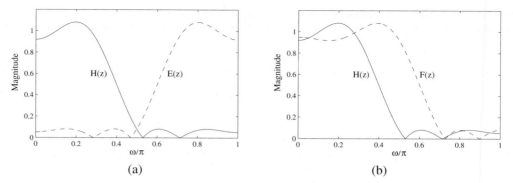

Figure 7.14: (a) Magnitude responses of the original Type 1 FIR lowpass transfer function $H(z)$ (solid line) and the transformed highpass transfer function $E(z)$, generated using Eq. (7.50a) (dashed line), and (b) the magnitude responses of the original Type 1 FIR lowpass transfer function $H(z)$ (solid line) and the transformed lowpass transfer function $F(z)$, generated using Eq. (7.50b)

$$\breve{H}(\omega) = 2 \sum_{n=1}^{(N+1)/2} h\left[\tfrac{N+1}{2} - n\right] \cos\left(\omega(n - \tfrac{1}{2})\right). \tag{7.52}$$

Type 3 FIR Transfer Function

The impulse response of a Type 3 FIR filter is of odd length; that is, the degree N is even and satisfies the antisymmetric condition of Eq. (7.48). Here the amplitude response $\breve{H}(\omega)$ is of the form

$$\breve{H}(\omega) = 2 \sum_{n=1}^{N/2} h\left[\tfrac{N}{2} - n\right] \sin(\omega n). \tag{7.53}$$

Type 4 FIR Transfer Function

The impulse response of a Type 4 FIR filter is of even length; that is, the degree N is odd and satisfies the antisymmetric condition of Eq. (7.48). The amplitude response in this case is of the form

$$\breve{H}(\omega) = 2 \sum_{n=1}^{(N+1)/2} h\left[\tfrac{N+1}{2} - n\right] \sin\left(\omega(n - \tfrac{1}{2})\right). \tag{7.54}$$

General Form of Frequency Response

In each of the four types of causal linear-phase FIR filters, the frequency response $H(e^{j\omega})$ is of the form

$$H(e^{j\omega}) = e^{-jN\omega/2} e^{j\beta} \breve{H}(\omega), \tag{7.55}$$

where β is either 0 or π, for the case of a symmetric impulse response, or $\pm\pi/2$, for the case of an antisymmetric impulse response. Examples 7.6 and 7.7 demonstrate the above form.

EXAMPLE 7.6 Type 1 FIR Transfer Functions with Identical Magnitude Responses but Different Phase Responses

Consider the causal Type 1 FIR transfer function:

$$H_1(z) = -1 + 2z^{-1} - 3z^{-2} + 6z^{-3} - 3z^{-4} + 2z^{-5} - z^{-6}. \qquad (7.56)$$

Its amplitude response is given by

$$\breve{H}_1(\omega) = 6 - 6\cos(\omega) + 4\cos(2\omega) - 2\cos(3\omega),$$

a plot of which is shown in Figure 7.15(a). The phase response of $H_1(z)$ is given by

$$\theta_1(\omega) = -3\omega$$

and is shown in Figure 7.15(b). On the other hand, the amplitude response of the Type 1 FIR transfer function

$$H_2(z) = 1 - 2z^{-1} + 3z^{-2} - 6z^{-3} + 3z^{-4} - 2z^{-5} + z^{-6},$$

is given by

$$\breve{H}_2(\omega) = -\breve{H}_1(\omega).$$

Its phase response is therefore given by

$$\theta_2(\omega) = -3\omega + \pi$$

and is shown in Figure 7.15(c).

EXAMPLE 7.7 Type 3 FIR Transfer Functions with Identical Magnitude Responses but Different Phase Responses

The causal Type 3 FIR transfer function

$$H_3(z) = 1 - 2z^{-1} + 3z^{-2} - 3z^{-4} + 2z^{-5} - z^{-6} \qquad (7.57)$$

has an amplitude response given by

$$\breve{H}_3(\omega) = -6\sin(\omega) + 4\sin(2\omega) + 2\sin(3\omega),$$

whose plot is shown in Figure 7.16(a). The phase response of $H_3(z)$ is given by

$$\theta_3(\omega) = -3\omega + \frac{\pi}{2},$$

where the factor $\frac{\pi}{2}$ is due to the presence of j in the expression for the frequency response given in Eq. (7.45). A plot of the phase response is shown in Figure 7.16(b). On the other hand, the amplitude response of the Type 3 FIR transfer function

$$H_4(z) = -1 + 2z^{-1} - 3z^{-2} + 3z^{-4} - 2z^{-5} + z^{-6}$$

is given by

$$\breve{H}_4(\omega) = -\breve{H}_3(\omega),$$

and its phase response is therefore given by

$$\theta_4(\omega) = -3\omega - \frac{\pi}{2}$$

and is shown in Figure 7.16(c).

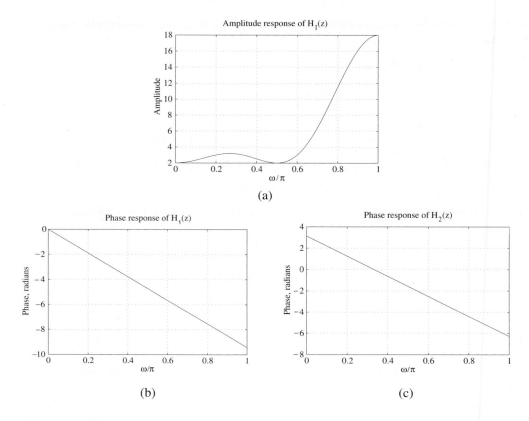

Figure 7.15: Example 7.6: (a) Amplitude response of $H_1(z)$, (b) unwrapped phase response of $H_1(z)$, and (c) unwrapped phase response of $H_2(z)$.

.

In general, the phase responses of any linear-phase FIR filter are given by

$$\theta(\omega) = -\frac{N\omega}{2} + \beta, \qquad (7.58)$$

where β is either 0 or π for either Type 1 or Type 2 filters, and $\pi/2$ or $3\pi/2$ for either Type 3 or Type 4 filters.

The magnitude response of all four types of linear-phase FIR filters is

$$|H(e^{j\omega})| = |\breve{H}(\omega)|, \qquad (7.59)$$

whereas the group delay is

$$\tau_g(\omega) = \frac{N}{2}. \qquad (7.60)$$

Note that, even though the group delay is constant, since in general $|H(e^{j\omega})|$ is not a constant, the output waveform is not a replica of the input waveform.

Zero-Phase FIR Filter

An FIR filter with a frequency response that is a real function of ω is often called a *zero-phase filter*. Such a filter must have a noncausal impulse response. A Type 1 or Type 3 linear-phase causal FIR filter with

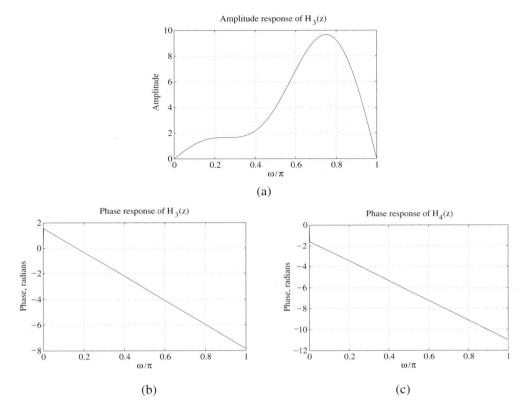

Figure 7.16: Example 7.7: (a) Amplitude response of $H_3(z)$, (b) unwrapped phase response of $H_3(z)$, and (c) unwrapped phase response of $H_4(z)$.

.

an impulse response $\{h[n]\}$ of length $N+1$ can be converted into a zero-phase FIR filter with an impulse response $\{h_{zp}[n]\}$ by shifting $\{h[n]\}$ by $N/2$ samples; that is,

$$h_{zp}[n] = h[n + \tfrac{N}{2}], \qquad -\tfrac{N}{2} \le n \le \tfrac{N}{2}.$$

7.3.1 Zero Locations of Linear-Phase FIR Transfer Functions

The symmetry or the antisymmetry of the impulse response of a linear-phase FIR transfer function imposes restrictions on the locations of the zeros of the transfer function, which, in turn, determines the types of filters these transfer functions can implement. It is thus of interest to investigate the restrictions on the zero locations of a linear-phase FIR transfer function.

Mirror-Image and Antimirror-Image Polynomials

Consider first an FIR filter with a symmetric impulse response. Its transfer function $H(z)$ can be written as

$$H(z) = \sum_{n=0}^{N} h[n]\, z^{-n} = \sum_{n=0}^{N} h[N-n]\, z^{-n}, \qquad (7.61)$$

using the symmetry condition of Eq. (7.44). By making a change of variable $m = N - n$, we can rewrite the right-most expression in Eq. (7.61) as

$$H(z) = \sum_{m=0}^{N} h[m]\, z^{-N+m} = z^{-N} \sum_{m=0}^{N} h[m]\, z^{m} = z^{-N} H(z^{-1}). \qquad (7.62)$$

Similarly, the transfer function $H(z)$ of an FIR filter with an antisymmetric impulse response satisfying the condition of Eq. (7.48) can be expressed as

$$H(z) = \sum_{n=0}^{N} h[n]\, z^{-n} = -\sum_{n=0}^{N} h[N-n]\, z^{-n} = -z^{-N} H(z^{-1}). \qquad (7.63)$$

A real-coefficient polynomial $H(z)$ satisfying the condition of Eq. (7.62) is called a *mirror-image polynomial* (MIP), whereas a real-coefficient polynomial $H(z)$ satisfying the condition of Eq. (7.63) is called an *antimirror-image polynomial* (AIP). An example of a mirror-image polynomial is the transfer function $H_1(z)$ of Eq. (7.56). Likewise, an example of an antimirror-image polynomial is the transfer function $H_3(z)$ of Eq. (7.57).

Zero Locations

For a linear-phase FIR filter with a transfer function that is either a mirror-image or an antimirror-image polynomial, it follows from Eqs. (7.62) and (7.63) that if $z = \xi_0$ is a zero of $H(z)$, so is $z = 1/\xi_0$. Moreover, for an FIR filter with a real impulse response, the zeros occur in complex conjugate pairs. Hence, a zero at $z = \xi_0$ is associated with a zero at $z = \xi_0^*$. Therefore, a complex zero that is not on the unit circle is associated with a set of four zeros given by

$$z = r e^{\pm j\phi}, \qquad z = \frac{1}{r} e^{\pm j\phi}.$$

For a zero on the unit circle, its reciprocal is also its complex conjugate. Hence, in this case, the zeros appear as a complex-conjugate pair

$$z = e^{\pm j\phi}.$$

A real zero is paired with its reciprocal zero appearing at

$$z = \rho, \qquad z = \frac{1}{\rho}.$$

Thus, the zeros of a linear-phase FIR filter exhibit a *mirror-image symmetry* with respect to the unit circle.

The principal difference among the four types of linear-phase FIR filters is with regard to the number of zeros at $z = 1$ and $z = -1$. If the Type 1 FIR transfer function has a zero at $z = 1$ or at $z = -1$, then because of its order being even and the symmetry of its impulse response, there must be an even number of zeros at $z = 1$ or at $z = -1$ or at both locations. On the other hand, the remaining three types of linear-phase FIR filters must have zeros at one of the above locations. We examine each one of the remaining FIR filters next.

A Type 2 FIR filter must have a zero at $z = -1$. To show this, we observe from Eq. (7.62) that

$$H(-1) = (-1)^{N} H(-1) = -H(-1),$$

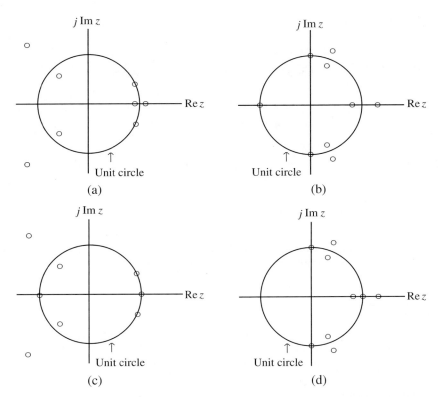

Figure 7.17: Examples of zero locations of linear-phase FIR transfer functions: (a) Type 1, (b) Type 2, (c) Type 3, and (d) Type 4 filters.

since N is odd, implying $H(-1) = 0$. In the case of a Type 3 or 4 FIR filter, Eq. (7.63) implies that $H(1) = -H(1)$, indicating that the filter must have a zero at $z = 1$. On the other hand, only a Type 3 FIR filter, for which N is even, is restricted to have a zero at $z = -1$, since here

$$H(-1) = -(-1)^N H(-1) = -H(-1),$$

forcing $H(-1) = 0$. Figure 7.17 shows some examples of zero locations of all four types of FIR filters.

Summarizing, we conclude that

(a) Type 1 FIR Filter: Either an even number or no zeros at $z = 1$ and $z = -1$.

(b) Type 2 FIR Filter: Either an even number or no zeros at $z = 1$ and an odd number of zeros at $z = -1$.

(c) Type 3 FIR Filter: An odd number of zeros at $z = 1$ and $z = -1$.

(d) Type 4 FIR Filter: An odd number of zeros at $z = 1$ and either an even number or no zeros at $z = -1$.

The presence of zeros at $z = \pm 1$ leads to the following limitations on the use of these linear-phase FIR filters for designing filters. For example, since the Type 2 FIR filter always has a zero at $z = -1$, it cannot be used to design a highpass filter. Likewise, the Type 3 FIR filter has zeros at both $z = 1$ and $z = -1$ and, as a result, cannot be used to design either a lowpass or a highpass or a bandstop filter. Similarly, the Type 4 FIR filter is not appropriate to design a lowpass filter due to the presence of a zero at $z = 1$. Finally, the Type 1 FIR filter has no such restrictions and can be used to design almost any type of filter.

EXAMPLE 7.8   Determination of a Linear-Phase FIR Transfer Function from Its Zero Locations

A length-9 Type 1 real coefficient FIR filter has the following zeros: $z_1 = -0.5$, $z_2 = 0.3 + j\,0.5$, $z_3 = -\frac{1}{2} + j\frac{\sqrt{3}}{2}$. We determine the locations of the remaining zeros and the expression for the transfer function $H(z)$ of the FIR filter.

Since the order of the transfer function is 8, there are 8 zeros. The real zero z_1 is paired with its reciprocal zero at $z_4 = \frac{1}{z_1} = -2$. The complex zero at z_2 is paired with its complex conjugate zero at $z_5 = z_2^* = 0.3 - j\,0.5$, and the reciprocal zeros at $z_6 = \frac{1}{z_2} = 0.12 - j\,0.1993$ and $z_7 = z_6^* = 0.12 + j\,0.1993$. Finally, the complex zero at z_3 is on the unit circle, and it is paired with its complex-conjugate zero at $z_8 = z_3^* = -\frac{1}{2} - j\frac{\sqrt{3}}{2}$. The associated transfer function is therefore given by $H(z) = \prod_{i=1}^{8}(1 - z_i z^{-1}) = 1 - 1.1353\,z^{-1} + 0.5635\,z^{-2} + 5.6841\,z^{-3} + 4.9771\,z^{-4} + 5.6841\,z^{-5} + 0.5635\,z^{-6} - 1.1353\,z^{-7} + z^{-8}$.

7.4   Simple Digital Filters

In Chapters 9 and 10, we outline several methods of designing frequency-selective filters satisfying prescribed specifications. In this section, we describe several low-order FIR and IIR digital filters with reasonable selective frequency responses that often are satisfactory in a number of applications.

7.4.1   Simple FIR Digital Filters

FIR digital filters considered here have very simple impulse response coefficients. These filters are employed in a number of practical applications, primarily because of their simplicity, which makes them amenable for inexpensive hardware implementation.

Lowpass FIR Digital Filters

The simplest lowpass FIR filter is the moving-average filter of Eq. (3.97) with $M = 2$, which has a transfer function

$$H_0(z) = \frac{1}{2}(1 + z^{-1}) = \frac{z + 1}{2z}. \tag{7.64}$$

The above transfer function has a zero at $z = -1$ and a pole at $z = 0$. It follows from our discussion in Section 6.7.4 that the pole vector has a magnitude of unity, the radius of the unit circle, for all values of ω. On the other hand, as ω increases from 0 to π, the magnitude of the zero vector decreases from a value of 2, the diameter of the unit circle, to zero. Hence, the magnitude response $|H_0(e^{j\omega})|$ is a monotonically decreasing function of ω from $\omega = 0$ to $\omega = \pi$. The maximum value of the magnitude function is unity at $\omega = 0$, and the minimum value is zero at $\omega = \pi$; that is,

$$|H_0(e^{j0})| = 1, \qquad |H_0(e^{j\pi})| = 0.$$

From Eq. (7.64), it follows that the frequency response of the above filter is given by

$$H_0(e^{j\omega}) = e^{-j\omega/2} \cos\left(\tfrac{\omega}{2}\right), \tag{7.65}$$

whose magnitude response is given by $\cos(\omega/2)$, which is seen to be a monotonically decreasing function of ω [see Figure 7.18(a)]. The frequency $\omega = \omega_c$ at which $|H_0(e^{j\omega_c})| = \frac{1}{\sqrt{2}}|H_0(e^{j0})|$ is of practical interest as here the gain $\mathcal{G}(\omega_c)$ in dB is given by

$$\mathcal{G}(\omega_c) = 20 \log_{10} |H_0(e^{j\omega_c})| = 20 \log_{10} |H_0(e^{j0})| - 20 \log_{10} \sqrt{2}$$
$$= 0 - 3.0103 \cong -3.0 \, \text{dB},$$

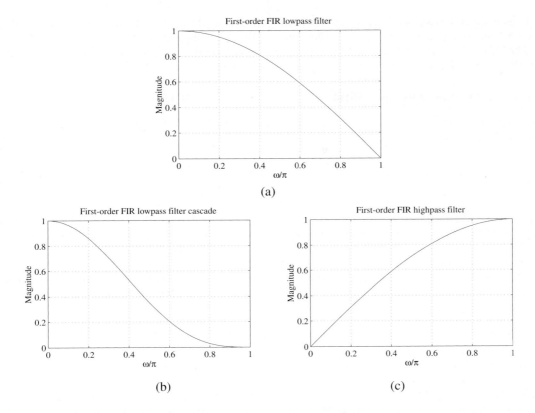

Figure 7.18: Magnitude responses of simple FIR filters: (a) first-order lowpass FIR filter $H_0(z)$ of Eq. (7.64), (b) cascade of 3 first-order lowpass FIR filters, and (c) first-order highpass filter $H_1(z)$ of Eq. (7.67).

since the dc gain $\mathcal{G}(0) = 20 \log_{10} |H_0(e^{j0})| = 0$. Thus, the gain $\mathcal{G}(\omega)$ at $\omega = \omega_c$ is approximately 3 dB less than that at zero frequency. As a result, ω_c is called the *3-dB cutoff frequency*. To determine the expression for ω_c, we set $|H_0(e^{j\omega_c})|^2 = \cos^2(\omega_c/2) = 1/2$, which yields $\omega_c = \pi/2$. This result checks with that given in the plot of Figure 7.18(a). The 3-dB cutoff frequency ω_c can be considered as the passband edge frequency, and as a result, for this filter, the passband width is approximately $\pi/2$. The stopband is here from $\pi/2$ to π. Note from Eq. (7.64), the transfer function $H_0(z)$ has a zero at $z = -1$ or $\omega = \pi$, which is in the stopband of the filter.

A cascade of the simple FIR filters of Eq. (7.64) results in an improved lowpass frequency response, as illustrated in Figure 7.18(b) for a cascade of three sections. The 3-dB cutoff frequency of a cascade of M sections of the lowpass filter of Eq. (7.64) can be shown to be given by (Problem 7.33)

$$\omega_c = 2 \cos^{-1}(2^{-1/2M}). \tag{7.66}$$

For $M = 3$, the above expression yields $\omega_c = 0.3002\pi$, which also checks with the plot in Figure 7.18(b). Thus, the cascade of first-order sections yields a sharper magnitude response but at the expense of a decrease in the passband width.

A better approximation to the ideal lowpass filter is given by the higher-order moving-average filter of Eq. (3.97). Signals with rapid fluctuations in sample values are generally associated with high-frequency components that are essentially eliminated by a moving-average filter of the type of Eq. (3.97), resulting in a much smoother output waveform, as illustrated earlier in Example 2.13.

The moving-average filter is often employed as the basic building block in the design of lowpass filters used in sampling rate alteration and is considered in Section 15.11.

Highpass FIR Digital Filters

The simplest highpass FIR filter is obtained by replacing z with $-z$ in Eq. (7.64), resulting in a transfer function

$$H_1(z) = \tfrac{1}{2}(1 - z^{-1}). \tag{7.67}$$

The corresponding frequency response is given by

$$H_1(e^{j\omega}) = je^{-j\omega/2}\sin(\tfrac{\omega}{2}), \tag{7.68}$$

whose magnitude response is shown in Figure 7.18(c). The monotonically increasing behavior of the magnitude function can again be demonstrated by examining the pole-zero pattern of the transfer function. The 3-dB cutoff frequency of this highpass filter is also at $\pi/2$. The transfer function $H_1(z)$ has a zero at $z = 1$ or $\omega = 0$, which is in the stopband of the filter.

Improved highpass frequency response can be obtained by cascading several sections of the simple highpass filter of Eq. (7.67). Alternatively, a higher-order highpass filter of the form

$$H_1(z) = \frac{1}{M}\sum_{n=0}^{M-1}(-1)^n z^{-n}, \tag{7.69}$$

obtained by replacing z with $-z$ in the expression for the transfer function of a moving-average lowpass filter, yields a sharper magnitude response.

An application of the FIR highpass filters is in moving-target-indicator (MTI) radars. In these radars, interfering signals, called *clutters,* are generated from fixed objects in the path of the radar beam [Sko62]. The clutter, generated mainly from ground echoes and weather returns, has frequency components primarily near zero frequency (dc) and can be removed by filtering the radar return signal through a *two-pulse canceler,* which is the first-order highpass filter of Eq. (7.67). Often, the frequency components of the clutter occupy a small band near dc, and for a more effective removal, it is necessary to use a highpass filter with a sharper magnitude response and a slightly broader stopband. To this end, a cascade of two two-pulse cancelers, called a *three-pulse canceler,* provides an improved performance. Problem 7.40 and Exercise M7.3 describe two simple IIR highpass filters proposed for clutter rejection.

7.4.2   Simple IIR Digital Filters

Causal FIR filters have their poles at the origin, and as a result, the shape of their frequency responses are determined only from the locations of the zeros. On the other hand, IIR filters allow the poles to move inside the unit circle, permitting them to contribute more heavily to the shape of their frequency responses. We now describe several simple IIR digital filters with first-order and second-order transfer functions and sketch their corresponding frequency responses. In many applications, use of such filters provides satisfactory results. Often, more complex frequency responses can be achieved by cascading these simple transfer functions.

Lowpass IIR Digital Filters

The first-order IIR transfer function of Eq. (7.3), considered in Example 7.1, has a lowpass magnitude response for $\alpha > 0$, as can be seen from Figure 7.2(a). A slightly improved magnitude response can be obtained by adding a factor $(1 + z^{-1})$ to the numerator of the transfer function, which forces the magnitude function to have a zero at $\omega = \pi$ in the stopband of the filter. On the other hand, the first-order IIR transfer function of Eq. (7.3) has a highpass magnitude response for $\alpha < 0$, as demonstrated in Figure 7.2(b). However, the modified transfer function obtained with the addition of a factor $(1 + z^{-1})$ to the numerator also exhibits a lowpass magnitude response.

The modified lowpass transfer function for both positive and negative values of α is then given by

$$H_{LP}(z) = \frac{K(1 + z^{-1})}{1 - \alpha z^{-1}}. \tag{7.70}$$

As ω increases from 0 to π, the magnitude of the zero vector decreases from a value of 2 to 0, whereas the magnitude of the pole vector increases from a value of $1 - \alpha$ to $1 + \alpha$. The maximum value of the magnitude function is $2K/(1 - \alpha)$ at $\omega = 0$, and the minimum value is zero at $\omega = \pi$; that is,

$$|H_{LP}(e^{j0})| = \frac{2K}{1 - \alpha}, \qquad |H_{LP}(e^{j\pi})| = 0.$$

Therefore, $|H_{LP}(e^{j\omega})|$ is a monotonically decreasing function of ω from $\omega = 0$ to $\omega = \pi$ (see Figure 7.19).

For most applications, it is usual to have a dc gain of 0 dB, that is, to have the maximum magnitude equal to 1. To this end, we can choose $K = (1 - \alpha)/2$, resulting in the scaled transfer function

$$H_{LP}(z) = \frac{1 - \alpha}{2} \cdot \frac{1 + z^{-1}}{1 - \alpha z^{-1}}, \tag{7.71}$$

where $0 < |\alpha| < 1$ for stability.

Plots of the magnitude responses of the above lowpass transfer function are sketched in Figure 7.19 for several positive and negative values of α between -1 and $+1$. For $0 < \alpha < 1$, $H_{LP}(z)$ has a pole on the positive real axis and a zero at $z = -1$. As a result, the magnitude function here is expected to have a narrow passband, as can be seen from Figure 7.19(a). On the other hand, for $-1 < \alpha < 0$, $H_{LP}(z)$ has a pole on the negative real axis and a zero at $z = -1$. Hence, in this case, the magnitude function is expected to have a wide passband, as indicated in Figure 7.19(b).

From Eq. (7.71), the squared magnitude function can be easily derived:

$$|H_{LP}(e^{j\omega})|^2 = \frac{(1 - \alpha)^2(1 + \cos \omega)}{2(1 + \alpha^2 - 2\alpha \cos \omega)}. \tag{7.72}$$

The derivative of $|H_{LP}(e^{j\omega})|^2$ with respect to ω is given by

$$\frac{d|H_{LP}(e^{j\omega})|^2}{d\omega} = \frac{-(1 - \alpha)^2(1 + 2\alpha + \alpha^2) \sin \omega}{2(1 + \alpha^2 - 2\alpha \cos \omega)^2},$$

which is always nonpositive in the range $0 \leq \omega \leq \pi$, verifying again the monotonically decreasing behavior of the magnitude function. The passband of the lowpass filter is usually defined by the frequency range from $\omega = 0$ to the frequency ω_c, called the *3-dB cutoff frequency*, where the gain is 3-dB below the gain at $\omega = 0$. To determine ω_c, we set $|H_{LP}(e^{j\omega_c})|^2 = 1/2$ in Eq. (7.72) and arrive at the equation

$$\frac{(1 - \alpha)^2(1 + \cos \omega_c)}{2(1 + \alpha^2 - 2\alpha \cos \omega_c)} = \frac{1}{2} \quad \text{or} \quad (1 - \alpha)^2(1 + \cos \omega_c) = 1 + \alpha^2 - 2\alpha \cos \omega_c,$$

which when solved yields

$$\cos \omega_c = \frac{2\alpha}{1 + \alpha^2}. \tag{7.73a}$$

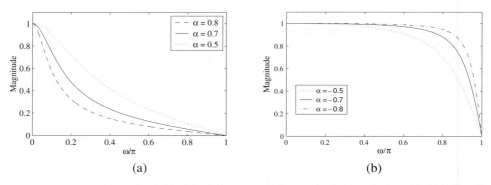

Figure 7.19: Magnitude responses of the first-order IIR lowpass filter of Eq. (7.71): (a) For three positive values of α and (b) for three negative values of α.

Equation (7.73a) can be solved for α yielding two solutions. The solution resulting in a stable transfer function $H_{LP}(z)$ is given by

$$\alpha = \frac{1 - \sin \omega_c}{\cos \omega_c}. \tag{7.73b}$$

It follows from Eq. (7.72) that the first-order lowpass transfer function $H_{LP}(z)$ of Eq. (7.71) is a bounded real (BR) function for $|\alpha| < 1$.

Highpass IIR Digital Filters

The first-order IIR transfer function of Eq. (7.3) of Example 7.1 has a highpass magnitude response for $\alpha < 0$, as shown in Figure 7.2(b). Slightly improved magnitude response is obtained by adding a factor $(1 - z^{-1})$ to the numerator of the transfer function, which forces the magnitude function to have a zero at $\omega = 0$ in the stopband of the filter. The magnitude response of the modified transfer function with a factor $(1 - z^{-1})$ added to the numerator also exhibits a highpass characteristics for $\alpha > 0$. For a 0 dB gain at $\omega = \pi$, we can choose $K = (1 + \alpha)/2$, resulting in the modified first-order highpass transfer function $H_{HP}(z)$, given by

$$H_{HP}(z) = \frac{1 + \alpha}{2} \cdot \frac{1 - z^{-1}}{1 - \alpha z^{-1}}, \tag{7.74}$$

where $|\alpha| < 1$ for stability. Its 3-dB cutoff frequency ω_c is also given by Eq. (7.73a).

Plots of the magnitude responses of the above highpass transfer function for several positive and negative values of α are shown in Figure 7.20. For $0 < \alpha < 1$, $H_{HP}(z)$ has a pole on the positive real axis and a zero at $z = 1$. Hence, the magnitude function is expected to have a wide passband as can be seen from Figure 7.20(a). On the other hand, for $-1 < \alpha < 0$, $H_{HP}(z)$ has a pole on the negative real axis and a zero at $z = 1$. In the latter case, the magnitude function is expected to have a narrow passband, as indicated in Figure 7.20(b).

A first-order highpass filter of Eq. (7.74) with a positive value of α close to 1 is an important structure in the digital waveguide modeling based approach to musical sound synthesis. It is used to block the dc component of a signal circulating in a delay-line loop and has been called a *dc blocker*. It is also finds applications in multitrack recording to eliminate the overflow of the mix caused by the addition of the dc components in the various tracks.

It can be shown that the first-order highpass transfer function of Eq. (7.74) is also a BR function for $|\alpha| < 1$.

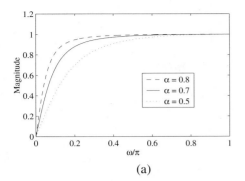 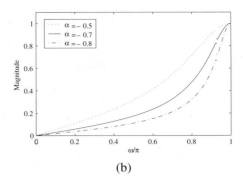

Figure 7.20: Magnitude responses of the first-order IIR highpass filter of Eq. (7.74): (a) For three positive values of α and (b) for three negative values of α.

EXAMPLE 7.9   Design of a First-Order Highpass Digital Filter

Design a first-order highpass digital filter with a 3-dB cutoff frequency of 0.8π. Now, $\sin(\omega_c) = \sin(0.8\pi) = 0.58778525$, and $\cos(\omega_c) = \cos(0.8\pi) = -0.80902$, which, when substituted in Eq. (7.73b), yields $\alpha = -0.50952449$. Therefore, from Eq. (7.74), the transfer function of the desired first-order highpass filter is given by

$$H_{HP}(z) = 0.245237755\left(\frac{1 - z^{-1}}{1 + 0.50952449z^{-1}}\right).$$

Bandpass IIR Digital Filters

As indicated by Figure 7.1(c), the passband of a bandpass filter is in the midband of the frequency range. Such a response cannot be generated with a first-order real-coefficient transfer function. The lowest order bandpass transfer function thus must have a pair of complex conjugate poles. In addition, improved magnitude response is obtained by ensuring the transfer function has a zero at $z = 1$ and a zero at $z = -1$, which forces the magnitude function to go to 0 at $\omega = 0$ and $\omega = \pi$, respectively. A second-order bandpass digital filter is thus described by the transfer function

$$H_{BP}(z) = \frac{K(1 - z^{-2})}{1 - \beta(1 + \alpha)z^{-1} + \alpha z^{-2}}.$$

The above transfer function has a pair of complex conjugate poles at $z = r\,e^{\pm j\phi}$, where $r = \sqrt{\alpha}$ and $\phi = \cos^{-1}\left(\beta(1 + \alpha)/2\sqrt{\alpha}\right)$. Now, for stability, $r < 1$, implying $|\alpha| < 1$. We shall show later in Section 7.9.1 that, for stability, $|\alpha| < 1$ and $|\beta| < 1$.

Its squared magnitude function is given by

$$
\begin{aligned}
|H_{BP}(e^{j\omega})|^2 &= \frac{4K^2(1 - \cos 2\omega)}{[1 + \beta^2(1 + \alpha)^2 + \alpha^2 - 2\beta(1 + \alpha)^2\cos\omega + 2\alpha\cos 2\omega]} \\
&= \frac{4K^2\sin^2\omega}{(1 + \alpha)^2(\beta - \cos\omega)^2 + (1 - \alpha)^2\sin^2\omega},
\end{aligned}
\tag{7.75}
$$

which goes to zero at $\omega = 0$ and at $\omega = \pi$. It assumes a maximum value of $2K/(1 - \alpha)$ at $\omega = \omega_o$ for which $\beta = \cos\omega_o$, or

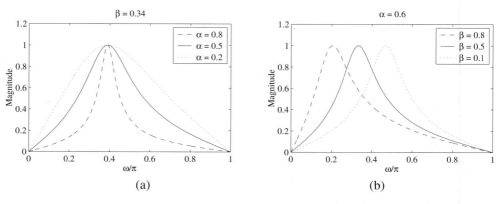

Figure 7.21: Magnitude response of the second-order IIR bandpass filter of Eq. (7.77): (a) three specific values of α with $\beta = 0.34$ and (b) three specific values of β with $\alpha = 0.6$.

$$\omega_o = \cos^{-1}(\beta). \qquad (7.76)$$

ω_o is called the *center frequency* of the bandpass filter.

To make the maximum value of the magnitude function equal to 1, we choose $K = (1-\alpha)/2$, resulting in the transfer function

$$H_{BP}(z) = \frac{1-\alpha}{2} \cdot \frac{1 - z^{-2}}{1 - \beta(1+\alpha)z^{-1} + \alpha z^{-2}}. \qquad (7.77)$$

The frequencies ω_{c1} and ω_{c2}, where the squared magnitude response goes to $1/2$, are called the *3-dB cutoff frequencies,* and their difference, B_w, assuming $\omega_{c2} > \omega_{c1}$, called the *3-dB bandwidth,* is given by

$$B_w = \omega_{c2} - \omega_{c1} = \cos^{-1}\left(\frac{2\alpha}{1+\alpha^2}\right). \qquad (7.78)$$

The *quality factor Q* of the gain response of the bandpass filter is given by

$$Q = \frac{\omega_o}{B_w}.$$

Plots of the magnitude response of the bandpass filter of Eq. (7.77) are given in Figure 7.21 for several values of α and β. It can be seen from Figure 7.21(a) that for a given center frequency ω_o, the 3-dB bandwidth decreases with an increase in α, bringing the poles closer to the unit circle.

It can be shown that the bandpass transfer function of Eq. (7.77) is a BR function for $|\alpha| < 1$ and $|\beta| < 1$.

EXAMPLE 7.10   Design of a Second-Order Bandpass Digital Filter

Design a second-order bandpass digital filter with center frequency at 0.4π and a 3-dB bandwidth of 0.1π. From Eq. (7.76),

$$\beta = \cos(\omega_o) = \cos(0.4\pi) = 0.309016994,$$

and from Eq. (7.78),

$$\frac{2\alpha}{1+\alpha^2} = \cos(B_w) = \cos(0.1\pi) = 0.951056516.$$

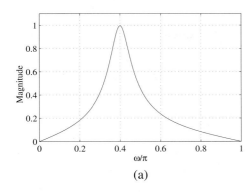 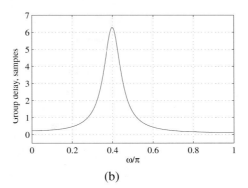

(a)   (b)

Figure 7.22: Magnitude response and the group delay of the IIR bandpass transfer function of Eq. (7.79).

The solution of the above quadratic equation yields two values for α: 1.37638192 and 0.726542528. The stable second-order bandpass transfer functions is obtained for $\alpha = 0.726542528$ and is given by

$$H_{BP}(z) = 0.136728736 \frac{1 - z^{-2}}{1 - 0.53353098z^{-1} + 0.726542528z^{-2}}. \tag{7.79}$$

Here, the magnitudes of both α and β are less than 1, implying the stability of the above transfer function.

Figure 7.22 shows the plots of the magnitude function and the group delay of $H_{BP}(z)$. The group delay has been computed using the M-file `grpdelay`.

The second-order IIR bandpass filter of Eq. (7.77) can be designed in MATLAB using the M-file `iirpeak`. We illustrate its use in Example 7.11.

EXAMPLE 7.11   **Direct Design of a Second-Order IIR Bandpass Filter**

We repeat the design of Example 7.10. The code fragment used is

```
[b,a] = iirpeak(0.4,0.1);
```

The numerator and denominator coefficient vectors obtained after running the above program are as follows:

```
b =
0.13672873599732     0   -0.13672873599732


a =
1.00000000000000   -0.53353098266474     0.72654252800536
```

which are seen to be the same as those derived in Example 7.10.

Bandstop IIR Digital Filters

We observe from Figure 7.1(d) that the stopband of a bandstop filter is in the midband of the frequency range. Such a response cannot also be generated with a first-order real-coefficient transfer function. We can place a zero of the transfer function at ω_o in the mid-frequency range by choosing a second-order factor of the form of $(1 - 2\cos\omega_o z^{-1} + z^{-2})$ in the numerator, which will force the magnitude function to go to 0 at ω_o. The frequency ω_o is called the *notch frequency*. Thus, a second-order bandstop transfer function is of the form

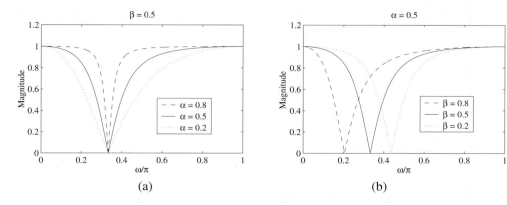

Figure 7.23: Frequency response of the second-order IIR bandstop filter of Eq. (7.80). (a) Three specific values of α with $\beta = 0.5$, and (b) three specific values of β with $\alpha = 0.5$.

$$H_{BS}(z) = \frac{K(1 - 2\beta z^{-1} + z^{-2})}{1 - \beta(1 + \alpha)z^{-1} + \alpha z^{-2}},$$

where $\beta = \cos \omega_o$. The maximum value of the magnitude function of the above transfer function is equal to $2K/(1 + \alpha)$ at both $\omega = 0$ and $\omega = \pi$. By choosing $K = (1 + \alpha)/2$, we can make the maximum values of the magnitude function to be equal to 1, resulting in the transfer function:

$$H_{BS}(z) = \frac{1 + \alpha}{2} \cdot \frac{1 - 2\beta z^{-1} + z^{-2}}{1 - \beta(1 + \alpha)z^{-1} + \alpha z^{-2}}. \tag{7.80}$$

Here, also for stability, $|\alpha| < 1$, and $|\beta| < 1$. The magnitude response of the above bandstop transfer function is plotted in Figure 7.23 for various values of α and β. A digital filter with a transfer function given by Eq. (7.80) is more commonly called a *notch filter* [Con69]. The *3-dB notch bandwidth B_w* is given by Eq. (7.78).

The *quality factor Q* of the gain response of the bandstop filter is given by

$$Q = \frac{\omega_o}{B_w}.$$

The bandstop transfer function of Eq. (7.80) is again a BR function for $|\alpha| < 1$ and $|\beta| < 1$.

The second-order bandstop filter of Eq. (7.80) can be directly designed using the M-file `iirnotch`.

Higher-Order IIR Digital Filters

By cascading the simple digital filters described above, we can implement digital filters with sharper magnitude responses. For example, for a cascade of K first-order lowpass sections characterized by the transfer function of Eq. (7.71), the overall structure has a transfer function $G_{LP}(z)$ given by

$$G_{LP}(z) = \left(\frac{1 - \alpha}{2} \cdot \frac{1 + z^{-1}}{1 - \alpha z^{-1}} \right)^K. \tag{7.81}$$

Using Eq. (7.72), we obtain the corresponding squared-magnitude function of the cascaded filter as

$$|G_{LP}(e^{j\omega})|^2 = \left[\frac{(1-\alpha)^2(1+\cos\omega)}{2(1+\alpha^2-2\alpha\cos\omega)} \right]^K. \tag{7.82}$$

To determine the relation between its 3-dB cutoff frequency ω_c and the parameter α, we set

$$\left[\frac{(1-\alpha)^2(1+\cos\omega_c)}{2(1+\alpha^2-2\alpha\cos\omega_c)} \right]^K = \frac{1}{2}, \tag{7.83}$$

which when solved for α yields, for a stable $G_{LP}(z)$,

$$\alpha = \frac{1+(1-C)\cos\omega_c - (\sqrt{2C-C^2})\sin\omega_c}{1-C+\cos\omega_c}, \tag{7.84}$$

where

$$C = 2^{(K-1)/K}. \tag{7.85}$$

It should be noted that Eq. (7.84) reduces to Eq. (7.73b) for $K = 1$.

EXAMPLE 7.12   Design of a Lowpass Filter Using a Single-Stage and a Multistage Cascade Structure

Design a lowpass filter with a 3-dB cutoff frequency at $\omega_c = 0.4\pi$ using a single-stage realization and a cascade of four first-order lowpass sections, and compare their gain responses.

From Eq. (7.73b) or, equivalently from Eq. (7.84), with $K = 1$, that is, $C = 1$, and $\omega_c = 0.4\pi$, we arrive at $\alpha = 0.1584$ for a single-stage design. Likewise, from Eq. (7.85), $C = 1.6818$ for $K = 4$, which when substituted in Eq. (7.84) along with $\omega_c = 0.4\pi$, yields $\alpha = -0.251$. Figure 7.24 shows the gain responses of a single first-order lowpass filter (plot marked $K = 1$) and a cascade of four identical first-order lowpass filters (plot marked $K = 4$). As can be seen from this figure, cascading has resulted in a sharper roll-off in the gain response, as expected.

Likewise, a cascade of first-order highpass sections results in a highpass filter with a sharper roll-off in the gain response.

Figure 7.25 shows the magnitude responses of a single second-order bandpass filter (plot marked 1), a cascade of two identical second-order bandpass filters (plot marked 2), and a cascade of three identical second-order bandpass filters (plot marked 3). All bandpass sections are characterized by $\alpha = 0.2$ and $\beta = 0.34$. Since the parameter β of all second-order sections is identical, the center frequency of the cascade structure is the same as that of the single section. However, the 3-dB bandwidth decreases with an increase in the number of sections. Similarly, a cascade of identical second-order bandstop filters results in a higher-order bandstop filter with an identical center frequency but with a narrower 3-dB bandwidth.

As pointed out earlier, the 3-dB bandwidth of a second-order bandpass filter can be made smaller by increasing α, which, in turn, brings the poles closer to the unit circle. We shall show later in Chapter 11 that as poles get closer to the unit circle, the performance of a digital filter implemented with finite precision decreases significantly. In such cases, it may be more attractive to design a digital filter by cascading a number of identical sections with poles further away from the unit circle.

7.4.3 Comb Filters

The simple filters of the previous two sections are characterized either by a single passband and/or a single stopband. There are a number of applications where filters with multiple passbands and stopbands are required. The *comb filter* is an example of this latter type of filter. In its most general form, a comb

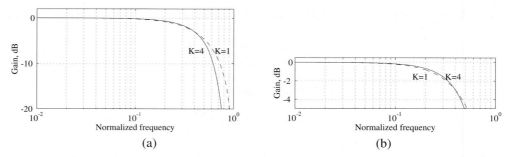

Figure 7.24: (a) Gain responses of a single first-order IIR lowpass filter ($K = 1$) and a cascade of four identical first-order IIR lowpass filters ($K = 4$) with a 3-dB cutoff frequency of $\omega_c = 0.4\pi$. (b) Passband details.

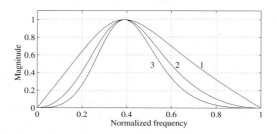

Figure 7.25: Gain responses of a single second-order IIR bandpass filter, a cascade of two identical second-order IIR bandpass filters, and a cascade of three identical second-order IIR bandpass filters. All sections characterized by $\alpha = 0.2$ and $\beta = 0.34$.

filter has a frequency response that is a periodic function of ω with a period $2\pi/L$, where L is a positive integer. If $H(z)$ is a filter with a monotonically decreasing or increasing magnitude response in the interval $[0, \pi]$, a comb filter can easily be generated from it by replacing each delay in its realization with L delays, resulting in a structure with a transfer function given by $G(z) = H(z^L)$. If the magnitude response $|H(e^{j\omega})|$ exhibits a peak at $\omega_p = \pi/2$, then the magnitude response of $|G(e^{j\omega})|$ will exhibit L peaks at $\pi(1 + 4k)/2L$, $1 \le k \le L$. Likewise, if the magnitude response $|H(e^{j\omega})|$ has a notch at $\omega_o = \pi/2$, then the magnitude response of $|G(e^{j\omega})|$ will have L notches at $\pi(1 + 4k)/2L$, $1 \le k \le L$. It should be noted that a comb filter can be generated from either an FIR or an IIR prototype filter.

FIR Comb Filters

To illustrate the generation of the FIR comb filter, consider the prototype lowpass FIR filter of Eq. (7.64), which has a lowpass magnitude response as indicated in Figure 7.18(a). The comb filter generated from $H_0(z)$ has a transfer function

$$G_0(z) = H_0(z^L) = \tfrac{1}{2}(1 + z^{-L}), \tag{7.86}$$

with a corresponding magnitude response as sketched in Figure 7.26(a) for $L = 5$. The new filter is essentially a notch filter with L notch frequencies in the range $0 \le \omega < 2\pi$ located at $\omega = (2k + 1)\pi/L$ and has L peaks in its magnitude response located at $\omega = 2\pi k/L$, $0 \le k \le L - 1$.

The comb filter $G_1(z)$ generated from the prototype highpass filter of Eq. (7.67) has a transfer function given by

$$G_1(z) = H_1(z^L) = \tfrac{1}{2}(1 - z^{-L}), \tag{7.87}$$

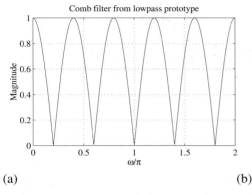 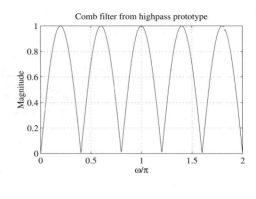

(a)　　　　　　　　　　　　　　　　　　(b)

Figure 7.26: Magnitude responses of FIR comb filters: (a) generated from a prototype lowpass filter of Eq. (7.64) with $L = 5$, and (b) generated from a prototype highpass filter of Eq. (7.67) with $L = 5$.

with a magnitude response as indicated in Figure 7.26(b) for $L = 5$. This comb filter again has L notch frequencies in the range $0 \leq \omega < 2\pi$ located at $\omega = 2\pi k/L$, $0 \leq k \leq L - 1$, which are exactly at the locations of the peaks of the comb filter $G_0(z)$ of Eq. (7.86). Likewise, its magnitude response has L peaks at $\omega = (2k + 1)\pi/L$, $0 \leq k \leq L - 1$, which are also precisely the locations of the notch frequencies of $G_0(z)$.

Depending on the application, comb filters with other types of periodic magnitude responses can be easily generated by appropriately choosing the prototype filter. For example, the moving-average filter of Eq. (6.80),

$$H(z) = \frac{1 - z^{-M}}{M(1 - z^{-1})},$$

can be used as the prototype. The above filter has a peak magnitude at $\omega = 0$ and $M - 1$ notches at $\omega = 2\pi\ell/M$, $1 \leq \ell \leq M - 1$. The comb filter generated from this prototype has a transfer function

$$G(z) = H(z^L) = \frac{1 - z^{-LM}}{M(1 - z^{-L})},$$

whose magnitude response has L peaks at $\omega = 2\pi k/L$, $0 \leq k \leq L - 1$, and $L(M - 1)$ notches at $\omega = 2\pi k/LM$, $1 \leq k \leq L(M - 1)$. By choosing L and M appropriately, peaks and notches can be created at desired locations. In ionospheric measurements of electron concentration, the weak lunar spectral components are usually masked by the strong solar spectral components. These two spectral components have been separated by using two such comb filters [Ber76].

One of the applications of a comb filter considered in Section 15.5.1 is in creating special audio effects in musical sound processing, where both FIR and IIR prototypes are employed. Comb filters with multiple notch frequencies find applications in the cancellation of periodic interference. The comb filter is also employed in LORAN navigation systems for the suppression of cross-rate interferences [Jac96].

An interesting application of the comb filters $G_0(z)$ and $G_1(z)$ of Eqs. (7.86) and (7.87), respectively, is in digital color television receivers for separating the luminance component containing the intensity information and the chrominance components containing the color information from the composite video signal [Orf96]. The basic structure for this purpose is shown in Figure 7.27, where the delay chain is chosen to provide a line delay, that is, the time to scan one horizontal line. However, a complete separation of the two components is not possible with this structure. Moreover, the filter $G_0(z)$ acts as a lowpass filter by averaging two successive horizontal lines of the video signal which blurs the luminance component. On

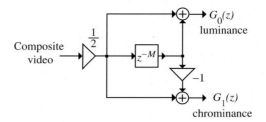

Figure 7.27: Filter structure for the separation of the luminance and chrominance components of a composite video signal.

the other hand, improved separation of the two components can be achieved by the structure of Figure 7.27 when the delay chain is chosen to provide a frame delay. Unfortunately, the separation fails if there is some motion from frame to frame.

IIR Comb Filters

The transfer functions of the simplest forms of the prototype IIR filter are given by

$$H_0(z) = K \frac{1 - z^{-1}}{1 - \alpha z^{-1}}, \tag{7.88a}$$

$$H_1(z) = K \frac{1 + z^{-1}}{1 - \alpha z^{-1}}, \tag{7.88b}$$

where $|\alpha| < 1$ for stability. Note that $H_0(z)$ is a first-order highpass filter with a zero at $z = 1$. Likewise, $H_1(z)$ is a first-order lowpass filter with a zero at $z = -1$. For a maximum gain of 0 dB, the scale factor K of $H_0(z)$ should be set equal to $(1 + \alpha)/2$ and the the scale factor K of $H_1(z)$ should be set equal to $(1 - \alpha)/2$. The transfer functions of the comb filters of order L generated from the prototype transfer functions of Eqs. (7.88a) and (7.88b) are thus

$$G_0(z) = H_0(z^L) = K \frac{1 - z^{-L}}{1 - \alpha z^{-L}}, \tag{7.89a}$$

$$G_0(z) = H_1(z^L) = K \frac{1 + z^{-L}}{1 - \alpha z^{-L}}. \tag{7.89b}$$

The IIR comb filters of Eqs. (7.89a) and (7.89b) can be designed using the M-file `iircomb`, which is available with several options. We illustrate the application of this M-file in Example 7.13.

EXAMPLE 7.13   Design of an IIR Comb Filter

Design an IIR comb filter of order 8 with a 3 dB bandwidth of 0.2π from a prototype IIR lowpass and an IIR highpass filter. The code fragments used for the design are

```
[b,a] = iircomb(8, 0.2,'type');
```

where `'type'` is `'notch'` for a lowpass prototype and is `'peak'` for a highpass prototype.indexM-file!iircomb Figure 7.28 shows the gain response plots of the designed comb filters.

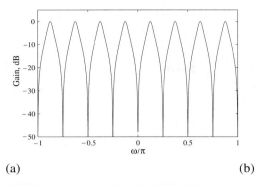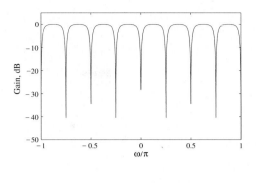

Figure 7.28: Gain responses of IIR comb filters: (a) generated from a prototype highpass IIR filter of Eq. (7.89a) with $L = 8$ and (b) generated from a prototype lowpass IIR filter of Eq. (7.89b) with $L = 8$.

7.5   Complementary Transfer Functions

A set of digital transfer functions with complementary characteristics often finds useful applications in practice, such as efficient realizations of the transfer functions, low sensitivity realizations, and filter bank design and implementation. We describe next four useful complementary relations and indicate some of their applications.

7.5.1   Delay-Complementary Transfer Functions

A set of L transfer functions $\{H_0(z), H_1(z), \ldots, H_{L-1}(z)\}$ is defined to be *delay-complementary* of each other if the sum of their transfer functions is equal to some integer multiple of the unit delay [Vai93]; that is,

$$\sum_{k=0}^{L-1} H_k(z) = \gamma z^{-n_0}, \quad \gamma \neq 0, \tag{7.90}$$

where n_0 is a nonnegative integer.

Consider a Type 1 linear-phase FIR transfer function $H_0(z)$ of even order N. From Eq. (7.55), its frequency response is of the form

$$H_0(e^{j\omega}) = e^{-j((\omega N/2)-\beta)} \breve{H}_0(\omega), \tag{7.91}$$

where $\breve{H}_0(\omega)$ is the *amplitude response*. Assume in the passband, $1 - \delta_p \leq \breve{H}_0(\omega) \leq 1 + \delta_p$, and in the stopband, $-\delta_s \leq \breve{H}_0(\omega) \leq \delta_s$. The delay-complementary transfer function $H_1(z)$ of $H_0(z)$ is then given by

$$H_1(z) = z^{-N/2} - H_0(z). \tag{7.92}$$

Note that Eq. (7.92) satisfies Eq. (7.90) with $\gamma = 1$ and $n_0 = N/2$. Now, $H_1(z)$ has a frequency response given by

$$H_1(e^{j\omega}) = e^{-j(n_0\omega-\beta)} \breve{H}_1(\omega) = e^{-j(n_0\omega-\beta)}[1 - \breve{H}_0(\omega)]. \tag{7.93}$$

It follows therefore from Eq. (7.93), in the stopband of $H_1(z)$, $-\delta_p \leq \breve{H}_1(\omega) \leq \delta_p$, and in the passband of $H_1(z)$, $1 - \delta_s \leq \breve{H}_1(\omega) \leq 1 + \delta_s$. As a result, $H_1(z)$ has a magnitude response characteristic that is complementary to that of $H_0(z)$ with a stopband exactly identical to the passband of $H_0(z)$ and a passband exactly equal to the stopband of $H_0(z)$. For example, if $H_0(z)$ is a lowpass filter, $H_1(z)$ will be a highpass

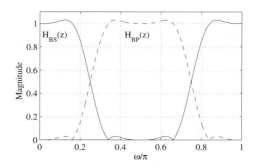

Figure 7.29: Delay-complementary linear-phase bandstop and bandpass FIR filters of Eqs. (7.94a) and (7.94b).

filter, and vice versa. Likewise, if $H_0(z)$ is a bandpass filter, $H_1(z)$ will be a bandstop filter, and vice versa. At the frequency ω_o for which $\breve{H}_0(\omega_o) = \breve{H}_1(\omega_o) = 0.5$, the gain responses of both filters are 6 dB below their maximum values. As a result, ω_o is called the *6-dB crossover frequency*.

EXAMPLE 7.14   An Example of a Delay-Complementary FIR Filter Pair

Consider the Type 1 bandstop transfer function $H_{BS}(z)$ given below [Aca83]:

$$H_{BS}(z) = \tfrac{1}{64}(1 + z^{-2})^4(1 - 4z^{-2} + 5z^{-4} + 5z^{-8} - 4z^{-10} + z^{-12}).\qquad(7.94a)$$

Its delay-complementary bandpass transfer function is given by

$$H_{BP}(z) = z^{-10} - H_{BS}(z) = -\tfrac{1}{64}(1 - z^{-2})^4(1 + 4z^{-2} + 5z^{-4} + 5z^{-8} + 4z^{-10} + z^{-12}).\qquad(7.94b)$$

Figure 7.29 shows the plots of the magnitude responses of $H_{BS}(z)$ and $H_{BP}(z)$.

The delay-complementary transfer function can be determined using the MATLAB code fragment d = firlp2hp(b, 'wide'), where d and b are the coefficient vectors of the delay-complementary transfer function and the parent Type 1 FIR transfer function, respectively.

An interesting application of the delay-complementary FIR transfer function pair is in digital television receivers [Orf96]. It has been pointed out earlier that the structure of Figure 7.27 for the separation of the luminance and chrominance components of the composite video signal tends to blur the luminance output, resulting in the loss of vertical details. The low-frequency vertical details can be recovered from the output of the comb filter $G_1(z)$ of Figure 7.27 by a lowpass filter and added to the output of the comb filter $G_0(z)$. The vertical details can be removed from the output of $G_1(z)$ by filtering it through a filter that is delay-complementary to the lowpass filter. In practice, the lowpass filter $H_{LP}(z)$ employed is a bandstop filter whose low-frequency passband coincides with the frequency range of the desired vertical details, while its delay-complementary filter $H_{HP}(z)$ is a bandpass transfer function. The overall structure is thus as shown in Figure 7.30, where the delay-chain of length K in the top path is chosen to equalize the total delay in both the top and bottom paths.

One set of delay-complementary linear-phase bandpass/bandstop filters proposed for use in Figure 7.30 is the one given in Eqs. (7.94a) and (7.94b) for which $K = 10$. Other linear-phase bandstop FIR transfer functions suitable for vertical details recovery are given in Problem 7.71.

Delay-complementary filter sets can also be used as crossover filters for separating the digital audio input signals into two or three subsignals occupying different frequency bands, which are then used to drive the appropriate speakers of a loudspeaker system [Orf96]. Design of such delay-complementary

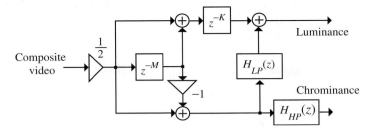

Figure 7.30: Luminance and chrominance components separation filter structure with improved vertical details.

crossover filters is considered in Exercises M10.20 and M10.21. Another important application of the delay-complementary property is in the realization of low-sensitivity FIR digital filters, discussed in Section 12.9.3.

7.5.2 Allpass-Complementary Transfer Functions

A set of M digital transfer functions $\{H_i(z)\}$, $0 \le i \le M - 1$, is defined to be *allpass-complementary* of each other if the sum of their transfer functions is equal to an allpass function $A(z)$ [Neu84a]; that is,

$$\sum_{i=0}^{M-1} H_i(z) = A(z). \tag{7.95}$$

EXAMPLE 7.15   An Allpass-Complementary IIR Filter Pair

Consider the first-order lowpass transfer function of Eq. (7.71) and the first-order highpass transfer function of Eq. (7.74). By adding these two transfer functions, we arrive at

$$H_{LP}(z) + H_{HP}(z) = \frac{1-\alpha}{2} \cdot \frac{1+z^{-1}}{1-\alpha z^{-1}} + \frac{1+\alpha}{2} \cdot \frac{1-z^{-1}}{1-\alpha z^{-1}}$$

$$= \frac{(1-\alpha)(1+z^{-1}) + (1+\alpha)(1-z^{-1})}{2(1-\alpha z^{-1})}$$

$$= \frac{2(1-\alpha z^{-1})}{2(1-\alpha z^{-1})} = 1,$$

which is a zeroth-order allpass transfer function. Hence, $H_{LP}(z)$ and $H_{HP}(z)$ are allpass-complementary pair.

It can be shown that the second-order bandpass transfer function $H_{BP}(z)$ of Eq. (7.77) and the second-order bandstop transfer function $H_{BS}(z)$ of Eq. (7.80) are also allpass-complementary pair with respect to a zeroth order allpass function.

7.5.3 Power-Complementary Transfer Functions

A set of M stable digital transfer functions $\{H_i(z)\}$, $0 \le i \le M - 1$, is defined to be *power-complementary* of each other if the sum of the squares of their magnitude responses is equal to a constant K for all values of ω [Neu84a], [Vai93]; that is,

$$\sum_{i=0}^{M-1} |H_i(e^{j\omega})|^2 = K, \qquad \text{for all } \omega, \tag{7.96a}$$

where $K > 0$ is a constant. By analytic continuation,[6] the above property is equivalent to

$$\sum_{i=0}^{M-1} H_i(z^{-1})H_i(z) = K, \qquad \text{for all } z, \tag{7.96b}$$

for real-coefficient transfer functions $H_i(z)$. Usually, by scaling the transfer functions, the power-complementary property is defined with respect to $K = 1$. For a pair of power-complementary transfer functions, $H_0(z)$ and $H_1(z)$, the frequency ω_o where $|H_0(e^{j\omega_o})|^2 = |H_1(e^{j\omega_o})|^2 = \frac{1}{2}$, is called the *crossover frequency*. At this frequency, the gain responses of both filters are 3 dB below their maximum values. As a result, ω_o is also called the *3-dB cutoff frequency* of both filters.

Consider two transfer functions $H_0(z)$ and $H_1(z)$ described by

$$H_0(z) = \tfrac{1}{2}[\mathcal{A}_0(z) + \mathcal{A}_1(z)], \tag{7.97a}$$

$$H_1(z) = \tfrac{1}{2}[\mathcal{A}_0(z) - \mathcal{A}_1(z)], \tag{7.97b}$$

$\mathcal{A}_0(z)$ and $\mathcal{A}_1(z)$ are stable allpass transfer functions. It follows from the above that the sum of the two transfer functions is an allpass function $A_0(z)$, and hence, $H_0(z)$ and $H_1(z)$ of Eqs. (7.97a) and (7.97b) are an *allpass-complementary* pair. It can be easily shown that the two filters described by Eq. (7.97a) and (7.97b) also form a *power-complementary* pair (Problem 7.73). It can also be easily shown that the transfer functions $H_0(z)$ and $H_1(z)$ of Eqs. (7.97a) and (7.97b) are also bounded real transfer functions (Problem 7.74).

The IIR power-complementary transfer function of a specified IIR transfer function can be determined using the M-file `iirpowcomp`. Its use is illustrated in Example 7.16.

EXAMPLE 7.16   Determination of the Power-Complementary Transfer Function Using MATLAB

Consider the fifth-order Type 2 Chebyshev lowpass transfer function with a minimum stopband attenuation of 50 dB at $\omega_s = 0.6\pi$:[7]

$$H_0(z) = \frac{0.0455846(1 + 3.090382\,z^{-1} + 5.068672\,z^{-2} + 5.068672\,z^{-3} + 3.090382\,z^{-4} + z^{-5})}{1 - 0.906343\,z^{-1} + 0.97418\,z^{-2} - 0.337845\,z^{-3} + 0.1138565\,z^{-4} - 0.008823\,z^{-5}}.$$

Its power-complementary transfer function obtained using the function `iirpowcomp` is given by

$$H_1(z) = \frac{0.1044077(1 - 5z^{-1} + 10z^{-2} - 10z^{-3} + 5z^{-4} - z^{-5})}{1 - 0.906343\,z^{-1} + 0.97418\,z^{-2} - 0.337845\,z^{-3} + 0.1138565\,z^{-4} - 0.008823\,z^{-5}}.$$

The plots of the gain responses of the two filters are shown in Figure 7.31.

7.5.4 Doubly-Complementary Transfer Functions

A set of M stable digital transfer functions satisfying both the allpass-complementary property of Eq. (7.95) and the power-complementary property of Eq. (7.96a) is known as a *doubly-complementary* set [Neu84a].

---

[6]See Section 6.7.3.

[7]This transfer function has been obtained using MATLAB. For details, see Section 9.6.

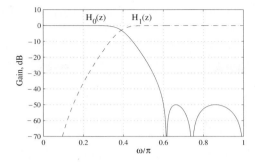

Figure 7.31: Gain responses of power-complementary IIR transfer functions of Example 7.16.

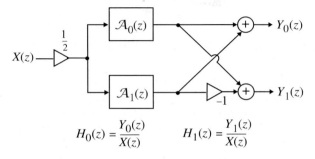

$$H_0(z) = \frac{Y_0(z)}{X(z)} \qquad H_1(z) = \frac{Y_1(z)}{X(z)}$$

Figure 7.32: Parallel allpass realization of doubly-complementary IIR transfer functions.

A pair of doubly-complementary IIR transfer functions, $H_0(z)$ and $H_1(z)$, with a sum of allpass decomposition in the form of Eqs. (7.97a) and (7.97b) can be simply realized by a parallel connection of the constituent allpass filters, as indicated in Figure 7.32. We shall demonstrate in Section 12.9.2 that such realizations also ensure low sensitivity in the passband with respect to the multiplier coefficients.

EXAMPLE 7.17   An Example of a Doubly-Complementary Filter Pair

The first-order lowpass transfer function of Eq. (7.71) can be expressed as

$$H_{LP}(z) = \frac{1}{2}\left[\frac{1 - \alpha + z^{-1} - \alpha z^{-1}}{1 - \alpha z^{-1}} \right]$$

$$= \frac{1}{2}\left[1 + \frac{-\alpha + z^{-1}}{1 - \alpha z^{-1}} \right] = \frac{1}{2}[\mathcal{A}_0(z) + \mathcal{A}_1(z)], \tag{7.98}$$

where

$$\mathcal{A}_0(z) = 1, \qquad \mathcal{A}_1(z) = \frac{-\alpha + z^{-1}}{1 - \alpha z^{-1}}. \tag{7.99}$$

From Eq. (7.97b), its power-complementary highpass transfer function is thus given by

$$H_{HP}(z) = \frac{1}{2}[\mathcal{A}_0(z) - \mathcal{A}_1(z)] = \frac{1}{2}\left[1 - \frac{-\alpha + z^{-1}}{1 - \alpha z^{-1}} \right] = \frac{1 + \alpha}{2} \cdot \frac{1 - z^{-1}}{1 - \alpha z^{-1}}, \tag{7.100}$$

which is precisely the first-order highpass transfer function of Eq. (7.74).

It can be easily shown that the bandpass transfer function $H_{BP}(z)$ of Eq. (7.77) and the bandstop transfer function $H_{BS}(z)$ of Eq. (7.80) form a doubly-complementary pair (Problem 7.76).

7.5.5   Power-Symmetric Filters and Conjugate Quadrature Filters

A real-coefficient causal digital filter with a transfer function $H(z)$ is said to be a *power-symmetric filter* if it satisfies the condition [Vai93]

$$H(z)H(z^{-1}) + H(-z)H(-z^{-1}) = K, \qquad (7.101)$$

where $K > 0$ is a constant. It can be shown that the gain function $\mathcal{G}(\omega)$ of a power-symmetric transfer function at $\omega = \pi/2$ is given by $10 \log_{10} K - 3$ dB (Problem 7.77).

If we define $G(z) = H(-z)$, then it follows from Eq. (7.101) that $H(z)$ and $G(z)$ are power-complementary as

$$H(z)H(z^{-1}) + G(z)G(z^{-1}) = \text{a constant.} \qquad (7.102)$$

If $H(z)$ of Eq. (7.101) is the transfer function of an FIR digital filter of order N, then the FIR digital filter with a transfer function

$$G(z) = z^{-N}H(-z^{-1}) \qquad (7.103)$$

is called the *conjugate quadratic filter* of $H(z)$, and vice versa [Vai88a]. Note that by definition, $G(z)$ is also a power-symmetric causal filter. It follows from Eqs. (7.101) and (7.103) that a pair of conjugate quadratic FIR filters $H(z)$ and $G(z)$ with real coefficients are also power-complementary as they satisfy Eq. (7.102).

EXAMPLE 7.18   An Example of a Power-Symmetric FIR Transfer Function

Consider $H(z) = 1 - 2z^{-1} + 6z^{-2} + 3z^{-3}$. We form

$$H(z)H(z^{-1}) + H(-z)H(-z^{-1}) = (1 - 2z^{-1} + 6z^{-2} + 3z^{-3})(1 - 2z + 6z^2 + 3z^3)$$

$$+ (1 + 2z^{-1} + 6z^{-2} - 3z^{-3})(1 + 2z + 6z^2 - 3z^3)$$

$$= (3z^3 + 4z + 50 + 4z^{-1} + 3z^{-3})$$

$$+ (-3z^3 - 4z + 50 - 4z^{-1} - 3z^{-3}) = 100.$$

Hence, $H(z)$ is a power-symmetric transfer function.

7.6   Inverse Systems

As indicated in Section 2.6.1, two LTI causal discrete-time systems with impulse responses $h_1[n]$ and $h_2[n]$ are inverses of each other if

$$h_1[n] \circledast h_2[n] = \delta[n]. \qquad (7.104)$$

An application of the inverse system design, as pointed out earlier, is in the recovery of a signal $x[n]$ that has been transmitted through an imperfect transmission channel. The received signal $y[n]$, in general, will be different from $x[n]$ because it will be distorted by the impulse response $h_1[n]$ of the channel. To recover the original signal $x[n]$, we need to pass $y[n]$ through a system with an impulse response $h_2[n]$, which is the inverse of the channel's impulse response (see Figure 7.33). The output $v[n]$ of the inverse system will be identical to the desired input $x[n]$.

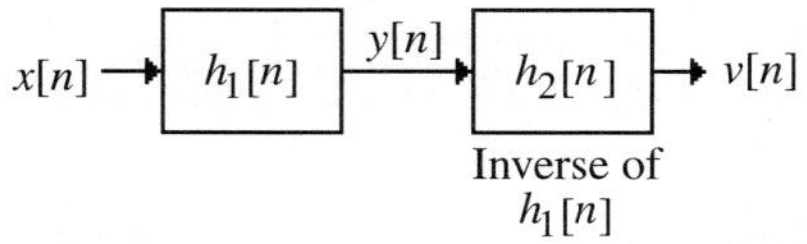

Figure 7.33: Cascade of a discrete-time system $h_1[n]$ with its inverse discrete-time system $h_2[n]$.

7.6.1   Representation in the z-Domain

It is easy to characterize the inverse system in the z-domain. Taking the z-transform of both sides of Eq. (7.104), we get

$$H_1(z)\,H_2(z) = 1, \qquad (7.105)$$

where $H_1(z)$ and $H_2(z)$ are the z-transforms of $h_1[n]$ and $h_2[n]$, respectively. From Eq. (7.105) it follows that the transfer function $H_2(z)$ of the inverse system is simply the reciprocal of $H_1(z)$; that is,

$$H_2(z) = \frac{1}{H_1(z)}. \qquad (7.106)$$

For a rational transfer function $H_1(z)$

$$H_1(z) = \frac{P(z)}{D(z)}, \qquad (7.107)$$

the transfer function $H_2(z)$ of the inverse system is then given by

$$H_2(z) = \frac{D(z)}{P(z)}. \qquad (7.108)$$

It follows from Eqs. (7.107) and (7.108) that the poles (zeros) of the inverse system $H_2(z)$ are the zeros (poles) of the system $H_1(z)$.

EXAMPLE 7.19   Determination of the Inverse of a Causal Stable Minimum-Phase LTI IIR Filter

Determine the inverse of the causal stable LTI system characterized by the transfer function:

$$H_1(z) = \frac{\left(z - \frac{1}{4}\right)\left(z + \frac{1}{5}\right)}{\left(z + \frac{1}{2}\right)\left(z - \frac{1}{3}\right)}, \qquad |z| > \frac{1}{2}.$$

Note that $H_1(z)$ is a stable system. Using Eq. (7.106), we arrive at the transfer function of the inverse system:

$$H_2(z) = \frac{\left(z + \frac{1}{2}\right)\left(z - \frac{1}{3}\right)}{\left(z - \frac{1}{4}\right)\left(z + \frac{1}{5}\right)}.$$

There are three possible ROCs of the above inverse system:

$$\mathcal{R}_1: \quad |z| > \tfrac{1}{4},$$

$$\mathcal{R}_2: \quad \tfrac{1}{5} < |z| < \tfrac{1}{4},$$

$$\mathcal{R}_3: \quad |z| < \tfrac{1}{5}.$$

If we take $\mathcal{R}_1$ as the ROC, then the inverse system is stable and has a causal impulse response. If we instead choose $\mathcal{R}_3$ as the ROC, then the inverse system is unstable and has an anticausal impulse response. If, on the other hand, $\mathcal{R}_2$ is selected as the ROC, the inverse system is unstable and has a two-sided impulse response.

It follows from Example 7.19 that to obtain a unique inverse, the ROC needs to be known a priori. To this end, it is a usual practice to look for a causal inverse of a causal system. The causal inverse of the parent causal system with a minimum-phase transfer function is always stable. However, the inverse of a nonminimum-phase system is unstable if causality is imposed.

7.6.2   Equalization of a Nonminimum-Phase Channel

An inverse system is often needed for the equalization of the frequency response of a communication channel so that the input signal applied to the channel can be recovered without any distortion after passing through the channel. The recovery algorithm is simpler if the channel can be represented by a minimum-phase transfer function, in which case the channel can be cascaded with an inverse system with a causal stable transfer function that is simply the reciprocal of the channel's transfer function. However, in the case of a channel modelled by a causal stable nonminimum-phase transfer function, the recovery of the input signal requires fairly complex algorithms. To get around this problem in the latter case, one solution is to equalize the channel by cascading it with a discrete-time system so that the overall transfer function is minimum-phase. Example 7.20 illustrates this approach.

EXAMPLE 7.20   Frequency Response Equalization of a Nonminimum-Phase Channel

Let the transmission channel be modeled by the causal stable nonminimum-phase transfer function

$$H_1(z) = \frac{(z-4)(z+5)}{\left(z+\frac{1}{2}\right)\left(z-\frac{1}{3}\right)}, \qquad |z| > \frac{1}{2}.$$

The above transfer function can be expressed in the form

$$H_1(z) = H_{\min}(z)A_1(z),$$

where

$$H_{\min}(z) = \frac{(4z-1)(5z+1)}{\left(z+\frac{1}{2}\right)\left(z-\frac{1}{3}\right)}, \qquad |z| > \frac{1}{2},$$

is a causal stable minimum-phase transfer function, and

$$A_1(z) = \frac{(z-4)(z+5)}{(4z-1)(5z+1)}, \qquad |z| > \frac{1}{4},$$

is a causal stable allpass transfer function.

To compensate for the magnitude response of the channel, we can cascade it with a causal stable filter with a transfer function that is reciprocal of $H_{\min}(z)$; that is,

$$H_2(z) = \frac{\left(z+\frac{1}{2}\right)\left(z-\frac{1}{3}\right)}{(4z-1)(5z+1)} \qquad |z| > \frac{1}{4}.$$

The transfer function of the cascaded system is then given by

$$H_{\text{overall}}(z) = H_1(z)H_2(z) = A_1(z).$$

On the other hand, to recover the input signal fully without any distortion, the channel must be cascaded with its exact inverse system with a transfer function given by

$$A_2(z) = \frac{(4z+1)(5z+1)}{(z-4)(z+5)},$$

which is seen to be an unstable transfer function if causality is imposed, implying a ROC given by $|z| > 5$. However, the above allpass section $A_2(z)$ has a stable noncausal impulse response if the ROC is chosen as $|z| < 4$. Hence, by operating the noncausal allpass section $A_2(z)$ in reversed time, the cascaded system will have a minimum-phase impulse response [Abr97].

7.6.3 Deconvolution

If the parent causal system has a known impulse response $h[n]$ and is excited by a causal input signal $x[n]$, then knowing the output signal $y[n]$ for $n \geq 0$, we can determine the samples of the input signal using a recursive relation without determining the inverse system. To develop this relation, we recall that the input–output relation in the time domain is given by

$$y[n] = x[n] \circledast h[n] = \sum_{k=0}^{n} x[k]\, h[n-k], \qquad n \geq 0. \tag{7.109}$$

From Eq. (7.109), for $n = 0$ we have

$$y[0] = x[0]\, h[0],$$

which leads to

$$x[0] = \frac{y[0]}{h[0]}. \tag{7.110}$$

To determine $x[n]$ for $n \geq 1$, we rewrite Eq. (7.109) as

$$y[n] = x[n]\, h[0] + \sum_{k=0}^{n-1} x[k]\, h[n-k],$$

which yields

$$x[n] = \frac{y[n] - \sum_{k=0}^{n-1} x[k]\, h[n-k]}{h[0]}, \qquad n \geq 1, \tag{7.111}$$

provided $h[0] \neq 0$. The process of determining the sequence $x[n]$ from the convolution sum given in Eq. (7.109) using Eqs. (7.110) and (7.111) is called *deconvolution*. The M-file deconv of MATLAB can be employed to carry out the deconvolution algorithm, as illustrated by Example 7.21.

EXAMPLE 7.21   Deconvolution Using MATLAB

Let the sequences $\{y[n]\}$ and $\{h[n]\}$ be causal finite-length sequences given by

$$y[n] = \{-2 \quad -4 \quad 1 \quad 3 \quad 1 \quad 5 \quad 1 \quad -3\},$$

$$h[n] = \{1 \quad 2 \quad 0 \quad -1\}.$$

Using MATLAB, we deconvolve the sequence $\{y[n]\}$ by the sequence $\{h[n]\}$ to determine the sequence $\{x[n]\}$. To this end, Program 7_1 can be used. When the program is run, the sequences $\{y[n]\}$ and $\{h[n]\}$ are entered as vectors, with the sample at time index $n = 0$ being the first sample. The output data generated by running the above program are given by

```
Sequence x[n]
-2     0     1    -1     3

Remainder Sequence r[n]
 0     0     0     0     0     0     0     0
```

To verify that the above result is correct, we convolve the sequence $\{x[n]\}$ generated with the sequence $\{h[n]\}$ using the M-file conv, which results in the output sequence:

```
 2    -4     1     3     1     5     1    -3
```

which is seen to be identical to the sequence $\{y[n]\}$ specified earlier.

Program 7_1.m

An alternate interpretation of the deconvolution algorithm given by these two equations can be established by treating the convolution sum as a polynomial multiplication in the z-domain. If $Y(z)$, $X(z)$, and $H(z)$ denote the z-transforms of the causal sequences $y[n]$, $x[n]$, and $h[n]$, respectively, then in the z-domain, the convolution sum of Eq. (7.109) can be written as

$$Y(z) = X(z)H(z). \tag{7.112}$$

Hence, $X(z)$ can be found by dividing the polynomial $Y(z)$ by the polynomial $H(z)$; that is,

$$X(z) = \frac{Y(z)}{H(z)}. \tag{7.113}$$

Now, the z-transforms $Y(z)$, $X(z)$, and $H(z)$ are polynomials in z^{-1}. Hence, $X(z)$ can be found by a long division of $Y(z)$ by $H(z)$, as indicated in Example 6.19 for the inverse z-transform computation with no carries performed out of a column [Pie96]. We illustrate this approach in Example 7.22.

EXAMPLE 7.22   Deconvolution Using Polynomial Division

We deconvolve $\{y[n]\} = \{6, 10, 3, -2, 5, -6\}, 0 \le n \le 5$, by the sequence $\{h[n]\} = \{2, 4, 1, -3\}, 0 \le n \le 3$. To this end, we divide $\{y[n]\}$ by $\{h[n]\}$, in a manner similar to the long division algorithm without any carries out of a column. The process is indicated below:

				n:	0	1	2	3	4	5
				$x[n]$:	3	−1	2			
2	4	1	−3		6	10	3	−2	5	−6
					6	12	3	−9		
					−2	0	7	5	−6	
					−2	−4	−1	3		
						4	8	2	−6	
									0	

The sequence $\{x[n]\}$ obtained is thus $x[n] = \{3 \quad -1 \quad 2\}$. To verify the above result, we carry out the convolution of the sequences $\{x[n]\}$ and $\{h[n]\}$ using the M-file conv, which results in the sequence as given below

```
 6     10     3     -2     5     -6
```

which is seen to be identical to the sequence $\{y[n]\}$ given above.

7.7 System Identification

There are applications where the objective is to determine either the impulse response $h[n]$ or the transfer function $H(z)$ of an unknown initially relaxed causal LTI system by exciting it with a known input sequence $x[n]$ and observing the corresponding output $y[n]$. The system identification is thus dual to the problem of determining the input $x[n]$ knowing the impulse response $h[n]$ and the output $y[n]$ described in the previous section. In the time domain, the system identification problem can be solved by interchanging the roles of the input and the impulse response in the method outlined above.

The recursive relation for computing the impulse response samples $h[n]$ of a causal LTI system from the specified causal input sequence $x[n]$ and the observed output sequence $y[n]$ is therefore as follows:

$$h[0] = \frac{y[0]}{x[0]}, \qquad h[n] = \frac{y[n] - \sum_{k=0}^{n-1} h[k]\,x[n-k]}{x[0]}, \qquad n \geq 1,$$

provided that $x[0] \neq 0$. The above process can be implemented in MATLAB by a simple modification of Program 7_1.

Alternately, $h[n]$ can be determined in the z-domain by dividing the z-transform $Y(z)$ of $y[n]$ by the z-transform $X(z)$ of $x[n]$. To this end, again the function deconv can be employed.

If the causal LTI system has a rational transfer function $H(z)$ of known order M, the numerator and the denominator coefficients of the transfer function can be determined from the first $2M + 1$ impulse response coefficients. The method to determine $H(z)$ is described in Section 11.1.4.

A second method of system identification is based on computing the energy density spectrum $S_{xx}(e^{j\omega})$ of the input signal $x[n]$ and the cross-energy density spectrum $S_{yx}(e^{j\omega})$ of the input signal $x[n]$ and the output signal $y[n]$. These spectrums can be evaluated by taking the DTFTs of the autocorrelation sequence $r_{xx}[\ell]$ of $x[n]$ and the cross-correlation sequence $r_{yx}[\ell]$ of $y[n]$ and $x[n]$. Consider a stable LTI discrete-time system with an impulse response $h[n]$. The input–output relation of this system is given by

$$y[n] = \sum_{k=-\infty}^{\infty} h[k]x[n-k]. \tag{7.114}$$

We assume that the autocorrelation sequence $r_{xx}[\ell]$ of the input is known. The cross-correlation sequence $r_{yx}[\ell]$ is defined by

$$r_{yx}[\ell] = \sum_{n=-\infty}^{\infty} y[n]x[n-\ell]. \tag{7.115}$$

Substituting Eq. (7.114) in Eq. (7.115), we get

$$r_{yx}[\ell] = \sum_{n=-\infty}^{\infty} \left(\sum_{k=-\infty}^{\infty} h[k]x[n-k] \right) x[n-\ell]$$

$$= \sum_{k=-\infty}^{\infty} h[k] \left(\sum_{n=-\infty}^{\infty} x[n-k]x[n-\ell] \right) = \sum_{k=-\infty}^{\infty} h[k]r_{xx}[\ell-k]. \tag{7.116}$$

If the system is assumed to have a causal finite-length impulse response of length N, then Eq. (7.116) reduces to

$$r_{yx}[\ell] = \sum_{k=0}^{N-1} h[k]r_{xx}[\ell-k]. \tag{7.117}$$

In the z-domain, Eq. (7.116) is equivalent to $S_{yx}(z) = H(z)S_{xx}(z)$, where $S_{xx}(z)$ and $S_{yx}(z)$ are the z-transforms of $r_{xx}[\ell]$ and $r_{yx}[\ell]$, respectively, and $H(z)$ is the transfer function of the LTI system. On the unit circle, Eq. (7.117) reduces to

$$S_{yx}(e^{j\omega}) = H(e^{j\omega})S_{xx}(e^{j\omega}) = H(e^{j\omega})|X(e^{j\omega})|^2, \qquad (7.118)$$

using Eq. (3.69). From Eq. (7.118), it follows that the frequency response of the LTI system can be expressed as

$$H(e^{j\omega}) = \frac{S_{yx}(e^{j\omega})}{S_{xx}(e^{j\omega})}. \qquad (7.119)$$

If $x[n]$ is selected to have a constant energy spectrum for all values of ω, that is, $S_{xx}(e^{j\omega}) = 1/K, 0 \le |\omega| \le \pi$, then Eq. (7.119) reduces to

$$H(e^{j\omega}) = K S_{yx}(e^{j\omega}).$$

It follows from Eq. (7.119) that the frequency response $H(e^{j\omega})$ of an LTI discrete-time system can be determined by taking the ratio of the cross-energy spectrum of the output and the input sequences, and the energy spectrum of the input. The frequency response is proportional to the cross-energy spectrum if the input has a constant energy spectrum.

In some applications, where the input signal $x[n]$ is not known, the system can be identified by computing the autocorrelation of the output signal $y[n]$, which is defined by

$$r_{yy}[\ell] = \sum_{n=-\infty}^{\infty} y[n]y[n - \ell]. \qquad (7.120)$$

Substituting Eq. (7.114) in Eq. (7.120) we arrive at

$$\begin{aligned}
r_{yy}[\ell] &= \sum_{n=-\infty}^{\infty} \left(\sum_{k=-\infty}^{\infty} h[k]x[n - k] \right) \left(\sum_{m=-\infty}^{\infty} h[m - \ell]x[n - m + \ell] \right) \\
&= \sum_{k=-\infty}^{\infty} h[k] \sum_{m=-\infty}^{\infty} h[m - \ell] \left(\sum_{n=-\infty}^{\infty} x[n - k]x[n - m + \ell] \right) \\
&= \sum_{k=-\infty}^{\infty} h[k] \sum_{m=-\infty}^{\infty} h[m - \ell]r_{xx}[m - \ell - k] = h[\ell] \circledast h[-\ell] \circledast r_{xx}[\ell]. \qquad (7.121)
\end{aligned}$$

In the z-domain, Eq. (7.121) becomes

$$S_{yy}(z) = H(z)H(z^{-1})S_{xx}(z),$$

where $S_{yy}(z)$ is the z-transform of $r_{yy}[\ell]$. On the unit circle, the above equation reduces to

$$S_{yy}(e^{j\omega}) = |H(e^{j\omega})|^2 S_{xx}(e^{j\omega}).$$

For an input signal with a flat energy density spectrum, we then have

$$S_{yy}(e^{j\omega}) = K|H(e^{j\omega})|^2,$$

or equivalently in the z-domain, with $K = 1$, this becomes

$$S_{yy}(z) = H(z)H(z^{-1}). \qquad (7.122)$$

If the system is characterized by a rational transfer function $H(z) = P(z)/D(z)$, then

$$S_{yy}(z) = \frac{A(z)}{B(z)} = \frac{P(z)P(z^{-1})}{D(z)D(z^{-1})},$$

indicating that the numerator and the denominator polynomials of $S_{yy}(z)$ exhibit mirror-image symmetry. To determine $H(z)$, one can determine the roots of the polynomials $A(z)$ and $B(z)$ and associate appropriate factors of these polynomials with the numerator and the denominator of $H(z)$.

It should be noted from the above discussion that the autocorrelation of the output signal can provide only the magnitude response of the system but not the phase response. A solution of Eq. (7.122) thus leads to many possible answers. A single solution can, however, be obtained by imposing some additional constraints on the phase response of the system.

EXAMPLE 7.23   Determination of the Transfer Function from Its Energy Density Spectrum

Let the output energy density spectrum $S_{yy}(e^{j\omega})$ of a causal stable LTI discrete-time system excited by an input sequence with a unity energy spectrum be given by

$$S_{yy}(e^{j\omega}) = \frac{1.04 + 0.4\cos\omega}{1.25 - \cos\omega}.$$

Using trigonometric identities, we rewrite the above as

$$S_{yy}(e^{j\omega}) = \frac{1.04 + 0.2(e^{j\omega} + e^{-j\omega})}{1.25 - 0.5(e^{j\omega} + e^{-j\omega})}.$$

Substituting $z = e^{j\omega}$ in the above and making use of Eq. (7.122), we arrive at

$$H(z)H(z^{-1}) = \frac{1.04 + 0.2(z + z^{-1})}{1.25 - 0.5(z + z^{-1})} = -0.4\left(\frac{z^2 + 5.2z + 1}{z^2 - 2.5z + 1}\right)$$

$$= -0.4\frac{(z+5)(z+0.2)}{(z-2)(z-0.5)} = \frac{(z^{-1} + 0.2)(1 + 0.2z^{-1})}{(z^{-1} - 0.5)(1 - 0.5z^{-1})}.$$

Therefore, for a minimum-phase system, the transfer function is given by

$$H(z) = \frac{1 + 0.2z^{-1}}{1 - 0.5z^{-1}},$$

whereas for a non-minimum-phase system, the transfer function is

$$H(z) = \frac{z^{-1} + 0.2}{1 - 0.5z^{-1}}.$$

7.8   Digital Two-Pairs

The LTI discrete-time systems considered so far are single-input, single-output structures characterized by a transfer function. Often, such a system can be efficiently realized by interconnecting two-input, two-output structures, more commonly called *two-pairs* [Mit73a]. Figure 7.34 shows two commonly used block diagram representations of a two-pair, with Y_1 and Y_2 denoting the two outputs and X_1 and X_2 denoting the two inputs, where the dependence on the variable z has been omitted for simplicity. We

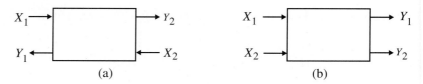

Figure 7.34: Two different representations of a digital two-pair.

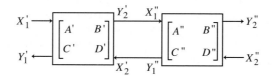

Figure 7.35: Γ-cascade connection of two-pairs.

consider here the transform-domain characterizations of such digital filter structures and discuss several two-pair interconnection schemes for the development of more complex structures. Later, we outline minimum-multiplier realizations of allpass transfer functions based on the two-pair representation.

7.8.1 Characterization

The input–output relation of a digital two-pair is given by

$$\begin{bmatrix} Y_1 \\ Y_2 \end{bmatrix} = \begin{bmatrix} t_{11} & t_{12} \\ t_{21} & t_{22} \end{bmatrix} \begin{bmatrix} X_1 \\ X_2 \end{bmatrix}, \tag{7.123}$$

In the above relation, the matrix $\boldsymbol{\tau}$ given by

$$\boldsymbol{\tau} = \begin{bmatrix} t_{11} & t_{12} \\ t_{21} & t_{22} \end{bmatrix} \tag{7.124}$$

is called the *transfer matrix* of the two-pair. From Eq. (7.123), it follows that the transfer parameters can be found as follows:

$$t_{11} = \left.\frac{Y_1}{X_1}\right|_{X_2=0}, \quad t_{12} = \left.\frac{Y_1}{X_2}\right|_{X_1=0}, \quad t_{21} = \left.\frac{Y_2}{X_1}\right|_{X_2=0}, \quad t_{22} = \left.\frac{Y_2}{X_2}\right|_{X_1=0}. \tag{7.125}$$

An alternative characterization of the two-pair is in terms of its chain parameters as

$$\begin{bmatrix} X_1 \\ Y_1 \end{bmatrix} = \begin{bmatrix} A & B \\ C & D \end{bmatrix} \begin{bmatrix} Y_2 \\ X_2 \end{bmatrix}, \tag{7.126}$$

where the matrix $\boldsymbol{\Gamma}$ given by

$$\boldsymbol{\Gamma} = \begin{bmatrix} A & B \\ C & D \end{bmatrix} \tag{7.127}$$

is called the *chain matrix* of the two-pair.

The relations between the transfer parameters and the chain parameters can be easily derived and are given as

$$t_{11} = \frac{C}{A}, \quad t_{12} = \frac{AD - BC}{A}, \quad t_{21} = \frac{1}{A}, \quad t_{22} = -\frac{B}{A}, \tag{7.128a}$$

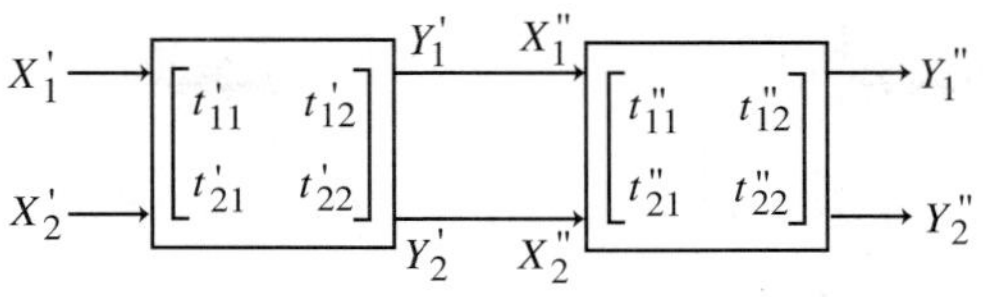

Figure 7.36: τ-cascade connection of two-pairs.

$$A = \frac{1}{t_{21}}, \quad B = -\frac{t_{22}}{t_{21}}, \quad C = \frac{t_{11}}{t_{21}}, \quad D = \frac{t_{12}t_{21} - t_{11}t_{22}}{t_{21}}. \tag{7.128b}$$

7.8.2 Two-Pair Interconnection Schemes

Two or more two-pairs can be connected in cascade in two different ways. The cascade connection of two-pairs shown in Figure 7.35 is called a Γ-*cascade*. We next determine the characterization of the overall two-pair. Let the individual two-pairs be characterized by their chain parameters as

$$\begin{bmatrix} X_1' \\ Y_1' \end{bmatrix} = \begin{bmatrix} A' & B' \\ C' & D' \end{bmatrix} \begin{bmatrix} Y_2' \\ X_2' \end{bmatrix}, \quad \begin{bmatrix} X_1'' \\ Y_1'' \end{bmatrix} = \begin{bmatrix} A'' & B'' \\ C'' & D'' \end{bmatrix} \begin{bmatrix} Y_2'' \\ X_2'' \end{bmatrix}. \tag{7.129}$$

However, from Figure 7.35, $X_1'' = Y_2'$, and $Y_1'' = X_2'$. Substituting these relations in the first equation of Eq. (7.129) and combining the two equations, we arrive at

$$\begin{bmatrix} X_1' \\ Y_1' \end{bmatrix} = \begin{bmatrix} A' & B' \\ C' & D' \end{bmatrix} \begin{bmatrix} A'' & B'' \\ C'' & D'' \end{bmatrix} \begin{bmatrix} Y_2'' \\ X_2'' \end{bmatrix} \tag{7.130}$$

Therefore, the chain matrix of the overall cascade is given by the product of the individual chain matrices:

$$\begin{bmatrix} A & B \\ C & D \end{bmatrix} = \begin{bmatrix} A' & B' \\ C' & D' \end{bmatrix} \begin{bmatrix} A'' & B'' \\ C'' & D'' \end{bmatrix}. \tag{7.131}$$

A second type of cascade connection, called the τ-*cascade,* is shown in Figure 7.36. If the individual two-pairs are described by their transfer matrices

$$\begin{bmatrix} Y_1' \\ Y_2' \end{bmatrix} = \begin{bmatrix} t_{11}' & t_{12}' \\ t_{21}' & t_{22}' \end{bmatrix} \begin{bmatrix} X_1' \\ X_2' \end{bmatrix}, \quad \begin{bmatrix} Y_1'' \\ Y_2'' \end{bmatrix} = \begin{bmatrix} t_{11}'' & t_{12}'' \\ t_{21}'' & t_{22}'' \end{bmatrix} \begin{bmatrix} X_1'' \\ X_2'' \end{bmatrix}, \tag{7.132}$$

the input–output relation of the cascaded two-pairs of Figure 7.36 is then

$$\begin{bmatrix} Y_1'' \\ Y_2'' \end{bmatrix} = \begin{bmatrix} t_{11}'' & t_{12}'' \\ t_{21}'' & t_{22}'' \end{bmatrix} \begin{bmatrix} t_{11}' & t_{12}' \\ t_{21}' & t_{22}' \end{bmatrix} \begin{bmatrix} X_1' \\ X_2' \end{bmatrix}. \tag{7.133}$$

It thus follows that the overall cascade is characterized by a transfer matrix that is the product of the transfer matrix of the constituent two-pairs; that is,

$$\begin{bmatrix} t_{11} & t_{12} \\ t_{21} & t_{22} \end{bmatrix} = \begin{bmatrix} t_{11}'' & t_{12}'' \\ t_{21}'' & t_{22}'' \end{bmatrix} \begin{bmatrix} t_{11}' & t_{12}' \\ t_{21}' & t_{22}' \end{bmatrix}. \tag{7.134}$$

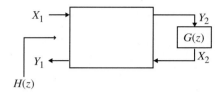

$H(z)$

Figure 7.37: A constrained two-pair.

Another useful interconnection is the constrained two-pair indicated in Figure 7.37. If we denote the transfer function Y_1/X_1 as $H(z)$, then it can be shown that $H(z)$ can be expressed in terms of the two parameters and the constraining transfer function $G(z)$ as

$$H(z) = \frac{Y_1}{X_1} = \frac{C + D \cdot G(z)}{A + B \cdot G(z)} \tag{7.135a}$$

$$= t_{11} + \frac{t_{12}t_{21}G(z)}{1 - t_{22}G(z)}. \tag{7.135b}$$

7.9 Algebraic Stability Test

We have shown in Section 6.7.5 that the BIBO stability of a causal rational transfer function requires that all its poles be inside the unit circle. For very-high-order transfer functions, it is difficult to determine the pole locations analytically, and the use of some type of root finding computer program is necessary. We outline here a simple algebraic stability test procedure that does not require the determination of the pole locations. The algorithm is based on the realization of an allpass transfer function with a denominator that is the same as that of the transfer function of interest.

7.9.1 The Stability Triangle

For a second-order transfer function, the stability can be checked easily by examining its denominator coefficients. Let

$$D(z) = 1 + d_1 z^{-1} + d_2 z^{-2} \tag{7.136}$$

denote the denominator of the transfer function. In terms of its poles, $D(z)$ can be expressed as

$$D(z) = (1 - \lambda_1 z^{-1})(1 - \lambda_2 z^{-1}) = 1 - (\lambda_1 + \lambda_2)z^{-1} + \lambda_1\lambda_2 z^{-2}. \tag{7.137}$$

Comparing Eqs. (7.136) and (7.137), we obtain

$$d_1 = -(\lambda_1 + \lambda_2), \quad d_2 = \lambda_1\lambda_2. \tag{7.138}$$

Now for stability of the transfer function, its poles must be inside the unit circle; that is,

$$|\lambda_1| < 1, \quad |\lambda_2| < 1.$$

Since from Eq. (7.136), the coefficient d_2 is given by the product of the poles, we must have

$$|d_2| < 1. \tag{7.139}$$

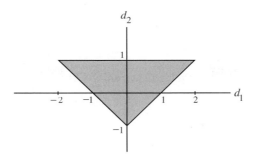

Figure 7.38: Stability triangle for a second-order IIR digital transfer function. Stability region is the shaded area.

Now the roots of the polynomial $D(z)$ are given by

$$\lambda_1 = -\frac{d_1 + \sqrt{d_1^2 - 4d_2}}{2},$$

$$\lambda_2 = -\frac{d_1 - \sqrt{d_1^2 - 4d_2}}{2}. \tag{7.140}$$

It can be shown (Problem 7.99) that a second coefficient condition is obtained using Eq. (7.140) and is given by

$$|d_1| < 1 + d_2. \tag{7.141}$$

The region in the (d_1, d_2)-plane where the two coefficient conditions of Eqs. (7.139) and (7.141) are satisfied is a triangle, as sketched in Figure 7.38, and is known as the *stability triangle* for a second-order digital transfer function [Jac96].

EXAMPLE 7.24   Illustration of Stability Testing of a Causal Second-Order IIR Transfer Function

For the second-order bandpass transfer function $H_{BP}(z)$ of Eq. (7.79) obtained using $\alpha = 0.726542528$, we observe that $d_1 = -0.53353098$ and $d_2 = 0.726542528$. In this case, $|d_2| < 1$, satisfying the condition of Eq. (7.139), and $|d_1| < 1 + d_2$, satisfying the condition of Eq. (7.141), indicating that $H_{BP}(z)$ is a stable transfer function.

On the other hand, the second-order bandpass transfer function obtained using $\alpha = 1.3768192$ is given by

$$H'_{BP}(z) = -0.18819096 \left(\frac{1 - z^{-2}}{1 - 0.7343423986z^{-1} + 1.37638192z^{-2}} \right),$$

where we have $d_1 = -0.7343423986$ and $d_2 = 1.37638192$. Note that here the condition of Eq. (7.141) is satisfied. However, the condition of Eq. (7.139) is not satisfied as $|d_2| > 1$, and hence, this transfer function is unstable.

7.9.2   A Stability Test Procedure

It is impossible to develop simple conditions on the coefficients of the denominator polynomial of a higher-order transfer function similar to that developed for a second-order function as given above. However, a number of methods have been proposed to determine the stability of an Mth-order transfer function $H(z)$ without factoring out the roots of its denominator polynomial $D_M(z)$. We describe below an alternate derivation of the *Schur-Cohn stability test method* [Sch18],[Coh22], which is based on an algebraic order reduction of an allpass function [Vai87e].

Let

$$D_M(z) = \sum_{i=0}^{M} d_i z^{-i}, \tag{7.142}$$

where we assume for simplicity $d_0 = 1$. We first form an Mth-order allpass transfer function from $D_M(z)$ as

$$A_M(z) = \frac{\tilde{D}_M(z)}{D_M(z)} = \frac{d_M + d_{M-1}z^{-1} + d_{M-2}z^{-2} + \cdots + d_1 z^{-(M-1)} + z^{-M}}{1 + d_1 z^{-1} + d_2 z^{-2} + \cdots + d_{M-1}z^{-(M-1)} + d_M z^{-M}}. \tag{7.143}$$

If we express

$$D_M(z) = \prod_{i=1}^{M}(1 - \lambda_i z^{-1}),$$

it then follows that the coefficient d_M is the product of all roots; that is,

$$d_M = (-1)^M \prod_{i=1}^{M} \lambda_i.$$

Now for stability, we must have $|\lambda_i| < 1$, which implies the condition $|d_M| < 1$. If we define

$$k_M = A_M(\infty) = d_M,$$

then a necessary condition for stability of $A_M(z)$ and, hence, the original transfer function $H(z)$, is given by

$$k_M^2 < 1. \tag{7.144}$$

Assume the above condition holds. Next, we form a new function $A_{M-1}(z)$ according to

$$A_{M-1}(z) = z\left[\frac{A_M(z) - k_M}{1 - k_M A_M(z)}\right] = z\left[\frac{A_M(z) - d_M}{1 - d_M A_M(z)}\right]. \tag{7.145}$$

Substituting in Eq. (7.145) the expression for $A_M(z)$ from Eq. (7.143), we arrive at

$$A_{M-1}(z) = \frac{\begin{array}{c}(d_{M-1} - d_M d_1) + (d_{M-2} - d_M d_2)z^{-1} + \cdots \\ + (d_1 - d_M d_{M-1})z^{-(M-2)} + (1 - d_M^2)z^{-(M-1)}\end{array}}{\begin{array}{c}(1 - d_M^2) + (d_1 - d_M d_{M-1})z^{-1} + \cdots \\ + (d_{M-2} - d_M d_2)z^{-(M-2)} + (d_{M-1} - d_M d_1)z^{-(M-1)}\end{array}}, \tag{7.146}$$

which is seen to be an $(M - 1)$th-order allpass function. $A_{M-1}(z)$ can be rewritten in the form

$$A_{M-1}(z) = \frac{d'_{M-1} + d'_{M-2}z^{-1} + \cdots + d'_1 z^{-(M-2)} + z^{-(M-1)}}{1 + d'_1 z^{-1} + \cdots + d'_{M-2}z^{-(M-2)} + d'_{M-1}z^{-(M-1)}}, \tag{7.147}$$

where

$$d'_i = \frac{d_i - d_M d_{M-i}}{1 - d_M^2}, \qquad i = 1, 2, \ldots, M - 1. \tag{7.148}$$

Now from Eq. (7.145), the poles λ_0 of $A_{M-1}(z)$ are given by the roots of the equation

$$A_M(\lambda_0) = \frac{1}{k_M}. \tag{7.149}$$

By assumption, Eq. (7.144) holds. Hence, Eq. (7.149) implies that

$$|A_M(\lambda_0)| > 1. \tag{7.150}$$

If $A_M(z)$ is a stable allpass function, then according to Eq. (7.20), $|A_M(z)| < 1$ for $|z| > 1$, $|A_M(z)| = 1$ for $|z| = 1$, and $|A_M(z)| > 1$ for $|z| < 1$. Therefore, if $A_M(z)$ is a stable allpass function, then $|\lambda_0| < 1$, or in other words, $A_{M-1}(z)$ is a stable allpass function. Thus, if $A_M(z)$ is a stable allpass function and $k_M^2 < 1$, then $A_{M-1}(z)$ is a stable allpass function.

We now prove the converse; that is, if $A_{M-1}(z)$ is a stable allpass function and $k_M^2 < 1$, then $A_M(z)$ is a stable allpass function. To this end, we invert the relation of Eq. (7.145) to arrive at

$$A_M(z) = \frac{k_M + z^{-1} A_{M-1}(z)}{1 + k_M z^{-1} A_{M-1}(z)}. \tag{7.151}$$

If ζ_0 is a pole of $A_M(z)$, then

$$\zeta_0^{-1} A_{M-1}(\zeta_0) = -\frac{1}{k_M}. \tag{7.152}$$

By assumption, Eq. (7.144) holds. Therefore, $|\zeta_0^{-1} A_{M-1}(\zeta_0)| > 1$; that is,

$$|A_{M-1}(\zeta_0)| > |\zeta_0|. \tag{7.153}$$

Assume $A_{M-1}(z)$ is a stable allpass function; then according to Eq. (7.20), $|A_{M-1}(z)| \le 1$ for $|z| \ge 1$. Now, if $|\zeta_0| \ge 1$, then from Eq. (7.20), $|A_{M-1}(\zeta_0)| \le 1$, which contradicts Eq. (7.153). On the other hand, if $|\zeta_0| < 1$, then from Eq. (7.20), $|A_{M-1}(\zeta_0)| > 1$, satisfying the condition of Eq. (7.153). Thus, if Eq. (7.144) holds and if $A_{M-1}(z)$ is a stable allpass function, then $A_M(z)$ is also a stable allpass function.

Summarizing, a necessary and sufficient set of conditions for the allpass function $A_M(z)$ to be stable is therefore

(a) $k_M^2 < 1$ and

(b) the allpass function $A_{M-1}(z)$ is stable.

Thus, once we have checked the condition $k_M^2 < 1$, we merely test for the stability of the lower-order allpass function $A_{M-1}(z)$. The process can now be repeated, generating a set of coefficients

$$k_M, k_{M-1}, \ldots, k_2, k_1$$

and a set of allpass functions of decreasing orders,

$$A_M(z), A_{M-1}(z), \ldots, A_2(z), A_1(z), A_0(z) = 1.$$

The allpass function $A_M(z)$ is stable if and only if $k_i^2 < 1$ for all i.

EXAMPLE 7.25   Illustration of the Algebraic Stability Testing Method

Let us test the stability of the transfer function

$$H(z) = \frac{1}{4z^4 + 3z^3 + 2z^2 + z + 1}. \tag{7.154}$$

From the above, we arrive at a fourth-order allpass function whose denominator is the same as that of $H(z)$ given above with the numerator being its mirror-image polynomial:

$$A_4(z) = \frac{\frac{1}{4}z^4 + \frac{1}{4}z^3 + \frac{1}{2}z^2 + \frac{3}{4}z + 1}{z^4 + \frac{3}{4}z^3 + \frac{1}{2}z^2 + \frac{1}{4}z + \frac{1}{4}} = \frac{d_4z^4 + d_3z^3 + d_2z^2 + d_1z + 1}{z^4 + d_1z^3 + d_2z^2 + d_3z + d_4}. \tag{7.155}$$

Note that $k_4 = A_4(\infty) = d_4 = 1/4$. Hence $|k_4| < 1$.

Using Eq. (7.148), we next determine the coefficients $\{d_i'\}$ of the third-order allpass transfer function $A_3(z)$ derived from $A_4(z)$:

$$d_i' = \frac{d_i - d_4 d_{4-i}}{1 - d_4^2}, \qquad i = 1, 2, 3. \tag{7.156}$$

Substituting the appropriate values of d_i from Eq. (7.155) in the above expression, we arrive at

$$A_3(z) = \frac{d_3'z^3 + d_2'z^2 + d_1'z + 1}{z^3 + d_1'z^2 + d_2'z + d_3'} = \frac{\frac{1}{15}z^3 + \frac{2}{5}z^2 + \frac{11}{15}z + 1}{z^3 + \frac{11}{15}z^2 + \frac{2}{5}z + \frac{1}{15}}.$$

It can be seen that $k_3 = A_3(\infty) = d_3' = 1/15$. Thus, $|k_3| < 1$.

Following the above procedure, we derive the next two lower-order allpass functions:

$$A_2(z) = \frac{\frac{79}{224}z^2 + \frac{159}{224}z + 1}{z^2 + \frac{159}{224}z + \frac{79}{224}},$$

$$A_1(z) = \frac{\frac{53}{101}z + 1}{z + \frac{53}{101}}.$$

Note from the above that $k_2 = A_2(\infty) = 79/224$, $k_1 = A_1(\infty) = 53/101$, implying that $|k_2| < 1$ and $|k_1| < 1$.

It should be noted that it is not necessary to derive $A_1(z)$ since $A_2(z)$ can be tested for stability using conditions of Eqs. (7.139) and (7.141). These two conditions are obviously satisfied for the $A_2(z)$ given above.

Since all of the stability test conditions are satisfied, $A_4(z)$ and, hence, $H(z)$ of Eq. (7.154) are stable transfer functions.

EXAMPLE 7.26   Illustration of Polynomial Root Location Testing

Determine whether all zeros of the denominator polynomial of a causal transfer function

$$D_4(z) = z^4 + 3.25z^3 + 3.75z^2 + 2.75z + 0.5$$

are inside the unit circle.

We form the allpass function

$$A_4(z) = \frac{\tilde{D}_4(z)}{D_4(z)} = \frac{0.5z^4 + 2.75z^3 + 3.75z^2 + 3.25z + 1}{z^4 + 3.25z^3 + 3.75z^2 + 2.75z + 0.5}.$$

Note from the above that $k_4 = A_4(\infty) = 0.5$, implying $|k_3| < 1$. Next, using Eq. (7.148), we derive the third-order allpass function $A_3(z)$, which can be shown to be equal to

$$A_3(z) = \frac{1.5z^3 + 2.5z^2 + 2.5z + 1}{z^3 + 2.5z^2 + 2.5z + 1.5}.$$

It can be seen from the above that $k_3 = A_3(\infty) = 1.5$, implying $|k_3| > 1$. Therefore, the allpass function $A_4(z)$ is unstable or, equivalently, not all zeros of $D_4(z)$ are inside the unit circle.

The M-file `poly2rc` can be employed to determine the stability test parameters $\{k_i\}$. We illustrate its application in Example 7.27.

EXAMPLE 7.27   Stability Testing Using MATLAB

We test the stability of the transfer function of Eq. (7.154) using MATLAB. To this end we make use of Program 7.2. The input data is the vector den of the coefficients of the denominator polynomial entered inside a square bracket in descending powers of z as indicated below:

Program 7.2.m

```
den = [4   3   2   1   1]
```

The output data are the stability test parameters $\{k_i\}$. The program also has a logical output, which is stable = 1 if the transfer function is stable; otherwise, stable = 0.

The output data generated for the transfer function of Eq. (7.154) are as follows:

```
The stability test parameters are
      0.2500    0.0667    0.3527    0.5248

stable = 1
```

Note that the stability test parameters are identical to those computed in Example 7.25.

7.10 Summary

The concept of filtering is introduced, and several ideal filters are defined. Several simple approximations to the ideal filters are next introduced. In addition, various special types of transfer functions that are often encountered in practice are reviewed. The concept of complementary transfer functions relating a set of transfer functions is discussed, and several types of complementary conditions are introduced.

The inverse system design is encountered in estimating the unknown input of a discrete-time system from its known output. The determination of the transfer function of the inverse of a causal LTI discrete-time system with a rational transfer function is outlined. The recursive computation of the unknown causal input signal from the impulse response of a causal LTI system and its known output is outlined. Next, two methods are outlined for the system identification problem. In one approach, a recursive algorithm is described for determining the impulse response of a causal initially relaxed system from its known input and output sequences. In the second method, the frequency response of the system is determined from the cross-energy spectrum of the output and the input signal and the energy spectrum of the input. Alternately, the square magnitude function of the system can be determined from the energy spectrum of the output and the input signals.

An important building block in the design of a single-input, single-output LTI discrete-time system is the digital two-pair, which is a two-input, two-output LTI discrete-time system. Characterizations of the digital two-pairs and their interconnections are discussed. A very simple algebraic procedure for testing the stability of a causal LTI transfer function is then introduced.

7.11 Problems

7.1 Show that the transfer function of the M-point moving-average filter of Eq. (3.97) is a BR function.

7.2 Consider the first-order causal and stable allpass transfer function given by

$$A_1(z) = \frac{1 - d_1^* z}{z - d_1}.$$

Determine the expression for $\left(1 - |A_1(z)|^2\right)$, and then show that

$$\left(1 - |A_1(z)|^2\right) \begin{cases} < 0, & \text{for } |z|^2 < 1, \\ = 0, & \text{for } |z|^2 = 1, \\ > 0, & \text{for } |z|^2 > 1. \end{cases}$$

Now, using the above approach, show that Property 2 given by Eq. (7.20) holds for any arbitrary causal stable allpass transfer function.

7.3 Derive Property 3 of a stable allpass transfer function given by Eq. (7.21).

7.4 A noncausal LTI FIR discrete-time system is characterized by an impulse response $h[n] = a_1\delta[n-2] - a_2\delta[n-1] - a_3\delta[n] + a_4\delta[n+1] - a_5\delta[n+2]$. For what values of the impulse response samples will its frequency response $H(e^{j\omega})$ have a zero phase?

7.5 Let a causal LTI discrete-time system be characterized by a real impulse response $h[n]$ with a DTFT $H(e^{j\omega})$. Consider the system of Figure P7.1, where $x[n]$ is a finite-length sequence. Determine the frequency response of the overall system $G(e^{j\omega})$ in terms of $H(e^{j\omega})$, and show that it has a zero-phase response.

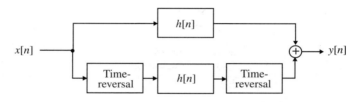

Figure P7.1

7.6 Show that an Mth-order causal complex coefficient allpass transfer function is of the form

$$A_M(z) = \pm\frac{d_M^* + d_{M-1}^* z^{-1} + \cdots + d_1^* z^{-M+1} + z^{-M}}{1 + d_1 z^{-1} + \cdots + d_{M-1} z^{-M+1} + d_M z^{-M}}.$$

7.7 Determine all possible causal stable transfer functions $H(z)$ with a square-magnitude function given by

$$|H(e^{j\omega})|^2 = \frac{9(1.0625 + 0.5\cos\omega)(1.49 - 1.4\cos\omega)}{(1.36 + 1.2\cos\omega)(1.64 + 1.6\cos\omega)}.$$

7.8 Consider the following five FIR transfer functions:

(i) $H_1(z) = 2 + 1.4z^{-1} - 0.9z^{-2} - 0.158z^{-3} + 0.4104z^{-4} + 0.0294z^{-5} - 0.0668z^{-6}$,

(ii) $H_2(z) = 1 + 2.7z^{-1} - 12.61z^{-2} - 24.757z^{-3} + 66.301z^{-4} + 62.072z^{-5} - 126.786z^{-6}$,

(iii) $H_3(z) = 0.2 - 0.26z^{-1} + 1.934z^{-2} + 10.413z^{-3} + 1.934z^{-4} - 0.26z^{-5} + 0.2z^{-6}$,

(iv) $H_4(z) = 1.25 + 0.5z^{-1} - 2.1z^{-2} - 2.1z^{-4} + 0.5z^{-5} + 1.25z^{-6}$,

(v) $H_5(z) = 1.1 + 3.12z^{-1} - 2.5z^{-2} + 0.6z^{-3} + 0.5z^{-4} + 0.06z^{-5} + z^{-6}$.

Using the M-file `zplane`, determine the zero locations of each, and then answer the following questions:

(a) Which one of the above FIR filters have a linear-phase response?

(b) Which one of the above FIR filters have a minimum-phase response?

(c) Which one of the above FIR filters have a maximum-phase response?

7.9 A third-order FIR filter has a transfer function given by

$$G_1(z) = (2 + 3.4z^{-1} - 4z^{-2})(3 - 1.5z^{-1}).$$

(a) Determine the transfer functions of all other FIR filters whose magnitude responses are identical to that of $G_1(z)$.

(b) Which one of these filters has a minimum-phase transfer function, and which one has a maximum-phase transfer function?

(c) If $g_k[n]$ denotes the impulse response of the kth FIR filter determined in Part (a), compute the partial energy of the impulse response given by

$$\mathcal{E}_k[n] = \sum_{m=0}^{n} g_k[m]^2, \qquad 0 \le n \le 3,$$

for all values of k, and show that

$$\sum_{m=0}^{n} |g_k[m]|^2 \le \sum_{m=0}^{n} |g_{\min}[m]|^2,$$

and

$$\sum_{m=0}^{\infty} |g_k[m]|^2 = \sum_{m=0}^{\infty} |g_{\min}[m]|^2,$$

for all values of k, and where $g_{\min}[n]$ is the impulse response of the minimum-phase FIR filter determined in Part (a).

7.10 The transfer functions of five FIR filters with identical magnitude responses are given below:

$$H_1(z) = 1 - 0.5z^{-1} + 0.8z^{-2} - 0.4z^{-3} + 0.25z^{-4} - 0.125z^{-5} + 0.2z^{-6} - 0.1z^{-7},$$
$$H_2(z) = 0.5 + 0.25z^{-1} + 0.4z^{-2} - 0.425z^{-3} + 0.75z^{-4} - 0.75z^{-5} + 0.6z^{-6} - 0.2z^{-7},$$
$$H_3(z) = -0.25 + 0.25z^{-1} + 0.175z^{-2} + 0.7z^{-3} - 0.45z^{-4} + 0.9z^{-5} - 0.6z^{-6} + 0.4z^{-7},$$
$$H_4(z) = -0.5 + z^{-1} - 0.4z^{-2} + 0.8z^{-3} - 0.125z^{-4} + 0.25z^{-5} - 0.1z^{-6} + 0.2z^{-7},$$
$$H_5(z) = -0.1 + 0.2z^{-1} - 0.125z^{-2} + 0.25z^{-3} - 0.4z^{-4} + 0.8z^{-5} - 0.5z^{-6} + z^{-7}.$$

Which transfer has all its zeros outside the unit circle? Which one has all its zeros inside the unit circle? How many other length-8 FIR filters exist that have the same magnitude response as that of the above transfer functions?

7.11 A causal LTI FIR discrete-time system is characterized by an impulse response $h[n] = a_1\delta[n] + a_2\delta[n-1] + a_3\delta[n-2] + a_4\delta[n-3] + a_5\delta[n-4] + a_6\delta[n-5]$. For what values of the impulse response samples will its frequency response $H(e^{j\omega})$ have a constant group delay?

7.12 An FIR LTI discrete-time system is described by the difference equation

$$y[n] = a_1x[n+k] - a_2x[n+k-1] + a_2x[n+k-3] - a_1x[n+k-4],$$

where $y[n]$ and $x[n]$ denote, respectively, the output and the input sequences. Determine the expression for its frequency response $H(e^{j\omega})$. For what values of the constant k will the system have a frequency response $H(e^{j\omega})$ that is a real function of ω?

7.13 Consider the cascade of two causal LTI systems: $h_1[n] = \alpha\delta[n] + \beta\delta[n-1]$ and $h_2[n] = \gamma^n\mu[n]$, $|\beta| < 1$. Determine the frequency response $H(e^{j\omega})$ of the overall system. For what values of α, β and γ will $|H(e^{j\omega})| = K$, where K is a real constant?

7.14 The input–output relation of a nonlinear discrete-time system in the frequency domain is given by

$$Y(e^{j\omega}) = |X(e^{j\omega})|^{\alpha} e^{j \arg X(e^{j\omega})}, \tag{7.157}$$

where $0 < \alpha \leq 1$ and $X(e^{j\omega})$ and $Y(e^{j\omega})$ denote the DTFTs of the input and output sequences, respectively. Determine the expression for its frequency response $H(e^{j\omega}) = Y(e^{j\omega})/X(e^{j\omega})$, and show that it has zero phase. The nonlinear algorithm described by Eq. (7.157) is known as the *alpha-rooting* method and has been used in image enhancement [Jai89].

7.15 An FIR filter of length 3 is defined by a symmetric impulse response; i.e., $h[0] = h[2]$. Let the input to this filter be a sum of two cosine sequences of angular frequencies 0.3 rad/samples and 0.6 rad/samples, respectively. Determine the impulse response coefficients so that the filter passes only the low-frequency component of the input.

7.16 (a) Design a length-5 FIR bandpass filter with an antisymmetric impulse response $h[n]$, i.e., $h[n] = -h[4 - n]$, $0 \leq n \leq 4$, satisfying the following magnitude response values: $|H(e^{j0.3\pi})| = 0.3$ and $|H(e^{j0.6\pi})| = 0.8$.
  (b) Determine the exact expression for the frequency response of the filter designed, and plot its magnitude and phase responses using MATLAB.

7.17 (a) Design a length-4 FIR highpass filter with an antisymmetric impulse response $h[n]$, i.e., $h[n] = -h[3 - n]$, $0 \leq n \leq 3$, satisfying the following magnitude response values: $|H(e^{j0.25\pi})| = 0.2$ and $|H(e^{j0.8\pi})| = 0.8$.
  (b) Determine the exact expression for the frequency response of the filter designed, and plot its magnitude and phase responses using MATLAB.

7.18 An FIR filter of length 5 is defined by a symmetric impulse response; i.e., $h[n] = h[4 - n]$, $0 \leq n \leq 4$. Let the input to this filter be a sum of three cosine sequences of angular frequencies: 0.3 rad/samples, 0.5 rad/samples, and 0.8 rad/samples, respectively. Determine the impulse response coefficients so that the filter blocks only the midfrequency component of the input.

7.19 The frequency response $H(e^{j\omega})$ of a length-4 FIR filter with a real impulse response has the following specific values: $H(e^{j0}) = 13$, $H(e^{j3\pi/4}) = -3 - j4$, and $H(e^{j\pi}) = -3$. Determine $H(z)$.

7.20 The frequency response $H(e^{j\omega})$ of a length-4 FIR filter with a real and antisymmetric impulse response has the following specific values: $H(e^{j\pi}) = 20$ and $H(e^{j3\pi/4}) = -5 - j5$. Determine $H(z)$.

7.21 Consider the two LTI causal digital filters with impulse responses given by

$$h_A[n] = 0.5\delta[n] - \delta[n - 1] + 0.5\delta[n - 2],$$
$$h_B[n] = 0.5\delta[n] + \delta[n - 1] + 0.5\delta[n - 2].$$

  (a) Sketch the magnitude responses of the two filters and compare their characteristics.
  (b) Let $h_A[n]$ be the impulse response of a causal digital filter with a frequency response $H_A(e^{j\omega})$. Define another digital filter whose impulse response $h_C[n]$ is given by

$$h_C[n] = (-1)^n h_A[n], \qquad \text{for all } n.$$

What is the relation between the frequency response $H_C(e^{j\omega})$ of this new filter and the frequency response $H_A(e^{j\omega})$ of the parent filter?

7.22 We have shown that a real-coefficient FIR transfer function $H(z)$ with a symmetric impulse response has a linear-phase response. As a result, the all-pole IIR transfer function $G(z) = 1/H(z)$ will also have a linear-phase response. What are the practical difficulties in implementing $G(z)$? Justify your answer mathematically.

7.23 Prove Eqs. (7.36) and (7.37) of the impulse response coefficients of a causal minimum-phase transfer function.

7.24 Check the stability of each of the following causal IIR transfer functions. If they are not stable, find a stable transfer function with an identical magnitude function. Are there any other transfer functions having the same magnitude response as those shown below?

(a) $H_1(z) = \frac{z^3+3z^2+2z+7}{(2z+3)(z^2+0.5z+0.8)}$, (b) $H_2(z) = \frac{4z^3-2z^2+5z-6}{(1.5z^2+3z-5)(z^2-0.3z+0.7)}$.

7.25 The *notch filter* is used to suppress a particular sinusoidal component of frequency ω_0 of an input signal $x[n]$ and has a transfer function with zeros at $z = e^{\pm j\omega_0}$. For each filter given below, (i) determine the notch frequency ω_0, (ii) show the form of the corresponding sinusoidal sequence to be suppressed, and (iii) verify by computing the output $y[n]$ by convolution that in the steady state, $y[n] = 0$ when the sinusoidal sequence is applied at the input of the filter.

(a) $H_1(z) = 1 - z^{-1} + z^{-2}$, (b) $H_2(z) = 1 - 0.8z^{-1} + z^{-2}$, (c) $H_3(z) = 1 - 1.6z^{-1} + z^{-2}$.

7.26 Let $G_L(z)$ and $G_H(z)$ represent ideal lowpass and highpass filters with magnitude responses as sketched in Figure P7.2(a). Determine the transfer functions $H_k(z) = Y_k(z)/X(z)$ of the discrete-time system of Figure P7.2(b), $k = 0, 1, 2, 3$, and sketch their magnitude responses.

7.27 Let $H_{LP}(z)$ denote the transfer function of an ideal real coefficient lowpass filter with a cutoff frequency of ω_p. Sketch the magnitude response of $H_{LP}(-z)$, and show that it is a highpass filter. Determine the relation between the cutoff frequency of this highpass filter in terms of ω_p and its impulse response in terms of the impulse response $h_{LP}[n]$ of the parent lowpass filter.

7.28 Let $H_{LP}(z)$ denote the transfer function of an ideal real coefficient lowpass filter having a cutoff frequency of ω_p, with $\omega_p < \pi/2$. Consider the complex coefficient transfer function $H_{LP}(e^{j\omega_0}z)$, where $\omega_p < \omega_o < \pi - \omega_p$. Sketch its magnitude response for $-\pi \leq \omega \leq \pi$. What type of filter does it represent? Now consider the transfer function $G(z) = H_{LP}(e^{j\omega_0}z) + H_{LP}(e^{-j\omega_0}z)$. Sketch its magnitude response for $-\pi \leq \omega \leq \pi$. Show that $G(z)$ is a real-coefficient bandpass filter with a passband centered at ω_o. Determine the width of its passband in terms of ω_p and its impulse response $g[n]$ in terms of the impulse response $h_{LP}[n]$ of the parent lowpass filter.

7.29 Let $H_{LP}(z)$ denote the transfer function of an ideal real coefficient lowpass filter with a cutoff frequency of ω_p with $0 < \omega_p < \pi/3$. Show that the transfer function $F(z) = H_{LP}(e^{j\omega_0}z) + H_{LP}(e^{-j\omega_0}z) + H_{LP}(z)$, where $\omega_o = \pi - \omega_p$, is a real-coefficient bandstop filter with the stopband centered at $\omega_o/2$. Determine the width of its stopband in terms of ω_p and its impulse response $f[n]$ in terms of the impulse response $h_{LP}[n]$ of the parent lowpass filter.

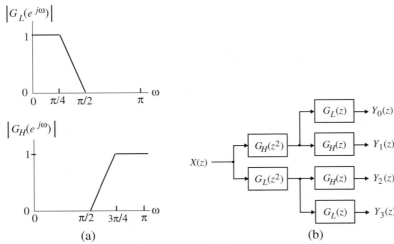

(a)

(b)

Figure P7.2

7.30 Show that the structure shown in Figure P7.3 implements the highpass filter of Problem 7.27.

7.31 Show that the structure shown in Figure P7.4 implements the bandpass filter of Problem 7.28.

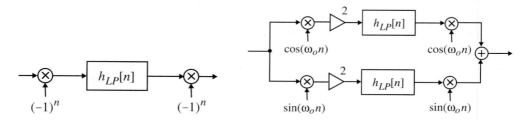

Figure P7.3   **Figure P7.4**

7.32 Let $H(z)$ be an ideal real-coefficient lowpass filter with a cutoff at ω_c, where $\omega_c = \pi/M$. Figure P7.5 shows a single-input, M-output filter structure, called an M-band analysis filter bank, where $H_k(z) = H(ze^{-j2\pi k/M})$, $k = 0, 1, \ldots, M - 1$. Sketch the magnitude response of each filter, and describe the operation of the filter bank.

7.33 Consider a cascade of M sections of the first-order FIR lowpass filter of Eq. (7.64). Show that its 3-dB cutoff frequency is given by Eq. (7.66).

7.34 Consider a cascade of M sections of the first-order FIR highpass filter of Eq. (7.67). Develop the expression for its 3-dB cutoff frequency.

7.35 Verify that the value of α given by Eq. (7.73b) ensures that the transfer function $H_{LP}(z)$ of Eq. (7.71) is stable.

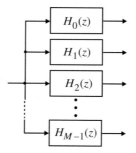

Figure P7.5

7.36 Show by trigonometric manipulation that Eq. (7.73a) can be alternately expressed as

$$\tan\left(\frac{\omega_c}{2}\right) = \frac{1 - \alpha}{1 + \alpha}. \tag{7.158}$$

Next show that the transfer function $H_{LP}(z)$ of Eq. (7.71) is stable for a value of α given by

$$\alpha = \frac{1 - \tan(\omega_c/2)}{1 + \tan(\omega_c/2)}. \tag{7.159}$$

7.37 Design a first-order lowpass IIR digital filter for each of the following normalized 3-dB cutoff frequencies: (a) 0.6 rad/samples, (b) 0.45π.

7.38 Show that the 3-dB cutoff frequency ω_c of the first-order highpass IIR digital filter of Eq. (7.74) is given by Eq. (7.73a).

7.39 Design a first-order highpass IIR digital filter for each of the following normalized 3-dB cutoff frequencies: (a) 0.6 rad/samples, (b) 0.55π.

7.40 The following first-order IIR transfer function has been proposed for clutter removal in MTI radars [Urk58]:

$$H(z) = \frac{1 - z^{-1}}{1 - kz^{-1}}.$$

Determine the magnitude response of the above transfer function and show that it has a highpass response. Scale the transfer function so that it has a 0-dB gain at $\omega = \pi$. Sketch the magnitude responses for $k = 0.95, 0.9$, and -0.5, respectively.

7.41 Show that the center frequency ω_o and the 3-dB bandwidth B_w of the second-order IIR bandpass filter of Eq. (7.75) are given by Eqs. (7.76) and (7.78), respectively.

7.42 Design a second-order bandpass IIR digital filter for each of the following specifications: (a) $\omega_o = 0.55\pi$, $B_w = 0.25\pi$, (b) $\omega_o = 0.3\pi$, $B_w = 0.3\pi$.

7.43 Show that the notch frequency ω_o and the 3-dB notch bandwidth B_ω of the second-order IIR bandstop filter of Eq. (7.80) are given by Eqs. (7.76) and (7.78), respectively.

7.44 Design a second-order bandstop IIR digital filter for each of the following specifications: (a) $\omega_o = 0.35\pi$, $B_w = 0.2\pi$, (b) $\omega_o = 0.6\pi$, $B_w = 0.15\pi$.

7.45 Consider a cascade of K identical first-order lowpass digital filters with a transfer function given by Eq. (7.71). Show that the coefficient α of the first-order section is related to the 3-dB cutoff frequency ω_c of the cascade according to Eq. (7.84), with the parameter C given by Eq. (7.85).

7.46 Consider a cascade of K identical first-order highpass digital filters with a transfer function given by Eq. (7.74). Express the coefficient α of the first-order section in terms of the 3-dB cutoff frequency ω_c of the cascade.

7.47 The filter structures shown in Figure P7.6, where $A_1(z)$ is a stable first-order allpass filter, can be used as low-frequency shelving filters in digital audio equalization [Zöl97] (see Section 15.5.2). Determine the transfer function of each structure.

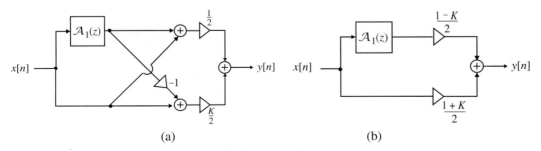

(a)   (b)

Figure P7.6

7.48 The filter structure shown in Figure P7.7(a), where $\mathcal{A}_1(z)$ is a stable first-order allpass filter, can be used as a low-frequency shelving filter in digital audio equalization [Zöl97] (see Section 15.5.2). Likewise, the filter structure shown in Figure P7.7(b), where $\mathcal{A}_1(z)$ is a stable first-order allpass filter, can be used as a high-frequency shelving filter in digital audio equalization [Zöl97]. Determine the transfer function of each structure.

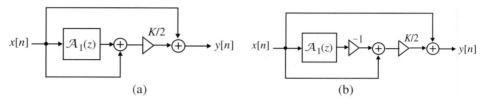

(a)   (b)

Figure P7.7

7.49 If $H(z)$ is a bandpass filter with passband edges at ω_{p1} and ω_{p2}, and stopband edges at ω_{s1} and ω_{s2}, with $\omega_{s1} < \omega_{p1} < \omega_{p2} < \omega_{s2}$, what type of filter is $H(-z)$? Determine the locations of the bandedges of $H(-z)$ in terms of the bandedges of $H(z)$.

7.50 Using the method of Problem 7.27, develop the transfer function $G_{HP}(z)$ of a first-order IIR highpass filter from the transfer function $H_{LP}(z)$ of the first-order IIR lowpass filter given by Eq. (7.71). Is it the same as that of the highpass transfer function of Eq. (7.74)? If not, determine the location of its 3-dB cutoff frequency as a function of the parameter α.

7.51 Let $H(z)$ be an ideal lowpass filter with a cutoff frequency at $\pi/3$. Sketch the magnitude responses of the following systems: (a) $H(z^3)$, (b) $H(z)H(z^3)$, (c) $H(-z)H(z^3)$, and (d) $H(z)H(-z^3)$.

7.52 Show that the amplitude response $\breve{H}(\omega)$ of Type 1 and Type 3 linear-phase FIR transfer functions is a periodic function of ω with a period 2π and the amplitude response $\breve{H}(\omega)$ of Type 2 and Type 4 linear-phase FIR transfer functions is a periodic function of ω with a period 4π.

7.53 A length-13 Type 1 real-coefficient FIR filter has the following zeros: $z_1 = 0.8$, $z_2 = -j$, $z_3 = 2 - j2$, $z_4 = -0.5 + j0.3$. (a) Determine the locations of the remaining zeros. (b) What is the transfer function $H_1(z)$ of the filter?

7.54 A length-12 Type 2 real-coefficient FIR filter has the following zeros: $z_1 = 3.1$, $z_2 = -2 + j4$, $z_3 = 0.8 + j0.4$. (a) Determine the locations of the remaining zeros. (b) What is the transfer function $H_2(z)$ of the filter?

7.55 A length-13 Type 3 real-coefficient FIR filter has the following zeros: $z_1 = 0.1 - j0.599$, $z_2 = -0.3 + j0.4$, $z_3 = 2$. (a) Determine the locations of the remaining zeros. (b) What is the transfer function $H_3(z)$ of the filter?

7.56 A length-12 Type 4 real-coefficient FIR filter has the following zeros: $z_1 = 2.2 + j3.4$, $z_2 = 0.6 + j0.9$, $z_3 = -0.5$. (a) Determine the locations of the remaining zeros. (b) What is the transfer function $H_4(z)$ of the filter?

7.57 Let $H(z)$ be a lowpass filter with unity passband magnitude, a passband edge at ω_p, and a stopband edge at ω_s, as shown in Figure P7.8.

(a) Sketch the magnitude response of the digital filter $G_1(z) = H(z^M)F_1(z)$, where $F_1(z)$ is a lowpass filter with unity passband magnitude, a passband edge at ω_p/M, and a stopband edge at $(2\pi - \omega_s)/M$. What are the bandedges of $G_1(z)$?

(b) Sketch the magnitude response of the digital filter $G_2(z) = H(z^M)F_2(z)$, where $F_2(z)$ is a bandpass filter with unity passband magnitude, and with passband edges at $(2\pi - \omega_p)/M$ and $(2\pi + \omega_p)/M$ and stopband edges at $(2\pi - \omega_s)/M$ and $(2\pi + \omega_s)/M$, respectively. What are the bandedges of $G_2(z)$?

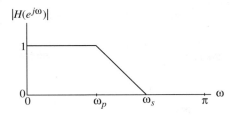

Figure P7.8

7.58 Show analytically that an FIR filter with a constant group delay must have either a symmetric or an antisymmetric impulse response.

7.59 Let the first five impulse response samples of a causal linear-phase FIR filter be given by $h[0] = a$, $h[1] = -b$, $h[2] = -c$, $h[3] = d$, and $h[4] = e$. Determine the remaining impulse response samples of the transfer function of lowest order for each type of linear-phase filter.

7.60 The first five samples of the impulse response of an FIR filter $H(z)$ are given by $h[0] = 1, h[1] = -3, h[2] = -4, h[3] = 6$, and $h[4] = 8$. Determine the remaining impulse response samples of $H(z)$ of lowest order for each type of linear-phase filter. Using zplane, determine the zero locations for $H(z)$ for each type of linear-phase filter. Does $H(z)$ have a zero at $z = 1$ and/or $z = -1$? Do the zeros on the unit circle appear in complex conjugate pairs? Do the zeros not on the unit circle appear in mirror-image symmetry? Justify your answers.

7.61 Let $H_1(z)$, $H_2(z)$, $H_3(z)$, and $H_4(z)$ be, respectively, Type 1, Type 2, Type 3, and Type 4 linear-phase FIR filters. Are the following filters composed of a cascade of the above filters' linear phase? If they are, what are their types?

  (a) $G_a(z) = H_1(z)H_1(z)$,   (b) $G_b(z) = H_1(z)H_2(z)$,   (c) $G_c(z) = H_1(z)H_3(z)$,
  (d) $G_d(z) = H_1(z)H_4(z)$,   (e) $G_e(z) = H_2(z)H_2(z)$,   (f) $G_f(z) = H_3(z)H_3(z)$,
  (g) $G_g(z) = H_4(z)H_4(z)$,   (h) $G_h(z) = H_2(z)H_3(z)$,   (i) $G_i(z) = H_3(z)H_4(z)$.

7.62 Consider a linear-phase FIR transfer function given by $H(z) = F_1(z)F_2(z)$. Determine the factor $F_2(z)$ of lowest order for each of the following choices for $F_1(z)$:

  (a) $F_1(z) = 2.1 - 3.5z^{-1} + 4.2z^{-2}$,   (b) $F_1(z) = 1.4 + 5.2z^{-1} - 2.2z^{-2} + 3.3z^{-3}$.

7.63 Consider a causal FIR transfer function given by

$$H(z) = K\left(1 + h[1]z^{-1} + h[2]z^{-2} + \cdots + h[N]z^{-N}\right) = K\prod_{i=1}^{N}(1 - \lambda_i z^{-1}), \qquad (7.160)$$

where K is a constant. The *root moment* of $H(z)$ is defined by [Fot2001]

$$S_m = \sum_{i=0}^{N} \lambda_i^m, \qquad 1 \le m \le N, \qquad (7.161)$$

where m is the degree of the moment. Prove the *Newton Identities* given by

$$S_m + h[1]S_{m-1} + h[2]S_{m-2} + \cdots + mh[m] = 0, \qquad 1 \le m \le N. \qquad (7.162)$$

Note: The above identities can be iteratively solved for all N root moments.

7.64 Assume that the causal FIR transfer function $H(z)$ of Eq. (7.160) has M_α roots $\{\alpha_i\}$ inside the unit circle, and M_β roots $\{\beta_i\}$ outside the unit circle, where $M_\alpha + M_\beta = N$. We can thus rewrite $H(z)$ as

$$H(z) = K \prod_{i=1}^{M_\alpha}(1 - \alpha_i z^{-1}) \prod_{i=1}^{M_\beta}(1 - \beta_i z^{-1}) = K G_\alpha(z) G_\beta(z), \tag{7.163}$$

where $G_\alpha(z) = \prod_{i=1}^{M_\alpha}(1 - \alpha_i z^{-1})$ is the minimum-phase factor and $G_\beta(z) = \prod_{i=1}^{M_\beta}(1 - \beta_i z^{-1})$ is the maximum-phase factor of $H(z)$.

(a) Show that the root moments of a $H(z)$ with real coefficients are real.

(b) Show that the root moments of a minimum-phase $H(z)$ decrease exponentially with increasing m.

(c) If $\check{H}(\omega)$ and $\theta(\omega)$ denote, respectively, the amplitude and phase responses of $H(z)$, show then [Fot2001]

$$\ln \check{H}(\omega) = \ln(K_1) - \sum_{m=1}^{\infty} \frac{S_m^\alpha - S_{-m}^\beta}{m} \cos(m\omega), \tag{7.164a}$$

$$\theta(\omega) = -\omega M_\beta + \sum_{m=1}^{\infty} \frac{S_m^\alpha - S_{-m}^\beta}{m} \sin(m\omega), \tag{7.164b}$$

where K_1 is an appropriate real constant, $\{S_m^\alpha\}$ are the root moments of the minimum-phase factor $G_\alpha(z)$, and $\{S_{-m}^\beta\}$ are the inverse root moments of the maximum-phase factor $G_\beta(z)$.

(d) Show that for the transfer function $H(z)$ to have linear phase, it must have zeros located outside the unit circle, and determine their number in relation to the number of zeros inside the unit circle.

7.65 (a) Show that the phase delay $\tau_p(\omega)$ of the first-order allpass transfer function

$$A_1(z) = \frac{d_1 + z^{-1}}{1 + d_1 z^{-1}},$$

is given by $\tau_p(\omega) \cong (1 - d_1)/(1 + d_1) = \delta$ [Ste96].

(b) Design a first-order allpass filter with a phase delay of $\delta = 0.5$ sample and operating at a sampling rate of 20 kHz. Determine the error in samples at 1 kHz in the phase delay from its design value of 0.5 sample.

7.66 Consider the second-order allpass transfer function

$$A_2(z) = \frac{d_2 + d_1 z^{-1} + z^{-2}}{1 + d_1 z^{-1} + d_2 z^{-2}}.$$

If δ denotes the desired low-frequency approximate value of the phase delay $\tau_p(\omega) = -\theta(\omega)/\omega$, show that [Fet72]

$$d_1 = 2\left(\frac{2 - \delta}{1 + \delta}\right), \qquad d_2 = \frac{(2 - \delta)(1 - \delta)}{(2 + \delta)(1 + \delta)}.$$

7.67 Let $G(z)$ be a causal stable nonminimum-phase transfer function, and let $H(z)$ denote another causal stable transfer function that is minimum-phase with $\left|G(e^{j\omega})\right| = \left|H(e^{j\omega})\right|$. Show that $G(z) = H(z)A(z)$, where $A(z)$ is a stable causal allpass transfer function.

7.68 A typical transmission channel is characterized by a causal transfer function

$$H(z) = \frac{(3z - 2.1)(z^2 + 2.5z + 5)}{(z - 0.65)(z + 0.48)}.$$

In order to correct for the magnitude distortion introduced by the channel on a signal passing through it, we wish to connect a causal stable digital filter characterized by a transfer function $G(z)$ at the receiving end. Determine $G(z)$.

7.69 Let $H(z)$ be a causal stable minimum-phase transfer function, and let $G(z)$ denote another causal stable transfer function that is nonminimum-phase with $\left|G(e^{j\omega})\right| = \left|H(e^{j\omega})\right|$. If $h[n]$ and $g[n]$ denote their respective impulse responses, show that

(a) $|g[0]| \leq |h[0]|$,

(b) $\sum_{\ell=0}^{n} |g[\ell]|^2 \leq \sum_{\ell=0}^{n} |h[\ell]|^2$.

7.70 Is the transfer function

$$H(z) = \frac{(2z+3)(4z-1)}{(z+0.4)(z-0.6)}$$

minimum-phase? If it is not minimum-phase, then construct a minimum-phase transfer function $G(z)$ such that $\left|G(e^{j\omega})\right| = \left|H(e^{j\omega})\right|$. Determine their corresponding unit sample responses, $g[n]$ and $h[n]$, for $n = 0, 1, 2, 3, 4$. For what values of m is $\sum_{n=0}^{m} |g[n]|^2$ bigger than $\sum_{n=0}^{m} |h[n]|^2$?

7.71 The following bandstop FIR transfer functions $H_{BS}(z)$ have also been proposed for the recovery of vertical details in the structure of Figure 7.30 employed for the separation of the luminance and the chrominance components [Aca83], [Pri80], [Ros75]:

(a) $H_{BS}(z) = \frac{1}{4}(1 + z^{-2})^2$,

(b) $H_{BS}(z) = \frac{1}{16}(1 + z^{-2})^2(-1 + 6z^{-2} - z^{-4})$,

(c) $H_{BS}(z) = \frac{1}{32}(1 + z^{-2})^2(-3 + 14z^{-2} - 3z^{-4})$.

Develop their delay-complementary transfer functions $H_{BP}(z)$.

7.72 Let $A_0(z)$ and $A_1(z)$ be two causal stable allpass transfer functions. Define two causal stable IIR transfer functions as follows:

$$H_0(z) = A_0(z) + A_1(z), \qquad H_1(z) = A_0(z) - A_1(z).$$

Show that the numerators of $H_0(z)$ and $H_1(z)$ are, respectively, a symmetric and an antisymmetric polynomial.

7.73 Show that the two transfer functions of Eqs. (7.97a) and (7.97b) are a power-complementary pair.

7.74 Show that the two transfer functions of Eqs. (7.97a) and (7.97b) are each a BR function.

7.75 Consider the transfer function $H(z)$ given by

$$H(z) = \frac{1}{M} \sum_{k=0}^{M-1} A_k(z),$$

where $A_k(z)$ are stable real-coefficient allpass functions. Show that $H(z)$ is a BR function.

7.76 Show that the bandpass transfer function $H_{BP}(z)$ of Eq. (7.77) and the bandstop transfer function $H_{BS}(z)$ of Eq. (7.80) form a doubly-complementary pair.

7.77 Show that the value of the gain function $\mathcal{G}(\omega)$ of a power-symmetric transfer function defined by Eq. (7.101) at $\omega = \pi/2$ is given by $10\log_{10} K - 3$ dB.

7.78 Consider the real-coefficient stable IIR transfer function $H(z) = A_0(z^2) + z^{-1}A_1(z^2)$, where $A_0(z)$ and $A_1(z)$ are stable allpass transfer functions. Show that $H(z)$ is a power-symmetric transfer function.

7.79 Show that

$$H(z) = \frac{-0.1 + 0.5z^{-1} + 0.05z^{-2} + 0.05z^{-3} + 0.5z^{-4} - 0.1z^{-5}}{1 + 0.1z^{-2} - 0.2z^{-4}}$$

is a power-symmetric IIR transfer function.

7.80 Show that the following causal FIR transfer functions satisfy the power-symmetric condition:

(a) $H_a(z) = 1 - 2z^{-1} + 4.5z^{-2} + 6z^{-3} + z^{-4} + 0.5z^{-5}$,

(b) $H_b(z) = 1 + \frac{1}{2}z^{-1} + \frac{15}{4}z^{-2} - z^{-4} + 2z^{-5}$.

7.81 Let $H(z) = a(1 + bz^{-1})$, where a and b are constants. Then $H(z)H(z^{-1})$ is of the form $cz + d + cz^{-1}$. Determine the condition on c and d so that $H(z)$ is a power-symmetric FIR transfer function with $K = 1$. Show that $a = 1/2$ and $b = 1$ satisfy the power-symmetric condition. Determine two other possible sets of values for a and b to ensure the power-symmetric condition. Using MATLAB, show that $H(z)$ and $G(z) = -z^{-1}H(-z^{-1})$ are power-complementary for the above values of the constants a and b.

7.82 Let $H(z) = a(1 + bz^{-1})(1 + d_1z^{-1} + d_2z^{-2})$, where $a, b, d_1,$ and d_2 are constants. Then $H(z)H(z^{-1})$ is of the form $(cz + d + cz^{-1})[d_2z^2 + d_1(1 + d_2)z + (1 + d_1^2 + d_2^2) + d_1(1 + d_2)z^{-1} + d_2z^{-2}]$. Determine the condition on c and d in terms of d_1 and d_2 so that $H(z)$ is a power-symmetric FIR transfer function with $K = 1$. For $d_1 = d_2 = 1$, evaluate the constraint on c and d, and using it, determine one realizable set of values for a and b. Using MATLAB, show that $H(z)$ and $G(z) = -z^{-3}H(-z^{-1})$ are power-complementary for these values of the constants a and b.

7.83 A set of M digital filters $\{G_i(z)\}$, $i = 0, 1, \ldots, M - 1$, is defined to be *magnitude-complementary* of each other if the sum of their magnitude responses is equal to a constant [Reg87c]; that is,

$$\sum_{i=0}^{M-1} |G_i(e^{j\omega})| = \beta, \qquad \text{for all } \omega, \tag{7.165}$$

where β is a positive nonzero constant. Consider two real-coefficient doubly-complementary transfer functions $H_0(z)$ and $H_1(z)$ that are related according to Eqs. (7.97a) and (7.97b). Define $G_0(z) = H_0^2(z)$ and $G_1(z) = -H_1^2(z)$. Show that $G_0(z)$ and $G_1(z)$ are a pair of magnitude-complementary transfer functions.

7.84 Show analytically that the following causal FIR transfer functions are BR functions:

(a) $H_1(z) = \frac{1}{4}\left(1 + 3z^{-1}\right)$,   (b) $H_2(z) = \frac{1}{2.2}\left(1 - 1.2z^{-1}\right)$,

(c) $H_3(z) = \frac{(1+\alpha z^{-1})(1-\beta z^{-1})}{(1+\alpha)(1+\beta)}$, $\alpha > 0$, $\beta > 0$,

(d) $H_4(z) = \frac{1}{2.34}(1 - 0.3z^{-1})(1 + 0.2z^{-1})(1 - 0.5z^{-1})$.

7.85 Show analytically that the following causal IIR transfer functions are BR functions:

(a) $H_1(z) = \dfrac{2.6 + 2.6z^{-1}}{4.2 + z^{-1}}$,   (b) $H_2(z) = \dfrac{1.6 - 1.6z^{-1}}{4.2 + z^{-1}}$,   (c) $H_3(z) = \dfrac{0.1(1 - z^{-2})}{1 + 0.4z^{-1} + 0.8z^{-2}}$,

(d) $H_4(z) = \dfrac{4.5 + 2z^{-1} + 4.5z^{-2}}{5 + 2z^{-1} + 4z^{-2}}$.

7.86 If $A_1(z)$ and $A_2(z)$ are two LBR functions, show that $A_1(1/A_2(z))$ is also an LBR function.

7.87 Let $G(z)$ be an LBR function of order N. Define

$$F(z) = z\left(\frac{G(z) + \alpha}{1 + \alpha G(z)}\right),$$

where $|\alpha| < 1$. Show that $F(z)$ is also LBR. What is the order of $F(z)$? Develop a realization of $G(z)$ in terms of $F(z)$.

7.88 If $G(z)$ and $A(z)$ are, respectively, a BR function and and an LBR function, show that $G(1/A(z))$ is a BR function.

7.89 Show analytically that each of the following pairs of transfer functions are doubly-complementary:

(a) $H(z) = \dfrac{2.6(1 + z^{-1})}{4.2 + z^{-1}}$, $G(z) = \dfrac{1.6(1 - z^{-1})}{4.2 + z^{-1}}$,

(b) $H(z) = \dfrac{0.1(1 - z^{-2})}{1 + 0.4z^{-1} + 0.8z^{-2}}$, $G(z) = \dfrac{0.9 + 0.4z^{-1} + 0.9z^{-2}}{1 + 0.4z^{-1} + 0.8z^{-2}}$.

7.90 Determine analytically the power-complementary transfer function of each of the following BR transfer functions:

(a) $H_a(z) = \dfrac{2(2 + z^{-1} + 2z^{-2})}{5 + 2z^{-1} + 3z^{-2}}$,　(b) $H_b(z) = \dfrac{3 + 7.5z^{-1} + 7.5z^{-2} + 3z^{-3}}{8 + 8z^{-1} + 4z^{-2} + z^{-3}}$.

7.91 Verify the relations between the transfer parameters and the chain parameters of a two-pair given in Eqs. (7.128a) and (7.128b).

7.92 A two-pair is said to be *reciprocal* if $t_{12} = t_{21}$ [Mit73b]. Show that for a reciprocal two-pair, $AD - BC = 1$.

7.93 Consider the Γ-cascade of Figure P7.9(a), where the two two-pairs are described by the transfer matrices

$$\tau_1 = \begin{bmatrix} k_1 & (1 - k_1^2)z^{-1} \\ 1 & -k_1 z^{-1} \end{bmatrix}, \qquad \tau_2 = \begin{bmatrix} k_2 & (1 - k_2^2)z^{-1} \\ 1 & -k_2 z^{-1} \end{bmatrix}.$$

Determine the transfer matrix of the cascade.

7.94 Consider the τ-cascade of Figure P7.9(b), where the two two-pairs are described by the chain matrices

$$\Gamma_1 = \begin{bmatrix} 1 & k_1 z^{-1} \\ k_1 & z^{-1} \end{bmatrix}, \qquad \Gamma_2 = \begin{bmatrix} 1 & k_2 z^{-1} \\ k_2 & z^{-1} \end{bmatrix}.$$

Determine the chain matrix of the cascade.

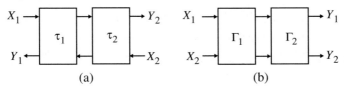

(a)　(b)

Figure P7.9

7.95 Determine the transfer parameters and the chain parameters of the digital two-pairs of Figure P7.10.

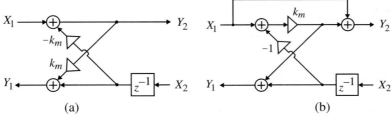

(a)　(b)

Figure P7.10

7.96 A transfer function $H(z)$ is realized in the form of Figure 7.37, where the constraining transfer function is given by $G(z)$. If the relation between $H(z)$ and $G(z)$ is of the form

$$G(z) = \frac{H(z) - k_m}{z^{-1}[1 - k_m H(z)]},$$

determine the transfer matrix and the chain matrix parameters of the two-pair of Figure 7.37.

7.97 A transfer function $H(z)$ is realized in the form of Figure 7.37, where the constraining transfer function is given by $G(z)$. The relation between $H(z)$ and $G(z)$ is of the form

$$H(z) = \frac{\alpha + z^{-1}G(z)}{1 + \alpha z^{-1}G(z)},$$

with α real and $|\alpha| < 1$.

(a) Determine the chain parameters of the two-pair of Figure 7.37.

(b) If $|G(z)| < 1$ for $|z| = 1$, show that $|H(z)|$ cannot have a maximum value greater than unity.

7.98 Determine chain parameters of the cascade of three lattice two-pairs of Figure P7.11. Using these chain parameters, determine the expression for the transfer function $A_3(z)$.

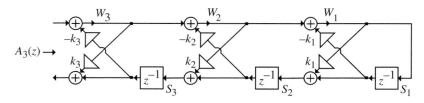

Figure P7.11

7.99 Derive the inequality of Eq. (7.141).

7.100 Determine by inspection which one of the following second-order polynomials has both roots inside the unit circle:

(a) $D_a(z) = 4 + 3z^{-1} + 2z^{-2}$,   (b) $D_b(z) = 2 + z^{-1} + z^{-2}$,

(c) $D_c(z) = 3 + 4z^{-1} - 4z^{-2}$,   (d) $D_d(z) = 3 - 0.5z^{-1} - z^{-2}$.

7.101 Test analytically the BIBO stability of the following causal IIR transfer functions:

(a) $H_a(z) = \dfrac{2z^2 + 3.75z + 10}{z^3 + 0.75z^2 + 0.5z + 0.25}$,   (b) $H_b(z) = \dfrac{0.1z^2 - 0.57z - 1.78}{3z^3 + 2z^2 - 2z - 1}$,

(c) $H_c(z) = \dfrac{8.7z^3 - 15.53z^2 + 33.1z + 12.5}{3z^4 + 2z^3 + 2z^2 + 1.5z - 0.5}$,   (d) $H_d(z) = \dfrac{1}{5 + 4z^{-1} + 3z^{-2} + 2z^{-3} + z^{-4}}$,

(e) $H_e(z) = \dfrac{1}{10 + 7z^{-1} + 5z^{-2} + 3z^{-3} + 2z^{-4} + z^{-5}}$.

7.102 Determine analytically whether all roots of the following polynomials are inside the unit circle:

(a) $D_a(z) = 10 + 8z^{-1} + 6z^{-2} + 4z^{-3} + 2z^{-4} + z^{-5}$,

(b) $D_b(z) = 4z^5 + 3.5z^4 + 3z^3 + 2.5z^2 + 2z - 1$.

7.12 MATLAB Exercises

M 7.1 Write a MATLAB program to simulate the filter designed in Problem 7.15, and verify its filtering operation.

M 7.2 Write a MATLAB program to simulate the filter designed in Problem 7.18, and verify its filtering operation.

M 7.3 The following third-order IIR transfer function has been proposed for clutter rejection in MTI radar [Whi58]:

$$H(z) = \frac{z^{-1}(1 - z^{-1})^2}{(1 - 0.4z^{-1})(1 - 0.88z^{-1} + 0.61z^{-2})}.$$

Using MATLAB, determine and plot its gain response, and show that it has a highpass response.

M 7.4 Show that for each case listed below, $H(z)$ and $H(-z)$ are power-complementary.

(a) $H(z) = \dfrac{1 - 2z^{-1} + 3.5z^{-2} + 3.5z^{-3} - 2z^{-4} + z^{-5}}{8 - 2z^{-2} - z^{-4}}$,

(b) $H(z) = \dfrac{1 + 1.5z^{-1} + 5.25z^{-2} + 7.25z^{-3} + 7.25z^{-4} + 5.25z^{-5} + 1.5z^{-6} + z^{-7}}{12 + 13z^{-2} + 4.5z^{-4} + 0.5z^{-6}}$.

To verify the power-complementary property, write a MATLAB program to evaluate $H(z)H(z^{-1}) + H(-z)H(-z^{-1})$, and show that this expression is equal to unity for each of the transfer functions given above.

M 7.5 Plot the magnitude and phase responses of the causal IIR digital transfer function

$$H(z) = \frac{0.2031(1 - z^{-1})(1 - 0.2743z^{-1} + z^{-2})}{(1 + 0.2695z^{-1})(1 + 0.4109z^{-1} + 0.6758z^{-2})}.$$

What type of filter does this transfer function represent? Determine the difference equation representation of the above transfer function.

M 7.6 Plot the magnitude and phase responses of the causal IIR digital transfer function

$$H(z) = \frac{0.2031(1 - z^{-1})(1 - 0.2743z^{-1} + z^{-2})}{(1 + 0.1532z^{-1} + 0.8351z^{-2})(1 + 0.487z^{-1} + 0.84z^{-2})}.$$

What type of filter does this transfer function represent? Determine the difference equation representation of the above transfer function.

M 7.7 Design an FIR lowpass filter with a 3-dB cutoff frequency at 0.45π using a cascade of five first-order lowpass filters of Eq. (7.71). Plot its gain response.

M 7.8 Using the result of Problem 7.46, design an FIR highpass filter with a 3-dB cutoff frequency at 0.4π using a cascade of six first-order highpass filters of Eq. (7.74). Plot its gain response.

M 7.9 Design a first-order IIR lowpass and a first-order IIR highpass filter with a 3-dB cutoff frequency of 0.6π. Using MATLAB, plot their magnitude responses on the same figure. Using MATLAB, show that these filters are both allpass-complementary and power-complementary.

M 7.10 Design a second-order IIR bandpass and a second-order IIR notch filter with a center (notch) frequency $\omega_o = 0.4\pi$ and a 3-dB bandwidth B_w (notch width) of 0.25π. Using MATLAB, plot their magnitude responses on the same figure. Using MATLAB, show that these filters are both allpass-complementary and power-complementary.

M 7.11 Design a stable second-order IIR bandpass filter with a center frequency at 0.6π and a 3-dB bandwidth of 0.2π. Plot its gain response.

M 7.12 Design a stable second-order IIR notch filter with a center frequency at 0.6π and a 3-dB bandwidth of 0.2π. Plot its gain response.

M 7.13 Using MATLAB, show that the transfer function pairs of Exercises M7.11 and M7.12 are both allpass-complementary and power-complementary.

M 7.14 Using MATLAB, show that the transfer function pairs of Problem 7.89 are doubly-complementary.

M 7.15 Develop the pole-zero plots of the transfer functions of Problem 7.90 using the function zplane of MATLAB, and show that they are stable. Next, plot the magnitude response of each transfer function using MATLAB, and show that it satisfies the bounded real property.

M 7.16 Using MATLAB, determine the power-complementary transfer function of each of the transfer functions of Problem 7.90.

M 7.17 Develop the pole-zero plots of the transfer functions of Problem 7.101 using the function zplane of MATLAB, and then test their stability.

M 7.18 Using Program 7_2, test the stability of the transfer functions of Problem 7.101.

M 7.19 Using Program 7_2, determine whether the roots of the polynomials of Problem 7.102 are inside the unit circle or not.

M 7.20 The FIR digital filter structure of Figure P7.12 is used for aperture correction in television to compensate for high-frequency losses [Dre90]. A cascade of two such circuits is used, with one correcting the vertical aperture and the other correcting the horizontal aperture. In the former case, the delay z^{-1} is a line delay, whereas in the latter case, it is 70 ns for the CCIR standard, and the weighting factor k provides an adjustable amount of correction. Determine the transfer function of this circuit, and plot its magnitude response using MATLAB for two different values of k.

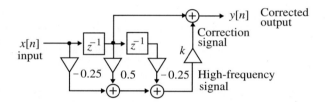

Figure P7.12

M 7.21 An improved aperture correction circuit for digital television is the FIR digital filter structure of Figure P7.13, where the delay z^{-1} is 70 ns for the CCIR standard, and the two weighting factors k_1 and k_2 provide an adjustable amount of correction with $k_1 > 0$ and $k_2 < 0$ [Dre90]. Determine the transfer function of this circuit, and plot its magnitude response using MATLAB for two different values of k_1 and k_2.

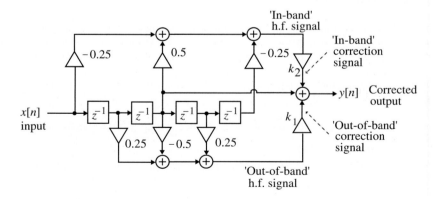

Figure P7.13

8 Digital Filter Structures

The description of the discrete-time system of Eq. (2.73a) or (2.73b) expresses the nth output sample, $y[n]$, as a convolution sum of the input, $x[n]$, with the impulse response of the system, $h[n]$, and in some sense, it is the most fundamental characterization of an LTI digital system. The convolution sum, in principle, can be employed to implement a digital filter with a known impulse response, and the implementation involves addition, multiplication, and delay, which are fairly simple operations. For an LTI system with an infinite-length impulse response, the approach is not practical. However, for the infinite impulse response (IIR) LTI digital filter described by a constant coefficient difference equation of the form of Eq. (2.91) and for the finite impulse response (FIR) LTI digital filter described by Eq. (2.118), the input–output relation involves a finite sum of products, and a direct implementation based on these equations is quite practical. In this text, we consider only these two types of LTI digital filters.

The actual implementation of a digital filter could be either in software or hardware, depending on applications. In both types of implementation, the signal variables and the filter coefficients cannot be represented with infinite precision. As a result, a direct implementation of a digital filter based on either Eq. (2.91) or Eq. (2.118) may not provide satisfactory performance due to the use of finite precision arithmetic. It is thus of interest to develop alternative realizations based on other types of time-domain representations with input–output relations equivalent to either Eq. (2.91) or Eq. (2.118), depending on the type of digital filter being implemented, and choose the realization that provides satisfactory performance using finite precision arithmetic.

In this chapter, we consider the realization problem of causal IIR and FIR transfer functions and outline realization methods based on both the time-domain and the transform-domain representations. In Chapter 12, we develop the methods for the analysis of such structures when implemented with finite precision arithmetic and present additional realizations that have been developed to minimize the effects of finite wordlength.

A structural representation using interconnected basic building blocks is the first step in the hardware or software implementation of an LTI digital filter. The structural representation provides the relations between some pertinent internal variables with the input and the output that, in turn, provide the keys to the implementation. There are various forms of the structural representation of a digital filter. We review in this chapter two such representations and then describe some popular schemes for the realization of real causal IIR and FIR digital filters. In addition, we outline a method for the realization of IIR digital filter structures that can be used for the generation of a pair of orthogonal sinusoidal sequences.

8.1 Block Diagram Representation

As indicated earlier, the input–output relations of an LTI digital filter can be expressed in various ways. In the time domain, it is given by the convolution sum

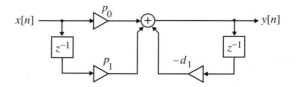

Figure 8.1: A first-order LTI digital filter.

$$y[n] = \sum_{k=-\infty}^{\infty} h[k]x[n-k], \tag{8.1}$$

or by the linear constant coefficient difference equation

$$y[n] = -\sum_{k=1}^{N} d_k y[n-k] + \sum_{k=0}^{M} p_k x[n-k]. \tag{8.2}$$

A digital filter can be implemented on a general-purpose digital computer in software form or with special-purpose hardware. To this end, it is necessary to describe the input–output relationship by means of a computational algorithm. To illustrate what we mean by a computational algorithm, consider a first-order causal LTI IIR digital filter described by

$$y[n] = -d_1 y[n-1] + p_0 x[n] + p_1 x[n-1]. \tag{8.3}$$

Using Eq. (8.3), we can compute $y[n]$ for $n = 0, 1, 2, \ldots$, knowing the initial condition $y[-1]$ and the input $x[n]$ for $n = -1, 0, 1, 2, \ldots$

$$y[0] = -d_1 y[-1] + p_0 x[0] + p_1 x[-1],$$
$$y[1] = -d_1 y[0] + p_0 x[1] + p_1 x[0],$$
$$y[2] = -d_1 y[1] + p_0 x[2] + p_1 x[1],$$
$$\vdots$$

We can continue this calculation for any value of n we desire. Each step of the calculation process requires a knowledge of the previously calculated value of the output sample (delayed value of the output), the present value of the input sample, and the previous value of the input sample (delayed value of the input). Knowing the data values, we multiply each of them appropriately with coefficients $-d_1$, p_0, and p_1 and then sum the products to compute the present value of the output. As a result, the difference equation of Eq. (8.3) can be interpreted as a valid computational algorithm.

8.1.1 Basic Building Blocks

The computational algorithm of an LTI digital filter can be conveniently represented in block diagram form using the basic building blocks representing the unit delay, the multiplier, the adder, and the pick-off node as shown earlier in Figure 2.5. Thus, a block diagram representation of the first-order digital filter described by Eq. (8.3) is as indicated in Figure 8.1. The adder shown here is a three-input adder. Note also the pick-off nodes at the input and the output.

There are several advantages in representing the digital filter in block diagram form: (1) it is easy to write down the computational algorithm by inspection, (2) it is easy to analyze the block diagram to determine the explicit relation between the output and the input, (3) it is easy to manipulate a block diagram to derive other "equivalent" block diagrams yielding different computational algorithms, (4) it is easy to determine the hardware requirements, and finally, (5) it is easy to develop block diagram representations from the transfer function directly leading to a variety of "equivalent" representations.

8.1.2 Analysis of Block Diagrams

Digital filter structures represented in block diagram form can often be analyzed by writing down the expressions for the output signals of each adder as a sum of its input signals, thereby developing a set of equations relating the filter input and output signals in terms of all internal signals. Eliminating the unwanted internal variables then results in the expression for the output signal as a function of the input signal and the filter parameters that are the multiplier coefficients.

Example 8.1 illustrates the analysis approach.

EXAMPLE 8.1   **Analysis of a Cascaded Lattice Digital Filter Structure**

To analyze the cascaded lattice digital filter structure of Figure 8.2, we first write down the expressions for the output signals of the four adders as given below:

$$W_1 = X - \alpha S_2, \tag{8.4a}$$
$$W_2 = W_1 - \delta S_1, \tag{8.4b}$$
$$W_3 = S_1 + \varepsilon W_2, \tag{8.4c}$$
$$Y = \beta W_1 + \gamma S_2. \tag{8.4d}$$

But, from the figure, $S_2 = z^{-1} W_3$ and $S_1 = z^{-1} W_2$. Substituting these relations in Eqs. (8.4a) to (8.4d), we obtain

$$W_1 = X - \alpha z^{-1} W_3, \tag{8.5a}$$
$$W_2 = W_1 - \delta z^{-1} W_2, \tag{8.5b}$$
$$W_3 = z^{-1} W_2 + \varepsilon W_2, \tag{8.5c}$$
$$Y = \beta W_1 + \gamma z^{-1} W_3. \tag{8.5d}$$

From Eq. (8.5b), $W_2 = W_1/(1 + \delta z^{-1})$, and from Eq. (8.5c), $W_3 = (\varepsilon + z^{-1}) W_2$. Combining these two equations, we get

$$W_3 = \frac{\varepsilon + z^{-1}}{1 + \delta z^{-1}} W_1. \tag{8.6}$$

Substituting Eq. (8.6) in Eqs. (8.5a) and (8.5d), we finally obtain, after some algebra, the expression for the transfer function $H(z)$ of the digital filter structure as

$$H(z) = \frac{Y}{X} = \frac{\beta + (\beta\delta + \gamma\varepsilon)z^{-1} + \gamma z^{-2}}{1 + (\delta + \alpha\varepsilon)z^{-1} + \alpha z^{-2}}. \tag{8.7}$$

8.1.3 The Delay-Free Loop Problem

For physical realizability of the digital filter structure, it is necessary that the block diagram representation contains no delay-free loops, that is, feedback loops without any delay elements. Figure 8.3 illustrates a typical delay-free loop that may appear unintentionally in a specific structure. Analysis of this structure

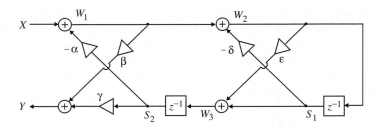

Figure 8.2: A cascaded lattice digital filter structure.

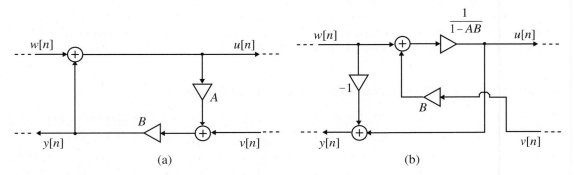

Figure 8.3: (a) An example of a delay-free loop and (b) its equivalent realization with no delay-free loop.

yields

$$y[n] = B\big(A\,(w[n] + y[n]) + v[n]\big).$$

The above expression implies that the determination of the current value of $y[n]$ requires the knowledge of $y[n]$. This is physically impossible to achieve due to the finite time required to carry out all arithmetic operations in a digital machine.

A simple graph-theoretic–based method has been proposed to detect the presence of delay-free loops in an arbitrary digital filter structure, along with the methods to locate and remove these loops without altering the overall input–output relations [Szc75b]. The removal is achieved by replacing the portion of the structure containing the delay-free loops by an equivalent realization with no delay-free loops. For example, Figure 8.3(b) shows an equivalent realization of the block diagram of Figure 8.3(a) without the delay-free loop.

8.1.4 Canonic and Noncanonic Structures

A digital filter structure is said to be *canonic* if the number of delays in the block diagram representation is equal to the order of the difference equation (i.e., the order of the transfer function). Otherwise, it is a *noncanonic* structure. For example, the structure of Figure 8.1 is a noncanonic realization since it employs two delays to realize the first-order difference equation of Eq. (8.3).

8.2 Equivalent Structures

Our main objective in this chapter is to develop various realizations of a given transfer function. We define two digital filter structures to be *equivalent* if they have the same transfer function. We outline in this

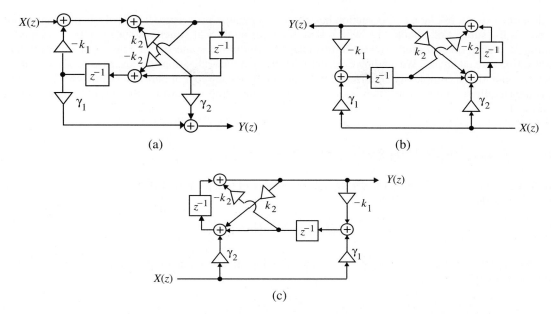

Figure 8.4: (a) The original digital filter structure, (b) its transposed form, and (c) redrawn transposed structure.

chapter and in Chapter 12 a number of methods for the generation of equivalent structures. However, a fairly simple way to generate an equivalent structure from a given realization is via the *transpose operation,* which is as follows [Jac70a]:

(i) reverse all paths,

(ii) replace pick-off nodes by adders, and vice versa, and

(iii) interchange the input and the output nodes.

We illustrate the application of the transpose operation in Example 8.2.

EXAMPLE 8.2   Illustration of the Generation of an Equivalent Structure Using the Transpose Operation

Consider the digital filter structure of Figure 8.4(a). The structure obtained by applying the transpose operation is shown in Figure 8.4(b). The redrawn transposed structure is shown in Figure 8.4(c).

All other methods for developing equivalent structures are based on a specific algorithm for each structure. There are literally an infinite number of equivalent structures realizing the same transfer function, and it is impossible to develop all such realizations. Moreover, a large variety of algorithms have been advanced by various authors, and space limitations prevent us from reviewing each method in this text. We therefore restrict ourselves to a discussion of some commonly used structures.

It should be noted that under infinite precision arithmetic, any given realization of a digital filter behaves identically to any other equivalent structure. However, in practice, due to the finite wordlength limitations, a specific realization behaves differently from its other equivalent realizations. Hence, it is important to choose a structure that has good quantization properties under a finite wordlength implementation. One way to arrive at such a structure is to determine a large number of equivalent structures, analyze the finite wordlength effects of each one, and then select the one that is least sensitive. In certain cases, it is possible to develop a structure that by construction has good quantization properties. The analysis of quantization effects is the subject of Chapter 12, which also describes additional structures specifically developed to minimize certain quantization effects. In this chapter, we discuss some simple realizations that in many applications are adequate. We do, however, compare each of the realizations discussed here with regard to their computational complexity determined in terms of the total number of multipliers and the total number of adders required for their implementation. This latter issue is important where the cost of implementation is critical.

8.3   Basic FIR Digital Filter Structures

We first consider the realization of FIR digital filters. Recall that a causal FIR filter of order N is characterized by a transfer function $H(z)$,

$$H(z) = \sum_{k=0}^{N} h[k]z^{-k}, \tag{8.8}$$

which is a polynomial in z^{-1} of degree N. In the time domain, the input–output relation of the above FIR filter is given by

$$y[n] = \sum_{k=0}^{N} h[k]x[n-k], \tag{8.9}$$

where $y[n]$ and $x[n]$ are the output and input sequences, respectively. Since FIR filters can be designed to provide exact linear phase over the whole frequency range and are always BIBO stable independent of the filter coefficients, such filters are often preferred in many applications. We now outline several realization methods for such filters.

8.3.1   Direct Forms

An FIR filter of order N is characterized by $N + 1$ coefficients and, in general, requires $N + 1$ multipliers and N two-input adders for implementation. Structures in which the multiplier coefficients are precisely the coefficients of the transfer function are called *direct form* structures. A direct form realization of an FIR filter can be readily developed from Eq. (8.9), as indicated in Figure 8.5(a) for $N = 4$. An analysis of this structure yields

$$y[n] = h[0]x[n] + h[1]x[n-1] + h[2]x[n-2] + h[3]x[n-3] + h[4]x[n-4],$$

which is precisely of the form of Eq. (8.9).

   The transpose of the structure of Figure 8.5(a), shown in Figure 8.5(b), is the second direct form structure. Both direct form structures are canonic with respect to delays.

8.3.2   Cascade Form

A higher-order FIR transfer function can also be realized as a cascade of FIR sections, with each section characterized by either a first-order or a second-order transfer function. To this end, we factor the FIR transfer function $H(z)$ of Eq. (8.8) and write it in the form

$$H(z) = h[0] \prod_{k=1}^{K} \left(1 + \beta_{1k}z^{-1} + \beta_{2k}z^{-2}\right), \tag{8.10}$$

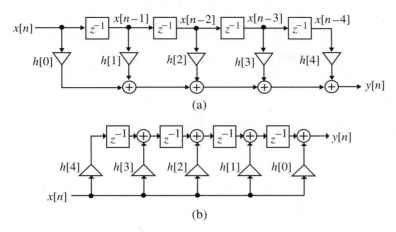

Figure 8.5: Direct form FIR structures.

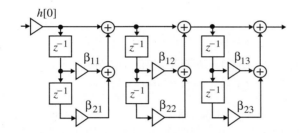

Figure 8.6: Cascade form FIR structure for a sixth-order FIR filter.

where $K = N/2$ if N is even and $K = (N+1)/2$ if N is odd, with $\beta_{2K} = 0$. A cascade realization of Eq. (8.10) requiring three second-order sections for $N = 6$ is shown in Figure 8.6. Each second-order section in Figure 8.6 can, of course, also be realized in the transposed direct form. Note that the cascade form is canonic and also employs N two-input adders and $N+1$ multipliers for an Nth-order FIR transfer function.

8.3.3 Polyphase Realization

Another interesting realization of an FIR filter is based on the polyphase decomposition of its transfer function and results in a parallel structure [Bel76]. To illustrate this approach, consider for simplicity a causal FIR transfer function $H(z)$ of length 9:

$$H(z) = h[0] + h[1]z^{-1} + h[2]z^{-2} + h[3]z^{-3} + h[4]z^{-4}$$
$$+ h[5]z^{-5} + h[6]z^{-6} + h[7]z^{-7} + h[8]z^{-8}. \tag{8.11}$$

The above transfer function can be expressed as a sum of two terms, with one term containing the even-indexed coefficients and the other containing the odd-indexed coefficients, as indicated below:

$$H(z) = \left(h[0] + h[2]z^{-2} + h[4]z^{-4} + h[6]z^{-6} + h[8]z^{-8} \right)$$

$$+ \left(h[1]z^{-1} + h[3]z^{-3} + h[5]z^{-5} + h[7]z^{-7} \right)$$

$$= \left(h[0] + h[2]z^{-2} + h[4]z^{-4} + h[6]z^{-6} + h[8]z^{-8} \right)$$

$$+ z^{-1} \left(h[1] + h[3]z^{-2} + h[5]z^{-4} + h[7]z^{-6} \right). \tag{8.12}$$

By using the notation

$$E_0(z) = h[0] + h[2]z^{-1} + h[4]z^{-2} + h[6]z^{-3} + h[8]z^{-4},$$

$$E_1(z) = h[1] + h[3]z^{-1} + h[5]z^{-2} + h[7]z^{-3}, \tag{8.13}$$

we can rewrite Eq. (8.12) as

$$H(z) = E_0(z^2) + z^{-1}E_1(z^2). \tag{8.14}$$

In a similar manner, by grouping the terms of Eq. (8.11) differently, we can reexpress it in the form

$$H(z) = E_0(z^3) + z^{-1}E_1(z^3) + z^{-2}E_2(z^3), \tag{8.15}$$

where now

$$E_0(z) = h[0] + h[3]z^{-1} + h[6]z^{-2},$$

$$E_1(z) = h[1] + h[4]z^{-1} + h[7]z^{-2}, \tag{8.16}$$

$$E_2(z) = h[2] + h[5]z^{-1} + h[8]z^{-2}.$$

The decomposition of $H(z)$ in the form of Eqs. (8.14) and (8.15) is more commonly known as the *polyphase decomposition*. In the general case, an L-branch polyphase decomposition of the transfer function of Eq. (8.8) of order N is of the form

$$H(z) = \sum_{m=0}^{L-1} z^{-m} E_m(z^L), \tag{8.17}$$

where[1]

$$E_m(z) = \sum_{n=0}^{\lfloor (N+1)/L \rfloor} h[Ln + m]z^{-n}, \qquad 0 \le m \le L - 1, \tag{8.18}$$

with $h[n] = 0$ for $n > N$. A realization of $H(z)$ based on the decomposition of Eq. (8.17) is called a *polyphase realization*. Figure 8.7 shows the four-branch, the three-branch, and the two-branch polyphase realizations of a transfer function. As indicated in Eqs. (8.13) and (8.16), the expression for the transfer function $E_0(z)$ is different for each structure and so are the expressions for $E_1(z)$, and so on.

The subfilters $E_m(z^L)$ in the polyphase realization of an FIR transfer function are also FIR filters and can be realized using any of the methods described earlier. However, to obtain a canonic realization of the overall structure, the delays in all subfilters must be shared. Figure 8.8 illustrates a canonic polyphase realization of a length-9 FIR transfer function obtained by delay-sharing. It should be noted that in developing this realization, we have used the transpose of the structure of Figure 8.7(b). Other canonic polyphase realizations can be similarly derived.

The polyphase structures find applications in multirate digital signal processing for computationally efficient realizations (see Section 13.4.2).

---

[1] $\lfloor x \rfloor$ is the integer part of x.

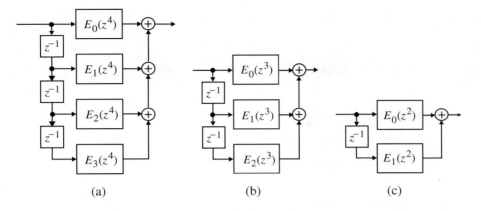

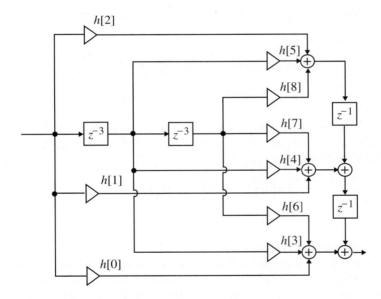

Figure 8.7: Polyphase realizations of an FIR transfer function.

Figure 8.8: Canonic three-branch polyphase realization of a length-9 FIR filter.

8.3.4 Linear-Phase FIR Structures

We showed in Section 7.3 that a linear-phase FIR filter of order N is either characterized by a symmetric impulse response

$$h[n] = h[N - n] \tag{8.19}$$

or by an antisymmetric impulse response

$$h[n] = -h[N - n]. \tag{8.20}$$

The symmetry (or antisymmetry) property of a linear-phase FIR filter can be exploited to reduce the total number of multipliers into almost half of that in the direct form implementations of the transfer

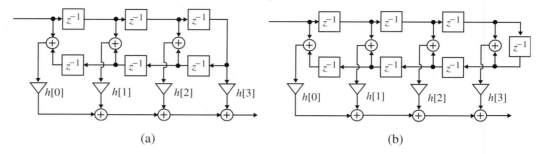

Figure 8.9: Linear-phase FIR structures: (a) Type 1 and (b) Type 2.

function. To this end, consider the realization of a length-7 Type 1 FIR transfer function with a symmetric impulse response

$$H(z) = h[0] + h[1]z^{-1} + h[2]z^{-2} + h[3]z^{-3} + h[2]z^{-4} + h[1]z^{-5} + h[0]z^{-6},$$

which can be rewritten in the form

$$
\begin{aligned}
H(z) = {} & h[0]\left(1 + z^{-6}\right) + h[1]\left(z^{-1} + z^{-5}\right) \\
& + h[2]\left(z^{-2} + z^{-4}\right) + h[3]z^{-3}.
\end{aligned}
\tag{8.21}
$$

A realization of $H(z)$ based on the decomposition of Eq. (8.21) is shown in Figure 8.9(a). A similar decomposition can be applied for the realization of a Type 2 FIR transfer function. For example, for a length-8 Type 2 FIR transfer function, the pertinent decomposition is given by

$$
\begin{aligned}
H(z) = {} & h[0]\left(1 + z^{-7}\right) + h[1]\left(z^{-1} + z^{-6}\right) \\
& + h[2]\left(z^{-2} + z^{-5}\right) + h[3]\left(z^{-3} + z^{-4}\right),
\end{aligned}
\tag{8.22}
$$

leading to the realization shown in Figure 8.9(b).

It should be noted that the structure of Figure 8.9(a) requires 4 multipliers, whereas a direct form realization of the original length-7 FIR filter would require 7 multipliers. Likewise, the structure of Figure 8.9(b) requires 4 multipliers, compared to 8 multipliers in the direct form realization of the original length-8 FIR filter. A similar savings occurs in the case of an FIR filter with an antisymmetric impulse response.

8.3.5 Tapped Delay Line

In some applications, such as musical sound processing, FIR filter structures of the form of Figure 8.10 are employed. The structure consists of a chain of $M_1 + M_2 + M_3$ unit delays with taps at the input, at the end of first M_1 delays, at the end of next M_2 delays, and at the output, respectively. Signals at these taps are then multiplied by constants α_0, α_1, α_2, and α_3 and added to form the output. Such a structure is usually referred to as the *tapped delay line.* The direct form FIR structure of Figure 8.5 can be seen to be a special case of a tapped delay line, where there is a tap after each unit delay.

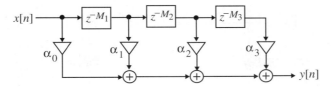

Figure 8.10: A tapped delay line.

Figure 8.11: A possible IIR filter realization scheme.

8.4 Basic IIR Digital Filter Structures

The causal IIR digital filters we are concerned with in this text are characterized by a real rational transfer function of the form of Eq. (6.13) or, equivalently, by the constant coefficient difference equation of Eq. (2.91). From the difference equation representation, it can be seen that the computation of the nth output sample requires the knowledge of several past samples of the output sequence, or in other words, the realization of a causal IIR filter requires some form of feedback. We outline here several simple and straightforward realizations of IIR filters.

8.4.1 Direct Forms

An Nth-order IIR digital filter transfer function is characterized by $2N + 1$ unique coefficients and, in general, requires $2N + 1$ multipliers and $2N$ two-input adders for implementation. As in the case of FIR filter realization, IIR filter structures in which the multiplier coefficients are precisely the coefficients of the transfer function are called *direct form* structures. We now describe the development of these structures.

Consider for simplicity a third-order IIR filter characterized by a transfer function

$$H(z) = \frac{P(z)}{D(z)} = \frac{p_0 + p_1 z^{-1} + p_2 z^{-2} + p_3 z^{-3}}{1 + d_1 z^{-1} + d_2 z^{-2} + d_3 z^{-3}}, \tag{8.23}$$

which we can implement as a cascade of two filter sections as shown in Figure 8.11, where

$$H_1(z) = \frac{W(z)}{X(z)} = P(z) = p_0 + p_1 z^{-1} + p_2 z^{-2} + p_3 z^{-3} \tag{8.24a}$$

and

$$H_2(z) = \frac{Y(z)}{W(z)} = \frac{1}{D(z)} = \frac{1}{1 + d_1 z^{-1} + d_2 z^{-2} + d_3 z^{-3}}. \tag{8.24b}$$

The filter section $H_1(z)$ of Eq. (8.24a) is seen to be an FIR filter and can be realized as shown in Figure 8.12(a). We next consider the realization of $H_2(z)$ given by Eq. (8.24b). Note that a time-domain representation of this transfer function is given by

$$y[n] = w[n] - d_1 y[n-1] - d_2 y[n-2] - d_3 y[n-3], \tag{8.25}$$

resulting in the realization indicated in Figure 8.12(b).

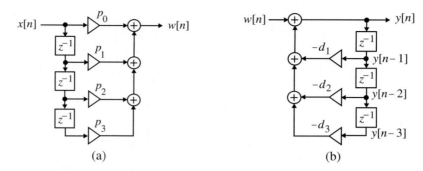

Figure 8.12: (a) Realization of the transfer function $H_1(z) = W(z)/X(z)$ and (b) realization of the transfer function $H_2(z) = Y(z)/W(z)$.

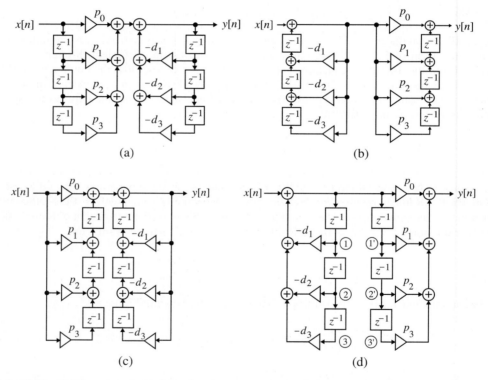

Figure 8.13: (a) Direct form I, (b) direct form I_t, (c) and (d) additional noncanonic direct form structures.

A cascade of the structures of Figure 8.12(a) and (b) as indicated in Figure 8.11 leads to a realization of the original IIR transfer function $H(z)$ of Eq. (8.23). The resulting structure is sketched in Figure 8.13(a) and is commonly known as the *direct form I* structure. Note that the overall realization is noncanonic since it employs six delays to implement a third-order transfer function. The transpose of this structure is sketched in Figure 8.13(b). Various other noncanonic direct forms can be derived by simple block diagram manipulations. Two such realizations are shown in Figure 8.13(c) and (d).

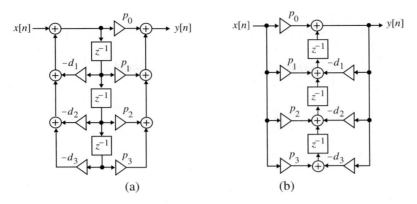

Figure 8.14: Direct form II and II$_t$ structures.

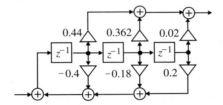

Figure 8.15: Direct form II realization of the IIR transfer function of Eq. (8.26).

To derive a canonic realization, we observe that in Figure 8.13(d), the signal variables at nodes ① and ①′are the same, and hence, the two top delays can be shared. Likewise, the signal variables at nodes ② and ②′ are the same, which permits the sharing of the two middle delays. Following the same argument, we can share the two delays at the bottom, leading to the final canonic structure shown in Figure 8.14(a), which is called the *direct form II* realization. The transpose of this is indicated in Figure 8.14(b).

The structures for direct form I and direct form II realizations of an Nth-order IIR transfer function should be evident from the third-order structures of Figures 8.13 and 8.14.

EXAMPLE 8.3   Illustration of Direct Form Realization

Consider the third-order IIR transfer function:

$$H(z) = \frac{0.44z^2 + 0.362z + 0.02}{z^3 + 0.4z^2 + 0.18z - 0.2} = \frac{0.44z^{-1} + 0.362z^{-2} + 0.02z^{-3}}{1 + 0.4z^{-1} + 0.18z^{-2} - 0.2z^{-3}}. \qquad (8.26)$$

A direct form II realization of the above transfer function is shown in Figure 8.15.

8.4.2 Cascade Realizations

By expressing the numerator and the denominator polynomials of the transfer function $H(z)$ as a product of polynomials of lower degree, a digital filter is often realized as a cascade of low-order filter sections. Consider, for example, $H(z) = P(z)/D(z)$, expressed as

$$H(z) = \frac{P_1(z)P_2(z)P_3(z)}{D_1(z)D_2(z)D_3(z)}. \qquad (8.27)$$

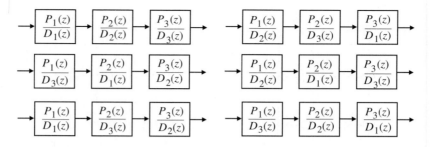

Figure 8.16: Examples of different equivalent cascade realizations obtained by different pole-zero pairings.

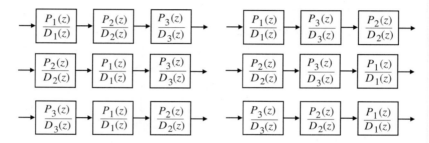

Figure 8.17: Different cascade realizations obtained by changing the ordering of the sections.

Various different cascade realizations of $H(z)$ can be obtained by different pole-zero polynomial pairings. Some examples of such realizations are shown in Figure 8.16. Additional cascade realizations are obtained by simply changing the ordering of the sections.[2] Figure 8.17 illustrates examples of different structures obtained by changing the ordering of the sections. There are altogether a total of 36 cascade realizations for the factored form indicated in Eq. (8.27) based on pole-zero pairings and ordering (Problem 8.18). In practice, due to the finite wordlength effects, each such cascade realization behaves differently from others.

Usually, the polynomials are factored into a product of first-order and second-order polynomials. In this case, $H(z)$ is expressed as

$$H(z) = p_0 \prod_k \left(\frac{1 + \beta_{1k} z^{-1} + \beta_{2k} z^{-2}}{1 + \alpha_{1k} z^{-1} + \alpha_{2k} z^{-2}} \right). \tag{8.28}$$

In the above, for a first-order factor, $\alpha_{2k} = \beta_{2k} = 0$. A possible realization of a third-order transfer function

$$H(z) = p_0 \left(\frac{1 + \beta_{11} z^{-1}}{1 + \alpha_{11} z^{-1}} \right) \left(\frac{1 + \beta_{12} z^{-1} + \beta_{22} z^{-2}}{1 + \alpha_{12} z^{-1} + \alpha_{22} z^{-2}} \right)$$

is shown in Figure 8.18.

EXAMPLE 8.4   Illustration of the Cascade Form Realization

We develop the cascade realizations of the third-order IIR transfer function of Eq. (8.26). By factoring the numerator and the denominator polynomials of $H(z)$ as given in Eq. (8.28), we obtain

---

[2]See Section 12.7.7.

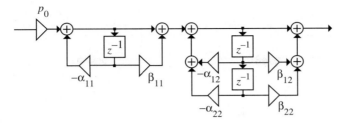

Figure 8.18: Cascade realization of a third-order IIR transfer function.

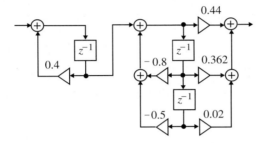

Figure 8.19: A cascade realization of the IIR transfer function of Eq. (8.26) based on direct form II realization of each section.

$$H(z) = \frac{0.44z^2 + 0.362z + 0.02}{(z^2 + 0.8z + 0.5)(z - 0.4)}$$

$$= \left(\frac{0.44 + 0.362z^{-1} + 0.02z^{-2}}{1 + 0.8z^{-1} + 0.5z^{-2}} \right) \left(\frac{z^{-1}}{1 - 0.4z^{-1}} \right).$$

From the above, we arrive at a cascade realization of $H(z)$, as shown in Figure 8.19. Another cascade realization is obtained by using a different pole-zero pairing:

$$H(z) = \left(\frac{z^{-1}}{1 + 0.8z^{-1} + 0.5z^{-2}} \right) \left(\frac{0.44 + 0.362z^{-1} + 0.02z^{-2}}{1 - 0.4z^{-1}} \right), \qquad (8.29)$$

whose realization is left as an exercise (Problem 8.23). However, it should be noted that the realization based on the above factored form will be noncanonic since it would require four delays instead of three delays employed in Figure 8.19. The total number of multipliers in both realizations remains the same.

8.4.3 Parallel Realizations

An IIR transfer function can be realized in a parallel form by making use of the partial-fraction expansion of the transfer function. A partial-fraction expansion of the transfer function in the form of Eq. (6.39) leads to the *parallel form I*. Thus, assuming simple poles, $H(z)$ is expressed in the form

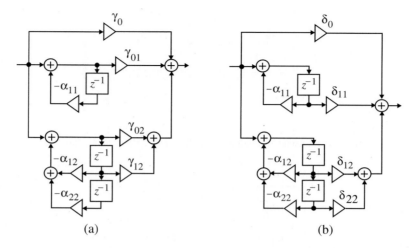

Figure 8.20: Parallel realizations of a third-order IIR transfer function: (a) parallel form I and (b) parallel form II.

$$H(z) = \gamma_0 + \sum_k \left(\frac{\gamma_{0k} + \gamma_{1k}z^{-1}}{1 + \alpha_{1k}z^{-1} + \alpha_{2k}z^{-2}} \right). \tag{8.30}$$

In the above, for a real pole, $\alpha_{2k} = \gamma_{1k} = 0$.

A direct partial-fraction expansion of the transfer function $H(z)$ expressed as a ratio of polynomials in z leads to the second basic form of the parallel structure, called the *parallel form II* [Mit77c]. Assuming simple poles, we arrive at

$$H(z) = \delta_0 + \sum_k \left(\frac{\delta_{1k}z^{-1} + \delta_{2k}z^{-2}}{1 + \alpha_{1k}z^{-1} + \alpha_{2k}z^{-2}} \right). \tag{8.31}$$

Here, for a real pole, $\alpha_{2k} = \delta_{2k} = 0$.

The two basic parallel realizations of a third-order IIR transfer function are sketched in Figure 8.20.

EXAMPLE 8.5   Illustration of the Parallel Form Realizations

We develop two different parallel realizations of the third-order IIR transfer function of Eq. (8.26). To this end, we make a partial-fraction expansion of $H(z)$ of the form of Eq. (6.39) expressed as a ratio of polynomials in z^{-1}, resulting in

$$H(z) = -0.1 + \frac{0.6}{1 - 0.4z^{-1}} + \frac{-0.5 - 0.2z^{-1}}{1 + 0.8z^{-1} + 0.5z^{-2}},$$

which leads to the parallel form I realization indicated in Figure 8.21(a).

Finally, a direct partial-fraction expansion of $H(z)$ expressed as a ratio of polynomials in z is given by

$$H(z) = \frac{0.24}{z - 0.4} + \frac{0.2z + 0.25}{z^2 + 0.8z + 0.5} = \frac{0.24z^{-1}}{1 - 0.4z^{-1}} + \frac{0.2z^{-1} + 0.25z^{-2}}{1 + 0.8z^{-1} + 0.5z^{-2}},$$

resulting in the parallel form II realization sketched in Figure 8.21(b).

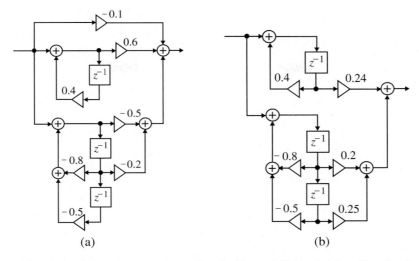

Figure 8.21: Example 8.5: (a) Parallel form I realization and (b) parallel form II realization

8.5 Realization of Basic Structures Using MATLAB

The basic FIR and IIR structures described in Sections 8.3 and 8.4 can be easily developed using MATLAB. In this section, we describe this approach.

8.5.1 Cascade Realization

The cascade realization of an FIR transfer function $H(z)$ involves its factorization in the form of Eq. (8.10). Likewise, the cascade realization of an IIR transfer function $H(z)$ involves its factorization in the form of Eq. (8.28). The factorization of a polynomial can be carried out in MATLAB using the function `factor`, which determines the second-order factors directly from the specified transfer function $H(z)$. We illustrate this approach in Examples 8.6 and 8.7.

factor.m

Program 8_1.m

EXAMPLE 8.6   Factorization of an FIR Transfer Function Using MATLAB

We develop the second-order factors of the sixth-order FIR transfer function:

$$H(z) = 50.4 + 28.02z^{-1} + 13.89z^{-2} + 7.42z^{-3} + 6.09z^{-4} + 3z^{-5} + z^{-6}.$$

To this end, we can use Program 8_1. The output data generated by running this program are as follows:

```
Factors
1.00000000000000   -0.70000000000000   0.33333333333333
1.00000000000000    0.38095238095238   0.23809523809524
1.00000000000000    0.87500000000000   0.25000000000000
```

The factorized form of $H(z)$ is therefore given by

$$H(z) = 50.4(1 - 0.7z^{-1} + 0.33333z^{-2})(1 + 0.38095z^{-1} + 0.2381z^{-2})(1 + 0.875z^{-1} + 0.25z^{-2}).$$

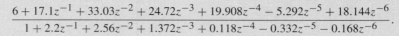

EXAMPLE 8.7   Factorization of an IIR Transfer Function Using MATLAB

Consider the sixth-order IIR transfer function:

$$\frac{6 + 17.1z^{-1} + 33.03z^{-2} + 24.72z^{-3} + 19.908z^{-4} - 5.292z^{-5} + 18.144z^{-6}}{1 + 2.2z^{-1} + 2.56z^{-2} + 1.372z^{-3} + 0.118z^{-4} - 0.332z^{-5} - 0.168z^{-6}}.$$

Program 8_2.m

The factors of this transfer function can be obtained using Program 8_2. The output data generated by running this program are

```
Numerator Factors
1.00000000000000     2.55474649598276     3.05353099104732
1.00000000000000     1.08357385033545     2.07384073616306
1.00000000000000    -0.78832034631821     0.47753373093274


Denominator Factors
1.00000000000000    -0.45677643628300                    0
1.00000000000000     0.79999999999998     0.69999999999999
1.00000000000000     1.20000000000001     0.80000000000002
1.00000000000000     0.65677643628301                    0
```

From the above, we arrive at the factored form of the specified IIR transfer function as given below:

$$H(z) = \frac{6(1 + 2.555z^{-1} + 3.0535z^{-2})(1 + 1.0836z^{-1} + 2.0738z^{-2})(1 - 0.788z^{-1} + 0.4775z^{-2})}{(1 - 0.4568z^{-1})(1 + 0.8z^{-1} + 0.7z^{-2})(1 + 1.2z^{-1} + 0.8z^{-2})(1 + 0.6568z^{-1})},$$

obtained by factoring out 6 from the first numerator factor.

8.5.2   Parallel Realization

The two parallel realizations described above can be developed readily in MATLAB using functions `residue` and `residuez`. We illustrate their use in Example 8.8.

Program 8_3.m

EXAMPLE 8.8   Parallel Realizations of an IIR Transfer Function Using MATLAB

We repeat the parallel realizations of Example 8.5 using Program 8_3. The input data requested by the program are the vectors num and den, containing the numerator and denominator coefficients, respectively. The output data generated by this program are

```
Parallel Form I
Residues are
 -0.2500 - 0.0000i
 -0.2500 + 0.0000i
  0.6000

Poles are at
 -0.4000 + 0.5831i
 -0.4000 - 0.5831i
  0.4000

Constant value
 -0.1000
```

```
        Parallel Form II
        Residues are
           0.1000 - 0.1458i
           0.1000 + 0.1458i
           0.2400

        Poles are at
          -0.4000 + 0.5831i
          -0.4000 - 0.5831i
           0.4000

        Constant value
```

The complex conjugate pairs in the partial-fraction expansion for parallel form I can be combined in MATLAB using the statement [b1,a1] = residuez(R1, P1, 0), where R1 is the two-element vector of the complex conjugate residues and P1 is the two-element vector of the complex conjugate poles. For this example, we use R1 = [r1(1) r1(2)] and P1 = [p1(1) p1(2)], resulting in

```
    b1 =
         -0.5000    -0.2000              0

    a1 =
          1.0000     0.8000     0.5000
```

verifying the correctness of the results obtained in Example 8.5.

Likewise, the complex conjugate pairs in the partial-fraction expansion for parallel form II can be combined in MATLAB using the statement [b2,a2] = residue(R2, P2, 0), where now R2 is the two-element vector of the complex conjugate residues and P2 is the two-element vector of the complex conjugate poles. For this example, we use R2 = [r2(1) r2(2)] and P2 = [p2(1) p2(2)], which yields

```
    b2 =
          0.2000     0.2500

    a2 =
          1.0000     0.8000     0.5000
```

The above results agree with those derived in Example 8.5.

8.6 Allpass Filters

We now turn our attention to the realization of a very special type of IIR transfer function, the allpass function, introduced earlier in Section 7.1.3. We recall from its definition in Eq. (7.7) that an IIR allpass transfer function $A(z)$ has a unity magnitude response for all frequencies, that is, $|A(e^{j\omega})| = 1$ for all values of ω. The digital allpass filter is a versatile building block for signal processing applications. Some of its possible applications have already been pointed out earlier. For example, as indicated in Section 7.1.3, it is used often as a delay equalizer, in which case it is designed such that when cascaded with another discrete-time system, the overall cascade has a constant group delay in the frequency range of interest. Another application, described in Section 7.5.4, is in the efficient implementation of a set of transfer functions satisfying certain complementary properties. For example, a pair of power-complementary first-order lowpass and highpass transfer functions can be implemented simultaneously, employing only a single

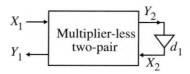

Figure 8.22: A multiplier-less two-pair constrained by a single multiplier.

first-order allpass filter. Likewise, a pair of power-complementary second-order bandpass and bandstop transfer functions can be implemented simultaneously, employing only a single second-order allpass filter. In fact, we shall demonstrate later in Section 8.10 that a large class of power-complementary transfer function pairs can be realized as a parallel connection of two allpass filters. It is thus of interest to develop techniques for the computationally efficient realization of allpass transfer functions, which is the subject of this section.

An Mth-order real-coefficient allpass transfer function is of the form of Eq. (7.8), with the numerator being the mirror-image polynomial of the denominator. A direct realization of an Mth-order allpass transfer function requires $2M$ multipliers. Since an Mth-order allpass transfer function is characterized by M unique coefficients, our objective here is to develop realization methods requiring only M multipliers. We outline here two different approaches to the minimum-multiplier realizations of allpass transfer functions.

8.6.1   Realization Based on the Multiplier Extraction Approach

Since an arbitrary allpass transfer function can be expressed as the product of second-order and/or first-order allpass transfer functions, we consider the realization of these lower-order transfer functions here. Even though an allpass transfer function can be realized using any of the methods discussed in this chapter, our objective here is to develop structures that remain allpass despite changes in the multiplier coefficients that may occur due to coefficient quantization [Mit74a].

Consider first the realization of a first-order allpass transfer function given by

$$A_1(z) = \frac{d_1 + z^{-1}}{1 + d_1 z^{-1}}. \tag{8.32}$$

Since the above transfer function is uniquely characterized by a single constant d_1, we attempt to realize it using a structure containing a single multiplier d_1 in the form of Figure 8.22. Substituting $G(z) = d_1$ in Eq. (7.135b), we express the input transfer function $A_1(z) = Y_1/X_1$ in terms of the transfer parameters of the two-pair as

$$A_1(z) = t_{11} + \frac{t_{12}t_{21}d_1}{1 - d_1 t_{22}} = \frac{t_{11} - d_1(t_{11}t_{22} - t_{12}t_{21})}{1 - d_1 t_{22}}. \tag{8.33}$$

A comparison of Eqs. (8.32) and (8.33) yields

$$t_{11} = z^{-1}, \qquad t_{22} = -z^{-1}, \tag{8.34}$$

$$t_{11}t_{22} - t_{12}t_{21} = -1. \tag{8.35}$$

Substituting Eq. (8.34) in Eq. (8.35), we arrive at

$$t_{12}t_{21} = 1 - z^{-2},$$

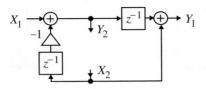

Figure 8.23: Development of the Type 1A first-order allpass structure.

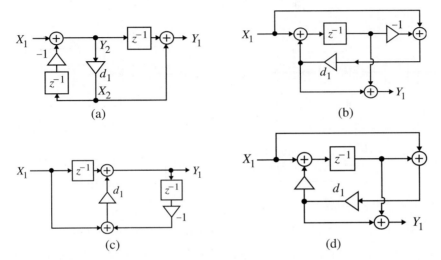

Figure 8.24: First-order single-multiplier allpass structures: (a) Type 1A, (b) Type 1B, (c) Type 1A$_t$, (d) Type 1B$_t$.

which leads to four possible solutions:

$$\text{Type 1A}: t_{11} = z^{-1}, \ t_{22} = -z^{-1}, \ t_{12} = 1 - z^{-2}, \ t_{21} = 1; \tag{8.36a}$$
$$\text{Type 1B}: t_{11} = z^{-1}, \ t_{22} = -z^{-1}, \ t_{12} = 1 + z^{-1}, \ t_{21} = 1 - z^{-1}; \tag{8.36b}$$
$$\text{Type 1A}_t: t_{11} = z^{-1}, \ t_{22} = -z^{-1}, \ t_{12} = 1, \ t_{21} = 1 - z^{-2}; \tag{8.36c}$$
$$\text{Type 1B}_t: t_{11} = z^{-1}, \ t_{22} = -z^{-1}, \ t_{12} = 1 - z^{-1}, \ t_{21} = 1 + z^{-1}. \tag{8.36d}$$

We now outline the development of the two-pair structure implementing the transfer parameters of Eq. (8.36a). From these equations, we arrive at

$$Y_2 = X_1 - z^{-1}X_2,$$
$$Y_1 = z^{-1}X_1 + (1 - z^{-2})X_2 = z^{-1}Y_2 + X_2.$$

A realization of the above is sketched in Figure 8.23. By constraining the $X_2 \, Y_2$ terminal-pair with the multiplier d_1, we arrive at the single multiplier realization of the first-order allpass transfer function, as sketched in Figure 8.24(a). In a similar fashion, the other three single-multiplier allpass structures can be realized from their transfer parameter descriptions of Eqs. (8.36b) to (8.36d) (Problem 8.33). The final realizations are indicated in Figure 8.24(b) to (d). These structures have been called the *Type 1 allpass networks*. It should be noted that the structures of Figure 8.24(c) and (d) are, respectively, the transpose of the structures of Figures 8.24(a) and (b). The structures of Figures 8.24(b) and (d) are both canonic with respect to the delays.

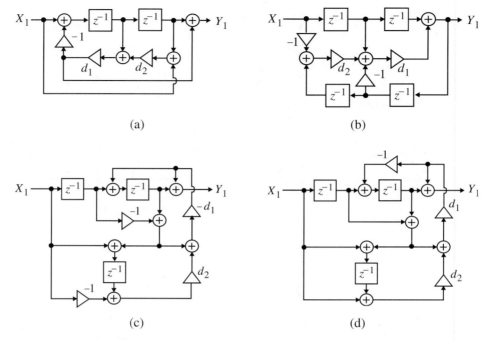

Figure 8.25: Second-order two-multiplier Type 2 allpass structures. (a) Type 2A, (b) Type 2D, (c) Type 2B, and (d) Type 2C.

We next consider the realization of a second-order allpass transfer function. Such a function is characterized by two unique coefficients and, hence, can be realized by a structure with two multipliers with multiplier constants d_1 and d_2. Various forms of the second-order allpass transfer functions exist. Networks realizing the transfer function of the form

$$A_2(z) = \frac{d_1 d_2 + d_1 z^{-1} + z^{-2}}{1 + d_1 z^{-1} + d_1 d_2 z^{-2}} \tag{8.37}$$

are called the *Type 2 allpass networks* and are sketched in Figure 8.25. Additional Type 2 allpass structures can be derived by transposing the structures of Figure 8.25.[3] The structure of Figure 8.25(a) is canonic with respect to the delays.

Another form of the second-order allpass transfer function is given by

$$A_2(z) = \frac{d_2 + d_1 z^{-1} + z^{-2}}{1 + d_1 z^{-1} + d_2 z^{-2}}. \tag{8.38}$$

The corresponding circuits are called the *Type 3 allpass structures* and are illustrated in Figure 8.26.[4] The structure of Figure 8.26(a) is canonic with respect to the delays.

A higher-order allpass transfer function can be realized as a cascade of first- and/or second-order allpass sections as illustrated in Example 8.9.

---

[3]The labeling of the structures here is as given in [Mit74a].
[4]The labeling of the structures here is as given in [Mit74a].

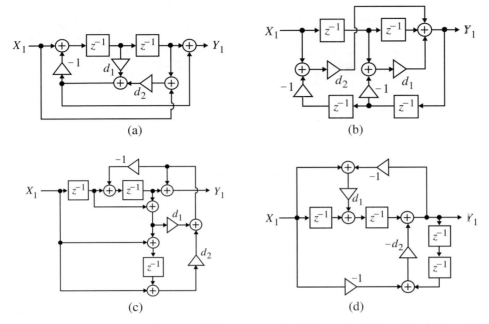

Figure 8.26: Second-order two-multiplier Type 3 allpass structures. (a) Type 3A, (b) Type 3D, (c) Type 3C, and (d) Type 3B.

EXAMPLE 8.9   Cascade Realization of an Allpass Transfer Function

Realize the third-order allpass transfer function

$$A_3(z) = \frac{-0.2 + 0.18z^{-1} + 0.4z^{-2} + z^{-3}}{1 + 0.4z^{-1} + 0.18z^{-2} - 0.2z^{-3}}. \tag{8.39}$$

By factoring the numerator and the denominator, we rewrite the above as

$$A_3(z) = \frac{(-0.4 + z^{-1})(0.5 + 0.8z^{-1} + z^{-2})}{(1 - 0.4z^{-1})(1 + 0.8z^{-1} + 0.5z^{-2})}. \tag{8.40}$$

A three-multiplier realization of the above allpass transfer function is shown in Figure 8.27, where the first-order allpass section has been implemented using the structure of Figure 8.24(b) and the second-order allpass section has been implemented using the structure of Figure 8.26(a).

8.6.2 Realization Based on the Two-Pair Extraction Approach

The stability test algorithm described in Section 7.9.2 also leads to an elegant realization of an Mth-order allpass transfer function $A_M(z)$ in the form of a cascaded two-pair [Vai87e]. This algorithm is based on the development of a series of $(m - 1)$th-order allpass transfer functions $A_{m-1}(z)$ from an mth-order allpass transfer function $A_m(z)$:

$$A_m(z) = \frac{d_m + d_{m-1}z^{-1} + d_{m-2}z^{-2} + \cdots + d_1 z^{-(m-1)} + z^{-m}}{1 + d_1 z^{-1} + d_2 z^{-2} + \cdots + d_{m-1}z^{-(m-1)} + d_m z^{-m}}, \tag{8.41}$$

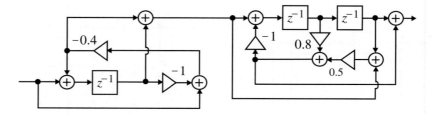

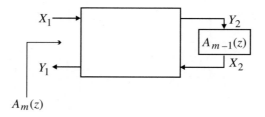

Figure 8.27: A three-multiplier realization of the allpass transfer function of Eq. (8.40).

Figure 8.28: Realization of $A_m(z)$ by two-pair extraction.

using the recursion

$$A_{m-1}(z) = z \left[\frac{A_m(z) - k_m}{1 - k_m A_m(z)} \right], \qquad m = M, M-1, \ldots 1, \tag{8.42}$$

where $k_m = A_m(\infty) = d_m$. It has been shown in Section 7.9.2 that $A_M(z)$ is stable if and only if

$$k_m^2 < 1, \quad \text{for } m = M, M-1, \ldots, 1. \tag{8.43}$$

If the allpass transfer function $A_{m-1}(z)$ is expressed in the form

$$A_{m-1}(z) = \frac{d'_{m-1} + d'_{m-2} z^{-1} + \cdots + d'_1 z^{-(m-2)} + z^{-(m-1)}}{1 + d'_1 z^{-1} + \cdots + d'_{m-2} z^{-(m-2)} + d'_{m-1} z^{-(m-1)}}, \tag{8.44}$$

then the coefficients of $A_{m-1}(z)$ are simply related to the coefficients of $A_m(z)$ through the expression

$$d'_i = \frac{d_i - d_m d_{m-i}}{1 - d_m^2}, \qquad i = 1, 2, \ldots, m-1. \tag{8.45}$$

To develop a realization of $A_m(z)$ using the above algorithm, we rewrite Eq. (8.42) as

$$A_m(z) = \frac{k_m + z^{-1} A_{m-1}(z)}{1 + k_m z^{-1} A_{m-1}(z)} \tag{8.46}$$

and realize $A_m(z)$ by extracting a two-pair constrained by $A_{m-1}(z)$ in the form of Figure 8.28.
Now from Eq. (7.135b), $A_m(z)$ can be expressed in terms of the transfer parameters of the two-pair as

$$A_m(z) = \frac{t_{11} - (t_{11} t_{22} - t_{12} t_{21}) A_{m-1}(z)}{1 - t_{22} A_{m-1}(z)}. \tag{8.47}$$

Comparing Eqs. (8.46) and (8.47), we readily obtain

$$t_{11} = k_m, \qquad t_{22} = -k_m z^{-1}, \tag{8.48a}$$

$$t_{11}t_{22} - t_{12}t_{21} = -z^{-1}. \tag{8.48b}$$

Substituting Eq. (8.48a) in Eq. (8.48b), we get

$$t_{12}t_{21} = (1 - k_m^2)z^{-1}. \tag{8.49}$$

As can be seen from the above, there are a number of solutions for t_{12} and t_{21} leading to different realizations for the two-pair. Some possible solutions are as indicated below:

$$t_{11} = k_m, \ t_{22} = -k_m z^{-1}, \ t_{12} = (1 - k_m^2)z^{-1}, \ t_{21} = 1, \tag{8.50a}$$

$$t_{11} = k_m, \ t_{22} = -k_m z^{-1}, \ t_{12} = (1 - k_m)z^{-1}, \ t_{21} = (1 + k_m), \tag{8.50b}$$

$$t_{11} = k_m, \ t_{22} = -k_m z^{-1}, \ t_{12} = \sqrt{1 - k_m^2}\,z^{-1}, \ t_{21} = \sqrt{1 - k_m^2}, \tag{8.50c}$$

$$t_{11} = k_m, \ t_{22} = -k_m z^{-1}, \ t_{12} = z^{-1}, \ t_{21} = (1 - k_m^2). \tag{8.50d}$$

The input–output relations of the two-pair described by Eq. (8.50a) are given by

$$Y_1 = k_m X_1 + (1 - k_m^2)z^{-1}X_2, \tag{8.51a}$$

$$Y_2 = X_1 - k_m z^{-1}X_2. \tag{8.51b}$$

A direct realization of the above equations leads to the three-multiplier two-pair shown in Figure 8.29(a). In a similar manner, direct realizations based on Eqs. (8.50b) and (8.50c) result in the four-multiplier structures of Figure 8.29(b) and (c), respectively.[5] A direct realization of Eq. (8.50d) results in a three-multiplier structure and is left as an exercise (Problem 8.40).

A two-multiplier realization can be derived by manipulating the input–output relations of Eqs. (8.51a) and (8.51b). Using Eq. (8.51b), we can rewrite Eq. (8.51a) as

$$Y_1 = k_m Y_2 + z^{-1}X_2. \tag{8.52}$$

A realization of the two-pair based on Eqs. (8.51b) and (8.52) is given in Figure 8.30(a). The two-pair of Figure 8.30(a) is often referred to as a *lattice structure*. A two-multiplier lattice realization of Eq. (8.50d) can be derived accordingly and is left as an excercise (Problem 8.41) [Lar99].

The two-pair described by Eq. (8.50b) can be realized using a single multiplier. To this erd, we first write its input–output relation from Eq. (8.50b) as

$$Y_1 = k_m X_1 + (1 - k_m)z^{-1}X_2, \tag{8.53a}$$

$$Y_2 = (1 + k_m)X_1 - k_m z^{-1}X_2. \tag{8.53b}$$

Defining

$$V_1 = k_m(X_1 - z^{-1}X_2), \tag{8.54}$$

we can rewrite Eqs. (8.53a) and (8.53b) as

$$Y_1 = V_1 + z^{-1}X_2, \tag{8.55a}$$

$$Y_2 = X_1 + V_1. \tag{8.55b}$$

---

[5]The structure of Figure 8.29(b) is called the *Kelly-Lochbaum form* [Kel62].

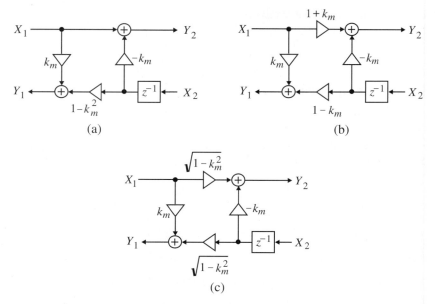

Figure 8.29: Direct realization of the two-pairs described by Eq. (8.50a) through (8.50c). (a) The two-pair described by Eq. (8.50a), (b) the two-pair described by Eq. (8.50b), and (c) the two-pair described by Eq. (8.50c).

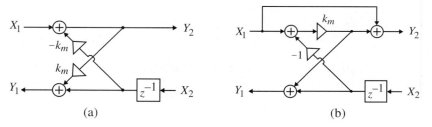

Figure 8.30: (a) A two-multiplier realization of the two-pair described by Eq. (8.50a) and (b) a one-multiplier realization of the two-pair described by Eq. (8.50b).

A realization based on Eqs. (8.54), (8.55a), and (8.55b) leads to the single-multiplier two-pair of Figure 8.30(b). Note that the two-input adder with an incoming multiplier with a coefficient -1 is implemented as a subtractor.

The realization of the mth-order allpass transfer function $A_m(z)$ is therefore obtained by constraining any one of the two-pairs of Figures 8.29 and 8.30 by the $(m - 1)$th-order allpass transfer function $A_{m-1}(z)$. For example, Figure 8.31(a) shows the realization of $A_m(z)$ by extracting the lattice two-pair of Figure 8.30(a).

Following the above algorithm, we can next realize $A_{m-1}(z)$ as a lattice two-pair constrained by the allpass transfer function $A_{m-2}(z)$. This process is repeated until the constraining transfer function is $A_0(z) = 1$. The complete realization of the original allpass transfer function $A_M(z)$ based on the extraction of the lattice two-pair of Figure 8.30(a) is therefore as indicated in Figure 8.31(b). It follows from our discussion in Section 7.9.2 that $A_M(z)$ is stable if the magnitudes of all multiplier coefficients in the realization are less than unity; that is, $|k_m| < 1$ for $m = M, M - 1, \ldots, 1$.

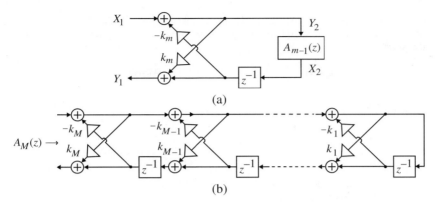

(a)

(b)

Figure 8.31: (a) Realization of $A_m(z)$ by extracting the lattice two-pair of Figure 8.30(a) and (b) cascaded lattice realization of $A_M(z)$.

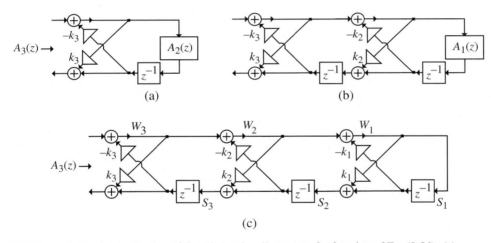

(a)   (b)

(c)

Figure 8.32: Cascaded lattice realization of the third-order allpass transfer function of Eq. (8.56): (a) $k_3 = d_3 = -0.2$, (b) $k_2 = d_2' = 0.2708333$, and (c) $k_1 = d_1'' = 0.3573771$.

Note that the above allpass structure requires $2M$ multipliers, which is twice that needed in the realization of an Mth-order allpass transfer function. However, a realization of $A_M(z)$ with M multipliers is obtained by extracting the two-pair of Figure 8.30(b). Here, also, the stability of $A_M(z)$ is ensured if $|k_m| < 1$ for $m = M, M - 1, \ldots, 1$.

EXAMPLE 8.10   Cascaded Lattice Realization of an IIR Transfer Function

Realize the third-order allpass transfer function $A_3(z)$ of Example 8.9 by means of a cascaded lattice structure. Here,

$$A_3(z) = \frac{d_3 + d_2 z^{-1} + d_1 z^{-2} + z^{-3}}{1 + d_1 z^{-1} + d_2 z^{-2} + d_3 z^{-3}}$$

$$= \frac{-0.2 + 0.18 z^{-1} + 0.4 z^{-2} + z^{-3}}{1 + 0.4 z^{-1} + 0.18 z^{-2} - 0.2 z^{-3}}. \tag{8.56}$$

Using the method outlined above, we realize $A_3(z)$ first in the form of a lattice two-pair characterized by the multiplier coefficient $k_3 = d_3 = -0.2$ and constrained by an allpass $A_2(z)$, as indicated in Figure 8.32(a). The allpass $A_2(z)$ is of the form

$$A_2(z) = \frac{d_2' + d_1' z^{-1} + z^{-2}}{1 + d_1' z^{-1} + d_2' z^{-2}}, \tag{8.57}$$

whose coefficients are obtained using Eq. (8.45) and are given by

$$d_1' = \frac{d_1 - d_3 d_2}{1 - d_3^2} = \frac{0.4 - (-0.2)(0.18)}{1 - (-0.2)^2} = 0.4541667, \tag{8.58a}$$

$$d_2' = \frac{d_2 - d_3 d_1}{1 - d_3^2} = \frac{0.18 - (-0.2)(0.4)}{1 - (-0.2)^2} = 0.2708333. \tag{8.58b}$$

Next, the allpass $A_2(z)$ is realized as a lattice two-pair characterized by the multiplier coefficient $k_2 = d_2' = 0.2708333$ and constrained by an allpass $A_1(z)$, as indicated in Figure 8.32(b). The allpass $A_1(z)$ is of the form

$$A_1(z) = \frac{d_1'' + z^{-1}}{1 + d_1'' z^{-1}}, \tag{8.59}$$

where

$$d_1'' = \frac{d_1' - d_2' d_1'}{1 - (d_2')^2} = \frac{d_1'}{1 + d_2'}$$

$$= \frac{0.4541667}{1.2708333} = 0.3573771. \tag{8.60}$$

Finally, the allpass $A_1(z)$ is realized as a lattice structure characterized by the multiplier coefficient $k_1 = d_1'' = 0.3573771$, resulting in the complete realization of the allpass $A_3(z)$, as shown in Figure 8.32(c). Note from Figure 8.32(c) that all multiplier coefficients have magnitudes less than unity. Therefore, the allpass transfer function of Eq. (8.56) is stable.

The M-file `poly2rc` in MATLAB can be employed to realize an allpass transfer function in the cascaded lattice form. We illustrate its application in Example 8.11.

Program 8_4.m

EXAMPLE 8.11   Cascaded Lattice Realization of an IIR Transfer Function Using MATLAB

We consider the development of the cascaded lattice realization of the allpass transfer function given in Eq. (8.39) of Example 8.9. To this end, we can use Program 8_4. Note the similarity between this program and Program 7_2 used for the stability test. The program requests the input data, which for this example is the vector of denominator coefficients in descending powers of z. It then prints the lattice section multiplier coefficients as given below:

```
The lattice section multipliers are
 -0.2    0.27083333    0.35737705
```

These values are seen to be identical to those derived in Example 8.10.

8.7   Tunable IIR Digital Filters

In Section 7.4.2, we described two first-order and two second-order IIR digital transfer functions with tunable frequency response characteristics. As we show next, these transfer functions can be realized easily

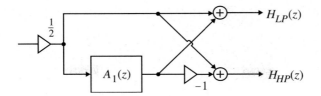

Figure 8.33: Allpass-based realization of the doubly-complementary first-order lowpass and highpass filters.

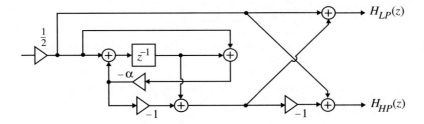

Figure 8.34: A tunable first-order lowpass/highpass filter structure.

using allpass structures, and the resulting realizations provide independent tuning of the filter parameters such as the cutoff frequency and bandwidth [Mit90a].

8.7.1 Tunable Lowpass and Highpass First-Order Filters

In Example 7.17, we showed that the first-order lowpass transfer function $H_{LP}(z)$ of Eq. (7.71) and the first-order highpass transfer function $H_{HP}(z)$ of Eq. (7.74) are a doubly-complementary pair and can be expressed as

$$H_{LP}(z) = \frac{1}{2}\left[\frac{1 - \alpha + z^{-1} - \alpha z^{-1}}{1 - \alpha z^{-1}}\right] = \frac{1}{2}\left[1 + \frac{-\alpha + z^{-1}}{1 - \alpha z^{-1}}\right]$$

$$= \frac{1}{2}[1 + A_1(z)], \tag{8.61}$$

$$H_{HP}(z) = \frac{1}{2}\left[\frac{1 + \alpha - z^{-1} - \alpha z^{-1}}{1 - \alpha z^{-1}}\right] = \frac{1}{2}\left[1 - \frac{-\alpha + z^{-1}}{1 - \alpha z^{-1}}\right]$$

$$= \frac{1}{2}[1 - A_1(z)], \tag{8.62}$$

where

$$A_1(z) = \frac{-\alpha + z^{-1}}{1 - \alpha z^{-1}} \tag{8.63}$$

is a first-order allpass transfer function. A combined realization of $H_{LP}(z)$ and $H_{HP}(z)$ based on the decompositions of Eqs. (8.61) and (8.62) is as indicated in Figure 8.33, in which the allpass filter given by Eq. (8.63) can be realized using any one of the four single-multiplier allpass structures of Figure 8.24. Figure 8.34 shows one such realization in which the 3-dB cutoff frequency of both filters can be simultaneously varied by changing the multiplier coefficient α. Figure 8.35 shows the composite magnitude responses of the two filters for two different values of α.

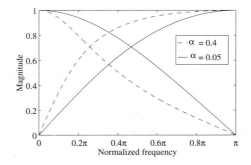

Figure 8.35: Magnitude responses of the lowpass/highpass filter structure of Figure 8.34 for two different values of the parameter α.

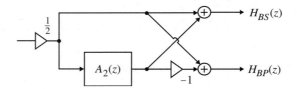

Figure 8.36: Allpass-based realization of doubly-complementary second-order bandpass/bandstop filter.

8.7.2   Tunable Bandpass and Bandstop Second-Order Filters

The second-order bandpass transfer function $H_{BP}(z)$ of Eq. (7.77) and the second-order bandstop transfer function $H_{BS}(z)$ of Eq. (7.80) also form a doubly-complementary pair and can be expressed as

$$H_{BP}(z) = \tfrac{1}{2}\left[1 - A_2(z)\right], \tag{8.64}$$

$$H_{BS}(z) = \tfrac{1}{2}\left[1 + A_2(z)\right], \tag{8.65}$$

where $A_2(z)$ is a second-order allpass transfer function given by

$$A_2(z) = \frac{\alpha - \beta(1+\alpha)z^{-1} + z^{-2}}{1 - \beta(1+\alpha)z^{-1} + \alpha z^{-2}}. \tag{8.66}$$

Therefore, the bandpass transfer function $H_{BP}(z)$ of Eq. (7.77) and the bandstop transfer function $H_{BS}(z)$ of Eq. (7.80) can be realized together, as indicated in Figure 8.36, where the allpass filter $A_2(z)$ is given by Eq. (8.66). A tunable bandpass/bandstop filter structure is then obtained by realizing the allpass $A_2(z)$ by means of a cascaded lattice structure described in Section 8.6.2, with the lattice two-pair realized using its single-multiplier equivalent of Figure 8.30(a). The final structure is indicated in Figure 8.37. Note that in the structure of Figure 8.37, the multiplier β controls the center frequency and the multiplier α controls the 3-dB bandwidth. Figure 8.38 illustrates the parametric tuning property of the bandpass/bandstop filter structure of Figure 8.37.

8.8   IIR Tapped Cascaded Lattice Structures

Several simple straightforward realizations of FIR and IIR transfer functions have been outlined in Sections 8.3 and 8.4, respectively. In most applications, these realizations work reasonably well, even under

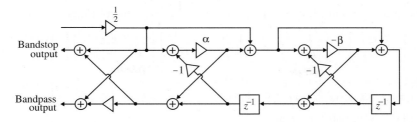

Figure 8.37: Tunable second-order bandpass/bandstop filter structure.

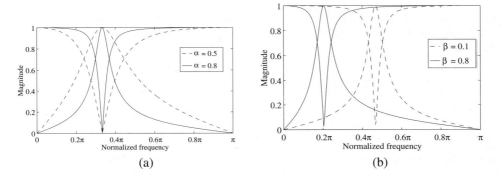

Figure 8.38: Magnitude responses of the bandpass/bandstop filter structure of Figure 8.37 for different values of the parameters α and β. (a) $\beta = 0.5$ and (b) $\alpha = 0.8$.

finite wordlength constraints. However, in some cases, digital filters with more robust properties are needed to provide satisfactory performances. We describe in this section two such realizations.

The cascaded lattice structure of Figure 8.31(b) can also realize an all-pole IIR digital filter, which finds application in the power spectrum estimation of random signals (see Section 15.4.2). It also forms the basis of an often used method for the realization of an arbitrary Mth-order transfer function $H(z)$ originally proposed by Gray and Markel [Gra73]. We first demonstrate the realization of an all-pole IIR transfer function and then outline the Gray and Markel realization method. We also provide a MATLAB program implementing both IIR structures.

8.8.1 Realization of an All-Pole IIR Transfer Function

We now show that the transfer function $W_1(z)/X_1(z)$ of the cascaded lattice structure of Figure 8.32(c) is an all-pole transfer function with the same denominator as the allpass transfer function $A_3(z)$ of Eq. (8.56). We first observe that a typical lattice two-pair here is described by the equations

$$W_i(z) = W_{i+1}(z) - k_i z^{-1} S_i(z),$$
$$S_{i+1}(z) = k_i W_i(z) + z^{-1} S_i(z).$$

From the above, we obtain the chain matrix description of the two-pair as

$$\begin{bmatrix} W_{i+1}(z) \\ S_{i+1}(z) \end{bmatrix} = \begin{bmatrix} 1 & k_i z^{-1} \\ k_i & z^{-1} \end{bmatrix} \begin{bmatrix} W_i(z) \\ S_i(z) \end{bmatrix}. \tag{8.67}$$

Thus, we can express the chain matrix of the cascaded lattice structure of Figure 8.32(b) as

$$
\begin{bmatrix} X_1(z) \\ Y_1(z) \end{bmatrix} = \begin{bmatrix} 1 & k_3 z^{-1} \\ k_3 & z^{-1} \end{bmatrix} \begin{bmatrix} 1 & k_2 z^{-1} \\ k_2 & z^{-1} \end{bmatrix} \begin{bmatrix} 1 & k_1 z^{-1} \\ k_1 & z^{-1} \end{bmatrix} \begin{bmatrix} W_1(z) \\ S_1(z) \end{bmatrix}
$$

$$
= \begin{bmatrix} 1 + k_2(k_1+k_3)z^{-1} + k_3 k_1 z^{-2} & k_1 z^{-1} + k_2(1+k_1 k_3)z^{-2} + k_3 z^{-3} \\ k_3 + k_2(1+k_1 k_3)z^{-1} + k_1 z^{-2} & k_1 k_3 z^{-1} + k_2(k_1+k_3)z^{-2} + z^{-3} \end{bmatrix} \begin{bmatrix} W_1(z) \\ S_1(z) \end{bmatrix}. \tag{8.68}
$$

From Eq. (8.68), we finally arrive at

$$
X_1(z) = \left(1 + [k_1(1+k_2) + k_2 k_3]z^{-1} + [k_2 + k_1 k_3(1+k_2)]z^{-2} + k_3 z^{-3} \right) W_1(z),
$$

where we have used the relation $S_1(z) = W_1(z)$. The transfer function $W_1(z)/X_1(z)$ is thus an all-pole transfer function with the same denominator polynomial as the third-order allpass transfer function of Eq. (8.56); that is,

$$
\frac{W_1(z)}{X_1(z)} = \frac{1}{1 + [k_1(1+k_2) + k_2 k_3]z^{-1} + [k_2 + k_1 k_3(1+k_2)]z^{-2} + k_3 z^{-3}} \tag{8.69}
$$

$$
= \frac{1}{1 + d_1 z^{-1} + d_2 z^{-2} + d_3 z^{-3}}, \tag{8.70}
$$

where we have used $k_1 = d_1''$, $k_2 = d_2'$, and $k_3 = d_3$ and Eqs. (8.58a), (8.58b), and (8.60).

It follows from the above discussion that, in the general case, the cascaded lattice structure of Figure 8.31(b) realizes an Mth-order all-pole transfer function with the same denominator as $A_M(z)$ if the output is tapped from the input of the right-most delay. Such an all-pole IIR structure results from modelling an autoregressive process, as described in Section 15.4.2. The multiplier coefficients $\{k_i\}$ of the cascaded lattice structure are called *reflection coefficients*.

8.8.2 Gray-Markel Method

The method of realizing an IIR transfer function $H(z) = P_M(z)/D_M(z)$ consists of two steps. In the first step, an intermediate allpass transfer function $A_M(z) = z^{-M} D_M(z^{-1})/D_M(z) = \tilde{D}_M(z)/D_M(z)$ is realized in the form of a cascaded lattice structure.[6] A set of independent variables of this structure are then summed in the second step, with appropriate weights to yield the desired numerator $P_M(z)$.

To illustrate the method of realizing the numerator, we consider for simplicity the implementation of a third-order IIR transfer function

$$
H(z) = \frac{P_3(z)}{D_3(z)} = \frac{p_0 + p_1 z^{-1} + p_2 z^{-2} + p_3 z^{-3}}{1 + d_1 z^{-1} + d_2 z^{-2} + d_3 z^{-3}}. \tag{8.71}
$$

In the first step, we form an intermediate allpass transfer function $A_3(z) = Y_1(z)/X_1(z) = \tilde{D}_3(z)/D_3(z)$. Realization of $A_3(z)$ has been illustrated in Example 8.10, resulting in the structure of Figure 8.32(c). Our objective is to sum the linearly independent signal variables Y_1, S_1, S_2, and S_3 with weights $\{\alpha_i\}$, as shown in Figure 8.39, to arrive at the desired numerator $P_3(z)$. To this end, we need to analyze the digital filter structure of Figure 8.32(c) and determine the transfer functions $S_1(z)/X_1(z)$, $S_2(z)/X_1(z)$, and $S_3(z)/X_1(z)$.

From Eq. (8.69), we have

$$
\frac{S_1(z)}{X_1(z)} = \frac{1}{D_3(z)}. \tag{8.72}
$$

---

[6] $\tilde{D}_M(z)$ denotes the mirror-image polynomial of $D_M(z)$ of degree M; that is, $\tilde{D}_M(z) = z^{-M} D_M(z^{-1})$.

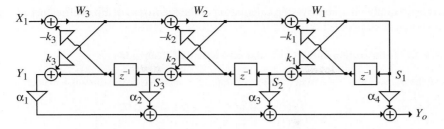

Figure 8.39: The Gray-Markel structure for a third-order transfer function.

Next, we observe from Figure 6.38 that $S_2(z) = (d_1'' + z^{-1})S_1(z)$ and, hence, from the above, we get

$$\frac{S_2(z)}{X_1(z)} = \frac{d_1'' + z^{-1}}{D_3(z)}. \tag{8.73}$$

Finally, from Figure 8.39, we have $S_3(z) = d_2'W_2(z) + z^{-1}S_2(z)$ and $S_1(z) = W_2(z) - d_1''z^{-1}S_1(z)$. From these equations, the relation $S_2(z) = (d_1'' + z^{-1})S_1(z)$, and Eqs. (8.58a), (8.58b), and (8.60), we arrive at $S_3(z) = (d_2' + d_1'z^{-1} + z^{-2})S_1(z)$, yielding

$$\frac{S_3(z)}{X_1(z)} = \frac{d_2' + d_1'z^{-1} + z^{-2}}{D_3(z)}. \tag{8.74}$$

It should be noted that the numerator of the transfer function $S_i(z)/X_1(z)$ is precisely the numerator of the allpass transfer function $S_i(z)/W_i(z)$.

We now form

$$\frac{Y_o(z)}{X_1(z)} = \alpha_1 \frac{Y_1(z)}{X_1(z)} + \alpha_2 \frac{S_3(z)}{X_1(z)} + \alpha_3 \frac{S_2(z)}{X_1(z)} + \alpha_4 \frac{S_1(z)}{X_1(z)}. \tag{8.75}$$

Substituting Eqs. (8.72) to (8.74) and the expression for $Y_1(z)/X_1(z) = A_3(z)$ as given by Eq. (8.56) in Eq. (8.75), we arrive at

$$\frac{Y_o(z)}{X_1(z)} = \frac{\begin{array}{c}\alpha_1(d_3 + d_2z^{-1} + d_1z^{-2} + z^{-3}) \\ + \alpha_2(d_2' + d_1'z^{-1} + z^{-2}) + \alpha_3(d_1'' + z^{-1}) + \alpha_4\end{array}}{D_3(z)}. \tag{8.76}$$

Comparing the numerators of the right-hand sides of Eqs. (8.71) and (8.76), and by equating the coefficients of like powers of z^{-1}, we thus obtain

$$\alpha_1 d_3 + \alpha_2 d_2' + \alpha_3 d_1'' + \alpha_4 = p_0,$$
$$\alpha_1 d_2 + \alpha_2 d_1' + \alpha_3 = p_1,$$
$$\alpha_1 d_1 + \alpha_2 = p_2,$$
$$\alpha_1 = p_3.$$

Solving the above, we get

$$\alpha_1 = p_3,$$
$$\alpha_2 = p_2 - \alpha_1 d_1,$$
$$\alpha_3 = p_1 - \alpha_1 d_2 - \alpha_2 d_1',$$
$$\alpha_4 = p_0 - \alpha_1 d_3 - \alpha_2 d_2' - \alpha_3 d_1'', \tag{8.77}$$

which is in a form suitable for step-by-step calculation of the feedforward tap coefficients α_i. We illustrate the above method in Example 8.12.

EXAMPLE 8.12   Illustration of the Gray-Markel Method of Realization

Let us realize the transfer function of Eq. (8.26) in Example 8.3, repeated below for convenience:

$$H(z) = \frac{P_3(z)}{D_3(z)} = \frac{0.44z^{-1} + 0.362z^{-2} + 0.02z^{-3}}{1 + 0.4z^{-1} + 0.18z^{-2} - 0.2z^{-3}}, \tag{8.78}$$

using the Gray-Markel method.

We form the intermediate allpass transfer function $A_3(z)$ having the same denominator as $H(z)$ in Eq. (8.78). Its realization has been carried out in Example 8.10 and is shown in Figure 8.32(c), where

$$d_3 = -0.2, \quad d_2' = 0.2708333, \quad \text{and} \quad d_1'' = 0.3573771.$$

To realize the numerator, we compute the tap coefficients $\{\alpha_i\}$ using Eq. (8.77). Substituting the values of

$$d_1 = 0.4, \quad d_2 = 0.18, \quad d_3 = -0.2, \quad d_1' = 0.4541667, \quad d_1'' = 0.3573771, \quad d_2' = 0.2708333,$$

and the values of the numerator coefficients,

$$p_0 = 0, \quad p_1 = 0.44, \quad p_2 = 0.36, \quad \text{and} \, p_3 = 0.02$$

in Eq. (8.77), we arrive at

$$\alpha_1 = 0.02, \quad \alpha_2 = 0.352, \quad \alpha_3 = 0.2765333, \quad \alpha_4 = -0.19016.$$

The final realization is thus as indicated in Figure 8.39, with the multiplier values as given above.

8.8.3   Realization Using MATLAB

Both the pole-zero and the all-pole IIR cascaded lattice structures can be developed from their specified transfer functions in MATLAB using the M-file `tf2latc`. The function `latc2tf` implements the reverse process and can be used to verify the realization.

We illustrate the use of the function `tf2latc` in Examples 8.13 and 8.14.

Program 8_5.m

EXAMPLE 8.13   Implementation of the Gray-Markel Method Using MATLAB

Using MATLAB, we determine the lattice and the feedforward parameters of the Gray-Markel structure for the IIR transfer function of Eq. (8.78). To this end, Program 8_5 can be employed. The input data called by this program are

```
num = [0   0.44   0.36    0.02]
den = [1   0.4    0.18   -0.2]
```

which results in the following output data generated by the program:

```
Lattice parameters are
    0.35737704918033    0.27083333333333   -0.2

Feedforward multipliers are
    0.02    0.352    0.27653333333333   -0.19016
```

which are seen to be identical to those derived in Example 8.9.

Program 8_5 can also be employed to develop the cascaded lattice realization of an all-pole IIR transfer function as illustrated in Example 8.14.

EXAMPLE 8.14   Cascaded Lattice Realization of an All-Pole IIR Transfer Function Using MATLAB

Program 8_5.m

We now realize the all-pole transfer function

$$H(z) = \frac{1}{1 + 0.5\,z^{-1} + 0.4\,z^{-2} + 0.3\,z^{-3} + 0.2\,z^{-4}}.$$

The output data generated by running Program 8_5 for the above transfer function are

```
Lattice parameters are
   0.325581395   0.24863884   0.20833333   0.20000000

Feedforward multipliers are
   0      0      0      0      1
```

Since the absolute values of all lattice parameters are less than 1, $H(z)$ is guaranteed stable.

Program 8_6 can be used to verify the cascaded lattice structure developed using Program 8_5, as illustrated in Example 8.15.

EXAMPLE 8.15   Determination of the Transfer Function of the Gray–Markel Structure Using MATLAB

Program 8_6.m

We verify the cascaded lattice structure generated in Example 8.13. The input data called by Program 8_6 are

```
k1 = [0.35737704918033    0.27083333333333   -0.2]
alpha = [0.02    0.352    0.27653333333333   -0.19016]
```

The program then generates the following output data,

```
Numerator coefficients are
   -0.00    0.44    0.36    0.02

Denominator coefficients are
    1.00    0.40    0.180    -0.20
```

which are seen to be the same as that given in Eq. (8.78).

8.9   FIR Cascaded Lattice Structures

There are two types of cascaded lattice structures for the realization of FIR transfer functions. In this section, we describe methods for their realizations.

8.9.1   Realization of Arbitrary FIR Transfer Functions

The cascaded lattice structure that can be used to realize an arbitrary FIR transfer function is shown in Figure 8.40 [Mak75]. The realization algorithm is developed first, followed by its MATLAB implementation.

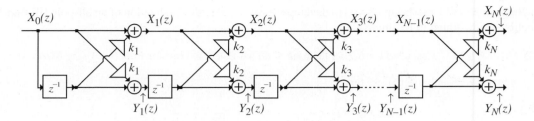

Figure 8.40: The FIR cascaded lattice structure.

Realization Method

For the cascaded lattice realization, the Nth-order FIR transfer function is assumed to be of the form

$$H_N(z) = 1 + \sum_{n=1}^{N} p_n z^{-n}. \tag{8.79}$$

An arbitrary FIR transfer function $H(z)$ of the form of Eq. (8.8) can be expressed in the form $H(z) = h[0]H_N(z)$, where $p_n = h[n]/h[0]$, $0 \leq n \leq N$. Hence, by placing a multiplier $h[0]$ at the input of the structure realizing $H_N(z)$, we obtain a realization of $H(z)$. The normalized transfer function $H_N(z)$ is realized in the form of Figure 8.40. To develop the realization algorithm, we first analyze it to determine the relations between the input $X(z) = X_0(z)$ and the intermediate variables $X_i(z)$ and $Y_i(z)$, that is, between the series of transfer functions $H_i(z) = X_i(z)/X_0(z)$ and $G_i(z) = Y_i(z)/X_0(z)$, $1 \leq i \leq N$. Note that $X_N(z)$ and $Y_N(z)$ are the output variables.

From Figure 8.40, we obtain

$$X_1(z) = X_0(z) + k_1 z^{-1} X_0(z), \tag{8.80a}$$

$$Y_1(z) = k_1 X_0(z) + z^{-1} X_0(z). \tag{8.80b}$$

The corresponding transfer functions are given by

$$H_1(z) = \frac{X_1(z)}{X_0(z)} = 1 + k_1 z^{-1}, \tag{8.81a}$$

$$G_1(z) = \frac{Y_1(z)}{X_0(z)} = k_1 + z^{-1}, \tag{8.81b}$$

which are seen to be first-order FIR transfer functions. Moreover, it can be seen from Eqs. (8.81a) and (8.81b) that

$$\tilde{H}_1(z) = z^{-1} H_1(z^{-1}) = z^{-1} + k_1 = G_1(z),$$

indicating that the FIR transfer function $G_1(z)$ of Eq. (8.81b) is the mirror image of the FIR transfer function $H_1(z)$ of Eq. (8.81a).

Next, we observe from Figure 8.40 that

$$X_2(z) = X_1(z) + k_2 z^{-1} Y_1(z), \tag{8.82a}$$

$$Y_2(z) = k_2 X_1(z) + z^{-1} Y_1(z). \tag{8.82b}$$

The corresponding transfer functions are given by

$$H_2(z) = \frac{X_2(z)}{X_0(z)} = H_1(z) + k_2 z^{-1} G_1(z), \tag{8.83a}$$

$$G_2(z) = \frac{Y_2(z)}{X_0(z)} = k_2 H_1(z) + z^{-1} G_1(z). \tag{8.83b}$$

Substituting Eqs. (8.81a) and (8.81b) in the above, we observe that $H_2(z)$ and $G_2(z)$ are second-order FIR transfer functions. Moreover, since $G_1(z) = \tilde{H}_1(z)$, it follows from the above that $G_2(z) = z^{-2} H_2(z^{-1}) = \tilde{H}_2(z)$, implying that $G_2(z)$ is the mirror image of $H_2(z)$.

Now, the input–output relation of the ith lattice section is given by

$$X_i(z) = X_{i-1}(z) + k_i z^{-1} Y_{i-1}(z), \tag{8.84a}$$

$$Y_i(z) = k_i X_{i-1}(z) + z^{-1} Y_{i-1}(z). \tag{8.84b}$$

The corresponding transfer functions, $H_i(z) = X_i(z)/X_0(z)$ and $G_i(z) = Y_i(z)/X_0(z)$, are related through

$$H_i(z) = H_{i-1}(z) + k_i z^{-1} G_{i-1}(z), \tag{8.85a}$$

$$G_i(z) = k_i H_{i-1}(z) + z^{-1} G_{i-1}(z), \tag{8.85b}$$

where $H_{i-1}(z) = X_{i-1}(z)/X_0(z)$ and $G_{i-1}(z) = Y_{i-1}(z)/X_0(z)$. It should be noted that the relations of Eqs. (8.85a) and (8.85b) hold for all $1 \leq i \leq N$. If we assume that $H_{i-1}(z)$ and $G_{i-1}(z)$ are $(i-1)$th-order FIR transfer functions, then it follows from Eqs. (8.85a) and (8.85b) that $H_i(z)$ and $G_i(z)$ are ith-order FIR transfer functions. Moreover, if $G_{i-1}(z) = \tilde{H}_{i-1}(z)$, then it also follows from Eqs. (8.85a) and (8.85b) that $G_i(z) = \tilde{H}_i(z)$. We have already shown that these two observations hold for $i = 1$ and $i = 2$. Therefore, by induction, they also hold for all $1 \leq i \leq N$.

To develop the realization algorithm, the above process is reversed. That is, given the expressions for $H_i(z)$ and $\tilde{H}_i(z)$, we find the expressions for $H_{i-1}(z)$ and $\tilde{H}_{i-1}(z)$ for $i = N, N - 1, \ldots, 2, 1$. We now develop a recursion relation to compute these intermediate transfer functions.

Solving Eqs. (8.85a) and (8.85b) for $i = N$, we obtain

$$H_{N-1}(z) = \frac{1}{(1 - k_N^2)z^{-1}} \left\{ z^{-1} H_N(z) - k_N z^{-1} \tilde{H}_N(z) \right\}, \tag{8.86}$$

$$\tilde{H}_{N-1}(z) = \frac{1}{(1 - k_N^2)z^{-1}} \left\{ -k_N H_N(z) + \tilde{H}_N(z) \right\}. \tag{8.87}$$

It can be easily verified that $\tilde{H}_{N-1}(z)$ of Eq. (8.87) is indeed the mirror image of $H_{N-1}(z)$ of Eq. (8.86). Substituting the expression for $H_N(z)$ of Eq. (8.79) in Eq. (8.86), we get

$$H_{N-1}(z) = \frac{1}{(1 - k_N^2)} \left\{ (1 - k_N p_N) + (p_1 - k_N p_{N-1})z^{-1} + \cdots \right.$$

$$\left. + (p_{N-1} - k_N p_1)z^{-(N-1)} + (p_N - k_N)z^{-N} \right\}. \tag{8.88}$$

If we choose $k_N = p_N$, $H_{N-1}(z)$ reduces to an FIR transfer function of degree $N - 1$ and can be written in the form

$$H_{N-1}(z) = 1 + \sum_{n=1}^{N-1} p'_n z^{-n}, \tag{8.89}$$

where

$$p'_n = \frac{p_n - k_N p_{N-n}}{1 - k_N^2}, \qquad 1 \le n \le N - 1. \tag{8.90}$$

Continuing the above recursion algorithm, all the multiplier coefficients of the structure of Figure 8.40 can be computed. Example 8.16 illustrates the procedure.

EXAMPLE 8.16   Cascaded Lattice Realization of an FIR Transfer Function

Realize the fourth-order FIR transfer function

$$H_4(z) = 1 + 1.2z^{-1} + 1.12z^{-2} + 0.12z^{-3} - 0.08z^{-4}. \tag{8.91}$$

From the above, the multiplier coefficient of the right-most lattice is given by $k_4 = p_4 = -0.08$. Using Eq. (8.90), we next determine the coefficients p'_n of the third-order transfer function $H_3(z)$ as

$$p'_3 = \frac{p_3 - k_4 p_1}{1 - k_4^2} = \frac{0.12 - (-0.08)(1.2)}{1 - (-0.08)^2} = 0.2173913,$$

$$p'_2 = \frac{p_2 - k_4 p_2}{1 - k_4^2} = \frac{1.12}{1 - 0.08} = 1.2173913,$$

$$p'_1 = \frac{p_1 - k_4 p_3}{1 - k_4^2} = \frac{1.2 - (-0.08)(0.12)}{1 - (-0.08)^2} = 1.2173913.$$

As a result,

$$H_3(z) = 1 + 1.2173913z^{-1} + 1.2173913z^{-2} + 0.2173913z^{-3}.$$

Thus, the multiplier coefficient of the third lattice two-pair is given by $k_3 = p'_3 = 0.2173913$.

We continue the above recursion process to determine the coefficients p''_n of the second-order transfer function $H_2(z)$, which are given by

$$p''_2 = \frac{p'_2 - k_3 p'_1}{1 - k_3^2} = \frac{p'_2}{1 + k_3} = \frac{1.2173913}{1.2173913} = 1.0,$$

$$p''_1 = \frac{p'_1 - k_3 p'_2}{1 - k_3^2} = \frac{p'_1}{1 + k_3} = 1.0,$$

leading to

$$H_2(z) = 1 + z^{-1} + z^{-2}.$$

Consequently, the multiplier coefficient of the second lattice two-pair is given by $k_2 = p''_2 = 1.0$.

The last recursion yields the multiplier coefficient of the first lattice two-pair:

$$k_1 = \frac{p''_1}{1 + k_2} = \frac{1.0}{2.0} = 0.5.$$

The final realization of the FIR transfer function of Eq. (8.91) is thus as shown in Figure 8.41.

Realization Using MATLAB

The function `tf2latc` in MATLAB can again be employed to compute the multiplier coefficients of the cascade lattice structure of Figure 8.40. To this end, we can make use of Program 8_7. The input data

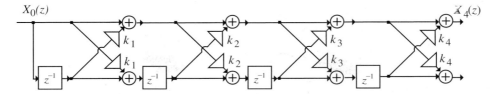

Figure 8.41: Cascaded lattice realization of the FIR transfer function of Eq. (8.91): $k_1 = 0.5$, $k_2 = 1.0$, $k_3 = 0.2173913$, and $k_4 = -0.08$.

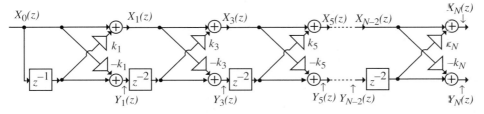

Figure 8.42: Power-symmetric FIR cascaded lattice structure.

called by the program is the vector num of transfer function coefficients entered in ascending powers of z^{-1}.

We illustrate its application in Example 8.17.

EXAMPLE 8.17   Cascaded Lattice Realization of an FIR Transfer Function Using MATLAB

We consider the realization of the FIR transfer function of Eq. (8.91) using Program 8_7. The output data obtained is the vector of lattice coefficients given below.

Program 8_7.m

```
Lattice coefficients are
 -0.08      0.2173913      1.0      0.5
```

which are seen to be identical to those computed in Example 8.11.

The coefficients of the FIR cascaded lattice can also be computed using the M-file poly2rc. The FIR cascaded lattice structure developed using Program 8_7 can also be verified using the function latc2tf.

8.9.2 Power-Symmetric FIR Cascaded Lattice Structure

Another type of a cascaded lattice structure for the realization of a real-coefficient FIR transfer function $H_N(z)$ of order N is shown in Figure 8.42 [Vai86b]. However, for realizability, the transfer function must satisfy the *power-symmetric condition* given by Eq. (7.101) and repeated below for convenience:

$$H_N(z)H_N(z^{-1}) + H_N(-z)H_N(-z^{-1}) = K_N, \qquad (8.92)$$

where K_N is a constant. We first analyze the structure of Figure 8.42 and then develop the synthesis procedure.

We define $H_i(z) = X_i(z)/X_0(z)$ and $G_i(z) = Y_i(z)/X_0(z)$. From Figure 8.42, we observe that

$$X_1(z) = X_0(z) + k_1 z^{-1} X_0(z),$$
$$Y_1(z) = -k_1 X_0(z) + z^{-1} X_0(z). \qquad (8.93)$$

Therefore,

$$H_1(z) = 1 + k_1 z^{-1},$$

$$G_1(z) = -k_1 + z^{-1}. \tag{8.94}$$

It can be easily verified that

$$G_1(z) = z^{-1} H_1(-z^{-1}). \tag{8.95}$$

Next, from Figure 8.42, it follows that the transfer functions $H_i(z)$ and $G_i(z)$ can be expressed in terms of $H_{i-2}(z)$ and $G_{i-2}(z)$ as

$$H_i(z) = H_{i-2}(z) + k_i z^{-2} G_{i-2}(z),$$

$$G_i(z) = -k_i H_{i-2}(z) + z^{-2} G_{i-2}(z). \tag{8.96}$$

It can be easily shown that

$$G_i(z) = z^{-i} H_i(-z^{-1}), \tag{8.97}$$

provided

$$G_{i-2}(z) = z^{-(i-2)} H_{i-2}(-z^{-1}).$$

However, as can be seen from Eq. (8.95), Eq. (8.97) holds for $i = 1$. Hence, the relation of Eq. (8.97) holds for all odd integer values of i. This result also implies that N must be an odd integer.

It is a simple exercise to show that both $H_i(z)$ and $G_i(z)$ satisfy the power-symmetry condition of Eq. (8.92). In addition, $H_i(z)$ and $G_i(z)$ are power-complementary; that is,

$$\left| H_i(e^{j\omega}) \right|^2 + \left| G_i(e^{j\omega}) \right|^2 = C_i, \tag{8.98}$$

where C_i is a constant.

To develop the synthesis equation, we invert Eq. (8.96) to arrive at

$$(1 + k_i^2) H_{i-2}(z) = H_i(z) - k_i G_i(z),$$

$$(1 + k_i^2) z^{-2} G_{i-2}(z) = k_i H_i(z) + G_i(z). \tag{8.99}$$

Note that at the ith step, the multiplier coefficient k_i is chosen to eliminate the coefficient of z^{-i}, the highest power of z^{-1}, in $H_i(z) - k_i G_i(z)$. For this choice of k_i, the coefficient of z^{-i+1} also vanishes, making $H_{i-2}(z)$ a polynomial of degree $i - 2$.

We start the process by setting $i = N$, and compute $G_N(z)$ using Eq. (8.97). Next, we determine the transfer functions $H_{N-2}(z)$ and $G_{N-2}(z)$, using the above recurrence relations. This process is repeated until all coefficients of the lattice have been determined.

We illustrate the above synthesis method in Example 8.18.

EXAMPLE 8.18   Cascaded Lattice Realization of a Power-Symmetric FIR Transfer Function

Let us realize the FIR transfer function

$$H_5(z) = 1 + 0.3z^{-1} + 0.2z^{-2} - 0.376z^{-3} - 0.06z^{-4} + 0.2z^{-5},$$

in the form of the cascaded lattice structure of Figure 8.42.

It can be easily verified that the above transfer function satisfies the power-symmetric condition of Eq. (8.92). Next, we form

$$G_5(z) = z^{-5}H_5(-z^{-1}) = -0.2 - 0.06z^{-1} + 0.376z^{-2} + 0.2z^{-3} - 0.3z^{-4} + z^{-5}.$$

To determine $H_3(z)$, we first form

$$H_5(z) - k_5 G_5(z) = (1 + 0.2k_5) + (0.3 + 0.06k_5)z^{-1} + (0.2 - 0.376k_5)z^{-2}$$
$$+ (-0.376 - 0.2k_5)z^{-3} + (-0.06 + 0.3k_5)z^{-4} + (0.2 - k_5)z^{-5}.$$

To cancel the coefficient of z^{-5} in the above expression, we choose

$$k_5 = 0.2.$$

Then, from Eq. (8.99) and the above equation, we get

$$H_3(z) = \tfrac{1}{1.04}\left(1.04 + 0.312z^{-1} + 0.1248z^{-2} - 0.416z^{-3}\right)$$
$$= 1 + 0.3z^{-1} + 0.12z^{-2} - 0.4z^{-3}.$$

We now compute

$$0.2H_5(z) + G_5(z) = (0.2 - 0.2) + (0.06 - 0.06)z^{-1} + (0.04 + 0.376)z^{-2}$$
$$+ (-0.0752 + 0.2)z^{-3} + (-0.012 - 0.3)z^{-4} + 1.04z^{-5}.$$

Therefore, from Eq. (8.99) and the above equation, we get

$$G_3(z) = \tfrac{1}{1.04}\left(0.416 + 0.1248z^{-1} - 0.312z^{-2} + 1.04z^{-3}\right)$$
$$= 0.4 + 0.12z^{-1} - 0.3z^{-2} + z^{-3}.$$

Continuing the above process, we arrive at the remaining two multiplier coefficients:

$$k_3 = -0.4, \qquad k_1 = 0.3.$$

8.10   Parallel Allpass Realization of IIR Transfer Functions

In Section 8.7.1, we have demonstrated that a pair of doubly-complementary first-order lowpass and highpass transfer functions can be realized as shown in Figure 8.33. Likewise, in Section 8.7.2, we have shown that a pair of doubly-complementary second-order bandpass and bandstop transfer functions can be realized as indicated in Figure 8.36. These two structures can be considered as special cases of a structure composed of a parallel connection of two stable allpass filters of the form of Figure 8.43, where one of the allpass sections is a zeroth-order transfer function. We consider in this section the realization of an Nth-order transfer function $G(z)$ in the form of Figure 8.43, along with its power-complementary transfer function $H(z)$ [Vai86a]. As we shall point out later, such structures have a number of very attractive properties from an implementation point of view.

From Figure 8.43, we observe that

$$G(z) = \tfrac{1}{2}\{A_0(z) + A_1(z)\}, \tag{8.100a}$$

$$H(z) = \tfrac{1}{2}\{A_0(z) - A_1(z)\}. \tag{8.100b}$$

It is a simple exercise to show that a necessary condition for $G(z)$ and $H(z)$ to have the sum of allpass decompositions of the form of Eqs. (8.100a) and (8.100b) is that each be a bounded real (BR) IIR transfer

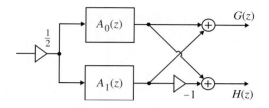

Figure 8.43: Realization of a doubly-complementary transfer function pair using a parallel allpass structure.

function (Problem 7.74).[7] The BR condition can be easily satisfied by any stable transfer function by a simple scaling. Let

$$G(z) = \frac{P(z)}{D(z)} = \frac{p_0 + p_1 z^{-1} + \cdots + p_N z^{-N}}{1 + d_1 z^{-1} + \cdots + d_N z^{-N}} \tag{8.101}$$

be an Nth-order real-coefficient BR transfer function with its power-complementary transfer function given by

$$H(z) = \frac{Q(z)}{D(z)} = \frac{q_0 + q_1 z^{-1} + \cdots + q_N z^{-N}}{1 + d_1 z^{-1} + \cdots + d_N z^{-N}}. \tag{8.102}$$

The power-complementary property implies that

$$\left| G(e^{j\omega}) \right|^2 + \left| H(e^{j\omega}) \right|^2 = 1. \tag{8.103}$$

From Eq. (8.100a), it follows that $G(z)$ must have a symmetric numerator, that is,

$$p_n = p_{N-n}, \tag{8.104}$$

and from Eq. (8.100b), it follows that $H(z)$ must have an antisymmetric numerator, that is,

$$q_n = -q_{N-n}. \tag{8.105}$$

We develop next the procedure to identify the two allpass transfer functions from a specified transfer function $G(z)$. Now, the symmetric property of the numerator of $G(z)$ as given by Eq. (8.104) implies that

$$P(z^{-1}) = z^N P(z). \tag{8.106}$$

Likewise, the antisymmetric property of the numerator of $H(z)$ as given by Eq. (8.105) implies that

$$Q(z^{-1}) = -z^N Q(z). \tag{8.107}$$

By analytic continuation, we can rewrite Eq. (8.103) as

$$G(z^{-1})G(z) + H(z)H(z^{-1}) = 1. \tag{8.108}$$

Substituting $G(z) = P(z)/D(z)$ and $H(z) = Q(z)/D(z)$ in Eq. (8.108), and making use of Eqs. (8.106) and (8.107), we arrive at

$$[P(z) + Q(z)][P(z) - Q(z)] = z^{-N}D(z^{-1})D(z). \tag{8.109}$$

---

[7]See Section 7.1.2 for the definition of a bounded real transfer function.

From the relations of Eqs. (8.106) and (8.107), we can write

$$P(z) - Q(z) = z^{-N}[P(z^{-1}) + Q(z^{-1})].$$ (8.110)

If we denote the zeros of $[P(z) + Q(z)]$ as $z = \xi_k$, $1 \le k \le N$, then it follows from Eq. (8.109) that $z = 1/\xi_k$, $1 \le k \le N$, are the zeros of $[P(z) - Q(z)]$; that is, $[P(z) - Q(z)]$ is the mirror-image polynomial of $[P(z) + Q(z)]$. From Eq. (8.110), it also follows that the zeros of $[P(z) + Q(z)]$ inside the unit circle are zeros of $D(z)$, whereas the zeros of $[P(z) + Q(z)]$ outside the unit circle are zeros of $D(z^{-1})$, since $G(z)$ and $H(z)$ have been assumed to be stable transfer functions. Let $z = \xi_k$, $1 \le k \le r$, be the r zeros of $[P(z) + Q(z)]$ inside the unit circle, and the remaining $N - r$ zeros, $z = \xi_k$, $r + 1 \le k \le N$, be outside the unit circle. Hence, it can be seen from Eq. (8.110) that the N zeros of $D(z)$ are given by

$$z = \begin{cases} \xi_k, & 1 \le k \le r, \\ \frac{1}{\xi_k}, & r + 1 \le k \le N. \end{cases}$$

All we need to do now is to identify the above zeros with the appropriate allpass transfer functions $A_0(z)$ and $A_1(z)$. To this end, we obtain from Eqs. (8.100a) and (8.100b)

$$A_0(z) = G(z) + H(z) = \frac{P(z) + Q(z)}{D(z)},$$ (8.111a)

$$A_1(z) = G(z) - H(z) = \frac{P(z) - Q(z)}{D(z)}.$$ (8.111b)

Therefore, the two allpass transfer functions can be expressed as

$$A_0(z) = \prod_{k=r+1}^{N} \frac{z^{-1} - (\xi_k^*)^{-1}}{1 - \xi_k^{-1} z^{-1}},$$ (8.112a)

$$A_1(z) = \prod_{k=1}^{r} \frac{z^{-1} - \xi_k^*}{1 - \xi_k z^{-1}}.$$ (8.112b)

In order to arrive at the above expressions for the two allpass transfer functions, it is necessary to determine the transfer function $H(z)$ that is power-complementary to $G(z)$. Denoting the $2N$th-degree polynomial $Q^2(z) = P^2(z) - z^{-N}D(z^{-1})D(z)$ as $U(z)$, we can rewrite Eq. (8.109) as

$$Q^2(z) = U(z) = \sum_{n=0}^{2N} u_n z^{-n}.$$ (8.113)

Solving the above equation for the coefficients q_k of $Q(z)$, we finally arrive at

$$q_0 = \sqrt{u_0}, \quad q_1 = \frac{u_1}{2q_0},$$ (8.114a)

$$q_k = -q_{N-k} = \frac{u_k - \sum_{\ell=1}^{k-1} q_\ell q_{n-\ell}}{2q_0}, \quad k \ge 2,$$ (8.114b)

where we have used the antisymmetric property of the coefficients. After $Q(z)$ has been determined, we form the polynomial $[P(z) + Q(z)]$, find its zeros $z = \xi_k$, and then determine the two allpass transfer functions using Eqs. (8.112a) and (8.112b).

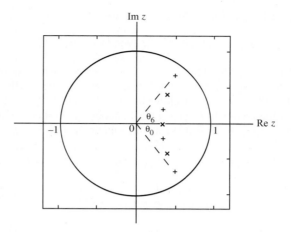

Figure 8.44: Pole interlacing property in the case of a seventh-order digital Butterworth lowpass filter. The poles marked "+" belong to the allpass function $A_0(z)$, and the poles marked "×" belong to the allpass function $A_1(z)$.

It can be shown that IIR digital transfer functions derived from the analog Butterworth, Chebyshev, and elliptic filters[8] via the bilinear transformation approach discussed in Section 9.2.1 can be decomposed in the form of Eqs. (8.100a) and (8.100b) [Vai86a]. Moreover, for lowpass–highpass filter pairs, the order N of the transfer function must be odd, with the orders of $A_0(z)$ and $A_1(z)$ differing by 1. Likewise, for bandpass–bandstop filter pairs, the order N of the transfer function must be even, with the orders of $A_0(z)$ and $A_1(z)$ differing by 2 [Vai86a].

In the case of odd-order digital Butterworth, Chebyshev, and elliptic lowpass or highpass transfer functions, there is a simple approach to identify the poles of the allpass transfer functions $A_0(z)$ and $A_1(z)$ from the poles λ_k, $0 \le k \le N - 1$, of the parent lowpass transfer function $G(z)$ or $H(z)$. Let θ_k denote the angle of the pole λ_k. If we assume that the poles are numbered such that $\theta_k < \theta_{k+1}$, then the poles of $A_0(z)$ are given by λ_{2k}, and the poles of $A_1(z)$ are given by λ_{2k+1} [Gaz85]. Figure 8.44 illustrates this *pole interlacing property* of the two allpass transfer functions.

EXAMPLE 8.19   Parallel Allpass Realization of a Lowpass Transfer Function

Consider the transfer function of a seventh-order Butterworth lowpass digital filter with a 3-dB cutoff frequency of 0.3π:[9]

$$G(z) = \frac{0.000963(1 + 7z^{-1} + 21z^{-2} + 35z^{-3} + 35z^{-4} + 21z^{-5} + 7z^{-6} + z^{-7})}{1 - 2.7825z^{-1} + 3.9668z^{-2} - 3.4051z^{-3} + 1.8757z^{-4}} \tag{8.115}$$
$$- 0.6509z^{-5} + 0.1308z^{-6} - 0.01166z^{-7}$$

Its seven poles are located at (numbered according to increasing angles)

$$\xi_0 = 0.4981133 - j0.668405, \quad \xi_1 = 0.3907071 - j0.4204395, \quad \xi_2 = 0.3399766 - j0.2030305,$$
$$\xi_3 = 0.32491969623291, \quad \xi_4 = 0.3399766 + j0.2030305, \quad \xi_5 = 0.3907071 + j0.4204395,$$
$$\xi_6 = 0.4981133 + j0.668405.$$

The corresponding z-plane pole-plot is indicated in Figure 8.44. Making use of the pole interlacing property, we identify the two allpass transfer functions as

---

[8] See Section 4.4 for a review of these analog filter approximation techniques.

[9] This transfer function has been obtained using MATLAB. For details, see Section 9.6.

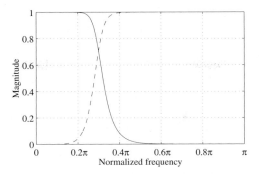

Figure 8.45: Magnitude responses of a pair of power-complementary seventh-order Butterworth lowpass and highpass transfer functions with a 3-dB cutoff frequency at 0.3π.

$$A_0(z) = \frac{(z^{-1} - \xi_0)(z^{-1} - \xi_2)(z^{-1} - \xi_4)(z^{-1} - \xi_6)}{(1 - \xi_0 z^{-1})(1 - \xi_2 z^{-1})(1 - \xi_4 z^{-1})(1 - \xi_6 z^{-1})}$$

$$= \frac{(0.69488 - 0.99623 z^{-1} + z^{-2})(0.15681 - 0.67995 z^{-1} + z^{-2})}{(1 - 0.99623 z^{-1} + 0.69488 z^{-2})(1 - 0.67995 z^{-1} + 0.15681 z^{-2})}, \qquad (8.116a)$$

$$A_1(z) = \frac{(z^{-1} - \xi_1)(z^{-1} - \xi_3)(z^{-1} - \xi_5)}{(1 - \xi_1 z^{-1})(1 - \xi_3 z^{-1})(1 - \xi_5 z^{-1})}$$

$$= \frac{(-0.32492 + z^{-1})(0.32942 - 0.781414 z^{-1} + z^{-2})}{(1 - 0.32492 z^{-1})(1 - 0.781414 z^{-1} + 0.32942 z^{-2})}. \qquad (8.116b)$$

It can be easily verified that by substituting the above two allpass transfer functions in Eq. (8.100a), we indeed arrive at the transfer function $G(z)$ of Eq. (8.115). In addition, making use of these in Eq. (8.100b), we also get its power-complementary highpass transfer function $H(z)$, given by

$$H(z) = \frac{0.108(1 - 7z^{-1} + 21z^{-2} - 35z^{-4} - 21z^{-5} + 7z^{-6} - z^{-7})}{1 - 2.7825 z^{-1} + 3.9668 z^{-2} - 3.4051 z^{-3} + 1.8757 z^{-4}} . \qquad (8.117)$$
$$- 0.6509 z^{-5} + 0.1308 z^{-6} - 0.01166 z^{-7}$$

Figure 8.45 shows the magnitude response plots of $G(z)$ of Eq. (8.115) and $H(z)$ of Eq. (8.117) demonstrating their power-complementary property.

The parallel allpass decomposition of a stable real coefficient transfer function can be easily carried out on MATLAB using the M-file tf2ca. The use of this function is illustrated in Example 8.20.

EXAMPLE 8.20   Direct Parallel Allpass Realization Using MATLAB

We repeat here Example 8.19. The code fragments used are

```
num = 0.000962*[1 7 21 35 35 21 7 1];
den = [1.0 -2.7825  3.9668  -3.4051  1.8758  -0.6509  0.1308  -0.01166];
[d1,d2] = tf2ca(num,den);
```

The denominator coefficient vectors of the two allpass transfer functions are

```
d0 =
Columns 1 through 5
1.0   -1.6761797957    1.5290748603    -0.6287009790     0.10896129785


d1 =
1.0   -1.1063340046    0.5833183258    -0.1070355097
```

which can be shown to be identical to that given in Eqs. (8.116a) and (8.116b), respectively.

The M-file `ca2tf` can be used to determine the transfer function $G(z)$ from the parallel allpass decomposition of Eq. (8.100a) or the transfer function $H(z)$ from the parallel allpass decomposition of Eq. (8.100b). We illustrate the use of this function in Example 8.21.

EXAMPLE 8.21   Reconstruction of Transfer Function from Its Parallel Allpass Decomposition

We reconstruct the transfer function $G(z)$ and its power-complementary transfer function $H(z)$ from its parallel allpass decomposition given in Example 8.20. The code fragments used are

```
d0 = [1.0  -1.6761797957    1.5290748603    -0.6287009790 ...
0.10896129785];
d1 = [1.0  -1.1063340046    0.5833183258    -0.1070355097];
[num,den,pcnum] = ca2tf(d0,d1);
```

The numerator and the denominator of $G(z)$ are given by

```
num =
Columns 1 through 4
0.0009628940     0.0067402583     0.0202207750     0.0337012917
Columns 5 through 8
0.0337012917     0.0202207750     0.0067402583     0.0009628940


den =
Columns 1 through 4
1.0              -2.7825138003    3.9668078919     -3.4051503945
Columns 5 through 8
1.8758627160    -0.6509456985     0.13085245168    -0.0116627280
```

which are seen to be identical to that given in Eq. (8.115). The numerator of $H(z)$ is given by

```
pcnum =
Columns 1 through 4
0.1079984037    -0.7559888263     2.2679664789     -3.7799441315
Columns 5 through 8
3.7799441315    -2.2679664789     0.7559888263     -0.1079984037
```

which can be shown to be identical to the numerator coefficients given in Eq. (8.117).

If the two allpass sections are to be realized in the cascaded lattice of Figure 8.31, then one can use the M-file `tf2cl` to determine the lattice parameters k_i of the allpass sections.

The parallel allpass structure has two other very attractive properties. As indicated in Section 8.6, an Mth-order allpass transfer function can be realized using only M multipliers. Therefore, the realization of an Nth-order IIR transfer function $G(z)$ based on the allpass decomposition of Eq. (8.100a) requires only $r + (N - r) = N$ multipliers. On the other hand, in general, a direct form realization of an Nth-order IIR transfer function uses $2N + 1$ multipliers. Moreover, with an additional adder and no additional multipliers, one can easily implement its power-complementary transfer function $H(z)$, also of Nth-order, as indicated in Figure 8.43. As a result, the parallel allpass realization of an IIR transfer function, if it exists, is computationally very efficient. We shall show later in Section 12.9.2, that the parallel allpass realization also has very low passband sensitivity with respect to multiplier coefficients if the allpass sections are realized in a structurally lossless form.

8.11 Digital Sine-Cosine Generator

We now consider the realization of digital sinusoidal oscillators. In particular, we consider here the design of digital sine-cosine generators that produce two sinusoidal sequences that are exactly 90 degrees out of phase with each other [Mit75]. These circuits have a number of applications, such as the computation of the discrete Fourier transform and certain digital communication systems. In some applications, it is more convenient to use numerical algorithms for the computation of the sine or cosine functions. These are discussed in Section 11.12.1.

Let $s_1[n]$ and $s_2[n]$ denote the two outputs of a digital sine-cosine generator given by

$$s_1[n] = \alpha \sin(\omega_o n), \tag{8.118a}$$

$$s_2[n] = \beta \cos(\omega_o n). \tag{8.118b}$$

From the above, we arrive at

$$s_1[n + 1] = \alpha \sin(\omega_o(n + 1))$$
$$= \alpha \sin(\omega_o n) \cos \omega_o + \alpha \cos(\omega_o n) \sin \omega_o, \tag{8.119a}$$

$$s_2[n + 1] = \beta \cos(\omega_o(n + 1))$$
$$= \beta \cos(\omega_o n) \cos \omega_o - \beta \sin(\omega_o n) \sin \omega_o. \tag{8.119b}$$

Making use of Eqs. (8.118a) and (8.118b), we can rewrite Eqs. (8.119a) and (8.119b) in a matrix form as

$$\begin{bmatrix} s_1[n + 1] \\ s_2[n + 1] \end{bmatrix} = \begin{bmatrix} \cos \omega_o & \frac{\alpha}{\beta} \sin \omega_o \\ -\frac{\beta}{\alpha} \sin \omega_o & \cos \omega_o \end{bmatrix} \begin{bmatrix} s_1[n] \\ s_2[n] \end{bmatrix}. \tag{8.120}$$

To generate $s_1[n]$ from $s_1[n + 1]$, we need a delay and, similarly, to generate $s_2[n]$ from $s_2[n + 1]$, we need a delay. Thus, such a structure must have at least two delays.

In order to arrive at a realization of the sine-cosine generator, we need to compare Eq. (8.120) with the equivalent expression of a general second-order structure with no delay-free loops. Such a structure is characterized by the following equations,

$$\begin{bmatrix} s_1[n + 1] \\ s_2[n + 1] \end{bmatrix} = \begin{bmatrix} 0 & A \\ 0 & 0 \end{bmatrix} \begin{bmatrix} s_1[n + 1] \\ s_2[n + 1] \end{bmatrix} + \begin{bmatrix} C & D \\ E & F \end{bmatrix} \begin{bmatrix} s_1[n] \\ s_2[n] \end{bmatrix}, \tag{8.121}$$

and can be implemented using five multipliers, as indicated in Figure 8.46.

From Figure 8.46, we readily arrive at the time-domain description of this structure:

$$\begin{bmatrix} s_1[n + 1] \\ s_2[n + 1] \end{bmatrix} = \begin{bmatrix} AE + C & AF + D \\ E & F \end{bmatrix} \begin{bmatrix} s_1[n] \\ s_2[n] \end{bmatrix}. \tag{8.122}$$

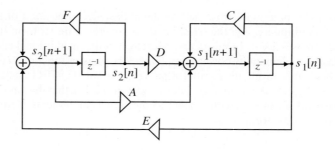

Figure 8.46: A general second-order digital filter structure with no input node.

Comparing Eqs. (8.120) and (8.122), we get

$$E = -\frac{\beta}{\alpha}\sin\omega_o, \qquad F = \cos\omega_o, \tag{8.123a}$$

$$AE + C = \cos\omega_o, \qquad AF + D = \frac{\alpha}{\beta}\sin\omega_o. \tag{8.123b}$$

From the above two equations, we obtain, after some algebra,

$$C\cos\omega_o + D\frac{\beta}{\alpha}\sin\omega_o = 1. \tag{8.123c}$$

Equations (8.123a) and (8.123c) ensure that the structure of Figure 8.46 will be a sine-cosine generator.

Expressing the multiplier constants A and D as a function of the multiplier constant C, we obtain, after some simple manipulations, the five-multiplier characterization of the sine-cosine generator as

$$\begin{bmatrix} s_1[n+1] \\ s_2[n+1] \end{bmatrix} = \begin{bmatrix} 0 & \frac{\alpha(C - \cos\omega_o)}{\beta\sin\omega_o} \\ 0 & 0 \end{bmatrix} \begin{bmatrix} s_1[n+1] \\ s_2[n+1] \end{bmatrix}$$

$$+ \begin{bmatrix} C & \frac{\alpha(1 - C\cdot\cos\omega_o)}{\beta\sin\omega_o} \\ -\frac{\beta}{\alpha}\sin\omega_o & \cos\omega_o \end{bmatrix} \begin{bmatrix} s_1[n] \\ s_2[n] \end{bmatrix}. \tag{8.124}$$

To reduce the total number of multipliers to four, we can choose the following specific values for the multiplier constant C: $\cos\omega_o$, 0, $+1$, and -1. For example, if $C = \cos\omega_o$, then Eq. (8.124) reduces to

$$\begin{bmatrix} s_1[n+1] \\ s_2[n+1] \end{bmatrix} = \begin{bmatrix} \cos\omega_o & \frac{\alpha}{\beta}\sin\omega_o \\ -\frac{\beta}{\alpha}\sin\omega_o & \cos\omega_o \end{bmatrix} \begin{bmatrix} s_1[n] \\ s_2[n] \end{bmatrix}, \tag{8.125}$$

which for $\beta = \pm\alpha$ leads to the four-multiplier sine-cosine generator described by Gold and Rader [Gol69b]. On the other hand, if we set $\alpha\sin\omega_o = \pm\beta$, then Eq. (8.125) leads to a three-multiplier realization. Another three-multiplier sine-cosine generator is obtained if we set $\alpha = \pm\beta\sin\omega_o$ in Eq. (8.125) (Problem 8.52).

A single-multiplier structure can be derived by setting $\beta\sin\omega_o/\alpha = 1 - \cos\omega_o$ or, equivalently, $\beta = \alpha\tan(\omega_o/2)$, in Eq. (8.125), resulting in

$$\begin{bmatrix} s_1[n+1] \\ s_2[n+1] \end{bmatrix} = \begin{bmatrix} \cos\omega_o & \cos\omega_o + 1 \\ \cos\omega_o - 1 & \cos\omega_o \end{bmatrix} \begin{bmatrix} s_1[n] \\ s_2[n] \end{bmatrix}, \tag{8.126}$$

leading to the realization of Figure 8.47. Another single-multiplier realization can be derived by setting $-\beta\sin\omega_o/\alpha = (1 + \cos\omega_o)$ or, equivalently, $\alpha = \beta\tan(\omega_o/2)$, in Eq. (8.125) (Problem 8.53). Various other single-multiplier sine-cosine generators can be developed for $C = 0$ and $C = \pm1$ (Problem 8.54).

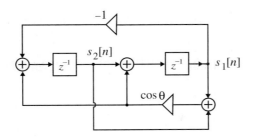

Figure 8.47: A single multiplier sine-cosine generator.

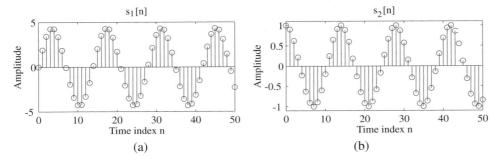

Figure 8.48: Sine-cosine sequences generated by the structure of Figure 8.47 for $\cos\theta = 0.9$.

It should be noted that to start the generation of the sinusoidal sequences, at least one of the signal variables $s_1[n]$ and $s_2[n]$ should have initially a nonzero value. Moreover, the actual amplitudes and the relative phases of the cosinusoidal and the sinusoidal sequences generated by the sine-cosine generator depend on the initial values chosen for the signal variables $s_1[n]$ and $s_2[n]$.

Figure 8.48 shows the plots of the sequences $s_1[n]$ and $s_2[n]$ generated by simulating Eq. (8.126) in MATLAB for $\cos\omega_o = 0.9$. As can be seen from this figure, $s_1[n]$ and $s_2[n]$ are, respectively, cosinusoidal and sinusoidal sequences. Note that the maximum value of the amplitudes of the two sequences can be made equal by scaling one of the sequences appropriately.

The single-multiplier structures retain their characteristic roots on the unit circle under finite wordlength constraints. On the other hand, in other realizations of the sine-cosine generators, roots may go inside or outside the unit circle due to the quantization of the multiplier coefficients, causing the oscillations to decay or to build up as n increases. In addition, due to product round-off errors, the sequences generated by the sine-cosine generator may not retain their sinusoidal behaviors, even in the case of a single-multiplier generator. It is therefore advisable to reset the variables $s_1[n]$ and $s_2[n]$ after some iterations at prescribed time instants so that the accumulated errors do not become unacceptable.

8.12 Computational Complexity of Digital Filter Structures

The computational complexity of a digital filter structure is given by the total number of multipliers and the total number of two-input adders required for its implementation, which roughly provides an indication of its cost of implementation. We summarize this measure here for various realizations discussed in this chapter. It should be emphasized, though, that the computational complexity measure is not the only criterion that should be used in selecting a particular structure for implementation of a given transfer

Table 8.1: Computational complexity comparison of various realizations of an FIR filter of order N.

Structure	No. of multipliers	No. of two-input adders
Direct form	$N+1$	N
Cascade form	$N+1$	N
Polyphase	$N+1$	N
Cascaded lattice	$2(N+1)$	$2N+1$
Linear phase	$\left\lfloor \dfrac{N+2}{2} \right\rfloor$	N

Table 8.2: Computational complexity comparison of various realizations of an IIR filter of order N.

Structure	No. of multipliers	No. of two-input adders
Direct form II and II$_t$	$2N+1$	$2N$
Cascade form	$2N+1$	$2N$
Parallel form	$2N+1$	$2N$
Gray-Markel 1	$3N+1$	$3N$
Gray-Markel 2	$2N+1$	$4N$
Parallel allpass	$N+1$	$3N+2$

Note: Gray-Markel 1 is based on the realization using the two-multiplier two-pair of Figure 8.30(a), whereas Gray-Markel 2 is based on the realization using the one-multiplier two-pair of Figure 8.30(b). Each allpass section of the parallel allpass structure has been assumed to be realized using the one-multiplier two-pair of Figure 8.30(b).

function. The performances of all equivalent realizations under finite wordlength constraints should also be considered together with the costs of implementation in selecting a structure.

Tables 8.1 and 8.2 show the computational complexity measures of all realizations discussed in this chapter. As can be seen from these tables, in general, the direct form realizations for the case of both the FIR and the IIR transfer functions require the least number of multipliers and two-input adders. However, in the case of the IIR transfer function, the parallel allpass realization is the most efficient if such a realization exists. Likewise, it is possible to realize certain special types of FIR transfer functions with many fewer multipliers and adders than that indicated in Table 8.1.

8.13 Summary

This chapter considered the realization of a causal digital transfer function. Such realizations, called structures, are usually represented in block diagram forms that are formed by interconnecting adders, multipliers, and unit delays. A digital filter structure represented in block diagram form can be analyzed to develop its input–output relationship either in the time domain or in the transform domain. Often, for analysis purposes, it is convenient to represent the digital filter structure as a signal flow-graph.

Several basic FIR and IIR digital filter structures are then reviewed. These structures, in most cases, can be developed from the transfer function description of the digital filter essentially by inspection.

The digital allpass filter is a versatile building block and has a number of attractive digital signal processing applications. Since the numerator and the denominator polynomials of a digital transfer function exhibit mirror-image symmetry, an Mth-order digital allpass filter can be realized with only M distinct multipliers. Two different approaches to the minimum-multiplier realization of a digital allpass transfer function are described. One approach is based on a realization in the form of a cascade of first- and second-order allpass filters. The second approach results in a cascaded lattice realization. The final realizations in both cases remain allpass independent of the actual values of the multiplier coefficients and are thus less sensitive to multiplier coefficient quantization. An elegant application of the first- and second-order minimum multiplier allpass structures, considered here, is in the implementation of some simple transfer functions with parametrically tunable properties.

The Gray-Markel method, to realize any arbitrary transfer function using the cascaded lattice form of allpass structure, is outlined. The realization of a large class of arbitrary Nth-order transfer functions using a parallel connection of two allpass filters is described. The final structure is shown to require only N multipliers. In Section 12.9.2, we demonstrate the low passband sensitivity property of these structures to small changes in the multiplier coefficients.

The cascaded lattice realization of an FIR transfer function is considered. The realization of a digital sine-cosine generator is then described, and various oscillator structures are systematically developed. The chapter concludes with a comparison of the computational complexities of FIR and IIR digital filter structures.

The digital filter realization methods outlined in this chapter assume the existence of the corresponding transfer function. Chapter 9 considers the development of such transfer functions meeting the given frequency response specifications. The analysis of the finite wordlength effects on the performance of the digital filter structures is treated in Chapter 12.

8.14 Problems

8.1 The digital filter structure of Figure P8.1 has a delay-free loop and is therefore unrealizable. Determine a realizable equivalent structure with identical input–output relations and without any delay-free loop. (*Hint:* Express the output variables $y[n]$ and $w[n]$ in terms of the input variables $x[n]$ and $u[n]$ only, and develop the corresponding block diagram representation from these input–output relations.)

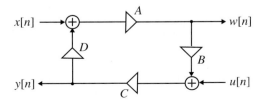

Figure P8.1

8.2 Determine by inspection whether or not the digital filter structures in Figure P8.2 have delay-free loops. Identify these loops if they exist. Develop equivalent structures without delay-free loops.

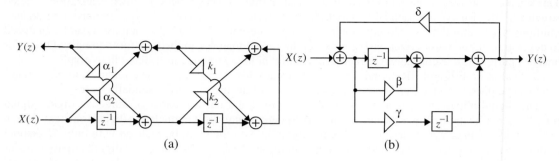

(a)   (b)

Figure P8.2

8.3 Figure P8.3 shows a typical closed-loop discrete-time feedback control system in which $G(z)$ is the plant and $C(z)$ is the compensator. If $G(z) = \frac{2}{1+3z^{-1}}$ and $C(z) = K$, determine the range of values of K for which the feedback structure is stable.

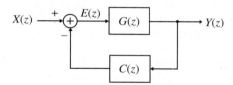

Figure P8.3

8.4 Figure P8.3 shows a typical closed-loop discrete-time feedback control system in which $G(z)$ is the plant and $C(z)$ is the compensator. If

$$G(z) = \frac{0.2(1 + 0.4z^{-1})}{1 + 0.9z^{-1} + 0.6z^{-2}},$$

determine the transfer function $C(z)$ of the compensator so that the overall closed-loop transfer function of the feedback system is

$$H(z) = \frac{0.2z^{-1} + 0.28z^{-2} + 0.18z^{-3} + 0.04z^{-4}}{1 + 0.5z^{-1} + 0.74z^{-2} + 0.18z^{-3} + 0.28z^{-4}}.$$

Check the stability of $G(z)$, $C(z)$, and $H(z)$ using the M-file `zplane`.

8.5 Analyze the digital filter structure of Figure P8.4, and determine its transfer function $H(z) = Y(z)/X(z)$. (a) Is this a canonic structure? (b) What should be the value of the multiplier coefficient K so that $H(z)$ has a unity gain at $\omega = 0$? (c) What should be the value of the multiplier coefficient K so that $H(z)$ has a unity gain at $\omega = \pi$? (d) Is there a difference between these two values of K? If not, why not?

8.6 Analyze the digital filter structure of Figure P8.5, where all multiplier coefficients are real, and determine the transfer function $H(z) = Y(z)/X(z)$. What are the range of values of the multiplier coefficients for which the filter is BIBO stable?

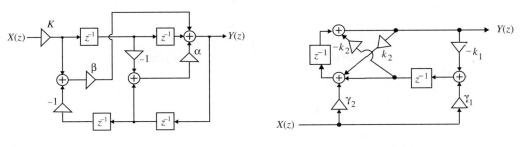

Figure P8.4   **Figure P8.5**

8.7 Determine the transfer function $H(z) = Y(z)/X(z)$ of the digital filter structure of Figure P8.6.

8.8 Determine the transfer function $H(z) = Y(z)/X(z)$ of the digital filter structure of Figure P8.7.

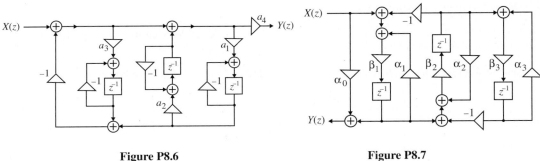

Figure P8.6   **Figure P8.7**

8.9 Determine the transfer function of the digital filter structure of Figure P8.9 [Kin72].

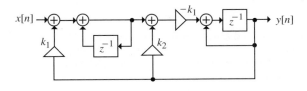

Figure P8.9

8.10 Realize the FIR transfer function

$$H(z) = (1 - 0.6z^{-1})^6 = 1 - 3.6z^{-1} + 5.4z^{-2} - 4.32z^{-3} + 1.944z^{-4} - 0.4666z^{-5} + 0.0467z^{-6}$$

in the following forms: (a) two different direct forms, (b) cascade of six first-order sections, (c) cascade of three second-order sections,(d) cascade of two third-order sections, and (e) cascade of two second-order sections and two first-order sections.

Compare the computational complexity of each of the above realizations.

8.11 Consider a length-10 FIR transfer function given by

$$H(z) = h[0] + h[1]z^{-1} + h[2]z^{-2} + h[3]z^{-3} + h[4]z^{-4}$$
$$+ h[5]z^{-5} + h[6]z^{-6} + h[7]z^{-7} + h[8]z^{-8} + h[9]z^{-9}.$$

(a) Develop a four-branch polyphase realization of $H(z)$ in the form of Figure 8.7(a), and determine the expressions for the polyphase transfer functions $E_0(z)$, $E_1(z)$, $E_2(z)$, and $E_3(z)$.

(b) From this realization, develop a canonic four-branch polyphase realization.

8.12 (a) Develop a three-branch polyphase realization of $H(z)$ of Problem 8.11 in the form of Figure 8.7(b), and determine the expressions for the polyphase transfer functions $E_0(z)$, $E_1(z)$, and $E_2(z)$.

(b) From this realization, develop a canonic three-branch polyphase realization of $H(z)$.

8.13 (a) Develop a two-branch polyphase realization of $H(z)$ of Problem 8.11 in the form of Figure 8.7(c), and determine the expressions for the polyphase transfer functions $E_0(z)$ and $E_1(z)$.

(b) From this realization, develop a canonic two-branch polyphase realization of $H(z)$.

8.14 Develop a minimum-multiplier realization of a length-7 Type 3 FIR transfer function.

8.15 Develop a minimum-multiplier realization of a length-8 Type 4 FIR transfer function.

8.16 Let $H(z)$ be a Type 1 linear-phase FIR filter of order N, with $G(z)$ denoting its delay-complementary filter. Develop a realization of both filters using only N delays and $(N + 2)/2$ multipliers.

8.17 Show that a Type 1 linear-phase FIR transfer function $H(z)$ of length $2M + 1$ can be expressed as

$$H(z) = z^{-M} \left[h[M] + \sum_{n=1}^{M} h[M - n] \left(z^n + z^{-n} \right) \right].   \tag{8.127}$$

By using the relation

$$z^{\ell} + z^{-\ell} = 2T_{\ell} \left(\frac{z + z^{-1}}{2} \right),$$

where $T_{\ell}(x)$ is the ℓth-order Chebyshev polynomial[10] in x, express $H(z)$ in the form

$$H(z) = z^{-M} \sum_{n=0}^{M} a[n] \left(\frac{z + z^{-1}}{2} \right)^n.   \tag{8.128}$$

Determine the relation between $a[n]$ and $h[n]$. Develop a realization of $H(z)$ based on Eq. (8.128) in the form of Figure P8.10, where $F_1(z^{-1})$ and $F_2(z^{-1})$ are causal structures. Determine the form of $F_1(z^{-1})$ and $F_2(z^{-1})$. The structure of Figure P8.9 is called the *Taylor structure* for linear-phase FIR filters [Sch72].

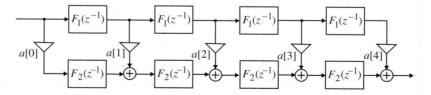

Figure P8.10   The Taylor structure shown for $M = 4$.

---

[10]For a definition of the Chebyshev polynomial and a recursive equation for generating such a polynomial, see Section 4.4.3.

8.18 Show that there are 36 distinct cascade realizations of the transfer function $H(z)$ of Eq. (8.27) obtained by different pole-zero pairings and different orderings of the individual sections.

8.19 Consider a real coefficient IIR transfer function $H(z)$ with its numerator and denominator expressed as a product of polynomials,

$$H(z) = \prod_{i=1}^{K} \frac{P_i(z)}{D_i(z)},$$

where $P_i(z)$ and $D_i(z)$ are either first-order or second-order polynomials with real coefficients. Determine the total number of distinct cascade realizations that can be obtained by different pole-zero pairings and different orderings of the individual sections.

8.20 Develop a canonic direct form realization of the transfer function

$$H(z) = \frac{2 - 5z^{-2} + 3z^{-3}}{1 - 4z^{-1} + 3z^{-2} + 6z^{-4}},$$

and then determine its transpose configuration.

8.21 Develop two different cascade canonic realizations of the following causal IIR transfer functions:

(a) $H_1(z) = \dfrac{(0.2z^{-1} + 0.6z^{-2})(3 - 2.4z^{-1})}{(2 - 3.2z^{-1} + 4.2z^{-2})(1 - 0.75z^{-1})}$,

(b) $H_2(z) = \dfrac{(1.7 + 6.2z)(z - 4.5)(3z^2 - 0.5)}{(2z - 0.3)(z + 0.2)(z^2 + 0.5z + 0.1)}$,

(c) $H_3(z) = \dfrac{(6 + 2.8z^{-1})(5.2 - 8.4z^{-1} + 7z^{-2})}{(4 + 12.4z^{-1})(1 - 2.4z^{-1} + 0.76z^{-2})}$.

8.22 Realize the transfer functions of Problem 8.21 in parallel forms I and II.

8.23 Develop a cascade realization of the transfer function of Eq. (8.26) using the factorization given in Eq. (8.29). Compare the computational complexity of this realization with the one shown in Figure 8.19(b).

8.24 Consider the cascade of three causal first-order LTI discrete-time systems shown in Figure P8.11, where

$$H_1(z) = \frac{2 - 0.3z^{-1}}{1 + 0.5z^{-1}}, \quad H_2(z) = \frac{0.4 + z^{-1}}{1 + 0.4z^{-1}}, \quad H_3(z) = \frac{3}{1 + 0.5z^{-1}}.$$

(a) Determine the transfer function of the overall system as a ratio of two polynomials in z^{-1}.
(b) Determine the difference equation characterizing the overall system.
(c) Develop the realization of the overall system with each section realized in direct form II.
(d) Develop a parallel form I realization of the overall system.
(e) Determine the impulse response of the overall system in closed form.

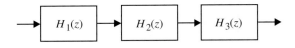

Figure P8.11

8.25 A causal LTI discrete-time system develops an output $y[n] = (0.4)^n \mu[n] - 0.3(0.4)^{n-1} \mu[n-1]$, for an input $x[n] = (0.2)^n \mu[n]$.

(a) Determine the transfer function of the system.
(b) Determine the difference equation characterizing the system.
(c) Develop a canonic direct form realization of the system with no more than three multipliers.
(d) Develop a parallel form I realization of the system.
(e) Determine the impulse response of the system in closed form.
(f) Determine the output $y[n]$ of the system for an input $x[n] = (0.3)^n \mu[n] - 0.4(0.3)^{n-1} \mu[n-1]$.

8.26 The structure shown in Figure P8.12 was developed in the course of a realization of the IIR digital transfer function

$$H(z) = \frac{2z^2 - 3.2z - 18.6}{z^2 - 2.6z - 1.2}.$$

However, by a mistake in the labeling, two of the multiplier coefficients in this structure have incorrect values. Find these two multipliers, and determine their correct values.

8.27 Figure P8.13 shows an incomplete realization of the causal IIR transfer function

$$H(z) = \frac{4z^2 - 5.6z}{z^2 + 0.2z - 0.08}.$$

Determine the values of the multiplier coefficients A and B.

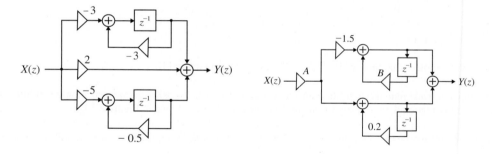

Figure P8.12   **Figure P8.13**

8.28 Develop a two-multiplier canonic realization of each of the following second-order transfer functions where the multiplier coefficients are α_1 and $-\alpha_2$, respectively [Hir73]:

(a) $H_1(z) = \dfrac{(1 - \alpha_1 + \alpha_2)(1 + z^{-1})^2}{1 - \alpha_1 z^{-1} + \alpha_2 z^{-2}}$,   (b) $H_2(z) = \dfrac{(1 - \alpha_2)(1 - z^{-2})}{1 - \alpha_1 z^{-1} + \alpha_2 z^{-2}}.$

8.29 Develop a four-multiplier canonic realization with multiplier coefficients $\alpha_1, \alpha_2, \alpha_3$ and α_4 for each one of the following second-order transfer functions [Szc75a]:

(a) $H_a(z) = \dfrac{1 - \alpha_3 z^{-1} + (\alpha_2 - \alpha_4) z^{-2}}{1 - (\alpha_1 + \alpha_3) z^{-1} - \alpha_4 z^{-2}}$,   (b) $H_b(z) = \dfrac{1 - \alpha_3 z^{-1} + \alpha_2 z^{-2}}{1 - (\alpha_1 + \alpha_3) z^{-1} - \alpha_4 z^{-2}}$,

8.30 In this problem, we develop an alternative cascaded lattice realization of an Nth-order IIR transfer function [Mit77b]

$$H_N(z) = \frac{p_0 + p_1 z^{-1} + \cdots + p_{N-1} z^{-N+1} + p_N z^{-N}}{1 + d_1 z^{-1} + \cdots + d_{N-1} z^{-N+1} + d_N z^{-N}}. \tag{8.129}$$

The first stage of the realization process is shown in Figure P8.14.

(a) Show that if the two-pair chain parameters are chosen as

$$A = 1, \qquad B = d_N z^{-1}, \qquad C = p_0, \qquad D = p_N z^{-1}, \tag{8.130}$$

then $H_{N-1}(z)$ is an $(N-1)$th-order IIR transfer function of the form

$$H_{N-1}(z) = \frac{p_0' + p_1' z^{-1} + \cdots + p_{N-1}' z^{-N+1}}{1 + d_1' z^{-1} + \cdots + d_{N-1}' z^{-N+1}}, \tag{8.131}$$

with coefficients given by

$$p_k' = \frac{p_0 d_{k+1} - p_{k+1}}{p_0 d_N - p_N}, \qquad k = 0, 1, \ldots, N-1, \tag{8.132a}$$

$$d_k' = \frac{p_k d_N - p_N d_k}{p_0 d_N - p_N}, \qquad k = 1, 2, \ldots, N-1. \tag{8.132b}$$

(b) Develop a lattice realization of the two-pair.

(c) Continuing the above process, we can realize $H_N(z)$ as a cascade connection of N lattice sections constrained by a transfer function $H_0(z) = 1$. What are the total number of multipliers and two-input adders in the final realization of $H_N(z)$?

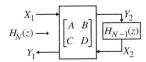

Figure P8.14

8.31 Realize the transfer functions of Problem 8.21 using the cascaded lattice realization method of Problem 8.30.

8.32 Show that the cascaded lattice realization method of Problem 8.30 results in the cascaded lattice structure described in Section 8.6.2 when $H_N(z)$ is an allpass transfer function.

8.33 Develop the structures of Types 1B, $1A_t$, and $1B_t$ first-order allpass transfer functions shown in Figure 8.24(b), (c), and (d), from Eqs. (8.50b), (8.50c), and (8.50d), respectively.

8.34 (a) Develop a cascade realization of the fourth-order allpass transfer function

$$A(z) = \left(\frac{a + z^{-1}}{1 + az^{-1}}\right) \left(\frac{b + z^{-1}}{1 + bz^{-1}}\right) \left(\frac{c + z^{-1}}{1 + cz^{-1}}\right) \left(\frac{d + z^{-1}}{1 + dz^{-1}}\right),$$

with each allpass section realized in Type 1A form. By sharing the delays between adjacent allpass sections, show that the total number of delays in the overall structure can be reduced from 8 to 6 [Mit74a].

(b) Repeat Part (a) with each allpass section realized in Type $1A_t$ form.

8.35 Analyze the digital filter structure of Figure P8.15, and show that it is a first-order allpass filter [Sto54].

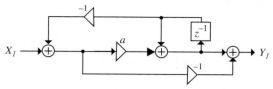

Figure P8.15

8.36 Develop formally the realizations of a second-order Type 2 allpass transfer function of Eq. (8.37), shown in Figure 8.25, using the multiplier extraction approach. Are there other Type 2 allpass structures?

8.37 Show that a cascade of two Type 2D second-order allpass structures can be realized with six delays by sharing the delays between adjacent sections. What is the minimum number of delays needed to implement a cascade of M Type 2D second-order allpass structures?

8.38 Develop formally the realizations of a second-order Type 3 allpass transfer function of Eq. (8.38), shown in Figure 8.26, using the multiplier extraction approach. Are there other Type 3 allpass structures?

8.39 Show that a cascade of two Type 3B second-order allpass structures can be realized with six delays by sharing the delays between adjacent sections. What is the minimum number of delays needed to implement a cascade of M Type 3H second-order allpass structures?

8.40 Develop a three-multiplier realization of the two-pair described by Eq. (8.50d).

8.41 Develop a lattice realization of the two-pair given by Eq. (8.50d). Determine the transfer function of an all-pole second-order cascaded lattice filter realized using this lattice structure and the transfer function of an all-pole second-order cascaded lattice filter realized using the lattice structure of Figure 8.30(a). Evaluate the approximate expressions for the gain of both second-order filters at resonance when the poles are close to the unit circle. Show that the gain of the first all-pole filter is approximately independent of the pole radius, whereas that of the second filter is not [Lar99].

8.42 Realize each of the following IIR transfer functions in the Gray-Markel form, and check the BIBO stability of each transfer function:

$$\text{(a)} H_1(z) = \frac{3.9 + 2.3z^{-1} + z^{-2}}{1 + 0.3z^{-1} + 0.5z^{-2}}, \qquad \text{(b)} \ H_2(z) = \frac{2.6 + 0.74z^{-1} + 3z^{-2}}{1 - 0.42z^{-1} + 0.4z^{-2}},$$

$$\text{(c)} \ H_3(z) = \frac{-3 + 5.192z^{-1} - 3.56z^{-2} + 2z^{-3}}{1 - 0.28z^{-1} + 0.056z^{-2} + 0.4z^{-3}}, \qquad \text{(d)} \ H_4(z) = \frac{1.6z^3 + 3.112z^2 + 0.84z + 2}{z^3 + 0.42z^2 + 0.056z - 0.3},$$

$$\text{(e)} \ H_5(z) = \frac{2z(z + 0.3)(z^2 - 0.5)}{(z + 0.5)(z - 0.4)(z^2 - 0.2z + 0.4)}.$$

8.43 Realize each of the IIR transfer functions of Problem 8.21 in the Gray-Markel form, and check their BIBO stability.

8.44 Realize the IIR transfer function

$$H(z) = \frac{0.2545(1 + z^{-1})(1 + 0.6985z^{-1} + z^{-2})}{(1 - 0.2169z^{-1})(1 + 0.0278z^{-1} + 0.7258z^{-2})}$$

in the following forms: (a) direct canonic form, (b) cascade form, (c) Gray-Markel form, and (d) cascaded lattice structure described in Problem 8.30. Compare their hardware requirements.

8.45 In this problem, the realization of an even-order real-coefficient transfer function using complex arithmetic is illustrated [Reg87a]. Let $G(z)$ be an Nth-order-real coefficient transfer function with simple poles in complex-conjugate pairs and with numerator degree less than or equal to that of the denominator.
 (a) Show that $G(z)$ can be expressed as a sum of two complex coefficient transfer functions of order $N/2$,

$$G(z) = H(z) + H^*(z^*),\qquad (8.133)$$

where the coefficients of $H^*(z^*)$ are complex conjugate of their corresponding coefficients of $H(z)$.
 (b) Generalize the above decomposition to the case when $G(z)$ has one or more simple real poles.
 (c) Consider a realization of $H(z)$ that has one real input $x[n]$ and a complex output $y[n]$. Show that the transfer function from the input to the real part of the output is simply $G(z)$. [*Hint:* Use a partial-fraction expansion to obtain the decomposition of Eq. (8.133).]

8.46 Develop a realization of a first-order complex coefficient transfer function $H(z)$ given by

$$H(z) = \frac{A + jB}{1 + (\alpha + j\beta)z^{-1}},$$

where A, B, α, and β are real constants. Show the real and imaginary parts of all signal variables separately. Determine the transfer functions from the input to the real and imaginary parts of the output.

8.47 Develop a cascaded lattice realization of an Nth-order complex coefficient allpass transfer function $A_N(z)$.

8.48 Realize the following transfer functions in the form of a parallel connection of two allpass filters:

(a) $H_1(z) = 4\left(\dfrac{1 + z^{-1}}{5 + 3z^{-1}}\right)$,　　(b) $H_2(z) = \dfrac{5 - z^{-1}}{6 - 4z^{-1}}$,

(c) $H_3(z) = \dfrac{0.5414(1 - z^{-1})(1 - 0.0757z^{-1} + z^{-2})}{(1 - 0.1768z^{-1})(1 - 0.004z^{-1} + 0.9061z^{-2})}$,

(d) $H_4(z) = \dfrac{0.4547(1 + z^{-1})(1 - 0.2859z^{-1} + z^{-2})}{(1 + 0.0712z^{-1})(1 - 0.0377z^{-1} + 0.8482z^{-2})}$.

8.49 Consider a causal length-6 FIR filter described by the convolution sum

$$y[n] = \sum_{k=0}^{5} h[k]x[n - k],\qquad n \geq 0,$$

where $y[n]$ and $x[n]$ denote, respectively, the output and the input sequences. .
 (a) Let the output and input sequences be blocked into length-2 vectors

$$\mathbf{Y}_\ell = \begin{bmatrix} y[2\ell] \\ y[2\ell + 1] \end{bmatrix},\quad \mathbf{X}_\ell = \begin{bmatrix} x[2\ell] \\ x[2\ell + 1] \end{bmatrix}.$$

Show that the above FIR filter can be equivalently described by a block convolution sum given by

$$\mathbf{Y}_\ell = \sum_{r=0}^{3} \mathbf{H}_r \mathbf{X}_{\ell-r},$$

where $\mathbf{H}_r$ is a 2×2 matrix composed of the impulse response coefficients. Determine the block convolution matrices $\mathbf{H}_r$. An implementation of the FIR filter based on the above block convolution sum is shown in Figure P8.16, where the block labeled "S/P" is a serial-to-parallel converter and the block marked "P/S" is a parallel-to-serial converter.
 (b) Develop the block convolution sum description of the above FIR filter for a block length of 3 for the input and the output.
 (c) Develop the block convolution sum description of the above FIR filter for a block length of 4 for the input and the output.

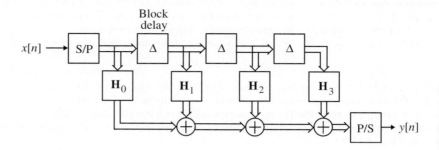

Figure P8.16

8.50 Consider a causal IIR filter described by a difference equation

$$\sum_{k=0}^{4} d_k y[n-k] = \sum_{k=0}^{4} p_k x[n-k], \qquad n \geq 0,$$

where $y[n]$ and $x[n]$ denote, respectively, the output and the input sequences.

(a) Let the output and input sequences be blocked into length-2 vectors

$$\mathbf{Y}_\ell = \begin{bmatrix} y[2\ell] \\ y[2\ell+1] \end{bmatrix}, \mathbf{X}_\ell = \begin{bmatrix} x[2\ell] \\ x[2\ell+1] \end{bmatrix}.$$

Show that the above IIR filter can be equivalently described by a block difference equation given by [Bur72]

$$\sum_{r=0}^{2} \mathbf{D}_r \mathbf{Y}_{\ell-r} = \sum_{r=0}^{2} \mathbf{P}_r \mathbf{X}_{\ell-r},$$

where $\mathbf{D}_r$ and $\mathbf{P}_r$ are 2×2 matrices composed of the difference equation coefficients $\{d_k\}$ and $\{p_k\}$, respectively. Determine the block difference equation matrices $\mathbf{D}_r$ and $\mathbf{P}_r$. An implementation of the IIR filter based on the above block difference equation is shown in Figure P8.17.

(b) Develop the block difference equation description of the above IIR filter for a block length of 3 for the input and the output.

(c) Develop the block difference equation description of the above IIR filter for a block length of 4 for the input and the output.

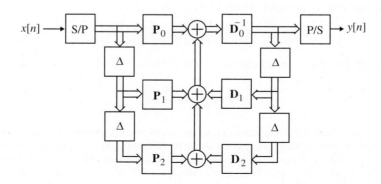

Figure P8.17

8.51 Develop a canonic realization of the block digital filter of Figure P8.17 employing only two block delays.

8.52 Develop the three-multiplier structure of a digital sine-cosine generator obtained by setting $\alpha = \pm\beta \sin \omega_o$ in Eq. (8.125).

8.53 Develop a one-multiplier structure of a digital sine-cosine generator obtained by setting $-\beta \sin \omega_o/\alpha = 1+\cos \omega_o$ in Eq. (8.125).

8.54 Develop a one-multiplier structure of a digital sine-cosine generator obtained by setting $C = -1$ in Eq. (8.124) and then choosing α and β properly. Show the final structure.

8.55 Signals generated by multiple sources or multiple sensors, called *multichannel signals*, are usually transmitted through independent channels in close proximity to each other. As a result, each component of the multichannel signal often gets corrupted by signals from adjacent channels during transmission, resulting in *cross talk*. Separation of the multichannel signal at the receiver is thus of practical interest. A model representing the cross talk between a pair of channels for a two-channel signal is depicted in Figure P8.18(a), and the corresponding discrete-time system for channel separation is as shown in Figure P8.18(b) [Yel96]. Determine two possible sets of conditions for perfect channel separation.

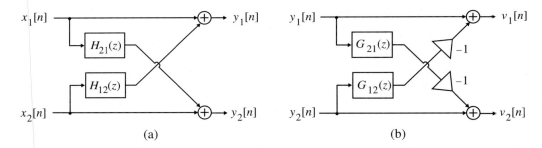

Figure P8.18

8.15 MATLAB Exercises

M 8.1 Using MATLAB, develop a cascade realization of each of the following linear-phase FIR transfer functions:
(a) $H_1(z) = -0.24 + 0.184z^{-1} + 0.4448z^{-2} + 1.296z^{-3} + 0.4448z^{-4} + 0.184z^{-5} - 0.24z^{-6}$,
(b) $H_2(z) = 4 - 13.6z^{-1} - 25.08z^{-2} + 77.2z^{-3} - 25.08z^{-4} - 13.6z^{-5} + 4z^{-6}$,
(c) $H_3(z) = -0.24 + 0.184z^{-1} + 0.4448z^{-2} - 0.4448z^{-4} - 0.184z^{-5} + 0.24z^{-6}$,
(d) $H_4(z) = 4 - 13.6z^{-1} - 25.08z^{-2} + 25.08z^{-4} + 13.6z^{-5} - 4z^{-6}$.

M 8.2 Consider the fourth-order IIR transfer function

$$G(z) = \frac{0.3901 + 0.6426z^{-1} + 0.8721z^{-2} + 0.6426z^{-3} + 0.3901z^{-4}}{1 + 0.5038z^{-1} + 0.8923z^{-2} + 0.3844z^{-3} + 0.1569z^{-4}}.$$

(a) Using MATLAB, express $G(z)$ in factored form.
(b) Develop two different cascade realizations of $G(z)$.
(c) Develop two different parallel form realizations of $G(z)$.
Realize each second-order section in direct form II.

M 8.3 Consider the fourth-order IIR transfer function given below:

$$H(z) = \frac{0.3549 + 0.2002z^{-1} + 0.7031z^{-2} + 0.2002z^{-3} + 0.3549z^{-4}}{1 + 1.2522z^{-1} + 1.9448z^{-2} + 0.9774z^{-3} + 0.5595z^{-4}}.$$

(a) Using MATLAB, express $H(z)$ in factored form.
(b) Develop two different cascade realizations of $H(z)$.
(c) Realize $H(z)$ in parallel forms I and II.
Realize each second-order section in direct form II.

M 8.4 Using Program 8_5, develop a Gray-Markel tapped cascaded lattice realization of the IIR transfer function $G(z)$ of Problem M8.2.

Program 8_5.m

M 8.5 Using Program 8_5, develop a Gray-Markel tapped cascaded lattice realization of the IIR transfer function $H(z)$ of Problem M8.3.

M 8.6 Using Program 8_7, develop a cascaded lattice realization of each of the FIR transfer functions of Problem

Program 8_7.m   M8.1.

M 8.7 (a) Realize the following IIR lowpass transfer function $G(z)$ in the form of a parallel allpass structure:

$$G(z) = \frac{0.2801(1 - 0.6006z^{-1} + 1.0338z^{-2} + 1.0338z^{-3} - 0.6006z^{-4} + z^{-5})}{1 - 1.9607z^{-1} + 2.9395z^{-2} - 2.14486z^{-3} + 1.165z^{-4} - 0.1962z^{-5}}.$$

(b) From the allpass decomposition, determine its power-complementary transfer function $H(z)$.
(c) Plot the square of the magnitude responses of the original transfer function $G(z)$ and its power-complementary transfer function $H(z)$ derived in Part (b), and verify that their sum is equal to 1 at all frequencies.

M 8.8 (a) Realize the following IIR highpass transfer function $G(z)$ in the form of a parallel allpass structure:

$$G(z) = \frac{0.2876(1 + 0.1318z^{-1} + 1.1861z^{-2} - 1.1861z^{-3} - 0.1318z^{-4} - z^{-5})}{1 + 1.57274z^{-1} + 2.712z^{-2} + 1.9431z^{-3} + 1.2979z^{-4} + 0.3018z^{-5}}.$$

(b) From the allpass decomposition, determine its power-complementary transfer function $H(z)$.
(c) Plot the square of the magnitude responses of the original transfer function $G(z)$ and its power-complementary transfer function $H(z)$ derived in Part (b), and verify that their sum is equal to 1 at all frequencies.

M 8.9 (a) Realize the following IIR bandpass transfer function $G(z)$ in the form of a parallel allpass structure.

$$G(z) = \frac{0.3082(1 - 1.9622z^{-1} + 2.3876z^{-2} - 1.9622z^{-3} + z^{-4})}{1 - 2.0936z^{-1} + 2.5697z^{-2} - 1.7322z^{-3} + 0.7076z^{-4}}.$$

(b) From the allpass decomposition, determine its power-complementary transfer function $H(z)$.
(c) Plot the square of the magnitude responses of the original transfer function $G(z)$ and its power-complementary transfer function $H(z)$ derived in Part (b), and verify that their sum is equal to 1 at all frequencies.

M 8.10 Using MATLAB, simulate the single-multiplier sine-cosine generator of Problem 8.53 with $\cos \omega_o = 0.8$, and plot the first 50 samples of its two output sequences. Scale the outputs so that they both have a maximum amplitude of ± 1. What is the effect of initial values of the variables $s_i[n]$?

M 8.11 Using MATLAB, simulate the single-multiplier sine-cosine generator of Problem 8.54 with $\cos \omega_o = 0.8$, and plot the first 50 samples of its two output sequences. Scale the outputs so that they both have a maximum amplitude of ± 1. What is the effect of initial values of the variables $s_i[n]$ on the outputs?

9 IIR Digital Filter Design

An important step in the development of a digital filter is the determination of a realizable transfer function $G(z)$ approximating the given frequency response specifications. If an IIR filter is desired, it is also necessary to ensure that $G(z)$ is stable. The process of deriving the transfer function $G(z)$ is called *digital filter design*. After $G(z)$ has been obtained, the next step is to realize it in the form of a suitable filter structure. In Chapter 8, we outlined a variety of basic structures for the realization of FIR and IIR transfer functions. In this chapter, we consider the IIR digital filter design problem. The design of FIR digital filters is treated in Chapter 10.

First we review some of the issues associated with the filter design problem. A widely used approach to IIR filter design based on the conversion of a prototype analog transfer function to a digital transfer function is discussed next. Typical design examples are included to illustrate this approach. We then consider the transformation of one type of IIR filter transfer function into another type, which is achieved by replacing the complex variable z by a function of z. Four commonly used transformations are summarized. Finally, we consider the computer-aided design of IIR digital filters. To this end, we restrict our discussion to the use of MATLAB in determining the transfer functions.

9.1 Preliminary Considerations

There are two major issues that need to be answered before one can develop the digital transfer function $G(z)$. The first and foremost issue is the development of a reasonable filter frequency response specification from the requirements of the overall system in which the digital filter is to be employed. The second issue is to determine whether an FIR or an IIR digital filter is to be designed. In this section, we examine these two issues first. Next we review the basic analytical approach to the design of IIR digital filters and then consider the determination of the filter order that meets the prescribed specifications. We also discuss appropriate scaling of the transfer function.

9.1.1 Digital Filter Specifications

As in the case of the design of analog filters, either the magnitude and/or the phase (delay) response is specified for the design of a digital filter for most applications. In some situations, the desired unit sample response or the step response may be specified. In most practical applications, the problem of interest is the development of a realizable approximation to a given magnitude response specification. As indicated in Section 7.1.3, the phase response of the designed filter can be corrected by cascading it with an allpass section. We describe in Section 9.7.2 a method for the design of allpass phase equalizers.

We restrict our attention in this chapter to the magnitude approximation problem only. We pointed out in Section 7.1.1 that there are four basic types of filters, whose magnitude responses are shown in Figure 7.1. Since the impulse response corresponding to each of these is noncausal and of infinite length,

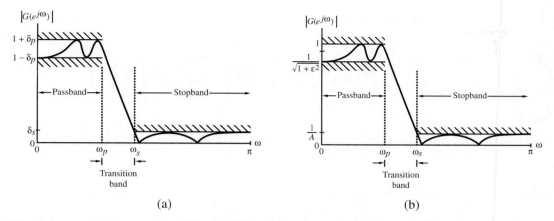

Figure 9.1: (a) Typical magnitude specifications for a digital lowpass filter and (b) normalized magnitude specifications for a digital lowpass filter.

these ideal filters are not realizable. One way of developing a realizable approximation to these filters would be to truncate the impulse response as indicated in Eq. (7.30) for a lowpass filter. The magnitude response of the FIR lowpass filter obtained by truncating the impulse response of the ideal lowpass filter does not have a sharp transition from passband to stopband but, rather, exhibits a gradual "roll-off."

Thus, as in the case of the analog filter design problem outlined in Section 4.4, the magnitude response specifications of a digital filter in the passband and in the stopband are given with some acceptable tolerances. In addition, a transition band is specified between the passband and the stopband to permit the magnitude to drop off smoothly. For example, the desired magnitude $\left|G(e^{j\omega})\right|$ of a lowpass filter may be given as shown in Figure 9.1(a). As indicated in the figure, in the *passband* defined by $0 \leq \omega \leq \omega_p$, we require that the magnitude approximates unity with an error of $\pm\delta_p$; that is,

$$1 - \delta_p \leq \left|G(e^{j\omega})\right| \leq 1 + \delta_p, \qquad \text{for } |\omega| \leq \omega_p. \tag{9.1}$$

In the *stopband*, defined by $\omega_s \leq \omega \leq \pi$, we require that the magnitude approximates zero with an error bound of δ_s; that is,

$$\left|G(e^{j\omega})\right| \leq \delta_s, \qquad \text{for } \omega_s \leq |\omega| \leq \pi. \tag{9.2}$$

The frequencies ω_p and ω_s are, respectively, called the *passband edge frequency* and the *stopband edge frequency*. The limits of the tolerances in the passband and stopband, δ_p and δ_s, are usually called the *peak ripple values*. Note that the frequency response $G(e^{j\omega})$ of a digital filter is a periodic function of ω, and the magnitude response of a real-coefficient digital filter is an even function of ω. As a result, the digital filter specifications are given only for the range $0 \leq |\omega| \leq \pi$.

Digital filter specifications are often given in terms of the loss function, $\mathcal{A}(\omega) = -20 \log_{10} \left|G(e^{j\omega})\right|$, in dB. Here the *peak passband ripple* α_p and the *minimum stopband attenuation* α_s are given in dB; that is, the loss specifications of a digital filter are given by

$$\alpha_p = -20 \log_{10}(1 - \delta_p) \text{ dB}, \tag{9.3}$$

$$\alpha_s = -20 \log_{10}(\delta_s) \text{ dB}. \tag{9.4}$$

EXAMPLE 9.1   Computation of Peak Passband and Stopband Ripples

The peak passband ripple α_p and the minimum stopband attenuation α_s of a digital filter are, respectively, 0.1 dB and 35 dB. Determine their corresponding peak ripple values δ_p and δ_s. From Eqs. (9.3) and (9.4), we obtain

$$\delta_p = 1 - 10^{-\alpha_p/20} = 1 - 10^{-0.005} = 0.01144690,$$

$$\delta_s = 10^{-\alpha_s/20} = 10^{-1.75} = 0.01778279.$$

As in the case of an analog lowpass filter, the magnitude response specifications for a digital lowpass filter may alternatively be given in a normalized form, as in Figure 9.1(b). Here the maximum value of the magnitude in the passband is assumed to be unity, and the maximum passband deviation, denoted as $1/\sqrt{1+\varepsilon^2}$, is given by the minimum value of the magnitude in the passband. The maximum stopband magnitude is denoted by $1/A$.[1]

For the normalized specification, the maximum value of the gain function or the minimum value of the loss function is therefore 0 dB. The quantity $\alpha_{\max}$ given by

$$\alpha_{\max} = 20 \log_{10} \left(\sqrt{1+\varepsilon^2} \right) \text{ dB} \tag{9.5}$$

is called the *maximum passband attenuation*. For $\delta_p \ll 1$, as is typically the case, it can be shown that

$$\alpha_{\max} \cong -20 \log_{10}(1 - 2\delta_p) \cong 2\alpha_p. \tag{9.6}$$

The passband and stopband edge frequencies, in most applications, are specified in Hz, along with the sampling rate of the digital filter. Since all filter design techniques are developed in terms of normalized angular frequencies ω_p and ω_s, the specified critical frequencies need to be normalized before a specific filter design algorithm can be applied. Let F_T denote the sampling frequency in Hz, and F_p and F_s denote, respectively, the passband and stopband edge frequencies in Hz. Then, the normalized angular edge frequencies in radians are given by

$$\omega_p = \frac{\Omega_p}{F_T} = \frac{2\pi F_p}{F_T} = 2\pi F_p T, \tag{9.7}$$

$$\omega_s = \frac{\Omega_s}{F_T} = \frac{2\pi F_s}{F_T} = 2\pi F_s T. \tag{9.8}$$

EXAMPLE 9.2   Conversion of Bandedge Frequencies to Normalized Digital Angular Frequencies

Let the specified passband and stopband edge frequencies of a digital highpass filter operating at a sampling rate of 25 kHz be 7 kHz and 3 kHz, respectively. Using Eqs. (9.7) and (9.8), we determine the corresponding normalized angular bandedge frequencies as

$$\omega_p = \frac{2\pi(7 \times 10^3)}{25 \times 10^3} = 0.56\pi,$$

$$\omega_s = \frac{2\pi(3 \times 10^3)}{25 \times 10^3} = 0.24\pi.$$

---

[1] The minimum stopband attenuation is therefore $20 \log_{10}(A)$.

9.1.2   Selection of the Filter Type

The second issue of interest is the selection of the digital filter type, that is, whether an IIR or an FIR digital filter is to be employed. The objective of digital filter design is to develop a causal transfer function $H(z)$ meeting the frequency response specifications. For IIR digital filter design, the transfer function is a real rational function of z^{-1}:

$$H(z) = \frac{p_0 + p_1 z^{-1} + p_2 z^{-2} + \cdots + p_M z^{-M}}{d_0 + d_1 z^{-1} + d_2 z^{-2} + \cdots + d_N z^{-N}}. \tag{9.9}$$

Moreover, $H(z)$ must be a stable transfer function, and for reduced computational complexity, it must be of lowest order N. On the other hand, for FIR filter design, the transfer function is a polynomial in z^{-1}:

$$H(z) = \sum_{n=0}^{N} h[n] z^{-n}. \tag{9.10}$$

For reduced computational complexity, the degree N of $H(z)$ must be as small as possible. In addition, if a linear phase is desired, then the FIR filter coefficients must satisfy the constraint:

$$h[n] = \pm h[N - n]. \tag{9.11}$$

There are several advantages in using an FIR filter, since it can be designed with exact linear phase and the filter structure is always stable with quantized filter coefficients. However, in most cases, the order N_{FIR} of an FIR filter is considerably higher than the order N_{IIR} of an equivalent IIR filter meeting the same magnitude specifications. In general, the implementation of the FIR filter requires approximately N_{FIR} multiplications per output sample, whereas the IIR filter requires $2N_{\text{IIR}} + 1$ multiplications per output sample. In the former case, if the FIR filter is designed with a linear phase, then the number of multiplications per output sample reduces to approximately $(N_{\text{FIR}} + 1)/2$. Likewise, most IIR filter designs result in transfer functions with zeros on the unit circle, and the cascade realization of an IIR filter of order N_{IIR} with all of the zeros on the unit circle requires $\lfloor (3N_{\text{IIR}} + 3)/2 \rfloor$ multiplications per output sample.[2] It has been shown that for most practical filter specifications, the ratio $N_{\text{FIR}}/N_{\text{IIR}}$ is typically of the order of tens or more and, as a result, the IIR filter usually is computationally more efficient [Rab75a]. However, if the group delay of the IIR filter is equalized by cascading it with an allpass equalizer, then the savings in computation may no longer be that significant [Rab75a]. In many applications, the linearity of the phase response of the digital filter is not an issue, making the IIR filter preferable because of the lower computational requirements.

9.1.3   Basic Approach to IIR Digital Filter Design

In the case of IIR filter design, the most common practice is to convert the digital filter specifications into analog lowpass prototype filter specifications, to determine the analog lowpass filter transfer function meeting these specifications, and then to transform it into the desired digital filter transfer function. This approach has been widely used for many reasons:

(a) Analog approximation techniques are highly advanced.

(b) They usually yield closed-form solutions.

(c) Extensive tables are available for analog filter design.

---

[2] $\lfloor x \rfloor$ denotes the integer part of x.

(d) Many applications require the digital simulation of analog filters.

In the following, we denote an analog transfer function as

$$H_a(s) = \frac{P_a(s)}{D_a(s)}, \tag{9.12}$$

where the subscript "a" specifically indicates the analog domain. The digital transfer function derived from $H_a(s)$ is denoted by

$$G(z) = \frac{P(z)}{D(z)}. \tag{9.13}$$

The basic idea behind the conversion of an analog prototype transfer function $H_a(s)$ into a digital IIR transfer function $G(z)$ is to apply a mapping from the s-domain to the z-domain so that the essential properties of the analog frequency response are preserved. This implies that the mapping function should be such that

(a) the imaginary ($j\Omega$) axis in the s-plane be mapped onto the unit circle of the z-plane,

(b) a stable analog transfer function be transformed into a stable digital transfer function.

To this end, the most widely used transformation is the bilinear transformation, described in Section 9.2.

9.1.4 IIR Digital Filter Order Estimation

After the type of the digital filter has been selected, the next step in the filter design process is to estimate the filter order N that is needed to meet the given filter specifications. For reduced computational complexity, the filter order should be the smallest integer greater than or equal to the estimated value.

For the design of a lowpass IIR digital filter $G(z)$ based on the conversion of an analog lowpass filter $H_a(s)$, the filter order of $H_a(s)$ is first estimated from its specifications using the appropriate formula given in Eq. (4.35), (4.43), or (4.54), depending on whether a Butterworth, Chebyshev, or equiripple filter approximation is desired. The order of $G(z)$ is then determined automatically from the transformation being used to convert $H_a(s)$ into $G(z)$. There are several M-files in MATLAB that can be used to directly estimate the minimum order of an IIR digital transfer function meeting the filter specifications for the class of approximations discussed in Section 4.4. These are discussed in Section 9.6.

9.1.5 Scaling the Digital Transfer Function

After a digital filter has been designed following any one of the techniques outlined in this chapter, the corresponding transfer function $G(z)$ has to be scaled in magnitude before it can be implemented. In magnitude scaling, the transfer function is multiplied by a scaling constant K so that the maximum magnitude of the scaled transfer function $G_t(z) = K\,G(z)$ in the passband is unity; that is, the scaled transfer function has a maximum gain of 0 dB. For a stable transfer function $G(z)$ with real coefficients, the scaled transfer function $K\,G(z)$ is then a bounded real (BR) function.[3]

For a frequency selective transfer function $G(z)$, if $G_{\max}$ is the maximum value of $|G(e^{j\omega})|$ in the frequency range $0 \le \omega \le \pi$, then $K = 1/G_{\max}$, which results in a maximum gain of 0 dB in the passband of the scaled transfer function. For example, in the case of a lowpass transfer function with a maximum magnitude at dc, it is usual practice to use $K = 1/G(1)$, implying a dc gain of 0 dB for the scaled transfer function. Likewise, in the case of a highpass transfer function with a maximum magnitude at $\omega = \pi$, K is set to $1/G(-1)$, yielding a gain of 0 dB at $\omega = \pi$ for the scaled transfer function. For a bandpass transfer function, it is common to use K equal to $1/|G(e^{j\omega_c})|$, where ω_c is the center frequency of the passband.

---

[3] See Section 7.1.2 for a definition of a bounded real (BR) function.

9.2   Bilinear Transformation Method of IIR Filter Design

A number of transformations have been proposed to convert an analog transfer function $H_a(s)$ into a digital transfer function $G(z)$ so that essential properties of the analog transfer function in the s-domain are preserved for the digital transfer function in the z-domain. Of these, the bilinear transformation is more commonly used to design IIR digital filters based on the conversion of analog prototype filters.

9.2.1   The Bilinear Transformation

The bilinear transformation from the s-plane to the z-plane is given by [Kai66]

$$s = \frac{2}{T}\left(\frac{1 - z^{-1}}{1 + z^{-1}}\right). \tag{9.14}$$

The above transformation is a one-to-one mapping; that is, it maps a single point in the s-plane to a unique point in the z-plane, and vice versa. The relation between the digital transfer function $G(z)$ and the parent analog transfer function $H_a(s)$ is then given by

$$G(z) = H_a(s)\Big|_{s=\frac{2}{T}\left(\frac{1-z^{-1}}{1+z^{-1}}\right)}. \tag{9.15}$$

The bilinear transformation is derived by applying the trapezoidal numerical integration approach to the differential equation representation of $H_a(s)$ that leads to the difference equation representation of $G(z)$ (see Example 2.46). The parameter T represents the step size in the numerical integration.

Let us now examine the above transformation. For $s = \sigma_o + j\Omega_o$,

$$z = \frac{1 + \frac{T}{2}(\sigma_o + j\Omega_o)}{1 - \frac{T}{2}(\sigma_o + j\Omega_o)} = \frac{\left(1 + \frac{T}{2}\sigma_o\right) + j\frac{T}{2}\Omega_o}{\left(1 - \frac{T}{2}\sigma_o\right) - j\frac{T}{2}\Omega_o}. \tag{9.16}$$

Therefore,

$$|z|^2 = \frac{\left(1 + \frac{T}{2}\sigma_o\right)^2 + \left(\frac{T}{2}\Omega_o\right)^2}{\left(1 - \frac{T}{2}\sigma_o\right)^2 + \left(\frac{T}{2}\Omega_o\right)^2}. \tag{9.17}$$

Thus, a point on the $j\Omega$-axis in the s-plane ($\sigma_o = 0$) is mapped onto a point on the unit circle in the z-plane as $|z| = 1$. A point in the left-half s-plane with $\sigma_o < 0$ is mapped onto a point inside the unit circle in the z-plane as $|z| < 1$. Likewise, a point in the right-half s-plane with $\sigma_o > 0$ is mapped onto a point outside the unit circle in the z-plane, as $|z| > 1$. Any point in the s-plane is mapped onto a unique point in the z-plane, and vice versa. The mapping of the s-plane into the z-plane via the bilinear transformation is illustrated in Figure 9.2 and is seen to have all the desired properties. Also, there is no aliasing due to the one-to-one mapping.

The exact relation between the imaginary axis in the s-plane ($s = j\Omega$) and the unit circle in the z-plane ($z = e^{j\omega}$) is of interest. From Eq. (9.14), it follows that

$$j\Omega = \frac{2}{T}\left(\frac{1 - e^{-j\omega}}{1 + e^{-j\omega}}\right) = j\frac{2}{T}\tan\left(\frac{\omega}{2}\right),$$

or

$$\Omega = \frac{2}{T}\tan\left(\frac{\omega}{2}\right), \tag{9.18}$$

which has been plotted in Figure 9.3. Note from this plot that the positive (negative) imaginary axis in the s-plane is mapped into the upper (lower) half of the unit circle in the z-plane. However, it is clear that

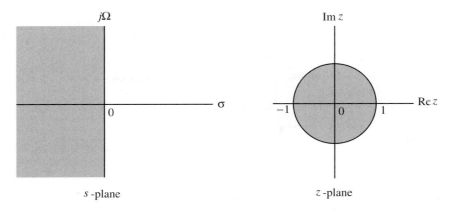

Figure 9.2: The bilinear transformation mapping.

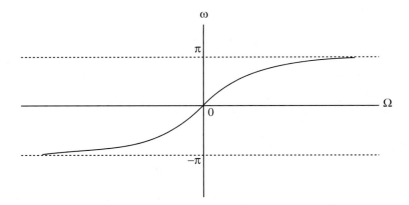

Figure 9.3: Mapping of the angular analog frequencies Ω to the angular digital frequencies ω via the bilinear transformation.

the mapping is highly nonlinear since the complete negative imaginary axis in the s-plane from $\Omega = -\infty$ to $\Omega = 0$ is mapped into the lower half of the unit circle from $\omega = -\pi$ (i.e., $z = -1$) to $\omega = 0$ (i.e., $z = +1$) and the complete positive imaginary axis in the s-plane from $\Omega = 0$ to $\Omega = +\infty$ is mapped into the upper half of the unit circle from $\omega = 0$ (i.e., $z = +1$) to $\omega = +\pi$ (i.e., $z = -1$). This introduces a distortion in the frequency axis called *frequency warping*. The effect of warping is more evident in Figure 9.4, which shows the transformation of a typical analog filter magnitude response to a digital filter magnitude response derived via the bilinear transformation. Thus, to develop a digital filter meeting a specified magnitude response, we must first prewarp the critical bandedge frequencies (ω_p and ω_s) to find their analog equivalents (Ω_p and Ω_s) using the relation of Eq. (9.18), design the analog prototype $H_a(s)$ using the prewarped critical frequencies, and then transform $H_a(s)$ using the bilinear transformation to obtain the desired digital filter transfer function $G(z)$.

It should be noted that the bilinear transformation preserves the magnitude response of an analog filter only if the specification requires piecewise constant magnitude. However, the phase response of the analog filter is not preserved after transformation. Hence, the transformation can be used only to design digital filters with prescribed magnitude response with piecewise constant values.

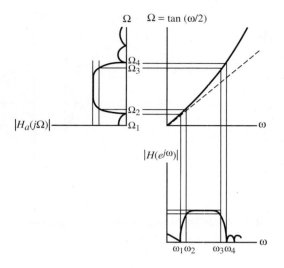

Figure 9.4: Illustration of the frequency warping effect.

The bilinear transformation of an analog transfer function can be carried out in MATLAB using the M-file `bilinear`. There are three forms of this function, each of which accepts an additional argument in the input specifying prewarping. The use of this function is illustrated in Examples 9.3, 9.4, and 9.5.

9.2.2 Design of Low-Order Digital Filters

We consider now the design of low-order digital filters by applying the bilinear transformation to the transfer functions of corresponding low-order analog filters. An application of these low-order digital filters are as equalizers in digital audio.[4] In the following section, we consider the design of higher-order digital filters.

First-Order Butterworth Lowpass and Highpass Digital Filters

It follows from Eqs. (4.36) and (4.37) that the transfer function of a first-order Butterworth lowpass analog filter with a 3-dB cutoff frequency at Ω_c is given by

$$H_{LP}(s) = \frac{\Omega_c}{s + \Omega_c}. \tag{9.19}$$

Applying a bilinear transformation to Eq. (9.19), we arrive at the expression for the transfer function $G(z)$ of a first-order Butterworth digital lowpass filter:

$$G_{LP}(z) = \left. \frac{\Omega_c}{s + \Omega_c} \right|_{s = \frac{2}{T}\left(\frac{1-z^{-1}}{1+z^{-1}}\right)} = \frac{\frac{\Omega_c T}{2}(1 + z^{-1})}{(1 - z^{-1}) + \frac{\Omega_c T}{2}(1 + z^{-1})}. \tag{9.20}$$

Rearranging terms, we can rewrite Eq. (9.20) as

---
[4]See Section 15.5.2.

$$G_{LP}(z) = \frac{1-\alpha}{2}\left(\frac{1+z^{-1}}{1-\alpha z^{-1}}\right), \tag{9.21}$$

where

$$\alpha = \frac{1 - \frac{\Omega_c T}{2}}{1 + \frac{\Omega_c T}{2}}. \tag{9.22}$$

The 3-dB cutoff frequency ω_c of the digital transfer function of Eq. (9.21) is related to the 3-dB cutoff frequency Ω_c of the analog transfer function through Eq. (9.18). Using Eq. (9.18) in Eq. (9.22), we can express α as a function of ω_c, which is thus given by

$$\alpha = \frac{1 - \tan(\omega_c T/2)}{1 + \tan(\omega_c T/2)}. \tag{9.23}$$

It should be noted that the first-order lowpass digital transfer function of Eq. (9.21) is identical to that of Eq. (7.71), where it was introduced without any derivation.

Applying a lowpass-to-highpass transformation of Eq. (4.62) to the lowpass transfer function of Eq. (9.19), we arrive at the transfer function of a first-order analog highpass Butterworth filter with a 3-dB cutoff frequency at Ω_c :

$$H_{HP}(s) = \frac{s}{s + \Omega_c}. \tag{9.24}$$

The transfer function of the first-order highpass digital Butterworth filter obtained by applying the bilinear transformation to the highpass analog transfer function $H_{HP}(s)$ of Eq. (9.24) is then given by

$$G_{HP}(z) = \left.\frac{s}{s + \Omega_c}\right|_{s = \frac{2}{T}\left(\frac{1-z^{-1}}{1+z^{-1}}\right)} = \frac{1+\alpha}{2}\left(\frac{1-z^{-1}}{1-\alpha z^{-1}}\right), \tag{9.25}$$

where α is as given in Eq. (9.23). It should be noted again that the first-order highpass transfer function of Eq. (9.25) is identical to that of Eq. (7.74), where it was also introduced without any derivation.

Second-Order Bandpass and Bandstop Digital Filters

Applying a lowpass-to-bandpass transformation of Eq. (4.64) to the lowpass transfer function of Eq. (9.19), we arrive at the transfer function $H_{BP}(s)$ of a second-order analog bandpass filter:

$$H_{BP}(s) = \frac{Bs}{s^2 + Bs + \Omega_o^2}, \tag{9.26}$$

where Ω_o is the angular frequency, called the *center frequency,* at which the magnitude response takes a maximum value of 1 and B is the *3-dB bandwidth* of the passband. The magnitude response goes to zero value at $\Omega = 0$ and $\Omega = \infty$; that is, $|H_{BP}(j\Omega_o)| = 1$ and $|H_{BP}(j0)| = |H_{BP}(j\infty)| = 0$. By substituting $B = \Omega_o/Q_o$, where Q_o is called the *quality factor* or simply the *Q-factor* of the bandpass filter, the above transfer function can be alternately expressed in the form

$$H_{BP}(s) = \frac{\frac{\Omega_o}{Q_o}s}{s^2 + \frac{\Omega_o}{Q_o}s + \Omega_o^2}. \tag{9.27}$$

The transfer function of a second-order digital bandpass filter with tunable center frequency and 3-dB bandwidth is obtained by applying the bilinear transformation to the analog bandpass transfer function $H_{BP}(s)$ of Eq. (9.26) and is given by

$$G_{BP}(z) = \frac{1-\alpha}{2} \left(\frac{1-z^{-2}}{1 - \beta(1+\alpha)z^{-1} + \alpha z^{-2}} \right), \tag{9.28}$$

where

$$\alpha = \frac{1 - \frac{BT}{2} + \frac{T^2\Omega_o^2}{4}}{1 + \frac{BT}{2} + \frac{T^2\Omega_o^2}{4}}, \tag{9.29a}$$

$$\beta = \frac{1 - \frac{T^2\Omega_o^2}{4}}{1 + \frac{T^2\Omega_o^2}{4}}. \tag{9.29b}$$

The parameters α and β are related to the *center frequency* ω_o and the *3-dB bandwidth* B_w of the digital bandpass filter through

$$\alpha = \frac{1 - \tan(B_w T/2)}{1 + \tan(B_w T/2)}, \tag{9.30a}$$

$$\beta = \cos(\omega_o T). \tag{9.30b}$$

The bandpass transfer function of Eq. (9.28) is the same as that of Eq. (7.77) introduced without any derivation.

A second-order analog bandstop filter has a transfer function given by

$$H_{BS}(s) = \frac{s^2 + \Omega_o^2}{s^2 + Bs + \Omega_o^2}. \tag{9.31}$$

Its magnitude response approaches unity values, that is, a gain of 0 dB, at $\Omega = 0$ and ∞. The magnitude has a zero value at the *notch frequency* $\Omega = \Omega_o$. If Ω_1 and Ω_2, $\Omega_2 > \Omega_1$, denote the frequencies at which the gain is down by -3 dB, it can be shown that the *3-dB notch bandwidth* defined by $(\Omega_2 - \Omega_1)$ is equal to B.

Applying a bilinear transformation to $H_{BS}(s)$ of Eq. (9.31), we arrive at

$$G_{BS}(z) = \frac{1+\alpha}{2} \left(\frac{1 - 2\beta z^{-1} + z^{-2}}{1 - \beta(1+\alpha)z^{-1} + \alpha z^{-2}} \right), \tag{9.32}$$

where the parameters α and β are again given by Eqs. (9.29a) and (9.29b), respectively. They are related to the *center frequency* ω_o and the *3-dB bandwidth* B_w of the digital bandpass filter through Eqs. (9.30a) and (9.30b), respectively.

It should be noted that Eq. (9.32) is precisely the transfer function of the second-order notch filter given in Eq. (7.80) and introduced without any derivation.

Simplified Bilinear Transformation

As we shall see later in the following section, the digital filter design procedure consists of two steps: first, the inverse bilinear transformation is applied to the digital filter specifications to arrive at the specifications of the analog filter prototype; then the bilinear transformation of Eq. (9.14) is employed to obtain the desired digital transfer function $G(z)$ from the analog transfer function $H_a(s)$ designed to meet the analog filter specifications. As a result, the parameter T has no effect on the expression for $G(z)$, and we shall choose, for convenience, $T = 2$ to simplify the design procedure.

The corresponding inverse transformation for $T = 2$ is given by

$$z = \frac{1+s}{1-s}. \tag{9.33}$$

9.3 Design of Lowpass IIR Digital Filters

We illustrate now the development of a lowpass IIR digital transfer function meeting given specifications using the bilinear transformation method. To this end, we first obtain the specifications for a prototype lowpass analog filter from the specifications of the lowpass digital filter using the inverse transformation. We then determine the analog transfer function $H_a(s)$ meeting the specifications of the prototype analog filter. Finally, the analog transfer function $H_a(s)$ is transformed into a digital transfer function $G(z)$ using the bilinear transformation.

Specifically, consider the design of a lowpass IIR digital filter $G(z)$ with a maximally flat magnitude characteristic. The passband edge frequency ω_p is 0.25π, with a passband ripple not exceeding 0.5 dB. The minimum stopband attenuation at the stopband edge frequency ω_s of 0.55π is 15 dB. Thus, if $|G(e^{j0})| = 1$, then we require that

$$20 \log_{10} \left| G(e^{j0.25\pi}) \right| \geq -0.5 \text{ dB}, \tag{9.34a}$$

$$20 \log_{10} \left| G(e^{j0.55\pi}) \right| \leq -15 \text{ dB}. \tag{9.34b}$$

We first prewarp the digital bandedge frequencies to obtain the corresponding analog bandedge frequencies. From Eq. (9.18), the pertinent analog bandedge frequencies Ω_p and Ω_s corresponding to the two digital bandedge frequencies ω_p and ω_s are given by

$$\Omega_p = \tan \left(\frac{\omega_p}{2} \right) = \tan \left(\frac{0.25\pi}{2} \right) = 0.4142136,$$

$$\Omega_s = \tan \left(\frac{\omega_s}{2} \right) = \tan \left(\frac{0.55\pi}{2} \right) = 1.1708496.$$

From Eq. (4.31), the inverse transition ratio is

$$\frac{1}{k} = \frac{\Omega_s}{\Omega_p} = \frac{1.1708496}{0.4142135} = 2.8266809.$$

From the specified passband ripple of 0.5 dB, we obtain $\varepsilon^2 = 0.1220185$, and from the minimum stopband attenuation of 15 dB, we obtain $A^2 = 31.622777$. Therefore, from Eq. (4.32), the inverse discrimination ratio is

$$\frac{1}{k_1} = \frac{\sqrt{A^2 - 1}}{\varepsilon} = 15.841979.$$

Substituting these values in Eq. (4.35), we obtain the filter order N as

$$N = \frac{\log_{10}(1/k_1)}{\log_{10}(1/k)} = \frac{\log_{10}(15.841979)}{\log_{10}(2.8266814)} = 2.6586997.$$

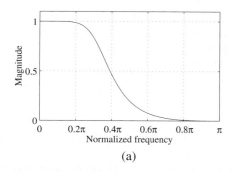 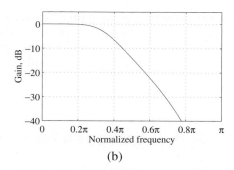

(a)   (b)

Figure 9.5: Magnitude and gain responses of the lowpass filter design based on the bilinear transformation method.

The nearest higher integer 3 is thus taken as the filter order.

The filter order is used next to determine the 3-dB cutoff frequency Ω_c. To this end, either Eq. (4.34a) or Eq. (4.34b) can be used. If Ω_c is determined using Eq. (4.34b), then the stopband specification at Ω_s is satisfied exactly, while the passband specification is exceeded providing a safety margin at Ω_p. However, if Eq. (4.34a) is used to solve for Ω_c, then the passband specification at Ω_p is met exactly, while the stopband specification at Ω_s is exceeded. Substituting the values of ε^2, Ω_p, and N in Eq. (4.34a), we arrive at

$$\Omega_c = 1.419915(\Omega_p) = 1.419915 \times 0.4142135 = 0.588148.$$

Using the function `buttap` of MATLAB,[5] we obtain the third-order normalized lowpass Butterworth transfer function as

$$H_{an}(s) = \frac{1}{(s+1)(s^2+s+1)},$$

which has a 3-dB frequency at $\Omega = 1$ and therefore has to be denormalized to bring the 3-dB frequency to $\Omega_c = 0.588148$. The denormalized transfer function is given by

$$H_a(s) = H_{an}\left(\frac{s}{0.588148}\right) = \frac{0.203451}{(s+0.588148)(s^2+0.588148s+0.345918)}.$$

Applying the bilinear transformation to the above, we finally arrive at the desired expression for the digital lowpass transfer function:

$$\begin{aligned} G(z) &= H_a(s)|_{s=(1-z^{-1})/(1+z^{-1})} \\ &= \frac{0.0662272(1+z^{-1})^3}{(1-0.2593284z^{-1})(1-0.6762858z^{-1}+0.3917468z^{-2})}. \end{aligned} \tag{9.35}$$

The corresponding magnitude and gain responses are plotted in Figure 9.5.

The above digital lowpass filter can be designed directly in the z-domain using the M-files `buttord` and `butter`.

9.4   Design of Highpass, Bandpass, and Bandstop IIR Digital Filters

In the previous section, we outlined the design of lowpass IIR digital filters. We now consider the design of the other three types of IIR digital filters. To this end, two approaches can be followed.

---

[5]See Section 4.4.6.

The first approach consists of the following steps:

Step 1: Prewarp the specified digital frequency specifications of the desired digital filter $G_D(z)$ using Eq. (9.18) to arrive at the frequency specifications of an analog filter $H_D(s)$ of the same type.

Step 2: Convert the frequency specifications of $H_D(s)$ into those of a prototype analog lowpass filter $H_{LP}(s)$ using an appropriate frequency transformation discussed in Section 4.5.

Step 3: Design the analog lowpass filter $H_{LP}(s)$ using the methods described in Section 4.4.

Step 4: Convert the transfer function $H_{LP}(s)$ into $H_D(s)$ using the inverse of the frequency transformation used in Step 2.

Step 5: Transform the transfer function $H_D(s)$ using the bilinear transformation of Eq. (9.14) to arrive at the desired digital IIR transfer function $G_D(z)$.

The second approach consists of the following steps:

Step 1: Prewarp the specified digital frequency specifications of the desired digital filter $G_D(z)$ using Eq. (9.18) to arrive at the frequency specifications of an analog filter $H_D(s)$ of the same type.

Step 2: Convert the frequency specifications of $H_D(s)$ into that of a prototype analog lowpass filter $H_{LP}(s)$ using an appropriate frequency transformation discussed in Section 4.5.

Step 3: Design the analog lowpass filter $H_{LP}(s)$ using the methods described in Section 4.4.

Step 4: Convert the transfer function $H_{LP}(s)$ into the transfer function $G_{LP}(z)$ of an IIR digital filter using the bilinear transformation of Eq. (9.14).

Step 5: Transform $G_{LP}(z)$ into the desired digital transfer function $G_D(z)$ using the appropriate spectral transformation discussed in Section 9.5.

We illustrate the first approach in this section with the aid of examples.

Design of Highpass IIR Digital Filter

We consider the design of a Type 1 Chebyshev IIR digital highpass filter in Example 9.3.

EXAMPLE 9.3   Design of a Highpass IIR Digital Filter

The specifications of the highpass filter are as follows: passband edge $F_p = 700\,\text{Hz}$, stopband edge $F_s = 500\,\text{Hz}$, passband ripple $\alpha_p = 1\,\text{dB}$, minimum stopband attenuation $\alpha_s = 32\,\text{dB}$, and sampling frequency $F_T = 2\,\text{kHz}$. Using Eqs. (9.7) and (9.8), we first determine the normalized angular bandedge frequencies as

$$\omega_p = \frac{2\pi F_p}{F_T} = \frac{2\pi(700)}{2000} = 0.7\pi,$$

$$\omega_s = \frac{2\pi F_s}{F_T} = \frac{2\pi(500)}{2000} = 0.5\pi.$$

We then prewarp the above digital edge frequencies using Eq. (9.18) to arrive at the following angular edge frequencies of the analog highpass filter:

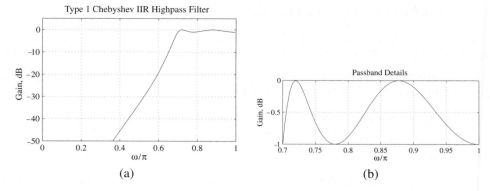

Figure 9.6: IIR Type 1 Chebyshev highpass filter of Example 9.3: (a) gain response and (b) passband details.

$$\hat{\Omega}_p = \tan\left(\frac{\omega_p}{2}\right) = \tan\left(\frac{0.7\pi}{2}\right) = 1.9626105,$$

$$\hat{\Omega}_s = \tan\left(\frac{\omega_s}{2}\right) = \tan\left(\frac{0.5\pi}{2}\right) = 1.0.$$

For the prototype analog lowpass filter, we choose the normalized passband edge to be $\Omega_p = 1$. From Eq. (4.63), the normalized stopband edge of the lowpass filter is thus $\Omega_s = 1.9626105$. The specifications of the analog lowpass filter are therefore as follows: passband edge at 1 rad/sec, stopband edge at 1.9626105 rad/sec, passband ripple of 1 dB, and a minimum stopband attenuation of 32 dB.

Using the M-file `cheb1ord`, the order N and the passband edge Wn of the lowpass filter $H_{LP}(s)$ are determined. Then using the M-file `cheby1`, the lowpass prototype filter $H_{LP}(s)$ is designed. Next, using the M-file `lp2hp`, the analog highpass filter $H_{HP}(s)$ is designed by applying the lowpass-to-highpass transformation of Eq. (4.62) to $H_{LP}(s)$. Finally, using the M-file `bilinear`, the desired digital IIR highpass filter $G_{HP}(z)$ is designed by applying the bilinear transformation of Eq. (9.14) to $H_{HP}(s)$. The code fragments used are

```
[N,Wn] = cheb1ord(1,1.9626105, 1, 32,'s');
[B,A]  = cheby1(N,1,Wn,'s');
[BT,AT] = lp2hp(B,A,1.9626105);
[num,den] = bilinear(BT,AT,0.5);
```

The transfer function $H_{LP}(s)$ of the prototype analog lowpass filter can be obtained by displaying the numerator coefficient vector B and the denominator coefficient vector A. Likewise, the transfer function $H_{HP}(s)$ of the analog highpass filter can be obtained by displaying the numerator coefficient vector BT and the denominator coefficient vector AT. Similarly, the transfer function $G_{HP}(z)$ of the desired digital IIR highpass filter can be obtained by displaying the numerator coefficient vector num and the denominator coefficient vector den. The gain response of the designed IIR digital highpass filter is shown in Figure 9.6.

The above IIR digital highpass filter can be directly designed in the z-domain using only the M-files `cheb1ord` and `cheby1`, as illustrated in Example 9.15.

Design of Bandpass IIR Digital Filter

The design of a Butterworth bandpass IIR digital filter is treated in Example 9.4.

EXAMPLE 9.4   Design of a Bandpass IIR Digital Filter

The desired specifications of the digital bandpass filter are as follows: normalized passband edges at $\omega_{p1} = 0.45\pi$ and $\omega_{p2} = 0.65\pi$, normalized stopband edges at $\omega_{s1} = 0.3\pi$ and $\omega_{s2} = 0.75\pi$, passband ripple of 1 dB, and a minimum stopband attenuation of 40 dB.

We first prewarp the digital edge frequencies using Eq. (9.18) to arrive at the following angular edge frequencies of the analog bandpass filter:

$$\hat{\Omega}_{p1} = \tan\left(\frac{\omega_{p1}}{2}\right) = \tan\left(\frac{0.45\pi}{2}\right) = 0.8540807,$$

$$\hat{\Omega}_{p2} = \tan\left(\frac{\omega_{p2}}{2}\right) = \tan\left(\frac{0.65\pi}{2}\right) = 1.6318517,$$

$$\hat{\Omega}_{s1} = \tan\left(\frac{\omega_{s1}}{2}\right) = \tan\left(\frac{0.3\pi}{2}\right) = 0.5095254,$$

$$\hat{\Omega}_{s2} = \tan\left(\frac{\omega_{s2}}{2}\right) = \tan\left(\frac{0.75\pi}{2}\right) = 2.41421356.$$

The width of the passband of the bandpass filter is $B_w = \hat{\Omega}_{p2} - \hat{\Omega}_{p1} = 0.777771$. The product of the two passband edge frequencies is given by $\hat{\Omega}_o^2 = 1.393733$, and the product of the two stopband edge frequencies is 1.23010325. It is a usual practice not to change the specified width B_w of the passband but to adjust the value of the lower stopband edge frequency so that the two stopband edge frequencies exhibit geometric symmetry with respect to $\hat{\Omega}_o = 1.1805647$. To this end, we modify the lower stopband edge frequency and set $\hat{\Omega}_{s1} = 0.577327$.

For the prototype analog lowpass filter, we choose the normalized passband edge frequency $\Omega_p = 1$. From Eq. (4.65), we get the stopband edge frequency of the lowpass filter to be

$$\Omega_s = \frac{1.393733 - 0.3332788}{0.5773031 \times 0.777771} = 2.3617627.$$

The specifications of the analog Butterworth lowpass filter are thus given by the following: normalized passband edge at 1 rad/sec, normalized stopband edge at 2.3617627 rad/sec, passband ripple of 1 dB, and minimum stopband attenuation of 40 dB.

We first use the M-file buttord to determine the filter order N and the passband edge angular frequency Wn of the prototype analog lowpass filter $H_{LP}(s)$. We then use the M-file butter to design the prototype analog lowpass filter $H_{LP}(s)$. Next, using the M-file lp2bp, we apply the lowpass-to-bandpass transformation of Eq. (9.64) to $H_{LP}(s)$ to design the analog bandpass filter $H_{BP}(s)$. Finally, using the M-file bilinear, we apply the bilinear transformation of Eq. (9.14) to $H_{BP}(s)$ to arrive at the transfer function $G_{BP}(z)$ of the desired bandpass IIR digital filter. The code fragments used are

```
[N,Wn]   = buttord(1,2.3617627, 1, 40,'s');
[B,A]    = butter(N,Wn,'s');
[BT,AT]  = lp2bp(B,A,1.1805647,0.777771);
[num,den] = bilinear(BT,AT,0.5);
```

The transfer function $H_{LP}(s)$ of the prototype analog lowpass filter can be obtained by displaying the numerator coefficient vector B and the denominator coefficient vector A. Likewise, the transfer function $H_{BP}(s)$ of the analog bandpass filter can be obtained by displaying the numerator coefficient vector BT and the denominator coefficient vector AT. Similarly, the transfer function $G_{BP}(z)$ of the desired Butterworth bandpass IIR digital filter can be obtained by displaying the numerator coefficient vector num and the denominator coefficient vector den. The gain response of the designed Butterworth bandpass IIR digital filter is shown in Figure 9.7.

The above IIR digital bandpass filter can be directly designed in the z-domain using only the M-files buttord and butter, as illustrated in Example 9.16.

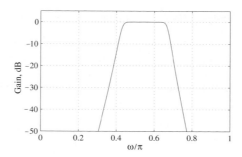

Figure 9.7: Gain response of the Butterworth bandpass IIR digital filter of Example 9.4.

Design of Bandstop IIR Digital Filter

We consider next the design of an elliptic bandstop IIR digital filter.

EXAMPLE 9.5   Design of a Bandstop IIR Digital Filter

The specifications of the bandstop filter are as follows: normalized stopband edges at $\omega_{s1} = 0.45\pi$ and $\omega_{s2} = 0.65\pi$, normalized passband edges at $\omega_{p1} = 0.3\pi$ and $\omega_{p2} = 0.75\pi$, passband ripple of 1 dB, and a minimum stopband attenuation of 40 dB.

As the bandedges of the above digital bandstop filter are the same as those of the digital bandpass filter of Example 9.4, the prewarped angular edge frequencies of the analog bandstop filter are obtained directly from Example 9.4:

$$\hat{\Omega}_{s1} = 0.8540807, \qquad \hat{\Omega}_{s2} = 1.6318517, \qquad \hat{\Omega}_{p1} = 0.5095254, \qquad \hat{\Omega}_{p2} = 2.41421356.$$

For this example, the width of the stopband of the bandstop filter is $B_w = \hat{\Omega}_{s2} - \hat{\Omega}_{s1} = 0.777771$. The product of the two stopband edge frequencies is given by $\hat{\Omega}_o^2 = 1.393733$, and the product of the two passband edge frequencies is 1.230103. In this case, we keep the width B_w of the stopband as specified and adjust the value of the lower passband edge frequency so that the two passband edge frequencies exhibit geometric symmetry with respect to $\hat{\Omega}_o = 1.1805647$. To this end, we modify the lower passband edge frequency and set $\hat{\Omega}_{p1} = 0.577303$. The normalized stopband edge frequency of the prototype analog lowpass filter is set at $\Omega_s = 1$. From Eq. (4.68), we get the passband edge frequency of the lowpass filter to be

$$\Omega_p = \frac{0.5095254 \times 0.777771}{1.393733 - 0.3332787} = 0.4234126.$$

The specifications of the analog elliptic lowpass filter are thus given by the following: normalized passband edge at 0.3494297 rad/sec, normalized stopband edge at 1 rad/sec, passband ripple of 1 dB, and minimum stopband attenuation of 40 dB.

The code fragments used to design the desired digital bandstop filter are

```
[N,Wn] = ellipord(0.4234126,1, 1, 40,'s');
[B,A] = ellip(N,1, 40,Wn,'s');
[BT,AT] = lp2bs(B,A,1.1805647,0.777771);
[num,den] = bilinear(BT,AT,0.5);
```

The gain response of the designed elliptic bandstop IIR digital filter is indicated in Figure 9.8. The transfer function $G_{BS}(z)$ of the elliptic bandstop IIR digital filter can be obtained by displaying the numerator coefficient vector num and the denominator coefficient vector den.

As in the case of Examples 9.3 and 9.4, the above bandstop filter can also be designed directly in the z-domain using only the M-files ellipord and ellip.

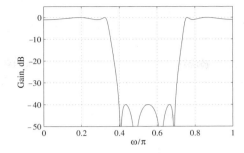

Figure 9.8: Gain response of the elliptic bandstop IIR digital filter of Example 9.5

9.5 Spectral Transformations of IIR Filters

Often, in practice, it is necessary to modify the characteristics of a filter to meet the new specifications without repeating the filter design procedure. For example, after a lowpass filter with a passband edge at 2 kHz has been designed, it may be required to move the passband edge to 2.1 kHz. It is possible to design a digital filter with highpass or bandpass or bandstop characteristics by transforming a given digital lowpass filter. We describe here the spectral transformations that can be used to transform a given lowpass digital IIR transfer function $G_L(z)$ to another digital transfer function $G_D(z)$ that could be a lowpass, highpass, bandpass, or bandstop filter [Con70]. Figure 7.1 shows the magnitude responses of these four types of ideal filters.

To eliminate the confusion between the complex variable z of the lowpass transfer function $G_L(z)$ and that of the desired transfer function $G_D(z)$, we shall use different symbols. Thus, we shall use z^{-1} to denote the unit delay in the prototype lowpass digital filter $G_L(z)$ and $\hat{z}^{-1}$ to denote the unit delay in the transformed filter $G_D(\hat{z})$. The unit circles in the z- and $\hat{z}$-planes are defined by

$$z = e^{j\omega}, \qquad \hat{z} = e^{j\hat{\omega}}.$$

We denote the transformation from the z-domain to the $\hat{z}$-domain as

$$z = F(\hat{z}). \tag{9.36}$$

Then, $G_L(z)$ is transformed to $G_D(\hat{z})$ through

$$G_D(\hat{z}) = G_L\{F(\hat{z})\}. \tag{9.37}$$

To transform a rational $G_L(z)$ into a rational $G_D(\hat{z})$, $F(\hat{z})$ must be a rational function of $\hat{z}$. In addition, to guarantee the stability of $G_D(\hat{z})$, the transformation should be such that the inside of the unit circle of the z-plane is mapped into the inside of the unit circle of the $\hat{z}$-plane. Finally, to ensure that a lowpass magnitude response is mapped into one of the four basic types of magnitude responses, points on the unit circle of the z-plane should be mapped to points on the unit circle of the $\hat{z}$-plane.

Now, in the z-plane, a point on the unit circle is characterized by $|z| = 1$, a point inside the unit circle is given by $|z| < 1$, and a point outside the unit circle is defined by $|z| > 1$. Thus, from Eq. (9.36), $|F(\hat{z})| = |z|$, and, therefore,

$$|F(\hat{z})| \begin{cases} > 1, & \text{if } |z| > 1, \\ = 1, & \text{if } |z| = 1, \\ < 1, & \text{if } |z| < 1. \end{cases} \tag{9.38}$$

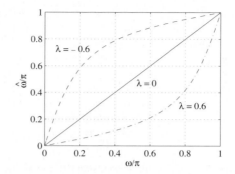

Figure 9.9: Mapping of the angular frequencies in the lowpass-to-lowpass transformation for three different values of the parameter λ.

From the above and Eq. (7.20), it follows that $1/F(\hat{z})$ is a stable allpass function. From Eq. (7.10), we observe that the most general form of $F(\hat{z})$ with real coefficients is given by

$$F(\hat{z}) = \pm \prod_{\ell=1}^{L} \left(\frac{\hat{z} - \lambda_\ell}{1 - \lambda_\ell^* \hat{z}} \right), \tag{9.39}$$

where λ_ℓ is either real or occurs in complex conjugate pairs with $|\lambda_\ell| < 1$ for stability.

9.5.1 Lowpass-to-Lowpass Transformation

To transform a prototype lowpass filter $G_L(z)$ with a cutoff frequency ω_c to another lowpass filter $G_D(\hat{z})$ with a cutoff frequency $\hat{\omega}_c$, we use the transformation

$$z^{-1} = F^{-1}(\hat{z}) = \frac{1 - \lambda \hat{z}}{\hat{z} - \lambda}, \tag{9.40}$$

with λ real. On the unit circle, the above transformation reduces to

$$e^{-j\omega} = \frac{e^{-j\hat{\omega}} - \lambda}{1 - \lambda e^{-j\hat{\omega}}},$$

from which we arrive at

$$\tan\left(\frac{\omega}{2}\right) = \left(\frac{1+\lambda}{1-\lambda}\right) \tan\left(\frac{\hat{\omega}}{2}\right). \tag{9.41}$$

A plot of the relation between ω and $\hat{\omega}$ is given in Figure 9.9 for three values of λ. Note that the mapping is nonlinear except for $\lambda = 0$, resulting in a warping of the frequency scale for nonzero values of λ. However, if $G_L(z)$ is a piecewise constant lowpass magnitude response, then the transformed filter $G_D(\hat{z})$ will likewise have a similar piecewise constant lowpass magnitude response due to the monotonicity of the transformation of Eq. (9.41). The relation between the cutoff frequency ω_c of $G_L(z)$ with the cutoff frequency $\hat{\omega}_c$ of $G_D(\hat{z})$ follows from Eq. (9.41):

$$\tan\left(\frac{\omega_c}{2}\right) = \left(\frac{1+\lambda}{1-\lambda}\right) \tan\left(\frac{\hat{\omega}_c}{2}\right),$$

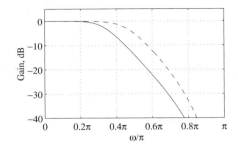

Figure 9.10: Gain responses of the prototype lowpass filter (solid line) and the transformed lowpass filter (dashed line) of Example 9.6.

which can be solved for λ yielding:

$$\lambda = \frac{\tan(\omega_c/2) - \tan(\hat{\omega}_c/2)}{\tan(\omega_c/2) + \tan(\hat{\omega}_c/2)} = \frac{\sin\left(\frac{\omega_c - \hat{\omega}_c}{2}\right)}{\sin\left(\frac{\omega_c + \hat{\omega}_c}{2}\right)}. \tag{9.42}$$

EXAMPLE 9.6   Changing the Passband Edge Frequency of a Lowpass IIR Digital Filter

Consider the third-order lowpass digital transfer function $G(z)$ of Eq. (9.35), which has a passband from dc to 0.25π with a 0.5-dB ripple, and a stopband from 0.55π to π with an attenuation greater than 15 dB.

We redesign this lowpass filter by applying the lowpass-to-lowpass transformation of Eq. (9 40) so that its passband edge moves from 0.25π to 0.35π. Here we have $\omega_c = 0.25\pi$ and $\hat{\omega}_c = 0.35\pi$. Substituting these values in Eq. (9.42), we get

$$\lambda = \frac{\sin\left(\frac{0.25\pi - 0.35\pi}{2}\right)}{\sin\left(\frac{0.25\pi + 0.35\pi}{2}\right)} = -\frac{\sin(0.05\pi)}{\sin(0.3\pi)} = -0.1933636.$$

Thus, the desired lowpass transfer function $G_D(\hat{z})$ is given by

$$G_D(\hat{z}) = G(z)\Big|_{z^{-1} = \frac{\hat{z}^{-1} + 0.1933636}{1 + 0.1933636\hat{z}^{-1}}}$$

$$= \frac{0.13402309(1 + \hat{z}^{-1})^3}{(1 - 0.0694472\hat{z}^{-1})(1 - 0.1848053\hat{z}^{-1} + 0.337566\hat{z}^{-2})}. \tag{9.43}$$

A plot of the gain responses of the original lowpass filter $G(z)$ of Eq. (9.35) and the new lowpass filter $G_D(\hat{z})$ of Eq. (9.43) is shown in Figure 9.10. Note from Figure 9.9 that the mapping ensures that the transformed lowpass transfer function $G_D(\hat{z})$ will have a 0-dB dc gain.

The M-file `allpasslp2lp` can be used to determine the allpass function needed for the lowpass-to-lowpass spectral transformation. The basic form of this function is

```
[AllpassNum, AllpassDen] = allpasslp2lp(wold, wnew)
```

where `wold` is the specified angular bandedge frequency of the original lowpass filter and `wnew` is the desired angular bandedge frequency of the transformed filter.

It should be noted that the lowpass-to-lowpass transformation can also be used to transform a highpass filter with a cutoff at ω_c to another highpass filter with a cutoff at $\hat{\omega}_c$ (Problem 9.25), a bandpass filter with a center frequency at ω_o to another bandpass filter with a center frequency at $\hat{\omega}_o$ (Problem 9.26), and

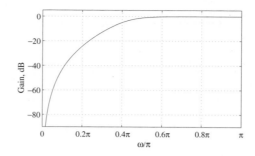

Figure 9.11: Gain response of the highpass filter of Example 9.7.

a bandstop filter with a center frequency at ω_o to another bandstop filter with a center frequency at $\hat{\omega}_o$ (Problem 9.27).

9.5.2 Other Transformations

Table 9.1 lists other useful transformations such as the lowpass-to-highpass, lowpass-to-bandpass, and lowpass-to-bandstop transformations, in addition to the lowpass-to-lowpass transformation discussed previously. It should be noted that these spectral transformations can be used only to map one frequency point ω_c in the magnitude response of the lowpass prototype filter into a new position $\hat{\omega}_c$; with the same magnitude response value for the transformed lowpass and highpass filters; or into two new positions, $\hat{\omega}_{c1}$ and $\hat{\omega}_{c2}$, with the same magnitude response values for the transformed bandpass and bandstop filters. Hence, it is possible only to map either the passband edge or the stopband edge of the lowpass prototype filter onto the desired position(s), but not both.

EXAMPLE 9.7   **Design of a Highpass IIR Digital Filter from a Lowpass IIR Digital Filter**

We now consider the design of a highpass filter by applying a spectral transformation to the third-order lowpass digital transfer function $G(z)$ of Eq. (9.35). The desired passband edge $\hat{\omega}_c$ of the highpass filter is 0.55π, whereas the passband edge of the prototype lowpass filter is at $\omega_c = 0.25\pi$. Substituting these values in the lowpass-to-highpass transformation given in Table 9.1, we arrive at

$$\lambda = -\frac{\cos\left(\frac{0.55\pi + 0.25\pi}{2}\right)}{\cos\left(\frac{0.55\pi - 0.25\pi}{2}\right)} = -\frac{\cos(0.4\pi)}{\cos(0.15\pi)} = -0.3468179.$$

Thus, from Table 9.1, the desired lowpass-to-highpass transformation is given by

$$z^{-1} = -\frac{\hat{z}^{-1} - 0.3468179}{1 - 0.3468179\hat{z}^{-1}}.$$

Using the above transformation, we obtain from Eqs. (9.35) and (9.37) the desired highpass transfer function

$$G_D(\hat{z}) = G(z)\big|_{z^{-1} = -\frac{\hat{z}^{-1} - 0.3468179}{1 - 0.3468179\hat{z}^{-1}}}$$

$$= \frac{0.218791(1 - \hat{z}^{-1})^3}{(1 - 0.0961359188872\hat{z}^{-1})(1 - 0.255685283\hat{z}^{-1} + 0.341493\hat{z}^{-2})}.$$

The gain response of the above transfer function has been plotted in Figure 9.11.

Table 9.1: Spectral transformations of a lowpass filter with a cutoff frequency ω_c.

Filter type	Spectral transformation	Design parameters
Lowpass	$z^{-1} = \dfrac{\hat{z}^{-1} - \lambda}{1 - \lambda \hat{z}^{-1}}$	$\lambda = \dfrac{\sin\left(\frac{\omega_c - \hat{\omega}_c}{2}\right)}{\sin\left(\frac{\omega_c + \hat{\omega}_c}{2}\right)}$ $\hat{\omega}_c$ = desired cutoff frequency
Highpass	$z^{-1} = -\dfrac{\hat{z}^{-1} + \lambda}{1 + \lambda \hat{z}^{-1}}$	$\lambda = -\dfrac{\cos\left(\frac{\omega_c + \hat{\omega}_c}{2}\right)}{\cos\left(\frac{\omega_c - \hat{\omega}_c}{2}\right)}$ $\hat{\omega}_c$ = desired cutoff frequency
Bandpass	$z^{-1} = -\dfrac{\hat{z}^{-2} - \frac{2\lambda\rho}{\rho+1}\hat{z}^{-1} + \frac{\rho-1}{\rho+1}}{\frac{\rho-1}{\rho+1}\hat{z}^{-2} - \frac{2\lambda\rho}{\rho+1}\hat{z}^{-1} + 1}$	$\lambda = \dfrac{\cos\left(\frac{\hat{\omega}_{c2} + \hat{\omega}_{c1}}{2}\right)}{\cos\left(\frac{\hat{\omega}_{c2} - \hat{\omega}_{c1}}{2}\right)}$ $\rho = \cot\left(\dfrac{\hat{\omega}_{c2} - \hat{\omega}_{c1}}{2}\right)\tan\left(\dfrac{\omega_c}{2}\right)$ $\hat{\omega}_{c2}, \hat{\omega}_{c1}$ = desired upper and lower cutoff frequencies
Bandstop	$z^{-1} = \dfrac{\hat{z}^{-2} - \frac{2\lambda}{1+\rho}\hat{z}^{-1} + \frac{1-\rho}{1+\rho}}{\frac{1-\rho}{1+\rho}\hat{z}^{-2} - \frac{2\lambda}{1+\rho}\hat{z}^{-1} + 1}$	$\lambda = \dfrac{\cos\left(\frac{\hat{\omega}_{c2} + \hat{\omega}_{c1}}{2}\right)}{\cos\left(\frac{\hat{\omega}_{c2} - \hat{\omega}_{c1}}{2}\right)}$ $\rho = \tan\left(\dfrac{\hat{\omega}_{c2} - \hat{\omega}_{c1}}{2}\right)\tan\left(\dfrac{\omega_c}{2}\right)$ $\hat{\omega}_{c2}, \hat{\omega}_{c1}$ = desired upper and lower cutoff frequencies

The M-file `allpasslp2hp` can be used to determine the allpass function needed for the lowpass-to-highpass spectral transformation. The basic form of this function is

```
[AllpassNum, AllpassDen] = allpasslp2hp(wold, wnew)
```

where `wold` is the specified angular bandedge frequency of the original lowpass filter and `wnew` is the desired angular bandedge frequency of the highpass filter.

For determining the allpass functions for the lowpass-to-bandpass and lowpass-to-bandstop spectral transformations, one can use the M-files `allpasslp2bp` and `allpasslp2bs`, respectively. ,

The lowpass-to-bandpass transformation given in Table 9.1 can be simplified if we consider the case when the bandwidth of the passband for the prototype lowpass filter is the same as that of the transformed bandpass filter; that is, $\omega_c = \hat{\omega}_{c2} - \hat{\omega}_{c1}$. Applying this constraint to the respective spectral transformation in Table 9.1, we observe that $\rho = 1$, and as a result, the modified spectral transformation is given by

$$z^{-1} = -\hat{z}^{-1} \frac{\hat{z}^{-1} - \lambda}{1 - \lambda \hat{z}^{-1}}. \tag{9.44}$$

The parameter λ is determined from the desired location of the center frequency $\hat{\omega}_o$ of the bandpass filter derived via the transformation of Eq. (9.44), which maps the zero frequency of the lowpass filter; that is, $\omega = 0$, to $\hat{\omega}_o$. From Eq. (9.44), we get

$$e^{-j\omega} = -e^{-j\hat{\omega}} \frac{e^{-j\hat{\omega}} - \lambda}{1 - \lambda e^{-j\hat{\omega}}}. \tag{9.45}$$

Substituting $\omega = 0$ and $\hat{\omega} = \hat{\omega}_o$ in Eq. (9.45), we arrive at

$$\lambda = \cos \hat{\omega}_o. \tag{9.46}$$

EXAMPLE 9.8   Development of a Second-Order Bandpass IIR Filter from a Lowpass IIR Filter

We develop the transfer function of a second-order bandpass filter by applying the transformation of Eq. (9.44) to the first-order lowpass transfer function of Eq. (9.21). The desired transfer function is thus given by

$$G_D(\hat{z}) = G(z)|_{z^{-1} = -\hat{z}^{-1}(\hat{z}^{-1} - \lambda)/(1 - \lambda \hat{z}^{-1})}$$

$$= \frac{1 - \alpha}{2} \left[\frac{1 - \hat{z}^{-2}}{1 - \lambda(1 + \alpha)\hat{z}^{-1} + \alpha \hat{z}^{-2}} \right], \tag{9.47}$$

whose passband center frequency is given by $\lambda = \cos \hat{\omega}_o$, and the 3-dB passband bandwidth $\hat{B}_w$ is given by $\alpha = [1 - \tan(\hat{B}_w)]/[1 + \tan(\hat{B}_w)]$. It should be noted that the above bandpass transfer function is precisely the same as in Eq. (7.77) if we replace λ with β, where it was introduced without any derivation. The relation between the transfer function parameters λ and α with the passband center frequency $\hat{\omega}_o$ and the 3-dB bandwidth $\hat{B}_w$ given above are identical to those given in Eqs. (7.76) and (7.78), respectively.

It should be noted that the lowpass-to-highpass transformation can also be used to transform a highpass filter with a cutoff at ω_c to a lowpass filter with a cutoff at $\hat{\omega}_c$ (Problem 9.28).

9.5.3 Spectral Transformation Using MATLAB

The M-files `iirlp2lp`, `iirlp2hp`, `iirlp2bp`, and `iirlp2bs` of MATLAB can be used to carry out the desired spectral transformations. We illustrate the use of the first three functions in Examples 9.9 through 9.11.

EXAMPLE 9.9   Illustration of Lowpass-to-Lowpass Transformation

To repeat Example 9.6 using the function `iirlp2lp` one can use the code fragments

```
b = 0.0662272*[1 3 3 1];
a = conv([1 -0.2593284],[1 -0.6762858 0.3917468]);
[num,den,allpassnum,allpassden] = iirlp2lp(b,a,0.25,0.35)
```

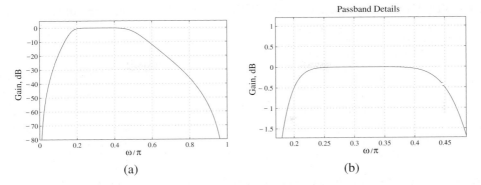

Figure 9.12: (a) Gain response of the bandpass filter and (b) passband details.

It can be shown that the allpass transfer function needed to carry out the spectral transformation has a numerator and a denominator that are the same as that given in Example 9.6. Likewise, the numerator and the denominator coefficients of the transformed filter obtained using the above program are also identical to those derived in Example 9.6.

EXAMPLE 9.10   Illustration of Lowpass-to-Highpass Transformation

To repeat Example 9.7 using the function `iirlp2hp` one can use the code fragments

```
b = 0.0662272*[1 3 3 1];
a = conv([1 -0.2593284],[1 -0.6762858 0.3917468]);
[num,den,allpassnum,allpassden]=iirlp2hp(b,a,0.25,0.55)
```

It can be shown that the allpass transfer function obtained using the above program is the same as that given in Example 9.7. Likewise, the numerator and the denominator coefficients of the highpass filter obtained using the above program are also identical to those derived in Example 9.7.

EXAMPLE 9.11   Illustration of Lowpass-to-Bandpass Transformation

We consider the design of a bandpass filter with passband edges at 0.2π and 0.45π by applying a lowpass-to-bandpass transformation to the lowpass filter $G(z)$ of Eq. (9.35). To this end, the code fragments used are

```
b = 0.0662272*[1 3 3 1];
a = conv([1 -0.2593284],[1 -0.6762858 0.3917468]);
[num,den,allpassnum,allpassden]=iirlp2bp(b,a,0.25,[0.2 0.45]);
```

A plot of the gain response of the bandpass filter is shown in Figure 9.12. The allpass function used for the spectral transformation can be found by printing `allpassnum` and `allpassden`. The numerator and denominator coefficient of the transfer function of the bandpass filter are given in num and den, respectively.

9.6   IIR Digital Filter Design Using MATLAB

The *Signal Processing Toolbox* of MATLAB includes a variety of M-files for the design of both IIR and FIR digital filters. We illustrate the use of some of these functions in this section.

The IIR digital filter design process involves two steps. In the first step, the filter order N and the frequency scaling factor Wn are determined from the given specifications. Using these parameters and the specified ripples, the coefficients of the transfer function are then determined in the next step. We describe the MATLAB implementation of these two steps below.

Order Estimation

For the estimation of the order of the IIR digital filter to be designed using the bilinear transformation method, the M-files to be used are as follows: buttord for the Butterworth filters, cheb1ord for the Type 1 Chebyshev filters, cheb2ord for the Type 2 Chebyshev filters, and ellipord for the elliptic filters. The use of the M-files for order estimation is illustrated in Examples 9.12 and 9.13.

EXAMPLE 9.12   Minimum Order of a Type 2 Chebyshev Highpass IIR Digital Filter

Determine the minimum order of the transfer function of a Type 2 Chebyshev digital highpass filter operating at a sampling rate of 4 kHz with the following specifications: passband edge at 1000 Hz, stopband edge at 600 Hz, passband ripple of 1 dB, and minimum stopband attenuation of 40 dB.

The normalized passband edge Wp is $2 \times 1000/4000 = 0.5$ and the normalized stopband edge Ws is $2 \times 600/4000 = 0.3$. For the design using the bilinear transformation method, we use the statement

```
[N,Wn] = cheb2ord(0.5,0.3,1,40);
```

which yields N = 5 and Wn = 0.3224.

EXAMPLE 9.13   Minimum Order of an Elliptic Bandpass IIR Digital Filter

Determine the minimum order of an elliptic bandpass filter operating at a sampling rate of 1600 Hz with the following specifications: passband edges at 200 Hz and 280 Hz, stopband edges at 160 Hz and 300 Hz, passband ripple of 0.1 dB, and minimum stopband attenuation of 70 dB.

The normalized passband edges are $2(200)/1600 = 0.25$ and $2(280)/1600 = 0.35$. Hence, Wp = [0.25 0.35]. Likewise, the normalized stopband edges are at $2(160)/1600 = 0.2$ and $2(300)/1600 = 0.375$, implying Ws = [0.2   0.375]. For the design using the bilinear transformation method, we use the statement

```
[N,Wn] = ellipord([0.25 0.35], [0.2 0.375],0.1, 70);
```

to determine the minimum order of the filter. Execution of the above statement results in N = 8 and Wn = [0.25 0.35]. Note that the order N = 8 here is the order of the prototype lowpass filter. The order of the bandpass filter is $2N = 16$.

Filter Design

For IIR filter design based on the bilinear transformation, the *Signal Processing Toolbox* of MATLAB includes functions for each one of the four magnitude approximation techniques. Specifically, the following M-files are available: butter for the design of butterworth filters, cheby1 for the design of Type 1 Chebyshev filters, cheby2 for the design of Type 2 Chebyshev filters, and ellip for the design of elliptic filters. The output files could be either the numerator and denominator coefficient vectors or the vector of zeros, the vector of poles, and the scalar gain factor. The numerator and denominator coefficients of the transfer function can be determined from the latter data using the function zp2tf. Alternately, the function zp2sos can be used to find the second-order factors of the numerator and the denominator of the transfer function.[6]

---

[6]It is preferable to use the function zp2sos which avoids the numerical problems of convolution to form the transfer function.

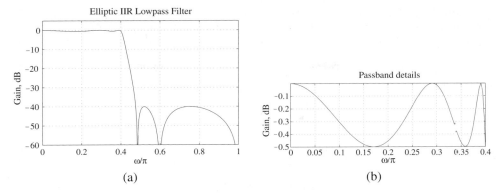

Figure 9.13: IIR elliptic lowpass filter of Example 9.14: (a) gain response and (b) passband details.

After the transfer function has been computed, the frequency response can be computed using the M-file `freqz`.

We illustrate the design of a digital lowpass filter using MATLAB in Example 9.14.

EXAMPLE 9.14   Design of an Elliptic IIR Lowpass Filter Using MATLAB

Determine the transfer function and plot the gain response of an elliptic IIR lowpass filter with the following specifications: passband edge $F_p = 800$ Hz, stopband edge $F_s = 1$ kHz, passband ripple of 0.5 dB, minimum stopband attenuation of 40 dB, and sampling rate $F_T = 4$ kHz.

From Eqs. (9.7) and (9.8), we arrive at the normalized bandedges given by $\omega_p = 2\pi F_p/F_T = 0.4\pi$ and $\omega_s = 2\pi F_s/F_T = 0.5\pi$. Program 9_1 can be used to design the above filter. As the program is run, it asks for the filter specifications to be typed in. It first computes the minimum order N of the filter and the desired cutoff frequency Wn necessary to meet the given specifications. In the elliptic filter case, Wn = Wp = 0.4. The gain response plot generated is shown in Figure 9.13. The numerator and the denominator coefficients of the transfer function can be obtained by displaying b and a.

Other types of digital filters can be designed by simple modifications of the filter function commands given in Program 9_1. Examples 9.15 and 9.16 illustrate the design of a highpass and a bandpass digital filter, respectively.

Program 9_1.m

EXAMPLE 9.15   Design of a Type 1 Chebyshev IIR Highpass Filter Using MATLAB

We repeat the design of the Type 1 Chebyshev IIR highpass filter of Example 9.3. Program 9_2 can be used to design such a filter. During execution, the program asks for the filter specifications. It first determines the minimum filter order N and the cutoff frequency Wn of the highpass filter needed to meet the specifications. In this case, Wn = Wp = 0.7.

The gain response generated by this program is the same as that shown in Figure 9.6. The numerator and the denominator coefficients of the transfer function are given by the vectors b and a, respectively.

Program 9_2.m

EXAMPLE 9.16   Design of a Butterworth IIR Bandpass Filter Using MATLAB

We repeat the design of the IIR Butterworth bandpass filter of Example 9.4. The MATLAB program for designing such a filter is Program 9_3. The input data are the vector of passband edges Wp = [0.45 0.65], the vector of stopband edges Ws = [0.3 0.75], the passband ripple Rp = 1, and the minimum stopband attenuation Rs = 40. The gain response generated by this program is the same as that shown in Figure 9.7. The transfer function

Program 9_3.m

coefficients can be obtained by displaying the numerator coefficient vector b and the denominator coefficient vector a. It should be noted that the parameter N used here is half the filter order. The filter designed above has N = 6; that is, it is a 12th-order bandpass filter.

For the design of higher-order filters, the functions computing the zeros and poles of the transfer function are more accurate than the functions computing the transfer function coefficients. However, the accuracy of the transfer function can be improved by forming second-order sections directly from the zeros and poles using the M-file `zp2sos`.

9.7 Computer-Aided Design of IIR Digital Filters

The IIR digital filter design algorithms described in Sections 9.2, 9.4, and 9.5 are based on the design of a prototype analog filter followed by its transformation to an IIR digital filter. These algorithms are used in applications requiring filters with a frequency-selective magnitude response with a lowpass, highpass, bandpass, or bandstop characteristics. In applications requiring IIR digital filters with other types of frequency responses, filter design algorithms rely on some type of iterative optimization techniques that are used to minimize the error between the desired frequency response and that of the computer-generated filter. In this section, we first review the basic idea behind the computer-based iterative design techniques and then outline a specific application for the group delay equalization of IIR digital filters.

9.7.1　Basic Idea

Let $H(e^{j\omega})$ denote the frequency response of the digital transfer function $H(z)$ to be designed so that it approximates the desired frequency response $D(e^{j\omega})$, given as a piecewise linear function of ω, in some sense. The objective is to determine iteratively the coefficients of the transfer function so that the difference between $H(e^{j\omega})$ and $D(e^{j\omega})$ for all values of ω over closed subintervals of $0 \leq \omega \leq \pi$ is minimized. Usually, this difference is specified as a weighted error function $\mathcal{E}(\omega)$, given by

$$\mathcal{E}(\omega) = W(e^{j\omega}) \left[H(e^{j\omega}) - D(e^{j\omega}) \right], \tag{9.48}$$

where $W(e^{j\omega})$ is some user-specified positive weighting function.

A commonly used approximation measure, called the *Chebyshev* or *minimax criterion*, is to minimize the peak absolute value of the weighted error $\mathcal{E}(\omega)$,

$$\varepsilon = \max_{\omega \in R} |\mathcal{E}(\omega)|, \tag{9.49}$$

where R is the set of disjoint frequency bands in the range $0 \leq \omega \leq \pi$, on which the desired frequency response is defined. In filtering applications, R is composed of the passbands and stopbands of the filter to be designed. For example, for a lowpass filter design, R is the disjoint union of the frequency ranges $[0, \omega_p]$ and $[\omega_s, \pi]$, where ω_p and ω_s are, respectively, the passband edge and the stopband edge.

A second approximation measure, called the *least-p criterion*, is to minimize the integral of pth power of the weighted error function $\mathcal{E}(\omega)$:

$$\varepsilon = \int_{\omega \in R} \left| W(e^{j\omega}) \left(H(e^{j\omega}) - D(e^{j\omega}) \right) \right|^p d\omega, \tag{9.50}$$

over the specified frequency range R with p a positive integer. The *least-squares criterion* obtained from Eq. (9.50) with $p = 2$ is used often for simplicity. If the weighting function $W(e^{j\omega})$ is 1 over the frequency range $[0, \pi]$, we shall show in Section 10.2.1 that the FIR filter designed by simply truncating the Fourier series representation of the desired amplitude response $D(e^{j\omega})$ has the least integral-squared error. However, the resulting FIR filter exhibits large peak errors near the bandedges due to the Gibbs phenomenon.[7] Hence, $W(e^{j\omega}) = 1$ is usually not used.

It can be shown that as $p \to \infty$, the least pth solution approaches the minimax solution. In practice, the integral error measure of Eq. (9.50) is approximated by a finite sum given by

$$\varepsilon = \sum_{i=1}^{K} \left| W(e^{j\omega_i}) \left(H(e^{j\omega_i}) - D(e^{j\omega_i}) \right) \right|^p , \tag{9.51}$$

where ω_i, $1 \leq i \leq K$, is a suitably chosen dense grid of digital angular frequencies. The *least-squares criterion* obtained from Eq. (9.51) with $p = 2$ is used often for simplicity.

In the case of IIR filter design, $H(e^{j\omega})$ and $D(e^{j\omega})$ are replaced with their magnitude functions. Moreover, the desired transfer function $H(z)$ is assumed to be a real, rational function of z with fixed orders of the numerator and the denominator polynomials. The adjustable filter parameters are either the coefficients of the numerator and the denominator polynomials or the poles and zeros of the transfer function. The design objective is thus to iteratively adjust the filter parameters so that ε defined by either Eq. (9.49) or Eq. (9.51) is a minimum. The function iirlpnorm designs optimal IIR filters directly in the digital domain by solving this problem.

9.7.2 Group Delay Equalization of IIR Digital Filters

As pointed out in Section 7.2.2, for a distortion-free transmission of an input signal in a prescribed frequency range through a digital filter, the transfer function of the filter should exhibit a unity magnitude response and a linear-phase response in the band of interest. We outline in Sections 10.2 and 10.3 methods for the design FIR digital filter with an exact linear-phase response. However, the IIR digital filter design methods described in Sections 9.2 through 9.5 lead to transfer functions with nonlinear phase responses resulting in group delays that are not constant in the passbands of the filters. Thus, to arrive at a frequency selective IIR digital filter with a constant group delay, a practical approach often followed is to first design an IIR digital filter meeting the magnitude response specifications and then design an allpass section so that the overall group delay of the IIR digital filter in cascade with the allpass section has a constant group delay in the passband. An example of this type of delay equalization was pointed out in Section 7.1.3. The allpass delay equalizer is usually designed using a computer-aided optimization method.

In this section, we outline the basic concepts behind this latter approach [Ko97]. Let $H(z)$ be the transfer function of the IIR digital filter, with a group delay given by $\tau_H(\omega)$. Our objective is to design a stable allpass section with a transfer function

$$A(z) = \prod_{\ell=1}^{M} \frac{d_{2,\ell} + d_{1,\ell}z^{-1} + z^{-2}}{1 + d_{1,\ell}z^{-1} + d_{2,\ell}z^{-2}}, \tag{9.52}$$

with a group delay $\tau_A(\omega)$ so that the overall group delay

$$\tau_{\text{overall}}(\omega) = \tau_H(\omega) + \tau_A(\omega)$$

of the cascaded system is approximately constant in the passband of the filter. Moreover, to guarantee the stability of the allpass section, we need to ensure that the coefficients of the allpass transfer function satisfy the constraints[8]

$$|d_{2,\ell}| < 1, \qquad |d_{1,\ell}| < 1 + d_{2,\ell}, \tag{9.53}$$

---

[7]See Section 10.2.3.
[8]See Section 7.9.1.

for $1 \le \ell \le M$. The allpass delay equalizer design can be formulated as a minimax optimization problem in which we minimize the peak absolute value of the error

$$\mathcal{E}(\omega) = \tau_{\text{overall}}(\omega) - \tau_o, \qquad (9.54)$$

in the passband of the filter. The adjustable parameters in the optimization procedure are the desired delay τ_o and the coefficients $d_{2,\ell}, d_{1,\ell}$ of the allpass transfer function [Cha80].

The M-file `iirgrpdelay` can be used to design the allpass delay equalizer which is availble with several versions. We illustrate its use in Example 9.17.

EXAMPLE 9.17   Delay Equalizer Design Using MATLAB

We design an allpass section of eighth order to equalize group delay in the passband of a fourth-order elliptic lowpass filter with a passband edge at 0.3π, passband ripple of 1 dB and a minimum stopband attenuation of 30 dB. To this end, we make use of Program 9_4. The group delays of the lowpass filter and the overall cascade are shown in Figure 7.8. The numerator and the denominator coefficients of the allpass section are given in `num` and `den`. It can be shown using the statement `poly2rc(den)` that the designed allpass is a stable transfer function as all eight reflection coefficients are of magnitude less than 1.

Program 9_4.m

9.8   Summary

The digital filter design problem is concerned with the development of a suitable transfer function meeting the frequency response specifications, which, in this chapter, is restricted to magnitude (or, equivalently, gain) response specifications. These specifications are usually given in terms of the desired passband edge and stopband edge frequencies and the allowable deviations from the desired passband and stopband magnitude (gain) levels. This chapter considered the design of causal, stable infinite impulse response (IIR) digital filters.

IIR filter design is usually carried out by transforming a prototype analog transfer function by means of a suitable mapping of the complex frequency variable s into the complex variable z. The widely used bilinear transform method, discussed in this chapter, is based on this approach.

The chapter then discusses some of the algorithms for the design of IIR digital filters that are available in the *Signal Processing Toolbox* of MATLAB as functions. In particular, it includes the design of IIR digital filters with Butterworth, Chebyshev, and elliptic magnitude responses.

Finally, the chapter reviews the basic idea behind the design of IIR digital filters using computer-aided iterative techniques and outlines a specific application of this approach to the design of group delay equalizers.

9.9   Problems

9.1 Determine the peak ripple values δ_p and δ_s for each of the following sets of peak passband ripple α_p and minimum stopband attenuation α_s:

    (a) $\alpha_p = 0.21$ dB, $\alpha_s = 53$ dB,   (b) $\alpha_p = 0.17$ dB, $\alpha_s = 78$ dB.

9.2 Determine the peak passband ripple α_p and minimum stopband attenuation α_s in dB for each of the following sets of peak ripple values δ_p and δ_s:

    (a) $\delta_p = 0.02$, $\delta_s = 0.03$,   (b) $\delta_p = 0.055$, $\delta_s = 0.033$.

9.3 Let $H(z)$ be the transfer function of a lowpass digital filter with a passband edge at ω_p, stopband edge at ω_s, passband ripple of δ_p, and stopband ripple of δ_s, as indicated in Figure 9.1. Consider a cascade of two identical filters with a transfer function $H(z)$. What are the passband and stopband ripples of the cascade at ω_p and ω_s, respectively? Generalize the results for a cascade of M identical sections.

9.4 Let $H_{LP}(z)$ denote the transfer function of a real-coefficient lowpass filter with a passband edge at ω_p, stopband edge at ω_s, passband ripple of δ_p, and stopband ripple of δ_s, as indicated in Figure 9.1. Sketch the magnitude response of the highpass transfer function $H_{LP}(-z)$ for $-\pi \le \omega < \pi$, and determine its passband and stopband edges in terms of ω_p and ω_s.

9.5 Consider the transfer function $G(z) = H_{LP}(e^{j\omega_o}z)$, where $H_{LP}(z)$ is the lowpass transfer function of Problem 9.4. Sketch its magnitude response for $-\pi \le \omega \le \pi$, and determine its passband and stopband edge frequencies in terms of ω_p, ω_s, and ω_o.

9.6 The impulse invariance method is another approach to the design of a causal IIR digital filter $G(z)$ based on the transformation of a prototype causal analog transfer function $H_a(s)$. If $h_a(t)$ is the impulse response of $H_a(s)$, in the impulse invariance method, we require that the unit sample response $g[n]$ of $G(z)$ be given by the sampled version of $h_a(t)$ sampled at uniform intervals of T seconds; that is,

$$g[n] = h_a(nT), \qquad n = 0, 1, 2, \ldots.$$

(a) Show that $G(z)$ and $H_a(s)$ are related through

$$G(z) = \mathcal{Z}\{g[n]\} = \mathcal{Z}\{h_a(nT)\}$$
$$= \frac{1}{T} \sum_{k=-\infty}^{\infty} H_a\left(s + j\frac{2\pi k}{T}\right)\Bigg|_{s=(1/T)\ln z}. \qquad (9.55)$$

(b) Show that the transformation

$$s = \frac{1}{T}\ln z, \qquad (9.56)$$

has the desirable properties enumerated in Section 9.1.3.

(c) Develop the condition under which the frequency response $G(e^{j\omega})$ of $G(z)$ will be a scaled replica of the frequency response $H_a(j\Omega)$ of $H_a(s)$.

(d) Show that the normalized digital angular frequency ω is related to the analog angular frequency Ω as

$$\omega = \Omega T. \qquad (9.57)$$

9.7 Show that the digital transfer function $G(z)$ obtained from an arbitrary rational analog transfer function $H_a(s)$ with simple poles via the impulse invariance method is given by

$$G(z) = \sum_{\substack{\text{all poles of} \\ H_a(s)}} \text{Residues}\left[\frac{H_a(s)}{1 - e^{sT}z^{-1}}\right]. \qquad (9.58)$$

9.8 Using Eq. (9.58), develop the expression for the causal digital transfer function $G(z)$ obtained from the causal analog transfer function $H(s) = A/(s + \alpha)$ via the impulse invariance method.

9.9 Determine the digital transfer functions obtained by transforming the following causal analog transfer functions using the impulse invariance method. Assume $T = 0.3$ sec.

(a) $H_a(s) = \dfrac{4(3s + 7)}{(s + 2)(s^2 + 4s + 5)}$, (b) $H_b(s) = \dfrac{8s^2 + 37s + 56}{(s^2 + 2s + 10)(s + 4)}$, (c) $H_c(s) = \dfrac{s^3 + s^2 + 6s - 14}{(s^2 + 2s + 5)(s^2 - s + 4)}$.

9.10 The following causal IIR digital transfer functions were designed using the impulse invariance method with $T = 0.2$ sec. Determine their respective parent causal analog transfer functions.

$$\text{(a) } G_a(z) = \frac{3z}{z - e^{-1.5}} + \frac{4z}{z - e^{-1.8}}, \quad \text{(b) } G_b(z) = \frac{z\,e^{-1.2}\sin(1.5)}{z^2 - 2z\,e^{-1.2}\cos(1.5) + e^{-2.4}}.$$

9.11 The following causal IIR digital transfer functions were designed using the bilinear transformation method with $T = 0.5$. Determine their respective parent causal analog transfer functions.

$$\text{(a) } G_a(z) = \frac{4(z^2 + 3z + 4)}{10z^2 + 4z + 6}, \quad \text{(b) } G_b(z) = \frac{54z^3 + 62z^2 + 26z + 18}{(3z + 1)(12z^2 - 4z + 8)}.$$

9.12 An IIR digital lowpass filter is to be designed by transforming an analog lowpass filter with a passband edge frequency F_p at 0.45 kHz using the impulse invariance method with $T = 0.3$ ms. What is the normalized passband edge angular frequency ω_p of the digital filter if there is no aliasing? What would be the normalized passband edge angular frequency ω_p of the digital filter if it is designed using the bilinear transformation with $T = 0.3$ ms?

9.13 An IIR lowpass digital filter has a normalized passband edge frequency $\omega = 0.56\pi$. What is the passband edge frequency in Hz of the prototype analog lowpass filter if the digital filter has been designed using the impulse invariance method with $T = 0.2$ ms? What is the passband edge frequency in Hz of the prototype analog lowpass filter if the digital filter has been designed using the bilinear transformation method with $T = 0.2$ ms?

9.14 Design an IIR lowpass digital filter $G(z)$ with a maximally flat magnitude response and meeting the specifications given by Eqs. (9.34a) and (9.34b) using the impulse invariance method. How does this filter compare with that designed via the bilinear transformation method in Section 9.2?

9.15 An LTI continuous-time system described by a linear constant coefficient differential equation is often solved numerically by developing an equivalent linear constant coefficient difference equation by replacing the derivative operators in the differential equation by their approximate difference equation representation. A commonly used difference equation representation of the first derivative at time $t = nT$ is given by

$$\left.\frac{d\,y(t)}{dt}\right|_{t=nT} \cong \frac{1}{T}(y[n] - y[n-1]),$$

where T is the sampling period and $y[n] = y(nT)$. The corresponding mapping from the s-domain to the z-domain is obtained by replacing s with the *backward difference operator* $\frac{1}{T}(1 - z^{-1})$. Investigate the above mapping and its properties. Does a stable $H_a(s)$ result in a stable $H(z)$? How useful is this mapping for digital filter design?

9.16 This problem illustrates how aliasing can be suitably exploited in order to realize interesting frequency response characteristics. An ideal causal analog lowpass filter with an impulse response $h_a(t)$ has a frequency response given by

$$H_a(j\Omega) = \begin{cases} 1, & |\Omega| < \Omega_c, \\ 0, & \text{otherwise.} \end{cases}$$

Let $H_1(e^{j\omega})$ and $H_2(e^{j\omega})$ be the frequency responses of digital filters obtained by sampling $h_a(t)$ at $t = nT$, where $T = 3\pi/2\Omega_c$ and π/Ω_c, respectively. Assume the transfer functions are later normalized so that $H_1(e^{j0}) = H_2(e^{j0}) = 1$.

(a) Sketch the frequency responses $G_1(e^{j\omega})$ and $G_2(e^{j\omega})$ of the two digital filter structures shown in Figure P9.1.

(b) What type of filters are $G_1(z)$ and $G_2(z)$ (lowpass, highpass, etc.)?

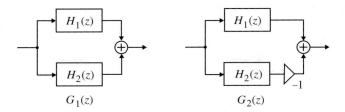

Figure P9.1

9.17 Let $H_a(s)$ be a real-coefficient causal and stable analog transfer function with a magnitude response bounded above by unity. Show that the digital transfer function $G(z)$ obtained by a bilinear transformation of $H_a(s)$ is a BR function.

9.18 We have shown in Section 8.7.2 that the transfer function $G(z)$ of a second-order IIR notch filter as given in Eq. (9.32) can be expressed in the form $G(z) = \frac{1}{2}[1 + A_2(z)]$, where $A_2(z)$ is a second-order allpass transfer function given by Eq. (8.66). Consider a notch filter with a notch frequency at $\omega = \pi/2$. Show that a notch filter with multiple notch frequencies is obtained if z^{-1} is replaced with z^{-N} [Reg88]. What are the locations of the new notch frequencies?

9.19 A notch filter with N notch frequencies can be realized by replacing the allpass filter $A_2(z)$ in Problem 9.18 with a cascade of N second-order allpass filters [Jos99]. In this problem, we consider the design of a notch filter with two notch frequencies ω_1, ω_2, and corresponding 3-dB notch bandwidths B_1, B_2. We thus replace $A_2(z)$ with a fourth-order allpass transfer function $A_4(z)$,

$$A_4(z) = \left(\frac{\alpha_1 - \beta_1(1+\alpha_1)z^{-1} + z^{-2}}{1 - \beta_1(1+\alpha_1)z^{-1} + \alpha_1 z^{-2}} \right) \left(\frac{\alpha_2 - \beta_2(1+\alpha_2)z^{-1} + z^{-2}}{1 - \beta_2(1+\alpha_2)z^{-1} + \alpha_2 z^{-2}} \right),$$

obtained by cascading two second-order allpass filters. The constants α_1 and α_2 are chosen as

$$\alpha_i = \frac{1 - \tan(B_i/2)}{1 + \tan(B_i/2)}, \quad i = 1, 2.$$

The transfer function of the modified structure is now given by $H(z) = \frac{1}{2}[1 + A_4(z)] = N(z)/D(z)$

(a) Show that $N(z)$ is a mirror-image polynomial of the form $a(1 + b_1 z^{-1} + b_2 z^{-2} + b_1 z^{-3} + z^{-4})$, and express the constants b_1 and b_2 in terms of the coefficients of $A_4(z)$.

(b) Show that $a = (1 + \alpha_1\alpha_2)/2$.

(c) By setting $N(e^{j\omega_i}) = 0$, $i = 1, 2$, solve for the constants b_1 and b_2 in terms of ω_1 and ω_2. From the equations in Parts (a) and (b), determine the expressions for the coefficients β_1 and β_2.

(d) Using the design equations derived above, design a double notch filter with the following specifications: $\omega_1 = 0.2\pi$, $\omega_2 = 0.6\pi$, $B_1 = 0.2\pi$, and $B_2 = 0.25\pi$. Using MATLAB, plot the magnitude response of the designed notch filter.

9.20 Let $H_{LP}(z)$ be an IIR lowpass transfer function with a zero (pole) at $z = z_k$. Let $H_D(\hat{z})$ denote the lowpass transfer function obtained by applying the lowpass-to-lowpass transformation given in Table 9.1, which moves the zero (pole) at $z = z_k$ of $H_{LP}(z)$ to a new location at $\hat{z} = \hat{z}_k$. Express $\hat{z}_k$ in terms of z_k. If $H_{LP}(z)$ has a zero at $z = -1$, show that $H_D(\hat{z})$ also has a zero at $z = -1$.

9.21 Let $H_{LP}(z)$ be an IIR lowpass transfer function with a zero (pole) at $z = z_k$. Let $H_D(\hat{z})$ denote the bandpass transfer function obtained by applying the lowpass-to-bandpass transformation given in Table 9.1, which moves the zero (pole) at $z = z_k$ of $H_{LP}(z)$ to a new location at $\hat{z} = \hat{z}_k$. Express $\hat{z}_k$ in terms of z_k. If $H_{LP}(z)$ has a zero at $z = -1$, show that $H_D(\hat{z})$ also has a zero at $z = \pm 1$.

9.22 A second-order lowpass IIR digital filter with a 3-dB cutoff frequency at $\omega_c = 0.55\pi$ has a transfer function

$$G_{LP}(z) = \frac{0.3404(1 + z^{-1})^2}{1 + 0.1842z^{-1} + 0.1776z^{-2}}. \tag{9.59}$$

Design a second-order lowpass filter $H_{LP}(z)$ with a 3-dB cutoff frequency at $\hat{\omega}_c = 0.42\pi$ by transforming the above lowpass transfer function using a lowpass-to-lowpass spectral transformation. Using MATLAB, plot the gain responses of the two lowpass filters on the same figure.

9.23 Design a second-order highpass filter $H_{HP}(z)$ with a 3-dB cutoff frequency at $\hat{\omega}_c = 0.47\pi$ by transforming the lowpass transfer function of Eq. (9.59) using a lowpass-to-highpass spectral transformation. Using MATLAB, plot the gain responses of the highpass and the lowpass filters on the same figure.

9.24 A second-order lowpass Type 2 Chebyshev IIR digital filter $G_{LP}(z)$ with a minimum attenuation of 20-dB at $\omega_c = 0.56\pi$ has a transfer function

$$G_{LP}(z) = \frac{0.1944(1 + 0.9802z^{-1} + z^{-2})}{1 - 0.7016z^{-1} + 0.281z^{-2}}. \tag{9.60}$$

Design a fourth-order bandpass filter $H_{BP}(z)$ with a center frequency at $\hat{\omega}_c = 0.41\pi$ by transforming the above lowpass transfer function using a lowpass-to-bandpass spectral transformation. Using MATLAB, plot the gain responses of the lowpass and the bandpass filters on the same figure.

9.25 A third-order elliptic highpass filter with a passband edge at $\omega_p = 0.52\pi$ has a transfer function

$$G_{HP}(z) = \frac{0.2397(1 - 1.5858z^{-1} + 1.5858z^{-2} - z^{-3})}{1 + 0.3272z^{-1} + 0.7459z^{-2} + 0.179z^{-3}}. \tag{9.61}$$

Design a highpass filter $H_{HP}(z)$ with a passband edge at $\omega_p = 0.45\pi$ by transforming the above highpass transfer function using the lowpass-to-lowpass spectral transformation. Using MATLAB, plot the gain responses of the two highpass filters on the same figure.

9.26 Design a second-order bandpass filter with a center frequency at $\omega_o = 0.5\pi$ by transforming the bandpass transfer function of Eq. (7.79) using the lowpass-to-lowpass spectral transformation. Using MATLAB, plot the gain responses of the two bandpass filters on the same figure.

9.27 Design a second-order notch filter operating at a sampling rate of 400 Hz with a notch frequency at 80 Hz and a 3-dB bandwidth of 5 Hz. By applying the lowpass-to-lowpass spectral transformation to this filter, design a notch filter with a notch frequency at 50 Hz. Using MATLAB, plot the gain responses of the two notch filters on the same figure.

9.28 Design a lowpass filter with a cutoff at $\omega_p = 0.45\pi$ by transforming the highpass transfer function of Eq. (9.61) using the lowpass-to-highpass spectral transformation. Using MATLAB, plot the gain responses of the highpass and the lowpass filters on the same figure.

9.29 A maximally flat group delay IIR allpass filter can be designed to approximate a fractional delay z^{-D}:

$$z^{-D} \simeq \frac{d_N + d_{N-1}z^{-1} + \cdots + d_1 z^{-(N-1)} + z^{-N}}{1 + d_1 z^{-1} + d_2 z^{-2} + \cdots + d_{N-1}z^{-(N-1)} + d_N z^{-N}}.$$

By expressing the desired positive delay as $D = N + \delta$, where N is a positive integer and δ a fractional number, it can be shown that the coefficient $\{d_k\}$ of the allpass filter is given by [Fet71]

$$d_k = (-1)^k C_k^N \prod_{n=0}^{N} \frac{D - N + n}{D - N + k + n},$$

where $C_k^N = N!/k!(N-k)!$ is a binomial coefficient. Design an allpass fractional delay filter of order 11 with a delay of 90/13 samples. Plot using MATLAB, the group delay response of the designed filter along with that of the ideal fractional delay filter. Comment on your results.

9.30 The desired frequency response of an ideal *integrator* is given by

$$H_{\text{int}}(e^{j\omega}) = \frac{1}{j\omega}. \tag{9.62}$$

Determine the transfer function $H_R(z)$ of an IIR integrator derived via the rectangular numerical integration method given by Eq. (2.141) and the transfer function $H_T(z)$ of an IIR integrator derived via the trapezoidal numerical integration method given by Eq. (2.119). Using MATLAB, plot the magnitude responses of $H_{\text{int}}(z)$, $H_R(z)$, and $H_T(z)$ for $T = 1$. Comment on your results.

9.31 An improved IIR digital integrator can be obtained by interpolating the rectangular and the trapezoidal integrators according to [Ala93]

$$H_N(z) = \tfrac{3}{4}H_R(z) + \tfrac{1}{4}H_T(z).$$

Using MATLAB, plot the magnitude responses of $H_N(z)$, $H_R(z)$, and $H_T(z)$ for $T = 1$. Comment on your results.

9.32 Develop an IIR digital differentiator by inverting the IIR digital integrator of Problem 9.31 [Ala93]. Is this a stable transfer function? If not, develop a stable equivalent. Using MATLAB, plot the magnitude responses of the ideal differentiator and the digital differentiator designed here. Comment on your results.

9.10 MATLAB Exercises

M 9.1 Design a digital Butterworth lowpass filter operating at a sampling rate of 100 kHz with a 0.4-dB cutoff frequency at 10 kHz and a minimum stopband attenuation of 50 dB at 30 kHz using the bilinear transformation method. Determine the order of the analog filter prototype using the formula given in Eq. (4.35), and then design the analog prototype filter using the M-file `buttap` of MATLAB. Transform the analog filter transfer function to the desired digital transfer function using the M-file `bilinear`. Plot the gain and phase responses using MATLAB. Show all steps used in the design.

M 9.2 Modify Program 9_3 to design a digital Butterworth lowpass filter using the bilinear transformation method. The input data required by the modified program should be the desired passband and stopband edges and maximum passband deviation and the minimum stopband attenuation in dB. Using the modified program, design the digital Butterworth lowpass filter of Exercise M9.1.

M 9.3 Design a digital filter by an impulse invariant transformation of a fifth-order analog Bessel transfer function for the following values of sampling frequencies: (a) $F_T = 1$ Hz and (b) $F_T = 2$ Hz. Plot the gain and the group delay responses of both designs using MATLAB, and compare these responses with that of the original Bessel transfer function. Comment on your results.

Program 9_3.m

M 9.4 Using the M-file `impinvar`, design the digital Butterworth lowpass filter of Exercise M9.1. Use the analog prototype filter order determined using the formula given in Eq. (4.35).

M 9.5 Design a digital Type 1 Chebyshev lowpass filter operating at a sampling rate of 100 kHz with a passband edge frequency at 10 kHz, a passband ripple of 0.4 dB, and a minimum stopband attenuation of 50 dB at 30 kHz using the impulse invariance method and the bilinear transformation method. Determine the order of the analog filter prototype using the formula given in Eq. (4.43), and then design the analog prototype filter using the M-file `cheb1ap` of MATLAB. Transform the analog filter transfer function to the desired digital transfer function using the M-file `bilinear`. Plot the gain and phase responses of both designs using MATLAB. Compare the performances of the two filters. Show all steps used in the design.

M 9.6 Modify Program 9_2 to design a digital Type 1 Chebyshev lowpass filter using the bilinear transformation method. The input data required by the modified program should be the desired passband and stopband edges and maximum passband deviation and the minimum stopband attenuation in dB. Using the modified program, design the digital Type 1 Chebyshev lowpass filter of Exercise M9.5.

Program 9_2.m   **M 9.7** Using the M-file impinvar, write a MATLAB program to design a digital Type 1 Chebyshev lowpass filter using the impulse invariance method. The input data required by the modified program should be the desired passband and stopband edges and maximum passband deviation and the minimum stopband attenuation in dB. Using your program, design the digital Type 1 Chebyshev lowpass filter of Exercise M9.5.

M 9.8 Design a digital elliptic lowpass filter operating at a sampling rate of 100 kHz with a passband edge frequency at 10 kHz, a stopband edge frequency at 30 kHz, passband ripple of 0.4 dB, and a stopband ripple of 50 dB using the impulse invariance method and the bilinear transformation method. Determine the order of the analog filter prototype using the formula given in Eq. (4.54), and then design the analog prototype filter using the M-file ellipap of MATLAB. Transform the analog filter transfer function to the desired digital transfer function using the M-file bilinear. Plot the gain and phase responses of both designs using MATLAB. Compare the performances of the two filters. Show all steps used in the design.

Program 9_3.m   **M 9.9** Modify Program 9_3 to design a digital elliptic lowpass filter using the bilinear transformation method. The input data required by the modified program should be the desired passband and stopband edges and the maximum passband deviation and the minimum stopband attenuation in dB. Using the modified program, design the digital elliptic lowpass filter of Exercise M9.8.

M 9.10 Using the bilinear transformation method, design a digital Butterworth highpass filter operating at a sampling rate of 1.5 MHz with the following specifications: passband edge at 600 kHz, stopband edge at 210 kHz, peak passband ripple of 0.4 dB, and minimum stopband attenuation of 45 dB. (a) What are the specifications of the analog highpass filter? (b) What are the specifications of the analog prototype lowpass filter? (c) Show all pertinent transfer functions. Plot the gain responses of the prototype analog lowpass filter, analog highpass filter, and desired digital highpass filter. Show all steps.

M 9.11 Using the bilinear transformation method, design a digital Type 1 Chebyshev bandpass filter operating at a sampling rate of 9 kHz with the following specifications: passband edges at 1.2 kHz and 2.2 kHz, stopband edges at 650 Hz and 3 kHz, peak passband ripple of 0.8 dB, and minimum stopband attenuation of 31 dB. (a) What are the specifications of the analog bandpass filter? (b) What are the specifications of the analog prototype lowpass filter? (c) Show all pertinent transfer functions. Plot the gain responses of the prototype analog lowpass filter, the analog bandpass filter, and desired digital bandpass filter. Show all steps.

M 9.12 Using the bilinear transformation method, design a digital elliptic bandstop filter operating at a sampling rate of 8 kHz with the following specifications: passband edges at 0.9 kHz and 2.1 kHz, stopband edges at 0.6 kHz and 3 kHz, peak passband ripple of 1.5 dB, and minimum stopband attenuation of 30 dB. (a) What are the specifications of the analog bandstop filter? (b) What are the specifications of the analog prototype lowpass filter? (c) Show all pertinent transfer functions. Plot the gain responses of the prototype analog lowpass filter, the analog bandstop filter, and desired digital bandstop filter. Show all steps.

M 9.13 Using the M-file iirgrpdelay, design an allpass section to equalize group delay in the passband of the Type 1 Chebyshev IIR highpass filter of Example 9.15.

M 9.14 Using the M-file iirgrpdelay, design an allpass section to equalize group delay in the passband of the Butterworth IIR bandpass filter of Example 9.16.

10 FIR Digital Filter Design

In Chapter 9 we considered the design of IIR digital filters. For such filters, it is also necessary to ensure that the derived transfer function $G(z)$ is stable. On the other hand, in the case of FIR digital filter design, the stability is not a design issue as the transfer function is a polynomial in z^{-1} and is thus always guaranteed stable. In this chapter, we consider the FIR digital filter design problem.

Unlike the IIR digital filter design problem, it is always possible to design FIR digital filters with exact linear-phase. First, we describe a popular approach to the design of FIR digital filters with linear-phase. We then consider the computer-aided design of linear-phase FIR digital filters. To this end, we restrict our discussion to the use of MATLAB in determining the transfer functions. Since the order of the FIR transfer function is usually much higher than that of an IIR transfer function meeting the same frequency response specifications, we outline two methods for the design of computationally efficient FIR digital filters requiring fewer multipliers than a direct form realization. Finally, we present a method of designing a minimum-phase FIR digital filter that leads to a transfer function with smaller group delay than that of a linear-phase equivalent. The minimum-phase FIR digital filter is thus attractive in applications where the linear-phase requirement is not an issue.

10.1 Preliminary Considerations

In this section, we first review some basic approaches to the design of FIR digital filters and the determination of the filter order to meet the prescribed specifications.

10.1.1 Basic Approaches to FIR Digital Filter Design

Unlike IIR digital filter design, FIR filter design does not have any connection with the design of analog filters. The design of FIR filters is therefore based on a direct approximation of the specified magnitude response, with the often added requirement that the phase response be linear. Recall a causal FIR transfer function $H(z)$ of length $N + 1$ is a polynomial in z^{-1} of degree N:

$$H(z) = \sum_{n=0}^{N} h[n]z^{-n}. \tag{10.1}$$

The corresponding frequency response is given by

$$H(e^{j\omega}) = \sum_{n=0}^{N} h[n]e^{-j\omega n}. \tag{10.2}$$

It has been shown in Section 5.3.1 that any finite duration sequence $x[n]$ of length $N + 1$ is completely characterized by $N + 1$ samples of its discrete-time Fourier transform $X(e^{j\omega})$. As a result, the design of an FIR filter of length $N + 1$ can be accomplished by finding either the impulse response sequence $\{h[n]\}$ or $N + 1$ samples of its frequency response $H(e^{j\omega})$. Also, to ensure a linear-phase design, the condition

$$h[n] = \pm h[N - n],$$

must be satisfied. Two direct approaches to the design of FIR filters are the windowed Fourier series approach and the frequency sampling approach. We describe the former approach in Section 10.2. The second approach is treated in Problems 10.31 and 10.32. In Section 10.3, we outline computer-based digital filter design methods.

10.1.2   Estimation of the Filter Order

After the type of the digital filter has been selected, the next step in the filter design process is to estimate the filter order N that is needed to meet the given filter specifications. For reduced computational complexity, the filter order should be the smallest integer greater than or equal to the estimated value.

FIR Digital Filter Order Estimation

For the design of lowpass FIR digital filters, several authors have advanced formulas for estimating the minimum value of the filter order N directly from the digital filter specifications: normalized passband edge angular frequency ω_p, normalized stopband edge angular frequency ω_s, peak passband ripple δ_p, and peak stopband ripple δ_s. We review three such formulas.

Kaiser's Formula.   A rather simple formula developed by Kaiser [Kai74] is given by

$$N \cong \frac{-20 \log_{10}\left(\sqrt{\delta_p \delta_s}\right) - 13}{14.6(\omega_s - \omega_p)/2\pi}. \tag{10.3}$$

We illustrate the application of the above formula in Example 10.1.

EXAMPLE 10.1   FIR Digital Filter Order Estimation Using Kaiser's Formula

Estimate the order of a linear-phase lowpass FIR filter with the following specifications: passband edge $F_p = 1.8$ kHz, stopband edge $F_s = 2$ kHz, peak passband ripple $\alpha_p = 0.1$ dB, minimum stopband attenuation $\alpha_s = 35$ dB, and sampling rate $F_T = 12$ kHz.

From Example 9.1, we get $\delta_p = 0.0114469$ and $\delta_s = 0.01778279$. Substituting these values along with the bandedge frequencies in Eq. (10.3), we get

$$N \cong \frac{-20 \log_{10}(\sqrt{0.00020355796}) - 13}{14.6(2000 - 1800)/12{,}000}$$

$$= \frac{23.913119}{0.2433333} = 98.2730.$$

Since the order N must be an integer, we round up the above value yielding $N = 99$.

As the order is odd, a Type 2 FIR filter can be designed to meet the specifications. A Type 1 FIR filter can be designed by increasing the order by 1 to 100.

Bellanger's Formula.   Another simple formula advanced by Bellanger is given by [Bel81]

$$N \cong -\frac{2 \log_{10}(10 \, \delta_p \delta_s)}{3 \, (\omega_s - \omega_p)/2\pi} - 1. \tag{10.4}$$

Its application is considered in Example 10.2.

EXAMPLE 10.2   FIR Digital Filter Order Estimation Using Bellanger's Formula

Estimate the order of a linear-phase FIR transfer function with the specifications given in Example 10.1 using Eq. (10.4)

Substituting the values of the specified transition bandwidth $(\omega_s - \omega_p)/2\pi = (F_s - F_p)/F_T = 0.017$, and the ripples $\delta_p = 0.0114469$ and $\delta_s = 0.01778279$ in Eq. (10.4), we get

$$N \cong 106.6525.$$

We choose the next higher integer 107 as the filter order.

Hermann's Formula.   The formula due to Hermann et al. [Her73] gives a slightly more accurate value for the order and is given by [1]

$$N \cong \frac{D_\infty(\delta_p, \delta_s) - F(\delta_p, \delta_s) \left[(\omega_s - \omega_p)/2\pi\right]^2}{(\omega_s - \omega_p)/2\pi}, \tag{10.5}$$

where

$$D_\infty(\delta_p, \delta_s) = \left[a_1 (\log_{10} \delta_p)^2 + a_2 (\log_{10} \delta_p) + a_3\right] \log_{10} \delta_s$$
$$- \left[a_4 (\log_{10} \delta_p)^2 + a_5 (\log_{10} \delta_p) + a_6\right], \tag{10.6a}$$

and

$$F(\delta_p, \delta_s) = b_1 + b_2 \left[\log_{10} \delta_p - \log_{10} \delta_s\right], \tag{10.6b}$$

with

$$a_1 = 0.005309, \qquad a_2 = 0.07114, \qquad a_3 = -0.4761, \tag{10.6c}$$

$$a_4 = 0.00266, \qquad a_5 = 0.5941, \qquad a_6 = 0.4278, \tag{10.6d}$$

$$b_1 = 11.01217, \qquad b_2 = 0.51244. \tag{10.6e}$$

The formula given in Eq. (10.5) is valid for $\delta_p \geq \delta_s$. If $\delta_p < \delta_s$, then the filter order formula to be used is obtained by interchanging δ_p and δ_s in Eq. (10.6a) and (10.6b).

For small values of δ_p and δ_s, all of the above formulas provide reasonably close and accurate results. On the other hand, when the values of δ_p and δ_s are large, Eq. (10.5) yields a more accurate value for the order.

EXAMPLE 10.3   FIR Digital Filter Order Estimation Using Hermann's Formula

Estimate the order of a linear-phase FIR transfer function with the specifications given in Example 10.1 using Eq. (10.5).

Substituting the values of the parameters a_i given in Eqs. (10.6c) and (10.6d) and the values of the specified ripples $\delta_p = 0.0114469$ and $\delta_s = 0.01778279$ in Eq. (10.6a), we get $D_\infty(\delta_p, \delta_s) = 1.7553533358$. Next, substituting the values of b_1 and b_2 from Eq. (10.6e) and the ripples in Eq. (10.6b), we get $F(\delta_p, \delta_s) = 10.9141341$.

---

[1] It has been obtained by carrying out a least-squares approximation to N as a function of the bandedges and ripples.

Now, $(\omega_s - \omega_p)/2\pi = (F_s - F_p)/F_T = 0.017$. Finally, using the computed values of $D_\infty(\delta_p, \delta_s)$, $F(\delta_p, \delta_s)$, and $(\omega_s - \omega_p)/2\pi$ in Eq. (10.5), we arrive at

$$N \cong 105.139.$$

We choose the next higher integer 106 as the estimated filter order.

A Comparison of FIR Filter Order Formulas

Note that the filter orders computed in Examples 10.1, 10.2, and 10.3, using Eqs. (10.3), (10.3), and (10.5), respectively, are all different. Each of these three formulas provide only an estimate of the required filter order. The frequency response of the FIR filter designed using this estimated order may or may not meet the given specifications. If the specifications are not met, it is recommended that the filter order be gradually increased until the specifications are met. Estimation of the FIR filter order using MATLAB is discussed in Section 10.5.1.

An important property of each of the above three formulas is that the estimated filter order N of the FIR filter is inversely proportional to the transition band width $(\omega_s - \omega_p)$ and does not depend on the actual location of the transition band. This implies that a sharp cutoff FIR filter with a narrow transition band would be of very high order, whereas an FIR filter with a wide transition band will have a very low order.

Another interesting property of Kaiser's and Bellanger's formulas is that the order depends on the product $\delta_p\delta_s$. This implies that if the values of δ_p and δ_s are interchanged, the order remains the same.

To compare the accuracy of the the above formulas, we estimate using each formula the order of three linear-phase lowpass FIR filters of known order, bandedges, and ripples. The specifications of the three filters are as follows:

Filter No. 1: $\omega_p = 0.10625\pi$, $\omega_s = 0.14375\pi$, $\delta_p = 0.0224$, $\delta_s = 0.000112$.

Filter No. 2: $\omega_p = 0.2075\pi$, $\omega_s = 0.2875\pi$, $\delta_p = 0.017$, $\delta_s = 0.034$.

Filter No. 3: $\omega_p = 0.345\pi$, $\omega_s = 0.575\pi$, $\delta_p = 0.0411$, $\delta_s = 0.0137$.

The results are given in Table 10.1.

Each one of the three formulas given above can also be used to estimate the order of highpass, bandpass, and bandstop FIR filters. In the case of the bandpass and bandstop filters, there are two transition bands. It has been found that here the filter order basically depends on the transition band with the smallest width. We illustrate the use of the Kaiser's formula in estimating the order of a linear-phase bandpass FIR filter in Example 10.4.

EXAMPLE 10.4   Estimation of the Order of a Linear-Phase Bandpass FIR Filter

Estimate the order of a linear-phase bandpass FIR filter with the following specifications: passband edges $F_{p1} = 0.35$ kHz and $F_{p2} = 1$ kHz, stopband edges $F_{s1} = 0.3$ kHz and $F_{s2} = 1.1$ kHz , passband ripple $\delta_p = 0.002$, stopband ripple $\delta_s = 0.001$, and sampling rate $F_T = 10$ kHz.

Since the widths of the two transition bands are not equal, we use the width of the smallest transition band to compute the order N. Substituting the appropriate values in Eq. (10.3), we get

$$N \cong \frac{-20\log_{10}(\sqrt{0.00004}) - 13}{14.6(350 - 300)/10{,}000}$$

$$= \frac{30.9794}{0.0730} = 424.3753.$$

The order of the FIR filter is therefore 425, which corresponds to a Type 2 filter. As before, by increasing the order by 1, we can design a Type 1 FIR filter to meet the specifications.

Table 10.1: Comparison of FIR filter orders.

Filter No.	Actual Order	Kaiser's Formula	Bellanger's Formula	Hermann's Formula
1	159	158	163	151
2	38	34	37	37
3	14	12	13	12

10.2 FIR Filter Design Based on Windowed Fourier Series

We now turn our attention to the design of real-coefficient FIR filters. These filters are described by a transfer function that is a polynomial in z^{-1} and therefore require different approaches for their design.

A variety of approaches has been proposed for the design of FIR digital filters. A direct and straightforward method is based on truncating the Fourier series representation of the prescribed frequency response and is discussed in this section. The second method is based on the observation that for a length-$(N + 1)$ FIR digital filter, $(N + 1)$ distinct equally spaced frequency samples of its frequency response constitute the $(N + 1)$-point DFT of its impulse response, and hence, the impulse response sequence can be readily computed by applying an inverse DFT to these frequency samples (Problems 10.31 and 10.32).

10.2.1 Least Integral-Squared Error Design of FIR Filters

Let $H_d(e^{j\omega})$ denote the desired frequency response function. Since $H_d(e^{j\omega})$ is a periodic function of ω with a period 2π, it can be expressed as a Fourier series,

$$H_d(e^{j\omega}) = \sum_{n=-\infty}^{\infty} h_d[n]e^{-j\omega n}, \tag{10.7}$$

where the Fourier coefficients $\{h_d[n]\}$ are precisely the corresponding impulse response samples and are given by

$$h_d[n] = \frac{1}{2\pi} \int_{-\pi}^{\pi} H_d(e^{j\omega})e^{j\omega n}d\omega, \qquad -\infty < n < \infty. \tag{10.8}$$

Thus, given a frequency response specification $H_d(e^{j\omega})$, we can compute $h_d[n]$ using Eq. (10.8) and, hence, determine the transfer function $H_d(z)$. However, for most practical applications, the desired frequency response is piecewise constant with sharp transitions between bands, in which case, the corresponding impulse response sequence $\{h_d[n]\}$ is of infinite length and noncausal.

Our objective is to find a finite-duration impulse response sequence $\{h_t[n]\}$ of length $2M + 1$ whose DTFT $H_t(e^{j\omega})$ approximates the desired DTFT $H_d(e^{j\omega})$ in some sense. One commonly used approximation criterion is to minimize the integral-squared error

$$\Phi_R = \frac{1}{2\pi} \int_{-\pi}^{\pi} \left| H_t(e^{j\omega}) - H_d(e^{j\omega}) \right|^2 d\omega, \tag{10.9}$$

where

$$H_t(e^{j\omega}) = \sum_{n=-M}^{M} h_t[n]e^{-j\omega n}. \tag{10.10}$$

Using Parseval's relation (Table 5.3), we can rewrite Eq. (10.9) as

$$\Phi_R = \sum_{n=-\infty}^{\infty} |h_t[n] - h_d[n]|^2$$

$$= \sum_{n=-M}^{M} |h_t[n] - h_d[n]|^2 + \sum_{n=-\infty}^{-M-1} h_d^2[n] + \sum_{n=M+1}^{\infty} h_d^2[n]. \tag{10.11}$$

It is evident from Eq. (10.11) that the integral-squared error is minimum when $h_t[n] = h_d[n]$ for $-M \leq n \leq M$, or, in other words, the best finite-length approximation to the ideal infinite-length impulse response in the mean-square error sense is simply obtained by truncation.

A causal FIR filter with an impulse response $h[n]$ can be derived from $h_t[n]$ by delaying the latter sequence by M samples, that is, by forming

$$h[n] = h_t[n - M]. \tag{10.12}$$

Note that the causal filter $h[n]$ has the same magnitude response as that of the noncausal filter $h_t[n]$ and its phase response has a linear phase shift of ωM radians with respect to that of the noncausal filter.

10.2.2 Impulse Responses of Ideal Filters

Four commonly used frequency selective filters are the lowpass, highpass, bandpass, and bandstop filters whose ideal frequency responses are shown in Figure 7.1. It is straightforward to develop their corresponding impulse responses. For example, the ideal *lowpass filter* of Figure 7.1(a) has a zero-phase frequency response

$$H_{LP}(e^{j\omega}) = \begin{cases} 1, & |\omega| \leq \omega_c, \\ 0, & \omega_c < |\omega| \leq \pi. \end{cases} \tag{10.13}$$

The corresponding impulse response coefficients were determined in Example 3.8 and are given by

$$h_{LP}[n] = \frac{\sin \omega_c n}{\pi n}, \qquad -\infty < n < \infty. \tag{10.14}$$

As can be seen from Eq. (10.14), the impulse response of an ideal lowpass filter is doubly infinite, not absolutely summable, and therefore unrealizable. By setting all impulse response coefficients outside the range $-M \leq n \leq M$ equal to zero, we arrive at a finite-length noncausal approximation of length $2M + 1$, which when shifted to the right yields the coefficients of a causal FIR lowpass filter:

$$\hat{h}_{LP}[n] = \begin{cases} \frac{\sin(\omega_c(n-M))}{\pi(n-M)}, & 0 \leq n \leq N - 1, \\ 0, & \text{otherwise.} \end{cases} \tag{10.15}$$

It should be noted that the above expression also holds for even length, in which case M is a fraction.

Likewise, the impulse response coefficients $h_{HP}[n]$ of the ideal *highpass filter* of Figure 7.1(b) are given by

$$h_{HP}[n] = \begin{cases} 1 - \frac{\omega_c}{\pi}, & \text{for } n = 0, \\ -\frac{\sin(\omega_c n)}{\pi n}, & \text{for } |n| > 0. \end{cases} \tag{10.16}$$

Correspondingly, the impulse response coefficients $h_{BP}[n]$ of the ideal *bandpass filter* of Figure 7.1(c) with cutoffs at ω_{c1} and ω_{c2} are given by

$$h_{BP}[n] = \frac{\sin(\omega_{c2}n)}{\pi n} - \frac{\sin(\omega_{c1}n)}{\pi n}, \qquad |n| \geq 0, \tag{10.17}$$

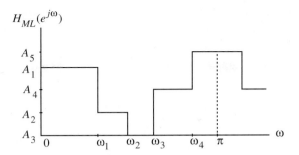

Figure 10.1: A typical zero-phase multilevel frequency response.

and those of the ideal *bandstop filter* of Figure 7.1(d) with cutoffs at ω_{c1} and ω_{c2} are given by

$$
h_{BS}[n] = \begin{cases} 1 - \frac{(\omega_{c2} - \omega_{c1})}{\pi}, & \text{for } n = 0, \\ \frac{\sin(\omega_{c1}n)}{\pi n} - \frac{\sin(\omega_{c2}n)}{\pi n}, & \text{for } |n| > 0. \end{cases} \tag{10.18}
$$

All the above impulse response formulas are for single passband or single stopband filters with two magnitude levels. However, it is quite straightforward to generalize the method to the design of multilevel FIR filters and obtain the expression for the impulse response coefficients. The zero-phase frequency response of an ideal L-band digital filter $H_{ML}(z)$ is given by

$$
H_{ML}(e^{j\omega}) = A_k, \quad \text{for } \omega_{k-1} \leq \omega \leq \omega_k, \quad k = 1, 2, \ldots, L, \tag{10.19}
$$

where $\omega_0 = 0$ and $\omega_L = \pi$. Figure 10.1 shows the zero-phase frequency response of a typical multilevel filter, for brevity only over the frequency range $[0, \pi]$. Its impulse response $h_{ML}[n]$ is given by

$$
h_{ML}[n] = \sum_{\ell=1}^{L} (A_\ell - A_{\ell+1}) \cdot \frac{\sin(\omega_\ell n)}{\pi n}, \tag{10.20}
$$

with $A_{L+1} = 0$.

Two other types of FIR digital filters that find applications are the discrete-time *Hilbert transformer* and the *differentiator*. The ideal Hilbert transformer, also called a *90-degree phase shifter*, is characterized by a frequency response

$$
H_{HT}(e^{j\omega}) = \begin{cases} j, & -\pi < \omega < 0, \\ -j, & 0 < \omega < \pi. \end{cases} \tag{10.21}
$$

The Hilbert transformer finds application in the generation of analytic signals (see Section 15.7.1). The impulse response $h_{HT}[n]$ of the Hilbert transformer is obtained by computing the inverse discrete-time Fourier transform of Eq. (10.21) and is given by

$$
h_{HT}[n] = \begin{cases} 0, & \text{for } n \text{ even,} \\ \frac{2}{\pi n}, & \text{for } n \text{ odd.} \end{cases} \tag{10.22}
$$

The relation between an ideal lowpass filter and an ideal Hilbert transformer is considered in Problem 10.7.

The ideal discrete-time differentiator is employed to perform the differentiation operation in discrete-time on the sampled version of a continuous-time signal. It is characterized by a frequency response given by

$$
H_{DIF}(e^{j\omega}) = j\omega, \quad 0 \leq |\omega| \leq \pi. \tag{10.23}
$$

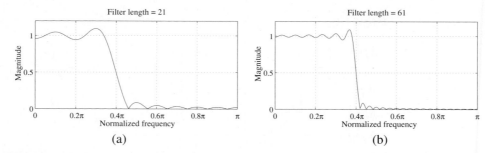

Figure 10.2: Magnitude responses of lowpass filters designed using the truncated impulse response of Eq. (10.15): (a) length $N = 21$ and (b) length $N = 61$.

The impulse response $h_{DIF}[n]$ of the ideal discrete-time differentiator is determined by an inverse discrete-time Fourier transform of Eq. (10.23) and is given by

$$h_{DIF}[n] = \begin{cases} 0, & n = 0, \\ \frac{\cos \pi n}{n}, & |n| > 0. \end{cases} \qquad (10.24)$$

Like the ideal lowpass filter, all the above five ideal filters are also characterized by doubly infinite impulse responses that are not absolutely summable, making them unrealizable. They can be made realizable by truncating the impulse response sequences to finite lengths and shifting the truncated coefficients to the right appropriately.

10.2.3 Gibbs Phenomenon

The causal FIR filters obtained by simply truncating the impulse response coefficients of the ideal filters given in the previous section exhibit an oscillatory behavior in their respective magnitude responses, which is more commonly referred to as the *Gibbs phenomenon*. We illustrate here the occurrence of the Gibbs phenomenon by considering the design of lowpass filters. Figure 10.2 shows the magnitude responses of a lowpass filter with a cutoff at $\omega_c = 0.4\pi$, designed using the formula of Eq. (10.15) for two different values of filter lengths. The oscillatory behavior of the magnitude response on both sides of the cutoff frequency is clearly visible in both cases. Moreover, as the length of the filter is increased, the number of ripples in both passband and stopband increases, with a corresponding decrease in the widths of the ripples. However, the heights of the largest ripples, which occur on both sides of the cutoff frequency, remain the same, independent of the filter length, and are approximately 11% of the difference between the passband and stopband magnitudes of the ideal filter [Par87].

A similar oscillatory behavior is also observed in the frequency responses of the truncated versions of the impulse responses of other types of ideal filters described in the previous section (also, see Exercises M10.1 to M10.3). For example, Figure 10.3 shows the magnitude response of a length-51 differentiator designed by truncating the impulse response coefficients of Eq. (10.24).

The reason behind the Gibbs phenomenon can be explained by considering the truncation operation as multiplication by a finite-length window sequence $w[n]$ and by examining the windowing process in the frequency domain. Thus, the FIR filter obtained by truncation can be alternatively expressed as

$$h_t[n] = h_d[n] \cdot w[n]. \qquad (10.25)$$

From the modulation theorem of Table 5.3, the Fourier transform of Eq. (10.25) is given by

$$H_t(e^{j\omega}) = \frac{1}{2\pi} \int_{-\pi}^{\pi} H_d(e^{j\varphi}) \Psi(e^{j(\omega-\varphi)}) \, d\varphi, \qquad (10.26)$$

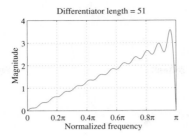

Figure 10.3: Magnitude response of a length-51 differentiator designed by truncating the impulse response of Eq. (10.24).

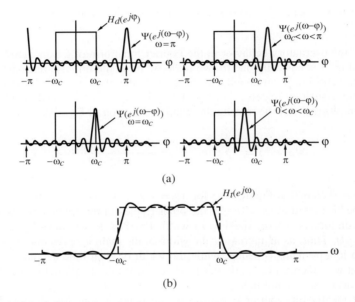

Figure 10.4: Illustration of the effect of windowing in the frequency domain.

where $H_t(e^{j\omega})$ and $\Psi(e^{j\omega})$ are the Fourier transforms of $h_t[n]$ and $w[n]$, respectively. Equation (10.26) implies that $H_t(e^{j\omega})$ is obtained by a periodic continuous convolution of the desired frequency response $H_d(e^{j\omega})$ with the Fourier transform $\Psi(e^{j\omega})$ of the window. The process is illustrated in Figure 10.4 with all Fourier transforms shown as real functions for convenience.

From Eq. (10.26), it follows that if $\Psi(e^{j\omega})$ is a very narrow pulse centered at $\omega = 0$ (ideally a delta function) compared to variations in $H_d(e^{j\omega})$, then $H_t(e^{j\omega})$ will approximate $H_d(e^{j\omega})$ very closely. This implies that the length $2M + 1$ of the window function $w[n]$ should be very large. On the other hand, the length $2M + 1$ of $h_t[n]$, and hence that of $w[n]$, should be as small as possible to make the computational complexity of the filtering processes as small as possible.

Now, the window used to achieve simple truncation of the ideal infinite-length impulse response is called a *rectangular window* and is given by

$$w_R[n] = \begin{cases} 1, & -M \le n \le M, \\ 0, & \text{otherwise.} \end{cases} \tag{10.27}$$

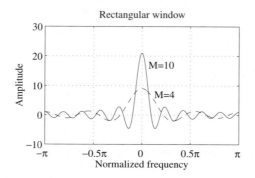

Figure 10.5: Frequency responses of the rectangular window for $M = 4$ and $M = 10$.

The presence of the oscillatory behavior in the Fourier transform of a truncated Fourier series representation of an ideal filter is basically due to two reasons. First, the impulse response of an ideal filter is infinitely long and not absolutely summable, and as a result, the filter is unstable. Second, the rectangular window has an abrupt transition to zero. The oscillatory behavior can be easily explained by examining the Fourier transform $\Psi_R(e^{j\omega})$ of the rectangular window function of Eq. (10.27):

$$\Psi_R(e^{j\omega}) = \sum_{n=-M}^{M} e^{-j\omega n} = \frac{\sin([2M+1]\omega/2)}{\sin(\omega/2)}. \tag{10.28}$$

A plot of the above is sketched in Figure 10.5 for $M = 4$ and 10. The frequency response $\Psi_R(e^{j\omega})$ has a narrow "main lobe" centered at $\omega = 0$. All the other ripples in the frequency response are called the "sidelobes." The main lobe is characterized by its width $4\pi/(2M+1)$ defined by the first zero crossings on both sides of $\omega = 0$. Thus, as M increases, the width of the main lobe decreases as desired. However, the area under each lobe remains constant, while the width of each lobe decreases with an increase in M. This implies that with increasing M, ripples in $H_t(e^{j\omega})$ around the point of discontinuity occur more closely but with no decrease in amplitude.

Recall also that ideally the Fourier transform of the window function should closely resemble an impulse function centered at $\omega = 0$, with its length $2M + 1$ being as small as possible to reduce the computational complexity of the FIR filter. An increase in the length of the rectangular window function reduces the main lobe width but unfortunately increases the computational complexity.

The rectangular window has an abrupt transition to zero outside the range $-M \leq n \leq M$, which is the reason behind the appearance of the Gibbs phenomenon in the magnitude response of the windowed ideal filter impulse response sequence. The Gibbs phenomenon can be reduced by either using a window that tapers smoothly to zero at each end or by providing a smooth transition from the passband to the stopband. Use of a tapered window causes the height of the sidelobes to diminish, with a corresponding increase in the main lobe width resulting in a wider transition at the discontinuity. We review in the next two sections a few of these windows and study their properties. Elimination of the Gibbs phenomenon by introducing a smooth transition in the filter specifications is considered in Section 10.2.6.

10.2.4   Fixed Window Functions

Many tapered windows have been proposed by various authors. A discussion of all these suggested windows is beyond the scope of this text. We restrict our discussion to the four commonly used tapered

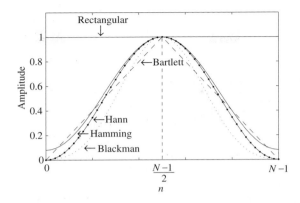

Figure 10.6: Plots of the fixed windows shown with solid lines for clarity.

windows of length $N = 2M + 1$ listed below [Sar93]:[2]

$$\textit{Bartlett: } w[n] = 1 - \frac{|n|}{M+1}, \qquad\qquad -M \leq n \leq M, \qquad\qquad (10.29)$$

$$\textit{Hann: } w[n] = \frac{1}{2}\left[1 + \cos\left(\frac{2\pi n}{2M+1}\right)\right], \qquad -M \leq n \leq M, \qquad\qquad (10.30)$$

$$\textit{Hamming: } w[n] = 0.54 + 0.46\cos\left(\frac{2\pi n}{2M+1}\right), \qquad -M \leq n \leq M, \qquad\qquad (10.31)$$

$$\textit{Blackman: } w[n] = 0.42 + 0.5\cos\left(\frac{2\pi n}{2M+1}\right)] + 0.08\cos\left(\frac{4\pi n}{2M+1}\right),$$

$$-M \leq n \leq M. \qquad\qquad (10.32)$$

In the literature, the Hann window is sometimes called the *Hanning window* or the *von Hann window*. Figure 10.6 shows these windows plotted as continuous functions for clarity.[3] It should be noted that except for the rectangular and the Hamming windows, the end samples of the three other windows are zeros; that is, $w[0] = w[N-1] = 0$. Hence, strictly speaking, these three windows are of length $N-2$.

A plot of the gain response of each of the above windows is shown in Figure 10.7 for $M = 25$. As can be seen from these plots, the magnitude spectrum of each window is characterized by a large main lobe centered at $\omega = 0$, followed by a series of sidelobes with decreasing amplitudes. Two parameters that somewhat predict the performance of a window in FIR filter design are its *main lobe width* and the *relative sidelobe level*. The main lobe width Δ_{ML} is the distance between the nearest zero crossings on both sides of the main lobe, and the relative sidelobe level $A_{s\ell}$ is the difference in dB between the amplitudes of the largest sidelobe and the main lobe.

To understand the effect of the window function on FIR filter design, we show in Figure 10.8 a typical relation among $H_t(e^{j\omega})$, $\Psi(e^{j\omega})$, and $H_d(e^{j\omega})$, the frequency responses of the windowed lowpass filter, the window function, and the desired ideal lowpass filter, respectively [Sar93]. Since the corresponding

---

[2]The expressions for the window functions given here are slightly different from that given in the literature.

[3]These window functions have been generated using the M-files `bartlett`, `hann`, `hamming` and `blackman` (see Section 10.5.3).

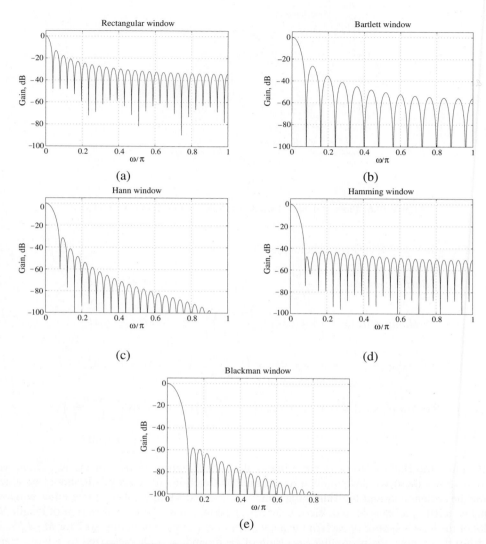

Figure 10.7: Gain responses of the fixed window functions.

impulse responses are symmetric with respect to $n = 0$, the frequency responses are of zero-phase. From this figure, we observe that for the windowed filter, $H_t(e^{j(\omega_c+\omega)}) + H_t(e^{j(\omega_c-\omega)}) \cong 1$, around the cutoff frequency ω_c. As a result, $H_t(e^{j\omega_c}) \cong 0.5$. Moreover, the passband and stopband ripples are the same. In addition, the distance between locations of the maximum passband deviation and the minimum stopband value is approximately equal to the width Δ_{ML} of the main lobe of the window, with the center at ω_c. The width of the transition band, defined by $\Delta\omega = \omega_s - \omega_p$, is less than Δ_{ML}. Therefore, to ensure a fast transition from the passband to the stopband, the window should have a very small main lobe width. On the other hand, to reduce the passband and stopband ripple δ, the area under the sidelobes should be very small. Unfortunately, these two requirements are contradictory.

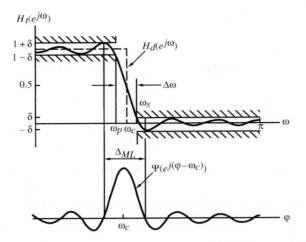

Figure 10.8: Relations among the frequency responses of an ideal lowpass filter, a typical window, and the windowed filter.

Table 10.2: Properties of some fixed window functions.[4]

Type of Window	Main Lobe Width Δ_{ML}	Relative Sidelobe Level $A_{s\ell}$	Minimum Stopband Attenuation	Transition Bandwidth $\Delta\omega$
Rectangular	$4\pi/(2M+1)$	13.3 dB	20.9 dB	$0.92\pi/M$
Barlett	$4\pi/(M+1)$	26.5 dB	See text	See text
Hann	$8\pi/(2M+1)$	31.5 dB	43.9 dB	$3.11\pi/M$
Hamming	$8\pi/(2M+1)$	42.7 dB	54.5 dB	$3.32\pi/M$
Blackman	$12\pi/(2M+1)$	58.1 dB	75.3 dB	$5.56\pi/M$

Table 10.2 summarizes the essential properties of the above window functions, except the Bartlett window. For the latter window, the stopband edge is difficult to determine as the frequency response of the filter designed using this window has no unit circle zeros, and as a result, the value of the stopband attenuation and the expression for the transition bandwidth are not precisely known. The Bartlett window finds applications in spectral estimation.

In the case of the window functions of Eqs. (10.27) and (10.30) to (10.31), the value of the ripple δ does not depend on the filter length, or the cutoff frequency ω_c, and is essentially constant. In addition, the transition bandwidth is approximately given by

$$\Delta\omega \approx \frac{c}{M},\tag{10.33}$$

where c is a constant for most practical purposes and is determined from Table 10.2 after the specific window has been selected [Sar93].

Example 10.5 illustrates the effect of each of the above windows on the frequency response of an FIR lowpass filter designed using the windowed Fourier series approach.

---

[4]This table has been adapted from [Sar93], with the values shown in the table for $\omega_c = 0.4\pi$ and $M = 128$.

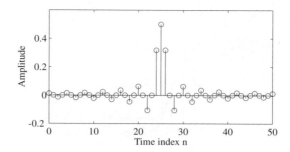

Figure 10.9: Impulse response of the truncated ideal lowpass FIR filter of length 51 with a cutoff at $\pi/2$.

EXAMPLE 10.5   **Effect of Windows on the Frequency Response of an Ideal Lowpass Filter**

The impulse response $h_{LP}[n]$ of the ideal lowpass filter is given in Eq. (10.14). To create a finite-duration zero-phase FIR filter of length N, we form $h_t[n] = h_{LP}[n] \cdot w[n]$, where $M = (N-1)/2$. Figure 10.9 shows the impulse response samples $h_t[n]$ for $N = 51$ and $\omega_c = \pi/2$. The gain response of each of the filters obtained by windowing the above impulse-response samples with various fixed window functions is sketched in Figure 10.10. It should be noted from this figure that the increase in the main lobe width of the window functions is clearly associated with an increase in the transition width. Likewise, a decrease in the sidelobe amplitude results in an increase in the stopband attenuation as expected.

For designing an FIR filter using one of the above windows, first the cutoff frequency ω_c is determined from the specified passband and stopband edge frequencies, ω_p and ω_s, by setting $\omega_c = (\omega_p + \omega_s)/2$. Next, M is estimated using Eq. (10.33), where the value of the constant c is obtained from Table 10.2 for the window chosen.

EXAMPLE 10.6   **Filter Length Estimation for Window-Based Design**

The desired specifications for the lowpass filter are as follows: passband edge $\omega_p = 0.3\pi$, stopband edge $\omega_s = 0.5\pi$, and minimum stopband attenuation $\alpha_s = 40$ dB. The cutoff frequency is thus given by $\omega_c = (\omega_p + \omega_s)/2 = 0.4\pi$. The normalized transition bandwidth is $\Delta\omega = \omega_s - \omega_p = 0.2\pi$.

We next observe from Table 10.2 that the desired minimum stopband attenuation α_s of 40 dB can be achieved using Hann, Hamming, and Blackman windows. Using Eq. (10.33) and Table 10.2, we arrive at the value of the parameter M as given below for the three windows:

$$\text{Hann}: \qquad M = \frac{3.11\pi}{0.2\pi} = 15.55,$$

$$\text{Hamming}: \qquad M = \frac{3.32\pi}{0.2\pi} = 16.6,$$

$$\text{Blackman}: \qquad M = \frac{55.6\pi}{0.2\pi} = 27.8.$$

It follows from the above that the FIR filter designed using the Hann window has the smallest length given by $\lceil 2M + 1 \rceil = 32$.

10.2.5   Adjustable Window Functions

As indicated above, the ripple δ of the filter designed using any one of the fixed window functions is fixed. Several windows have been developed that provide control over δ by means of an additional parameter characterizing the window. We describe here two such windows.

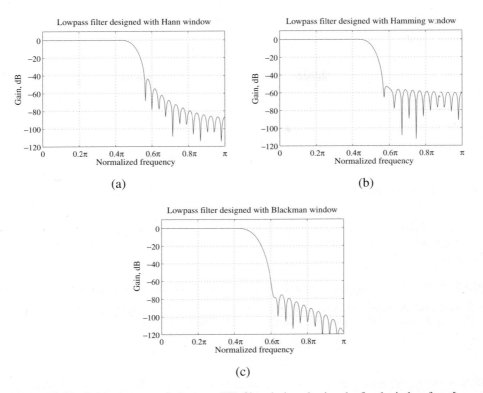

Figure 10.10: Gain responses of a lowpass FIR filter designed using the fixed window functions.

The *Dolph-Chebyshev window* of length $2M + 1$ is defined by [Hel68]

$$w[n] = \frac{1}{2M + 1}\left[\frac{1}{\gamma} + 2\sum_{k=1}^{M} T_{2M}\left(\beta\cos\frac{k\pi}{2M + 1}\right)\cos\frac{2nk\pi}{2M + 1}\right], \quad -M \le n \le M, \quad (10.34)$$

where γ is the relative sidelobe amplitude expressed as a fraction,

$$\gamma = \frac{\text{sidelobe amplitude}}{\text{main lobe amplitude}}, \quad (10.35)$$

$$\beta = \cosh\left(\frac{1}{2M}\cosh^{-1}\frac{1}{\gamma}\right), \quad (10.36)$$

and $T_{\ell}(x)$ is the ℓth-order Chebyshev polynomial in x defined by

$$T_{\ell}(x) = \begin{cases} \cos(\ell\cos^{-1}x), & \text{for } |x| \le 1, \\ \cosh(\ell\cosh^{-1}x), & \text{for } |x| > 1. \end{cases} \quad (10.37)$$

The Dolph-Chebyshev window can be designed with any specified relative sidelobe level and, as in the case of the other windows, its main lobe width can be adjusted by choosing the length appropriately. The filter order $N = 2M$ is estimated using the formula [Sar93]

$$N = \frac{2.056\alpha_s - 16.4}{2.285(\Delta\omega)}, \quad (10.38)$$

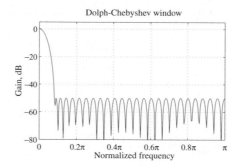

Figure 10.11: Gain response of a length-51 Dolph-Chebyshev window with a relative sidelobe level of 50 dB.

where α_s is the minimum stopband attenuation in dB and $\Delta\omega$ is the normalized transition bandwidth. In the case of a lowpass filter with normalized angular passband and stopband edge frequencies ω_p and ω_s, respectively, $\Delta\omega = \omega_s - \omega_p$.

EXAMPLE 10.7   Order Estimation of an FIR Filter for Design Using the Dolph-Chebyshev Window

We consider the determination of the filter order of the FIR filter to be designed using the Dolph-Chebyshev window to meet the specifications of Example 10.6. Using Eq. (10.38), the filter order N here is given by

$$N = \left\lceil \frac{2.056 \times 40 - 16.4}{2.285 \times 0.2\pi} \right\rceil = \lceil 45.8589 \rceil = 46.$$

It should be noted that the order of the filter designed using the Hamming window is 64, which is much higher than the above filter order.

Figure 10.11 shows the gain response of a Dolph-Chebyshev window for $M = 25$, that is, a window length of 51, and a relative sidelobe level of 50 dB.[5] As can be seen from this plot, all sidelobes are of equal height. As a result, the stopband approximation error of filters designed using this window has essentially an equiripple behavior. Another interesting property of this window is that for a given window length, it has the smallest main lobe width compared to other windows, resulting in filters with the smallest transition band.

The most widely used adjustable window is the *Kaiser window*, given by [Kai74]

$$w[n] = \frac{I_0\left\{\beta\sqrt{1 - (n/M)^2}\right\}}{I_0(\beta)}, \qquad -M \leq n \leq M, \tag{10.39}$$

where β is an adjustable parameter and $I_0(u)$ is the modified zeroth-order Bessel function, which can be expressed in a power series form

$$I_0(u) = 1 + \sum_{r=1}^{\infty} \left[\frac{(u/2)^r}{r!} \right]^2, \tag{10.40}$$

which is seen to be positive for all real values of u. In practice, it is sufficient to keep only the first 20 terms in the summation of Eq. (10.40) to arrive at a reasonably accurate value of $I_0(u)$.

The parameter β controls the minimum attenuation $\alpha_s = -20\log_{10}(\delta_s)$ in the stopband of the windowed filter response. Formulas for estimating β and the filter order $N = 2M$, for specified α_s and normalized

---

[5]The coefficients of the window have been computed using the M-file `chebwin` (see Section 10.5.4).

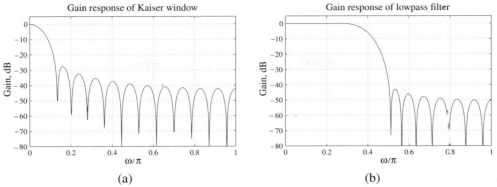

Figure 10.12: (a) Gain response of the Kaiser window of Example 10.8 and (b) gain response of the lowpass filter designed using this window.

transition bandwidth $\Delta\omega$, have been developed by Kaiser [Kai74]. The parameter β is computed from [6]

$$\beta = \begin{cases} 0.1102(\alpha_s - 8.7), & \text{for } \alpha_s > 50, \\ 0.5842(\alpha_s - 21)^{0.4} + 0.07886(\alpha_s - 21), & \text{for } 21 \leq \alpha_s \leq 50, \\ 0, & \text{for } \alpha_s < 21. \end{cases} \tag{10.41}$$

The filter order N is estimated using the formula

$$N = \frac{\alpha_s - 8}{2.285(\Delta\omega)}, \tag{10.42}$$

where $\Delta\omega$ is the normalized transition bandwidth. It should be noted that the Kaiser window provides no independent control over the passband ripple δ_p. However, in practice, δ_p is approximately equal to δ_s.

We illustrate the design of a linear-phase lowpass filter using the Kaiser window in Example 10.8.

EXAMPLE 10.8   Design of a Lowpass FIR Filter Using the Kaiser Window

We now consider the design of the FIR filter of Example 10.6.

For the specified minimum stopband attenuation, the peak stopband ripple δ_s is computed using Eq. (9.4) and is found to be $\delta_s = 0.01$. Using Eq. (10.41), the parameter β is next calculated and is determined to be $\beta = 3.3953$. The normalized transition bandwidth $\Delta\omega$ is given by

$$\Delta\omega = \omega_s - \omega_p = 0.5\pi - 0.3\pi = 0.2\pi.$$

Applying Eq. (10.42), we next determine the filter order N to be 22.2886, which when rounded up to the next higher integer is 23, implying $M = 11$. From Eqs. (10.14) and (10.39), we then arrive at the impulse response coefficients of the FIR filter obtained by windowing as

$$h_t[n] = \frac{\sin \omega_c n}{\pi n} \cdot w[n], \quad -M \leq n \leq M,$$

where $M = 11$, $\omega_c = 0.4\pi$, and $w[n]$ is the nth coefficient of a length-23 Kaiser window. It should be noted that the above filter is noncausal and can be converted into a causal filter by delaying the filter coefficients by M samples. Since N is odd for our example, the delayed filter here is a Type 2 FIR linear-phase filter.

The coefficients of the Kaiser window can be determined in MATLAB using the M-file kaiser.[7] The gain responses of the length-23 Kaiser window and the windowed filter are sketched in Figure 10.12.

---

[6]Determined empirically.

[7]See Section 10.5.4.

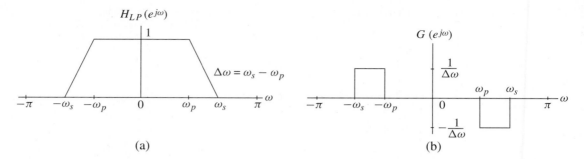

Figure 10.13: (a) Lowpass filter specification with a transition region and (b) the specification of its derivative function.

10.2.6 Impulse Responses of FIR Filters with a Smooth Transition

We showed earlier that the FIR filter obtained by truncating the infinite-length impulse response of a digital filter developed from a frequency response specification with sharp discontinuities exhibits oscillatory behavior, called the Gibbs phenomenon, in its frequency response. One way of reducing the height of the ripples to acceptable values is to truncate the infinite-length impulse response by tapered window functions. Another approach to eliminate the Gibbs phenomenon is by modifying the frequency response specification of the digital filter to have a transition band between the passband and the stopband and to provide a smooth transition between the bands [Orm61]. We now discuss this second approach for lowpass filter design. Similar modifications can be carried out for the design of the other types of filters.

The simplest modification to the zero-phase lowpass filter specification is to provide a transition band between the passband and the stopband responses and to connect these two with a first-order spline function (straight line), as indicated in Figure 10.13(a). An inverse Fourier transform of the modified frequency response $H_{LP}(e^{j\omega})$ leads to the expression for its corresponding impulse response coefficients $h_{LP}[n]$. However, as illustrated in the following, a simpler method is to compute $h_{LP}[n]$ from the inverse Fourier transform of the derivative of the specified frequency response $H_{LP}(e^{j\omega})$.

Let $G(e^{j\omega})$ denote $dH_{LP}(e^{j\omega})/d\omega$ with a corresponding inverse Fourier transform $g[n]$. Its specification will be thus as indicated in Figure 10.13(b). It follows from the differentiation-in-frequency property of the Fourier transform given in Table 3.4, $h_{LP}[n] = jg[n]/n$. From the inverse Fourier transform $g[n]$ of $G(e^{j\omega})$ given in Figure 10.13(b), we arrive at the impulse response of the modified lowpass filter [Par87]:

$$h_{LP}[n] = \begin{cases} \dfrac{\omega_c}{\pi}, & n = 0, \\[2mm] \dfrac{2\sin(\Delta\omega n/2)}{\Delta\omega n} \cdot \dfrac{\sin(\omega_c n)}{\pi n}, & |n| > 0, \end{cases} \tag{10.43}$$

where

$$\Delta\omega = \omega_s - \omega_p \quad \text{and} \quad \omega_c = \frac{\omega_p + \omega_s}{2}.$$

A still smoother transition between the passband and the stopband of the lowpass filter can be provided by specifying the transition function by a higher-order spline. The corresponding impulse response for a Pth-order spline as the transition function is given by [Bur92], [Par87]

$$h_{LP}[n] = \begin{cases} \dfrac{\omega_c}{\pi}, & n = 0, \\[2mm] \left(\dfrac{\sin(\Delta\omega n/2P)}{\Delta\omega n/2P}\right)^P \cdot \dfrac{\sin(\omega_c n)}{\pi n}, & |n| > 0. \end{cases} \tag{10.44}$$

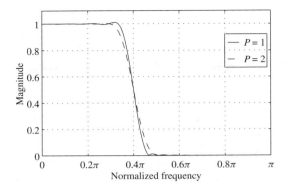

Figure 10.14: Magnitude responses of FIR lowpass filters with smooth transition between bands obtained using spline transition functions.

EXAMPLE 10.9   Design of a Lowpass FIR Filter Using Spline Transition Functions

Design a lowpass filter with a passband edge at 0.35π and a stopband edge at 0.45π using a first-order and a second-order spline as the transition functions.

The magnitude response of a length-41 filter obtained by using a first-order spline tranistion function ($P = 1$) and then truncating the impulse response of Eq. (10.43) is shown by the solid line in Figure 10.14. The magnitude response of a length-61 FIR filter derived by using a second-order spline tranistion function ($P = 2$) and then truncating the impulse response of Eq. (10.44) is shown in Figure 10.14 by the dashed line.

As Example 10.9 points out, the effect of P on the frequency response of the truncated filter is not obvious. For a given filter order $2M$ and transition bandwidth $\Delta\omega$, the optimum value of P minimizing the integral-squared approximation error has been shown to be given by [Bur92]

$$P = 1.248(\Delta\omega)M. \tag{10.45}$$

Various other transition functions have been investigated to reduce the ripple heights. For example, the raised cosine function has been proposed as a transition function. (Problem 10.18) [Par87]. These transition functions can also be employed for the design of other types of filters. So far, no specific guidelines have been advanced to select the optimum transition function for a given filter design problem. As a result, it should be selected by a trial-and-error procedure.

10.3   Computer-Aided Design of Equiripple Linear-Phase FIR Filters

In this section, we consider the application of computer-aided optimization techniques for the design of FIR filters. As described in Section 9.7, the basic idea behind the computer-based technique is to minimize iteratively an error measure that is a function of the difference between the desired frequency response $D(e^{j\omega})$ and the frequency response $H(e^{j\omega})$ of the filter being designed. Commonly used error measures are also defined in this section.

In the case of linear-phase FIR filter design, $H(e^{j\omega})$ and $D(e^{j\omega})$ are zero-phase frequency responses. On the other hand, for IIR filter design, these functions are replaced with their magnitude functions. The design objective here is to iteratively adjust the filter parameters so that ε defined by either Eq. (9.49) or Eq. (9.51) is a minimum. The linear-phase FIR filter obtained by minimizing the peak absolute value of the weighted error ε given by Eq. (9.49) is usually called the *equiripple FIR filter*, since here, after ε has been minimized, the weighted error function $\mathcal{E}(\omega)$ exhibits an equiripple behavior in the frequency range of interest. In this section, we first briefly outline the *weighted-Chebyshev approximation method* advanced by Parks and McClellan for designing equiripple linear-phase FIR filters [Par72]. This widely used method is more commonly known as the *Parks–McClellan algorithm*. We then mention an improved formulation of the optimization algorithm that has been found to eliminate some of the difficulties encountered in practice with the Parks–McClellan algorithm [Shp90].

The Parks–McClellan Algorithm

In Section 7.3, we defined the four types of linear-phase FIR filters. The general form of the frequency response $H(e^{j\omega})$ of a causal linear-phase FIR filter of length $N + 1$ is given by

$$H(e^{j\omega}) = e^{-jN\omega/2}e^{j\beta}\breve{H}(\omega), \tag{10.46}$$

where the amplitude response $\breve{H}(\omega)$ is a real function of ω. The weighted error function in this case involves the amplitude response and is given by

$$\mathcal{E}(\omega) = W(\omega)\left[\breve{H}(\omega) - D(\omega)\right], \tag{10.47}$$

where $D(\omega)$ is the desired amplitude response and $W(\omega)$ is a positive weighting function. $W(\omega)$ is chosen to control the relative size of the peak errors in the specified frequency bands. The Parks–McClellan algorithm is based on iteratively adjusting the coefficients of the amplitude response until the peak absolute value of $\mathcal{E}(\omega)$ is minimized.

If the minimum value of the peak absolute value of $\mathcal{E}(\omega)$ in a band $\omega_a \leq \omega \leq \omega_b$ is ε_o, then the absolute error satisfies

$$|\breve{H}(\omega) - D(\omega)| \leq \frac{\varepsilon_o}{|W(\omega)|}, \qquad \omega_a \leq \omega \leq \omega_b.$$

In typical filter design applications, the desired amplitude response is given by

$$D(\omega) = \begin{cases} 1, & \text{in the passband,} \\ 0, & \text{in the stopband,} \end{cases}$$

and the amplitude response $\breve{H}(\omega)$ is required to satisfy the above desired response with a ripple of $\pm\delta_p$ in the passband and a ripple of δ_s in the stopband. As a result, it is evident from Eq. (10.47) that the weighting function can be chosen either as

$$W(\omega) = \begin{cases} 1, & \text{in the passband,} \\ \delta_p/\delta_s, & \text{in the stopband,} \end{cases}$$

or as

$$W(\omega) = \begin{cases} \delta_s/\delta_p, & \text{in the passband,} \\ 1, & \text{in the stopband.} \end{cases}$$

By a clever manipulation, the expression for the amplitude response for each of the four types of linear-phase FIR filters can be expressed in the same form and, as a result, basically the same algorithm can be adapted to design any one of the four types of filters. To develop this general form for the amplitude response expression, we consider each of the four types of filters separately.

For the Type 1 linear-phase FIR filter, the amplitude response given by Eq. (7.49) can be rewritten using the notation $N = 2M$ in the form

$$\breve{H}(\omega) = \sum_{k=0}^{M} a[k] \cos(\omega k), \tag{10.48}$$

where

$$a[0] = h[M], \qquad a[k] = 2\,h[M-k], \qquad 1 \le k \le M. \tag{10.49}$$

For the Type 2 linear-phase FIR filter, the amplitude response given by Eq. (7.52) can be rewritten in the form

$$\breve{H}(\omega) = \sum_{k=1}^{(2M+1)/2} b[k] \cos\left(\omega(k - \tfrac{1}{2})\right), \tag{10.50}$$

where

$$b[k] = 2h\left[\tfrac{2M+1}{2} - k\right], \qquad 1 \le k \le \tfrac{2M+1}{2}. \tag{10.51}$$

Equation (10.50) can be expressed in the form

$$\breve{H}(\omega) = \cos\left(\tfrac{\omega}{2}\right) \sum_{k=0}^{(2M-1)/2} \tilde{b}[k] \cos(\omega k), \tag{10.52}$$

where

$$
\begin{aligned}
b[1] &= \tfrac{1}{2}\left(\tilde{b}[1] + 2\tilde{b}[0]\right), \\
b[k] &= \tfrac{1}{2}\left(\tilde{b}[k] + \tilde{b}[k-1]\right), \qquad 2 \le k \le \tfrac{2M-1}{2}, \\
b[\tfrac{2M+1}{2}] &= \tfrac{1}{2}\,\tilde{b}\left[\tfrac{2M-1}{2}\right].
\end{aligned}
\tag{10.53}
$$

The amplitude response for the case of the Type 3 linear-phase FIR filter given by Eq. (7.53) can be rewritten in the form

$$\breve{H}(\omega) = \sum_{k=1}^{M} c[k] \sin(\omega k), \tag{10.54}$$

where

$$c[k] = 2\,h[M-k], \qquad 1 \le k \le M. \tag{10.55}$$

Equation (10.54) can be expressed in the form

$$\breve{H}(\omega) = \sin(\omega) \sum_{k=0}^{M-1} \tilde{c}[k] \cos(\omega k), \tag{10.56}$$

where

$$
\begin{aligned}
c[1] &= \tilde{c}[0] - \tfrac{1}{2}\tilde{c}[1], \\
c[k] &= \tfrac{1}{2}\left(\tilde{c}[k-1] - \tilde{c}[k]\right), \qquad 2 \le k \le M-1, \\
c[M] &= \tfrac{1}{2}\tilde{c}[M-1].
\end{aligned}
\tag{10.57}
$$

Likewise, the amplitude response for the case of the Type 4 linear-phase FIR filter given by Eq. (7.54) can be rewritten in the form

$$\breve{H}(\omega) = \sum_{k=1}^{(2M+1)/2} d[k] \sin \omega(k - \tfrac{1}{2}), \tag{10.58}$$

where

$$d[k] = 2\, h[\tfrac{2M+1}{2} - k], \qquad 1 \le k \le \tfrac{2M+1}{2}. \tag{10.59}$$

Equation (10.58) can be expressed in the form

$$\breve{H}(\omega) = \sin(\tfrac{\omega}{2}) \sum_{k=0}^{(2M-1)/2} \tilde{d}[k] \cos(\omega k), \tag{10.60}$$

where

$$d[1] = \tilde{d}[0] - \tfrac{1}{2}\tilde{d}[1], \tag{10.61}$$

$$d[k] = \tfrac{1}{2}\left(\tilde{d}[k-1] - \tilde{d}[k]\right), \qquad 2 \le k \le \tfrac{2M-1}{2},$$

$$d[\tfrac{2M+1}{2}] = \tilde{d}[\tfrac{2M-1}{2}].$$

If we now examine Eqs. (10.48), (10.52), (10.56), and (10.60), we observe that the amplitude response for all four types of linear-phase FIR filters can be expressed in the form

$$\breve{H}(\omega) = Q(\omega)\, A(\omega), \tag{10.62}$$

where the first factor $Q(\omega)$ is given by

$$Q(\omega) = \begin{cases} 1, & \text{for Type 1,} \\ \cos(\omega/2), & \text{for Type 2,} \\ \sin(\omega), & \text{for Type 3,} \\ \sin(\omega/2), & \text{for Type 4,} \end{cases} \tag{10.63}$$

and the second factor $A(\omega)$ is of the form

$$A(\omega) = \sum_{k=0}^{L} \tilde{a}[k] \cos(\omega k), \tag{10.64}$$

where

$$\tilde{a}[k] = \begin{cases} a[k], & \text{for Type 1,} \\ \tilde{b}[k], & \text{for Type 2,} \\ \tilde{c}[k], & \text{for Type 3,} \\ \tilde{d}[k], & \text{for Type 4,} \end{cases} \tag{10.65}$$

with

$$L = \begin{cases} M, & \text{for Type 1,} \\ \tfrac{2M-1}{2}, & \text{for Type 2,} \\ M - 1, & \text{for Type 3,} \\ \tfrac{2M-1}{2}, & \text{for Type 4.} \end{cases} \tag{10.66}$$

Substituting Eq. (10.62) in Eq. (10.47), we arrive at a modified form of the weighted approximation function given by

$$\mathcal{E}(\omega) = W(\omega)\left[Q(\omega)A(\omega) - D(\omega)\right]$$

$$= W(\omega)Q(\omega)\left[A(\omega) - \frac{D(\omega)}{Q(\omega)}\right]. \tag{10.67}$$

Using the notation $\tilde{W}(\omega) = W(\omega)Q(\omega)$ and $\tilde{D}(\omega) = D(\omega)/Q(\omega)$, we can rewrite Eq. (10.67) as

$$\mathcal{E}(\omega) = \tilde{W}(\omega)\left[A(\omega) - \tilde{D}(\omega)\right]. \tag{10.68}$$

The optimization problem now becomes the determination of the coefficients $\tilde{a}[k]$, $0 \leq k \leq L$, which minimize the peak absolute value ε of the weighted approximation error $\mathcal{E}(\omega)$ of Eq. (10.68) over the specified frequency bands $\omega \in R$. After the coefficients $\{\tilde{a}[k]\}$ have been determined, the corresponding coefficients of the original amplitude response are computed from which the filter coefficients are then obtained. For example, if the filter being designed is of Type 2, we observe from Eq. (10.65) that $\tilde{b}[k] = \tilde{a}[k]$ and from Eq. (10.66) $M = (2L + 1)/2$. Knowing $\tilde{b}[k]$ and M, we determine next $b[k]$ using Eq. (10.53). Substituting these values of $b[k]$ in Eq. (10.51), we finally arrive at the filter coefficients $h[n]$. In a similar manner, the filter coefficients for the other three types of FIR filters can be determined from $\tilde{a}[k]$.

Parks and McClellan solved the above problem applying the following theorem from the theory of Chebyshev approximation [Par72]:

Alternation Theorem. The amplitude function $A(\omega)$ of Eq. (10.64) is the best unique approximation of the desired amplitude response obtained by minimizing the peak absolute value ε of $\mathcal{E}(\omega)$ given by Eq. (10.67) if and only if there exist at least $L + 2$ extremal angular frequencies, $\omega_0, \omega_1, \ldots, \omega_{L+1}$, in a closed subset R of the frequency range $0 \leq \omega \leq \pi$ such that $\omega_0 < \omega_1 < \cdots < \omega_L < \omega_{L+1}$ and $\mathcal{E}(\omega_i) = -\mathcal{E}(\omega_{i+1})$, with $|\mathcal{E}(\omega_i)| = \varepsilon$ for all i in the range $0 \leq i \leq L + 1$.

Let us examine the behavior of the amplitude response for a Type 1 equiripple lowpass FIR filter whose approximation error $\mathcal{E}(\omega)$ satisfies the condition of the above theorem. The peaks of $\mathcal{E}(\omega)$ are at $\omega = \omega_i$, $0 \leq i \leq L + 1$, where

$$\frac{d\mathcal{E}(\omega)}{d\omega} = 0.$$

Since in the passband and the stopband, $\tilde{W}(\omega)$ and $\tilde{D}(\omega)$ are piecewise constant, it follows from Eq. (10.68) that

$$\left.\frac{d\mathcal{E}(\omega)}{d\omega}\right|_{\omega=\omega_i} = \left.\frac{dA(\omega)}{d\omega}\right|_{\omega=\omega_i} = 0,$$

or, in other words, the amplitude response $A(\omega)$ also has peaks at $\omega = \omega_i$. Using the relation

$$\cos(\omega k) = T_k(\cos\omega),$$

where $T_k(x)$ is the kth order Chebyshev polynomial defined by

$$T_k(x) = \cos(k \cos^{-1} x),$$

the amplitude response $A(\omega)$ given by Eq. (10.64) can be expressed as a power series in $\cos\omega$.

$$A(\omega) = \sum_{k=0}^{L} \alpha[k](\cos\omega)^k,$$

which is seen to be an Lth order polynomial in $\cos\omega$. As a result, $A(\omega)$ can have at most $L-1$ local minima and maxima inside the specified passband and stopband. Moreover, at the bandedges, $\omega = \omega_p$ and $\omega = \omega_s$, $|\mathcal{E}(\omega)|$ is a maximum, and hence, $A(\omega)$ has extrema at these angular frequencies. In addition, $A(\omega)$ may also have extrema at $\omega = 0$ and $\omega = \pi$. Therefore, there are, at most, $L+3$ extremal frequencies of $\mathcal{E}(\omega)$. Similarly, in the case of a linear-phase FIR filter with K specified bandedges and designed using the Remez algorithm, there can be at most $L + K + 1$ extremal frequencies. An equiripple linear-phase FIR filter with more than $L + 2$ extremal frequencies has been called an *extra-ripple filter*.

To arrive at the optimum solution, we need to solve the set of $L + 2$ equations

$$\tilde{W}(\omega_i)[A(\omega_i) - \tilde{D}(\omega_i)] = (-1)^i \varepsilon, \qquad 0 \le i \le L+1, \tag{10.69}$$

for the unknowns $\tilde{a}[i]$ and ε, provided the $L + 2$ extremal angular frequencies are known. To this end, Eq. (10.69) is rewritten in a matrix form as

$$\begin{bmatrix} 1 & \cos(\omega_0) & \cdots & \cos(L\omega_0) & -1/\tilde{W}(\omega_0) \\ 1 & \cos(\omega_1) & \cdots & \cos(L\omega_1) & 1/\tilde{W}(\omega_1) \\ \vdots & \vdots & \ddots & \vdots & \vdots \\ 1 & \cos(\omega_L) & \cdots & \cos(L\omega_L) & (-1)^{L-1}/\tilde{W}(\omega_L) \\ 1 & \cos(\omega_{L+1}) & \cdots & \cos(L\omega_{L+1}) & (-1)^L/\tilde{W}(\omega_{L+1}) \end{bmatrix} \begin{bmatrix} \tilde{a}[0] \\ \tilde{a}[1] \\ \vdots \\ \tilde{a}[L] \\ \varepsilon \end{bmatrix}$$

$$= \begin{bmatrix} \tilde{D}(\omega_0) \\ \tilde{D}(\omega_1) \\ \vdots \\ \tilde{D}(\omega_L) \\ \tilde{D}(\omega_{L+1}) \end{bmatrix}, \tag{10.70}$$

which can, in principle, be solved for the unknowns if the locations of the $L + 2$ extremal frequencies are known a priori. The *Remez exchange algorithm,* a highly efficient iterative procedure, is used to determine the locations of the extremal frequencies and consists of the following steps at each iteration stage:

Step 1: A set of initial values for the extremal frequencies are either chosen or are available from the completion of the previous iteration.

Step 2: The value ε is then computed by solving Eq. (10.70), resulting in the expression

$$\varepsilon = \frac{c_0\tilde{D}(\omega_0) + c_1\tilde{D}(\omega_1) + \cdots + c_{L+1}\tilde{D}(\omega_{L+1})}{\dfrac{c_0}{\tilde{W}(\omega_0)} - \dfrac{c_1}{\tilde{W}(\omega_1)} + \cdots + \dfrac{(-1)^{L+1}c_{L+1}}{\tilde{W}(\omega_{L+1})}}, \tag{10.71}$$

where the constant c_n is given by

$$c_n = \prod_{\substack{i=0 \\ i \ne n}}^{L+1} \frac{1}{\cos(\omega_n) - \cos(\omega_i)}. \tag{10.72}$$

Step 3: The values of the amplitude response $A(\omega)$ at $\omega = \omega_i$ are then computed using

$$A(\omega_i) = \frac{(-1)^i \varepsilon}{\tilde{W}(\omega_i)} + \tilde{D}(\omega_i), \qquad 0 \le i \le L+1.$$

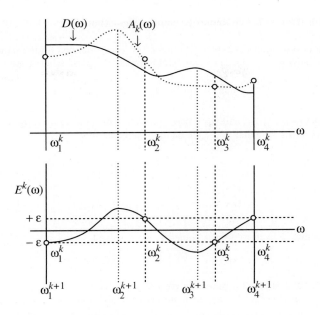

Figure 10.15: Plots of the desired response $D(\omega)$, the amplitude response $A_k(\omega)$, and the error $E^k(\omega)$ at the end of the kth iteration. The locations of the new extremal frequencies are given by ω_i^{k+1}.

Step 4: The polynomial $A(\omega)$ is determined by interpolating the above values at the $L+2$ extremal frequencies using the Lagrange interpolation formula:

$$A(\omega) = \sum_{i=0}^{L+1} A(\omega_i)\, P_i(\cos\omega),$$

where

$$P_i(\cos\omega) = \prod_{\substack{\ell=0,\\ \ell\neq i}}^{L+1} \left(\frac{\cos\omega - \cos\omega_\ell}{\cos\omega_i - \cos\omega_\ell} \right), \qquad 0 \leq i \leq L+1.$$

Step 5: The new weighted error function $\mathcal{E}(\omega)$ of Eq. (10.68) is computed at a dense set S ($S \geq L$) of frequencies. In practice, $S = 16L$ is adequate. Determine the $L+2$ new extremal frequencies from the values of $\mathcal{E}(\omega)$ evaluated at the dense set of frequencies.

Step 6: If the peak values ε of $\mathcal{E}(\omega)$ are equal in magnitude, the algorithm has converged. Otherwise, go back to Step 2.

Figure 10.15 demonstrates how the weighted error function changes from one iteration stage to the next and how the new extremal frequencies are determined. Finally, the iteration process is stopped after the difference between the value of the peak error ε calculated at any stage and that at the previous stage is below a preset threshold value, such as 10^{-6}. In practice, the process converges after very few iterations. The basic principle of the Remez exchange algorithm is illustrated in Example 10.10 [Par87].

EXAMPLE 10.10   Illustration of the Remez Exchange Algorithm

Let the desired function be a quadratic function $D(x) = 1.1x^2 - 0.1$, defined for the range $0 \le x \le 2$. We wish to approximate $D(x)$ by a linear function $a_1 x + a_0$ by minimizing the peak value of the absolute error; that is,

$$\max_{x \in [0,2]} \left| 1.1x^2 - 0.1 - a_0 - a_1 x \right|. \tag{10.73}$$

Since there are three unknowns, a_0, a_1, and ε, we need three extremal points on x, which we arbitrarily choose as $x_1 = 0$, $x_2 = 0.5$, and $x_3 = 1.5$. We then solve the three linear equations

$$a_0 + a_1 x_\ell - (-1)^\ell \varepsilon = D(x_\ell), \qquad \ell = 1, 2, 3, \tag{10.74}$$

which leads to

$$\begin{bmatrix} 1 & 0 & 1 \\ 1 & 0.5 & -1 \\ 1 & 1.5 & 1 \end{bmatrix} \begin{bmatrix} a_0 \\ a_1 \\ \varepsilon \end{bmatrix} = \begin{bmatrix} -0.1 \\ 0.175 \\ 2.375 \end{bmatrix}, \tag{10.75}$$

for the given extremal points. The above matrix equation, when solved, yields

$$a_0 = -0.375, \qquad a_1 = 1.65, \qquad \varepsilon = 0.275. \tag{10.76}$$

Figure 10.16(a) shows the plot of the corresponding error $\mathcal{E}_1(x) = 1.1x^2 - 1.65x + 0.275$, along with the values of error at the chosen extremal points. As expected, at the extremal points, the errors are equal in magnitude and alternate in sign.

   The next set of extremal points are those points where $\mathcal{E}_1(x)$ assumes its maximum absolute values. These extremal points are given by $x_1 = 0$, $x_2 = 0.75$, and $x_3 = 2$. The new values of the unknowns are obtained by solving

$$\begin{bmatrix} 1 & 0 & 1 \\ 1 & 0.75 & -1 \\ 1 & 2 & 1 \end{bmatrix} \begin{bmatrix} a_0 \\ a_1 \\ \varepsilon \end{bmatrix} = \begin{bmatrix} -0.1 \\ 0.5188 \\ 4.3 \end{bmatrix}, \tag{10.77}$$

which yields

$$a_0 = -0.6156, \quad a_1 = 2.2, \quad \varepsilon = 0.5156. \tag{10.78}$$

The plot of the new error $\mathcal{E}_2(x) = 1.1x^2 - 2.2x + 0.5156$ and the values of the error at the new set of extremal points are shown in Figure 10.16(b).

   We continue the process and determine the new set of extremal points, where $\mathcal{E}_2(x)$ takes on the maximum absolute values. These are now given by $x_1 = 0$, $x_2 = 1$, and $x_3 = 2$. Using these extremal points, we determine the next set of values for the unknowns by solving

$$\begin{bmatrix} 1 & 0 & 1 \\ 1 & 1 & -1 \\ 1 & 2 & 1 \end{bmatrix} \begin{bmatrix} a_0 \\ a_1 \\ \varepsilon \end{bmatrix} = \begin{bmatrix} -0.1 \\ 1.0 \\ 4.3 \end{bmatrix}, \tag{10.79}$$

which results in

$$a_0 = -0.65, \quad a_1 = 2.2, \quad \varepsilon = 0.55. \tag{10.80}$$

Figure 10.16(c) depicts the new error $\mathcal{E}_3(x) = 1.1x^2 - 2.2x + 0.55$ and the values of the error at the new extremal points. This time the algorithm converged as ε is also the maximum value of the absolute error.

The Shpak–Antoniou Algorithm

The weighted-Chebyshev method employed in the Parks–McClellan algorithm is often not very reliable and robust when the number of specified bands is more than two, resulting in either a very slow convergence or with filters exhibiting large undesirable magnitude response behavior in the transition regions. An example

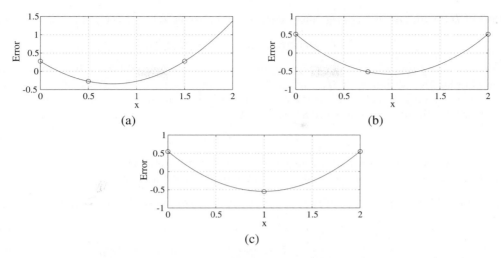

Figure 10.16: Illustration of the Remez algorithm: (a) $\mathcal{E}_1(x)$, (b) $\mathcal{E}_2(x)$, and (c) $\mathcal{E}_3(x)$.

of the latter behavior can be found in Example 10.17. These problems has been shown to be caused by an underdetermined approximating polynomial [Rab74c]. A generalized method proposed by Shpak and Antoniou has been shown to eliminate the above problems [Shp90]. The improved method can be used to design *maximum-ripple* and extra-ripple FIR filters in addition to the weighted-Chebyshev FIR filters.[8] The approximating function used in their algorithm is either fully determined in the case of the maximum ripple filters or is less underdetermined in the case of the extra-ripple filters than that encountered in the weighted-Chebyshev approach.

10.4 Design of Minimum-Phase FIR Filters

Linear-phase FIR filters with narrow transition bands are of very high order and as a result have a very long group delay that is half the filter order. Moreover, making use of the filter coefficient symmetry to compute the output sample may not be possible. As a result, in certain applications, the use of a linear-phase FIR filter may not be very attractive. By relaxing the linear phase requirement, it is possible to design an FIR filter of lower order thus reducing the overall group delay and the computational cost. We outline a very simple and straightforward method to design a minimum-phase FIR filter [Her70].

To understand the basic idea behind this method, consider an arbitrary FIR transfer function of degree N:

$$H(z) = \sum_{n=0}^{N} h[n]\, z^{-n} = h[0] \prod_{k=1}^{N}(1 - \xi_k z^{-1}). \tag{10.81}$$

The mirror-image polynomial to $H(z)$ is the transfer function given by

$$\hat{H}(z) = z^{-N} H(z^{-1})$$
$$= \sum_{n=0}^{N} h[N-n]\, z^{-n} = h[N] \prod_{k=1}^{N}(1 - z^{-1}/\xi_k). \tag{10.82}$$

---

[8]A maximum ripple filter has the maximum number of ripples in its magnitude response.

The zeros of $\hat{H}(z)$ are thus at $z = 1/\xi_k$, that is, are reciprocal to the zeros of $H(z)$ at $z = \xi_k$. As a result,

$$G(z) = H(z)\hat{H}(z) = z^{-N} H(z)H(z^{-1}), \tag{10.83}$$

has zeros exhibiting mirror-image symmetry in the z-plane and is thus a Type 1 linear-phase transfer function of order $2N$. Moreover, if $H(z)$ has a zero on the unit circle, $\hat{H}(z)$ will also have a zero on the unit circle at the conjugate reciprocal position, implying that the unit circle zeros of real-coefficient $G(z)$ occur in pairs. On the unit circle, Eq. (10.83) reduces to

$$|H(e^{j\omega})|^2 = \breve{G}(\omega). \tag{10.84}$$

The amplitude response $\breve{G}(\omega)$ is thus nonnegative. Moreover, it has double zeros in the frequency range $[0, \pi]$. The minimum-phase FIR filter design procedure of Herrmann and Schüssler is thus as follows.

Step 1: Design a Type I linear-phase FIR transfer function $F(z)$ of degree $2N$ satisfying the specifications:

$$1 - \delta_p^{(F)} \leq \breve{F}(\omega) \leq 1 + \delta_p^{(F)}, \qquad \text{for } \omega \in [0, \omega_p], \tag{10.85a}$$

$$-\delta_s^{(F)} \leq \breve{F}(\omega) \leq \delta_s^{(F)}, \qquad \text{for } \omega \in [\omega_s, \pi]. \tag{10.85b}$$

Note that $F(z)$ has single-unit circle zeros.

Step 2: Determine the linear-phase transfer function

$$G(z) = \delta_s^{(F)} z^{-N} + F(z).$$

Its amplitude response $\breve{G}(\omega)$ satisfies

$$1 + \delta_s^{(F)} - \delta_p^{(F)} \leq \breve{G}(\omega) \leq 1 + \delta_s^{(F)} + \delta_p^{(F)}, \qquad \text{for } \omega \in [0, \omega_p],$$

$$0 \leq \breve{G}(\omega) \leq 2\delta_s^{(F)}, \qquad \text{for } \omega \in [\omega_s, \pi].$$

Note that $G(z)$ has double zeros in the stopband with all other zeros situated with a mirror-image symmetry. It can thus be expressed in the form

$$G(z) = z^{-N} H_m(z)H_m(z^{-1}), \tag{10.86}$$

where $H_m(z)$ is a minimum-phase FIR transfer function containing the zeros of $G(z)$ that are inside the unit circle and one each of the double zeros of $G(z)$ on the unit circle.

Step 3: Determine $H_m(z)$ from $G(z)$ by applying a *spectral factorization*.

The passband ripple $\delta_p^{(F)}$ and the stopband ripple $\delta_s^{(F)}$ of $F(z)$ must be chosen to ensure that the specified passband ripple δ_p and the stopband ripple δ_s of $H_m(z)$ are satisfied. To this end, it can be shown that (Problem 10.26)

$$\delta_p^{(F)} = \sqrt{1 + \frac{\delta_p}{1 + \delta_s}} - 1, \tag{10.87a}$$

$$\delta_s^{(F)} = \sqrt{\frac{2\delta_s}{1 + \delta_s}}. \tag{10.87b}$$

An estimate of the order N of $H_m(z)$ can be found by first estimating the order of $F(z)$ using either Eq. (10.3) or (10.4) or (10.5) and then dividing it by 2. If the estimated order of $F(z)$ is an odd integer, it should be increased by 1 to make it even.

10.5   FIR Digital Filter Design Using MATLAB

The *Signal Processing Toolbox* of MATLAB includes a variety of M-files for the design of FIR digital filters. We illustrate the use of some of these functions in this section.

10.5.1   FIR Digital Filter Order Estimation Using MATLAB

As in the case of IIR digital filter design, the FIR digital filter design process also consists of two steps. In the first step, the filter order is estimated from the given specifications. In the second step, the coefficients of the transfer function are determined using the estimated order and the filter specifications. We consider in this section the order estimation problem. In the next two sections, we treat the FIR filter design problem.

Either one of the three formulas of Eqs. (10.3), (10.4), and (10.5) can be used to estimate the order of an FIR filter.

The M-file `kaiord` determines the FIR filter order using Kaiser's formula of Eq. (10.3). Example 10.11 illustrates the application of the above function.

kaiord.m

EXAMPLE 10.11   **FIR Filter Order Estimation Using Kaiser's Formula**

We recompute the order of the linear-phase FIR filter of Example 10.1 using the function `kaiord`. Here, $\delta_p = 0.0114469$, $\delta_s = 0.01778279$, $Fp = 1800$, $Fs = 2000$, and $FT = 12,000$. The code fragment used is as follows:

```
N = kaiord(0.0114469, 0.01778279, 1800, 2000, 12000)
```

The data displayed by running this code is N = 99, which is the same as that derived in Example 10.1.

The M-file `bellangord` which determines the FIR filter order using Bellanger's formula of Eq. (10.4). Example 10.12 illustrates the application of the above function.

bellangord.m

EXAMPLE 10.12   **FIR Filter Order Estimation Using Bellanger's Formula**

We recompute the order of the linear-phase FIR filter of Example 10.2. Here, $\delta_p = 0.0114469$, $\delta_s = 0.01778279$, $Fp = 1800$, $Fs = 2000$, and $FT = 12,000$. The code fragment used is as follows:

```
N = bellangord(0.0114469, 0.01778279, 1800, 2000, 12000)
```

The data displayed by running this code is N = 107, which is the same as that derived in Example 10.2.

The *Signal Processing Toolbox* of MATLAB includes the M-file `remezord`, which determines the FIR filter order employing the formula of Eq. (10.5). It is available with two different options. We illustrate its use in Example 10.13.

Program 10_1.m

EXAMPLE 10.13   **FIR Filter Order Estimation Using Hermann's Formula**

Estimate the order of a linear-phase FIR transfer function with the specifications given in Example 10.1 using `remezord`. To this end, we can use Program 10_1. The output data generated by this program is given by

```
Filter order is 106
```

which is seen to be the same as that derived in Example 10.3.

For FIR filter design using the Kaiser window, the window order should be estimated using the formula of Eq. (10.42). To this end, the M-file `kaiserord` in the *Signal Processing Toolbox* of MATLAB can be employed. This function is available with three options. We demonstrate its application in Example 10.14.

> **EXAMPLE 10.14   Estimation of Filter Order for Kaiser Window-Based Design**
>
> Using MATLAB, we determine the order N and the parameter β of the Kaiser window needed to design the lowpass filter of Example 10.8. Using the program statement
>
> ```
> [N,Wn,beta,ftype] = kaiserord([0.3 0.5],[1 0],[0.01 0.01])
> ```
>
> we get $N = 23$ and $\beta = 3.3953$, which are seen to be the same as those derived in Example 10.8.

The function `kaiserord`, in some cases, can generate a value for N that is either greater or smaller than the required minimum value. If the FIR filter designed using N does not meet the specifications, the order should be either gradually increased or decreased by 1 until the specifications are met. The filter order N estimated using Eq. (10.42) has been found to be within ± 2 of the required value for a broad range of filter specifications.

The output data generated by this function can be directly used to design the FIR filter based on the windowed Fourier series approach using the function `fir1` discussed later in this section.

10.5.2   Equiripple Linear-Phase FIR Filter Design Using MATLAB

For designing an equiripple linear-phase FIR filter employing the Parks–McClellan algorithm (Section 10.3), we can use the MATLAB M-file `remez`,[9] which is available with several options. This function can design any type of multiband linear-phase filters.

Incorrect use of `remez` results in self-explanatory diagnostic messages. The use of the function `remez` is illustrated next for the design of frequency selective filters, differentiators, and Hilbert transformers.

Lowpass FIR Filter Design Examples

We consider first the design of a lowpass linear-phase filter using the function `remez`. Recall from our discussion in Section 7.3.1 that the transfer function of a Type 3 FIR filter has zeros at both $z = 1$ and $z = -1$, and the transfer function of a Type 4 FIR filter has a zero at $z = 1$. These filters cannot be used to design a lowpass filter. On the other hand, the transfer function of a Type 2 FIR filter has a zero at $z = -1$, and hence, it can be used to design a lowpass filter along with the Type 1 FIR filter that has no restrictions.

Example 10.15 illustrates the design of a lowpass FIR filter using `remez`.

Program 10_2.m

> **EXAMPLE 10.15   Design of an Equiripple Linear-Phase FIR Lowpass Filter Using Remez Algorithm**
>
> Design an equiripple linear-phase FIR lowpass filter with the same specifications as given in Example 9.14; that is, passband edge $F_p = 800$ Hz, stopband edge $F_s = 1000$ Hz, passband ripple $\alpha_p = 0.5$ dB, minimum stopband attenuation $\alpha_s = 40$ dB, and sampling frequency $F_T = 4000$ Hz. From Eqs. (9.3) and (9.4), we thus obtain $\delta_p = 0.0559$, and $\delta_s = 0.01$.
>
> This filter can be designed using Program 10_2. From the output data generated by this program it can be shown that the filter coefficients satisfy the symmetry constraint $h[k] = h[N - k]$. Since the filter length is 29 (the filter order $N = 28$), it is a Type 1 FIR filter. The computed gain response is plotted in Figure 10.17(a), with passband details in Figure 10.17(b).
>
> The passband ripple and the minimum stopband attenuation of the designed filter are 0.6 dB and 38.7 dB, respectively, which do not meet the given specifications. We next increase the filter order by 1 and find that for a

---

[9]Renamed `firpm` in Release 14 of MATLAB.

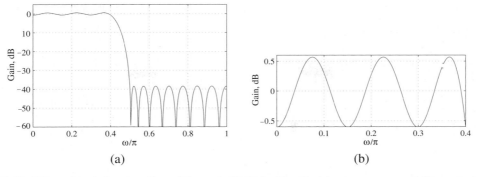

Figure 10.17: FIR equiripple lowpass filter of Example 10.15 for $N = 28$: (a) gain response and (b) passband details.

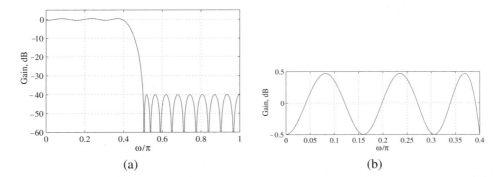

Figure 10.18: FIR equiripple lowpass filter of Example 10.15 for $N = 30$: (a) gain response and (b) passband details.

filter order of $N = 30$, the passband ripple and the minimum stopband attenuation are now 0.5 dB and 40.02 dB, respectively, which are now satisfactory. The gain response of this filter along with the passband details are shown in Figure 10.18.

Bandpass FIR Filter Design Examples

In this case, all four types of linear-phase FIR filters can be used. Example 10.16 considers the design of a bandpass linear-phase FIR filter using MATLAB and investigates the relation between the filter order and the ripple ratio on the frequency response of the filter.

EXAMPLE 10.16   Design of an Equiripple Linear-Phase FIR Bandpass Filter Using Remez Algorithm

Consider the design of a linear-phase FIR bandpass filter of order 26 with a passband from 0.3 to 0.5 and stopbands from 0 to 0.25 and from 0.55 to 1.0. The desired passband and stopband magnitudes are, respectively, 1 and 0. Assume equal weights in the passband and the stopbands.

We modify Program 10.2 by replacing 5 statements below `format long` with the following statements:

```
N = input('Filter order = ');
fpts = input('Normalized band edges = ');
mag = input('Desired magnitude values in each band = ');
wt = input('Desired weights for each band = ');
```

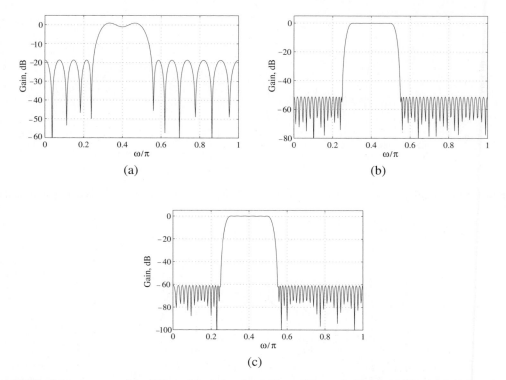

(a)

(b)

(c)

Figure 10.19: Gain response of the FIR equiripple bandpass filter of Example 10.16: (a) $N = 26$, weight ratio = 1; (b) $N = 110$, weight ratio = 1; and (c) $N = 110$, weight ratio = 1/10.

The computed gain response obtained using the modified program is plotted in Figure 10.19(a). The passband ripple and the minimum stopband attenuation of the designed filter are about 1 dB and 18.7 dB, respectively.

We next repeat the design with the order increased to 110. The gain response obtained is shown in Figure 10.19(b). The new filter has now a passband ripple of 0.024 dB, which is much smaller than that of the previous design, and a minimum stopband attenuation of 51.2 dB, which is larger than that obtained earlier. Hence, an increase in the filter order improves the frequency response characteristics at the expense of an increasing computational complexity.

It should be noted that the minimum stopband attenuation can be increased at the expense of a larger passband ripple, while keeping the order as given earlier by decreasing the relative weight ratio $W(\omega) = \delta_p/\delta_s$. Figure 10.19(c) shows the gain response of the bandpass filter of order 110 obtained with a weight vector [1, 0.1, 1]. This weight ratio puts more weight on the stopband response relative to the passband response. As expected, this unequal weighting has increased the passband ripple to 0.076 dB, while increasing the minimum stopband attenuation to 60.86 dB.

The absolute error for each of the above three bandpass filter design examples has been plotted in Figure 10.20. As can be seen from Figure 10.20(a), the absolute error, as expected, has the same peak value in all three bands. Since $L = 13$, and there are four bandedges, there can be at most $L - 1 + 6 = 18$ extrema in this design. The error plot exhibits 17 extrema. In Figure 10.20(b), we again observe the error plot has the same peak value in each band. On the other hand, as can be seen from Figure 10.20(c), the peak absolute value of the error in the passband is 10 times the peak absolute error in the stopbands as expected.

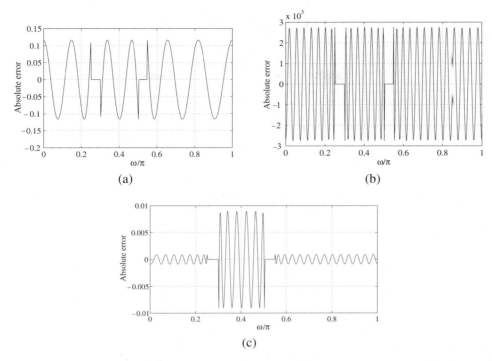

Figure 10.20: Absolute error of the FIR equiripple bandpass filter of Example 10.16: (a) $N = 26$, weight ratio = 1; (b) $N = 110$, weight ratio = 1; and (c) $N = 110$, weight ratio = 1/10.

The Parks–McClellan algorithm can create some unusual results in the design of linear-phase FIR filters with more than two bands. We investigate this problem in Example 10.17.

EXAMPLE 10.17   Unusual Results Resulting from the Use of Parks–McClellan Algorithm

Consider now the design of a linear-phase FIR bandpass filter of order 60 with a normalized passband from 0.3π to 0.5π and stopbands from 0 to 0.25π and from 0.6π to π. The desired passband and stopband magnitudes are, respectively, 1 and 0. Assume equal weights in the passband and the low-frequency stopband and a weight in the high-frequency stopband that is 30 % of that in the other two bands.

From the computed gain response shown in Figure 10.21(a), obtained using the modified program of Example 10.16, we observe that the response in the second transition band is nonmonotonic and shows a peak with a value higher than that in the passband. This indicates the presence of extremal frequencies in the second transition band. However, this result does not contradict the alternation theorem, which is concerned with extremal frequencies only in the passbands and the stopbands. As the FIR filter order is 60, $M = 30$. Thus, according to the alternation theorem, there must be at least $M + 2 = 32$ extremal frequencies. We note that the plot of the absolute error given in Figure 10.21(b) also shows the presence of 32 extremal frequencies.

Even though the Parks–McClellan algorithm guarantees equiripple error in the specified bands, as the response in a transition band is left unspecified, it cannot ensure that the gain response of a filter with more than two bands will decrease monotonically from its value in the passband. If the gain response of the filter designed exhibits a nonmonotonic behavior, it is recommended that either the filter order or the bandedges or the weighting function be adjusted until a satisfactory gain response is obtained. For example, the bandpass filter of Example 10.17 has an acceptable gain response, as shown in Figure 10.21(c), if the second stopband edge is moved from 0.6π to 0.55π, with all other parameters kept at their original values.

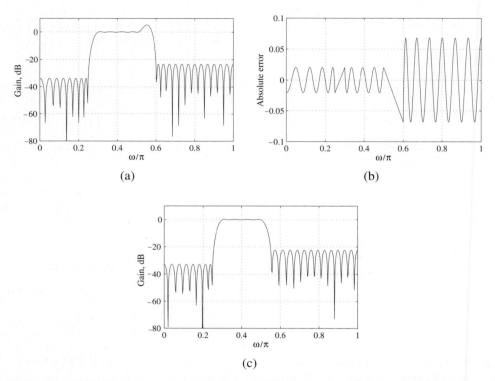

Figure 10.21: FIR equiripple bandpass filter of Example 10.17: (a) gain response with original bandedge specifications, (b) absolute error, and (c) gain response with a slightly modified stopband edge.

FIR Differentiator Design Examples

Now, from Eq. (10.24), we observe that an ideal differentiator is characterized by an antisymmetric impulse response, which implies that either a Type 3 or a Type 4 FIR filter can be used for its realization. However, from Eq. (10.23), for an ideal differentiator, $\breve{H}(\pi) = \pi$. Hence, a Type 3 FIR filter cannot be used for its realization as its transfer function has a zero at $z = -1$, which forces the amplitude response to go to zero at $\omega = \pi$. As a result, only a Type 4 FIR filter can be used for its design.

In most practical applications, signals of interest are in a frequency range $0 \leq \omega \leq \omega_p$. Consequently, a lowpass differentiator with a bandlimited frequency response

$$H_{DIF}(e^{j\omega}) = \begin{cases} j\omega, & 0 \leq |\omega| \leq \omega_p, \\ 0, & \omega_s \leq |\omega| \leq \pi, \end{cases} \tag{10.88}$$

can be employed. In Eq. (10.88), $0 \leq \omega \leq \omega_p$ and $\omega_s \leq \omega \leq \pi$ represent, respectively, the passband and the stopband of the differentiator. The frequency ω_p is usually called its bandwidth. Hence, both Type 3 and Type 4 FIR filters can be used to design a lowpass differentiator. For the design phase, we choose the weighting function as

$$W(\omega) = \frac{1}{\omega}, \qquad 0 \leq \omega \leq \omega_p,$$

and the desired amplitude response as

$$D(\omega) = 1, \qquad 0 \leq \omega \leq \omega_p.$$

The function `remezord` cannot be used to estimate the order of an FIR differentiator, as the formula of Eq. (10.5) has been developed for conventional filters with two or more bands with constant gain levels. However, the function `remez` can be employed to design an equiripple FIR differentiator, as demonstrated in Examples 10.18 and 10.19.

EXAMPLE 10.18   Design of a Full-Band Differentiator Using Parks–McClellan Algorithm

Consider the design of a full-band differentiator of order 11. Since the length is even, we can use a Type 4 FIR filter for its design. We use the program statement

```
b = remez(11,[0 1],[0 pi],'differentiator');
```

The plots of the amplitude response of the differentiator designed and the absolute error between the amplitude response of the differentiator and the desired amplitude response $[A(\omega) - \omega]$ are shown in Figure 10.22. We observe from Figure 10.22(b) that the absolute error increases as ω increases from 0 to π. This is expected as the application of the Parks–McClellan algorithm results in an equiripple error of the function $[\frac{A(\omega)}{\omega} - 1]$. The filter coefficients given by b can be shown to satisfy the antisymmetric condition.

EXAMPLE 10.19   Design of a Lowpass Differentiator Using Parks–McClellan Algorithm

We now consider the design of a lowpass differentiator with the following specifications: passband edge $\omega_p = 0.4\pi$, stopband edge $\omega_s = 0.45\pi$. Here, we can use either a Type 3 or a Type 4 FIR filter for the implementation.

We consider here the design using a Type 4 filter of order 50. Figure 10.23 shows the magnitude response of the FIR lowpass differentiator and the resulting absolute error obtained using the program statement

```
b = remez(50,[0 0.4 0.45 1],[0 0.4*pi 0 0],'differentiator');
```

Rabiner and Schafer [Rab74a] have investigated the relations between the filter order N, the bandwidth ω_p, and the peak absolute error ε of a lowpass differentiator through extensive designs. Their results, available in the form of design charts, can be used to estimate the filter order for a specified bandwidth and the peak absolute error in dB.

FIR Hilbert Transformer Design Examples

Like the ideal differentiator, as can be seen from Eq. (10.22), the ideal Hilbert transformer also has an antisymmetric impulse response, implying either a Type 3 or a Type 4 FIR filter for its realization. However, from Eq. (10.23), we observe that the magnitude response of an ideal Hilbert transformer is unity for all ω, which cannot be satisfied by a Type 3 FIR filter whose magnitude response has a zero at $\omega = 0$ or by a Type 4 FIR filter whose magnitude response has a zero at both $\omega = 0$ and $\omega = \pi$. In practice, the signals of interest are in a finite range $\omega_L \leq |\omega| \leq \omega_H$, and as a consequence, the Hilbert transformer can be designed with a bandpass amplitude response given by

$$D(\omega) = 1, \qquad \omega_L \leq |\omega| \leq \omega_H, \tag{10.89}$$

with the weighting function $P(\omega)$ set to unity in the band of interest.

As indicated by Eq. (10.22), the impulse response samples of an ideal Hilbert transformer satisfy the condition

$$h_{HT}[n] = 0, \qquad \text{for } n \text{ even.} \tag{10.90}$$

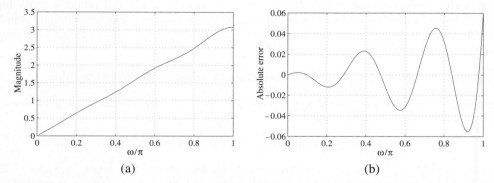

Figure 10.22: FIR equiripple differentiator of Example 10.18 of length 11: (a) magnitude response and (b) absolute error.

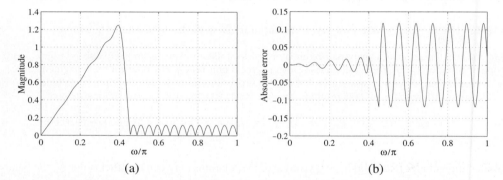

Figure 10.23: FIR equiripple lowpass differentiator of Example 10.19 of length 51: (a) magnitude response and (b) absolute error.

It can be shown that the above attractive property can be maintained by a Type 3 linear-phase FIR filter if the desired amplitude response $\breve{H}(\omega)$ is symmetric with respect to $\pi/2$; that is,

$$\breve{H}(\omega) = \breve{H}(\pi - \omega).$$

However, the condition of Eq. (10.90) cannot be met by a Type 4 linear-phase filter with an antisymmetric amplitude response (Problem 10.39).

As in the case of the FIR differentiator design, the function `remezord` cannot be used to estimate the order of an FIR Hilbert transformer. Rabiner and Schafer [Rab74b] have investigated the relations between the order N, passband ripple δ_p, and the normalized transition bandwidth ω_L of a bandpass Hilbert transformer using extensive designs. Based on their investigations, the following formula has been developed to estimate the order of the Hilbert transformer:

$$N \cong -\frac{3.833 \, \log_{10} \delta_p}{\omega_L}.$$

The design of an equiripple FIR bandpass Hilbert transformer using `remez` is demonstrated in Example 10.20.

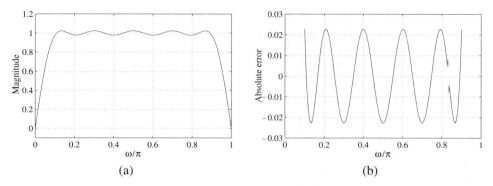

Figure 10.24: FIR equiripple bandpass Hilbert transformer of Example 10.20 of length 21: (a) magnitude response and (b) absolute error.

EXAMPLE 10.20   Design of an Equiripple FIR Bandpass Hilbert Transformer Using Remez Algorithm

Design a linear-phase bandpass FIR Hilbert transformer of order 20 having a magnitude of unity in the passband, and passband edges at 0.1π and 0.9π, respectively. The magnitude response of the Type 3 FIR bandpass Hilbert transformer and its absolute error obtained using the program statement

```
b = remez(20,[0.1 0.9],[1 1],'Hilbert');
```

are shown in Figure 10.24. It can be shown that all even indexed samples of the impulse response vector b are zero, as expected, because of the symmetry imposed on the desired amplitude response. It can be shown that if the desired amplitude response does not satisfy the symmetry condition, the even-indexed samples of the impulse response of the Hilbert transformer designed will not have zero values. To ensure zero values for the even-indexed samples, an elegant solution has been provided by Vaidyanathan and Nguyen [Vai87b]. We describe their technique in Section 13.6.3.

Generalized Remez FIR Filter Design

The M-file `gremez`[10] is a generalized version of the M-file `remez` and can be used to design a variety of real coefficient FIR filters with an equiripple error characteristic [Shp90]. There are various forms of this function. In its simplest form, `b = gremez(N,fpts,mag,wt)`, it designs an order N linear-phase FIR filter b having the best approximation in the minimax sense to the desired frequency response defined by `fpts`, `mag`, and `wt`, where these parameter vectors are as defined for the M-file `remez`. With an additional string `'Hilbert'` or `'differentiator'`, this function can also be used to design either a Hilbert transformer or a differentiator.

An interesting variation of the above function is `b = gremez('order',fpts,mag,dev)`, where `'order'` is either the string `'minorder'` or `'mineven'` or `'minodd'`. If `'minorder'` is used, it designs filters repeatedly until a filter with the least order meeting the specifications is found. In the other two cases, it finds the filter with a minimum even or odd order, depending on the string used. The peak ripple per frequency band is specified by the vector `dev`. An alternate version of the above variation is `b = gremez({'order',NI},fpts,mag,dev)`, where NI is the initial estimate of the filter order. This version is useful in designing either the Hilbert transformer or the differentiator for which `remezord` cannot be used to estimate the order.

---

[10]Renamed `firgr` in Release 14 of MATLAB.

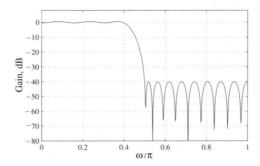

Figure 10.25: Gain response of the FIR equiripple lowpass filter of Example 10.21.

There are many other versions of the function `gremez`. We illustrate the use of the function `gremez` in Example 10.21.

EXAMPLE 10.21   Design of a Minimum Order Equiripple FIR Lowpass Filter

In Example 10.15, we observed that the function `remezord` underestimated the order of the lowpass filter and the stopband attenuation was met by increasing the estimated order by 2. We repeat the filter design of this example using the function `gremez` with the option `'minorder'`. The code fragment used is

```
b = gremez('minorder',[0 0.4 0.5 1],[1 1 0 0],[0.0559 0.01]);
```

The gain response of the resulting FIR filter is shown in Figure 10.25 and can be seen to meet the stopband specifications. The length of the filter designed can be determined using the statement `length(b)` and is found to be 31. Thus, the order of the filter meeting the specifications is 30.

10.5.3   Minimum-Phase FIR Filter Design Using MATLAB

The design of a minimum-phase FIR filter $H_m(z)$ of order N as outlined in Section 10.4 involves several steps. First, a Type 1 linear-phase FIR filter $F(z)$ of order $2N$ with specifications derived from the specifications of the desired minimum-phase FIR filter is designed. Then, an intermediate linear-phase FIR filter $G(z)$ with a nonnegative amplitude response is developed by adding a constant equal to the peak stopband ripple to the amplitude response of $H(z)$. Finally, a spectral-factorization of $G(z)$ yields

$$G(z) = z^{-N} H_m(z) H_m(z^{-1}),$$

where $H_m(z)$ contains all zeros of $G(z)$ that are inside the unit circle and one of each double zeros of $G(z)$ on the unit circle. Several methods to determine the spectral factor $H_m(z)$ without determining the roots of $G(z)$ have been advanced [Boi81], [Mia82]. We describe below the spectral factorization method recently advanced by Orchard and Willson [Orc2003].

The Spectral Factorization Method

Without any loss of generality, consider the minimum-phase sepectral factorization of a sixth-order linear-phase FIR filter $G(z)$ with a nonnegative amplitude response:

$$G(z) = g_3 + g_2 z^{-1} + g_1 z^{-2} + g_0 z^{-3} + g_1 z^{-4} + g_2 z^{-5} + g_3 z^{-6}. \tag{10.91}$$

Our objective is to express the above $G(z)$ in the form

$$G(z) = z^{-3} H_m(z) H_m(z^{-1}), \qquad (10.92)$$

where

$$H_m(z) = a_0 + a_1 z^{-1} + a_2 z^{-2} + a_3 z^{-3}, \qquad (10.93)$$

is the minimum-phase factor.

Expressing $G(z)$ in terms of the coefficients of $H_m(z)$, we get

$$G(z) = (a_0 + a_1 z^{-1} + a_2 z^{-2} + a_3 z^{-3})(a_3 + a_2 z^{-1} + a_1 z^{-2} + a_0 z^{-3}). \qquad (10.94)$$

Forming the product of the two polynomials in Eq. (10.94) and comparing the coefficients of like powers of z^{-1} of the product with the coefficients of $G(z)$ given in Eq. (10.91), we arrive at the following set of nonlinear equations:

$$\begin{aligned} g_0 &= a_0^2 + a_1^2 + a_2^2 + a_3^2, \\ g_1 &= a_0 a_1 + a_1 a_2 + a_2 a_3, \\ g_2 &= a_0 a_2 + a_1 a_3, \\ g_3 &= a_0 a_3. \end{aligned} \qquad (10.95)$$

The above set of equations is then solved iteratively using the Newton–Raphson method. First, the initial values of the coefficients a_i are chosen to ensure that $H_m(z)$ has all zeros strictly inside the unit circle. Then, the coefficients a_i are changed by adding the corrections e_i so that the modified values $c_i + e_i$ satisfy Eq. (10.95) better. The process is repeated until the iteration converges, yielding the desired values for the coefficients of $H_m(z)$. Substituting $a_i + e_i$ in Eq. (10.95) and expanding the products, a set of linear equations are obtained by eliminating all quadratic terms in e_i from the expansion. In matrix form, these equations can be written as

$$\mathbf{Ae} = \mathbf{b}, \qquad (10.96)$$

where

$$\mathbf{A} = \begin{bmatrix} 2a_0 & 2a_1 & 2a_2 & 2a_3 \\ a_1 & a_0 + a_2 & a_3 + a_1 & a_2 \\ a_2 & a_3 & a_0 & a_1 \\ a_3 & 0 & 0 & a_0 \end{bmatrix}, \qquad \mathbf{e} = \begin{bmatrix} e_0 \\ e_1 \\ e_2 \\ e_3 \end{bmatrix},$$

and

$$\mathbf{b} = \begin{bmatrix} g_0 - a_0^2 - a_1^2 - a_2^2 - a_3^2 \\ g_1 - a_0 a_1 - a_1 a_2 - a_2 a_3 \\ g_2 - a_0 a_2 - a_1 a_3 \\ g_3 - a_0 a_3 \end{bmatrix}.$$

The matrix $\mathbf{A}$ can be expressed as a sum of two triangular matrices:

$$\mathbf{A} = \begin{bmatrix} a_0 & a_1 & a_2 & a_3 \\ a_1 & a_2 & a_3 & 0 \\ a_2 & a_3 & 0 & 0 \\ a_3 & 0 & 0 & 0 \end{bmatrix} + \begin{bmatrix} a_0 & a_1 & a_2 & a_3 \\ 0 & a_0 & a_1 & a_2 \\ 0 & 0 & a_0 & a_1 \\ 0 & 0 & 0 & a_0 \end{bmatrix}.$$

The iteration convergence is checked at each step by evaluating the error term $\sum_{i=0}^{3} e_i^2$. The error first decreases monotonically, and the iteration is stopped when the error starts increasing.

The M-file `minphase`[11] implements the above spectral factorization method.   minphase.m

---

[11] Reproduced with permission of the authors [Orc2003].

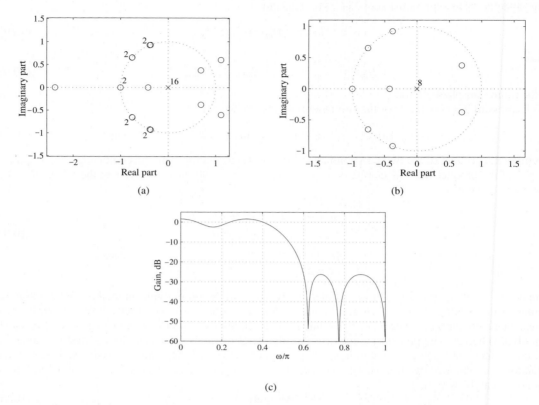

Figure 10.26: (a) Plot of zeros of $G(z)$, (b) plot of zeros of $H_m(z)$, and (c) gain response of $H_m(z)$.

Design Example

Program 10_3.m

We illustrate the application of the above spectral factorization method in Example 10.22.

> **EXAMPLE 10.22   Design of a Minimum-Phase Lowpass FIR Digital Filter**
>
> Design a minimum-phase lowpass FIR filter with a passband edge at $\omega_p = 0.45\pi$, stopband edge at $\omega_s = 0.6\pi$, passband ripple of $R_p = 2$ dB, and a minimum stopband attenuation of $R_s = 26$ dB.
>   Program 10_3 can be used to design the above filter. It should be noted that the actual value of the peak stopband ripple was added to the middle impulse response coefficient of $H(z)$ to ensure that the linear-phase filter $G(z)$ has a nonnegative amplitude response. The plots of the zeros of $G(z)$ and $H_m(z)$ generated by the above program are shown in Figure 10.26(a) and (b), respectively. The gain response of the designed minimum-phase filter $H_m(z)$ is shown in Figure 10.26(c).

Direct Design Using MATLAB

A minimum-phase FIR filter can also be designed using the M-file `firminphase`. We illustrate its use in Example 10.23.

> **EXAMPLE 10.23**   **Direct Design of the Minimum-Phase Spectral Factor Using MATLAB**
>
> We determine the minimum-phase spectral factor of the linear-phase FIR filter c of Example 10.22. The code fragment used is
>
> ```
> h = firminphase(c)
> ```
>
> A plot of the zeros of the minimum-phase factor is shown in Figure 10.27. It should be noted that the minimum-phase factor obtained using the above program statement is nearly identical to that derived in Example 10.22.

Minimum-Phase FIR Filter Design Using GREMEZ

The M-file gremez can also be used to design a minimum-phase FIR filter by adding the argument 'minphase', as shown below:

```
b = gremez(....,'minphase')
```

Maximum-Phase FIR Filter Design

The maximum-phase spectral factor g of a linear-phase FIR filter with an impulse response b of even order with a nonnegative zero-phase response can be determined by first computing its minimum-phase spectral factor h and then using the statement

```
g = fliplr{h}
```

A maximum-phase FIR filter can also be designed directly using the M-file gremez. To this end, an additional argument is added to the statement as indicated below:

```
b = gremez(....,'maxphase)
```

10.5.4 Window-Based FIR Filter Design Using MATLAB

The window-based FIR filter design process involves three steps. In the first step, the type of window to be used is determined to meet the desired stopband attenuation. In the second step, the order of the FIR filter to be designed is estimated using either the data given in Table 10.2 or Eq. (10.38) if using a Dolph-Chebyshev window or Eq. (10.42) if using a Kaiser window. Finally, in the last step, the desired impulse response of the ideal filter is computed, which is then multiplied by the window coefficients generated in the first step to yield the coefficients of the FIR filter.

FIR filter order estimation using MATLAB has been described in Section 10.5.1. We now discuss the window generation using MATLAB.

Window Generation

The *Signal Processing Toolbox* of MATLAB includes the following functions for generating the windows discussed earlier in Section 10.2:

```
w = blackman(L),    w = hamming(L),    w = hann(L)
w = chebwin(L,Rs),  w = kaiser (L,beta)
```

The above functions generate a vector w of window coefficients of odd length L.[12] The parameter *beta* in the Kaiser window is the same as the parameter β of Eq. (10.39) and can be computed using Eq. (10.41).

---

[12]It should be noted that the Hann and Blackman windows generated by MATLAB have zero-valued first and last coefficients. As a result, to keep the length the same for all windows, these two windows should be generated using a value of L that is greater by 2 than the value of L used for the Hamming and the Kaiser windows.

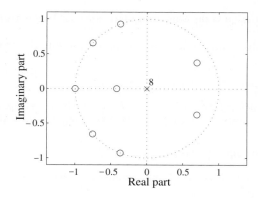

Figure 10.27: Plot of the zeros of the minimum-phase factor.

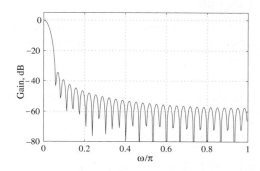

Figure 10.28: Gain response of the Kaiser window generated in Example 10.24.

Example 10.24 illustrates the generation of the coefficients of the Kaiser window.

EXAMPLE 10.24   Kaiser Window Generation Using MATLAB

Consider the determination of the coefficients of a Kaiser window to be used for the design of a lowpass FIR filter. The filter specifications are $\omega_p = 0.3\pi$, $\omega_s = 0.4\pi$, and $\alpha_s = 50$ dB. Substituting the value of α_s in Eq. (9.4), we arrive at the value of the peak stopband ripple $\delta_s = 0.003162$.

The gain response of the window obtained using Program 10_4 given is shown in Figure 10.28.

Program 10_4.m

Filter Design

The functions available in MATLAB for the design of FIR filters using the windowed Fourier series approach are fir1 and fir2. The function fir1 is used to design conventional lowpass, highpass, bandpass, bandstop, and multiband FIR filters, whereas the function fir2 is employed to design FIR filters with arbitrarily shaped magnitude response.

We illustrate next the use of the above two functions in designing linear-phase FIR filters. Without any loss of generality, we restrict our attention to filter design using the Kaiser window. Example 10.25 considers the design of a lowpass filter.

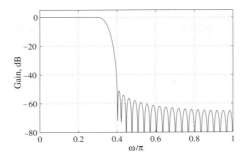

Figure 10.29: Gain response of the lowpass filter of Example 10.25.

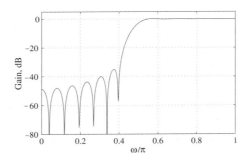

Figure 10.30: Gain response of the highpass filter of Example 10.26.

EXAMPLE 10.25   Lowpass FIR Filter Design Using the Kaiser Window

We continue here with the remaining steps in the design of the lowpass FIR filter of Example 10.24 using the Kaiser window. The desired specifications for the filter are $\omega_p = 0.3\pi$, $\omega_s = 0.4\pi$, and $\delta_s = 0.003\,62$. The MATLABprogram used for this filter design problem is Program 10_5. The input data for this program are the same as in Example 10.24. The gain response of the resulting lowpass filter is shown in Figure 10.29. The filter order computed by the above program is N = 59. The filter designed is therefore of Type 2.

Program 10_5.m

Example 10.26 considers the design of a highpass filter. It should be noted that for highpass and bandstop filter design, the function `fir1` requires that the order of the filter N be even. If the order generated using `kaiserord` is not even, it should be increased by 1 before the window and filter coefficients are generated.

EXAMPLE 10.26   Highpass FIR Filter Design Using the Kaiser Window

A highpass FIR filter is to be designed meeting the following specifications: passband edge $\omega_p = 0.55\pi$, stopband edge $\omega_s = 0.4\pi$, and stopband ripple $\delta_s = 0.02$. To design a highpass filter (or a bandstop filter), we modify Program 10_5 by changing the statement b = fir1(N,Wn, kw); to b = fir1(N,Wn,'highpass', kw);. With this modification, we arrive at the computed gain response indicated in Figure 10.30 for the given specifications.

In Example 10.27 we consider the application of the function `fir2` of MATLAB in designing an FIR filter with passbands having different gain levels.

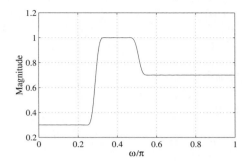

Figure 10.31: Magnitude response of the multilevel filter of Example 10.27.

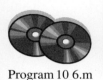

Program 10_6.m

> **EXAMPLE 10.27   Design of Multiband FIR Filter Using Hamming Window**
>
> Consider the design of an FIR filter of order 100 with three different constant magnitude levels: 0.3 in the frequency range 0 to 0.28, 1.0 in the frequency range 0.3 to 0.5, and 0.7 in the frequency range 0.52 to 1.0. To this end, we use Program 10_6. The computed magnitude response is indicated in Figure 10.31.

10.6   Design of Computationally Efficient FIR Digital Filters

We have indicated earlier in Section 9.1.2, in general, the order N of an FIR transfer function is considerably higher than the order of an equivalent IIR transfer function meeting the same magnitude response specifications. However, an FIR filter is often attractive in many applications as it is always stable and can be designed with linear-phase. Moreover, as indicated by Eqs. (10.3) and (10.5), the order of an FIR filter is inversely proportional to the width $\Delta\omega$ of the transition band. Hence, in the case of an FIR filter with a very sharp transition, the order of the filter is extremely high. This is particularly critical in designing very narrow-band or very wide-band FIR filters.

The computational complexity of a digital filter is basically determined by the total number of multipliers and adders needed to implement the filter. The direct form implementation of a linear-phase FIR transfer function of order N requires, in general, $\lfloor \frac{N+1}{2} \rfloor$ multipliers and N two-input adders. Hence, it is important to develop realization techniques that require significantly fewer number of these arithmetic units than that needed in the direct form realization.

In this section, we outline two methods of realizing computationally efficient linear-phase FIR filters. The basic building block in both realizations is an FIR subfilter structure with a periodic amplitude response.

10.6.1   The Periodic Filter Section

To understand the principles behind both design methods, we first introduce the concept of the periodic filter section and examine its properties. Consider a Type 1 linear-phase FIR filter $F(z)$ of even degree N:

$$F(z) = \sum_{n=0}^{N} f[n]z^{-n}.$$

(10.97)

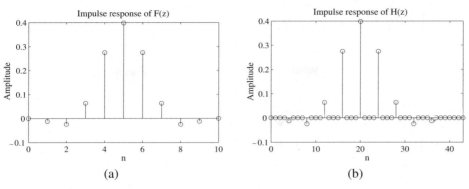

Figure 10.32: (a) Impulse response of $F(z)$ and (b) impulse response of $H(z) = F(z^L)$ for $L = 4$.

Its delay-complementary filter $E(z)$ is given by

$$E(z) = z^{-N/2} - F(z) = z^{-N/2} - \sum_{n=0}^{N} f[n]z^{-n}$$

$$= (1 - f[N/2])z^{-N/2} - \sum_{\substack{n=0 \\ n \neq N/2}}^{N} f[n]z^{-n}. \qquad (10.98)$$

The transfer function $H(z)$ obtained by replacing z^{-1} in $F(z)$ of Eq. (10.97) with z^{-L}, with L being a positive integer, is given by

$$H(z) = F(z^L) = \sum_{n=0}^{N} f[n]z^{-nL}. \qquad (10.99)$$

The order of the transfer function $H(z)$ is thus NL. A direct realization of $H(z)$ is obtained simply by replacing each unit delay in the realization of $F(z)$ by L unit delays, while keeping the number of multipliers and the adders the same. The transfer function $H(z)$ has a sparse impulse response of length $NL + 1$, with $L - 1$ zero-valued samples inserted between every consecutive pair of impulse response samples of $F(z)$, as shown in Figure 10.32 for a typical case. The parameter L is referred to as the *sparsity factor.*

The relations between the amplitude responses of these two filters is then given by

$$\check{H}(\omega) = \check{F}(L\omega). \qquad (10.100)$$

It follows from the above that the amplitude response $\check{H}(\omega)$ is a periodic function of ω with a period $2\pi/L$. One period of $\check{H}(\omega)$ is obtained by compressing the amplitude response $\check{F}(\omega)$ in the interval $[0, 2\pi]$ to the interval $[0, 2\pi/L]$.[13] A transfer function $H(z)$ with a frequency response that is a periodic function of ω with a period $2\pi/L$ is called a *periodic filter.*

If $F(z)$ is a lowpass filter with a single passband and a single stopband, $H(z)$ will be a multiband filter with $\lfloor \frac{L}{2} \rfloor + 1$ passbands and $\lceil \frac{L}{2} \rceil$ stopbands, as shown in Figure 10.33 for $L = 4$. Let $F(z)$ be a lowpass filter with a passband edge and stopband edge at $\omega_p^{(F)}$ and $\omega_s^{(F)}$, respectively, where $\omega_s^{(F)} < \pi$. Then the passband

---

[13]This approach is used in designing a comb filter, as described in Section 7.4.3.

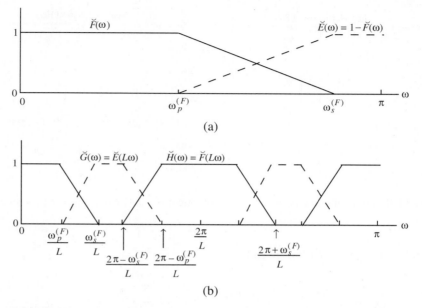

Figure 10.33: (a) Amplitude response of $F(z)$ (solid line) and $E(z)$ (dashed line) and (b) amplitude responses of $H(z) = F(z^L)$ (solid line) and $G(z) = E(z^L)$ (dashed line) for $L = 4$.

edge and the stopband edge of the first band of $H(z) = F(z^L)$ are at $\omega_p^{(F)}/L$ and $\omega_s^{(F)}/L$, respectively. The passband edges and the stopband edges of the second band are, respectively, at $(2\pi \pm \omega_p^{(F)})/L$ and $(2\pi \pm \omega_s^{(F)})/L$, and so on. The width of the transition bands of $H(z)$ are $(\omega_s^{(F)} - \omega_p^{(F)})/L$, which is $\frac{1}{L}$-th of that of $F(z)$.

Likewise, the transfer function $G(z)$ obtained by replacing z^{-1} in E(z) of Eq. (10.98) with z^{-L}, with L being a positive integer, is given by

$$G(z) = E(z^L) = z^{-NL/2} - F(z^L)$$

$$= z^{-NL/2} - \sum_{n=0}^{N} f[n]z^{-nL}. \tag{10.101}$$

The amplitude response of $G(z)$ is related to that of $H(z)$ and $F(z)$, as indicated below:

$$\breve{G}(\omega) = 1 - \breve{H}(\omega) = 1 - \breve{F}(L\omega). \tag{10.102}$$

Figure 10.33 also shows the amplitude responses of the delay-complementary filter $E(z)$ and the periodic filter $G(z) = E(z^L)$.

10.6.2   Interpolated FIR Filter

The main idea behind this approach is to design the overall filter $H_{\text{IFIR}}(z)$ as a cascade of a linear-phase FIR filter $F(z^L)$ with a periodic amplitude response and another linear-phase FIR filter $I(z)$ that suppresses the undesired passbands of the periodic filter section, as indicated in Figure 10.34 [Neu84b]. This approach leads to a single passband filter with a sharper transition band and a narrower passband than the parent FIR

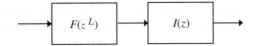

Figure 10.34: IFIR filter structure.

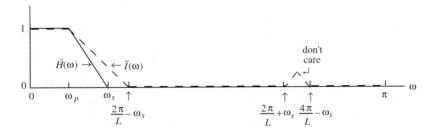

Figure 10.35: Amplitude responses of $H(z)$ (solid line) and $I(z)$ (dashed line).

filter $F(z)$. The widths of the transition band and the passband of the overall cascade are $1/L$-th of those of $F(z)$. It should be noted that the filter $F(z^L)$ has a sparse impulse response, with $L - 1$ zero-valued samples between every consecutive pair of impulse response samples of $F(z)$.

The cascaded structure is called the *interpolated finite impulse response* (IFIR) filter, as the missing samples of the sparse impulse response of the periodic filter section $F(z^L)$ are being interpolated by the second section $I(z)$, called the *interpolator*. As the filter $F(z)$ determines approximately the shape of the amplitude response of the IFIR filter, it is called the *shaping filter*.

The Design Steps

Given the specifications of the overall filter $H_{\text{IFIR}}(z)$, the design problem is then to determine the specifications of the shaping filter $F(z)$ and the interpolator $I(z)$. We outline below a straightforward approach to the design of a lowpass IFIR filter [Sar93].

Let the desired passband and stopband edges of the IFIR lowpass filter $H_{\text{IFIR}}(z)$ be ω_p and ω_s, respectively, with the specified passband and stopband ripples given by δ_p and δ_s, respectively. The passband and stopband edges of the shaping filter $F(z)$ are then given by $\omega_p^{(F)} = L\omega_p$ and $\omega_s^{(F)} = L\omega_s$. The interpolator $I(z)$ has to be designed to preserve the passband of $F(z^L)$ in the frequency range $[0, \omega_p]$ and mask the amplitude response of $F(z^L)$ in the frequency range $[\omega_s, \pi]$ where the periodic subfilter has unwanted passbands and transition bands. It follows from Figure 10.33(b) that this latter region is defined by

$$R_\omega = \bigcup_{k=1}^{\lfloor L/2 \rfloor} \left[\frac{2\pi k}{L} - \omega_s, \min\left(\frac{2\pi k}{L} + \omega_s, \pi \right) \right]. \tag{10.103}$$

The transition band of the interpolator is the frequency range $[\omega_p, \frac{2\pi}{L} - \omega_s]$. Figure 10.35 shows the amplitude responses of $H(z)$ and $I(z)$.

Next we need to determine the appropriate values of the passband and stopband ripples of the periodic subfilter and the interpolator so that the cascade of these two sections meet the ripple specifications of $H_{\text{IFIR}}(z)$. Let $\delta_p^{(F)}$ and $\delta_s^{(F)}$ denote, respectively, the passband and stopband ripples of the shaping filter

$F(z)$. Likewise, let $\delta_p^{(I)}$ and $\delta_s^{(I)}$ denote, respectively, the passband and stopband ripples of the interpolator $I(z)$. To ensure that the passband ripples of the cascade meet the specification, we choose

$$\delta_p^{(F)} + \delta_p^{(I)} = \delta_p.$$

One simple choice for the passband ripples for the two filter sections are

$$\delta_p^{(F)} = \frac{\delta_p}{2}, \qquad \delta_p^{(I)} = \frac{\delta_p}{2}. \qquad (10.104)$$

On the other hand, it is safe to choose

$$\delta_s^{(F)} = \delta_s^{(I)} = \delta_s, \qquad (10.105)$$

as it provides extra attenuation in the stopband of the cascade than that required.

Summarizing, the design specifications for the shaping filter $F(z)$ and the interpolator $I(z)$ can be as follows:

$$1 - \delta_p^{(F)} \leq \check{F}(\omega) \leq 1 + \delta_p^{(F)}, \qquad \text{for } \omega \in [0, L\omega_p], \qquad (10.106a)$$

$$-\delta_s^{(F)} \leq \check{F}(\omega) \leq \delta_s^{(F)}, \qquad \text{for } \omega \in [L\omega_s, \pi], \qquad (10.106b)$$

$$1 - \delta_p^{(I)} \leq \check{I}(\omega) \leq 1 + \delta_p^{(I)}, \qquad \text{for } \omega \in [0, \omega_p], \qquad (10.106c)$$

$$-\delta_s^{(I)} \leq \check{I}(\omega) \leq \delta_s^{(I)}, \qquad \text{for } \omega \in R_\omega. \qquad (10.106d)$$

The two linear-phase FIR filters $F(z)$ and $I(z)$ can then be designed using the Parks–McClellan algorithm to meet the specifications given by Eqs. (10.106a)–(10.106d), with the passband and stopband ripples chosen according to Eq. (10.104) and (10.105).

EXAMPLE 10.28   Design of a Linear-Phase Lowpass IFIR Digital Filter

We consider the design of a linear-phase FIR lowpass filter $H_{\text{IFIR}}(z)$ in the form of $F(z^L)I(z)$ using the IFIR filter design approach. The specifications for the filter $H_{\text{IFIR}}(z)$ are as follows: $\omega_p = 0.15\pi$, $\omega_s = 0.2\pi$, $\delta_p = 0.002$, and $\delta_s = 0.001$.

We first determine the specification for the shaping filter $F(z)$ and the interpolator $I(z)$. Now, the passband and stopband edges of $F(z)$ are given by

$$\omega_p^{(F)} = L\omega_p = 0.15\pi L,$$

$$\omega_s^{(F)} = L\omega_s = 0.2\pi L.$$

It follows from Figure 10.33 that to ensure no overlaps between adjacent passbands of $F(z^L)$, we should choose the sparsity factor L to satisfy the condition

$$\frac{\omega_s^{(F)}}{L} < \frac{2\pi - \omega_s^{(F)}}{L},$$

or, equivalently,

$$0.2\pi < \frac{2\pi}{L} - 0.2\pi, \quad \text{implying} \quad L < 5.$$

Hence, the largest value of the sparsity factor that can be used for the design is $L = 4$, yielding an IFIR structure requiring the least number of multipliers. As a result, the specifications for the shaping filter $F(z)$ and the interpolator $I(z)$ are as follows:

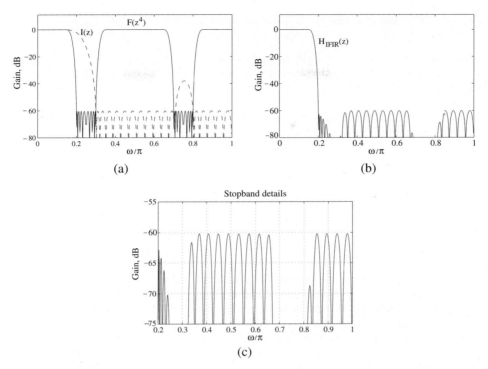

Figure 10.36: Gain responses of (a) $F(z^4)$ and $I(z)$, (b) $H_{\text{IFIR}}(z)$, and (c) stopband details

$$F(z) : \omega_p^{(F)} = 0.6\pi, \qquad \omega_s^{(F)} = 0.8\pi, \quad \delta_p^{(F)} = 0.001, \quad \delta_s^{(F)} = 0.001, \qquad (10.107\text{a})$$

$$I(z) : \omega_p^{(I)} = 0.15\pi, \qquad \omega_s^{(I)} = 0.3\pi, \quad \delta_p^{(I)} = 0.001, \quad \delta_s^{(I)} = 0.001. \qquad (10.107\text{b})$$

The filter orders of $F(z)$ and $I(z)$ obtained using the function `remezord` are as follows:

$$\text{Order of } F(z) = 32, \qquad (10.108\text{a})$$

$$\text{Order of } I(z) = 43. \qquad (10.108\text{b})$$

It can be shown that the the shaping filter $F(z)$ and the interpolator $I(z)$ designed using `remez` with the above orders do not lead to an IFIR filter $H_{IFIR}(z)$ meeting the stopband specification of a minimum attenuation of 60 dB. To meet the stopband specification, the orders of $F(z)$ and $I(z)$ need to be increased to 33 and 46, respectively. The pertinent gain responses of the redesigned IFIR filter are indicated in Figure 10.36.

The number of multipliers needed to implement $F(z)$ and, hence, $F(z^4)$ is

$$\mathcal{R}_F = \left\lceil \frac{33 + 1}{2} \right\rceil = 17.$$

Likewise, the number of multipliers needed to implement $I(z)$ is

$$\mathcal{R}_I = \left\lceil \frac{46 + 1}{2} \right\rceil = 24.$$

As a result, the total number of multipliers needed to implement $H_{\text{IFIR}}(z) = F(z^4)I(z)$ is
$$\mathcal{R}_{\text{IFIR}} = 17 + 24 = 41.$$

On the other hand, we note that the order of the direct single-stage realization of the filter obtained using the function `remezord` is 122, resulting in a structure requiring $\lfloor(122 + 1)/2\rfloor = 62$ multipliers. Hence, the IFIR realization results in a saving of about 34% in the number of multipliers.

An improved algorithm that results in a much lower order interpolator $I(z)$ is based on designing iteratively the shaping filter $F(z)$ and the interpolator $I(z)$ using the Parks–McClellan algorithm to meet the specifications [Sar88]:

$$1 - \delta_p \leq \breve{F}(\omega)\breve{I}(\omega L) \leq 1 + \delta_p, \qquad \text{for } \omega \in [0, L\omega_p], \qquad (10.109a)$$

$$-\delta_s \leq \breve{F}(\omega)\breve{I}(\omega L) \leq \delta_s, \qquad \text{for } \omega \in [L\omega_s, \pi], \qquad (10.109b)$$

$$I(0) = 1, \qquad (10.109c)$$

$$-\delta_s \leq \breve{F}(L\omega)\breve{I}(\omega) \leq \delta_s, \qquad \text{for } \omega \in R_\omega, \qquad (10.109d)$$

until the difference between successive designs is below some acceptable tolerance.

In many cases, particularly for very narrow-band filter design, simple FIR filter sections singly or in cascade can be used as interpolators to mask unwanted passbands of $F(z^4)$, yielding IFIR realizations with considerably fewer number of multipliers than that obtained using the method outlined above [Neu84b]. For masking a passband centered at $\omega = \pi$, one can use the first-order lowpass section $I_1(z) = \frac{1}{2}(1 + z^{-1})$, which places a zero at $z = -1$. Likewise, to mask a passband centered at $\omega = 0$, the first-order highpass section $I_2(z) = \frac{1}{2}(1 - z^{-1})$, which places a zero at $z = 1$, can be employed. For the masking of a passband centered at ω_o, a second-order section $I_3(z) = \frac{1}{K}(1 + 2\cos\omega_o z^{-1} + z^{-2})$, with a pair of zeros on the unit circle at $\pi \pm \omega_o$, can be used. For higher attenuation of the unwanted passbands, these simple interpolators can be used more than once.

A narrow-band highpass FIR filter can also be designed using the above approach. Also, a wide-band lowpass (highpass) FIR filter $H(z)$ with an amplitude response $\breve{H}(\omega)$ can be designed by first designing a narrow-band highpass (lowpass) FIR filter $G(z)$ with an amplitude response given by

$$\breve{G}(\omega) = 1 - \breve{H}(\omega)$$

and then implementing its delay-complementary filter.

IFIR Filter Design Using MATLAB

The M-file `ifir` can be used to design the interpolated FIR filter. We illustrate the use of this function in Example 10.29.

EXAMPLE 10.29   Direct Design of a Linear-Phase Lowpass IFIR Digital Filter

We consider again the design of the FIR filter of Example 10.28. The code fragment used here is

```
[F,I]=ifir(4,'low',[0.15,0.2],[0.002,0.001]);
```

Figure 10.37 shows the gain responses of $F(z^4)$ and $I(z)$ and $H_{\text{IFIR}}(z)$. The length of the impulse responses of $F(z^4)$ and $I(z)$ are, respectively, 133 and 46. The orders of $F(z)$ and $I(z)$ are therefore, respectively, 33 and 45. Hence, the total number of multipliers needed to implement the IFIR filter is 78.

For a more computationally efficient realization, we use the `'advanced'` option. The code fragment to be used then is

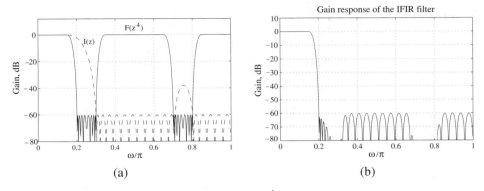

Figure 10.37: Gain responses of (a) $F(z^4)$ and $I(z)$ and (b) $H_{\text{IFIR}}(z)$.

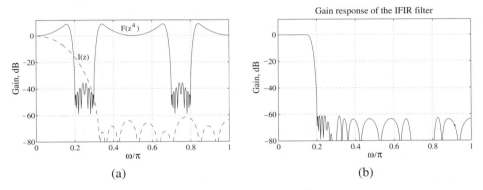

Figure 10.38: Gain responses of the redesigned (a) $F(z^4)$ and $I(z)$ and (b) $H_{\text{IFIR}}(z)$.

```
[F,I]=ifir(4,'low',[0.15,0.2],[0.002,0.001]),'advanced';
```

Figure 10.38 shows the gain responses of $F(z^4)$ and $I(z)$ and $H_{\text{IFIR}}(z)$ obtained using the above approach. The orders of $F(z)$ and $I(z)$ are now, respectively, 31 and 16. As a result, the total number of multipliers needed to implement the IFIR filter is now only 47.

10.6.3 Frequency-Response Masking Approach

This approach makes use of the relation between a periodic filter $H(z) = F(z^L)$ generated from a Type 1 linear-phase FIR filter $F(z)$ of even degree N and its delay-complementary filter $G(z)$ as given in Eq. (10.101).

The amplitude responses of $F(z)$ and its delay-complementary filter $E(z)$ are shown in Figure 10.33(a) for a typical case. The amplitude responses of the periodic filter $H(z)$ and its delay-complementary periodic filter $G(z)$ are shown in Figure 10.33(b) for $L = 4$. By selectively masking out the unwanted passbands of both $H(z)$ and $G(z)$ by cascading each with appropriate masking filters $I_1(z)$ and $I_2(z)$ and connecting the resulting cascades in parallel, one can design a large class of FIR filters with sharper transition bands

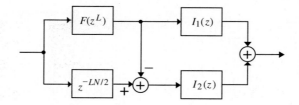

Figure 10.39: Overall filter structure for the frequency-response masking based approach.

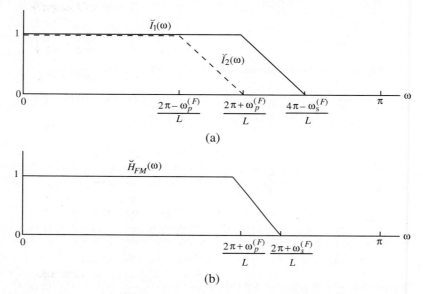

Figure 10.40: Case A: (a) amplitude response of $I_1(z)$ and $I_2(z)$ and (b) amplitude responses of the overall filter $H_{FM}(z)$.

[Lim86]. The overall structure is then realized as indicated in Figure 10.39, whose transfer function is given by

$$H_{FM}(z) = H(z)I_1(z) + G(z)I_2(z)$$
$$= F(z^L)I_1(z) + [z^{-LN/2} - F(z^L)]I_2(z). \qquad (10.110)$$

It should be noted that the delay block $z^{-LN/2}$ can be realized by tapping the FIR structure implementing $F(z^L)$. Similarly, $I_1(z)$ and $I_2(z)$ can share the same delay-chain if they are realized using the transposed direct-form structure shown in Figure 8.5(b), provided their orders are the same. From Eq. (10.110), it follows that the amplitude response of the overall filter is of the form

$$\breve{H}_{FM}(\omega) = \breve{F}(L\omega)\breve{I}_1(\omega) + [1 - \breve{F}(L\omega)]\breve{I}_2(\omega). \qquad (10.111)$$

The overall computational complexity here is given by the complexities of the shaping filter $F(z)$ and the two masking filters $I_1(z)$ and $I_2(z)$. As in the case of the IFIR filter structure, all these three filters have wide transition bands and, in general, require considerably fewer multipliers and adders than that required in a direct design of the desired sharp cutoff filter.

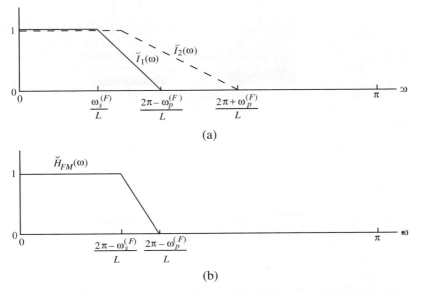

Figure 10.41: Case B: (a) amplitude response of $I_1(z)$ and $I_2(z)$ and (b) amplitude responses of the overall filter $H_{FM}(z)$.

The Design Steps

Given the specifications of $H_{FM}(z)$, the design problem here is to determine the specifications of the subfilters $F(z)$, $I_1(z)$, and $I_2(z)$. We consider here the design of a lowpass filter. However, the procedure can be easily modified to design other types of filters. As indicated next for the lowpass filter design, two different situations may arise depending on how the transition band of $H_{FM}(z)$ is created.

If the transition band is from one of the transition bands of $H(z)$, to be referred to as *Case A*, the bandedges of $H_{FM}(z)$ are related to the bandedges of $F(z)$ as follows (Figure 10.40):

$$\omega_p = \frac{2\ell\pi + \omega_s^{(F)}}{L}, \qquad \omega_s = \frac{2\ell\pi + \omega_p^{(F)}}{L}, \qquad (10.112)$$

where $\omega_p^{(F)}$ and $\omega_s^{(F)}$ are, respectively, the passband and stopband edges of $F(z)$ and $2 \leq \ell \leq L - 1$. If, on the other hand, the transition band is from one of the transition bands of $G(z)$, to be referred to as *Case B*, the bandedges of $H_{FM}(z)$ are related to the bandedges of $F(z)$ as follows (Figure 10.41):

$$\omega_p = \frac{2\ell\pi - \omega_p^{(F)}}{L}, \qquad \omega_s = \frac{2\ell\pi - \omega_s^{(F)}}{L}. \qquad (10.113)$$

We illustrate the design of an FIR filter using the above approach in Example 10.30.

EXAMPLE 10.30   Design of a Lowpass FIR Digital Filter Using the Frequency-Response Masking Approach

We design a linear-phase FIR lowpass filter $H_{FM}(z)$ using the frequency-response masking approach. The specifications for the filter $H_{FM}(z)$ are as follows: $\omega_p = 0.4\pi$, $\omega_s = 0.402\pi$, $\delta_p = 0.01$, and $\delta_s = 0.0001$ [Sar93].

The order N of the direct single-stage realization of the lowpass filter obtained using the function *remezord* is 3093, requiring 1548 multipliers if coefficient symmetry is exploited. For designing $H_{FM}(z)$, the optimum value of the parameter L is in the range $3 \leq L \leq 29$. To determine the optimum value of L resulting in a structure with the least number of multipliers, we calculate the values of the multipliers to realize the shaping filter $F(z)$, and the masking filters, $I_1(z)$, and $I_2(z)$, for all values of L in the above range. Table 10.3 lists for each value of L, the value of ℓ, the passband and stopband edges of the shaping filter $F(z)$, and N_F, N_{I_1}, and N_{I_2}, the estimated orders of $F(z)$, $I_1(z)$, and $I_2(z)$, respectively, obtained using the function *remezord*. This table also includes the total number of multipliers $\mathcal{R}_M$ needed to implement $H_{FM}(z)$, determined using

$$\mathcal{R}_M = \frac{N_F}{2} + 1 + \left\lfloor \frac{N_{I_1}+2}{2} \right\rfloor + \left\lfloor \frac{N_{I_2}+2}{2} \right\rfloor. \tag{10.114}$$

In choosing the optimum value of the sparsity factor L, there is a tradeoff between the orders of the shaping and the masking filters and, hence, in the computational complexities of these filters. An increase in the value of L leads to a decrease in the number of multipliers in the shaping filter $F(z)$, but an increase in the number of multipliers in the masking filters $I_1(z)$ and $I_2(z)$. If there are several values of L with each giving equal number of multipliers in the overall filter structure, then the value of the L chosen should be the one for which the number of multipliers in $F(z)$ is less than the total number of multipliers of $I_1(z)$ and $I_2(z)$ [Sar93]. From Table 10.3, we observe that the optimum value of L is 16 for which the total number of multipliers is $\mathcal{R}_M = 229$, which is about 15% of the total number required in the direct single-stage realization. However, the order of the overall filter has increased to 3664, which represents a 14% increase in the filter order.

The parameters listed in Table 10.3 were found using the function *remez*. Since this function uses the remez exchange algorithm, adjustments to the weight and gain parameters are necessary to ensure that the filter specifications are met [Lim86]. After adjusting the shaping filter length and the weights used in the function *remez*, we arrive at the following values for the orders of the shaping filter and the two masking filters: $N_F = 240$, $N_{I_1} = 96$, and $N_{I_1} = 136$. In the final design, the total number of multipliers has increased to 240, while the order of the shaping filter is 3840. The gain response of the final filter $H_{FM}(z)$ is shown in Figure 10.42. It should be noted that the order of the FIR filter for a direct design obtained using the function *remez* has increased to 3892.

10.7   Summary

This chapter considered the design of finite impulse response (FIR) causal digital filters. The FIR filter design is carried out directly from the digital filter specifications. The method outlined in this chapter is based on the truncated Fourier series expansion of the desired frequency response. Formulas for computing the Fourier series coefficients of some ideal frequency response specifications are included. To reduce the effect of the Gibbs phenomenon, the truncation can be carried out by applying a suitable window function. Some commonly used window functions are reviewed, along with their properties. The effect of the Gibbs phenomenon can also be reduced by providing a smooth transition between the passband and the stopband.

Next, the theory behind the popular Parks–McClellan algorithm was described. This algorithm develops a linear-phase FIR transfer function with equiripple passband and stopband magnitude responses and makes use of the Remez optimization algorithm. A simple algorithm was then outlined for the design of a minimum-phase FIR filter as it is usually of lower order and, hence, smaller group delay than an equivalent linear-phase FIR filter.

The chapter then discussed some of the M-files for the design of FIR digital filters that are available in the *Signal Processing Toolbox* of MATLAB as functions. It includes functions for the design of linear-phase filters based on the windowed Fourier series approach and the Parks–McClellan algorithm, in addition to the design of minimum-phase FIR filters. Finally, two methods for the design of computationally efficient FIR filters with very few multipliers were outlined.

Table 10.3: Total number of multipliers as a function of L.

L	ℓ	Case	$\omega_p^{(F)}$	$\omega_s^{(F)}$	N_F	N_{I_1}	N_{I_2}	$\mathcal{R}_M$
3	1	B	0.794	0.800	1166	51	11	616
4	1	B	0.392	0.400	876	22	36	470
6	1	A	0.400	0.412	584	35	51	337
7	1	A	0.800	0.814	500	127	29	330
8	2	B	0.784	0.800	438	135	35	306
9	2	B	0.382	0.400	390	51	81	263
11	2	A	0.400	0.422	318	65	95	241
12	2	A	0.800	0.824	292	223	51	285
13	3	B	0.774	0.800	270	214	58	274
14	3	B	0.372	0.400	250	80	128	232
16	3	A	0.400	0.432	220	96	136	229
17	3	A	0.800	0.834	206	325	73	304
18	4	B	0.764	0.800	196	289	81	285
19	4	B	0.362	0.400	184	107	175	235
21	4	A	0.400	0.442	168	127	175	237
22	4	A	0.800	0.844	160	433	95	346
23	5	B	0.754	0.800	152	361	105	311
24	5	B	0.352	0.400	146	135	223	254
26	5	A	0.400	0.452	136	159	215	257
27	5	A	0.800	0.854	130	546	114	398
28	6	B	0.744	0.800	126	430	128	345
29	6	B	0.342	0.400	122	161	275	281

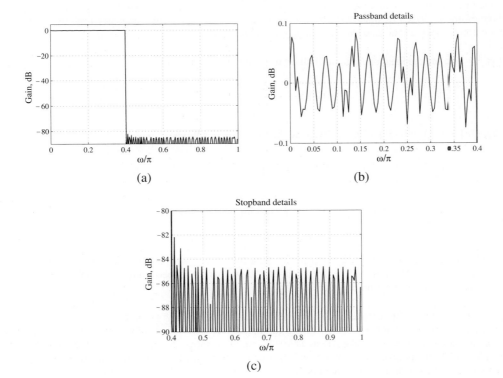

Figure 10.42: (a) Gain response of $H_{FM}(z)$, (b) passband details, and (c) stopband details.

The *Filter Design Toolbox* of MATLAB has a large number of M-files to aid in the design of FIR digital filters. This toolbox may be useful in many practical applications. The article by Losada in the Mathworks website is recommended to the reader interested in finding out about some of the practical aspects of FIR filter design.[14]

10.8 Problems

10.1 Verify the FIR filter orders given in Table 10.1.

10.2 A lowpass FIR filter of order $N = 75$ is to be designed with a transition band given by $\omega_s - \omega_p = 0.05\pi$ using the Parks–McClellan method. Determine the approximate value of the stopband attenuation α_s in dB and the corresponding stopband ripple δ_s of the designed filter if the filter order is estimated using each of the following formulas: (a) Kaiser's formula of Eq. (10.3), (b) Bellanger's formula of Eq. (10.4), and (c) Hermann's formula of Eq. (10.5). Assume the passband and stopband ripples to be the same.

10.3 Repeat Problem 10.2 if the filter is designed using the Kaiser's window-based method.

10.4 Verify the expression for the impulse response coefficients $h_{ML}[n]$ given in Eq. (10.20) for the zero-phase multiband filter with a frequency response $H_{ML}(e^{j\omega})$ defined in Eq. (10.19) and shown in Figure 10.1.

10.5 Show that the ideal Hilbert transformer with a frequency response $H_{HT}(e^{j\omega})$ defined in Eq. (10.21) has an impulse response $h_{HT}[n]$ as given in Eq. (10.22). Since the impulse response is doubly infinite, the ideal discrete-time Hilbert transformer is not realizable. To make it realizable, the impulse response has to be truncated to $|n| \le M$. What type of linear-phase FIR filter is the truncated impulse response? Plot the frequency response of the truncated approximation for various values of M. Comment on your results.

10.6 Let $\mathcal{H}\{\cdot\}$ denote the ideal operation of Hilbert transformation defined by

$$\mathcal{H}\{x[n]\} = \sum_{\ell=-\infty}^{\infty} h_{HT}[n - \ell]x[\ell],$$

where $h_{HT}[n]$ is as given in Eq. (10.22). Evaluate the following quantities:

$$\text{(a) } \mathcal{H}\{\mathcal{H}\{\mathcal{H}\{\mathcal{H}\{x[n]\}\}\}\}, \quad \text{(b) } \sum_{\ell=-\infty}^{\infty} x[\ell]\mathcal{H}\{x[\ell]\}.$$

10.7 Let $h_{LP}[n]$ denote the impulse response of an ideal lowpass filter with a cutoff at $\omega_c = \pi/2$. Show that

$$h_{HT}[n] = (-1)^n 2h_{LP}[2n]$$

is the impulse response of an ideal Hilbert transformer [Che2001]. If $h_{LP}[n]$ is the impulse response of a causal Type 1 FIR lowpass filter of order N with $M = N/2$ odd, then show that the Hilbert transformer obtained using the above relation is a causal Type 3 FIR filter of order M.

10.8 Show that the ideal differentiator with a frequency response $H_{DIF}(e^{j\omega})$ defined in Eq. (10.23) has an impulse response $h_{DIF}[n]$ as given in Eq. (10.24). Since the impulse response is doubly infinite, the ideal discrete-time differentiator is not realizable. To make it realizable, the impulse response has to be truncated to $|n| \le M$. What type of linear-phase FIR filter is the truncated impulse response? Plot the frequency response of the truncated approximation for various values of M. Comment on your results.

---

[14]See the white paper by R. Losada in the website **www.mathworks.com/products/filterdesign/**.

10.9 Develop the expression for the impulse response $\hat{h}_{HP}[n]$ of a causal highpass FIR filter of length $N = 2M + 1$ obtained by truncating and shifting the impulse response $h_{HP}[n]$ of the ideal highpass filter given by Eq. (10.16). Show that the causal lowpass FIR filter $\hat{h}_{LP}[n]$ of Eq. (10.15) and $\hat{h}_{HP}[n]$ are a delay-complementary pair.

10.10 Determine the impulse response $h_{LLP}[n]$ of a zero-phase ideal linear passband lowpass filter characterized by a frequency response shown in Figure P10.1(a).

10.11 Determine the impulse response $h_{BLDIF}[n]$ of a zero-phase ideal band-limited differentiator characterized by a frequency response shown in Figure P10.1(b).

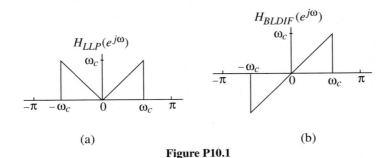

(a)   (b)

Figure P10.1

10.12 The magnitude response of an ideal *notch filter* $H_{\text{notch}}(z)$ is defined by

$$|H_{\text{notch}}(e^{j\omega})| = \begin{cases} 0, & |\omega| = \omega_o, \\ 1, & \text{otherwise,} \end{cases} \tag{10.115}$$

where ω_o is the *notch frequency*. Determine its impulse response $h_{\text{notch}}[n]$ [Yu90].

10.13 Let $h_d[n], -\infty < n < \infty$, denote the impulse response samples of a zero-phase filter with a frequency response $H_d(e^{j\omega})$. We have shown in Section 10.2.1 that the frequency response $H_t(e^{j\omega})$ of the zero-phase FIR filter $h_t[n], -M \leq n \leq M$, obtained by multiplying $h_d[n]$ with a rectangular window $w_R[n], -M \leq n \leq M$, has the least integral-squared error Φ_R defined in Eq. (10.9). Let Φ_{Hann} denote the integral-squared error if a length-$2M + 1$ Hann window is used to develop the FIR filter. Determine an expression for the excess error $\Phi_{excess} = \Phi_R - \Phi_{Hann}$.

10.14 Repeat Problem 10.13 if a Hamming window is used instead.

10.15 For each of the lowpass filter specifications given below, design an FIR filter with the smallest length meeting the specifications using the window-based approach, and plot its magnitude response using MATLAB:
   (a) $\omega_p = 0.47\pi, \omega_s = 0.59\pi, \delta_p = 0.001, \delta_s = 0.007$, (b) $\omega_p = 0.61\pi, \omega_s = 0.78\pi, \delta_p = 0.001, \delta_s = 0.002$.

10.16 Design a bandpass FIR filter with the smallest length using the window-based approach and meeting the following specifications: $\omega_{p1} = 0.45\pi, \omega_{p2} = 0.65\pi, \omega_{s1} = 0.3\pi, \omega_{s2} = 0.8\pi, \delta_p = 0.01, \delta_{s1} = 0.008$, and $\delta_{s2} = 0.05$, where δ_{s1} and δ_{s2} are, respectively, the ripple in the lower and upper stopbands. Plot the magnitude response of the filter designed using MATLAB.

10.17 Design a bandstop FIR filter with the smallest length using the window-based approach and meeting the following specifications: $\omega_{p1} = 0.3\pi, \omega_{p2} = 0.8\pi, \omega_{s1} = 0.45\pi, \omega_{s2} = 0.65\pi, \delta_{p1} = 0.05, \delta_{p2} = 0.009$, and $\delta_s = 0.02$, where δ_{p1} and δ_{p2} are, respectively, the ripple in the lower and upper passbands. Plot the magnitude response of the filter designed using MATLAB.

10.18 The frequency response of a zero-phase lowpass filter with a passband edge at ω_p, a stopband edge at ω_s, and a raised cosine transition function is given by [Bur92], [Par87]

$$H_{LP}(e^{j\omega}) = \begin{cases} 1, & 0 \le |\omega| < \omega_p, \\ \frac{1}{2}\left(1 + \cos\left(\frac{\pi(\omega - \omega_p)}{\omega_s - \omega_p}\right)\right), & \omega_p \le |\omega| \le \omega_s, \\ 0, & \omega_s < |\omega| < \pi. \end{cases} \tag{10.116}$$

Show that its impulse response is of the form

$$h_{LP}[n] = \left[\frac{\cos(\Delta\omega n/2)}{1 - (\Delta\omega/\pi)^2 n^2}\right] \cdot \frac{\sin(\omega_c n)}{\pi n}, \tag{10.117}$$

where $\Delta\omega = \omega_s - \omega_p$ and $\omega_c = (\omega_p + \omega_s)/2$.

10.19 The length-$(2M + 1)$ Hann, Hamming, and Blackman window sequences given in Eqs. (10.30) to (10.32) are all of the form of raised cosine windows and can be expressed as

$$w_{GC}[n] = \left[\alpha + \beta\cos\left(\frac{2\pi n}{2M + 1}\right) + \gamma\cos\left(\frac{4\pi n}{2M + 1}\right)\right] w_R[n], \tag{10.118}$$

where $w_R[n]$ is a length-$(2M+1)$ rectangular window sequence. Express the Fourier transform of the above generalized cosine window in terms of the Fourier transform of the rectangular window $\Psi_R(e^{j\omega})$. From this expression, determine the Fourier transform of the Hann, Hamming, and Blackman window sequences.

10.20 In this problem, we consider the design of an FIR digital filter approximating a fractional delay z^{-D}:

$$z^{-D} \cong \sum_{n=0}^{N} h[n]z^{-n},$$

where the delay D is a positive real rational number.

(a) Show that the filter coefficients obtained using the Lagrange interpolation method [15] are given by [Laa96]

$$h[n] = \prod_{\substack{k=0 \\ k \ne n}}^{N} \frac{D - k}{n - k}, \qquad 0 \le n \le N.$$

(b) Design a length-21 FIR fractional delay filter with a delay of 90/13 samples. Using MATLAB, plot the group delay response of the designed filter along with that of the ideal fractional delay filter. Comment on your results.

10.21 An ideal zero-phase comb filter with notches at a fundamental frequency ω_o and its harmonics has a frequency response given by

$$H_{\text{comb}}(e^{j\omega}) = \begin{cases} 0, & \omega = k\omega_o, \ 0 \le k \le M, \\ 1, & \text{otherwise.} \end{cases} \tag{10.119}$$

If the input to the comb filter is of the form $x[n] = s[n] + r[n]$, where $s[n]$ is the desired signal and $r[n] = \sum_{k=0}^{M} A_k \sin(k\omega_o n + \phi_k)$ is the harmonic interference with a fundamental frequency ω_o, the comb filter suppresses the interference and generates $s[n]$ as its output. Let $D = 2\pi/\omega_o$ denote the fractional sample delay.

(a) Show that $r[n - D] = r[n]$.

(b) Next, by computing the output $y[n]$ of a filter $H(z) = 1 - z^{-D}$ whose input is $x[n]$, show that $y[n]$ does not contain any harmonic interference.

---

[15]See Section 13.5.2.

(c) Even though the filter $H(z) = 1 - z^{-D}$ eliminates the harmonic interference completely, it does not have a unity magnitude at frequencies $\omega \neq k\omega_o$, thus introducing signal distortion at its output. The distortion in the passband of $H(z)$ can be eliminated by modifying the filter according to [Pei98]

$$H_c(z) = \frac{1 - z^{-D}}{1 - \rho^D z^{-D}},$$

where $0 < \rho < 1$. In practice, ρ should be close to 1. Using MATLAB, plot the magnitude response of $H_c(z)$ for $\omega_o = 0.22\pi$ and $\rho = 0.99$.

(d) Develop an efficient realization of the improved comb filter $H_c(z)$.

10.22 Using the method of Problem 10.21 and the FIR fractional delay filter design method of Problem 10.20, design a comb filter of order 18 for $\omega_o = 0.18\pi$ and $\rho = 0.98$. Using MATLAB, plot the magnitude response of the designed filter.

10.23 Using the method of Problem 10.21 and the allpass fractional delay filter design method of Problem 9.29, design an IIR comb filter of order 11 for $\omega_o = 0.18\pi$ and $\rho = 0.98$. Using MATLAB, plot the magnitude response of the designed filter.

10.24 By computing the inverse discrete-time Fourier transform of the frequency response $H_{LP}(e^{j\omega})$ of the zero-phase modified lowpass filter of Figure 10.13(a) with a first-order spline as the transition function, verify the expression for its impulse response $h_{LP}[n]$ as given in Eq. (10.43). Show that $h_{LP}[n]$ of Eq. (10.43) can also be derived by computing the inverse discrete-time Fourier transform of the derivative function $G(e^{j\omega})$ of Figure 10.13(b) and then using the differentiation-in-frequency property of the discrete-time Fourier transform given in Table 3.3.

10.25 Show that the impulse response $h_{LP}[n]$ of the zero-phase modified lowpass filter with a Pth-order spline as the transition function is given by Eq. (10.44).

10.26 Prove Eqs. (10.87a) and (10.87b).

10.27 Many applications require the fitting of a set of $2L+1$ equally spaced data samples $x[n]$ by a smooth polynomial $x_a(t)$ of degree N where $N < 2L$. In the least-squares fitting approach, the polynomial coefficients $\alpha_i, i = 0, 1, \ldots, N$, are determined so that the mean-square error

$$\varepsilon(\alpha_i) = \sum_{k=-L}^{L} \{x[k] - x_a(k)\}^2 \tag{10.120}$$

is a minimum [Ham89]. In smoothing a very long data sequence $x[n]$ based on the least-squares fitting approach, the central sample in a set of consecutive $2L + 1$ data samples is replaced by the polynomial coefficient minimizing the corresponding mean-square error.

(a) Develop the smoothing algorithm for $N = 1$ and $L = 5$, and show that it is a moving average FIR filter of length 5.

(b) Develop the smoothing algorithm for $N = 2$ and $L = 5$. What type of digital filter is represented by this algorithm?

(c) By comparing the frequency responses of the previous two FIR smoothing filters, select the filter that provides better smoothing.

10.28 An improved smoothing algorithm is *Spencer's* 15-point smoothing formula given by [Ham89]

$$\begin{aligned}
y[n] = \tfrac{1}{320} \{ &-3x[n-7] - 6x[n-6] - 5x[n-5] + 3x[n-4] \\
&+ 21x[n-3] + 46x[n-2] + 67x[n-1] + 74x[n] \\
&+ 67x[n+1] + 46x[n+2] + 21x[n+3] + 3x[n+4] \\
&- 5x[n+5] - 6x[n+6] - 3x[n+7] \}. \tag{10.121}
\end{aligned}$$

Evaluate its frequency response and, comparing it with that of the two smoothing filters of Problem 10.27, show why Spencer's formula yields the better result.

10.29 In Problem 9.3, we considered filtering by a cascade of a number of identical filters. While the cascade provides more stopband attenuation than that obtained by a single filter section, it also increases the passband ripple or, in effect, decreases the passband width for a given maximum passband deviation. In the case of an FIR filter $H(z)$ with a symmetric impulse response, improved passband and stopband performances can be achieved by employing the *filter sharpening approach* [Kai77] in which the overall system $G(z)$ is implemented as

$$G(z) = \sum_{\ell=1}^{L} \alpha_\ell [H(z)]^\ell, \tag{10.122}$$

where $\{\alpha_\ell\}$ are real constants. In this problem, we outline the method of selecting the weighting coefficients $\{\alpha_\ell\}$ for a given L. It follows from the above that $G(z)$ is also an FIR filter with a symmetric impulse response. Let x denote a specific value of the amplitude response of $H(z)$ at a given angular frequency ω. If we denote the value of the amplitude response of $G(z)$ at this value of ω as $P(x)$, then it is related to x through

$$P(x) = \sum_{\ell=1}^{L} \alpha_\ell x^\ell. \tag{10.123}$$

$P(x)$ is called the *amplitude change function*. For a BR transfer function $H(z)$, $0 \le x \le 1$, where $x = 0$ is in the stopband and $x = 1$ is in the passband. If we further desire $G(z)$ to be a BR transfer function, then the amplitude change function must satisfy the two basic properties $P(0) = 0$ and $P(1) = 1$. Additional conditions on the amplitude change function are obtained by constraining the behavior of its slope at $x = 0$ and $x = 1$. To improve the performance of $G(z)$ in the stopband, we need to ensure

$$\left. \frac{d^k P(x)}{dx^k} \right|_{x=0} = 0, \quad k = 1, 2, \ldots, n, \tag{10.124}$$

and to improve the performance of $G(z)$ in the passband, we need to ensure

$$\left. \frac{d^k P(x)}{dx^k} \right|_{x=1} = 0, \quad k = 1, 2, \ldots, m, \tag{10.125}$$

where $m + n = L - 1$. Determine the coefficients $\{\alpha_\ell\}$ for $L = 3, 4$, and 5.

10.30 Consider a Type 3 linear-phase FIR filter with an amplitude response as given in Eq. (10.54). Show that if the amplitude response is symmetric, that is, $\breve{H}(\omega) = \breve{H}(\pi - \omega)$, then it is possible to choose the parameters $c[k]$ of Eq. (10.54) so that the even-indexed impulse response samples $h[n]$ are zero.

10.31 In the frequency sampling approach of FIR filter design, the specified frequency response $H_d(e^{j\omega})$ is first uniformly sampled at M equally spaced points $\omega_k = 2\pi k/M$, $0 \le k \le M - 1$, providing M *frequency samples* $H[k] = H_d(e^{j\omega_k})$. These M frequency samples constitute an M-point DFT $H[k]$, whose M-point inverse-DFT thus yields the impulse response coefficients $h[n]$ of the FIR filter of length M [Gol69a]. The basic assumption here is that the specified frequency response is uniquely characterized by the M frequency samples and, hence, can be fully recovered from these samples.

(a) Show that the transfer function $H(z)$ of the FIR filter can be expressed as

$$H(z) = \frac{1 - z^{-M}}{M} \sum_{k=0}^{M-1} \frac{H[k]}{1 - W_M^{-k} z^{-1}}.$$

(b) Develop a realization of the FIR filter based on the above expression.

(c) Show that the frequency response $H(e^{j\omega})$ of the FIR filter designed via the frequency sampling-based approach has exactly the specified frequency samples $H(e^{j\omega_k}) = H[k]$ at $\omega_k = 2\pi k/M$, $0 \le k \le M - 1$.

10.32 Let $|H_d(e^{j\omega})|$ denote the desired magnitude response of a real linear-phase FIR filter of length M.

(a) For M odd (Type 1 FIR filter), show that the DFT samples $H[k]$ needed for a frequency sampling-based design are given by

$$H[k] = \begin{cases} |H_d\left(e^{j2\pi k/M}\right)|e^{-j2\pi k(M-1)/2M}, & k = 0, 1, \ldots, \frac{M-1}{2}, \\ |H_d\left(e^{j2\pi k/M}\right)|e^{j2\pi(M-k)(M-1)/2M}, & k = \frac{M+1}{2}, \ldots, M-1. \end{cases} \tag{10.126}$$

(b) For M even (Type 2 FIR filter), show that the DFT samples $H[k]$ needed for a frequency sampling-based design are given by

$$H[k] = \begin{cases} |H_d\left(e^{j2\pi k/M}\right)|e^{-j2\pi k(M-1)/2M}, & k = 0, 1, \ldots, \frac{M}{2} - 1, \\ 0, & k = \frac{M}{2}, \\ |H_d\left(e^{j2\pi k/M}\right)|e^{j2\pi(M-k)(M-1)/2M}, & k = \frac{M}{2} + 1, \ldots, M-1. \end{cases} \tag{10.127}$$

10.33 Design a linear-phase FIR lowpass filter of length 19 with a passband edge at $\omega_p = 0.55\pi$ using the frequency sampling approach. Assume an ideal brickwall characteristic for the desired magnitude response.

(a) Using Eq. (10.126), develop the exact values for the desired frequency samples.

(b) Using MATLAB, plot the magnitude response of the designed filter.

10.34 Design a linear-phase FIR lowpass filter of length 39 with a passband edge at $\omega_p = 0.35\pi$ using the frequency sampling approach. Assume an ideal brickwall characteristic for the desired magnitude response.

(a) Using Eq. (10.126), develop the exact values for the desired frequency samples.

(b) Using MATLAB, plot the magnitude response of the designed filter.

10.35 By solving Eq. (10.70), derive the value of ε given by Eq. (10.71).

10.36 Determine the weighting function $W(\omega)$ that is to be used to design a Type 1 linear-phase FIR lowpass filter using the Parks–McClellan method to meet the following specifications: $\omega_p = 0.45\pi$, $\omega_s = 0.6\pi$, $\delta_p = 0.2043$, and $\delta_s = 0.0454$.

10.37 Determine the weighting function $W(\omega)$ that is to be used to design a Type 1 linear-phase FIR highpass filter using the Parks–McClellan method to meet the following specifications: $\omega_p = 0.7\pi$, $\omega_s = 0.55\pi$ $\delta_p = 0.03808$, and $\delta_s = 0.0112$.

10.38 Determine the weighting function $W(\omega)$ that is to be used to design a Type 1 linear-phase FIR bandpass filter using the Parks–McClellan method to meet the following specifications: $\omega_{p1} = 0.55\pi$, $\omega_{p2} = 0.7\pi$, $\omega_{s1} = 0.44\pi$, $\omega_{s2} = 0.82\pi$, $\delta_p = 0.01$, $\delta_{s1} = 0.007$, and $\delta_{s2} = 0.002$, where δ_{s1} and δ_{s2} are, respectively, the ripples in the lower and upper stopbands.

10.39 Show that the condition of Eq. (10.90) on the impulse response samples $h_{HT}[n]$ of an ideal Hilbert transformer cannot be met by a Type 4 linear-phase FIR filter.

10.40 The *warped discrete Fourier transform* (WDFT) can be employed to determine the N frequency samples of the z-transform $X(z)$ of a length-N sequence $x[n]$ at a warped frequency scale on the unit circle. The N-point WDFT $\check{X}[k]$ of $x[n]$ is given by the N equally spaced frequency samples on the unit circle of the modified z-transform $X(\check{z})$ obtained by applying an allpass first-order spectral transformation to $X(z)$ [Mak2001]:

$$X(\check{z}) = X(z)\Big|_{z^{-1} = \frac{-\alpha + \check{z}^{-1}}{1 - \alpha\check{z}^{-1}}} = \frac{P(\check{z})}{D(\check{z})}, \tag{10.128}$$

where $|\alpha| < 1$. Thus, the N-point WDFT $\check{X}[k]$ of $x[n]$ is given by

$$\check{X}[k] = X(\check{z})\Big|_{\check{z} = e^{j2\pi k/N}}, \quad 0 \leq k \leq N-1. \tag{10.129}$$

(a) Develop the expressions for $P(\breve{z})$ and $D(\breve{z})$.

(b) If we denote

$$P(\breve{z}) = \sum_{n=0}^{N-1} p[n]\breve{z}^{-n} \text{ and } D(\breve{z}) = \sum_{n=0}^{N-1} d[n]\breve{z}^{-n},$$

show that $\breve{X}[k] = P[k]/D[k]$, where $P[k]$ and $D[k]$ are, respectively, the N-point DFTs of the sequences $p[n]$ and $d[n]$.

(c) If we denote $\mathbf{P} = [p[0]\ p[1]\ \cdots\ p[N-1]]^T$, and $\mathbf{X} = [x[0]\ x[1]\ \cdots x[N-1]]^T$, show that $\mathbf{P} = \mathbf{Q} \cdot \mathbf{X}$, where $\mathbf{Q} = [q_{r,s}]$ is a real $N \times N$ matrix whose first row is given by $q_{0,s} = \alpha^s$, first column is given by $q_{r,0} = {}^{N-1}C_r\alpha^r$, and remaining elements $q_{r,s}$ can be derived using the recursion relation

$$q_{r,s} = q_{r-1,s-1} + \alpha q_{r,s-1} - \alpha q_{r-1,s}.$$

10.9   MATLAB Exercises

M 10.1 Plot the magnitude response of a linear-phase FIR highpass filter by truncating the impulse response $h_{HP}[n]$ of the ideal highpass filter of Eq. (10.16) to length $N = 2M + 1$ for two different values of M, and show that the truncated filter exhibits oscillatory behavior on both sides of the cutoff frequency.

M 10.2 Plot the magnitude response of a linear-phase FIR bandpass filter by truncating the impulse response $h_{BP}[n]$ of the ideal bandpass filter of Eq. (10.17) to length $N = 2M + 1$ for two different values of M, and show that the truncated filter exhibits oscillatory behavior on both sides of the cutoff frequency.

M 10.3 Plot the magnitude response of a linear-phase FIR Hilbert transformer by truncating the impulse response $h_{HT}[n]$ of the ideal Hilbert transformer of Eq. (10.22) to length $N = 2M + 1$ for two different values of M, and show that the truncated filter exhibits oscillatory behavior near $\omega = 0$ and $\omega = \pi$.

M 10.4 The impulse response obtained by convolving K rectangular windows of length N each approximates an ideal zero-mean FIR Gaussian filter. Using the M-file `firgauss`, generate several such filters, and show that the approximation gets better with an increase in K or N.

M 10.5 Write a MATLAB program to design a linear-phase FIR notch filter by windowing the impulse response of the ideal notch filter of Problem 10.12. Using this program, design an FIR notch filter of order 36 operating at a 500-Hz sampling rate with a notch frequency of 50 Hz.

M 10.6 Determine a linear approximation $a_0 + a_1 x$ to the quadratic function $D(x) = 3.2x^2 + 4.05x - 5.5$ defined for the range $-3 \le x \le 2$ by minimizing the peak value of the absolute error $|D(x) - a_0 - a_1 x|$, i.e.,

$$\max_{-3 \le x \le 2} |D(x) - a_0 - a_1 x|,$$

using the Remez algorithm. Plot the error function after convergence of the algorithm.

M 10.7 Determine a quadratic approximation $a_0 + a_1 x + a_2 x^2$ to the cubic function $D(x) = -5x^3 - 0.2x^2 + 8x + 5.5$ defined for the range $-2 \le x \le 2$ by minimizing the peak value of the absolute error $\left| D(x) - a_0 - a_1 x - a_2 x^2 \right|$, i.e.,

$$\max_{-2 \le x \le 2} \left| D(x) - a_0 - a_1 x - a_2 x^2 \right|,$$

using the Remez algorithm. Plot the error function after convergence of the algorithm.

M 10.8 Using the windowed Fourier series approach, design a linear-phase FIR lowpass filter with the following specifications: passband edge at 4 rad/sec, stopband edge at 6 rad/sec, maximum passband attenuation of 0.2 dB, minimum stopband attenuation of 42 dB, and a sampling frequency of 18 rad/sec. Use each of the following windows for the design: Hamming, Hann, and Blackman. Show the impulse response coefficients, and plot the gain response of the designed filters for each case. Comment on your results. Do not use the M-file `fir1`.

M 10.9 Repeat Exercise M10.8 using the Kaiser window. Do not use the M-file `fir1`.

M 10.10 Using the windowed Fourier series approach, design a linear-phase FIR lowpass filter of lowest order with the following specifications: passband edge at 0.4π, stopband edge at 0.6π, and minimum stopband attenuation of 42 dB. Which window function is appropriate for this design? Show the impulse response coefficients and plot the gain response of the designed filter. Comment on your results. Do not use the M-file `fir1`.

M 10.11 Repeat Exercise M10.10 using the Dolph-Chebyshev window. Do not use the M-file `fir1`. Compare your results with that obtained in Exercise M10.10.

M 10.12 Repeat Exercise M10.10 using the M-file `fir1`. Compare your results with that obtained in Exercise M10.10.

M 10.13 Design a linear-phase highpass FIR filter of length 36 with a passband edge at $\omega_p = 0.45\pi$ using the frequency sampling approach. Show the impulse response coefficients, and plot the magnitude response of the designed filter using MATLAB.

M 10.14 Design a linear-phase bandpass FIR filter of order 45 with passband edges at $\omega_{p1} = 0.5\pi$ and $\omega_{p2} = 0.7\pi$ using the frequency sampling approach. Show the impulse response coefficients, and plot the magnitude response of the designed filter using MATLAB.

M 10.15 Using the frequency sampling-based approach, redesign the linear-phase lowpass filter of Problem 10.34 by including a transition band with one frequency sample of magnitude 1/2. Plot the magnitude response of the new filter using MATLAB, and compare it with that obtained in Problem 10.34.

M 10.16 Repeat Exercise M10.15 by including a transition band with two frequency samples of magnitude 2/3 and 1/3, respectively.

M 10.17 Design the linear-phase FIR lowpass filter of Exercise M10.8 using the function `fir1` of MATLAB. Use each of the following windows for the design: Hamming, Hann, Blackman, and Kaiser. Show the impulse response coefficients, and plot the gain response of the designed filters for each case. Compare your results with those obtained in Exercises M10.8 and M10.9.

M 10.18 Using the M-file `fir1`, design a linear-phase FIR highpass filter with the following specifications: stopband edge at 0.4π, passband edge at 0.55π, maximum passband attenuation of 0.1 dB, and minimum stopband attenuation of 42 dB. Use each of the following windows for the design: Hamming, Hann, Blackman, and Kaiser. Show the impulse response coefficients, and plot the gain response of the designed filters for each case. Comment on your results.

M 10.19 Using the M-file `fir1`, design a linear-phase FIR bandpass filter with the following specifications: stopband edges at 0.55π and 0.75π, passband edges at 0.65π and 0.85π, maximum passband attenuation of 0.2 dB, and minimum stopband attenuation of 42 dB. Use each of the following windows for the design: Hamming, Hann, Blackman, and Kaiser. Show the impulse response coefficients, and plot the gain response of the designed filters for each case. Comment on your results.

M 10.20 Design a two-channel crossover FIR lowpass and highpass filter pair for digital audio applications. The lowpass and the highpass filters are of length 31 and have a crossover frequency of 15 kHz operating at a sampling rate of 70 kHz. Use the function `fir1` with a Hamming window to design the lowpass filter, while the highpass filter is derived from the lowpass filter using the delay-complementary property. Plot the gain responses of both filters on the same figure. What is the minimum number of delays and multipliers needed to implement the crossover network?

M 10.21 Design a three-channel crossover FIR filter system for digital audio applications. All filters are of length 33 and operate at a sampling rate of 44.1 kHz. The two crossover frequencies are at 5.5 kHz and 12 kHz, respectively. Use the function `fir1` with a Hann window to design the lowpass and the highpass filters, while the bandpass filter is derived from the lowpass and highpass filters using the delay-complementary property. Plot the gain responses of all filters on the same figure. What is the minimum number of delays and multipliers needed to implement the crossover network?

M 10.22 The M-file `fir2` is employed to design FIR filters with arbitrarily shaped magnitude responses. Using this function, design an FIR filter of order 70 with three different constant magnitude levels: 0.2 in the frequency range 0 to 0.35, 1.0 in the frequency range 0.4 to 0.7, and 0.6 in the frequency range 0.72 to 1.0. Plot the gain response of the designed filter.

M 10.23 Design the linear-phase FIR lowpass filter of Problem 10.36 using the function `remez` and plot its magnitude response.

M 10.24 Design the linear-phase FIR highpass filter of Problem 10.37 using the function `remez` and plot its magnitude response.

M 10.25 Design the linear-phase FIR bandpass filter of Problem 10.38 using the function `remez` and plot its magnitude response.

M 10.26 Design a length-30 discrete-time FIR differentiator using the function `remez` and plot its magnitude response.

M 10.27 Design a 30th-order FIR Hilbert transformer using the function `remez`. The passband is from 0.07π to 0.95π. The two stopbands are from 0.02π to 0.05π, and from 0.97π to π. Plot its magnitude response.

M 10.28 Design a minimum-phase lowpass FIR filter with the passband edge at $\omega_p = 0.35\pi$, stopband edge at $\omega_s = 0.5\pi$, passband ripple of $R_p = 1$ dB, and a minimum stopband attenuation of $R_s = 28$ dB.

M 10.29 Determine the minimum-phase spectral factor of the polynomial

$$Q(z) = 2.4 + 6.76z^{-1} + 26.15z^{-2} + 68.43z^{-3} + 186.83z^{-4} + 326.51z^{-5} + 565.53z^{-6} + 678.95z^{-7} + 805.24z^{-8}$$
$$+ 678.95z^{-9} + 565.53z^{-10} + 326.51z^{-11} + 186.83z^{-12} + 68.43z^{-13} + 26.15z^{-14} + 6.76z^{-15} + 2.4z^{-16}.$$

M 10.30 Design a linear-phase narrow-band FIR lowpass filter using the interpolated FIR filter design approach to meet the following specifications: $\omega_p = 0.1\pi$, $\omega_s = 0.15\pi$, $\delta_p = 0.001$, and $\delta_s = 0.001$.

M 10.31 Design a linear-phase narrow-band FIR highpass filter using the interpolated FIR filter design approach to meet the following specifications: $\omega_p = 0.9\pi$, $\omega_s = 0.95\pi$, $\delta_p = 0.002$, and $\delta_s = 0.004$.

M 10.32 Another approach to the design of a computationally efficient FIR filter is the *prefilter-equalizer method* [Ada83]. In this method, first, a computationally efficient FIR prefilter $H(z)$ with a frequency response reasonably close to the desired response is selected. Next, an FIR equalizer $F(z)$ is designed so that the cascade of the prefilter and the equalizer meets the desired specifications. An attractive prefilter structure for the design of a lowpass FIR filter is the recursive running sum (RRS) FIR filter of order N, which has a transfer function

$$H(z) = \frac{1 - z^{-(N+1)}}{1 - z^{-1}}.$$

The first null of the frequency response of the RRS filter is at $\omega = 2\pi/(N+1)$. Thus, if the desired stopband edge is at ω_s, the order of the RRS filter should be chosen as $N \cong 2\pi/\omega_s$. If N is a fraction, then both the integer values nearest to $2\pi/\omega_s$ are good candidates for the order of the RRS filter. The Park–McClellan algorithm can be modified to incorporate the frequency response of the RRS filter in the weighting function of the error function $W(\omega)$ of Eq. (9.48). Using the prefilter-equalizer approach, design a computationally efficient narrow-band FIR lowpass filter with the following specifications: $\omega_p = 0.05\pi$, $\omega_s = 0.15\pi$, $\alpha_p = 0.15$ dB, and $\alpha_s = 40$ dB.

11 DSP Algorithm Implementation

There are basically two types of digital signal processing (DSP) algorithms: filtering algorithms and signal analysis algorithms. These algorithms can be based on either the difference equation, both recursive and nonrecursive, or the discrete Fourier transform (DFT) and can be implemented in any one of the following forms: hardware, firmware, and software. In the hardware approach, the algorithm can be implemented using digital circuitry, such as a shift register to provide the delaying operation, a digital multiplier, and a digital adder. Alternatively, a special-purpose VLSI chip can be designed and fabricated to implement a specific filtering algorithm. In the firmware approach, the algorithm is implemented on a read-only-memory (ROM) chip. Additional control circuitry, and storage registers, are usually needed in the final hardware or firmware realization. Finally, in the software approach, the algorithm is implemented as a computer program on a general-purpose computer such as a workstation, a minicomputer, a personal computer, or a programmable DSP chip. This chapter is concerned with the implementation aspects of DSP algorithms. We first examine two major issues concerning all the above types of approaches to implementation. We then discuss the software implementation of digital filtering and DFT algorithms on a computer using MATLAB to illustrate the main points. This section is followed by a review of various schemes for the representation of numbers and signal variables on a digital machine. The number representation scheme is basic to the development of methods for the analysis of finite wordlength effects discussed in Chapter 12. Next, we review algorithms that are employed to implement addition and multiplication, the two key arithmetic operations in digital signal processing. We then briefly review operations developed to handle overflow. Finally, two general methods for the design and implementation of tunable digital filters are outlined, followed by a discussion of algorithms for the approximation of certain special functions. A detailed discussion of hardware, firmware, and DSP chip implementations is beyond the scope of this book. Information on programming a range of DSP chips can be found in the books and application notes published by the manufacturers of these chips.

11.1 Basic Issues

We examine first two specific problems that generally are encountered before a digital filter is actually implemented. The first problem is concerned with the computability of the equations describing the filter structure, and the second problem is concerned with the verification of the structure developed to realize a prescribed transfer function.

11.1.1 Matrix Representation of the Digital Filter Structure

As indicated in Chapter 8, a digital filter can be described in the time domain by a set of equations relating the output sequence to the input sequence and, in some cases, one or more internally generated sequences. The ordering of these equations in computing the output samples is important, as discussed next.

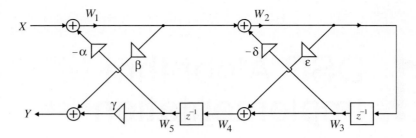

Figure 11.1: A cascaded lattice digital filter structure.

Consider the digital filter structure of Figure 11.1. We can describe this structure by the following set of equations relating the signal variables $W_k(z)$, the output $Y(z)$, and the input $X(z)$:

$$W_1(z) = X(z) - \alpha W_5(z), \tag{11.1a}$$

$$W_2(z) = W_1(z) - \delta W_3(z), \tag{11.1b}$$

$$W_3(z) = z^{-1} W_2(z), \tag{11.1c}$$

$$W_4(z) = W_3(z) + \varepsilon W_2(z), \tag{11.1d}$$

$$W_5(z) = z^{-1} W_4(z), \tag{11.1e}$$

$$Y(z) = \beta W_1(z) + \gamma W_5(z). \tag{11.1f}$$

In the time domain, the above set of equations is equivalent to

$$w_1[n] = x[n] - \alpha w_5[n], \tag{11.2a}$$

$$w_2[n] = w_1[n] - \delta w_3[n], \tag{11.2b}$$

$$w_3[n] = w_2[n - 1], \tag{11.2c}$$

$$w_4[n] = w_3[n] + \varepsilon w_2[n], \tag{11.2d}$$

$$w_5[n] = w_4[n - 1], \tag{11.2e}$$

$$y[n] = \beta w_1[n] + \gamma w_5[n]. \tag{11.2f}$$

The above set of equations does not describe a valid computational algorithm since the equations cannot be implemented in the order shown, with each variable on the left side computed before the variable below is computed. For example, computation of $w_1[n]$ in the first step requires the knowledge of $w_5[n]$ that is computed in the fifth step. Likewise, the computation of $w_2[n]$ in the second step requires the knowledge of $w_3[n]$ that is computed in the following step. We call the ordered set of equations of Eqs. (11.2a) to (11.2f) *noncomputable*.

Suppose we reorder the above equations and write them in the form

$$w_3[n] = w_2[n - 1], \tag{11.3a}$$

$$w_5[n] = w_4[n - 1], \tag{11.3b}$$

$$w_1[n] = x[n] - \alpha w_5[n], \tag{11.3c}$$

$$w_2[n] = w_1[n] - \delta w_3[n], \tag{11.3d}$$

$$y[n] = \beta w_1[n] + \gamma w_5[n], \tag{11.3e}$$

$$w_4[n] = w_3[n] + \varepsilon w_2[n]. \tag{11.3f}$$

It can be seen that this ordered set of equations now describes a valid computational algorithm since the equations can be implemented in the sequential order shown, with each variable on the left side computed before the variable below is computed.

In most practical applications, the equations characterizing the digital filter can be put into a computable order by inspection. It is, however, instructive to examine the computability of the equations describing a digital filter in a more formal fashion, which is described next [Cro75].

To this end, we write the equations of the digital filter in matrix form. Thus, a matrix representation of Eqs. (11.2a) to (11.2f) is given by

$$
\begin{bmatrix} w_1[n] \\ w_2[n] \\ w_3[n] \\ w_4[n] \\ w_5[n] \\ y[n] \end{bmatrix} = \begin{bmatrix} x[n] \\ 0 \\ 0 \\ 0 \\ 0 \\ 0 \end{bmatrix} + \begin{bmatrix} 0 & 0 & 0 & 0 & -\alpha & 0 \\ 1 & 0 & -\delta & 0 & 0 & 0 \\ 0 & 0 & 0 & 0 & 0 & 0 \\ 0 & \varepsilon & 1 & 0 & 0 & 0 \\ 0 & 0 & 0 & 0 & 0 & 0 \\ \beta & 0 & 0 & 0 & \gamma & 0 \end{bmatrix} \begin{bmatrix} w_1[n] \\ w_2[n] \\ w_3[n] \\ w_4[n] \\ w_5[n] \\ y[n] \end{bmatrix}
$$

$$
+ \begin{bmatrix} 0 & 0 & 0 & 0 & 0 & 0 \\ 0 & 0 & 0 & 0 & 0 & 0 \\ 0 & 1 & 0 & 0 & 0 & 0 \\ 0 & 0 & 0 & 0 & 0 & 0 \\ 0 & 0 & 0 & 1 & 0 & 0 \\ 0 & 0 & 0 & 0 & 0 & 0 \end{bmatrix} \begin{bmatrix} w_1[n-1] \\ w_2[n-1] \\ w_3[n-1] \\ w_4[n-1] \\ w_5[n-1] \\ y[n-1] \end{bmatrix}, \qquad (11.4)
$$

which we can write compactly as

$$
\mathbf{y}[n] = \mathbf{x}[n] + \mathbf{F}\mathbf{y}[n] + \mathbf{G}\mathbf{y}[n-1], \qquad (11.5)
$$

where

$$
\mathbf{y}[n] = \begin{bmatrix} w_1[n] \\ w_2[n] \\ w_3[n] \\ w_4[n] \\ w_5[n] \\ y[n] \end{bmatrix}, \qquad \mathbf{x}[n] = \begin{bmatrix} x[n] \\ 0 \\ 0 \\ 0 \\ 0 \\ 0 \end{bmatrix}, \qquad (11.6a)
$$

$$
\mathbf{F} = \begin{bmatrix} 0 & 0 & 0 & 0 & -\alpha & 0 \\ 1 & 0 & -\delta & 0 & 0 & 0 \\ 0 & 0 & 0 & 0 & 0 & 0 \\ 0 & \varepsilon & 1 & 0 & 0 & 0 \\ 0 & 0 & 0 & 0 & 0 & 0 \\ \beta & 0 & 0 & 0 & \gamma & 0 \end{bmatrix}, \qquad \mathbf{G} = \begin{bmatrix} 0 & 0 & 0 & 0 & 0 & 0 \\ 0 & 0 & 0 & 0 & 0 & 0 \\ 0 & 1 & 0 & 0 & 0 & 0 \\ 0 & 0 & 0 & 0 & 0 & 0 \\ 0 & 0 & 0 & 1 & 0 & 0 \\ 0 & 0 & 0 & 0 & 0 & 0 \end{bmatrix}. \qquad (11.6b)
$$

If we examine Eq. (11.4), we observe that, for the computation of the present value of a particular signal variable, the nonzero entries in the corresponding rows of the matrices $\mathbf{F}$ and $\mathbf{G}$ determine the variables whose present and previous values are needed. If the diagonal element in $\mathbf{F}$ is nonzero, then the computation of the present value of the corresponding variable requires the knowledge of its present value, indicating the presence of a delay-free loop, making the structure totally noncomputable. Any nonzero entries in the same row above the diagonal of $\mathbf{F}$ imply that the computation of the present value of the corresponding variable requires the present values of other variables that have not yet been computed, thus making the set of equations noncomputable.

It follows, therefore, for computability, all elements of the $\mathbf{F}$ matrix on the diagonal and above the diagonal must be zeros.

In the $\mathbf{F}$ matrix of this example, the diagonal elements are all zeros, indicating that there are no simple delay-free loops in the structure. However, there are nonzero entries above the diagonal in the first and second rows of $\mathbf{F}$, indicating that the set of equations of Eqs. (11.2a) to (11.2f) is not in proper order for computation.

On the other hand, the matrix representation of Eqs. (11.3a) to (11.3f) results in

$$
\begin{bmatrix} w_3[n] \\ w_5[n] \\ w_1[n] \\ w_2[n] \\ y[n] \\ w_4[n] \end{bmatrix} = \begin{bmatrix} 0 \\ 0 \\ x[n] \\ 0 \\ 0 \\ 0 \end{bmatrix} + \begin{bmatrix} 0 & 0 & 0 & 0 & 0 & 0 \\ 0 & 0 & 0 & 0 & 0 & 0 \\ 0 & -\alpha & 0 & 0 & 0 & 0 \\ -\delta & 0 & 1 & 0 & 0 & 0 \\ 0 & \gamma & \beta & 0 & 0 & 0 \\ 1 & 0 & 0 & \varepsilon & 0 & 0 \end{bmatrix} \begin{bmatrix} w_3[n] \\ w_5[n] \\ w_1[n] \\ w_2[n] \\ y[n] \\ w_4[n] \end{bmatrix}
$$

$$
+ \begin{bmatrix} 0 & 0 & 0 & 1 & 0 & 0 \\ 0 & 0 & 0 & 0 & 0 & 1 \\ 0 & 0 & 0 & 0 & 0 & 0 \\ 0 & 0 & 0 & 0 & 0 & 0 \\ 0 & 0 & 0 & 0 & 0 & 0 \\ 0 & 0 & 0 & 0 & 0 & 0 \end{bmatrix} \begin{bmatrix} w_3[n-1] \\ w_5[n-1] \\ w_1[n-1] \\ w_2[n-1] \\ y[n-1] \\ w_4[n-1] \end{bmatrix}, \qquad (11.7)
$$

for which the $\mathbf{F}$ matrix satisfies the computability condition, and thus, the equations describing the filter are in proper order.

11.1.2 Precedence Graph

We now describe a simple algorithm for testing the computability of a digital filter structure and for developing the proper ordering sequence for a set of equations describing a computable structure. To this end, we first form a signal flow-graph description of the digital filter structure. In a signal flow-graph [?], the dependent and the independent signal variables are represented by *nodes,* whereas the multiplier and the delay units are represented by *directed branches.* In the latter case, the directed branch has an attached symbol denoting the *branch-gain* or the *transmittance,* which, for a multiplier branch, is the multiplier coefficient value and for a delay branch is simply z^{-1}. For example, the signal flow-graph representation of the digital filter structure of Figure 11.1 is as shown in Figure 11.2.

As the output of the delay branches can always be computed at any instant since they are the delayed values of their respective input signals computed at the previous instant, we remove all delay branches from the complete signal flow-graph of the digital filter structure. Similarly, all branches coming out of

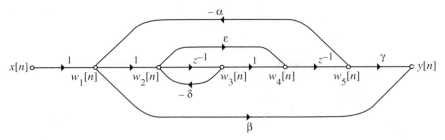

Figure 11.2: Signal flow-graph representation of the digital filter structure of Figure 11.1.

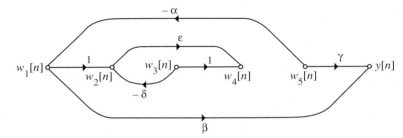

Figure 11.3: Reduced signal flow-graph obtained by removing the branches going out of the input node and the delay branches from the signal flow-graph of Figure 11.2.

the input node are also removed since the input variables are always available at each instant. For our example, the resulting reduced signal flow-graph is as shown in Figure 11.3.

We now group the remaining nodes in the reduced signal flow-graph as follows. All nodes with only outgoing branches are grouped into one set labeled $\{\mathcal{N}_1\}$. Next, we form the set $\{\mathcal{N}_2\}$ containing nodes that have branches coming in only from one or more of the nodes in the set $\{\mathcal{N}_1\}$ and have outgoing branches to the other nodes. We then form a set $\{\mathcal{N}_3\}$ containing nodes that have branches coming in only from one or more of the nodes in the sets $\{\mathcal{N}_1\}$ and $\{\mathcal{N}_2\}$ and have outgoing branches to the other nodes. This process is continued until we have a set $\{\mathcal{N}_f\}$ containing nodes with only incoming branches.

Since the signal variables belonging to the set $\{\mathcal{N}_1\}$ do not depend on the present values of the other signal variables, these variables should be computed first. Next, the signal variables belonging to the set $\{\mathcal{N}_2\}$ can be computed since they depend on the present values of the signal variables contained in the set $\{\mathcal{N}_1\}$ that have already been computed. This is followed by the computation of the signal variables in the sets $\{\mathcal{N}_3\}$, $\{\mathcal{N}_4\}$, and so on. Finally, in the last step, the signal variables belonging to the set $\{\mathcal{N}_f\}$ are computed. This process of sequential computation ensures the development of a valid computational algorithm. However, if there is no final set $\{\mathcal{N}_f\}$ containing only incoming branches, the signal flow-graph is noncomputable. The rearranged signal flow-graph without the delay branches and with nodes grouped as indicated above is called a *precedence graph* [Cro75].

For our example, the pertinent groupings of node variables according to their precedence relations are as follows:

$$\{\mathcal{N}_1\} = \{w_3[n], \ w_5[n]\},$$
$$\{\mathcal{N}_2\} = \{w_1[n]\},$$
$$\{\mathcal{N}_3\} = \{w_2[n]\},$$
$$\{\mathcal{N}_4\} = \{w_4[n], \ y[n]\}.$$

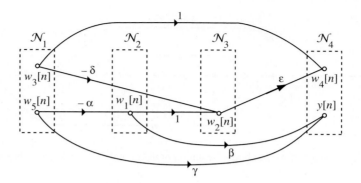

Figure 11.4: Precedence graph of Figure 11.3 redrawn with signal variables grouped according to their precedence relations.

The precedence graph redrawn according to the above groupings is indicated in Figure 11.4. Since the final node set $\{\mathcal{N}_4\}$ has only incoming nodes, the structure of Figure 11.1 has no delay-free loops. Therefore, for our example structure, we can compute first the signal variables $w_3[n]$ and $w_5[n]$ in any order, then compute the signal variable $w_1[n]$, followed by the computation of the signal variable $w_2[n]$, and finally, compute the signal variables $w_4[n]$ and $y[n]$ in any order to arrive at a valid computational algorithm.

P_G_A.m

Computability Test Using MATLAB

The M-file P_G_A[1] can be used to check the computability of a digital filter structure. In order to apply this M-file to check the computability of a digital filter structure with D delays and M multipliers, the multiplier coefficients should be assigned numerical values. The input data are the vectors, defining the locations of each delay and the locations and coefficient values of each multiplier. The delay vector delay is entered as a $2 \times D$ matrix where the elements of the ith row are the input and output nodes of the ith delay. The multiplier vector mult is a $3 \times M$ matrix whose ith row contains the input node number, the output node number, and the coefficient value of the ith multiplier. A direct connection from one node to another is considered as a multiplier with a unity coefficient.

We illustrate the application of the above M-file in Example 11.1.

EXAMPLE 11.1   Illustration of the Computability Test Using MATLAB

In this example, we check the computability of the digital filter structure of Figure 11.1. The internal nodes of this structure are labeled as shown by the signal flow-graph in Figure 11.2. We label the output node as 6. The values assigned arbitrarily to the multipliers shown in the graph are as follows:

$$\alpha = 5, \ \beta = 4, \ \gamma = 3, \ \delta = 2, \ \epsilon = 1.$$

The two direct paths between nodes 1 and 2 and between nodes 3 and 4 are assumed to contain multipliers with unity coefficients. The code fragments used to test the structure are as given in the following:

```
delay = [4 5; 2 3];
mult = [1 2 1; 1 6 4; 2 4 1; 3 4 1; 5 1 -5; 5 6 3; 3 2 -2];
[F,G,NN,F_new,G_new] = P_G_A(delay, mult)
```

---

[1]Reproduced with permission from Dr. Hugo Tassignon, KHBO, Ostende, Belgium.

The output data generated by the above program is given below:

```
NN =
    3     5     1     2     4     6

F =
    0     0     0     0    -5     0
    1     0    -2     0     0     0
    0     0     0     0     0     0
    0     1     1     0     0     0
    0     0     0     0     0     0
    4     0     0     0     3     0

G =
    0     0     0     0     0     0
    0     0     0     0     0     0
    0     1     0     0     0     0
    0     0     0     0     0     0
    0     0     0     1     0     0
    0     0     0     0     0     0

F_new =
    0     0     0     0     0     0
    0     0     0     0     0     0
    0    -5     0     0     0     0
   -2     0     1     0     0     0
    1     0     0     1     0     0
    0     3     4     0     0     0

G_new =
    0     0     0     1     0     0
    0     0     0     0     1     0
    0     0     0     0     0     0
    0     0     0     0     0     0
    0     0     0     0     0     0
    0     0     0     0     0     0
```

Note that the F and G matrices are precisely those given in Eq. (11.6b). The new ordering of the nodes as given above is nearly the same as shown in Eqs. (11.3a)–(11.3f), except the the ordering of the last two equations has been reversed. It follows from the precedence graph given in Figure 11.4 that the last node set $\mathcal{N}_4$ contains nodes 4 and 6. Hence, it does not matter which node variable is computed first.

It should be noted that if the digital filter structure is noncomputable, the M-file P_G_A will give the error message

STRUCTURE NOT COMPUTABLE

11.1.3 Direct Analysis of Digital Filter Structures Using MATLAB

In many applications, it is often useful to determine explicitly the expression of the transfer function of a digital filter structure from the set of equations describing the structure. This can be achieved in one of two ways. In the first approach, the set of equations described in the z-domain is solved using standard matrix algebra. In the second approach, the digital filter structure is simulated on a computer by implementing the set of equations in the time domain after the proper ordering of the equations has been made to ensure computability, and then the transfer function is determined using the structure verification method described in Section 11.1.4. We describe here the first approach that is based on a method of symbolic analysis of analog circuits [Vla83]. The second approach is illustrated later in Section 11.2.

Without any loss of generality, consider the analysis of the structure of Figure 11.1. A matrix representation of Eqs. (11.1a) to (11.1f) leads to

$$\mathbf{T} \cdot \mathbf{W} = \mathbf{X}, \tag{11.8}$$

where

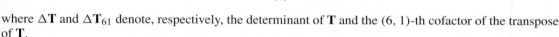

$$\mathbf{T} = \begin{bmatrix} 1 & 0 & 0 & 0 & \alpha & 0 \\ -1 & 1 & \delta & 0 & 0 & 0 \\ 0 & -z^{-1} & 1 & 0 & 0 & 0 \\ 0 & -\epsilon & 1 & 1 & 0 & 0 \\ 0 & 0 & -z^{-1} & 1 & 0 \\ -\beta & 0 & 0 & 0 & -\gamma & 1 \end{bmatrix}, \quad \mathbf{W} = \begin{bmatrix} W_1(z) \\ W_2(z) \\ W_3(z) \\ W_4(z) \\ W_5(z) \\ Y(z) \end{bmatrix}, \quad \mathbf{X} = \begin{bmatrix} X(z) \\ 0 \\ 0 \\ 0 \\ 0 \\ 0 \end{bmatrix}. \tag{11.9}$$

A solution of Eq. (11.8) yields the desired transfer function given by

$$H(z) = \frac{Y(z)}{X(z)} = \frac{\Delta \mathbf{T}_{61}}{\Delta \mathbf{T}}, \tag{11.10}$$

sym_di_f.m

where $\Delta \mathbf{T}$ and $\Delta \mathbf{T}_{61}$ denote, respectively, the determinant of $\mathbf{T}$ and the (6, 1)-th cofactor of the transpose of $\mathbf{T}$.

Since the symbolic matrix analysis using the *Symbolic Analysis Toolbox* of MATLAB is limited to lower order matrices, operations on the transfer matrix $\mathbf{T}$ of a structure with a large number of components very quickly becomes impractical. We instead make use of the M-file `sym_di_f` [2], which is based on a semisymbolic analysis method. To use this function, numerical values to the multiplier coefficients are first assigned. Next, the transfer function is numerically evaluated at N_d equidistant points on the unit circle symbol. Note that the number of coefficients in the numerator and denominator polynomials of the transfer function is N_d each, where $N_d - 1$ is the total number of unit delays in the structure.[3] Hence, the transfer function is evaluated at $\omega = e^{-j2\pi n/N_d}$, $0 \leq n \leq N_d - 1$. As a result, the transfer matrix $\mathbf{T}$ has numerical values only at each point on the unit circle, and its determinant $\Delta \mathbf{T}$ and the adjoint $\Delta \mathbf{T}_{61}$ can be easily calculated. The coefficients of the numerator and the denominator polynomials of $H(z)$ can be simply determined by applying an inverse DFT to the frequency samples of the numerator and the denominator polynomials.

We illustrate the application of the function `sym_di_f` in Example 11.2.

EXAMPLE 11.2   Direct Analysis of a Digital Filter Structure Using MATLAB

We determine the transfer function of the structure of Figure 11.1. We number the nodes and assign values to the multipliers as in Example 11.1. For our example, there are 2 unit delays in the structure, and hence, we evaluate the transfer function at 3 points on the unit circle.

---

[2]Reproduced with permission from Dr. Hugo Tassignon, KHBO, Ostende, Belgium.
[3]It is tacitly assumed here that the structure is canonic with respect to the delays.

The code fragments used to analyze the structure are as given below:

```
delay = [4 5; 2 3];
mult = [1 2 1; 1 6 4; 2 4 1; 3 4 1; 5 1 -5; 5 6 3; 3 2 -2];
in_node = 1; out_node = 6;
[Num,Den] = sym_di_f(delay,mult,in_node,out_node)
```

The output data generated by the above program are given by

```
Num =
4.0000    11.0000    3.0000

Den =
1.0000    7.0000    5.0000
```

Thus, the transfer function of the digital filter structure of Figure 11.1 for the multiplier values given in Example 11.1 is

$$H(z) = \frac{4 + 11\,z^{-1} + 3\,z^{-2}}{1 + 7\,z^{-1} + 5\,z^{-2}}.$$

The above result can be easily verified by substituting the multiplier values in the expression for the transfer function of the structure of Figure 11.1.

11.1.4 Structure Verification

An important step that needs to be considered in the hardware or software implementation of a digital transfer function is to ensure that no computational and/or other errors have taken place during the course of the realization process and that the structure obtained is indeed characterized by the prescribed transfer function $H(z)$. A simple technique to verify the structure is outlined next [Mit77a].

Without any loss of generality, consider a causal LTI digital filter structure characterized by a fourth-order transfer function

$$H(z) = \frac{P(z)}{D(z)} = \frac{p_0 + p_1 z^{-1} + p_2 z^{-2} + p_3 z^{-3} + p_4 z^{-4}}{1 + d_1 z^{-1} + d_2 z^{-2} + d_3 z^{-3} + d_4 z^{-4}}. \tag{11.11}$$

If $\{h[n]\}$ denotes the unit sample response corresponding to $H(z)$, then

$$H(z) = \sum_{n=0}^{\infty} h[n]z^{-n}. \tag{11.12}$$

From Eqs. (11.11) and (11.12), it follows that

$$P(z) = H(z)D(z) \tag{11.13}$$

or, equivalently, in the time domain by the convolution sum

$$p_n = h[n] \circledast d_n, \tag{11.14}$$

which explicitly shows the relation between the numerator and the denominator coefficients of the transfer function $H(z)$ of Eq. (11.11) and its impulse response samples. Since the total number of transfer function coefficients is nine, we need only any consecutive nine equations of the set of Eq. (11.14) to have unique relations between the transfer function coefficients and the impulse response samples. Writing out Eq. (11.14) explicitly for $n = 0, 1, 2, \ldots, 8$, we obtain

$$
\begin{aligned}
p_0 &= h[0], \\
p_1 &= h[1] + h[0]d_1, \\
p_2 &= h[2] + h[1]d_1 + h[0]d_2, \\
p_3 &= h[3] + h[2]d_1 + h[1]d_2 + h[0]d_3, \\
p_4 &= h[4] + h[3]d_1 + h[2]d_2 + h[1]d_3 + h[0]d_4, \\
0 &= h[5] + h[4]d_1 + h[3]d_2 + h[2]d_3 + h[1]d_4, \\
0 &= h[6] + h[5]d_1 + h[4]d_2 + h[3]d_3 + h[2]d_4, \\
0 &= h[7] + h[6]d_1 + h[5]d_2 + h[4]d_3 + h[3]d_4, \\
0 &= h[8] + h[7]d_1 + h[6]d_2 + h[5]d_3 + h[4]d_4.
\end{aligned}
$$

In matrix form, the above equations can be rewritten as

$$
\begin{bmatrix} p_0 \\ p_1 \\ p_2 \\ p_3 \\ p_4 \\ 0 \\ 0 \\ 0 \\ 0 \end{bmatrix} =
\left[\begin{array}{ccccc}
h[0] & 0 & 0 & 0 & 0 \\
h[1] & h[0] & 0 & 0 & 0 \\
h[2] & h[1] & h[0] & 0 & 0 \\
h[3] & h[2] & h[1] & h[0] & 0 \\
h[4] & h[3] & h[2] & h[1] & h[0] \\
\cdots & \cdots & \cdots & \cdots & \cdots & \cdots \\
h[5] & \vdots & h[4] & h[3] & h[2] & h[1] \\
h[6] & \vdots & h[5] & h[4] & h[3] & h[2] \\
h[7] & \vdots & h[6] & h[5] & h[4] & h[3] \\
h[8] & \vdots & h[7] & h[6] & h[5] & h[4]
\end{array} \right]
\begin{bmatrix} 1 \\ d_1 \\ d_2 \\ d_3 \\ d_4 \end{bmatrix}.
\tag{11.15}
$$

In partitioned form, Eq. (11.15) can be reexpressed as

$$
\begin{bmatrix} \mathbf{p} \\ \cdots \\ \mathbf{0} \end{bmatrix} =
\begin{bmatrix} & \mathbf{H}_1 & \\ \cdots & \cdots & \cdots \\ \mathbf{h} & \vdots & \mathbf{H}_2 \end{bmatrix}
\begin{bmatrix} 1 \\ \cdots \\ \mathbf{d} \end{bmatrix}
\tag{11.16}
$$

where

$$
\mathbf{p} = \begin{bmatrix} p_0 \\ p_1 \\ p_2 \\ p_3 \\ p_4 \end{bmatrix}, \quad
\mathbf{d} = \begin{bmatrix} d_1 \\ d_2 \\ d_3 \\ d_4 \end{bmatrix}, \quad
\mathbf{0} = \begin{bmatrix} 0 \\ 0 \\ 0 \\ 0 \end{bmatrix},
\tag{11.17a}
$$

$$
\mathbf{H}_1 = \begin{bmatrix}
h[0] & 0 & 0 & 0 & 0 \\
h[1] & h[0] & 0 & 0 & 0 \\
h[2] & h[1] & h[0] & 0 & 0 \\
h[3] & h[2] & h[1] & h[0] & 0 \\
h[4] & h[3] & h[2] & h[1] & h[0]
\end{bmatrix}, \quad
\mathbf{h} = \begin{bmatrix} h[5] \\ h[6] \\ h[7] \\ h[8] \end{bmatrix},
$$

$$
\mathbf{H}_2 = \begin{bmatrix}
h[4] & h[3] & h[2] & h[1] \\
h[5] & h[4] & h[3] & h[2] \\
h[6] & h[5] & h[4] & h[3] \\
h[7] & h[6] & h[5] & h[4]
\end{bmatrix}.
\tag{11.17b}
$$

Equation (11.15) therefore can be written as two matrix equations:

$$\mathbf{p} = \mathbf{H}_1 \begin{bmatrix} 1 \\ \mathbf{d} \end{bmatrix}, \tag{11.18}$$

$$\mathbf{0} = [\,\mathbf{h} \quad \mathbf{H}_2\,] \begin{bmatrix} 1 \\ \mathbf{d} \end{bmatrix}. \tag{11.19}$$

Solving Eq. (11.19), we obtain the vector $\mathbf{d}$ composed of the denominator coefficients:

$$\mathbf{d} = -\mathbf{H}_2^{-1}\mathbf{h}. \tag{11.20}$$

Substituting Eq. (11.20) in Eq. (11.18), we arrive at the vector $\mathbf{p}$ containing the numerator coefficients:

$$\mathbf{p} = \mathbf{H}_1 \begin{bmatrix} 1 \\ -\mathbf{H}_2^{-1}\mathbf{h} \end{bmatrix}. \tag{11.21}$$

In the general case of an Nth-order IIR transfer function, knowing the first $2N + 1$ impulse response samples is sufficient to determine the transfer function coefficients. Here, the vector $\mathbf{p}$ is of length $N + 1$, the vector $\mathbf{d}$ is of length N, the vector $\mathbf{h}$ is of length N, the matrix $\mathbf{H}_1$ is of size $(N + 1) \times (N + 1)$, and the matrix $\mathbf{H}_2$ is of size $N \times N$.

We illustrate the above approach for the reconstruction of the transfer function of a causal IIR filter from its impulse response coefficients in Example 11.3.

EXAMPLE 11.3   Reconstruction of a Transfer Function from Its Impulse Response Coefficients

Consider the causal IIR transfer function

$$H(z) = \frac{2 + 6z^{-1} + 3z^{-2}}{1 + z^{-1} + 2z^{-2}}.$$

Since $H(z)$ is of second-order, we need to compute only the first five impulse response samples. Using long division, we determine the first five terms in the series expansion of $H(z)$:

$$H(z) = 2 + 4z^{-1} - 5z^{-2} - 3z^{-3} + 13z^{-4} + \cdots.$$

Therefore, here,

$$h[0] = 2, \qquad h[1] = 4, \qquad h[2] = -5, \qquad h[3] = -3, \qquad h[4] = 13.$$

The equation corresponding to Eq. (11.15) is given by

$$\begin{bmatrix} p_0 \\ p_1 \\ p_2 \\ 0 \\ 0 \end{bmatrix} = \begin{bmatrix} 2 & 0 & 0 \\ 4 & 2 & 0 \\ -5 & 4 & 2 \\ -3 & -5 & 4 \\ 13 & -3 & -5 \end{bmatrix} \begin{bmatrix} 1 \\ d_1 \\ d_2 \end{bmatrix}.$$

Therefore, from Eq. (11.20),

$$\begin{bmatrix} d_1 \\ d_2 \end{bmatrix} = -\begin{bmatrix} -5 & 4 \\ -3 & -5 \end{bmatrix}^{-1} \begin{bmatrix} -3 \\ 13 \end{bmatrix} = -\frac{1}{37}\begin{bmatrix} -5 & -4 \\ 3 & -5 \end{bmatrix} \begin{bmatrix} -3 \\ 13 \end{bmatrix} = \begin{bmatrix} 1 \\ 2 \end{bmatrix}.$$

Next, from Eq. (11.21),

$$\begin{bmatrix} p_0 \\ p_1 \\ p_2 \end{bmatrix} = \begin{bmatrix} 2 & 0 & 0 \\ 4 & 2 & 0 \\ -5 & 4 & 2 \end{bmatrix} \begin{bmatrix} 1 \\ 1 \\ 2 \end{bmatrix} = \begin{bmatrix} 2 \\ 6 \\ 3 \end{bmatrix}.$$

This gives us a straightforward means of finding the transfer function of any discrete-time structure knowing the first $2M + 1$ samples of $\{h[n]\}$, where M is the order of the transfer function $H(z)$. This approach to structure verification is illustrated later in Examples 11.6 and 11.11. It can also be used to determine the effect of coefficient quantization by computing the transfer function realized with the multiplier coefficients quantized to the desired number of bits. This actual transfer function is then used to compute the frequency response, to determine the actual pole locations, and so on. Another application of the above method is in the determination of noise transfer functions for computing the output noise power due to product round-offs in fixed-point digital filter implementations considered in Section 12.6 and in the determination of scaling transfer functions needed for dynamic range scaling discussed in Section 12.7. In most cases, we can assume that the denominator coefficients are known; therefore, the solution of Eq. (11.20) requiring a matrix inversion can be avoided. The numerator coefficients are easily found from the first $M + 1$ samples of $\{h[n]\}$ using Eq. (11.21).

11.2 Structure Simulation and Verification Using Matlab

As indicated earlier, we concentrate in this book only on software implementation of DSP algorithms. In this section, we consider the implementation of digital filtering algorithms. The following section is devoted to the implementation of discrete Fourier transform algorithms.

The software implementation of a digital filtering algorithm on a computer is often carried out before the algorithm is implemented in a hardware form to verify that the chosen algorithm does indeed meet the goals of the application on hand. Moreover, such an implementation is adequate if the application under consideration does not require real-time signal processing.

For computer simulation, we basically describe the structure in the form of a set of equations. These equations must be ordered properly to ensure computability. For simplicity, the procedure is to express the output variable of each adder and the filter output variable in terms of all incoming signal variables. For example, for the structure of Figure 11.1, a valid computational algorithm involving the least number of equations is

$$w_1[n] = x[n] - \alpha w_4[n - 1],$$
$$w_2[n] = w_1[n] - \delta w_2[n - 1],$$
$$w_4[n] = w_2[n - 1] + \varepsilon w_2[n],$$
$$y[n] = \beta w_1[n] + \gamma w_4[n - 1].$$

The above set of equations is evaluated for increasing values of n, starting at $n = 0$. At the beginning, the initial conditions $w_2[-1]$ and $w_4[-1]$ can be set to any desired values, which are typically zero. After the computation of the last equation at time instant n, the computed values of $w_2[n]$ and $w_4[n]$ replace the values of $w_2[n - 1]$ and $w_4[n - 1]$ before the set of equations are evaluated for the next time instant $n + 1$.

In Chapter 8, we outlined a number of methods for the realization of both FIR and IIR digital transfer functions, resulting in a variety of structures. We restrict our attention to some of these structures to demonstrate the simulation of digital filters using Matlab. The structure being simulated can be verified by computing its transfer function using the method described in Section 11.1.4. To this end, the M-file `strucver` can be used.

strucver.m

11.2.1 Simulation of Direct Form IIR Filters

The M-file `filter` in the *Signal Processing Toolbox* of Matlab basically implements the IIR digital filter in the transposed direct form II structure shown in Figure 11.5 for a third-order filter.[4] As indicated in

---

[4]See Section 8.4.1 for the development of IIR direct form structures.

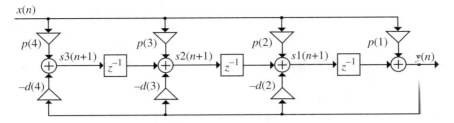

Figure 11.5: Transposed direct form II IIR structure.

this figure, $d(1)$ has been assumed to be equal to 1. If $d(1) \neq 1$, the program automatically normalizes all filter coefficients in p and d to make $d(1) = 1$.[5] The basic forms of this function are as follows:

```
y = filter(num,den,x)
[y,sf] = filter(num,den,x,si)
```

The numerator and the denominator coefficients are contained in the vectors num and den, respectively. These vectors do not have to be of the same size. The input vector is x, while the output vector generated by the filtering algorithm is y.

As indicated in Figure 11.5, the function filter simulates the filtering operation in the time domain in accordance with the following representation of the digital filter:

```
s3(n+1) = p(4)x(n) - d(4)y(n),
s2(n+1) = p(3)x(n) - d(3)y(n) + s3(n),
s1(n+1) = p(2)x(n) - d(2)y(n) + s2(n),
   y(n) = p(1)x(n) + s1(n).
```

In the second form of the function filter, the initial conditions of the delay (state) variables, sk(n), $k = 1, 2, \ldots$, can be specified through the argument si. Moreover, the function filter can return the final values of the delay (state) variables through the output vector sf. The size of the initial (final) condition vector si (sf) is one less than the maximum of the sizes of the filter coefficient vectors num and den. The final values of the state variables given as vector sf are useful if the input vector to be processed is very long and needs to be segmented into small blocks of data for processing in stages. In such a situation, after the ith block of input data has been processed, the final state vector sf is fed as the initial state vector si for the processing of the $(i + 1)$th block of input data, and so on.

For simulating a causal IIR filter realized in the direct form II structure, the function direct2 can be employed.

Example 11.4 illustrates the application of the M-file filter in generating the impulse response coefficients of a causal IIR digital filter.

direct2.m

EXAMPLE 11.4   Determination of Impulse Response from Digital Filter Simulation

Determine and plot the first 25 samples of the impulse response sequence, from $n = 0$ to $n = 24$, of the IIR lowpass digital filter described by the transfer function of Eq. (9.35), repeated in the following in a form suitable for direct form implementation:

---

[5]It should be noted that in Figure 11.5, we have used the MATLAB notations for vector elements instead of the notations used elsewhere in the text for representing filter coefficient and signal variables.

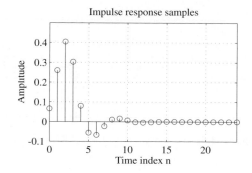

Figure 11.6: Impulse response samples of the IIR digital filter of Eq. (11.22).

$$H(z) = \frac{0.0662272(1 + 3z^{-1} + 3z^{-2} + z^{-3})}{1 - 0.9356142z^{-1} + 0.5671269z^{-2} - 0.10159107z^{-3}}. \tag{11.22}$$

Program 11_1 can be used to compute the impulse response. During execution, the program first requests the length of the impulse response to be computed. The program then requests the numerator and denominator vectors to be typed in. These vectors must be entered using the square brackets. After execution, it plots the impulse response as shown in Figure 11.6 for the IIR filter of Eq. (11.22).

Program 11_1.m
Program 11_2.m

The digital filtering application of the M-file `filter` is considered in Examples 11.5 to 11.7.

EXAMPLE 11.5   Illustration of Filtering by a Lowpass IIR Filter

We illustrate in this example the filtering of a signal composed of the sum of two sinusoids of normalized angular frequencies, 0.1π and 0.8π, using the lowpass IIR digital filter of Eq. (11.22). Program 11_2 can be utilized for this purpose.

Figure 11.7 shows the plots generated by this program, verifying the blocking of the high-frequency sinusoidal sequence by the lowpass filter. Note also the transients generated by the zero initial conditions at the beginning of the output sequence.

11.2.2   Simulation of Cascade Form IIR Filters

In Example 11.6, we illustrate the simulation of cascade form realizations of IIR transfer functions. We first verify the simulation using the method described in Section 11.1.4.

EXAMPLE 11.6   Cascade Realization of IIR Filters and Structure Verification

We repeat now Example 11.5, using instead a cascade realization of the digital filter of Eq. (11.22). From Eq. (9.35), we arrive at one possible realization using a cascade of a first-order transfer function $H_1(z)$ and a second-order transfer function $H_2(z)$ given by

$$H_1(z) = \frac{0.0662272(1 + z^{-1})}{1 - 0.2593284z^{-1}} \tag{11.23a}$$

and

$$H_2(z) = \frac{1 + 2z^{-1} + z^{-2}}{1 - 0.6762858z^{-1} + 0.3917468z^{-2}}. \tag{11.23b}$$

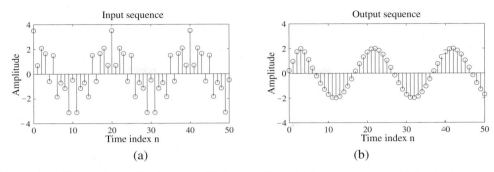

Figure 11.7: Illustration of filtering by an IIR lowpass filter: (a) input sequence and (b) output sequence.

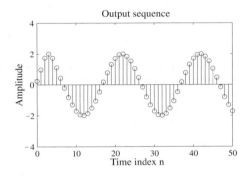

Figure 11.8: Output of the cascade realization of the lowpass filter of Eq. (11.22).

Program 11_3 for the simulation of the cascade realization computes the first seven impulse response samples. From these impulse response samples, the transfer function of the simulated filter is determined using the function `strucver` to verify the correctness of the simulation. The output data generated by running this program are identical to those given in Eq. (11.22).

Program 11_3.m
Program 11_4.m

EXAMPLE 11.7   Illustration of Lowpass Filtering Using the Cascade Realization

We next modify Program 11_3 to demonstrate the lowpass filtering using the cascade realization of the IIR transfer function of Eq. (11.22). The modified version is Program 11_4. The output generated by the above program is shown in Figure 11.8.

11.2.3 Simulation of Overlap-Add Filtering Method

As indicated in Section 5.10.2, a long input sequence can be filtered using the overlap-add method in which the input sequence is segmented into a set of contiguous short input blocks. Each block is then filtered separately, and the overlaps in the output blocks are added appropriately to generate the long output sequence. This method of filtering can be implemented on MATLAB using the M-file `fftfilt`. It can also be easily implemented using the second form of the M-file `filter`. Here, the final values of the internal variable vector `sf` at any stage of filtering is fed back in the following stage of filtering as the initial condition vector `si`.

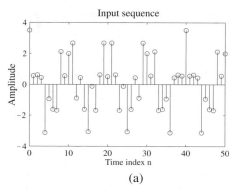
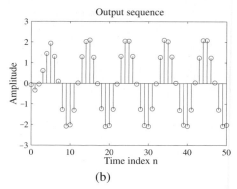

Figure 11.9: Illustration of filtering by an FIR lowpass filter: (a) input sequence and (b) output sequence.

Program 11_5.m

EXAMPLE 11.8   Illustration of Overlap-Add Filtering Approach

We use Program 11_5 to illustrate the method by repeating the lowpass filtering in Example 11.7. The output plot generated by the above program is identical to the one shown in Figure 11.8.

11.2.4   Simulation of Direct Form FIR Filters

Both `filter` and `direct2` can be used also to simulate direct form FIR filter structures by setting the denominator vector `den` equal to 1. Example 11.9 illustrates the use of the former M-file.

EXAMPLE 11.9   Illustration of Filtering by a Lowpass FIR Filter

We consider in this example the filtering of a signal formed by the sum of two sinusoidal sequences of normalized angular frequencies, 0.2π and 0.75π, by an equiripple FIR lowpass filter. The coefficients are generated using the function `remez`. For simplicity, we consider the design of a ninth-order Type 1 FIR filter. Program 11_6 has been utilized to illustrate the filtering process.

Program 11_6.m

Figure 11.9 shows the input and the output sequences generated by this program. These plots verify the lowpass filtering operation. The transients generated by the zero initial conditions at the beginning of the output sequence are again clearly visible.

11.2.5   Zero-Phase Filtering

The *Signal Processing Toolbox* of MATLAB includes the M-file `filtfilt`, which implements the zero-phase filtering scheme discussed in Section 7.2.1. In its basic form, `y = filtfilt(p,d,x)` implements zero-phase filtering by performing both forward and time-reversed processing operations. The resulting filter thus has double the order of the filter characterized by the coefficients p and d. If $H(z)$ denotes the transfer function of the original filter, `filtfilt` then realizes a filter with a zero-phase frequency response given by $|H(e^{j\omega})|^2$ and therefore has a passband ripple in dB and a minimum stopband attenuation in dB that are twice those of the original filter, respectively. We illustrate one possible application of the function `filtfilt` in Example 11.10.

Program 11_7.m

EXAMPLE 11.10   Illustration of Zero-Phase Digital Filtering

We have indicated in Section 9.1.2 that IIR digital filters are, in general, computationally more efficient than equivalent FIR digital filters meeting the same magnitude specifications. On the other hand, IIR filters cannot

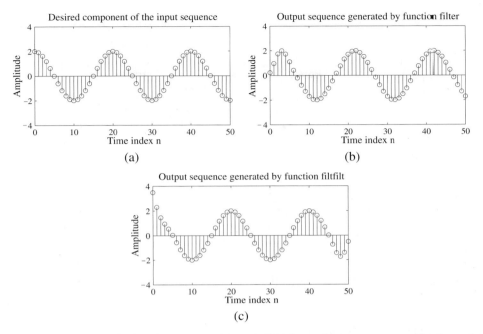

Figure 11.10: Example 11.10: (a) low-frequency component of the input, (b) output generated by forward-only filtering, and (c) output generated by both forward and time-reversed filtering.

be designed with linear phase, whereas it is much easier to design FIR filters with linear-phase. One way to implement an IIR filter with linear-phase is to use both forward and time-reversed filtering operations as discussed in Section 7.2.1. To this end we repeat the lowpass filtering problem of Example 11.5 and compare the results of the functions `filter` and `filtfilt` using Program 11_7.

Figure 11.10 depicts the low-frequency component of the input, the output generated by forward-only filtering, and the output generated by both forward and time-reversed filtering. Several comments are in order here. Both outputs show the effect of the transients. In the case of forward-only filtering, as shown in Figure 11.10(b), only the first few output samples are corrupted by the transients generated by the initial conditions. On the other hand, in the case of both forward and time-reversed filtering indicated in Figure 11.10(c), effects of transients can be seen at both the beginning and the end. Next, as expected, both outputs are nearly identical to the input except for the transient part, but the output in the first case is delayed with respect to the input, whereas the output in the second case is not.

11.2.6 Simulation of Cascaded Lattice Filter Structures

The function `latcfilt` in the *Signal Processing Toolbox* can be used to simulate the IIR and the FIR cascaded lattice filter structure of Sections 8.8 and 8.9.1, respectively. There are three basic forms of this function. In the first form, `[f,g] = latcfilt(k,x)` simulates an FIR cascaded lattice filter structure with lattice filter coefficients given by vector `k` and generates the forward output vector `f` and the backward output vector `g` for an input vector `x`. The second form, `[f,g] = latcfilt(k,alpha,x)`, simulates an IIR tapped cascaded lattice structure with lattice coefficient vector `k` and the feedforward multiplier vector `alpha`. The last form, `[f,g] = latcfilt(k,1,x)`, simulates an all-pole IIR cascaded lattice filter structure.

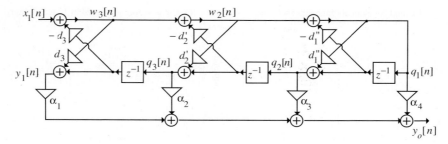

Figure 11.11: Cascaded lattice realization of the IIR transfer function of Eq. (11.24).

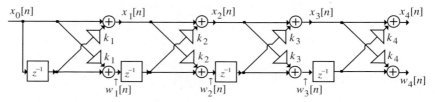

Figure 11.12: Cascaded lattice realization of the FIR transfer function of Eq. (11.25): $k_1 = 0.5$, $k_2 = 1.0$, $k_3 = 0.2173913$, and $k_4 = -0.08$.

We illustrate in Examples 11.11 and 11.12 the simulation of both IIR and FIR cascaded lattice filter structures.

EXAMPLE 11.11   **Simulation of IIR Cascaded Lattice Filter Structure**

Consider the tapped cascaded lattice realization of the third-order IIR transfer function of Example 8.3, repeated below for convenience:

$$H(z) = \frac{0.44z^{-1} + 0.36z^{-2} + 0.02z^{-3}}{1 + 0.4z^{-1} + 0.18z^{-2} - 0.2z^{-3}}. \tag{11.24}$$

The realization derived in this example is depicted in Figure 11.11. The multiplier coefficients as derived in Examples 8.10 and 8.12 are given by

$$d_3 = -0.2, \qquad d_2' = 0.270833, \qquad d_1'' = 0.357377,$$
$$\alpha_1 = 0.02, \qquad \alpha_2 = 0.352, \qquad \alpha_3 = 0.276533, \qquad \alpha_4 = -0.19016.$$

Program 11_8.m

Program 11_8 can be used for the simulation of an IIR cascaded lattice structure. The input vector x consists of the first $2N + 1$ coefficients of a unit sample sequence where N is the order of the filter. The output vector f contains the first $2N + 1$ impulse response coefficients, which are then used to determine the numerator and the denominator coefficients of the actual transfer function implemented using the method described in Section 11.1.4 The resulting transfer function coefficients generated by running this program are identical to those given in Eq. (11.24), thus verifying the structure of Figure 11.11.

EXAMPLE 11.12   **Simulation of FIR Cascaded Lattice Filter Structure**

We consider the simulation of the FIR cascaded lattice structure of Example 8.16. The pertinent transfer function is given by:

$$H_4(z) = 1 + 1.2z^{-1} + 1.12z^{-2} + 0.12z^{-3} - 0.08z^{-4}. \tag{11.25}$$

The realization derived in this example is shown in Figure 11.12. The multiplier coefficients as determined in Example 8.16 are given by

$$k_1 = 0.5, \qquad k_2 = 1, \qquad k_3 = 0.2173913, \qquad k_4 = -0.08.$$

Program 11_9 can be used for the simulation of an FIR transfer function of order N as a FIR cascaded lattice structure of the form of Figure 11.12. The input vector x0 contains the first $N + 1$ samples of a unit sample sequence, while the output vector f contains the first $N + 1$ samples of the impulse response sequence. The program also computes the mirror-image output vector g.

The output data displayed are the coefficients of the transfer function $H_4(z)$ and the coefficients of its mirror-image transfer function $\tilde{H}_4(z)$. The coefficients of $H_4(z)$ are identical to those given in Eq. (11.25).

Program 11_9.m

11.3 Computation of the Discrete Fourier Transform

The discrete Fourier transform (DFT) is another widely used DSP algorithm. As indicated earlier, it can be employed to implement the linear convolution of two sequences, a key digital filtering operation. It is also used for the spectral analysis of signals, discussed in Sections 14.2 to 14.4. Because of the widespread use of the DFT, it is of interest to investigate efficient implementation methods, which are discussed in this section.

In Section 5.2, we introduced the concept of the N-point DFT $X[k]$ of a sequence $x[n]$ of length N as the N samples of its Fourier transform, $X(e^{j\omega})$, evaluated uniformly on the ω-axis at $\omega_k = 2\pi k/N$, $0 \le k \le N - 1$; that is,

$$X[k] = X(e^{j\omega})\Big|_{\omega=2\pi k/N} = \sum_{n=0}^{N-1} x[n]e^{-j2\pi kn/N}, \qquad 0 \le k \le N - 1. \qquad (11.26)$$

Since a finite-length sequence is always absolutely summable, the DFT is thus the samples of its z-transform $X(z)$ evaluated on the unit circle at N equally spaced points $z = W_N^{-k}$:

$$X[k] = X(z)|_{z=e^{j2\pi k/N}}, \qquad 0 \le k \le N - 1. \qquad (11.27)$$

As can be seen from Eq. (11.26), the computation of each sample of the DFT sequence requires N complex multiplications and $N - 1$ complex additions. Hence, the computation of the N-point DFT sequence requires N^2 complex multiplications and $(N - 1)N$ complex additions. In the case of a sequence of length N, it can be shown that the computation of its N-point DFT sequence requires $4N^2$ real multiplications and $(4N - 2)N$ real additions (Problem 11.12). As a result, the total number of computations to compute an N-point DFT increases very rapidly as N increases. For large N, the number of complex multiplications and additions is approximately equal to N^2. Hence, it is of practical interest to develop more efficient or fast algorithms for computing the DFT.

11.3.1 Goertzel's Algorithm

An elegant approach to computing a single sample of the DFT is to use a recursive computation scheme. One such scheme, which is quite popular, is the *Goertzel's algorithm*, derived next [Goe58]. This algorithm makes use of the identity

$$W_N^{-kN} = 1, \qquad (11.28)$$

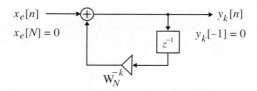

Figure 11.13: Recursive computation of the kth DFT sample.

obtained using the periodicity of W_N^{-kn}. Using the identity of Eq. (11.28), we can rewrite Eq. (11.26) as

$$X[k] = \sum_{\ell=0}^{N-1} x[\ell]W_N^{k\ell} = W_N^{-kN} \sum_{\ell=0}^{N-1} x[\ell]W_N^{k\ell} = \sum_{\ell=0}^{N-1} x[\ell]W_N^{-k(N-\ell)}. \qquad (11.29)$$

The above expression can be expressed in the form of a convolution. To this end, we define a new sequence

$$y_k[n] = \sum_{\ell=0}^{n} x_e[\ell]W_N^{-k(n-\ell)}, \qquad (11.30)$$

which is a direct convolution of the causal sequence $x_e[n]$ defined by

$$x_e[n] = \begin{cases} x[n], & 0 \le n \le N-1, \\ 0, & n < 0, n \ge N, \end{cases} \qquad (11.31a)$$

with a causal infinite-length sequence

$$h_k[n] = \begin{cases} W_N^{-kn}, & n \ge 0, \\ 0, & n < 0. \end{cases} \qquad (11.31b)$$

It follows from Eqs. (11.29) and (11.30) that

$$X[k] = y_k[n]\big|_{n=N} .$$

By taking the z-transform of both sides of Eq. (11.30), we arrive at

$$Y_k(z) = \frac{X_e(z)}{1 - W_N^{-k}z^{-1}}, \qquad (11.32)$$

where $1/(1 - W_N^{-k}z^{-1})$ is the z-transform of the causal sequence $h_k[n]$ and $X_e(z)$ is the z-transform of $x_e[n]$. Equation 11.32 implies that $y_k[n]$ is the output of an initially relaxed LTI digital filter with a transfer function

$$H_k(z) = \frac{1}{1 - W_N^{-k}z^{-1}}, \qquad (11.33)$$

with an input $x[n]$, as indicated in Figure 11.13. When $n = N$, the output of the filter $y_k[N]$ is precisely $X[k]$.

From Figure 11.13, the DFT computation algorithm is given by

$$y_k[n] = x_e[n] + W_N^{-k}y_k[n-1] \qquad 0 \le n \le N, \qquad (11.34)$$

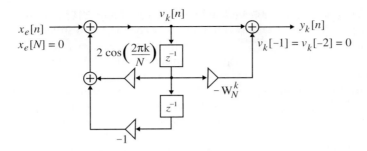

Figure 11.14: Modified approach to the recursive computation of the kth DFT sample.

with $y_k[-1] = 0$ and $x_e[N] = 0$. Since a complex multiplication can be implemented using four real multiplications and two real additions, computation of each new value of $y_k[n]$ thus, in general, requires four real multiplications and four real additions.[6] As a result, the computation of $X[k] = y_k[N]$ involves $4N$ real multiplications and $4N$ real additions, resulting in $4N^2$ real multiplications and $4N^2$ real additions for the computation of all N DFT samples.

Hence, the above algorithm, in comparison to the direct DFT computation, requires the same number of real multiplications but $2N$ more real additions. As a result, it is computationally slightly more inefficient than the direct approach. The advantage of the recursive algorithm, however, is that the N complex coefficients W_N^{kn} required to compute $X[k]$ do not have to be either computed or stored in advance, but are computed recursively as needed.

The algorithm can be made more efficient by observing that $H_k(z)$ of Eq. (11.33) can be rewritten as

$$H_k(z) = \frac{1}{1 - W_N^{-k} z^{-1}} = \frac{1 - W_N^k z^{-1}}{(1 - W_N^{-k} z^{-1})(1 - W_N^k z^{-1})}$$

$$= \frac{1 - W_N^k z^{-1}}{1 - 2\cos\left(\frac{2\pi}{N} k\right) z^{-1} + z^{-2}},$$

(11.35)

resulting in the new realization shown in Figure 11.14.

The DFT computation equations are now given by

$$v_k[n] = x_e[n] + 2\cos\left(\frac{2\pi k}{N}\right) v_k[n-1] - v_k[n-2], \quad 0 \le n \le N,$$ (11.36a)

$$X[k] = y_k[N] = v_k[N] - W_N^k v_k[N-1].$$ (11.36b)

Note that the computation of each sample of the intermediate variable $v_k[n]$ involves only two real multiplications and four real additions.[7] The complex multiplication by the constant W_N^k needs to be performed only once at $n = N$. Thus, to compute one sample $X[k]$ of the N-point DFT, we need $(2N + 4)$ real multiplications and $(4N + 4)$ real additions. As a result, the modified Goertzel's algorithm for computing the N-point DFT requires $2(N + 2)N$ real multiplications and $4(N + 1)N$ real additions.

Further savings in computational requirements can be obtained by comparing the realization of $H_{N-k}(z)$ with that of $H_k(z)$, shown in Figure 11.14. In the case of the former, the multiplier in the feedback path is $2\cos(2\pi(N - k)/N) = 2\cos(2\pi k/N)$, which is the same as in Figure 11.14. Hence, $v_{N-k}[n] = v_k[n]$,

---

[6]It can be shown that a simple modification of the complex multiplication algorithm can reduce the number of real multiplications to 3, while increasing the number of real additions to 5 (see Problem 11.13).

[7]In general, $x[n]$ and $v_k[n]$ are complex sequences.

indicating that the intermediate variables computed for determining $X[k]$ need no longer be computed again for determining $X[N - k]$. The only difference between these two structures is in the feedforward path, where the multiplier is instead $W_N^{N-k} = W_N^{-k}$, which is the complex conjugate of the coefficient W_N^k used in Figure 11.14. Thus, the computation of the two samples of the N-point DFT, $X[k]$ and $X[N - k]$, requires $2(N + 4)$ real multiplications and $4(N + 2)$ real additions. Or, in other words, all N samples of the DFT can be determined using approximately N^2 real multiplications and approximately $2N^2$ real additions. The number of real multiplications is thus about one-fourth and the number of real additions is about one-half of those needed in the direct DFT computation.

The M-file `goertzel` implements the modified Goertzel's algorithm.

Goertzel's algorithm is attractive in applications requiring the computation of a few samples of the DFT. One such example is the *dual-tone multifrequency* (DTMF) signal detection in the TOUCH-TONE® telephone dialing system, discussed in Section 15.1. In such an application, the input $x[n]$ is a real sequence, and the square magnitude of the DFT sample, $|X[k]|^2$, is of interest. Since $x[n]$ is a real sequence, the intermediate sequence $v_k[n]$ generated in the modified Goertzel's algorithm is also a real sequence. As a result, from Eq. (11.36b) we obtain

$$|X[k]|^2 = |y_k[N]|^2 = v_k^2[N] + v_k^2[N - 1] - 2\cos(2\pi k/N)v_k[N]v_k[N - 1]. \qquad (11.37)$$

The above scheme uses only real multiplications, avoiding the complex multiplication required for the computation of $y_k[N]$ as indicated in Eq. (11.36b).

We next describe two fast algorithms for the computation of all samples of the DFT when the length N is a composite number that is expressible as a product of integers. As we shall demonstrate, in the case of N that is a power of 2, the total number of computations in the new algorithms can be made proportional to $N \log_2(N)$, and as a result, these algorithms are highly attractive in applications requiring the computation of all DFT samples.

11.3.2   Cooley–Tukey FFT Algorithms

The basic idea behind all fast algorithms for computing the discrete Fourier transform (DFT), commonly called the *fast Fourier transform* (FFT) algorithms, is to decompose successively the N-point DFT computation into computations of smaller-size DFTs and to take advantage of the periodicity and symmetry properties of the complex number W_N^{kn}. Such decompositions, if properly carried out, can result in a significant savings in the computational complexity given by the total number of multiplications and the total number of additions needed to compute all N DFT samples. There are various versions of the FFT algorithms. We review here the main concepts behind the two most basic FFT algorithms [Coo65].

Decimation-in-Time FFT Algorithm

Consider a sequence $x[n]$ of length N that is assumed to be a power of 2. Using a two-band polyphase decomposition[8] of $x[n]$, we can express its z-transform $X(z)$ as

$$X(z) = X_0(z^2) + z^{-1}X_1(z^2), \qquad (11.38)$$

where

$$X_0(z) = \sum_{n=0}^{\frac{N}{2}-1} x_0[n]z^{-n} = \sum_{n=0}^{\frac{N}{2}-1} x[2n]z^{-n}, \qquad (11.39a)$$

---

[8]See Section 8.3.3 for a discussion on the polyphase decomposition.

$$X_1(z) = \sum_{n=0}^{\frac{N}{2}-1} x_1[n]z^{-n} = \sum_{n=0}^{\frac{N}{2}-1} x[2n+1]z^{-n}. \qquad (11.39\text{b})$$

Thus, $X_0(z)$ is the z-transform of the $(N/2)$-length sequence $x_0[n] = x[2n]$ formed from the even-indexed samples of $x[n]$, while $X_1(z)$ is the z-transform of the $(N/2)$-length sequence $x_1[n] = x[2n+1]$ formed from the odd-indexed samples of $x[n]$.

Evaluating $X(z)$ on the unit circle at N equally spaced points, $z = W_N^{-k}$, from Eq. (11.38), we arrive at the N-point DFT of $x[n]$ given by[9]

$$X[k] = X_0[\langle k \rangle_{N/2}] + W_N^k X_1[\langle k \rangle_{N/2}], \qquad 0 \le k \le N-1, \qquad (11.40)$$

where $X_0[k]$ and $X_1[k]$ are the $(N/2)$-point DFTs of the $(N/2)$-length sequences $x_0[n]$ and $x_1[n]$, respectively; that is,

$$X_0[k] = \sum_{r=0}^{\frac{N}{2}-1} x_0[r]W_{N/2}^{rk} = \sum_{r=0}^{\frac{N}{2}-1} x[2r]W_{N/2}^{rk}, \qquad 0 \le k \le \frac{N}{2}-1, \qquad (11.41\text{a})$$

$$X_1[k] = \sum_{r=0}^{\frac{N}{2}-1} x_1[r]W_{N/2}^{rk} = \sum_{r=0}^{\frac{N}{2}-1} x[2r+1]W_{N/2}^{rk}, \qquad 0 \le k \le \frac{N}{2}-1. \qquad (11.41\text{b})$$

It should be noted that in deriving Eq. (11.40) from Eq. (11.38), we have made use of the fact that the $N/2$-points DFTs, $X_0[k]$ and $X_1[k]$, are periodic sequences in k with a period $N/2$.

It is instructive at this point to examine a block diagram interpretation of the modified DFT computation scheme of Eq. (11.40) that computes an N-point DFT of the original length-N sequence $x[n]$ by forming a weighted sum of two $(N/2)$-point DFTs of two $(N/2)$-length subsequences formed from the even-indexed samples $x_0[n] = x[2n]$ and the odd-indexed samples $x_1[n] = x[2n+1]$. To this end, we need the *down-sampler*, introduced earlier in Section 2.1.2, to develop the two subsequences $x_0[n]$ and $x_1[n]$ from $x[n]$. It follows from the definition of Eq. (2.21), if the input $x[n]$ to a factor-of-2 down-sampler is a length-N sequence defined for $0 \le n \le N-1$, its output $x_0[n]$ is a length-$(N/2)$ sequence defined for $0 \le n \le (N/2)-1$ and is composed of the even-indexed samples of $x[n]$; that is, $x_0[n] = x[2n]$, as shown in Figure 11.15(a). To generate the subsequence $x_1[n]$ composed of the odd-indexed samples of $x[n]$; that is, $x_1[n] = x[2n+1]$, $0 \le n \le (N/2)-1$, we need to pass $x[n+1]$ through a factor-of-2 down-sampler. The sequence $x[n+1]$ can be developed from the sequence $x[n]$ by means of an advance operation. The process is illustrated in Figure 11.15(b).

From the above discussion, it follows then that a block diagram interpretation of the DFT computation scheme of Eq. (11.40) is as indicated in Figure 11.16. Figure 11.17 shows its flow-graph representation for the case of $N = 8$.

Before proceeding further, let us evaluate the computational requirements for computing an N-point DFT using two $(N/2)$-point DFTs based on the decomposition of Eq. (11.40). Now a direct computation of an N-point DFT requires N^2 complex multiplications and $N^2 - N \cong N^2$ complex additions. On the other hand, computation of an N-point DFT using the decomposition of Eq. (11.40) requires the computations of two $(N/2)$-point DFTs that need to be combined with N complex multiplications and N complex additions, resulting in a total of $N + (N^2/2)$ complex multiplications and approximately $N + (N^2/2)$ complex additions. It can easily be verified that for $N \ge 3$, $N + (N^2/2) < N^2$.

We can continue the above process by expressing each of the two $(N/2)$-point DFTs, $X_0[k]$ and $X_1[k]$, as a weighted combination of two $(N/4)$-point DFTs since, by assumption, $N/2$ is even. For example, we

---

[9] $\langle k \rangle_{N/2} = k$ modulo $(N/2)$. For an explanation of the modulo operation, see Section 5.4.1.

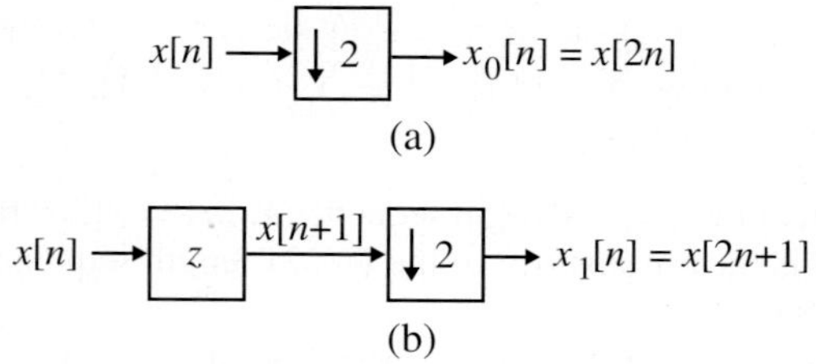

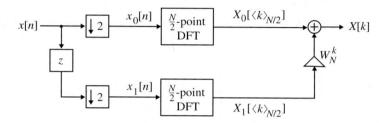

$x[n] \longrightarrow \boxed{\downarrow 2} \longrightarrow x_0[n] = x[2n]$

(a)

$x[n] \longrightarrow \boxed{z} \xrightarrow{x[n+1]} \boxed{\downarrow 2} \longrightarrow x_1[n] = x[2n+1]$

(b)

Figure 11.15: (a) Generation of a subsequence containing even-indexed input samples and (b) generation of a subsequence containing odd-indexed input samples.

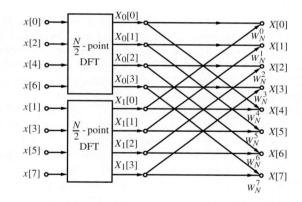

Figure 11.16: Structural interpretation of the DFT decomposition scheme of Eq. (11.40).

Figure 11.17: Flow-graph of the first stage in the decimation-in-time FFT algorithm for $N = 8$.

can express $X_0[k]$ as

$$X_0[k] = X_{00}[\langle k \rangle_{N/4}] + W_{N/2}^k X_{01}[\langle k \rangle_{N/4}], \quad 0 \le k \le \tfrac{N}{2} - 1, \tag{11.42}$$

where $X_{00}[k]$ and $X_{01}[k]$ are the $(N/4)$-point DFTs of the $(N/4)$-length sequences, $x_{00}[n]$ and $x_{01}[n]$, generated from the even and odd samples of $x_0[n]$, respectively. Likewise, we can express $X_1[k]$ as

$$X_1[k] = X_{10}[\langle k \rangle_{N/4}] + W_{N/2}^k X_{11}[\langle k \rangle_{N/4}], \quad 0 \le k \le \tfrac{N}{2} - 1, \tag{11.43}$$

where $X_{10}[k]$ and $X_{11}[k]$ are the $(N/4)$-point DFTs of the $(N/4)$-length sequences, $x_{10}[n]$ and $x_{11}[n]$, generated from the even and odd samples of $x_1[n]$, respectively.

Substituting Eqs. (11.42) and (11.43) in Eq. (11.40) and making use of the identity $W_{N/2}^k = W_N^{2k}$, we then arrive at the two-stage decomposition of an N-point DFT computation in terms of four $(N/4)$-point

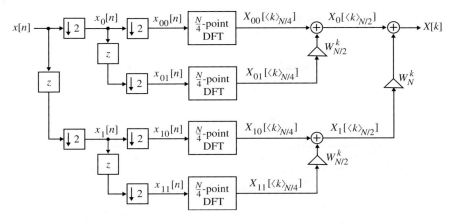

Figure 11.18: Structural interpretation of the two-stage DFT decomposition scheme.

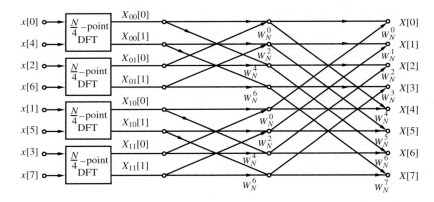

Figure 11.19: Flow-graph of the second stage in the decimation-in-time FFT algorithm for $N = 8$.

DFTs, as indicated by the block diagram of Figure 11.18. The corresponding flow-graph representation is shown in Figure 11.19 for $N = 8$. In the case of the 8-point DFT computation illustrated in Figure 11.20, the $(N/4)$-point DFT is a 2-point DFT, and no further decomposition is possible. The 2-point DFTs, $X_{00}[k]$, $X_{01}[k]$, $X_{10}[k]$, and $X_{11}[k]$, can be easily computed. For example, for the computation of $X_{00}[k]$, the pertinent expressions are

$$X_{00}[k] = \sum_{n=0}^{1} x_{00}[n] W_2^{nk} = x[0] + W_2^k x[4], \quad k = 0, 1. \tag{11.44}$$

The corresponding flow-graph is indicated in Figure 11.20, where we have used the identity $W_2^k = W_N^{(N/2)k}$. Replacing each of the 2-point DFTs in Figure 11.19 with their respective flow-graph representations, we finally arrive at the complete flow-graph of the basic decimation-in-time DFT algorithm, as shown in Figure 11.21.

If we examine the flow-graph of Figure 11.21, we note that it consists of three stages. The first stage computes the four 2-point DFTs, the second stage computes the two 4-point DFTs, and finally, the last stage computes the desired 8-point DFT. Moreover, the number of complex multiplications and additions

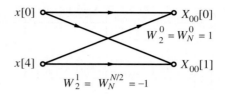

Figure 11.20: Flow-graph of the 2-point DFT.

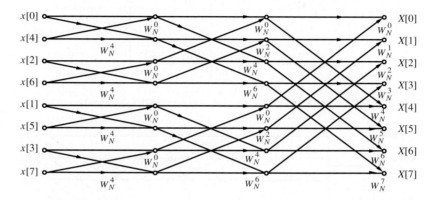

Figure 11.21: Complete flow-graph of the basic decimation-in-time FFT algorithm for $N = 8$.

performed at each stage is equal to 8, the size of the transform. As a result, the total number of complex multiplications and additions in computing all 8 DFT samples is equal to $3 \times 8 = 24$.

It follows from the above observation that in the general case when $N = 2^\mu$, the number of stages of computation of the (2^μ)-point DFT in the fast algorithm will be $\mu = \log_2 N$. Therefore, the total number of complex multiplications and additions in computing all N DFT samples is equal to $N(\log_2 N)$. In developing this count, we have, for the present, considered multiplications with $W_N^0 = 1$ and $W_N^{N/2} = -1$ to be complex. In addition, we have not taken advantage of the symmetry property $W_N^{(N/2)+k} = -W_N^k$. These properties can be made use of in reducing the computational complexity further.

Computational Considerations

Examination of the flow-graph of Figure 11.21 also reveals that each stage of the DFT computation process employs the same basic computational module in which two output variables are generated from a weighted combination of two input variables. To see this aspect more clearly, let us label the N-input and the N-output variables in the rth stage of the DFT computation as $\Psi_r[m]$ and $\Psi_{r+1}[m]$, respectively, with $r = 1, 2, \ldots, \mu$, and $m = 0, 1, \ldots, N-1$. According to this labeling, in the case of Figure 11.21 for the 8-point DFT computation, $\Psi_1[0] = x[0]$, $\Psi_1[1] = x[4]$, and so on. Similarly, here, $\Psi_4[0] = X[0]$, $\Psi_4[1] = X[1]$, and so on. Based on this labeling scheme, it can be easily verified that the basic computational module is represented by the flow-graph of Figure 11.22 and described by the following input–output relations:

$$\Psi_{r+1}[\alpha] = \Psi_r[\alpha] + W_N^\ell \Psi_r[\beta], \tag{11.45a}$$

$$\Psi_{r+1}[\beta] = \Psi_r[\alpha] + W_N^{\ell+(N/2)} \Psi_r[\beta]. \tag{11.45b}$$

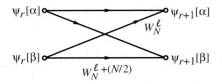

Figure 11.22: Flow-graph of the basic computational module in the decimation-in-time FFT method.

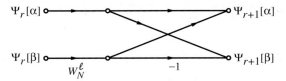

Figure 11.23: Flow-graph of the modified butterfly computational module.

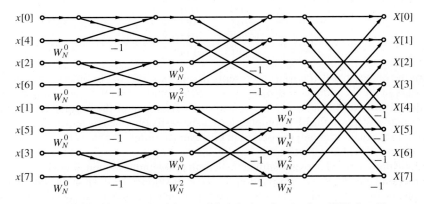

Figure 11.24: Flow-graph of the modified decimation-in-time FFT algorithm.

Because of its shape, the basic computational module of Figure 11.22 is referred to in the literature as the *butterfly computation.*

Substituting $W_N^{\ell+(N/2)} = -W_N^\ell$ in Eq. (11.45b), we can rewrite this equation as

$$\Psi_{r+1}[\beta] = \Psi_r[\alpha] - W_N^\ell \Psi_r[\beta]. \tag{11.45c}$$

The modified butterfly computation is thus as indicated in Figure 11.23 and requires only one complex multiplication. Use of this modified butterfly computational module in the FFT computation leads to a reduction in the total number of complex multiplications by 50%, as can be seen from the new flow-graph for the $N = 8$ case, illustrated in Figure 11.24. Further savings in the computational complexity arise by taking into consideration that multiplications by $W_N^0 = 1$, $W_N^{N/2} = -1$, $W_N^{N/4} = j$, and $W_N^{3N/4} = -j$ can be avoided in the DFT computation process.

Another attractive feature of the FFT algorithm described above is with regard to memory requirements. Since each stage employs the same butterfly computation to compute the two output variables $\Psi_{r+1}[\alpha]$ and $\Psi_{r+1}[\beta]$ from the input variables $\Psi_r[\alpha]$ and $\Psi_r[\beta]$, after $\Psi_{r+1}[\alpha]$ and $\Psi_{r+1}[\beta]$ have been determined, they can be stored in the same memory locations where $\Psi_r[\alpha]$ and $\Psi_r[\beta]$ were previously stored. Thus, at the end of the computation at any stage, the output variables $\Psi_{r+1}[m]$ can be stored in the same registers previously occupied by the corresponding input variables $\Psi_r[m]$. This type of memory location sharing feature is commonly known as the *in-place computation,* resulting in a significant saving in the overall memory requirements.

It should be noted, however, from Figure 11.24, that while the DFT samples $X[k]$ appear at the output in a sequential order, the input time-domain samples $x[n]$ appear in a different order. Thus, a sequentially ordered input $x[n]$ must be reordered appropriately before the FFT algorithm described by the above structure can begin. To understand the basic scheme, in the input reordering scheme, consider the 8-point DFT computation illustrated in Figure 11.24. If we represent the arguments of the input samples $x[n]$ and their sequentially ordered new representations $\Psi_1[m]$ in binary forms, we arrive at the following relations between m and n:

m	n
000	000
001	100
010	010
011	110
100	001
101	101
110	011
111	111

It follows from the above that if $(b_2b_1b_0)$ represents the index n of $x[n]$ in binary form, then the sample $x[b_2b_1b_0]$ appears in the location $m = b_0b_1b_2$ as $\Psi[b_0b_1b_2]$ before the DFT computation is started, or, in other words, the location of $\Psi_1[m]$ is in *bit-reversed order* from that of the original input array $x[n]$.

Various alternative forms of the FFT computation can be easily obtained by reordering the computations, such as the input in normal order and the output in bit-reversed order (Problem 11.17), and both input and output in normal order (Problem 11.18).

The FFT algorithm outlined above assumed the length of the input sequence $x[n]$ to be a power of 2. If it is not, we can extend the length by zero-padding $x[n]$ (see Section 5.2.3) so that the length of the extended sequence is a power of 2.[10] Even after zero-padding, the DFT computation based on the fast algorithm derived above may be computationally more efficient than the direct DFT computation of the original shorter sequence. Alternatively, we can develop fast algorithms that make use of polyphase decomposition with more than two subsequences. To illustrate this modification to the basic FFT algorithm, consider a sequence $x[n]$ of a length that is a power of 3. Here, the DFT samples in the first stage are computed using a three-band polyphase decomposition of $X(z)$,

$$X(z) = X_0(z^3) + z^{-1}X_1(z^3) + z^{-2}X_2(z^3), \tag{11.46}$$

resulting in

$$X[k] = X_0[\langle k \rangle_{N/3}] + W_N^k X_1[\langle k \rangle_{N/3}] + W_N^{2k} X_2[\langle k \rangle_{N/3}], \quad 0 \le k \le N-1, \tag{11.47}$$

where $X_0[k]$, $X_1[k]$, and $X_2[k]$ are now $(N/3)$-point DFTs. This process can be repeated to compute each of the $(N/3)$-point DFTs in terms of three $(N/9)$-point DFTs, and so on, until the smallest computational module is a 3-point DFT and no further decomposition is possible.

The FFT computation schemes described above are called *decimation-in-time (DIT) FFT algorithms,* since here the input sequence $x[n]$ is first decimated to form a set of subsequences before the DFT is

---

[10]It should be noted that zero-padding increases the effective length of the original sequence, and hence, the longer-length DFT samples are different frequency samples of the frequency response and are more closely spaced on the unit circle than that of the original shorter-length DFT samples.

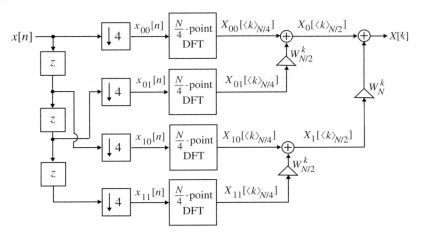

Figure 11.25: First stage in a radix-4 decimation-in-time FFT algorithm.

computed. For example, the relation between the input sequence $x[n]$ and its even and odd parts, $x_0[n]$ and $x_1[n]$, respectively, generated by the first stage of the DIT algorithm, shown in Figure 11.16, is as follows:

$x[n]$:	$x[0]$	$x[1]$	$x[2]$	$x[3]$	$x[4]$	$x[5]$	$x[6]$	$x[7]$
$x_0[n]$:	$x[0]$		$x[2]$		$x[4]$		$x[6]$	
$x_1[n]$:		$x[1]$		$x[3]$		$x[5]$		$x[7]$

Likewise, the relation between the input sequence $x[n]$ and the sequences $x_{00}[n]$, $x_{01}[n]$, $x_{10}[n]$, and $x_{11}[n]$, generated by the two-stage decomposition of the DIT algorithm and illustrated in Figure 11.19, is given by

$x[n]$:	$x[0]$	$x[1]$	$x[2]$	$x[3]$	$x[4]$	$x[5]$	$x[6]$	$x[7]$
$x_{00}[n]$:	$x[0]$				$x[4]$			
$x_{01}[n]$:	$x[2]$				$x[6]$			
$x_{10}[n]$:	$x[1]$				$x[5]$			
$x_{11}[n]$:	$x[3]$				$x[7]$			

Or, in other words, the subsequences $x_{00}[n]$, $x_{01}[n]$, $x_{10}[n]$, and $x_{11}[n]$ can be generated directly by a factor-of-4 decimation process, leading to the single-stage decomposition given in Figure 11.25.

If at each stage the decimation is by a factor of R, the resulting FFT algorithm is called a *radix-R* FFT algorithm. Thus, Figure 11.24 shows a radix-2 DIT FFT algorithm. Likewise, Figure 11.25 illustrates the first stage of a radix-4 DIT FFT algorithm. It also follows from the above discussion that, depending on the value of N, various combinations of decompositions of $X[k]$ can be used to develop different types of DIT FFT algorithms. If the scheme uses a mixture of decimations by different factors, it is called a *mixed radix* FFT algorithm.

Modified Goertzel's Algorithm Based on DIT Decomposition

In some applications, it may be convenient to apply the Goertzel's algorithm in computing the smaller-size DFTs after one or two or more stages of decimation-in-time decompositions of the input sequence $x[n]$. For example, the four $(N/4)$-DFTs in Figure 11.18 can be computed using Goertzel's algorithm. This approach reduces the overall computational complexity of the direct Goertzel's algorithm, while still permitting the computation of a few DFT samples.

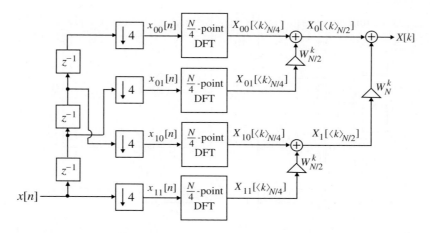

Figure 11.26: Modified structure for the first stage of a radix-4 decimation-in-time FFT algorithm.

It should be noted, however, that the structural interpretations of the DIT FFT algorithms given in Figures 11.16, 11.18, and 11.25 make use of the advance block with a transfer function z that is not physically realizable. If a hardware implementation of these structures is desired, we can insert an appropriate amount of delays at the input and then move them through the chain of advance blocks to make the overall structure physically realizable. Figure 11.26 shows the realizable version of Figure 11.25. Moreover, the input sequence $x[n]$ should be delayed by three sample periods before the down-sampling operations are carried out. The relations between the subsequences $x_{00}[n]$, $x_{01}[n]$, $x_{10}[n]$, and $x_{11}[n]$ and the original sequence $x[n]$ are now

$$
\begin{array}{llllllll}
x[n]: & x[0] & x[1] & x[2] & x[3] & x[4] & x[5] & x[6] & x[7] \\
x_{00}[n]: & & & & x[0] & & & & x[4] \\
x_{01}[n]: & & & & x[2] & & & & x[6] \\
x_{10}[n]: & & & & x[1] & & & & x[5] \\
x_{11}[n]: & & & & x[3] & & & & x[7]
\end{array}
$$

Decimation-in-Frequency FFT Algorithm

The basic idea behind the decimation-in-time FFT algorithm is to decompose sequentially the N-point sequence $x[n]$ into sets of smaller and smaller subsequences and then form a weighted combination of the DFTs of these subsequences. The same idea can be applied to the N-point DFT sequence $X[k]$ to decompose it sequentially into sets of smaller and smaller subsequences. This approach leads to another class of DFT computation schemes collectively known as the *decimation-in-frequency (DIF) FFT algorithm*.

To illustrate the basic difference between the above two decomposition schemes, we develop below the first stage of the DIF FFT algorithm for the case when N is a power of 2. We first express the z-transform $X(z)$ of $x[n]$ as

$$
X(z) = X_0(z) + z^{-N/2} X_1(z), \tag{11.48}
$$

where

$$
X_0(z) = \sum_{n=0}^{(N/2)-1} x[n] z^{-n}, \quad X_1(z) = \sum_{n=0}^{(N/2)-1} x\left[\frac{N}{2} + n\right] z^{-n}. \tag{11.49}
$$

Evaluating $X(z)$ on the unit circle at $z = W_N^{-k}$, we get from Eqs. (11.48) and (11.49),

$$X[k] = \sum_{n=0}^{(N/2)-1} x[n] W_N^{nk} + W_N^{(N/2)k} \sum_{n=0}^{(N/2)-1} x\left[\tfrac{N}{2} + n\right] W_N^{nk}. \tag{11.50}$$

Equation (11.50) can be rewritten as

$$X[k] = \sum_{n=0}^{(N/2)-1} \left(x[n] + (-1)^k x\left[\tfrac{N}{2} + n\right]\right) W_N^{nk}, \tag{11.51}$$

where we have used the identity $W_N^{(N/2)k} = (-1)^k$. Two different forms of Eq. (11.51) are obtained, depending on whether k is even or odd:

$$X[2\ell] = \sum_{n=0}^{(N/2)-1} \left(x[n] + x\left[\tfrac{N}{2} + n\right]\right) W_N^{2n\ell}$$

$$= \sum_{n=0}^{(N/2)-1} \left(x[n] + x\left[\tfrac{N}{2} + n\right]\right) W_{N/2}^{n\ell}, \qquad 0 \le \ell \le \tfrac{N}{2} - 1, \tag{11.52a}$$

$$X[2\ell + 1] = \sum_{n=0}^{(N/2)-1} \left(x[n] - x\left[\tfrac{N}{2} + n\right]\right) W_N^{n(2\ell+1)}$$

$$= \sum_{n=0}^{(N/2)-1} \left(x[n] - x\left[\tfrac{N}{2} + n\right]\right) W_N^{n} W_{N/2}^{n\ell}, \qquad 0 \le \ell \le \tfrac{N}{2} - 1. \tag{11.52b}$$

The above two expressions represent the $\tfrac{N}{2}$-point DFTs of the following two $\tfrac{N}{2}$-point sequences:

$$x_0[n] = \left(x[n] + x\left[\tfrac{N}{2} + n\right]\right),$$
$$x_1[n] = \left(x[n] - x\left[\tfrac{N}{2} + n\right]\right) W_N^{n}, \qquad 0 \le n \le \tfrac{N}{2} - 1, \tag{11.53}$$

respectively. The flow-graph of the first stage of the DFT computation scheme defined by Eqs. (11.52a) and (11.52b) is shown in Figure 11.27 for $N = 8$. As can be seen from this figure, here the input samples are in a sequential order, while the output DFT samples appear in a decimated form, with the even-indexed samples appearing as the outputs of one $(N/2)$-point DFT and the odd-indexed samples appearing as the outputs of the other $(N/2)$-point DFT.

We can continue the above decomposition process by expressing the even- and odd-indexed samples of each one of the two $(N/2)$-point DFTs as a sum of two $(N/4)$-point DFTs. This process can continue until the smallest DFTs are 2-point DFTs. The complete flow-graph of the decimation-in-frequency DFT computation scheme for the $N = 8$ case is shown in Figure 11.28.

It can be seen from Figure 11.28 that in the DIF FFT algorithm, the input $x[n]$ appears in the normal order, while the output $X[k]$ appears in the bit-reversed order. Just as in the case of the radix-2 DIT FFT algorithm, the total number of complex multiplications per stage in the radix-2 DIF FFT algorithm is $N/2$. Hence, the total number of complex multiplications for computing the N-point DFT samples here is also equal to $(N/2) \log_2 N$, ignoring the fact that multiplications by $W_N^{(N/4)\ell}$, $\ell = 0, 1, 2, 3, 4$, can be avoided. As before, various forms of the DIF FFT algorithm can be generated.

The DIT and DIF FFT algorithms described here are often referred to as *Cooley–Tukey* FFT algorithms.

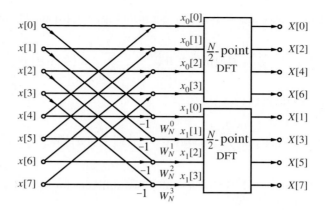

Figure 11.27: Flow-graph of the first stage of the decimation-in-frequency FFT algorithm for $N = 8$.

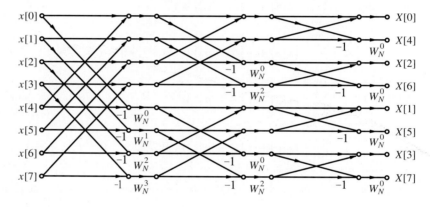

Figure 11.28: Complete flow-graph of the decimation-in-frequency FFT algorithm for $N = 8$.

An examination of the flow-graph of the first stage of the radix-2 DIF FFT algorithm given in Figure 11.27 reveals that the even- and odd-indexed samples of $X[k]$ can be computed independently of each other. It has been shown by Duhamel [Duh86] that significant reduction in the computational complexity can be obtained by using a radix-2 DIF algorithm to compute the even-indexed DFT samples and a radix-4 DIF FFT algorithm to compute the odd-indexed DFT samples. This type of computational scheme has been called a *split-radix FFT algorithm* (Problem 11.32).

11.3.3   Inverse DFT Computation

An FFT algorithm for computing the DFT samples can also be used to calculate efficiently the inverse DFT (IDFT). To show this, consider an N-point sequence $x[n]$ with an N-point DFT $X[k]$. The sequence $x[n]$ is related to the samples $X[k]$ through

$$x[n] = \frac{1}{N} \sum_{k=0}^{N-1} X[k] W_N^{-nk}. \tag{11.54}$$

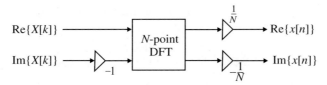

Figure 11.29: Inverse DFT computation via DFT.

If we multiply both sides of the above equation by N and take the complex conjugate, we arrive at

$$Nx^*[n] = \sum_{k=0}^{N-1} X^*[k] W_N^{nk}. \tag{11.55}$$

The right-hand side of the above expression can be recognized as the N-point DFT of a sequence $X^*[k]$ and can be computed using any one of the FFT algorithms discussed earlier. The desired IDFT $x[n]$ is then obtained as

$$x[n] = \frac{1}{N} \left\{ \sum_{k=0}^{N-1} X^*[k] W_N^{nk} \right\}^*. \tag{11.56}$$

In summary, given an N-point DFT $X[k]$, we first form its complex conjugate sequence $X^*[k]$, then compute the N-point DFT of $X^*[k]$, form the complex conjugate of the DFT computed, and, finally, divide each sample by N. The inverse DFT computation process is illustrated in Figure 11.29. Two other approaches to inverse DFT computation are described in Problems 11.26 and 11.27.

11.4 Fast DFT Algorithms Based on Index Mapping

The fast DFT algorithms described in the previous section are for sequences of length N that is a power-of-2 integer. We indicated here that if the length N is not a power-of-2, it can be modified by zero-padding to arrive at a modified sequence of length that is a power-of-2 integer. However, the zero-padding also increases the DFT length and, as a result, the computational complexity. For the case when the length N is a composite number that is expressible as a product of integers, it is possible to develop computationally fast DFT computation algorithms via an index mapping approach in which the sample indices n and k are mapped into two-dimensional indices [Coo65], [Bur77]. The corresponding algorithms also compute the length-N DFT through a series of smaller length DFTs. In this section, we first generalize the Cooley–Tukey FFT algorithms for composite N and then describe a more efficient algorithm for N that is expressible as a product of prime numbers.

11.4.1 General Form of Cooley–Tukey FFT Algorithms

Consider a sequence $x[n]$ of length N that is a product of two integers, N_1 and N_2:

$$N = N_1 N_2. \tag{11.57}$$

The time-domain index n can be represented as a function of two indices, n_1 and n_2, as follows:

$$n = n_1 + N_1 n_2, \qquad \begin{cases} 0 \leq n_1 \leq N_1 - 1, \\ 0 \leq n_2 \leq N_2 - 1. \end{cases} \tag{11.58}$$

Likewise, the frequency-domain index k of the DFT sequence $X[k]$ can also be represented as a function of two indices, k_1 and k_2, as follows:

$$k = N_2 k_1 + k_2, \qquad \begin{cases} 0 \leq k_1 \leq N_1 - 1, \\ 0 \leq k_2 \leq N_2 - 1. \end{cases} \tag{11.59}$$

It follows from Eq. (11.58) that each value of the index n in the range from 0 to $N - 1$ is represented by a unique pair of indices n_1 and n_2. The index mapping of Eq. (11.58) effectively maps the one-dimensional (1-D) sequence $x[n]$ of length-N into a two-dimensional (2-D) sequence $X[n_1, n_2]$ of size $N_1 \times N_2$, containing N_1 rows and N_2 columns. Similarly, it follows from Eq. (11.59) that each value of the index k in the range from 0 to $N - 1$ is represented by an unique pair of indices k_1 and k_2. Likewise, the index mapping of Eq. (11.59) effectively maps the 1-D sequence $X[k]$ of length-N into a 2-D sequence $X[k_1, k_2]$ of size $N_1 \times N_2$, containing N_1 rows and N_2 columns.

Using the above index mappings, we can express the DFT samples

$$X[k] = \sum_{n=0}^{N-1} x[n] W_N^{nk}, \qquad 0 \leq k \leq N - 1, \tag{11.60}$$

as

$$X[k] = X[N_2 k_1 + k_2] = \sum_{n_1=0}^{N_1-1} \sum_{n_2=0}^{N_2-1} x[n_1 + N_1 n_2] W_N^{[n_1+N_1 n_2][N_2 k_1 + k_2]}$$

$$= \sum_{n_1=0}^{N_1-1} \sum_{n_2=0}^{N_2-1} x[n_1 + N_1 n_2] W_N^{n_1 N_2 k_1} W_N^{n_1 k_2} W_N^{N_1 n_2 N_2 k_1} W_N^{N_1 n_2 k_2}, \tag{11.61}$$

where $0 \leq k_1 \leq N_1$ and $0 \leq k_2 \leq N_2$. Now $W_N^{n_1 N_2 k_1} = W_{N_1 N_2}^{n_1 N_2 k_1} = W_{N_1}^{n_1 k_1}$, $W_N^{N_1 n_2 N_2 k_1} = W_{N_1 N_2}^{N_1 n_2 N_2 k_1} = 1$, and $W_N^{N_1 n_2 k_2} = W_{N_1 N_2}^{N_1 n_2 k_2} = W_{N_2}^{n_2 k_2}$. Hence, Eq. (11.61) can be rewritten as

$$X[N_2 k_1 + k_2] = \sum_{n_1=0}^{N_1-1} \sum_{n_2=0}^{N_2-1} x[n_1 + N_1 n_2] W_{N_1}^{n_1 k_1} W_N^{n_1 k_2} W_{N_2}^{n_2 k_2}$$

$$= \sum_{n_1=0}^{N_1-1} \left[\left(\sum_{n_2=0}^{N_2-1} x[n_1 + N_1 n_2] W_{N_2}^{n_2 k_2} \right) W_N^{n_1 k_2} \right] W_{N_1}^{n_1 k_1}, \begin{cases} 0 \leq k_1 \leq N_1 - 1, \\ 0 \leq k_2 \leq N_2 - 1. \end{cases} \tag{11.62}$$

Define

$$G[n_1, k_2] = \sum_{n_2=0}^{N_2-1} x[n_1 + N_1 n_2] W_{N_2}^{n_2 k_2}, \qquad 0 \leq k_2 \leq N_2 - 1. \tag{11.63}$$

It should be noted that for each value of the index n_1, the quantity $G[n_1, k_2]$ can be considered as an N_2-point DFT of the length-N_2 sequence in the row numbered n_1 of the 2-D array defined by $x[[n_1 + N_1 n_2]$. Substituting Eq. (11.63) in Eq. (11.62), we arrive at

$$X[k_1 + N_1 k_2] = \sum_{n_1=0}^{N_1-1} \left(G[n_1, k_2] W_N^{n_1 k_2} \right) W_{N_1}^{n_1 k_1} = \sum_{n_1=0}^{N_1-1} \hat{G}[n_1, k_2] W_{N_1}^{n_1 k_1}, \tag{11.64}$$

where

$$\hat{G}[n_1, k_2] = G[n_1, k_2] W_N^{n_1 k_2}. \tag{11.65}$$

It follows from Eq. (11.65) that the set of N_1 sequences $\hat{G}[n_1, k_2]$ of length N_2 each have been obtained by multiplying the row DFTs $G[n_1, k_2]$ by the twiddle factors $W_N^{n_1 k_2}$. For each value of k_2, the right-hand side of Eq. (11.64) defines an N_1-point DFT of the length N_1 sequence in the column numbered k_2 of the 2-D array defined by $\hat{G}[n_1, k_2]$. It should be further noted that the row DFTs of Eq. (11.63), the column DFTs of Eq. (11.64), and the multiplication by the twiddle factors given by Eq. (11.65) can all be performed in-place.

EXAMPLE 11.13   Illustration of Index Mapping for FFT Computation

Let $N = 15$. Choose $N_1 = 3$, and $N_2 = 5$. In this case, the index mappings are given by

$$n = n_1 + 3n_2, \qquad \begin{cases} 0 \leq n_1 \leq 2, \\ 0 \leq n_2 \leq 4, \end{cases} \quad \text{and} \quad k = 5k_1 + k_2, \qquad \begin{cases} 0 \leq k_1 \leq 2, \\ 0 \leq k_2 \leq 4. \end{cases}$$

Here, Eq. (11.62) reduces to

$$X[5k_1 + k_2] = \sum_{n_1=0}^{2} \left[\left(\sum_{n_2=0}^{4} x[n_1 + 3n_2] W_5^{n_2 k_2} \right) W_{15}^{n_1 k_2} \right] W_3^{n_1 k_1}, \quad \begin{cases} 0 \leq k_1 \leq 2, \\ 0 \leq k_2 \leq 4. \end{cases} \qquad (11.66)$$

The input index mapping results in the 2-D representation of the input sequence as indicated in the following:

n_1 \ n_2	0	1	2	3	4
0	$x[0]$	$x[3]$	$x[6]$	$x[9]$	$x[12]$
1	$x[1]$	$x[4]$	$x[7]$	$x[10]$	$x[13]$
2	$x[2]$	$x[5]$	$x[8]$	$x[11]$	$x[14]$

The 5-point DFTs of each one of the three rows given above results in the 2-D array as shown below:

n_1 \ k_2	0	1	2	3	4
0	$G[0,0]$	$G[0,1]$	$G[0,2]$	$G[0,3]$	$G[0,4]$
1	$G[1,0]$	$G[1,1]$	$G[1,2]$	$G[1,3]$	$G[1,4]$
2	$G[2,0]$	$G[2,1]$	$G[2,2]$	$G[2,3]$	$G[2,4]$

The above 5-point DFTs $G[n_1, k_2]$ are next multiplied by the twiddle factors $W_{15}^{n_1 k_2}$, leading to the 2-D array indicated below:

n_1 \ k_2	0	1	2	3	4
0	$\hat{G}[0,0]$	$\hat{G}[0,1]$	$\hat{G}[0,2]$	$\hat{G}[0,3]$	$\hat{G}[0,4]$
1	$\hat{G}[1,0]$	$\hat{G}[1,1]$	$\hat{G}[1,2]$	$\hat{G}[1,3]$	$\hat{G}[1,4]$
2	$\hat{G}[2,0]$	$\hat{G}[2,1]$	$\hat{G}[2,2]$	$\hat{G}[2,3]$	$\hat{G}[2,4]$

Finally, the 3-point DFTs of each column of the array $\hat{G}[n_1, k_2]$ are carried out, leading to the desired 15-point DFT $X[k]$ given by

k_1 \ k_2	0	1	2	3	4
0	$X[0]$	$X[1]$	$X[2]$	$X[3]$	$X[4]$
1	$X[5]$	$X[6]$	$X[7]$	$X[8]$	$X[9]$
2	$X[10]$	$X[11]$	$X[12]$	$X[13]$	$X[14]$

The flow-graph representation of the above 15-point FFT algorithm is shown in Figure 11.30.

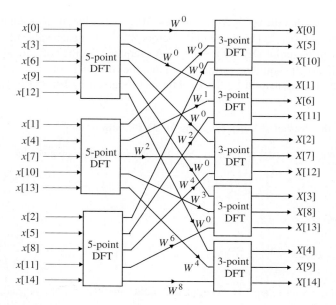

Figure 11.30: Flow-graph representation of the 15-point DFT computation using the Cooley–Tukey algorithm.

It is interesting to point out the relation between the FFT algorithm developed here and the Cooley–Tukey FFT algorithms described in the previous section. For example, if $N_1 = 2$ and $N_2 = N/2$, the index mappings of Eqs. (11.58) and (11.59) lead to the first-stage of the decimation-in-time FFT algorithm, whereas if $N_1 = N/2$ and $N_2 = 2$, the index mappings of Eqs. (11.58) and (11.59) lead to the first-stage of the decimation-in-frequency FFT algorithm.

An alternate index mappings is obtained by reversing the mappings of Eqs. (11.58) and (11.59) as indicated below:

$$n = N_2 n_1 + n_2, \qquad \begin{cases} 0 \le n_1 \le N_1 - 1, \\ 0 \le n_2 \le N_2 - 1, \end{cases} \tag{11.67}$$

$$k = k_1 + N_1 k_2, \qquad \begin{cases} 0 \le k_1 \le N_1 - 1, \\ 0 \le k_2 \le N_2 - 1. \end{cases} \tag{11.68}$$

In this case, if $N_1 = 2$ and $N_2 = N/2$, the above index mappings lead to the first-stage of the decimation-in-frequency FFT algorithm, whereas if $N_1 = N/2$ and $N_2 = 2$, the index mappings lead to the first-stage of the decimation-in-time FFT algorithm.

In either one of the above two FFT algorithms based on index mapping, the process can be repeated to implement the smaller size DFTs until no further decompositions are possible. For N, which is a composite number expressible in the form of a product of integers,

$$N = r_1 \cdot r_2 \cdots r_\nu,$$

it can be shown that the total number of complex multiplications (additions) in a Cooley–Tukey type FFT algorithm based on a ν-stage decomposition is given by (Problem 11.19)

$$\text{No. of multiply (add) operations} = \left(\sum_{i=1}^{\nu} r_i - \nu \right) N. \tag{11.69}$$

11.4.2 Prime Factor Algorithms

More computationally efficient DFT algorithms could be developed if the multiplication by the twiddle factors, as indicated by Eq. (11.65), can be eliminated by choosing appropriately the index mappings. One such index mapping is given by [Bur77]

$$n = \langle An_1 + Bn_2 \rangle_N, \qquad \begin{cases} 0 \leq n_1 \leq N_1 - 1, \\ 0 \leq n_2 \leq N_2 - 1, \end{cases} \tag{11.70}$$

$$k = \langle Ck_1 + Dk_2 \rangle_N, \qquad \begin{cases} 0 \leq k_1 \leq N_1 - 1, \\ 0 \leq k_2 \leq N_2 - 1, \end{cases} \tag{11.71}$$

where $\langle \ell \rangle_N$ denotes ℓ modulo N. For $A = D = 1$, $B = N_1$, and $C = N_2$, the above index mappings reduce to the mappings given in Eqs. (11.58) and (11.59). Likewise, the index mappings of Eqs. (11.67) and (11.68) are obtained for $A = N_2$, $B = C = 1$, and $D = N_1$. For either of these choices, the values of the indices n and k remain within the range $(0, N - 1)$, and as a result, the modulo operation is not needed.

Using the above index mappings, we rewrite the DFT samples of Eq. (11.60):

$$X[k] = X[\langle Ck_1 + Dk_2 \rangle_N]$$

$$= \sum_{n_1=0}^{N_1-1} \sum_{n_2=0}^{N_2-1} x[\langle An_1 + Bn_2 \rangle_N] W_N^{[\langle An_1+Bn_2 \rangle_N][\langle Ck_1+Dk_2 \rangle_N]}, \qquad \begin{cases} 0 \leq k_1 \leq N_1 - 1, \\ 0 \leq k_2 \leq N_2 - 1, \end{cases} \tag{11.72}$$

Now,

$$W_N^{[\langle An_1+Bn_2 \rangle_N][\langle Ck_1+Dk_2 \rangle_N]} = W_N^{\langle ACn_1k_1 \rangle_N} W_N^{\langle ADn_1k_2 \rangle_N} W_N^{\langle BCn_2k_1 \rangle_N} W_N^{\langle BDn_2k_2 \rangle_N}. \tag{11.73}$$

The twiddle factors can be eliminated only when N_1 and N_2 are relatively prime[11] and the constants in Eqs. (11.70) and (11.71) are chosen appropriately. For example, if we choose the constants so that

$$\langle AC \rangle_N = N_2, \qquad \langle BD \rangle_N = N_1, \qquad \langle AD \rangle_N = \langle BC \rangle_N = 0, \tag{11.74}$$

then the factors on the right-hand side of Eq. (11.73) simplify to

$$W_N^{\langle ACn_1k_1 \rangle_N} = W_{N_1 N_2}^{N_2 n_1 k_1} = W_{N_1}^{n_1 k_1},$$

$$W_N^{\langle ADn_1k_2 \rangle_N} = W_N^0 = 1,$$

$$W_N^{\langle BCn_2k_1 \rangle_N} = W_N^0 = 1,$$

$$W_N^{\langle BDn_2k_2 \rangle_N} = W_{N_1 N_2}^{N_1 n_2 k_2} = W_{N_2}^{n_2 k_2},$$

and as a result, Eq. (11.73) reduces to

$$W_N^{[\langle An_1+Bn_2 \rangle_N][\langle Ck_1+Dk_2 \rangle_N]} = W_{N_1}^{k_1 n_1} W_{N_2}^{k_2 n_2},$$

thus eliminating the twiddle factors. One possible choice for the constants satisfying Eq. (11.74) is given by [Bur77]

$$A = N_2, \qquad B = N_1, \tag{11.75a}$$

$$C = N_2 \langle N_2^{-1} \rangle_{N_1}, \qquad D = N_1 \langle N_1^{-1} \rangle_{N_2}, \tag{11.75b}$$

where $\langle N_a^{-1} \rangle_{N_b}$ denotes the *multiplicative inverse* of N_a reduced modulo N_b. Note that if $\langle N_1^{-1} \rangle_{N_2} = \alpha$, then $\langle N_1 \alpha \rangle_{N_2} = 1$. Or, in other words, $N_1 \alpha = N_2 \beta + 1$, where β is any integer. Likewise, if $\langle N_2^{-1} \rangle_{N_1} = \gamma$, then $\langle N_2 \gamma \rangle_{N_1} = 1$. Hence, $N_2 \gamma = N_1 \delta + 1$, where δ is any integer.

---

[11] The integers N_1 and N_2 are relatively prime if they have no common factors.

EXAMPLE 11.14   Determination of the Multiplicative Inverses of Two Relatively Prime Factors

Let $N_1 = 3$ and $N_2 = 5$, then $\langle 3^{-1} \rangle_5 = 2$ as $\langle 2 \cdot 3 \rangle_5 = 1$. Likewise, $\langle 5^{-1} \rangle_3 = 2$ as $\langle 2 \cdot 5 \rangle_3 = 1$.

There are several other choices for the constants A, B, C, and D. Determination of these choices requires the application of the *Chinese Remainder Theorem*, a discussion of which is beyond the scope of this book. For the choice given by Eqs. (11.75a) and (11.75b), we have

$$\langle AC \rangle_N = \langle N_2 \cdot (N_2 \langle N_2^{-1} \rangle_{N_1}) \rangle_N = \langle N_2 (N_1 \delta + 1) \rangle_N = \langle N_2 N_1 \delta + N_2 \rangle_N = N_2,$$

$$\langle BD \rangle_N = \langle N_1 \cdot (N_1 \langle N_1^{-1} \rangle_{N_2}) \rangle_N = \langle N_1 (N_2 \beta + 1) \rangle_N = \langle N_1 N_2 \beta + N_1 \rangle_N = N_1,$$

$$\langle AD \rangle_N = \langle N_2 \cdot (N_1 \langle N_1^{-1} \rangle_{N_2}) \rangle_N = \langle N \alpha \rangle_N = 0,$$

$$\langle BC \rangle_N = \langle N_1 \cdot (N_2 \langle N_2^{-1} \rangle_{N_1}) \rangle_N = \langle N \gamma \rangle_N = 0.$$

Making use of the results in Eq (11.72), we arrive at

$$
\begin{aligned}
X[k] &= X[\langle Ck_1 + Dk_2 \rangle_N] \\
&= \sum_{n_1=0}^{N_1-1} \sum_{n_2=0}^{N_2-1} x[\langle An_1 + Bn_2 \rangle_N] W_N^{N_2 n_1 k_1} W_N^{N_1 n_2 k_2}, \\
&= \sum_{n_1=0}^{N_1-1} \left(\sum_{n_2=0}^{N_2-1} x[\langle An_1 + Bn_2 \rangle_N] W_{N_2}^{n_2 k_2} \right) W_{N_1}^{n_1 k_1}, \qquad \begin{cases} 0 \le k_1 \le N_1 - 1, \\ 0 \le k_2 \le N_2 - 1. \end{cases}
\end{aligned}
\tag{11.76}
$$

Substituting

$$G[n_1, k_2] = \sum_{n_2=0}^{N_2-1} x[\langle An_1 + Bn_2 \rangle_N] W_{N_2}^{n_2 k_2}, \qquad \begin{cases} 0 \le n_1 \le N_1 - 1, \\ 0 \le k_2 \le N_2 - 1 \end{cases} \tag{11.77}$$

in Eq. (11.76), we get

$$X[k] = \sum_{n_1=0}^{N_1-1} G[n_1, k_2] W_{N_1}^{n_1 k_1}, \qquad \begin{cases} 0 \le k_2 \le N_2 - 1, \\ 0 \le k \le N. \end{cases} \tag{11.78}$$

It should be noted that Eq. (11.77) defines a N_2-point DFT of the n_1-th row of the 2-D array $x[\langle An_1 + Bn_2 \rangle_N]$. Likewise, Eq. (11.78) defines a N_1-point DFT of the k_2-th column of the 2-D array $G[n_1, k_2]$. As indicated above, the N-point DFT $X[k]$ can be computed by first computing the row DFTs and then computing the column DFTs. Since no twiddle factors are involved, the computation of the 1-D DFT $X[k]$ now involves the computation of a 2-D DFT.

We can also rewrite Eq. (11.76) in the form

$$X[k] = \sum_{n_2=0}^{N_2-1} \left(\sum_{n_1=0}^{N_1-1} x[\langle An_1 + Bn_2 \rangle_N] W_{N_1}^{n_1 k_1} \right) W_{N_2}^{n_2 k_2}, \qquad \begin{cases} 0 \le k_1 \le N_1 - 1, \\ 0 \le k_2 \le N_2 - 1. \end{cases} \tag{11.79}$$

The quantity inside the parentheses on the right-hand side of Eq. (11.79) represents an N_1-point DFT $H[k_1, n_2]$ of the n_2-th column of the 2-D array $x[\langle An_1 + Bn_2 \rangle_N]$:

$$H[k_1, n_2] = \sum_{n_1=0}^{N_1-1} x[\langle An_1 + Bn_2 \rangle_N] W_{N_1}^{n_1 k_1}, \qquad \begin{cases} 0 \le k_1 \le N_1 - 1, \\ 0 \le n_2 \le N_2 - 1. \end{cases}$$

Making use of $H[k_1, n_2]$ in Eq. (11.79), we then arrive at

$$X[k] = \sum_{n_2=0}^{N_2-1} H[k_1, n_2] W_{N_2}^{n_2 k_2}, \qquad \begin{cases} 0 \le k_1 \le N_1 - 1, \\ 0 \le k \le N, \end{cases}$$

which represents a N_2-point DFT of the k_1-th row of the 2-D array $H[k_1, n_2]$. Thus, the 1-D DFT $X[k]$ can also be computed by first computing the column DFTs and then the row DFTs. The above class of FFT algorithms is more commonly known as the *prime factor algorithm,* as the factors forming the composite N are relatively prime [Kol77]. In practice, computationally efficient short prime-length DFT algorithms are employed to compute the row and column short-length DFTs (butterflies) [Bur77].

EXAMPLE 11.15   15-point DFT Computation Using the Prime Factor Algorithm

Let $N = 15$. We choose $N_1 = 3$ and $N_2 = 5$, which are seen to be relatively prime. From Example 11.14, we observe that

$$\langle 3^{-1} \rangle_5 = 2, \qquad \langle 5^{-1} \rangle_3 = 2.$$

Using these values in Eqs. (11.75a) and (11.75b), we get

$$A = 5, \qquad B = 3, \qquad C = 5\langle 5^{-1} \rangle_3 = 10, \qquad D = 3\langle 3^{-1} \rangle_5 = 6.$$

Substituting these values of the constants in Eqs. (11.70) and (11.71), we arrive at the index maps

$$n = \langle 5n_1 + 3n_2 \rangle_{15}, \qquad \begin{cases} 0 \le n_1 \le 2, \\ 0 \le n_2 \le 4, \end{cases} \tag{11.80a}$$

$$k = \langle 10k_1 + 6k_2 \rangle_{15}, \qquad \begin{cases} 0 \le k_1 \le 2, \\ 0 \le k_2 \le 4. \end{cases} \tag{11.80b}$$

The corresponding expression for the DFT is now given by

$$X[\langle 10k_1 + 6k_2 \rangle_{15}] = \sum_{n_2=0}^{4} \left(\sum_{n_1=0}^{2} x[\langle 5n_1 + 3n_2 \rangle_{15}] W_3^{k_1 n_1} \right) W_5^{n_2 k_2}, \quad \begin{cases} 0 \le k_1 \le 2, \\ 0 \le k_2 \le 4. \end{cases} \tag{11.81}$$

The index map of Eq. (11.80a) develops the 2-D array representation of the input as

n_1 \ n_2	0	1	2	3	4
0	$x[0]$	$x[3]$	$x[6]$	$x[9]$	$x[12]$
1	$x[5]$	$x[8]$	$x[11]$	$x[14]$	$x[2]$
2	$x[10]$	$x[13]$	$x[1]$	$x[4]$	$x[7]$

The 5-point DFTs of each one of the three rows given above results in the 2-D array given below:

n_1 \ k_2	0	1	2	3	4
0	$G[0, 0]$	$G[0, 1]$	$G[0, 2]$	$G[0, 3]$	$G[0, 4]$
1	$G[1, 0]$	$G[1, 1]$	$G[1, 2]$	$G[1, 3]$	$G[1, 4]$
2	$G[2, 0]$	$G[2, 1]$	$G[2, 2]$	$G[2, 3]$	$G[2, 4]$

Finally, the 3-point DFTs of each column of the array $G[n_1, k_2]$ are carried out, leading to the desired 15-point DFT $X[k]$ given by

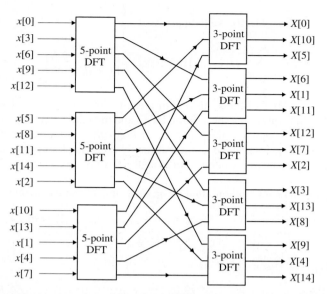

Figure 11.31: Flow-graph representation of the 15-point DFT computation using the prime factor algorithm.

k_1 \ k_2	0	1	2	3	4
0	$X[0]$	$X[6]$	$X[12]$	$X[3]$	$X[9]$
1	$X[10]$	$X[1]$	$X[7]$	$X[13]$	$X[4]$
2	$X[5]$	$X[11]$	$X[2]$	$X[8]$	$X[14]$

The flow-graph representation of the above 15-point prime factor algorithm is shown in Figure 11.31.

There are two major differences between the prime factor algorithm and the Cooley–Tukey radix-2 or radix-4 FFT algorithms. First, the former requires a more complex indexing of the input data than that employed in the latter algorithms. Second, the butterfly structure in the prime factor algorithm is different for different factors, whereas a single butterfly requiring no multiplications is employed in the Cooley–Tukey radix-2 or radix-4 FFT algorithms.

11.5 DFT and IDFT Computation Using MATLAB

The M-files `fft` and `ifft` can be used for the computation of the DFT and the IDFT. Since vectors in MATLAB are indexed from 1 to N instead of 0 to $N - 1$, the DFT and the IDFT computed in the above MATLAB functions make use of the expressions

$$X[k] = \sum_{n=1}^{N} x[n] W_N^{(n-1)(k-1)}, \qquad 1 \le k \le N, \tag{11.82a}$$

$$x[n] = \frac{1}{N} \sum_{k=1}^{N} X[k] W_N^{-(n-1)(k-1)}, \qquad 1 \le n \le N. \tag{11.82b}$$

Examples 11.16 and 11.17 illustrate two different applications of the DFT.

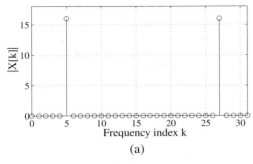
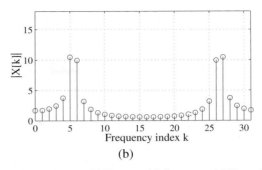

(a)   (b)

Figure 11.32: The magnitudes of a 32-point DFT of a sinusoid sampled at a 64-Hz rate: (a) frequency 10 Hz and (b) frequency 11 Hz.

EXAMPLE 11.16   Spectral Analysis of a Finite-Length Sinusoidal Sequence

We consider here the computation of the DFT of a sinusoidal sequence of finite length. Program 11_10 can be used for this purpose.

Program 11_10.m

During execution, the program requests the following input data: the length N of the DFT, the sampling frequency F_T in Hz, and the frequency of the sinusoid in Hz. We compute here the 32-point DFT $X[k]$ of a length-32 sinusoidal sequence $x[n]$ of frequency 10 Hz with a sampling rate of 64 Hz. Note that the sampling frequency of 64 Hz is considerably higher than 20 Hz, the Nyquist frequency, and as a result, no aliasing distortion will occur after sampling. The magnitudes of the DFT samples generated by this program are plotted in Figure 11.32(a).

Several comments are in order here. Since the time-domain sequence $x[n]$ is a pure sinusoid, its discrete-time Fourier transform $X(e^{j2\pi f})$ contains two impulses at $f = \pm 10$ Hz and is zero everywhere else. Its 32-point DFT is obtained by sampling $X(e^{j2\pi f})$ at

$$f = \frac{64k}{32} = 2k \text{ Hz} \qquad 0 \le k \le 31.$$

As a result, the impulse at $f = 10$ Hz appears at the DFT frequency bin location $k = 5$ as $X[5]$, while the impulse at $f = -10$ Hz appears at the bin location $k = 32 - 5 = 27$ as $X[27]$. Thus, the DFT has correctly identified the frequency of the sinusoid in this case. Note also that the first half of the DFT samples for $k = 0$ to $k = (N/2) - 1$ corresponds to the positive frequency axis from $f = 0$ to $f = F_T/2$, excluding the point $F_T/2$, while the second half for $k = N/2$ to $k = N - 1$ corresponds to the negative frequency axis from $f = -F_T/2$ to $f = 0$, excluding the point $f = 0$.

We next compute the 32-point DFTs $X[k]$ of a sinusoid $x[n]$ of frequency 11 Hz with a sampling frequency of 64 Hz. Since the time-domain sequence $x[n]$ is a pure sinusoid, its discrete-time Fourier transform $X(e^{j2\pi f})$ contains two impulses at $f = \pm 11$ Hz and is zero everywhere else. However, these frequency locations are between the bins $k = 5$ and $k = 6$, and $k = 26$ and $k = 27$. From the computed magnitudes of the DFT samples, shown in Figure 11.32(b), it can be seen that the spectrum contains frequency components at all bins, with two strong components at bins $k = 5$ and $k = 6$ and with two strong components at bins $k = 26$ and $k = 27$. This type of phenomenon is called the *leakage*. It is further discussed in detail in Section 15.2.

EXAMPLE 11.17   Computation of Linear Convolution Using the DFT

Program 11_11.m

In this example, we illustrate the application of the fft and ifft functions of MATLAB in performing the linear convolution of two finite-length sequences. To this end, Program 11_11 can be employed. We illustrate its use in developing the linear convolution of the two length-4 sequences of Example 5.7: $g[n] = \{1 \quad 2 \quad 0 \quad 1\}$ and $h[n] = \{2 \quad 2 \quad 1 \quad 1\}$. The result obtained using Program 11_11 is identical to that determined in Example 5.9.

11.6 Sliding Discrete Fourier Transform

In some applications involving a very long sequence, it is of interest to determine the spectral properties of a subset consisting of a fixed number of consecutive samples at successive values of the time instant n. For example, if N is the length of the subset, we compute the N-point DFT of the finite-length segment, $\{x[n], x[n-1], \ldots, x[n-N+1]\}$, inside a length-$N$ sliding window. The DFT computation is repeated for each increasing value of n by advancing the window one sample for each computation.[12] Such spectral measurements are particularly of interest in the case of signals with time-varying spectrums. The above type of DFT computation is called a *sliding discrete Fourier transform* or *running discrete Fourier transform*. We outline here a computationally efficient method of evaluating the sliding DFT [Rab75a].

To indicate the time dependency of the DFT samples, we denote the k-th DFT sample at time instant n as $S_k[n]$, where

$$S_k[n] = \sum_{\ell=n-N+1}^{n} x[\ell] W_N^{(\ell-n+N-1)}. \tag{11.83}$$

It can easily be shown that $S_k[n]$ can be expressed in terms of $S_k[n-1]$, the k-th DFT sample at the previous time instant $(n-1)$, as

$$S_k[n] = W_N^{-k} \left(S_k[n-1] + x[n] - x[n-N] \right). \tag{11.84}$$

As a result, once the k-th DFT sample at any time instant has been computed, the values of this DFT sample at all successive time instants can be computed recursively, requiring only a single complex multiplication by W_N^{-k} for each new sample.

A block diagram representation of Eq. (11.84) is shown in Figure 11.33. The transfer function of the digital filter structure of Figure 11.33 is given by

$$H_{\mathrm{SDFT}}^{(k)}(z) = \frac{W_N^{-k}(1 - z^{-N})}{1 - W_N^{-k} z^{-1}}, \tag{11.85}$$

which is seen to have a complex pole on the unit circle at $z = W_N^{-k}$. Because of the unavoidable quantization effects, the system can become unstable if the pole moves outside the unit circle. To get around this possible instability problem, it has been recommended that the coefficient W_N^{-k} in Eq. (11.85) be replaced with $r W_N^{-k}$, where r is slightly smaller than 1, to force the pole to remain strictly inside the unit circle. However, such modification to the transfer function, or equivalently, to the recursion of Eq. (11.84) also introduces an error in the DFT sample being computed. Alternative stabilization method and other practical details can be found in [Jac2003].

11.7 DFT Computation Over a Narrow Frequency Band

In applications requiring the computation of the DFT samples over a specified frequency range, the FFT algorithms described earlier are computationally not very attractive, in particular for sequences of very large length, as they compute all samples of the DFT. Such applications typically occur when the sequences are zero-padded to increase the frequency resolution of the DFT samples, and for very large length, the cost of computing all DFT samples may be prohibitively high. In this section, we describe two computationally efficient algorithms for the calculation of a subset of the DFT samples.

---

[12]In practice, the window is advanced by more than one sample, e.g., 1/4-th the window length.

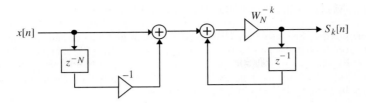

Figure 11.33: Recursive implementation of the sliding DFT.

11.7.1 Zoom FFT

The *zoom FFT* can be used to compute the samples of an N-point DFT $X[k]$ of a length-N sequence $x[n]$ in a small range of values of the frequency index k, $i \leq k \leq i + K - 1$, where $0 \leq i \leq N - K + 1$ [Por97]. It is assumed that N is an integer multiple of K, that is, $N = KR$, which can be always satisfied by zero-padding $x[n]$ if necessary.

To develop the appropriate expression for the samples of the DFT in a limited range, we apply an R-band polyphase decomposition to the sequence $x[n]$, which develops R sub-sequences

$$x_r[m] = x[r + mR], \qquad 0 \leq r \leq R - 1,$$

of length K each. Note that $x[r + mR]$ is basically the r-th sub-sequence of length-k obtained by down-sampling $x[n]$ by a factor R. If $X(z)$ denotes the z-transform of $x[n]$, then using a R-band polyphase decomposition, we can express $X(z)$ as

$$X(z) = \sum_{r=0}^{R-1} z^{-r} X_r(z^R), \tag{11.86}$$

where

$$X_r(z) = \sum_{n=0}^{K-1} x[r + nR]z^{-n}. \tag{11.87}$$

Evaluating Eq. (11.86) on the unit circle at K equally-spaced points, $z = W_N^{-k}, i \leq k \leq i + K - 1$, we arrive at K samples of the N-point DFT $X[k]$ of $x[n]$ given by

$$X[k] = \sum_{r=0}^{R-1} W_N^{kr} X_r[\langle k \rangle_K], \qquad i \leq k \leq i + K - 1, \tag{11.88}$$

where $X_r[\ell]$ is the K-point DFT of the sub-sequence $x[r + mR]$:

$$X_r[\ell] = \sum_{m=0}^{K-1} x[r + mR]W_K^{\ell m}, \qquad 0 \leq \ell \leq K - 1. \tag{11.89}$$

The above K-point DFT is periodic in ℓ with a period K. It can be seen that Eq. (11.88) basically represents the first stage of a radix-R decimation-in-time Cooley–Tukey FFT algorithm.

Thus to compute $X[k]$ in the given range of k, we first compute the K-point DFTs $X_r[\ell]$ of each of the R sub-sequences $x_r[m]$ obtained by down-sampling $x[n]$ by a factor of R. Then, a weighted combination of these K-point DFTs $X_r[\ell]$, as indicated in Eq. (11.88), yields the desired $X[k]$. Note that the short-length K-point DFTs $X_r[\ell]$, for example, can be computed using the radix-2 Cooley–Tukey FFT algorithm if K is a power-of-2.

zoomfft.m
Program 11_12.m

EXAMPLE 11.18　　MATLAB Implementation of the Zoom FFT Algorithm

The computation of selected samples of the DFT using the zoom FFT algorithm can be carried out using the function `zoomfft`. We illustrate the use of this function by computing the eight samples of the DFT of the second length-32 sinusoid of Example 11.16 with a frequency resolution of $2\pi/512$ in the neighborhood of the two peaks shown in Figure 11.32(b). The frequency bins are from 85 to 92. To this end, we make use of Program 11_12. The output data generated by running the above program are given by

```
DFT Samples
14.6025    15.2441    15.6965    15.9501    16.0000    15.8458
15.4919    14.9470
```

The above result can be verified by computing the 512-point FFT of the sinusoid using the command `XF = fft(x,512);` and displaying the eight samples for the frequency indices from k = 85 to k = 92 using the command `disp(abs(XF(85:92)))`.

11.7.2　Chirp Fourier Transform

There are applications where it is required to compute a limited number of samples of the z-transform of a finite-length sequence at points equally placed in angle over a portion of a spiral contour in the z-plane. Thus, if $x[n]$ is a length-N sequence with a z-transform $X(z)$,

$$X(z) = \sum_{n=0}^{N-1} x[n]\, z^{-n}, \tag{11.90}$$

the objective is to evaluate $X(z)$ at K points z_k, $(K < N)$, in the z-plane given by

$$z_k = r_k e^{j\omega_k} = A\, V^{-k}, \qquad 0 \le k \le K-1, \tag{11.91}$$

where

$$V = V_0 e^{-j\phi_0}, \tag{11.92a}$$
$$A = A_0 e^{j\theta_0}. \tag{11.92b}$$

In the above equations, the constants V_0 and A_0 are positive real numbers. The points z_k defined by Eq. (11.91) are thus located on a spiral contour, with $z_0 = A$ being the starting point on the contour. The K samples of the z-transform $X(z)$ to be computed are therefore given by

$$X(z_k) = \sum_{n=0}^{N-1} x[n]\, z_k^{-n}$$
$$= \sum_{n=0}^{N-1} x[n]\, A^{-n} V^{nk}, \qquad 0 \le k \le K-1, \tag{11.93}$$

and are called the *chirp z transform* (CZT) of the sequence $x[n]$, which can be computed efficiently using a special algorithm [Rab69]. In most applications, this algorithm may also be computationally more efficient than computing the samples directly.

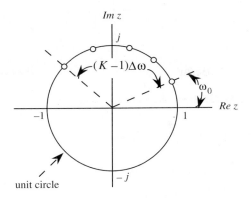

Figure 11.34: Locations of frequencies for the CFT.

If $A_0 = V_0 = 1$, the contour is a portion of the unit circle, and the points z_k are equally spaced points on the unit circle. Substituting $V_0 = 1$ and $\phi_0 = \Delta\omega$ in Eqs. (11.92a) and substituting $A_0 = 1$ and $\theta_0 = \omega_0$ in (11.92b), the K points on the unit circle are then given by

$$\omega_k = \omega_0 + k(\Delta\omega), \qquad 0 \le k \le K - 1, \tag{11.94}$$

where ω_0 is the starting frequency point and $\Delta\omega$ is the desired frequency resolution. The frequency points on the unit circle for computing the CFT samples are shown in Figure 11.34 for a typical case. In this case, the $X(z_k)$ of Eq. (11.93) are simply the samples of Fourier transform $X(e^{j\omega})$:

$$X(e^{j\omega_k}) = \sum_{n=0}^{N-1} x[n] e^{-j\omega_k n}$$

$$= \sum_{n=0}^{N-1} x[n] e^{-j[\omega_0 + k(\Delta\omega)n]}, \qquad 0 \le k \le K - 1, \tag{11.95}$$

of the sequence $x[n]$ and will be referred to as the *chirp Fourier transform* (CFT). We develop here the CFT algorithm.

Making use of the notation of Eq. (11.92a) in the above equation, we get

$$X(e^{j\omega_k}) = \sum_{n=0}^{N-1} x[n] e^{-j\omega_0 n} V^{nk}. \tag{11.96}$$

Replacing the exponent nk in the above equation with the identity

$$nk = \tfrac{1}{2}[n^2 + k^2 - (k - n)^2], \tag{11.97}$$

we arrive at

$$X(e^{j\omega_k}) = \sum_{n=0}^{N-1} x[n] e^{-j\omega_0 n} V^{[n^2+k^2-(k-n)^2]/2}$$

$$= V^{k^2/2} \sum_{n=0}^{N-1} x[n] e^{-j\omega_0 n} V^{n^2/2} V^{-(k-n)^2/2}. \tag{11.98}$$

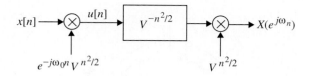

Figure 11.35: Block-diagram representation of Eq. (11.102).

The right-hand side of Eq. (11.98) can be expressed in the form of a convolution. To this end, we define a new sequence

$$u[n] = x[n]e^{-j\omega_0 n}V^{n^2/2},\tag{11.99}$$

which, when substituted in Eq. (11.98), results in

$$X(e^{j\omega_k}) = V^{n^2/2}\sum_{n=0}^{N-1}u[n]V^{-(k-n)^2/2}, \qquad 0 \le k \le K-1.\tag{11.100}$$

Interchanging the variables k and n, we rewrite Eq. (11.100) as

$$X(e^{j\omega_n}) = V^{n^2/2}\sum_{k=0}^{N-1}u[k]V^{-(n-k)^2/2}, \qquad 0 \le n \le K-1.\tag{11.101}$$

The complex exponential sequence

$$V^{-n^2/2} = e^{j(\Delta\omega)n^2/2}$$

has a linearly increasing frequency and is often referred to as a *chirp signal*. This has led to the name *chirp Fourier transform* to the scheme of Eq. (11.101) for the computation of the Fourier transform samples. The right-hand side of Eq. (11.101) can be interpreted as the convolution of the sequence $u[n]$ with the sequence $V^{-n^2/2}$ multiplied by the factor $V^{n^2/2}$; that is,

$$X(e^{j\omega_n}) = V^{n^2/2}\left(u[n]\circledast V^{-n^2/2}\right).\tag{11.102}$$

A block-diagram representation of Eq. (11.102) is indicated in Figure 11.35.

Now the complex exponential sequence $V^{-n^2/2}$ is of infinite length, whereas $u[n]$ is a finite-length sequence of length N. Hence, the convolution sum of Eq. (11.102) is carried out over a finite number of terms. It is thus more convenient to evaluate the CFT samples by replacing the infinite length sequence $V^{-n^2/2}$ in Eq. (11.102) with a finite-length noncausal sequence $g[n]$ defined by

$$g[n] = \begin{cases} V^{-n^2/2}, & -(N-1) \le n \le K-1, \\ 0, & \text{otherwise,} \end{cases}\tag{11.103}$$

which is an LTI finite-impulse response (FIR) filter. The modified CFT algorithm is thus given by

$$y[n] = X(e^{j\omega_n}) = V^{n^2/2}(u[n]\circledast g[n]),\tag{11.104}$$

whose block-diagram representation is shown in Figure 11.36.

Several comments are in order here. First, by delaying $g[n]$ by $(N-1)$ samples, the LTI system of Figure 11.36 can be made causal. Second, the convolution sum can be implemented using the FFT-based

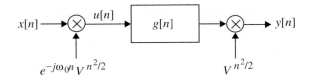

Figure 11.36: Block-diagram representation of the CFT algorithm.

method of Section 5.10.1. Third, the two parameters, ω_0 and $\Delta\omega$, can be chosen arbitrarily unlike those for the DFT. In fact, in some applications, the discrete-time Fourier transform sample locations need not correspond to DFT bin locations.

The chirp-z transform can be computed using the MATLAB M-file $\mathtt{czt}$. The basic form of this M-file is $\mathtt{Y = czt(x,K,V,A)}$, which computes a length-$\mathtt{K}$ CZT of the sequence $\mathtt{x}$ on a spiral contour defined by $\mathtt{V}$ and $\mathtt{A}$. Here, $\mathtt{V}$ is as defined in Eq. (11.92a), and $\mathtt{A}$ is the complex-valued starting point on the contour defined in Eq. (11.92b).

We illustrate the application of the above M-file in Example 11.19.

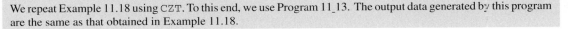

EXAMPLE 11.19   DFT Sample Computation Using the Chirp z-Transform

We repeat Example 11.18 using $\mathtt{CZT}$. To this end, we use Program 11_13. The output data generated by this program are the same as that obtained in Example 11.18.

Program 11_13.m

11.8 Number Representation

The binary representation is used to represent numbers (and signal variables) in most digital computers and special-purpose digital signal processors used for implementing digital filtering algorithms. In this form, the number is represented using the symbols 0 and 1, called *bits*,[13] with the *binary point* separating the integer part from the fractional part. For example, the binary representation of the decimal number 11.625 is given by

$$1011_\triangle 101,$$

where $\triangle$ denotes the binary point. The four bits, 1011, to the left of the binary point form the integer part, and the three bits, 101, to the right of the binary point represent the fractional part. In general, the decimal equivalent of a binary number η consisting of B integer bits and b fractional bits,

$$a_{B-1}a_{B-2}\cdots a_1 a_0 {}_\triangle a_{-1}a_{-2}\cdots a_{-b},$$

is given by

$$\sum_{i=-b}^{B-1} a_i 2^i,$$

where each bit a_i is either a 0 or a 1. The leftmost bit, a_{B-1}, is called the *most significant bit* (*MSB*), and the rightmost bit, a_{-b}, is called the *least significant bit* (*LSB*).

To avoid the confusion between a decimal number containing the digits 1 and 0 and a binary number containing the bits 1 and 0, we shall include a subscript 10 to the right of the least significant digit to indicate a decimal number and a subscript 2 to the right of the least significant bit to indicate a binary number. Thus, for example, 1101_{10} represents a decimal number, whereas 1101_2 represents a binary number whose decimal equivalent is 13_{10}. If there is no ambiguity in the representation, then the subscript is dropped.

The block of bits representing a number is called a *word*, and the number of bits in the word is called the *wordlength* or *word size*. The wordlength is typically a positive integer power of 2, such as 8 or 16 or

---
[13]**bit** is an abbreviated form of **b**inary dig**it**.

32. The word size is often expressed in units of eight bits called a *byte*. For example, a 4-byte word is equivalent to a 32-bit-size word.

Digital circuits implementing the arithmetic operations, addition and multiplication of two binary numbers, are specifically designed to develop the results, the sum and the product, respectively, in binary form, with their binary points in the assumed locations. There are two basic types of binary representations of numbers, fixed-point and floating-point, as discussed below.

11.8.1 Fixed-Point Representation

In this type of representation, the binary point is assumed to be fixed at a specific location, and the hardware implementation of the arithmetic circuits takes into account the fixed location in performing the arithmetic operations. Implementation of the addition operation carried out by a digital adder circuit is independent of the location of the binary points in the two numbers being added as long as they are in the same location for both numbers. On the other hand, it is not simple to locate the binary point in the product of two binary numbers unless they are both integers or are both fractions. In the case of the multiplication of two integers by the multiplier circuit, the result is also an integer. Likewise, multiplying two fractions results in a fraction. In digital signal processing applications, therefore, fixed-point numbers are always represented as fractions.

The range of nonnegative integers η that can be represented by B bits in a fixed-point representation is given by

$$0 \leq \eta \leq 2^B - 1. \tag{11.105}$$

Similarly, the range of positive fractions η that can be represented by B bits in a fixed-point representation is given by

$$0 \leq \eta \leq 1 - 2^{-B}. \tag{11.106}$$

In either case, the range is fixed. If $\eta_{\max}$ and $\eta_{\min}$ denote, respectively, the maximum and the minimum values of the numbers that can be represented in a B-bit fixed-point representation, then the dynamic range R of the numbers that can be represented with B bits is given by $R = \eta_{\max} - \eta_{\min}$, and the *resolution* of the representation is defined by

$$\delta = \frac{R}{2^B - 1}, \tag{11.107}$$

where δ is also known as the *quantization level*.

11.8.2 Floating-Point Representation

In the normalized floating-point representation, a positive number η is represented using two parameters, the *mantissa M* and the *exponent* or the *characteristic E* in the form

$$\eta = M \cdot 2^E, \tag{11.108}$$

where the mantissa M is a binary fraction restricted to lie in the range

$$\tfrac{1}{2} \leq M < 1, \tag{11.109}$$

and the exponent E is either a positive or a negative binary integer.

The floating-point system provides a variable resolution for the range of numbers being represented. The resolution increases exponentially as the magnitude of the number being represented increases. Floating-point numbers are stored in a register by assigning B_E bits of the register to the exponent and the remaining B_M bits to the mantissa. If the same number of total bits is used in both floating-point and fixed-point

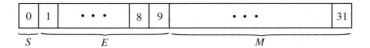

Figure 11.37: IEEE 32-bit floating-point format.

representations, that is, $B = B_M + B_E$, then the former provides a larger dynamic range than the latter (Problem 11.45).

The most widely followed floating-point formats for 32-bit and 64-bit representations are those given by the ANSI/IEEE Standard 754-1985 [IEEE85]. In this format, a 32-bit number is divided into fields. The exponent field is of 8-bit length, the mantissa[14] field is of 23-bit length, and 1 bit is assigned for the sign, as indicated in Figure 11.37. The exponent is coded in a biased form as $E - 127$. Thus, a floating-point number η under this scheme is represented as

$$\eta = (-1)^S \cdot 2^{E-127}(M), \tag{11.110}$$

with the mantissa in the range

$$0 \le M < 1. \tag{11.111}$$

The following conventions are followed in interpreting the representation of Eq. (11.110):

1. If $E = 255$ and $M \neq 0$, then η is not a number (abbreviated as *NaN*),

2. If $E = 255$ and $M = 0$, then $\eta = (-1)^S \cdot \infty$,

3. If $0 < E < 255$, then $\eta = (-1)^S \cdot 2^{E-127}(1_\Delta M)$,

4. If $E = 0$ and $M \neq 0$, then $\eta = (-1)^S \cdot 2^{-126}(0_\Delta M)$,

5. If $E = 0$ and $M = 0$, then $\eta = (-1)^S \cdot 0$,

where $1_\Delta M$ is a number with one integer bit and 23 fractional bits and $0_\Delta M$ is a fraction. The range of a 32-bit floating-point number in the above format is from 1.18×10^{-38} to 3.4×10^{38} (Problem 11.46). The IEEE 32-bit standard has been adopted for number representation in almost all commercial floating-point digital signal processor chips.

11.8.3   Representation of Negative Numbers

To accommodate the representation of both positive and negative b-bit fractions, an additional bit, called the *sign bit*, is placed at the leading position of the register to indicate the sign of the number (Figure 11.38). Independent of the scheme being used to represent the negative number, except for the offset-binary representation, the *sign bit* is 0 for a positive number and 1 for a negative number. On the other hand, for the offset-binary representation, the *sign bit* is 1 for a positive number and 0 for a negative number.

A fixed-point negative number is represented in one of three different forms. In the *sign-magnitude* format, if the sign bit $s = 0$, the b-bit fraction is a positive number with a magnitude $\sum_{i=1}^{b} a_{-i} 2^{-i}$, and if $s = 1$, the b-bit fraction is a negative number with a magnitude $\sum_{i=1}^{b} a_{-i} 2^{-i}$.

In the *ones'-complement* form, a positive fraction is represented as in the sign-magnitude form, while its negative is represented by complementing each bit of the binary representation of the positive fraction.

---

[14]In the IEEE floating-point standard, the mantissa is called the *significand*.

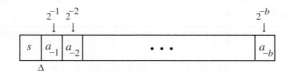

Figure 11.38: Representation of a general signed b-bit fixed-point fraction.

Table 11.1: Binary number representations.

Decimal Equivalent	Sign-Magnitude	Ones'-Complement	Two's-Complement	Offset Binary
7/8	$0_\Delta 111$	$0_\Delta 111$	$0_\Delta 111$	$1_\Delta 111$
6/8	$0_\Delta 110$	$0_\Delta 110$	$0_\Delta 110$	$1_\Delta 110$
5/8	$0_\Delta 101$	$0_\Delta 101$	$0_\Delta 101$	$1_\Delta 101$
4/8	$0_\Delta 100$	$0_\Delta 100$	$0_\Delta 100$	$1_\Delta 100$
3/8	$0_\Delta 011$	$0_\Delta 011$	$0_\Delta 011$	$1_\Delta 011$
2/8	$0_\Delta 010$	$0_\Delta 010$	$0_\Delta 010$	$1_\Delta 010$
1/8	$0_\Delta 001$	$0_\Delta 001$	$0_\Delta 001$	$1_\Delta 001$
0	$0_\Delta 000$	$0_\Delta 000$	$0_\Delta 000$	$1_\Delta 000$
-0	$1_\Delta 000$	$1_\Delta 111$	N/A	N/A
$-1/8$	$1_\Delta 001$	$1_\Delta 110$	$1_\Delta 111$	$0_\Delta 111$
$-2/8$	$1_\Delta 010$	$1_\Delta 101$	$1_\Delta 110$	$0_\Delta 110$
$-3/8$	$1_\Delta 011$	$1_\Delta 100$	$1_\Delta 101$	$0_\Delta 101$
$-4/8$	$1_\Delta 100$	$1_\Delta 011$	$1_\Delta 100$	$0_\Delta 100$
$-5/8$	$1_\Delta 101$	$1_\Delta 010$	$1_\Delta 011$	$0_\Delta 011$
$-6/8$	$1_\Delta 110$	$1_\Delta 001$	$1_\Delta 010$	$0_\Delta 010$
$-7/8$	$1_\Delta 111$	$1_\Delta 000$	$1_\Delta 001$	$0_\Delta 001$
$-8/8$	N/A	N/A	$1_\Delta 000$	$0_\Delta 000$

In this representation, the decimal equivalent of a positive or a negative fraction is thus given by $-s(1 - 2^{-b}) + \sum_{i=1}^{b} a_{-i}2^{-i}$.

Finally, in the *two's-complement* representation, the positive fraction is represented as in the sign-magnitude form, while its negative is represented by complementing each bit (i.e., by replacing each 0 with a 1, and vice-versa) of the binary representation of the positive fraction and adding a 1 to the LSB, the bth bit. In this form, the decimal equivalent of a positive or a negative fraction is thus given by $-s + \sum_{i=1}^{b} a_{-i}2^{-i}$.

Table 11.1 illustrates the above three representations for a 4-bit number (3-bit fraction and a sign bit).

EXAMPLE 11.20   Binary Representations of a Negative Decimal Fraction

We determine the 5-bit binary representations of the decimal fraction -0.625_{10}. Now, the sign-magnitude representation of 0.625_{10} is given by $0_\triangle 1010 (= 2^{-1} + 2^{-3} = 0.5 + 0.125)$. Therefore, the sign-magnitude representation of -0.625_{10} is simply $1_\triangle 1010$. The ones'-complement representation of -0.625_{10} is obtained by complementing each bit of $0_\triangle 1010$ individually, leading to $1_\triangle 0101$. Finally, the two's-complement representation of -0.625_{10} is obtained by adding a 1 to the LSB of the ones'-complement representation, resulting in $1_\triangle 0110$.

11.8.4   Offset Binary Representation

In the offset binary representation, used primarily in bipolar digital-to-analog conversion, a b-bit fraction with an additional sign bit is considered as a $(b + 1)$-bit number representing 2^{b+1} decimal numbers. About half of these numbers represent the negative fractions, and the remaining half represent the positive fractions, as illustrated in Table 11.1 for $b = 3$. It should be noted that the two's-complement representation can be converted to the offset binary representation by simply complementing the sign bit.

11.8.5   Signed Digit Representation

The radix-2 signed digit (SD) format is a 3-valued representation of a radix-2 number and employs three digit values, 0, 1, and $\bar{1}$ (with the last symbol representing -1). In many instances, the SD representation of a binary number requires fewer nonzero digits and has been exploited in developing an algorithm for a faster hardware implementation of the multiplication operation (see Section 11.9.2). A simple algorithm for the conversion of a radix-2 binary number into an equivalent SD representation is as follows [Boo51]. Let $a_{-1}a_{-2} \cdots a_b$ denote a binary number with its equivalent SD representation denoted by $c_{-1}c_{-2} \cdots c_b$. The bits c_{-i} of the SD number are determined through the relation

$$c_{-i} = a_{-i-1} - a_{-i}, \qquad i = b, b-1, \ldots, 1,$$

where $a_{-b-1} = 0$. Example 11.21 illustrates the radix-2 SD representation.

EXAMPLE 11.21   Signed-Digit Representations of a Binary Fraction

Consider the binary number $\eta = 0_\triangle 0111_2$, whose decimal equivalent is 0.4375_{10}. Its various equivalent SD representations are indicated as follows, along with their decimal equivalents:

$$0_\triangle 100\bar{1} = (0.5)_{10} - (0.0625)_{10}$$

$$0_\triangle 10\bar{1}1 = (0.5625)_{10} - (0.125)_{10}$$

$$0_\triangle 1\bar{1}11 = (0.6875)_{10} - (0.25)_{10}$$

As can be seen from Example 11.21, the SD representation is not unique, and a representation with the fewest number of nonzero digits is called a *minimal* SD representation. A minimal SD representation containing no adjacent nonzero digits is called a *canonic signed-digit* (CSD) representation. An algorithm to derive the CSD representation of a binary number is given in Hwang [Hwa79].

11.9   Arithmetic Operations

The two basic arithmetic operations in the implementation of digital filtering algorithms are addition (subtraction) and multiplication illustrated next for binary numbers.

11.9.1 Fixed-Point Addition

The procedure for the addition of two binary numbers represented in the signed-magnitude representation is similar to that used in the addition of two decimal numbers. The serial addition of two positive binary fractions

$$0_\triangle a_{-1}a_{-2}\cdots a_{-b} \quad \text{and} \quad 0_\triangle d_{-1}d_{-2}\cdots d_{-b}$$

is carried out in b steps. At the first step, we add the two LSBs a_{-b} and d_{-b}, whose sum will consist of two bits; the first bit is a carry bit c_{-b}, and the second bit is the LSB s_{-b} of the final sum. At all other steps, we add three bits, a_{-i}, d_{-i}, and c_{-i-1}, where c_{-i-1} is the carry bit generated at the previous step. Their sum again will consist of a carry bit c_i and a sum bit s_i. If the addition of the two positive binary fractions is still a binary fraction, the above process yields a correct result. This happens if at the bth step, the addition of a_{-1}, d_{-1}, and c_{-2} does not create a carry bit 1. However, if their sum is no longer a fraction, an *overflow* is said to have occurred, and the carry bit $c_1 = 1$ generated is used to indicate the overflow.

We illustrate the addition process in Example 11.22.

EXAMPLE 11.22   Binary Addition of Two Positive Binary Fractions

Consider the binary addition of the two positive binary fractions $0_\triangle 1001$ and $0_\triangle 0101$, whose decimal equivalents are, respectively, 0.5625_{10} and 0.3125_{10}. The serial addition of these two numbers is shown below and is carried out in four steps:

Note that the sum of these two numbers is the positive binary fraction $0_\triangle 1110$, whose decimal equivalent is 0.875_{10}, and no overflow has occurred.

Now, consider the binary addition of the two positive binary fractions $0_\triangle 1101$ and $0_\triangle 0111$, whose decimal equivalents are, respectively, 0.8175_{10} and 0.4375_{10}. The serial addition of these two numbers is illustrated below:

In this case, the addition of two positive fractions has resulted in a sum whose decimal equivalent, 1.25_{10}, is no longer a fraction and thus appears as a negative fraction. The appearance of a sign bit of 1 in the sum of two positive fractions indicates an overflow.

The subtraction of a positive binary number from another positive binary number, or, equivalently, the addition of a positive binary number to a negative binary number in signed-magnitude form, is slightly more complicated since here we may have to introduce a borrowing process. Example 11.23 illustrates the method.

EXAMPLE 11.23   Subtraction of a Binary Number from Another Binary Number

Consider the subtraction of $0_\triangle 0111$ (called the *subtrahend*) from $0_\triangle 1101$ (called the *minuend*). The subtraction is carried out sequentially in four steps as shown on next page.

$$
\begin{array}{ccccccl}
-1 & & -1 & & & & \leftarrow \text{ borrow} \\
0_\Delta 1 & 11 & 10 & 1 & & & \leftarrow \text{ minuend} \\
- \ 0_\Delta 0 & 1 & 1 & 1 & & & \leftarrow \text{ subtrahend} \\
\hline
0_\Delta 0 & 1 & 1 & 0 & & & \leftarrow \text{ difference}
\end{array}
$$

In the first step, the LSB of the subtrahend is subtracted from the LSB of the minuend. Since both LSBs are 1, their difference is 0. Next, the difference of the two bits to the left of the LSB is formed. Note that here the bit in the minuend is 0, whereas the corresponding bit in the subtrahend is 1. Thus, to form the difference, we borrow a 1 from the bit to its left in the minuend, making its effective value 10. Subtracting a 1 from it results in a difference of 1. This process is continued until the difference of the MSBs is computed.

Note that in Example 11.23, the subtrahend is smaller in value than the minuend, resulting in the correct difference. However, in the opposite case, the arithmetic operation will cause an *overflow*.

As illustrated above, the algorithms for the addition and subtraction operations for binary numbers in the signed-magnitude form are different, and, therefore, separate circuits are needed in their hardware implementations. On the other hand, both operations can be carried out using essentially the same algorithm for binary numbers in either the ones'-complement or the two's-complement form. The basic difference in the arithmetic operations between the two representations is in the handling of the carry bit resulting in the last step. Examples 11.24 and 11.25 illustrate the respective algorithms.

EXAMPLE 11.24   Subtraction Using the Two's-Complement Representation of the Negative Number

Consider the computation of the difference $0_\Delta 1101 - 0_\Delta 0111$. We can treat the subtraction operation as an addition of a positive number $0_\Delta 1101$ to a negative number $-0_\Delta 0111$. The two's-complement representation of $-0_\Delta 0111$ is simply $1_\Delta 1001$. In the addition of the positive binary fraction $0_\Delta 1101$ and the two's-complement negative binary fraction $1_\Delta 1001$, we treat the sign bits as the rest of the number as indicated below.

$$
\begin{array}{crcccc}
& 0_\Delta 1 & 1 & 0 & 1 \\
+ & 1_\Delta 1 & 0 & 0 & 1 \\
\hline
1 & 0_\Delta 0 & 1 & 1 & 0 \\
\uparrow & & & & \\
\text{drop} & & & &
\end{array}
$$

By dropping the carry bit generated at the fifth step, we arrive at the correct result given by $0_\Delta 0110$.

EXAMPLE 11.25   Subtraction Using the Ones'-Complement Representation of the Negative Number

We repeat the problem of Example 11.24 by representing the negative number $-0_\Delta 0111$ in its ones'-complement form, which is given by $1_\Delta 1000$. The addition of $0_\Delta 1101$ and $1_\Delta 1000$ is shown below.

$$
\begin{array}{crcccc}
& 0_\Delta 1 & 1 & 0 & 1 \\
+ & 1_\Delta 1 & 0 & 0 & 0 \\
\hline
1 & 0_\Delta 0 & 1 & 0 & 1 \\
+ & & & & 1 & \leftarrow \text{ end around carry} \\
\hline
& 0_\Delta 0 & 1 & 1 & 0 \\
\end{array}
$$

Here also, the addition operation has created an extra bit 1 to the left of the sign bit due to the carry bit 1 generated at the last step. This bit is next added to the LSB position of the intermediate sum, resulting in the correct sum $0_\Delta 0110$.

11.9.2 Fixed-Point Multiplication

The multiplication of two b-bit binary fractions, $A = a_s {}_\triangle a_{-1} a_{-2} \cdots a_{-b}$ (called the *multiplicand*) and $D = d_s {}_\triangle d_{-1} d_{-2} \cdots d_{-b}$ (called the *multiplier*), in sign-magnitude form is carried out by forming the product $P^{(b)}$ of their respective magnitudes first and then assigning the appropriate sign to the product from the signs of the multiplier and the multiplicand. The product of the magnitudes can be implemented serially in b steps, where at the ith step, the partial product $P^{(i)}$ is determined as follows:

$$P^{(i)} = \left(P^{(i-1)} + d_{-b+i-1} \cdot A \right) \cdot 2^{-1}, \qquad i = 1, 2, \ldots, b, \tag{11.112}$$

with $P^{(0)} = 0$. It follows from Eq. (11.112) that if $d_{-b+i-1} = 0$, the new partial product is obtained by simply shifting the previous partial product to the right by one bit position. On the other hand, if $d_{-b+i-1} = 1$, the multiplicand A is added to the previous partial product and then shifted to the right by one bit position to arrive at the new partial product. The final product is a $2b$-bit fraction. Example 11.26 illustrates the algorithm.

EXAMPLE 11.26   Fixed-Point Multiplication of Two Binary Numbers

Consider the multiplication of the two 3-bit binary fractions, $a_s {}_\triangle 110$ and $d_s {}_\triangle 101$. We first form the product of their magnitudes using the recursive algorithm of Eq. (11.112), as illustrated below. The final answer in sign-magnitude form is $p_s {}_\triangle 011110$, where the sign of the product $p_s = 0$ if $a_s = d_s$; otherwise it is 1. Note that the product is a 6-bit fraction.

$$
\begin{array}{rcccccll}
        &   & 1 & 1 & 0 &   &   & = & A \\
\times &   & 1 & 0 & 1 &   &   & = & D \\
\hline
        &   & 0 & 0 & 0 &   &   & = & P^{(0)} \\
+   &   & 1 & 1 & 0 &   &   &   & \\
\hline
        &   & 1 & 1 & 0 &   &   &   & \\
\hline
        & 0 & 1 & 1 & 0 &   &   &   & P^{(1)} \\
\hline
        & 0 & 0 & 1 & 1 & 0 &   &   & P^{(2)} \\
+   & 1 & 1 & 0 &   &   &   &   & \\
\hline
        & 1 & 1 & 1 & 1 & 0 &   &   & \\
\hline
        & 0 & 1 & 1 & 1 & 1 & 0 & = & P^{(3)}
\end{array}
$$

If the multiplicand is a negative fraction in either two's-complement form or in ones'-complement form and the multiplier is a positive fraction, the algorithm of Eq. (11.112) can be followed without any change, except that the addition of the negative multiplicand is carried out according to the method outlined in Section 11.9.1, and after shifting the sum, the sign bit of the sum is left as is. On the other hand, if the multiplier is a negative fraction, a correction step is needed at the end to arrive at the correct result [Kor93].

In Booth's multiplication algorithm, employed in most DSP chips, the serial multiplication process described above is implemented, with the multiplier recoded in the SD form resulting, in general, in a faster operation [Boo51]. For example, a multiplier $0_\triangle 011001111_2$ requires six additions, whereas its SD representation would require four add/subtract operations. Booth's algorithm generates the correct product if both the multiplicand and the multiplier are represented in the two's-complement form, provided the sign bit of the multiplicand is employed in determining whether to perform an add or a subtract operation at the last step.

EXAMPLE 11.27   Illustration of Booth's Multiplication Algorithm

Let $A = 1_\Delta 101 = -0.375_{10}$ and $D = 1_\Delta 011 = -0.625_{10}$, represented in two's-complement form. The SD representation of D is given by $0_\Delta 10\bar{1}$. The multiplication of $1_\Delta 101$ with $0_\Delta 10\bar{1}$ using Booth's algorithm is illustrated below:

```
sign bit ⌐
              1   1   0   1                    = A = -0.375₁₀
         ×    0   1̄   0   1̄
              0   0   0   0                     = P⁽⁰⁾
         +    0   0   1   1                     Add  −A
              0   0   1   1
              0   0   0   1   1                 = P⁽¹⁾
              0   0   0   0   1   1             = P⁽²⁾
         +    0   0   1   1                     Add  −A
              0   0   1   1   1   1
           0 ∆ 0  0   1   1   1   1            = P⁽³⁾
```

The final product is a positive binary fraction $0_\Delta 001111$ of value 0.234375_{10}, as expected.

It should be noted that the multiplication of two fixed-point binary numbers may result in more bits for the product than those in the two numbers being multiplied. As a result, when the product is stored back in the memory, the product must either be truncated or rounded to fit the memory wordlength.

11.9.3   Floating-Point Arithmetic

Addition of two floating-point binary numbers can be carried out easily, if their exponents are equal. Thus, in a floating-point adder, the mantissa of the smaller number is shifted to the right by an appropriate number of bits to make its exponent equal to that of the larger number, and then the two mantissas are added. If the sum of the two mantissas is within the range of Eq. (11.107), then no additional normalization is necessary; otherwise, a normalization is carried out to bring it to the proper range, with a corresponding adjustment to the exponent.

EXAMPLE 11.28   Addition of Two Floating-Point Numbers

Consider the addition of the two floating-point numbers

$$\eta_1 = (0_\Delta 1110)2^{01},$$
$$\eta_2 = (0_\Delta 1101)2^{10},$$

whose decimal equivalents are, respectively, 1.75_{10} and 3.25_{10}. We first rewrite the floating-point representation of the smaller number η_1 so that its characteristic is equal to that of the larger number η_2:

$$\eta_1 = (0_\Delta 0111)2^{10}.$$

Then, the sum of these two numbers is given by

$$\eta_1 + \eta_2 = (0_\Delta 1101 + 0_\Delta 0111)2^{10} = (1_\Delta 0100)2^{10}.$$

Since the mantissa of the sum is greater than 1, its characteristic is therefore increased by one unit, and the mantissa is shifted to the right by one bit, resulting in

$$\eta_1 + \eta_2 = (0_\triangle 1010)2^{11},$$

whose decimal equivalent is 5_{10}, as expected.

In a floating-point multiplier, on the other hand, the multiplication is carried out by multiplying the two mantissas and adding their corresponding exponents. Since the range of the product of the mantissas is now between $\frac{1}{4}$ and 1, a normalization of the product is carried out if it is less than $\frac{1}{2}$, along with a corresponding adjustment to the new exponent.

EXAMPLE 11.29   **Multiplication of Two Floating-Point Numbers**

Consider the multiplication of the two floating-point numbers

$$\eta_1 = (0_\triangle 1010)2^{01},$$
$$\eta_2 = (0_\triangle 1001)2^{10},$$

whose decimal equivalents are, respectively, 1.25_{10} and 2.25_{10}. Their product is thus given by

$$\eta_1 \eta_2 = (0_\triangle 1010)(0_\triangle 1001)2^{01+10} = (0_\triangle 0101101)2^{11}.$$

As the mantissa of the product is less than $\frac{1}{2}$, the characteristic is reduced by one unit, and the mantissa is shifted to the left by one bit, resulting in the correct floating-point representation

$$\eta_1 \eta_2 = (0_\triangle 1011010)2^{10},$$

whose decimal equivalent is 2.8125_{10}, as expected.

11.10   Handling of Overflow

In Section 11.9.1, we pointed out that the results of the addition of two fixed-point fractions can result in a sum exceeding the dynamic range of the register storing the result of the addition, thus resulting in an overflow. Occurrence of overflow leads to severe output distortion and can often result in large-amplitude oscillations at the filter output (see Section 12.11.2). Therefore, the sum should be substituted with another number that is within the dynamic range. Two widely used schemes for handling the overflow are described next.

Let η denote the sum; then in either of the two schemes, if η exceeds the dynamic range $[-1, 1)$, it is substituted with a number ξ, which is within the range. In the *saturation overflow* scheme shown in Figure 11.39(a), if $\eta \geq 1$, it is replaced with $1 - 2^{-b}$, where the number is assumed to be a b-bit fraction with an additional bit for the sign, and if $\eta \leq -1$, it is replaced with -1. On the other hand, in the *two's-complement overflow* scheme shown in Figure 11.39(b), whenever η is outside the range $[-1, 1)$, it is replaced with $\xi = \langle \eta + 1 \rangle_2$. Basically, here, when η is outside the range, the bits to the left of the sign bit (overflow bits) are ignored. The second scheme is usually implemented in nonrecursive digital filters employing two's-complement arithmetic. In most applications, the first scheme is usually preferred.

We shall discuss in Section 12.7 the dynamic range scaling of the digital filter structure to either eliminate completely or reduce the probability of overflow. The effects of the two overflow handling schemes on the performance of the digital filter are also considered in Section 12.7.

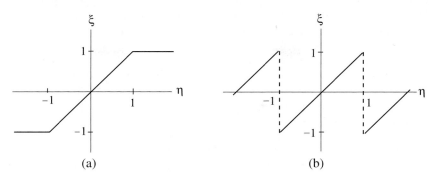

Figure 11.39: (a) Saturation overflow and (b) two's-complement overflow.

11.11 Tunable Digital Filters

Many applications require the use of digital filters with easily tunable characteristics. In Section 8.7, we outlined the design of tunable first-order and second-order digital filters that may provide adequate solutions in some applications. There are other applications where use of higher-order tunable digital filters may be necessary. The design of such tunable filters is the subject of this section.

11.11.1 Tunable IIR Digital Filters

The basis for the design of tunable digital filters is the spectral transformation discussed in Section 9.5, which can be used to tune a given digital filter realization with a specified cutoff frequency to another realization with a different cutoff frequency. Thus, if $G_{old}(z)$ is the transfer function of the original realization, the transfer function of the new structure is $G_{new}(z)$, where

$$G_{new}(z) = G_{old}(z)|_{z^{-1}=F^{-1}(z)}, \qquad (11.113)$$

in which $F^{-1}(z)$ is a stable allpass function of the form given in Table 9.1, with the parameters of the transformation being the tuning parameters. One straightforward way to implement this transformation would be to replace each delay block in the realization of $G_{old}(z)$ with an allpass structure realizing $F^{-1}(z)$. However, such an approach leads, in general, to a structure realizing $G_{new}(z)$ with delay-free loops and cannot be implemented, as explained in Section 8.1.3.

We describe now a very simple practical modification to the above approach that does not result in a structure with delay-free loops [Mit90b]. In Section 8.10, we outlined a method of realization of a large class of stable IIR transfer functions $G(z)$ in the form [Vai86a]

$$G(z) = \tfrac{1}{2}\{A_0(z) + A_1(z)\}, \qquad (11.114)$$

where $A_0(z)$ and $A_1(z)$ are stable allpass filters. The conditions for the realization of $G(z)$ as a parallel connection of two allpass sections are that $G(z)$ be a bounded real transfer function with a symmetric numerator and that it have a power complementary transfer function $H(z)$ with an antisymmetric numerator. These conditions are satisfied by all odd-order lowpass Butterworth, Chebyshev, and elliptic transfer functions.

The allpass filters $A_0(z)$ and $A_1(z)$ can be realized using any one of the approaches discussed in Section 8.6. We consider here their realizations as a cascade of first-order and second-order sections. As indicated in Section 8.6, there are a large variety of structurally lossless canonic structures realizing the first-order and second-order allpass transfer functions [Mit74a], [Szc88]. These structures use only one multiplier and one delay for the realization of a first-order allpass function and two multipliers and two delays for the realization of a second-order allpass function.

Consider first the tuning of the cutoff frequency of a lowpass IIR filter realized by a parallel allpass structure. From Section 9.5, we note that the lowpass-to-lowpass transformation is given by

$$z^{-1} \rightarrow F^{-1}(z^{-1}) = \frac{z^{-1} - \alpha}{1 - \alpha z^{-1}}, \tag{11.115}$$

where the parameter α is related to the old and new cutoff frequencies, ω_c and $\hat{\omega}_c$, respectively, through

$$\alpha = \frac{\sin[(\omega_c - \hat{\omega}_c)/2]}{\sin[(\omega_c + \hat{\omega}_c)/2]}. \tag{11.116}$$

Substituting the transformation of Eq. (11.115) on a Type 1 first-order allpass transfer function

$$a_1(z) = \frac{d_1 + z^{-1}}{1 + d_1 z^{-1}}, \tag{11.117}$$

we obtain the new first-order allpass transfer function

$$\hat{a}_1(z) = a_1(z)|_{z^{-1}=(z^{-1}-\alpha)/(1-\alpha z^{-1})} = \frac{d_1 + \left(\frac{z^{-1}-\alpha}{1-\alpha z^{-1}}\right)}{1 + d_1 \left(\frac{z^{-1}-\alpha}{1-\alpha z^{-1}}\right)}$$

$$= \frac{(d_1 - \alpha) + (1 - \alpha d_1)z^{-1}}{(1 - \alpha d_1) + (d_1 - \alpha)z^{-1}} = \frac{\left(\frac{d_1 - \alpha}{1 - \alpha d_1}\right) + z^{-1}}{1 + \left(\frac{d_1 - \alpha}{1 - \alpha d_1}\right)z^{-1}}. \tag{11.118}$$

If α is very small, we can make a Taylor series expansion of the coefficient $(d_1 - \alpha)/(1 - \alpha d_1)$ of the allpass function $\hat{a}_1(z)$ of Eq. (11.118) and arrive at an approximation

$$\hat{a}_1(z) \cong \frac{\left[d_1 + \alpha(d_1^2 - 1)\right] + z^{-1}}{1 + \left[d_1 + \alpha(d_1^2 - 1)\right]z^{-1}}, \tag{11.119}$$

which is seen to be a Type 1 first-order allpass transfer function with a coefficient that is now a linear function of α. The approximated allpass section $\hat{a}_1(z)$ can be simply implemented by replacing each multiplier d_1 in Figure 8.24 with a parallel connection of two multipliers, as indicated in Figure 11.40(a). Note that for $\alpha = 0$, $\hat{a}_1(z)$ of Eq. (11.119) reduces to $a_1(z)$ of Eq. (11.117), as expected.

Applying a similar procedure to the Type 3 second-order allpass transfer function

$$a_2(z) = \frac{d_2 + d_1 z^{-1} + z^{-2}}{1 + d_1 z^{-1} + d_2 z^{-2}}, \tag{11.120}$$

we arrive at

$$\hat{a}_2(z) = a_2(z)|_{z^{-1}=(z^{-1}-\alpha)/(1-\alpha z^{-1})} = \frac{d_2 + d_1 \left(\frac{z^{-1}-\alpha}{1-\alpha z^{-1}}\right) + \left(\frac{z^{-1}-\alpha}{1-\alpha z^{-1}}\right)^2}{1 + d_1 \left(\frac{z^{-1}-\alpha}{1-\alpha z^{-1}}\right) + d_2 \left(\frac{z^{-1}-\alpha}{1-\alpha z^{-1}}\right)^2}. \tag{11.121}$$

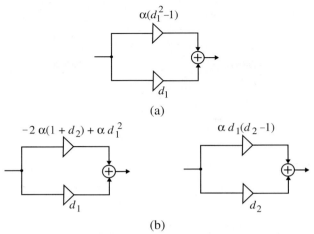

(a)

(b)

Figure 11.40: Multiplier replacement schemes in the constituent allpass sections for designing a tunable IIR filter: (a) Type 1 allpass network and (b) Type 3 allpass network.

Again, if we assume α to be very small, we can rewrite the above expression as

$$\hat{a}_2(z) \cong \frac{(d_2 - \alpha d_1) + (d_1 - 2\alpha[1 + d_2])\, z^{-1} + (1 - \alpha d_1)z^{-2}}{(1 - \alpha d_1) + (d_1 - 2\alpha[1 + d_2])\, z^{-1} + (d_2 - \alpha d_1)z^{-2}}$$

$$= \frac{\left(\frac{d_2 - \alpha d_1}{1 - \alpha d_1}\right) + \left(\frac{d_1 - 2\alpha[1 + d_2]}{1 - \alpha d_1}\right) z^{-1} + z^{-2}}{1 + \left(\frac{d_1 - 2\alpha[1 + d_2]}{1 - \alpha d_1}\right) z^{-1} + \left(\frac{d_2 - \alpha d_1}{1 - \alpha d_1}\right) z^{-2}}, \qquad (11.122)$$

by neglecting coefficients containing α^2. Next, an approximation based on the Taylor series expansion of the last term in Eq. (11.122) results in

$$\hat{a}_2(z) \cong \frac{[d_2 + \alpha d_1(d_2 - 1)] + \left[d_1 - 2\alpha(1 + d_2) + \alpha d_1^2\right] z^{-1} + z^{-2}}{1 + \left[d_1 - 2\alpha(1 + d_2) + \alpha d_1^2\right] z^{-1} + [d_2 + \alpha d_1(d_2 - 1)]\, z^{-2}}, \qquad (11.123)$$

which is seen to be a Type 3 second-order allpass transfer function with coefficients that are a linear function of α. The approximated allpass section $\hat{a}_2(z)$ can be simply implemented by replacing the multipliers d_1 and d_2 in Figure 8.26 with a parallel connection of two multipliers, as indicated in Figure 11.40(b). Note that for $\alpha = 0$, $\hat{a}_2(z)$ of Eq. (11.123) reduces to $a_2(z)$ of Eq. (11.120), as expected.

Note that in the case of both the first-order and the second-order allpass filters, the tuning rule is now a linear function of α. Even though this tuning algorithm is approximate and has been derived for small values of α, in practice, tuning ranges of several octaves have been observed to hold in the case of narrow-band lowpass elliptic filters [Mit90b]. Figure 11.41 shows the gain responses of a fifth-order tunable elliptic lowpass filter for three values of the tuning parameter α. The prototype filter ($\alpha = 0$) has a passband edge at $\omega_p = 0.4\pi$, passband ripple of 0.5 dB, and minimum stopband attenuation of 40 dB.

By applying a lowpass-to-bandpass transformation

$$z^{-1} \rightarrow F^{-1}(z^{-1}) = -z^{-1}\frac{z^{-1} + \beta}{1 + \beta z^{-1}} \qquad (11.124)$$

to a tunable lowpass IIR filter, we can design a tunable bandpass filter whose center frequency ω_0 is tuned by adjusting the parameter $\beta = \cos\omega_0$, and the bandwidth is tuned by changing α [Mit90b]. Unlike

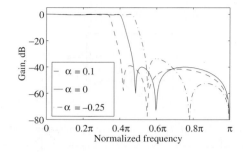

Figure 11.41: Gain responses of a fifth-order tunable elliptic lowpass filter for three values of α.

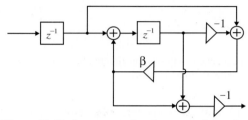

Figure 11.42: Lowpass-to-bandpass transformation.

the lowpass-to-lowpass transformation of Eq. (11.115), the transformation of Eq. (11.124) can be directly implemented on the structure realizing the tunable lowpass filter by replacing each delay with the structure shown in Figure 11.42 to arrive at a tunable bandpass filter structure.

11.11.2 Tunable FIR Digital Filters

The spectral transformation approach of Section 9.5 can also be applied to an FIR filter to develop a filter structure with tunable characteristics. However, the resulting structure is no longer FIR since the replacement of the delays in the prototype FIR structure by allpass sections implementing the spectral transformation makes it an IIR filter. We outline below a straightforward method for designing tunable linear-phase lowpass FIR filters and later show how to modify the method to the design of other types of tunable FIR filters [Jar88]. The method discussed preserves the FIR structure and permits easy tuning of the cutoff frequency.

The basic idea behind the tuning procedure is the observation that, for an ideal lowpass FIR filter with a zero-phase response given by

$$H_d(e^{j\omega}) = \begin{cases} 1, & \text{for } 0 \le |\omega| \le \omega_c, \\ 0, & \text{for } \omega_c < |\omega| \le \pi, \end{cases} \tag{11.125}$$

the impulse response coefficients are given by

$$h_d[n] = \frac{\sin(\omega_c n)}{\pi n}, \quad 0 \le |n| < \infty, \tag{11.126}$$

as derived in Example 3.6. We can truncate the above expression and obtain the coefficients of a realizable approximation given by

$$h_{LP}[n] = \begin{cases} c[n]\omega_c, & \text{for } n = 0, \\ c[n]\sin(\omega_c n), & \text{for } 1 \le |n| \le N, \\ 0, & \text{otherwise,} \end{cases} \tag{11.127}$$

where ω_c is the 6-dB cutoff frequency, and

$$c[n] = \begin{cases} 1/\pi, & \text{for } n = 0, \\ 1/\pi n, & \text{for } 1 \leq |n| \leq N. \end{cases} \tag{11.128}$$

It follows from the above that once an FIR lowpass filter has been designed for a given cutoff frequency, it can be tuned simply by changing ω_c and recomputing the filter coefficients according to the above expression. It can be shown that Eq. (11.127) can also be used to design a tunable FIR lowpass filter by equating the coefficients of a prototype filter developed using any of the FIR filter design methods outlined in Chapter 7 with those of Eq. (11.127) and solving for $c[n]$ [Jar88]. Thus, if $h_{LP}[n]$ denotes the coefficients of the prototype lowpass filter designed for a cutoff frequency ω_c, from Eq. (11.127), the constants $c[n]$ are given by

$$c[0] = \frac{h_{LP}[0]}{\omega_c}, \tag{11.129a}$$

$$c[n] = \frac{h_{LP}[n]}{\sin(\omega_c n)}, \quad 1 \leq |n| \leq N. \tag{11.129b}$$

Then, the coefficients $\hat{h}_{LP}[n]$ of the transformed FIR filter with a cutoff frequency $\hat{\omega}_c$ are given by

$$\hat{h}_{LP}[0] = c[0]\hat{\omega}_c = \left(\frac{\hat{\omega}_c}{\omega_c}\right) h_{LP}[0], \tag{11.130a}$$

$$\hat{h}_{LP}[n] = c[n]\sin(\hat{\omega}_c n) = \left(\frac{\sin(\hat{\omega}_c n)}{\sin(\omega_c n)}\right) h_{LP}[n], \quad 1 \leq |n| \leq N. \tag{11.130b}$$

This tuning procedure has been recommended for filters with equal passband and stopband ripples. It has also been recommended that the prototype filter be designed such that its coefficients have values not too close to zero.

We illustrate the design of a tunable FIR filter in Example 11.30.

EXAMPLE 11.30   Design of a Tunable Lowpass FIR Filter

The prototype filter is a linear-phase lowpass FIR filter of length 51 with a passband edge ω_p at 0.36π and a stopband edge ω_s at 0.46π and a transition bandwidth of 0.1π. The cutoff frequency ω_c is thus at 0.41π. It has been designed using the function remez of MATLAB, assuming equal weights to the passband and the stopband. Using the method discussed above, the filter coefficients for a cutoff at 0.31π and at 0.51π have been computed. Figure 11.43 shows the plots of the gain responses of the three FIR filters.

For the design of tunable FIR filters with unequal passband and stopband ripples, the following modification is used. The impulse response coefficients of the tunable filter of odd length are now given by

$$h_{LP}[n] = \begin{cases} c[n]\omega_c + d[n], & \text{for } n = 0, \\ c[n]\sin(\omega_c n) + d[n]\cos(\omega_c n), & \text{for } 1 \leq |n| \leq N. \end{cases} \tag{11.131}$$

The constants $c[n]$ and $d[n]$ are determined from two different optimal prototype filters, substituting them in Eq. (11.131), and then solving for these constants. For example, if the passband weight W_p is greater than the stopband weight W_s, the cutoff frequencies ω_{ca} and ω_{cb} of the two filters are chosen as

$$\omega_{ca} = 0.8 - \frac{0.25}{N}, \quad \omega_{cb} = 0.8 + \frac{0.25}{N}.$$

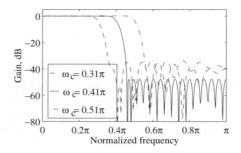

Figure 11.43: Tunable lowpass FIR filter of length 51.

On the other hand, if the passband weight W_p is less than the stopband weight W_s, the cutoff frequencies ω_{ca} and ω_{cb} of the two filters are chosen as

$$\omega_{ca} = 0.2 - \frac{0.25}{N}, \quad \omega_{cb} = 0.2 + \frac{0.25}{N}.$$

The two optimal filters $h_a[n]$ and $h_b[n]$ are then designed using the Parks–McClellan algorithm. As before, the prototype filters should be designed with nonzero coefficients.

The above modification is not recommended for the design of tunable wide-band or very narrow-band filters. Figure 11.44 shows the gain responses of a lowpass FIR filter of length 41 and a transition width of 0.1π, with a variable 6-dB cutoff frequency for various values of passband and stopband weights. The cutoff frequencies of the filters vary from 0.2 to 0.9, as indicated in Figure 11.44.

The above methods can be directly applied to the design of a tunable highpass FIR filter from a highpass FIR prototype filter. Alternatively, the latter can be designed as a delay-complementary tunable Type 1 FIR lowpass filter.[15]

To develop the methods for designing a tunable bandpass FIR filter, we observe that a filter $H_{BP}(z)$ with a symmetrical bandpass magnitude response can be derived from a lowpass prototype filter $H_{LP}(z)$ by applying a frequency translation to its frequency response. This process results in

$$H_{BP}(e^{j\omega}) = H_{LP}(e^{j(\omega+\omega_0)}) + H_{LP}(e^{j(\omega-\omega_0)}), \tag{11.132}$$

where ω_0 is the desired center frequency of the bandpass filter. The relation between the impulse responses of the bandpass and the prototype lowpass filters is given by

$$h_{BP}[n] = \left(e^{-j\omega_0 n} + e^{j\omega_0 n}\right) h_{LP}[n] = 2\cos(\omega_0 n) h_{LP}[n]. \tag{11.133}$$

If δ_p and δ_s denote the passband and stopband ripples of the lowpass prototype, the corresponding ripples of the bandpass filter are $\delta_p + \delta_s$ for the passband and $2\delta_s$ for the stopband. Equation (11.133) forms the basis for designing a tunable bandpass FIR filter with adjustable center frequency. A similar approach can be followed for the design of a tunable bandstop FIR filter, in which case the prototype is a highpass FIR filter.

It should be noted that the tunable FIR filters designed using the methods suggested above have the same hardware requirements as that of their prototypes. They also have linear-phase if the prototype filter has linear-phase.

---

[15]See Section 7.5.1.

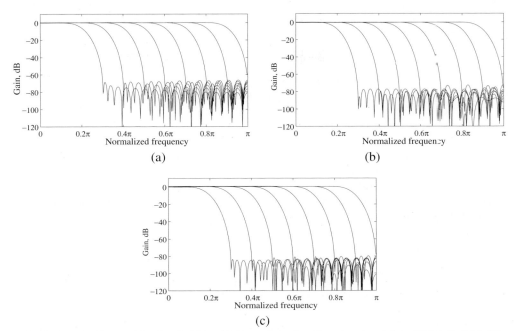

Figure 11.44: Length 41, transition bandwidth of 0.1π, and passband/stopband weights of (a) $1/1$, (b) $1/10$, and (c) $1/50$.

11.12 Function Approximation

Often there is a need to use transcendental functions and random numbers in certain DSP applications. For example, the FFT computations require the generation of the complex exponential sequences. Certain digital communication systems require sinusoidal sequences. Both of these sequences can, of course, be generated by second-order digital filter structures, described in Section 8.11. We outline below an alternative approach to the generation of transcendental functions based on truncated polynomial expansions. These types of expansions are often used in DSP chip implementations [Mar92].

11.12.1 Trigonometric Function Approximation

The sine of a number x can be approximated using the expansion [Abr72]

$$\sin(x) \cong x - 0.166667x^3 + 0.008333x^5 - 0.0001984x^7 + 0.0000027x^9, \tag{11.134}$$

where the argument x is in radians and its range is restricted to the first quadrant, that is, from 0 to $\pi/2$. If x is outside this range, its sine can be computed by making use of the identities $\sin(-x) = -\sin(x)$, and $\sin[(\pi/2) + x] = \sin[(\pi/2) - x]$. Figure 11.45 shows the plots of the sine approximation computed using Eq. (11.134) and the error due to the approximation.

For computing the arctangent of a number x where $-1 \le x \le 1$, a recommended expansion is given by [Abr72]

$$\tan^{-1}(x) \cong 0.999866x - 0.3302995x^3 + 0.180141x^5 - 0.085133x^7 + 0.0208351x^9. \tag{11.135}$$

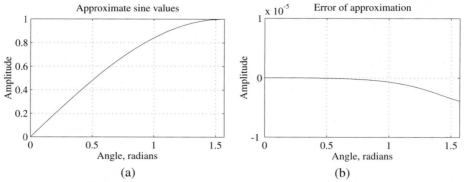

Figure 11.45: (a) Plot of the sine value computed using Eq. (11.134) and (b) plot of the error between the actual sine value and the approximation.

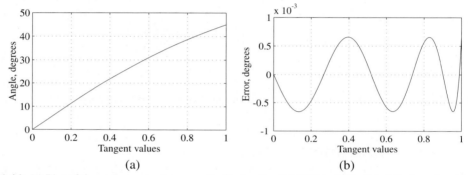

Figure 11.46: (a) Plot of the arctangent value computed using Eq. (11.135) and (b) plot of the error between the actual arctangent value and the approximation.

Figure 11.46 shows the plots of the arctangent approximation computed using Eq. (11.135) and the error due to the approximation. If $x \geq 1$, then its arcangent can be computed by making use of the identity $\tan^{-1}(x) = (\pi/2) - \tan^{-1}(1/x)$.

Exercises M11.10 and M11.11 list several polynomial approximations for computing these trigonometric functions. Various trigonometric functions can also be computed from Eq. (11.134) using trigonometric identities.

11.12.2 Square-Root and Logarithm Approximation

The square root of a positive number x in the range $0.5 \leq x \leq 1$ can be evaluated using the truncated polynomial approximation [Mar92]:

$$\sqrt{x} \cong 0.2075806 + 1.454895x - 1.34491x^2 + 1.106812x^3$$
$$- 0.536499x^4 + 0.1121216x^5. \tag{11.136}$$

Figure 11.47 shows the plot of the error resulting from the approximation given by Eq. (11.136). If x is outside the range from 0.5 to 1, it can be multiplied by a binary constant K^2 to bring the product $x' = K^2 x$ into the desirable range, compute $\sqrt{x'}$ using Eq. (11.136), and then determine $\sqrt{x} = \sqrt{x'}/K$.

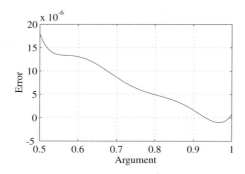

Figure 11.47: Plot of the error between the actual square-root value and the approximation given by Eq. (11.136).

Polynomial expansions have also been advanced for the approximate computation of both the logarithm (base 10) and natural logarithm (base e) of any number x between one and two. Two such expressions are given by [Bur73], [Mar92]

$$\log_{10}(x) \cong 0.4339142(x-1) - 0.21278385(x-1)^2 + 0.1240692(x-1)^3$$
$$- 0.05778505(x-1)^4 + 0.0136261(x-1)^5, \tag{11.137}$$
$$\log_e(x) \cong 0.999115(x-1) - 0.4899597(x-1)^2 + 0.2856751(x-1)^3$$
$$- 0.13305665(x-1)^4 + 0.031372071(x-1)^5. \tag{11.138}$$

For the calculation of the logarithm of a number x that is outside the range from 1 to 2, the number x must be scaled by an appropriate factor K with known logarithm to bring the product $x' = Kx$ to this range, and then the logarithm of the product x' must be computed. From the logarithm of the product x', the logarithm of K is subtracted to get the logarithm of x.

11.12.3   Random Number Generation

Random number generation is used in a number of applications. For example, it is used as a training signal for the adaptive equalizer in high-speed modems [Mar92]. Another application is in the elimination of limit cycles in IIR digital filter structures using the random rounding method (Section 12.11.4). A variety of approaches has been proposed for the generation of uniformly distributed random numbers [Knu69]. One rather simple method is based on the linear congruence method and is given by the recursive equation

$$x[n+1] = \langle \alpha x[n] + \beta \rangle_M, \tag{11.139}$$

where the modulus M is a positive integer. The rules for the selection of the constants α and β are given in [Knu69]. Equation 11.139 generates a periodic pseudo-random sequence, having a period M with proper choice of α and β fairly independent of the *seed value* $x[0]$. The samples of the sequence generated by Eq. (11.139) within a period are uniformly distributed integers from 0 to $M-1$. A family of such random sequences can be generated by choosing different seed values for each sequence. To ensure the randomness of each sequence generated, M should be chosen as large as possible.

The M-file `rand` can be used to generate random numbers and matrices with elements uniformly distributed in the interval (0, 1). Likewise, the M-file `randn` can be used to generate random numbers and matrices with elements that are normally (Gaussian) distributed with zero mean and unity variance.

Both functions are available with several options. The function `rand` was used in Programs 2_3 and 2_4 of Example 2.14 to generate a random sequence.

11.13   Summary

Some of the common factors in the implementation of DSP algorithms, which are independent of the type of implementation being carried out, are discussed in this chapter. The implementation is usually carried out by a sequential implementation of the set of equations describing the algorithm. These equations can be derived directly from the structure realizing the algorithm. The computability condition of these equations is derived, and an algorithm for testing the computability is described. A simple algebraic technique to verify the structure from the input and output samples is also outlined.

MATLAB-based software implementation of digital filtering algorithms and the discrete Fourier transform (DFT) are then considered. The basic ideas behind the fast Fourier transform (FFT) used for faster computation of the DFT samples are explained, and several commonly used FFT algorithms are derived.

Various schemes for the binary representation of the numbers and the signal variables that are employed in the digital computers and special-purpose DSP chips are reviewed, followed by a discussion of the algorithms used for the implementation of the addition and the multiplication operation. In certain cases, the result of an addition of numbers may cause an overflow of the dynamic range of the register storing the sum. Two methods of physically preventing the overflow are outlined.

Methods for the design of IIR and FIR digital filters with tunable characteristics are introduced next. The chapter concludes with a discussion on the approximation of certain functions that are needed in a number of applications requiring implementation using DSP chips. In particular, the approximation of trigonometric functions, square roots, logarithms, and the generation of random numbers are considered.

11.14   Problems

11.1 Develop a set of time-domain equations describing the digital filter structure of Figure P11.1 in terms of the input $x[n]$, output $y[n]$, and the intermediate variables $w_k[n]$ in a sequential order. Does this set describe a valid computational algorithm? Justify your answer by developing a matrix representation of the digital filter structure and by examining the matrix **F**.

11.2 Develop a computable set of time-domain equations describing the digital filter structure of Figure P11.1. Verify the computability condition by forming an equivalent matrix representation and by examining the matrix **F**.

11.3 Develop a set of time-domain equations describing the digital filter structure of Figure P11.2 in terms of the input $x[n]$, output $y[n]$, and the intermediate variables $w_k[n]$ in a sequential order. Does this set of equations describe a valid computational algorithm? Justify your answer by developing a matrix representation of the digital filter structure and by examining the matrix **F**.

11.4 Develop the precedence graph of the digital filter structure of Figure P11.1, and investigate its realizability. If the structure is found to be realizable, then from the precedence graph, determine a valid computational algorithm describing the structure.

11.5 Develop the precedence graph of the digital filter structure of Figure P11.2, and investigate its realizability. If the structure is found to be realizable, then from the precedence graph, determine a valid computational algorithm describing the structure.

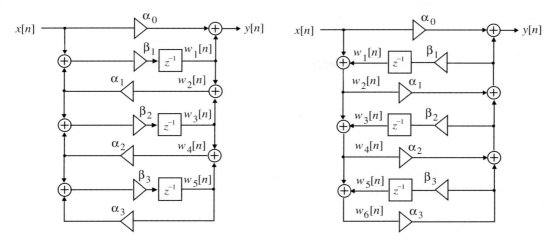

Figure P11.1   Figure P11.2

11.6 (a) Write down the time-domain equations relating the node variables $w_i[n]$, $s_i[n]$, $y[n]$ and the input $x[n]$ of the digital filter structure of Figure P11.3. Check formally the computability of the set of equations if the equations are ordered sequentially with increasing values of the node indices.

(b) Develop a signal flow-graph representation of this digital filter structure, and then determine its precedence graph. From the precedence graph, develop a set of computable equations describing the structure, and show formally that these equations are indeed computable.

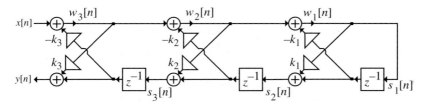

Figure P11.3

11.7 The measured impulse response samples of a causal second-order IIR digital filter with a transfer function $H(z) = P(z)/(1 - 3z^{-1} + 1.5z^{-2})$ are given by

$$h[0] = 3.2, \ h[1] = 5.6, \ h[2] = 7.0, \ h[3] = 12.6, \ h[4] = 27.3, \ldots$$

Determine the numerator $P(z)$ of the transfer function $H(z)$.

11.8 The first five impulse response samples of a causal second-order IIR digital filter are given by

$$h[0] = 2, \ h[1] = -2, \ h[2] = 4, \ h[3] = 8, \ h[4] = -12.$$

Determine the transfer function $H(z)$.

11.9 Determine the transfer function of a third-order causal IIR digital filter whose first 10 impulse response samples are given by

$$\{h[n]\} = \{2 \quad -4 \quad 8 \quad -8 \quad 12 \quad 16 \quad -8 \quad 168 \quad 12.48 \quad 746.048\}.$$

11.10 The first four impulse response samples of a causal third-order IIR digital filter with a transfer function $G(z) = P(z)/(1 - 0.6z^{-1} + 0.2z^{-2} + 1.8z^{-3})$ are given by

$$g[0] = 2, \ g[1] = -4, \ g[2] = 4, \ g[3] = -6.$$

Determine the numerator polynomial $P(z)$ of the transfer function.

11.11 The first 11 impulse response samples of a causal IIR transfer function

$$H(z) = \frac{p_0 + p_1 z^{-1} + p_2 z^{-2} + p_3 z^{-3} + p_4 z^{-4}}{1 + \frac{2}{3}z^{-1} - 2z^{-3} - \frac{5}{3}z^{-4}}$$

are given by

$$\{h[n]\} = \{2 \quad -2 \quad 4 \quad 6 \quad -8 \quad 10 \quad 12 \quad -14 \quad 16 \quad 30 \quad -28\}.$$

Determine its numerator coefficients $\{p_i\}$.

11.12 Show that the direct computation of the N-point DFT of a length-N sequence requires $4N^2$ real multiplications and $(4N - 2)N$ real additions.

11.13 Develop an algorithm for the complex multiplication of two complex numbers using only three real multiplications and five real additions.

11.14 The twiddle factors in the implementation of the FFT and inverse FFT algorithms are complex numbers with magnitude one. Let $W_N = c + js$ be a complex twiddle factor. Show that the multiplication of a complex signal Ψ_r by the twiddle factor W_N can be implemented using three real multiplications, as indicated in Figure P11.4 [Ora2002]. Develop the structure for implementing the multiplication by the twiddle factor W_N^{-1} using three real multiplications.

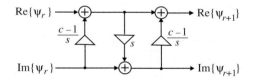

Figure P11.4

11.15 A digital oscillator working at a sampling rate of 2500 Hz can generate any one of four sinusoidal signals of frequencies 150 Hz, 375 Hz, 620 Hz, and 850 Hz. We would like to detect the tone frequency of the signal being generated by computing only one sample of the four N-point DFTs using Goertzel's method. What is the smallest value of the DFT length N so that the four tone frequencies fall as close as possible to four DFT bin indices to make the leakage to adjacent bins as small as possible? The DFT length N should be the same in all four cases.

11.16 Let $H_k(z)$ denote the transfer function of the kth filter used in the Goertzel's algorithm to calculate the N-point DFT $X[k]$ of a length-N sequence $x[n]$. Consider the following input sequence applied to $H_k(z)$:

$$x[n] = \{1, \ 0, \ 0, \ldots, \ 1, \ 0, \ 0, \ldots\},$$
$$\quad\quad\uparrow \quad\quad\quad\quad\quad \uparrow$$
$$\quad\quad 0 \quad\quad\quad\quad\quad N/2$$

where $N/2$ is assumed to be divisible by k. What is the output sequence $y[n]$ for $k = 1$? What is the output sequence $y[n]$ for $k = N/2$? Give a qualitative sketch.

11.17 Develop the flow-graph for the FFT algorithm from the radix-2 DIT FFT algorithm of Figure 11.24 for the $N = 8$ case in which the input is in normal order and the output is in the bit-reversed order.

11.18 Develop the flow-graph for the FFT algorithm from the radix-2 DIT FFT algorithm of Figure 11.24 for the $N = 8$ case in which both the input and the output are in normal order.

11.19 Derive Eq. (11.69).

11.20 As indicated in Section 11.3.2, the total number of complex multiplications $\mathcal{R}(\nu)$ needed in the implementation of the radix-2 DIT and DIF FFT algorithms can be made smaller than $\frac{1}{2} N \log_2 N$ if the multiplications by ± 1 can be avoided. Develop an exact expression for $\mathcal{R}(\nu)$ that includes only multiplications by complex twiddle factors with nonunity magnitudes.

11.21 If M DFT samples of the N-point DFT of a length-N are required with $M \leq N$, what is the smallest value of M for which the N-point radix-2 FFT algorithm is computationally more efficient than a direct computation of the M DFT samples? What are the values of M for the following values of N: (a) $N = 512$, (b) $N = 1024$, and (c) $N = 2048$.

11.22 Develop the structural interpretation of the first stage of the radix-3 DIT FFT algorithm.

11.23 Develop the flow-graph for the radix-3 DIT FFT algorithm for the $N = 9$ case in which the input is in digit-reversed order and the output is in the normal order.

11.24 Develop the flow-graph for a mixed-radix DIT FFT algorithm for the $N = 15$ case in which the input is in digit-reversed order and the output is in the normal order.

11.25 Form the transposed graph of the flow-graph of Figure 11.24. Replace each complex multiplicand W_N^r in the transposed flow-graph with $\frac{1}{2} W_N^{-r}$. Show that the final flow-graph implements an inverse DFT if the input is $X[k]$.

11.26 A second approach to the inverse DFT computation using a DFT algorithm is illustrated in Figure P11.5. Let $X[k]$ be the N-point DFT of a length-N sequence $x[n]$. Define a length-N time-domain sequence $q[n]$ as

$$\text{Re}\{q[n]\} = \text{Im}\, X[k]|_{k=n}, \quad \text{Im}\{q[n]\} = \text{Re}\, X[k]|_{k=n},$$

with $Q[k]$ denoting its N-point DFT. Show that

$$\text{Re}\{x[n]\} = \frac{1}{N} \cdot \text{Im}\, Q[k]\bigg|_{k=n}, \quad \text{Im}\{x[n]\} = \frac{1}{N} \cdot \text{Re}\, Q[k]\bigg|_{k=n}.$$

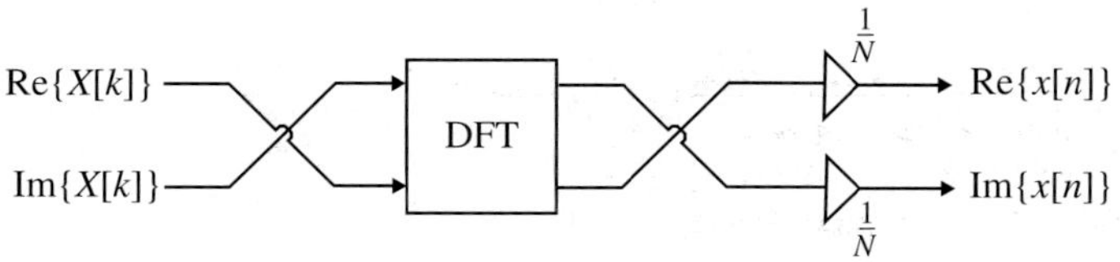

Figure P11.5

11.27 A third approach to the inverse DFT computation using a DFT algorithm is described below. Let $X[k]$ be the N-point DFT of a length-N sequence $x[n]$. Define a length-N time-domain sequence $r[n]$ as

$$r[n] = X[k]|_{k=\langle -n \rangle_N},$$

with $R[k]$ denoting its N-point DFT. Show that

$$x[n] = \frac{1}{N} \cdot R[k]\bigg|_{k=n}.$$

11.28 We wish to determine the sequence $y[n]$ generated by a linear convolution of a length-16 real sequence $x[n]$ and a length-9 real sequence $h[n]$. To this end, we can follow one of the following methods:

Method # 1. Direct computation of the linear convolution.

Method # 2. Computation of the linear convolution via a single circular convolution.

Method # 3. Computation of the linear convolution using radix-2 FFT algorithms.

Determine the least number of real multiplications needed in each of the above methods. For the radix-2 FFT algorithm, do not include in the count multiplications by ± 1, $\pm j$, and W_N^0.

11.29 Repeat Problem 11.28 for the computation of the linear convolution of a length-16 sequence $x[n]$ with a length-10 sequence $h[n]$.

11.30 An input sequence $x[n]$ of length 2048 is to be filtered using a linear-phase FIR filter $h[n]$ of length 72. This filtering process involves the linear convolution of two finite-length sequences and can be computed using the overlap-add algorithm discussed in Section 5.10.2, where the short linear convolutions are performed using the DFT-based approach of Figure 5.13 with the DFTs implemented by the Cooley–Tukey FFT algorithm.

(a) Determine the appropriate power-of-2 transform length that would result in a minimum number of multiplications, and calculate the total number of multiplications that would be required.

(b) What would be the total number of multiplications if the direct convolution method is used?

11.31 We pointed out in Eq. (5.25) that the vector $\mathbf{X}$ of DFT samples can be expressed as the product of the DFT matrix $\mathbf{D}_N$ and the vector $\mathbf{x}$ of input samples, where $\mathbf{D}_N$ is given by Eq. (5.28).

(a) Show that the 8-point DIT FFT algorithm shown in Figure 11.21 is equivalent to expressing the DFT matrix as a product of four matrices as indicated below:

$$\mathbf{D}_N = \mathbf{V}_8 \mathbf{V}_4 \mathbf{V}_2 \mathbf{E}. \tag{11.140}$$

Determine the matrices given above and show that multiplication by each matrix $\mathbf{V}_k$, $k = 8, 4, 2$, requires at most eight complex multiplications.

(b) Since the DFT matrix $\mathbf{D}_N$ is its own transpose, i.e., $\mathbf{D}_N = \mathbf{D}_N^T$, another FFT algorithm is readily obtained by forming the transpose of the right-hand side of Eq. (11.140), resulting in a factorization of $\mathbf{D}_N$ given by

$$\mathbf{D}_N = \mathbf{E}^T \mathbf{V}_2^T \mathbf{V}_4^T \mathbf{V}_8^T. \tag{11.141}$$

Show that the flow-graph representation of the above factorization is precisely the 8-point DIF FFT algorithm of Figure 11.28.

11.32 The basic idea behind the split-radix FFT (SRFFT) algorithm is explored in this problem [Duh86]. As in the case of the decimation-in-frequency FFT algorithm, in the SRFFT algorithm, the even-indexed and the odd-indexed samples of the DFT are computed separately. For the computation of the even-indexed samples, we write, as in the DIF FFT approach,

$$X[2\ell] = \sum_{n=0}^{N-1} x[n] W_N^{2\ell n}$$

$$= \sum_{n=0}^{(N/2)-1} x[n] W_N^{2\ell n} + \sum_{n=N/2}^{N-1} x[n] W_N^{2\ell n}, \quad \ell = 0, 1, \ldots, \tfrac{N}{2} - 1. \tag{11.142}$$

Show that the above can be reexpressed as an $\frac{N}{2}$-point DFT in the form

$$X[2\ell] = \sum_{n=0}^{(N/2)-1} \left(x[n] + x\left[n + \tfrac{N}{2}\right] \right) W_{N/2}^{\ell n}, \quad \ell = 0, 1, \ldots, \tfrac{N}{2} - 1. \tag{11.143}$$

For the computation of the odd-indexed samples, we write two different expressions, depending on whether the frequency index k can be expressed as $4\ell + 1$ or $4\ell + 3$:

$$X[4\ell + 1] = \sum_{n=0}^{(N/4)-1} x[n]W_N^{(4\ell+1)n} + \sum_{n=\frac{N}{4}}^{(N/2)-1} x[n]W_N^{(4\ell+1)n}$$

$$+ \sum_{n=N/2}^{(3N/4)-1} x[n]W_N^{(4\ell+1)n} + \sum_{n=3N/4}^{N-1} x[n]W_N^{(4\ell+1)n},$$

$$\ell = 0, 1, \ldots, \frac{N}{4} - 1, \tag{11.144a}$$

$$X[4\ell + 3] = \sum_{n=0}^{(N/4)-1} x[n]W_N^{(4\ell+3)n} + \sum_{n=N/4}^{(N/2)-1} x[n]W_N^{(4\ell+3)n}$$

$$+ \sum_{n=N/2}^{(3N/4)-1} x[n]W_N^{(4\ell+3)n} + \sum_{n=3N/4}^{N-1} x[n]W_N^{(4\ell+3)n},$$

$$\ell = 0, 1, \ldots, \frac{N}{4} - 1. \tag{11.144b}$$

Show that Eqs. (11.144a) and (11.144b) can be rewritten as two $(N/4)$-point DFTs of the form

$$X[4\ell + 1] = \sum_{m=0}^{(N/4)-1} \left\{ \left(x[m] - x\left[m + \frac{N}{2}\right] \right) \right.$$

$$\left. - j\left(x\left[m + \frac{N}{4}\right] - x\left[m + \frac{3N}{4}\right] \right) \right\} W_N^m W_{N/4}^{\ell m}, \tag{11.145a}$$

$$X[4\ell + 3] = \sum_{m=0}^{(N/4)-1} \left\{ \left(x[m] - x\left[m + \frac{N}{2}\right] \right) \right.$$

$$\left. + j\left(x\left[m + \frac{N}{4}\right] - x\left[m + \frac{3N}{4}\right] \right) \right\} W_N^{3m} W_{N/4}^{\ell m}, \tag{11.145b}$$

where $\ell = 0, 1, \ldots, \frac{N}{4} - 1$. Sketch the flow-graph of a typical butterfly in the above algorithm for computing two even-numbered and two odd-numbered points, and show that the split-radix FFT algorithm requires only two complex multiplications per butterfly.

11.33 Develop the flow-graph for the computation of an 8-point DFT based on the split-radix algorithm described in Problem 11.32. What is the total number of real multiplications needed to implement this algorithm? How does this number compare with that required in a radix-2 DIF FFT algorithm? Ignore multiplications by ± 1 and $\pm j$.

11.34 Develop the flow-graph for the computation of a 16-point DFT based on the split-radix algorithm described in Problem 11.32. What is the total number of real multiplications needed to implement this algorithm? How does this number compare with that required in a radix-2 DIF FFT algorithm? Ignore multiplications by ± 1 and $\pm j$.

11.35 Develop the index mapping for implementing an N-point DFT $X[k]$ of a length-N sequence $x[n]$ using the Cooley–Tukey FFT algorithm for (a) $N = 12$, (b) $N = 15$, (c) $N = 21$, and (d) $N = 35$.

11.36 Develop the index mapping for implementing an N-point DFT $X[k]$ of a length-N sequence $x[n]$ using the prime factor algorithm for (a) $N = 12$, (b) $N = 15$, (c) $N = 21$, and (d) $N = 35$.

11.37 Consider the computation of a 12-point DFT $X[k]$ of a length-N sequence $x[n]$. The index mapping to be used can be either

$$n = \langle An_1 + Bn_2 \rangle_{12}, \quad k = \langle Ck_1 + Dk_2 \rangle_{12},$$

or

$$n = \langle Cn_1 + Dn_2 \rangle_{12}, \quad k = \langle Ak_1 + Bk_2 \rangle_{12},$$

where A, B, C, and D are appropriately chosen constants. Use of the mapping $g[n_1, n_2] = x[\langle An_1 + Bn_2 \rangle_{12}]$ results in

$$G[k_1, k_2] = \sum_{n_1=0}^{2} \sum_{n_2=0}^{3} g[n_1, n_2] W_4^{n_2 k_2} W_3^{n_1 k_1}, \tag{11.146}$$

$$X[\langle Ck_1 + Dk_2 \rangle_{12}] = G[k_1, k_2]. \tag{11.147}$$

If, on the other hand, the index mapping used is $h[n_1, n_2] = x[\langle Cn_1 + Dn_2 \rangle_{12}]$, we obtain

$$H[k_1, k_2] = \sum_{n_1=0}^{2} \sum_{n_2=0}^{3} h[n_1, n_2] W_4^{n_2 k_2} W_3^{n_1 k_1}, \tag{11.148}$$

$$Y[\langle Ak_1 + Bk_2 \rangle_{12}] = H[k_1, k_2]. \tag{11.149}$$

What is the relation between $X[k]$ and $Y[k]$?

11.38 Develop the flow-graph for the computation of an N-point DFT based on the prime factor algorithm for the following values of N: (a) $N = 6$, (b) $N = 10$, (c) $N = 12$, and (d) $N = 15$.

11.39 Develop a scheme to compute the 1536-point DFT of a sequence of length 1536 using 258-point FFT modules and complex multiplications and additions. Show the scheme in block-diagram form. How many FFT modules and complex multiplications and additions are needed for the overall computation?

11.40 A 512-point DFT of a length-498 sequence $x[n]$ is to be computed. How many zero-valued samples should be appended to $x[n]$ prior to the computation of the DFT? What are the total number of complex multiplications and additions needed for the direct evaluation of all DFT samples? What are the total number of complex multiplications and additions needed if a Cooley–Tukey type FFT is used to compute the DFT samples?

11.41 We wish to compute the L-point chirp-z transform (CZT) samples $X(z_\ell)$, $\ell = 0, 1, 2, \ldots, L-1$, of a length-$N$ sequence $x[n]$ according to Eq. (11.93), where $z_\ell = AV^{-\ell}$ with $A = A_o e^{j\theta_o}$ and $V = V_o e^{j\phi_o}$. What are the values of A_o, θ_o, V_o, and ϕ_o if the CZT needs to be calculated at points $\{z_\ell\}$ on the real axis in the z-plane such that $z_\ell = \alpha^\ell$, $0, \le \ell \le L-1$, for α real and $\alpha \ne \pm 1$?

11.42 Let $H(z) = \sum_{n=0}^{N-1} h[n]z^{-n}$ and $X(z) = \sum_{n=0}^{N-1} x[n]z^{-n}$ be two real polynomials of degree $(N-1)$. Their product $Y(z)$ is then a polynomial of degree $(2N-2)$ and a direct evaluation of $Y(z)$ requires N^2 multiplications and $N+1$ additions. Note also that the coefficients $y[n]$ of $Y(z)$ are the same as that obtained by a linear convolution of the two length-N sequences $h[n]$ and $x[n]$. The number of multiplications in computing the product $H(z)X(z)$ can be reduced but with an increase in the number of additions by making use of the Cook–Toom algorithm [?], [Knu69], which is studied in this problem.

Let z_k, $k = 0, 1, \ldots, 2N-1$, be $2N-1$ distinct points in the z-plane. Then $\hat{Y}[z_k] = H(z_k)X(z_k)$ represent the $(2N-1)$-point NDFT of $y[n]$, and from these $2N-1$ samples, we can uniquely determine $Y(z)$ using the Lagrange interpolation formula as discussed in Problem 6.29. By choosing the values of z_k appropriately, $\hat{Y}[z_k]$ can be evaluated with only $2N-1$ multiplications if we ignore multiplications (or divisions) by a power-of-2 integer that can be implemented using simple shifts in a binary representation.

(a) We first develop the Cook–Toom algorithm for $N = 2$. Here,

$$Y(z) = y[0] + y[1]z^{-1} + y[2]z^{-2} = \left(h[0] + h[1]z^{-1}\right)\left(x[0] + x[1]z^{--}\right).$$

Using the Lagrange interpolation formula, express $Y(z)$ in terms of its 3-point NDFT samples evaluated at $z_0 = -1$, $z_1 = \infty$, and $z_2 = +1$. Develop the expression for $y[0]$, $y[1]$, and $y[2]$ in terms of the parameters $X(z_k)$ and $H(z_k)$, and show that the computations of $\{y[n]\}$ require only a total of three multiplications. Determine the total number of additions required by this algorithm. Note that in many applications, $h[n]$ represents a fixed FIR filter, and, hence, the multiplications by integers needed to evaluate $y[n]$ can be included in the constants $H(z_k)$ to eliminate them from future computations.

(b) Develop the Cook–Toom algorithm for the linear convolution of two length-3 sequences.

11.43 Let $H(z) = h[0] + h[1]z^{-1}$ and $X(z) = x[0] + x[1]z^{-1}$. Show that $Y(z) = H(z)X(z) = y[0] + y[1]z^{-1} + y[2]z^{-2}$ can be written as [Jen91]

$$Y(z) = h[0]x[0] + [(h[0] + h[1])(x[0] + x[1]) - (h[0]x[0] + h[1]x[1])]z^{-1} + h[1]x[1]z^{-2}.$$

The evaluation of the product of the two first-order real polynomials thus can be carried out using three multiplications instead of four as required in the direct product. Equivalently, a linear convolution of two length-2 sequences can thus be implemented using only three multiplications.

11.44 Develop a multistage algorithm to compute the linear convolution of two length-N real sequences based on the scheme outlined in Problem 11.43 for the case when N is a power of 2 [Jen91]. What is the least number of multiplications required to compute the convolution using the multistage algorithm?

11.45 Let a 32-bit register be used to represent a floating-point number, with E bits assigned for the exponent and M bits plus a sign bit for the mantissa. Determine the approximate dynamic range of this floating-point representation for the following pairs of bit assignments for the exponent and the mantissa by evaluating the values of the smallest and the largest numbers that can be represented by the floating-point representation: (a) $E = 6$ and $M = 25$, (b) $E = 7$ and $M = 24$, and (c) $E = 8$ and $M = 23$. Determine the dynamic range of a 32-bit fixed-point representation of a signed integer. Show that the floating-point representation provides a larger dynamic range than the fixed-point representation.

11.46 Show that the range of a 32-bit floating-point number in the IEEE standard is from 1.18×10^{-38} to 3.4×10^{38}.

11.47 Show that the decimal equivalent of a positive or a negative binary fraction given by $s_\Delta a_{-1} a_{-2} \cdots a_{-b}$ in two's-complement form is $-s + \sum_{i=1}^{b} a_{-i} 2^{-i}$.

11.48 Show that the decimal equivalent of a positive or a negative binary fraction given by $s_\Delta a_{-1} a_{-2} \cdots a_{-b}$ in ones'-complement form is $-s(1 - 2^{-b}) + \sum_{i=1}^{b} a_{-i} 2^{-i}$.

11.49 Consider the binary addition of the following three numbers represented in two's-complement form with five bits: $\eta_1 = 0.78125_{10}$, $\eta_2 = 0.6875_{10}$, and $\eta_3 = -0.53125_{10}$. The addition is carried out in two steps: first we form the sum $\eta_1 + \eta_2$, and then we form the sum $(\eta_1 + \eta_2) + \eta_3$. Since the magnitudes of η_1 and η_2 are both greater than 0.5, their sum will lead to overflow, indicated by a 1 in the sign bit of the sum. Ignore this overflow by keeping all bits in the partial sum and add η_3. Show that the final sum is correct in spite of the overflow generated by the first addition.

11.50 The Taylor structure of Figure P11.6(a), developed for the realization of a Type 1 linear-phase FIR transfer function in Problem 8.17, has been proposed for designing tunable FIR filters [Cro76b], [Opp76]. From Eq. (8.128), observe that the zero-phase frequency response, also called the amplitude response, of a Type 1 FIR transfer function of length $2M + 1$ is given by

$$\breve{H}(\omega) = \sum_{n=0}^{M} a[n] (\cos \omega)^n. \tag{11.150}$$

Let $\hat{\omega}$ denote the angular frequency variable of the transformed FIR filter $\hat{H}(z)$. Show that a lowpass-to-lowpass transformation can be achieved by substituting

$$\cos \omega = \alpha + \beta \cos \hat{\omega} \qquad (11.151)$$

in Eq. (11.150). Show that this transformation can be implemented by replacing each block with a transfer function $(1 + z^{-2})/2$ in Figure P11.5 by a block with a transfer function

$$\alpha z^{-1} + \frac{\beta}{2}(1 + z^{-2}). \qquad (11.152)$$

Let ω_c and $\hat{\omega}_c$ denote, respectively, the cutoff frequency of the prototype filter and the desired cutoff frequency of the transformed filter. Show that if $\hat{\omega}_c < \omega_c$, it is convenient to choose $\beta = 1 - \alpha$, with $0 \le \alpha < 1$. Sketch the mapping from $\cos \omega_c$ to $\cos \hat{\omega}_c$ for this case. On the other hand, if $\hat{\omega}_c > \omega_c$, show that it is convenient to choose $\beta = 1 + \alpha$, with $-1 < \alpha \le 0$. Sketch the mapping from $\cos \omega_c$ to $\cos \hat{\omega}_c$ for this second case.

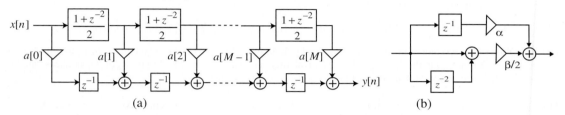

(a)   (b)

Figure P11.6

11.51 The *warped discrete Fourier transform* (WDFT) can be employed to determine the N frequency samples of the z-transform $X(z)$ of a length-N sequence $x[n]$ at a warped frequency scale on the unit circle. The N-point WDFT $\check{X}[k]$ of $x[n]$ is given by the N equally spaced frequency samples on the unit circle of the modified z-transform $X(\check{z})$, obtained by applying an allpass first-order spectral transformation to $X(z)$ [Mak2001]:

$$X(\check{z}) = X(z)\big|_{z^{-1} = \frac{-\alpha + \check{z}^{-1}}{1 - \alpha \check{z}^{-1}}} = \frac{P(\check{z})}{D(\check{z})}, \qquad (11.153)$$

where $|\alpha| < 1$. Thus, the N-point WDFT $\check{X}[k]$ of $x[n]$ is given by

$$\check{X}[k] = X(\check{z})\big|_{\check{z} = e^{j2\pi k/N}}, \quad 0 \le k \le N - 1. \qquad (11.154)$$

(a) Develop the expressions for $P(\check{z})$ and $D(\check{z})$.
(b) If we denote

$$P(\check{z}) = \sum_{n=0}^{N-1} p[n]\check{z}^{-n} \text{ and } D(\check{z}) = \sum_{n=0}^{N-1} d[n]\check{z}^{-n},$$

show that $\check{X}[k] = P[k]/D[k]$, where $P[k]$ and $D[k]$ are, respectively, the N-point DFTs of the sequences $p[n]$ and $d[n]$.

(c) If we denote $\mathbf{P} = [p[0]\ p[1]\ \cdots p[N-1]]^T$, and $\mathbf{X} = [x[0]\ x[1]\ \cdots x[N-1]]^T$, show that $\mathbf{P} = \mathbf{Q} \cdot \mathbf{X}$, where $\mathbf{Q} = [q_{r,s}]$ is a real $N \times N$ matrix, whose first row is given by $q_{0,s} = \alpha^s$, first column is given by $q_{r,0} = {}^{N-1}C_r \alpha^r$, and remaining elements $q_{r,s}$ can be derived using the recursion relation

$$q_{r,s} = q_{r-1,s-1} + \alpha q_{r,s-1} - \alpha q_{r-1,s}.$$

11.52 The transfer function of a third-order highpass elliptic IIR filter with a passband edge at $\omega_p = 0.6\pi$ is given by

$$H(z) = \frac{0.1676(1 - 1.5708z^{-1} + 1.5708z^{-2} - z^{-3})}{1 + 0.7048z^{-1} + 0.7608z^{-2} + 0.1941z^{-3}}.$$

Determine the transfer function of a highpass filter with a tunable passband edge frequency from the above prototype filter.

11.53 The transfer function of a third-order lowpass Butterworth IIR filter with a 3-dB cutoff frequency at 0.4π is given by

$$H(z) = \frac{0.0985(1 + 3z^{-1} + 3z^{-2} + z^{-3})}{1 - 0.5772z^{-1} + 0.4218z^{-2} - 0.0563z^{-3}}.$$

Determine the transfer function of a bandpass filter with a tunable center frequency from the above prototype filter.

11.15 MATLAB Exercises

M 11.1 Using Program 11_1, determine the first 30 samples of the impulse response coefficients of the following filters:
- (a) Fifth-order elliptic lowpass filter developed in Example 9.14,
- (b) Fourth-order Type 1 Chebyshev highpass filter developed in Example 9.15,
- (c) Eighth-order Butterworth bandpass filter developed in Example 9.16.

Program 11_1.m

Program 11_2.m

M 11.2 Modify Program 11_2 to demonstrate filtering of a sum of two sinusoidal sequences by an arbitrary IIR causal digital filter. The input data to the modified program should be the angular frequencies of the sinusoidal sequences and the numerator and denominator coefficients of the transfer function of the IIR digital filter. Apply the modified program to filter a sum of two sinusoidal sequences of angular frequencies 0.3π and 0.6π by the filter developed in Example 9.14, and verify its lowpass filtering property.

M 11.3 Apply the modified program developed in Problem M11.2 to filter a sum of two sinusoidal sequences of angular frequencies 0.3π and 0.6π by the filter developed in Example 9.15, and verify its highpass filtering property.

Program 11_3.m

M 11.4 Modify Program 11_3 to demonstrate the filtering of a sum of two sinusoidal sequences by an arbitrary IIR causal digital filter implemented in a cascade form. The individual sections in the cascade are either second-order or first-order with real coefficients. The input data to the modified program should be the angular frequencies of the sinusoidal sequences, the numerator and denominator coefficients of the transfer function of the individual sections in the cascade. Apply the modified program to filter a sum of two sinusoidal sequences of angular frequencies 0.3π and 0.6π by the filter developed in Example 9.14, and verify its lowpass filtering property. The output plot generated by the modified Program 11_3 should be identical to that generated in Problem M11.2.

M 11.5 Apply the modified program developed in Problem M11.4 to filter a sum of two sinusoidal sequences of angular frequencies 0.3π and 0.6π by the filter developed in Example 9.15, and verify its highpass filtering property. The output plot generated by the modified Program 8_3 should be identical to that generated in Problem M11.3.

direct2.m

M 11.6 Write a MATLAB program using the function direct2 to simulate the direct form II structure and demonstrate the filtering of input sequences. Apply this program to filter a sum of two sinusoidal sequences of angular frequencies 0.3π and 0.6π by the filter developed in Example 9.14, and verify its lowpass filtering property. The output plot generated by your new program should be identical to that generated in Problem M11.2.

M 11.7 Using the function goertzel for computing a single DFT sample by Goertzel's algorithm, write a MATLAB program to compute the N-point DFT of an arbitrary length-N sequence, and compare the DFT samples generated with those obtained using the M-file function fft. Verify the DFT computation of several sequences of lengths $N = 8, 12$, and 16.

M 11.8 Write a MATLAB program to verify the plots given in Figure 11.43.

M 11.9 Write a MATLAB program to verify the plots given in Figure 11.44.

M 11.10 A polynomial expansion suggested for the approximation of the sine of a number x is given by [Mar92]

$$\sin(\pi x) \cong 3.140625x + 0.02026367x^2 - 5.325196x^3 + 0.5446778x^4 + 1.800293x^5, \qquad (11.155)$$

where x is normalized by π, and its range is restricted to the first quadrant given by $0 < x < 0.5$. (*Note*: 0.5 is the normalized value of $\pi/2$.) Using MATLAB, compute and plot the values of $\sin(\pi x)$ given by Eq. (11.155) and the error due to the above approximation. Compare this approximation with that given by Eq. (11.134).

M 11.11 A polynomial expansion suggested for the approximation of the arctangent of a number x in the range $-1 \le x \le 1$ is given by [Mar92]

$$\tan^{-1}(x/\pi) \cong 0.318253x + 0.003314x^2 - 0.130908x^3 + 0.068542x^4 - 0.009159x^5. \qquad (11.156)$$

Using MATLAB, compute and plot the values of $\tan^{-1}(x)$ given by Eq. (11.156) and the error due to the above approximation. Compare this approximation with that given by Eq. (11.135).

12 Analysis of Finite Wordlength Effects

So far, we have assumed that we are dealing with discrete-time systems characterized by linear difference equations with constant coefficients, where both the coefficients and the signal variables have infinite precision taking any value between $-\infty$ and ∞. However, when implemented in either software form on a general-purpose computer or in special-purpose hardware form, the system parameters, along with the signal variables, can take only discrete values within a specified range since the registers of the digital machine where they are stored are of finite length. The discretization process results in nonlinear difference equations characterizing the discrete-time systems. These nonlinear equations, in principle, are almost impossible to analyze and deal with exactly. Fortunately, if the quantization amounts are small compared to the values of the signal variables and filter constants, a simpler approximate theory based on a statistical model can be applied, and it is possible to derive the effects of discretization and develop results that can be verified experimentally.

To illustrate the various sources of errors arising from the discretization process in the implementation of a digital filter, consider for simplicity, the causal first-order IIR digital filter of Figure 12.1, defined by the linear constant coefficient difference equation

$$y[n] = \alpha y[n-1] + x[n], \tag{12.1}$$

where $y[n]$ and $x[n]$ are the output and the input signal variables, respectively. The corresponding transfer function describing the above digital filter is given by

$$H(z) = \frac{1}{1 - \alpha z^{-1}} = \frac{z}{z - \alpha}, \qquad |z| > |\alpha|. \tag{12.2}$$

When implemented on a digital machine, the filter coefficient α can assume only certain discrete values $\hat{\alpha}$ and, in general, can only approximate the original design value of α. As a result, the actual transfer function implemented is given by

$$\hat{H}(z) = \frac{z}{z - \hat{\alpha}}, \tag{12.3}$$

which can be different from the desired transfer function $H(z)$ of Eq. (12.2). Therefore, the actual frequency response can be quite different from the desired frequency response. This coefficient quantization problem is similar to the sensitivity problem encountered in analog filter implementation.

If we assume that the input sequence $x[n]$ has been obtained by sampling an analog signal $x_a(t)$, it is discretized by the A/D converter being employed to convert the output of the sample-and-hold to digital samples. If we represent the output of the A/D converter as $\hat{x}[n]$, then the actual input to the digital filter of Figure 12.1 is given by

$$\hat{x}[n] = x[n] + e[n], \tag{12.4}$$

where $e[n]$ is the A/D conversion error generated by the input quantization process.

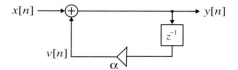

Figure 12.1: A first-order IIR digital filter.

The quantization of arithmetic operations leads to another source of errors. In the case of our simple digital filter of Eq. (12.1), the output of the multiplier $v[n]$ generated by multiplying the signal $y[n-1]$ with α,

$$v[n] = \alpha y[n-1], \tag{12.5}$$

is quantized to fit the register containing the product. The quantized signal $\hat{v}[n]$ can be represented as

$$\hat{v}[n] = v[n] + e_\alpha[n], \tag{12.6}$$

where $e_\alpha[n]$ is the error sequence generated by the product quantization process. The properties of this type of round-off error are somewhat similar to those of the A/D conversion error.

In addition to the above sources of errors, another type of error occurs in digital filters due to the nonlinearity caused by the quantization of arithmetic operations. These errors manifest themselves in the form of oscillations, called limit cycles, at the output of the filter, usually in the absence of input or sometimes in the presence of constant input signals or sinusoidal input signals. In this chapter, we analyze the effects of the above sources of quantization errors and then describe structures that are less sensitive to these effects.

12.1   The Quantization Process and Errors

There are two basic types of the binary representations of data, fixed-point and floating-point formats, as described in Section 11.8. In each of these formats, a negative number can be represented in one of three different forms. The arithmetic operations involved in digital signal processing are the addition (subtraction) and the multiplication operations, discussed in Section 11.9. Various problems can arise in the digital implementation of the arithmetic operations involving the binary data due to the finite wordlength limitations of the registers storing the numbers and the results of the arithmetic operations. For example, in fixed-point arithmetic, as demonstrated in Example 11.26, the product of two b-bit numbers is $2b$ bits long, which has to be quantized to b bits to fit the prescribed wordlength of the registers. Moreover, in fixed-point arithmetic, the addition operation can result in a sum exceeding the register wordlength, causing an overflow, as illustrated in Example 11.22. On the other hand, there is essentially no overflow in a floating-point addition. However, the results of both addition and multiplication may have to be quantized to fit the prescribed wordlength of the registers.

An analysis of the various quantization effects on the performance of a digital filter in practice depends on whether the numbers are in fixed-point or floating-point format, the type of representation for the negative numbers being used, the quantization method being employed to quantize the data, and the digital filter structure being used for implementation. Since the number of all possible combinations of the type of arithmetic, type of quantization method, and digital filter structure (of which there are literally thousands) is very large, we consider in this chapter analysis of quantization effects in some selected practical cases. However, the analysis presented can easily be extended to other cases.

We now describe the three different types of quantization that are generally used. As indicated in Section 11.8, it is a common practice in digital signal processing applications to represent data in a digital

Figure 12.2: A general $(b + 1)$-bit fixed-point fraction.

Figure 12.3: The quantization process model.

machine either as a fixed-point fraction or as a floating-point binary number with the mantissa as a binary fraction. We assume the available wordlength is $(b+1)$ bits, with the *most significant bit (MSB)* representing the sign of the number. Consider first the data to be a $(b + 1)$-bit fixed-point fraction with the binary point just to the right of the sign bit, as indicated in Figure 12.2. The smallest positive number that can be represented in this format will have a *least significant bit (LSB)* of 1, with the remaining bits being all 0's. Its decimal equivalent is 2^{-b}. Numbers represented with $(b + 1)$ bits are thus quantized in steps of 2^{-b}, which is called the *quantization step* or the *width of quantization*.

Before quantization, the wordlength is much larger than that indicated above. Assume that the original data x is represented as a $(\beta + 1)$-bit fraction with $\beta >> b$. To convert it into a $(b + 1)$-bit fraction, to be denoted as $Q(x)$, we can employ either *truncation* or *rounding*. In either case, the quantization process can be modeled as shown in Figure 12.3. Since the representation of a positive binary fraction is the same, independent of the format being used to represent the negative binary fraction, the effect of quantization of a positive fraction remains unchanged. However, the effect on negative fractions is not the same for the three different types of representations.

12.2 Quantization of Fixed-Point Numbers

To truncate a fixed-point number from $(\beta + 1)$ bits to $(b + 1)$ bits, we simply discard the least significant $(\beta - b)$ bits, as indicated in Figure 12.4. Let ε_t denote the truncation error defined by

$$\varepsilon_t = Q(x) - x. \tag{12.7}$$

For a positive number x, the magnitude of the number $Q(x)$ obtained after truncation is less than or equal to the magnitude of x. Therefore, $\varepsilon_t \leq 0$ for a positive x. The error ε_t is equal to zero if all bits being discarded are 0's and is largest if all bits being discarded are 1's. In the latter case, the decimal equivalent of the portion being discarded is equal to $2^{-b} - 2^{-\beta}$. Hence the range of the error ε_t in the case of truncation of a positive number x is given by

$$-(2^{-b} - 2^{-\beta}) \leq \varepsilon_t \leq 0. \tag{12.8}$$

For a negative number x, each one of the three different representations needs to be examined individually. For a negative fraction in sign-magnitude form, the magnitude of the truncated number $Q(x)$ is smaller than that of the unquantized negative number x. Thus, it follows from the definition of the quantization error ε_t given by Eq. (12.7) that here

$$0 \leq \varepsilon_t \leq 2^{-b} - 2^{-\beta}. \tag{12.9}$$

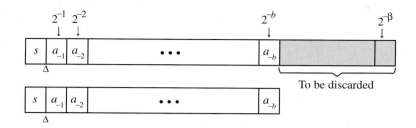

Figure 12.4: Illustration of the truncation operation.

For a negative fraction x in ones'-complement form $1_\triangle a_{-1}a_{-2}\ldots a_{-\beta}$, the numerical value is $-(1 - 2^{-\beta}) + \sum_{i=1}^{\beta} a_{-i}2^{-i}$. The numerical value of its quantized version $Q(x)$ is thus $-(1 - 2^{-b}) + \sum_{i=1}^{b} a_{-i}2^{-i}$, and hence, the error ε_t is given by

$$\varepsilon_t = Q(x) - x = -(1 - 2^{-b}) + \sum_{i=1}^{b} a_{-i}2^{-i} + (1 - 2^{-\beta}) - \sum_{i=1}^{\beta} a_{-i}2^{-i}$$

$$= (2^{-b} - 2^{-\beta}) - \sum_{i=b+1}^{\beta} a_{-i}2^{-i}. \tag{12.10}$$

The truncation error for this representation is always positive and has a range

$$0 \le \varepsilon_t \le 2^{-b} - 2^{-\beta}. \tag{12.11}$$

Now, consider a negative fraction x given in the two's-complement format $1_\triangle a_{-1}a_{-2}\ldots a_{-\beta}$. Its numerical value is given by $(-1 + \sum_{i=1}^{\beta} a_{-i}2^{-i})$. The representation of $Q(x)$, obtained after truncating x, is given by $1_\triangle a_{-1}a_{-2}\ldots a_{-b}$, with a numerical value $(-1 + \sum_{i=1}^{b} a_{-i}2^{-i})$. Therefore,

$$\varepsilon_t = Q(x) - x = \left(-1 + \sum_{i=1}^{b} a_{-i}2^{-i}\right) - \left(-1 + \sum_{i=1}^{\beta} a_{-i}2^{-i}\right) = -\sum_{i=b+1}^{\beta} a_{-i}2^{-i}. \tag{12.12}$$

The truncation error is here always negative and has a range

$$-(2^{-b} - 2^{-\beta}) \le \varepsilon_t \le 0. \tag{12.13}$$

In the case of rounding, the number is quantized to the nearest quantization level. We assume that a number exactly halfway between two quantization levels is rounded up to the nearest higher level. Therefore, if the bit $a_{-(b+1)}$ is 0, rounding is equivalent to truncation, and if this bit is 1, then 1 is added to the LSB position of the truncated number. It should be noted that the rounding error ε_r does not depend on the format being used to represent the negative fraction since the operation is solely based on the magnitude of the number. To determine the range of ε_r, we observe that the quantization step after rounding has a value 2^{-b}. The maximum rounding error ε_r therefore has a magnitude $(2^{-b})/2$. As a result, the range of ε_r is given by

$$-\tfrac{1}{2}(2^{-b} - 2^{-\beta}) < \varepsilon_r \le \tfrac{1}{2}(2^{-b} - 2^{-\beta}). \tag{12.14}$$

In practice, $\beta \gg b$. For example, the wordlength of a product is typically twice that of the numbers being multiplied. Hence, we can set $2^{-\beta} \cong 0$ in the inequalities of Eqs. (12.9), (12.11), (12.13), and

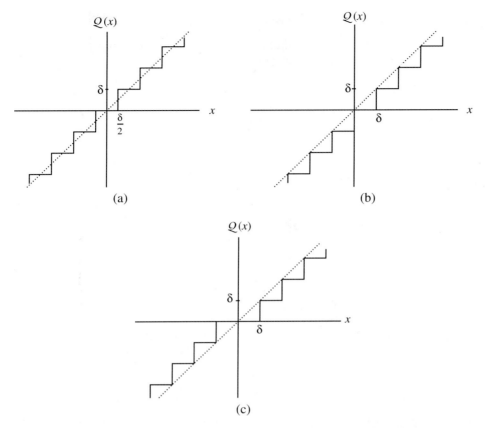

Figure 12.5: Input-output relationships of the quantizer: (a) rounding, (b) two's complement truncation, and (c) ones'-complement and sign-magnitude truncation. Here $\delta = 2^{-b}$.

(12.14), leading to the simpler inequalities shown in Table 12.1, where we have set the quantization step as $\delta = 2^{-b}$. Plots of the input–output characteristics of the quantizer for the three different representations and the two different quantization methods are shown in Figure 12.5.

12.3 Quantization of Floating-Point Numbers

In the case of floating-point numbers, quantization is carried out only on the mantissa. As a result, here it is more relevant to consider the relative error caused by the quantization process. To this end, we define the relative error ε in terms of the numerical values of quantized floating-point number $Q(x) = 2^E Q(M)$ and the unquantized number $x = 2^E M$ as

$$\varepsilon = \frac{Q(x) - x}{x} = \frac{Q(M) - M}{M}. \tag{12.15}$$

It can be shown (Problem 12.1) that the range of the relative errors for the different representations of a floating-point binary number are as indicated in Table 12.2, assuming $2^{-\beta} << 2^{-b}$.

Table 12.1: Range of quantization error.

Type of Quantization	Number Representation	Range of error $\mathcal{Q}(x) - x$
Truncation	Positive number Two's-complement negative number	$-\delta < \varepsilon_t \leq 0$
Truncation	Sign-magnitude negative number Ones'-complement negative number	$0 \leq \varepsilon_t < \delta$
Rounding	All positive and negative numbers	$-\frac{\delta}{2} < \varepsilon_r \leq \frac{\delta}{2}$

Note: $\delta = 2^{-b}$.

Table 12.2: Range of relative error $\varepsilon = (\mathcal{Q}(x) - x)/x$.

Type of Quantization	Number Representation	Range of Relative Error
Truncation	Two's-complement	$-2\delta < \varepsilon_t \leq 0, \quad x > 0$ $0 \leq \varepsilon_t < 2\delta, \qquad x < 0$
Truncation	Sign-magnitude Ones'-complement	$-2\delta < \varepsilon_t \leq 0$
Rounding	All numbers	$-\delta < \varepsilon_r \leq \delta$

Note: $\delta = 2^{-b}$.

A discussion of the analysis of quantization effects of digital filters implemented using floating-point arithmetic is beyond the scope of this text. Interested readers are referred to the publications listed at the end of this book [Kan71], [Liu69], [Opp75], [San67], [Wei69a], [Wei69b]. We consider here the fixed-point implementation case.

12.4   Analysis of Coefficient Quantization Effects

The effect of multiplier coefficient quantization on digital filters is similar to that observed in analog filters. The transfer function $\hat{H}(z)$ of the digital filter implemented in hardware or software form with quantized coefficients is different from the desired transfer function $H(z)$. The main effect of the coefficient quantization is therefore on the poles and zeros that move to different locations from the original desired locations. As a result, the actual frequency response $\hat{H}(e^{j\omega})$ is different from the desired frequency response $H(e^{j\omega})$ and may not be acceptable to the user. Moreover, the poles can move outside the unit circle, causing the implemented digital filter to become unstable even though the original transfer function with unquantized coefficients is stable, as illustrated earlier in Example 6.38. However, it can be shown using the results of Example 6.41 that the same transfer function, when implemented in a cascade form, remained stable after coefficient quantization.

12.4.1 Analysis Using MATLAB

Before we study the effect of the coefficient quantization on the performance of a digital filter analytically, it is instructive to investigate the effect on a computer using MATLAB.

Since MATLAB uses decimal numbers and arithmetic, to study the quantization effects on the digital filter implemented using binary numbers and arithmetic, we need to develop the decimal equivalents of the quantized representations of binary numbers and signals. The latter can be quantized using either truncation or rounding. The two functions, a2dT and a2dR,, develop the decimal equivalent beq of the binary representation of a vector d of decimal numbers with N bits for the magnitude part by truncation and rounding, respectively.

To illustrate the effect of coefficient quantization on the frequency response and the pole-zero locations of a digital filter, we need to evaluate these characteristics with both infinite and finite precision for the filter coefficients. Since MATLAB uses double-precision decimal numbers and arithmetic, the filter coefficients and signals generated using MATLAB can be considered to be of infinite precision for all practical purposes. To evaluate the effect of quantized binary representation, we can use either the function a2dT or the function a2dR given above to develop the decimal equivalent of the quantized binary numbers and signals.

a2dT.m
a2dR.m

We first illustrate the effect of coefficient quantization of an IIR digital filter implemented in direct form using Program 12_1. In it present form, the program evaluates the frequency response of a fifth-order elliptic lowpass digital filter with a cutoff at 0.4π, a passband ripple of 0.4 dB, and a minimum stopband attenuation of 50 dB. With simple modifications, the program can be used to study the effect on other types of filters with different specifications. For truncating the transfer function coefficients, this program uses the function a2dT.

Program 12_1.m

The above program uses the function plotzp to plot the locations of the poles and zeros of the transfer function, with the quantized numerator and denominator coefficients, bq and aq, respectively, using different markers.

plotzp.m

Figure 12.6(a) and (b) shows the gain response of the ideal filter with infinite precision coefficients (shown with a solid line) and the gain response obtained when the transfer function coefficients are truncated to 5-bit length (shown with a dashed line). It can be seen from this figure that the effect of the coefficient quantization is more severe around the bandedges with a higher passband ripple and a smaller transition band. The minimum stopband attenuation has also become smaller. Moreover, the transmission zeros closest to the stopband edge have moved closer to the passband edge.

Figure 12.6(c) shows the locations of the poles and zeros of the original elliptic lowpass filter transfer function with unquantized coefficients and of the transfer function of the elliptic filter implemented with quantized coefficients. As can be seen from this plot, coefficient quantization can cause substantial displacement of the poles and zeros from their desired nominal locations. In this example, the zero closest to the pole has moved the farthest from its original location and has moved closer to the new location of its nearest pole, which is now much closer to the unit circle.

It is of interest to compare the performance of the direct form realization of an IIR transfer function with that of a cascade realization when implemented with quantized coefficients. Program 12_2 can be used to evaluate the effect of quantization of the transfer function coefficients of each section of a cascade form realization of the above elliptic lowpass filter. However, this program can be easily modified to study the effect on other types of filters with different specifications.

Program 12_2.m

Figure 12.7 shows the fullband gain response and the passband gain response details of the ideal cascade for realization with infinite precision coefficients (shown with a solid line) and the gain response obtained when the transfer function coefficients of each section are truncated to 5-bit length (shown with a dashed line). It can be seen from this figure that the effect of the coefficient quantization here is not as severe as

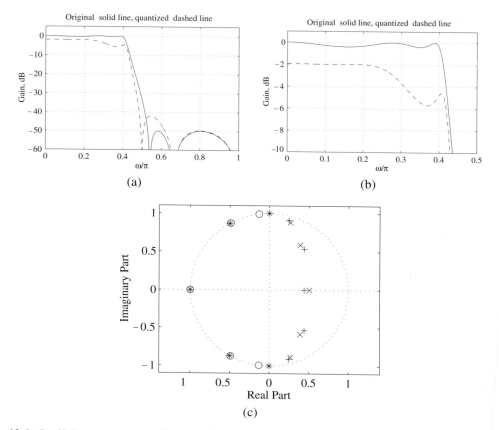

Figure 12.6: Coefficient quantization effects on a fifth-order IIR elliptic lowpass filter implemented in direct form:
(a) fullband gain responses with unquantized (shown with solid line) and quantized coefficients (shown with dashed
line), (b) passband details, and (c) pole-zero movements. Pole and zero locations of the filter with unquantized
coefficients denoted by "x" and "o", respectively, and pole and zero locations of the filter with quantized coefficients
denoted by "+" and "*", respectively.

Program 12_3.m

in the previous case. A flat loss has been added to the passband response with an increase in the passband
ripple. Here each complex zero-pair is realized by a second-order section, and hence, the zeros remain on
the unit circle. In fact, all of the zeros remain pretty much at their original locations. The overall effect on
the stopband response is minimal.

In general, a higher-order IIR transfer function should never be realized as a single direct form structure,
but instead should be realized as a cascade of second-order and first-order sections to minimize the effect
of coefficient quantization.

The above two programs can easily be modified to study the effect of coefficient quantization on the
performance of an FIR digital filter. Program 12_3 can be employed to study the effect of coefficient
quantization on the frequency response of a lowpass equiripple FIR digital filter implemented in direct
form.

Figure 12.8 shows the gain responses of the FIR filter generated by the above program. As can be
seen from this figure, the effect of coefficient quantization on an FIR filter implemented in direct form is
to reduce the passband width, increase the passband ripple, increase the transition band, and reduce the

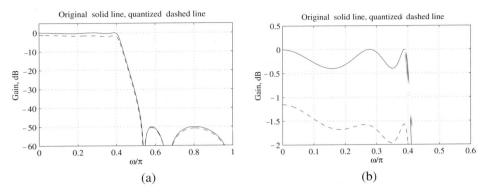

Figure 12.7: Coefficient quantization effects on a fifth-order IIR elliptic lowpass filter implemented in a cascade form: (a) fullband gain responses with unquantized (shown with solid line) and quantized coefficients (shown with dashed line), and (b) passband details.

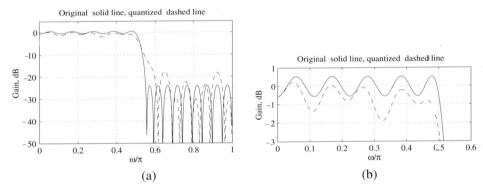

Figure 12.8: Coefficient quantization effects on a 39th-order FIR equiripple lowpass filter implemented in direct form: (a) fullband gain responses with unquantized (shown with solid line) and quantized coefficients (shown with dashed line), and (b) passband details.

minimum stopband attenuation.

12.4.2 Pole Sensitivity Analysis of Second-Order Structures

In many applications, a higher-order IIR digital transfer function is realized either in the form of a cascade of second-order sections or in the form of a parallel connection of second-order sections. It is thus of interest to study the sensitivity properties of the multiplier coefficients of a second-order digital filter structure. Since the poles of the transfer function are more critical in determining the frequency response of the filter, in this section we investigate the movement of poles caused by quantization of the multiplier coefficients. If a pole remains close to the original location after coefficient quantization, the structure is expected to exhibit low pole sensitivity. On the other hand, if the pole moves further away from the original location after coefficient quantization, the structure is expected to exhibit high pole sensitivity.

We restrict our attention here to two second-order digital filter structures and investigate their pole distributions when their respective multiplier coefficients take all possible quantized values within the specified wordlength. A similar procedure can be followed to investigate the pole sensitivity properties

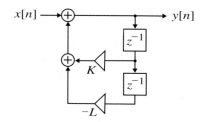

Figure 12.9: Direct form II realization of the second-order IIR transfer function of Eq. (12.16).

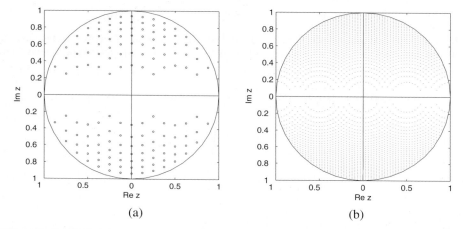

Figure 12.10: Pole distribution of a second-order direct form IIR structure with the multiplier coefficients in sign-magnitude format: (a) 4-bit wordlength and (b) 6-bit wordlength.

of other second-order digital filter structures. We shall demonstrate the validity of our conclusions in the following sections by carrying out an exact sensitivity analysis.

Consider first the second-order direct form II digital filter structure of Figure 12.9. Its transfer function is given by

$$H(z) = \frac{z^2}{z^2 - Kz + L}.$$ (12.16)

The two poles of the above transfer function are at

$$z_1 = re^{j\theta} = \frac{K + \sqrt{K^2 - 4L}}{2}, \qquad z_2 = re^{-j\theta} = \frac{K - \sqrt{K^2 - 4L}}{2}.$$ (12.17)

The pole distribution plot obtained with the multiplier coefficients K and L represented in a sign-magnitude format with a 4-bit wordlength (3 bits plus the sign bit) is shown in Figure 12.10(a). For example, if $K = 0_\Delta 101 = 0.625_{10}$ and $L = 0_\Delta 111 = 0.875_{10}$, from Eq. (12.17), the complex-conjugate pole-pair is at

$$z_1 = 0.3125 + j0.8817, \qquad z_2 = 0.3125 - j0.8817.$$

As can be seen from this plot, the poles appear close to each other around $z = \pm j$ and further apart around the real axis.

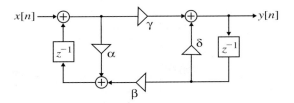

Figure 12.11: A second-order coupled form structure.

The pole distribution plot of the direct form II structure for a 6-bit wordlength (5 bits plus the sign bit) is shown in Figure 12.10(b), which clearly indicates a higher density of poles around $z = \pm j$ and a sparser density of poles around the real axis.

We can thus conclude that the direct form structure exhibits high pole sensitivity in the case of transfer functions with poles closer to the real axis and low pole sensitivity in the case of transfer functions with poles closer to $z = \pm j$. As a result, the direct form structure is not suitable for implementing lowpass or highpass transfer functions for which the poles are closer to the real axis.

Now consider the coupled-form structure of Figure 12.11, whose transfer function is given by

$$H(z) = \frac{\gamma z^2}{z^2 - (\alpha + \delta)z + (\alpha\delta - \beta\gamma)}. \tag{12.18}$$

If $\alpha = \delta = r\cos\theta$ and $\beta = -\gamma = r\sin\theta$, then the transfer function of Eq. (12.18) becomes

$$H(z) = \frac{\gamma z^2}{z^2 - 2r\cos\theta z + r^2}. \tag{12.19}$$

Comparing the denominator polynomial of Eq. (12.18) with that of Eq. (12.16), we observe

$$K = \alpha + \delta = 2\alpha, \tag{12.20a}$$
$$L = \alpha\delta - \beta\gamma = \alpha^2 + \beta^2. \tag{12.20b}$$

The pole distribution plot obtained for this structure with the multiplier coefficients α and β represented in a sign-magnitude format with a 4-bit wordlength (3 bits plus the sign bit) is shown in Figure 12.12(a). Figure 12.12(b) shows the pole distribution plot when the multiplier coefficients α and β are in a sign-magnitude format with a 6-bit wordlength (5 bits plus the sign bit). As can be seen from these two plots, in this case, all poles are distributed with a uniform density. Hence, the coupled form structure is more suitable for implementing any type of second-order transfer functions, particularly the lowpass and highpass functions.

12.4.3 Estimation of Pole-Zero Displacements

A measure of the coefficient quantization effects on the performance of a digital filter is given by the pole-zero displacements from their original positions. These displacements can be evaluated analytically as described next [Mit74c].

Consider an Nth-degree polynomial $B(z)$ with simple roots:

$$B(z) = \sum_{i=0}^{N} b_i z^i = \prod_{k=1}^{N} (z - z_k), \tag{12.21}$$

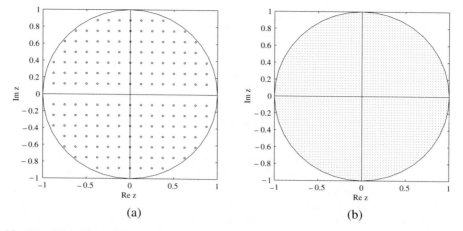

(a)   (b)

Figure 12.12: Pole distribution of a second-order coupled form IIR structure with multiplier coefficients in sign-magnitude format: (a) 4-bit wordlength and (b) 6-bit wordlength.

where $b_N = 1$. The roots z_k of $B(z)$ are given by

$$z_k = r_k e^{j\theta_k}. \tag{12.22}$$

Note that $B(z)$ can either be the denominator polynomial or the numerator polynomial of the digital transfer function. The effect of coefficient quantization is manifested by the change of the polynomial coefficients from b_i to $b_i + \Delta b_i$, and as a result, the polynomial $B(z)$ changes to a new polynomial $\hat{B}(z)$ given by

$$\hat{B}(z) = \sum_{i=0}^{N} (b_i + \Delta b_i)z^i = B(z) + \sum_{i=0}^{N-1} (\Delta b_i)z^i = \prod_{k=1}^{N} (z - \hat{z}_k), \tag{12.23}$$

with $\hat{z}_k$ denoting the roots of the polynomial $\hat{B}(z)$. Note that $\hat{z}_k$ are the new locations to which the roots z_k of $B(z)$ have moved. For small changes, $\hat{z}_k$ will be close to z_k and can be expressed as

$$\hat{z}_k = (r_k + \Delta r_k)e^{j(\theta_k + \Delta\theta_k)}, \tag{12.24}$$

where Δr_k and $\Delta\theta_k$ represent the changes in the radius and the angle of the kth root due to coefficient quantization. Our aim is to develop simple expressions for estimating Δr_k and $\Delta\theta_k$, knowing the changes Δb_i in the coefficients of the polynomial $B(z)$. If we assume the change Δb_i to be very small, the changes in the radius and the angle, Δr_k and $\Delta\theta_k$, of the kth root can be also considered to be very small, and we can rewrite the expression for $\hat{z}_k$ in Eq. (12.24) as

$$\hat{z}_k = (r_k + \Delta r_k)e^{j\Delta\theta_k}e^{j\theta_k} \cong (r_k + \Delta r_k)(1 + j\Delta\theta_k)e^{j\theta_k}$$
$$\cong r_k e^{j\theta_k} + (\Delta r_k + jr_k\Delta\theta_k)e^{j\theta_k}, \tag{12.25}$$

neglecting higher-order terms. The root displacement can now be expressed as

$$\hat{z}_k - z_k = \Delta z_k \cong (\Delta r_k + jr_k\Delta\theta_k)e^{j\theta_k}. \tag{12.26}$$

Now consider the rational function $1/B(z)$. Its partial-fraction expansion is given by

$$\frac{1}{B(z)} = \sum_{k=1}^{N} \frac{\rho_k}{z - z_k}, \tag{12.27}$$

where ρ_k is the residue of $1/B(z)$ at the pole $z = z_k$; that is,

$$\rho_k = \left.\frac{(z - z_k)}{B(z)}\right|_{z=z_k} = R_k + jX_k. \qquad (12.28)$$

If we assume that $\hat{z}_k$ is very close to z_k, then we can write

$$\frac{1}{B(\hat{z}_k)} \cong \frac{\rho_k}{\hat{z}_k - z_k}, \qquad (12.29)$$

or

$$\Delta z_k = \rho_k \cdot B(\hat{z}_k). \qquad (12.30)$$

However,

$$\hat{B}(\hat{z}_k) = 0 = B(\hat{z}_k) + \sum_{i=0}^{N-1} (\Delta b_i)(\hat{z}_k)^i. \qquad (12.31)$$

Therefore, from Eqs. (12.30) and (12.31), we arrive at

$$\Delta z_k = -\rho_k \left\{ \sum_{i=0}^{N-1} (\Delta b_i)(\hat{z}_k)^i \right\} \cong -\rho_k \left\{ \sum_{i=0}^{N-1} (\Delta b_i)(z_k)^i \right\}, \qquad (12.32)$$

assuming that $\hat{z}_k$ is very close to z_k. Rewriting Eq. (12.32), we obtain

$$(\Delta r_k + jr_k \Delta\theta_k)e^{j\theta_k} = -(R_k + jX_k)\left\{ \sum_{i=0}^{N-1} (\Delta b_i)(r_k e^{j\theta_k})^i \right\}. \qquad (12.33)$$

Equating real and imaginary parts of the above, we arrive at, after some algebra,

$$\Delta r_k = (-R_k \mathbf{P}_k + X_k \mathbf{Q}_k) \cdot \Delta\mathbf{B} = \mathbf{S}_b^{r_k} \cdot \Delta\mathbf{B}, \qquad (12.34a)$$

$$\Delta\theta_k = -\frac{1}{r_k}(X_k \mathbf{P}_k + R_k \mathbf{Q}_k) \cdot \Delta\mathbf{B} = \mathbf{S}_b^{\theta_k} \cdot \Delta\mathbf{B}, \qquad (12.34b)$$

where

$$\mathbf{P}_k = \left[\cos\theta_k \quad r_k \quad r_k^2 \cos\theta_k \cdots r_k^{N-1}\cos(N-2)\theta_k \right], \qquad (12.35a)$$

$$\mathbf{Q}_k = \left[-\sin\theta_k \quad 0 \quad r_k^2 \sin\theta_k \cdots r_k^{N-1}\sin(N-2)\theta_k \right], \qquad (12.35b)$$

$$\Delta\mathbf{B} = \left[\Delta b_0 \ \Delta b_1 \ \Delta b_2 \cdots \Delta b_{N-1} \right]^T. \qquad (12.35c)$$

It should be noted that the sensitivity vectors $\mathbf{S}_b^{r_k}$ and $\mathbf{S}_b^{\theta_k}$ depend on only $B(z)$ and are independent of $\Delta\mathbf{B}$. Hence, once these vectors have been calculated, pole-zero displacements for any sets of $\Delta\mathbf{B}$ can be rapidly calculated using Eqs. (12.34a) and (12.34b). Moreover, the elements of $\Delta\mathbf{B}$ are multiplier coefficient changes only for the direct form realization.

We illustrate in Example 12.1 the application of Eqs. (12.34a) and (12.34b) in computing the pole displacements of a second-order direct form IIR digital filter structure due to coefficient quantization.

EXAMPLE 12.1   Coefficient Sensitivity Analysis of a Second-Order Direct Form II Structure

Consider the second-order direct form II structure of Figure 12.9. From Eq. (12.16), the denominator polynomial of its transfer function is given by

$$B(z) = z^2 - Kz + L = z^2 - 2r\cos\theta z + r^2 = (z - z_1)(z - z_2), \tag{12.36}$$

where

$$z_1 = re^{j\theta}, \qquad z_2 = re^{-j\theta}. \tag{12.37}$$

From Eqs. (12.16) and (12.36), we get

$$K = 2r\cos\theta, \qquad L = r^2. \tag{12.38}$$

The residue ρ_1 of $1/B(z)$ at $z = z_1$ is given by

$$\rho_1 = \left.\frac{z - z_1}{B(z)}\right|_{z=z_1} = -\frac{j}{2r\sin\theta}. \tag{12.39}$$

Therefore, for the IIR filter of Figure 12.9, we have

$$\Delta\mathbf{B} = [\Delta L \quad -\Delta K]^T, \tag{12.40a}$$

$$\mathbf{Q}_1 = [-\sin\theta \quad 0], \tag{12.40b}$$

$$\mathbf{P}_1 = [\cos\theta \quad r]. \tag{12.40c}$$

Substituting Eqs. (12.40a) to (12.40c) in Eqs. (12.34a) and (12.34b), we arrive at

$$\Delta r = X_1\mathbf{Q}_1\Delta\mathbf{B} = \frac{1}{2r}\Delta L, \tag{12.41a}$$

$$\Delta\theta = -\frac{1}{r}(X_1\mathbf{P}_1\Delta\mathbf{B}) = \frac{\Delta L}{2r^2\tan\theta} - \frac{\Delta K}{2r\sin\theta}. \tag{12.41b}$$

It is evident from the above expressions that the second-order direct form IIR structure is highly sensitive to coefficient quantizations for transfer functions with poles close to $\theta = 0$ or π, that is, for narrow bandwidth lowpass and highpass filters, verifying the conclusions made in the previous section from the pole distribution plots of Figure 12.10.

We now extend the results of Eqs. (12.34a) and (12.34b) to an arbitrary structure with R multipliers given by α_k, $k = 1, 2, \ldots, R$. Due to coefficient quantization, these coefficients change to $\alpha_k + \Delta\alpha_k$. Since the coefficients b_i of the polynomial $B(z)$ are multilinear functions of the multiplier coefficients α_i, we can express the change Δb_i in the transfer function coefficient b_i to the change $\Delta\alpha_k$ in the multiplier coefficient α_k through

$$\Delta b_i = \sum_{k=1}^{R} \frac{\partial b_i}{\partial\alpha_k}\Delta\alpha_k, \qquad i = 0, 1, \ldots, N - 1. \tag{12.42}$$

In matrix form, the above can be written as

$$\Delta\mathbf{B} = \mathbf{C} \cdot \Delta\boldsymbol{\alpha}, \tag{12.43}$$

where

$$\mathbf{C} = \begin{bmatrix} \dfrac{\partial b_0}{\partial \alpha_1} & \dfrac{\partial b_0}{\partial \alpha_2} & \cdots & \dfrac{\partial b_0}{\partial \alpha_R} \\[2mm] \dfrac{\partial b_1}{\partial \alpha_1} & \dfrac{\partial b_1}{\partial \alpha_2} & \cdots & \dfrac{\partial b_1}{\partial \alpha_R} \\[2mm] \vdots & \vdots & \ddots & \vdots \\[2mm] \dfrac{\partial b_{N-1}}{\partial \alpha_1} & \dfrac{\partial b_{N-1}}{\partial \alpha_2} & \cdots & \dfrac{\partial b_{N-1}}{\partial \alpha_R} \end{bmatrix}, \tag{12.44a}$$

$$\Delta \boldsymbol{\alpha} = [\Delta \alpha_1 \quad \Delta \alpha_2 \cdots \Delta \alpha_R]^T. \tag{12.44b}$$

Substituting Eq. (12.43) in Eqs. (12.34a) and (12.34b), we arrive at the desired result:

$$\Delta r_k = \mathbf{S}_b^{r_k} \cdot \mathbf{C} \cdot \Delta \boldsymbol{\alpha}, \tag{12.45a}$$

$$\Delta \theta_k = \mathbf{S}_b^{\theta_k} \cdot \mathbf{C} \cdot \Delta \boldsymbol{\alpha}, \tag{12.45b}$$

where the sensitivity vectors are as given in Eqs. (12.34a) and (12.34b). It should be noted that $\mathbf{C}$ depends on the structure but has to be computed only once.

The application of Eqs. (12.45a) and (12.45b) in computing the pole displacements of a second-order IIR digital filter structure is treated in Example 12.2.

EXAMPLE 12.2   Coefficient Sensitivity Analysis of the Coupled-Form Structure

Consider the coupled-form structure [Gol69b] of Figure 12.11 for which the transfer function is given by Eq. (12.19). The relations between the multiplier coefficients α and β and the coefficients K and L of the direct form II structure are given by Eqs. (12.20a) and (12.20b), from which we arrive at

$$\begin{bmatrix} \Delta L \\ \Delta K \end{bmatrix} = \begin{bmatrix} 2r \cos \theta & 2r \sin \theta \\ 2 & 0 \end{bmatrix} \begin{bmatrix} \Delta \alpha \\ \Delta \beta \end{bmatrix}. \tag{12.46}$$

Making use of Eqs. (12.41a) and (12.41b) in the above, we then get

$$\begin{bmatrix} \Delta r \\ \Delta \theta \end{bmatrix} = \begin{bmatrix} \dfrac{1}{2r} & 0 \\[2mm] \dfrac{1}{2r^2 \tan \theta} & -\dfrac{1}{2r \sin \theta} \end{bmatrix} \begin{bmatrix} 2r \cos \theta & 2r \sin \theta \\ 2 & 0 \end{bmatrix} \begin{bmatrix} \Delta \alpha \\ \Delta \beta \end{bmatrix}$$

$$= \begin{bmatrix} \cos \theta & \sin \theta \\ -\dfrac{1}{r} \sin \theta & \dfrac{1}{r} \cos \theta \end{bmatrix} \begin{bmatrix} \Delta \alpha \\ \Delta \beta \end{bmatrix}. \tag{12.47}$$

As can be seen from the above, the coupled-form structure is less sensitive to multiplier coefficient quantization than the direct form structure supporting the conclusions made in the previous section from the pole distribution plots of Figure 12.12. Several other low-sensitivity second-order structures are considered in Problems 12.4 and 12.5.

12.4.4 Analysis of Coefficient Quantization Effects in FIR Filters

The analysis of the displacement of the roots of a polynomial due to coefficient quantization outlined in the previous section can of course be applied to an FIR transfer function to determine the sensitivity of its zeros to changes in the coefficients. A more meaningful analysis is obtained by examining the changes in the frequency response due to coefficient quantization as described next.

Consider an Nth-order FIR transfer function:

$$H(z) = \sum_{n=0}^{N} h[n] z^{-n}. \tag{12.48}$$

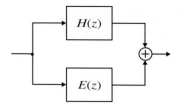

Figure 12.13: Model of the FIR filter with quantized coefficients.

Quantization of the filter coefficients results in a new transfer function

$$\hat{H}(z) = \sum_{n=0}^{N} \hat{h}[n]z^{-n} = \sum_{n=0}^{N} (h[n] + e[n]) z^{-n}, \tag{12.49}$$

which can be rewritten as

$$\hat{H}(z) = H(z) + E(z), \tag{12.50}$$

where

$$E(z) = \sum_{n=0}^{N} e[n]z^{-n}. \tag{12.51}$$

Thus, the FIR filter with quantized coefficients can be modeled as a parallel connection of two FIR filters, $H(z)$ and $E(z)$, as shown in Figure 12.13, where $H(z)$ represents the desired FIR filter with unquantized coefficients and $E(z)$ is the FIR filter representing the error in the transfer function due to coefficient quantization.

Without any loss of generality, assume the FIR filter $H(z)$ to be a Type 1 linear-phase filter of order N with impulse response $h[n]$. Hence, $E(z)$ is also a Type 1 linear-phase FIR transfer function. The frequency response of $H(z)$ can be expressed as

$$H(e^{j\omega}) = e^{-j\omega N/2} \left(h\left[\tfrac{N}{2}\right] + \sum_{n=0}^{(N-2)/2} 2h[n] \cos\left[\left(\tfrac{N}{2} - n\right)\omega\right] \right). \tag{12.52}$$

The frequency response of the actual FIR filter with quantized coefficients $\hat{h}[n]$ can be expressed as

$$\hat{H}(e^{j\omega}) = H(e^{j\omega}) + E(e^{j\omega}), \tag{12.53}$$

where $E(e^{j\omega})$ represents the error in the desired frequency response $H(e^{j\omega})$:

$$E(e^{j\omega}) = \sum_{n=0}^{N} e[n]e^{-j\omega n}, \tag{12.54}$$

with $e[n] = \hat{h}[n] - h[n]$. The error in the frequency response is thus bounded by

$$\left| E(e^{j\omega}) \right| = \left| \sum_{n=0}^{N} e[n]e^{-j\omega n} \right| \leq \sum_{n=0}^{N} |e[n]| \left| e^{-j\omega n} \right| \leq \sum_{n=0}^{N} |e[n]|. \tag{12.55}$$

Assume each impulse response coefficient $h[n]$ is a $(b + 1)$-bit signed fraction. In this case, the range of coefficient quantization error $e[n]$ is exactly the same as that indicated in Table 12.1. Using the data given in this table, an upper bound on $|E(e^{j\omega})|$ can be derived from Eq. (12.55). For example, for rounding, $|e[n]| \leq \delta/2$, where $\delta = 2^{-b}$ is the quantization step. As a result,

$$|E(e^{j\omega})| \leq \frac{(N + 1)\delta}{2}. \tag{12.56}$$

The above bound is rather conservative and can be reached only if all errors in Eq. (12.55) are of the same sign and have the maximum value in the range. A more realistic bound can be derived assuming $e[n]$ are statistically independent random variables [Cha73]. From Eqs. (12.52) to (12.54), we obtain

$$E(e^{j\omega}) = e^{-j\omega N/2} \left(e\left[\tfrac{N}{2}\right] + \sum_{n=0}^{(N-2)/2} 2e[n] \cos\left[\left(\tfrac{N}{2} - n\right)\omega\right] \right). \tag{12.57}$$

From the above, we observe that $E(e^{j\omega})$ is a sum of independent random variables. If we denote the variance of $e[n]$ as σ_e^2, then the variance of $E(e^{j\omega})$ is simply given by

$$\sigma_E^2(\omega) = \sigma_e^2 \left(1 + 4 \sum_{n=1}^{N/2} \cos^2(\omega n) \right) = \sigma_e^2 \left(N + \frac{\sin(N + 1)\omega}{\sin \omega} \right). \tag{12.58}$$

Using the notation

$$W_N(\omega) = \left[\frac{1}{2N + 1} \left(N + \frac{\sin(N + 1)\omega}{\sin \omega} \right) \right]^{1/2}, \tag{12.59}$$

the standard deviation of $E(e^{j\omega})$ can be expressed as

$$\sigma_E(\omega) = \sigma_e \left(\sqrt{2N + 1} \right) W_N(\omega). \tag{12.60}$$

For uniformly distributed $e[n]$, $\sigma_e = \delta/\sqrt{12} = 2^{-b-1}/\sqrt{3}$.

A plot of a typical weighting function $W_N(\omega)$ is sketched in Figure 12.14. In fact, it can be shown that $W_N(\omega)$ is in the range $(0, 1)$, and hence, the standard deviation $\sigma_E(\omega)$ is bounded by

$$\sigma_E(\omega) \leq \delta \sqrt{\frac{2N + 1}{12}}. \tag{12.61}$$

Based on the above bound, Chan and Rabiner [Cha73] have advanced a method to estimate the wordlength of the FIR filter coefficients to meet the prescribed filter specifications.

12.5 A/D Conversion Noise Analysis

In many applications, digital signal processing techniques are employed to process continuous-time (analog) band-limited signals that are either voltage or current waveforms. These analog signals must be converted into digital form before they can be processed digitally. According to the sampling theorem given in Section 4.2.1, an analog band-limited signal $x_a(t)$ can be represented uniquely by its sampled version $x_a(nT)$ if the sampling frequency $\Omega_T = 2\pi/T$ is greater than twice the highest frequency Ω_m contained in $x_a(t)$. In order to ensure that the sampling frequency chosen does satisfy this condition, an

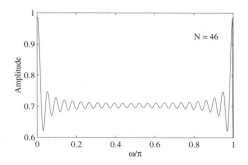

Figure 12.14: Plot of a typical $W_N(\omega)$.

Figure 12.15: Model of a practical A/D conversion system.

anti-aliasing filter is used to band-limit the analog signal to half of the sampling frequency. The discrete-time sequence $x_a(nT) = x[n]$ is then converted into a digital sequence for digital signal processing. As indicated by Figure 4.1, the conversion of an analog signal into a digital sequence is implemented in practice by a cascade of two devices, a sample-and-hold (S/H) circuit followed by an analog-to-digital (A/D) converter.

The digital samples produced by the A/D converter are usually represented in a binary form. As indicated in Section 11.8, there are several different forms of binary representations of which the two's-complement representation is usually employed in digital signal processing for convenience in the implementation of the arithmetic operations. Consequently, the A/D converters used for the digital signal processing of analog signals in general are those that employ the two's-complement fixed-point representation to represent the digital equivalent of the input analog signal. Moreover, for the processing of bipolar analog signals, the A/D converter generates a bipolar output represented as a fixed-point signed binary fraction.

12.5.1   Quantization Noise Model

The digital sample generated by the A/D converter is the binary representation of the quantized version of that produced by an ideal sampler with infinite precision. Because of the finite wordlength of the output register, the digital equivalent can take a value from a finite set of discrete values within the dynamic range of the register. For example, if the output word is of length $(b + 1)$ bits including the sign bit, the total number of discrete levels available for representation is 2^{b+1}. The dynamic range of the output register depends on the binary number representation selected for the A/D converter. The operation of a practical analog-to-digital conversion system consisting of a sample-and-hold circuit followed by an A/D converter therefore can be modelled as shown in Figure 12.15. The quantizer maps the input analog sample $x[n]$ into $\hat{x}[n]$, one of a set of discrete values, and the coder determines its binary equivalent $\hat{x}_{eq}[n]$ based on the binary representation scheme adopted by the A/D converter.

The quantization process employed by the quantizer in the A/D converter can be either rounding or truncation. Assuming rounding is used, the input–output characteristic of a 3-bit A/D converter with the output in two's-complement form is as shown in Figure 12.16. This figure also shows the binary equivalents of the quantized samples.

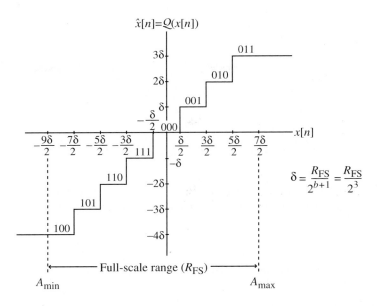

Figure 12.16: Input–output characteristic of a 3-bit bipolar A/D converter with two's-complement representation.

As indicated earlier and shown explicitly in Figure 12.16, for a two's-complement binary representation, the binary equivalent $\hat{x}_{eq}[n]$ of the quantized input analog sample $\hat{x}[n]$ is a binary fraction in the range

$$-1 \leq \hat{x}_{eq}[n] < 1. \tag{12.62}$$

It is related to the quantized sample $\hat{x}[n]$ through

$$\hat{x}_{eq}[n] = \frac{2\hat{x}[n]}{R_{FS}}, \tag{12.63}$$

where R_{FS} denotes the *full-scale range* of the A/D converter. We shall assume that the input signal has been scaled to be in the range of ± 1 by dividing its amplitude by $R_{FS}/2$, as is usually the case. Then the decimal equivalent of $\hat{x}_{eq}[n]$ is equal to $\hat{x}[n]$, and we will not differentiate between these two numbers.

For a $(b+1)$-bit bipolar A/D converter, the total number of quantization levels is 2^{b+1}, and the full-scale range R_{FS}, usually given in volts or amperes, is given by

$$R_{FS} = 2^{b+1}\delta, \tag{12.64}$$

where δ is the quantization step size, also called the quantization width, in volts or amperes. If the input signal is in the range given by Eq. (12.62), $R_{FS} = 2$, and then $\delta = 2^{-b}$. For the 3-bit A/D converter depicted in Figure 12.16, the total number of levels is $2^3 = 8$, and the full-scale range is $R_{FS} = 8\delta$, with a maximum value of $A_{max} = 7\delta/2$ and a minimum value of $A_{min} = -9\delta/2$. If the input analog sample $x_a(nT)$ is within the full-scale range,

$$-\frac{9\delta}{2} < x_a(nT) \leq \frac{7\delta}{2}, \tag{12.65}$$

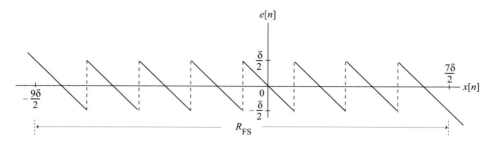

Figure 12.17: Quantization error as a function of the input for a 3-bit bipolar A/D converter.

it is quantized to one of the 8 discrete values indicated in Figure 12.16. In general, for a $(b + 1)$-bit wordlength A/D converter employing two's-complement representation, the full-scale range is given by

$$-(2^{b+1} + 1)\frac{\delta}{2} < x_a(nT) \le (2^{b+1} - 1)\frac{\delta}{2}. \tag{12.66}$$

If we denote the difference between the quantized value $\mathcal{Q}(x[n]) = \hat{x}[n]$ and the input sample $x_a(nT) = x[n]$ as the quantization error $e[n]$,

$$e[n] = \mathcal{Q}(x[n]) - x[n] = \hat{x}[n] - x[n], \tag{12.67}$$

it follows from Figure 12.16 that $e[n]$ is within the range

$$-\frac{\delta}{2} < e[n] \le \frac{\delta}{2}, \tag{12.68}$$

assuming that a sample exactly halfway between two levels is rounded up to the nearest higher level and assuming that the analog input is within the A/D converter full-scale range as given by Eq. (12.62). In this case, the quantization error $e[n]$, called *granular noise*, is bounded in magnitude according to Eq. (12.68). A plot of the quantization error $e[n]$ of the above 3-bit A/D converter as a function of the input sample $x[n]$ is given in Figure 12.17.

As can be seen from this figure, when the input analog sample is outside the A/D converter full-scale range, the magnitude of the error $e[n]$ increases linearly with an increase in the magnitude of the input. In the latter situation, the A/D converter error $e[n]$ is called the *saturation error* or the *overload noise* as the A/D converter output, independent of the actual value of the input, is "clipped" to the maximum value $(1 - 2^{-b})$ if the analog input is positive or the minimum value -1 if the analog input is negative. A clipping of the A/D converter output causes signal distortion with highly undesirable effects and must be avoided by scaling down the analog input $x_a(nT)$ to ensure that it remains within the A/D converter full-scale range.

In order to develop the necessary mathematical model for analyzing the effect of the finite wordlength of the A/D converter output, we assume that analog input samples are within the full-scale range, and as a result, there is no saturation error at the converter output. Since the input–output characteristic of an A/D converter is nonlinear and the analog input signal, in most practical cases, is not known a priori, it is reasonable to assume for analysis purposes that the quantization error $e[n]$ is a random signal and to use a statistical model of the quantizer operation as indicated in Figure 12.18. Furthermore, for simplified analysis, we make the following assumptions:

(a) The error sequence $\{e[n]\}$ is a sample sequence of a wide-sense stationary (WSS) white noise process, with each sample $e[n]$ being uniformly distributed over the range of the quantization error, as indicated in Figure 12.19, where δ is the quantization step.

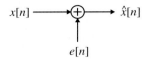

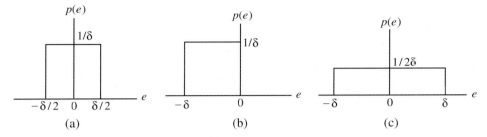

Figure 12.18: A statistical model of the A/D quantizer.

Figure 12.19: Quantization error probability density functions: (a) rounding, (b) two's-complement truncation, and (c) ones'-complement truncation.

(b) The error sequence is uncorrelated with its corresponding input sequence $\{x[n]\}$.

(c) The input sequence is a sample sequence of a stationary random process.

These assumptions hold in most practical situations for input signals whose samples are large and change in amplitude very rapidly in time relative to the quantization step (in a somewhat random fashion) and have been verified to be valid assumptions experimentally [Ben48], [Wid56], [Wid61] and by computer simulations [DeF88]. The statistical model also makes the analysis of A/D conversion noise more tractable, and results derived have been found to be useful for most applications. It should be pointed out that in the case of an A/D converter employing ones'-complement or sign-magnitude truncation, the quantization error is correlated to the input signal since here the sign of each error sample $e[n]$ is exactly opposite to the sign of the corresponding input sample $x[n]$. As a result, practical A/D converters use either rounding or two's-complement truncation. The mean and the variance of a general uniformly distributed random variable defined in Eq. (A.5) are given by Eqs. (A.7a) and (A.7c), respectively. Using these results, we observe that the mean and variance of the error sample in the case of rounding (Figure 12.19(a)) are given by

$$m_e = \frac{(\delta/2) - (\delta/2)}{2} = 0, \tag{12.69}$$

$$\sigma_e^2 = \frac{((\delta/2) - (-\delta/2))^2}{12} = \frac{\delta^2}{12}. \tag{12.70}$$

The corresponding parameters for the two's-complement truncation (Figure 12.19(b)) are as follows:

$$m_e = \frac{0 - \delta}{2} = -\frac{\delta}{2}, \tag{12.71}$$

$$\sigma_e^2 = \frac{(0 - \delta)^2}{12} = \frac{\delta^2}{12}. \tag{12.72}$$

12.5.2   Signal-to-Quantization Noise Ratio

Based on the model of Figure 12.17, we can evaluate the effect of additive quantization noise $e[n]$ on the input signal $x[n]$ by computing the *signal-to-quantization noise ratio* (SNR$_{A/D}$) in dB defined by

$$\text{SNR}_{A/D} = 10 \log_{10} \left(\frac{\sigma_x^2}{\sigma_e^2} \right) \text{dB}, \tag{12.73}$$

where σ_x^2 is the input signal variance representing the signal power and σ_e^2 is the noise variance representing the quantization noise power. For rounding, the quantization error is uniformly distributed in the range $(-\delta/2, \delta/2)$, and for two's-complement truncation, the quantization error is uniformly distributed in the range $(-\delta, 0)$, as indicated in Figure 12.18(a) and (b), respectively. In the case of a bipolar $(b + 1)$-bit A/D converter, $\delta = 2^{-(b+1)} R_{FS}$, and hence,

$$\sigma_e^2 = \frac{2^{-2b}(R_{FS})^2}{48}. \tag{12.74}$$

Substituting Eq. (12.74) in Eq. (12.73), we arrive at

$$\text{SNR}_{A/D} = 10 \log_{10} \left(\frac{48\sigma_x^2}{2^{-2b}(R_{FS})^2} \right)$$

$$= 6.02b + 16.81 - 20 \log_{10} \left(\frac{R_{FS}}{\sigma_x} \right) \text{dB}. \tag{12.75}$$

The above expression is used to determine the minimum A/D converter wordlength needed to meet a specified signal-to-quantization noise ratio. As can be seen from this expression, the SNR increases by approximately 6 dB for each bit added to the wordlength. For a given wordlength, the actual SNR depends on the last term in Eq. (12.75), which, in turn, depends on σ_x, the rms value of the input signal amplitude, and the full-scale range R_{FS} of the converter, as illustrated by Example 12.3.

EXAMPLE 12.3   Signal-to-Quantization Noise Ratio in the Digital Equivalent of an Analog Sample

Determine the signal-to-quantization noise ratio in the digital equivalent of an analog sample $x[n]$ with a zero-mean Gaussian distribution using a $(b + 1)$-bit A/D converter having a full-scale range $R_{FS} = K\sigma_x$. From Eq. (12.75), we obtain

$$\text{SNR}_{A/D} = 6.02\,b + 16.81 - 20 \log_{10}(K). \tag{12.76}$$

Table 12.3 shows the computed values of the SNR for various values of b and K.

Now, the probability of a particular input analog sample with a zero-mean Gaussian distribution staying within the full-scale range $K\sigma_x$ is given by [Par60]

$$2\Phi(K/2) - 1 = \sqrt{\frac{2}{\pi}} \int_0^{K/2} e^{-y^2/2} \, dy. \tag{12.77}$$

For example, for $K = 4$, the probability of an input analog sample staying within the full-scale range of $4\sigma_x$ is 0.9544. This implies that, on average, about 456 samples out of 10,000 samples will fall outside the range and be clipped. If we increase the full-scale range to $6\sigma_x$, the probability of an input analog sample staying within the expanded full-scale range increases to 0.9974, in which case, on average, about

Table 12.3: Signal-to-quantization noise ratio of an A/D converter as a function of wordlength and full-scale range.

	$b = 7$	$b = 9$	$b = 11$	$b = 13$	$b = 15$
$K = 4$	46.91	58.95	70.99	83.04	95.08
$K = 6$	43.39	55.43	67.47	79.51	91.56
$K = 8$	40.89	52.93	64.97	77.01	89.05

26 samples out of 10,000 samples will now be outside the range. In most applications, a full-scale range of $8\sigma_x$ is more than adequate to ensure no clipping in conversion.

12.5.3 Effect of Input Scaling on SNR

Now, consider the effect of scaling of the input on the SNR. Let the input scaling factor be A with $A > 0$. Since the variance of the scaled input $Ax[n]$ is $A^2\sigma_x^2$, the signal-to-quantization noise expression of Eq. (12.76) changes to

$$\mathrm{SNR}_{\mathrm{A/D}} = 6.02\,b + 16.81 - 20\log_{10}(K) + 20\log_{10}(A). \tag{12.78}$$

For a given b, the SNR can also be increased by scaling up the input analog signal by making $A > 1$. However, this process also increases the probability of some of the input analog samples being outside the full-scale range R_{FS}, and as a result, Eq. (12.78) no longer holds. Furthermore, the output is clipped, causing severe distortion in the digital representation of the analog input. On the other hand, a scaling down of the input analog signal by making $A < 1$ decreases the SNR. It is therefore necessary to ensure that the analog sample range matches as closely as possible the full-scale range of the A/D converter to get the maximum possible SNR without any signal distortion.

It should be noted here that the above analysis assumed an ideal A/D conversion. However, as pointed out in Section 4.8.5, a practical A/D converter is a nonideal device exhibiting a variety of errors, resulting in the actual signal-to-quantization noise ratio being smaller than that predicted by Eq. (12.75). Hence, the effective wordlength of the A/D converter, in general, is less than the wordlength computed using Eq. (12.75) by 1 to 2 bits. This factor should be taken into consideration in selecting an appropriate A/D converter for a given application.

12.5.4 Propagation of Input Quantization Noise to Digital Filter Output

In most applications, the quantized signal $\hat{x}[n]$ generated by the A/D converter is processed by a linear time-invariant discrete-time system $H(z)$. It is thus of interest to determine how the input quantization noise propagates to the output of the digital filter. In determining the noise at the filter output generated by the input noise, we can assume that the digital filter is implemented using infinite precision. However, as we shall point out later, in practice, the quantization of the arithmetic operations generates errors inside the digital filter structure, which also propagate to the output and appear as noise. These noise sources are assumed to be independent of the input quantization noise, and their effects can be analyzed separately and added to that due to the input noise.

As indicated in Figure 12.18, the quantized signal $\hat{x}[n]$ can be considered as a sum of two sequences: the unquantized input $x[n]$ and the input quantization noise $e[n]$. Because of the linearity property and the assumption that $x[n]$ and $e[n]$ are uncorrelated, the output $\hat{y}[n]$ of the LTI system can thus be expressed as a sum of two sequences: $y[n]$ generated by the unquantized input $x[n]$ and $v[n]$ generated by the error sequence $e[n]$, as shown in Figure 12.20. As a result, we can compute the output noise component $v[n]$ as a linear convolution of $e[n]$ with the impulse response sequence $h[n]$ of the LTI system:

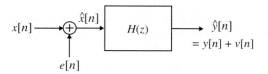

Figure 12.20: Model for the analysis of the effect of processing a quantized input by an LTI discrete-time system.

$$v[n] = \sum_{m=-\infty}^{\infty} e[m]h[n-m].$$ (12.79)

From Eq. (A.48), the mean m_v of the output noise $v[n]$ is given by

$$m_v = m_e H(e^{j0}),$$ (12.80)

and, from Eq. (A.67), its variance σ_v^2 is given by

$$\sigma_v^2 = \frac{\sigma_e^2}{2\pi} \int_{-\pi}^{\pi} \left| H(e^{j\omega}) \right|^2 d\omega.$$ (12.81)

The output noise power spectrum is given by

$$P_{vv}(\omega) = \sigma_e^2 \left| H(e^{j\omega}) \right|^2.$$ (12.82)

The normalized output noise variance is given by

$$\sigma_{v,n}^2 = \frac{\sigma_v^2}{\sigma_e^2} = \frac{1}{2\pi} \int_{-\pi}^{\pi} \left| H(e^{j\omega}) \right|^2 d\omega$$ (12.83a)

$$= \| H \|_2^2,$$ (12.83b)

where $\| H \|_2$ denotes the $\mathcal{L}_2$-norm of $H(e^{j\omega})$.[1]
    An equivalent representation of the normalized output noise variance of Eq. (12.83b) is given by

$$\sigma_{v,n}^2 = \frac{1}{2\pi j} \oint_C H(z)H(z^{-1})z^{-1} \, dz,$$ (12.83c)

where C is a counterclockwise contour in the ROC of $H(z)H(z^{-1})$. An alternate equivalent expression for Eq. (12.83c) in the time domain obtained using Eq. (A.69) is given by

$$\sigma_{v,n}^2 = \sum_{n=-\infty}^{\infty} |h[n]|^2.$$ (12.83d)

12.5.5   Algebraic Computation of Output Noise Variance

We now outline a simple algebraic approach for computing the normalized output noise variance using Eq. (12.83c) [Mit74b]. Several other algebraic approaches have been suggested for computing integrals of the form of Eq. (12.83c) [Chu95], [Dug80], [Dug82].

---

[1] See Eq. (3.53) for a definition of the $\mathcal{L}_2$-norm.

In general, $H(z)$ is a causal stable real rational function with all poles inside the unit circle in the z-plane. It can be expressed in a partial-fraction form as

$$H(z) = \sum_{i=1}^{R} H_i(z), \tag{12.84}$$

where $H_i(z)$ is a low-order real rational causal stable transfer function. Substituting the above in Eq. (12.83c), we arrive at

$$\sigma_{v,n}^2 = \frac{1}{2\pi j} \sum_{k=1}^{R} \sum_{\ell=1}^{R} \oint_C H_k(z) H_\ell(z^{-1}) z^{-1} \, dz. \tag{12.85}$$

Since $H_k(z)$ and $H_\ell(z)$ are stable transfer functions, it can be shown that

$$\oint_C H_k(z) H_\ell(z^{-1}) z^{-1} \, dz = \oint_C H_\ell(z) H_k(z^{-1}) z^{-1} \, dz. \tag{12.86}$$

As a result, Eq. (12.85) can be rewritten as

$$\sigma_{v,n}^2 = \frac{1}{2\pi j} \left\{ \sum_{k=1}^{R} \oint_C H_k(z) H_k(z^{-1}) z^{-1} \, dz \right.$$
$$\left. + 2 \sum_{k=1}^{R-1} \sum_{\ell=k+1}^{R} \oint_C H_k(z) H_\ell(z^{-1}) z^{-1} \, dz \right\}. \tag{12.87}$$

In most practical cases, $H(z)$ has only simple poles, with $H_k(z)$ being either a first-order or a second-order transfer function. As a result, each of the above contour integrations is much simpler to perform using Cauchy's residue theorem, and the results are tabulated for easy reference. Typical terms in the partial-fraction expansion of $H(z)$ are as follows:[2]

$$A, \qquad \frac{B_k}{z - a_k}, \qquad \frac{C_k z + D_k}{z^2 + b_k z + d_k}. \tag{12.88}$$

Let us denote a typical contour integral in Eq. (12.87) as

$$I_i = \frac{1}{2\pi j} \oint_C H_k(z) H_\ell(z^{-1}) z^{-1} \, dz. \tag{12.89}$$

The expressions obtained after performing the contour integrations for different I_i are listed in Table 12.4.

EXAMPLE 12.4   Output Noise Variance of a First-Order Digital Filter Due to A/D Quantization Noise

Let us determine the variance of the noise generated by the A/D quantization noise at the output of a first-order digital filter with a transfer function

---

[2]It should be noted that $H(z)$ is expressed as a ratio of polynomials in z before a direct partial-fraction expansion is carried out to arrive at the terms given in Eq. (12.88).

$$H(z) = \frac{1}{1 - \alpha z^{-1}}. \tag{12.90}$$

We rewrite $H(z)$ as

$$H(z) = \frac{z}{z - \alpha},$$

whose partial-fraction expansion is given by

$$H(z) = 1 + \frac{\alpha}{z - \alpha}. \tag{12.91}$$

The two terms in the partial-fraction expansion are

$$H_1(z) = 1, \qquad H_2(z) = \frac{\alpha}{z - \alpha}.$$

Therefore, using the results of Table 12.4, we finally obtain the expression for the normalized output noise variance as

$$\sigma_{v,n}^2 = 1^2 + \frac{\alpha^2}{1 - \alpha^2} = \frac{1}{1 - \alpha^2}. \tag{12.92}$$

If the pole is close to the unit circle, we can write $|\alpha| = 1 - \varepsilon$, where $\varepsilon \cong 0$. In which case, we can rewrite Eq. (12.92) as

$$\sigma_{v,n}^2 = \frac{1}{1 - (1 - \varepsilon)^2} \cong \frac{1}{2\varepsilon}, \tag{12.93}$$

indicating that as the pole gets closer to the unit circle, the output noise increases rapidly to very high values approaching infinity. Thus, for high Q realizations, the wordlengths of the registers storing the signal variables should be of longer length to keep the output round-off noise below a prescribed level.

The exact value of the output noise variance can be obtained from Eq. (12.92) by multiplying it with the variance of the A/D conversion noise σ_e^2.

12.5.6 Computation of Output Noise Variance Using MATLAB

The algebraic method of calculating the output noise variance developed in the previous section can be carried out much more easily using MATLAB. To this end, it is more convenient to develop a partial-fraction expansion of the real coefficient transfer function $H(z)$ using the function `residue`, which results in terms of the form A and $B_k/(z - a_k)$ only, where the residues B_k and the poles a_k are either real or complex numbers. Thus, for the variance calculation, only the terms I_1 and I_2 of Table 12.4 are employed. Program 12_4 is based on the approach of Section 12.5.5. We illustrate its use in Example 12.5.

Program 12_4.m

EXAMPLE 12.5   Computation of the Output Noise Variance of a Digital Filter Using MATLAB

Using Program 12_4, determine the output noise variance due to the input quantization of a causal fourth-order elliptic lowpass digital filter with a transfer function given by

$$H(z) = \frac{\begin{matrix} 0.06891875 + 0.13808186z^{-1} + 0.18636107z^{-2} \\ + 0.13808186z^{-3} + 0.06891875z^{-4} \end{matrix}}{\begin{matrix} 1 - 1.30613249z^{-1} + 1.48301305z^{-2} \\ - 0.77709026z^{-3} + 0.2361457z^{-4} \end{matrix}}. \tag{12.94}$$

The output generated by the program is

```
Output Noise Variance = 0.40263012267534
```

Table 12.4: Expressions for typical contour integrals in the output noise variance calculation.

$H_k(z)$	A	$\dfrac{B_\ell}{z^{-1} - a_\ell}$	$\dfrac{C_\ell z^{-1} + D_\ell}{z^{-2} + b_\ell z^{-1} + d_\ell}$
			$H_\ell(z^{-1})$
A	I_1	0	0
$\dfrac{B_k}{z - a_k}$	0	I_2	I_4'
$\dfrac{C_k z + D_k}{z^2 + b_k z + d_k}$	0	I_4	I_3

$I_1 = A^2$

$I_2 = \dfrac{B_k B_\ell}{1 - a_k a_\ell}$

$I_3 = \dfrac{(C_k C_\ell + D_k D_\ell)(1 - d_k d_\ell) - (D_k C_\ell - C_k D_\ell d_k)b_\ell - (C_k D_\ell - D_k C_\ell d_\ell)b_k}{(1 - d_k d_\ell)^2 + d_k b_\ell^2 + d_\ell b_k^2 - (1 + d_k d_\ell)b_k b_\ell}$

$I_4 = \dfrac{B_\ell(C_k + D_k a_\ell)}{1 + b_k a_\ell + d_k a_\ell^2}$

$I_4' = \dfrac{B_k(C_\ell + D_\ell a_k)}{1 + b_\ell a_k + d_\ell a_k^2}$

The normalized output noise variance can also be computed in MATLAB, using the M-file `filternorm`, described earlier in Section 3.2.5 (Exercise M12.4).

An alternative, fairly simple and straightforward computer-based method for the computation of the approximate value of the normalized output noise makes use of its equivalent expression given by Eq. (12.83d). For a causal and stable digital filter, the impulse response decays rapidly to zero values, and hence, Eq. (12.83d) can be approximated as a finite sum

$$\sigma_{v,n}^2 = S_L \cong \sum_{n=0}^{L} |h[n]|^2. \tag{12.95}$$

To determine $\sigma_{v,n}^2$, we can iteratively compute the above partial sum for $L = 1, 2, \ldots,$ and stop the computation when the difference $S_L - S_{L-1}$ becomes smaller than a specified value κ, which is typically chosen as 10^{-8}.

Example 12.6 illustrates this approach.

Program 12_5.m

EXAMPLE 12.6 **Computation of the Approximate Value of the Normalized Output Noise**

We determine again the normalized round-off noise variance of the fourth-order digital filter of Example 12.5 due to input quantization. To this end, we make use of Program 12_5. The normalized output noise variance of the transfer function of Eq. (12.94) obtained using this program is

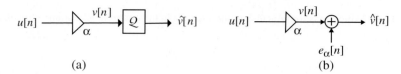

Figure 12.21: (a) Product quantization process and (b) its statistical model for the product round-off error analysis.

```
            Output Noise Variance = 0.40254346745459
```

The approximate value computed using the method of Eq. (12.95) is seen to be close to the actual value computed in Example 12.5.

12.6 Analysis of Arithmetic Round-Off Errors

As illustrated earlier in Section 11.9.2, in the fixed-point implementation of a digital filter, the result of only the multiplication operation is quantized. In this section, we develop the tools for the analysis of product round-off errors. Figure 12.21(a) shows the representation of a practical multiplier with the quantizer at its output. Its statistical model is indicated in Figure 12.21(b), which will be used to develop the error analysis methods. Here, the output $v[n]$ of the ideal multiplier is quantized to a value $\hat{v}[n]$, where $\hat{v}[n] = v[n] + e_\alpha[n]$. For analysis purposes, we again make assumptions similar to those made for the A/D conversion error analysis; that is,

(a) the error sequence $\{e_\alpha[n]\}$ is a sample sequence of a stationary white noise process, with each sample $e_\alpha[n]$ being uniformly distributed over the range of the quantization error;

(b) the quantization error sequence $\{e_\alpha[n]\}$ is uncorrelated with the sequence $\{v[n]\}$, the input sequence $\{x[n]\}$ to the digital filter, and all other quantization noise sources.

Recall that the assumption of $\{e_\alpha[n]\}$ being uncorrelated with $\{v[n]\}$ holds only for rounding and two's-complement truncation. The range of the error sample $e_\alpha[n]$ for these two cases are as given in Table 12.1. The mean and variance of the error sample for rounding are given by Eqs. (12.69) and (12.70), respectively, while those for the two's-complement truncation are given by Eqs. (12.71) and (12.72), respectively.

Using the above model for each multiplier, the representation of a digital filter to determine the effect of product quantizations at the output of the digital filter is as indicated in Figure 12.22(a), which explicitly shows the ℓth adder with an output $v_\ell[n]$ summing the quantized outputs of the k_ℓ multipliers at its input. This figure also shows the internal rth branch node associated with the signal variable $u_r[n]$ that needs to be scaled to prevent overflow at this node. These nodes are typically the inputs to the multipliers, as indicated in Figure 12.23. In digital filters employing two's-complement arithmetic, these nodes are outputs of adders forming sums of products, since here the sums will still have the correct values even though some of the products and/or the partial sums overflow (Problem 11.49). If we assume the error sources are statistically independent of each other, then each error source develops a round-off noise at the output of the digital filter. An equivalent statistical model is then as shown in Figure 12.22(b).

Let the impulse response from the digital filter input to the rth branch node be denoted as $f_r[n]$ and the impulse response from the input of the ℓth adder to the digital filter output be denoted as $g_\ell[n]$, with their corresponding z-transforms denoted by $F_r(z)$ and $G_\ell(z)$, respectively. $F_r(z)$ is called the *scaling transfer function* and plays a role in the dynamic range scaling schemes employed in fixed-point digital

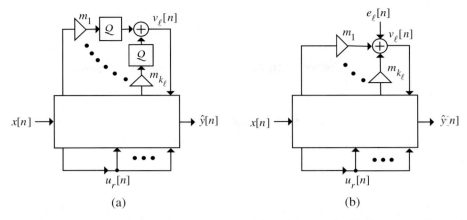

Figure 12.22: (a) Representation of a digital filter structure with product round-off before summation and (b) its statistical model.

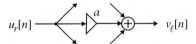

Figure 12.23: A typical multiplier branch with input as a branch node and output feeding into an adder.

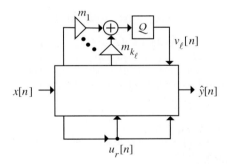

Figure 12.24: Digital filter structure with product round-off after summation.

filter structures, to be discussed in Section 12.7. $G_\ell(z)$ is called the *noise transfer function*, which is used in computing the noise power at the filter output due to product round-off as described next.

If σ_0^2 denotes the variance of each individual noise source at the output of each multiplier, the variance of $e_\ell[n]$ in Figure 12.22(b) is simply $k_\ell \sigma_0^2$ since we have assumed each noise source to be statistically independent of all others. The variance of the output noise caused by $e_\ell[n]$ is then given by

$$\sigma_0^2 \left[k_\ell \left(\frac{1}{2\pi j} \oint_C G_\ell(z) G_\ell(z^{-1}) z^{-1} dz \right) \right] = \sigma_0^2 \left[k_\ell \left(\frac{1}{2\pi} \int_{-\pi}^{\pi} \left| G_\ell(e^{j\omega}) \right|^2 d\omega \right) \right]. \qquad (12.96)$$

If there are L such adders in the digital filter structure, the total output noise power due to all product round-offs is given by

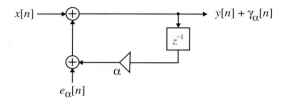

Figure 12.25: Model for the arithmetic round-off error analysis of Figure 12.1.

$$\sigma_\gamma^2 = \sigma_0^2 \sum_{\ell=1}^{L} k_\ell \left(\frac{1}{2\pi j} \oint_C G_\ell(z) G_\ell(z^{-1}) z^{-1} \, dz \right). \tag{12.97}$$

In many hardware implementation schemes, such as those employing DSP chips, the multiplication operation is carried out as a multiply–add operation with the result stored in a double-precision register. In such cases, if the signal variable being generated by a sum of product operations, the quantization operation can be carried out after all the multiply–add operations have been completed, reducing the number of quantization error sources to one for each such sum of product operations, as indicated in Figure 12.24. In such a case, the statistical model of Figure 12.22(b) can still be used, except the variance of the noise source $e_\ell[n]$ is now σ_0^2, thus resulting in considerably lower noise at the digital filter output.

We illustrate the product round-off error analysis method for two simple digital filter structures in Examples 12.7 and 12.8.

EXAMPLE 12.7   Output Noise Variance of a First-Order IIR Digital Filter Due to Product Round-Off

Consider first the filter of Figure 12.1. The model for the arithmetic round-off error analysis of this structure is as given in Figure 12.25, where the output is now given by a sum of two sequences, the desired output $y[n]$ generated by the input sequence $x[n]$ and the round-off noise $\gamma_\alpha[n]$ generated by the error sequence $e_\alpha[n]$. It follows from the figure that the noise transfer function $G_\alpha(z)$ is the same as the filter transfer function $H(z)$ as given by Eq. (12.2),

$$G_\alpha(z) = \frac{z}{z - \alpha}, \tag{12.98}$$

and thus the normalized output noise variance due to product round-off is the same as that due to input quantization given in Eq. (12.92). The actual output noise variance is

$$\sigma_\gamma^2 = \frac{\sigma_0^2}{1 - \alpha^2}, \tag{12.99}$$

where σ_0^2 is the variance of the error sequence $e_\alpha[n]$. The quantity $\sigma_\gamma^2 / \sigma_0^2$ is called the *noise gain* or the *normalized round-off noise variance* and is denoted as $\sigma_{\gamma,n}^2$.

EXAMPLE 12.8   Output Noise Variance of a Second-Order IIR Digital Filter Due to Product Round-Off

Consider next the causal second-order digital filter structure of Figure 12.26(a). Its transfer function is given by

$$H(z) = \frac{Y(z)}{X(z)} = \frac{z^2}{z^2 + \alpha_1 z + \alpha_2}. \tag{12.100}$$

A model for the analysis of the product round-off error is as indicated in Figure 12.26(b), where we have assumed that each product is rounded first before addition. It follows from this figure that the two noise transfer functions $G_{\alpha_1}(z)$ and $G_{\alpha_2}(z)$ are the same as the transfer function $H(z)$ of the digital filter given by Eq. (12.100). Thus, the total normalized variance $\sigma_{\gamma,n}^2$ of the output noise generated by the error sources $e_{\alpha_1}[n]$ and $e_{\alpha_2}[n]$ is given by

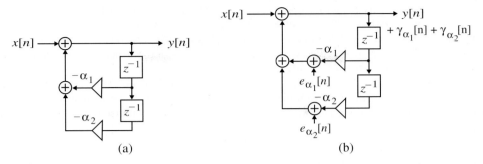

Figure 12.26: (a) A second-order digital filter structure, and (b) its model for product round-off error analysis.

$$\sigma_{\gamma,n}^2 = 2\left[\frac{1}{2\pi j}\oint_C H(z)H(z^{-1})z^{-1}\,dz\right], \tag{12.101}$$

where $H(z)$ is as given in Eq. (12.100).

Now a direct partial-fraction expansion of $H(z)$ is given by

$$H(z) = 1 + \frac{-\alpha_1 z - \alpha_2}{z^2 + \alpha_1 z + \alpha_2}. \tag{12.102}$$

Applying the technique of Section 12.5.5, we arrive at the expression for the normalized output noise variance,

$$\sigma_{\gamma,n}^2 = 2\left[1 + \frac{(\alpha_1^2 + \alpha_2^2)(1 - \alpha_2^2) - 2(\alpha_1\alpha_2 - \alpha_1\alpha_2^2)\alpha_1}{(1 - \alpha_2^2)^2 + 2\alpha_2\alpha_1^2 - (1 + \alpha_2^2)\alpha_1^2}\right], \tag{12.103}$$

which, after some algebra, simplifies to

$$\sigma_{\gamma,n}^2 = 2\left(\frac{1 + \alpha_2}{1 - \alpha_2}\right)\left(\frac{1}{1 + \alpha_2^2 + 2\alpha_2 - \alpha_1^2}\right). \tag{12.104}$$

It is instructive to express the multiplier coefficients in terms of the pole positions $re^{\pm j\theta}$, in the above expression. In this case, $\alpha_1 = -2r\cos\theta$ and $\alpha_2 = r^2$. Substituting these values in Eq. (12.104), we arrive, after some simple manipulations, at

$$\sigma_{\gamma,n}^2 = 2\left(\frac{1 + r^2}{1 - r^2}\right)\left(\frac{1}{1 + r^4 - 2r^2\cos 2\theta}\right). \tag{12.105}$$

If the pole is close to the unit circle, we can replace r in Eq. (12.105) with $1 - \varepsilon$, where ε is a very small positive number. This leads to

$$\sigma_{\gamma,n}^2 \cong \frac{1}{2\sin^2\theta}\left(\frac{1 - \varepsilon}{\varepsilon}\right)\left(\frac{1}{1 - 2\varepsilon}\right). \tag{12.106}$$

It can be seen that as the poles get closer to the unit circle, $\varepsilon \to 0$, and as a result, the total output noise variance increases rapidly.

12.7 Dynamic Range Scaling

In a digital filter implemented using fixed-point arithmetic, overflow can occur at certain internal nodes, such as inputs to multipliers and/or the adder outputs that can lead to large amplitude oscillations at the filter output, causing unsatisfactory operation, as discussed in Section 12.11.2. The probability of overflow can be minimized significantly by properly scaling the internal signal levels with the aid of scaling multipliers inserted at appropriate points in the digital filter structure. In many cases, most of these scaling multipliers can be absorbed with existing multipliers in the structure, thus reducing the total number needed to implement the scaled filter.

To understand the basic concepts involved in scaling, consider again the digital filter structure of Figure 12.22, showing explicitly the rth node variable $u_r[n]$ that needs to be scaled. We assume that all fixed-point numbers are represented as binary fractions and the input sequence of the filter is bounded by unity; that is,

$$|x[n]| \leq 1, \qquad \text{for all values of } n. \tag{12.107}$$

The objective of scaling is to ensure that

$$|u_r[n]| \leq 1, \qquad \text{for all } r \text{ and for all values of } n. \tag{12.108}$$

We now derive three different conditions to ensure that $u_r[n]$ satisfies the above bound.

12.7.1 An Absolute Bound

The inverse z-transform of the scaling transfer function $F_r(z)$ is the impulse response $f_r[n]$ from the filter input to the rth node. Therefore, $u_r[n]$ can be expressed as the linear convolution of $f_r[n]$ and the input $x[n]$:

$$u_r[n] = \sum_{k=-\infty}^{\infty} f_r[k]x[n-k]. \tag{12.109}$$

From Eq. (12.109), it follows then that

$$|u_r[n]| = \left| \sum_{k=-\infty}^{\infty} f_r[k]x[n-k] \right| \leq \sum_{k=-\infty}^{\infty} |f_r[k]|.$$

Thus, Eq. (12.108) is satisfied if

$$\sum_{k=-\infty}^{\infty} |f_r[k]| \leq 1, \qquad \text{for all } r. \tag{12.110}$$

The above condition is both necessary and sufficient to guarantee no overflow [Jac70a].[3] If it is not satisfied in the unscaled realization, we scale the input signal with a multiplier K of value

$$K = \frac{1}{\max_r \sum_{k=-\infty}^{\infty} |f_r[k]|}. \tag{12.111}$$

It should be noted that the above scaling rule is based on a worst case bound and does not fully utilize the dynamic range of all adder output registers and, as a result, reduces the SNR significantly. Even though it is difficult to compute analytically the value of K, an approximate value can be computed on a computer using an approach similar to that given by Eq. (12.95).

More practical and easy to use scaling rules can be derived in the frequency domain if some information about the input signals is known a priori [Jac70a]. The bounds to be derived assume that the input $x[n]$ to the digital filter is a deterministic signal with a Fourier transform $X(e^{j\omega})$. The bounds are given in terms of $\mathcal{L}_p$-norms as defined in Eq. (3.53).

---

[3]It also follows from the BIBO stability requirements discussed in Section 2.5.3 and Eq. (2.78).

12.7.2 $\mathcal{L}_\infty$-Bound

By taking the Fourier transform of both sides of Eq. (12.109), we get

$$U_r(e^{j\omega}) = F_r(e^{j\omega})X(e^{j\omega}), \tag{12.112}$$

where $U_r(e^{j\omega})$ and $F_r(e^{j\omega})$ are the Fourier transforms of $u_r[n]$ and $f_r[n]$, respectively. The inverse Fourier transform of Eq. (12.112) yields

$$u_r[n] = \frac{1}{2\pi} \int_{-\pi}^{\pi} F_r(e^{j\omega})X(e^{j\omega})e^{j\omega n} d\omega. \tag{12.113}$$

As a result,

$$
\begin{aligned}
|u_r[n]| &\leq \frac{1}{2\pi} \int_{-\pi}^{\pi} \left| F_r(e^{j\omega}) \right| \cdot \left| X(e^{j\omega}) \right| d\omega \\
&\leq \left\| F_r(e^{j\omega}) \right\|_\infty \left[\frac{1}{2\pi} \int_{-\pi}^{\pi} \left| X(e^{j\omega}) \right| d\omega \right] \\
&\leq \| F_r \|_\infty \cdot \| X \|_1 .
\end{aligned}
\tag{12.114}
$$

If $\| X \|_1 \leq 1$, then the dynamic range constraint of Eq. (12.108) is satisfied if

$$\left\| F_r(e^{j\omega}) \right\|_\infty \leq 1. \tag{12.115}$$

Thus, if the mean absolute value of the input spectrum is bounded by unity, then there will be no adder overflow if the peak gains from the filter input to all adder outputs are scaled satisfying the bound of Eq. (12.115). In general, this scaling rule is rarely used since, with most input signals encountered in practice, $\| X \|_1 \leq 1$ does not hold!

12.7.3 $\mathcal{L}_2$-Bound

Applying the Schwartz inequality to Eq. (12.113), we arrive at [Jac70a]

$$|u_r[n]|^2 \leq \left(\frac{1}{2\pi} \int_{-\pi}^{\pi} \left| F_r(e^{j\omega}) \right|^2 d\omega \right) \cdot \left(\frac{1}{2\pi} \int_{-\pi}^{\pi} \left| X(e^{j\omega}) \right|^2 d\omega \right), \tag{12.116}$$

or, equivalently,

$$|u_r[n]| \leq \| F_r \|_2 \cdot \| X \|_2 . \tag{12.117}$$

Thus, if the input to the filter has finite energy bounded by unity, that is, $\| X \|_2 \leq 1$, then the adder overflow can be prevented by scaling the filter such that the rms values of the transfer functions from the input to all adder outputs are bounded by unity:

$$\| F_r \|_2 \leq 1, \qquad r = 1, 2, \ldots, R. \tag{12.118}$$

12.7.4 A General Scaling Rule

A more general scaling rule obtained using Holder's inequality is given by [Jac70a]

$$|u_r[n]| \leq \| F_r \|_p \cdot \| X \|_q , \tag{12.119}$$

for all $p, q \geq 1$ satisfying $(1/p) + (1/q) = 1$. Note that for the $\mathcal{L}_\infty$-bound of Eq. (12.114), $p = \infty$ and $q = 1$, and for the $\mathcal{L}_2$-bound of Eq. (12.117), $p = q = 2$. Another useful scaling rule, $\mathcal{L}_1$-bound, is obtained for $p = 1$ and $q = \infty$.

After scaling, the scaling transfer functions become $\bar{F}_r(z)$, and the scaling constants should be chosen such that

$$\left\| \bar{F}_r \right\|_p \leq 1, \qquad r = 1, 2, \ldots, R. \tag{12.120}$$

In many structures, all scaling multipliers can be absorbed into the existing feedforward multipliers without any increase in the total number of multipliers and, hence, noise sources. However, in some cases, the scaling process may introduce additional multipliers in the system. If all scaling multipliers are regular b-bit units, then Eq. (12.120) can be satisfied with an equality sign, providing a full utilization of the dynamic range of each adder output and, thus, yielding a maximum SNR. An attractive option from a hardware point of view, and preferred in cases where scaling introduces new multipliers, is to make as many scaling multiplier coefficients as possible in the scaled structure take values that are powers of 2 [Jac70a]. In this case, these multipliers can be implemented simply by a shift operation. Now, the norm of the scaling transfer function satisfies

$$\tfrac{1}{2} < \left\| \bar{F}_r \right\|_p \leq 1 \tag{12.121}$$

with a slight decrease in the SNR.

It should be pointed out here that the output round-off noise should be always computed after the digital filter structure has been scaled. For the scaled structure, the expression for the output round-off noise of Eq. (12.96) thus changes to

$$\sigma_\gamma^2 = \sigma_0^2 \sum_{\ell=1}^{L} \bar{k}_\ell \left(\frac{1}{2\pi j} \oint_C \bar{G}_\ell(z) \bar{G}_\ell(z^{-1}) z^{-1} \, dz \right), \tag{12.122}$$

where $\bar{k}_\ell$ is the total number of multipliers feeding the ℓth adder with $\bar{k}_\ell \geq k_\ell$, and $\bar{G}_\ell(z)$ is the modified noise transfer function from the input of the ℓth adder to the filter output.

We illustrate next the application of the above method in the scaling of a cascade realization of an IIR transfer function [Jac70b].

12.7.5 Scaling of a Cascade Form IIR Digital Filter Structure

Consider the unscaled structure of Figure 12.27 consisting of a cascade of R second-order IIR sections realized in direct form II. Its transfer function is given by

$$H(z) = K \prod_{i=1}^{R} H_i(z), \tag{12.123}$$

where

$$H_i(z) = \frac{B_i(z)}{A_i(z)} = \frac{1 + b_{1i} z^{-1} + b_{2i} z^{-2}}{1 + a_{1i} z^{-1} + a_{2i} z^{-2}}. \tag{12.124}$$

The branch nodes to be scaled are marked by $(*)$ in the figure, which are seen to be the inputs to the multipliers in each second-order section. The transfer functions from the input to these branch nodes are the scaling transfer functions, which are given by

$$F_r(z) = \frac{K}{A_r(z)} \prod_{\ell=1}^{r-1} H_\ell(z), \qquad r = 1, 2, \ldots, R. \tag{12.125}$$

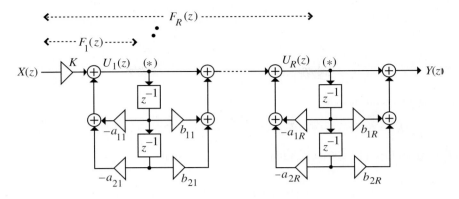

Figure 12.27: An unscaled cascade realization of second-order IIR sections.

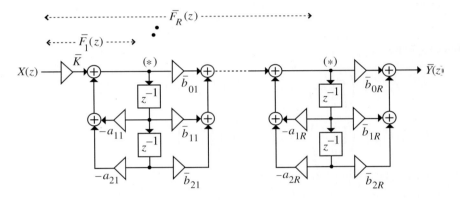

Figure 12.28: The scaled cascade realization.

The scaled version of the above structure is shown in Figure 12.28, with new values for the feedforward multipliers. Note that the scaling process has introduced a new multiplier $\bar{b}_{0\ell}$ in each second-order section. If the zeros of the transfer function are on the unit circle, as is usually the case in practice, then $b_{2\ell} = \pm 1$, in which case, we can choose $\bar{b}_{0\ell} = \bar{b}_{2\ell} = 2^{-\beta}$ to reduce the total number of actual multiplications in the final realization.

It can be seen from Figure 12.28 that

$$\bar{F}_r(z) = \frac{\bar{K}}{A_r(z)} \prod_{\ell=1}^{r-1} \bar{H}_\ell(z), \tag{12.126a}$$

$$\bar{H}(z) = \bar{K} \prod_{\ell=1}^{R} \bar{H}_\ell(z), \tag{12.126b}$$

where

$$\bar{H}_\ell(z) = \frac{\bar{b}_{0\ell} + \bar{b}_{1\ell}z^{-1} + \bar{b}_{2\ell}z^{-2}}{1 + a_{1\ell}z^{-1} + a_{2\ell}z^{-2}}. \tag{12.127}$$

Denote

$$\|F_r\|_p \overset{\Delta}{=} \alpha_r, \qquad r = 1, 2, \ldots, R, \tag{12.128a}$$

$$\|H\|_p \overset{\Delta}{=} \alpha_{R+1}, \tag{12.128b}$$

and choose the scaling constants as

$$\bar{K} = \beta_0 K; \qquad \bar{b}_{\ell r} = \beta_r b_{\ell r}, \qquad \ell = 0, 1, 2; \quad r = 1, 2, \ldots, R. \tag{12.129}$$

Now, it follows from Eqs. (12.129), (12.126a), and (12.126b) that

$$\bar{F}_r(z) = \left(\prod_{i=0}^{r-1} \beta_i\right) F_r(z), \qquad r = 1, 2, \ldots, R, \tag{12.130a}$$

$$\bar{H}(z) = \left(\prod_{i=0}^{R} \beta_i\right) H(z). \tag{12.130b}$$

After scaling, we require

$$\|\bar{F}_r\|_p = \left(\prod_{i=0}^{r-1} \beta_i\right) \|F_r\|_p = \alpha_r \left(\prod_{i=0}^{r-1} \beta_i\right) = 1, \quad r = 1, 2, \ldots, R, \tag{12.131a}$$

$$\|\bar{H}\|_p = \left(\prod_{i=0}^{R} \beta_i\right) \|H\|_p = \alpha_{R+1} \left(\prod_{i=0}^{R} \beta_i\right) = 1. \tag{12.131b}$$

Solving the above for the scaling constants, we arrive at

$$\beta_0 = \frac{1}{\alpha_1}, \tag{12.132a}$$

$$\beta_r = \frac{\alpha_r}{\alpha_{r+1}}, \qquad r = 1, 2, \ldots, R. \tag{12.132b}$$

12.7.6 Dynamic Range Scaling Using MATLAB

The dynamic range scaling using the $\mathcal{L}_2$-norm rule can be easily performed using MATLAB by actual simulation of the digital filter structure. If we denote the impulse response from the input of the digital filter to the output of the rth branch node as $\{f_r[n]\}$ and assume that the branch nodes have been ordered in accordance to their precedence relations with increasing i (see Section 11.1.2), then we can compute the $\mathcal{L}_2$-norm $\|F_1\|_2$ of $\{f_1[n]\}$ first and divide the input by a multiplier $k_1 = \|F_1\|_2$. Next, we compute the $\mathcal{L}_2$-norm $\|F_2\|_2$ of $\{f_2[n]\}$ and scale the multipliers feeding into the second adder by dividing with a constant $k_2 = \|F_2\|_2$. This process is continued until the output node has been scaled to yield an $\mathcal{L}_2$-norm of unity.

We illustrate the method by means of Example 12.9.

EXAMPLE 12.9   Dynamic Range Scaling of a Cascaded Digital Filter Structure

Consider the cascade realization of the third-order lowpass digital filter of Eq. (11.22). The transfer functions of the two sections in the cascade are given by Eqs. (11.23a) and (11.23b), repeated below for convenience:

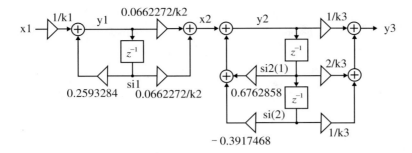

Figure 12.29: Cascade realization of the transfer functions of Eqs. (12.133a) and (12.133b) in direct form II.

$$H_1(z) = \frac{0.0662272(1 + z^{-1})}{1 - 0.2593284z^{-1}}, \tag{12.133a}$$

$$H_2(z) = \frac{1 + 2z^{-1} + z^{-2}}{1 - 0.6762858z^{-1} + 0.3917468z^{-2}}. \tag{12.133b}$$

Their realization in the direct form II structure is as shown in Figure 12.29. The MATLAB program simulating this structure is given by Program 12_6. The program is first run with all scaling constants set to unity; that is, $k1 = k2 = k3 = 1$. In the line indicating the approximate $\mathcal{L}_2$-norm calculation, the output variable is chosen as $y1$. The program computes the square of the $\mathcal{L}_2$-norm of the impulse response at node $y1$ as 1.07210002757252, which is used to set $k1 = \sqrt{1.07210002757252}$, with other scaling constants still set to unity. A second run of the program shows the $\mathcal{L}_2$-norm of the impulse response at node $y1$ as 1.00000000000000, verifying the success of scaling the input. In the second step, in the line indicating the approximate $\mathcal{L}_2$-norm calculation, the output variable is chosen as $y2$. The program then yields the square of the $\mathcal{L}_2$-norm of the impulse response at node $y2$ as 0.02679820762398, which is used to set $k2 = \sqrt{0.02679820762398}$, with $k3$ still set to unity. This process is repeated for node $y3$, resulting in $k3 = \sqrt{11.96975400608943}$, with the final value of the $\mathcal{L}_2$-norm of the impulse response at node $y3$ as 0.99999683131540. The program listing given is after all adder outputs have been scaled.

Program 12_6 can be easily modified to compute the product round-off noise variance at the output of the scaled structure. To this end, we set the digital filter input to zero and apply an impulse at the input of the first adder. This is equivalent to setting $x1 = 1$ in Program 12_6. The normalized output noise variance due to a single error source is computed as 1.07209663042567. Next, we apply an impulse at the input of the second adder, with the digital filter input set to zero. This is achieved by replacing $x2$ in the calculation of $y2$ as $x1$. The program yields the normalized output noise variance due to a single error source at the second adder as 1.26109014071707. The total normalized output round-off noise, assuming all products to be quantized before addition, is then equal to

$$2 \times 1.07209663042567 + 4 \times 1.26109014071707 + 3 = 10.18855382371962.$$

On the other hand, if we assume quantization after addition of all products at each adder, the total normalized output round-off noise becomes

$$1.07209663042567 + 1.26109014071707 + 1 = 3.33318677114274.$$

EXAMPLE 12.10   **Effect of Ordering on Output Round-Off Noise of a Cascaded Digital Filter Structure**

Program 12_6.m

We repeat the problem of Example 12.9 by interchanging the locations of the two sections $G_1(z)$ and $G_2(z)$ in the cascade connection, as indicated in Figure 12.30. By modifying Program 12_6, we arrive at the following values for the scaling constants:

$$k1 = \sqrt{1.54643736440090}, \qquad k2 = \sqrt{14.44567893169408}, \qquad k3 = \sqrt{0.01539326750782}.$$

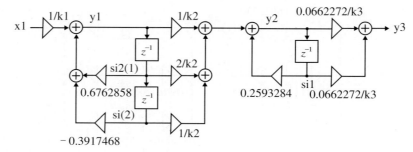

Figure 12.30: Alternate cascade realization of the transfer functions of Eqs. (12.133a) and (12.133b) in direct form II.

The total normalized output round-off noise assuming all products to be quantized before addition in this case is equal to

$$3 \times 1.5465220947704 + 4 \times 0.76938948942268 + 2 = 9.717124200195,$$

whereas the total normalized output round-off noise assuming quantization after addition is given by

$$1.5465220947704 + 0.76938948942268 + 1 = 3.31591158419309.$$

Note from above that the second cascade connection of Figure 12.30 yields a slightly lower round-off noise than the first of Figure 12.29.

12.7.7 Optimum Ordering and Pole-Zero Pairing of the Cascade Form IIR Digital Filter

As indicated in Section 8.4.2 and illustrated in Figures 8.16 and 8.17, there are many possible cascade realizations of a higher-order IIR transfer function obtained by different pole-zero pairings and ordering. In fact, for a cascade of R second-order sections, there are $(R!)^2$ different possible realizations (Problem 12.31). Each one of these realizations will have different output noise power, as illustrated in Examples 12.9 and 12.10, and hence, it is of interest to determine the cascade realization with the lowest output noise power.

A fairly simple heuristic set of rules for determining an optimum pole-zero pairing and ordering of sections in a cascade realization has been advanced by Jackson [Jac70b]. To understand the reasoning behind these rules, we first develop the expression for the output noise variance due to product round-off in a cascade IIR structure implemented in fixed-point arithmetic.

Expression for the Output Noise Variance

To determine expression for the output noise variance, we make use of the noise model of the scaled cascade structure of Figure 12.28 as shown in Figure 12.31. It follows from Figure 12.31 and Eqs. (12.127) and (12.129) that the scaled noise transfer functions are given by

$$\bar{G}_\ell(z) = \prod_{i=\ell}^{R} \bar{H}_i(z) = \left(\prod_{i=\ell}^{R} \beta_i \right) G_\ell(z), \qquad \ell = 1, 2, \ldots, R, \tag{12.134a}$$

$$\bar{G}_{R+1}(z) = 1, \tag{12.134b}$$

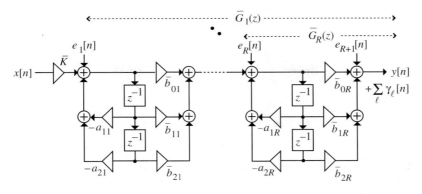

Figure 12.31: The noise model of the scaled cascade realization of Figure 12.27.

where $\bar{H}_i(z)$ is as given in Eq. (12.127) and the unscaled noise transfer function is given by

$$G_\ell(z) = \prod_{i=\ell}^{R} H_i(z).$$

(12.135)

Hence, the output noise power spectrum due to product round-off is given by

$$P_{\gamma\gamma}(\omega) = \sigma_0^2 \left[\sum_{\ell=1}^{R+1} k_\ell \left| \bar{G}_\ell(e^{j\omega}) \right|^2 \right],$$

(12.136)

and the output noise variance is thus

$$\sigma_\gamma^2 = \sigma_0^2 \left[\sum_{\ell=1}^{R+1} k_\ell \left(\frac{1}{2\pi} \int_{-\pi}^{\pi} \left| \bar{G}_\ell(e^{j\omega}) \right|^2 d\omega \right) \right]$$

$$= \sigma_0^2 \left[\sum_{\ell=1}^{R+1} k_\ell \left\| \bar{G}_\ell \right\|_2^2 \right],$$

(12.137)

where we have used the fact that the quantity inside the parentheses in the middle expression is the square of the $\mathcal{L}_2$-norm of $\bar{G}_\ell(e^{j\omega})$ as defined in Eq. (3.53). In Eqs. (12.136) and (12.137), k_ℓ is the total number of multipliers connected to the ℓth adder. If all products are rounded before summation, then

$$k_1 = k_{R+1} = 3,$$

(12.138a)

$$k_\ell = 5, \qquad \text{for } \ell = 2, 3, \ldots, R.$$

(12.138b)

On the other hand, if all products are rounded after summation, then

$$k_\ell = 1, \qquad \text{for } \ell = 1, 2, \ldots, R + 1.$$

(12.139)

From Eqs. (12.132a) and (12.132b),

$$\prod_{i=\ell}^{R} \beta_i = \frac{\alpha_\ell}{\alpha_{R+1}} = \frac{\|F_\ell\|_p}{\|H\|_p}.$$

(12.140)

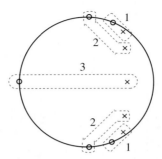

Figure 12.32: Illustration of the pole-zero pairing and ordering rules.

Substituting Eq. (12.140) in Eq. (12.136), we obtain for the output noise power spectrum

$$P_{\gamma\gamma}(\omega) = \frac{\sigma_0^2}{\|H\|_p^2}\left[k_{R+1}\|H\|_p^2 + \sum_{\ell=1}^{R} k_\ell\|F_\ell\|_p^2\left|G_\ell(e^{j\omega})\right|^2\right]. \tag{12.141}$$

Correspondingly, the output noise variance is given by

$$\sigma_\gamma^2 = \frac{\sigma_0^2}{\|H\|_p^2}\left[k_{R+1}\|H\|_p^2 + \sum_{\ell=1}^{R} k_\ell\|F_\ell\|_p^2\|G_\ell\|_2^2\right]. \tag{12.142}$$

Now, the scaling transfer function $F_\ell(z)$ contains a product of section transfer functions, $H_i(z)$, $i = 1, 2, \ldots, \ell-1$, whereas the noise transfer function $G_\ell(z)$ contains the product of section transfer functions, $H_i(z)$, $i = \ell, \ell+1, \ldots, R$. Thus, every term in the sum in Eqs. (12.141) and (12.142) includes the transfer function of all R sections in the cascade realization. This implies that to minimize the output noise power, the norms of $H_i(z)$ should be minimized for all values of i by appropriately pairing the poles and zeros. To this end, Jackson has proposed the following rule [Jac70b]:

Pole-Zero Pairing Rule

First, the complex pole-pair closest to the unit circle should be paired with the nearest complex zero-pair. Next, the complex pole-pair that is closest to the previous set of poles should be matched with its nearest complex zero-pair. This process should be continued until all poles and zeros have been paired.

The above type of pairing of poles and zeros is likely to lower the peak gain of the section characterized by the paired poles and zeros. Lowering of the peak gain, in turn, reduces the possibility of overflow and attenuates the round-off noise. The pole-zero pairing rule is illustrated in Figure 12.32 for a fifth-order elliptic lowpass IIR digital filter with passband edge at 0.3π, passband ripple of 0.5 dB, and minimum stopband attenuation of 40 dB.

Once the appropriate pole-zero pairings have been made, the next question that needs to be answered is how to order the sections in the cascade structure [Jac70b]. A section in the front part of the cascade has its transfer function $H_i(z)$ appearing more frequently in the scaling transfer function expressions in Eqs. (12.141) and (12.142), whereas a section near the output end of the cascade has its transfer function $H_i(z)$ appearing more frequently in the noise transfer function expressions. The best location for $H_i(z)$ obviously depends on the type of norms being applied to the scaling and noise transfer functions.

To understand the reasoning behind the optimum ordering of sections, consider a sixth order IIR transfer function with its poles and zeros ordered according to the rules given above. In this case, the expression for the output power spectrum of Eq. (12.141) reduces to

$$P_{\gamma\gamma}(\omega) = \frac{\sigma_0^2}{\|H\|_p^2} \Big[k_1 \|F_1\|_p^2 \Big| G_1(e^{j\omega}) \Big|^2 + k_2 \|F_2\|_p^2 \Big| G_2(e^{j\omega}) \Big|^2$$
$$+ k_3 \|F_3\|_p^2 \Big| G_3(e^{j\omega}) \Big|^2 + k_4 \|H\|_p^2 \Big].$$

(12.143)

Likewise, the expression for the output noise variance of Eq. (12.142) reduces to

$$\sigma_\gamma^2 = \frac{\sigma_0^2}{\|H\|_p^2} \Big[k_1 \|F_1\|_p^2 \|G_1\|_2^2 + k_2 \|F_2\|_p^2 \|G_2\|_2^2 + k_3 \|F_3\|_p^2 \|G_3\|_2^2 + k_4 \|H\|_p^2 \Big].$$

(12.144)

A careful examination of Eq. (12.144) reveals that if the $\mathcal{L}_2$-scaling is used, then each term inside the square brackets on the right-hand side contains the $\mathcal{L}_2$-norms of all three sections of the cascade, and hence, ordering of paired sections does not influence too much the output noise power. This fact is evident from the results of Examples 12.9 and 12.10, where the total normalized output round-off noise power of the cascade realization of Figure 12.29 is seen to be quite close to that of the cascade realization of Figure 12.30. If, however, an $\mathcal{L}_\infty$-scaling is being employed, then in the expression on the right-hand side of Eq. (12.144), the transfer function $H_3(z)$ appears only once, transfer function $H_2(z)$ appears twice, and the transfer function $H_1(z)$ appears three times. Hence, the section with the poles closest to the unit circle exhibit the most peaking magnitude response and should be chosen as $H_3(z)$ and placed at the output end. The section $H_2(z)$ should be the section with the next most peaking magnitude response, and $H_1(z)$ should be the section with the least-peaked magnitude response. The ordering rule in the general case is therefore to place the least-peaked section to the most-peaked section, starting at the input end.

On the other hand, the ordering scheme is exactly opposite if the objective is to minimize the peak noise $\|P_{\gamma\gamma}(\omega)\|_\infty$ and an $\mathcal{L}_2$-scaling is used. However, the ordering has no effect on the peak noise if an $\mathcal{L}_\infty$-scaling is used.

Pole-Zero Pairing and Ordering Using MATLAB

The MATLAB *Signal Processing Toolbox* includes the function zp2sos that can be employed to determine the optimum pole-zero pairing and ordering according to the above discussed rules. Its basic form is sos = zp2sos(z,p,k). This function generates a matrix sos containing the coefficients of each second-order section of the equivalent transfer function $H(z)$ determined from the specified zero-pole form. sos is an $L \times 6$ matrix of the form

$$\text{sos} = \begin{bmatrix} b_{01} & b_{11} & b_{21} & a_{01} & a_{11} & a_{21} \\ b_{02} & b_{12} & b_{22} & a_{02} & a_{12} & a_{22} \\ \vdots & \vdots & \vdots & \vdots & \vdots & \vdots \\ b_{0L} & b_{1L} & b_{2L} & a_{0L} & a_{1L} & a_{2L} \end{bmatrix},$$

whose ith row contains the coefficients, $\{b_{i\ell}\}$ and $\{a_{i\ell}\}$, of the numerator and denominator polynomials of the ith second-order section, with L denoting the total number of sections in the cascade. The form of the overall transfer function is thus given by

$$H(z) = \prod_{i=1}^{L} H_i(z) = \prod_{i=1}^{L} \frac{b_{0i} + b_{1i}z^{-1} + b_{2i}z^{-2}}{a_{0i} + a_{1i}z^{-1} + a_{2i}z^{-2}}.$$

The rows are ordered so that the first row ($i = 1$) contains the coefficients of the pole pair farthest from the unit circle and its nearest zero pair. If a reverse ordering is desired, with the first row containing the coefficients of the pole pair closest to the unit circle and its nearest zero pair, then the statement `sos = zp2sos(z,p,k,'down')` should be used.

EXAMPLE 12.11   Optimum Pairing and Ordering of a Fifth-Order Lowpass Filter

We develop the optimum pairing and ordering of the fifth-order elliptic lowpass filter whose pole-zero plot is shown in Figure 12.31. The filter specifications are $\alpha_p = 0.5$ dB, $\alpha_s = 40$ dB, and $\omega_p = 0.3\pi$.
    To this end, we first determine the zeros, poles, and the gain constant of the desired lowpass filter using the command `[z,p,k] = ellip(5,0.5,40,0.3)` and then employ the command `sos = zp2sos(z,p,k)` to determine the coefficients of the second-order sections. The transfer functions of the sections are given by

$$H_1(z) = \left[\frac{0.03 + 0.03z^{-1}}{1 - 0.6135686z^{-1}} \right],$$

$$H_2(z) = \left[\frac{1 - 0.0861z^{-1} + z^{-2}}{1 - 1.152863z^{-1} + 0.6116484z^{-2}} \right],$$

$$H_3(z) = \left[\frac{1 - 0.76382924z^{-1} + z^{-2}}{1 - 1.0983341z^{-1} + 0.899076z^{-2}} \right].$$

As indicated above, the first section in the cascade is $H_1(z)$, followed by $H_2(z)$, and the last section is $H_3(z)$. This pairing is identical to that displayed in Figure 12.31, but the ordering is reversed. This can be easily verified by using the function `[z,p,k] = sos2zp(sos)`, which computes the zeros and poles of each individual second-order section defined by the matrix `sos`.

    The numerator and the denominator polynomials of the transfer function can be determined from the matrix `sos` using the M-file `sos2tf`.

12.8   Signal-to-Noise Ratio in Low-Order IIR Filters

The output round-off noise variances of unscaled digital filters do not provide a realistic picture of the performances of these structures in practice since introduction of scaling multipliers can increase the number of error sources and the gain for the noise transfer functions. It is therefore important to scale the digital filter structure first before its round-off noise performance is analyzed. In most applications, the round-off noise variance by itself is not sufficient, and a more meaningful result is obtained by computing instead the expression for the signal to round-off noise ratio (SNR) for performance evaluation. We illustrate such computation for the first-order and the second-order IIR structures considered earlier in Examples 12.7 and 12.8, respectively [Vai87c]. Most conclusions derived from the detailed analysis of these simple structures outlined here are also valid in the case of more complex structures. Moreover, the methods followed here can be easily extended to the general case.

12.8.1   First-Order Section

Consider first the causal unscaled first-order IIR filter of Figure 12.1. Its round-off noise variance was computed in Example 12.7 and is given by Eq. (12.8). To determine its signal-to-noise ratio, assume the input $x[n]$ to be a wide-sense stationary (WSS) random signal with a uniform probability density function and a variance σ_x^2. The variance σ_y^2 of the output signal $y[n]$ generated by this input is then given by

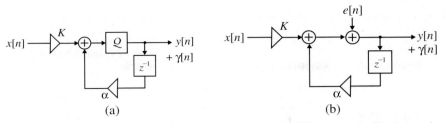

Figure 12.33: (a) Scaled first-order section with quantizer and (b) its round-off noise model.

$$\sigma_y^2 = \sigma_x^2 \left(\sum_{n=0}^{\infty} h^2[n] \right) = \frac{\sigma_x^2}{1 - \alpha^2}. \tag{12.145}$$

Taking the ratio of Eqs. (12.145) and (12.99), we arrive at the expression for the signal-to-noise ratio of the unscaled section as

$$\text{SNR} = \frac{\sigma_y^2}{\sigma_\gamma^2} = \frac{\sigma_x^2}{\sigma_0^2}, \tag{12.146}$$

implying that the SNR is independent of α. However, this is not a valid result since the adder is likely to overflow in an unscaled structure. It is therefore important first to scale the structure and then to compute the SNR to arrive at a more meaningful result.

The scaled structure is as indicated in Figure 12.33, along with its round-off error analysis model, assuming quantization after addition of all products. With the scaling multiplier present, the output signal power now becomes

$$\sigma_y^2 = \frac{K^2 \sigma_x^2}{1 - \alpha^2}, \tag{12.147}$$

modifying the signal-to-noise ratio to

$$\text{SNR} = \frac{K^2 \sigma_x^2}{\sigma_0^2}. \tag{12.148}$$

Since the scaling multiplier coefficient K depends on the pole location and the type of scaling rule being followed, the SNR will thus reflect this dependence. The scaling parameter K can be chosen according to the rules outlined in Section 12.7. To this end, we need first to determine the scaling transfer function $F(z)$ of the unscaled structure. It follows from Figure 12.33(a) that

$$F(z) = H(z) = \frac{1}{1 - \alpha z^{-1}}, \tag{12.149}$$

with a corresponding impulse response

$$f[n] = \mathcal{Z}^{-1}\{F(z)\} = \alpha^n \mu[n]. \tag{12.150}$$

Now, as indicated in Section 12.7, there are different scaling rules. To guarantee no overflow, we can use the scaling rule of Eq. (12.111) and arrive at

$$K = \frac{1}{\sum_{n=0}^{\infty} |f[n]|} = 1 - |\alpha|. \tag{12.151}$$

To evaluate the SNR, we need to know the type of input $x[n]$ being applied. If $x[n]$ is uniformly distributed with $|x[n]| \leq 1$, its variance is given by

$$\sigma_x^2 = \tfrac{1}{3}. \tag{12.152}$$

Table 12.5: Signal-to-noise ratio of first-order IIR digital filters for different inputs. Adapted from [Vai87c].

Scaling rule	Input type	SNR	Typical SNR, dB $(b = 12, |\alpha| = 0.95)$
No overflow	WSS, white uniform density	$\dfrac{(1 - |\alpha|)^2}{3\sigma_0^2}$	52.24
No overflow	WSS, white Gaussian density $(\sigma_x^2 = 1/9)$	$\dfrac{(1 - |\alpha|)^2}{9\sigma_0^2}$	47.97
No overflow	Sinusoid, known frequency	$\dfrac{(1 - |\alpha|)^2}{2\sigma_0^2}$	69.91

Substituting Eqs. (12.151) and (12.152) in Eq. (12.148), we arrive at

$$\text{SNR} = \frac{(1 - |\alpha|)^2}{3\sigma_0^2}. \tag{12.153}$$

For a $(b + 1)$-bit signed fraction with round-off or two's-complement truncation, $\sigma_0^2 = 2^{-2b}/12$. Substituting this figure in Eq. (12.153), we obtain the signal-to-noise ratio in dB as

$$\text{SNR}_{\text{dB}} = 20 \log_{10}(1 - |\alpha|) + 6.02 + 6.02\, b. \tag{12.154}$$

As a result, for a given transfer function, the SNR increases by 6 dB for each additional bit added to the register storing the product.

The above analysis can be carried out for other types of input and different scaling rules. We summarize in Table 12.5 the results for three different types of input for the scaling ensuring no overflow. Several conclusions can be made by examining this table. Independent of the type of input being applied, the SNR decreases rapidly as the pole moves closer to the unit circle. For a given transfer function, knowing the type of input being applied, the internal wordlength can be computed to achieve a desired SNR.

12.8.2 Second-Order Section

Consider next the causal unscaled second-order IIR filter of Figure 12.26(a). Its scaled version is indicated in Figure 12.34, along with the round-off noise analysis model, assuming again quantization after addition of all products. Now for a WSS input with a uniform probability density function and a variance σ_x^2, the signal power at the output is given by

$$\sigma_y^2 = \sigma_x^2 \left(K^2 \sum_{n=0}^{\infty} h^2[n] \right), \tag{12.155}$$

while the round-off noise power at the output is given by

$$\sigma_\gamma^2 = \sigma_0^2 \left(\sum_{n=0}^{\infty} h^2[n] \right). \tag{12.156}$$

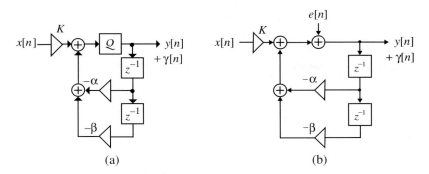

Figure 12.34: (a) Scaled second-order section with quantizer and (b) its round-off noise model.

Therefore, the signal-to-noise ratio of the scaled structure is of the form

$$\text{SNR} = \frac{\sigma_y^2}{\sigma_\gamma^2} = \frac{K^2 \sigma_x^2}{\sigma_0^2}. \tag{12.157}$$

To determine the appropriate value of the scaling multiplier K, we need to compute the scaling transfer function $F(z)$. It can be seen from Figure 12.34(a) that $F(z)$ is identical to the filter transfer function $H(z)$. If the poles of the transfer function are at $z = re^{-j\theta}$, then both transfer functions take the form

$$F(z) = H(z) = \frac{1}{1 + \alpha z^{-1} + \beta z^{-2}} = \frac{1}{1 - 2r \cos\theta z^{-1} + r^2 z^{-2}}. \tag{12.158}$$

The corresponding impulse response is obtained by taking the inverse z-transform of the above, resulting in (Problem 6.22)

$$f[n] = h[n] = \frac{r^n \sin(n+1)\theta}{\sin\theta} \cdot \mu[n]. \tag{12.159}$$

To eliminate completely the overflow at the output in Figure 12.34, the scaling multiplier K must be chosen as

$$K = \frac{1}{\sum_{n=0}^{\infty} |h[n]|}. \tag{12.160}$$

The summation in the denominator of Eq. (12.160) with $h[n]$ given by Eq. (12.159) is difficult to compute analytically. However, it is possible to establish some bounds on the summation to provide a reasonable estimate of the value of the scaling multiplier coefficient K [Opp75]. To this end, note that the amplitude of the response of the unscaled second-order section of Figure 12.26(a) to a sinusoidal input at the resonant frequency $\omega = \theta$; that is, $x[n] = \cos(\theta n)$, is given by

$$\left| H(e^{j\theta}) \right| = \left[\frac{1}{(1-r)^2(1 - 2r\cos\theta + r^2)} \right]^{1/2}. \tag{12.161}$$

However, $\left| H(e^{j\theta}) \right|$ cannot be greater than $\sum_{n=0}^{\infty} |h[n]|$ since the latter is the largest possible value of output $y[n]$ for an input $x[n]$ with $|x[n]| \le 1$. Moreover,

$$\sum_{n=0}^{\infty} |h[n]| = \frac{1}{\sin\theta} \sum_{n=0}^{\infty} r^n |\sin(n+1)\theta|$$

$$\leq \frac{1}{\sin\theta} \sum_{n=0}^{\infty} r^n = \frac{1}{(1-r)\sin\theta}. \tag{12.162}$$

A tighter upper bound on $\sum_{n=0}^{\infty} |h[n]|$ has been derived in [Ünv75] and is given by

$$\sum_{n=0}^{\infty} |h[n]| \leq \frac{4}{\pi(1-r^2)\sin\theta}. \tag{12.163}$$

From Eqs. (12.160), (12.161), and (12.163), we therefore arrive at

$$(1-r)^2(1-2r\cos\theta+r^2) \geq K^2 \geq \frac{\pi^2(1-r^2)^2\sin^2\theta}{16}. \tag{12.164}$$

Following the technique carried out for the first-order section, we can derive the bounds on the SNR for the all-pole second-order section from Eqs. (12.157) and (12.164) for various types of inputs. As in the case of the first-order section, as the poles move closer to the unit circle ($r \to 1$), the gain of the filter increases, causing the input signal to be scaled down significantly to avoid the overflow, while at the same time boosting the output round-off noise. This type of interplay between round-off noise and dynamic range is a characteristic of all fixed-point digital filters [Jac70a].

In certain cases, it is possible to develop digital filter structures that have inherently the least quantization effects. In the following section, we consider these structures.

12.9   Low-Sensitivity Digital Filters

In Section 12.2, we considered the effect of multiplier coefficient quantization on the performance of a digital filter. One major consequence is that the frequency response of the digital filter with quantized multiplier coefficients is different from that of the desired digital filter with unquantized coefficients, and this difference may be significant enough to make the practical digital filter unsuitable in most applications. It is thus of interest to develop digital filter structures that are inherently less sensitive to coefficient quantization. To this end, the first approach advanced is based on the conversion of an inherently low sensitivity analog network composed of inductors, capacitors, and resistors to a digital filter structure by replacing each analog network component and their interconnections to a corresponding digital filter equivalent such that the overall structure "simulates" the analog prototype [Fet71], [Fet86]. The resulting digital filter structure is called a *wave digital filter*, which also shares some additional properties of its analog prototype. An alternative approach is to determine directly the conditions for low coefficient sensitivity to be satisfied by the digital filter structure and to develop realization methods that ensure that the final structure indeed satisfies these conditions [Vai84]. In this section, we examine the latter approach.

12.9.1   Requirements for Low Coefficient Sensitivity

Let the prescribed transfer function $H(z)$ be a *bounded real* (BR) function, as defined in Section 7.1.2. That is, $H(z)$ is a causal stable real-coefficient function characterized by a magnitude response $|H(e^{j\omega})|$ bounded above by unity; that is,

$$\left| H(e^{j\omega}) \right| \leq 1. \tag{12.165}$$

Assume that $H(z)$ is such that at a set of frequencies ω_k, the magnitude is exactly equal to 1:

$$\left| H(e^{j\omega_k}) \right| = 1. \tag{12.166}$$

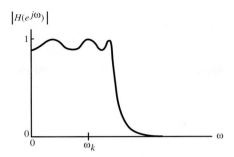

Figure 12.35: A typical magnitude response of a bounded real transfer function.

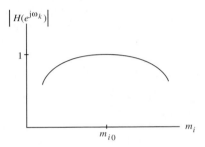

Figure 12.36: Illustration of the zero-sensitivity property.

Since the magnitude function is bounded above by unity, the frequencies ω_k must be in the passband of the filter. A typical frequency response satisfying the above conditions is shown in Figure 12.35. Note that any causal stable transfer function can be scaled to satisfy conditions of Eqs. (12.165) and (12.166).

Let the digital filter structure $\mathcal{N}$ realizing $H(z)$ be characterized by a set of R multipliers with coefficients m_i. Moreover, let the nominal values of these multiplier coefficients, assuming infinite precision realization, be m_{i0}. Assume that the structure $\mathcal{N}$ is such that, regardless of actual values of the multiplier coefficients m_i in the immediate neighborhood of their design values m_{i0}, its transfer function remains bounded real, satisfying the condition of Eq. (12.165). Now, consider $\left|H(e^{j\omega_k})\right|$, the value of the magnitude response at $\omega = \omega_k$, which is equal to 1 when the multiplier values have infinite precision. Because of the assumption on $\mathcal{N}$ if a multiplier coefficient m_i is quantized, then $\left|H(e^{j\omega_k})\right|$ can only become less than 1 because of the BR condition of Eq. (12.165). Thus, a plot of $\left|H(e^{j\omega_k})\right|$ as a function of m_i will appear as indicated in Figure 12.36, with a zero-valued slope at $m_i = m_{i0}$,

$$\left. \frac{\partial \left|H(e^{j\omega_k})\right|}{\partial m_i} \right|_{m_i=m_{i0}} = 0, \tag{12.167}$$

implying that the first-order sensitivity of the magnitude function $\left|H(e^{j\omega})\right|$ with respect to each multiplier coefficient m_i is zero at all frequencies ω_k where $\left|H(e^{j\omega})\right|$ assumes its maximum value of unity. Since all frequencies ω_k, where the magnitude function is exactly equal to unity, are in the passband of the filter, and if these frequencies are closely spaced, we expect the sensitivity of the magnitude function to be very low at other frequencies in the passband.

A digital filter structure satisfying the above conditions for low passband sensitivity is called a *structurally bounded* system. Since the output energy of such a structure is less than the input energy for all

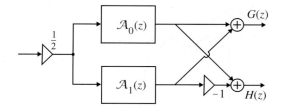

Figure 12.37: Parallel allpass realization of a power-complementary pair of transfer functions.

finite energy inputs [see Eq. (7.6)], it is also called a *structurally passive* system. If Eq. (12.165) holds with an equality sign, the transfer function $H(z)$ is called a *lossless bounded real* (LBR) function, that is, a stable allpass function. An allpass realization satisfying the LBR condition is therefore a *structurally lossless* or *LBR* system, implying that the structure remains allpass under coefficient quantization.

We now outline methods for the realization of low passband sensitivity digital filters.

12.9.2 Low Passband Sensitivity IIR Digital Filter

In Section 8.10, we outlined a method for the realization of a large class of stable IIR transfer functions $G(z)$ in the form of a parallel allpass structure:

$$G(z) = \tfrac{1}{2}\{\mathcal{A}_0(z) + \mathcal{A}_1(z)\}, \tag{12.168}$$

where $\mathcal{A}_0(z)$ and $\mathcal{A}_1(z)$ are stable allpass transfer functions. Such a realization is possible if $G(z)$ is a BR function with a symmetric numerator and has a power-complementary BR transfer function $H(z)$ with an antisymmetric numerator [Vai86a]. In this case, $H(z)$ can be expressed as

$$H(z) = \tfrac{1}{2}\{\mathcal{A}_0(z) - \mathcal{A}_1(z)\}. \tag{12.169}$$

The allpass decompositions of Eqs. (12.168) and (12.169) permit the realization of the power-complementary pair $\{G(z), H(z)\}$, as indicated in Figure 12.37.

Now, on the unit circle, Eq. (12.168) becomes

$$G(e^{j\omega}) = \frac{1}{2}\left\{e^{j\theta_0(\omega)} + e^{j\theta_1(\omega)}\right\}, \tag{12.170}$$

since $\mathcal{A}_0(z)$ and $\mathcal{A}_1(z)$ are allpass functions with unity magnitude responses. Thus, if the allpass transfer functions are realized in structurally lossless forms, $|G(e^{j\omega})|$ will remain bounded above by unity. Moreover, at frequencies ω_k, where $|G(e^{j\omega_k})| = 1$, $\theta_0(\omega_k) = \langle\theta_1(\omega_k)\rangle_{2\pi}$, and $|e^{j\theta_0(\omega_k)} + e^{j\theta_1(\omega_k)}| = 2$. Or in other words, the realization of Figure 12.37 is structurally passive, implying low passband sensitivity with respect to multiplier coefficient quantization.

Section 8.10 describes a method for determining the two allpass transfer functions $\mathcal{A}_0(z)$ and $\mathcal{A}_1(z)$ for a given BR transfer function $H(z)$ satisfying the conditions given above. Once these allpass transfer functions have been determined, they can then be realized in structurally lossless forms using one of the two methods discussed in Section 8.6.

Example 12.12 illustrates the low passband sensitivity characteristics of a parallel allpass structure.

EXAMPLE 12.12   Low Passband Sensitivity of a Parallel Allpass Structure

Consider the realization of the fifth-order elliptic IIR lowpass filter of Example 9.14. The filter specifications are

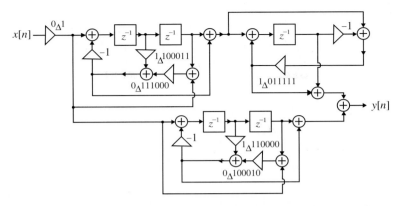

Figure 12.38: Parallel allpass realization of the lowpass filter using structurally lossless allpass sections with quantized multiplier coefficients.

passband ripple of 0.5 dB, minimum stopband attenuation of 40 dB, and passband edge at 0.4. Using a modified version of Program 9_1, we arrive at the desired transfer function given by

$$G(z) = \frac{\begin{aligned}&0.0528168 + 0.0797082z^{-1} + 0.1294975z^{-2} + 0.1294975z^{-3} \\ &\quad + 0.0797082z^{-4} + 0.0528168z^{-5}\end{aligned}}{\begin{aligned}&(1 - 0.4908777z^{-1})(1 - 0.7624319z^{-1} + 0.5390008z^{-2}) \\ &\quad \times (1 - 0.5573823z^{-1} + 0.8828417z^{-2})\end{aligned}}.$$

Its parallel allpass decomposition is given by

$$G(z) = \frac{1}{2}\left[\frac{(0.4908777 - z^{-1})(0.8828417 - 0.5573823z^{-1} + z^{-2})}{(1 - 0.4908777z^{-1})(1 - 0.5573823z^{-1} + 0.8828417z^{-2})} \right.$$
$$\left. + \frac{0.5390008 - 0.7624319z^{-1} + z^{-2}}{1 - 0.7624319z^{-1} + 0.5390008z^{-2}} \right].$$

A 5-multiplier realization of the above decomposition using a signed 7-bit fractional sign-magnitude representation of each multiplier coefficient is shown in Figure 12.38. The gain responses of the filter with infinite precision multiplier coefficients and that with the quantized coefficients are shown in Figure 12.39. The low passband sensitivity properties of the parallel allpass structure are evident from this figure. Figure 12.40 depicts the passband details of the gain response of a direct form realization of the above fifth-order lowpass filter with infinite precision multiplier coefficients and that with the quantized coefficients. The poor passband sensitivity of the direct form realization is clearly seen in this plot.

It should be noted that the power-complementary transfer function $H(z)$ realized in the form of Eq. (12.169) also remains BR if the allpass filters are structurally LBR. Hence, it exhibits low sensitivity in its passband [which is the stopband of $G(z)$]. However, low passband sensitivity of $H(z)$ does not imply low stopband sensitivity of $G(z)$ (Problem 12.36).

A number of other methods have been advanced for the realization of BR IIR digital transfer functions [Dep80], [Hen83], [Rao84], [Vai84], [Vai85a].

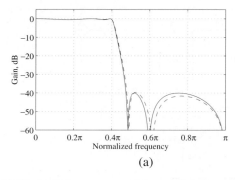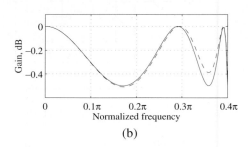

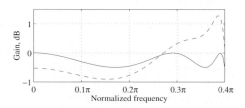

Figure 12.39: (a) Gain responses of the parallel allpass realization with infinite precision coefficients (shown with solid line) and with quantized coefficients (dashed line) and (b) passband details.

Figure 12.40: Passband responses of the direct form realization with infinite precision coefficients (shown with solid line), and with quantized coefficients (dashed line).

12.9.3 Low Passband Sensitivity FIR Digital Filter

In many applications linear-phase FIR filters are preferred. As indicated in Section 7.3, there are four types of linear-phase FIR filters of which the Type 1 filter is the most general and can realize any type of frequency response. We therefore restrict our attention here to a BR Type 1 FIR transfer function and outline a simple approach to realize it in a structurally passive form, thus ensuring low passband sensitivity to multiplier coefficients [Vai85b].

Now, from Eq. (7.39), the frequency response of a Type 1 FIR filter of order N can be expressed as

$$H(e^{j\omega}) = e^{-j\omega N/2}\breve{H}(\omega) \tag{12.171}$$

where $\breve{H}(\omega)$, a real function of ω, is its amplitude response and given by Eq. (7.49). Since $H(z)$ is a BR function, $\breve{H}(\omega) \leq 1$. Its delay-complementary filter $G(z)$ defined by (see Section 7.5.1)

$$G(z) = z^{-N/2} - H(z) \tag{12.172}$$

has a frequency response given by

$$G(e^{j\omega}) = e^{-j\omega N/2}\left[1 - \breve{H}(\omega)\right] = e^{-j\omega N/2}\breve{G}(\omega), \tag{12.173}$$

where $\breve{G}(\omega) = 1 - \breve{H}(\omega)$ is its amplitude response. Amplitude responses of a typical delay-complementary FIR filter pair are depicted in Figure 12.41. It follows from this figure that at $\omega = \omega_k$, where $\left|H(e^{j\omega_k})\right| = 1$, $\breve{G}(\omega)$ has double zeros. Thus, $G(z)$ can be expressed as

$$G(z) = G_a(z)\prod_{k=1}^{L}\left(1 - 2\cos\omega_k z^{-1} + z^{-2}\right)^2 = G_a(z)G_b(z). \tag{12.174}$$

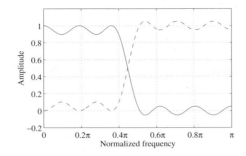

Figure 12.41: Amplitude responses of typical delay-complementary Type 1 FIR filters.

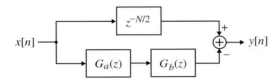

Figure 12.42: Low passband sensitivity realization of a Type 1 FIR filter $H(z)$.

A delay-complementary implementation of $H(z)$ based on Eq. (12.172) is sketched in Figure 12.42, where the FIR filter $G(z)$ is realized according to Eq. (12.174) as a cascade of $G_a(z)$ and L fourth-order FIR sections, with the kth section having a transfer function $(1 - 2\cos\omega_k z^{-1} + z^{-2})^2$. Now if the multiplier coefficient $2\cos\omega_k$ of the kth section is quantized, its zeros are still double and remain on the unit circle. As a result, quantization of the coefficients in $G_b(z)$ does not change the sign of the amplitude response $\breve{G}(\omega)$, and in the passband of $H(z)$, $\breve{G}(\omega) \geq 0$. Moreover, $G_a(z)$ has no zeros on the unit circle, and quantization of its coefficients also does not affect the sign of $\breve{G}(\omega)$. Hence, $\breve{H}(\omega)$ continues to remain bounded above by unity, i.e., the realization of $H(z)$ as indicated in Figure 12.41 remains structurally BR or structurally passive with regard to all multiplier coefficients, resulting in a low passband sensitivity realization.

Program 10_5.m

EXAMPLE 12.13   Low Passband Sensitivity Realization of a Lowpass FIR Filter

We consider here the low passband sensitivity realization of a lowpass FIR filter using the above method. The filter specifications are length 13 with a normalized passband edge at 0.5π and a normalized stopband edge at 0.6π with equal weights to passband and stopband ripples. Using Program 10_5, we first design the lowpass filter $H(z)$ and then form its delay-complementary filter as $G(z) = z^{-6} - H(z)$. Using the function roots in MATLAB, we next determine the roots of $G(z)$, which has six zeros on the unit circle: two at $z = 1$, a pair of complex conjugate zeros at $z = -0.26463064626566 \pm j0.96434984370664$, and a pair of complex conjugate zeros at $z = -0.27683551142484 \pm j0.96091732194510$. These unit circle zeros constitute the transfer function

$$G_b(z) = (1 - z^{-1})^2(1 - 0.52926129253132z^{-1} + z^{-2})$$
$$\times (1 - 0.55367102284968z^{-1} + z^{-2}).$$

By factoring out $G_b(z)$ from $G(z)$ using the function deconv in MATLAB, we arrive at the remaining term $G_a(z)$:

$$G_a(z) = 0.04107997195619 + 0.05197154435126z^{-1} - 0.12094731168744z^{-2} - 0.30704562224137z^{-3}$$
$$- 0.12094731168744z^{-4} + 0.05197154435126z^{-5} + 0.04107997195619z^{-6}.$$

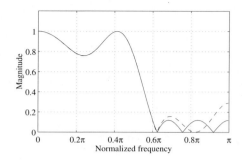

Figure 12.43: Magnitude responses of the original FIR lowpass filter (solid line) and its low sensitivity realization (dashed line).

Next, we quantize the coefficients of $G_a(z)$ and $G_b(z)$ by rounding the fractional part to two decimal digits, and from the quantized $G(z)$ determine again the delay complementary filter, as indicated in Figure 12.42. Plots of the frequency responses of the original transfer function $H(z)$ and its realization as in Figure 12.42 with quantized coefficients are shown in Figure 12.43, verifying the low passband sensitivity behavior of the proposed realization. Another novel approach to the design of low passband sensitivity FIR filters is described in [Vai86b].

12.10   Reduction of Product Round-Off Noise Using Error Feedback

We have indicated earlier in Section 12.6 that in digital filters implemented using fixed-point arithmetic, the quantization of multiplication operations can be treated as a round-off noise at the output of the filter structure and can be analyzed using a statistical model of the quantization process. In many applications, this noise may decrease the output signal-to-noise ratio to an unacceptable level. It is thus of interest to investigate techniques that can reduce the output round-off noise. In this section, we describe two possible solutions that require additional hardware. In critical applications, the cost of the additional hardware may be justified.

The basic idea behind the error-feedback approach is to make use of the difference between the unquantized and quantized signal in reducing the round-off noise. This difference, called the error, is fed back to the digital filter structure in such a way as to not change the transfer function originally implemented by the structure, while effectively decreasing the noise power [Cha81], [Hig84], [Mun81], [Thô77]. We illustrate the approach for a first-order and a second-order filter structure. The error-feedback approach is often used in designing high-precision oversampling A/D converters (see Section 15.11).

12.10.1   First-Order Error-Feedback Structure

Consider again the scaled first-order section of Figure 12.33(a). We also assume that all multiplier coefficients are signed $(b + 1)$-bit fractions. The quantization error signal is then given by

$$e[n] = y[n] - v[n]. \tag{12.175}$$

We now modify the structure of Figure 12.33(a) as shown in Figure 12.44, where now the error signal is being fed back to the system through a delay and a multiplier with a coefficient β. In practice, the coefficient β is chosen to be a simple integer or fraction, such as ± 1, ± 2, or ± 0.5, so that the multiplication can be simply performed using a shift operation and will not introduce an additional quantization error.

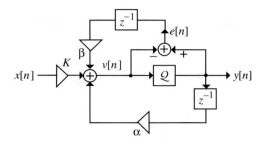

Figure 12.44: A first-order digital filter structure with error feedback.

Analyzing Figure 12.44, we arrive at the expression for the transfer function of the digital filter structure with $y[n]$ as the output as

$$H(z) = \frac{Y(z)}{X(z)}\bigg|_{E(z)=0} = \frac{K}{1 - \alpha z^{-1}}. \tag{12.176}$$

The noise transfer function $G(z)$ with the error feedback, with $y[n]$ as the output, is given by

$$G(z) = \frac{Y(z)}{E(z)}\bigg|_{X(z)=0} = \frac{1 + \beta z^{-1}}{1 - \alpha z^{-1}}, \tag{12.177}$$

and without the error feedback ($\beta = 0$) is given by

$$G(z) = \frac{1}{1 - \alpha z^{-1}}. \tag{12.178}$$

Following steps similar to that outlined in Example 9.7, we arrive at the output noise variance of the error-feedback structure as

$$\sigma_\gamma^2 = \frac{1 + 2\alpha\beta + \beta^2}{1 - \alpha^2}\sigma_0^2, \tag{12.179}$$

where σ_0^2 is the variance of the noise source $e[n]$. Note from the above that the output noise variance is a minimum when $\beta = -\alpha$. However, in practice, $|\alpha| < 1$, and hence, this choice for β will introduce an additional quantization noise source, making the analysis resulting in Eq. (12.179) invalid. Thus, from a practical point of view, it is more attractive to reduce the output round-off noise variance by choosing β as an integer with a value close to that of $-\alpha$, in which case, the only noise source is due to the quantization of $\alpha y[n-1]$.

For $|\alpha| < 0.5$, $\beta = 0$, implying no error feedback. However, in this case, the pole of $H(z)$ is far from the unit circle, and as a result, the noise variance is not high. For $|\alpha| \geq 0.5$, we choose $\beta = (-1)\text{sgn}(\alpha)$.[4] Substituting this value of β in Eq. (12.179), we arrive at

$$\sigma_\gamma^2 = \frac{2}{1 + |\alpha|}\sigma_0^2. \tag{12.180}$$

Comparing Eqs. (12.180) and (12.179) with $\beta = 0$, we note that the introduction of error feedback has increased the SNR by a factor of $-10\log_{10}[2(1 - |\alpha|)]$. The above increase in SNR is quite significant if

---

[4]$\text{sgn}(\alpha) = +1$ for $\alpha \geq 0$, and $\text{sgn}(\alpha) = -1$ for $\alpha < 0$.

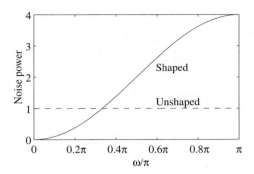

Figure 12.45: The normalized error power spectrum of the first-order section with and without error feedback.

the pole is closer to the unit circle. For example, if $|\alpha| = 0.99$, the improvement is about 17 dB, which is equivalent to about 3 bits of increased accuracy compared to the case without error feedback. The additional hardware requirements for the structure of Figure 12.44 are two new adders and an additional storage register.

Comparing the two expressions of Eqs. (12.177) and (12.178), we conclude that the noise transfer function with error feedback is given by that without the error feedback multiplied by the expression $(1 + \beta z^{-1})$. Equivalently, the error-feedback circuit is *shaping* the error spectrum by modifying the input quantization noise $E(z)$ to $E_s(z) = (1 + \beta z^{-1})E(z)$. The output noise is generated by passing $E_s(z)$ through the usual noise transfer function of Eq. (12.178).

To examine the effect of noise spectrum shaping, consider the case of a narrowband lowpass first-order filter with $\alpha \to 1$. In this case, we choose $\beta = -1$, and as a result, $E_s(z)$ has a zero at $z = 1$ (i.e., $\omega = 0$). The power spectral density of the unshaped quantization noise $E(z)$ is σ_0^2, a constant. The power spectral density of the shaped noise source $E_s(z)$ is given by $4\sin^2(\omega/2)\sigma_0^2$ and is plotted in Figure 12.45, along with that of the unshaped case. As can be seen from this figure, the noise shaping basically redistributes the noise so as to move it mostly into the stopband of the lowpass filter, thus reducing the noise variance. Because of the noise redistribution caused by the error-feedback circuit, this method of round-off noise reduction has also been called the *error-spectrum shaping* approach in the literature [Hig84].

12.10.2 Second-Order Error-Feedback Structure

The error-feedback approach for round-off noise reduction has also been applied to second-order IIR digital filter structures [Hig84], [Mun81]. One proposed structure obtained by modifying Figure 12.34(a) is indicated in Figure 12.46. The transfer function $H(z)$ of this structure is given by Eq. (12.158). It should be noted that the inclusion of the error-feedback circuit to the structure of Figure 12.34(a) does not affect either $H(z)$ or the scaling transfer function $F(z)$. Analyzing Figure 12.46, we arrive at the expression for the noise transfer function

$$G(z) = \frac{1 + \beta_1 z^{-1} + \beta_2 z^{-2}}{1 + \alpha_1 z^{-1} + \alpha_2 z^{-2}}. \tag{12.181}$$

The output round-off noise variance for $\mathcal{L}_2$-scaling is given by

$$\sigma_\gamma^2 = (\|G\|_2)^2 \sigma_0^2. \tag{12.182}$$

Note that a choice of $\beta_1 = \alpha_1$ and $\beta_2 = \alpha_2$ makes $\|G\|_2 = 1$, yielding $\sigma_\gamma^2 = \sigma_0^2$, an apparent optimal solution. However, this choice for the multiplier coefficients in the error-feedback path introduces additional

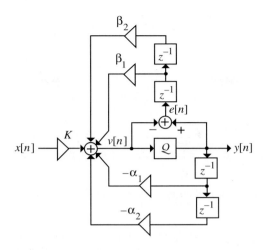

Figure 12.46: A second-order digital filter with error feedback.

quantization noise sources that were not taken into account in the above analysis. As in the case of the first-order section with error feedback, a more attractive solution is to make β_1 and β_2 integers with values close to α_1 and α_2, respectively. For example, for a narrowband lowpass transfer function, the poles are close to the unit circle and to the real axis; that is, $r \approx 1$ and $\theta \approx 0$. In this case, α_1 is close to -2, and α_2 is close to 1. We therefore choose here $\beta_1 = -2$ and $\beta_2 = 1$, resulting in a noise transfer function

$$G(z) = \frac{1 - 2z^{-1} + z^{-2}}{1 + \alpha_1 z^{-1} + \alpha_2 z^{-2}}. \tag{12.183}$$

It has been shown by Vaidyanathan [Vai87c] that for a very narrowband lowpass filter with $r = 0.995$, $\theta = 0.07\pi$, and $b = 16$, the second-order error-feedback structure has an SNR that is approximately 25 dB higher than that without the error feedback. A detailed comparison of the second-order section with and without the error feedback for various types of inputs is left as an exercise (Problem 12.37).

It should be noted that here also the error-feedback circuit provides noise shaping, as in the first-order case. In fact, for the parameters indicated above, the noise transfer function with error feedback is given by that without error-feedback multiplied by the expression $(1 - z^{-1})^2$. Or in other words, the error-feedback circuit is *shaping* the error spectrum by modifying the input quantization noise $E(z)$ to $E_s(z) = (1 - z^{-1})^2 E(z)$. The output noise is generated by passing $E_s(z)$ through the usual noise transfer function of Eq. (12.181), with $\beta_1 = \beta_2 = 0$. The power spectral density of the shaped noise source $E_s(z)$ is given by $16 \sin^4(\omega/2)\sigma_0^2$, whereas that of the unshaped case is simply σ_0^2. These power spectral densities have been plotted in Figure 12.47. Here also the error feedback lowers the noise in the passband by pushing it into the stopband of the filter.

12.11   Limit Cycles in IIR Digital Filters

So far we have treated the analysis of finite wordlength effects using a linear model of the system. However, a practical digital filter is a nonlinear system caused by the quantization of the arithmetic operations. Such nonlinearities may cause an IIR filter, which is stable under infinite precision, to exhibit an unstable behavior under finite precision arithmetic for specific input signals, such as zero or constant inputs. This type of instability usually results in an oscillatory periodic output called a *limit cycle,* and the system remains

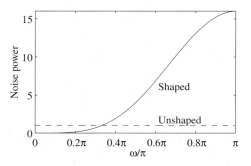

Figure 12.47: Normalized error power spectrum of the second-order section with and without error feedback.

in this condition until an input of sufficiently large amplitude is applied to move the system into a more conventional operation.

In applications where the digital filter is operative at all times, oscillatory output in the absence of an input is highly undesirable. In particular, limit cycles with a frequency of oscillation in the audio frequency range can be quite annoying to the listener in audio and musical sound processing applications.

It should be noted that limit cycles occur in IIR filters due to the presence of a feedback path. Such oscillations are absent in FIR structures, which do not have any feedback path.

There are basically two types of limit cycles: (1) *granular* and (2) *overflow*. The former type of limit cycle is usually of low amplitude, whereas overflow oscillations have large amplitudes. We examine both types of limit cycles in this section.

12.11.1 Granular Limit Cycles

Two types of granular limit cycles have been observed in IIR digital filters: *inaccessible* and *accessible limit cycles* [Cla73]. The former type can appear only if the initial conditions of the digital filter at the time of starting pertain to that limit cycle, whereas in the second case, the limit cycle condition can be reached by starting the digital filter with initial conditions not pertaining to that limit cycle. We illustrate the generation of the limit cycle by analyzing the nonlinear behavior of a first-order and a second-order IIR digital filter.

EXAMPLE 12.14   Granular Limit Cycles in a First-Order IIR Digital Filter

Consider the causal first-order digital filter of Figure 12.1 characterized by the difference equation of Eq. (12.1), where for stability $|\alpha| < 1$. We assume that the quantization operation being performed is rounding and is carried out on the result of the multiplication, as indicated in Figure 12.48. In this case, Eq. (12.1) becomes the nonlinear difference equation

$$\hat{y}[n] = \mathcal{Q}\left(\alpha \hat{y}[n-1]\right) + x[n], \tag{12.184}$$

where $\hat{y}[n]$ denotes the actual output of the filter. Without any loss of generality, we assume the digital filter is being implemented using a signed 6-bit fractional arithmetic with a quantization step of $\delta = 2^{-5}$. Table 12.6 shows the first seven output samples for two different pole positions for an impulse input with $x[0] = 0_\Delta 1101$ and $x[n] = 0$ for $n > 0$, and an initial condition $\hat{y}[-1] = 0$. Observe that the steady-state output in the first case is a nonzero constant, that is, periodic with a period of 1, whereas in the second case, it is periodic with a period of 2. On the other hand, with infinite precision, the ideal output goes to zero exponentially as $n \to \infty$.

The above types of oscillations at the output are called *zero-input limit cycles*, and the amplitude ranges of the oscillations are often called *dead bands* [Bla65]. A digital filter exhibiting limit cycles at its output can be modelled

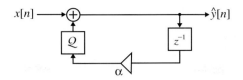

Figure 12.48: A first-order IIR filter with a quantizer.

Table 12.6: Limit cycle behavior of the first-order IIR digital filter.

n	$\alpha = 0_\Delta 1011,\ \hat{y}[-1] = 0$		$\alpha = 1_\Delta 1011,\ \hat{y}[-1] = 0$	
	$\alpha\hat{y}[n-1]$	$\hat{y}[n]$	$\alpha\hat{y}[n-1]$	$\hat{y}[n]$
0	0	$0_\Delta 1101$	0	$0_\Delta 1101$
1	$0_\Delta 10001111$	$0_\Delta 1001$	$1_\Delta 10001111$	$1_\Delta 1001$
2	$0_\Delta 01100011$	$0_\Delta 0110$	$0_\Delta 01100011$	$0_\Delta 0110$
3	$0_\Delta 01000010$	$0_\Delta 0100$	$1_\Delta 01000010$	$1_\Delta 0100$
4	$0_\Delta 00101100$	$0_\Delta 0011$	$0_\Delta 00101100$	$0_\Delta 0011$
5	$0_\Delta 00100001$	$0_\Delta 0010$	$1_\Delta 00100001$	$1_\Delta 0010$
6	$0_\Delta 00010110$	$0_\Delta 0001$	$0_\Delta 00010110$	$0_\Delta 0001$
7	$0_\Delta 00001011$	$0_\Delta 0001$	$1_\Delta 00001011$	$1_\Delta 0001$
8	$0_\Delta 00001011$	$0_\Delta 0001$	$0_\Delta 00001011$	$0_\Delta 0001$

by a linear system with its poles on the unit circle [Jac69]. Using this representation, it is possible to determine the range of the dead band in low-order filters. For example, in the case of the first-order IIR filter considered above, we assume that the system under the limit cycle condition has an effective pole at $z = 1$ when $\alpha > 0$ and at $z = -1$ when $\alpha < 0$. This implies that effectively

$$\mathcal{Q}\left(\alpha\hat{y}[n-1]\right) = \begin{cases} \hat{y}[n-1], & \alpha > 0, \\ -\hat{y}[n-1], & \alpha < 0. \end{cases} \tag{12.185}$$

In addition, the quantization error due to rounding is bounded by $\pm\delta/2$, where δ is the quantization step; that is,

$$\left|\mathcal{Q}\left(\alpha\hat{y}[n-1]\right) - \alpha\hat{y}[n-1]\right| \le \frac{\delta}{2}. \tag{12.186}$$

From Eqs. (12.185) and (12.186), we arrive at the dead band range of the first-order IIR filter as

$$\left|\hat{y}[n-1]\right| \le \frac{\delta}{2\left(1 - |\alpha|\right)}. \tag{12.187}$$

As a result, if for any value of n, the output $\hat{y}[n-1]$ of the delay unit is in the above range with the input set to zero, then the system gets trapped into a limit cycle mode. For our numerical example cited above, the dead band range is $|\hat{y}[n-1]| \le 0.1$, which definitely is satisfied by the entries in Table 12.6.

The limit cycle generation can be easily illustrated on a computer. MATLAB Program 12_7 can be used to study the granular limit cycle process. This program uses the function a2dR of Section 12.4.1 to develop the decimal equivalent of the binary representation of the filter coefficient with N bits for the

Program 12_7.m

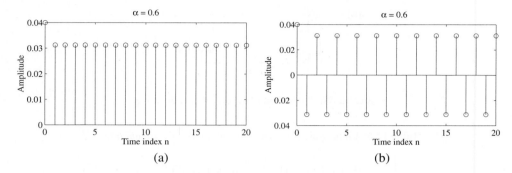

Figure 12.49: Illustration of limit cycles in a first-order IIR digital filter: (a) $\alpha = 0.6$ and (b) $\alpha = -0.6$.

magnitude after rounding. Figure 12.49 shows the plots of the first 21 samples of the output response of the first-order IIR digital filter of Eq. (12.1) implemented with filter coefficients rounded to 6 bits. The input is $x[n] = 0.04\delta[n]$ and the filter coefficients are $\alpha = \pm 0.6$. The initial condition $y[-1]$ is set to 0.

EXAMPLE 12.15   Granular Limit Cycles in a Second-Order IIR Digital Filter

In the case of a second-order IIR filter, a similar situation exists, except there are various types of limit cycles here. Consider the all-pole second-order IIR digital filter of Figure 12.26(a). It is described by the difference equation

$$y[n] = -\alpha_1 y[n-1] - \alpha_2 y[n-2] + x[n]. \tag{12.188}$$

The poles of this system are at

$$z = -\frac{\alpha_1}{2} - \frac{\sqrt{\alpha_1^2 - 4\alpha_2}}{2}, \tag{12.189}$$

which are complex for $\alpha_1^2 < 4\alpha_2$ and on the unit circle if $\alpha_2 = 1$. If we assume the products are first quantized by rounding, as indicated in Figure 12.50, the difference equation of Eq. (12.188) becomes

$$\hat{y}[n] = \mathcal{Q}\left(\alpha_1 \hat{y}[n-1]\right) - \mathcal{Q}\left(\alpha_2 \hat{y}[n-2]\right) + x[n], \tag{12.190}$$

with $\hat{y}[n]$ denoting the actual output. One way the system of Figure 12.50 could be in the limit cycle mode with zero input is if its effective pole is on the unit circle. In this case,

$$\mathcal{Q}\left(\alpha_2 \hat{y}[n-2]\right) = \hat{y}[n-2]. \tag{12.191}$$

Moreover, because of rounding, we have

$$\left|\mathcal{Q}\left(\alpha_2 \hat{y}[n-2]\right) - \alpha_2 \hat{y}[n-2]\right| \le \frac{\delta}{2}. \tag{12.192}$$

From Eqs. (12.191) and (12.192), we arrive at the dead band region governing the limit cycle mode:

$$\left|\hat{y}[n-2]\right| \le \frac{\delta}{2\left|1+\alpha_2\right|}, \tag{12.193}$$

with α_1 determining the frequency of oscillation.

Limit cycles can also occur in second-order IIR filters if the system with quantizers has effective poles at $z = \pm 1$. In this mode, the dead band is bounded by $2\delta/(1 - |\alpha_1| + \alpha_2)$ [Jac69].

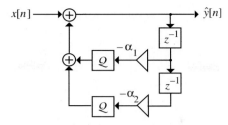

Figure 12.50: A second-order IIR section with quantizers after the multipliers.

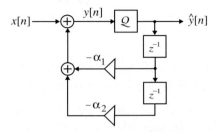

Figure 12.51: A second-order IIR section with a quantizer after the accumulator.

It should be noted that the limit cycles occurring with the amplitude bound of Eq. (12.193) are basically inaccessible limit cycles, and it is highly unlikely, in practice, that the digital filter will start with initial conditions pertaining to these limit cycles [Cla73]. On the other hand, for the second-order IIR filter structure of Figure 12.50, with arbitrary initial conditions, accessible limit cycles are highly likely to occur, and their amplitude can only be bounded from below:

$$\left| \hat{y}[n-2] \right| \geq \frac{\delta}{2\,|1+\alpha_2|}. \tag{12.194}$$

Even though we have considered here the zero-input limit cycle generation in very simple IIR digital filter structures, limit cycles also occur in higher-order structures. However, their analysis is almost impossible, except for the determination of the bounds on the amplitudes of the limit cycles [Lon73]. In some structures, periodic limit cycles have been observed with nonzero constant amplitudes and also for constant amplitude sinusoidal inputs.

12.11.2 Overflow Limit Cycles

As indicated earlier, limit-cycle-like oscillations can also result from overflow in digital filters implemented with finite precision arithmetic. The amplitude of the overflow oscillations can cover the whole dynamic range of the register experiencing the overflow and are much more serious in nature than the granular type. We illustrate the generation of an overflow oscillation on a computer in Example 12.16.

EXAMPLE 12.16   Overflow Limit Cycles in a Second-Order IIR Digital Filter

We consider the causal all-pole second-order IIR digital filter of Figure 12.26(a). We assume its implementation using sign-magnitude arithmetic with a rounding of the sum of products by a single quantizer, as indicated in Figure 12.51. In this case, the linear difference equation of Eq. (12.188) describing the ideal filter reduces to the nonlinear difference equation given by

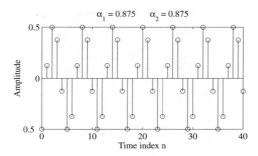

Figure 12.52: Overflow limit cycles of Figure 12.51.

$$\hat{y}[n] = \mathcal{Q}\left(-\alpha_1\hat{y}[n-1] - \alpha_2\hat{y}[n-2] + x[n]\right), \tag{12.195}$$

where $\mathcal{Q}(\cdot)$ represents the rounding operation and $\hat{y}[n]$ denotes the actual output of the filter. All numbers are assumed to be signed 4-bit fractions.

Let the filter coefficients be given by $\alpha_1 = 1_\triangle001 = -0.875_{10}$ and $\alpha_2 = 0_\triangle111 = 0.875_{10}$. The initial conditions are assumed to be $\hat{y}[-1] = -0.625_{10}$ and $\hat{y}[-2] = -0.125_{10}$. We consider the zero-input case; that is, $x[n] = 0$ for $n \geq 0$.

Program 12_8 can be used to illustrate the overflow limit cycle generation. It uses the function a2dR of Section 12.4.1 to perform the rounding operation on the sum of products, as indicated in Eq. (12.195).

Figure 12.52 shows the output generated by the above program demonstrating the generation of overflow limit cycles with zero input. It should be noted that, as shown in the following section, the structure of Figure 12.51 does not exhibit overflow limit cycles if sign-magnitude truncation is used to quantize the sum of products. This property can be easily demonstrated by replacing the function a2dR in Program 12_8 with the function a2dT of Section 12.4.1.

Program 12_8.m

For stability, we have shown in Section 7.9.1 that the filter coefficients of the second-order direct form IIR structure of Figure 12.51 must remain inside the stability triangle of Figure 12.53. However, the structure can still get into a zero-input overflow oscillation mode for a large range of values of the filter constants, satisfying this stability constraint when implemented using two's-complement arithmetic with rounding. It has been shown that overflow limit cycles under zero-input cannot occur if the filter coefficients lie in the shaded region inside the stability triangle, as indicated in Figure 12.53 [Ebe69]. This region is defined by the relation

$$|\alpha_1| + |\alpha_2| < 1. \tag{12.196}$$

As the above condition is quite restrictive, we now examine other IIR structures that do not sustain limit cycles with less severe constraint on the pole locations.

12.11.3 Limit Cycle Free Structures

A number of authors have advanced digital filter structures that are limit cycle free when implemented using specific arithmetic schemes. The most general approach to the development of such structures is based on the state-space representation [Mil78]. In Chapter 2, we considered the time-domain description of an LTI discrete-time system in terms of the convolution sum and the linear constant coefficient difference equation relating the input and the output signals. Another time-domain representation for a causal LTI discrete-time system is in terms of internal variables called the state variables, which are usually the output variables of all unit delays. For a second-order causal LTI discrete-time system, the state-space representation relating the output sequence $y[n]$ to the input sequence $x[n]$ is given by

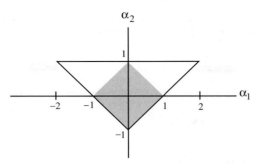

Figure 12.53: Coefficient ranges of a second-order direct form IIR filter to guarantee no overflow oscillations (shown by shaded region).

$$\begin{bmatrix} s_1[n+1] \\ s_2[n+1] \end{bmatrix} = \begin{bmatrix} a_{11} & a_{12} \\ a_{21} & a_{22} \end{bmatrix} \begin{bmatrix} s_1[n] \\ s_2[n] \end{bmatrix} + \begin{bmatrix} b_1 \\ b_2 \end{bmatrix} x[n], \tag{12.197}$$

$$y[n] = [\,c_1 \quad c_2\,] \begin{bmatrix} s_1[n] \\ s_2[n] \end{bmatrix} + dx[n]. \tag{12.198}$$

Denoting[5]

$$\mathbf{s}[n] = [s_1[n] \quad s_2[n]]^t, \tag{12.199}$$

$$\mathbf{A} = \begin{bmatrix} a_{11} & a_{12} \\ a_{21} & a_{22} \end{bmatrix}, \qquad \mathbf{B} = \begin{bmatrix} b_1 \\ b_2 \end{bmatrix}, \qquad \mathbf{C} = [\,c_1 \quad c_2\,], \tag{12.200}$$

we can rewrite Eqs. (12.197) and (12.198) in a compact form as

$$\mathbf{s}[n+1] = \mathbf{A}\,\mathbf{s}[n] + \mathbf{B}x[n], \tag{12.201}$$

$$y[n] = \mathbf{C}\,\mathbf{s}[n] + dx[n]. \tag{12.202}$$

In the above equations, $\mathbf{s}[n]$, given by Eq. (12.199), is called the *state vector*, with its elements $s_i[n]$ known as the *state variables*. The matrix $\mathbf{A}$ given in Eq. (12.200) is called the *state transition matrix*.

Even though in an actual implementation, the right-hand sides of both Eqs. (12.201) and (12.202) are quantized, the quantization errors caused by the quantization of the right-hand side of Eq. (12.201) go through the feedback loop and are responsible for the generation of limit cycles. We assume that the variables $s_1[n+1]$ and $s_2[n+1]$ are quantized, and the delayed versions of these quantized signals are the state variables $s_1[n]$ and $s_2[n]$, respectively.

We define a quantizer to be *passive* if

$$|\mathcal{Q}(x)| \leq |x|, \qquad \text{for all } x. \tag{12.203}$$

If x is inside the given dynamic range of the system, then for magnitude truncation (Section 12.2), it is evident that Eq. (12.203) holds; that is, the quantizer is passive. If x is outside the dynamic range caused,

---

[5]In this text, row and column vectors, in general, are indicated by boldface lowercase letters, while the matrices are represented by boldface uppercase letters.

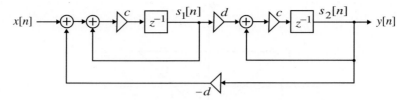

Figure 12.54: A second-order IIR section.

for example, by overflow, it must be brought back to the range by following either the saturation arithmetic scheme or the two's-complement overflow scheme discussed in Section 11.10. As a result, magnitude truncation followed by either of the above two overflow handling schemes is again a passive quantizer.

A digital filter structure with a state transition matrix satisfying

$$\mathbf{A}^t \mathbf{A} = \mathbf{A} \mathbf{A}^t \tag{12.204}$$

has been called a *normal form structure.* It has been shown that such a structure with passive quantizers does not support zero-input limit cycles of either type [Mil78]. The matrix $\mathbf{A}$ satisfying the above condition and $\|\mathbf{A}\|_2 < 1$ is called a *normal matrix.*

EXAMPLE 12.17   A Second-Order Digital Filter Structure With No Zero-Input Limit Cycles

Consider the digital filter structure of Figure 12.54 [Yan82]. Analysis yields

$$s_1[n + 1] = cs_1[n] - cds_2[n] + cx[n], \tag{12.205a}$$
$$s_2[n + 1] = cds_1[n] + cs_2[n]. \tag{12.205b}$$

From the above, we easily identify the state transition matrix as

$$\mathbf{A} = \begin{bmatrix} c & -cd \\ cd & c \end{bmatrix}. \tag{12.206}$$

Comparing the determinant of $(z\mathbf{I} - \mathbf{A})$ with the denominator of a second-order IIR transfer function with poles at $z = re^{\pm j\theta}$ (with $r < 1$ for stability), we obtain

$$c = r\cos\theta, \qquad d = \tan\theta. \tag{12.207}$$

Substituting the above values in Eq. (12.206), we arrive at

$$\mathbf{A} = \begin{bmatrix} r\cos\theta & -r\sin\theta \\ r\sin\theta & r\cos\theta \end{bmatrix}. \tag{12.208}$$

Note that $\mathbf{A}^t \mathbf{A} = \mathbf{A} \mathbf{A}^t = r^2 \mathbf{I}$. Since $r < 1$, we have $\|\mathbf{A}\|_2 = r < 1$. Therefore, the filter of Figure 12.54 is a normal form structure and will not exhibit zero-input limit cycles of either type.

12.11.4   Random Rounding

A conceptually simple technique to suppress zero-input granular limit cycles in second-order IIR sections is called *random rounding* [Büt77], [Law78]. To explain the operation of this method, let the register storing the signal $v[n]$ prior to the quantization be as indicated in Figure 12.55. In the random rounding method, an uncorrelated binary random sequence $x[n]$ is generated externally by a random number generator, and

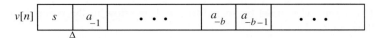

Figure 12.55: Register storing the signal to be quantized.

the $(b + 1)$th bit a_{-b-1} of $v[n]$ is replaced with $x[n]$. The modified signal $v[n]$ is then passed through a passive quantizer. In addition to the increase in hardware complexity, this approach also results in a slightly higher quantization noise that is, however, distributed over the whole frequency range.

12.12   Round-Off Errors in FFT Algorithms

Since FFT is often employed in a number of digital signal processing applications, it is of interest to analyze the effects of finite wordlengths in FFT computations. The most critical error in the computation is that due to the arithmetic round-off errors. As in the earlier sections, we assume that the DFT computations are being carried out using fixed-point arithmetic, and we thus restrict our analysis to the effect of product round-off errors in the DFT computation via the FFT algorithms and to compare them with the errors generated in the direct DFT computation [Opp89], [Pro96], [Wel69]. The model for the round-off error analysis to be employed here is the same as in the case of LTI digital filters and is shown in Figure 12.20.

Now the multiplier coefficients W_N^{kn} and the signals in the DFT computation are, in general, complex numbers resulting in complex multiplications. Since each complex multiplication usually requires four real multiplications, there are four quantization errors per multiplication. If the DFT computation requires K complex multiplications, there are $4K$ sources of quantization errors in the computational structure. We make the usual assumptions about the statistical properties of the noise sources:

(a) All $4K$ errors are uncorrelated with each other and uncorrelated with the input sequence.

(b) The quantization errors are random variables uniformly distributed with a variance $\sigma_0^2 = 2^{-2b}/12$, assuming a signed b-bit fractional fixed-point arithmetic.

12.12.1   Direct DFT Computation

Recall that the N-point DFT $X[k]$ of a length-N complex sequence $x[n]$ is given by

$$X[k] = \sum_{n=0}^{N-1} x[n] W_N^{nk}, \qquad 0 \leq k \leq N - 1. \tag{12.209}$$

Thus, the computation of a single DFT sample requires N complex multiplications, and hence, the total number of real multiplications for the computation of a single DFT sample is $4N$. As a result, there are $4N$ quantization error sources. The variance of the error in the computation of one DFT sample is therefore[6]

$$\sigma_\gamma^2 = 4N\sigma_0^2 = \frac{2^{-2b}N}{3}, \tag{12.210}$$

indicating that the output round-off error is proportional to the DFT length.

---

[6]Strictly speaking, the output noise variance will be less than the value here since multiplications with W_N^0 and $W_N^{N/2}$ do not develop any error.

Now, the input sequence $x[n]$ must be scaled to avoid overflow in the computation of $X[k]$. From Eq. (12.209), it follows that

$$|X[k]| \leq \sum_{n=0}^{N-1} |x[n]| < N, \qquad (12.211)$$

assuming that the input samples satisfy the dynamic range constraint $|x[n]| \leq 1$. To prevent overflow, we need to ensure that

$$|X[k]| < 1, \qquad (12.212)$$

which can be guaranteed by dividing each sample $x[n]$ in the input sequence by N.

To analyze the effect of the above scaling, assume the input to be a white noise sequence, with each sample uniformly distributed in the range $(-1/N, 1/N)$ [Pro96]. The input signal power is then given by

$$\sigma_x^2 = \frac{(2/N)^2}{12} = \frac{1}{3N^2}. \qquad (12.213)$$

The corresponding output signal power is

$$\sigma_X^2 = N\sigma_x^2 = \frac{1}{3N}. \qquad (12.214)$$

As a result, the signal-to-noise ratio is

$$\text{SNR} = \frac{\sigma_X^2}{\sigma_\gamma^2} = \frac{2^{2b}}{N^2}. \qquad (12.215)$$

The above expression indicates that the SNR has been reduced by a factor of N^2 due to scaling and the round-off error. The wordlength size needed to compute a DFT of given length with a desired SNR can be determined using Eq. (12.215) (Problem 12.44).

12.12.2 DFT Computation via FFT Algorithm

We now consider the round-off error analysis of the DFT computation based on an FFT algorithm. Without any loss of generality, we analyze the decimation-in-time radix-2 FFT algorithm. However, the results derived can be easily extended to other types of fast DFT algorithms.

From the flow-graph of the DIT FFT algorithm given in Figure 11.24, it can be seen that the DFT samples are computed by a series of butterfly computations with a single complex multiplication per butterfly module. Some of the butterfly computations require multiplications by ± 1 or $\pm j$ that we do not treat separately here to simplify the analysis.

Consider now the computation of a single DFT sample, as indicated in Figure 12.56. It follows from this figure that the computation of a single DFT sample involves $\nu = \log_2 N$ stages. The number of butterflies in a particular stage depends on the stage's location in the computational chain with $N/2^r = 2^{\nu-r}$ butterflies in the rth stage, where $r = 1, 2, \ldots, \nu$. The total number of butterflies involved per DFT sample is therefore

$$1 + 2 + 2^2 + \cdots + 2^{\nu-2} + 2^{\nu-1} = 2^\nu - 1 = N - 1. \qquad (12.216)$$

It also follows from Figure 12.56 that the quantization errors introduced at the rth stage appear at the output after propagating through $(r - 1)$ stages, while getting multiplied by the twiddle factors at each subsequent stage. Since the magnitude of the twiddle factors is always unity, the variances of the quantization errors do not change while propagating to the output. The total number of error sources

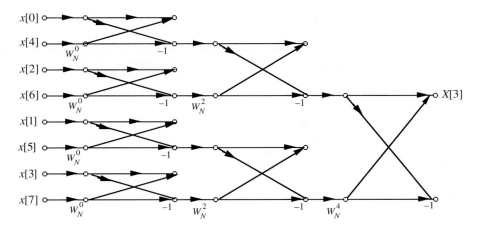

Figure 12.56: Reduced flow-graph for the computation of $X[3]$.

contributing to the output round-off error is $4(N-1)$. Assuming that the quantization errors introduced in each butterfly are uncorrelated with those generated at other butterflies, the variance of the output round-off error is then

$$\sigma_\gamma^2 = 4(N-1)\frac{2^{-2b}}{12} \cong \frac{2^{-2b}N}{3}. \tag{12.217}$$

It should be noted that the above expression for σ_γ^2 is identical to that derived for the direct DFT computation case given by Eq. (12.210). This is to be expected since the FFT algorithm does not alter the total number of complex multiplications to compute a single DFT sample; rather, it organizes the computations more efficiently so that the number of multiplications to compute all N DFT samples is reduced.

Now, to prevent overflow at the output if we scale the input samples to satisfy the condition $|x[n]| < 1/N$, the SNR obtained in the FFT algorithm remains the same as given in Eq. (12.215). However, in the case of the FFT algorithm, the SNR can be improved by following a different scaling rule, as described below.

Instead of scaling the input samples by $1/N$, we can scale the input signals at each stage by $1/2$. This scaling rule also guarantees that the output DFT samples are scaled by a factor of $(1/2)^\nu = 1/N$ as desired. However, each scaling by a factor of $1/2$ reduces the round-off noise variance by a factor of $1/4$. As a result, the round-off noise variances of the $4(2^{\nu-r})$ noise sources at the rth stage are reduced by a factor of $1/4^{r-1}$, while the noise propagates to the output. It can be shown that the total round-off noise variance at the output is now given by

$$\sigma_\gamma^2 = \tfrac{2}{3} \cdot 2^{-2b} \left(1 - \tfrac{1}{N}\right) \cong \tfrac{2}{3} \cdot 2^{-2b}, \tag{12.218}$$

assuming N to be large enough so that $\frac{1}{N} \ll 1$ (Problem 12.45). The SNR in this case reduces to

$$\frac{\sigma_X^2}{\sigma_\gamma^2} = \frac{2^{2b}}{2N}. \tag{12.219}$$

Hence, distribution of the scaling into each stage has increased the SNR by a factor of N. The formula given in Eq. (12.219) can be used to determine the wordlength required to achieve a specified SNR for computing DFT of a given length N (Problem 12.46).

12.13   Summary

This chapter is concerned with the effects of the finite wordlengths caused by the actual implementation of a digital signal processing algorithm that causes the results of the algorithm to be different, in general, from the desired ones obtained in the ideal case with infinite precision wordlengths. To develop the appropriate models for the analysis of these effects prior to actual implementation, we first review the quantization process used to fit the infinite precision data into a finite wordlength register and the resulting errors caused by it. The quantization of both fixed-point and floating-point numbers are considered. However, the discussion in the rest of the chapter is restricted to fixed-point implementations.

The effect of the quantization of the multiplier coefficients in the implementation of a digital filter is considered next as a coefficient sensitivity problem. Simple formulas are derived for such sensitivity analysis for the infinite impulse response (IIR) filter and the finite impulse response (FIR) filter.

In the digital processing of a continuous-time signal, the latter is sampled periodically by a sample-and-hold device and then converted into a digital form by an analog-to-digital (A/D) converter. A statistical model is developed for the analysis of the input quantization error caused by the A/D conversion and is used to derive the expression for the signal-to-quantization ratio as a function of the A/D converter wordlength. The A/D quantization error propagates to the output of the digital filter processing the digitized continuous-time signal, appearing as a noise added to the desired output, and methods for the statistical analysis of the output error are provided.

The effect of product round-off in the fixed-point implementation of a digital filter is then analyzed using a statistical model. A statistical analysis of the output noise caused by the propagation of these internally generated errors also to the output of the digital filter is provided. An overflow may occur at certain internal nodes in a digital filter implemented in fixed-point arithmetic, which can result in a large amplitude oscillation at the filter output. Methods of scaling the internal signal variables with the aid of suitably placed scaling multipliers to minimize the probability of overflow are discussed. The performance of the cascade form of digital filters is particularly examined in depth to illustrate the effect of pole-zero pairing and the ordering of the low-order sections. A detailed analysis of the output signal-to-noise ratios of scaled first-order and second-order IIR digital filter sections is then provided.

Conditions for the low passband sensitivity realizations of IIR and FIR digital filters are next derived, and a method for such low sensitivity realization for each case is outlined. Two approaches to the reduction of round-off errors in IIR digital filter structures are then described.

The discretization process can also cause the occurrence of periodic oscillation, called limit cycles, at the output of an IIR digital filter under certain conditions. These limit cycles are difficult to analyze in the general case. Their existence is demonstrated here only for the first-order and second-order IIR digital filters. Conditions for the limit cycle free operation of a state-space structure is derived.

The chapter concludes with an analysis of round-off errors in the implementation of DFT and FFT algorithms.

12.14   Problems

12.1 Derive the ranges of the relative quantization errors of Table 12.2.

12.2 Compute the pole sensitivity of the first-order digital filter structure of Figure 8.34 with respect to the coefficient α.

12.3 Compute the pole sensitivities of the second-order digital filter structure of Figure 8.37 with respect to the coefficients α and β.

12.4 The digital filter structure of Figure 12.54 is more commonly known as a *modified coupled-form structure* [Yan82]. Another modified coupled-form structure is shown in Figure P12.1. Determine the transfer function of both structures, and then compute their respective pole sensitivities. Compare these sensitivities with those of the coupled-form structure of Figure 12.11 as given in Eq. (12.19).

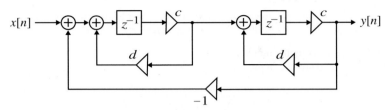

Figure P12.1

12.5 Determine the transfer function of the second-order digital filter structure shown in Figure P12.2, and compute its pole sensitivities [Aga75]. Compare these sensitivities with those of the structures of Figure 12.11 and Figure P12.1.

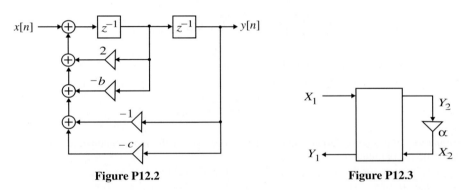

Figure P12.2   **Figure P12.3**

12.6 A third-order elliptic lowpass transfer function

$$H(z) = \frac{0.339(z + 1)(z^2 + 1.1883z + 1)}{(z - 0.0606)(z^2 + 0.5625z + 0.7386)}$$

is realized in (1) direct form and (2) cascade form. Compute the pole sensitivities of each structure.

12.7 Consider the digital filter structure of Figure P12.3 characterized by a transfer function $H(z) = Y_1/X_1$. Show that the sensitivity of $H(z)$ with respect to the multiplier coefficient α defined by $\partial H(z)/\partial \alpha$ is given by

$$\frac{\partial H(z)}{\partial \alpha} = F_\alpha(z) \cdot G_\alpha(z),$$

where $F_\alpha(z)$ is the scaling transfer function from the input node X_1 to the input node Y_2 of the multiplier α and $G_\alpha(z)$ is the noise transfer function from the output node X_2 of the multiplier α to the filter output node Y_1.

12.8 Show that the function $W_N(\omega)$ defined in Eq. (12.59) satisfies the following properties:
   (a) $0 < W_N(\omega) \leq 1$,
   (b) $W_N(0) = W_N(\pi) = 1,$ for all N,
   (c) $\lim_{N \to \infty} W_N(\omega) = \frac{1}{\sqrt{2}},$ $0 < \omega < \pi.$

12.9 Verify the SNR values given in Table 12.3.

12.10 An alternate approach to the algebraic calculation of output round-off noise variance is outlined in this problem [Pat80]. Let the partial-fraction expansion in z of an Nth-order real rational noise transfer function $H(z)$ with simple poles be given by

$$H(z) = \sum_{k=1}^{N} \frac{A_k}{z + a_k},$$

where A_k and a_k are, in general, complex numbers.

(a) Show that $H(z)$ can be expressed in the form

$$H(z) = \sum_{k=1}^{N} C_k \left(\frac{a_k z + 1}{z + a_k} \right) + B,$$

where C_k and B are constants. Determine the expressions for C_k and B.

(b) Show that the normalized output round-off noise variance σ_0^2 can be expressed in the form

$$\sigma_0^2 = \sum_{k=1}^{N} \frac{A_k^2}{1 - a_k^2} + 2 \sum_{k=1}^{N-1} \sum_{\ell=k+1}^{N} \frac{A_k A_\ell}{1 - a_k a_\ell}.$$

(c) Show that the above expression can be further simplified as

$$\sigma_0^2 = \sum_{k=1}^{N} \sum_{\ell=1}^{N} \frac{A_k A_\ell}{1 - a_k a_\ell}.$$

12.11 Determine the output noise variance due to the propagation of the input quantization noise for each of the following causal IIR digital filters:

(a) $H_1(z) = \dfrac{(z - 2)(z + 4)}{(z - 0.2)(z - 0.6)}$,   (b) $H_2(z) = \dfrac{2(3z - 1)(2z^2 - 4z + 1)}{(2z - 1)(z + 0.5)(z^2 + 0.8z + 0.5)}$,

(c) $H_3(z) = \dfrac{(z + 1)^2}{(z^2 + 0.6z + 0.4)}$,   (d) $H_4(z) = \dfrac{(z - 2.5)(z + 1.25)(z + 1)}{(z - 0.4)(z + 0.8)(z + 0.5)}$.

12.12 Determine the expression for the normalized output noise variance due to input quantization of the following digital filter realized in a parallel form

$$H(z) = A + \frac{B}{z + \beta} + \frac{C}{z + \gamma}.$$

Each section in the parallel structure is realized in direct form II. What is the value of the variance for $\beta = 0.6$, $\gamma = -0.8$, $A = 3$, $B = -4$, $C = 2$?

12.13 Realize the following transfer function

$$H(z) = \frac{p_0 + p_1 z^{-1}}{1 + d_1 z^{-1}}$$

in direct form II and direct form II$_t$ structures.

(a) Show the noise model for each unscaled structure for the computation of the product round-off noise at the output, assuming quantization of products before addition and assuming fixed-point implementation with either rounding or two's-complement truncation. Compute the expression for the normalized output round-off noise variance for each realization.

(b) Repeat Part (a), assuming quantization after addition of product.

12.14 Realize the transfer function
$$G(z) = \frac{(z + 0.4)(z - 0.3)}{(z + 0.2)(z - 0.6)}$$
in four different cascade forms, with each first-order stage implemented in direct form II.

(a) Show the noise model for each unscaled structure for the computation of the product round-off noise at the output, assuming quantization of products before addition and assuming fixed-point implementation with either rounding or two's-complement truncation. Compute the normalized output round-off noise variance for each realization. Which cascade realization has the lowest round-off noise?

(b) Repeat Part (a), assuming quantization after addition of product.

12.15 Realize the transfer function of Problem 12.14 in two different parallel forms, with each first-order stage implemented in direct form II.

(a) Show the noise model for each unscaled structure for the computation of the product round-off noise at the output, assuming quantization of products before addition and assuming fixed-point implementation with either rounding or two's-complement truncation. Compute the normalized output round-off noise variance for each realization. Which parallel realization has the lowest round-off noise?

(b) Repeat Part (a), assuming quantization after addition of products.

12.16 Realize the following second-order transfer function
$$H(z) = \frac{2 - 0.4z^{-1} - 0.6z^{-2}}{1 + 0.2z^{-1} - 0.15z^{-2}}$$
in (1) direct form, (2) cascade form, and (3) parallel form. Each section in the cascade and parallel structures is realized in direct form II. Show the noise model for each unscaled structure for the computation of the product round-off noise at the output, assuming quantization of products before addition and assuming fixed-point implementation with either rounding or two's-complement truncation. Compute the product round-off noise variance for each realization. Which realization has the lowest round-off noise? *Note*: There are four cascade and four parallel realizations.

12.17 Realize the transfer function of Problem 12.16 in the Gray–Markel form.

(a) Show the noise model for the unscaled structure for the computation of the product round-off noise variance at the output, assuming quantization of products before addition, and compute its product round-off noise variance.

(b) Repeat Part (a), assuming quantization after addition of products.

12.18 The structure of Figure P12.4 is a second-order allpass filter with a transfer function
$$A(z) = \frac{d_1 d_2 + d_1 z^{-1} + z^{-2}}{1 + d_1 z^{-1} + d_1 d_2 z^{-2}}.$$

Derive the expression for the normalized steady-state output noise variance due to product round-off, assuming fixed-point implementation with either rounding or two's-complement truncation.

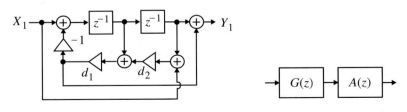

Figure P12.4   **Figure P12.5**

12.19 Develop the noise model for the product round-off noise analysis of the following second-order coupled-form structures: (a) Figure 12.11 and (b) Figure P12.1. Determine the normalized output round-off noise variances of each of the above structures due to product round-off before summation and after summation.

12.20 Develop the noise model for the product round-off noise analysis of the second-order Kingsbury structure of Figure P8.8 of Problem 8.9. Determine the normalized output round-off noise variances due to product round-off before summation and after summation.

12.21 The allpass section of Figure P12.4 is employed to alter the phase response of the structure realizing the transfer function $G(z)$ of Problem 12.14 as indicated in Figure P12.5. The allpass equalizer of Figure P12.5 has a transfer function

$$A(z) = \frac{-0.5 + 0.2z^{-1} + z^{-2}}{1 + 0.2z^{-1} - 0.5z^{-2}}.$$

Compute the normalized steady-state output noise variance due to product round-off of the phase-equalized structure of Figure P12.5 if $G(z)$ is realized in a cascade form with the lowest product round-off noise.

12.22 Consider the digital filter structure of Figure P12.6 that is assumed to be implemented using 9-bit signed two's-complement fixed-point arithmetic with all products quantized before additions. Draw the linear noise model of the unscaled system and compute its total output noise power.

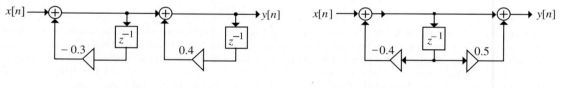

Figure P12.6   **Figure P12.7**

12.23 Scale the first-order digital filter structure of Figure P12.7 using the $\mathcal{L}_2$-norm scaling rule.

12.24 Scale the second-order digital filter structure of Figure P12.8 using the $\mathcal{L}_2$-norm scaling rule.

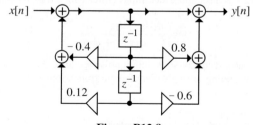

Figure P12.8

12.25 Scale the structure realized in Problem 12.12 using the $\mathcal{L}_2$-norm scaling rule and then compute its output noise variance due to product round-off, assuming quantization of products before addition.

12.26 Scale the structures realized in Problem 12.14 using the $\mathcal{L}_2$-norm scaling rule and then compute the output noise variances due to product round-off, assuming quantization of products before addition. What would be the output noise variances if quantization is carried out after addition?

12.27 Scale the structures realized in Problem 12.15 using the $\mathcal{L}_2$-norm scaling rule and then compute the output noise variances due to product round-off, assuming quantization of products before addition. What would be the output noise variances if quantization is carried out after addition?

12.28 Scale the structures realized in Problem 12.16 using the $\mathcal{L}_2$-norm scaling rule and then compute the output noise variance due to product round-off, assuming quantization of products before addition. What would be the output noise variance if quantization is carried out after addition?

12.29 (a) Calculate the output noise variance of the digital filter structure of Figure P12.9(a) due to product round-off before addition. Assume all numbers are fractions and represented in a two's-complement fixed-point representation with a wordlength of $b + 1$ bits. Note that the multiplier "-1" does not generate noise.
   (b) Now consider the configuration of Figure P12.9(b), where the filters $A_1(z)$, $A_2(z)$, and $A_3(z)$ are implemented as in Figure P12.9(a), with the multiplier coefficients d_i replaced with d_1, d_2, and d_3, respectively. Calculate the output noise variance of this new digital filter structure due to product round-off before addition.

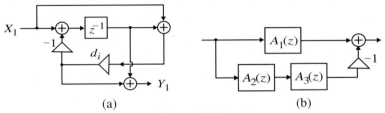

(a)   (b)

Figure P12.9

12.30 Consider the digital filter structure of Figure P12.5, where $A(z)$ is a first-order allpass filter with a transfer function

$$A(z) = \frac{-0.4 + z^{-1}}{1 - 0.4z^{-1}}.$$

Let σ_0^2 represent the total output noise variance due to product round-off in $G(z)$ and $A(z)$. If each delay in the realization of $G(z)$ is replaced with two delays, i.e., z^{-1} replaced with z^{-2}, but $A(z)$ is left unchanged, calculate the output noise variance due to product round-off in terms of σ_0.

12.31 Show that there are $(R!)^2$ different possible realizations of a cascade of R second-order sections

12.32 (a) What is the optimum pole-zero pairing and ordering of each of the following transfer functions for obtaining the smallest peak output noise due to product round-off under an $\mathcal{L}_2$-scaling rule?
   (b) Repeat Part (a) if the objective is to minimize the output noise power due to product round-off under an $\mathcal{L}_\infty$-scaling rule.
   (i) $H_1(z) = \dfrac{0.0959(z^2 + 1.0094z + 1)(z^2 + 1.2914z + 1)(z^2 + 1.8606z + 1)}{(z^2 - 0.4159z + 0.2066)(z^2 + 0.6081z + 0.9208)(z^2 + 0.2749z + 0.6673)}$,
   (ii) $H_2(z) = \dfrac{0.1331(z^2 - 1.1141z + 1)(z^2 - 0.8508z + 1)(z^2 - 1.7988z + 1)}{(z^2 - 0.6120z + 0.9405)(z^2 + 0.2993z + 0.2035)(z^2 - 0.3474z + 0.7127)}$.

12.33 Derive the expressions for the SNR given in Table 12.5.

12.34 Show that the realization of $H_{LP}(z)$ in the form of Figure 8.33 exhibits low sensitivity in the passband at $\omega = 0$ with respect to the multiplier coefficient α, i.e.,

$$\left. \frac{\partial \left| H_{LP}(e^{j\omega}) \right|}{\partial \alpha} \right|_{\omega=0} = 0.$$

Likewise, show that the realization of $H_{HP}(z)$ in the form of Figure 8.33 exhibits low sensitivity in the passband at $\omega = \pi$ with respect to the multiplier coefficient α, i.e.,

$$\frac{\partial \left| H_{HP}(e^{j\omega}) \right|}{\partial \alpha} \Bigg|_{\omega=\pi} = 0.$$

12.35 Show that the realization of $H_{BP}(z)$ in the form of Figure 8.36 exhibits low sensitivity in the passband at the center frequency ω_0 with respect to the multiplier coefficients α and β, i.e.,

$$\frac{\partial \left| H_{BP}(e^{j\omega}) \right|}{\partial \alpha} \Bigg|_{\omega=\omega_0} = 0, \qquad \frac{\partial \left| H_{BP}(e^{j\omega}) \right|}{\partial \beta} \Bigg|_{\omega=\omega_0} = 0.$$

Likewise, show that the realization of $H_{BS}(z)$ in the form of Figure 8.36 exhibits low sensitivity in the passband at $\omega = 0$ and $\omega = \pi$ with respect to the multiplier coefficients α and β, i.e.,

$$\frac{\partial \left| H_{BS}(e^{j\omega}) \right|}{\partial \alpha} \Bigg|_{\omega=0} = 0, \qquad \frac{\partial \left| H_{BS}(e^{j\omega}) \right|}{\partial \beta} \Bigg|_{\omega=0} = 0,$$

$$\frac{\partial \left| H_{BS}(e^{j\omega}) \right|}{\partial \alpha} \Bigg|_{\omega=\pi} = 0, \qquad \frac{\partial \left| H_{BS}(e^{j\omega}) \right|}{\partial \beta} \Bigg|_{\omega=\pi} = 0.$$

12.36 In the parallel allpass realization of a bounded real (BR) transfer function $G(z)$, the passband of $G(z)$ is the stopband of its power-complementary transfer function $H(z)$. Show why low passband sensitivity of $G(z)$ does not imply low stopband sensitivity of $H(z)$.

12.37 Develop the expressions for the SNR of the second-order IIR filter structure of Figure 12.45 with and without error feedback for the following types of inputs: (a) WSS, white uniform density; (b) WSS, white Gaussian density $(\sigma_x = 1/3)$; and (c) sinusoid with known frequency. The poles are at $z = (1 - \varepsilon)e^{\pm j\theta}$, with $\varepsilon \to 0$, $\theta \to 0$, and $\theta \gg \varepsilon$. Assume a $(b + 1)$-bit signed representations for the number representation.

12.38 Modify the coupled-form structure of Figure 12.11 to include error feedback, and determine the normalized output round-off noise power of the modified structure. What are the appropriate values of the multiplier coefficients in the error-feedback loops to minimize the output round-off noise power? What is the expression for the output round-off noise power in this case?

12.39 Modify the Kingsbury structure of Figure P8.9 to include error feedback, and determine the normalized output round-off noise power of the modified structure. What are the appropriate values of the multiplier coefficients in the error-feedback loops to minimize the output round-off noise power? What is the expression for the output round-off noise power in this case?

12.40 Show that the coupled-form structure of Figure 12.11 will not support zero-input limit cycles under magnitude quantization.

12.41 Show that the modified coupled-form structure of Figure P12.1 will not support zero-input limit cycles under magnitude quantization.

12.42 Digital filter structures realized using the *delta operator* have been shown to yield improved finite word-length performances [Mid86]. The delta operator is defined as

$$\delta = \frac{z-1}{\Gamma}, \tag{12.220}$$

where z is the unit advance operator and Γ is a free parameter that can be selected to improve the finite word-length properties of the filter structure. Develop the equivalent delta-domain transfer function $H(\delta)$ of a second-order z-domain transfer function

$$H(z) = \frac{\beta_0 + \beta_1 z^{-1} + \beta_2 z^{-2}}{1 + \alpha_1 z^{-1} + \alpha_2 z^{-2}},$$

by replacing the unit delay operator z^{-1} with the inverse delta operator δ^{-1} [Kau98]. Show explicitly the expressions for the coefficients of $H(\delta)$.

12.43 Determine the expression for the output round-off noise variance in the computation of a single sample of a length-N DFT using Goertzel's algorithm implemented in a fractional signed $(b+1)$-bit fixed-point arithmetic.

12.44 Determine the number of bits b to compute a single sample of a 256-point DFT of an input sequence of length 256 by direct computation with an SNR of 30 dB.

12.45 Derive Eq. (12.218).

12.46 Determine the number of bits needed to compute a single sample of a 256-point DFT of an input sequence of length 256 using a radix-2 decimation-in-time FFT algorithm with an SNR of 30 dB. Assume a distributed scaling to prevent overflow at the output.

12.15 MATLAB Exercises

M 12.1 Write a MATLAB program to plot the pole distribution of a second-order two-multiplier structure with multiplier coefficients represented in sign-magnitude form with b bits.

M 12.2 Using the program developed in Exercise M12.1, plot the pole distributions of the direct form, the coupled form of Figure 9.9, and the Kingsbury structure of Figure P8.9 for a 4-bit wordlength. Comment on the coefficient sensitivity of each of these structures from the pole distribution plots.

Program 12_1.m

M 12.3 Modify Program 12_1 to design an elliptic highpass IIR filter with the following specifications: passband edge at 0.7π, stopband edge at 0.6π, passband ripple of 0.02 dB, and minimum stopband attenuation of 55 dB. Quantize the transfer function coefficients to 6 bits using the M-function a2dT. Plot the magnitude responses and the pole-zero plots of the two transfer functions. Comment on your results.

M 12.4 Using the function filternorm, write a MATLAB program to compute the normalized output noise variance due to input quantization noise for a stable IIR digital filter. The input data for this program should be the numerator and denominator coefficients of the filter transfer function. Using this program, determine the normalized output noise variance of the transfer function of Eq. (12.94) and compare it with that obtained in Example 12.5.

M 12.5 Determine the factored form of an elliptic lowpass transfer function with the following specifications: passband edge at 0.45π, stopband edge at 0.55π, passband ripple of 0.2 dB, and a minimum stopband attenuation of 50 dB. From its pole-zero locations, determine the optimum pole-zero pairing and their ordering to minimize the output noise power under an $\mathcal{L}_\infty$-scaling rule. Verify your result using MATLAB.

M 12.6 Realize the transfer function of Exercise M12.5 as a parallel connection of two allpass filters and determine its power-complementary highpass transfer function. Plot on the same figure the gain responses of the two filters. Verify the low passband sensitivity of the parallel allpass realization by making a 1% change in the filter coefficients.

M 12.7 Write a MATLAB program to simulate a fourth-order transfer function of the form

$$H(z) = \frac{(1 + b_1 z^{-1} + b_2 z^{-2})(1 + b_3 z^{-1} + b_4 z^{-2})}{(1 + a_1 z^{-1} + a_2 z^{-2})(1 + a_3 z^{-1} + a_4 z^{-2})} \tag{12.221}$$

in a cascade form, with each second-order section realized in direct form II. The input data to your program are the numerator coefficients $\{b_i\}$ and the denominator coefficients $\{a_i\}$ of the second-order sections. Using this program, simulate two different cascade realizations of the following transfer function

$$H(z) = \frac{(1 - 0.7589 z^{-1} + z^{-2})(1 + 0.6564 z^{-1} + z^{-2})}{(1 + 1.0462 z^{-1} + 0.8385 z^{-2})(1 + 1.0657 z^{-1} + 0.4046 z^{-2})}. \tag{12.222}$$

Using this program, determine the impulse response of each pertinent scaling transfer functions and their approximate $\mathcal{L}_2$-norms. Based on this information, scale each realization using an $\mathcal{L}_2$-scaling rule, and compute the total product round-off noise variance for each scaled structure, assuming rounding before addition.

M 12.8 Write a MATLAB program to simulate a fourth-order transfer function of the form given in Eq. (12.221) in parallel forms I and II. The input data to your program are the numerator coefficients $\{b_i\}$ and the denominator coefficients $\{a_i\}$. Using this program, simulate the two different parallel realizations of the transfer function of Eq. (12.222). Determine the impulse response of each pertinent scaling transfer functions and their approximate $\mathcal{L}_2$-norms. Based on this information, scale each realization using an $\mathcal{L}_2$-scaling rule, and compute the total product round-off noise variance for each scaled structure, assuming rounding before addition. Compare the output round-off noises of the two parallel structures with those of the cascade realizations of Exercise M12.7.

M 12.9 Write a MATLAB program to simulate a fourth-order transfer function of the form given in Eq. (12.221) using the Gray–Markel method. The input data to your program are the numerator coefficients $\{b_i\}$ and the denominator coefficients $\{a_i\}$. Using this program, simulate the Gray–Markel realization of the transfer function of Eq. (12.222). Determine the impulse response of each pertinent scaling transfer function and its approximate $\mathcal{L}_2$-norm. Based on this information, scale the realization using an $\mathcal{L}_2$-scaling rule, and compute the total product round-off noise variance for the scaled structure, assuming rounding before addition. Compare the output round-off noise of the Gray–Markel realization with those of the cascade realizations of Exercise M12.7 and the two parallel structures of Exercise M12.8.

M 12.10 Using Program 12_7, investigate the granular limit cycles of Figure 12.47 for the following sets of values of the coefficient, the initial condition, and the scale factor of the input impulse: (a) $\alpha = 0.6$, $y[-1] = 0.3$, $x[0] = 0.08$; (b) $\alpha = 0.6$, $y[-1] = 0.9$, $x[0] = 0.03$; and (c) $\alpha = 0.6$, $y[-1] = 8$, $x[0] = 5$. Comment on your results.

M 12.11 Modify Program 12_8 by replacing the M-function a2dR with the function a2dT and then demonstrate by running the modified program that the structure of Figure 12.49 does not exhibit overflow limit cycles if sign-magnitude truncation is used to truncate the sum of products of Eq. (12.195).

13 Multirate Digital Signal Processing Fundamentals

The digital signal processing structures discussed so far in this text belong to the class of single-rate systems since the sampling rates at the input, at the output, and at all internal nodes are the same. There are many applications where the signal at a given sampling rate needs to be converted into another signal with a different sampling rate. For example, in digital audio, three different sampling rates are presently employed: 32 kHz in broadcasting, 44.1 kHz in digital CD, and 48 kHz in digital audio tape (DAT) and other applications [Lag82]. Conversion of sampling rates of audio signals among these three different rates is necessary in many situations. Another example is pitch control of audio recordings usually performed by varying tape recorder speed. However, such an approach changes the sampling frequency of the digital signal and, as a result, conversion to the original sampling rate is needed [Lag82]. In video applications, the sampling rates of NTSC (National Television Systems Committee) and PAL (Phase Alternate Line) composite video signals are, respectively, 14.3181818 MHz and 17.734475 MHz, whereas the sampling rates of the digital component video signal are 13.5 MHz and 6.75 MHz for the luminance and the color-difference signals, respectively [Lut91].[1] There are other applications where it is convenient (and often judicious) to have unequal rates of sampling at the filter input and output and at internal nodes. Examples of such sampling rate alterations are the oversampling A/D and D/A converters discussed in Sections 4.8.4 and 4.9.3, respectively, and analyzed in detail in Sections 15.11 and 15.12, respectively. Additional applications of sampling rate alterations are also given in Sections 15.8 through 15.10.

To change the sampling rate of a digital signal, multirate digital signal processing systems use a *down-sampler* and an *up-sampler*, the two basic sampling rate alteration devices in addition to conventional elements such as an adder, a multiplier, and a delay. Discrete-time systems with unequal sampling rates at various parts of the system are called *multirate systems* and are the subject of discussion of this chapter.

We first examine the input–output relations of an up-sampler and a down-sampler both in the time domain and the transform domain. As in many applications, cascade connections of the basic sampling rate alteration devices and digital filters are employed; some basic cascade equivalences are then reviewed. For sampling rate alterations, the basic sampling rate alteration devices are invariably employed together with appropriate lowpass digital filters. The frequency response specifications of these filters are developed next. A computationally more efficient approach to sampling rate alteration, based on a multistage implementation, is illustrated by means of a specific design problem. The polyphase decomposition of a sequence is reexamined next in the framework of multirate theory, and its application to developing computationally efficient sampling rate alteration systems is illustrated. A review of the design of sampling rate converters with arbitrary conversion rate based on the Lagrange and spline interpolation algorithms is then discussed. The chapter concludes with a discussion on Lth band filters and their design.

Chapter 14 considers the analysis and design of multirate filter banks.

---

[1]CCIR Recommendation No. 601.

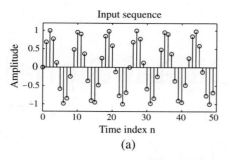
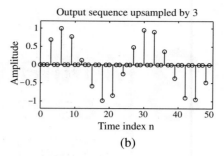

Figure 13.1: Illustration of the up-sampling process.

$$x[n] \longrightarrow \boxed{\uparrow L} \longrightarrow x_u[n]$$

Figure 13.2: Block diagram representation of an up-sampler.

13.1 The Basic Sampling Rate Alteration Devices

The two basic components in sampling rate alteration are the up-sampler and the down-sampler, introduced earlier in Section 2.1.2 where we examined their input–output relations in the time domain. However, it is also instructive to analyze their operations in the frequency domain. This will point out why these devices must be used with additional filters. In addition, a frequency-domain analysis provides the basic foundation for analyzing more complex multirate systems introduced in the latter parts of the chapter.

13.1.1 Time-Domain Characterization

We reexamine the time-domain characterizations of the two basic sampling rate alteration devices. An up-sampler with an up-sampling factor L, where L is a positive integer, develops an output sequence $x_u[n]$ with a sampling rate that is L times larger than that of the input sequence $x[n]$. The up-sampling operation is implemented by inserting $L - 1$ equidistant zero-valued samples between two consecutive samples of the input sequence $x[n]$ according to the relation

$$x_u[n] = \begin{cases} x[n/L], & n = 0, \pm L, \pm 2L, \ldots, \\ 0, & \text{otherwise.} \end{cases} \tag{13.1}$$

The up-sampling operation is illustrated in Example 13.1.

Program
13_1.m

> **EXAMPLE 13.1   Illustration of Up-Sampling Operation**
>
> Program 13_1 can be used to study up-sampling of a sinusoidal input sequence. Its input data are the length of the input sequence, the up-sampling factor, and the frequency of the sinusoid in Hz. It then plots the input sequence and its up-sampled version. Figure 13.1 shows the result obtained for a length-50 sinusoidal sequence with a frequency of 0.12 Hz and with an up-sampling factor of 3.

The block diagram representation of the up-sampler, also called a *sampling rate expander* or simply an *expander,* is shown in Figure 13.2.

In practice, the zero-valued samples inserted by the up-sampler are interpolated using some type of filtering process in order that the new higher-rate sequence has no unnecessary spectral components. This process, called *interpolation,* is discussed later in this chapter.

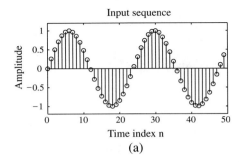

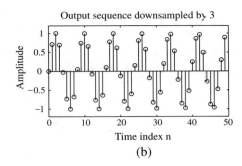

Figure 13.3: Illustration of the down-sampling process.

$$x[n] \longrightarrow \boxed{\downarrow M} \longrightarrow y[n]$$

Figure 13.4: Block diagram representation of a down-sampler.

On the other hand, a down-sampler with a down-sampling factor M, where M is a positive integer, creates an output sequence $y[n]$ with a sampling rate that is $(1/M)$th of that of the input sequence $x[n]$. The down-sampling operation is implemented by keeping every Mth sample of the input sequence and removing the $M-1$ in-between samples to generate the output sequence according to the relation

$$y[n] = x[nM]. \tag{13.2}$$

As a result, all input samples with indices equal to an integer multiple of M are retained at the output and all others are discarded, as illustrated in Example 13.2.

EXAMPLE 13.2   Illustration of Down-Sampling Operation

We investigate down-sampling of a sinusoidal input sequence using MATLAB. To this end, we utilize Program 13_2. Its input data are the length of the input sequence, the down-sampling factor, and the frequency of the sinusoid in Hz. It then plots the input sequence and its down-sampled version. The result obtained for a length-50 sinusoidal sequence with a frequency 0.042 Hz and with a down-sampling factor of 3 are shown in Figure 13.3.

Program
13_2.m

The block diagram representation of the *down-sampler* or *sampling rate compressor* is shown in Figure 13.4.

The sampling periods involved have not been explicitly shown in Figures 13.2 and 13.4. This is in the interest of simplicity and in view of the fact that the mathematical theory of multirate systems can be understood without bringing T or the sampling frequency F_T into the picture. It is instructive in the beginning to explicitly see the time dimensions at various stages in the sampling rate alteration process, as indicated in Figure 13.5. In the remainder of this book, however, the explicit appearance of T or the sampling frequency F_T is not shown unless their actual values are relevant.

The up-sampler and the down-sampler building blocks of Figures 13.2 and 13.4 are often used together in a number of applications involving multirate signal processing and are discussed in more detail in this chapter and Chapter 14. For example, one application using both types of sampling rate alteration devices is to achieve a sampling rate change by a rational number rather than an integer value. Example 13.3 illustrates another application.

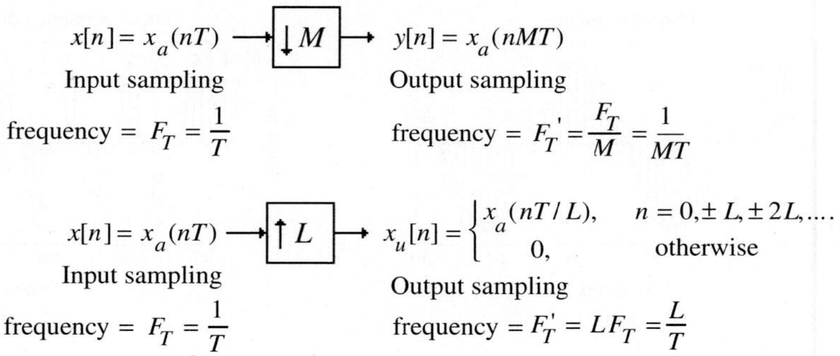

$$x[n] = x_a(nT) \longrightarrow \boxed{\downarrow M} \longrightarrow y[n] = x_a(nMT)$$

Input sampling   Output sampling

$$\text{frequency} = F_T = \frac{1}{T} \qquad\qquad \text{frequency} = F_T' = \frac{F_T}{M} = \frac{1}{MT}$$

$$x[n] = x_a(nT) \longrightarrow \boxed{\uparrow L} \longrightarrow x_u[n] = \begin{cases} x_a(nT/L), & n = 0, \pm L, \pm 2L, \dots \\ 0, & \text{otherwise} \end{cases}$$

Input sampling   Output sampling

$$\text{frequency} = F_T = \frac{1}{T} \qquad\qquad \text{frequency} = F_T' = LF_T = \frac{L}{T}$$

Figure 13.5: The sampling rate alteration building blocks with sampling rates explicitly shown.

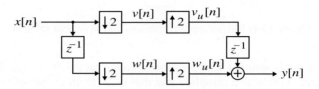

Figure 13.6: A simple multirate system.

EXAMPLE 13.3   Analysis of a Simple Multirate Structure

Consider the multirate system of Figure 13.6. Its operation can be analyzed by writing down the relations between various signal variables and the input, as depicted below:

n:	0	1	2	3	4	5	6	7	8
$x[n]$:	$x[0]$	$x[1]$	$x[2]$	$x[3]$	$x[4]$	$x[5]$	$x[6]$	$x[7]$	$x[8]$
$v[n]$:	$x[0]$	$x[2]$	$x[4]$	$x[6]$	$x[8]$	$x[10]$	$x[12]$	$x[14]$	$x[16]$
$w[n]$:	$x[-1]$	$x[1]$	$x[3]$	$x[5]$	$x[7]$	$x[9]$	$x[11]$	$x[13]$	$x[15]$
$v_u[n]$:	$x[0]$	0	$x[2]$	0	$x[4]$	0	$x[6]$	0	$x[8]$
$w_u[n]$:	$x[-1]$	0	$x[1]$	0	$x[3]$	0	$x[5]$	0	$x[7]$
$v_u[n-1]$:	0	$x[0]$	0	$x[2]$	0	$x[4]$	0	$x[6]$	0
$y[n]$:	$x[-1]$	$x[0]$	$x[1]$	$x[2]$	$x[3]$	$x[4]$	$x[5]$	$x[6]$	$x[7]$

It can be seen from the above that the output $y[n]$ of the multirate system of Figure 13.6 is given by $y[n] = v_u[n-1] + w_u[n]$, which is simply $x[n-1]$. A formal proof of the perfect reconstruction property of the multirate structure of Figure 13.6 is left as an exercise (Problem 13.3).

A multirate structure in which the output sequence is a delayed and scaled replica of the input sequence is called a *perfect reconstruction multirate system*. The structure of Figure 13.6 thus is a very simple example of such a system.

It can be shown that the up-sampler and the down-sampler are linear and time-varying discrete-time systems (Problems 13.1 and 13.2).

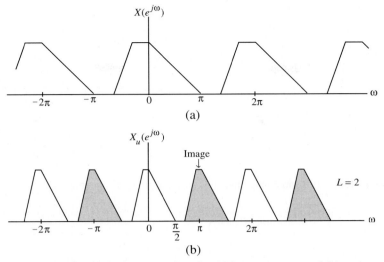

Figure 13.7: Effects of up-sampling in the frequency domain: (a) input spectrum and (b) output spectrum for $L = 2$.

13.1.2 Frequency-Domain Characterization

We first derive the relations between the spectrums of the input and the output of a factor-of-2 up-sampler. From the input–output relation of the factor-of-L up-sampler given by Eq. (13.1), we arrive at the corresponding relation for the factor-of-2 up-sampler:

$$x_u[n] = \begin{cases} x[n/2], & n = 0, \pm 2, \pm 4, \ldots, \\ 0, & \text{otherwise.} \end{cases} \tag{13.3}$$

In terms of the z-transform, the input–output relation is then given by

$$X_u(z) = \sum_{n=-\infty}^{\infty} x_u[n]z^{-n} = \sum_{\substack{n=-\infty \\ n \text{ even}}}^{\infty} x[n/2]z^{-n}$$

$$= \sum_{m=-\infty}^{\infty} x[m]z^{-2m} = X(z^2). \tag{13.4}$$

In a similar manner, we can show that for the factor-of-L up-sampler,

$$X_u(z) = X(z^L). \tag{13.5}$$

Let us examine the implication of the above relation on the unit circle. For $z = e^{j\omega}$, the above equation becomes $X_u(e^{j\omega}) = X(e^{j\omega L})$. Figure 13.7(a) shows the DTFT $X(e^{j\omega})$ that has been assumed to be a real function of ω for convenience. Moreover, the DTFT $X(e^{j\omega})$ shown is not an even function of ω, implying that $x[n]$ is a complex sequence. The asymmetric response has been purposely chosen to illustrate more clearly the effect of up-sampling.

As shown in Figure 13.7(b), a factor-of-2 sampling rate expansion thus leads to a 2-fold repetition of $X(e^{j\omega})$, indicating that the Fourier transform is compressed by a factor of 2. This process is called *imaging* because we get an additional "image" of the input spectrum. In the case of a factor-of-L sampling rate expansion, there will be $L - 1$ additional images of the input spectrum in the baseband. Thus, a

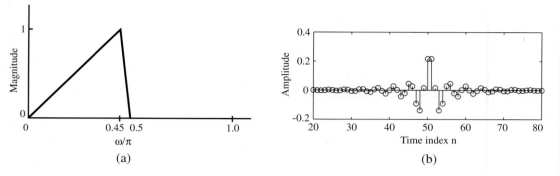

Figure 13.8: (a) Desired magnitude response and (b) corresponding time sequence.

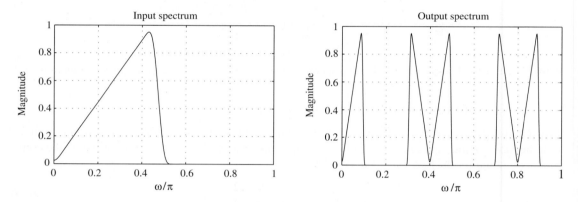

Figure 13.9: MATLAB-generated input and output spectrum of a factor-of-5 up-sampler.

spectrum $X(e^{j\omega})$ band-limited to the low-frequency region does not look like a low-frequency spectrum after up-sampling because of the insertion of zero-valued samples between the nonzero samples of $x_u[n]$. Lowpass filtering of $x_u[n]$ removes the $L-1$ images and, in effect, "fills in" on the zero-valued samples in $x_u[n]$ with interpolated sample values.

Example 13.4 illustrates the frequency-domain properties of the up-sampler.

Program
13_3.m

EXAMPLE 13.4   Illustration of the Frequency-Domain Properties of the Up-Sampler Using MATLAB

To investigate the effect of up-sampling, we use Program 13_3. The input is a causal finite-length sequence with a band-limited frequency response generated using the M-file `fir2`. The input to `fir2` is as follows: length of the sequence is `100`, the desired magnitude response vector `mag = [0 1 0 0]`, and the vector of frequency points `freq = [0 0.45 0.5 1]`. The desired magnitude response is thus as indicated in Figure 13.8(a). A plot of the middle 61 samples of the signal generated is shown in Figure 13.8(b).

The program determines the output of the up-sampler and then plots the input and output spectrums, as indicated in Figure 13.9 for `L = 5`. It can be seen from this figure that, as expected, the output spectrum consists of a factor-of-5 compressed version of the input spectrum followed by `L-1 = 4` images.

We now derive the relations between the spectrums of the input and the output of a down-sampler. Applying the z-transform to the input–output relation given in Eq. (13.2), we arrive at

$$Y(z) = \sum_{n=-\infty}^{\infty} x[Mn]z^{-n}. \tag{13.6}$$

The expression on the right-hand side of Eq. (13.6) cannot be directly expressed in terms of $X(z)$. To get around this problem, we define an intermediate sequence $x_{\text{int}}[n]$ whose sample values are the same as that of $x[n]$ at the values of n that are multiples of M and are zeros at other values of n:

$$x_{\text{int}}[n] = \begin{cases} x[n], & n = 0, \pm M, \pm 2M, \ldots, \\ 0, & \text{otherwise.} \end{cases} \tag{13.7}$$

Then,

$$Y(z) = \sum_{n=-\infty}^{\infty} x[Mn]z^{-n} = \sum_{n=-\infty}^{\infty} x_{\text{int}}[Mn]z^{-n}$$

$$= \sum_{k=-\infty}^{\infty} x_{\text{int}}[k]z^{-k/M} = X_{\text{int}}(z^{1/M}). \tag{13.8}$$

Now $x_{\text{int}}[n]$ can be formally related to $x[n]$ through $x_{\text{int}}[n] = c[n]x[n]$, where $c[n]$ is defined by

$$c[n] = \begin{cases} 1, & n = 0, \pm M, \pm 2M, \ldots, \\ 0, & \text{otherwise.} \end{cases} \tag{13.9}$$

A convenient representation of $c[n]$ is given by (Problem 13.4)

$$c[n] = \frac{1}{M} \sum_{k=0}^{M-1} W_M^{kn}, \tag{13.10}$$

where $W_M = e^{-j2\pi/M}$ is the quantity defined in Eq. (5.12). Substituting $x_{\text{int}}[n] = c[n]x[n]$ and making use of Eq. (13.10) in the z-transform of $x_{\text{int}}[n]$, we obtain

$$X_{\text{int}}(z) = \sum_{n=-\infty}^{\infty} c[n]x[n]z^{-n} = \frac{1}{M} \sum_{n=-\infty}^{\infty} \left(\sum_{k=0}^{M-1} W_M^{kn} \right) x[n]z^{-n}$$

$$= \frac{1}{M} \sum_{k=0}^{M-1} \left(\sum_{n=-\infty}^{\infty} x[n]W_M^{kn}z^{-n} \right) = \frac{1}{M} \sum_{k=0}^{M-1} X\left(zW_M^{-k}\right). \tag{13.11}$$

The desired input–output relation in the transform domain for a factor-of-M down-sampler is then obtained by substituting Eq. (13.11) in Eq. (13.8), resulting in

$$Y(z) = \frac{1}{M} \sum_{k=0}^{M-1} X(z^{1/M} W_M^{-k}). \tag{13.12}$$

To understand the implication of the above relation, consider a factor-of-2 down-sampler with an input $x[n]$ whose spectrum is as shown in Figure 13.10(a). As before, for convenience, we assume again $X(e^{j\omega})$ to be a real function with an asymmetric frequency response. From Eq. (13.12), we get

$$Y(e^{j\omega}) = \tfrac{1}{2}\left\{X(e^{j\omega/2}) + X(-e^{j\omega/2})\right\}. \tag{13.13}$$

The plot of $\tfrac{1}{2}X(e^{j\omega/2})$ is shown by the solid line in Figure 13.10(b). To determine the relation of the second term in Eq. (13.13) with respect to the first, we observe next that

$$X(-e^{j\omega/2}) = X(e^{j(\omega-2\pi)/2}), \tag{13.14}$$

indicating that the second term $X(e^{-j\omega/2})$ in Eq. (13.13) is obtained simply by shifting the first term $X(e^{j\omega/2})$ to the right by an amount 2π, as shown by the dotted lines in Figure 13.10(b). The plots of the two terms in Eq. (13.13) have an overlap, and hence, in general, the original "shape" of $X(e^{j\omega})$ is lost when $x[n]$ is down-sampled. This overlap causes the *aliasing* that takes place due to undersampling (i.e., down-sampling). There is no overlap, that is, no aliasing, only if $X(e^{j\omega})$ is zero for $|\omega| \geq \pi/2$. Note that $Y(e^{j\omega})$ in Eq. (13.13) is indeed periodic with a period 2π, even though the stretched version $X(e^{j\omega})$ is periodic with a period 4π. For the general case, the situation is essentially the same, and the relation between the Fourier transform of the output and the input of the factor-of-M down-sampler is given by

$$Y(e^{j\omega}) = \frac{1}{M}\sum_{k=0}^{M-1} X(e^{j(\omega-2\pi k)/M}). \tag{13.15}$$

The above relation implies that $Y(e^{j\omega})$ is a sum of M uniformly shifted and stretched versions of $X(e^{j\omega})$, scaled by a factor $1/M$. Aliasing due to a factor-of-M down-sampling is absent if and only if the signal $x[n]$ is band-limited to $\pm\pi/M$, as shown in Figure 13.11 for $M = 2$.

We illustrate the aliasing effect caused by down-sampling in Example 13.5.

Program
13_4.m

EXAMPLE 13.5   **Illustration of the Frequency-Domain Properties of the Down-Sampler Using MATLAB**

To investigate the effect of down-sampling, we use Program 13_4. The input signal is again the signal generated using `fir2` with a triangular magnitude response, as in Figure 13.8(a). However, here the frequency vector has been selected to be `freq = [0 0.42 0.48 1]` to ensure that there are no appreciable signal components above the normalized frequency of 0.5.

The plots generated by Program 13_4 are shown in Figure 13.12. The input spectrum is shown in Figure 13.12(a). Since the input signal is band-limited to $\pi/2$, the output spectrum for a down-sampling factor of M = 2 shown in Figure 13.12(b) is nearly of the same shape as the input spectrum, except it has been stretched by a factor of 2 in frequency and its magnitude is reduced by one-half as predicted by the factor 1/2 in Eq. (13.13). On the other hand, the output spectrum for a down-sampling factor of M = 3, shown in Figure 13.12(c), shows a severe distortion caused by the aliasing.

Any linear discrete-time multirate system can be analyzed in the transform domain by using the input–output relations of the up-sampler and the down-sampler given by Eqs. (13.5) and (13.12), respectively. We illustrate the applications of these relations in the latter parts of this chapter.

13.1.3   Cascade Equivalences

As we shall observe later, a complex multirate system is formed by an interconnection of the basic sampling rate alteration devices and the components of an LTI digital filter. In many applications, these devices appear in a cascade form. An interchange of the positions of the branches in a cascade often can lead to a computationally efficient realization. We investigate certain specific cascade connections and their equivalences, which leave input–output relations invariant.

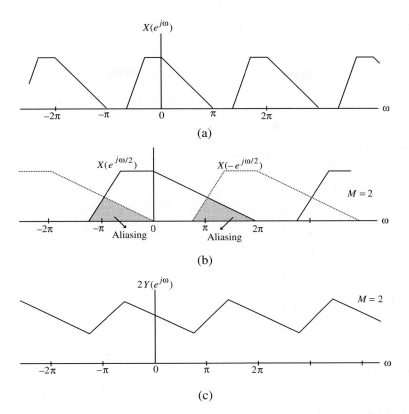

Figure 13.10: Illustration of the aliasing effect in the frequency domain caused by down-sampling.

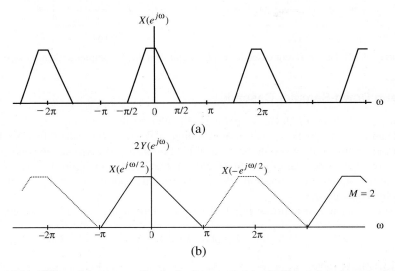

Figure 13.11: Effect of down-sampling in the frequency domain illustrating absence of aliasing.

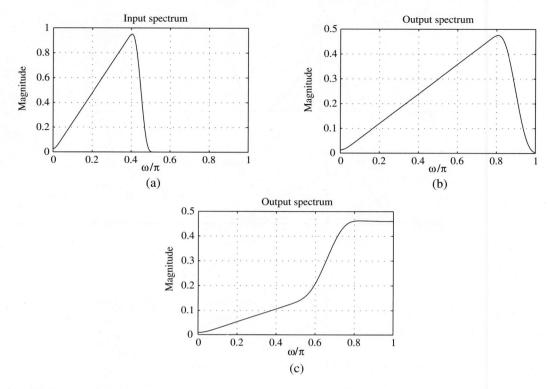

Figure 13.12: (a) Input spectrum, (b) output spectrum for a down-sampling factor of $M = 2$, and (c) output spectrum for a down-sampling factor of $M = 3$.

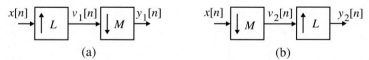

Figure 13.13: Two different cascade arrangements of a down-sampler and an up-sampler.

Up-Sampler and Down-Sampler Cascade

The basic sampling rate alteration devices can be used to change the sampling rate of a signal only by an integer factor. Therefore, to implement a fractional change in the sampling rate, it follows that a cascade of a down-sampler and an up-sampler should be used. It is of interest to determine the condition under which a cascade of a factor-of-M down-sampler with a factor-of-L up-sampler (Figure 13.13) is *interchangeable,* with no change in the input–output relation. We determine this condition next by developing the input–output relations for both cascades.

Consider first the cascade structure of Figure 13.13(a). Here, using Eq. (13.5), we get

$$V_1(z) = X(z^L),$$

and using Eq. (13.12), we have

$$Y_1(z) = \frac{1}{M} \sum_{k=0}^{M-1} V_1(z^{1/M} W_M^{-k}).$$

Combining the last two equations, we arrive at

$$Y_1(z) = \frac{1}{M} \sum_{k=0}^{M-1} X(z^{L/M} W_M^{-kL}).$$

In the case of the cascade structure Figure 13.13(b), likewise, we have

$$V_2(z) = \frac{1}{M} \sum_{k=0}^{M-1} X(z^{1/M} W_M^{-k}),$$

$$Y_2(z) = V_2(z^L).$$

Combining the above two equations, we obtain

$$Y_2(z) = \frac{1}{M} \sum_{k=0}^{M-1} X(z^{L/M} W_M^{-k}).$$

It follows from the above that $Y_1(z) = Y_2(z)$ if

$$\sum_{k=0}^{M-1} X(z^{L/M} W_M^{-kL}) = \sum_{k=0}^{M-1} X(z^{L/M} W_M^{-k}).$$

The above equality holds if and only if M and L are *relatively prime*, that is, M and L do not have a common factor that is an integer $r > 1$, as then W_M^{-k} and W_M^{-kL} take the same set of values for $k = 0, 1, \ldots, M - 1$ [Vai90] (Problem 13.5).

We discuss fractional changes in the sampling rate in Section 13.2.

Noble Identities

Two other simple cascade equivalence relations are depicted in Figure 13.14. Verification of these two cascade equivalences can be shown by proving $y_1[n] = y_2[n]$ (Problem 13.6). These noble identities enable us to move the basic sampling rate alteration devices in multirate networks to more advantageous positions. Such rules are extremely useful in the design and analysis of more complicated multirate systems, as we shall demonstrate later.

13.2 Multirate Structures for Sampling Rate Conversion

The process of reducing the sampling rate of a sequence is more commonly known as *decimation,* and the multirate structure used for decimation is called a *decimator.* Likewise, the reverse process of increasing the sampling rate is called *interpolation,* and the multirate structure used for interpolation is called an *interpolator.* From the sampling theorem introduced in Section 4.2.1, it is known that the sampling rate of a critically sampled discrete-time signal with a spectrum occupying the full Nyquist range cannot be

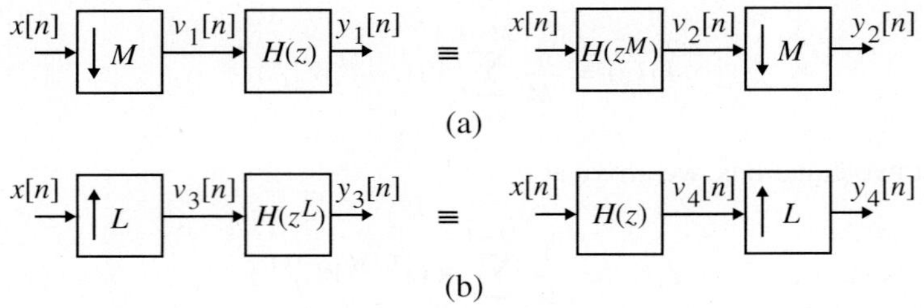

Figure 13.14: Cascade equivalences: (a) equivalence #1 and (b) equivalence #2.

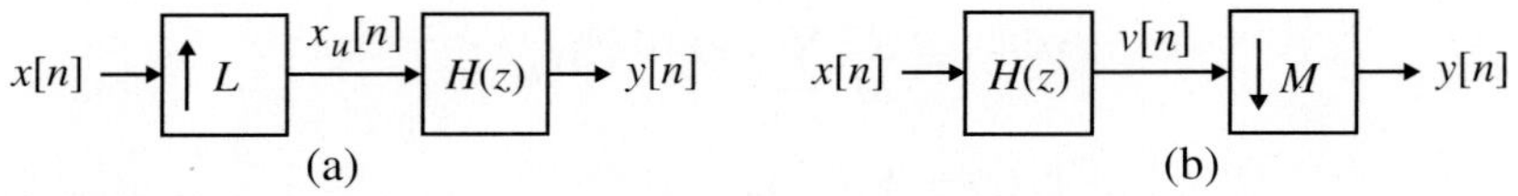

Figure 13.15: Sampling rate alteration systems for integer valued conversion factors: (a) interpolator and (b) decimator.

reduced any further, since such a reduction will introduce aliasing. Hence, the bandwidth of a critically sampled signal must first be reduced by lowpass filtering before its sampling rate is reduced by a down-sampler. Likewise, the zero-valued samples introduced by an up-sampler must be interpolated for an effective sampling rate increase. As we shall show next, this interpolation can be achieved simply by digital lowpass filtering. Since a fractional-rate sampling rate converter with a rational conversion factor can be realized by cascading an interpolator with a decimator, filters are also needed in the design of such sampling multirate systems. We consider in this section some of the issues concerning these lowpass filters. We first develop the input–output relations of the multirate structures used for sampling rate conversion. We then develop the frequency response specifications of the lowpass filter employed in these multirate structures. Next, we illustrate the decimation and interpolation of sequences using MATLAB. Finally, we investigate the computational complexity issues of the required lowpass digital filters.

13.2.1 Basic Structures

Since up-sampling by an integer factor L causes periodic repetition of the basic spectrum (Figure 13.7), the basic interpolator structure for integer-valued sampling rate increase consists of an up-sampler followed by a lowpass filter $H(z)$ with a cutoff at π/L, as indicated in Figure 13.15(a). The lowpass filter $H(z)$, called the *interpolation filter,* removes the $L - 1$ unwanted images in the spectra of the up-sampled signal $x_u[n]$. On the other hand, as down-sampling by an integer factor M may result in aliasing, the basic decimator structure for integer-valued sampling rate decrease consists of a lowpass filter $H(z)$ with a cutoff at π/M, followed by the down-sampler, as indicated in Figure 13.15(b). Here, the lowpass filter $H(z)$, called the *decimation filter,* band-limits the input signal $v[n]$ to $|\omega| < \pi/M$ prior to down-sampling, to ensure no aliasing.[2] It can be shown that the transpose of a factor-of-M decimator is a factor-of-M interpolator and vice-versa (Problem 13.12).

---

[2]It has been tacitly assumed here that the discrete-time signal to be interpolated or decimated has a lowpass frequency response and, as a result, the desired interpolation or the decimation filter is a lowpass filter. However, if the discrete-time signal to be interpolated or decimated has a highpass (bandpass) frequency response, then the desired interpolation or the decimation filter is a highpass (bandpass) filter.

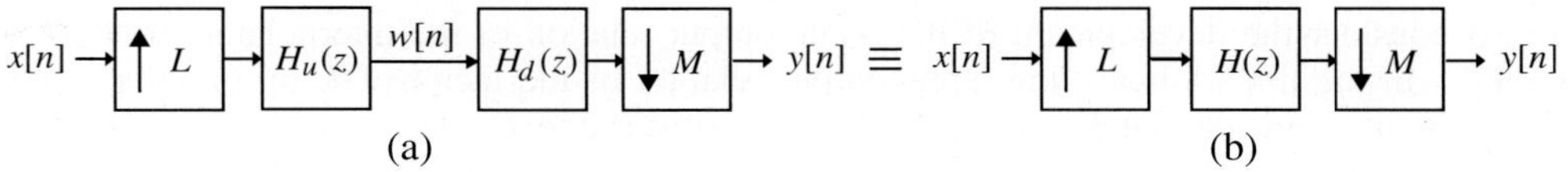

Figure 13.16: (a) General scheme for increasing the sampling rate by L/M and (b) an efficient implementation.

A fractional change in the sampling rate by a rational factor can be achieved by cascading a factor-of-L interpolator with a factor-of-M decimator, where L and M are positive integers. The interpolator must precede the decimator, as indicated in Figure 13.16(a), to ensure that the baseband of $w[n]$ is greater than or equal to that of $x[n]$ or $y[n]$ [Cro83]. As both the interpolation filter $H_u(z)$ and the decimation filter $H_d(z)$ in Figure 13.16(a) operate at the same sampling rate, they can be replaced with a single filter designed to avoid aliasing that may be caused by down-sampling and eliminate images resulting from up-sampling, resulting in the computationally efficient structure of Figure 13.16(b).

13.2.2 Input–Output Relations

We first develop the input–output relation of the decimator structure of Figure 13.15(b). Let $h[n]$ denote the impulse response of the decimation filter $H(z)$. Then, the output $v[n]$ of the filter is given by the convolution sum of the $h[n]$ and the input $x[n]$:

$$v[n] = \sum_{\ell=-\infty}^{\infty} h[n - \ell]x[\ell].$$

Using Eq. (13.2), we can express the output $y[n]$ of the factor-of-M down-sampler to its input $v[n]$ through

$$y[n] = v[Mn].$$

Combining the last two equations, we then have the desired time-domain input–output relation of the decimator structure as

$$y[n] = v[Mn] = \sum_{\ell=-\infty}^{\infty} h[Mn - \ell]x[\ell]. \tag{13.16}$$

In the z-domain, the input–output relation of the decimation filter is

$$V(z) = H(z)X(z),$$

where $V(z)$ and $X(z)$ are the z-transforms of $v[n]$ and $x[n]$, respectively. From Eq. (13.12), the z-domain input–output relation of the down-sampler is

$$Y(z) = \frac{1}{M} \sum_{k=0}^{M-1} V(z^{1/M} W_M^{-k}).$$

Combining these two equations, we arrive at the z-domain input–output relation of the decimator as

$$Y(z) = \frac{1}{M} \sum_{k=0}^{M-1} H(z^{1/M} W_M^{-k}) X(z^{1/M} W_M^{-k}). \tag{13.17}$$

We next consider the development of the input–output relation of the interpolator structure of Figure 13.15(a). In the time-domain, the input–output relation of the factor-of-L up-sampler is given by Eq. (13.1), which can be alternately written by a change of variable as

$$x_u[Lm] = x[m], \qquad m = 0, \pm 1, \pm 2, \dots.$$

Substituting the above equation in the input–output relation of the interpolation filter

$$y[n] = \sum_{\ell=-\infty}^{\infty} h[n-\ell] x_u[\ell],$$

and with a change of variable, we arrive at the desired time-domain input–output relation of the interpolator as

$$y[n] = \sum_{m=-\infty}^{\infty} h[n-Lm] x[m]. \tag{13.18}$$

In the z-domain, the input–output relation of the up-sampler is given by Eq. (13.5), which combined with the input–output relation of the interpolation filter yields the desired result for the interpolator as

$$Y(z) = H(z) X(z^L). \tag{13.19}$$

For the fractional-rate sampling rate converter of Figure 13.16(b), the input–output relation can be derived in a similar manner. In the time domain, it is given by

$$y[n] = \sum_{m=-\infty}^{\infty} h[Mn - Lm] x[m], \tag{13.20}$$

whereas in the z-domain, it is given by

$$Y(z) = \frac{1}{M} \sum_{k=0}^{M-1} H(z^{1/M} W_M^{-k}) X(z^{L/M} W_M^{-kL}). \tag{13.21}$$

13.2.3 Filter Specifications

The specifications for the lowpass filters in Figures 13.15 and 13.16(b) can now be derived. We first develop the specifications for the interpolation filter. Assume $x[n]$ has been obtained by sampling a band-limited continuous-time signal $x_a(t)$ at the Nyquist rate. If $X_a(j\Omega)$ and $X(e^{j\omega})$ denote the Fourier transforms of $x_a(t)$ and $x[n]$, respectively, from Eq. (4.15a), it follows that these Fourier transforms are related through

$$X(e^{j\omega}) = \frac{1}{T_o} \sum_{k=-\infty}^{\infty} X_a\left(\frac{j\omega - j2\pi k}{T_o}\right), \tag{13.22}$$

where T_o is the sampling period. Since the sampling is being done at the Nyquist rate, there is no overlap between the shifted spectras of $X_a(j\omega/T_o)$. If we instead sample $x_a(t)$ at a much higher rate, $1/T = L/T_o$ yielding $y[n]$, its Fourier transform $Y(e^{j\omega})$ is related to $X_a(j\Omega)$ through

$$Y(e^{j\omega}) = \frac{1}{T} \sum_{k=-\infty}^{\infty} X_a\left(\frac{j\omega - j2\pi k}{T}\right) = \frac{L}{T_o} \sum_{k=-\infty}^{\infty} X_a\left(\frac{j\omega - j2\pi k}{T_o/L}\right). \tag{13.23}$$

On the other hand, if we pass $x[n]$ through a factor-of-L up-sampler generating $x_u[n]$, from Eq. (13.5), the relation between the Fourier transforms $X_u(e^{j\omega})$ and $X(e^{j\omega})$ is given by

$$X_u(e^{j\omega}) = X(e^{j\omega L}).\tag{13.24}$$

It follows from Eqs. (13.22)-(13.24) that if $x_u[n]$ is passed through an ideal lowpass filter with a cutoff at π/L and a gain of L, the output of the filter will be precisely $y[n]$.

In practice, a transition band is provided to ensure the realizability and stability of the lowpass interpolation filter $H(z)$. Hence, the desired lowpass filter should have a stopband edge at $\omega_s = \pi/L$ and a passband edge ω_p close to ω_s to reduce the distortion of the spectrum of the signal $x[n]$.[3] If ω_c denotes the highest frequency that needs to be preserved in the signal to be interpolated, the passband edge ω_p of the lowpass filter should be at $\omega_p = \omega_c/L$. Summarizing, the specifications for the lowpass interpolation filter are thus given by

$$\left|H(e^{j\omega})\right| = \begin{cases} L, & |\omega| \leq \omega_c/L, \\ 0, & \text{otherwise.} \end{cases}\tag{13.25}$$

It should be noted that in many interpolation applications, one requirement is to ensure that the input samples are not changed at the output. This requirement can be satisfied by a Nyquist interpolation filter discussed in Section 13.6.

In a similar manner, we can develop the specifications for the lowpass decimation filter that are given by

$$\left|H(e^{j\omega})\right| = \begin{cases} 1, & |\omega| \leq \omega_c/M, \\ 0, & \text{otherwise,} \end{cases}\tag{13.26}$$

where ω_c denotes the highest frequency that needs be preserved in the decimated signal.

The effects of decimation and interpolation in the frequency domain are illustrated in Figures 13.17 and 13.18, respectively, for $M = 2$ and $L = 2$. Figure 13.17(a) shows the spectrum $X(e^{j\omega})$ of the input signal $x[n]$ and the spectrum $H(e^{j\omega})$ of the decimation filter. The passband edge and the stopband edge of the decimation filters are, respectively, $\omega_c/2$ and $\pi/2$, where ω_c is the highest frequency in the input signal that is to be preserved at the output of the decimator. From Eq. (13.13), the frequency response $Y(e^{j\omega})$ of the decimator output $y[n]$ is now given by

$$Y(e^{j\omega}) = \tfrac{1}{2}\left\{V(e^{j\omega/2}) + V(-e^{j\omega/2})\right\},$$

where $V(e^{j\omega})$ is the frequency response of the filter output $v[n]$. Due to prior filtering, there is no overlap in these two spectra, resulting in no aliasing at the decimator output with the output spectrum $Y(e^{j\omega})$, as sketched in Figure 13.17(b). However, it is not possible to recover the original input signal $x[n]$ from the decimated version $y[n]$. The filter $H(z)$ is used to preserve $X(e^{j\omega})$ in the range $-\omega_c/2 < \omega < \omega_c/2$, and one can reconstruct this portion exactly from $y[n]$. On the other hand, if there is no filtering of the input $x[n]$ prior to down-sampling, the two components $\tfrac{1}{2}X(e^{j\omega/2})$ and $\tfrac{1}{2}X(-e^{j\omega/2})$ will overlap, resulting in a severe aliasing at the decimator output, as indicated in Figure 13.17(c). We discuss fractional sampling rate alteration in Section 13.5.

In the case of the factor-of-2 interpolation, the spectrum $V(e^{j\omega})$ of the up-sampler output is given by $X(e^{j2\omega})$. Figure 13.18(b) shows $V(e^{j\omega})$ for an input spectrum indicated in Figure 13.18(a). The spectrum $Y(e^{j\omega})$ of the interpolator output obtained by filtering $v[n]$ is thus as sketched in Figure 13.18(c).

In the case of a fractional sampling rate converter, as indicated in Figure 13.16(a), only one of the filters, $H_u(z)$ or $H_d(z)$, is adequate to serve as both the interpolation filter and the decimation filter, depending on which one of the two stopband frequencies, π/L or π/M, is a minimum. Thus, the lowpass filter $H(z)$ has a normalized stopband cutoff frequency at [Cro83]

---

[3]The meanings of ω_p, ω_s, δ_p, and δ_s in this chapter are precisely as in Section 9.1.1.

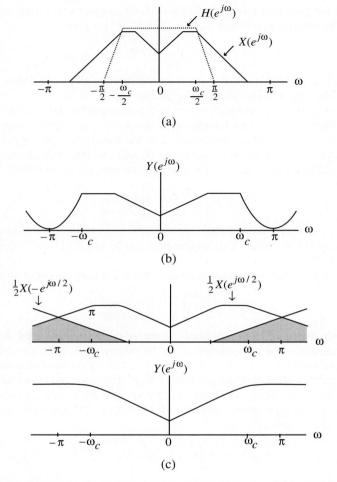

Figure 13.17: Spectrum of (a) the input $x[n]$, (b) the output $y[n]$ of the factor-of-2 decimator with $x[n]$ filtered, and (c) the output $y[n]$ of the factor-of-2 decimator with no filtering of $x[n]$ showing the effect of aliasing. The spectrum of the decimation filter $H(z)$ is shown in (a) with dotted lines. In the figures ω_c is the highest frequency of the input signal that is to be preserved.

$$\omega_s = \min\left(\frac{\pi}{L}, \frac{\pi}{M}\right),$$

which suppresses the imaging caused by the interpolator, while at the same time ensuring the absence of aliasing that would otherwise be caused by the decimator. Hence, the specifications of the filter $H(z)$ for a fractional sampling rate conversion are given by

$$\left|H(e^{j\omega})\right| = \begin{cases} L, & |\omega| \leq \min\left(\frac{\pi}{L}, \frac{\pi}{M}\right), \\ 0, & \text{otherwise.} \end{cases} \tag{13.27}$$

The design of $H(z)$ is a standard IIR or FIR lowpass filter design problem. Any of the techniques outlined in Chapter 9 can be applied here for the design of these lowpass filters.

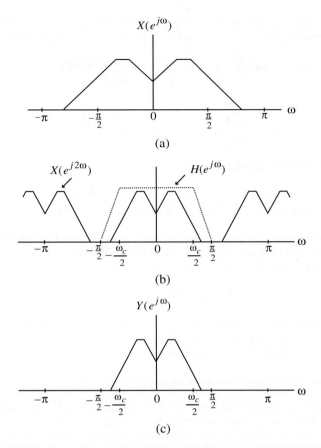

Figure 13.18: Spectrum of (a) the input $x[n]$, (b) the output $x[n]$ of the up-sampler, and (c) the output $y[n]$ of the factor-of-2 interpolator. The spectrum of the interpolation filter $H(z)$ is shown in (b) with dotted lines. In the figures ω_c is the highest frequency of the input signal that is to be preserved.

13.2.4 Computational Requirements

As indicated earlier, the lowpass decimation or interpolation filter can be designed either as an FIR or an IIR digital filter. In the case of single-rate digital signal processing, IIR filters are, in general, computationally more efficient than the FIR digital filters, and are therefore preferred in applications where computational cost needs to be minimized. This issue is not quite the same in the case of multirate digital signal processing, as elaborated next.

Consider the factor-of-M decimator structure of Figure 13.15(b). If the decimation filter $H(z)$ is an FIR filter of length N implemented in a direct form, then

$$v[n] = \sum_{m=0}^{N-1} h[m]x[n-m]. \qquad (13.28)$$

Now, the down-sampler keeps only every Mth sample of $v[n]$ at its output. As a result, it is sufficient to compute $v[n]$ using Eq. (13.28) only for values of n that are multiples of M and skip the computations of the in-between $M - 1$ samples. This leads to a factor-of-M savings in the computational complexity. If, on the other hand, $H(z)$ in Figure 13.15(b) is an IIR filter of order K with a transfer function

$$H(z) = \frac{V(z)}{X(z)} = \frac{P(z)}{D(z)}, \tag{13.29}$$

where

$$P(z) = \sum_{n=0}^{K} p_n z^{-n}, \qquad D(z) = 1 + \sum_{n=1}^{K} d_n z^{-n}, \tag{13.30}$$

its direct form implementation is given by

$$w[n] = -d_1 w[n-1] - d_2 w[n-2] - \cdots - d_K w[n-K] + x[n], \tag{13.31a}$$

$$v[n] = p_0 w[n] + p_1 w[n-1] + \cdots + p_K w[n-K]. \tag{13.31b}$$

Since $v[n]$ is being downsampled, it is sufficient to compute $v[n]$ using Eq. (13.31b) only for values of n that are integer multiples of M. However, the intermediate signal $w[n]$ in Eq. (13.31a) must be computed for all values of n. For example, in the computation of

$$v[M] = p_0 w[M] + p_1 w[M-1] + \cdots + p_K w[M-K],$$

$K + 1$ successive values of $w[n]$ are *still* required. As a result, the savings in the computations in the case of an IIR filter is going to be less than a factor of M.

Example 13.6 provides a more detailed comparison.

EXAMPLE 13.6   **Decimator Computational Complexity**

In this example, we compare the computational complexity of various implementations of the factor-of-M decimator of Figure 13.15(b).[4] Let the sampling frequency for the input $x[n]$ in Figure 13.15(b) be F_T. Then the number of multiplications per second, to be denoted as $\mathcal{R}_M$, are as follows for various computational schemes.

For an FIR $H(z)$ of length N,
$$\mathcal{R}_{M,\text{FIR}} = N \times F_T.$$

For an FIR $H(z)$ of length N followed by a factor-of-M down-sampler,

$$\mathcal{R}_{M,\text{FIR-DEC}} = N \times F_T / M.$$

For an IIR $H(z)$ of order K,
$$\mathcal{R}_{M,\text{IIR}} = (2K + 1) \times F_T.$$

For an IIR $H(z)$ of order K followed by a factor-of-M down-sampler,

$$\mathcal{R}_{M,\text{IIR-DEC}} = K \times F_T + (K + 1) \times F_T / M.$$

Thus, in the FIR case, we save computations by a factor of M. In the IIR case, we save by a factor of $M(2K + 1)/[(M + 1)K + 1]$, which is not significant for large K. For $M = 10$ and $K = 9$, the savings is only by a factor of 1.9. We shall point out later that in certain cases, the IIR filter can be computationally more efficient.

For the case of the interpolation filter in Figure 13.15(a), very similar arguments hold. If $H(z)$ is an FIR filter, then the computational savings are by a factor of L (since $v[n]$ has $L - 1$ zeros between its two

---

[4]In this chapter, the "computational complexity of an implementation" is taken to be equal to the number of multiplications required per second. Also, the symmetry of FIR impulse responses, which leads to about 50% computational savings, is ignored.

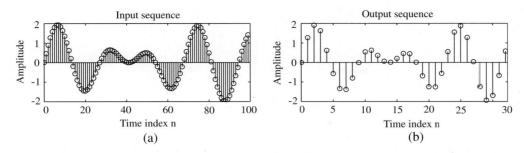

Figure 13.19: (a) The input and (b) output sequences of the factor-of-2 decimator of Example 13.7.

consecutive nonzero samples). On the other hand, computational savings are significantly less with IIR filters.

13.2.5 Sampling Rate Alteration Using MATLAB

Four specific functions available in MATLAB for sampling rate alteration are `decimate`, `interp`, `upfirdn`, and `resample`.

The function `decimate` can be employed to reduce the sampling rate of an input signal vector by an integer factor and is available with several options. We illustrate its application in Example 13.7.

EXAMPLE 13.7   Illustration of Decimation Operation Using MATLAB

We consider here the decimation of a sum of two sinusoidal sequences of normalized frequencies f_1 and f_2 by an arbitrary down-sampling factor M using Program 13.5. The input data to the program are the length N of the input sequence $x[n]$, the down-sampling factor M, and the two normalized frequencies in Hz. The program uses a 30-tap FIR lowpass decimation filter designed to have a stopband edge at π/M.[5] It then plots the input and the output sequences, as indicated in Figure 13.19, for $N = 100$, $M = 2$, $f_1 = 0.043$, and $f_2 = 0.031$.

Program
13_5.m

The function `interp`[6] can be employed to increase the sampling rate of an input signal by an integer factor and is available with several options. We consider its application in Example 13.8.

EXAMPLE 13.8   Illustration of Interpolation Operation Using MATLAB

To illustrate the interpolation process, we consider an input x that is given by a sum of two sinusoidal sequences. Program 13.6 is then used to find its interpolated output y for an up-sampling factor of L. The input data to the program are the length N of the input vector x, the up-sampling factor L, and the two normalized frequencies in Hz. The frequencies of the two sinusoids considered here are $f_1 = 0.043$ and $f_2 = 0.031$, while the interpolation factor is 2. The input length is 50. The output plots generated by this program are shown in Figure 13.20.

Program
13_6.m

The M-files `resample` and `upfirdn` can be employed to increase the sampling rate of an input vector by a ratio of two positive integers. Each of these functions is available with several options. Example 13.9 demonstrates the application of the function `resample`.

---

[5]The group delay of the default decimation filter is 14.5 samples. For an integer-valued group delay, a decimation filter of odd length should be used.

[6]`interp` is based on the interpolator design of Oetken et al. [Oet75].

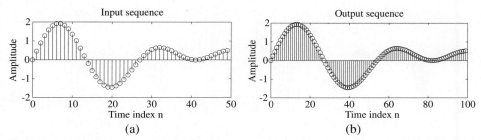

Figure 13.20: (a) The input and (b) output sequences of the factor-of-2 interpolator of Example 13.8.

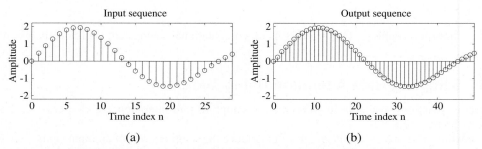

Figure 13.21: Illustration of sampling rate increase by a rational number 5/3.

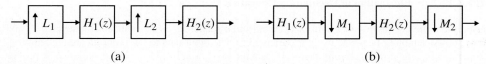

Figure 13.22: Two-stage implementation of sampling rate alteration systems: (a) interpolator and (b) decimator.

Program
13_7.m

EXAMPLE 13.9   Illustration of Fractional-Rate Interpolation Using MATLAB

In this example, we consider the sampling rate increase by a ratio of two positive integers of a signal composed of two sinusoids employing Program 13_7. The input data to the program are the length N of the input vector x, the up-sampling factor L, the down-sampling factor M, and the two frequencies. The up-sampling and down-sampling factors used here are 5 and 3, respectively. The frequencies of the two sinusoids considered here are $f_1 = 0.043$ and $f_2 = 0.031$. Figure 13.21 shows the output plots generated by this program.

13.3 Multistage Design of Decimator and Interpolator

The decimator and the interpolator of Figure 13.15 are single-stage structures since here the basic scheme for the implementation involves a single lowpass filter and a single sampling rate alteration device. If the interpolation factor L can be expressed as a product of two integers L_1 and L_2, the factor-of-L interpolator of Figure 13.15(a) can also be realized in two stages, as indicated in Figure 13.22(a). Likewise, the factor-of-M decimator of Figure 13.15(b) can be implemented in two stages, as shown in Figure 13.22(b), if the decimation factor M is a product of two integers M_1 and M_2. Of course, the design can involve more than two stages, depending on the number of factors used to express L and M, respectively. It turns out that, in general, the computational efficiency is improved significantly by designing the sampling rate alteration system as a cascade of several stages. We demonstrate this feature in Example 13.10.

EXAMPLE 13.10   Multistage Design of a Decimator

Consider the design of a decimator for the reduction of the sampling rate of a signal from 12 kHz to 400 Hz, as indicated in Figure 13.23. The desired decimation factor is thus $M = 12,000/400 = 30$. The specifications for the decimation filter $H(z)$ are as follows: passband edge $F_p = 180$ Hz, stopband edge $F_s = 200$ Hz, passband ripple $\delta_p = 0.002$, and stopband ripple $\delta_s = 0.001$. If $H(z)$ is required to be an equiripple linear-phase FIR lowpass filter, then its order N_H determined using `remezord` is given by

$$N_H = 1827. \tag{13.32}$$

Thus, the number of multiplications per second, ignoring the filter coefficient symmetry, in the single-stage implementation of the factor-of-30 decimator of Figure 13.23 is

$$\mathcal{R}_{M,H} = (N_H + 1) \times \frac{F_T}{M} = 1828 \times \frac{12,000}{30} = 730,800. \tag{13.33}$$

Let us now consider the implementation of $H(z)$ as a cascade realization in the form of $F(z^{15})I(z)$ using the interpolated FIR (IFIR) realization method discussed in Section 10.6.2. The shaping filter $F(z)$ should now have specifications as shown in Figure 13.24(a). This corresponds to stretching the specifications of $H(z)$ by 15. Figure 13.24(b) shows the magnitude response of $F(z^{15})$. The desired response of the interpolator $I(z)$ is also shown in the same figure. Note that $I(z)$ has a wider transition band as it takes into account the spectral gaps between the passbands of $F(z^{15})$.

The resulting IFIR implementation is shown in Figure 13.25(a). Because of the cascade connection, the overall ripple of the cascade in dB is given by the sum of the passband ripples of $I(z)$ and $F(z^{15})$ in dB. This can be compensated for by designing $I(z)$ and $F(z)$ to have a passband ripple of $\delta_p = 0.001$ (rather than 0.002) each. On the other hand, the cascade of $I(z)$ and $F(z^{15})$ has a stopband at least as good as $I(z)$ or $F(z^{15})$ individually, so the specification for δ_s for both $I(z)$ and $F(z)$ can be retained as $\delta_s = 0.001$. Therefore, the specifications for the IFIR approach are as follows:

$$F(z): \quad \omega_p^{(F)} = 2.7\,kHz, \quad \omega_s^{(F)} = 3\,kHz, \quad \delta_p^{(F)} = 0.001, \quad \delta_s^{(F)} = 0.001, \tag{13.34a}$$

$$I(z): \quad \omega_p^{(I)} = 180\,Hz, \quad \omega_s^{(I)} = 600\,Hz, \quad \delta_p^{(I)} = 0.001, \quad \delta_s^{(I)} = 0.001. \tag{13.34b}$$

The filter orders N_F and N_I of $F(z)$ and $I(z)$, respectively, obtained using `remezord` are as follows:

$$N_F = 130, \quad N_I = 93. \tag{13.35a}$$

It should be noted that Figure 13.25(a) has been obtained by replacing $H(z)$ in Figure 13.23 with $I(z)F(z^{15})$. The length of $H(z)$ for a direct implementation is 1828 from Eq. (13.32), whereas the length of $I(z)F(z^{15})$ is $93 + 15 \times 130 + 1 = 2044$. In other words, the length of the overall filter has been increased. However, the computational complexity (i.e., number of multiplications per second) of the new realization can be dramatically reduced by making use of the cascade equivalence #1 in Figure 13.14, as demonstrated next. To this end, we redraw Figure 13.25(a) as in Figure 13.25(b) by factoring the down-sampling factor M as $M_1 M_2$, with $M_1 = 15$ and $M_2 = 2$. Next, by invoking the equivalence of Figure 13.14(a), we replace it with Figure 13.25(c). From this figure, it can be seen that the implementation of $F(z)$ at the sampling rate 800 Hz, followed by a down-sampling by a factor of 2, requires

$$\mathcal{R}_{M,F} = 131 \times \frac{800}{2} = 52,400 \tag{13.36}$$

multiplications per second. The implementation of $I(z)$ followed by the down-sampler of factor 15 requires about

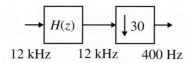

Figure 13.23: Block-diagram representation of the single-stage factor-of-30 decimator.

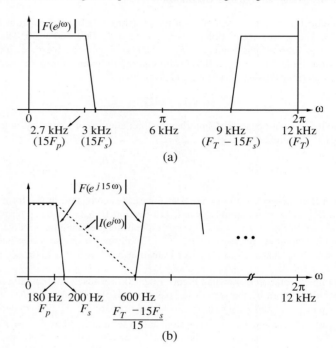

Figure 13.24: Decimation filter design based on the IFIR approach (frequency response plots not shown to scale).

$$\mathcal{R}_{M,I} = 94 \times \frac{12,000}{15} = 75,200 \tag{13.37}$$

multiplications per second. Thus, the interpolator $I(z)$ requires more multiplications per second than the shaping filter $F(z)$. The total complexity of the IFIR-based implementation of Figure 13.25(c) is therefore

$$52,400 + 75,200 = 127,600 \tag{13.38}$$

multiplications per second, which is about 5.73 times smaller than the complexity Eq. (13.33) of a direct implementation, as in Figure 13.23.

In the general case of a k-stage implementation, as indicated in Figure 13.26(a), the decimation factor is $M = M_1 M_2 \cdots M_k$. For a given M_k, there are $k-1$ quantities, $M_1, M_2, \ldots, M_{k-1}$, which can be chosen in various ways, with one combination leading to an optimum multistage realization of a decimator with the least computational complexity. The determination of an optimum realization depends on the selection of k and the "best" combination and ordering of $M_1, M_2, \ldots, M_k$ that minimizes the required number of multiplications per second [Cro83]. The corresponding multistage interpolator design is an analogous problem [Figure 13.26(b)] [Cro83].

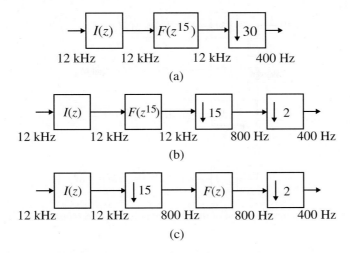

Figure 13.25: The steps in the two-stage realization of the decimator structure.

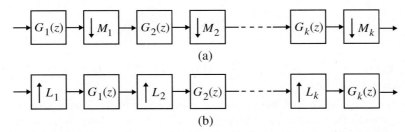

Figure 13.26: (a) A general multistage structure for decimation and (b) a general multistage structure for interpolation.

13.4 The Polyphase Decomposition

We have shown in Section 13.2.4 that a single-stage decimator or interpolator employing FIR lowpass filters can be computationally efficient since the necessary multiplications required to compute the output sample can be carried out only when needed. We demonstrated in the previous section that the computational requirements can be further decreased using a multistage design. Additional reduction in the computational complexity is possible by realizing the FIR filters using the polyphase decomposition described in Section 8.3.3. In certain cases, it is also possible to realize IIR decimation and interpolation filters in polyphase forms, resulting in reduced computational complexity realizations. We review the polyphase decomposition again here and illustrate its application in the efficient realization of the decimator and the interpolator.

13.4.1 The Decomposition

Consider an arbitrary sequence $\{x[n]\}$ with a z-transform $X(z)$:

$$X(z) = \sum_{n=-\infty}^{\infty} x[n]z^{-n}. \tag{13.39}$$

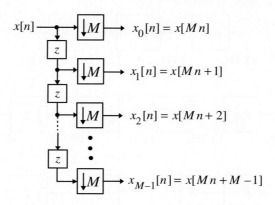

Figure 13.27: A structural interpretation of the M-band polyphase decomposition of a sequence $x[n]$.

We can rewrite $X(z)$ as

$$X(z) = \sum_{k=0}^{M-1} z^{-k} X_k(z^M), \tag{13.40}$$

where

$$X_k(z) = \sum_{n=-\infty}^{\infty} x_k[n]z^{-n} = \sum_{n=-\infty}^{\infty} x[Mn+k]z^{-n}, \qquad 0 \le k \le M-1. \tag{13.41}$$

The subsequences $\{x_k[n]\}$ are called the polyphase components of the parent sequence $x[n]$, and the functions $X_k(z)$, given by the z-transform of $\{x_k[n]\}$, are called the *polyphase components* of $X(z)$ [Bel76]. The relation between the subsequences $\{x_k[n]\}$ and the original sequence $\{x[n]\}$ is given by

$$x_k[n] = x[Mn+k], \qquad 0 \le k \le M-1. \tag{13.42}$$

Equation (13.40) can be written in matrix form as

$$X(z) = \begin{bmatrix} 1 & z^{-1} & \cdots & z^{-(M-1)} \end{bmatrix} \begin{bmatrix} X_0(z^M) \\ X_1(z^M) \\ \vdots \\ X_{M-1}(z^M) \end{bmatrix}. \tag{13.43}$$

A multirate structural interpretation of the polyphase decomposition is given in Figure 13.27.

     The polyphase decomposition of an FIR transfer function can be carried out by inspection as illustrated in Section 8.3.3, where we developed two-branch and three-branch polyphase decompositions of a length-9 FIR transfer function. Figure 8.7 illustrates the parallel realizations of an FIR transfer function based on the polyphase decomposition. In the following section, we consider several variations of the parallel realization of an FIR transfer function based on a polyphase decomposition.

     The polyphase decomposition of an IIR transfer function $H(z) = P(z)/D(z)$, on the other hand, is not that straightforward. One way to arrive at an M-branch polyphase decomposition of $H(z)$ is to express it in the form $P'(z)/D'(z^M)$ by multiplying the denominator $D(z)$ and the numerator $P(z)$ with an appropriately chosen polynomial and then applying an M-branch polyphase decomposition to $P'(z)$. This approach is illustrated in Example 13.11.

EXAMPLE 13.11   Polyphase Decomposition of an IIR Transfer Function

Consider an IIR transfer function

$$H(z) = \frac{1 - 2z^{-1}}{1 + 3z^{-1}}.$$

To obtain a two-band decomposition of the form of Eq. (13.41) with $M = 2$, we rewrite $H(z)$ as

$$H(z) = \frac{(1 - 2z^{-1})(1 - 3z^{-1})}{(1 + 3z^{-1})(1 - 3z^{-1})} = \frac{1 - 5z^{-1} + 6z^{-2}}{1 - 9z^{-2}} = \left(\frac{1 + 6z^{-2}}{1 - 9z^{-2}}\right) + z^{-1}\left(\frac{-5}{1 - 9z^{-2}}\right).$$

Therefore, $H(z) = E_0(z^2) + z^{-1}E_1(z^2)$, where

$$E_0(z) = \frac{1 + 6z^{-1}}{1 - 9z^{-1}}, \quad E_1(z) = \frac{-5}{1 - 9z^{-1}}.$$

Note that the above approach increases the overall order and the complexity of $H(z)$. However, when used in certain multirate structures, the approach may result in a more efficient structure. An alternative, more attractive approach is considered in Example 13.12.

EXAMPLE 13.12   Parallel Allpass Decomposition of an IIR Transfer Function

Consider the transfer function of a fifth-order Butterworth lowpass filter with a 3-dB cutoff frequency at 0.5π:

$$H(z) = \frac{0.0527864045(1 + z^{-1})^5}{1 + 0.633436854z^{-2} + 0.05572809z^{-4}}.$$

Using the pole-interlacing property outlined in Section 8.10, we can express $H(z)$ in the form

$$H(z) = \frac{1}{2}\left[\left(\frac{0.10557281 + z^{-2}}{1 + 0.10557281z^{-2}}\right) + z^{-1}\left(\frac{0.527864045 + z^{-2}}{1 + 0.527864045z^{-2}}\right)\right].$$

Therefore, we can express $H(z) = E_0(z^2) + z^{-1}E_1(z^2)$, where

$$E_0(z) = \frac{1}{2}\left(\frac{0.10557281 + z^{-1}}{1 + 0.10557281z^{-1}}\right), \quad E_1(z) = \frac{1}{2}\left(\frac{0.527864045 + z^{-1}}{1 + 0.527864045z^{-1}}\right).$$

Note that in the above polyphase-like decomposition, the branch transfer functions are stable allpass functions. Moreover, the decomposition has not increased the order of the overall transfer function $H(z)$.

13.4.2   FIR Filter Structures Based on the Polyphase Decomposition

We have illustrated in Section 8.3.3 that a parallel realization of an FIR filter transfer function $H(z)$ can be obtained using a polyphase decomposition. As we shall point out later in this chapter, such a realization often results in computationally efficient structures in certain multirate applications. We revisit the polyphase decomposition based FIR filter realization again here using the more commonly used notation for the polyphase components and develop several other alternate realizations.

Consider first an M-branch polyphase decomposition of $H(z)$ given by

$$H(z) = \sum_{k=0}^{M-1} z^{-k} E_k(z^M). \tag{13.44}$$

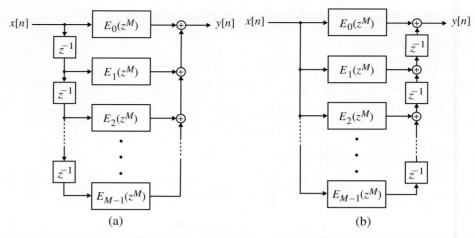

(a)   (b)

Figure 13.28: (a) Direct realization of an FIR filter based on a Type I polyphase decomposition and (b) its transpose.

A direct realization based on Eq. (13.44) is shown in Figure 13.28(a). The transpose of this realization is indicated in Figure 13.28(b). An alternative representation of the transpose structure of Figure 13.28(b) is obtained by using the notation

$$R_\ell(z^M) = E_{M-1-\ell}(z^M), \qquad 0 \le \ell \le M-1, \tag{13.45}$$

resulting in the structure of Figure 13.29. The corresponding polyphase decomposition is thus given by

$$H(z) = \sum_{\ell=0}^{M-1} z^{-(M-1-\ell)} R_\ell(z^M). \tag{13.46}$$

To differentiate between the two decompositions of Eqs. (13.44) and (13.46), the former is usually called the *Type I polyphase decomposition,* while the latter is called *Type II polyphase decomposition.*

13.4.3 Computationally Efficient Interpolator and Decimator Structures

Computationally efficient decimator and interpolator structures employing lowpass filters can be derived by applying a polyphase decomposition to their corresponding transfer functions. We demonstrate this property next.

   Consider first the use of the polyphase decomposition in the realization of the decimation filter of Figure 13.15(b). The overall decimator structure is of the form of Figure 13.30(a) when the lowpass filter $H(z)$ is realized as in Figure 13.28(a). An equivalent realization indicated in Figure 13.30(b), obtained using the cascade equivalence #1 of Figure 13.14(a), is computationally more efficient than the structure of Figure 13.30(a). To illustrate this point, assume that the decimation filter $H(z)$ of Figure 13.15(b) is a length-N FIR structure and the input sampling period $T = 1$. Since the decimator output $y[n]$ is obtained by downsampling the filter output $v_1[n]$ by a factor of M, it is necessary only to compute $v[n]$ at $n = \ldots, -2M, -M, 0, M, 2M, \ldots$. The computational requirements are therefore N multiplications and $(N-1)$ additions per output sample being computed. However, as n increases, the stored signals in the delay registers change. As a result, all computations need to be completed in one sampling period, and for the following $(M-1)$ sampling periods, the arithmetic units remain idle. Now consider the structure of Figure 13.30(b). If the lengths of the subfilter $E_k(z)$ is N_k, then $N = \sum_{k=0}^{M-1} N_k$. The computational

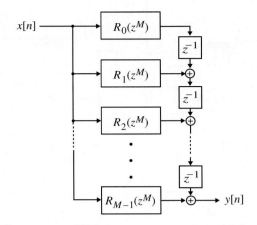

Figure 13.29: Realization of an FIR filter based on a Type II polyphase decomposition.

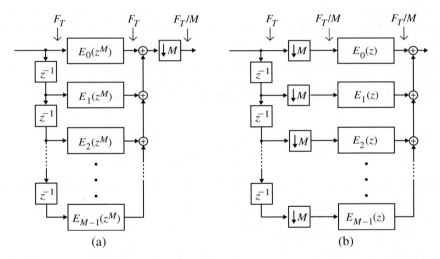

Figure 13.30: (a) Decimator implementation based on a Type I polyphase decomposition and (b) computationally efficient decimator structure. In the figures, the sampling rates of pertinent sequences are indicated by an arrow.

requirements of the kth subfilter are N_k multiplications and $N_k - 1$ additions per output sample, and that for the overall structure is therefore $\sum_{k=0}^{M-1} N_k = N$ multiplications and $\sum_{k=0}^{M-1}(N_k - 1) + (M - 1) = N - 1$ additions per decimator output sample. However, in the latter structure, the arithmetic units are operative at all instants of the output sampling period, which is M times that of the input sampling period.

Similar savings are also obtained in the case of the interpolator structure employing polyphase decomposition in the realization of computationally efficient interpolators. Figure 13.31(a) shows the interpolator structure derived from Figure 13.15(a) by making use of the L-band Type I polyphase decomposition of the interpolation filter $H(z)$ and the cascade equivalence #2 of Figure 13.14(b). An alternative realization obtained using the Type II polyphase decomposition of the interpolation filter $H(z)$ and the cascade equivalence #2 is shown in Figure 13.31(b).

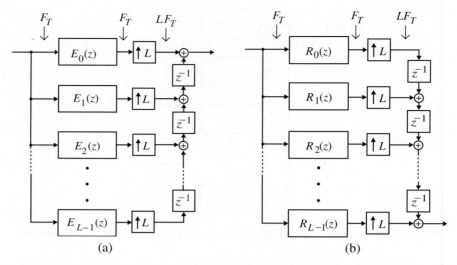

Figure 13.31: Computationally efficient interpolator structures: (a) Type I polyphase decomposition and (b) Type II polyphase decomposition. In the figures, the sampling rates of pertinent sequences are indicated by an arrow.

More efficient interpolator and decimator structures can be realized by exploiting the symmetry of the filter coefficients of $H(z)$ in the case of linear-phase filters. Consider, for example, the realization of a factor-of-3 ($M = 3$) decimator using a length-12 linear-phase FIR lowpass filter with a symmetric impulse response:

$$H(z) = h[0] + h[1]z^{-1} + h[2]z^{-2} + h[3]z^{-3} + h[4]z^{-4} + h[5]z^{-5} + h[5]z^{-6}$$
$$+ h[4]z^{-7} + h[3]z^{-8} + h[2]z^{-9} + h[1]z^{-10} + h[0]z^{-11}. \qquad (13.47)$$

A conventional polyphase decomposition of the above $H(z)$ yields the following subfilters:

$$E_0(z) = h[0] + h[3]z^{-1} + h[5]z^{-2} + h[2]z^{-3},$$
$$E_1(z) = h[1] + h[4]z^{-1} + h[4]z^{-2} + h[1]z^{-3}, \qquad (13.48)$$
$$E_2(z) = h[2] + h[5]z^{-1} + h[3]z^{-2} + h[0]z^{-3}.$$

Note that the subfilter $E_1(z)$ still has a symmetric impulse response, whereas the impulse response of $E_2(z)$ is the mirror image of that of $E_0(z)$. These relations can be made use of in developing a computationally efficient realization using only six multipliers and 11 two-input adders, as depicted in Figure 13.32.

13.4.4 Computationally Efficient Rational Sampling Rate Converter

The complexity of the design of the fractional sampling rate converter depends on the ratio of the sampling rates between the input and the output digital signals. For example, in digital audio applications, the three different sampling frequencies employed are 44.1 kHz, 32 kHz, and 48 kHz. As a consequence, there are three different values for the sampling rate conversion factor – 2:3 (or 3:2), 147:160 (or 160:147), and 320:441 (or 441:320). Likewise, in digital video applications, the sampling rates of composite video signals are 14.3181818 MHz and 17.734475 MHz, whereas the sampling rates of the digital component video signal are 13.5 MHz and 6.75 MHz for the luminance and the color-difference signals, respectively, for the NTSC and PAL systems. Here, the sampling rates for the component and the NTSC composite

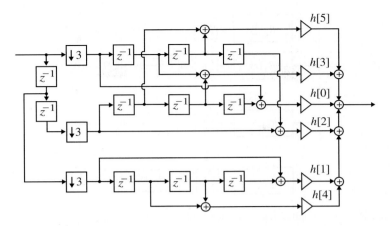

Figure 13.32: A computationally efficient realization of a factor-of-3 decimator exploiting the linear-phase symmetry of the length-12 decimation filter.

video signals are related by a ratio 35:33, whereas the sampling rates for the component and the PAL composite video signals are related by a ratio 709,379:540,000. There are applications, such as in the pitch control of audio signals, where the ratio is irrational, in which case there is no periodic relation between the sampling instants of the input and the output digital signals.

We consider here the implementation of a computationally efficient fractional rate converter with a rational conversion factor L/M, where L and M are mutually prime. In developing the computationally efficient structure, we invoke a result from number theory, which states that the two mutually prime integers L and M satisfy the relation

$$\mu M - \lambda L = 1, \tag{13.49}$$

for a set of unique distinct positive integers μ and λ. The general structure for a rational sampling rate converter of Figure 13.16(b) can be made computationally efficient by making use of one of the structures based on the polyphase decompositions, such as either Figure 13.29 or Figure 13.31. A more computationally efficient structure can be realized by combining both of these approaches, as described next [Hsi87], [Fli94].

Without any loss of generality, we assume the case when $L < M$. The structure for the case $L < M$ can be derived by applying the transpose operation.

To develop the computationally efficient structure for the case when $L < M$, first the cascade of the factor-of-L up-sampler and the filter $H(z)$ in Figure 13.16(b) is replaced with its equivalent Type II polyphase decomposition based realization shown in Figure 13.31(a), with the factor-of-M down-sampler moved to all L branches, resulting in the structure of Figure 13.33. Consider the kth polyphase branch in Figure 13.33, as shown in Figure 13.34(a). Now we can write

$$z^{-k} = z^{-k(\mu M - \lambda L)}, \qquad 0 \le k \le L - 1,$$

by making use of Eq. (13.49). Hence, we can replace the block of k unit delays in Figure 13.34(a) with a block of $k\mu M$ unit delays and a block of $k\lambda L$ unit advances, as indicated in Figure 13.34(b). This branch can be further redrawn, as shown in Figure 13.34(c), by invoking the noble identities of Figure 13.14. Since the up-sampling factor L and the down-sampling factor M are mutually prime, we can interchange their positions, resulting in the structure shown in Figure 13.34(d). As a result, the general rational sampling rate converter structure of Figure 13.33 can be redrawn as shown in Figure 13.35(a), whose equivalent realizable form is as indicated in Figure 13.35(b).

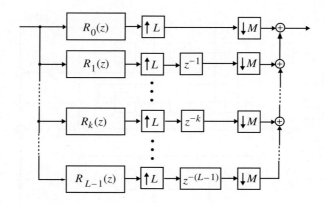

Figure 13.33: Rational sampling rate converter implementation based on a polyphase decomposition.

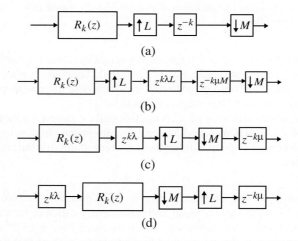

Figure 13.34: Various steps in the development of the kth branch.

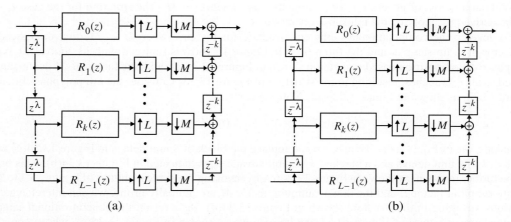

Figure 13.35: Equivalent realizations of the rational sampling rate converter of Figure 13.33.

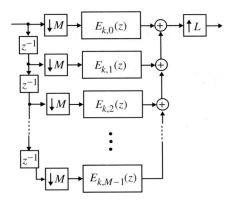

Figure 13.36: Computationally efficient realization of the kth polyphase branch of Figure 13.35(b).

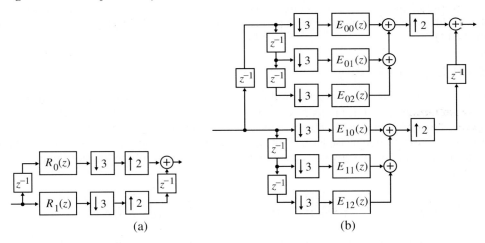

(a) (b)

Figure 13.37: (a) Original rational sampling rate converter with a conversion factor of 2/3 and (b) its computationally efficient realization.

Next, the cascade of the polyphase section $R_k(z)$ followed by the factor-of-M down-sampler can be replaced with a computationally efficient realization based on a Type I polyphase decomposition in the form of Figure 13.31(b), resulting in the structure shown in Figure 13.36. Finally, by combining all k branches as indicated in Figure 13.35(b), we arrive at a computationally efficient realization of a rational sampling rate converter.

We illustrate the above approach in Example 13.13.

EXAMPLE 13.13   Design of a Rational Sampling Rate Converter with a 2/3 Conversion Factor

A rational sampling rate interpolator with an interpolation factor of 2/3 is needed in the conversion of a digital audio signal of 48-kHz sampling rate to one of 32-kHz sampling rate. The basic form of the desired sampling rate converter is as indicated in Figure 13.16(a), where the lowpass filter $H(z)$ has a stopband edge at $\pi/3$. Note that in this structure, the filter $H(z)$ operates at the 96-kHz rate.

For this design, we have $L = 2$, $M = 3$, and $k = 1$. Equation (13.49) is seen to be satisfied with $\mu = \lambda = 1$. Hence, the structure of Figure 13.35(b) reduces to that shown in Figure 13.37(a). By realizing the filters $R_0(z)$ and $R_1(z)$ in Type I polyphase forms and then applying the cascade equivalence of Figure 13.14(a), we arrive at the final realization shown in Figure 13.37(b), in which all filters now operate at the 16-kHz rate.

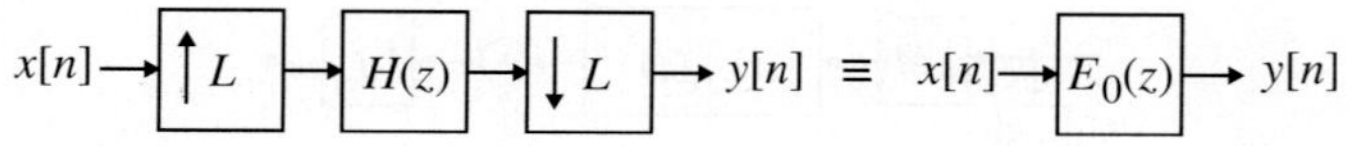

$$x[n] \longrightarrow \boxed{\uparrow L} \longrightarrow \boxed{H(z)} \longrightarrow \boxed{\downarrow L} \longrightarrow y[n] \quad \equiv \quad x[n] \longrightarrow \boxed{E_0(z)} \longrightarrow y[n]$$

(a)   (b)

Figure 13.38: (a) A cascade multirate structure and (b) its equivalent form.

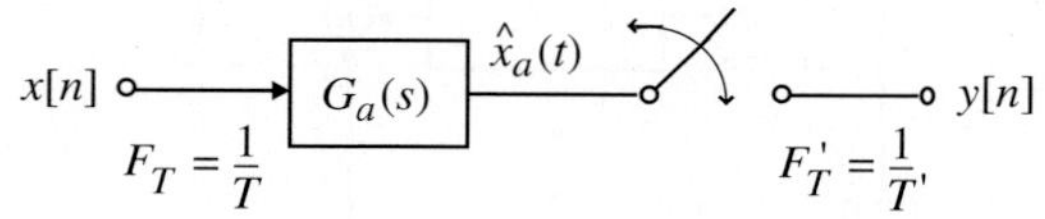

$$x[n] \circ \longrightarrow \boxed{G_a(s)} \xrightarrow{\hat{x}_a(t)} \circ \quad \circ \longrightarrow \circ \; y[n]$$

$$F_T = \frac{1}{T} \qquad\qquad\qquad F_T' = \frac{1}{T'}$$

Figure 13.39: Sampling rate alteration based on conversion of input digital signal to analog form, followed by a resampling at the desired output rate.

The transpose of the structure of Figure 13.37(b) yields the realization of a rational sampling rate interpolator with a conversion factor of 3/2 and can be used as the 32-48-kHz sampling rate converter (Problem 13.14).

13.4.5 A Useful Identity

The cascade multirate digital filter structure of Figure 13.38(a) appears in a number of applications. If we express the transfer function $H(z)$ in its L-term Type I polyphase form $\sum_{k=0}^{L-1} z^{-k} E_k(z^L)$, it can be easily shown that the structure of Figure 13.38(a) is equivalent to the time-invariant digital filter of Figure 13.38(b), where $E_0(z)$ is the zeroth polyphase term (Problem 13.28) [Vai93]. This equivalence can be exploited in simplifying complex multirate networks containing cascade structures of the form of Figure 13.38(a) for analysis purposes.

13.5 Arbitrary-Rate Sampling Rate Converter

There are many applications requiring the estimation of a discrete-time signal value at an arbitrary time instant between a consecutive pair of known samples. Applications include conversion between arbitrary sampling rates, timing adjustment in digital receivers, time-delay estimation, echo cancellation in modems, and beam steering and direction finding in antenna arrays.

The estimation problem can be solved by using some type of interpolation that basically forms an approximating continuous-time signal from a set of known consecutive samples of the given discrete-time signal and then evaluates the value of the continuous-time signal at the desired time instant. This interpolation process can be directly implemented by designing a digital interpolation filter. An all-digital design of a sampling rate converter with an arbitrary conversion factor is not simple. In particular, the design is quite difficult and expensive when the conversion factor is a ratio of two very large integers or an irrational number.

13.5.1 Ideal Sampling Rate Converter

In principle, a sampling rate conversion by an arbitrary conversion factor can be implemented simply by passing the input digital signal through an ideal analog reconstruction lowpass filter whose output is resampled at the desired output rate, as indicated in Figure 13.39 [Ram84]. If the impulse response of the analog lowpass filter is denoted by $g_a(t)$, the output of the filter is then given by

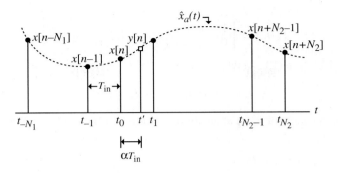

Figure 13.40: Interpolation by an arbitrary factor.

$$\hat{x}_a(t) = \sum_{\ell=-\infty}^{\infty} x[\ell] g_a(t - \ell T). \tag{13.50}$$

If the analog filter is chosen to band-limit its output to the frequency range $F_g < F_T'/2$, its output $\hat{x}_a(t)$ can then be resampled at the rate F_T', and, hence, the output $y[n]$ of the resampler at new time instants $t = nT'$ is given by

$$y[n] = \hat{x}_a(nT') = \sum_{\ell=-\infty}^{\infty} x[\ell] g_a(nT' - \ell T). \tag{13.51}$$

Since the impulse response $g_a(t)$ of an ideal lowpass analog filter is of infinite duration and the samples $g_a(nT' - \ell T)$ have to be computed at each output sampling instant, implementation of the ideal band-limited interpolation algorithm of Eq. (13.51) in exact form is not practical. Thus, approximations to this ideal interpolation algorithm are usually employed in practice.

The basic interpolation problem based on a finite weighted sum of input signal samples can then be stated as follows: Given $N_2 + N_1 + 1$ input signal samples, $x[k], k = -N_1, \ldots, N_2$, obtained by sampling an analog signal $x_a(t)$ at $t = t_k = t_0 + kT_{\text{in}}$, determine the sample value $x_a(t_0 + \alpha T_{\text{in}}) = y[\alpha]$ at time $t' = t_0 + \alpha T_{\text{in}}$, where $-N_1 \le \alpha \le N_2$. Figure 13.40 illustrates the interpolation process by an arbitrary factor.

We describe in this section two commonly employed interpolation algorithms based on a weighted sum of finite number of input samples.

13.5.2 Lagrange Interpolation Algorithm

In this approach, a polynomial approximation $\hat{x}_a(t)$ to $x_a(t)$ is defined as

$$\hat{x}_a(t) = \sum_{k=-N_1}^{N_2} P_k(t) x[n + k], \tag{13.52}$$

where $P_k(t)$ are the Lagrange polynomials given by

$$P_k(t) = \prod_{\substack{\ell=-N_1 \\ \ell \ne k}}^{N_2} \left(\frac{t - t_\ell}{t_k - t_\ell} \right), \qquad -N_1 \le k \le N_2. \tag{13.53}$$

Since

$$P_k(t_r) = \begin{cases} 1, & k = r \\ 0, & k \neq r \end{cases}, \qquad -N_1 \leq r \leq N_2, \tag{13.54}$$

it follows from Eqs. (13.52) to (13.54) that

$$\hat{x}_a(t_k) = x_a(t_k), \qquad -N_1 \leq k \leq N_2. \tag{13.55}$$

From Eq. (13.52), the value of $x_a(t)$ at an arbitrary value $t' = t_0 + \alpha T_{\text{in}}$ is given by

$$\hat{x}_a(t') = \hat{x}_a(t_0 + \alpha T_{\text{in}}) = y[n] = \sum_{k=-N_1}^{N_2} P_k(\alpha)x[n+k], \tag{13.56}$$

where

$$P_k(\alpha) = P_k(t_0 + \alpha T_{\text{in}}) = \prod_{\substack{\ell=-N_1 \\ \ell \neq k}}^{N_2} \left(\frac{\alpha - \ell}{k - \ell} \right), \qquad -N_1 \leq k \leq N_2. \tag{13.57}$$

We illustrate in Example 13.14 the application of the Lagrange interpolation algorithm in designing a fractional-rate sampling-rate converter.

EXAMPLE 13.14   Interpolation Using Lagrange Interpolation Algorithm

Consider the design of a fractional-rate interpolator with an interpolation factor of 3/2. For simplicity, we make use of a third-order polynomial approximation, with $N_1 = 2$ and $N_2 = 1$, in which case Eq. (13.56) reduces to

$$y[n] = P_{-2}(\alpha)x[n-2] + P_{-1}(\alpha)x[n-1] + P_0(\alpha)x[n] + P_1(\alpha)x[n+1]. \tag{13.58}$$

Here, the Lagrange polynomials $P_k(\alpha)$ are given by

$$P_{-2}(\alpha) = \frac{(\alpha+1)\alpha(\alpha-1)}{-6} = \tfrac{1}{6}(-\alpha^3 + \alpha), \tag{13.59a}$$

$$P_{-1}(\alpha) = \frac{(\alpha+2)\alpha(\alpha-1)}{2} = \tfrac{1}{2}(\alpha^3 + \alpha^2 - 2\alpha), \tag{13.59b}$$

$$P_0(\alpha) = \frac{(\alpha+2)(\alpha+1)(\alpha-1)}{-2} = -\tfrac{1}{2}(\alpha^3 + 2\alpha^2 - \alpha - 2), \tag{13.59c}$$

$$P_1(\alpha) = \frac{(\alpha+2)(\alpha+1)\alpha}{6} = \tfrac{1}{6}(\alpha^3 + 3\alpha^2 + 2\alpha), \tag{13.59d}$$

obtained from Eq. (13.53), with $N_1 = 2$ and $N_2 = 1$.

Figure 13.41 shows the locations of the samples of the input and the output sequences for an interpolator with a conversion factor of 3/2, where input sample locations are marked with a circle and output sample locations are marked with a square. The locations of the output samples $y[0]$, $y[1]$, and $y[2]$ in the input sample domain are marked with an arrow.

Consider the computations of $y[n]$, $y[n+1]$, and $y[n+2]$ using four input samples, $x[n-2]$ through $x[n+1]$. To this end, we require a third-order Lagrange polynomial; that is, $N_1 = 2$ and $N_2 = 1$ in Eq. (13.57). From Figure 13.41, it follows that, for the computation of $y[n]$, the value of α in Eq. (13.58), to be labeled α_0, is 0. Equation (13.58) then is of the form

$$y[n] = P_{-2}(\alpha_0)x[n-2] + P_{-1}(\alpha_0)x[n-1] + P_0(\alpha_0)x[n] + P_1(\alpha_0)x[n+1]. \tag{13.60}$$

Replacing α with $\alpha_0 = 0$ in Eqs. (13.59a) to (13.59d), we arrive at the desired values of the filter coefficients as

$$P_{-2}(\alpha_0) = 0, \qquad P_{-1}(\alpha_0) = 0, \tag{13.61a}$$

$$P_0(\alpha_0) = 1, \qquad P_1(\alpha_0) = 0. \tag{13.61b}$$

Next, we observe from Figure 13.41 that for the computation of $y[n+1]$, the value of α, to be labeled α_1, is equal to 2/3. The expression for $y[n+1]$ is then of the form

$$y[n+1] = P_{-2}(\alpha_1)x[n-2] + P_{-1}(\alpha_1)x[n-1] + P_0(\alpha_1)x[n] + P_1(\alpha_1)x[n+1]. \tag{13.62}$$

Using this value of α in Eqs. (13.59a) to (13.59d), we obtain the new set of values for the interpolator filter coefficients as

$$P_{-2}(\alpha_1) = 0.0617, \qquad P_{-1}(\alpha_1) = -0.2963, \tag{13.63a}$$

$$P_0(\alpha_1) = 0.7407, \qquad P_1(\alpha_1) = 0.4938. \tag{13.63b}$$

Finally, for the computation of $y[n+2]$, the value of α, to be labeled α_2, is 4/3. The corresponding expression for $y[n+2]$ is now of the form

$$y[n+2] = P_{-2}(\alpha_2)x[n-2] + P_{-1}(\alpha_2)x[n-1] + P_0(\alpha_2)x[n] + P_1(\alpha_2)x[n+1]. \tag{13.64}$$

For the above value of α_2, we arrive at the following filter coefficient values:

$$P_{-2}(\alpha_2) = -0.1728, \qquad P_{-1}(\alpha_2) = 0.7407, \tag{13.65a}$$

$$P_0(\alpha_2) = -1.2963, \qquad P_1(\alpha_2) = 1.7284. \tag{13.65b}$$

Combining Eqs. (13.60), (13.62), and (13.64) into matrix form, we arrive at the input–output relation of the interpolation filter as

$$\begin{bmatrix} y[n] \\ y[n+1] \\ y[n+2] \end{bmatrix} = \begin{bmatrix} P_{-2}(\alpha_0) & P_{-1}(\alpha_0) & P_0(\alpha_0) & P_1(\alpha_0) \\ P_{-2}(\alpha_1) & P_{-1}(\alpha_1) & P_0(\alpha_1) & P_1(\alpha_1) \\ P_{-2}(\alpha_2) & P_{-1}(\alpha_2) & P_0(\alpha_2) & P_1(\alpha_2) \end{bmatrix} \begin{bmatrix} x[n-2] \\ x[n-1] \\ x[n] \\ x[n+1] \end{bmatrix}$$

$$= \mathbf{H} \begin{bmatrix} x[n-2] \\ x[n-1] \\ x[n] \\ x[n+1] \end{bmatrix}, \tag{13.66}$$

where $\mathbf{H}$ is the block coefficient matrix, which for the above factor of 3/2 interpolator design is given by

$$\mathbf{H} = \begin{bmatrix} 0 & 0 & 1 & 0 \\ 0.0617 & -0.2963 & 0.7407 & 0.4938 \\ -0.1728 & 0.7407 & -1.2963 & 1.7284 \end{bmatrix}. \tag{13.67}$$

It should be evident from an examination of Figure 13.41 that the filter coefficients to compute $y[n+3]$, $y[n+4]$, and $y[n+5]$ are again given by the coefficients in Eqs. (13.61a) and (13.61b), Eqs. (13.63a) and (13.63b), and Eqs. (13.65a) and (13.65b), respectively. Or in other words, the desired interpolation filter is a time-varying filter with a period of three samples. A realization of the above factor-of-3/2 interpolator based on an implementation of Eq. (13.66) is indicated in Figure 13.42(a). Note that, in practice, the overall system delay will be three sample periods, and as a result, the output sample $y[n]$ actually will appear at the time index $n+3$.

An alternative realization of the fractional rate interpolator in the form of a time-varying FIR filter is indicated in Figure 13.42(b). The filter coefficients of the fifth-order time-varying FIR filter have a period of 3 and are assigned the values as indicated in Figure 13.42(c).

Another realization of the above fractional-rate interpolator is obtained by substituting the Lagrange polynomials of Eqs. (13.59a) to (13.59d) in Eq. (13.58), which yields

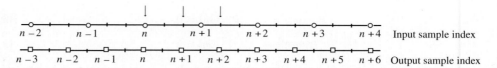

Figure 13.41: The input and output sample locations of an interpolator with an up-sampling factor of $3/2$.

$$y[n] = \alpha^3 \left(-\tfrac{1}{6}x[n-2] + \tfrac{1}{2}x[n-1] - \tfrac{1}{2}x[n] + \tfrac{1}{6}x[n+1] \right)$$

$$+ \alpha^2 \left(\tfrac{1}{2}x[n-1] - x[n] + \tfrac{1}{2}x[n+1] \right)$$

$$+ \alpha \left(\tfrac{1}{6}x[n-2] - x[n-1] + \tfrac{1}{2}x[n] + \tfrac{1}{3}x[n+1] \right) + x[n]. \tag{13.68}$$

A digital filter realization of the above equation leads to the *Farrow structure* of Figure 13.43 [Far88], where the transfer functions of the three FIR digital filters are given by

$$H_0(z) = -\tfrac{1}{6}z^{-2} + \tfrac{1}{2}z^{-1} - \tfrac{1}{2} + \tfrac{1}{6}z,$$
$$H_1(z) = \tfrac{1}{2}z^{-1} - 1 + \tfrac{1}{2}z,$$
$$H_2(z) = \tfrac{1}{6}z^{-2} - z^{-1} + \tfrac{1}{2} + \tfrac{1}{3}z.$$

Note that in this realization, only the value of the multiplier coefficient α is changed periodically, with the remaining digital filter structure kept unchanged.

Figure 13.44 shows the plots of the input and the output of the above fractional-rate interpolator for a sinusoidal input of frequency 0.05 Hz sampled at a 1-Hz rate, along with that of the error sequence defined by the sample-wise difference between the output sequence and the actual sine values. Note the beginning transient samples in Figure 13.44(c).

13.5.3 Spline Interpolation

In this approach, a polynomial approximation $\hat{x}_a(t)$ to $x_a(t)$ is made using the *B-spline functions* as the basis. The time instants t_k, $m \le k \le N+m$, at which the samples $x_a(t_k)$ of the signal $x_a(t)$ are known, are called *knots*. The Lth order B-spline $B_m^{(L)}(t)$ defined in the interval $[t_m, \ldots, t_{N+m}]$ is given by [Zöl97]

$$B_m^{(L)}(t) = \sum_{i=m}^{N+m} a_i \phi_i(t), \tag{13.69}$$

where $\phi_i(t)$, called the *truncated power functions,* are polynomials of degree L;

$$\phi_i(t) = (t - t_i)_+^L = \begin{cases} 0, & t < t_i, \\ (t - t_i)^L, & t \ge t_i. \end{cases} \tag{13.70}$$

The polynomial approximation $\hat{x}_a(t)$ to $x_a(t)$ is given by

$$\hat{x}_a(t) = \sum_{k=m}^{N+m} B_k^{(L)}(t) x_a(t_k). \tag{13.71}$$

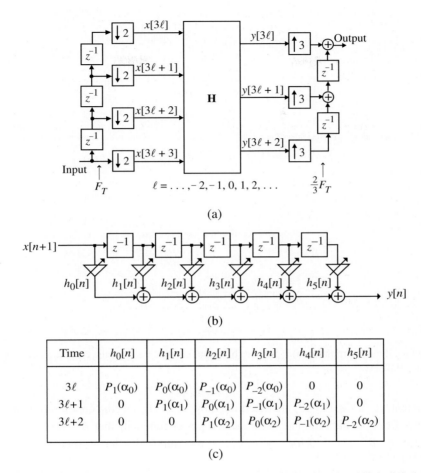

(a)

(b)

Time	$h_0[n]$	$h_1[n]$	$h_2[n]$	$h_3[n]$	$h_4[n]$	$h_5[n]$
3ℓ	$P_1(\alpha_0)$	$P_0(\alpha_0)$	$P_{-1}(\alpha_0)$	$P_{-2}(\alpha_0)$	0	0
$3\ell+1$	0	$P_1(\alpha_1)$	$P_0(\alpha_1)$	$P_{-1}(\alpha_1)$	$P_{-2}(\alpha_1)$	0
$3\ell+2$	0	0	$P_1(\alpha_2)$	$P_0(\alpha_2)$	$P_{-1}(\alpha_2)$	$P_{-2}(\alpha_2)$

(c)

Figure 13.42: Implementation of the fractional-rate interpolator of Example 13.14: (a) block digital filter implementation, (b) implementation using a time-varying FIR interpolation filter, and (c) coefficients of the time-varying filter as a function of sample index.

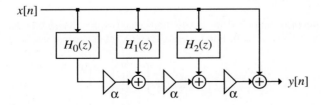

Figure 13.43: Implementation of the fractional-rate interpolator of Example 13.14 using the Farrow structure.

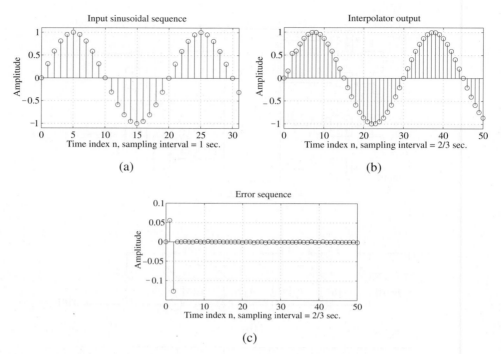

Figure 13.44: Plots of the interpolator (a) input and (b) output sequence and (c) the error sequence for a sinusoidal input of frequency 0.05 Hz.

The coefficients a_i in Eq. (13.69) are determined by imposing specific conditions at the knots t_m and t_{N+m}. It follows from the definition of the truncated power functions that $B_m^{(L)}(t) = 0$ for $t \leq t_m$. An additional condition, $B_m^{(L)}(t) = 0$ for $t \geq t_{N+m}$, is also imposed. Hence, for $t \geq t_{N+m}$, substituting Eq. (13.70) in Eq. (13.69), we get

$$\sum_{i=m}^{N+m} a_i (t - t_i)^L = 0. \tag{13.72}$$

The above set of homogeneous linear equations has nontrivial solutions for $N > L$. In particular, an elegant and simple solution exists for $N = L + 1$ as shown in Example 13.15.

EXAMPLE 13.15   Cubic B-Spline

In this example we develop the solution for the widely used cubic B-spline. Here $L = 3$ and therefore $N = 4$. For notational convenience, we choose $m = 0$. For this case, Eq. (13.72) in matrix form becomes

$$\begin{bmatrix} 1 & 1 & 1 & 1 & 1 \\ t_0 & t_1 & t_2 & t_3 & t_4 \\ t_0^2 & t_1^2 & t_2^2 & t_3^2 & t_4^2 \\ t_0^3 & t_1^3 & t_2^3 & t_3^3 & t_4^3 \end{bmatrix} \begin{bmatrix} a_0 \\ a_1 \\ a_2 \\ a_3 \\ a_4 \end{bmatrix} = \begin{bmatrix} 0 \\ 0 \\ 0 \\ 0 \end{bmatrix}. \tag{13.73}$$

Since the above matrix equation is an underdetermined system and all rows are linearly independent, assuming $t_i \neq t_j$ if $i \neq j$, we can choose any one coefficient as the free parameter and solve for the other four coefficients in terms of the free parameter. Considering a_4 to be the free parameter, we rewrite Eq. (13.73) as

$$
\begin{bmatrix}
1 & 1 & 1 & 1 \\
t_0 & t_1 & t_2 & t_3 \\
t_0^2 & t_1^2 & t_2^2 & t_3^2 \\
t_0^3 & t_1^3 & t_2^3 & t_3^3
\end{bmatrix}
\begin{bmatrix}
a_0 \\ a_1 \\ a_2 \\ a_3
\end{bmatrix}
= -a_4
\begin{bmatrix}
1 \\ t_4 \\ t_4^2 \\ t_4^3
\end{bmatrix}.
\tag{13.74}
$$

We can solve Eq. (13.74) for the coefficients a_i using Cramer's rule. For example, the coefficient a_0 is given by

$$
a_0 = -a_4 \frac{
\begin{vmatrix}
1 & 1 & 1 & 1 \\
t_4 & t_1 & t_2 & t_3 \\
t_4^2 & t_1^2 & t_2^2 & t_3^2 \\
t_4^3 & t_1^3 & t_2^3 & t_3^3
\end{vmatrix}
}{
\begin{vmatrix}
1 & 1 & 1 & 1 \\
t_0 & t_1 & t_2 & t_3 \\
t_0^2 & t_1^2 & t_2^2 & t_3^2 \\
t_0^3 & t_1^3 & t_2^3 & t_3^3
\end{vmatrix}
}.
\tag{13.75}
$$

The numerator and the denominator on the right-hand side of the above equation are determinants of Vandermonde matrices and have nonzero values if the knots t_i are distinct. It can be shown (Problem 13.32) that Eq. (13.75) reduces to

$$
\begin{aligned}
a_0 &= -a_4 \frac{(t_4 - t_1)(t_4 - t_2)(t_4 - t_3)(t_1 - t_2)(t_1 - t_3)(t_2 - t_3)}{(t_0 - t_1)(t_0 - t_2)(t_0 - t_3)(t_1 - t_2)(t_1 - t_3)(t_2 - t_3)} \\
&= -a_4 \frac{(t_4 - t_1)(t_4 - t_2)(t_4 - t_3)}{(t_0 - t_1)(t_0 - t_2)(t_0 - t_3)}.
\end{aligned}
\tag{13.76}
$$

Choosing the free parameter a_4 to be

$$
a_4 = \frac{1}{(t_4 - t_0)(t_4 - t_1)(t_4 - t_2)(t_4 - t_3)},
\tag{13.77}
$$

we get from Eq. (13.76)

$$
a_0 = \frac{1}{(t_0 - t_1)(t_0 - t_2)(t_0 - t_3)(t_0 - t_4)}.
\tag{13.78}
$$

In a similar manner the expressions for the other three coefficients can be derived and are given by

$$
a_1 = \frac{1}{(t_1 - t_0)(t_1 - t_2)(t_1 - t_3)(t_1 - t_4)},
\tag{13.79a}
$$

$$
a_2 = \frac{1}{(t_2 - t_0)(t_2 - t_1)(t_2 - t_3)(t_2 - t_4)},
\tag{13.79b}
$$

$$
a_3 = \frac{1}{(t_3 - t_0)(t_3 - t_1)(t_3 - t_2)(t_3 - t_4)}.
\tag{13.79c}
$$

In the general case, the coefficients a_i are given by

$$
a_i = \frac{(-1)^{L+1}}{\prod_{k=m,i \neq k}^{N+m}(t_i - t_k)}, \qquad m \leq i \leq N + m,
\tag{13.80}
$$

and the Lth-order B-spline function is then of the form

$$B_m^{(L)}(t) = (-1)^{L+1} \sum_{i=m}^{N+m} \frac{(t - t_i)_+^L}{\prod_{k=m, i \neq k}^{N+m} (t_i - t_k)}. \tag{13.81}$$

Since the maximum value of the B-spline decreases with increasing L, it is a common practice to use instead, a normalized form

$$\beta_m^{(L)}(t) = (t_{N+m} - t_m) B_m^{(L)}(t), \tag{13.82}$$

in the interpolation formula of Eq. (13.71).

In digital signal processing applications, the knots are uniformly spaced at the sampling instants, resulting in simpler values for the coefficients a_i, as illustrated in Example 13.16.

EXAMPLE 13.16   **Second-Order Normalized B-Spline**

Determine the values of the coefficients of a second-order normalized B-spline, and plot the spline and its corresponding power functions for $t_i = i$, $m \leq i \leq m + 3$. Here, $L = 2$ and hence, $N = 3$. Substituting these values in Eq. (13.80), we arrive at the coefficients as given below:

$$a_m = -\frac{1}{(m - m - 1)(m - m - 2)(m - m - 3)} = \frac{1}{6},$$

$$a_{m+1} = -\frac{1}{(m + 1 - m)(m + 1 - m - 2)(m + 1 - m - 3)} = -\frac{1}{2},$$

$$a_{m+2} = -\frac{1}{(m + 2 - m)(m + 2 - m - 1)(m + 2 - m - 3)} = \frac{1}{2},$$

$$a_{m+3} = -\frac{1}{(m + 3 - m)(m + 3 - m - 1)(m + 3 - m - 2)} = -\frac{1}{6}.$$

Next, from Eq. (13.69) and the above values for a_i, we arrive at the expression for the second-order B-spline, as indicated below:

$$B_m^{(2)}(t) = \begin{cases} 0, & t < m, \\ a_m(t - m)^2, & m \leq t < m + 1, \\ a_m(t - m)^2 + a_{m+1}(t - m - 1)^2, & m + 1 \leq t < m + 2, \\ a_m(t - m)^2 + a_{m+1}(t - m - 1)^2 + a_{m+2}(t - m - 2)^2, & m + 2 \leq t < m + 3, \\ 0, & t \geq m + 3. \end{cases} \tag{13.83}$$

As a result, the expression for the normalized second-order B-spline is of the form

$$\beta_m^{(2)}(t) = (m + 3 - m) B_m^{(2)}(t) = \begin{cases} 0, & t < m, \\ \frac{t^2}{2} - mt + \frac{m^2}{2}, & m \leq t < m + 1, \\ -t^2 + 3t + 2mt - 3m - m^2 - \frac{3}{2}, & m + 1 \leq t < m + 2, \\ \frac{t^2}{2} - 3t - mt + \frac{m^2}{2} + 3m + \frac{9}{2}, & m + 2 \leq t < m + 3, \\ 0, & t \geq m + 3. \end{cases} \tag{13.84}$$

A plot of the normalized second-order B-spline $\beta_m^{(2)}(t)$ and the corresponding power functions for several values of m are shown in Figure 13.45.

The interpolation formula is obtained by forming a linear combination of the normalized B-splines weighted by the known values of the function $x_a(t)$ at the knots $t_k = n + k$. The interpolated value at the time instant $t' = t_0 + \alpha T_{in}$ is then given by

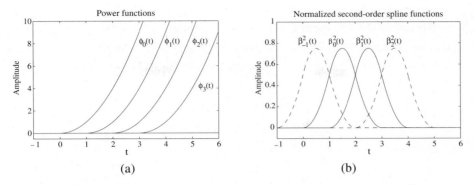

Figure 13.45: (a) Power functions and (b) normalized second-order spline functions.

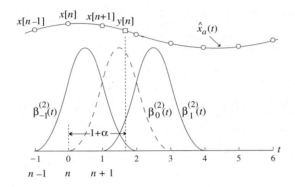

Figure 13.46: Interpolation using second-order B-spline.

$$\hat{x}_a(t') = \hat{x}_a(t_0 + \alpha T_{in})$$

$$= y[n] = \sum_{k=m}^{L+m+1} \beta_k^{(L)}(t_0 + \alpha T_{in})x[n+k]. \qquad (13.85)$$

It should be noted that, unlike the Lagrange interpolation algorithm, in the case of spline interpolation, $\hat{x}_a(t_k) \neq x_a(t_k)$. We illustrate the development of the interpolation formula using the normalized second-order B-spline in Example 13.17.

EXAMPLE 13.17   Interpolation Using Second-Order B-Spline

The interpolation process is illustrated in Figure 13.46. As can be seen from this figure, the position $t' = 1 + \alpha$ of the desired value $y[n]$ of $\hat{x}_a(t)$ is between the knots $t = 1$ and $t = 2$. Here, Eq. (13.85) reduces to

$$y[n] = \sum_{k=-1}^{2} \beta_k^{(2)}(\alpha)x[n+k], \qquad (13.86)$$

where $\beta_{-1}^{(2)}(\alpha)$ is computed using the fourth entry in Eq. (13.84), with $t = 1 + \alpha$ and $m = -1$; $\beta_0^{(2)}(\alpha)$ is computed using the third entry in Eq. (13.84), with $t = 1 + \alpha$ and $m = 0$; $\beta_1^{(2)}(\alpha)$ is computed using the second entry in Eq. (13.84), with $t = 1 + \alpha$ and $m = 1$; and $\beta_2^{(2)}(\alpha)$ is computed using the first entry in Eq. (13.84), with $t = 1 + \alpha$ and $m = 2$, resulting in

$$\beta_{-1}^{(2)}(\alpha) = \frac{\alpha^2}{2} - \alpha + \frac{1}{2}, \tag{13.87a}$$

$$\beta_0^{(2)}(\alpha) = -\alpha^2 + \alpha + \frac{1}{2}, \tag{13.87b}$$

$$\beta_1^{(2)}(\alpha) = \frac{\alpha^2}{2}, \tag{13.87c}$$

$$\beta_2^{(2)}(\alpha) = 0. \tag{13.87d}$$

Substituting the above in Eq. (13.86), we finally get

$$y[n] = \sum_{k=-1}^{1} \beta_k^{(2)}(\alpha)x[n + k]$$
$$= \left(\frac{\alpha^2}{2} - \alpha + \frac{1}{2}\right)x[n - 1] + \left(-\alpha^2 + \alpha + \frac{1}{2}\right)x[n] + \left(\frac{\alpha^2}{2}\right)x[n + 1]. \tag{13.88}$$

Rewriting Eq. (13.88) we arrive at

$$y[n] = \left(\frac{1}{2}x[n - 1] + \frac{1}{2}x[n]\right) + \alpha\left(-x[n - 1] + x[n]\right) + \alpha^2\left(\frac{1}{2}x[n - 1] - x[n] + \frac{1}{2}x[n + 1]\right), \tag{13.89}$$

which leads to the Farrow structure shown in Figure 13.47, where

$$H_0(z) = \frac{1}{2}z^{-1} - 1 + \frac{1}{2}z, \qquad H_1(z) = -z^{-1} + 1, \qquad H_2(z) = \frac{1}{2}z^{-1} + \frac{1}{2}. \tag{13.90}$$

In Example 13.18, we develop the interpolation formula using the cubic B-spline.

EXAMPLE 13.18   Interpolation Using Cubic B-Spline

The method for the derivation of the expressions for a_i and the cubic B-spline for uniformly spaced knots at $t_i = i, m \leq i \leq m + 4$ is similar to that outlined in Example 13.14 for the Lagrange interpolation. Here, now, $L = 3$ and $N = 4$. Following the procedures described in Examples 13.16 and 13.17, we arrive at the interpolation formula [Cuc91]:

$$y[n] = \sum_{k=-1}^{2} \beta_k^{(3)}(\alpha)x[n + k], \tag{13.91}$$

where

$$\beta_{-1}^{(3)}(\alpha) = -\frac{\alpha^3}{6} + \frac{\alpha^2}{2} - \frac{\alpha}{2} + \frac{1}{6}, \tag{13.92a}$$

$$\beta_0^{(3)}(\alpha) = \frac{\alpha^3}{2} - \alpha^2 + \frac{2}{3}, \tag{13.92b}$$

$$\beta_1^{(3)}(\alpha) = -\frac{\alpha^3}{2} + \frac{\alpha^2}{2} + \frac{\alpha}{2} + \frac{1}{6}, \tag{13.92c}$$

$$\beta_2^{(3)}(\alpha) = \frac{\alpha^3}{6}. \tag{13.92d}$$

The derivation of the normalized cubic B-splines of Eqs. (13.92a) - (13.92d) and the corresponding Farrow structure for interpolation [Doo99] are left as an exercise (Problem 13.34).

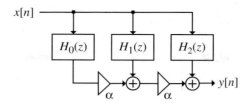

Figure 13.47: Implementation of the fractional-rate interpolator of Example 13.17 using the Farrow structure.

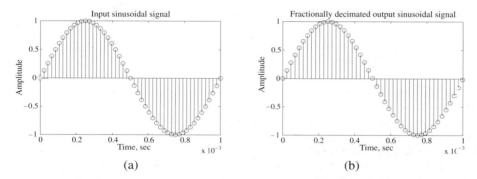

Figure 13.48: (a) Input sinusoidal sequence and (b) output sinusoidal sequence.

The M-file spline can be used to implement the interpolation using a more general spline function.

Design of Rational Sampling Rate Converter Using MATLAB

The M-file mfilt.firsrc can be used to design a direct form FIR polyphase fractional sampling rate converter. We illustrate the use of this function in Example 13.19.

EXAMPLE 13.19   Fractional Sampling-Rate Conversion Using MATLAB

We consider the design of a fractional sampling-rate converter to change the sampling rate from 48 kHz used in digital audio to 44.1 kHz used in compact disc. The desired sampling-rate conversion factor is

$$\frac{44.1}{48} = 0.91875 = \frac{147}{160}.$$

Hence, we need to first increase the sampling rate by an up-sampling factor $L = 147$ and then decrease it by a down-sampling factor of $M = 160$. The code fragment using the default FIR filter is

```
hm = mfilt.firsrc(147,160)
```

Figure 13.48 shows the plots of the input sinusoidal signal of frequency 1 kHz sampled at 48 kHz and the output sinusoidal signal with a sampling frequency of 44.1 kHz.

13.5.4 Practical Considerations

A direct design of a fractional sampling rate converter in most applications is impractical since the length of the time-varying filter needed is usually very long and the corresponding filter coefficient calculations in real time are thus nearly impossible. As a result, a fractional sampling rate converter is almost always

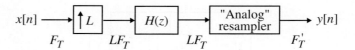

Figure 13.49: Hybrid form of a fractional-rate sampling converter.

realized in a hybrid form, consisting of a digital sampling rate converter with an integer-valued conversion factor, followed by an "analog" fractional rate converter, as indicated in Figure 13.49 [Ram84]. The digital sampling rate converter, of course, if need be, can be implemented in a multistage form (see Section 13.3.)

13.6   Nyquist Filters

In this section, we introduce a special type of lowpass filter with a transfer function that, by design, has certain zero-valued coefficients. Due to the presence of these zero-valued coefficients, these filters are, by nature, computationally more efficient than other lowpass filters of the same order. In addition, when used as interpolator filters, they preserve the nonzero samples of the up-sampler output at the interpolator output. These filters, called Lth-band filters or Nyquist filters and as discussed in this section, are often used both in single-rate and multirate signal processing. For example, they are usually preferred in decimator and interpolator design. Another application is in the design of a quadrature-mirror filter bank, discussed in Section 14.3. A third application, described in Section 15.7.1, is in the design of a Hilbert transformer, which is employed for the generation of analytic signals.

13.6.1 Lth-Band Filters

Consider the factor-of-L interpolator of Figure 13.15(a). The relation between the output and the input of the interpolator is given by

$$Y(z) = H(z)X(z^L). \tag{13.93}$$

If the interpolation filter $H(z)$ is realized in the L-band polyphase form, then we have

$$H(z) = E_0(z^L) + z^{-1}E_1(z^L) + z^{-2}E_2(z^L) + \cdots + z^{-(L-1)}E_{L-1}(z^L).$$

Assume that the kth polyphase component of $H(z)$ is a constant; that is, $E_k(z) = \alpha$:

$$\begin{aligned}
H(z) = E_0(z^L) + z^{-1}E_1(z^L) + \cdots + z^{-(k-1)}E_{k-1}(z^L) + \alpha z^{-k} \\
+ z^{-(k+1)}E_{k+1}(z^L) + \cdots + z^{-(L-1)}E_{L-1}(z^L).
\end{aligned} \tag{13.94}$$

Then, we can express $Y(z)$ as

$$Y(z) = \alpha z^{-k}X(z^L) + \sum_{\substack{\ell=0 \\ \ell \neq k}}^{L-1} z^{-\ell}E_\ell(z^L)X(z^L). \tag{13.95}$$

As a result, for a given k, $y[Ln + k] = \alpha x[n]$; that is, the input samples appear at the output without any distortion at time instants $Ln + k$, $-\infty < n < \infty$, whereas the in-between $(L-1)$ output samples are determined by interpolation.

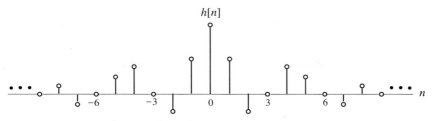

Figure 13.50: The impulse response of a typical third-band filter.

Figure 13.51: Frequency responses of $H(zW_L^k)$ for $k = 0, 1, \ldots, L-1$.

A filter with the above property is called a *Nyquist filter* or an *Lth-band filter* and its impulse response has many zero-valued samples, making it computationally very attractive. For example, the impulse response of the zero-phase Lth-band filter obtained for $k = 0$ satisfies the following condition:

$$h[Ln] = \begin{cases} \alpha, & n = 0, \\ 0, & \text{otherwise.} \end{cases} \tag{13.96}$$

Usually, in practice, $\alpha = 1$, to preserve the exact value of the corresponding input sample.

Figure 13.50 shows a typical impulse response of a zero-phase third-band filter ($L = 3$). If $H(z)$ is a zero-phase transfer function satisfying Eq. (13.94) with $k = 0$; that is, $E_0(z) = \alpha$, then it can be shown that (Problem 13.36)

$$\sum_{k=0}^{L-1} H(zW_L^k) = L\alpha = 1 \qquad \text{(assuming } \alpha = 1/L\text{).} \tag{13.97}$$

Since the frequency response of $H(zW_L^k)$ is the shifted version $H(e^{j(\omega - 2\pi k/L)})$ of $H(e^{j\omega})$, the sum of all of these L uniformly shifted versions of $H(e^{j\omega})$ add up to a constant (see Figure 13.51). The Lth-band filters can be either FIR or IIR filters.

13.6.2 Half-Band Filters

An Lth-band filter for $L = 2$ is called a *half-band filter*. From Eq. (13.94), the transfer function of a zero-phase half-band filter is thus given by

$$H(z) = \alpha + z^{-1}E_1(z^2), \tag{13.98}$$

with its impulse response satisfying Eq. (13.96) with $L = 2$. The condition of Eq. (13.97) on the frequency response of a zero-phase Lth-band filter reduces for $L = 2$ case to

$$H(z) + H(-z) = 1 \qquad \text{(assuming } \alpha = \tfrac{1}{2}\text{).} \tag{13.99}$$

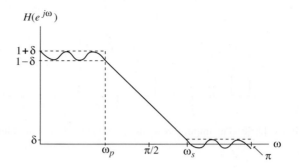

Figure 13.52: Frequency response of a zero-phase half-band filter.

If $H(z)$ has real coefficients, then $H(-e^{j\omega}) = H(e^{j(\pi-\omega)})$, and Eq. (13.99) leads to

$$H(e^{j\omega}) + H(e^{j(\pi-\omega)}) = 1. \tag{13.100}$$

The above equality implies that $H(e^{j(\pi/2-\theta)})$ and $H(e^{j(\pi/2+\theta)})$ add up to unity for all θ. In other words, $H(e^{j\omega})$ exhibits a symmetry with respect to the half-band frequency $\pi/2$, thus justifying the name "half-band filter." Figure 13.52 illustrates this symmetry for a half-band lowpass filter for which the passband and stopband ripples are equal; that is, $\delta_p = \delta_s$, and the passband and stopband edge frequencies are symmetric with respect to $\pi/2$; that is, $\omega_p + \omega_s = \pi$.

An important attractive property of the half-band filter is that about 50% of the coefficients of $h[n]$ are zero. This reduces the number of multiplications required in its implementation, making the filter computationally quite efficient. For example, if $N = 101$, an arbitrary Type 1 FIR transfer function requires about 50 multipliers, whereas a Type 1 half-band filter requires only about 25 multipliers.

An FIR half-band filter can be designed with linear phase. However, there is a constraint on its length. Consider a zero-phase real-coefficient half-band FIR filter for which $h[n] = h[-n]$. Let the highest nonzero coefficient be $h[R]$. Then R is odd as a result of the condition of Eq. (13.96). Therefore, $R = 2K + 1$ for some integer K. Thus, the length of the impulse response $h[n]$ is restricted to be of the form $2R + 1 = 4K + 3$ [unless $H(z)$ is a constant].

13.6.3   Design of Linear-Phase Lth-Band FIR Filters

A lowpass linear-phase Nyquist Lth-band FIR filter with a cutoff at $\omega_c = \pi/L$ and a good frequency response can be readily designed via the windowed Fourier series approach, described in Section 10.2. In this approach, the impulse response coefficients of the lowpass filter are chosen as

$$h[n] = h_{LP}[n] \cdot w[n], \tag{13.101}$$

where $h_{LP}[n]$ is the impulse response of an ideal lowpass filter with a cutoff at π/L and $w[n]$ is a suitable window function. If

$$h_{LP}[n] = 0 \qquad \text{for } n = \pm L, \pm 2L, \ldots, \tag{13.102}$$

then Eq. (13.96) is indeed satisfied.

Now the impulse response $h_{LP}[n]$ of an ideal Lth-band filter is obtained from Eq. (10.15) by substituting $\omega_c = \pi/L$ and is given by

$$h_{LP}[n] = \frac{\sin(\pi n/L)}{\pi n}, \qquad -\infty < n < \infty. \tag{13.103}$$

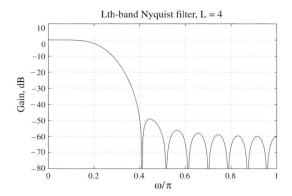

Figure 13.53: Gain response of the linear-phase fourth-band FIR filter of length-23 of Example 13.20.

It can be seen from the above that the impulse response coefficients do indeed satisfy the condition of Eq. (13.102). Hence, an Lth-band filter can be designed by applying a suitable window function to Eq. (13.103).

Likewise, an Lth-band filter with a frequency response as in Figure 10.13(a) can also be designed from Eq. (10.43) by replacing ω_c with π/L, resulting in an impulse response

$$h_{LP}[n] = \begin{cases} \frac{1}{L}, & n = 0, \\ \frac{2\sin(\Delta\omega n/2)}{\Delta\omega n} \cdot \frac{\sin(\pi n/L)}{\pi n}, & |n| > 0, \end{cases} \qquad (13.104)$$

which again is seen to satisfy the condition of Eq. (13.102). Other candidates for Lth-band filter design are the lowpass filters of Eq. (10.44) and the raised cosine filter of Eq. (10.116) in Problem 10.18.

We illustrate the lowpass half-band filter design in Example 13.20.

EXAMPLE 13.20   Window-Based Design of Lth-Band Filter

Design a fourth-band ($L = 4$) linear-phase lowpass filter of length-23 using the windowed Fourier series approach with a Hamming window. To this end, we use Program 13_8.

The program determines the impulse response coefficients of the Lth-band FIR filter using the expression of Eq. (13.103) and computes and plots the gain response of the designed filter as shown in Figure 13.53. It can be shown that the impulse response coefficients generated by this program are equal to zero for $n = \pm 4, \pm 8$.

Program 13_8.m

A lowpass Lth band FIR filter can also be designed using the M-file `firnyquist`. There are several options available with this function. We illustrate its use in Example 13.21.

EXAMPLE 13.21   Direct Design of Lth-Band Filter

We repeat the design of the fourth-band filter of Example 13.20 with two different roll-off factors, 0.1 and 0.4, respectively. The code fragments used are

```
b1 = firnyquist(22,4,0.4);
b2 = firnyquist(22,4,0.1);
```

The gain responses of the two fourth-band filters are shown in Figure 13.54. By printing the filter coefficients, it can be shown for both designs, the impulse response coefficients are equal to zero for $n = \pm 4, \pm 8$.

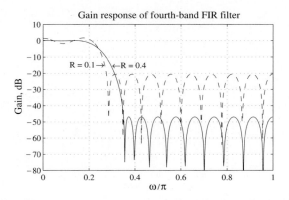

Figure 13.54: Gain response of the linear-phase fourth-band FIR filter of length-23 of Example 13.21.

An elegant method for the design of linear-phase half-band FIR filters is considered in the following section. Section 13.6.5 describes a method for the design of half-band IIR filters. Several other design approaches have been advanced in the literature [Ren87], [Vai87a].

13.6.4   Design of Linear-Phase Half-Band FIR Filters

The problem of designing a real-coefficient half-band FIR filter can be transformed into the design of a single passband FIR filter with no stopband, which can then be easily designed using the popular Parks–McClellan algorithm, described in Section 10.3 [Vai87b]. An inverse transformation on the realization of this wide-band filter then yields the implementation of the desired real-coefficient half-band filter.

Let the specifications of the real half-band lowpass filter $G(z)$ of order N be as follows: passband edge at ω_p, stopband edge at ω_s, passband ripple of δ_p, and stopband ripple of δ_s. As discussed in Section 13.6.2, the passband and stopband ripples of a linear-phase lowpass half-band FIR transfer function $G(z)$ are equal; that is, $\delta_p = \delta_s = \delta$, and the order N is even with $N/2$ odd. Moreover, the passband and stopband edge frequencies are related through $\omega_p + \omega_s = \pi$.

Now, consider a wide-band linear-phase filter $F(z)$ of degree $N/2$ with a passband from 0 to $2\omega_p$, a transition band from $2\omega_p$ to π, and a passband ripple of 2δ. Since $N/2$ is odd, $F(z)$ has a zero at $z = -1$. The wide-band filter $F(z)$ can be designed using the Parks–McClellan method. Define

$$G(z) = \tfrac{1}{2}\left[z^{-N/2} + F(z^2)\right]. \tag{13.105}$$

It follows from Eq. (13.105) that $G(z)$ is indeed the transfer function of a causal half-band lowpass filter and has an impulse response

$$g[n] = \begin{cases} \tfrac{1}{2}f\left[\tfrac{n}{2}\right], & n \text{ even}, \\ 0, & n \text{ odd}, n \neq \tfrac{N}{2} \\ \tfrac{1}{2}, & n = \tfrac{N}{2}, \end{cases} \tag{13.106}$$

where $f[n]$ is the impulse response of $F(z)$.

Example 13.22 illustrates the above design method.

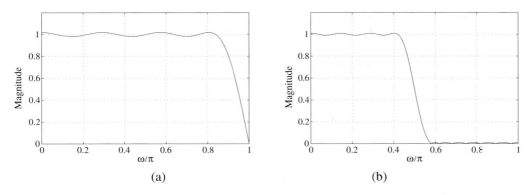

Figure 13.55: Magnitude responses of (a) wide-band lowpass filter and (b) desired lowpass half-band filter.

EXAMPLE 13.22   Design of Half-Band FIR Filters Using the Parks–McClellan Algorithm

Using MATLAB, we design a wide-band FIR filter $F(z)$ of degree 13 with a passband from 0 to 0.85π and an extremely small stopband from 0.9π to π. We use the function remez with a magnitude vector m = [1 1 0 0]. The weight vector used to weigh the passband and the stopband is wt = [2 0.05]. The magnitude responses of the wide-band filter $F(z)$ and the half-band filter $G(z)$ are shown in Figure 13.55.

A lowpass half-band FIR filter can also be designed using the M-file firhalfband. There are several options available with this function. Its use is illustrated in Example 13.23.

EXAMPLE 13.23   Direct Design of a Lowpass Half-Band FIR Filter

We design a minimum-order lowpass half-band FIR filter with a passband edge at 0.45π and a stopband deviation of 0.002; that is, -53.9794 dB. The code fragment used is

```
b = firhalfband('minorder',0.45, 0.002);
```

The gain response of the above filter is shown in Figure 13.56. The order of the filter designed is 58. It can be shown that all even-indexed filter coefficients are equal to zero except that at $n = 0$.

A highpass half-band FIR filter can be designed using b = firhalfband(....,'high').

13.6.5   Design of Half-Band IIR Filters

We now outline a method for the design of a half-band IIR filter. We have shown earlier in Section 8.10 that an odd-order bounded real lowpass IIR transfer function with a symmetric numerator, and with no common factors between its numerator and denominator, can be decomposed into a sum of two stable allpass transfer functions provided its power-complementary highpass IIR transfer function has an antisymmetric numerator. In the case of an odd-order bounded real (BR) lowpass transfer function $G(z) = P(z)/D(z)$ with a symmetric numerator and satisfying the power-symmetry condition

$$G(z)G(z^{-1}) + G(-z)G(-z^{-1}) = 1, \tag{13.107}$$

the sum of allpass decomposition is of the form

$$G(z) = \tfrac{1}{2}[\mathcal{A}_0(z^2) + z^{-1}\mathcal{A}_1(z^2)], \tag{13.108}$$

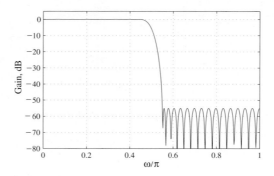

Figure 13.56: Gain response of the lowpass FIR half-band of Example 13.23.

where $\mathcal{A}_0(z)$ and $\mathcal{A}_1(z)$ are stable allpass transfer functions [Vai87f]. It follows from Eq. (13.107) that $G(z)G(z^{-1})$ is a half-band lowpass filter as it satisfies the condition of Eq. (13.99).

Since the passband and stopband ripples of the Chebyshev IIR filter are not equal, it is not possible to design a Chebyshev IIR half-band filter. We therefore consider here only the design of Butterworth and elliptic half-band filters. The former can be designed by first designing an odd-order analog Butterworth filter with a 3-dB cutoff frequency $\Omega_c = 1$ and applying a bilinear transformation to arrive at a Butterworth IIR digital filter $G(z)$, which can be shown to satisfy the power-symmetry property of Eq. (13.107) (Problem 14.12). The design of an elliptic IIR half-band filter is treated next.

It can be easily verified that the transfer function $H(z)$ of Example 13.12 satisfies the power-symmetry condition. It has been shown that any odd-order elliptic lowpass half-band filter $G(z)$ with a frequency response specification given by

$$1 - 2\delta_p \leq \left|G(e^{j\omega})\right| \leq 1, \quad \text{for } 0 \leq \omega \leq \omega_p, \tag{13.109a}$$

$$\left|G(e^{j\omega})\right| \leq \delta_s, \quad \text{for } \omega_s \leq \omega \leq \pi, \tag{13.109b}$$

and satisfying the conditions

$$\omega_p + \omega_s = \pi, \tag{13.110a}$$

$$\delta_s^2 = 4\delta_p(1 - \delta_p), \tag{13.110b}$$

is a power-symmetric transfer function and can always be expressed as in Eq. (13.108) [Vai87f]. Moreover, it can be shown that the poles of the elliptic filter satisfying the above two conditions lie on the imaginary axis (Problem 14.11). Using the pole-interlacing property outlined in Section 8.10, we can readily identify the expressions for the two allpass transfer functions $\mathcal{A}_0(z)$ and $\mathcal{A}_1(z)$.

The steps in the design of an IIR elliptic half-band filter are outlined next, which have been obtained by modifying the algorithm for designing digital elliptic filters outlined in [Ant93], [Vai93]. Since the two bandedges are related through Eq. (13.110a) and the two ripples are related through Eq. (13.110b), the filter specifications include only one of the bandedges and one of the ripples. Let the specified stopband edge and stopband ripple be ω_s and δ_s, respectively. The passband edge ω_p and the passband ripple δ_p are then determined from from Eqs. (13.110a) and (13.110b), respectively. The minimum stopband attenuation is therefore $A_s = -20 \log_{10} \delta_s$ dB, and the maximum passband ripple is $A_{max} = -20 \log_{10}(1 - 2\delta_p)$ dB.

Next, define the parameters

$$r = \frac{\tan(\omega_p/2)}{\tan(\omega_s/2)}, \tag{13.111a}$$

$$r' = \sqrt{1 - r^2}, \tag{13.111b}$$

$$q_0 = \frac{(1 - \sqrt{r'})}{2(1 + \sqrt{r'})}, \tag{13.111c}$$

and compute

$$q = q_0 + 2\,q_0^5 + 15\,q_0^9 + 150\,q_0^{13}, \tag{13.112a}$$

$$D = \left(\frac{1 - \delta_s^2}{\delta_s^2}\right)^2. \tag{13.112b}$$

Then estimate the order N of $G(z)$ by choosing the smallest odd integer satisfying

$$N \geq \frac{\log_{10}(16D)}{\log_{10}(1/q)}. \tag{13.113}$$

Now the integer value of N is almost always higher than the quantity on the right-hand side of Eq. (13.113). As a result, the corresponding stopband ripple δ_s will be smaller than the original specified value. To determine the actual value of δ_s, the actual value of the parameter D is first computed from

$$D = \frac{10^{N \log_{10}(1/q)}}{16}. \tag{13.114}$$

From this value of D, the actual value of δ_s is computed by solving Eq. (13.112b), and then the actual value of δ_p is obtained from Eq. (13.110b).

Next, the poles of the two allpass filters are computed as follows. Define for $1 \leq k \leq (N-1)/2$,

$$\lambda_k = \frac{2q^{1/4} \sum_{i=0}^{\infty}(-1)^i q^{i(i+1)} \sin\left((2i+1)k\pi/N\right)}{1 + 2\sum_{i=1}^{\infty}(-1)^i q^{i^2} \cos(2\pi ki/N)}, \tag{13.115a}$$

$$b_k = \sqrt{\left(1 - r\lambda_k^2\right)\left(1 - \frac{\lambda_k^2}{r}\right)}, \tag{13.115b}$$

$$c_k = \frac{2b_k}{1 + \lambda_k^2}, \tag{13.115c}$$

$$\alpha_{k-1} = \frac{2 - c_k}{2 + c_k}. \tag{13.115d}$$

In general, the two infinite sums in Eq. (13.115a) converge after the addition of five or six terms.

The poles of the two allpass sections are on the imaginary axis at $z = \pm j\sqrt{\alpha_k}$ and are inside the unit circle, as the parameters α_k determined using Eq. (13.115d) are distinct with magnitudes less than 1. Using the pole-interlacing property, then poles of $\mathcal{A}_0(z)$ and $\mathcal{A}_1(z)$ are selected. Their corresponding zeros are at the mirror-image locations.

Program 13_9 can be used to design an IIR lowpass half-band filter. We illustrate its use in Example 13.24.

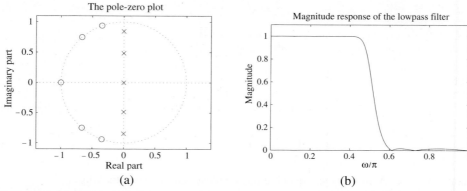

Figure 13.57: Example 13.24: (a) Pole-zero plot and (b) magnitude response of the fifth-order elliptic IIR half-band filter.

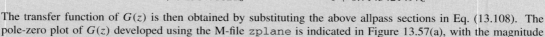

EXAMPLE 13.24   Design of Half-Band IIR Filters

The specified stopband edge and the stopband ripple are $\omega_s = 0.6\pi$ and $\delta_s = 0.016$, respectively. The desired minimum stopband attenuation is thus 35.9176 dB.

Using Program 13_9, we arrive at the transfer functions of the two allpass sections of the half-band filter $G(z)$:

$$\mathcal{A}_0(z^2) = \frac{0.236471021 + z^{-2}}{1 + 0.236471021z^{-2}}, \quad \mathcal{A}_1(z^2) = \frac{0.7145421497 + z^{-2}}{1 + 0.7145421497z^{-2}}. \quad (13.116)$$

Program 13_9.m

The transfer function of $G(z)$ is then obtained by substituting the above allpass sections in Eq. (13.108). The pole-zero plot of $G(z)$ developed using the M-file `zplane` is indicated in Figure 13.57(a), with the magnitude response of the half-band filter shown in Figure 13.57(b).

The adjusted stopband ripple δ_s is 0.01541972077441; that is, a minimum stopband attenuation of 36.23846 dB, which is slightly more than the specified value.

13.7 Summary

The basic theory of multirate digital signal processing is introduced in this chapter, along with the design of some useful multirate systems. The two basic sampling rate alteration devices are the up-sampler and the down-sampler. We first discuss the input–output relations of these devices, both in the time domain and in the frequency domain. We then describe several cascade equivalences that are used to develop computationally efficient implementation of a multirate system by permitting the movement of the up-sampler and down-sampler from one part of the system to another part.

The sampling rate converter is implemented using either the up-sampler or the down-sampler or both, and a lowpass digital filter. The design issues for the sampling rate conversion are discussed next. We demonstrate that a computationally efficient sampling rate converter can often be designed as a cascade of such converters. We also outline another approach to the implementation of a computationally efficient sampling rate converter that is based on the use of the polyphase decomposition of the lowpass digital filter.

We then describe the Lagrange and spline interpolation algorithms that can be used to design arbitrary rate sampling rate converters. The chapter concludes with a discussion of the design of Nyquist filters, which are computationally more efficient than other equivalent filters due to the presence of certain zero-valued coefficients, and when used as interpolation filters, they preserve the nonzero samples of the up-sampler at the interpolator output.

Chapter 14 considers the application of the basic concepts of multirate digital signal processing discussed here to the analysis and design of multirate filter banks.

For additional details on multirate digital signal processing, we refer the reader to the texts by Akansu and Haddad [Aka92], Crochiere and Rabiner [Cro83], Fliege [Fli94], Vaidyanathan [Vai93], and Vetterli and Kovacevic [Vet95].

13.8 Problems

13.1 Show that the up-sampler defined by Eq. (13.1) and the down-sampler defined by Eq. (13.2) are linear systems.

13.2 Show that the up-sampler defined by Eq. (13.1) and the down-sampler defined by Eq. (13.2) are time-varying systems.

13.3 Express the output $y[n]$ of Figure 13.6 as a function of the input $x[n]$. By simplifying the expression derived, show that $y[n] = x[n-1]$.

13.4 Prove the identity of Eq. (13.10).

13.5 (a) Show that W_M^{-k} and W_M^{-kL} take the same set of values for $k = 0, 1, \ldots, M-1$, with $L = 5$ and $M = 6$.
(b) Show that the above result holds in general for arbitrary mutually prime values of L and M.

13.6 Verify the cascade equivalences of Figure 13.14.

13.7 (a) Show that the structure of Figure P13.1 is a LTI system, and determine its overall transfer function $H(z) = Y(z)/X(z)$, where $Y(z)$ and $X(z)$ are, respectively, the z-transforms of $y[n]$ and $x[n]$.
(b) Show that the structure of Figure P13.1 is an identity system, i.e., $y[n] = x[n]$, if $\frac{1}{L} \sum_{k=0}^{L-1} G(z^{-1/L} W_L^k) = 1$.

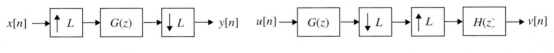

$$\text{Figure P13.1} \qquad\qquad\qquad \text{Figure P13.2}$$

13.8 Consider the structure of Figure P13.2 where the transfer functions $G(z)$ and $H(z)$ satisfy the condition $\frac{1}{L} \sum_{k=0}^{L-1} G(z^{1/L} W_L^k) H(z^{1/L} W_L^k) = 1$. Show that it is an identity system, i.e., $v[n] = u[n]$ [Vai2001].

13.9 Develop an expression for the output $y[n]$ as a function of the input $x[n]$ for the multirate structure of Figure P13.3.

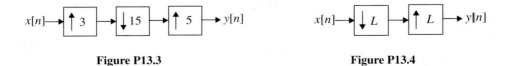

$$\text{Figure P13.3} \qquad\qquad\qquad \text{Figure P13.4}$$

13.10 Develop an expression for the output $y[n]$ as a function of the input $x[n]$ for the multirate structure of Figure P13.4.

13.11 Develop an expression for the output $y[n]$ as a function of the input $x[n]$ for the multirate structure of Figure P13.5.

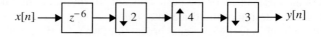

$$x[n] \longrightarrow \boxed{z^{-6}} \longrightarrow \boxed{\downarrow 2} \longrightarrow \boxed{\uparrow 4} \longrightarrow \boxed{\downarrow 3} \longrightarrow y[n]$$

Figure P13.5

13.12 Show that the transpose of a factor-of-M decimator is a factor-of-M interpolator if the transpose of a factor-of-M down-sampler is a factor-of-M up-sampler, and vice versa.

13.13 Derive the input–output relations of a fractional-rate sampling rate converter given by Eqs. (13.20) and (13.21), respectively.

13.14 Develop a computationally efficient structure for realizing a fractional-rate interpolator with an interpolation factor of 3/2 by taking the transpose of Figure 13.37(b).

13.15 Consider the fractional-rate converter of Figure 13.16(b) with $L = 8$ and $M = 15$. (a) If the sampling frequency F_T of the input signal $x[n]$ is 400 Hz, determine the sampling rate of the output signal $y[n]$. (b) What is the stopband edge frequency of the filter $H(z)$ to ensure no aliasing?

13.16 Repeat Problem 13.15 for the following values of the input sampling rate and up-sampling and down-sampling factors: $F_T = 650Hz$, $L = 5$, and $M = 9$.

13.17 Show that the transpose of an M-channel analysis filter bank is an M-channel synthesis bank.

13.18 Develop an alternate two-stage design of the decimator of Example 13.10 by designing the decimation filter in the form $H(z) = I(z)F(z^5)$. Compare its computational requirements with that of the design in Example 13.10.

13.19 Repeat Problem 13.18 for a filter of the form $H(z) = I(z)F(z^3)$. Compare its computational requirements with that of the designs in Example 13.10 and Problem 13.18.

13.20 (a) Determine the computational complexity of a single-stage decimator designed to reduce the sampling rate from 40 kHz to 2 kHz. The decimation filter is to be designed as an equiripple FIR filter with a passband edge at 800 Hz, a passband ripple of 0.02, and a stopband ripple of 0.01. Use the total multiplications per second as a measure of the computational complexity.

(b) The above decimator is to be designed as a two-stage structure. Develop an optimum design with the smallest computational complexity.

13.21 (a) Determine the computational complexity of a single-stage interpolator to be designed to increase the sampling rate from 480 Hz to 24 kHz. The interpolator is to be designed as an equiripple FIR filter with a passband edge at 190 Hz, a passband ripple of 0.002, and a stopband ripple of 0.002. Use Kaiser's formula, given in Eq. (10.3), to estimate the order of the FIR filter. The measure of computational complexity is given by the total number of multiplications per second.

(b) Develop a two-stage design of the above interpolator and compare its computational complexity with that of the single-stage design.

13.22 Develop a computationally efficient realization of a factor-of-3 interpolator employing a length-15 linear-phase FIR filter.

13.23 Develop a computationally efficient realization of a factor-of-4 decimator employing a length-14 linear-phase FIR filter.

13.24 Show that the *running-sum filter,* also called the *boxcar filter,* $H(z) = \sum_{i=0}^{N-1} z^{-i}$, can be expressed in the form

$$H(z) = (1 + z^{-1})(1 + z^{-2}) \cdots (1 + z^{-2^{K-1}}),$$

where $N = 2^K$. Develop a computationally efficient realization of a factor-of-16 interpolator using a length-16 boxcar filter.

13.25 The multirate system of Figure P13.6 is usually employed in exchanging discrete-time signals between two discrete-time systems with incommensurate sampling rates [Cro83]. Samples at the input of the digital sample-and-hold circuit may often be repeated or totally dropped, resulting in an error in the overall sampling rate conversion process. Let $\mathcal{E}$ denote the ratio of the energy in the sample-to-sample difference signal to that in the original signal $y[n]$ at the output of the factor-of-L interpolator, and let $\mathcal{C}$ denote the sample-to-sample correlation of the signal $y[n]$. Express $\mathcal{E}$ as a function of $\mathcal{C}$, and show that as L becomes large, $\mathcal{E}$ becomes small, i.e., the error in the overall sampling rate conversion process becomes small.

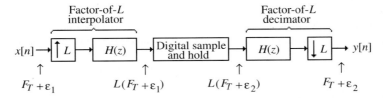

Figure P13.6

13.26 The multirate system of Figure P13.7 implements a fixed delay of L/M samples, where L and M are relatively prime integers [Cro83]. Let $H(z)$ be a Type 1 length-N linear-phase FIR lowpass filter with a cutoff at π/M and a passband magnitude approximately equal to M. Develop the relation between the DTFTs $Y(e^{j\omega})$ and $X(e^{j\omega})$ of the output $y[n]$ and the input $x[n]$, respectively, assuming $N = 2KM + 1$, where K is a positive integer.

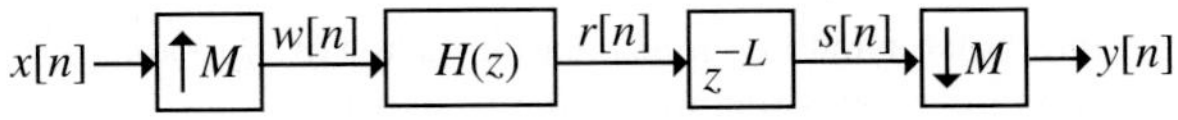

$$x[n] \rightarrow \boxed{\uparrow M} \xrightarrow{w[n]} \boxed{H(z)} \xrightarrow{r[n]} \boxed{z^{-L}} \xrightarrow{s[n]} \boxed{\downarrow M} \rightarrow y[n]$$

Figure P13.7

13.27 Let an ideal lowpass filter $H(z)$ with a cutoff at π/M be expressed as

$$H(z) = \sum_{k=0}^{M-1} z^{-k} H_k(z^M).$$

Show that each polyphase subfilter $H_k(z)$ is an allpass filter.

13.28 Prove the identity shown in Figure 13.38.

13.29 A generalization of the polyphase decomposition is considered in this problem. Let $H(z) = \sum_{n=0}^{N-1} h[n] z^{-n}$ be a causal FIR transfer function of degree $N - 1$, with N even.
(a) Show that $H(z)$ can be expressed in the form

$$H(z) = (1 + z^{-1}) H_0(z^2) + (1 - z^{-1}) H_1(z^2). \tag{13.117}$$

(b) Express $H_0(z)$ and $H_1(z)$ in terms of the coefficients of the polyphase components $E_0(z)$ and $E_1(z)$. The decomposition of Eq. (13.117) is an example of a generalized polyphase decomposition, called the *structural subband decomposition* [Mit93].

(c) Show that the decomposition of Eq. (13.117) can be expressed in the form

$$H(z) = [1 \quad z^{-1}] \begin{bmatrix} 1 & 1 \\ 1 & -1 \end{bmatrix} \begin{bmatrix} H_0(z^2) \\ H_1(z^2) \end{bmatrix}. \tag{13.118}$$

The 2×2 matrix in Eq. (13.118) is called a Hadamard matrix of order 2 and, is denoted by $\mathbf{R}_2$.

(d) Show that if $N = 2^L$, then $H(z)$ can be expressed in the form

$$H(z) = [1 \quad z^{-1} \ldots z^{-(L-1)}]\mathbf{R}_L \begin{bmatrix} H_0(z^L) \\ H_1(z^L) \\ \vdots \\ H_{L-1}(z^L) \end{bmatrix}, \tag{13.119}$$

where $\mathbf{R}_L$ is an $L \times L$ Hadamard matrix. Express the structural subband components $\{H_i(z)\}$ in terms of the polyphase components $\{E_i(z)\}$ of $H(z)$.

13.30 Develop a computationally efficient realization of a factor-of-4 interpolator employing a length-16 linear-phase FIR filter $H(z)$ and using a four-band structural subband decomposition of $H(z)$ in the form of Eq. (13.119).

13.31 Design a fractional-rate interpolator with an interpolation factor of 4/3 using the Lagrange interpolation algorithm. Use a fourth-order polynomial approximation. Develop realizations of the interpolator based on a block filtering approach and the Farrow structure.

13.32 Derive Eq. (13.76).

13.33 Derive Eqs. (13.79a) through (13.79c).

13.34 Derive Eqs. (13.92a) through (13.92d), and develop the corresponding Farrow structure for interpolation.

13.35 Design a fractional-rate interpolator with an interpolation factor of 4/3 using the cubic B-spline interpolation algorithm. Develop the corresponding Farrow structure for interpolation.

13.36 Derive Eq. (13.97).

13.37 Let $h[n]$ denote the impulse response of a lowpass half-band filter with a zero at $z = -1$. Show that

$$h[0] = \sum_{\substack{n=-\infty \\ n \neq 0}}^{\infty} h[n].$$

13.38 Show that the following FIR linear-phase transfer functions are lowpass half-band filters [Goo77]. Plot their magnitude responses using MATLAB.

(a) $H_1(z) = 1 + 2z^{-1} + z^{-2}$,

(b) $H_2(z) = -1 + 9z^{-2} + 16z^{-3} + 9z^{-4} - z^{-6}$,

(c) $H_3(z) = -3 + 19z^{-2} + 32z^{-3} + 19z^{-4} - 3z^{-6}$,

(d) $H_4(z) = 3 - 25z^{-2} + 150z^{-4} + 256z^{-5} + 150z^{-6} - 25z^{-8} + 3z^{-10}$,

(e) $H_5(z) = 9 - 44z^{-2} + 208z^{-4} + 346z^{-5} + 208z^{-6} - 44z^{-8} + 9z^{-10}$.

13.39 The general form of the frequency response of the zero-phase half-band maximally flat Daubechies FIR filter is given by [Dau88]

$$H(e^{j\omega}) = \left(\frac{1 + \cos \omega}{2}\right)^p \sum_{\ell=0}^{p-1} \binom{p-1+\ell}{\ell} \left(\frac{1 - \cos \omega}{2}\right)^\ell. \tag{13.120}$$

(a) Show that $H(e^{j\omega})$ has pth order zero at $\omega = 0$ and pth order zero at $\omega = \pi$.
(b) Show also that $H(e^{j\pi/2}) = \frac{1}{2}$.
(c) Determine the transfer function for $p = 6$, and using MATLAB, plot its magnitude response.

13.40 The structure of Figure P13.8 has been proposed for the computationally efficient implementation of FIR digital filters [Vet88].
(a) Show that the structure is alias-free, and determine the overall transfer function $T(z) = Y(z)/X(z)$ in terms of $H_0(z)$ and $H_1(z)$.
(b) Determine the expression for $T(z)$ if

$$H_0(z^2) = \tfrac{1}{2}\{H(z) + H(-z)\}, \qquad H_1(z^2) = \tfrac{1}{2}\{H(z) - H(-z)\}z.$$

(c) If $H(z)$ is a length-$2K$ FIR filter, what are the lengths of the filters $H_0(z)$ and $H_1(z)$?
(d) Determine the computational efficiency of this structure.

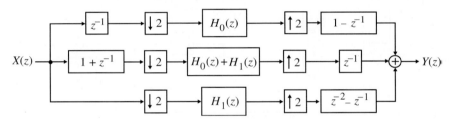

Figure P13.8

13.41 Analyze the structure of Figure P13.9 and determine its input–output relations. Comment on your results.

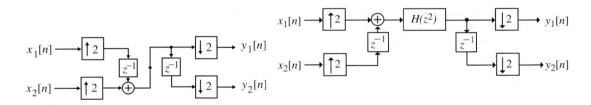

Figure P13.9 **Figure P13.10**

13.42 An efficient implementation of two separate single-input, single-output LTI discrete-time systems with an identical transfer function $H(z)$ by a single two-input, two-output multirate discrete-time system is obtained using the *pipelining/interleaving* (PI) technique, as shown in Figure P13.10 [Jia97]. Show that the system of Figure P13.10 is time-invariant, and determine the transfer functions from each input to each output.

13.43 Show that the multirate system of Figure P13.11 is time-invariant and determine its transfer function [Jia97].

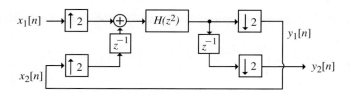

Figure P13.11

13.44 Problem 10.29 describes the filter sharpening approach [Kai77], which is used to improve the magnitude response of a filter $H(z)$ in both the passband and the stopband by employing multiple copies of the filter. For example, the *thricing method* of filter sharpening implements the transfer function

$$G(z) = z^{-2}[3H^2(z) - 2H^3(z)], \tag{13.121}$$

where $H(z)$ is the prototype zero-phase FIR filter. Show that the multirate structure of Figure P13.12 implements Eq. (13.121) using the PI technique for an appropriate value of the constant C [Jia97].

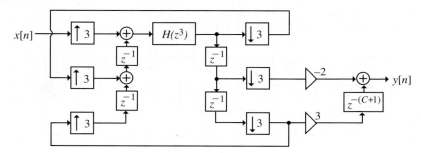

Figure P13.12

13.9 MATLAB Exercises

Program 13_1.m

M 13.1 (a) Modify Program 13_1 to study the operation of a factor-of-5 up-sampler on the following input sequences: (i) sum of two sinusoidal sequences of normalized frequencies 0.25 and 0.4 rad/sec, (ii) ramp sequence, and (iii) square wave sequence with various duty cycles. Choose the input length to be 50. Plot the input and the output sequences.

(b) Repeat Part (a) for a factor-of-6 up-sampler.

Program 13_2.m

M 13.2 (a) Modify Program 13_2 to study the operation of a factor-of-5 down-sampler on the following input sequences: (i) sum of two sinusoidal sequences of normalized frequencies 0.25 and 0.35 rad/sec, (ii) ramp sequence, and (iii) square wave sequence with various duty cycles. Choose the output length to be 50. Plot the input and the output sequences.

(b) Repeat Part (a) for a factor-of-6 down-sampler.

Program 13_3.m

M 13.3 (a) Use a vector of frequency points `freq = [0 0.96 0.99 1]` in Program 13_3 and run it with an up-sampling factor of $L = 4$. Comment on your results.

(b) Repeat Part (a) for $L = 7$. Comment on your results.

M 13.4 (a) Use a vector of frequency points `freq = [0 0.96 0.99 1]` and a magnitude response vector `mag = [1 0 0 0]` in Program 13_3 and run it with an up-sampling factor of $L = 5$. Comment on your results.

(b) Repeat Part (a) for $L = 7$. Comment on your results.

M 13.5 (a) Use a vector of frequency points `freq = [0 0.18 0.20 1]` and a magnitude response vector `mag = [1 0 0 0]` in Program 13_4 and run it with a down-sampling factor of $M = 4$. Comment on your results.
 (b) Repeat Part (a) for $M = 7$. Comment on your results.

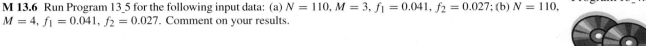

Program 13_4.m

M 13.6 Run Program 13_5 for the following input data: (a) $N = 110$, $M = 3$, $f_1 = 0.041$, $f_2 = 0.027$; (b) $N = 110$, $M = 4$, $f_1 = 0.041$, $f_2 = 0.027$. Comment on your results.

M 13.7 Run Program 13_6 for the following input data: (a) $N = 45$, $L = 3$, $f_1 = 0.041$, $f_2 = 0.027$; (b) $N = 35$, $L = 4$, $f_1 = 0.041$, $f_2 = 0.027$. Comment on your results.

Program 13_5.m

M 13.8 Run Program 13_7 for the following input data: (a) $N = 35$, $L = 2$, $M = 3$, $f_1 = 0.041$, $f_2 = 0.027$; (b) $N = 45$, $L = 5$, $M = 3$, $f_1 = 0.041$, $f_2 = 0.027$. Comment on your results.

M 13.9 Design a length-71 linear-phase FIR lowpass filter with a cutoff frequency at $\pi/5$ using the windowed Fourier series approach. Express the transfer function in a five-band polyphase decomposition. Using MATLAB, compute and plot the frequency responses of each polyphase component. Show that all polyphase components have constant magnitude responses.

Program 13_6.m

M 13.10 Design an Nth-order IIR half-band Butterworth lowpass filter and realize it using the least number of multipliers for the following values of N: (a) $N = 3$, (b) $N = 5$, and (c) $N = 7$. For each values of N, determine the power-complementary half-band highpass filter, and using MATLAB, plot the magnitude responses of the lowpass and highpass filters in the same figure.

Program 13_7.m

M 13.11 Using MATLAB, design a real-coefficient elliptic half-band IIR filter $H_0(z)$ of odd order with the following specifications: $\omega_s = 0.55\pi$, and $\delta_s = 0.001$. Note that the half-band filter constraint is satisfied if $\omega_p + \omega_s = \pi$ and $(1 - \delta_p)^2 + \delta_s^2 = 1$. Express $H_0(z)$ in the form

$$H_0(z) = \tfrac{1}{2}\{\mathcal{A}_0(z^2) + z^{-1}\mathcal{A}_1(z^2)\},$$

where $\mathcal{A}_0(z)$ and $\mathcal{A}_1(z)$ are stable allpass transfer functions. Plot the magnitude responses of $H_0(z)$ and its power-complementary transfer function $H_1(z)$ in the same figure.

M 13.12 Design a linear-phase fifth-band lowpass FIR filter of order 39 with $\omega_p = 0.18\pi$ and $\omega_s = 0.22\pi$ using the windowed Fourier series approach by modifying Program 13_8. Use the Hann window.

Program 13_8.m

14 Multirate Filter Banks and Wavelets

Many multirate systems employ a bank of filters with either a common input or a summed output. These filter banks are introduced next, and the design and computationally efficient implementation of a class of such filter banks are discussed. This is followed by a discussion of quadrature-mirror filter (QMF) banks that find applications in signal compression and other areas. The chapter concludes with a short review on the development of certain classes of discrete wavelet transforms from a tree-structure QMF bank. Several applications of the multirate filter banks and wavelets are discussed in Chapter 15.

14.1 Digital Filter Banks

In Chapter 13, we discussed mostly the design, realization, and applications of single-input, single-output digital filters. There are applications, as in the case of a spectrum analyzer, where it is desirable to separate a signal into a set of subband signals occupying, usually nonoverlapping, portions of the original frequency band. In other applications, it may be necessary to combine many such subband signals into a single composite signal occupying the whole Nyquist range. To this end, digital filter banks play an important role and are the subject of discussion in this section.

14.1.1 Definitions

A digital filter bank is a set of digital bandpass filters with either a common input or a summed output, as shown in Figure 14.1. The structure of Figure 14.1(a) is called an M-band *analysis filter bank,* with the subfilters $H_k(z)$ known as the *analysis filters*. It is used to decompose the input signal $x[n]$ into a set of M subband signals $v_k[n]$, with each subband signal occupying a portion of the original frequency band. (The signal is being "analyzed" by being separated into a set of narrow spectral bands.)

The *dual* of the analysis operation, whereby a set of subband signals $\hat{v}_k[n]$ (typically belonging to contiguous frequency bands) is combined into one signal $y[n]$ is called a *synthesis filter bank*. Figure 14.1(b) shows an L-band synthesis bank where each filter $F_k(z)$ is called a *synthesis filter*.

14.1.2 Uniform DFT Filter Banks

We now outline a simple technique for the design of a class of filter banks with equal passband widths. Let $H_0(z)$ represent a causal lowpass digital filter with a real impulse response $h_0[n]$:

$$H_0(z) = \sum_{n=0}^{\infty} h_0[n]z^{-n}, \tag{14.1}$$

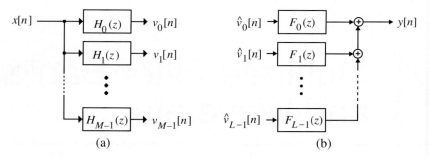

Figure 14.1: (a) Analysis filter bank and (b) synthesis filter bank.

which we assume to be an IIR filter without any loss of generality. Let us now assume that $H_0(z)$ has its passband edge ω_p and stopband edge ω_s around π/M, where M is some arbitrary integer, as indicated in Figure 14.2(a). Now, consider the causal impulse response $h_k[n], 0 \le k \le -1$, obtained by modulating $h_0[n]$ with the exponential sequence $e^{-j2\pi k/M}$:

$$h_k[n] = h_0[n]W_M^{-kn}, \tag{14.2}$$

where we have used the notation $W_M = e^{-j2\pi/M}$, as defined in Eq. (5.12). The corresponding transfer function is then given by

$$H_k(z) = \sum_{n=0}^{\infty} h_k[n]z^{-n} = \sum_{n=0}^{\infty} h_0[n]\left(zW_M^k\right)^{-n}, \qquad 0 \le k \le M-1; \tag{14.3}$$

that is,

$$H_k(z) = H_0\left(zW_M^k\right), \qquad 0 \le k \le M-1. \tag{14.4}$$

Its frequency response is

$$H_k(e^{j\omega}) = H_0(e^{j(\omega-2\pi k/M)}), \qquad 0 \le k \le M-1. \tag{14.5}$$

In other words, the frequency response of $H_k(z)$ is obtained by shifting the response of the lowpass filter $H_0(z)$ to the right, by an amount $2\pi k/M$. The responses of $H_1(z), H_2(z), \ldots, H_{M-1}(z)$ are shown in Figure 14.2(b). Note from Eq. (14.2) that the corresponding impulse responses $h_k[n]$ are, in general, complex, and hence, $|H_k(e^{j\omega})|$ does not necessarily exhibit symmetry with respect to zero frequency. Figure 14.2(b) therefore represents the responses of $M-1$ filters $H_1(z), H_2(z), \ldots, H_{M-1}(z)$, which are uniformly shifted versions of the response of the basic *prototype* filter $H_0(z)$ of Figure 14.2(a).

The M filters $H_k(z)$ defined by Eq. (14.4) could be used as the analysis filters in the analysis filter bank of Figure 14.1(a) or as the synthesis filters $F_k(z)$ in the synthesis filter bank of Figure 14.1(b).

Since the set of magnitude responses $|H_k(e^{j\omega})|, 1 \le k \le M-1$, are uniformly shifted versions of a basic prototype $|H_0(e^{j\omega})|$, that is,

$$\left|H_k(e^{j\omega})\right| = \left|H_0\left(e^{j(\omega-2\pi k/M)}\right)\right|, \tag{14.6}$$

the filter bank obtained is called a *uniform filter bank*.

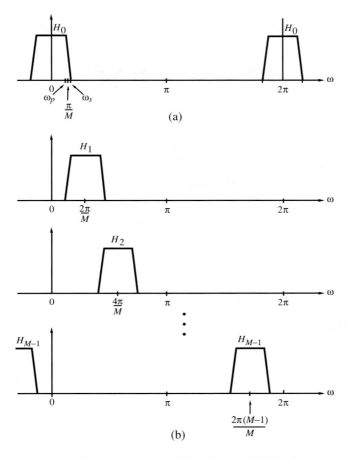

Figure 14.2: (a) Frequency response of prototype lowpass filter $H_0(z)$ and (b) the frequency responses of $M - 1$ bandpass filters $H_k(z)$, $1 \leq k \leq M - 1$.

14.1.3 Polyphase Implementations of Uniform Filter Banks

Let Figure 14.1(a) represent a uniform filter bank with the M analysis filters $H_k(z)$ related through Eq. (14.4). The impulse response sequences $\{h_k[n]\}$ of the analysis filters are accordingly related as in Eq. (14.2). Instead of realizing each analysis filter as a separate filter, it is possible to develop a computationally more efficient realization of the above uniform filter bank, which is described next.

Let the lowpass prototype transfer function $H_0(z)$ be represented in its M-band polyphase form,

$$H_0(z) = \sum_{\ell=0}^{M-1} z^{-\ell} E_\ell(z^M), \tag{14.7}$$

where $E_\ell(z)$ is the ℓth polyphase component of $H_0(z)$,

$$E_\ell(z) = \sum_{n=0}^{\infty} e_\ell[n] z^{-n} = \sum_{n=0}^{\infty} h_0[\ell + nM] z^{-n}, \qquad 0 \leq \ell \leq M - 1. \tag{14.8}$$

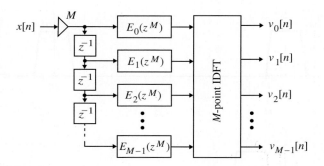

Figure 14.3: Polyphase implementation of a uniform DFT analysis filter bank where $H_k(z) = V_k(z)/X(z)$.

Substituting z with zW_M^k in Eq. (14.7), we arrive at the M-band polyphase decomposition of $H_k(z)$:

$$H_k(z) = \sum_{\ell=0}^{M-1} z^{-\ell} W_M^{-k\ell} E_\ell(z^M W_M^{kM}) = \sum_{\ell=0}^{M-1} z^{-\ell} W_M^{-k\ell} E_\ell(z^M), \qquad 0 \le k \le M-1, \qquad (14.9)$$

where we have used the identity $W_M^{kM} = 1$.

Note that Eq. (14.9) can be written in matrix form as

$$H_k(z) = \begin{bmatrix} 1 & W_M^{-k} & W_M^{-2k} & \cdots & W_M^{-(M-1)k} \end{bmatrix} \begin{bmatrix} E_0(z^M) \\ z^{-1} E_1(z^M) \\ z^{-2} E_2(z^M) \\ \vdots \\ z^{-(M-1)} E_{M-1}(z^M) \end{bmatrix}, \qquad (14.10)$$

for $k = 0, 1, \ldots, M-1$. All these M equations can be combined into a matrix equation as

$$\begin{bmatrix} H_0(z) \\ H_1(z) \\ H_2(z) \\ \vdots \\ H_{M-1}(z) \end{bmatrix} = M\mathbf{D}^{-1} \begin{bmatrix} E_0(z^M) \\ z^{-1} E_1(z^M) \\ z^{-2} E_2(z^M) \\ \vdots \\ z^{-(M-1)} E_{M-1}(z^M) \end{bmatrix}, \qquad (14.11)$$

where $\mathbf{D}$ denotes the DFT matrix:

$$\mathbf{D} = \begin{bmatrix} 1 & 1 & 1 & \cdots & 1 \\ 1 & W_M^1 & W_M^2 & \cdots & W_M^{(M-1)} \\ 1 & W_M^2 & W_M^4 & \cdots & W_M^{2(M-1)} \\ \vdots & \vdots & \vdots & \ddots & \vdots \\ 1 & W_M^{(M-1)} & W_M^{2(M-1)} & \cdots & W_M^{(M-1)^2} \end{bmatrix}. \qquad (14.12)$$

An efficient implementation of the M-band analysis filter bank based on Eq. (14.11) is thus as shown in Figure 14.3, where the prototype lowpass filter $H_0(z)$ has been implemented in a polyphase form. The structure of Figure 14.3 is more commonly known as the *uniform DFT analysis filter bank*.

The computational complexity of Figure 14.3 is much smaller than that of a direct implementation, as in Figure 14.1(a). For example, an M-band uniform DFT analysis filter bank based on an N-tap prototype

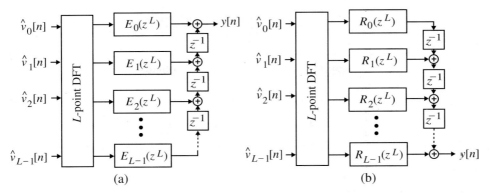

Figure 14.4: Uniform DFT synthesis filter bank: (a) realization based on Type I polyphase decomposition and (b) realization based on Type II polyphase decomposition.

lowpass FIR filter requires a total of $(M/2)\log_2 M + N$ multipliers, whereas a direct implementation requires NM multiplications.

Following a development similar to that outlined above, we can derive the structure for a *uniform DFT synthesis filter bank*. Efficient realizations based on the Types I and II polyphase decompositions of the prototype lowpass filter $H_0(z)$ are indicated in Figure 14.4.

EXAMPLE 14.1   Design of a Four-Channel Uniform DFT Filter Bank

We design a four-channel uniform DFT filter bank using the fourth-band linear-phase lowpass filter of Example 13.20. Here the four polyphase components are as follows:

$$E_0(z) = 0.00163694 - 0.01121888\,z^{-1} + 0.06311487\,z^{-2} + 0.22088513\,z^{-3}$$
$$-0.02725549\,z^{-4} + 0.00382693\,z^{-5},$$
$$E_1(z) = 0.00313959 - 0.025174873\,z^{-1} + 0.147532912\,z^{-2} + 0.147532912\,z^{-3}$$
$$-0.025174873\,z^{-4} + 0.00313959\,z^{-5},$$
$$E_2(z) = 0.00382693 - 0.02725549\,z^{-1} + 0.22088513\,z^{-2} + 0.06311487\,z^{-3}$$
$$-0.01121888\,z^{-4} + 0.00163694\,z^{-5},$$
$$E_3(z) = 0.25\,z^{-2}.$$

From Eq. (14.11), with $M = 4$, we then have

$$\begin{bmatrix} H_0(z) \\ H_1(z) \\ H_2(z) \\ H_3(z) \end{bmatrix} = \begin{bmatrix} 1 & 1 & 1 & 1 \\ 1 & j & -1 & -j \\ 1 & -1 & 1 & -1 \\ 1 & -j & -1 & j \end{bmatrix} \begin{bmatrix} E_0(z^4) \\ z^{-1}E_1(z^4) \\ z^{-2}E_2(z^4) \\ z^{-3}E_3(z^4) \end{bmatrix},$$

where the polyphase components are as given earlier. Hence, the four analysis filters are given by

$$H_0(z) = E_0(z^4) + z^{-1}E_1(z^4) + z^{-2}E_2(z^4) + z^{-3}E_3(z^4),$$
$$H_1(z) = E_0(z^4) + j\,z^{-1}E_1(z^4) - z^{-2}E_2(z^4) - j\,z^{-3}E_3(z^4),$$
$$H_2(z) = E_0(z^4) - z^{-1}E_1(z^4) + z^{-2}E_2(z^4) - z^{-3}E_3(z^4),$$
$$H_3(z) = E_0(z^4) - j\,z^{-1}E_1(z^4) - z^{-2}E_2(z^4) + j\,z^{-3}E_3(z^4).$$

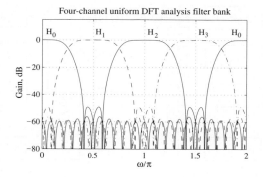

Figure 14.5: Gain responses of the filters of the four-channel uniform DFT analysis filter bank of Example 14.1.

A plot of the gain responses of the four analysis filters of the four-channel uniform DFT filter bank is shown in Figure 14.5.

It follows from Eq. (14.11) that the polyphase components $E_i(z^M)$ can be expressed in terms of the prototype transfer function $H_0(z)$ and its modulated versions $H_k(z) = H_0(zW_M^k)$ according to

$$
\begin{bmatrix} E_0(z^M) \\ z^{-1}E_1(z^M) \\ z^{-2}E_2(z^M) \\ \vdots \\ z^{-(M-1)}E_{M-1}(z^M) \end{bmatrix} = \frac{1}{M}\mathbf{D}\begin{bmatrix} H_0(z) \\ H_1(z) \\ H_2(z) \\ \vdots \\ H_{M-1}(z) \end{bmatrix} = \frac{1}{M}\mathbf{D}\begin{bmatrix} H_0(z) \\ H_0(zW_M) \\ H_0(zW_M^2) \\ \vdots \\ H_0(zW_M^{M-1}) \end{bmatrix}.
\tag{14.13}
$$

Equation 14.13 can be used to determine the polyphase components of an IIR transfer function. The method is illustrated in Example 14.2.

EXAMPLE 14.2   Three-Band Polyphase Decomposition of an IIR Transfer Function

We develop a three-band polyphase decomposition of the following first-order IIR transfer function:

$$
H(z) = \frac{a + bz^{-1}}{1 + cz^{-1}}, \qquad |c| < 1.
$$

From Eq. (14.13), we then have

$$
\begin{bmatrix} E_0(z^3) \\ z^{-1}E_1(z^3) \\ z^{-2}E_2(z^3) \end{bmatrix} = \frac{1}{3}\begin{bmatrix} 1 & 1 & 1 \\ 1 & W_3^1 & W_3^2 \\ 1 & W_3^2 & W_3^1 \end{bmatrix}\begin{bmatrix} H(z) \\ H(zW_3^1) \\ H(zW_3^2) \end{bmatrix}.
$$

Therefore,

$$
\begin{aligned}
E_0(z^3) &= \tfrac{1}{3}\left[H(z) + H(zW_3^1) + H(zW_3^2)\right] \\
&= \tfrac{1}{3}\left[\frac{a+bz^{-1}}{1+cz^{-1}} + \frac{a+be^{j2\pi/3}z^{-1}}{1+ce^{j2\pi/3}z^{-1}} + \frac{a+be^{j4\pi/3}z^{-1}}{1+ce^{j4\pi/3}z^{-1}}\right] \\
&= \frac{a + bc^2 z^{-3}}{1 + c^3 z^{-3}},
\end{aligned}
$$

$$z^{-1}E_1(z^3) = \frac{1}{3}\left[H(z) + W_3^1 H(z\,W_3^1) + W_3^2 H(z\,W_3^2)\right]$$

$$= \frac{1}{3}\left[\frac{a+b\,z^{-1}}{1+c\,z^{-1}} + e^{j2\pi/3}\left(\frac{a+b\,e^{j2\pi/3}z^{-1}}{1+c\,e^{j2\pi/3}z^{-1}}\right) + e^{j4\pi/3}\left(\frac{a+b\,e^{j4\pi/3}z^{-1}}{1+c\,e^{j4\pi/3}z^{-1}}\right)\right]$$

$$= z^{-1}\left(\frac{b-ac}{1+c^3 z^{-3}}\right),$$

$$z^{-2}E_2(z^3) = \frac{1}{3}\left[H(z) + W_3^2 H(z\,W_3^1) + W_3^1 H(z\,W_3^2)\right]$$

$$= \frac{1}{3}\left[\frac{a+b\,z^{-1}}{1+c\,z^{-1}} + e^{j4\pi/3}\left(\frac{a+b\,e^{j2\pi/3}z^{-1}}{1+c\,e^{j2\pi/3}z^{-1}}\right) + e^{j2\pi/3}\left(\frac{a+b\,e^{j4\pi/3}z^{-1}}{1+c\,e^{j4\pi/3}z^{-1}}\right)\right]$$

$$= z^{-2}\left(\frac{-bc+ac^2}{1+c^3 z^{-3}}\right).$$

Hence, the three polyphase components of $H(z)$ are given by

$$E_0(z) = \frac{a+bc^2 z^{-1}}{1+c^3 z^{-1}}, \qquad E_1(z) = \frac{b-ac}{1+c^3 z^{-1}}, \qquad E_2(z) = \frac{-bc+ac^2}{1+c^3 z^{-1}}.$$

14.2 Two-Channel Quadrature-Mirror Filter Bank

In many applications, a discrete-time signal $x[n]$ is first split into a number of subband signals $\{v_k[n]\}$ by means of an analysis filter bank; the subband signals are then processed and finally combined by a synthesis filter bank, resulting in an output signal $y[n]$. If the subband signals are band-limited to frequency ranges much smaller than that of the original input signal, they can be downsampled before processing. Because of the lower sampling rate, the processing of the down-sampled signals can be carried out more efficiently. After processing, these signals are upsampled before being combined by the synthesis bank into a higher-rate signal. The combined structure employed is called a *quadrature-mirror filter (QMF) bank*. If the down-sampling and the up-sampling factors are equal to or greater than the number of bands of the filter bank, then the output $y[n]$ can be made to retain some or all of the characteristics of the input $x[n]$ by properly choosing the filters in the structure. In case of equality, the filter bank is said to be a *critically sampled filter bank*. The most common application of this scheme is in the efficient coding of a signal $x[n]$ (see Section 15.8). Another possible application is in the design of an analog voice privacy system to provide secure telephone conversations [Cox83]. In this section, we study a two-channel QMF bank.

14.2.1 The Filter Bank Structure

Figure 14.6 shows the basic two-channel QMF bank–based subband codec (coder/decoder). Here, the input signal $x[n]$ is first passed through a two-band analysis filter bank containing the filters $H_0(z)$ and $H_1(z)$, which typically have lowpass and highpass frequency responses, respectively, with a cutoff frequency at $\pi/2$, as indicated in Figure 14.7. The subband signals $\{v_k[n]\}$ are then downsampled by a factor of 2. Each down-sampled subband signal is encoded by exploiting the special spectral properties of the signal, such as energy levels and perceptual importance (see Section 15.8). The coded subband signals are combined into one sequence by multiplexing and either stored for later retrieval or transmitted. At the receiving end, the coded subband signals are first recovered by demultiplexing, and decoders are used to produce approximations of the original down-sampled signals. The decoded signals are then upsampled by a factor of 2 and passed through a two-band synthesis filter bank composed of the filters $G_0(z)$ and $G_1(z)$, whose outputs are then added yielding $y[n]$. It follows from the figure that the sampling rates of the input signal $x[n]$ and output signal $y[n]$ are the same. The analysis and the synthesis filters in the QMF bank are chosen

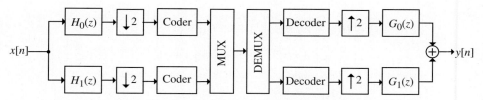

Figure 14.6: The two-channel filter bank based coder/decoder.

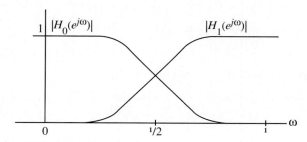

Figure 14.7: Typical frequency responses of the analysis filters.

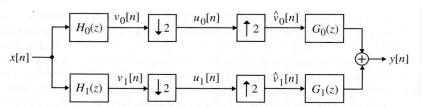

Figure 14.8: The two-channel quadrature-mirror filter (QMF) bank.

so as to ensure that the reconstructed output $y[n]$ is a reasonable replica of the input $x[n]$. Moreover, they are also designed to provide good frequency selectivity in order to ensure that the sum of the power of the subband signals is reasonably close to the input signal power.

In practice, various errors are generated in this scheme. In addition to the coding error and errors caused by transmission through the channel, the QMF bank itself introduces several errors due to the sampling rate alterations and imperfect filters. We ignore the coding and channel errors, and investigate only the errors generated by the down-samplers and up-samplers in the filter bank and their effects on the performance of the system. To this end, we consider the QMF bank structure without the coders and the decoders, as indicated in Figure 14.8 [Cro76a], [Est77].

14.2.2 Analysis of the Two-Channel QMF Bank

It is convenient to analyze the filter bank in the z-domain. To this end, we make use of the input–output relations of the up-sampler and the down-sampler derived earlier in Section 13.1 and given by Eqs. (13.5) and (13.12), respectively. The expressions for the z-transforms of various intermediate signals in Figure 14.8, obtained using Eqs. (13.5) and (13.12), are then given by

$$V_k(z) = H_k(z)X(z), \tag{14.14a}$$

$$U_k(z) = \tfrac{1}{2}\left\{V_k(z^{1/2}) + V_k(-z^{1/2})\right\}, \tag{14.14b}$$

$$\hat{V}_k(z) = U_k(z^2), \tag{14.14c}$$

for $k = 0, 1$. From Eqs. (14.14a) to (14.14c), we obtain, after some algebra

$$\hat{V}_k(z) = \tfrac{1}{2}\{V_k(z) + V_k(-z)\} = \tfrac{1}{2}\{H_k(z)X(z) + H_k(-z)X(-z)\}. \tag{14.15}$$

The reconstructed output of the filter bank is given by

$$Y(z) = G_0(z)\hat{V}_0(z) + G_1(z)\hat{V}_1(z). \tag{14.16}$$

Substituting Eq. (14.15) in Eq. (14.16), we obtain, after some rearrangement, the expression for the output of the filter bank as

$$
\begin{aligned}
Y(z) = &\tfrac{1}{2}\{H_0(z)G_0(z) + H_1(z)G_1(z)\} X(z) \\
&+ \tfrac{1}{2}\{H_0(-z)G_0(z) + H_1(-z)G_1(z)\} X(-z).
\end{aligned}
\tag{14.17}
$$

The second term in Eq. (14.17) is precisely due to the aliasing caused by sampling rate alteration. Equation (14.17) can be compactly expressed as

$$Y(z) = T(z)X(z) + A(z)X(-z), \tag{14.18}$$

where

$$T(z) = \tfrac{1}{2}\{H_0(z)G_0(z) + H_1(z)G_1(z)\} \tag{14.19}$$

is called the *distortion transfer function* and

$$A(z) = \tfrac{1}{2}\{H_0(-z)G_0(z) + H_1(-z)G_1(z)\} \tag{14.20}$$

is the aliasing term.

14.2.3   Linear Time-Invariant Two-Channel QMF Bank

As noted in Section 13.1, the up-sampler and the down-sampler are linear time-varying components and, as a result, in general, the QMF structure of Figure 14.8 is a linear time-varying (LTV) system. It can be shown also that it has a period of 2 (Problem 14.13). However, it is possible to choose the analysis and synthesis filters such that the aliasing effect is cancelled, resulting in a linear time-invariant (LTI) operation. In addition, by choosing these filters appropriately, the output of the QMF bank can be made to preserve certain characteristics of the input.

Aliasing Cancellation Condition

It follows from Eq. (14.18) that aliasing is cancelled if $A(z) = 0$. From Eq. (14.20), we therefore arrive at the aliasing cancellation condition

$$H_0(-z)G_0(z) + H_1(-z)G_1(z) = 0. \tag{14.21}$$

If the above relations hold, then Eq. (14.18) reduces to

$$Y(z) = T(z)X(z), \tag{14.22}$$

where $T(z)$ is as given in Eq. (14.19). On the unit circle, we have

$$Y(e^{j\omega}) = T(e^{j\omega})X(e^{j\omega}) = \left| T(e^{j\omega}) \right| e^{j\phi(\omega)} X(e^{j\omega}). \tag{14.23}$$

If $T(z)$ is an allpass function, that is, $\left| T(e^{j\omega}) \right| = d \neq 0$, then

$$\left| Y(e^{j\omega}) \right| = d \left| X(e^{j\omega}) \right|, \tag{14.24}$$

indicating that the output of the QMF bank has the same magnitude response as that of the input (scaled by d) but exhibits phase distortion, and the filter bank is said to be *magnitude preserving*. If $T(z)$ has linear phase, that is,

$$\theta(\omega) = \alpha\omega + \beta, \tag{14.25}$$

then

$$\arg\left\{ Y(e^{j\omega}) \right\} = \arg\left\{ X(e^{j\omega}) \right\} + \alpha\omega + \beta, \tag{14.26}$$

and the filter bank is said to be *phase preserving* but exhibits magnitude distortion.

Condition for Perfect Reconstruction

If an alias-free QMF bank has no amplitude and phase distortions, then it is called a *perfect reconstruction* (PR) QMF bank. In such a case, we must have

$$H_0(z)G_0(z) + H_1(z)G_1(z) = 2 z^{-\ell}, \tag{14.27}$$

that is, $T(z) = z^{-\ell}$, with ℓ a positive integer, resulting in

$$Y(z) = z^{-\ell}X(z), \tag{14.28}$$

which in the time domain is equivalent to

$$y[n] = x[n - \ell] \tag{14.29}$$

for all possible inputs, indicating that the reconstructed output $y[n]$ is a delayed replica of the input $x[n]$.

EXAMPLE 14.3   A Trivial Two-Channel Perfect Reconstruction QMF Bank

The multirate system of Figure 13.6 can be considered as a two-channel QMF bank. Comparing it with the filter bank structure of Figure 14.8, we conclude that the analysis and synthesis filters of Figure 13.6 are given by

$$H_0(z) = 1, \qquad H_1(z) = z^{-1}, \qquad G_0(z) = z^{-1}, \qquad G_1(z) = 1.$$

Substituting these values in Eqs. (14.19) and (14.20) we get

$$T(z) = \tfrac{1}{2}\left(z^{-1} + z^{-1} \right) = z^{-1},$$

$$A(z) = \tfrac{1}{2}\left(z^{-1} - z^{-1} \right) = 0,$$

confirming that the structure is an alias-free perfect reconstruction filter bank. However, the filters in the bank do not provide any frequency selectivity.

Product Filters

The perfect reconstruction condition of Eq. (14.27) involves the sum of two product filters [Str96]:

$$P_0(z) = H_0(z)G_0(z), \qquad P_1(z) = H_1(z)G_1(z). \tag{14.30}$$

The product filter $P_0(z)$ is a lowpass filter, as both $H_0(z)$ and $G_0(z)$ are lowpass filters. Likewise, the product filter $P_1(z)$ is a highpass filter, as both $H_1(z)$ and $G_1(z)$ are highpass filters.

14.2.4 A Simple Perfect Reconstruction QMF Bank

Consider a two-channel QMF bank with analysis filters given by

$$H_0(z) = \tfrac{1}{\sqrt{2}}(1 + z^{-1}), \qquad H_1(z) = \tfrac{1}{\sqrt{2}}(1 - z^{-1}), \tag{14.31}$$

and synthesis filters given by

$$G_0(z) = \tfrac{1}{\sqrt{2}}(1 + z^{-1}), \qquad G_1(z) = \tfrac{1}{\sqrt{2}}(-1 + z^{-1}). \tag{14.32}$$

It can be seen from our discussion in Section 7.4.1 that $H_0(z)$ and $G_0(z)$ are very simple first-order FIR lowpass filters, and $H_1(z)$ and $G_1(z)$ are very simple first-order FIR highpass filters. Substituting Eqs. (14.31) and (14.32) in Eq. (14.21) we get

$$H_0(-z)G_0(z) + H_1(-z)G_1(z) = \tfrac{1}{2}(1 - z^{-1})(1 + z^{-1}) + \tfrac{1}{2}(1 + z^{-1})(-1 + z^{-1})$$
$$= \tfrac{1}{2}(1 - z^{-2}) + \tfrac{1}{2}(-1 + z^{-2}) = 0.$$

Thus, the two-channel QMF bank with analysis and synthesis filters given by Eqs. (14.31) and (14.32), respectively, is an alias-free system.

We next substitute the expressions for the analysis and synthesis filters of Eqs. (14.31) and (14.32) in Eq. (14.19), resulting in

$$T(z) = \tfrac{1}{2}\{H_0(z)G_0(z) + H_1(z)G_1(z)\}$$
$$= \tfrac{1}{4}\{(1 + z^{-1})(1 + z^{-1}) + (1 - z^{-1})(-1 + z^{-1})\} = 2z^{-1}.$$

As a result, the two-channel QMF bank with analysis and synthesis filters given by Eqs. (14.31) and (14.32), respectively, is a perfect-reconstruction system.

In the time domain, the outputs of the analysis filters are related to the input $x[n]$ of the QMF bank (see Figure 14.8) as

$$u_0[n] = \tfrac{1}{\sqrt{2}}(x[n] + x[n-1]), \tag{14.33a}$$

$$u_1[n] = \tfrac{1}{\sqrt{2}}(x[n] - x[n-1]). \tag{14.33b}$$

It follows from the above that the samples $u_0[n]$ and $u_1[n]$ are simply the sum and difference of the input samples $x[n]$ and $x[n-1]$, scaled by $1/\sqrt{2}$.[1] Equations (14.33a) and (14.33b) can be written in matrix form as

$$\begin{bmatrix} u_0[n] \\ u_1[n] \end{bmatrix} = \mathbf{H}_2 \begin{bmatrix} x[n] \\ x[n-1] \end{bmatrix}, \tag{14.34}$$

---

[1] The scale factor $1/\sqrt{2}$ preserves the energy.

where

$$\mathbf{H}_2 = \frac{1}{\sqrt{2}} \begin{bmatrix} 1 & 1 \\ 1 & -1 \end{bmatrix} \tag{14.35}$$

is seen to be the 2×2 *Haar transform matrix* of Eq. (5.164) except for the scale factor $1/\sqrt{2}$. It should be noted that $\mathbf{H}_2$ is a symmetric orthonormal matrix, and hence, $\mathbf{H}_2^{-1} = \mathbf{H}_2^t = \mathbf{H}_2$. In matrix form, the reconstruction of the input data $x[n]$ and $x[n-1]$ from the output data $u_0[n]$ and $u_1[n]$ is then given by

$$\begin{bmatrix} x[n] \\ x[n-1] \end{bmatrix} = \mathbf{H}_2^t \begin{bmatrix} u_0[n] \\ u_1[n] \end{bmatrix}. \tag{14.36}$$

14.2.5 An Alias-Free Realization

The aliasing cancellation condition of Eq. (14.21) is satisfied if we choose

$$G_0(z) = H_1(-z), \qquad G_1(z) = -H_0(-z). \tag{14.37}$$

In this case, the product filter $P_1(z)$ of Eq. (14.30) becomes

$$P_1(z) = -H_0(-z)H_1(z) = -P_0(-z). \tag{14.38}$$

Using the above result in Eq. (14.27), the perfect reconstruction condition can be rewritten as [Str96]

$$P_0(z) - P_0(-z) = 2z^{-\ell}. \tag{14.39}$$

Since the even powers of z in $P_0(z)$ are cancelled by the even powers of z in $P_0(-z)$, ℓ must be an odd integer. Moreover, $P_0(z)$ cannot contain any terms with odd powers of z except the term involving $z^{-\ell}$, whose coefficient is simply 1.

It can be seen from the expressions for the analysis and synthesis filters given in Eqs. (14.31) and (14.32) that the transfer functions satisfy the condition of Eq. (14.37), ensuring alias-free operation of the two-channel QMF bank. Here, $P_0(z) = H_0(z)H_1(-z) = \frac{1}{2}(1 + 2z^{-1} + z^{-2})$. Thus, $P_0(z) - P_0(-z) = 2z^{-1}$, or in other words, the perfect reconstruction condition of Eq. (14.39) is satisfied with $\ell = 1$. It should be noted also that here $P_0(z)$, as expected, does not contain any terms with odd powers of z, except the term involving z^{-1}, whose coefficient is simply 1.

Under the alias-free condition of Eq. (14.37), the algorithm for the design of a perfect reconstruction two-channel QMF bank thus involves three steps: (1) the design of a lowpass product filter $P_0(z)$ satisfying the condition of Eq. (14.39), (2) factorization of $P_0(z)$ to determine the lowpass filters $H_0(z)$ and $G_0(z)$, and (3) determination of the filters $H_1(z)$ and $G_1(z)$ by using Eq. (14.37).

A number of algorithms have been proposed to design the lowpass product filter $P_0(z)$. We discuss a few of these methods in this chapter. Also, a number of methods have been advanced to determine the factors $H_0(z)$ and $G_0(z)$ from $P_0(z)$. We outline three different approaches to the factorization problem.

A more convenient form of the perfect reconstruction condition is obtained by multiplying both sides of Eq. (14.39) with z^ℓ, resulting in

$$z^\ell P_0(z) - z^\ell P_0(-z) = 2.$$

Define a normalized form of the lowpass product filter as

$$P(z) = z^\ell P_0(z). \tag{14.40}$$

From Eq. (14.40) we have $P(-z) = (-z)^\ell P_0(-z) = -z^\ell P_0(-z)$, as ℓ is an odd integer. Thus, the perfect reconstruction condition can be rewritten as

$$P(z) + P(-z) = 2. \tag{14.41}$$

Comparing Eq. (14.41) with Eq. (13.99) we observe that $P(z)$ is a zero-phase half-band lcwpass filter whose constant term, that is, the coefficient of z^0, is 1.

A very simple factorization of the product filter $P_0(z)$ is obtained by choosing [Cro76a]

$$G_0(z) = H_0(z). \tag{14.42}$$

From Eq. (14.37), we then get

$$H_1(z) = G_0(-z) = H_0(-z), \tag{14.43a}$$
$$G_1(z) = -H_1(z) = -H_0(-z). \tag{14.43b}$$

Equation (14.43a), in the case of a real coefficient filter, implies

$$\left| H_1(e^{j\omega}) \right| = \left| H_0(e^{j(\pi-\omega)}) \right|, \tag{14.44}$$

indicating that if $H_0(z)$ is a lowpass filter, then $H_1(z)$ is a highpass filter, and vice versa. In fact, Eq. (14.43a) indicates that $\left| H_1(e^{j\omega}) \right|$ is a mirror image of $\left| H_0(e^{j\omega}) \right|$ with respect to $\pi/2$, the *quadrature frequency*. This has given rise to the name quadrature-mirror filter bank.

Equations (14.42), (14.43a), and (14.43b) imply that the two analysis filters and the two synthesis filters in the QMF bank are essentially determined from one transfer function $H_0(z)$. The distortion transfer function $T(z)$ of Eq. (14.19) in this case reduces to

$$T(z) = \frac{1}{2}\left\{ H_0^2(z) - H_1^2(z) \right\} = \frac{1}{2}\left\{ H_0^2(z) - H_0^2(-z) \right\}. \tag{14.45}$$

A computationally efficient realization of the above alias-free two-channel QMF bank is obtained by realizing the analysis and the synthesis filters in polyphase form. Let the two-band Type I polyphase representation of $H_0(z)$ be given by

$$H_0(z) = E_0(z^2) + z^{-1}E_1(z^2). \tag{14.46a}$$

From Eq. (14.43b), it follows then that

$$H_1(z) = E_0(z^2) - z^{-1}E_1(z^2). \tag{14.46b}$$

In matrix form, Eqs. (14.46a) and (14.46b) can be expressed as

$$\begin{bmatrix} H_0(z) \\ H_1(z) \end{bmatrix} = \begin{bmatrix} 1 & 1 \\ 1 & -1 \end{bmatrix} \begin{bmatrix} E_0(z^2) \\ z^{-1}E_1(z^2) \end{bmatrix}. \tag{14.47}$$

Likewise, the synthesis filters, in matrix form, can be expressed as

$$[\, G_0(z) \quad G_1(z)\,] = [\, z^{-1}E_1(z^2) \quad E_0(z^2)\,] \begin{bmatrix} 1 & 1 \\ 1 & -1 \end{bmatrix}. \tag{14.48}$$

Using Eqs. (14.47) and (14.48), we can redraw the two-channel QMF bank as shown in Figure 14.9(a), which can be further simplified using the cascade equivalences of Figure 13.14, resulting in the computationally efficient realization of Figure 14.9(b).

The expression for the distortion transfer function in this case, obtained by substituting the expressions for the analysis filters of Eqs. (14.46a) and (14.46b) and the corresponding expressions for the synthesis filters obtained using Eqs. (14.42) and (14.43b) in Eq. (14.19), is then given by

$$T(z) = 2z^{-1}E_0(z^2)E_1(z^2). \tag{14.49}$$

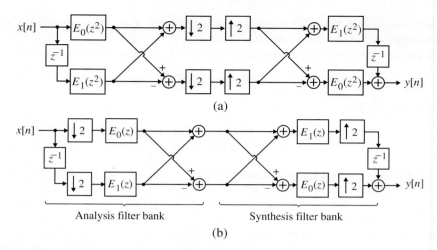

Figure 14.9: Polyphase realization of the two-channel QMF bank: (a) direct polyphase realization and (b) computationally efficient realization.

14.2.6   Alias-Free FIR QMF Bank

In Section 14.2.4 we presented a simple alias-free FIR QMF bank with first-order linear-phase FIR analysis and synthesis filters. We now consider the design of an alias-free linear-phase FIR QMF bank with higher-order filters. Let the prototype analysis filter be a linear-phase FIR filter of order N with a real coefficient transfer function $H_0(z)$ given by

$$H_0(z) = \sum_{n=0}^{N} h_0[n]z^{-n}. \tag{14.50}$$

Note that $H_0(z)$ can be either a Type 1 or Type 2 linear-phase function since it has to be a lowpass filter. As a result, its impulse-response coefficients must satisfy the condition $h_0[n] = h_0[N - n]$, in which case we can write

$$H_0(e^{j\omega}) = e^{-j\omega N/2}\, \breve{H}_0(\omega), \tag{14.51}$$

where $\breve{H}_0(\omega)$ is the amplitude function, a real function of ω. By making use of Eq. (14.51) in Eq. (14.45), along with the property that $|H_0(e^{j\omega})|$ is an even function of ω, we can express the frequency response of the distortion transfer function as

$$T(e^{j\omega}) = \frac{e^{-jN\omega}}{2}\left\{ \left| H_0(e^{j\omega}) \right|^2 - (-1)^N \left| H_0(e^{j(\pi-\omega)}) \right|^2 \right\}. \tag{14.52}$$

From Eq. (14.52) it can be seen that if N is even, then $T(e^{j\omega}) = 0$ at $\omega = \pi/2$, implying severe amplitude distortion at the output of the filter bank. As a result, N must be chosen to be odd, in which case Eq. (14.52) reduces to

$$\begin{aligned}
T(e^{j\omega}) &= \frac{e^{-jN\omega}}{2}\left\{ \left| H_0(e^{j\omega}) \right|^2 + \left| H_0(e^{j(\pi-\omega)}) \right|^2 \right\} \\
&= \frac{e^{-jN\omega}}{2}\left\{ \left| H_0(e^{j\omega}) \right|^2 + \left| H_1(e^{j\omega}) \right|^2 \right\}.
\end{aligned} \tag{14.53}$$

It follows from the above expression, the FIR two-channel filter bank with linear-phase analysis and synthesis filters will be of perfect reconstruction type if

$$\left|H_0(e^{j\omega})\right|^2 + \left|H_1(e^{j\omega})\right|^2 = 1, \tag{14.54}$$

that is, the two analysis filters are power-complementary. Except for the two trivial filter banks of Example 14.3 and Section 14.2.4, it can be shown that it is not possible to realize a perfect reconstruction two-channel filter bank with linear-phase power-complementary analysis filters [Vai85c].

As can be seen from Eq. (14.53), the QMF bank has no phase distortion, but will always exhibit amplitude distortion unless $|T(e^{j\omega})|$ is a constant for all values of ω. If $H_0(z)$ is a very good lowpass filter with $|H_0(e^{j\omega})| \cong 1$ in the passband and $|H_0(e^{j\omega})| \cong 0$ in the stopband, then $H_1(z)$ is a very good highpass filter, with its passband coinciding with the stopband of $H_0(z)$, and vice versa. As a result, $|T(e^{j\omega})| \cong 1/2$ in the passbands of $H_0(z)$ and $H_1(z)$. The amplitude distortion thus occurs primarily in the transition band of these filters, with the degree of distortion being determined by the amount of overlap between their squared-magnitude responses. This distortion can be minimized by controlling the overlap, which, in turn, can be controlled by appropriately choosing the passband edge of $H_0(z)$.

One way to minimize the amplitude distortion is to employ a computer-aided optimization method to iteratively adjust the filter coefficients $h_0[n]$ of $H_0(z)$ such that the constraint

$$\left|H_0(e^{j\omega})\right|^2 + \left|H_1(e^{j\omega})\right|^2 \cong 1 \tag{14.55}$$

is satisfied for all values of ω [Joh80]. To this end, the objective function Φ to be minimized can be chosen as a linear combination of two functions: (1) stopband attenuation of $H_0(z)$ and (2) the sum of the squared-magnitude responses of $H_0(z)$ and $H_1(z)$ as indicated in Eq. (14.55). One such objective function is given by

$$\Phi = \alpha\Phi_1 + (1-\alpha)\Phi_2, \tag{14.56}$$

where

$$\Phi_1 = \int_{\omega_s}^{\pi} \left|H_0(e^{j\omega})\right|^2 d\omega, \tag{14.57a}$$

$$\Phi_2 = \int_0^{\pi} \left(1 - \left|H_0(e^{j\omega})\right|^2 - \left|H_1(e^{j\omega})\right|^2\right)^2 d\omega, \tag{14.57b}$$

and $0 < \alpha < 1$ and $\omega_s = (\pi/2) + \varepsilon$ for some small $\varepsilon > 0$. Note that since $|T(e^{j\omega})|$ is symmetric with respect to $\pi/2$, the second integral in Eq. (14.57b) can be replaced with

$$2\int_0^{\pi/2} \left(1 - \left|H_0(e^{j\omega})\right|^2 - \left|H_1(e^{j\omega})\right|^2\right)^2 d\omega.$$

After Φ has been made very small by the minimization procedure, both Φ_1 and Φ_2 will also be very small. This, in turn, will make $H_0(z)$ have a magnitude response satisfying $|H_0(e^{j\omega})| \cong 1$ in its passband and $|H_0(e^{j\omega})| \cong 0$ in its stopband, as desired. Moreover, since the power-complementary condition of Eq. (14.55) will be satisfied approximately, the magnitude response of the power-complementary highpass filter $H_1(z)$ to the lowpass filter will satisfy $|H_1(e^{j\omega})| \cong 0$ in the passband of $H_0(z)$ and $|H_1(e^{j\omega})| \cong 1$ in the stopband of $H_0(z)$.

Using the above approach, Johnston has designed a large class of linear-phase FIR lowpass filters $H_0(z)$ meeting a variety of specifications and has tabulated their impulse response coefficients [Joh80], [Cro83], [Ans93]. Example 14.4 examines the performance of one such filter.

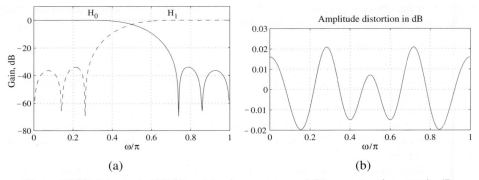

Figure 14.10: Johnston's 12B filter: (a) gain responses and (b) reconstruction error in dB.

EXAMPLE 14.4   Johnston's Optimized Filter Design Example

The impulse response coefficients of the length-12 linear-phase lowpass filter 12B of Johnston [Joh80] are given by

$$
\begin{aligned}
h_0[0] &= -0.006443977 &= h_0[11], & \quad h_0[1] &= 0.02745539 &= h_0[10], \\
h_0[2] &= -0.00758164 &= h_0[9], & \quad h_0[3] &= -0.0913825 &= h_0[8], \\
h_0[4] &= 0.09808522 &= h_0[7], & \quad h_0[5] &= 0.4807962 &= h_0[6].
\end{aligned}
$$

We use Program 14_1 to verify the performance of the lowpass filter 12B of Johnston. The input data requested by the program are the first half of the filter coefficients of $H_0(z)$. It determines the remaining half by using the function `fliplr`. The program computes the gain response of $H_0(z)$ and that of its complementary highpass filter $H_1(z) = H_0(-z)$, as indicated in Figure 14.10(a). It then computes the amplitude distortion function $|H_0(e^{j\omega})|^2 + |H_1(e^{j\omega})|^2$ in dB, as shown in Figure 14.10(b). From Figure 14.10(a), we note that the stopband edge frequency ω_s of filter 12B is about 0.71π, which corresponds to a transition bandwidth $(\omega_s - 0.5\pi)/2\pi = 0.105$. The minimum stopband attenuation is approximately 34 dB. We also observe from Figure 14.10(b) that the amplitude distortion function is indeed very close to 0 dB in both the passbands and the stopbands of the two filters, with a peak value of ± 0.02 dB.

Program 14_1.m

It should be noted that the two-channel QMF bank with the analysis and synthesis filters given by Eqs. (14.31) and (14.32), respectively, is the only example of a perfect reconstruction system with linear-phase FIR filters. A proof of this observation has been left as an exercise in Problem 14.23. An example of a two-channel perfect reconstruction QMF bank with nonminimum phase FIR filters is given in Problem 14.25.

14.2.7   Alias-Free IIR QMF Bank

We now consider the design of an alias-free QMF bank employing IIR analysis and synthesis filters. Under the alias-free conditions of Eqs. (14.37) with $H_1(z)$ chosen according to Eq. (14.43a), the distortion transfer function $T(z)$ of the two-channel QMF bank is given by $2z^{-1}E_0(z^2)E_1(z^2)$, as indicated in Eq. (14.45). If $T(z)$ is an allpass function, then its magnitude response is a constant, and, as a result, the corresponding QMF bank has no magnitude distortion [Vai87f]. Let the polyphase components $E_0(z)$ and $E_1(z)$ of $H_0(z)$ be expressed as

$$E_0(z) = \tfrac{1}{2}\mathcal{A}_0(z), \qquad E_1(z) = \tfrac{1}{2}\mathcal{A}_1(z), \qquad (14.58)$$

where $\mathcal{A}_0(z)$ and $\mathcal{A}_1(z)$ are stable allpass functions. Then $T(z) = \tfrac{1}{2}z^{-1}\mathcal{A}_0^2(z)\mathcal{A}_1^2(z)$, which is seen to be an IIR allpass function.

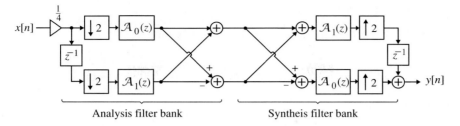

Analysis filter bank $\qquad$ Syntheis filter bank

Figure 14.11: Magnitude-preserving two-channel QMF bank.

Substituting Eq. (14.58) in Eqs. (14.46a) and (14.46b), we get

$$H_0(z) = \tfrac{1}{2}[\mathcal{A}_0(z^2) + z^{-1}\mathcal{A}_1(z^2)], \qquad (14.59a)$$

$$H_1(z) = \tfrac{1}{2}[\mathcal{A}_0(z^2) - z^{-1}\mathcal{A}_1(z^2)]. \qquad (14.59b)$$

Equations (14.59a) and (14.59b) can be written in matrix form as

$$\begin{bmatrix} H_0(z) \\ H_1(z) \end{bmatrix} = \frac{1}{2} \begin{bmatrix} 1 & 1 \\ 1 & -1 \end{bmatrix} \begin{bmatrix} \mathcal{A}_0(z^2) \\ z^{-1}\mathcal{A}_1(z^2) \end{bmatrix}. \qquad (14.60)$$

The corresponding synthesis filters are obtained from Eq. (14.48) and are given by

$$[G_0(z) \quad G_1(z)] = \tfrac{1}{2}\begin{bmatrix} z^{-1}\mathcal{A}_1(z^2) & \mathcal{A}_0(z^2) \end{bmatrix}\begin{bmatrix} 1 & 1 \\ 1 & -1 \end{bmatrix}, \qquad (14.61)$$

which yields

$$G_0(z) = \tfrac{1}{2}[\mathcal{A}_0(z^2) + z^{-1}\mathcal{A}_1(z^2)] = H_0(z), \qquad (14.62a)$$

$$G_1(z) = \tfrac{1}{2}[-\mathcal{A}_0(z^2) + z^{-1}\mathcal{A}_1(z^2)] = -H_1(z). \qquad (14.62b)$$

The realization of the *magnitude-preserving* two-channel QMF bank, shown in Figure 14.11 is obtained by making use of Eq. (14.58) in Figure 14.9(b) [Cro76a].

From Section 13.6.5, we observe that the lowpass transfer function $H_0(z)$ of Eq. (14.59a) satisfies the power-symmetry condition given by

$$H_0(z)H_0(z^{-1}) + H_0(-z)H_0(-z^{-1}) = 1 \qquad (14.63)$$

and can be designed either as an odd order Butterworth or elliptic lowpass half-band filter. By using the pole-interlacing property, the two allpass sections $\mathcal{A}_0(z)$ and $\mathcal{A}_1(z)$ can be determined from the poles of $H_0(z)$, as illustrated in Example 13.24.

14.3   Perfect Reconstruction Two-Channel FIR Filter Banks

A perfect reconstruction two-channel FIR filter bank with linear-phase FIR filters can be designed if the power-complementary requirement of Eq. (14.54) between the two analysis filters $H_0(z)$ and $H_1(z)$ is not imposed. We develop in this section the pertinent design equations.

14.3.1 Modulation Matrices

Observe from Eq. (14.17) that $Y(z)$ can be expressed in matrix form as

$$Y(z) = \tfrac{1}{2}[G_0(z) \quad G_1(z)] \begin{bmatrix} H_0(z) & H_0(-z) \\ H_1(z) & H_1(-z) \end{bmatrix} \begin{bmatrix} X(z) \\ X(-z) \end{bmatrix}. \tag{14.64}$$

From Eq. (14.64) we obtain

$$Y(-z) = \tfrac{1}{2}[G_0(-z) \quad G_1(-z)] \begin{bmatrix} H_0(z) & H_0(-z) \\ H_1(z) & H_1(-z) \end{bmatrix} \begin{bmatrix} X(z) \\ X(-z) \end{bmatrix}. \tag{14.65}$$

Combining Eqs. (14.64) and (14.65) we arrive at

$$\begin{bmatrix} Y(z) \\ Y(-z) \end{bmatrix} = \tfrac{1}{2} \begin{bmatrix} G_0(z) & G_1(z) \\ G_0(-z) & G_1(-z) \end{bmatrix} \begin{bmatrix} H_0(z) & H_0(-z) \\ H_1(z) & H_1(-z) \end{bmatrix} \begin{bmatrix} X(z) \\ X(-z) \end{bmatrix}$$

$$= \tfrac{1}{2} \mathbf{G}^{(m)}(z)\,[\mathbf{H}^{(m)}(z)]^t \begin{bmatrix} X(z) \\ X(-z) \end{bmatrix}, \tag{14.66}$$

where

$$\mathbf{G}^{(m)}(z) = \begin{bmatrix} G_0(z) & G_1(z) \\ G_0(-z) & G_1(-z) \end{bmatrix}, \qquad \mathbf{H}^{(m)}(z) = \begin{bmatrix} H_0(z) & H_1(z) \\ H_0(-z) & H_1(-z) \end{bmatrix}. \tag{14.67}$$

In Eq. (14.67) $\mathbf{H}^{(m)}(z)$ is called the *analysis modulation matrix* and $\mathbf{G}^{(m)}(z)$ is called the *synthesis modulation matrix*.

14.3.2 Perfect Reconstruction Condition

It follows from Eq. (14.66) that for perfect reconstruction, we must have $Y(z) = z^{-\ell} X(z)$ and, correspondingly, $Y(-z) = (-z)^{-\ell} X(-z)$. Substituting these relations in Eq. (14.66) we conclude that the perfect reconstruction condition is satisfied if

$$\mathbf{G}^{(m)}(z)\,[\mathbf{H}^{(m)}(z)]^t = 2 \begin{bmatrix} z^{-\ell} & 0 \\ 0 & (-z)^{-\ell} \end{bmatrix}. \tag{14.68}$$

Thus, knowing the analysis filters $H_0(z)$ and $H_1(z)$, the synthesis filters $G_0(z)$ and $G_1(z)$ can be determined from

$$\mathbf{G}^{(m)}(z) = 2 \begin{bmatrix} z^{-\ell} & 0 \\ 0 & (-z)^{-\ell} \end{bmatrix} \left([\mathbf{H}^{(m)}(z)]^t \right)^{-1},$$

which yields, after some algebra,

$$G_0(z) = \frac{2z^{-\ell}}{\det[\mathbf{H}^{(m)}(z)]} \cdot H_1(-z), \tag{14.69a}$$

$$G_1(z) = -\frac{2z^{-\ell}}{\det[\mathbf{H}^{(m)}(z)]} \cdot H_0(-z), \tag{14.69b}$$

where

$$\det[\mathbf{H}^{(m)}(z)] = H_0(z)H_1(-z) - H_0(-z)H_1(z), \tag{14.70}$$

and ℓ is an odd positive integer.

For FIR analysis filters $H_0(z)$ and $H_1(z)$, the synthesis filters $G_0(z)$ and $G_1(z)$ will also be FIR filters while ensuring perfect reconstruction if

$$\det[\mathbf{H}^{(m)}(z)] = c\,z^{-k}, \tag{14.71}$$

where c is a real number and k is a positive integer. In this case, the two synthesis filters are given by

$$G_0(z) = \tfrac{2}{c}z^{-(\ell-k)}H_1(-z), \tag{14.72a}$$
$$G_1(z) = -\tfrac{2}{c}z^{-(\ell-k)}H_0(-z). \tag{14.72b}$$

14.3.3 Biorthogonal Filter Banks

As pointed out by Eq. (14.41), for a perfect reconstruction filter bank, the normalized product filter $P(z)$ must be a zero-phase half-band lowpass filter. Thus, $P(z)$ is a symmetric polynomial of the form

$$P(z) = 1 + p_1(z + z^{-1}) + p_3(z^3 + z^{-3}) + p_5(z^5 + z^{-5}) + \cdots. \tag{14.73}$$

In signal compression using QMF banks, it is preferable to choose the lowpass filters $H_0(z)$ and $G_0(z)$ with a maximum number of zeros at $z = -1$, which are also zeros of $P_0(z)$ and, hence, of $P(z)$ [Dau88]. The general form of $P(z)$ is then given by [Vet95]

$$P(z) = (1 + z^{-1})^m(1 + z)^m R(z), \tag{14.74}$$

where $R(z)$ is a symmetric polynomial, that is, $R(z) = R(z^{-1})$, with $R(e^{j\omega}) \geq 0$. This class of half-band filters has been called the *binomial* or *maxflat filter*, as they have a frequency response $P(e^{j\omega})$ that is maximally flat at $\omega = 0$ and $\omega = \pi$ [Dau88], [Her71]. When $R(z)$ is of the form

$$R(z) = r_0 + \sum_{s=1}^{m-1} r_s\,(z^s + z^{-s}), \tag{14.75}$$

it is said to be of *minimal degree* and is of practical interest.

In the simplest case, $m = 1$ and $R(z) = \tfrac{1}{2}$, resulting in

$$P(z) = \tfrac{1}{2}(z + 2 + z^{-1})$$
$$= \tfrac{1}{2}z(1 + z^{-1})(1 + z^{-1}) = z^\ell H_0(z)G_0(z). \tag{14.76}$$

If we choose

$$H_0(z) = \tfrac{1}{\sqrt{2}}(1 + z^{-1}),$$

the lowest order $G_0(z)$ is obtained with $\ell = 1$ and is given by

$$G_0(z) = \tfrac{1}{\sqrt{2}}(1 + z^{-1}).$$

These lowpass filters are precisely the Haar filters of Eqs. (14.31) and (14.32), discussed in Section 14.2.4. The corresponding highpass filters are also given in these two equations.

For $m = 2$, $R(z)$ is of the form [Vet95]

$$R(z) = a\,z + b + a\,z^{-1},$$

and, as a result,

$$P(z) = (1 + z^{-1})^2(1 + z)^2(a\,z + b + a\,z^{-1})$$
$$= a\,z^3 + (4a + b)z^2 + (7a + 4b)z + (8a + 6b)$$
$$+ (7a + 4b)z^{-1} + (4a + b)z^{-2} + a\,z^{-3}. \tag{14.77}$$

Since even powers of $P(z)$ must be zeros and the coefficient of the z^0 be equal to 1, we must have

$$4a + b = 0, \qquad 8a + 6b = 1. \tag{14.78}$$

Solving Eq. (14.78), we get

$$a = -\tfrac{1}{16}, \qquad b = \tfrac{1}{4}. \tag{14.79}$$

Thus, in this case, we have

$$P(z) = [1 + \tfrac{1}{2}(z + z^{-1})]^2[1 - \tfrac{1}{4}(z + z^{-1})]$$
$$= \tfrac{1}{16}z^3(1 + 2z^{-1} + z^{-2})^2(-1 + 4z^{-1} - z^{-2}). \tag{14.80}$$

One possible factorization of $P(z)$, with $\ell = 3$, is given by [Kin2001]

$$H_0(z) = \tfrac{1}{2}(1 + 2z^{-1} + z^{-2}), \tag{14.81a}$$
$$G_0(z) = \tfrac{1}{8}(1 + 2z^{-1} + z^{-2})(-1 + 4z^{-1} - z^{-2})$$
$$= \tfrac{1}{8}(-1 + 2z^{-1} + 6z^{-2} + 2z^{-3} - z^{-4}). \tag{14.81b}$$

The corresponding highpass filters are given by

$$H_1(z) = G_0(-z) = \tfrac{1}{8}(-1 - 2z^{-1} + 6z^{-2} - 2z^{-3} - z^{-4}), \tag{14.82a}$$
$$G_1(z) = -H_0(-z) = -\tfrac{1}{2}(1 - 2z^{-1} + z^{-2}). \tag{14.82b}$$

A plot of the magnitude responses of the analysis filters of Eqs. (14.81a) and (14.82a) is shown in Figure 14.12(a). The above set of perfect reconstruction QMF filters were originally proposed by LeGall [LeG88]. Since the lowpass and the highpass analysis filters, $H_0(z)$ and $H_1(z)$, are, respectively, length-3 and length-5 linear-phase FIR filters, this set of filters is often referred to as *LeGall 3/5-tap filter pair.*

A different set of perfect reconstruction QMF filters are obtained by interchanging the factors assigned in the above two sets of equations [Kin2001]:

$$H_0(z) = \tfrac{1}{8}(-1 + 2z^{-1} + 6z^{-2} + 2z^{-3} - z^{-4}), \tag{14.83a}$$
$$G_0(z) = \tfrac{1}{2}(1 + 2z^{-1} + z^{-2}), \tag{14.83b}$$
$$H_1(z) = \tfrac{1}{2}(1 - 2z^{-1} + z^{-2}), \tag{14.83c}$$
$$G_1(z) = \tfrac{1}{8}(1 + 2z^{-1} - 6z^{-2} + 2z^{-3} + z^{-4}). \tag{14.83d}$$

The above set of perfect reconstruction linear-phase FIR analysis filters, $H_0(z)$ and $H_1(z)$, is known as the *LeGall 5/3-tap filter pair* [Dau88].

Another possible factorization of $P(z)$, with $\ell = 3$, leads to the lowpass filters

$$H_0(z) = \tfrac{1}{8}(1 + 3z^{-1} + 3z^{-2} + z^{-3}), \tag{14.84a}$$
$$G_0(z) = \tfrac{1}{2}(-1 + 3z^{-1} + 3z^{-2} - z^{-3}). \tag{14.84b}$$

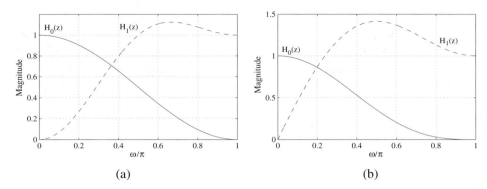

Figure 14.12: (a) Magnitude responses of LeGall 3/5-tap analysis filter pair and (b) magnitude responses of Daubechies 4/4-tap analysis filter pair.

The corresponding highpass filters are given by

$$H_1(z) = \tfrac{1}{2}(-1 - 3z^{-1} + 3z^{-2} + z^{-3}), \tag{14.85a}$$

$$G_1(z) = \tfrac{1}{8}(-1 + 3z^{-1} - 3z^{-2} + z^{-3}). \tag{14.85b}$$

The perfect reconstruction linear-phase FIR analysis filters of Eqs. (14.84a) and (14.85a) are more commonly known as the as the *Daubechies 4/4-tap filter pair* [Dau88]. A plot of the magnitude responses of these analysis filters is shown in Figure 14.12(b).

It is easy to show that the above choices for the analysis filters do satisfy the condition of Eq. (14.71) on the determinant of the analysis modulation matrix $\mathbf{H}^{(m)}(z)$. For example, for the Haar filters we have

$$
\begin{aligned}
\det[\mathbf{H}^{(m)}(z)] &= H_0(z)H_1(-z) - H_0(-z)H_1(z) \\
&= P_0(z) - P_0(-z) \\
&= \tfrac{1}{2}(1 + z^{-1})^2 - \tfrac{1}{2}(1 - z^{-1})^2 = 2\,z^{-1},
\end{aligned}
$$

which is seen to satisfy the condition with $c = 2$ and $k = 1$. Likewise, for the analysis filters generated from the lowpass product filter of Eq. (14.80), we get

$$
\begin{aligned}
\det[\mathbf{H}^{(m)}(z)] &= \tfrac{1}{16}(1 + 2z^{-1} + z^{-2})^2(-1 + 4z^{-1} - z^{-2}) \\
&\quad - \tfrac{1}{16}(1 - 2z^{-1} + z^{-2})^2(-1 - 4z^{-1} - z^{-2}) = 2\,z^{-3},
\end{aligned}
$$

which again satisfies the condition with $c = 2$ and $k = 3$.

Several other choices for the analysis filters are possible, with some not resulting in linear-phase filters.

In many applications, it is desirable to use a QMF bank in which the outputs of the analysis filters are represented as an orthonormal transformation of the input signal, while the reconstruction process implemented by the synthesis filters is represented by a transformation that is simply the transpose of the analysis filtering transform matrix. However, the subband filtering process described above cannot be represented as an orthonormal transformation of the input signal. QMF banks obtained using the perfect reconstruction linear-phase FIR filters and represented by nonorthonormal transformation matrices are often referred to as the *biorthogonal filter banks*.

We consider next the design of perfect reconstruction QMF banks with nonlinear phase FIR filters based on orthogonal transformation.

14.3.4　Orthogonal Filter Banks

These filter banks are designed using a different scheme for the factorization of the lowpass product filter $P(z)$ of order $2N$, which results in FIR analysis and synthesis filters that are no longer linear-phase. Now $P(z)$ being a symmetric polynomial of the form of Eq. (14.73), it has factors of the form $(\alpha z + 1)(1 + \alpha z^{-1})$. We can assign the subfactor $(1 + \alpha z^{-1})$ to $H_0(z)$ and the subfactor $z^{-1}(\alpha z + 1)$ to $G_0(z)$. This process is continued with respect to each factor of $P(z)$, leading to $H_0(z)$ and its mirror-image transfer function

$$G_0(z) = z^{-N} H_0(z^{-1}). \tag{14.86}$$

For real-coefficient filters, the above relation implies that

$$G_0(e^{j\omega}) = H_0(e^{-j\omega}),$$

implying that the magnitude responses of the two filters are the same, while the phase responses are opposite of each other.

As in the case of the biorthogonal filter banks, it is desirable to have filters with as many zeros as possible at $z = -1$, that is, to have maximum number of zeros of $P(z)$ at $z = -1$ [Dau88]. We can thus choose $P(z)$ to be of the form of Eq. (14.74). We can assign, for example, all factors of $P(z)$ having zeros inside the unit circle to $H_0(z)$ and assign all factors of $P(z)$ having zeros outside the unit circle to $G_0(z)$. Moreover, zeros of $P(z)$ on the unit circle are of even multiplicity, and hence, half of these zeros are assigned to $H_0(z)$ and the other half to $G_0(z)$. This assignment makes $G_0(z)$ the mirror-image of $H_0(z)$. In addition, $H_0(z)$ is then a minimum-phase filter, whereas $G_0(z)$ is a maximum-phase filter.

As an example, we consider the factorization of $P(z)$ of Eq. (14.80). Note that the factor $(1 - 4z^{-1} + z^{-2})$ has a zero inside the unit circle at $z = 2 - \sqrt{3}$ and another zero outside the unit circle at $z = 2 + \sqrt{3}$. The minimum-phase spectral factor of $P(z)$ thus yields the lowpass analysis filter:

$$H_0(z) = \tfrac{1}{4(\sqrt{3}-1)}(1 + z^{-1})^2 \left(1 - (2 - \sqrt{3})z^{-1}\right)$$
$$= 0.48296 + 0.83652z^{-1} + 0.224144z^{-2} - 0.12941z^{-3}, \tag{14.87}$$

whereas, the maximum-phase spectral factor yields the lowpass synthesis filter, which is the mirror-image polynomial of $H_0(z)$:

$$G_0(z) = z^{-3} H_0(z^{-1}) = -0.12941 + 0.224144z^{-1} + 0.83652z^{-2} + 0.48296z^{-3}. \tag{14.88}$$

The transfer functions of the highpass analysis and synthesis filter are then given by

$$H_1(z) = G_0(-z) = -0.12941 - 0.224144z^{-1} + 0.83652z^{-2} - 0.48296z^{-3}, \tag{14.89a}$$
$$G_1(z) = -H_0(-z) = -0.48296 + 0.83652z^{-1} - 0.224144z^{-2} - 0.12941z^{-3}. \tag{14.89b}$$

Figure 14.13 shows the magnitude responses of the above two analysis filters.

A general method for the design of a lowpass zero-phase half-band filter $P(z) = H_0(z)H_0(z^{-1})$ of order $2N$ with N odd is outlined next [Smi84], [Min85]. Let $H_0(z)$ be an FIR filter of odd order N satisfying the power-symmetric condition of Eq. (14.63). If we then choose

$$H_1(z) = z^{-N} H_0(-z^{-1}), \tag{14.90}$$

Eq. (14.70) reduces to

$$\det[\mathbf{H}^{(m)}(z)] = -z^{-N} \left(H_0(z)H_0(z^{-1}) + H_0(-z)H_0(-z^{-1})\right) = -z^{-N}. \tag{14.91}$$

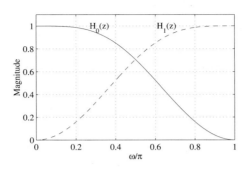

Figure 14.13: Magnitude responses of third-order maximally flat analysis filters of Eqs. (14.87) and (14.89a).

Comparing Eq. (14.91) with Eq. (14.71), we observe that $c = -1$ and $k = N$. Using Eqs. (14.90) and (14.91) in Eqs. (14.72a) and (14.72b), with $\ell = k = N$, we get

$$G_0(z) = z^{-N} H_0(z^{-1}), \qquad G_1(z) = z^{-N} H_1(z^{-1}). \tag{14.92}$$

It should be noted that if $H_0(z)$ is a causal FIR filter, the other three filters are also causal FIR filters. Moreover, from Eqs. (14.90) and (14.92), it follows that $|G_i(e^{j\omega})| = |H_i(e^{j\omega})|$, for $i = 1, 2$. In addition, $|H_1(e^{j\omega})| = |H_0(-e^{j\omega})|$, which for a real-coefficient transfer function implies that if $H_0(z)$ is a lowpass filter, then $H_1(z)$ is a highpass filter. A perfect reconstruction power-symmetric filter bank is also called an *orthogonal filter bank*.

The filter bank design problem thus reduces to the design of a power-symmetric lowpass filter $H_0(z)$. To this end, we can design an even order $P_0(z) = H_0(z)H_0(z^{-1})$ whose *spectral factorization* yields $H_0(z)$. Now, the power-symmetric condition of Eq. (14.63) implies that $P_0(z)$ be a zero-phase half-band lowpass filter with a nonnegative frequency response $P_0(e^{j\omega})$. Such a half-band filter can be designed using the minimum-phase FIR filter design method described in Section 10.5.3.

Several comments are in order here. First, as shown in Section 13.6.2, the order of the half-band filter $P_0(z)$ is of the form $4K + 2$, where K is a positive integer. This implies that the order of $H_0(z)$ is $N = 2K + 1$, which is odd as required. Second, the zeros of $P_0(z)$ appear with mirror-image symmetry in the z-plane, with the zeros on the unit circle being of multiplicity 2. Any appropriate half of these zeros can be grouped to form the spectral factor $H_0(z)$. For example, a minimum-phase $H_0(z)$ can be formed by grouping all the zeros inside the unit circle with half of the zeros on the unit circle. Likewise, a maximum-phase $H_0(z)$ can be formed by grouping all the zeros outside the unit circle with half of the zeros on the unit circle. However, it is not possible to form a spectral factor with a linear phase. Third, the stopband edge frequency is the same for $P_0(z)$ and $H_0(z)$. If the desired minimum stopband attenuation of $H_0(z)$ is α_s dB, the minimum stopband attenuation of $P_0(z)$ is approximately $2\alpha_s + 6.02$ dB.

We illustrate the power-symmetric filter bank design in Example 14.5.

EXAMPLE 14.5   Lowpass Power-Symmetric Filter Design

Design a lowpass real-coefficient power-symmetric filter $H_0(z)$ with the following specifications: stopband edge at $\omega_s = 0.63\pi$ and a minimum stopband attenuation of $\alpha_s = 17$ dB. The specifications of the corresponding zero-phase half-band filter $F(z)$ are therefore as follows: stopband edge at $\omega_s = 0.63\pi$ and a minimum stopband attenuation of $\alpha_s = 40$ dB. The desired stopband ripple is thus $\delta_s = 0.01$, which is also the passband ripple. The passband edge is at $\omega_p = \pi - 0.63\pi = 0.37\pi$. Using the function remezord we then estimate the order of $F(z)$, and using the function remez we next design $Q(z)$. To this end the code fragments used are

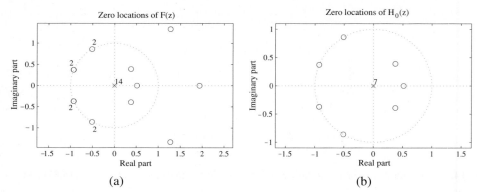

Figure 14.14: Zero locations of (a) the zero-phase half-band filter $P_0(z)$ and (b) its minimum-phase spectral factor $H_0(z)$ of Example 14.5.

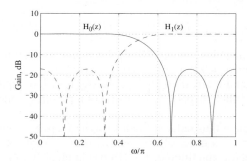

Figure 14.15: Gain responses of seventh-order power-symmetric analysis filters of Example 14.5.

minphase.m

```
[N,fpts,mag,wt] = remezord([0.37 0.63],[1 0], [0.01 0.01]);
[q,err] = remez(N,fpts,mag,wt);
```

The order of $P_0(z)$ is found to be 14, implying the order of $H_0(z)$ to be 7, which is odd, as desired. Note the parameter `err` provides the maximum value of the ripple. To determine the coefficients of the half-band filter $F(z)$, we add `err` to the central coefficient $q[7]$. The roots of $F(z)$ should theoretically exhibit a mirror-image symmetry with respect to the unit circle, with double roots on the unit circle. However, it is found that a slightly larger value than `err` should be added to ensure double zeros of $P_0(z)$ on the unit circle. Next, using the M-file `minphase`, we determine the minimum-phase spectral factor $H_0(z)$. Its corresponding highpass analysis filter $H_1(z)$ is then obtained using Eq. (14.90). Figure 14.14 shows the zero locations of $F(z)$ and $H_0(z)$. The gain responses of the two analysis filters $H_0(z)$ and $H_1(z)$ are shown in Figure 14.15.

In realizing the analysis filter bank, if the two filters $H_0(z)$ and $H_1(z)$ are implemented independently, the overall structure would require $2(N + 1)$ multipliers and $2N$ two-input adders. However, a computationally efficient realization requiring $N + 1$ multipliers and $2N$ two-input adders can be developed by exploiting the relation of Eq. (14.90) (Problem 14.22).

14.3.5   Design of Orthogonal Filter Banks Using MATLAB

The M-file `firpr2chfb` can be used to design a perfect reconstruction two-channel FIR filter bank. We illustrate its use in Example 14.6.

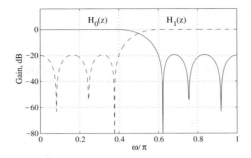

Figure 14.16: Gain responses of the two analysis filters of Example 14.6.

Figure 14.17: The paraunitary lattice structure.

EXAMPLE 14.6   Design of a Perfect Reconstruction FIR Orthogonal Filter Bank

We consider the design of two-channel perfect reconstruction FIR orthogonal filter bank with filters of order 11. The normalized angular passband edge frequency of the lowpass filters is 0.4. The code fragment used is

```
[h0,h1,g0,g1] = firpr2chfb(11,0.4);
```

The aliasing cancellation condition of Eq. (14.21) can be checked on MATLAB by using the statements

```
n = 0:11;
ac = conv(g0,((-1).^n.*h0))+ conv(g1,((-1).^n.*h1))
```

The aliasing condition condition is seen to be satisfied as the maximum absolute value of the sequence ac is 0.2776×10^{-16}. The perfect reconstruction condition of Eq. (14.27) can be checked on MATLAB by using the statements

```
n = 0:11;
pr = 0.5*conv(g0,h0)+ 0.5*conv(g1,h1);
```

The 12th sample of the sequence pr is 0.9985. The remaining samples of this sequence are either equal to zero or nearly equal to zero. A plot of the gain responses of the two analysis filters is shown in Figure 14.16.

14.3.6 Paraunitary Filter Banks

A p-input, q-output LTI discrete-time system with a transfer matrix $\mathbf{T}_{pq}(z)$ is called a *paraunitary system* if $\mathbf{T}_{pq}(z)$ is a *paraunitary matrix,* that is,

$$\tilde{\mathbf{T}}_{pq}(z)\mathbf{T}_{pq}(z) = c\mathbf{I}_p, \tag{14.93}$$

where $\tilde{\mathbf{T}}_{pq}(z)$ is the paraconjugate of $\mathbf{T}_{pq}(z)$ given by the transpose of $\mathbf{T}_{pq}(z^{-1})$ with each coefficient replaced by its conjugate, $\mathbf{I}_p$ is an $p \times p$ identity matrix, and c is a real constant. A causal, stable paraunitary system is also a lossless system.

It can be easily shown that the modulation matrix $\mathbf{H}^{(m)}(z)$ defined in Eq. (14.67) of a power-symmetric filter bank is a paraunitary matrix. Hence, a power-symmetric filter bank has also been referred to as a *paraunitary filter bank*.

Since the cascade of a paraunitary system with a transfer matrix $\mathbf{T}_{pq}^{(1)}(z)$ and a paraunitary system with a transfer matrix $\mathbf{T}_{qr}^{(2)}(z)$ is also paraunitary, it is easier to design a paraunitary filter bank without resorting to spectral factorization by cascading simpler paraunitary blocks. To this end, the cascaded FIR lattice structure introduced in Section 8.9.2 can be employed. The overall structure is lossless, as each lattice stage shown in Figure 14.17 is lossless, that is, paraunitary, causal, and stable [Vai86b], [Vai88a]. The synthesis procedure outlined in [Vai86b], [Vai88a] realizes both the power-symmetric transfer function $H_0(z)$ and its conjugate quadratic transfer function $H_1(z)$. Three important properties of the QMF lattice structure are structurally induced. First, the QMF lattice filter bank guarantees perfect reconstruction, independent of the lattice parameters. Second, it exhibits very small coefficient sensitivity to lattice parameters, as each stage remains lossless under coefficient quantization. Third, its computational complexity is about one-half that of any other realization, as it requires $(N-1)/2$ total numbers of multipliers for an order-N filter.

EXAMPLE 14.7   FIR Cascaded Lattice Analysis Filter Bank

Following the method outlined in Section 8.9.2, we develop the FIR cascaded lattice realization of the analysis filter bank designed in Example 14.5. The input multiplier here is $h[0] = 0.32308146$, which is used to normalize the analysis transfer function so that its constant coefficient is unity. The lattice coefficients obtained from the normalized analysis transfer function using Eq. (8.99) are given by

$$k_7 = -0.15165236, \quad k_5 = 0.23538743, \quad k_3 = -0.483934447, \quad k_1 = 1.6100196.$$

It should be noted that, because of the numerical accuracy problem, the coefficients of the spectral factor obtained in Example 14.5 are not very accurate. As a result, the coefficient of $z^{-(i-1)}$ of the transfer function $H_{i-2}(z)$ generated from the transfer function $H_i(z)$ using Eq. (8.99) is not exactly zero and has been set to zero at each iteration. Two interesting properties of the cascaded lattice filter bank can be seen from the above coefficient values. First, the signs of the lattice multiplier coefficients alternate between stages. Second, the values of the multiplier coefficients $\{k_i\}$ decrease with increasing i.

The QMF lattice structure can be used directly to design the power-symmetric analysis filter $H_0(z)$ using an iterative computer-aided optimization technique. The goal here is to determine the lattice parameters k_i by minimizing the energy in the stopband of $H_0(z)$. To this end, the objective function is given by

$$\Phi = \int_{\omega_s}^{\pi} \left| H_0(e^{j\omega}) \right|^2 d\omega. \tag{14.94}$$

It should be noted that the power-symmetric property ensures good passband response.

14.4   *L*-Channel QMF Banks

We now generalize the discussion of the previous section to the case of a QMF bank with more than two channels. The basic structure of the L-channel QMF bank is shown in Figure 14.18.

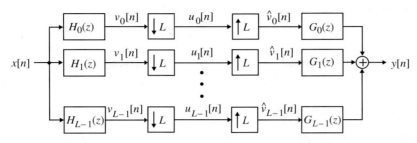

Figure 14.18: The basic L-channel QMF filter bank structure.

14.4.1 Analysis of the *L*-Channel Filter Bank

We analyze the operation of the L-channel QMF bank of Figure 14.18 in the z-domain. The expressions for the z-transforms of various intermediate signals in Figure 14.18 are given by

$$V_k(z) = H_k(z)X(z), \tag{14.95a}$$

$$U_k(z) = \frac{1}{L}\sum_{\ell=0}^{L-1} H_k(z^{1/L}W_L^{\ell})X(z^{1/L}W_L^{\ell}), \tag{14.95b}$$

$$\hat{V}_k(z) = U_k(z^L), \tag{14.95c}$$

where $0 \leq k \leq L - 1$.

Define the vector of down-sampled subband signals $U_k(z)$ as[2]

$$\mathbf{u}(z) = [\, U_0(z) \quad U_1(z) \quad \cdots \quad U_{L-1}(z)\,]^t, \tag{14.96}$$

the modulation vector of the input signals as

$$\mathbf{x}^{(m)}(z) = [\, X(z) \quad X(zW_L) \quad \cdots \quad X(zW_L^{L-1})\,]^t, \tag{14.97}$$

and the *analysis filter bank modulation matrix* as

$$\mathbf{H}^{(m)}(z) = \begin{bmatrix} H_0(z) & H_1(z) & \cdots & H_{L-1}(z) \\ H_0(zW_L^1) & H_1(zW_L^1) & \cdots & H_{L-1}(zW_L^1) \\ \vdots & \vdots & \ddots & \vdots \\ H_0(zW_L^{L-1}) & H_1(zW_L^{L-1}) & \cdots & H_{L-1}(zW_L^{L-1}) \end{bmatrix}. \tag{14.98}$$

Then, Eq. (14.95b) can be compactly expressed in the form

$$\mathbf{u}(z) = \tfrac{1}{L}[\mathbf{H}^{(m)}(z^{1/L})]^t \mathbf{x}^{(m)}(z^{1/L}). \tag{14.99}$$

The output of the QMF bank is given by

$$Y(z) = \sum_{k=0}^{L-1} G_k(z)\hat{V}_k(z), \tag{14.100}$$

---

[2]The superscript "*t*" denotes matrix transposition.

which can be expressed in a matrix form as

$$Y(z) = \mathbf{g}^t(z)\mathbf{u}(z^L), \tag{14.101}$$

where

$$\mathbf{g}(z) = [\, G_0(z) \quad G_1(z) \quad \cdots \quad G_{L-1}(z)\,]^t. \tag{14.102}$$

14.4.2   Alias-Free L-Channel Filter Bank

We now develop the condition for alias-free operation of the L-channel filter bank of Figure 14.18 [Fli94]. From Eq. (14.101), the modulated versions of the output signal are given by

$$Y(zW_L^k) = \mathbf{g}^t(zW_L^k)\mathbf{u}(z^L W_L^{kL}) = \mathbf{g}^t(zW_L^k)\mathbf{u}(z^L), \qquad 0 \le k \le L-1, \tag{14.103}$$

which can be expressed in a matrix form as

$$\mathbf{y}^{(m)}(z) = [\, Y(z) \quad Y(zW_L) \quad \cdots \quad Y(zW_L^{L-1})\,]^t. \tag{14.104}$$

Using Eqs. (14.101) and (14.102) in Eq. (14.104), we therefore obtain

$$\mathbf{y}^{(m)}(z) = \mathbf{G}^{(m)}(z)\mathbf{u}(z^L), \tag{14.105}$$

where

$$\mathbf{G}^{(m)}(z) = \begin{bmatrix} G_0(z) & G_1(z) & \cdots & G_{L-1}(z) \\ G_0(zW_L^1) & G_1(zW_L^1) & \cdots & G_{L-1}(zW_L^1) \\ \vdots & \vdots & \ddots & \vdots \\ G_0(zW_L^{L-1}) & G_1(zW_L^{L-1}) & \cdots & G_{L-1}(zW_L^{L-1}) \end{bmatrix}, \tag{14.106}$$

is the *synthesis filter bank modulation matrix*.

Combining Eqs. (14.99) and (14.105), we arrive at the input–output relationship of the L-channel filter bank as

$$\mathbf{y}^{(m)}(z) = \tfrac{1}{L}\mathbf{G}^{(m)}(z)[\mathbf{H}^{(m)}(z)]^t\mathbf{x}^{(m)}(z)$$
$$= \mathbf{T}(z)\mathbf{x}^{(m)}(z), \tag{14.107}$$

where $\mathbf{T}(z) = \tfrac{1}{L}\mathbf{G}^{(m)}(z)[\mathbf{H}^{(m)}(z)]^t$ is called the *transfer matrix,* relating the input signal $X(z)$ and its frequency-modulated versions $X(zW_L^k)$, $1 \le k \le L-1$, with the output signal $Y(z)$ and its frequency-modulated versions $Y(zW_L^k)$, $1 \le k \le L-1$. The filter bank is alias-free if the transfer matrix $\mathbf{T}(z)$ is a diagonal matrix of the form

$$\mathbf{T}(z) = \text{diag}\,[\, T(z) \quad T(zW_L) \quad \cdots \quad T(zW_L^{L-1})\,]. \tag{14.108}$$

The first element $T(z)$ of the above diagonal matrix is called the *distortion transfer function* of the L-channel filter bank. Substituting Eqs. (14.95a) to (14.95c) in Eq. (14.100), we arrive at

$$Y(z) = \sum_{\ell=0}^{L-1} a_\ell(z) X(zW_L^\ell), \tag{14.109}$$

where

$$a_\ell(z) = \frac{1}{L}\sum_{k=0}^{L-1} H_k(zW_L^\ell)G_k(z), \qquad 0 \le \ell \le L-1. \tag{14.110}$$

On the unit circle, the term $X(zW_L^\ell)$ becomes

$$X(e^{j\omega}W_L^\ell) = X(e^{j(\omega - 2\pi\ell/L)}). \tag{14.111}$$

Thus, from Eq. (14.109), the output spectrum $Y(e^{j\omega})$ is a weighted sum of $X(e^{j\omega})$ and its uniformly shifted versions $X(e^{j(\omega - 2\pi\ell/L)})$ for $\ell = 1, 2, \ldots, L - 1$, which are caused by the sampling rate alteration operations. The term $X(zW_L^\ell)$ is called the ℓth *aliasing term*, with $a_\ell(z)$ representing its *gain* at the output. In general, the QMF bank of Figure 14.18 is a linear, time-varying system with a period of L.

It follows from Eq. (14.109) that the aliasing effect can be completely eliminated at the output if and only if

$$a_\ell(z) = 0, \qquad 1 \le \ell \le L - 1, \tag{14.112}$$

for all possible inputs $x[n]$. If Eq. (14.112) holds, then the L-channel QMF bank of Figure 14.18 becomes a linear time-invariant system with an input–output relation given by

$$Y(z) = T(z)X(z), \tag{14.113}$$

where $T(z)$ is the distortion transfer function given by

$$T(z) = a_0(z) = \frac{1}{L}\sum_{k=0}^{L-1} H_k(z)G_k(z). \tag{14.114}$$

If $T(z)$ has a constant magnitude, then the system of Figure 14.18 is a *magnitude-preserving QMF bank*. If $T(z)$ has a linear phase, then the QMF bank has no phase distortion. Finally, if $T(z)$ is a pure delay, it is a *perfect reconstruction QMF bank*.

Using Eqs. (14.98) and (14.102), we can express Eq. (14.110) in a matrix form as

$$L \cdot \mathbf{A}(z) = \mathbf{H}^{(m)}(z)\mathbf{g}(z), \tag{14.115}$$

where

$$\mathbf{A}(z) = [a_0(z) \quad a_1(z) \quad \cdots \quad a_{L-1}(z)]^t. \tag{14.116}$$

The aliasing cancellation condition can now be rewritten as

$$\mathbf{H}^{(m)}(z)\mathbf{g}(z) = \mathbf{t}(z), \tag{14.117}$$

where

$$\mathbf{t}(z) = [L a_0(z) \quad 0 \quad \cdots \quad 0]^t = [L \cdot T(z) \quad 0 \quad \cdots \quad 0]^t. \tag{14.118}$$

From Eq. (14.117), it follows that by knowing the set of analysis filters $\{H_k(z)\}$, we can determine the desired set of synthesis filters $\{G_k(z)\}$ as

$$\mathbf{g}(z) = [\mathbf{H}^{(m)}(z)]^{-1}\mathbf{t}(z), \tag{14.119}$$

provided, of course, $[\det \mathbf{H}^{(m)}(z)] \neq 0$. Moreover, a perfect reconstruction QMF bank results if we set $T(z) = z^{-n_0}$ in the expression for $\mathbf{t}(z)$ in Eq. (14.118). In practice, the above approach is difficult to carry out for a number of reasons. A more practical solution to the design of perfect reconstruction QMF bank is obtained via the polyphase representation outlined next [Vai87d].

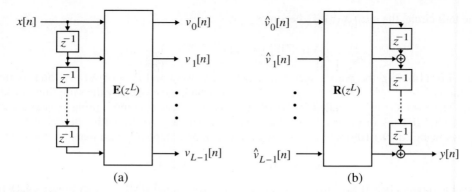

Figure 14.19: (a) Type I polyphase representation of the analysis filter bank and (b) Type II polyphase representation of the synthesis filter bank.

14.4.3   Polyphase Representation

Consider the Type I polyphase representation of the kth analysis filter $H_k(z)$:

$$H_k(z) = \sum_{\ell=0}^{L-1} z^{-\ell} E_{k\ell}(z^L), \qquad 0 \le k \le L-1. \tag{14.120}$$

A matrix representation of the above set of equations is given by

$$\mathbf{h}(z) = \mathbf{E}(z^L)\mathbf{e}(z), \tag{14.121}$$

where

$$\mathbf{h}(z) = \begin{bmatrix} H_0(z) & H_1(z) & \cdots & H_{L-1}(z) \end{bmatrix}^t, \tag{14.122a}$$

$$\mathbf{e}(z) = \begin{bmatrix} 1 & z^{-1} & \cdots & z^{-(L-1)} \end{bmatrix}^t, \tag{14.122b}$$

and

$$\mathbf{E}(z) = \begin{bmatrix} E_{00}(z) & E_{01}(z) & \cdots & E_{0,L-1}(z) \\ E_{10}(z) & E_{11}(z) & \cdots & E_{1,L-1}(z) \\ \vdots & \vdots & \ddots & \vdots \\ E_{L-1,0}(z) & E_{L-1,1}(z) & \cdots & E_{L-1,L-1}(z) \end{bmatrix}. \tag{14.122c}$$

The matrix $\mathbf{E}(z)$ defined above is called the *Type I polyphase component matrix*. Figure 14.19(a) shows the Type I polyphase representation of the analysis filter bank.

Likewise, we can represent the L synthesis filters in a Type II polyphase form:

$$G_k(z) = \sum_{\ell=0}^{L-1} z^{-(L-1-\ell)} R_{\ell k}(z^L), \qquad 0 \le k \le L-1. \tag{14.123}$$

In matrix form, the above set of L equations can be rewritten as

$$\mathbf{g}^t(z) = z^{-(L-1)} \tilde{\mathbf{e}}(z) \mathbf{R}(z^L), \tag{14.124}$$

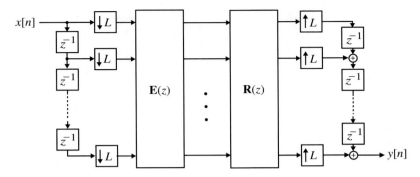

Figure 14.20: *L*-channel QMF bank structure based on the polyphase representations of the analysis ard synthesis filter banks.

where

$$\mathbf{g}(z) = \begin{bmatrix} G_0(z) & G_1(z) & \cdots & G_{L-1}(z) \end{bmatrix}^t, \tag{14.125a}$$

$$\tilde{\mathbf{e}}(z) = \begin{bmatrix} 1 & z & \cdots & z^{L-1} \end{bmatrix} = \mathbf{e}^t(z^{-1}), \tag{14.125b}$$

and

$$\mathbf{R}(z) = \begin{bmatrix} R_{00}(z) & R_{01}(z) & \cdots & R_{0,L-1}(z) \\ R_{10}(z) & R_{11}(z) & \cdots & R_{1,L-1}(z) \\ \vdots & \vdots & \ddots & \vdots \\ R_{L-1,0}(z) & R_{L-1,1}(z) & \cdots & R_{L-1,L-1}(z) \end{bmatrix}. \tag{14.125c}$$

The matrix $\mathbf{R}(z)$ defined above is called the *Type II polyphase component matrix*. Figure 14.19(b) shows the Type II polyphase representation of the synthesis filter bank.

Making use of the polyphase representations of Figure 14.19 in Figure 14.18 and the cascade equivalences of Figure 13.14, we arrive at an equivalent realization of the *L*-channel QMF bank shown in Figure 14.20.

The relation between the modulation matrix $\mathbf{H}^{(m)}(z)$ of Eq. (14.98) and the Type I polyphase component matrix $\mathbf{E}(z)$ of Eq. (14.122c) can be easily established. From Eqs. (14.98), (14.121), and (14.122a), we observe that

$$[\mathbf{H}^{(m)}(z)]^t = \begin{bmatrix} \mathbf{h}(z) & \mathbf{h}(zW_L^1) & \cdots & \mathbf{h}(zW_L^{L-1}) \end{bmatrix}$$
$$= \mathbf{E}(z^L) \begin{bmatrix} \mathbf{e}(z) & \mathbf{e}(zW_L^1) & \cdots & \mathbf{e}(zW_L^{L-1}) \end{bmatrix}. \tag{14.126}$$

Now, from Eq. (14.122b), it follows that

$$\mathbf{e}(zW_L^k) = \mathbf{\Delta}(z) \begin{bmatrix} 1 \\ W_L^{-k} \\ \vdots \\ W_L^{-k(L-1)} \end{bmatrix}, \tag{14.127}$$

where

$$\mathbf{\Delta}(z) = \operatorname{diag} \begin{bmatrix} 1 & z^{-1} & \cdots & z^{-(L-1)} \end{bmatrix}. \tag{14.128}$$

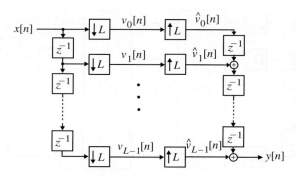

Figure 14.21: A simple L-channel perfect reconstruction multirate system.

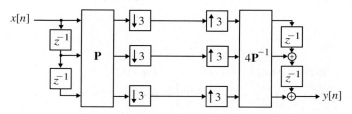

Figure 14.22: A three-channel analysis/synthesis filter bank.

Making use of Eq. (14.128) in Eq. (14.126), we arrive at the desired result, after some algebra:

$$\mathbf{H}(z) = \mathbf{D}^\dagger \mathbf{\Delta}(z)\mathbf{E}^t(z^L),   \tag{14.129}$$

where $\mathbf{D}$ is the $L \times L$ DFT matrix.

14.4.4 Condition for Perfect Reconstruction

If the polyphase component matrices of Figure 14.20 satisfy the relation

$$\mathbf{R}(z)\mathbf{E}(z) = c\mathbf{I},   \tag{14.130}$$

where $\mathbf{I}$ is an $L \times L$ identity matrix and c is a constant, the structure of Figure 14.20 reduces to the one shown in Figure 14.21. Comparing Figure 14.21 with Figure 14.18, we note that the former can be considered as a special case of an L-channel QMF bank if we set

$$H_k(z) = z^{-k}, \qquad G_k(z) = z^{-(L-1-k)}, \qquad 0 \le k \le L-1.   \tag{14.131}$$

Substituting Eq. (14.131) in Eq. (14.110), we arrive at

$$a_\ell(z) = \frac{1}{L} \sum_{k=0}^{L-1} z^{-k} W_L^{-\ell k} z^{-(L-1-k)} = z^{-(L-1)} \left(\frac{1}{L} \sum_{k=0}^{L-1} W_L^{-\ell k} \right).   \tag{14.132}$$

From Eqs. (13.9) and (13.10), it follows that $a_0(z) = 1$ and $a_\ell(z) = 0$ for $\ell \neq 0$. Hence, from Eq. (14.114), we note that $T(z) = z^{-(L-1)}$, or in other words, the structure of Figure 14.20 is a perfect reconstruction L-channel QMF bank if the condition of Eq. (14.130) is satisfied.

It should be noted that a more general condition for perfect reconstruction is given by

$$\mathbf{R}(z)\mathbf{E}(z) = cz^{-K}\mathbf{I}, \tag{14.133}$$

where K is a nonnegative integer.

The analysis and synthesis filters of a perfect reconstruction filter bank of the form of Figure 14.20 can be easily determined, as illustrated in Example 14.8.

EXAMPLE 14.8   A Simple Perfect Reconstruction Three-Channel Analysis/Synthesis Filter Bank

The three-channel analysis/synthesis filter bank of Figure 14.22 is by construction a perfect reconstruction filter bank. The output of the filter bank is simply $y[n] = dx[n - 2]$. Moreover, here, $\mathbf{E}(z^3) = \mathbf{P}$ and $\mathbf{R}(z^3) = d\mathbf{P}^{-1}$. Consider

$$\mathbf{P} = \begin{bmatrix} 1 & 1 & 1 \\ 1 & -1 & 1 \\ 1 & 0 & -1 \end{bmatrix}.$$

From Eqs. (14.121) and (14.122a) to (14.122c), we thus get

$$\begin{bmatrix} H_0(z) \\ H_1(z) \\ H_2(z) \end{bmatrix} = \begin{bmatrix} 1 & 1 & 1 \\ 1 & -1 & 1 \\ 1 & 0 & -1 \end{bmatrix} \begin{bmatrix} 1 \\ z^{-1} \\ z^{-2} \end{bmatrix},$$

resulting in the analysis filters

$$H_0(z) = 1 + z^{-1} + z^{-2}, \qquad H_1(z) = 1 - z^{-1} + z^{-2}, \qquad H_2(z) = 1 - z^{-2}.$$

Let $d = 4$. Then,

$$d\mathbf{P}^{-1} = \begin{bmatrix} 1 & 1 & 2 \\ 2 & -2 & 0 \\ 1 & 1 & -2 \end{bmatrix}.$$

Using Eqs. (14.124) and (14.125a) to (14.125c), we arrive at

$$\begin{bmatrix} G_0(z) \\ G_1(z) \\ G_2(z) \end{bmatrix} = \begin{bmatrix} z^{-2} & z^{-1} & 1 \end{bmatrix} \begin{bmatrix} 1 & 1 & 2 \\ 2 & -2 & 0 \\ 1 & 1 & -2 \end{bmatrix},$$

leading to the synthesis filters

$$G_0(z) = 1 + 2z^{-1} + z^{-2}, \quad G_1(z) = 1 - 2z^{-1} + z^{-2}, \quad G_2(z) = -2 + 2z^{-2}.$$

Now, for a given L-channel analysis filter bank, the polyphase matrix $\mathbf{E}(z)$ is known. From Eq. (14.130), it therefore follows that a perfect reconstruction L-channel QMF bank can be simply designed by constructing a synthesis filter bank with a polyphase matrix $\mathbf{R}(z) = [\mathbf{E}(z)]^{-1}$. In general, it is not easy to compute the inverse of a rational $L \times L$ matrix. An alternative elegant approach is to design the analysis filter bank with an invertible polyphase matrix. For example, $\mathbf{E}(z)$ can be chosen to be a paraunitary matrix satisfying the condition

$$\tilde{\mathbf{E}}(z)\mathbf{E}(z) = c\mathbf{I}, \text{ for all } z, \tag{14.134}$$

where $\tilde{\mathbf{E}}(z)$ is the paraconjugate of $\mathbf{E}(z)$ given by the transpose of $\mathbf{E}(z^{-1})$, with each coefficient replaced by its conjugate and choosing $\mathbf{R}(z) = \tilde{\mathbf{E}}(z)$.

For the design of a perfect reconstruction FIR L-channel QMF bank, the matrix $\mathbf{E}(z)$ can be expressed in a product form [Vai89]

$$\mathbf{E}(z) = \mathbf{E}_R(z)\mathbf{E}_{R-1}(z) \cdots \mathbf{E}_1(z)\mathbf{E}_0, \tag{14.135}$$

where $\mathbf{E}_0$ is a constant unitary matrix, and

$$\mathbf{E}_\ell(z) = \mathbf{I} - \mathbf{v}_\ell[\mathbf{v}_\ell^*]^t + z^{-1}\mathbf{v}_\ell[\mathbf{v}_\ell^*]^t, \tag{14.136}$$

in which $\mathbf{v}_\ell$ is a column vector of order L with unit norm, that is, $[\mathbf{v}_\ell^*]^T \mathbf{v}_\ell = 1$. With this constraint on $\mathbf{E}(z)$, one can set up an appropriate objective function that can then be minimized to arrive at a set of L analysis filters meeting the desired passband and stopband specifications. To this end, a suitable objective function is given by

$$\Phi = \sum_{k=0}^{L-1} \int_{k\text{th stopband}} \left| H_k(e^{j\omega}) \right|^2 d\omega. \tag{14.137}$$

The optimization parameters are the elements of $\mathbf{v}_\ell$ and $\mathbf{E}_0$.

EXAMPLE 14.9   Design of an FIR Perfect Reconstruction Three-Channel QMF Bank

We consider the design of a three-channel FIR perfect reconstruction QMF bank with a passband of width $\pi/3$ [Vai93]. The passband of the lowpass analysis filter $H_0(z)$ is from 0 to $\pi/3$, the passband of the bandpass analysis filter $H_1(z)$ is from $\pi/3$ to $2\pi/3$, and the passband of the highpass analysis filter $H_2(z)$ is from $2\pi/3$ to π. The objective function to be minimized here is thus of the form

$$\Phi = \int_{\frac{\pi}{3}+\varepsilon}^{\pi} \left| H_0(e^{j\omega}) \right|^2 d\omega + \int_0^{\frac{\pi}{3}-\varepsilon} \left| H_1(e^{j\omega}) \right|^2 d\omega + \int_{\frac{2\pi}{3}+\varepsilon}^{\pi} \left| H_1(e^{j\omega}) \right|^2 d\omega + \int_0^{\frac{2\pi}{3}-\varepsilon} \left| H_2(e^{j\omega}) \right|^2 d\omega.$$

The impulse response coefficients of the analysis filters $\{h_k[n]\}$, $k = 0, 1, 2$, of length 15 obtained by minimizing Φ are given in [Vai93]. The gain responses of these filters are plotted in Figure 14.23. It should be noted that the coefficients of the corresponding synthesis filters are given by $g_k[n] = h_k[14 - n]$, $k = 0, 1, 2$.

14.5   Multilevel Filter Banks

A multichannel analysis/synthesis filter bank can be developed by iterating a two-channel QMF bank. Moreover, if the two-band QMF bank is of the perfect reconstruction type, the generated multiband structure also exhibits the perfect reconstruction property (Problems 14.56 and 14.57). In this section, we consider this approach.

14.5.1   Filter Banks with Equal Passband Widths

By inserting a two-channel maximally decimated QMF bank in each channel of another two-channel maximally decimated QMF bank between the down-sampler and the up-sampler, we can generate a four-channel maximally decimated QMF bank, as shown in Figure 14.24. Since the analysis and the synthesis filter banks are formed like a tree, the overall system is usually called a *tree-structured filter bank*. It should be noted that in the four-channel tree-structured filter bank of Figure 14.24, the 2 two-channel QMF banks in the second level do not have to be identical. However, if they are different QMF banks with different analysis and synthesis filters, to compensate for the unequal gains and unequal delays of the 2 two-channel systems, additional delays of appropriate values need to be inserted at the middle to ensure perfect reconstruction of the overall four-channel system.

An equivalent representation of the four-channel QMF system of Figure 14.24 is shown in Figure 14.25. The analysis and synthesis filters in the equivalent representation are related to those of the parent two-level tree-structured filter bank as follows:

$$H_{00}(z) = H_0(z)H_0(z^2), \qquad H_{01}(z) = H_0(z)H_1(z^2), \tag{14.138a}$$

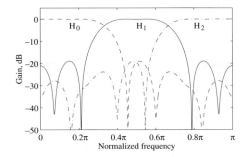

Figure 14.23: Gain responses of the three-channel FIR QMF bank of Example 14.9.

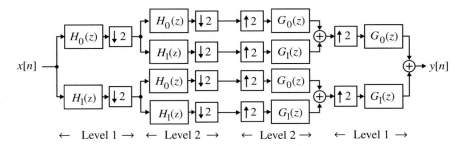

$$\leftarrow \text{ Level 1 } \rightarrow \quad \leftarrow \text{ Level 2 } \rightarrow \quad \leftarrow \text{ Level 2 } \rightarrow \quad \leftarrow \text{ Level 1 } \rightarrow$$

Figure 14.24: A two-level four-channel maximally decimated QMF structure.

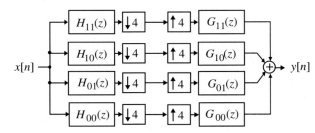

Figure 14.25: An equivalent representation of the four-channel QMF structure of Figure 14.24.

$$H_{10}(z) = H_1(z)H_0(z^2), \qquad H_{11}(z) = H_1(z)H_1(z^2), \qquad (14.138b)$$
$$G_{00}(z) = G_0(z)G_0(z^2), \qquad G_{01}(z) = G_0(z)G_1(z^2). \qquad (14.138c)$$
$$G_{10}(z) = G_1(z)G_0(z^2), \qquad G_{11}(z) = G_1(z)G_1(z^2). \qquad (14.138d)$$

EXAMPLE 14.10   Design of a Tree-Structured Four-Channel QMF Bank

We illustrate the design of a four-channel QMF bank by iterating the two-channel QMF bank based on the filter 12B of Johnston [Joh80].

From the filter's impulse response given in Example 14.4, we compute the impulse response of each of the four analysis filters using Eqs. (14.138a) and (14.138b) and then determine the gain responses of each using Program 14_1. Figure 14.26 shows the gain responses of the four analysis filters of the final four-channel QMF bank.

Program 14_1.m

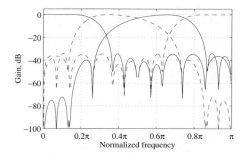

Figure 14.26: Gain responses of the four analysis filters of Example 14.10.

From Eqs. (14.138a) and (14.138b), it can be seen that each analysis filter $H_{r,s}(z)$ is a cascade of two filters, one with a single passband and a single stopband and the other with two passbands and two stopbands. The passband of the cascade is the frequency range where the passbands of the two filters overlap. On the other hand, the stopband of the cascade is formed from three different frequency ranges. In two of the frequency ranges, the passband of one coincides with the stopband of the other, while in the third range, the two stopbands overlap. As a result, the gain responses of the cascade in the three regions of the stopband are not equal, resulting in an uneven stopband attenuation characteristic. This type of behavior of the gain response can also be seen in Figure 14.26 and should be taken into account in the design of the tree-structured filter bank.

By continuing the process of inserting a two-channel maximally decimated QMF bank, QMF banks with more than four channels can be easily constructed. It should be noted that the number of channels resulting from this approach is restricted to a power of 2; that is, $L = 2^\nu$. In addition, as illustrated by Figure 14.26, the filters in the analysis (synthesis) branch have passbands of equal width, given by π/L. However, by a simple modification to the approach we can design QMF banks with analysis (synthesis) filters having passbands of unequal width, as described next.

14.5.2 Filter Banks with Unequal Passband Widths

Consider the two-channel maximally decimated QMF bank of Figure 14.27(a). By inserting another two-channel maximally decimated QMF bank in the top subband channel between the down-sampler and the up-sampler at the position marked by a $*$, we arrive at a three-channel maximally decimated QMF bank, as shown in Figure 14.27(b). The equivalent representation of the generated three-channel filter bank is indicated in Figure 14.28(a), where the analysis and synthesis filters are given by

$$H_{00}(z) = H_0(z)H_0(z^2), \qquad H_{01}(z) = H_0(z)H_1(z^2), \qquad H_1(z),$$
$$G_{00}(z) = G_0(z)G_0(z^2), \qquad G_{01}(z) = G_0(z)G_1(z^2), \qquad G_1(z). \qquad (14.139)$$

Typical magnitude responses of the analysis filters of the two-channel QMF bank of Figure 14.27(a) and that of the derived three-channel filter of Figure 14.27(b) are sketched in Figure 14.29(a) and (b), respectively.

We can continue this process and generate a four-channel QMF bank from the three-channel QMF bank of Figure 14.27(b) by inserting a two-channel QMF bank in the top subband channel at the position marked by a $*$, resulting in the structure of Figure 14.27(c). Its equivalent representation is indicated in Figure 14.28(b), where

$$H_{000}(z) = H_0(z)H_0(z^2)H_0(z^4), \qquad H_{001}(z) = H_0(z)H_0(z^2)H_1(z^4),$$

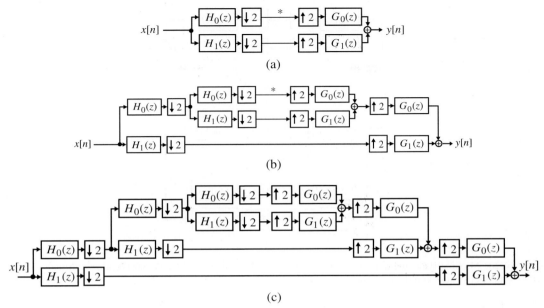

Figure 14.27: (a) A two-channel QMF bank, (b) a three-channel QMF bank derived from the two-channel QMF bank, and (c) a four-channel QMF bank derived from the three-channel QMF bank.

$$H_{01}(z) = H_0(z)H_1(z^2), \qquad H_1(z),$$
$$G_{000}(z) = G_0(z)G_0(z^2)G_0(z^4), \quad G_{001}(z) = G_0(z)G_0(z^2)G_1(z^4),$$
$$G_{01}(z) = G_0(z)G_1(z^2), \qquad G_1(z). \tag{14.140}$$

Figure 14.29(c) shows typical magnitude responses of the analysis (synthesis) filters of the four-channel QMF bank of Figure 14.27(c) derived from a parent two-channel QMF bank, with magnitude responses as indicated in Figure 14.29(a).

Because of the unequal passband widths of the analysis and synthesis filters, these structures belong to the class of nonuniform QMF banks. The tree-structured filter banks of Figure 14.27 are also referred to as *octave band QMF banks* [Fli94]. Various other types of nonuniform filter banks can be generated by iterating branches of a parent uniform two-channel QMF in different forms. Nonuniform filter banks have been used in speech and image coding applications.

14.6 Discrete Wavelet Transform[3]

There is a unique relation between the parameters of a class of perfect reconstruction octave band QMF banks and certain discrete wavelet transforms. We develop in this section this relation for a few well-known discrete wavelet transforms.

The function of an octave-band analysis filter bank with down-sampling, also called the *binary tree,* can be considered as a transformation of the input sequence to a set of sub-sequences at the output of the down-samplers. Consider for example a four-level binary tree as shown in Figure 14.30(a), where the parent analysis filters, $H_0(z)$ and $H_1(z)$, are the analysis filters of a perfect reconstruction two-channel FIR

---

[3]Portions of this section have been adapted from [Kin2001] by permission of the author.

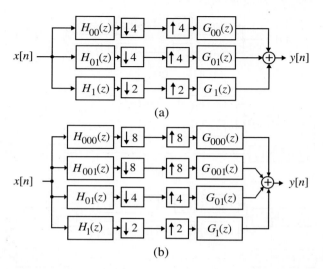

$$(a)$$

$$(b)$$

Figure 14.28: Maximally decimated QMF banks with unequal passband width analysis (synthesis) filters: (a) three-channel system and (b) four-channel system

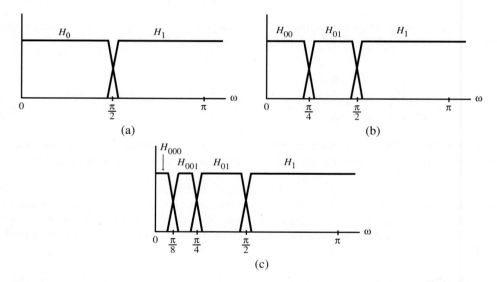

Figure 14.29: Magnitude responses of the analysis filters of a (a) two-channel QMF bank, (b) three-channel QMF bank derived from a two-channel QMF bank, and (c) four-channel QMF bank derived from a three-channel QMF bank.

QMF bank. If the input to this filter bank is a finite-length sequence $x[n]$ of length M, then the output $u_0[n]$ of the highpass analysis filter at the first level is of length $M/2$. Next, the output $u_1[n]$ of the highpass analysis filter at the second level is of length $M/4$. Continuing this process, we arrive at the outputs of the analysis filters at all levels as shown below along with their lengths.

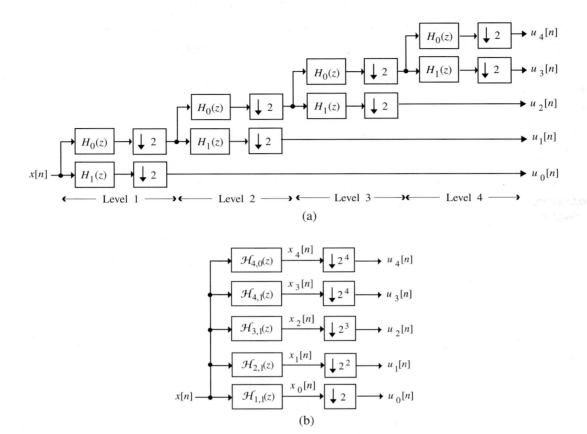

Figure 14.30: (a) A four-level octave-band analysis filter bank and its (b) equivalent representation.

Signal	No. of Samples	Approximate Bandwidth
$x[n]$	M	$[0, \pi]$
$u_0[n]$	$\frac{M}{2}$	$[\frac{\pi}{2}, \pi]$
$u_1[n]$	$\frac{M}{4}$	$[\frac{\pi}{4}, \frac{\pi}{2}]$
$u_2[n]$	$\frac{M}{8}$	$[\frac{\pi}{8}, \frac{\pi}{4}]$
$u_3[n]$	$\frac{M}{16}$	$[\frac{\pi}{16}, \frac{\pi}{8}]$
$u_4[n]$	$\frac{M}{16}$	$[0, \frac{\pi}{16}]$

It can be seen from the data given above, the total number of output samples is M, independent of the number of levels of the overall octave-band analysis filter bank. The lowpass analysis filter $H_0(z)$ has an approximate passband of width $[0, \frac{\pi}{2}]$, and the highpass analysis filter $H_1(z)$ has an approximate passband of width $[\frac{\pi}{2}, \pi]$. The lowpass and highpass filters at the second level have passbands of approximate widths given by $[0, \frac{\pi}{4}]$ and $[\frac{\pi}{4}, \frac{\pi}{2}]$, respectively. In a similar manner, the approximate widths of the analysis filters at each succeeding levels can be determined. The above table shows the approximate bandwidths of all signals at the outputs of the down-samplers of the four-level binary tree of Figure 14.30(a). An equivalent representation with a single down-sampler in each channel is as indicated in Figure 14.30(b). The transfer functions from the input of the four-level octave-band analysis filter bank to each output (before down-sampling) are given by

$$\mathcal{H}_{1,1}(z) = H_1(z),$$
$$\mathcal{H}_{2,1}(z) = H_0(z)H_1(z^2),$$
$$\mathcal{H}_{3,1}(z) = H_0(z)H_0(z^2)H_1(z^4),$$
$$\mathcal{H}_{4,1}(z) = H_0(z)H_0(z^2)H_0(z^4)H_1(z^8),$$
$$\mathcal{H}_{4,0}(z) = H_0(z)H_0(z^2)H_0(z^4)H_0(z^8). \tag{14.141}$$

In the general case, the transfer functions to the two outputs at level k of the binary tree before down-sampling are given by

$$\mathcal{H}_{k,1}(z) = \left[\prod_{i=0}^{k-2} H_0(z^{2^i})\right] H_1(z^{2^{k-1}}), \tag{14.142a}$$

$$\mathcal{H}_{k,0}(z) = \prod_{i=0}^{k-1} H_0(z^{2^i}). \tag{14.142b}$$

14.6.1   Wavelets

The process of generating the set of output sub-sequences $u_k[n]$, $0 \le k \le L$, from the finite-length input sequence $x[n]$ using the analysis filters of an L-level octave-band perfect reconstruction QMF bank is known as the *discrete wavelet transform* (DWT).

More precisely, the *wavelet* at level k is related to the impulse response $\hbar_{k,1}[n]$ of the filter $\mathcal{H}_{k,1}(z)$ of Eq. (14.142a) for the kth bandpass output. The frequency response of the wavelet at level k given by $\mathcal{H}_{k,1}(e^{j\omega})$. The basic impulse response $\hbar_{1,1}[n]$ given by the inverse-z transform of $\mathcal{H}_{1,1}(z) = H_1(z)$ is called the *mother wavelet*. The impulse response $\hbar_{k,0}[n]$ of the lowpass filter $\mathcal{H}_{k,0}(z)$ of Eq. (14.142b) is called the *scaling function* at level k.

The input–output of the kth channel before down-sampling in the z-domain of an L-level octave-band perfect reconstruction QMF bank is related through

$$X_k(z) = \mathcal{H}_{k,1}(z)X(z), \qquad 0 \le k \le L-1,$$
$$X_k(z) = \mathcal{H}_{L,0}(z)X(z), \qquad k = L.$$

or equivalently, in the time domain, through

$$x_k[n] = \sum_{m=-\infty}^{\infty} \hbar_{k,1}[n-m]x[m], \qquad 0 \le k \le L-1, \tag{14.143}$$

$$x_k[n] = \sum_{m=-\infty}^{\infty} \hbar_{k,0}[n-m]x[m], \qquad k = L. \tag{14.144}$$

The equivalent representation of the kth bandpass channel of the binary analysis tree, for $0 \le k \le L-1$, is a factor-of-2^{k+1} decimator with a decimation filter $\mathcal{H}_{k,1}(z)$ and that of the L-th lowpass channel is a factor-of-2^L decimator with a decimation filter $\mathcal{H}_{L,0}(z)$. Hence, using Eq. (13.16), we arrive at the DWT

$u_k[n], 0 \le k \le L$, of $x[n]$ given by [Vai93]

$$u_k[n] = \sum_{m=-\infty}^{\infty} \hbar_{k,1}[2^{k+1}n - m]x[m], \qquad 0 \le k \le L-1, \qquad (14.145a)$$

$$u_k[n] = \sum_{m=-\infty}^{\infty} \hbar_{k,0}[2^k n - m]x[m], \qquad k = L. \qquad (14.145b)$$

The reverse process of reconstructing the output sequence $y[n]$, which is a replica of the input $x[n]$, from the sub-sequences $u_k[n], 1 \le k \le L$, using the synthesis filters of the L-level octave-band perfect reconstruction QMF bank is the *inverse discrete wavelet transform* (IDWT). The synthesis binary tree corresponding to the analysis binary tree of Figure 14.30(a) is shown in Figure 14.31(a). An equivalent representation of Figure 14.31(a) is given in Figure 14.31(b). The transfer functions from each input (after up-sampling) of the L-level octave-band synthesis filter bank to the output are given by

$$\mathcal{G}_{1,1}(z) = G_1(z),$$
$$\mathcal{G}_{2,1}(z) = G_0(z)G_1(z^2),$$
$$\mathcal{G}_{3,1}(z) = G_0(z)G_0(z^2)G_1(z^4),$$
$$\mathcal{G}_{4,1}(z) = G_0(z)G_0(z^2)G_0(z^4)G_1(z^8),$$
$$\mathcal{G}_{4,0}(z) = G_0(z)G_0(z^2)G_0(z^4)G_0(z^8). \qquad (14.146)$$

In the general case, the transfer functions to the output from the two inputs at level k of the binary tree after up-sampling are given by

$$\mathcal{G}_{k,1}(z) = \left[\prod_{i=0}^{k-2} G_0(z^{2^i})\right] G_1(z^{2^{k-1}}), \qquad (14.147a)$$

$$\mathcal{G}_{k,0}(z) = \prod_{i=0}^{k-1} G_0(z^{2^i}). \qquad (14.147b)$$

Under the perfect reconstruction condition, the output of the L-channel synthesis filter bank is identical to the input of the analysis filter bank; that is., $y[n] = x[n]$. In the z-domain, we can express the output in terms of the inputs to the synthesis filter bank (after up-sampling) through [Vai93]

$$Y(z) = \mathcal{G}_{1,1}(z)U_0(z^2) + \mathcal{G}_{2,1}(z)U_1(z^{2^2}) + \cdots$$
$$+ \mathcal{G}_{L,1}(z)U_{L-1}(z^{2^L}) + \mathcal{G}_{L,0}(z)U_L(z^{2^L}). \qquad (14.148)$$

Here, the equivalent representation of the kth bandpass channel, $0 \le k \le L-1$, of the synthesis binary tree is a factor-of-2^{k+1} interpolator with an interpolation filter $\mathcal{G}_{k,1}(z)$ and that of the Lth lowpass channel is a factor-of-2^L interpolator with an interpolation filter $\mathcal{G}_{L,0}(z)$. Making use of Eq. (13.18), we can develop the inverse-z transform in Eq. (14.148) expressing the output $y[n]$ of the synthesis binary tree in the time domain as a function of its input $u_k[n]$ [Vai93]:

$$y[n] = \sum_{k=0}^{L-1} \sum_{m=-\infty}^{\infty} g_{k,1}[n - 2^{k+1}m]u_k[m] + \sum_{m=-\infty}^{\infty} g_{L,0}[n - 2^L m]u_L[m], \qquad (14.149)$$

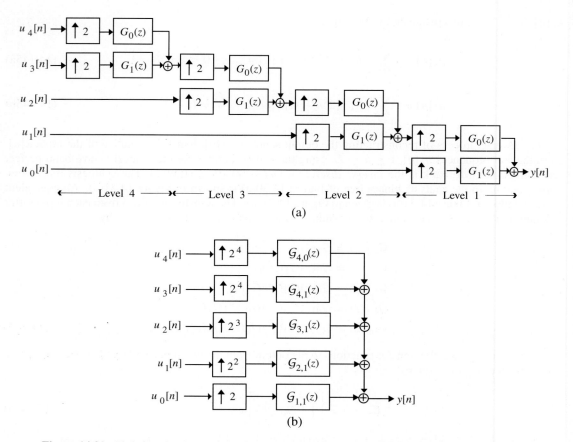

Figure 14.31: (a) A four-level octave-band synthesis filter bank and its (b) equivalent representation.

where $g_{k,\ell}[n]$ is the impulse response of $\mathcal{G}_{k,\ell}(z)$, where $\ell = 0, 1$. Using the notation

$$\eta_{k,m}[n] = g_{k,1}[n - 2^{k+1}m], \qquad 0 \le k \le L - 1, \tag{14.150a}$$

$$\eta_{L,m}[n] = g_{L,0}[n - 2^L m], \tag{14.150b}$$

we can rewrite Eq. (14.149) as

$$y[n] = \sum_{k=0}^{L} \sum_{m=-\infty}^{\infty} \eta_{k,m}[n] u_k[m], \tag{14.151}$$

which is the expression for the inverse discrete wavelet transform. In Eq. (14.151), $u_k[m]$ are the wavelet coefficients of $x[n]$ with respect to the basis functions $\eta_{k,m}[n]$.

We next review a few well-known wavelet transforms.

14.6.2 Biorthogonal Wavelets

The *biorthogonal wavelets* are generated from an octave-band filter bank designed from a two-channel biorthogonal perfect reconstruction QMF bank. The simplest type of biorthogonal wavelets are the *Haar*

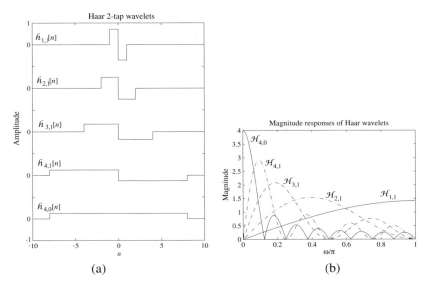

Figure 14.32: (a) Impulse responses and (b) magnitude responses of the four-level tree of Haar filters.

wavelets which are generated using the Haar filters of Eq. (14.31). In the case of a four-level octave-band analysis filter bank, the transfer functions of the analysis filters in the tree are given by

$$\mathcal{H}_{1,1}(z) = \frac{1}{\sqrt{2}}(1 - z^{-1}),$$

$$\mathcal{H}_{2,1}(z) = \frac{1}{2}(1 + z^{-1} - z^{-2} - z^{-3}),$$

$$\mathcal{H}_{3,1}(z) = \frac{1}{2\sqrt{2}}(1 + z^{-1} + z^{-2} + z^{-3} - z^{-4} - z^{-5} - z^{-6} - z^{-7}),$$

$$\mathcal{H}_{4,1}(z) = \frac{1}{4}(1 + z^{-1} + z^{-2} + z^{-3} + z^{-4} + z^{-5} + z^{-6} + z^{-7}$$
$$-z^{-8} - z^{-9} - z^{-10} - z^{-11} - z^{-12} - z^{-13} - z^{-14} - z^{-15}),$$

$$\mathcal{H}_{4,0}(z) = \frac{1}{4}(1 + z^{-1} + z^{-2} + z^{-3} + z^{-4} + z^{-5} + z^{-6} + z^{-7}$$
$$+z^{-8} + z^{-9} + z^{-10} + z^{-11} + z^{-12} + z^{-13} + z^{-14} + z^{-15}). \qquad (14.152)$$

The corresponding impulse responses are given in Table 14.1. Figure 14.32 shows the impulse responses (shown as continuous plots) and the magnitude responses of the four-level tree of Haar filters. It can be seen that the shape of the basic impulse response $h_{1,1}[n]$ is almost independent of scale and is the mother wavelet.

The Haar wavelets exhibit abrupt transitions at the middle and at both ends. These transitions often lead to visible blocking artifacts in the images reconstructed from their compressed versions. As a result, wavelets with gradual transitions are usually preferred for image compression applications. A few such wavelets are discussed in the following.

The *LeGall 3/5-tap wavelets* are generated from the analysis filters of Eqs. (14.81a) and (14.82a). The impulse responses (shown as continuous plots) and the magnitude responses of the four-level tree of LeGall 3/5-tap filters are shown in Figures 14.33 and 14.34, respectively. The scaling function $h_{4,0}[n]$, shown

Table 14.1: Haar wavelets.

n	$h_{1,1}[n]$	$h_{2,1}[n]$	$h_{3,1}[n]$	$h_{4,1}[n]$	$h_{4,0}[n]$
0	$\frac{1}{\sqrt{2}}$	$\frac{1}{2}$	$\frac{1}{2\sqrt{2}}$	$\frac{1}{4}$	$\frac{1}{4}$
1	$-\frac{1}{\sqrt{2}}$	$\frac{1}{2}$	$\frac{1}{2\sqrt{2}}$	$\frac{1}{4}$	$\frac{1}{4}$
2	0	$-\frac{1}{2}$	$\frac{1}{2\sqrt{2}}$	$\frac{1}{4}$	$\frac{1}{4}$
3	0	$-\frac{1}{2}$	$\frac{1}{2\sqrt{2}}$	$\frac{1}{4}$	$\frac{1}{4}$
4	0	0	$-\frac{1}{2\sqrt{2}}$	$\frac{1}{4}$	$\frac{1}{4}$
5	0	0	$-\frac{1}{2\sqrt{2}}$	$\frac{1}{4}$	$\frac{1}{4}$
6	0	0	$-\frac{1}{2\sqrt{2}}$	$\frac{1}{4}$	$\frac{1}{4}$
7	0	0	$-\frac{1}{2\sqrt{2}}$	$\frac{1}{4}$	$\frac{1}{4}$
8	0	0	$-\frac{1}{2\sqrt{2}}$	$\frac{1}{4}$	$\frac{1}{4}$
9	0	0	0	$-\frac{1}{4}$	$\frac{1}{4}$
10	0	0	0	$-\frac{1}{4}$	$\frac{1}{4}$
11	0	0	0	$-\frac{1}{4}$	$\frac{1}{4}$
12	0	0	0	$-\frac{1}{4}$	$\frac{1}{4}$
13	0	0	0	$-\frac{1}{4}$	$\frac{1}{4}$
14	0	0	0	$-\frac{1}{4}$	$\frac{1}{4}$
15	0	0	0	$-\frac{1}{4}$	$\frac{1}{4}$

in Figure 14.33, converges to a pure triangular pulse, after many levels in which case the wavelets are basically determined by the superposition of two triangular pulses.

Unlike the Haar wavelets, the LeGall 3/5-tap wavelets do not exhibit abrupt discontinuites. Also, the sidelobes of their frequency responses are much smaller and narrower. As a result, they are more appropriate for image compression applications than the Haar wavelets. However, their inverse wavelets, obtained using the synthesis filters of Eqs. (14.81b) and (14.82b), are not very good when used for image reconstruction from the compressed data. As can be seen from the plots of the impulse responses and magnitude responses of the *LeGall 5/3-tap wavelets* shown in Figures 14.35, these wavelets do not converge to a smooth function after many levels, and their corresponding frequency responses have much larger sidelobes. It has been suggested that a better approach would be to use the LeGall 5/3-tap wavelets in the analysis branch for the decomposition and use the smoother LeGall 3/5-tap wavelets for the reconstruction.

14.6.3 Orthogonal Wavelets

The *orthogonal wavelets* are generated from an octave-band filter bank designed from a two-channel orthogonal perfect reconstruction QMF bank. An example of orthogonal wavelets are the Daubechies 4/4-tap wavelets generated from the analysis filters of Eqs. (14.84a) and (14.85a). The impulse responses (shown as continuous plots) and the magnitude responses of the four-level tree of Daubeches 4/4-tap filters are shown in Figure 14.36.

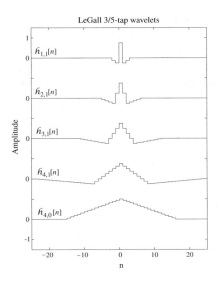

Figure 14.33: Impulse responses of the four-level tree of LeGall 3/5-tap filters.

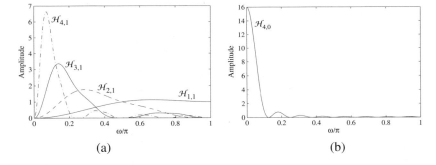

Figure 14.34: Magnitude responses of the four-level tree of LeGall 3/5-tap filters.

The Daubechies 4/4-tap wavelets and the scaling function have nonlinear phases and are not as smooth as the LeGall 3/5-tap wavelets. In addition, the magnitude responses of the former exhibit much higher sidelobes. Smoother orthogonal wavelets can be generated using higher-order Daubechies filters. However, these wavelets also have nonlinear phase and have been found not very suitable for image compression due to the artifacts in the reconstructed images, particularly around sharp edges.

14.7 Summary

This chapter considered the application of the concepts introduced in Chapter 13. In particular, it considered the analysis and design of analysis and synthesis banks and their polyphase representations. It then considered the analysis and design of the so-called quadrature-mirror filter (QMF) bank that is formed by a combination of an analysis filter bank with down-sampled outputs, followed by a synthesis filter bank with up-sampled inputs. Conditions satisfied by the analysis and synthesis filters for an alias-free operation of the QMF bank are derived. Several types of QMF banks are defined, and their design equations are

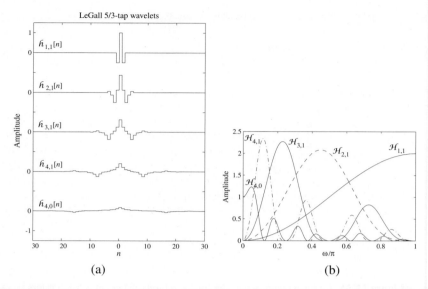

Figure 14.35: (a) Impulse responses and (b) magnitude responses of the four-level tree of LeGall 5/3-tap filters.

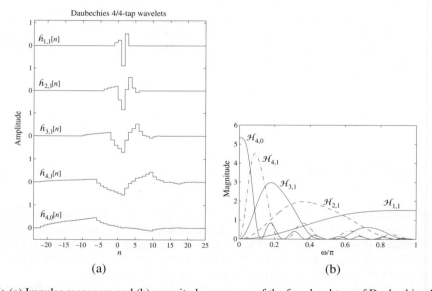

Figure 14.36: (a) Impulse responses and (b) magnitude responses of the four-level tree of Daubechies 4/4-tap filters.

developed. Finally, the concept of discrete wavelet transform was developed from the analysis of tree-structured multilevel QMF banks, and several popular discrete wavelets were developed.

Several other applications of multirate discrete-time systems are outlined in Chapter 15. These include transmultiplexers for signal conversion between frequency-division multiplex (FDM) to time-division multiplex (TDM) communication systems (Section 15.9), discrete multitone transmission (Section 15.10), oversampling A/D conversion (Section 15.11), and oversampling D/A conversion (Section 15.12).

14.8 Problems

14.1 By multiplying the numerator and the denominator polynomials of the transfer function $H(z) = \frac{a+bz^{-1}}{1+cz^{-1}}$, $|c| < 1$, with an appropriate polynomial $P(z)$, develop a two-band and a three-band polyphase decomposition of $H(z)$.

14.2 Using the method of Eq. (14.13), develop a two-band polyphase decomposition of each of the following IIR transfer functions:

(a) $H_a(z) = \dfrac{a + bz^{-1}}{1 + cz^{-1}}$, $|c| < 1$,   (b) $H_b(z) = \dfrac{3 - 4z^{-1} + 2.1z^{-2}}{1 - 0.8z^{-1} + 0.7z^{-2}}$,

(c) $H_c(z) = \dfrac{4 + 2.5z^{-1} - 3.5z^{-2} + 2z^{-3}}{(1 + 0.2z^{-1})(1 - 0.6z^{-1} + 0.9z^{-2})}$.

14.3 Using the method of Eq. (14.13), develop a three-band polyphase decomposition of each of the following IIR transfer functions:

(a) $H_1(z) = \dfrac{3 - 4z^{-1}}{1 - 0.5z^{-1}}$,   (b) $H_2(z) = \dfrac{4 + 2.1z^{-1} - 3.4z^{-2}}{1 - 0.8z^{-1} + 0.6z^{-2}}$.

14.4 Consider the two-channel synthesis filter bank structure of Figure P14.1, where $H_0(z)$ is a length-6 FIR filter with a transfer function given by

$$H_0(z) = h[0] + h[1]z^{-1} + h[2]z^{-2} + h[3]z^{-3} + h[4]z^{-4} + h[5]z^{-5}. \tag{14.153}$$

If $H_1(z) = H_0(-z)$, develop a realization of this filter bank with five delays and six multipliers.

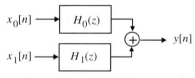

Figure P14.1

14.5 Consider the two-channel synthesis filter bank structure of Figure P14.1, where $H_0(z)$ is an FIR filter of the form of Eq. (14.153). If $H_1(z)$ has a transfer function that is the mirror image of $H_0(z)$, i.e., $H_1(z) = z^{-5}H_0(z^{-1})$, develop a realization of this filter bank with only six multipliers.

14.6 The four-channel synthesis filter bank of Figure P14.2(a), where $\mathbf{D}$ is a 4×4 DFT matrix, is characterized by the set of four transfer functions: $G_i(z) = Y(z)/X_i(z)$, $i = 0, 1, 2, 3$. Let the transfer functions of the four subfilters be given by $R_0(z) = 1 + 3.7z^{-1} + 1.7z^{-2}$, $R_1(z) = 4 - 0.9z^{-1} + 2.3z^{-2}$, $R_2(z) = 1 + 0.3z^{-1} - 0.8z^{-2}$, $R_3(z) = 2 - 1.5z^{-1} + 3.1z^{-2}$.

(a) Determine the expressions for the four transfer functions, $G_i(z)$, $0 \leq i \leq 3$.

(b) Assume that the synthesis filter $G_1(z)$ has a magnitude response as indicated in Figure P14.2(b). Sketch the magnitude responses of the other three analysis filters.

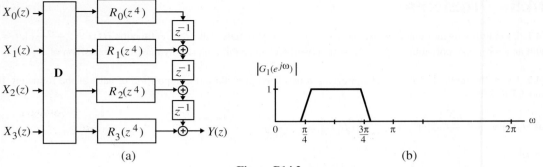

(a)   (b)

Figure P14.2

14.7 Consider the analysis–synthesis filter bank of Figure P14.3. Develop the input–output relation of this structure in the z-domain. Let $H_0(z) = (1 + z^{-1})/2$ and $H_1(z) = (1 - z^{-1})/2$. Determine the synthesis filters $G_0(z)$ and $G_1(z)$ so that the structure of Figure P14.3 is a perfect reconstruction filter bank.

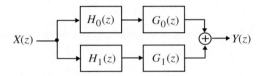

Figure P14.3

14.8 Let the analysis filters $H_0(z)$ and $H_1(z)$ of the structure of Figure P14.3 be power-complementary FIR filters of order N each.

(a) Show that this structure becomes a perfect reconstruction filter bank if the synthesis filters $G_0(z)$ and $G_1(z)$ are chosen as

$$G_0(z) = z^{-N} H_0(z^{-1}), \qquad G_1(z) = z^{-N} H_1(z^{-1}). \qquad (14.154)$$

(b) Show that the synthesis filters are causal FIR filters if the analysis filters are causal.

(c) Show that the analysis and synthesis filters satisfying the perfect reconstruction condition cannot all be of linear phase.

14.9 (a) Decompose the third-order IIR transfer function

$$G(z) = \frac{0.0985(1 + z^{-1})^3}{(1 - 0.1584z^{-1})(1 - 0.4189z^{-1} + 0.3554z^{-2})},$$

in the form

$$G(z) = \tfrac{1}{2}\{\mathcal{A}_0(z) + \mathcal{A}_1(z)\},$$

where $\mathcal{A}_0(z)$ and $\mathcal{A}_1(z)$ are stable allpass transfer functions.

(b) Realize $G(z)$ as a parallel connection of allpass filters with $\mathcal{A}_0(z)$ and $\mathcal{A}_1(z)$ realized with the fewest number of multipliers.

(c) Determine the transfer function $H(z)$, which is power-complementary to $G(z)$.

(d) Sketch the magnitude responses of $G(z)$ and $H(z)$.

14.10 Repeat Problem 14.9 for the following third-order IIR transfer function:

$$G(z) = \frac{0.1868(1 + 1.0902z^{-1} + 1.0902z^{-2} + z^{-3})}{(1 - 0.3628z^{-1})(1 - 0.5111z^{-1} + 0.7363z^{-2})}.$$

14.11 Show that the poles of a real-coefficient odd order bounded real elliptic lowpass transfer function with a symmetric numerator and satisfying the power-symmetric condition lie on the imaginary axis in the z-plane.

14.12 Let $G_a(s)$ denote the Nth-order analog lowpass Butterworth transfer function with a 3-dB cutoff frequency at 1 rad/sec with N odd. Show that the corresponding digital Butterworth transfer function $H_0(z)$ obtained by a bilinear transformation is a half-band lowpass filter expressible in the form

$$H_0(z) = \tfrac{1}{2}\{\mathcal{A}_0(z^2) + z^{-1}\mathcal{A}_1(z^2)\},$$

where $\mathcal{A}_0(z)$ and $\mathcal{A}_1(z)$ are stable allpass transfer functions.

14.13 Show that the two-channel QMF bank of Figure 14.8, in general, is a linear, time-varying system with a period of 2.

14.14 Using the method of Problem 14.12, develop the transfer function of a third-order lowpass half-band filter $H_0(z)$, and then determine its power-complementary transfer function $H_1(z)$. Develop a realization of a magnitude-preserving two-channel QMF bank whose analysis filters are $H_0(z)$ and $H_1(z)$, using no more than one multiplier for the analysis stage.

14.15 Using the method of Problem 14.12, develop the transfer function of a fifth-order lowpass half-band filter $H_0(z)$, and then determine its power-complementary transfer function $H_1(z)$. Develop a realization of a magnitude-preserving two-channel QMF bank whose analysis filters are $H_0(z)$ and $H_1(z)$, using no more than two multipliers for the analysis stage.

14.16 Let the order of the half-band lowpass IIR filter $H_0(z)$ of the QMF bank of Figure 14.8 be N, where N is odd.

(a) What is the total number of multipliers needed to implement the QMF bank of Figure 14.8? How many multiplications per second are needed to implement this structure?

(b) If the QMF bank is of the magnitude-preserving type, the analysis and synthesis filters can be realized as a sum of IIR allpass filters in the form of Figure 14.11. What is the total number of multipliers needed in the implementation of Figure 14.11? How many multiplications per second are needed to implement this structure?

14.17 The four filters of the two-channel QMF bank are given by $H_0(z) = 4z^{-2}$, $H_1(z) = z^{-1}$, $G_0(z) = 0.5z^{-1}$, and $G_1(z) = z^{-2}$. Show that the QMF bank is a perfect reconstruction system.

14.18 The transfer functions of the lowpass analysis and synthesis filters of a two-channel QMF bank are given by $H_0(z) = a + bz^{-1} + cz^{-2}$ and $G_0(z) = d + ez^{-1} + fz^{-2} + gz^{-3} + hz^{-4}$. Determine the expressions for the transfer functions of the highpass analysis and synthesis filters $H_1(z)$ and $G_1(z)$ to ensure aliasing cancellation.

14.19 Consider a two-channel perfect reconstruction filter bank with analysis filters $H_0(z)$ and $H_1(z)$ and synthesis filters $G_0(z)$ and $G_1(z)$. If the filters $G_0(z)$ and $G_1(z)$ are used as analysis filters and the filters $H_0(z)$ and $H_1(z)$ are used as synthesis filters, is the two-channel filter bank still a perfect reconstruction system? Justify your answer.

14.20 In a two-channel biorthogonal filter bank, can the linear-phase FIR filters $H_0(z)$ and $G_0(z)$ be of even and odd lengths, respectively? Can $H_0(z)$ and $G_0(z)$ be symmetric and antisymmetric, respectively? Justify your answers?

14.21 The general form of the expression for the polynomial $R(z)$ for a normalized zero-phase half-band lowpass product filter $P(z)$ of Eq. (14.74) for $m = 3$ is $R(z) = az^2 + bz + c + bz^{-1} + az^{-2}$. Determine the values of the constants a, b, and c. Using this polynomial, determine the expressions for the transfer functions of the *Daubechies 6/6-tap filter pairs* $H_0(z)$ and $H_1(z)$.

14.22 Consider a two-channel orthogonal filter bank structure of Figure 14.8, where the analysis filter $H_0(z)$ is a power-symmetric FIR transfer function of odd order N. If the second analysis filter $H_1(z)$ is chosen according to Eq. (14.90), show that both analysis filters can be realized using only $N + 1$ multipliers and $2N$ two-input adders.

14.23 Show that a two-channel perfect reconstruction, orthogonal filter bank cannot be designed with linear-phase real-coefficient FIR analysis and synthesis filters, except for the Haar filters.

14.24 The highpass filter $H_1(z)$ of a two-channel perfect reconstruction orthogonal filter bank is given by $H_1(z) = a + bz^{-1} + cz^{-2} + dz^{-3}$. Determine the expressions for the remaining three filters, $H_0(z)$, $G_0(z)$, and $G_1(z)$.

14.25 Show that the two-channel QMF bank of Figure 14.8 is a perfect reconstruction system for the following analysis and synthesis filters: $H_0(z) = 3 + 4z^{-1}$, $H_1(z) = 1 + 2z^{-1}$, $G_0(z) = -0.5 + z^{-1}$, and $G_1(z) = 1.5 - 2z^{-1}$.

14.26 The lowpass analysis filter of a two-channel QMF bank is given by $H_0(z) = a + bz^{-1} + cz^{-2} + dz^{-3} + ez^{-4} + fz^{-5}$. Determine the highpass analysis filter $H_1(z)$, and the two synthesis filters $G_0(z)$ and $G_1(z)$ so that the two-channel QMF bank is an orthogonal filter bank.

14.27 A cascaded lattice realization of the analysis filter part of a two-channel orthogonal QMF bank is as shown in Figure P14.4. Develop the cascaded lattice realization of its synthesis filter part.

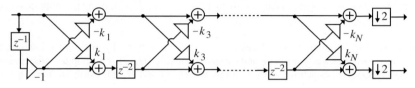

Figure P14.4

14.28 Develop a cascaded lattice realization of the orthogonal perfect reconstruction QMF bank characterized by Eqs. (14.87), (14.88), (14.89a), and (14.89b).

14.29 The analysis filter of a paraunitary perfect reconstruction QMF bank is given by $H_0(z) = 1 + 3z^{-1} + 14z^{-2} + 22z^{-3} - 12z^{-4} + 4z^{-5}$.
    (a) Determine the transfer function of the analysis filter $H_1(z)$.
    (b) Develop a cascaded lattice realization of the QMF bank.

14.30 Repeat Problem 14.29 for the analysis transfer function $H_0(z) = \frac{1}{2} - z^{-1} + \frac{21}{2}z^{-2} - \frac{27}{2}z^{-3} - 5z^{-4} - \frac{5}{2}z^{-5}$.

14.31 Consider the two-channel analysis filter bank shown in Figure P14.5, where the ℓth two-pair is characterized by a transfer matrix

$$\mathbf{T}_\ell = \begin{bmatrix} 1 & k_\ell \\ k_\ell & 1 \end{bmatrix}.$$

    (a) Develop a lattice realization of the above two-pair.
    (b) Let $P_m(z)$ and $Q_m(z)$ denote the transfer functions $R_m(z)/X(z)$ and $S_m(z)/X(z)$, respectively, for $0 \le m \le M$. Express transfer functions $P_m(z)$ and $Q_m(z)$ as functions of $P_{m-1}(z)$ and $Q_{m-1}(z)$.
    (c) If $Q_m(z) = z^{-(2m+1)} P_m(z^{-1})$, show that the two analysis transfer functions $H_0(z)$ and $H_1(z)$ are linear-phase.
    (d) Develop a lattice realization of the synthesis filter bank to ensure perfect reconstruction.

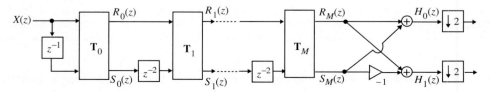

Figure P14.5

14.32 Develop the lattice analysis and synthesis filter structures for linear phase FIR two-channel QMF banks for each pair of analysis transfer functions that follow:

(a) $H_0(z) = 3 + 6z^{-1} + 27z^{-2} + 27z^{-3} + 6z^{-4} + 3z^{-5}$, $H_1(z) = -1 - 2z^{-1} - 3z^{-2} + 3z^{-3} + 2z^{-4} + z^{-5}$,

(b) $H_0(z) = -1 + 2.5z^{-1} + 4.5z^{-2} + 4.5z^{-3} + 2.5z^{-4} - z^{-5}$, $H_1(z) = 3 - 7.5z^{-1} - 31.5z^{-2} + 31.5z^{-3} + 7.5z^{-4} - 3z^{-5}$,

(c) $H_0(z) = 3 - 7.5z^{-1} + 6.75z^{-2} + 6.75z^{-3} - 7.5z^{-4} + 3z^{-5}$, $H_1(z) = -1 + 2.5z^{-1} - 5.25z^{-2} + 5.25z^{-3} - 2.5z^{-4} + z^{-5}$.

14.33 Show that, in general, the L-channel QMF bank of Figure 14.18 is a linear, time-varying system with a period of L.

14.34 Consider a two-channel QMF bank with a lowpass analysis filter $H_0(z) = E_0(z^2) + z^{-1}E_1(z^2)$ of degree N.

(a) Express the Type 1 polyphase component matrix $\mathbf{E}(z)$ as a function of the polyphase components $E_0(z)$ and $E_1(z)$ if the highpass analysis filter is given by $H_1(z) = H_0(-z)$.

(b) Repeat Part (a) if $H_1(z) = z^{-N}H_0(z^{-1})$.

14.35 Consider the two-channel QMF structure of Figure 14.11 where $\mathcal{A}_i(z), i = 0, 1$, are stable allpass transfer functions. Let $\mathbf{E}(z)$ and $\mathbf{R}(z)$ denote the polyphase matrices of the analysis and synthesis filter banks.

(a) Determine the expressions for $\mathbf{E}(z)$ and $\mathbf{R}(z)$ in terms of $\mathcal{A}_i(z)$.

(b) Is $\mathbf{E}(z)$ lossless?

(c) Express $\mathbf{R}(z)$ in terms of $\mathbf{E}(z)$.

(d) What is the product $\mathbf{R}(z)\mathbf{E}(z)$?

14.36 The Type I polyphase component matrix of the analysis filters of a two-channel QMF bank is given by

$$\mathbf{E}(z) = \begin{bmatrix} -2z^{-2} + 8z^{-1} - 6 & 2 - z^{-1} \\ -3 + 2z^{-1} & 1 \end{bmatrix}.$$

(a) Determine the transfer functions $H_0(z)$ and $H_1(z)$ of the two analysis filters and the transfer functions $G_0(z)$ and $G_1(z)$ of the two synthesis filters for designing a perfect reconstruction system.

(b) Derive the corresponding modulation matrix $\mathbf{H}^{(m)}(z)$.

14.37 The three-channel analysis/synthesis filter bank of Figure 14.22 is, by construction, a perfect reconstruction filter bank. For

$$\mathbf{P} = \begin{bmatrix} 2 & -2 & 1 \\ 3 & 1 & -2 \\ 2 & 4 & 2 \end{bmatrix},$$

determine the transfer functions of the analysis and synthesis filters.

14.38 Consider the QMF bank structure of Figure 14.20 with $L = 4$. Let the Type II polyphase component matrix be given by

$$\mathbf{R}(z) = \begin{bmatrix} 1 & -2 & 3 & -1 \\ 2 & 1 & 1 & 0 \\ 0 & 3 & -1 & 1 \\ 1 & 1 & -1 & 2 \end{bmatrix}.$$

Determine the Type I polyphase component matrix $\mathbf{E}(z)$ such that the four-channel QMF structure is a perfect reconstruction system with an input–output relation $y[n] = 4x[n-3]$.

14.39 The analysis filters of a three-channel QMF bank are given by

$$\begin{bmatrix} H_0(z) & H_1(z) & H_2(z) \end{bmatrix} = \begin{bmatrix} z^{-2} & z^{-1} & 1 \end{bmatrix} \begin{bmatrix} 1 & 2 & 2 \\ 3 & -3 & 1 \\ 2 & 1 & 2 \end{bmatrix}.$$

Determine the synthesis filters for implementing a perfect reconstruction filter bank.

14.40 Determine the synthesis filters of a four-channel perfect reconstruction QMF bank whose analysis filters are given by

$$
\begin{bmatrix} H_0(z) \\ H_1(z) \\ H_2(z) \\ H_3(z) \end{bmatrix} = \begin{bmatrix} 2 & 1 & 1 & 2 \\ 3 & 1 & -1 & -2 \\ 3 & 1 & 2 & -1 \\ 1 & 3 & 1 & 2 \end{bmatrix} \begin{bmatrix} 1 \\ z^{-1} \\ z^{-2} \\ z^{-3} \end{bmatrix}.
$$

14.41 Determine the analysis filters of a three-channel perfect reconstruction QMF bank with synthesis filters given by

$$
\begin{bmatrix} G_0(z) \\ G_1(z) \\ G_2(z) \end{bmatrix} = \begin{bmatrix} -2 & 3 & 2 \\ 3 & 1 & -2 \\ 2 & -1 & 1 \end{bmatrix} \begin{bmatrix} 1 \\ z^{-1} \\ z^{-2} \end{bmatrix}.
$$

14.42 Consider an alias-free L-channel maximally decimated QMF bank with $H_k(z)$ and $G_k(z)$, $0 \le k \le L-1$, denoting, respectively, the analysis and the synthesis filters. Let $T(z)$ denote its distortion transfer function. Show that if the analysis and the synthesis filters of each branch are interchanged, that is, $G_k(z)$ are now the analysis filters and $H_k(z)$ are the synthesis filters, the resulting system is still alias-free and has the same distortion transfer function.

14.43 Let the analysis filter bank of the L-channel QMF bank of Figure 14.20 be as shown in Figure P14.6, where the L-input, L-output blocks are characterized by constant $L \times L$ nonsingular matrices $\mathbf{S}_\ell$, $0 \le \ell \le K$. Determine the structure for the synthesis bank so that it results in a perfect reconstruction system.

14.44 Consider the analysis filter bank of Figure P14.6 with $L = 3$ and $K = 1$. The 3×3 constant matrices $\mathbf{S}_0$ and $\mathbf{S}_0$ are given by

$$
\mathbf{S_0} = \begin{bmatrix} 3 & -5 & 3 \\ -1 & 3 & -2 \\ 0 & -1 & 1 \end{bmatrix}, \qquad \mathbf{S_1} = \begin{bmatrix} 1 & -2 & 3 \\ 0 & 1 & -2 \\ 0 & 0 & 1 \end{bmatrix}.
$$

(a) Determine the transfer functions of the three analysis filters $H_0(z)$, $H_1(z)$, and $H_2(z)$ and the three synthesis filters $G_0(z)$, $G_1(z)$, and $G_2(z)$ for perfect reconstruction.

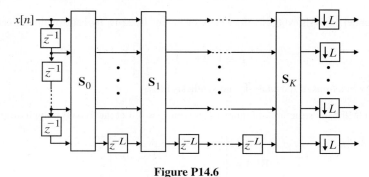

Figure P14.6

14.45 Consider the L-channel QMF structure of Figure 14.20. Denote $\mathbf{P}(z) = \mathbf{R}(z)\mathbf{E}(z)$. Show that this structure is time-invariant with no aliasing if and only if $\mathbf{P}(z)$ is a pseudocirculant matrix of the form [Vai88b]

$$
\mathbf{P} = \begin{bmatrix} P_0(z) & P_1(z) & \cdots & P_{L-2}(z) & P_{L-1}(z) \\ z^{-1}P_{L-1}(z) & P_0(z) & \cdots & P_{L-3}(z) & P_{L-2}(z) \\ \vdots & \vdots & \ddots & \vdots & \vdots \\ z^{-1}P_2(z) & z^{-1}P_3(z) & \cdots & P_0(z) & P_1(z) \\ z^{-1}P_1(z) & z^{-1}P_2(z) & \cdots & z^{-1}P_{L-1}(z) & P_0(z) \end{bmatrix}. \tag{14.155}
$$

14.46 Consider the two-channel QMF bank of Figure 14.8, with the analysis and synthesis filters chosen according to Eqs. (14.43a) and (14.43b) for alias-free realization. Determine its Type I and Type II polyphase component matrices $\mathbf{E}(z)$ and $\mathbf{R}(z)$. Show that $\mathbf{P}(z) = \mathbf{R}(z)\mathbf{E}(z)$ is a pseudocirculant matrix as defined in Eq. (14.155).

14.47 Consider the following power-symmetric FIR transfer functions.
  (a) $H_0(z) = 1 - 2z^{-1} + 4.5z^{-2} + 6z^{-3} + z^{-4} + 0.5z^{-5}$,
  (b) $H_0(z) = 1 + \frac{1}{2}z^{-1} + \frac{15}{4}z^{-2} - z^{-4} + 2z^{-5}$.
  Using $H_0(z)$ as one of the analysis filters, determine the remaining three filters of the corresponding two-channel orthogonal filter bank. In each case, show that the filter bank is alias-free and satisfies the perfect reconstruction condition.

14.48 The analysis filters of a biorthogonal two-channel filter bank are given by $H_0(z) = 1 + az^{-1} + z^{-2}$ and $H_1(z) = 1 + az^{-1} + bz^{-2} + az^{-3} + z^{-4}$ [Vet89]. Determine the two synthesis filters $G_0(z)$ and $G_1(z)$ using Eqs. (14.72a) and (14.72b), with $c = 2$ and $k = \ell$. Show that the two-channel filter bank is alias-free and satisfies the perfect reconstruction property with $a \neq 0$ and $b \neq 2$.

14.49 Design a two-channel perfect reconstruction filter bank such that the lowpass analysis filter $H_0(z)$ is of length 4 and has two zeros at $z = -1$. Is it possible to design the analysis filters so that they have linear phase in addition to having two zeros at $z = -1$?

14.50 The transfer function of a zero-phase *Lagrange half-band filter* with $4K - 1$ coefficients is given by [Ans91]

$$H_K(z) = \frac{1}{2} + \sum_{n=1}^{K} h_K[2n - 1]\left(z^{2n-1} + z^{-(2n-1)}\right), \tag{14.156}$$

where the coefficients are determined using the *Lagrange interpolation formula* and are given by

$$h_K[2n - 1] = \frac{(-1)^{n+K-1}\prod_{i=1}^{2K}(K - i + \frac{1}{2})}{(K - n)!(K - 1 + n)!(2n - 1)}, \qquad 1 \leq n \leq K. \tag{14.157}$$

These filters have zeros of multiplicity $2K$ at $z = -1$ and $2K - 2$ zeros situated in a mirror-image symmetry with respect to the unit circle. The frequency response of a Lagrange half-band filter is thus nonnegative and can be used to design perfect reconstruction two-channel FIR QMF banks by appropriately factorizing $H_K(z)$. As the transfer function has a large number of zeros at $z = -1$, it is highly suitable for the design of both biorthogonal and orthogonal wavelets.
  (a) Develop the expression for the transfer function $H_4(z)$.
  (b) Design a perfect reconstruction two-channel biorthogonal QMF bank by factorizing the above Lagrange half-band transfer function in the form of $H_4(z) = H_0(z)G_0(z)$, where $H_0(z)$ and $G_0(z)$ are linear-phase factors.
  (c) Design a perfect reconstruction two-channel orthogonal QMF bank by factorizing the above Lagrange half-band transfer function in the form of $H_4(z) = H_0(z)H_0(z^{-1})$.
  (d) Show that the above orthogonal solution is a scaled version of Daubechies 4-tap filter.

14.51 Given the length-11 zero-phase half-band maximally flat filter $P(z)$ of Problem 14.21,
  (a) find all factors of the filters $H_0(z)$ and $G_0(z)$ for designing a two-channel perfect reconstruction orthogonal filter bank,
  (b) find all factors of the filters $H_0(z)$ and $G_0(z)$ resulting in a symmetric design with even length of a two-channel perfect reconstruction filter bank,
  (c) find all factors of the filters $H_0(z)$ and $G_0(z)$ resulting in a symmetric design with odd length of a two-channel perfect reconstruction filter bank.

14.52 Show that the Haar wavelets at different scales are orthogonal and the scaling function is orthogonal to the wavelets.

14.53 Given the two-channel perfect reconstruction filter bank of Problem 14.51, find the wavelets and the scaling function for six scales. Discuss the smoothness of wavelet and scaling functions as the number of zeros at $\omega = \pi$ for $H_0(z)$ and $G_0(z)$.

14.54 We have demonstrated in Example 13.3 that the multirate structure of Figure 13.6 is a perfect reconstruction system, with the output $y[n]$ being a replica of the input $x[n]$ but delayed by one sample. Figure P14.7(a) shows the structure obtained from Figure 13.6 using the *lifting scheme* [Swe96]. Show that it is also a perfect reconstruction system.

14.55 The lifting scheme can be repeatedly applied to develop perfect reconstruction systems with more desired features. Figure P14.7(b) shows a structure derived from Figure P14.7(a) by applying the lifting scheme a second time. Show that this structure is also a perfect reconstruction system.

14.56 Show that the four-channel QMF bank of Figure 14.24 is a perfect reconstruction type if the two parent two-channel QMF banks are of the perfect reconstruction type.

14.57 Show that the three-channel and the four-channel QMF banks of Figure 14.27(b) and Figure 14.27(c), respectively, are of perfect reconstruction type if the two parent two-channel QMF banks of Figure 14.27(a) are of the perfect reconstruction type.

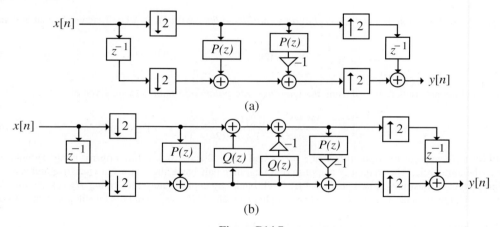

(a)

(b)

Figure P14.7

14.9 MATLAB Exercises

M 14.1 Using MATLAB, design a real-coefficient power-symmetric FIR lowpass filter $H_0(z)$ with a stopband edge at 0.6π and a minimum stopband attenuation of 30 dB. Design next a perfect reconstruction two-channel QMF bank based on $H_0(z)$. Show the transfer functions of all four filters. Plot the magnitude responses of the two analysis filters in the same figure.

M 14.2 Write a MATLAB program to design a two-channel QMF paraunitary lattice filter bank. Using this program design a lattice structure with filters of order 21 and a stopband edge at $\omega_s = 0.6\pi$. Plot the sum of the magnitude squares of the two analysis filters. What is the minimum stopband attenuation in dB of your filters? Plot the amplitude distortion in dB. Quantize the lattice coefficients to six decimal digits, and plot the gain responses of the two analysis filters along with those of the original filters on the same figure. Comment on your results.

M 14.3 Design and realize a four-channel uniform DFT analysis filter bank using a prototype linear-phase FIR filter of length 19. Design the prototype filter using the function `remez` of MATLAB. Assume a transition band of width 0.2π. Plot the magnitude responses of each filter on the same figure.

M 14.4 Design a three-channel QMF bank in the form of Figure 14.27(b) by iterating the two-channel QMF bank based on Filter 16A of Johnston [Ans93], [Cro83], [Joh80]. Plot the gain responses of the three analysis filters, $H_0(z)$, $H_1(z)$, and $H_2(z)$ on the same figure. Comment on your results.

M 14.5 From the analysis filter $H_0(z)$ of the two-channel orthogonal filter bank of Problem 14.51, determine and plot the impulse responses $h_{k,0}[n]$ and $h_{k,1}[n]$ of the four-level tree of Daubechies 6/6-tap wavelets. Also plot the magnitude responses of the transfer functions $\mathcal{H}_{k,0}(z)$ and $\mathcal{H}_{k,0}(z)$ of the four-level tree of Daubechies 6/6-tap wavelets.

15 Applications of Digital Signal Processing

As mentioned in Chapter 1, digital signal processing techniques are increasingly replacing conventional analog signal processing methods in many fields, such as speech analysis and processing, radar and sonar signal processing, biomedical signal analysis and processing, telecommunications, and geophysical signal processing. In this chapter, we include a few simple applications to provide a glimpse of the potential of DSP.

We first describe several applications of the discrete Fourier transform (DFT) introduced in Section 5.2. The first application considered is the detection of the frequencies of a pair of sinusoidal signals, called tones, employed in telephone signaling. Next, we discuss the use of the DFT in the determination of the spectral contents of a continuous-time signal. The effect of the DFT length and the windowing of the sequence are examined in detail here. In the following section, we introduce the concept of the short-time Fourier transform (STFT) and discuss its application in the spectral analysis of nonstationary signals. We then consider the spectral analysis of random signals using both nonparametric and parametric methods. Application of digital filtering methods to musical sound processing is considered next, and a variety of practical digital filter structures useful for the generation of certain audio effects, such as artificial reverberation, flanging, phasing, filtering, and equalization, are introduced. Generation of discrete-time analytic signals by means of a discrete-time Hilbert transformer is then considered, and several methods of designing these circuits are outlined along with an application. The basic concepts of signal compression are reviewed next, along with a technique for image compression based on Haar wavelets. The theory and design of transmultiplexers are discussed in the following section. One method of digital data transmission employing digital signal processing methods is then introduced. The basic concepts behind the design of the oversampling A/D and D/A converters are reviewed in the following two sections. Finally, we review the sparse antenna array design for ultrasound scanners.

15.1 Dual-Tone Multifrequency Signal Detection

Dual-tone multifrequency (DTMF) signaling, increasingly being employed worldwide with push-button telephone sets, offers a high dialing speed over the dial-pulse signaling used in conventional rotary telephone sets. In recent years, DTMF signaling has also found applications requiring interactive control, such as in voice mail, electronic mail (e-mail), telephone banking, and ATM machines.

A DTMF signal consists of a sum of two tones, with frequencies taken from two mutually exclusive groups of preassigned frequencies. Each pair of such tones represents a unique number or a symbol. Decoding of a DTMF signal thus involves identifying the two tones in that signal and determining their corresponding number or symbol. The frequencies allocated to the various digits and symbols of a push-button keypad are internationally accepted standards and are shown in Figure 1.35 [ITU84]. The four keys in the last column of the keypad, as shown in this figure, are not yet available on standard handsets and are reserved for future use. Since the signaling frequencies are all located in the frequency band used for speech transmission, this is an *in-band system*. Interfacing with the analog input and output devices is provided by *codec* (coder/decoder) chips or A/D and D/A converters (Sections 4.8 and 4.9).

Although a number of chips with analog circuitry are available for the generation and decoding of DTMF signals in a single channel, these functions can also be implemented digitally on DSP chips. Such a digital implementation surpasses analog equivalents in performance, since it provides better precision, stability, versatility, and reprogrammability to meet other tone standards and the scope for multichannel operation by time-sharing, leading to a lower chip count.

The digital implementation of a DTMF signal involves adding two finite-length digital sinusoidal sequences, with the latter simply generated by using look-up tables or by computing a polynomial expansion. The digital tone detection can be easily performed by computing the DFT of the DTMF signal and then measuring the energy present at the eight DTMF frequencies. The minimum duration of a DTMF signal is 40 ms. Thus, with a sampling rate of 8 kHz, there are at most $0.04 \times 8000 = 320$ samples available for decoding each DTMF digit. The actual number of samples used for the DFT computation is less than this number and is chosen so as to minimize the difference between the actual location of the sinusoid and the nearest integer value DFT index k.

The DTMF decoder computes the DFT samples closest in frequency to the eight DTMF fundamental tones and their respective second harmonics. In addition, a practical DTMF decoder also computes the DFT samples closest in frequency to the second harmonics corresponding to each of the fundamental tone frequencies. This latter computation is employed to distinguish between human voices and the pure sinusoids generated by the DTMF signal. In general, the spectrum of a human voice contains components at all frequencies including the second harmonic frequencies. On the other hand, the DTMF signal generated by the handset has negligible second harmonics. The DFT computation scheme employed is a slightly modified version of Goertzel's algorithm, as described in Section 11.3.1, for the computation of the squared magnitudes of the DFT samples that are needed for the energy computation.

The DFT length N determines the frequency spacing between the locations of the DFT samples and the time it takes to compute the DFT sample. A large N makes the spacing smaller, providing higher resolution in the frequency domain, but increases the computation time. The frequency f_k in Hz corresponding to the DFT index (bin number) k is given by

$$f_k = \frac{k F_T}{N}, \qquad k = 0, 1, \ldots, N - 1, \tag{15.1}$$

where F_T is the sampling frequency. If the input signal contains a sinusoid of frequency f_{in} different from that given above, its DFT will contain not only large-valued samples at values of k closest to $N f_{\text{in}}/F_T$ but also nonzero values at other values of k due to a phenomenon called leakage (see Example 11.16). To minimize the leakage, it is desirable to choose N appropriately so that the tone frequencies fall as close as possible to a DFT bin, thus providing a very strong DFT sample at this index value relative to all other values. For an 8-kHz sampling frequency, the best value of the DFT length N to detect the eight fundamental DTMF tones has been found to be 205 and that for detecting the eight second harmonics is 201 [Mar92]. Table 15.1 shows the DFT index values closest to each of the tone frequencies and their second harmonics for these two values of N, respectively. Figure 15.1 shows 16 selected DFT samples computed using a 205-point DFT of a length-205 sinusoidal sequence for each of the fundamental tone frequencies.

Program 15_1 can be used to demonstrate the DFT-based DTMF detection algorithm. The outputs generated by this program for the input symbol # are displayed in Figure 15.2.

Program 15_1.m

15.2   Spectral Analysis of Sinusoidal Signals

An important application of digital signal processing methods is in determining in the discrete-time domain the frequency contents of a continuous-time signal, more commonly known as *spectral analysis*. More

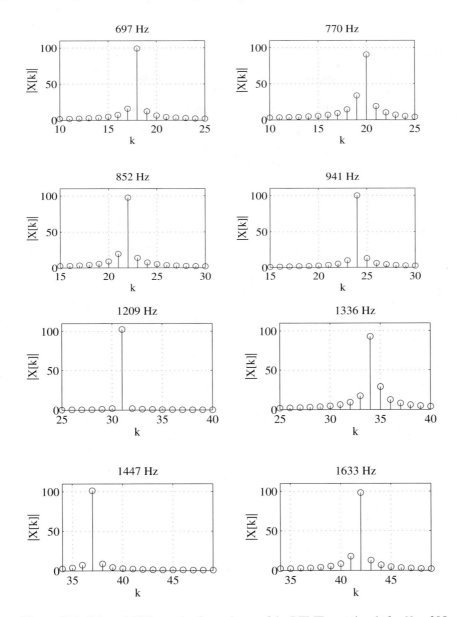

Figure 15.1: Selected DFT samples for each one of the DTMF tone signals for $N = 205$.

specifically, it involves the determination of either the energy spectrum or the power spectrum of the signal. Applications of digital spectral analysis can be found in many fields and are widespread. The spectral analysis methods are based on the following observation. If the continuous-time signal $g_a(t)$ is reasonably band-limited, the spectral characteristics of its discrete-time equivalent $g[n]$ should provide a good estimate of the spectral properties of $g_a(t)$. However, in most cases, $g_a(t)$ is defined for $-\infty < t < \infty$, and as a result, $g[n]$ is of infinite extent and defined for $-\infty < n < \infty$. Since it is difficult to evaluate the spectral

Table 15.1: DFT index values for DTMF tones for $N = 205$ and their second harmonics for $N = 201$ [Mar92].

Basic tone in Hz	Exact k value	Nearest integer k value	Absolute error in k
697	17.861	18	0.139
770	19.731	20	0.269
852	21.833	22	0.167
941	24.113	24	0.113
1209	30.981	31	0.019
1336	34.235	34	0.235
1477	37.848	38	0.152
1633	41.846	42	0.154
Second harmonic in Hz	**Exact k value**	**Nearest integer k value**	**Absolute error in k**
1394	35.024	35	0.024
1540	38.692	39	0.308
1704	42.813	43	0.187
1882	47.285	47	0.285
2418	60.752	61	0.248
2672	67.134	67	0.134
2954	74.219	74	0.219
3266	82.058	82	0.058

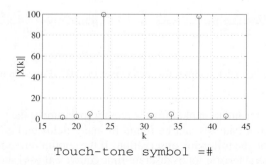

Touch-tone symbol =#

Figure 15.2: A typical output of Program 15_1.

parameters of an infinite-length signal, a more practical approach is as follows. First, the continuous-time signal $g_a(t)$ is passed through an analog anti-aliasing filter before it is sampled to eliminate the effect of aliasing. The output of the filter is then sampled to generate a discrete-time sequence equivalent $g[n]$. It is assumed that the anti-aliasing filter has been designed appropriately, and hence, the effect of aliasing can be ignored. Moreover, it is further assumed that the A/D converter wordlength is large enough so that the A/D conversion noise can be neglected.

This and the following two sections provide a review of some spectral analysis methods. In this section, we consider the Fourier analysis of a stationary signal composed of sinusoidal components. In Section 15.3, we discuss the Fourier analysis of nonstationary signals with time-varying parameters. Section 15.4 considers the spectral analysis of random signals. For a detailed exposition of spectral analysis and a concise review of the history of this area, see Kumaresan [Kum93].

For the spectral analysis of sinusoidal signals, we assume that the parameters characterizing the sinusoidal components, such as amplitudes, frequencies, and phase, do not change with time. For such a signal $g[n]$, the Fourier analysis can be carried out by computing its Fourier transform $G(e^{j\omega})$:

$$G(e^{j\omega}) = \sum_{n=-\infty}^{\infty} g[n]e^{-j\omega n}. \tag{15.2}$$

In practice, the infinite-length sequence $g[n]$ is first windowed by multiplying it with a length-N window $w[n]$ to make it into a finite-length sequence $\gamma[n] = g[n] \cdot w[n]$ of length N. The spectral characteristics of the windowed finite-length sequence $\gamma[n]$ obtained from its Fourier transform $\Gamma(e^{j\omega})$ then is assumed to provide a reasonable estimate of the Fourier transform $G(e^{j\omega})$ of the discrete-time signal $g[n]$. The Fourier transform $\Gamma(e^{j\omega})$ of the windowed finite-length segment $\gamma[n]$ is next evaluated at a set of $R(R \geq N)$ discrete angular frequencies equally spaced in the range $0 \leq \omega < 2\pi$ by computing its R-point discrete Fourier transform (DFT) $\Gamma[k]$. To provide sufficient resolution, the DFT length R is chosen to be greater than the window N by zero-padding the windowed sequence with $R - N$ zero-valued samples. The DFT is usually computed using an FFT algorithm.

We examine the above approach in more detail to understand its limitations so that we can properly make use of the results obtained. In particular, we analyze here the effects of windowing and the evaluation of the frequency samples of the Fourier transform via the DFT.

Before we can interpret the spectral content of $\Gamma(e^{j\omega})$, that is, $G(e^{j\omega})$, from $\Gamma[k]$, we need to reexamine the relations between these transforms and their corresponding frequencies. Now, the relation between the R-point DFT $\Gamma[k]$ of $\gamma[n]$ and its Fourier transform $\Gamma(e^{j\omega})$ is given by

$$\Gamma[k] = \Gamma(e^{j\omega})\Big|_{\omega=2\pi k/R}, \qquad 0 \leq k \leq R - 1. \tag{15.3}$$

The normalized discrete-time angular frequency ω_k corresponding to the DFT bin number k (DFT frequency) is given by

$$\omega_k = \frac{2\pi k}{R}. \tag{15.4}$$

Likewise, the continuous-time angular frequency Ω_k corresponding to the DFT bin number k (DFT frequency) is given by

$$\Omega_k = \frac{2\pi k}{RT}. \tag{15.5}$$

To interpret the results of the DFT-based spectral analysis correctly, we first consider the frequency-domain analysis of a sinusoidal sequence. Now an infinite-length sinusoidal sequence $g[n]$ of normalized angular frequency ω_o is given by

$$g[n] = \cos(\omega_o n + \phi). \tag{15.6}$$

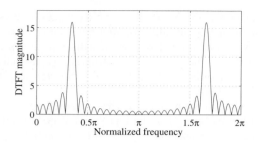

Figure 15.3: Fourier transform of a sinusoidal sequence windowed by a rectangular window.

By expressing the above sequence as

$$g[n] = \tfrac{1}{2}\left(e^{j(\omega_o n + \phi)} + e^{-j(\omega_o n + \phi)}\right) \tag{15.7}$$

and making use of Table 3.3, we arrive at the expression for its Fourier transform as

$$G(e^{j\omega}) = \pi \sum_{\ell=-\infty}^{\infty}\left(e^{j\phi}\delta(\omega - \omega_o + 2\pi\ell) + e^{-j\phi}\delta(\omega + \omega_o + 2\pi\ell)\right). \tag{15.8}$$

Thus, the Fourier transform is a periodic function of ω with a period 2π containing two impulses in each period. In the frequency range, $-\pi \le \omega < \pi$, there is an impulse at $\omega = \omega_o$ of complex amplitude $\pi e^{j\phi}$ and an impulse at $\omega = -\omega_o$ of complex amplitude $\pi e^{-j\phi}$.

To analyze $g[n]$ in the spectral domain using the DFT, we employ a finite-length version of the sequence given by

$$\gamma[n] = \cos(\omega_o n + \phi), \qquad 0 \le n \le N - 1. \tag{15.9}$$

The computation of the DFT of a finite-length sinusoid has been considered in Example 11.16. In this example, using Program 11_10, we computed the DFT of a length-32 sinusoid of frequency 10 Hz sampled at 64 Hz, as shown in Figure 11.32(a). As can be seen from this figure, there are only two nonzero DFT samples, one at bin $k = 5$ and the other at bin $k = 27$. From Eq. (15.5), bin $k = 5$ corresponds to frequency 10 Hz, while bin $k = 27$ corresponds to frequency 54 Hz, or equivalently, -10 Hz. Thus, the DFT has correctly identified the frequency of the sinusoid.

Next, using the same program, we computed the 32-point DFT of a length-32 sinusoid of frequency 11 Hz sampled at 64 Hz, as shown in Figure 11.32(b). This figure shows two strong peaks at bin locations $k = 5$ and $k = 6$, with nonzero DFT samples at other bin locations in the positive half of the frequency range. Note that the bin locations 5 and 6 correspond to frequencies 10 Hz and 12 Hz, respectively, according to Eq. (15.5). Thus the frequency of the sinusoid being analyzed is exactly halfway between these two bin locations.

The phenomenon of the spread of energy from a single frequency to many DFT frequency locations as demonstrated by Figure 11.32(b) is called *leakage*. To understand the cause of this effect, we recall that the DFT $\Gamma[k]$ of a length-N sequence $\gamma[n]$ is given by the samples of its discrete-time Fourier transform (Fourier transform) $\Gamma(e^{j\omega})$ evaluated at $\omega = 2\pi k/N, k = 0, 1, \ldots, N - 1$. Figure 15.3 shows the Fourier transform of the length-32 sinusoidal sequence of frequency 11 Hz sampled at 64 Hz. It can be seen that the DFT samples shown in Figure 11.32(b) are indeed obtained by the frequency samples of the plot of Figure 15.3.

To understand the shape of the Fourier transform shown in Figure 15.3, we observe that the sequence of Eq. (15.9) is a windowed version of the infinite-length sequence $g[n]$ of Eq. (15.6) obtained using a rectangular window $w[n]$:

$$w[n] = \begin{cases} 1, & 0 \leq n \leq N - 1, \\ 0, & \text{otherwise.} \end{cases} \tag{15.10}$$

Hence, the Fourier transform $\Gamma(e^{j\omega})$ of $\gamma[n]$ is given by the frequency-domain convolution of the Fourier transform $G(e^{j\omega})$ of $g[n]$ with the Fourier transform $\Psi_R(e^{j\omega})$ of the rectangular window $w[n]$:

$$\Gamma(e^{j\omega}) = \frac{1}{2\pi} \int_{-\pi}^{\pi} G(e^{j\varphi}) \Psi_R(e^{j(\omega-\varphi)}) \, d\varphi, \tag{15.11}$$

where

$$\Psi_R(e^{j\omega}) = e^{-j\omega(N-1)/2} \frac{\sin(\omega N/2)}{\sin(\omega/2)}. \tag{15.12}$$

Substituting $G(e^{j\omega})$ from Eq. (15.8) into Eq. (15.11), we arrive at

$$\Gamma(e^{j\omega}) = \tfrac{1}{2} e^{j\phi} \Psi_R(e^{j(\omega-\omega_o)}) + \frac{1}{2} e^{-j\phi} \Psi_R(e^{j(\omega+\omega_o)}). \tag{15.13}$$

As indicated by Eq. (15.13), the Fourier transform $\Gamma(e^{j\omega})$ of the windowed sequence $\gamma[n]$ is a sum of the frequency shifted and amplitude scaled Fourier transform $\Psi_R(e^{j\omega})$ of the window $w[n]$, with the amount of frequency shifts given by $\pm\omega_o$. Now, for the length-32 sinusoid of frequency 11 Hz sampled at 64 Hz, the normalized frequency of the sinusoid is $11/64 = 0.172$. Hence, its Fourier transform is obtained by frequency shifting the Fourier transform $\Psi_R(e^{j\omega})$ of a length-32 rectangular window to the right and to the left by the amount $0.172 \times 2\pi = 0.344\pi$, adding both shifted versions, and then amplitude scaling by a factor 1/2. In the normalized angular frequency range 0 to 2π, which is one period of the Fourier transform, there are two peaks, one at 0.344π and the other at $2\pi(1 - 0.172) = 1.656\pi$, as verified by Figure 15.3. A 32-point DFT of this Fourier transform is precisely the DFT shown in Figure 11.32(b). The two peaks of the DFT at bin locations $k = 5$ and $k = 6$ are frequency samples of the main lobe located on both sides of the peak at the normalized frequency 0.172. Likewise, the two peaks of the DFT at bin locations $k = 26$ and $k = 27$ are frequency samples of the main lobe located on both sides of the peak at the normalized frequency 0.828. All other DFT samples are given by the samples of the sidelobes of the Fourier transform of the window causing the leakage of the frequency components at $\pm\omega_o$ to other bin locations, with the amount of leakage determined by the relative amplitude of the main lobe and the sidelobes. Since the relative sidelobe level $A_{s\ell}$, defined by the ratio in dB of the amplitude of the main lobe to that of the largest sidelobe, of the rectangular window is very high, there is a considerable amount of leakage to the bin locations adjacent to the bins showing the peaks in Figure 11.32(b).

The above problem gets more complicated if the signal being analyzed has more than one sinusoid, as is typically the case. We illustrate the DFT-based spectral analysis approach by means of Examples 15.1 through 15.3. Through these examples, we examine the effects of the length R of the DFT, the type of window being used, and its length N on the results of spectral analysis.

EXAMPLE 15.1   Effect of the DFT Length on Spectral Analysis

The signal to be analyzed in the spectral domain is given by

$$x[n] = \tfrac{1}{2} \sin(2\pi f_1 n) + \sin(2\pi f_2 n), \qquad 0 \leq n \leq N - 1. \tag{15.14}$$

Let the normalized frequencies of the two length-16 sinusoidal sequences be $f_1 = 0.22$ and $f_2 = 0.34$. We compute the DFT of their sum $x[n]$ for various values of the DFT length R. To this end, we use Program 15_2 given in the Appendix whose input data are the length N of the signal, length R of the DFT, and the two frequencies f_1 and f_2. The program generates the two sinusoidal sequences, forms their sum, then computes the DFT of the sum

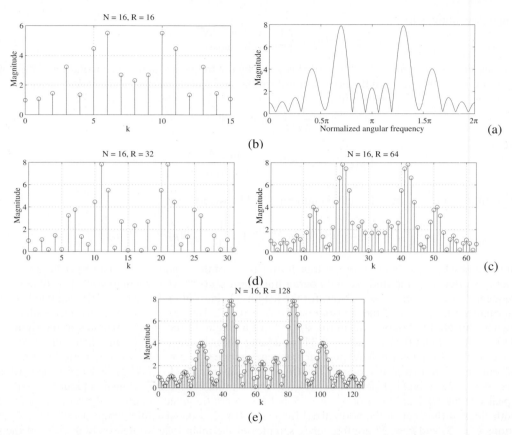

Figure 15.4: (a)–(e) DFT-based spectral analysis of a sum of two finite-length sinusoidal sequences of normalized frequencies 0.22 and 0.34, respectively, of length 16 each for various values of DFT lengths.

and plots the DFT samples. In this example, we fix $N = 16$ and vary the DFT length R from 16 to 128. Note that when $R > N$, the M-file fft(x,R) automatically zero-pads the sequence x with R−N zero-valued samples.

Figure 15.4(a) shows the magnitude $|X[k]|$ of the DFT samples of the signal $x[n]$ of Eq. (15.14) for $R = 16$. From the plot of the magnitude $|X(e^{j\omega})|$ of the Fourier transform given in Figure 15.4(b), it is evident that the DFT samples given in Figure 15.4(a) are indeed the frequency samples of the frequency response, as expected. As is customary, the horizontal axis in Figure 15.4(a) has been labeled in terms of the DFT frequency sample (bin) number k, where k is related to the normalized angular frequency ω through Eq. (15.4). Thus, $\omega = 2\pi \times 8/16 = \pi$ corresponds to $k = 8$, and $\omega = 2\pi \times 15/16 = 1.875\pi$ corresponds to $k = 15$.

From the plot of Figure 15.4(a), it is difficult to determine whether there is one or more sinusoids in the signal being examined and the exact locations of the sinusoids. To increase the accuracy of the locations of the sinusoids, we increase the size of the DFT to 32 and recompute the DFT, as indicated in Figure 15.4(c). In this plot, there appear to be some concentrations around $k = 7$ and around $k = 11$ in the normalized frequency range from 0 to 0.5. Figure 15.4(d) shows the DFT plot obtained for $R = 64$. In this plot, there are two clear peaks occurring at $k = 13$ and $k = 22$ that correspond to the normalized frequencies of 0.2031 and 0.3438, respectively. To improve further the accuracy of the peak location, we compute next a 128-point DFT, as indicated in Figure 15.4(e), in which the peak occurs around $k = 27$ and $k = 45$, corresponding to the normalized frequencies of 0.2109 and 0.3516, respectively. However, this plot also shows a number of minor peaks, and it is not clear by examining this DFT plot whether additional sinusoids of lesser strengths are present in the original signal or not.

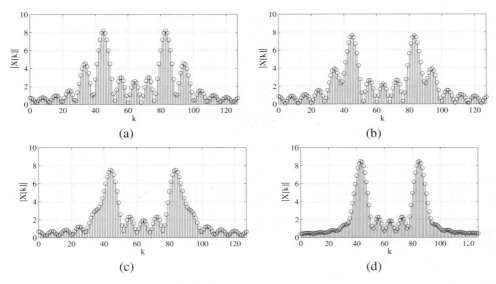

Figure 15.5: Illustration of the frequency resolution property: (a) $f_1 = 0.28$, $f_2 = 0.34$; (b) $f_1 = 0.29$, $f_2 = 0.34$; (c) $f_1 = 0.3$, $f_2 = 0.34$; and (d) $f_1 = 0.31$, $f_2 = 0.34$.

As Example 15.1 points out, in general, an increase in the DFT length improves the sampling accuracy of the Fourier transform by reducing the spectral separation of adjacent DFT samples.

EXAMPLE 15.2   Effect of Spectral Separation on the DFT of a Sum of Two Sinusoids

In this example, we compute the DFT of a sum of two finite-length sinusoidal sequences, as given by Eq. (15.14), with one of the sinusoids at a fixed frequency, while the frequency of the other sinusoid is varied. Specifically, we keep $f_2 = 0.34$ and vary f_1 from 0.28 to 0.31. The length of the signal being analyzed is 16, while the DFT length is 128.

Figure 15.5 shows the plots of the DFTs computed, along with the frequencies of the sinusoids obtained using Program 15.2. As can be seen from these plots, the two sinusoids are clearly resolved in Figures 15.5(a) and (b), while they cannot be resolved in Figures 15.5(c) and (d). The reduced resolution occurs when the difference between the two frequencies becomes less than 0.04.

As indicated by Eq. (15.11), the Fourier transform $\Gamma(e^{j\omega})$ of a length-N sinusoid of normalized angular frequency ω_1 is obtained by frequency translating the Fourier transform $\Psi_R(e^{j\omega})$ of a length-N rectangular window to the frequencies $\pm\omega_1$ and scaling their amplitudes appropriately. In the case of a sum of two length-N sinusoids of normalized angular frequencies ω_1 and ω_2, the Fourier transform is obtained by summing the Fourier transforms of the individual sinusoids. As the difference between the two frequencies becomes smaller, the main lobes of the Fourier transforms of the individual sinusoids get closer and eventually overlap. If there is a significant overlap, it will be difficult to resolve the peaks. It follows therefore that the frequency resolution is essentially determined by the main lobe Δ_{ML} of the Fourier transform of the window.

Program 15_2.m

Now from Table 10.2, the main lobe width Δ_{ML} of a length-N rectangular window is given by $4\pi/N$. In terms of normalized frequency, the main lobe width of a length-16 rectangular window is 0.125. Hence, two closely spaced sinusoids windowed with a rectangular window of length 16 can be clearly resolved if the difference in their frequencies is about half of the main lobe width, that is, 0.0625.

Even though the rectangular window has the smallest main lobe width, it has the largest relative sidelobe amplitude and, as a consequence, causes considerable leakage. As seen from Examples 15.1 and 15.2, the large amount of leakage results in minor peaks that may be falsely identified as sinusoids. We now study the effect of windowing the signal with a Hamming window.[1]

EXAMPLE 15.3   **Minimization of the Leakage Using a Tapered Window**

We compute the DFT of a sum of two sinusoids windowed by a Hamming window. The signal being analyzed is $x[n] \cdot w[n]$, where $x[n]$ is given by

$$x[n] = 0.85 \sin(2\pi f_1 n) + \sin(2\pi f_2 n),$$

and $w[n]$ is a Hamming window of length N. The two normalized frequencies are $f_1 = 0.22$ and $f_2 = 0.26$.

Figure 15.6(a) shows the 16-point DFT of the windowed signal with a window length of 16. As can be seen from this plot, the leakage has been reduced considerably, but it is difficult to resolve the two sinusoids. We next increase the DFT length to 64, while keeping the window length fixed at 16. The resulting plot is shown in Figure 15.6(b), indicating a substantial reduction in the leakage but with no change in the resolution. From Table 10.2, the main lobe width Δ_{ML} of a length-N Hamming window is $8\pi/N$. Thus, for $N = 16$, the normalized main lobe width is 0.25. Hence, with such a window, we can resolve two frequencies if their difference is of the order of half the main lobe width, that is, 0.125 or better. In our example, the difference is 0.04, which is considerably smaller than this value.

In order to increase the resolution, we increase the window length to 32, which reduces the main lobe width by half. Figure 15.6(c) shows its 32-point DFT. There now appears to be two peaks. Increasing the DFT size to 64 clearly separates the two peaks, as indicated in Figure 15.6(d). This separation becomes more visible with an increase in the DFT size to 256, as shown in Figure 15.6(e). Finally, Figure 15.6(f) shows the result obtained with a window length of 64 and a DFT length of 256.

It is clear from Examples 15.1 through 15.3 that performance of the DFT-based spectral analysis depends on several factors, the type of window being used and its length, and the size of the DFT. To improve the frequency resolution, one must use a window with a very small main lobe width, and to reduce the leakage, the window must have a very small relative sidelobe level. The main lobe width can be reduced by increasing the length of the window. Furthermore, an increase in the accuracy of locating the peaks is achieved by increasing the size of the DFT. To this end, it is preferable to use a DFT length that is a power of 2 so that very efficient FFT algorithms can be employed to compute the DFT. Of course, an increase in the DFT size also increases the computational complexity of the spectral analysis procedure.

15.3   Spectral Analysis of Nonstationary Signals

The discrete Fourier transform can be employed for the spectral analysis of a finite-length signal composed of sinusoidal components as long as the frequency, amplitude, and phase of each sinusoidal component are time-invariant and independent of the signal length. There are practical situations where the signal to be analyzed is instead nonstationary, for which these signal parameters are time-varying. An example of such a time-varying signal is the chirp signal given by

$$x[n] = A \cos(\omega_o n^2) \tag{15.15}$$

and shown in Figure 15.7 for $\omega_o = 10\pi \times 10^{-5}$. Note from Eq. (15.15), that the instantaneous frequency of $x[n]$ is given by $2\omega_o n$, which is not a constant but increases linearly with time. Speech, radar, and sonar

---

[1] For a review of some commonly used windows, see Sections 10.2.4 and 10.2.5.

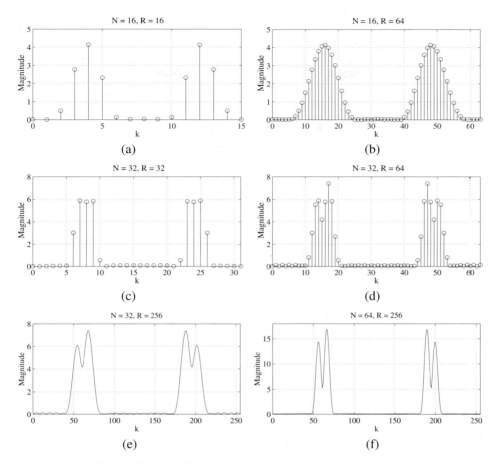

Figure 15.6: (a)–(f) Spectral analysis using a Hamming window.

signals are other examples of such nonstationary signals. A description of such signals in the frequency domain using a simple DFT of the complete signal will provide misleading results. To get around the time-varying nature of the signal parameters, an alternative approach would be to segment the sequence into a set of subsequences of short length, with each subsequence centered at uniform intervals of time and its DFT computed separately. If the subsequence length is reasonably small, it can be safely assumed to be stationary for practical purposes. As a result, the frequency-domain description of a long sequence is given by a set of short-length DFTs, that is, a time-dependent DFT.

To represent a nonstationary signal $x[n]$ in terms of a set of short-length subsequences, we can multiply it with a window $w[n]$ that is stationary with respect to time and move the signal through the window. For example, Figure 15.8 shows four segments of the chirp signal of Figure 15.7 as seen through a stationary rectangular window of length 200. As illustrated in this figure, the segments could be overlapping in time. A discrete-time Fourier transform of the short sequence obtained by windowing is called the short-term Fourier transform, which is thus a function of the location of the window relative to the original long sequence and the frequency. In this section, we review the basic concepts associated with this type of transform, study some of its properties, and point out one of its important applications. A detailed exposition of this subject can be found in [All77], [Naw88], [Opp89], and [Rab78].

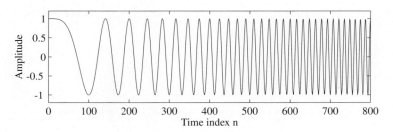

Figure 15.7: First 800 samples of a casual chirp signal $\cos(\omega_o n^2)$ with $\omega_o = 10\pi \times 10^{-5}$.

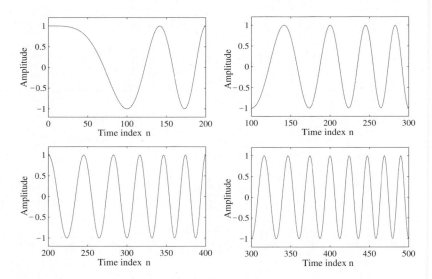

Figure 15.8: Examples of subsequences of the chirp signal of Figure 15.7 generated by a length-200 rectangular window.

15.3.1   Short-Time Fourier Transform

The *short-time Fourier transform* (STFT), also known as the *time-dependent Fourier transform,* of a sequence $x[n]$ is defined by

$$X_{\text{STFT}}(e^{j\omega}, n) = \sum_{m=-\infty}^{\infty} x[n - m]w[m]e^{-j\omega m}, \tag{15.16}$$

where $w[n]$ is a suitably chosen window sequence. It should be noted that the function of the window is to extract a finite-length portion of the sequence $x[n]$ such that the spectral characteristics of the section extracted are approximately stationary over the duration of the window for practical purposes.

   Note that if $w[n] = 1$, the definition of STFT given in Eq. (15.16) reduces to the conventional discrete-time Fourier transform (Fourier transform) of $x[n]$. However, even though the Fourier transform of $x[n]$ exists under certain well-defined conditions, the windowed sequence in Eq. (15.16), being finite in length, ensures the existence of the STFT for *any* sequence $x[n]$. It should be noted also that, unlike the conventional Fourier transform, the STFT is a function of two variables: the integer variable time

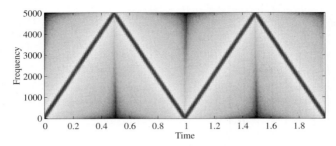

Figure 15.9: Spectrogram of a chirp signal.

index n and the continuous frequency variable ω. It also follows from the definition of Eq. (15.16) that $X_{\text{STFT}}(e^{j\omega}, n)$ is a periodic function of ω with a period 2π.

In most applications, the magnitude of the STFT is of interest. The display of the magnitude of the STFT is usually referred to as the *spectrogram*. However, since the STFT is a function of two variables, the display of its magnitude would normally require three dimensions. Often, it is plotted in two dimensions, with the magnitude represented by the darkness of the plot. Here, the white areas represent zero-valued magnitudes, while the gray areas represent nonzero magnitudes, with the largest magnitudes being shown in black. In the STFT magnitude display, the vertical axis represents the frequency variable (ω) and the horizontal axis represents the time index (n). The STFT can also be visualized as a mesh plot in a three-dimensional coordinate frame in which the STFT magnitude is a point in the z-direction above the x-y plane.

Figure 15.9 shows the STFT of the chirp sequence of Eq. (15.15), with $\omega_o = 10\pi \times 10^{-5}$ for a length of 20,000 samples computed using a Hamming window of length 200 in the above two forms. In Figure 15.9, the STFT for a given value of the time index n is essentially the DFT of a segment of a sinusoidal sequence. Recall from our discussion in Section 15.2 that the shape of the DFT of such a segment is similar to that shown in Figure 15.3, with large nonzero-valued DFT samples around the frequency of the sinusoid and smaller nonzero-valued DFT samples at other frequency points. In the spectrogram plot, the large-valued DFT samples show up as narrow, nearly black, very short vertical lines, while the other DFT samples show up as gray points. As the instantaneous frequency of the chirp signal increases linearly, the short black line moves up in the vertical direction, and eventually, because of aliasing, the black line starts moving down in the vertical direction. As a result, the spectrogram of the chirp signal essentially appears as a thick line in the form of a triangular shape.

15.3.2 Sampling in the Time and Frequency Dimensions

In practice, the STFT is computed at a finite set of discrete values of ω. Moreover, due to the finite length of the windowed sequence, the STFT is accurately represented by its frequency samples as long as the number of frequency samples is greater than the window length. In addition, the portion of the sequence $x[n]$ inside the window can be fully recovered from the frequency samples of the STFT.

To be more precise, let the window be of length R, defined in the range $0 \leq m \leq R - 1$. We sample $X_{\text{STFT}}(e^{j\omega}, n)$ at N equally spaced frequencies $\omega_k = 2\pi k/N$, with $N \geq R$ as indicated in the following:

$$X_{\text{STFT}}[k, n] = X_{\text{STFT}}(e^{j\omega}, n)\Big|_{\omega=2\pi k/N} = X_{\text{STFT}}(e^{j2\pi k/N}, n)$$

$$= \sum_{m=0}^{R-1} x[n - m]w[m]e^{-j2\pi km/N}, \quad 0 \leq k \leq N - 1. \qquad (15.17)$$

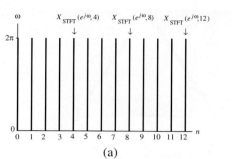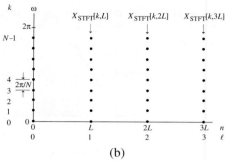

Figure 15.10: (a) Lines in the (ω, n)-plane for $X_{\text{STFT}}(e^{j\omega}, n)$ and (b) sampling grid in the (ω, n)-plane for $X_{\text{STFT}}[k, \ell L]$ for $N = 9$ and $L = 4$.

It follows from Eq. (15.17), assuming $w[m] \neq 0$, that $X_{\text{STFT}}[k, n]$ is simply the DFT of $x[n-m]w[m]$. Note that $X_{\text{STFT}}[k, n]$ is a two-dimensional sequence and is periodic in k with a period N. Applying the IDFT, we thus arrive at

$$x[n-m]w[m] = \frac{1}{N} \sum_{k=0}^{N-1} X_{\text{STFT}}[k, n]e^{j2\pi km/N}, \qquad 0 \leq m \leq R-1, \tag{15.18}$$

or in other words,

$$x[n-m] = \frac{1}{Nw[m]} \sum_{k=0}^{N-1} X_{\text{STFT}}[k, n]e^{j2\pi km/N}, \qquad 0 \leq m \leq R-1, \tag{15.19}$$

verifying that the sequence values inside the window can be fully recovered from $X_{\text{STFT}}[k, n]$ as long as the number N of frequency samples is greater than or equal to the window length R. It should be evident by now that $x[n]$ for $-\infty < n < \infty$ can be fully recovered if $X_{\text{STFT}}(e^{j\omega}, n)$ or $X_{\text{STFT}}[k, n]$ is sampled also in the time dimension. More precisely, if we set $n = n_o$ in Eq. (15.19), we recover the signal in the interval $n_o \leq n \leq n_o + R - 1$ from $X_{\text{STFT}}[k, n_o]$. Likewise, by setting $n = n_o + R$ in Eq. (15.19), we recover the signal in the interval $n_o + R \leq n \leq n_o + 2R - 1$ from $X_{\text{STFT}}[k, n_o + R]$, and so on.

The sampled STFT for a window defined in the region $0 \leq m \leq R - 1$ is given by

$$X_{\text{STFT}}[k, \ell L] = X_{\text{STFT}}(e^{j2\pi k/N}, \ell L) = \sum_{m=0}^{R-1} x[\ell L - m]w[m]e^{-j2\pi km/N}, \tag{15.20}$$

where ℓ and k are integers such that $-\infty < \ell < \infty$ and $0 \leq k \leq N - 1$. Figure 15.10 shows lines in the (ω, n)-plane corresponding to $X_{\text{STFT}}(e^{j\omega}, n)$ and the grid of sampling points in the (ω, n)-plane for the case $N = 9$ and $L = 4$. As we have shown, it is possible to uniquely reconstruct the original signal from such a 2-D discrete representation provided $N \geq R \geq L$.

15.3.3 Window Selection

As in the case of the DFT-based spectral analysis of deterministic signals discussed in Section 15.2, in the STFT analysis of nonstationary signals, the window also plays an important role. Both the length and shape of the window are critical issues that need to be examined carefully.

The function of the window $w[n]$ is to extract a portion of the signal for analysis and ensure that the extracted section of $x[n]$ is approximately stationary. To this end, the window length R should be small, in particular for signals with widely varying spectral parameters. A decrease in the window length increases the time-resolution property of the STFT, whereas the frequency-resolution property of the STFT increases with an increase in the window length. A shorter window thus provides a *wide-band spectrogram,* while a longer window results in a *narrow-band spectrogram.*

The two frequency-domain parameters characterizing the Fourier transform of a window are its main lobe width Δ_{ML} and the relative sidelobe amplitude $A_{s\ell}$. The former parameter determines the ability of the window to resolve two signal components in the vicinity of each other, while the latter controls the degree of leakage of one component into a nearby signal component. It thus follows that in order to obtain a reasonably good estimate of the frequency spectrum of a time-varying signal, the window should be chosen to have a very small relative sidelobe amplitude with a length chosen based on the acceptable accuracy of the frequency and time resolutions.

15.3.4   STFT Computation Using MATLAB

The *Signal Processing Toolbox* of MATLAB includes the function `specgram` for the computation of the STFT of a signal. There are a number of versions of this function. We illustrate its application in Section 15.3.5.

15.3.5   Analysis of Speech Signals Using STFT

The short-term Fourier transform is often used in the analysis of speech, since speech signals are generally nonstationary. As indicated in Section 1.3, the speech signal, generated by the excitation of the vocal tract, is composed of two types of basic waveforms: voiced and unvoiced sounds. A typical speech signal is shown in Figure 1.16. As can be seen from this figure, a speech segment over a small time interval can be considered as a stationary signal, and as a result, the DFT of the speech segment can provide a reasonable representation of the frequency domain characteristic of the speech in this time interval. However, in the STFT analysis, the size of the window is critical since a shorter window developing a wide-band spectrogram provides a better time resolution, whereas a longer window developing a narrow-band spectrogram results in an improved frequency resolution. In order to provide a reasonably good estimate of the changes in the vocal tract and the excitation, a wide-band spectrogram is preferable. To this end, the window size is selected to be approximately close to one pitch period, which is adequate for resolving the formants though not adequate to resolve the harmonics of the pitch frequencies. On the other hand, to resolve the harmonics of the pitch frequencies, a narrow-band spectrogram with a window size of several pitch periods is desirable.

Example 15.4 illustrates the STFT analysis of a speech signal.

EXAMPLE 15.4   Short-Time Fourier Transform Analysis of a Speech Signal

The `mtlb.mat` file in the *Signal Processing Toolbox* of MATLAB contains a speech signal of duration 4001 samples sampled at 7418 Hz. We compute its STFT using a Hamming window of length 256 with an overlap of 50 samples between consecutive windowed signals using Program 15.3. Figures 15.11(b) and (c) show, respectively, a narrow-band spectrogram and a wide-band spectrogram of the speech signal of Figure 15.11(a). The frequency and time resolution trade-off between the two spectrograms of Figure 15.11 should be evident.

Program 15_3.m

15.4   Spectral Analysis of Random Signals

As discussed in Section 15.2, in the case of a deterministic signal composed of sinusoidal components, a Fourier analysis of the signal can be carried out by taking the discrete Fourier transform (DFT) of a finite-

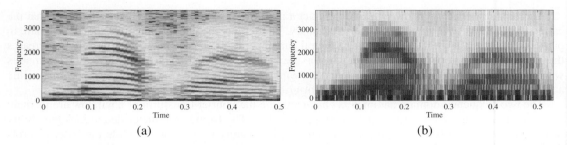

Figure 15.11: (a) Narrow-band spectrogram and (b) wide-band spectrogram of a speech signal.

length segment of the signal obtained by appropriate windowing, provided the parameters characterizing the components are time-invariant and independent of the window length. On the other hand, the Fourier analysis of nonstationary signals with time-varying parameters is best carried out using the short-time Fourier transform (STFT) described in Section 15.3.

Neither the DFT nor the STFT is applicable for the spectral analysis of naturally occurring random signals as here the spectral parameters are also random. These type of signals are usually classified as noiselike random signals such as the unvoiced speech signal generated when a letter such as "/f/" or "/s/" is spoken, and signal-plus-noise random signals, such as seismic signals and nuclear magnetic resonance signals [Rob82]. Spectral analysis of a noiselike random signal is usually carried out by estimating the power density spectrum using Fourier-analysis-based nonparametric methods, whereas a signal-plus-noise random signal is best analyzed using parametric-model-based methods in which the autocovariance sequence is first estimated from the model and then the Fourier transform of the estimate is evaluated. In this section, we review both of these approaches.

15.4.1   Nonparametric Spectral Analysis

Consider a wide-sense stationary (WSS) random signal $g[n]$ with zero mean. According to the Wiener–Khintchine theorem of Eq. (A.33) given in the Appendix, the power spectrum of $g[n]$ is given by

$$\mathcal{P}_{gg}(\omega) = \sum_{\ell=-\infty}^{\infty} \phi_{gg}[\ell] e^{-j\omega\ell}, \tag{15.21}$$

where $\phi_{gg}[\ell]$ is its autocorrelation sequence, which from Eq. (A.20b) is given by

$$\phi_{gg}[\ell] = E(g[n+\ell]g^*[n]). \tag{15.22}$$

In Eq. (15.22), $E(\cdot)$ denotes the expectation operator as defined in Eq. (A.4a).

Periodogram Analysis

Assume that the infinite-length random discrete-time signal $g[n]$ is windowed by a length-N window sequence $w[n]$, $0 \le n \le N - 1$, resulting in the length-N sequence $\gamma[n] = g[n] \cdot w[n]$. The Fourier transform $\Gamma(e^{j\omega})$ of $\gamma[n]$ is given by

$$\Gamma(e^{j\omega}) = \sum_{n=0}^{N-1} \gamma[n] e^{-j\omega n} = \sum_{n=0}^{N-1} g[n] \cdot w[n] e^{-j\omega n}. \tag{15.23}$$

The estimate $\hat{\mathcal{P}}_{gg}(\omega)$ of the power spectrum $\mathcal{P}_{gg}(\omega)$ is then obtained using

$$\hat{\mathcal{P}}_{gg}(\omega) = \frac{1}{CN}|\Gamma(e^{j\omega})|^2, \tag{15.24}$$

where the constant C is a normalization factor given by

$$C = \frac{1}{N}\sum_{n=0}^{N-1}|w[n]|^2 \tag{15.25}$$

and included in Eq. (15.24) to eliminate any bias in the estimate occurring due to the use of the window $w[n]$. The quantity $\hat{\mathcal{P}}_{gg}(e^{j\omega})$ defined in Eq. (15.24) is called the *periodogram* when $w[n]$ is a rectangular window and is called a *modified periodogram* for other types of windows.

In practice, the periodogram $\hat{\mathcal{P}}_{gg}(\omega)$ is evaluated at a discrete set of equally spaced R frequencies, $\omega_k = 2\pi k/R$, $0 \le k \le R-1$, by replacing the Fourier transform $\Gamma(e^{j\omega})$ with an R-point DFT $\Gamma[k]$ of the length-N sequence $\gamma[n]$:

$$\hat{\mathcal{P}}_{gg}[k] = \frac{1}{CN}|\Gamma[k]|^2. \tag{15.26}$$

As in the case of the Fourier analysis of sinusoidal signals discussed earlier, R is usually chosen to be greater than N to provide a finer grid of the samples of the periodogram.

It can be shown that the mean value of the periodogram $\hat{\mathcal{P}}_{gg}(\omega)$ is given by

$$E\left(\hat{\mathcal{P}}_{gg}(\omega)\right) = \frac{1}{2\pi CN}\int_{-\pi}^{\pi} \mathcal{P}_{gg}(v)|\Psi(e^{j(\omega-v)})|^2 \, dv, \tag{15.27}$$

where $\mathcal{P}_{gg}(\omega)$ is the desired power spectrum and $\Psi(e^{j\omega})$ is the Fourier transform of the window sequence $w[n]$. The mean value being nonzero for any finite-length window sequence, the power spectrum estimate given by the periodogram is said to be *biased*. By increasing the window length N, the bias can be reduced.

We illustrate the power spectrum computation in Example 15.5.

EXAMPLE 15.5   Power Spectrum of a Noise-Corrupted Sinusoidal Sequence

Let the random signal $g[n]$ be composed of two sinusoidal components of angular frequencies 0.06π and 0.14π radians, corrupted with a Gaussian distributed random signal of zero mean and unity variance, and windowed by a rectangular window of two different lengths: $N = 128$ and 1024. The random signal is generated using the M-file `randn`. Figures 15.12(a) and (b) show the plots of the estimated power spectrum for the two cases. Ideally the power spectrum should show four peaks at ω equal to 0.06, 0.14, 0.86, and 0.94, respectively, and a flat spectral density at all other frequencies. However, Figure 15.12(a) shows four large peaks and several other smaller peaks. Moreover, the spectrum shows large amplitude variations throughout the whole frequency range. As N is increased to a much larger value, the peaks get sharper due to increased resolution of the DFT, while the spectrum shows more rapid amplitude variations.

To understand the cause behind the rapid amplitude variations of the computed power spectrum encountered in Example 15.5, we assume $w[n]$ to be a rectangular window and rewrite the expression for the periodogram given in Eq. (15.24) using Eq. (15.23) as

$$\hat{\mathcal{P}}_{gg}(\omega) = \frac{1}{N}\sum_{n=0}^{N-1}\sum_{m=0}^{N-1} g[m]g^*[n]e^{-j\omega(m-n)}$$

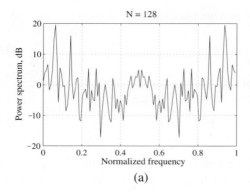
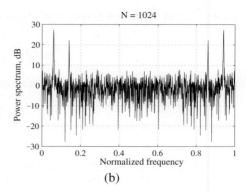

(a)   (b)

Figure 15.12: Power spectrum estimate of a signal containing two sinusoidal components corrupted with a white noise sequence of zero mean and unit variance Gaussian distribution: (a) Periodogram with a rectangular window of length $N = 128$ and (b) periodogram with a rectangular window of length $N = 1024$.

$$= \sum_{k=-N+1}^{N-1} \left(\frac{1}{N} \sum_{n=0}^{N-1-|k|} g[n+k]g^*[n] \right) e^{-j\omega k}$$

$$= \sum_{k=-N+1}^{N-1} \hat{\phi}_{gg}[k] e^{-j\omega k}. \tag{15.28}$$

Now $\hat{\phi}_{gg}[k]$ is the periodic correlation of $g[n]$ and is an estimate of the true correlation $\phi_{gg}[k]$. Hence, $\hat{\mathcal{P}}_{gg}(\omega)$ is actually the Fourier transform of $\hat{\phi}_{gg}[k]$. A few samples of $g[n]$ are used in the computation of $\hat{\phi}_{gg}[k]$ when k is near N, yielding a poor estimate of the true correlation. This, in turn, results in rapid amplitude variations in the periodogram estimate. A smoother power spectrum estimate can be obtained by the periodogram averaging method discussed next.

Periodogram Averaging

The power spectrum estimation method, originally proposed by Bartlett [Bar48] and later modified by Welch [Wel67], is based on the computation of the modified periodogram of R overlapping portions of length-N input samples and then averaging these R periodograms. Let the overlap between adjacent segments be K samples. Consider the windowed rth segment of the input data

$$\gamma^{(r)}[n] = g[n+rK]w[n], \qquad 0 \le n \le N-1, \qquad 0 \le r \le R-1, \tag{15.29}$$

with a Fourier transform given by $\Gamma^{(r)}(e^{j\omega})$. Its periodogram is given by

$$\hat{\mathcal{P}}_{gg}^{(r)}(\omega) = \frac{1}{CN} |\Gamma^{(r)}(e^{j\omega})|^2. \tag{15.30}$$

The Welch estimate is then given by the average of all R periodograms $\hat{\mathcal{P}}_{gg}^{(r)}(\omega), 0 \le r \le R-1$:

$$\hat{\mathcal{P}}_{gg}^W(\omega) = \frac{1}{R} \sum_{r=1}^{R-1} \hat{\mathcal{P}}_{gg}^{(r)}(\omega). \tag{15.31}$$

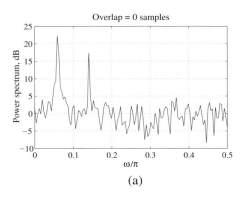
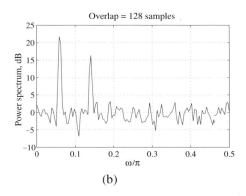

Figure 15.13: Power spectrum estimates: (a) Bartlett's method and (b) Welch's method.

It can be shown that the variance of the Welch estimate of Eq. (15.31) is reduced approximately by a factor R if the R periodogram estimates are assumed to be independent of each other. For a fixed-length input sequence, R can be increased by decreasing the window length N which in turn decreases the DFT resolution. On the other hand, an increase in the resolution is obtained by increasing N. Thus, there is a trade-off between resolution and the bias.

It should be noted that if the data sequence is segmented by a rectangular window into contiguous segments with no overlap, the periodiogram estimate given by Eq. (15.31) reduces to Barlett estimate [Bar48].

Periodogram Estimate Computation Using MATLAB

The *Signal Processing Toolbox* of MATLAB includes the M-file `psd` for modified periodogram estimate computation. It is available with several options. We illustrate its use in Example 15.6.

Program 15_4.m

> **EXAMPLE 15.6   Estimation of the Power Spectrum of a Noise-Corrupted Sinusoidal Sequence**
>
> We consider here the evaluation of the Bartlett and Welch estimates of the power spectrum of the random signal considered in Example 15.6. To this end, Program 15_4 can be used. This program is run first with no overlap and with a rectangular window generated using the function `boxcar`. The power spectrum computed by the above program is then the Bartlett estimate, as indicated in Figure 15.13(a). It is then run with an overlap of 128 samples and a Hamming window. The corresponding power spectrum is then the Welch estimate, as shown in Figure 15.13(b). It should be noted from Figure 15.13 that the Welch periodogram estimate is much smoother than the Bartlett periodogram estimate, as expected. Compared to the power spectrums of Figure 15.12, there is a decrease in the variance in the smoothed power spectrums of Figure 15.13, but the latter are still biased. Because of the overlap between adjacent data segments, Welch's estimate has a smaller variance than the others. It should be noted that both periodograms of Figure 15.13 show clearly two distinct peaks at 0.06 and 0.14.

15.4.2   Parametric Model-Based Spectral Analysis

In the model-based method, a causal LTI discrete-time system with a transfer function

$$H(z) = \sum_{n=0}^{\infty} h[n] z^{-n}$$

$$= \frac{P(z)}{D(z)} = \frac{\sum_{k=0}^{L} p_k z^{-k}}{1 + \sum_{k=1}^{M} d_k z^{-k}} \tag{15.32}$$

is first developed, whose output, when excited by a white noise sequence $e[n]$ with zero mean and variance σ_e^2, matches the specified data sequence $g[n]$ [Kum93]. An advantage of the model-based approach is that it can extrapolate a short-length data sequence to create a longer data sequence for improved power spectrum estimation. On the other hand, in nonparametric methods, spectral leakages limit the frequency resolution if the data length is short.

The model of Eq. (15.32) is called an *autoregressive moving-average* (ARMA) process of order (L, M) if $P(z) \neq 1$, an *all-pole* or *autoregressive* (AR) process of order M if $P(z) = 1$, and an all-zero or *moving-average* (MA) process of order L if $D(z) = 1$. For an ARMA or an AR model, for stability, the denominator $D(z)$ must have all its zeros inside the unit circle. In the time domain, the input–output relation of the model is given by

$$g[n] = -\sum_{k=1}^{M} d_k g[n - k] + \sum_{k=0}^{L} p_k e[n - k]. \tag{15.33}$$

As indicated in Section A.8 of the Appendix, the output $g[n]$ of the model is a WSS random signal. From Eq. (A.58), it follows that the power spectrum $\mathcal{P}_{gg}(\omega)$ of $g[n]$ can be expressed as

$$\mathcal{P}_{gg}(\omega) = \sigma_e^2 |H(e^{j\omega})|^2 = \sigma_e^2 \frac{|P(e^{j\omega})|^2}{|D(e^{j\omega})|^2}, \tag{15.34}$$

where $H(e^{j\omega}) = P(e^{j\omega})/D(e^{j\omega})$ is the frequency response of the model, and

$$P(e^{j\omega}) = \sum_{k=0}^{L} p_k e^{-j\omega k}, \qquad D(e^{j\omega}) = 1 + \sum_{k=1}^{M} d_k e^{-j\omega k}.$$

In the case of an AR or an MA model, the power spectrum is thus given by

$$\mathcal{P}_{gg}(\omega) = \begin{cases} \sigma_e^2 |P(e^{j\omega})|^2, & \text{for an MA model,} \\ \frac{\sigma_e^2}{|D(e^{j\omega})|^2}, & \text{for an AR model.} \end{cases} \tag{15.35}$$

The spectral analysis is carried out by first determining the model and then computing the power spectrum using either Eq. (15.34) for an ARMA model or using Eq. (15.35) for an MA or an AR model. To determine the model, we need to decide the type of the model (i.e., pole-zero IIR structure, all-pole IIR structure, or all-zero FIR structure) to be used; determine an appropriate order of its transfer function $H(z)$ (i.e., both L and M for an ARMA model or M for an AR model or L for an MA model); and then, from the specified length-N data $g[n]$, estimate the coefficients of $H(z)$. We restrict our discussion here to the development of the AR model, as it is simpler and often used. Applications of the AR model include spectral analysis, system identification, speech analysis and compression, and filter design. For a discussion on the development of the MA model and the ARMA model, see [Kum93].

Relation Between Model Parameters and the Autocorrelation Sequence

The model filter coefficients $\{p_k\}$ and $\{d_k\}$ are related to the autocorrelation sequence $\phi_{gg}[\ell]$ of the random signal $g[n]$. To establish this relation, we obtain from Eq. (15.33),

$$\phi_{gg}[\ell] = -\sum_{k=1}^{M} d_k \phi_{gg}[\ell - k] + \sum_{k=0}^{L} p_k \phi_{eg}[\ell - k], \qquad -\infty < \ell < \infty, \tag{15.36}$$

by multiplying both sides of the equation with $g^*[n - \ell]$ and taking the expected values. In Eq. (15.36), the cross-correlation $\phi_{eg}[\ell]$ between $g[n]$ and $e[n]$ can be written as

$$\phi_{eg}[\ell] = E(g^*[n]e[n + \ell])$$

$$= \sum_{k=0}^{\infty} h^*[k] \, E(e^*[n - k]e[n + \ell]) = \sigma_e^2 h^*[-\ell], \qquad (15.37)$$

where $h[n]$ is the causal impulse response of the LTI model as defined in Eq. (15.32) and σ_e^2 is the variance of the white noise sequence $e[n]$ applied to the input of the model.

For an AR model, $L = 0$, and hence Eq. (15.36) reduces to

$$\phi_{eg}[\ell] = \begin{cases} -\sum_{k=1}^{M} d_k \phi_{gg}[\ell - k], & \text{for } \ell > 0, \\ -\sum_{k=1}^{M} d_k \phi_{gg}[\ell - k] + \sigma_e^2, & \text{for } \ell = 0, \\ \phi_{gg}^*[-\ell], & \text{for } \ell < 0. \end{cases} \qquad (15.38)$$

From Eq. (15.38), we obtain for $1 \le \ell \le M$, a set of M equations,

$$\sum_{k=1}^{M} d_k \phi_{gg}[\ell - k] = -\phi_{eg}[\ell], \qquad 1 \le \ell \le M,$$

which can be written in matrix form as

$$\begin{bmatrix} \phi_{gg}[0] & \phi_{gg}[-1] & \cdots & \phi_{gg}[-M+1] \\ \phi_{gg}[1] & \phi_{gg}[0] & \cdots & \phi_{gg}[-M+2] \\ \vdots & \vdots & \ddots & \vdots \\ \phi_{gg}[M-1] & \phi_{gg}[M-2] & \cdots & \phi_{gg}[0] \end{bmatrix} \begin{bmatrix} d_1 \\ d_2 \\ \vdots \\ d_M \end{bmatrix} = - \begin{bmatrix} \phi_{gg}[1] \\ \phi_{gg}[2] \\ \vdots \\ \phi_{gg}[M] \end{bmatrix}. \qquad (15.39)$$

For $\ell = 0$, we also get from Eq. (15.38)

$$\phi_{gg}[0] + \sum_{k=1}^{M} d_k \phi_{gg}[-k] = \sigma_e^2. \qquad (15.40)$$

Combining Eq. (15.40) with Eq. (15.39) we arrive at

$$\begin{bmatrix} \phi_{gg}[0] & \phi_{gg}[-1] & \cdots & \phi_{gg}[-M] \\ \phi_{gg}[1] & \phi_{gg}[0] & \cdots & \phi_{gg}[-M+1] \\ \vdots & \vdots & \ddots & \vdots \\ \phi_{gg}[M] & \phi_{gg}[M-1] & \cdots & \phi_{gg}[0] \end{bmatrix} \begin{bmatrix} 1 \\ d_1 \\ d_2 \\ \vdots \\ d_M \end{bmatrix} = \begin{bmatrix} \sigma_e^2 \\ 0 \\ 0 \\ \vdots \\ 0 \end{bmatrix}. \qquad (15.41)$$

The matrix equation of Eq. (15.41) is more commonly known as the *Yule–Walker equation*. It can be seen from Eq. (15.41) that knowing the $M + 1$ autocorrelation samples $\phi_{xx}[\ell]$ for $0 \le \ell \le M$, we can determine the model parameters d_k for $1 \le k \le M$ by solving the matrix equation. The $(M + 1) \times (M - 1)$ matrix in Eq. (15.41) is a *Toeplitz matrix*.[2] Because of the structure of the Toeplitz matrix, the matrix equation of Eq. (15.41) can be solved using the fast *Levinson–Durbin algorithm* [Dur59], [Lev47]. The causal all-pole LTI system $H(z) = 1/D(z)$ resulting from the application of the Levinson–Durbin recursions is guaranteed to be BIBO stable. Moreover, the recursion automatically leads to a realization in the form of a cascaded FIR lattice structure, as shown in Figure 8.39.

---

[2]A Toeplitz matrix has the same element values along each negative-sloping diagonal.

Power Spectrum Estimation Using an AR Model

The AR model parameters can be determined using the *Yule–Walker method*, which makes use of the estimates of the autocorrelation sequence samples, as their actual values are not known a priori. The autocorrelation at lag ℓ is determined from the specified data samples $g[n]$ for $0 \leq n \leq N - 1$ using

$$\hat{\phi}_{gg}[\ell] = \frac{1}{N} \sum_{n=0}^{N-1-|\ell|} g^*[n]\, g[n + \ell], \qquad 0 \leq \ell \leq N - 1. \tag{15.42}$$

The above estimates are used in Eq. (15.39) in place of the true autocorrelation samples, with the AR model parameters d_k replaced with their estimates $\hat{d}_k$. The resulting equation is next solved using the Levinson–Durbin algorithm to determine the estimates of the AR model parameters $\hat{d}_k$. The power spectrum estimate is then evaluated using

$$\hat{\mathcal{P}}_{gg}(\omega) = \frac{\hat{\mathcal{E}}_M}{\left| 1 + \sum_{k=1}^{M} \hat{d}_k\, e^{-j\omega k} \right|^2}, \tag{15.43}$$

where $\hat{\mathcal{E}}_M$ is the prediction error for the Mth-order AR model:

$$\hat{\mathcal{E}}_M = \hat{\phi}_{gg}[0] \prod_{i=1}^{M} \left(1 - |\hat{K}_i|^2 \right). \tag{15.44}$$

The Yule–Walker method is related to the linear prediction problem. Here the problem is to predict the N-th sample $g[N]$ from the previous M data samples $g[n]$, $0 \leq n \leq M - 1$, with the assumption that data samples outside this range are zeros. The predicted value $\hat{g}[n]$ of the data sample $g[n]$ can be found by a linear combination of the previous M data samples as

$$\hat{g}[n] = -\sum_{k=1}^{M} \hat{d}_k g[n - k]$$
$$= g[n] - e[n], \tag{15.45}$$

where $e[n]$ is the prediction error. For the specified data sequence, Eq. (15.45) leads to $N + M$ prediction equations given by

$$g[n] + \sum_{k=1}^{M} g[n - k]\hat{d}_k = e[n], \qquad 0 \leq n \leq N + M - 1. \tag{15.46}$$

The optimum linear predictor coefficients $\hat{d}_k$ are obtained by minimizing the error

$$\frac{1}{N} \sum_{n=0}^{N+M-1} |e[n]|^2.$$

It can be shown that the solution of the minimization problem is given by Eq. (15.39). Thus, the best all-pole linear predictor filter is also the AR model resulting from the solution of Eq. (15.39).

It should be noted that the AR model is guaranteed stable. But the all-pole filter developed may not model an AR process exactly of the same order due to the windowing of the data sequence to a finite length, with samples outside the window range assumed to be zeros.

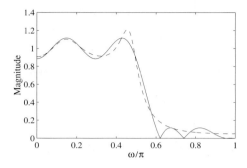

Figure 15.14: Magnitude response of the FIR filter (shown with solid line) and the all-pole IIR model (shown with dashed line).

The function `lpc` in MATLAB finds the AR model using the above method.[3] We illustrate its application in Example 15.7.

EXAMPLE 15.7   Development of an AR Model of an FIR Filter

We consider the approximation of an FIR digital filter of order 13 with an all-pole IIR digital filter of order 7. The coefficients of the FIR filter are obtained using the function `remez`, and the all-pole IIR filter is designed using the function `lpc`. Program 15_5 can be used for the design. The magnitude response plots generated by running this program are shown in Figure 15.14.

Several comments are in order here. First, the linear predictor coefficients $\{d_i\}$ match the power spectral densities of the all-pole model with that of the sequence $\{g_i\}$. Since, the sequence of the FIR filter coefficients $\{b_i\}$ is not a power signal, and to convert the energy spectrum of the sequence $\{b_i\}$ to a power spectrum, the sequence $\{b_i\}$ needs to be divided by its length N. Hence, to approximate the power spectrum density of the sequence $\{b_i\}$ with that of the AR model, we need to scale the ARMA filter transfer function with the factor $\sqrt{NE}$, where E is the variance of the prediction error. Second, it can be seen from this figure that the AR model has reasonably matched the passband response, with peaks occurring very close to the peaks of the magnitude response of the FIR system. However, there are no nulls in the stopband response of the AR model even though the stopband response of the FIR system has nulls. Since the nulls are generated by the zeros of the transfer function, an all-pole model cannot produce nulls.

Program 15_5.m

In order to apply the above method to power spectrum estimation, it is necessary to estimate first the model order M. A number of formulae have been advanced for order estimation [Kum93]. Unfortunately, none of the these formulae yields a really good estimate of the true model order in many applications.

15.5   Musical Sound Processing

Recall from our discussion in Section 1.4.1 that almost all musical programs are produced in basically two stages. First, sound from each individual instrument is recorded in an acoustically inert studio on a single track of a multitrack tape recorder. Then, the signals from each track are manipulated by the sound engineer to add special audio effects and are combined in a mix-down system to finally generate the stereo recording on a two-track tape recorder [Ble78], [Ear86]. The audio effects are artificially generated using various signal processing circuits and devices, and they are increasingly being performed using digital signal processing techniques [Ble78], [Orf96].

---

[3]The second to last line in the M-file `lpc.m` in the *Signal Processing Toolbox* should be changed to R = R./m; to get the correct AR model.

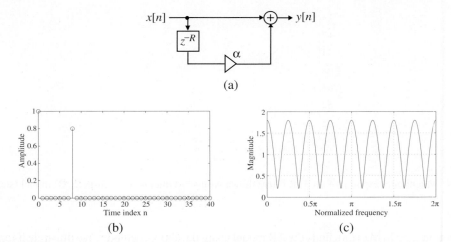

(a)

(b)   (c)

Figure 15.15: Single echo filter: (a) filter structure, (b) typical impulse response, and (c) magnitude response for $R = 8$ and $\alpha = 0.8$.

Some of the special audio effects that can be implemented digitally are reviewed in this section.

15.5.1 Time-Domain Operations

Commonly used time-domain operations carried on musical sound signals are echo generation, reverberation, flanging, chorus generation, and phasing. In each of these operations, the basic building block is a delay.

Single Echo Filter

Echoes are simply generated by delay units. For example, the direct sound and a single echo appearing R sampling periods later can be simply generated by the FIR filter of Figure 15.15(a), which is characterized by the difference equation

$$y[n] = x[n] + \alpha x[n - R], \qquad |\alpha| < 1, \tag{15.47}$$

or, equivalently, by the transfer function

$$H(z) = 1 + \alpha z^{-R}. \tag{15.48}$$

In Eqs. (15.47) and (15.48), the delay parameter R denotes the time the sound wave takes to travel from the sound source to the listener after bouncing back from the reflecting wall, whereas the parameter α, with $|\alpha| < 1$, represents the signal loss caused by propagation and reflection.

The impulse response of the single echo filter is sketched in Figure 15.15(b). The magnitude response of a single echo FIR filter for $\alpha = 0.8$ and $R = 8$ is shown in Figure 15.15(c). The magnitude response exhibits R peaks and R dips in the range $0 \le \omega < 2\pi$, with the peaks occurring at $\omega = 2\pi k/R$ and the dips occurring at $\omega = (2k+1)\pi/R$, $k = 0, 1, \ldots, R-1$. Because of the comblike shape of the magnitude response, such a filter is also known as a *comb filter*. The maximum and minimum values of the magnitude response are given by $1 + \alpha = 1.8$ and $1 - \alpha = 0.2$, respectively.

Program 15_6[4] can be used to investigate the effect of a single echo on the speech signal shown in Figure 1.16.

Program 15_6.m

[4]Reproduced with permission of Prof. Dale Callahan, University of Alabama, Birmingham, AL.

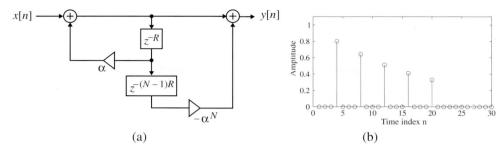

(a)   (b)

Figure 15.16: Multiple echo filter generating $N - 1$ echoes: (a) filter structure and (b) impulse response with $\alpha = 0.8$ for $N = 6$ and $R = 4$.

Multiple Echo Filter

To generate a fixed number of multiple echoes spaced R sampling periods apart with exponentially decaying amplitudes, one can use an FIR filter with a transfer function of the form

$$H(z) = 1 + \alpha z^{-R} + \alpha^2 z^{-2R} + \cdots + \alpha^{N-1} z^{-(N-1)R} = \frac{1 - \alpha^N z^{-NR}}{1 - \alpha z^{-R}}. \tag{15.49}$$

An IIR realization of this filter is sketched in Figure 15.16(a). The impulse response of a multiple echo filter with $\alpha = 0.8$ for $N = 6$ and $R = 4$ is shown in Figure 15.16(b).

An infinite number of echoes spaced R sampling periods apart with exponentially decaying amplitudes can be created by an IIR filter with a transfer function of the form

$$H(z) = 1 + \alpha z^{-R} + \alpha^2 z^{-2R} + \alpha^3 z^{-3R} + \cdots$$
$$= \frac{1}{1 - \alpha z^{-R}}, \qquad |\alpha| < 1. \tag{15.50}$$

Figure 15.17(a) shows one possible realization of the above IIR filter whose first 61 impulse response samples for $R = 4$ are indicated in Figure 15.17(b). The magnitude response of this IIR filter for $R = 7$ is sketched in Figure 15.17(c). The magnitude response exhibits R peaks and R dips in the range $0 \leq \omega < 2\pi$, with the peaks occurring at $\omega = 2\pi k/R$ and the dips occurring at $\omega = (2k+1)\pi/R$, $k = 0, 1, \ldots, R-1$. The maximum and minimum values of the magnitude response are given by $1/(1-\alpha) = 5$ and $1/(1+\alpha) = 0.5556$, respectively.

The *fundamental repetition frequency* of the IIR multiple echo filter of Eq. 15.50 is given by $F_R = F_T/R$ Hz, or $\omega_R = 2\pi/R$ radians. In practice, the repetition frequency F_R is often locked to the fundamental frequency of an accompanying musical instrument, such as the drum beat. For a specified F_R, the delay parameter R can be determined from $R = F_R/F_T$, resulting in a time delay of $RT = R/F_T$ seconds [Orf96].

Program 15_7[5] can be used to investigate the effect of multiple echos on the speech signal shown in Figure 1.16.

Program 15_7.m

Reverberation

As indicated in Section 1.4.1, the sound reaching the listener in a closed space, such as a concert hall, consists of several components: direct sound, early reflections, and reverberation. The early reflections are

---

[5]Reproduced with permission of Prof. Dale Callahan, University of Alabama, Birmingham, AL.

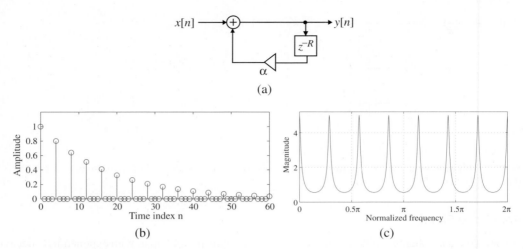

(a)

(b)   (c)

Figure 15.17: IIR filter generating an infinite number of echoes: (a) filter structure, (b) impulse response with $\alpha = 0.8$ for $R = 4$, and (c) magnitude response with $\alpha = 0.8$ for $R = 7$.

composed of several closely spaced echoes that are basically delayed and attenuated copies of the direct sound, whereas the reverberation is composed of densely packed echoes. The sound recorded in an inert studio is different from that recorded inside a closed space, and, as a result, the former does not sound "natural" to a listener. However, digital filtering can be employed to convert the sound recorded in an inert studio into a natural-sounding one by artificially creating the echoes and adding them to the original signal.

The IIR comb filter of Figure 15.17(a) by itself does not provide natural-sounding reverberations for two reasons [Sch62]. First, as can be seen from Figure 15.17(c), its magnitude response is not constant for all frequencies, resulting in a "coloration" of many musical sounds that are often unpleasant for listening purposes. Second, the output echo density, given by the number of echoes per second, generated by a unit impulse at the input, is much lower than that observed in a real room, thus causing a "fluttering" of the composite sound. It has been observed that approximately 1000 echoes per second are necessary to create a reverberation that sounds free of flutter [Sch62]. To develop a more realistic reverberation, a reverberator with an allpass structure, as indicated in Figure 15.17(a), has been proposed [Sch62].[6] Its transfer function is given by

$$H(z) = \frac{\alpha + z^{-R}}{1 + \alpha z^{-R}}, \qquad |\alpha| < 1. \tag{15.51}$$

In the steady state, the spectral balance of the sound signal remains unchanged due to the unity magnitude response of the allpass reverberator.

Program 15_8[7] can be used to investigate the effect of an allpass reverberator on the speech signal shown in Figure 1.16.

Program 15_8.m   The IIR comb filter of Figure 15.17(a) and the allpass reverberator of Figure 15.18(a) are basic reverberator units that are suitably interconnected to develop a natural-sounding reverberation. Figure 15.19 shows one such interconnection composed of a parallel connection of four IIR echo generators in cascade with two allpass reverberators [Sch62]. By choosing different values for the delays in each section (obtained by

---

[6]The structures shown here are the canonic single multiplier realization of a first-order allpass transfer function [Mit74a]. See also Section 8.6.1.

[7]Reproduced with permission of Prof. Dale Callahan, University of Alabama, Birmingham, AL.

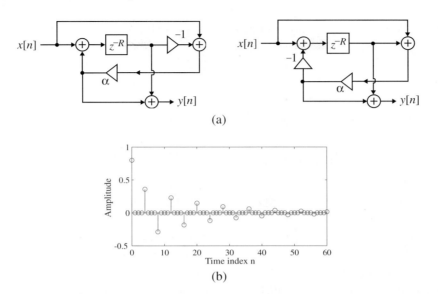

(a)

(b)

Figure 15.18: Allpass reverberator: (a) block diagram representation and (b) impulse response with $\alpha = 0.8$ for $R = 4$.

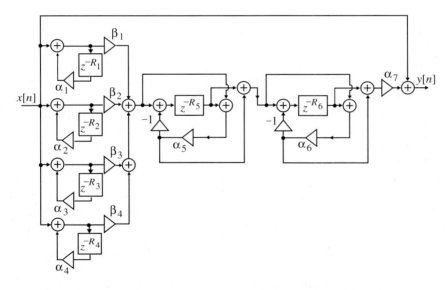

Figure 15.19: A proposed natural-sounding reverberator scheme.

adjusting R_i) and the multiplier constants α_i, it is possible to arrive at a pleasant-sounding reverberation, duplicating that occurring in a specific closed space, such as a concert hall.

Program 15_9[8] can be used to investigate the effect of the above natural-sounding reverberator on the speech signal shown in Figure 1.16.

Program 15_9.m

---

[8]Reproduced with permission of Prof. Dale Callahan, University of Alabama, Birmingham, AL.

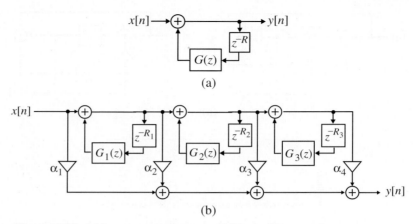

Figure 15.20: (a) Lowpass reverberator and (b) a multitap reverberator structure.

An interesting modification of the basic IIR comb filter of Figure 15.17(a) is obtained by replacing the multiplier α with a lowpass FIR or IIR filter $G(z)$, as indicated in Figure 15.20(a). It has a transfer function given by

$$H(z) = \frac{1}{1 - z^{-R}G(z)}, \tag{15.52}$$

obtained by replacing α in Eq. (15.50) with $G(z)$. This structure has been referred to as the *teeth filter* and has been introduced to provide a natural tonal character to the artificial reverberation generated by it [Egg84]. This type of reverberator should be carefully designed to avoid the stability problem. To provide a reverberation with a higher echo density, the teeth filter has been used as a basic unit in a more complex structure such as that indicated in Figure 15.18(b).

Additional details concerning these and other such composite reverberator structures can be found in [Sch62], [Moo79].

Flanging

There are a number of special sound effects that are often used in the mix-down process. One such effect is called *flanging*. Originally, it was created by feeding the same musical piece to two tape recorders and then combining their delayed outputs while varying the difference Δt between their delay times. One way of varying Δt is to slow down one of the tape recorders by placing the operator's thumb on the flange of the feed reel, which led to the name flanging [Ear86]. The FIR comb filter of Figure 15.15(a) can be modified to create the flanging effect. In this case, the unit generating the delay of R samples, or equivalently, a delay of RT seconds, where T is the sampling period, is made a time-varying delay $\beta(n)$, as indicated in Figure 15.21. The corresponding input–output relation is then given by

$$y[n] = x[n] + \alpha x[n - \beta(n)]. \tag{15.53}$$

Periodically varying the delay $\beta(n)$ between 0 and R with a low frequency ω_o such as

$$\beta(n) = \frac{R}{2}(1 - \cos(\omega_o n)) \tag{15.54}$$

generates a flanging effect on the sound. It should be noted that, as the value of $\beta(n)$ at an instant n, in general, has a noninteger value, in an actual implementation, the output sample value $y[n]$ should be computed using some type of interpolation method such as that outlined in Section 13.5.

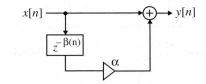

Figure 15.21: Generation of a flanging effect.

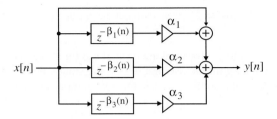

Figure 15.22: Generation of a chorus effect.

Program 15_10[9] can be used to investigate the effect of flanging on the musical sound signal dt.wav.

Chorus Generator

The *chorus* effect is achieved when several musicians are playing the same musical piece at the same time but with small changes in the amplitudes and small timing differences between their sounds. Such an effect can also be created synthetically by a *chorus generator* from the music of a single musician. A simple modification of the digital filter of Figure 15.21 leads to a structure that can be employed to simulate this sound effect. For example, the structure of Figure 15.22 can effectively create a chorus of four musicians from the music of a single musician. To achieve this effect, the delays $\beta_i(n)$ are randomly varied with very slow variations.

The *phasing* effect is produced by processing the signal through a narrowband notch filter with variable notch characteristics and adding a scaled portion of the notch filter output to the original signal as indicated in Figure 15.23 [Orf96]. The phase of the signal at the notch filter output can dramatically alter the phase of the combined signal, particularly around the notch frequency when it is varied slowly. The tunable notch filter can be implemented using the technique described in Section 8.7.2. The notch filter in Figure 15.23 can be replaced with a cascade of tunable notch filters to provide an effect similar to flanging. However, in flanging, the swept notch frequencies are always equally spaced, whereas in phasing, the locations of the notch frequencies and their corresponding 3-dB bandwidths are varied independently.

15.5.2 Frequency-Domain Operations

The frequency responses of individually recorded instruments or musical sounds of performers are frequently modified by the sound engineer during the mix-down process. These effects are achieved by passing the original signals through an equalizer, briefly reviewed in Section 1.4.1. The purpose of the equalizer is to provide "presence" by peaking the midfrequency components in the range of 1.5 to 3 kHz and to modify the bass–treble relationships by providing "boost" or "cut" to components outside this range. It is usually formed from a cascade of first-order and second-order filters with adjustable frequency responses. Many

Program 15_10.m

---

[9]Reproduced with permission of Prof. Dale Callahan, University of Alabama, Birmingham, AL.

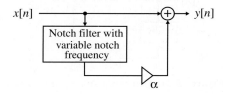

Figure 15.23: Generation of the phasing effect.

of the low-order digital filters employed for implementing these functions have been obtained by applying the bilinear transformation to analog filters. We first review the analog filters and then develop their digital equivalents. In addition, we describe some new structures with more flexible frequency responses.

Analog Filters

Simple lowpass and highpass analog filters with a Butterworth magnitude response are usually employed in analog mixers. The transfer functions of first-order analog lowpass and highpass Butterworth filters were given in Eq. (9.19) and (9.24), respectively. The transfer functions of higher-order lowpass and highpass analog Butterworth filters can be derived using the method outlined in Section 4.4.2. Also used in analog mixers are second-order analog bandpass and bandstop filters whose transfer functions were given in Eq. (9.26) and Eq. (9.31), respectively.

A first-order lowpass analog shelving filter for boost has a transfer function given by [Reg87b]

$$H_{LP}^{(B)}(s) = \frac{s + K\Omega_c}{s + \Omega_c}, \qquad K > 1. \tag{15.55}$$

It follows from Eq. (15.55) that

$$H_{LP}^{(B)}(0) = K, \qquad H_{LP}^{(B)}(\infty) = 1.$$

The transfer function $H_{LP}^{(B)}(s)$ of Eq. (15.55) can also be used for cut if $K < 1$. However, in this case, $H_{LP}^{(B)}(s)$ has a magnitude response that is not symmetrical to that for the case of $K > 1$ (boost) with respect to the frequency axis without changing the cutoff frequency [Reg87b]. The first-order lowpass analog shelving filter providing cut that retains the cutoff frequency has a transfer function given by [Zöl95]

$$H_{LP}^{(C)}(s) = \frac{s + \Omega_c}{s + \Omega_c/K}, \qquad K < 1, \tag{15.56}$$

for which

$$H_{LP}^{(C)}(0) = K, \qquad H_{LP}^{(C)}(\infty) = 1.$$

The first-order highpass analog shelving filter $H_{HP}^{(B)}(s)$ for the boost and cut can be derived by applying a lowpass-to-highpass transformation to the transfer functions of Eqs. (15.55) and (15.56), respectively. The transfer function for boost is given by

$$H_{HP}^{(B)}(s) = \frac{Ks + \Omega_c}{s + \Omega_c}, \qquad K > 1, \tag{15.57}$$

for which

$$H_{HP}^{(B)}(0) = 1, \qquad H_{HP}^{(B)}(\infty) = K.$$

Likewise, the transfer function for cut is given by

$$H_{HP}^{(C)}(s) = K \left(\frac{s + \Omega_c}{s + K\Omega_c} \right), \qquad K < 1, \tag{15.58}$$

for which

$$H_{HP}^{(C)}(0) = 1, \qquad H_{HP}^{(C)}(\infty) = K.$$

The *peak filter* is used for boost or cut at a finite frequency Ω_o. The transfer function of an analog second-order peak filter is given by

$$H_{BP}^{(BC)}(s) = \frac{s^2 + KBs + \Omega_o^2}{s^2 + Bs + \Omega_o^2}, \tag{15.59}$$

for which the maximum (minimum) value of the magnitude response, determined by K, occurs at the center frequency Ω_o. The above peak filter operates as a bandpass filter for $K > 1$ and as a bandstop filter for $K < 1$. The 3-dB bandwidth of the passband for a bandpass response and the 3-dB bandwidth of the stopband for a bandstop response is given by $B = \Omega_o/Q_o$.

First-Order Digital Filters and Equalizers

The analog filters can be converted into their digital equivalents by applying the bilinear transformation of Eq. (9.14) to their corresponding transfer functions. The design of first-order Butterworth digital lowpass and highpass filters derived via bilinear transformation of corresponding analog transfer functions has been treated in Section 9.2.2. The relevant transfer functions are given in Eqs. (9.21) and (9.25), respectively.

The transfer functions of the first-order digital lowpass and highpass filters given by Eqs. (9.21) and (9.25) can be alternatively expressed as[10]

$$G_{LP}(z) = \tfrac{1}{2} \left\{ 1 - \mathcal{A}_1(z) \right\}, \tag{15.60a}$$

$$G_{HP}(z) = \tfrac{1}{2} \left\{ 1 + \mathcal{A}_1(z) \right\}, \tag{15.60b}$$

where $\mathcal{A}_1(z)$ is a first-order allpass transfer function given by

$$\mathcal{A}_1(z) = \frac{\alpha - z^{-1}}{1 - \alpha z^{-1}}. \tag{15.61}$$

A composite realization of the above two transfer functions is sketched in Figure 15.24, where the first-order allpass digital transfer function $\mathcal{A}_1(z)$ can be realized using any one of the single multiplier structures of Figure 8.24. Note that in this structure, the 3-dB cutoff frequency ω_c of both digital filters is independently controlled by the multiplier constant α of the allpass section.

To derive the transfer function $G_{LP}^{(B)}(z)$ of a first-order digital low-frequency shelving filter for boost, we first observe that Eq. (15.57) can be rewritten as a sum of a first-order analog lowpass and a first-order analog highpass transfer function [Reg87b]:

$$H_{LP}^{(B)}(s) = K \left(\frac{\Omega_c}{s + \Omega_c} \right) + \left(\frac{s}{s + \Omega_c} \right). \tag{15.62}$$

---

[10]See Section 8.7.1.

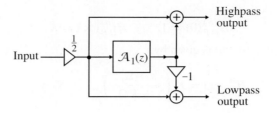

Figure 15.24: A parametrically tunable first-order lowpass/highpass filter.

Applying the bilinear transformation to the transfer function of Eq. (15.62) and making use of Eqs. (15.60a) and (15.60b), we arrive at

$$G_{LP}^{(B)}(z) = \tfrac{K}{2}\left[1 - \mathcal{A}_B(z)\right] + \tfrac{1}{2}\left[1 + \mathcal{A}_B(z)\right], \tag{15.63}$$

where, to emphasize the fact that the above shelving filter is for boost, we have replaced $\mathcal{A}_1(z)$ with $\mathcal{A}_B(z)$, with the latter rewritten as

$$\mathcal{A}_B(z) = \frac{\alpha_B - z^{-1}}{1 - \alpha_B z^{-1}}. \tag{15.64}$$

From Eq. (9.22) the tuning parameter α_B is given by

$$\alpha_B = \frac{1 - \tan(\omega_c T/2)}{1 + \tan(\omega_c T/2)}. \tag{15.65}$$

Likewise, the transfer function of a first-order digital low-frequency shelving filter for cut is obtained by applying the bilinear transformation to $H_{LP}^{(C)}(s)$ of Eq. (15.58) [Zöl97]. To this end, we first rewrite $H_{LP}^{(C)}(s)$ as a sum of a lowpass and a highpass transfer functions as indicated below:

$$H_{LP}^{(C)}(s) = \left(\frac{\Omega_c}{s + \Omega_c/K}\right) + \left(\frac{s}{s + \Omega_c/K}\right), \tag{15.66}$$

which, after a bilinear transformation, leads to the transfer function of a first-order low-frequency digital shelving filter for cut as given by

$$G_{LP}^{(C)}(z) = \tfrac{K}{2}[1 - \mathcal{A}_C(z)] + \tfrac{1}{2}[1 + \mathcal{A}_C(z)], \tag{15.67}$$

where

$$\mathcal{A}_C(z) = \frac{\alpha_C - z^{-1}}{1 - \alpha_C z^{-1}}, \tag{15.68}$$

with

$$\alpha_C = \frac{K - \tan(\omega_c T/2)}{K + \tan(\omega_c T/2)}. \tag{15.69}$$

It should be noted that $G_{LP}^{(C)}(z)$ of Eq. (15.67) is identical in form to $G_{LP}^{(B)}(z)$ of Eq. (15.63). Hence, the digital filter structure shown in Figure 15.25 can be used for both boost and cut, except for boost $\mathcal{A}_1(z) = \mathcal{A}_B(z)$ and for cut $\mathcal{A}_1(z) = \mathcal{A}_C(z)$. Two other possible realizations of the low-frequency shelving filter are considered in Problem 7.47.

Figures 15.26(a) and (b) show the gain responses of the first-order lowpass digital shelving filter obtained by varying the multiplier constant K and ω_c. Note that the parameter K controls the amount

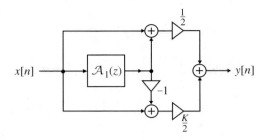

Figure 15.25: Low-frequency shelving filter where $\mathcal{A}_1(z) = \mathcal{A}_B(z)$ for boost and $\mathcal{A}_1(z) = \mathcal{A}_C(z)$ for cut.

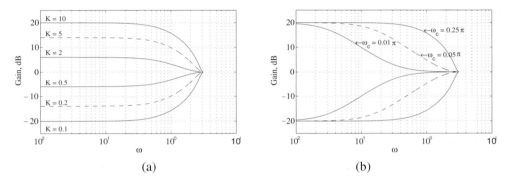

Figure 15.26: Gain responses of the low-frequency digital shelving filter (a) for six values of K with $\omega_c = 0.25\pi$ and $T = 1$ and (b) for three values of ω_c with $T = 1$ and $K = 10$ for boost and $K = 0.1$ for cut.

of boost or cut at low frequencies, while the parameters α_B and α_C control the boost bandwidth and cut bandwidth, respectively.

To derive the transfer function $G_{HP}^{(B)}(z)$ of a first-order high-frequency shelving filter for boost, we first express Eq. (15.57) as a sum of a first-order analog lowpass and highpass transfer function and then apply the bilinear transformation to the resulting expression, arriving at

$$G_{HP}^{(B)}(z) = \tfrac{1}{2}[1 - \mathcal{A}_B(z)] + \tfrac{K}{2}[1 + \mathcal{A}_B(z)], \tag{15.70}$$

where $\mathcal{A}_B(z)$ is as given by Eq. (15.64), with the multiplier constant α_B given by Eq. (15.65).

Likewise, the transfer function $G_{HP}^{(C)}(z)$ of a first-order high-frequency shelving filter for cut is obtained by expressing Eq. (15.58) as a sum of a first-order analog lowpass and highpass transfer function and then applying the bilinear transformation resulting in

$$G_{HP}^{(C)}(z) = \tfrac{1}{2}[1 - \mathcal{A}_C(z)] + \tfrac{K}{2}[1 + \mathcal{A}_C(z)], \tag{15.71}$$

where $\mathcal{A}_C(z)$ is as given by Eq. (15.67), with the multiplier constant α_C given by

$$\alpha_C = \frac{1 - K\tan(\omega_c T/2)}{1 + K\tan(\omega_c T/2)}. \tag{15.72}$$

As $G_{HP}^{(B)}(z)$ of Eq. (15.70) and $G_{HP}^{(C)}(z)$ of Eq. (15.71) are identical in form, the digital filter structure of Figure 15.27 can be employed for both boost and cut, except for boost $\mathcal{A}_1(z) = \mathcal{A}_B(z)$ and for cut $\mathcal{A}_1(z) = \mathcal{A}_C(z)$.

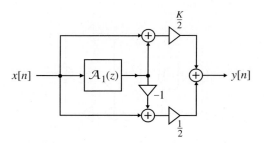

Figure 15.27: High-frequency shelving filter.

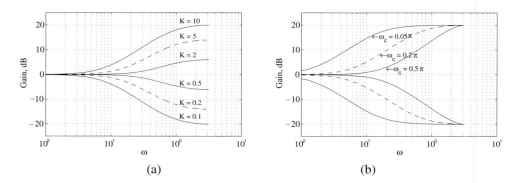

Figure 15.28: Gain responses of the high-frequency shelving filter of Figure 15.27 (a) for three values of the parameter K, with $\omega_c = 0.5\pi$ and $T = 1$, and (b) for three values of the parameter ω_c, with $K = 10$ and $T = 1$.

Figures 15.28(a) and (b) show the gain responses of the first-order high-frequency shelving filter obtained by varying the multiplier constant K and ω_c. Note that, as in the case of the low-frequency shelving filter, here the parameter K controls the amount of boost or cut at high frequencies, while the parameters α_B and α_C control the boost bandwidth and cut bandwidth, respectively.

Additional digital filter structures of the first-order lowpass and highpass shelving filters based on an alternate allpass decomposition are considered in Problem 7.48.

Second-Order Digital Filters and Equalizers

The design of second-order digital bandpass and bandstop filters derived via bilinear transformation of corresponding analog transfer functions has been treated in Section 9.2.2. The relevant transfer functions, given in Eqs. (9.28) and (9.32), can be alternatively expressed as[11]

$$G_{BP}(z) = \tfrac{1}{2}\left[1 - \mathcal{A}_2(z)\right], \tag{15.73a}$$

$$G_{BS}(z) = \tfrac{1}{2}\left[1 + \mathcal{A}_2(z)\right], \tag{15.73b}$$

where $\mathcal{A}_2(z)$ is a second-order allpass transfer function given by

$$\mathcal{A}_2(z) = \frac{\alpha - \beta(1 + \alpha)z^{-1} + z^{-2}}{1 - \beta(1 + \alpha)z^{-1} + \alpha z^{-2}}. \tag{15.74}$$

---

[11]See Section 8.7.2.

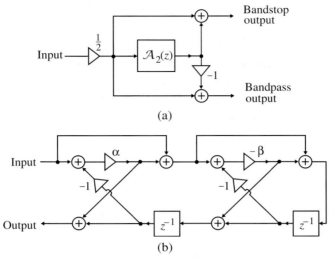

Figure 15.29: A parametrically tunable second-order digital bandpass/bandstop filter: (a) overall structure and (b) allpass section realizing $\mathcal{A}_2(z)$.

A composite realization of both filters is indicated in Figure 15.29(a), where the second-order allpass section is realized using the cascaded lattice structure of Figure 15.29(b) for independent tuning of the center (notch) frequency ω_o and the 3-dB bandwidth B_w.

The transfer function $G_{BP}^{(B)}(z)$ of a second-order peak filter for boost can be derived by applying the lowpass-to-bandpass spectral transformation of Eq. (9.44) to the lowpass shelving filter of Eq. (15.63), resulting in [Zöl95]

$$G_{BP}^{(B)}(z) = G_{LP}^{(B)}(z) \mid_{z^{-1} \to -z^{-1}\left(\frac{z^{-1}+\beta}{1+\beta z^{-1}}\right)} = \frac{K}{2}[1 - \mathcal{A}_{2B}(z)] + \frac{1}{2}[1 + \mathcal{A}_{2B}(z)], \qquad (15.75)$$

where

$$\beta = \cos(\omega_o), \qquad (15.76)$$

determines the center angular frequency ω_o where the bandpass response peaks and

$$\mathcal{A}_{2B}(z) = \frac{\alpha_B - \beta(1 + \alpha_B)z^{-1} + z^{-2}}{1 - \beta(1 + \alpha_B)z^{-1} + \alpha_B z^{-2}} \qquad (15.77)$$

is a second-order allpass transfer function obtained by applying the lowpass-to-bandpass transformation of Eq. (9.44) to the first-order allpass transfer function $\mathcal{A}_{(B)}(z)$ of Eq. (15.64). Here, the parameter α is now related to the 3-dB bandwidth B_w of the bandpass response through

$$\alpha_B = \frac{1 - \tan(B_w T/2)}{1 + \tan(B_w T/2)}. \qquad (15.78)$$

Likewise, the transfer function $G_{BP}^{(C)}(z)$ of a second-order peak filter for cut is obtained by applying the lowpass-to-bandpass transformation to the lowpass shelving filter for cut of Eq. (15.67), resulting in

$$G_{BP}^{(C)}(z) = G_{LP}^{(C)}(z) \Big|_{z^{-1} \to -z^{-1}\left(\frac{z^{-1}+\beta}{1+\beta z^{-1}}\right)} = \frac{K}{2}[1 - \mathcal{A}_{2C}(z)] + \frac{1}{2}[1 + \mathcal{A}_{2C}(z)]. \qquad (15.79)$$

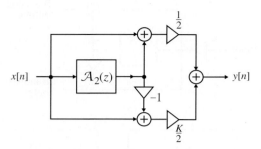

Figure 15.30: A parametrically tunable second-order peak filter for boost and cut.

In Eq. (15.79), the center angular frequency ω_o, where the bandstop response dips, is related to the parameter β through Eq. (15.76) and

$$\mathcal{A}_{2C}(z) = \frac{\alpha_C - \beta(1 + \alpha_C)z^{-1} + z^{-2}}{1 - \beta(1 + \alpha_C)z^{-1} + \alpha_C z^{-2}} \tag{15.80}$$

is a second-order allpass transfer function obtained by applying the lowpass-to-bandpass transformation of Eq. (9.44) to the first-order allpass transfer function $\mathcal{A}_{(C)}(z)$ of Eq. (15.67). Here, the parameter α_C is now related to the 3-dB bandwidth B_w of the bandstop response through

$$\alpha_C = \frac{K - \tan(B_w T/2)}{K + \tan(B_w T/2)}. \tag{15.81}$$

Since both $G_{BP}^{(B)}(z)$ and $G_{BP}^{(C)}(z)$ are identical in form, the digital filter structure of Figure 15.30 can be employed for both boost and cut, except for boost $\mathcal{A}_2(z) = \mathcal{A}_{2B}(z)$ and for cut $\mathcal{A}_2(z) = \mathcal{A}_{2C}(z)$.

It follows from the above discussion that the peak or the dip of the gain response occurs at the frequency ω_o, which is controlled independently by the parameter β according to Eq. (15.76), and the 3-dB bandwidth B_w of the gain response is determined solely by the parameter α_B of Eq. (15.78) for boost or by the parameter α_C of of Eq. (15.81) for cut. Moreover, the height of the peak of the magnitude response for boost is given by $K = G_{BP}^{(B)}(e^{j\omega_o})$ and the height of the dip of the magnitude response for cut is given by $K = G_{BP}^{(C)}(e^{j\omega_o})$. Figures 15.31(a), (b), and (c) show the gain responses of the second-order peak filter obtained by varying the parameters K, ω_o, and B_w.

Higher-Order Equalizers

A graphic equalizer with tunable gain response can be built using a cascade of first-order and second-order equalizers with external control of the maximum gain values of each section in the cascade. Figure 15.32(a) shows the block diagram of a cascade of one first-order and three second-order equalizers with nominal frequency response parameters as indicated. Figure 15.32(b) shows its gain response for some typical values of the parameter K (maximum gain values) of the individual sections.

15.6   Digital Music Synthesis

We indicated in Section 1.4.4 that there are basically four methods of musical sound synthesis: (1) *wavetable synthesis*, (2) *spectral modeling synthesis*, (3) *nonlinear synthesis*, and (4) *physical modeling synthesis* [Rab2001], [Smi91]. A detailed discussion of all these methods is beyond the scope of this book. In this

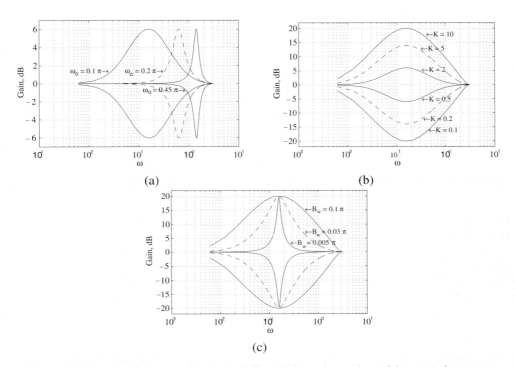

Figure 15.31: Gain responses of the second-order peak filter (a) for various values of the center frequency ω_o, with $B_w = 0.1\pi$, $T = 1$, and $K = 10$ for boost and $K = 0.1$ for cut; (b) for various values of the parameter K, with $\omega_o = 0.45\pi$, $B_w = 0.1\pi$, and $T = 1$; and (c) for various values of the parameter B_w, with $\omega_0 = 0.45\pi$, $T = 1$ and $K = 10$ for boost and $K = 0.1$ for cut.

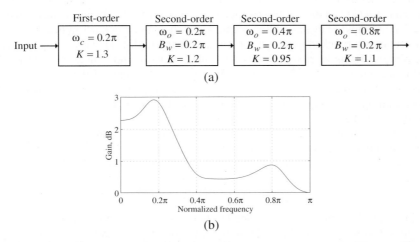

Figure 15.32: (a) Block diagram of a typical graphic equalizer and (b) its gain response for the section parameter values shown.

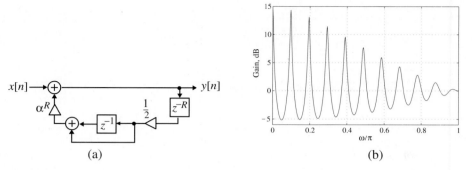

Figure 15.33: (a) Basic plucked-string filter structure and (b) its gain response for $R = 20$ and $\alpha = 0.99$.

section, we outline a simple wavetable synthesis-based method for generating the sounds of plucked-string instruments [Kar83].

The basic idea behind the wavetable synthesis method is to store one period of a desired musical tone and repeat it over and over to generate a periodic signal. Such a signal can be generated by running the IIR digital filter structure of Figure 15.17(a) with specified initial conditions, called *wavetable*, stored in the delay register z^{-R} and with no input. Mathematically, the generated periodic note can be expressed as

$$y[n] = y[n - R],$$

where R, called the *wavetable length*, is the period. The frequency of the tone is F_T / R, where F_T is the sampling frequency. Usually, samples of simple waveforms are used as initial conditions.

A simple modification of the algorithm has been used to generate plucked-string tones [Kar83]. The modified algorithm is given by

$$y[n] = \frac{\alpha^R}{2} (y[n - R] + y[n - R - 1]). \tag{15.82}$$

The corresponding *plucked-string filter* structure is shown in Figure 15.33(a). It should be noted that this structure has been derived from the IIR filter structure of Figure 15.17(a) by inserting a lowpass filter $G(z)$ consisting of a 2-point moving average filter in cascade with a gain block α^R in the feedback path.

The initial sound of a plucked guitar string contains many high-frequency components. To simulate this effect, the plucked-string filter structure is run with zero input and with zero-mean random numbers initially stored in the delay block z^{-R}. The high-frequency components of the stored data get repeatedly lowpass filtered by $G(z)$ as they circulate around the feedback loop of the filter structure of Figure 15.33(a) and decay faster than the low-frequency components. Since the 2-point moving average filter has a group delay of $\frac{1}{2}$ samples, the pitch period of the tone is $R + \frac{1}{2}$ samples.

It is instructive to examine the gain response of the plucked-string filter [Ste96]. The transfer function of the the filter structure of Fig. 15.33(a) is given by

$$H(z) = \frac{1}{1 - \frac{\alpha^R}{2}(1 + z^{-1})z^{-R}}. \tag{15.83}$$

As the loop delay is 20.5 samples, the resonance frequencies are expected to occur at integer multiples of the pitch frequency $F_T / 20.5$, where F_T is the sampling frequency. It can be seen from the gain response plot shown in Figure 15.33(b) for $R = 20$ and $\alpha = 0.99$, the resonance peaks occur at frequencies very close to the expected values. In addition, the amplitudes of the peaks decrease with increasing frequencies as desired. Moreover, the widths of the resonance peaks increase with increasing frequency, as expected.

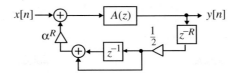

Figure 15.34: Modified plucked-string filter structure.

For better control of the pitch frequency, an allpass filter $A(z)$ is inserted in the feedback loop, as indicated in Figure 15.34 [Jaf83]. The fractional group delay of the allpass filter can be adjusted to tune the overall loop delay of the modified structure. A detailed discussion on the design of the modified plucked-string filter structure for the generation of a sound with a given fundamental frequency can be found in Steiglitz [Ste96].

15.7 Discrete-Time Analytic Signal Generation

As discussed in Section 1.2.3, an analytic continuous-time signal has a zero-valued spectrum for all negative frequencies. Such a signal finds applications in single-sideband analog communication systems and analog frequency-division multiplex systems. A discrete-time signal with a similar property finds applications in digital communication systems and is the subject of this section. We illustrate here the generation of an analytic signal $y[n]$ from a discrete-time real signal $x[n]$ and describe some of its applications.

Now, the Fourier transform $X(e^{j\omega})$ of a real signal $x[n]$, if it exists, is nonzero for both positive and negative frequencies. On the other hand, a signal $y[n]$ with a single-sided spectrum $Y(e^{j\omega})$ that is zero for negative frequencies must be a complex signal. Consider the complex analytic signal

$$y[n] = x[n] + j\hat{x}[n], \tag{15.84}$$

where $x[n]$ and $\hat{x}[n]$ are real. Its Fourier transform $Y(e^{j\omega})$ is given by

$$Y(e^{j\omega}) = X(e^{j\omega}) + j\hat{X}(e^{j\omega}), \tag{15.85}$$

where $\hat{X}(e^{j\omega})$ is the Fourier transform of $\hat{x}[n]$. Now, $x[n]$ and $\hat{x}[n]$ being real, their corresponding Fourier transforms are conjugate symmetric; that is, $X(e^{j\omega}) = X^*(e^{-j\omega})$ and $\hat{X}(e^{j\omega}) = \hat{X}^*(e^{-j\omega})$. Hence, from Eq. (15.85), we obtain

$$X(e^{j\omega}) = \tfrac{1}{2}\left[Y(e^{j\omega}) + Y^*(e^{-j\omega})\right], \tag{15.86a}$$

$$j\hat{X}(e^{j\omega}) = \tfrac{1}{2}\left[Y(e^{j\omega}) - Y^*(e^{-j\omega})\right]. \tag{15.86b}$$

Since, by assumption, $Y(e^{j\omega}) = 0$ for $-\pi \le \omega < 0$, we obtain from Eq. (15.86a)

$$Y(e^{j\omega}) = \begin{cases} 2X(e^{j\omega}), & 0 \le \omega < \pi, \\ 0, & -\pi \le \omega < 0. \end{cases} \tag{15.87}$$

Thus, the analytic signal $y[n]$ can be generated by passing $x[n]$ through a linear discrete-time system, with a frequency response $H(e^{j\omega})$ given by

$$H(e^{j\omega}) = \begin{cases} 2, & 0 \le \omega < \pi, \\ 0, & -\pi \le \omega < 0, \end{cases} \tag{15.88}$$

as indicated in Figure 15.35(a).

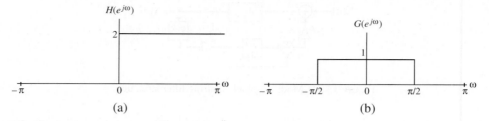

Figure 15.35: (a) Frequency response of the discrete-time filter generating an analytic signal and (b) half-band lowpass filter.

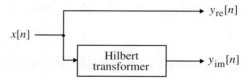

Figure 15.36: Generation of an analytic signal using a Hilbert transformer.

15.7.1   The Discrete-Time Hilbert Transformer

We now relate the imaginary part $\hat{x}[n]$ of the analytic signal $y[n]$ to its real part $x[n]$. From Eq. (15.86b), we get

$$\hat{X}(e^{j\omega}) = \tfrac{1}{2j}\left[Y(e^{j\omega}) - Y^*(e^{-j\omega})\right]. \tag{15.89}$$

For $0 \le \omega < \pi$, $Y(e^{-j\omega}) = 0$, and for $-\pi \le \omega < 0$, $Y(e^{j\omega}) = 0$. Using this property and Eq. (15.87) in Eq. (15.89), it can be easily shown that

$$\hat{X}(e^{j\omega}) = \begin{cases} -jX(e^{j\omega}), & 0 \le \omega < \pi, \\ jX(e^{j\omega}), & -\pi \le \omega < 0. \end{cases} \tag{15.90}$$

Thus, the imaginary part $\hat{x}[n]$ of the analytic signal $y[n]$ can be generated by passing its real part $x[n]$ through a linear discrete-time system, with a frequency response $H_{\mathrm{HT}}(e^{j\omega})$ given by

$$H_{\mathrm{HT}}(e^{j\omega}) = \begin{cases} -j, & 0 \le \omega < \pi, \\ j, & -\pi \le \omega < 0. \end{cases} \tag{15.91}$$

The linear system defined by Eq. (15.91) is usually referred to as the ideal *Hilbert transformer*. Its output $\hat{x}[n]$ is called the *Hilbert transform* of its input $x[n]$. The basic scheme for the generation of an analytic signal $y[n] = y_{\mathrm{re}}[n] + jy_{\mathrm{im}}[n]$ from a real signal $x[n]$ is thus as indicated in Figure 15.36. Observe that $\left|H_{\mathrm{HT}}(e^{j\omega})\right| = 1$ for all frequencies and has a -90-degree phase-shift for $0 \le \omega < \pi$ and a $+90$-degree phase-shift for $-\pi \le \omega < 0$. As a result, an ideal Hilbert transformer is also called a 90-*degree phase-shifter.*

The impulse response $h_{\mathrm{HT}}[n]$ of the ideal Hilbert transformer is obtained by taking the inverse Fourier transform of $H_{\mathrm{HT}}(e^{j\omega})$ and can be shown to be (Problem 10.5)

$$h_{\mathrm{HT}}[n] = \begin{cases} 0, & \text{for } n \text{ even}, \\ \dfrac{2}{\pi n}, & \text{for } n \text{ odd}. \end{cases} \tag{15.92}$$

Since the ideal Hilbert transformer has a two-sided infinite-length impulse response defined for $-\pi < n < \pi$, it is an unrealizable system. Moreover, its transfer function $H_{HT}(z)$ exists only on the unit circle. We describe later two approaches for developing a realizable approximation.

15.7.2   Relation with Half-Band Filters

Consider the filter with a frequency response $G(e^{j\omega})$ obtained by shifting the frequency response $H(e^{j\omega})$ of Eq. (15.88) by $\pi/2$ radians and scaling by a factor $\frac{1}{2}$ (see Figure 15.35):

$$G(e^{j\omega}) = \tfrac{1}{2}H(e^{j(\omega+\pi/2)}) = \begin{cases} 1, & 0 < |\omega| < \frac{\pi}{2}, \\ 0, & \frac{\pi}{2} < |\omega| < \pi. \end{cases} \tag{15.93}$$

From our discussion in Section 13.6.2, we observe that $G(e^{j\omega})$ is a half-band lowpass filter. Because of the relation between $H(e^{j\omega})$ of Eq. (15.88) and the real coefficient half-band lowpass filter $G(e^{j\omega})$ of Eq. (15.93), the filter $H(e^{j\omega})$ has been referred to as a *complex half-band filter* [Reg93].

15.7.3   Design of the Hilbert Transformer

It also follows from the above relation that a complex half-band filter can be designed simply by shifting the frequency response of a half-band lowpass filter by $\pi/2$ radians and then scaling by a factor 2. Equivalently, the relation between the transfer functions of a complex half-band filter $H(z)$ and a real half-band lowpass filter $G(z)$ is given by

$$H(z) = j2G(-jz). \tag{15.94}$$

Three methods of the design of the real half-band filter have been presented in Section 13.6. We adopt two of these methods here for the design of complex half-band filters.

FIR Complex Half-Band Filter

Let $G(z)$ be the desired FIR real half-band linear-phase lowpass filter of even degree N, with the passband edge at ω_p, stopband edge at ω_s, and passband and stopband ripples of δ_p, with $\omega_p + \omega_s = \pi$. The half-band filter $G(z)$ is then designed by first designing a wide-band linear-phase filter $F(z)$ of degree $N/2$ with a passband from 0 to $2\omega_p$, a transition band from $2\omega_p$ to π, and a passband ripple of 2δ. The desired half-band filter $G(z)$ is then obtained by forming

$$G(z) = \tfrac{1}{2}\left[z^{-N/2} + F(z^2)\right]. \tag{15.95}$$

Substituting Eq. (15.95) in Eq. (15.94), we obtain

$$H(z) = j\left[(-jz)^{-N/2} + F(-z^2)\right] = z^{-N/2} + jF(-z^2). \tag{15.96}$$

An FIR implementation of the complex half-band filter based on the above decomposition is indicated in Figure 15.37. The linear-phase FIR filter $F(-z^2)$ is thus an approximation to a Hilbert transformer.

We illustrate the above approach in Example 15.8.

EXAMPLE 15.8   FIR Complex Half-Band Filter Design

Using MATLAB, we design a wide-band FIR filter $F(z)$ of degree 13 with a passband from 0 to 0.35π and an extremely small stopband from 0.9π to π. We use the function `remez` with a magnitude vector `m = [1 1 0 0]`. The weight vector used to weigh the passband and the stopband is `wt = [2 0.05]`. The magnitude responses of the wide-band filter $F(z)$ and the Hilbert transformer $F(-z^2)$ are shown in Figures 15.38(a) and (b).

The FIR Hilbert transformer can be designed directly using the function `remez`. Example 15.9 illustrates this approach.

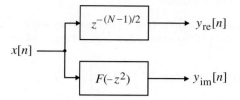

Figure 15.37: FIR realization of a complex half-band filter.

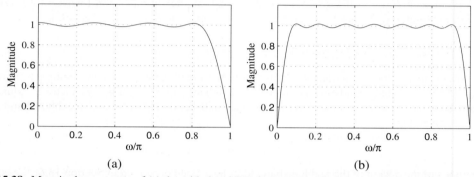

(a)   (b)

Figure 15.38: Magnitude responses of (a) the wide-band FIR filter $F(z)$ and (b) the approximate Hilbert transformer $F(-z^2)$.

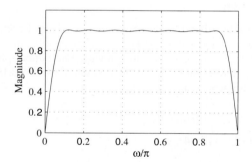

Figure 15.39: The magnitude response of the Hilbert transformer designed directly using MATLAB.

EXAMPLE 15.9   Direct Design of FIR Complex Half-Band Filter Using MATLAB

We design a 26th-order FIR Hilbert transformer with a passband from 0.1π to 0.9π. It should be noted that for the design of a Hilbert transformer, the first frequency point in the vector f containing the specified bandedges cannot be a 0. The magnitude response of the designed Hilbert transformer obtained using the program statement b = remez(26, [0.1 0.9], [1 1], 'Hilbert') is indicated in Figure 15.39. It should be noted that due to numerical round-off problems, unlike the design of Example 15.8, the odd impulse response coefficients of the Hilbert transformer here are not exactly zero.

IIR Complex Half-Band Filter

We outlined in Section 13.6.5 a method to design stable IIR real coefficient half-band filters of odd order in the form [Vai87f]

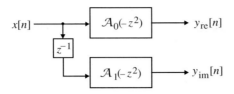

Figure 15.40: IIR realization of a complex half-band filter.

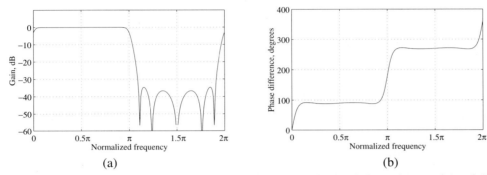

Figure 15.41: (a) Gain response of the complex half-band filter (normalized to 0 dB maximum gain) and (b) phase difference between the two allpass sections of the complex half-band filter.

$$G(z) = \tfrac{1}{2}[\mathcal{A}_0(z^2) + z^{-1}\mathcal{A}_1(z^2)], \tag{15.97}$$

where $\mathcal{A}_0(z)$ and $\mathcal{A}_1(z)$ are stable allpass transfer functions. Substituting Eq. (15.97) in Eq. (15.94), we therefore arrive at

$$H(z) = \mathcal{A}_0(-z^2) + jz^{-1}\mathcal{A}_1(-z^2). \tag{15.98}$$

A realization of the complex half-band filter based on the above decomposition is thus as shown in Figure 15.40.

We illustrate the above approach to Hilbert transformer design in Example 15.10.

EXAMPLE 15.10   IIR Complex Half-Band Filter Design

In Example 13.24, we designed a real half-band elliptic filter with the following frequency response specifications: $\omega_s = 0.6\pi$ and $\delta_s = 0.016$. The transfer function of the real half-band filter $G(z)$ can be expressed as in Eq. (15.97), where the transfer functions of the two allpass sections $\mathcal{A}_0(z^2)$ and $\mathcal{A}_1(z^2)$ are given in Eq. (13.116). The gain response of the complex half-band filter $H(z)$ obtained using Eq. (15.98) is sketched in Figure 15.41(a). Figure 15.41(b) shows the phase difference between the two allpass functions $\mathcal{A}_0(-z^2)$ and $z^{-1}\mathcal{A}_1(-z^2)$ of the complex half-band filter. Note that, as expected, the phase difference is 90 degrees for most of the positive frequency range and 270 degrees for most of the negative frequency range. In plotting the gain response of the complex half-band filter and the phase difference between its constituent two allpass sections, the M-file `freqz(num,den,n,'whole')` has been used to compute the pertinent frequency response values over the whole normalized frequency range from 0 to 2π.

15.7.4   Single-Sideband Modulation

For efficient transmission over long distances, a real low-frequency band-limited signal $x[n]$, such as speech or music, is modulated by a very high frequency sinusoidal carrier signal $\cos \omega_c n$, with the carrier

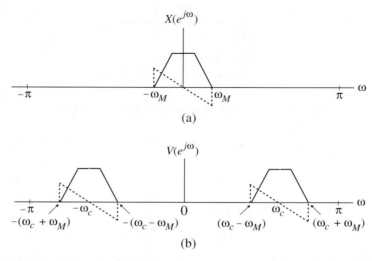

Figure 15.42: Spectra of a real signal and its modulated version. (Solid lines represent the real parts, and dashed lines represent the imaginary parts.)

frequency ω_c being less than half of the sampling frequency. The spectrum $V(e^{j\omega})$ of the resulting signal $v[n] = x[n] \cos \omega_c n$ is given by

$$V(e^{j\omega}) = \tfrac{1}{2} \left[X(e^{j(\omega - \omega_c)}) + X(e^{j(\omega + \omega_c)}) \right]. \tag{15.99}$$

As indicated in Figure 15.42, if $X(e^{j\omega})$ is band-limited to ω_M, the spectrum $V(e^{j\omega})$ of the modulated signal $v[n]$ has a bandwidth of $2\omega_M$ centered at $\pm\omega_c$. By choosing widely separated carrier frequencies, one can modulate a number of low-frequency signals to high-frequency signals, combine them by frequency-division multiplexing, and transmit over a common channel. The carrier frequencies are chosen appropriately to ensure that there is no overlap in the spectra of the modulated signals when combined by frequency-division multiplexing. At the receiving end, each of the modulated signals is then separated by a bank of bandpass filters of center frequencies corresponding to the different carrier frequencies.

It is evident from Figure 15.42 that, for a real low-frequency signal $x[n]$, the spectrum of its modulated version $v[n]$ is symmetric with respect to the carrier frequency ω_c. Thus, the portion of the spectrum in the frequency range from ω_c to $(\omega_c + \omega_M)$, called the *upper sideband*, has the same information content as the portion in the frequency range from $(\omega_c - \omega_M)$ to ω_c, called the *lower sideband*. Hence, for a more efficient utilization of the channel bandwidth, it is sufficient to transmit either the upper or the lower sideband signal. A conceptually simple way of eliminating one of the sidebands is to pass the modulated signal $v[n]$ through a sideband filter whose passband covers the frequency range of one of the sidebands.

An alternative, often preferred, approach for single-sideband signal generation is by modulating the analytic signal whose real and imaginary parts are, respectively, the real signal and its Hilbert transform. To illustrate this approach, let $y[n] = x[n] + j\hat{x}[n]$, where $\hat{x}[n]$ is the Hilbert transform of $x[n]$. Consider

$$
\begin{aligned}
s[n] = y[n]e^{j\omega_c n} &= (y_{\text{re}}[n] + jy_{\text{im}}[n])(\cos \omega_c n + j \sin \omega_c n) \\
&= \left(x[n] \cos \omega_c n - \hat{x}[n] \sin \omega_c n \right) \\
&\quad + j \left(x[n] \sin \omega_c n + \hat{x}[n] \cos \omega_c n \right).
\end{aligned} \tag{15.100}
$$

From Eq. (15.100), the real and imaginary parts of $s[n]$ are thus given by

$$s_{re}[n] = x[n] \cos \omega_c n - \hat{x}[n] \sin \omega_c n, \qquad (15.101a)$$

$$s_{im}[n] = x[n] \sin \omega_c n + \hat{x}[n] \cos \omega_c n. \qquad (15.101b)$$

Figure 15.43 shows the spectra of $x[n]$, $\hat{x}[n]$, $y[n]$, $s[n]$, $s_{re}[n]$, and $s_{im}[n]$. It therefore follows from these plots that a single-sideband signal can be generated using either one of the modulation schemes described by Eqs. (15.101a) and (15.101b), respectively. A block diagram representation of the scheme of Eq. (15.101a) is sketched in Figure 15.44.

15.8 Signal Compression

As mentioned earlier, signals carry information, and the objective of signal processing is to preserve the information contained in the signal and extract and manipulate it when necessary. Most digital signals encountered in practice contain a huge amount of data. For example, a gray-level image of size 512×512 with 8-bits per pixel contains $(512)^2 \cdot 8 = 2,097,152$ bits. A color image of the same size contains 3 times as many bits. For efficient storage of digital signals, it is often necessary to compress the data into a smaller size requiring significantly fewer number of bits. A signal in compressed form also requires less bandwidth for transmission. Roughly speaking, signal compression is concerned with the reduction of the amount of data, while preserving the information content of the signal with some acceptable fidelity.

Most practical signals exhibit *data redundancy,* as they contain some amount of data with no relevant information. Three types of data redundancy are usually encountered in practice: *coding redundancy*, *intersample redundancy*, and *psychovisual redundancy* [Gon2002]. Signal compression methods exploit one or more of these redundancies to achieve data reduction.

A signal coding system consists of an encoder and a decoder. The input to the encoder is the signal **x** to be compressed, and its output is the compressed bit stream **d**. The decoder performs the reverse operation. Its input is the compressed bit stream **d** developed by the encoder, and its output $\hat{x}$ is a reasonable replica of the original input signal of the encoder. The basic components of the encoder and the decoder are shown in Figure 15.45.

The *energy compression* block transforms the input sequence **x** into another sequence **y** with the same total energy, while packing most of the energy in very few of samples **y**. The *quantizer* block develops an approximate representation of **y** for a given level of accuracy in the form of an integer-valued sequence **q** by adjusting the quantizer step size to control the trade-off between distortion and bit rate. The *entropy coding* block uses variable-length entropy coding to encode the integers in the sequence **q** into a binary bitstream **d**, with the aim of minimizing the total number of bits in **d** by making use of the statistics of the class of samples in **q**.

The *entropy decoding* block regenerates the integer-valued sequence **q** from the binary bit stream **d**. The *inverse quantizer* develops $\hat{y}$, a best estimate of **y** from **q**. Finally, the *reconstruction* block develops $\hat{x}$, the best approximation of the original input sequence **x** from $\hat{y}$.

The signal compression methods can be classified into two basic groups: *lossless* and *lossy*. In the lossless compression methods, no information is lost due to compression, and the original signal can be recovered exactly from the compressed data by the decoder. On the other hand, in the lossy compression methods, some amount of information (usually less relevant) is lost, due to compression and the signal reconstructed by the decoder is not a perfect replica of the original signal but is still an acceptable approximation for the application at hand. Naturally, the latter method can result in a significant reduction in the number of bits necessary to represent the signal and is considered here. Moreover, for conciseness, we discuss image compression methods that exploit only the coding redundancy. A detailed exposition of compression methods exploiting all types of data redundancies is beyond the scope of this book.

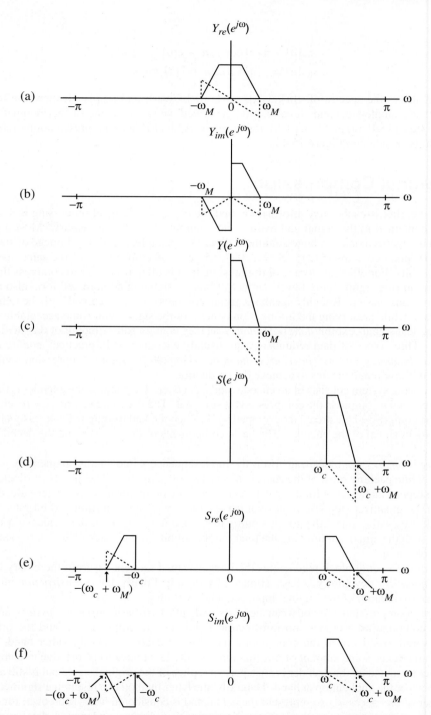

Figure 15.43: (a)–(f) Illustration of the generation of single-sideband signals via the Hilbert transform. (Solid lines represent the real parts, and dashed lines represent the imaginary parts.)

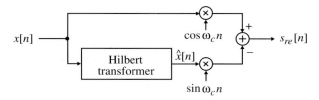

Figure 15.44: Schematic of the single-sideband generation scheme of Eq. (15.101a).

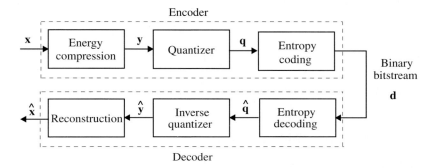

Figure 15.45: The block diagram representation of the signal compression system.

15.8.1 Coding Redundancy

We assume that each sample of the discrete-time signal $\{x[n]\}$ is a random variable r_i taking one of Q distinct values with a probability p_i, $0 \le i \le Q-1$, where $p_i \le 1$ and $\sum_{i=0}^{Q-1} p_i = 1$. Each possible value r_i is usually called a *symbol*. The probability p_i of each symbol r_i can be estimated from the histogram of the signal. Thus, if the signal contains a total of N samples, with m_i denoting the total number of samples taking the value r_i, then

$$p_i = \frac{m_i}{N}, \qquad 0 \le i \le Q-1. \tag{15.102}$$

Let b_i denote the length of the i-th codeword, that is, the total number of bits necessary to represent the value of the random variable r_i. A measure of the coding redundancy is then given by the average number of bits needed to represent each sample of the signal $\{x[n]\}$:

$$B_{\text{av}} = \sum_{i=0}^{Q-1} b_i\, p_i \ \text{bits}. \tag{15.103}$$

As a result, the total number of bits required to represent the signal is $N \cdot B_{\text{av}}$.

15.8.2 Entropy

The goal of the compression method is to reduce the volume of data while retaining the information content of the original signal with some acceptable fidelity. The information content represented by a symbol can be informally related to its unexpectedness; that is, if a symbol that arrives is the one that was expected, it does not convey very much information. On the other hand, if an unexpected symbol arrives, it conveys much more information. Thus, the information content of a particular symbol can be related to its probability of occurrence, as described next.

For a discrete-time sequence $\{x[n]\}$ with samples taking one of Q distinct symbols r_i with a probability p_i, $0 \leq i \leq Q - 1$, a measure of the information content I_i of the i-th symbol r_i is defined by [Jai89]

$$I_i = -\log_2 p_i. \tag{15.104}$$

It follows from the above definition that $I_i \geq 0$. Moreover, it also can be seen that I_i is very large when p_i is very small.

A measure of the average information content of the signal $\{x[n]\}$ is then given by its *entropy*, which is defined by

$$\mathcal{H}_x = \sum_{i=0}^{Q-1} p_i I_i = -\sum_{i=0}^{Q-1} p_i \log_2 p_i \text{ bits/symbol}. \tag{15.105}$$

The coding redundancy is defined as the difference between the actual data rate and the entropy of a data stream.

15.8.3   A Signal Compression Example[12]

We now consider the compression of a gray level image to illustrate the various concepts introduced earlier in this section. For the energy compression stage, we make use of the Haar wavelets of Section 14.6.1. Since the image is a two-dimensional sequence, the wavelet decomposition is first applied row-wise and then column-wise. Applying the Haar transform $\mathbf{H}$ to the input image $\mathbf{x}$, first row-wise and then column-wise, we get

$$\mathbf{y} = \mathbf{HxH}^T,$$

where

$$\mathbf{H} = \frac{1}{\sqrt{2}} \begin{bmatrix} 1 & 1 \\ 1 & -1 \end{bmatrix}.$$

To understand the effect of the decomposition on the image, consider a 2×2 two-dimensional sequence given by

$$\mathbf{x} = \begin{bmatrix} a & b \\ c & d \end{bmatrix}.$$

Then,

$$\mathbf{y} = \frac{1}{2} \begin{bmatrix} a+b+c+d & a-b+c-d \\ a+b-c-d & a-b-c+d \end{bmatrix}.$$

The element $a + b + c + d$ at the top left position of $\mathbf{y}$ is the 4-point average of $\mathbf{x}$ and therefore contains only the vertical and horizontal low-frequency components of $\mathbf{x}$. It is labeled as the **LL** part of $\mathbf{x}$. The element $a - b + c - d$ at the top right position of $\mathbf{y}$ is obtained by forming the differences of the horizontal components and the sum of the vertical components and hence contains the vertical low- and horizontal high-frequency components. It is labeled as the **HL** part of $\mathbf{x}$. The element $a + b - c - d$ at the bottom left position of $\mathbf{y}$ is obtained by forming the sum of the horizontal components and the differences of the vertical components and hence contains the horizontal low- and vertical high-frequency components. It is labeled as the **LH** part of $\mathbf{x}$. Finally, the element $a - b - c + d$ at the bottom right is obtained by forming the differences between both the horizontal and vertical components and therefore contains only the vertical and horizontal high-frequency components of $\mathbf{x}$. It is labeled as the **HH** part of $\mathbf{x}$.

---

[12]Portions of this section have been adapted from [Kin2001] by permission of the author.

(a)   (b)

Figure 15.46: (a) Original "Goldhill" image and (b) subimages after one-level Haar wavelet decomposition.

Applying a one-level Haar wavelet decomposition to the image "Goldhill" of Figure 15.46(a), down-sampling the outputs of all filters by a factor-of-2 in both horizontal and vertical directions, we arrive at the four subimages shown in Figure 15.46(b). The original image is of size 512×512 pixels. The subimages of Figure 15.46(b) are of size 256×256 pixels each. The total energies of the subimages and their percentages of the total energy of all subimages are now as follows:

LL:	**HL:**	**LH:**	**HH:**
3919.91×10^6	6.776×10^6	7.367×10^6	1.483×10^6
$99.603\,\%$	$0.172\,\%$	$0.187\,\%$	$0.038\,\%$

The sum of the energies of all subimages is equal to 3935.54×10^6, which is also the total energy of the original image of Figure 15.46(a). As can be seen from the above energy distribution data and Figure 15.46(b), the **LL** subimage contains most of the energy of the original image, whereas, the **HH** subimage contains least of the energy. Also, the **HL** subimage has mostly the near-horizontal edges, whereas the **LH** subimage has mostly the near-vertical edges.

To evaluate the entropies, we use uniform scalar quantizers for all signals with a quantization step size $Q = 15$. The entropy of the original image computed from its histogram, after compression with $Q = 15$, is $\mathcal{H}_x = 3.583$ bits/pixel. The entropies of the sub-images after a one-level Haar decomposition are as given below:

LL:	**HL:**	**LH:**	**HH:**
$4.549/4$	$1.565/4$	$1.375/4$	$0.574/4$
$= 1.1370$	$= 0.3911$	0.3438	0.1436

The entropy of the wavelet representation is $\mathcal{H}_y = 2.016$ bits/pixel, obtained by adding the entropies of the subimages given above. Hence, the compression ratio is 1.78-to-1.0. Figures 15.47(a) and (b) show, respectively, the reconstructed "Goldhill" image after direct quantization of the original pixels and after quantization of the wavelet coefficients.

A commonly used measure of the quality of the reconstructed image compared with the original image is the *peak-signal-to-noise ratio* (PSNR). Let $x[m, n]$ denote the (m, n)th pixel of an original image $\mathbf{x}$ of

(a)   (b)

Figure 15.47: Reconstructed "Goldhill" image: (a) after direct quantization of the original pixels and (b) after quantization of the wavelet coefficients.

size $M \times N$, and, $y[m, n]$ denote the (m, n)th pixel of the reconstructed image **y** of the same size, with 8-bits per pixel. Then, the PSNR is defined by

$$\text{PSNR} = 20 \log_{10} \left(\frac{255}{\text{RMSE}} \right) \text{ dB,} \tag{15.106}$$

where RMSE is the *root mean square error,* which is the square root of the *mean square error* (MSE), given by

$$\text{MSE} = \frac{\sum_{m=1}^{M} \sum_{n=1}^{N} (x^2[m, n] - y^2[m, n])}{MN}. \tag{15.107}$$

The PSNR of the reconstructed image of Figure 15.47(a) is 35.42 dB and that of Figure 15.47(b) is 35.72 dB.

We next apply a two-level Haar wavelet decomposition to "Goldhill" image. The process is equivalent to applying a one-level decomposition to the **LL**-subimage of Figure 15.47(b). Figure 15.48(a) shows the seven subimages. The subimage at the top left is of size 128×128 and contains only the low frequencies and is labeled **LLL**. The remaining 3 subimages obtained after the second-level decomposition are labeled accordingly.

The total energies of the four subimages of size 128×128 at the top left corner and their percentages of the total energy of all subimages are now as follows:

LLL:	**LHL:**	**LLH:**	**LHH:**
3898.26×10^6	9.412×10^6	10.301×10^6	1.940×10^6
$99.053\,\%$	$0.239\,\%$	$0.262\,\%$	$0.049\,\%$

The total energies of the remaining three subimages of size 256×256 at the top right and bottom left and right corners remain the same as given earlier. The sum of the energies of all subimages is again equal to 3935.54×10^6. The entropy of the wavelet representation after a two-level decomposition is now $\mathcal{H}_y = 1.620$ bits/pixel, obtained by adding the entropies of the subimages given above and is seen to be much smaller than that obtained after a one-level decomposition. The compression ratio is 2.2-to-1.0.

(a)   (b)

Figure 15.48: (a) Subimages after a two-level Haar wavelet decomposition and (b) reconstructed image after quantization of the two-level wavelet coefficients.

Figure 15.48(b) shows the reconstructed "Goldhill" image after quantization of the wavelet coefficients at the second level. The PSNR of the reconstructed image is now 35.75 dB.

The compression ratio advantage of the wavelet decomposition is a consequence of the entropy difference between the quantization of the image in the space domain and the separate quantization of each subimage. The wavelet decomposition allows for an entropy reduction since most of the signal energy is allocated to low-frequency subimages with a smaller number of pixels. If the quantization step does not force the entropy reduction to be very large, then the wavelet-reconstructed image can still have better quality than that obtained from space-domain coding, which can seen from the compression examples given in Figures 15.47 and 15.48.

In order to exploit the entropy of the image representation after quantization, a lossless source coding scheme such as Huffman or arithmetic coding[13] is required. The design of these codes goes beyond the scope of this book and is therefore not included here. It is enough to state that lossless codes allow the image in these examples to be compressed in practice at rates (number of bits/pixel) arbitrarily close to those expressed by the entropy values that have been shown.

The histograms of all subimages, except the one with lowest frequency content (i.e., the top left subimage), have only one mode and are centered at zero, with tails that usually decay with exponential behavior. This means that many of the subimage pixels are assigned to the quantized interval centered at zero. Run-length coding is a lossless coding scheme that encodes sequences of zeros by a special symbol denoting the beginning of such sequences, followed by the length of the sequence [Jay84]. This different representation allows for further reduction of the subimages entropy after the quantization, and thus, run-length coding can be used to improve, without any loss of quality, the compression ratios shown in the examples of this section.

15.9 Transmultiplexers

In the United States and most other countries, the telephone service employs two types of multiplexing schemes to transmit multiple low-frequency voice signals over a wide-band channel. In the *frequency-*

---

[13]For a discussion on Huffman and arithmetic coding, see [Jay84].

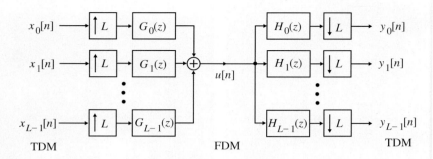

Figure 15.49: The basic L-channel transmultiplexer structure.

$$x_\ell[n] \rightarrow \boxed{\uparrow L} \rightarrow \boxed{G_\ell(z)} \rightarrow \boxed{H_k(z)} \rightarrow \boxed{\downarrow L} \rightarrow y_k[n] \quad \equiv \quad x_\ell[n] \rightarrow \boxed{F_{k\ell}(z)} \rightarrow y_k[n]$$

(a)   (b)

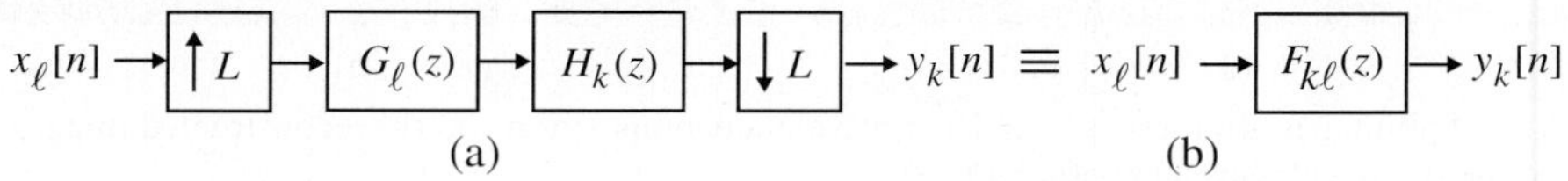

Figure 15.50: The k, ℓ-path of the L-channel transmultiplexer structure.

division multiplex (FDM) telephone system, multiple analog voice signals are first modulated by single-sideband (SSB) modulators onto several subcarriers, combined, and transmitted simultaneously over a common wide-band channel. To avoid *cross-talk*, the subcarriers are chosen to ensure that the spectra of the modulated signals do not overlap. At the receiving end, the modulated subcarrier signals are separated by analog bandpass filters and demodulated to reconstruct the individual voice signals. On the other hand, in the *time-division multiplex* (TDM) telephone system, the voice signals are first converted into digital signals by sampling and A/D conversion. The samples of the digital signals are time-interleaved by a digital multiplexer, and the combined signal is transmitted. At the receiving end, the digital voice signals are separated by a digital demultiplexer and then passed through a D/A converter and an analog reconstruction filter to recover the original analog voice signals.

The TDM system is usually employed for short-haul communication, while the FDM scheme is preferred for long-haul transmission. Until the telephone service becomes all digital, it is necessary to translate signals between the two formats. This is achieved by the transmultiplexer system discussed next.

The *transmultiplexer* is a multi-input, multi-output, multirate structure, as shown in Figure 15.49. It is exactly the opposite to that of the L-channel QMF bank of Figure 14.18 and consists of an L-channel synthesis filter bank at the input end, followed by an L-channel analysis filter bank at the output end. To determine the input–output relation of the transmultiplexer, consider one typical path from the kth input to the ℓth output as indicated in Figure 15.50(a) [Vai93]. A polyphase representation of the structure of Figure 15.49 is shown in Figure 15.51(a). Invoking the identity of Section 13.1.3, we note that the structure of Figure 15.51(a) is equivalent to that shown in Figure 15.51(b), consisting of an LTI branch with a transfer function $F_{k\ell}(z)$ that is the zeroth polyphase component of $H_k(z)G_\ell(z)$. The input–output relation of the transmultiplexer is therefore given by

$$Y_k(z) = \sum_{\ell=0}^{L-1} F_{k\ell}(z)X_\ell(z), \quad 0 \le k \le L - 1. \tag{15.108}$$

Denoting

$$\mathbf{Y}(z) = \begin{bmatrix} Y_0(z) & Y_1(z) & \cdots & Y_{L-1}(z) \end{bmatrix}^t, \tag{15.109a}$$

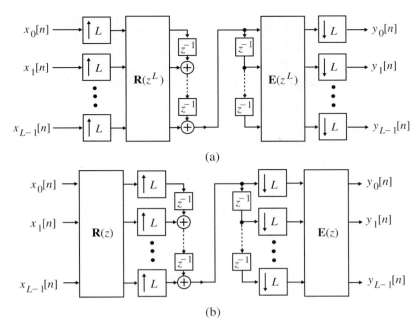

(a)

(b)

Figure 15.51: (a) Polyphase representation of the L-channel transmultiplexer and (b) its computationally efficient realization.

$$\mathbf{X}(z) = \begin{bmatrix} X_0(z) & X_1(z) & \cdots & X_{L-1}(z) \end{bmatrix}^t, \tag{15.109b}$$

we can rewrite Eq. (15.108) as

$$\mathbf{Y}(z) = \mathbf{F}(z)\mathbf{X}(z), \tag{15.110}$$

where $\mathbf{F}(z)$ is an $L \times L$ matrix whose (k, ℓ)th element is given by $F_{k\ell}(z)$. The objective of the transmultiplexer design is to ensure that $y_k[n]$ is a reasonable replica of $x_k[n]$. If $y_k[n]$ contains contributions from $x_r[n]$ with $r \neq n$, then there is *cross-talk* between these two channels. It follows from Eq. (15.110) that cross-talk is totally absent if $F(z)$ is a diagonal matrix, in which case Eq. (15.110) reduces to

$$Y_k(z) = F_{kk}(z)X_k(z), \quad 0 \le k \le L - 1. \tag{15.111}$$

As in the case of the QMF bank, we can define three types of transmultiplexer systems. It is a phase-preserving system if $F_{kk}(z)$ is a linear-phase transfer function for all values of k. Likewise, it is a magnitude-preserving system if $F_{kk}(z)$ is an allpass function. Finally, for a perfect reconstruction transmultiplexer,

$$F_{kk}(z) = \alpha_k z^{-n_k}, \quad 0 \le k \le L - 1, \tag{15.112}$$

where n_k is an integer and α_k is a nonzero constant. For a perfect reconstruction system, $y_k[n] = \alpha_k x_k[n - n_k]$.

The perfect reconstruction condition can also be derived in terms of the polyphase components of the synthesis and analysis filter banks of the transmultiplexer of Figure 15.49, as shown in Figure 15.51(a) [Koi91]. Using the cascade equivalences of Figure 13.14, we arrive at the equivalent representation indicated in Figure 15.51(b). Note that the structure in the center part of this figure is a special case of the system of Figure 15.49, where $G_\ell(z) = z^{-(L-1-\ell)}$ and $H_k(z) = z^{-k}$, with $\ell, k = 0, 1, \ldots, L - 1$. Here the zeroth polyphase component of $H_{\ell+1}(z)G_\ell(z)$ is z^{-1} for $\ell = 0, 1, \ldots, L - 2$, the zeroth polyphase

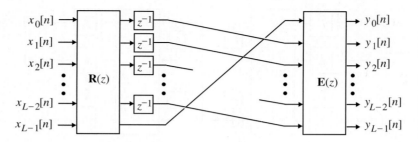

Figure 15.52: Simplified equivalent circuit of Figure 15.51.

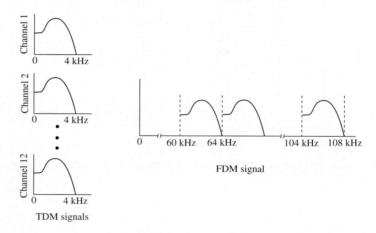

Figure 15.53: Spectrums of TDM signals and the FDM signal.

component of $H_0(z)G_{L-1}(z)$ is 1, and the zeroth polyphase component of $H_k(z)G_\ell(z)$ is 0 for all other cases. As a result, a simplified equivalent representation of Figure 15.51(b) is as shown in Figure 15.52.

The transfer matrix characterizing the transmultiplexer is thus given by

$$\mathbf{F}(z) = \mathbf{E}(z) \begin{bmatrix} \mathbf{0} & 1 \\ z^{-1}\mathbf{I}_{L-1} & \mathbf{0} \end{bmatrix} \mathbf{R}(z), \tag{15.113}$$

where $\mathbf{I}_{L-1}$ is an $(L-1) \times (L-1)$ identity matrix. Now, for a perfect reconstruction system, it is sufficient to ensure that

$$\mathbf{F}(z) = dz^{-n_o}\mathbf{I}_L, \tag{15.114}$$

where n_o is a positive integer. From Eqs. (15.113) and (15.114) we arrive at the condition for perfect reconstruction in terms of the polyphase components as

$$\mathbf{R}(z)\mathbf{E}(z) = dz^{-m_o} \begin{bmatrix} \mathbf{0} & \mathbf{I}_{L-1} \\ z^{-1} & 0 \end{bmatrix}, \tag{15.115}$$

where m_o is a suitable positive integer.

It is possible to develop a perfect reconstruction transmultiplexer from a perfect reconstruction QMF bank with analysis filters $H_\ell(z)$ and synthesis filters $G_\ell(z)$, with a distortion transfer function given by $T(z) = dz^{-K}$, where d is a nonzero constant and K is a positive integer. It can be shown that a perfect reconstruction transmultiplexer can then be designed using the analysis filters $H_\ell(z)$ and synthesis filters $z^{-R}G_\ell(z)$, where R is a positive integer less than L such that $R + K$ is a multiple of L [Koi91]. We illustrate this approach in Example 15.11.

EXAMPLE 15.11 Consider the perfect reconstruction analysis/synthesis filter bank of Example 14.8 with an input–output relation $y[n] = 2x[n - 2]$. In this case, the analysis and synthesis filters are given by

$$H_0(z) = 1 + z^{-1} + z^{-2}, \quad H_1(z) = 1 - z^{-1} + z^{-2}, \quad H_2(z) = 1 - z^{-2},$$
$$G_0(z) = 1 + 2z^{-1} + z^{-2}, \quad G_1(z) = 1 - 2z^{-1} + z^{-2}, \quad G_2(z) = -2 + 2z^{-2}.$$

Here, $d = 2$ and $K = 2$. We thus choose $R = 1$ so that $R + K = 3$. The synthesis filters of the transmultiplexer are thus given by $z^{-1}G_\ell(z)$.

We now examine the products $z^{-1}G_\ell(z)H_k(z)$, for $\ell, k = 0, 1, 2$, and determine their zeroth polyphase components. Thus,

$$z^{-1}G_0(z)H_0(z) = z^{-1} + 3z^{-2} + 4z^{-3} + 3z^{-4} + z^{-5},$$

whose zeroth polyphase component is given by $4z^{-1}$, and hence, $y_0[n] = 4x_0[n - 1]$. Likewise,

$$z^{-1}G_1(z)H_1(z) = z^{-1} - 3z^{-2} + 4z^{-3} - 3z^{-4} + z^{-5},$$

with a zeroth polyphase component $4z^{-1}$, resulting in $y_1[n] = 4x_1[n - 1]$. Similarly,

$$z^{-1}G_2(z)H_2(z) = -2z^{-1} + 4z^{-3} - 2z^{-5},$$

whose zeroth polyphase component is again $4z^{-1}$, implying $y_2[n] = 4x_2[n - 1]$. It can be shown that the zeroth polyphase components for all other products $z^{-1}G_\ell(z)H_k(z)$, with $\ell \neq k$, is 0, indicating a total absence of cross-talk between channels.

In a typical TDM-to-FDM format translation, 12 digitized speech signals are interpolated by a factor of 12, modulated by single-sideband modulation, digitally summed, and then converted into an FDM analog signal by D/A conversion. At the receiving end, the analog signal is converted into a digital signal by A/D conversion and passed through a bank of 12 single-sideband demodulators whose outputs are then decimated, resulting in the low-frequency speech signals. The speech signals have a bandwidth of 4 kHz and are sampled at an 8-kHz rate. The FDM analog signal occupies the band 60 kHz to 108 kHz, as illustrated in Figure 15.53. The interpolation and the single-sideband modulation can be performed by up-sampling and appropriate filtering. Likewise, the single-sideband demodulation and the decimation can be implemented by appropriate filtering and down-sampling.

15.10 Discrete Multitone Transmission of Digital Data

Binary data are normally transmitted serially as a pulse train, as indicated in Figure 15.54(a). However, in order to faithfully extract the information transmitted, the receiver requires complex equalization procedures to compensate for channel imperfection and to make full use of the channel bandwidth. For example, the pulse train of Figure 15.54(a) arriving at the receiver may appear as indicated in Figure 15.54(b). To alleviate the problems encountered with the transmission of data as a pulse train, frequency-division multiplexing with overlapping subchannels has been proposed. In such a system, each binary digit a_r, $r = 0, 1, 2, \ldots, N - 1$, modulates a subcarrier sinusoidal signal $\cos(2\pi rt/T)$, as indicated in Figure 15.54(c), for the transmission of the data of Figure 15.54(a), and then the modulated subcarriers are summed and transmitted as one composite analog signal. At the receiver, the analog signal is passed through a bank of coherent demodulators whose outputs are tested to determine the digits transmitted. This is the basic idea behind the multicarrier modulation/demodulation scheme for digital data transmission.

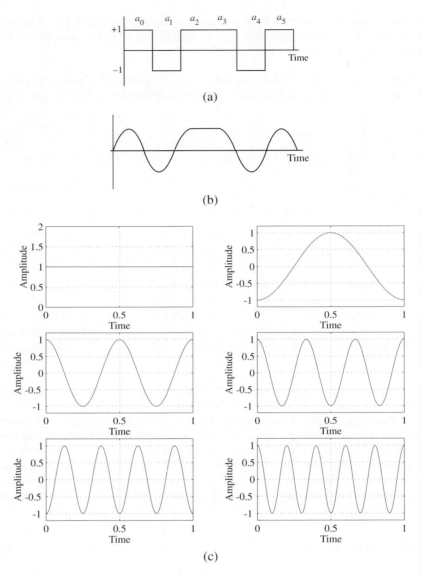

Figure 15.54: (a) Serial binary data stream, (b) baseband serially transmitted signal at the receiver, and (c) signals generated by modulating a set of subcarriers by the digits of the pulse train in (a).

A widely used form of the multicarrier modulation is the discrete multitone transmission (DMT) scheme in which the modulation and demodulation processes are implemented via the discrete Fourier transform (DFT), efficiently realized using fast Fourier transform (FFT) methods. This approach leads to an all-digital system, eliminating the arrays of sinusoidal generators and the coherent demodulators [Cio91], [Pel80].

We outline here the basic idea behind the DMT scheme. Let $\{a_k[n]\}$ and $\{b_k[n]\}$, $0 \leq k \leq M - 1$, be two $M - 1$ real-valued data sequences operating at a sampling rate of F_T that are to be transmitted. Define a new set of complex sequences $\{\alpha_k[n]\}$ of length $N = 2M$ according to

$$\alpha_k[n] = \begin{cases} a_0[n], & k = 0, \\ a_k[n] + jb_k[n], & 1 \le k \le \frac{N}{2} - 1, \\ b_0[n], & k = \frac{N}{2}, \\ a_{N-k}[n] - jb_{N-k}[n], & \frac{N}{2} + 1 \le k \le N - 1. \end{cases} \qquad (15.116)$$

We apply an inverse DFT, and the above set of N sequences is transformed into another new set of N signals $\{u_\ell[n]\}$, given by

$$u_\ell[n] = \frac{1}{N} \sum_{k=0}^{N-1} \alpha_k[n] W_N^{-\ell k}, \qquad 0 \le \ell \le N - 1, \qquad (15.117)$$

where $W_N = e^{-j2\pi/N}$. Note that the method of generation of the complex sequence set $\{\alpha_k[n]\}$ ensures that its IDFT $\{u_\ell[n]\}$ will be a real sequence. Each of these N signals is then upsampled by a factor of N and time-interleaved, generating a composite signal $\{x[n]\}$ operating at a rate of NF_T that is assumed to be equal to $2F_c$. The composite signal is converted into an analog signal $x_a(t)$ by passing it through a D/A converter, followed by an analog reconstruction filter. The analog signal $x_a(t)$ is then transmitted over the channel.

At the receiver, the received analog signal $y_a(t)$ is passed through an analog anti-aliasing filter and then converted into a digital signal $\{y[n]\}$ by an S/H circuit, followed by an A/D converter operating at a rate of $NF_T = 2F_c$. The received digital signal is then deinterleaved by a delay chain containing $N - 1$ unit delays, whose outputs are next down-sampled by a factor of N, generating the set of signals $\{v_\ell[n]\}$. Applying the DFT to these N signals, we finally arrive at N signals $\{\beta_k[n]\}$

$$\beta_k[n] = \sum_{\ell=0}^{N-1} v_\ell[n] W_N^{\ell k}, \qquad 0 \le k \le N - 1. \qquad (15.118)$$

Figure 15.55 shows schematically the overall DMT scheme. If we assume the frequency response of the channel to have a flat passband and assume the analog reconstruction and anti-aliasing filters to be ideal lowpass filters, then neglecting the nonideal effects of the D/A and the A/D converters, we can assume $y[n] = x[n]$. Hence, the interleaving circuit of the DMT structure at the transmitting end connected to the deinterleaving circuit at the receiving end is identical to the same circuit in the transmultiplexer structure of Figure 15.51(b) (with $L = N$). From the equivalent representation given in Figure 15.52, it follows that

$$v_k[n] = u_{k-1}[n - 1], \qquad 0 \le k \le N - 2,$$
$$v_0[n] = u_{N-1}[n], \qquad\qquad\qquad (15.119)$$

or, in other words,

$$\beta_k[n] = \alpha_{k-1}[n - 1], \qquad 0 \le k \le N - 2,$$
$$\beta_0[n] = \alpha_{N-1}[n]. \qquad\qquad\qquad (15.120)$$

Transmission channels, in general, have a bandpass frequency response $H_{\text{ch}}(f)$, with a magnitude response dropping to zero at some frequency F_c. In some cases, in the passband of the channel, the magnitude response, instead of being flat, drops very rapidly outside its passband, as indicated in Figure 15.56. For reliable digital data transmission over such a channel and its recovery at the receiving end, the channel's frequency response needs to be compensated by essentially a highpass equalizer at the receiver. However, such an equalization also amplifies high-frequency noise that is invariably added to the data signal as it passes through the channel.

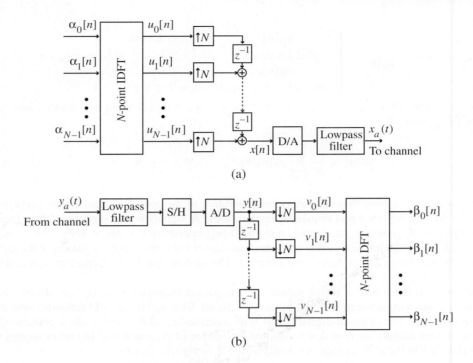

(a)

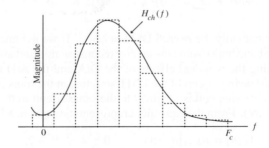

(b)

Figure 15.55: The DMT scheme: (a) transmitter and (b) receiver.

Figure 15.56: Frequency response of a typical band-limited channel.

For a large value of the DFT length N, the channel can be assumed to be composed of a series of contiguous narrow-bandwidth bandpass subchannels. If the bandwidth is reasonably narrow, the corresponding bandpass subchannel can be considered to have an approximately flat magnitude response, as indicated by the dotted lines in Figure 15.56, and the channel can be approximately characterized by a single complex number given by the value of its frequency response at $\omega = 2\pi k/N$. The values can be determined by first transmitting a known training signal of unmodulated carriers and generating the respective channel frequency response samples. The real data samples are then divided by these complex numbers at the receiver to compensate for channel distortion.

Further details on the performance of the above DMT scheme under nonideal conditions can be found in [Bin90], [Cio91], [She95].

15.11   Oversampling A/D Converter

For the digital processing of an analog continuous-time signal, the signal is first passed through a sample-and-hold circuit whose output is then converted into a digital form by means of an A/D converter. However, according to the sampling theorem, discussed in Section 4.2.1, a band-limited continuous-time signal with a lowpass spectrum can be fully recovered from its uniformly sampled version if it is sampled at a sampling frequency that is at least twice the highest frequency contained in the analog signal. If this condition is not satisfied, the original continuous-time signal cannot be recovered from its sampled version because of aliasing. To prevent aliasing, the analog signal is thus passed through an analog anti-aliasing lowpass filter prior to sampling, which enforces the condition of the sampling theorem. The passband cutoff frequency of the lowpass filter is chosen equal to the frequency of the highest signal frequency component that needs to be preserved at the output. The anti-aliasing filter also cuts off all out-of-band signal components and any high-frequency noise that may be present in the original analog signal, which otherwise would alias into the baseband after sampling. The filtered signal is then sampled at a rate that is at least twice that of the cutoff frequency.

Let the signal band of interest be the frequency range $0 \leq f \leq F_m$. Then, the Nyquist rate is given by $F_N = 2F_m$. Now, if the sampling rate F_T is the same as the Nyquist rate, we need to use before the sampler an anti-aliasing lowpass filter with a very sharp cutoff in its frequency response, satisfying the requirements as given by Eq. (4.69).[14] This requires the design of a very high-order anti-aliasing filter structure built with high-precision analog components, and it is usually difficult to implement such a filter in VLSI technology. Moreover, such a filter also introduces undesirable phase distortion in its output. An alternative approach mentioned in Section 4.8.4 is to sample the analog signal at a rate much higher than the Nyquist rate, use a fast low-resolution A/D converter, and then decimate the digital output of the converter to the Nyquist rate. This approach relaxes the sharp cutoff requirements of the analog anti-aliasing filter, resulting in a simpler filter structure that can be built using low-precision analog components while requiring fast, more complex digital signal processing hardware at later stages. The overall structure is not only amenable to VLSI fabrication but also can be designed to provide linear-phase response in the signal band of interest.

The oversampling approach is an elegant application of multirate digital signal processing and is increasingly being employed in the design of high-resolution A/D converters for many practical systems [Can92], [Fre94]. In this section, we analyze the quantization noise performance of the conventional A/D converter and show analytically how the oversampling approach decreases the quantization noise power in the signal band of interest [Fre94]. We then show that further improvement in the noise performance of an oversampling A/D converter can be obtained by employing a sigma-delta ($\Sigma\Delta$) quantization scheme. For simplicity, we restrict our discussion to the case of a basic first-order sigma-delta quantizer.

To illustrate the noise performance improvement property, consider a b-bit A/D converter operating at F_T Hz. Now, for a full-scale peak-to-peak input analog voltage of R_{FS}, the smallest voltage step represented by b bits is

$$\Delta V = \frac{R_{FS}}{2^b - 1} \cong \frac{R_{FS}}{2^b}. \tag{15.121}$$

From Eq. (12.70), the rms quantization noise power σ_e^2 of the error voltage, assuming a uniform distribution of the error between $-\Delta V/2$ and $\Delta V/2$, is given by

$$\sigma_e^2 = \frac{(\Delta V)^2}{12}. \tag{15.122}$$

The rms noise voltage, given by σ_e, therefore has a flat spectrum in the frequency range from 0 to $F_T/2$.

---

[14]Recall that $F_T = 1/T$, where T is the sampling period.

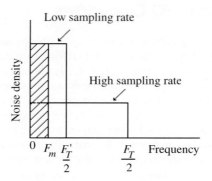

Figure 15.57: A/D converter noise density.

The noise power per unit bandwidth, called the *noise density*, is then given by

$$P_{e,n} = \frac{(\Delta V)^2/12}{F_T/2} = \frac{(\Delta V)^2}{6F_T}. \tag{15.123}$$

A plot of the noise densities for two different sampling rates is shown in Figure 15.57, where the shaded portion indicates the signal band of interest. As can be seen from this figure, the total amount of noise in the signal band of interest for the high sampling rate case is smaller than that for the low sampling rate case. The total noise in the signal band of interest, called the *in-band noise power*, is given by

$$P_{\text{total}} = \frac{(R_{\text{FS}}/2^b)^2}{12} \cdot \frac{F_m}{F_T/2}. \tag{15.124}$$

It is interesting to compute the needed wordlength β of the A/D converter operating at the Nyquist rate in order that its total noise in the signal band of interest be equal to that of a b-bit A/D converter operating at a higher rate. Substituting $F_T = 2F_m$ and replacing b with β in Eq. (15.124), we arrive at

$$P_{\text{total}} = \frac{(R_{\text{FS}}/2^\beta)^2}{12} = \frac{(R_{\text{FS}}/2^b)^2}{12} \cdot \frac{F_m}{F_T/2}, \tag{15.125}$$

which leads to the desired relation

$$\beta = b + \frac{1}{2} \log_2 M, \tag{15.126}$$

where $M = F_T/2F_m$ denotes the *oversampling ratio* (OSR). Thus, $\beta - b$ denotes the increase in the resolution of a b-bit converter whose oversampled output is filtered by an ideal brick-wall lowpass filter. A plot of the increase in resolution as a function of the oversampling ratio is shown in Figure 15.58. For example, for an OSR of $M = 1000$, an 8-bit oversampling A/D converter has an effective resolution equal to that of a 13-bit A/D converter operating at the Nyquist rate. Note that Eq. (15.126) implies that the increase in the resolution is $\frac{1}{2}$-bit per doubling of the OSR.

We now illustrate the improvement in the noise performance obtained by employing a sigma-delta ($\Sigma\Delta$) quantization scheme. The sigma-delta A/D converter was briefly introduced in Section 4.8.4 and is shown in block-diagram form in Figure 15.59 for convenience. This figure also indicates the sampling rates at various stages of the structure. It should be noted here that the 1-bit output samples of the quantizer after decimation become b-bit samples at the output of the sigma-delta A/D converter due to the filtering operations involving b-bit multiplier coefficients of the Mth-band digital lowpass filter.

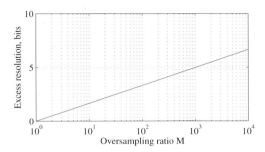

Figure 15.58: Excess resolution as a function of the oversampling ratio M.

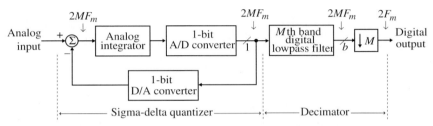

Figure 15.59: Oversampling sigma-delta A/D converter structure.

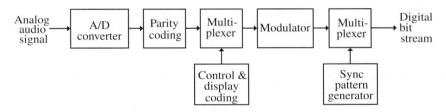

Figure 15.60: Compact disk encoding system for one channel of a stereo audio.

Since the oversampling ratio M is typically very large in practice, the sigma-delta A/D converter is most useful in low-frequency applications such as digital telephony, digital audio, and digital spectrum analyzers. For example, Figure 15.60 shows the block diagram of a typical compact disk encoding system used to convert the input analog audio signal into a digital bit stream that is then applied to generate the master disk [Hee82]. Here, the oversampling sigma-delta A/D converter employed has a typical input sampling rate of 3175.2 kHz and an output sampling rate of 44.1 kHz [Kam86].

To understand the operation of the sigma-delta A/D converter of Figure 15.59, we need to study the operation of the sigma-delta quantizer shown in Figure 15.61(a). To this end, it is convenient to use the discrete-time equivalent circuit of Figure 15.61(b), where the integrator has been replaced with an accumulator.[15] Here, the input $x[n]$ is a discrete-time sequence of analog samples developing an output sequence of binary-valued samples $y[n]$. From this diagram, we observe that, at each discrete instant of time, the circuit forms the difference (Δ) between the input and the delayed output, which is accumulated by a summer (Σ) whose output is then quantized by a one-bit A/D converter, that is., a comparator.

Even though the input–output relation of the sigma-delta quantizer is basically nonlinear, the low-

---

[15] In practice, the integrator is implemented as a discrete-time switched-capacitor circuit.

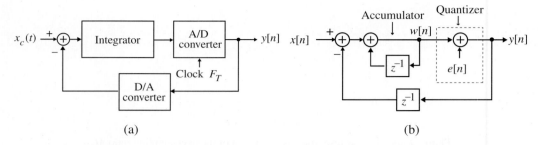

Figure 15.61: Sigma-delta quantization scheme: (a) quantizer and (b) its discrete-time equivalent.

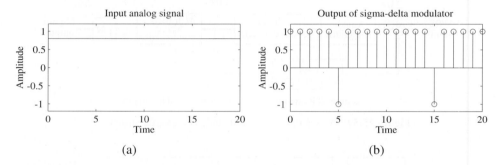

Figure 15.62: (a) Input and (b) output waveforms of the sigma-delta quantizer of Figure 15.61(a) for a constant input.

frequency content of the input $x_c(t)$ can be recovered from the output $y[n]$ by passing it through a digital lowpass filter. This property can be easily shown for a constant input analog signal $x_a(t)$ with a magnitude less than $+1$. In this case, the output $w[n]$ of the accumulator is a bounded sequence with sample values equal to either -1 or $+1$. This can happen only if the input to the accumulator has an average value of zero. Or in other words, the average value of $w[n]$ must be equal to the average value of the input $x[n]$ [Sch91]. Examples 15.12 and 15.13 illustrate the operation of a sigma-delta quantizer.

Program 15_11.m

EXAMPLE 15.12   Sigma-Delta Quantization of a Constant Amplitude Signal

We first consider the operation for the case of a constant input signal using MATLAB. To this end, we can use Program 15_11. The plots generated by this program are the input and output waveforms of the sigma-delta quantizer of Figure 15.61(a) and are shown in Figure 15.62. The program also prints the average value of the output as indicated below:

```
Average value of output is =
       0.8095
```

which is very close to the amplitude 0.8 of the constant input. It can be easily verified that the average value of the output gets closer to the amplitude of the constant input as the length of the input increases.

Program 15_12.m

EXAMPLE 15.13   Sigma-Delta Quantization of a Sinusoidal Signal

We now verify the operation of the sigma-delta A/D converter for a sinusoidal input of frequency 0.01 Hz using MATLAB. To this end, we make use of the Program 15_12. Because of the short length of the input sequence, the filtering operation is performed here in the DFT domain [Sch91]. Figure 15.63 shows the input and output

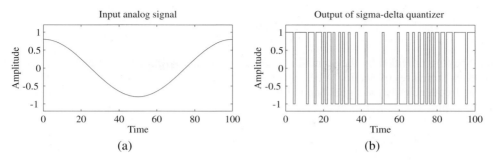

Figure 15.63: Input and output waveforms of the sigma-delta quantizer of Figure 15.61(a) with a sine wave input.

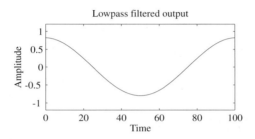

Figure 15.64: The lowpass filtered version of the waveform of Figure 15.63(b).

waveforms of the sigma-delta quantizer of Figure 15.61(a) for a sinusoidal input. Figure 15.64 depicts the lowpass filtered version of the output signal shown in Figure 15.63(b). As can be seen from these figures, the filtered output is nearly an exact replica of the input.

It follows from Figure 15.61(b) that the output $y[n]$ of the quantizer is given by

$$y[n] = w[n] + e[n], \tag{15.127}$$

where

$$w[n] = x[n] - y[n - 1] + w[n - 1]. \tag{15.128}$$

From Eqs. (15.127) and (15.128), we obtain, after some algebra,

$$y[n] = x[n] + (e[n] - e[n - 1]), \tag{15.129}$$

where the quantity inside the parentheses represents the noise due to sigma-delta modulation. The noise transfer function is simply $G(z) = (1 - z^{-1})$. The power spectral density of the modulation noise is therefore given by

$$P_y(f) = \left| G(e^{j2\pi fT}) \right|^2 P_e(f) = 4\sin^2\left(\frac{2\pi fT}{2}\right) P_e(f), \tag{15.130}$$

where we have assumed the power spectral density $P_e(\omega)$ of the quantization noise to be the one-sided power spectral density defined for positive frequencies only. For a random signal input $x[n]$, $P_e(f)$ is constant for all frequencies and is given by

$$P_e(f) = \frac{(\Delta V)^2/12}{F_T/2}. \tag{15.131}$$

Substituting Eq. (15.131) in Eq. (15.130), we arrive at the power spectral density of the output noise, given by

$$P_y(f) = \frac{2}{3} \frac{(\Delta V)^2}{F_T} \sin^2(\pi f T). \tag{15.132}$$

The noise-shaping provided by the sigma-delta quantizer is similar to that encountered in the first-order error-feedback structures of Section 9.10.1 and shown in Figure 9.42. For a very large OSR, as is usually the case, the frequencies in the signal band of interest are much smaller than F_T, the sampling frequency. Thus, we can approximate $P_y(f)$ of Eq. (15.132) as

$$P_y(f) \cong \frac{2}{3} \frac{(\Delta V)^2}{F_T} (\pi f T)^2 = \tfrac{2}{3}\pi^2 (\Delta V)^2 T^3 f^2, \qquad f << F_T. \tag{15.133}$$

From Eq. (15.133), the in-band noise power of the sigma-delta A/D converter is thus given by

$$P_{\text{total,sd}} = \int_0^{F_m} P_y(f)\, df = \tfrac{2}{3}\pi^2 (\Delta V)^2 T^3 \int_0^{F_m} f^2\, df = \frac{2}{9}\pi^2 (\Delta V)^2 T^3 (F_m)^3. \tag{15.134}$$

It is instructive to compare the noise performance of the sigma-delta A/D converter with that of a direct oversampling A/D converter operating at a sampling rate of F_T with a signal band of interest from dc to F_m. From Eq. (15.125), the in-band noise power of the latter is given by

$$P_{\text{total,os}} = \tfrac{1}{6}(\Delta V)^2 T F_m. \tag{15.135}$$

The improvement in the noise performance is therefore given by

$$10 \log_{10}\left(\frac{P_{\text{total,os}}}{P_{\text{total,sd}}}\right) = 10 \log_{10}\left(\frac{3M^2}{\pi^2}\right) = -5.1718 + 20 \log_{10}(M)\, \text{dB}, \tag{15.136}$$

where we have used $M = F_T/2F_m$ to denote the OSR. For example, for an OSR of $M = 1000$, the improvement in the noise performance using the sigma-delta modulation scheme is about 55 dB. In this case, the increase in the resolution is about 1.5 bits per doubling of the OSR.

The improved noise performance of the sigma-delta A/D converter results from the shape of $\left|G(e^{j2\pi f T})\right|$, which decreases the noise power spectral density in-band ($0 \leq f \leq F_m$), while increasing it outside the signal band of interest ($f > F_m$). Since this type of converter also employs oversampling, it requires a less stringent analog anti-aliasing filter.

The A/D converter of Figure 15.59 employs a single-loop feedback and is often referred to as a first-order sigma-delta converter. Multiple feedback loop modulation schemes have been advanced to reduce the in-band noise further. However, the use of more than two feedback loops may result in unstable operation of the system, and care must be taken in the design to ensure stable operation [Can92].

As indicated in Figure 15.59, the quantizer output is passed through an Mth-band lowpass digital filter whose output is then down-sampled by a factor of M to reduce the sampling rate to the desired Nyquist rate. The function of the digital lowpass filter is to eliminate the out-of-band quantization noise and the out-of-band signals that would be aliased into the passband by the down-sampling operation. As a result, the filter must exhibit a very sharp cutoff frequency response with a passband edge at F_m. This necessitates the use of a very high-order digital filter. In practice, it is preferable to use a filter with a transfer function having simple integer-valued coefficients to reduce the cost of hardware implementation and to permit all multiplication operations to be carried out at the down-sampled rate. In addition, most applications require the use of linear-phase digital filters, which can be easily implemented using FIR filters.

The simplest lowpass FIR filter is the moving-average filter of Eq. (2.61), repeated below for convenience:[16]

$$H(z) = 1 + z^{-1} + z^{-2} + \cdots + z^{-(N-1)}. \tag{15.137}$$

---

[16]For simplicity, we have ignored the scale factor of $1/N$, which is needed to provide a dc gain of 0 dB.

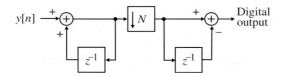

Figure 15.65: A very simple factor-of-N decimator structure.

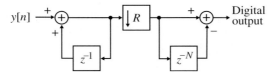

Figure 15.66: A two-stage CIC decimator structure.

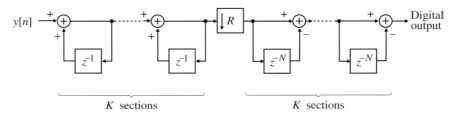

Figure 15.67: A CIC decimator structure with cascaded sections.

A more convenient form of the above transfer function for realization purposes is given by

$$H(z) = \frac{1 - z^{-N}}{1 - z^{-1}}, \tag{15.138}$$

also known as a *recursive running-sum filter* or a *boxcar filter*. A realization of a factor-of-N decimator based on the decimation filter of Eq. (15.138) is sketched in Figure 15.65.[17]

Since the decimator based on a running-sum filter does not provide sufficient out-of-band attenuation, often a multistage decimator formed by a cascade of the running-sum decimators, more commonly known as *cascaded integrator comb* (CIC) filters, is used in practice [Hog81]. The structure of a two-stage CIC decimator is shown in Figure 15.66. It can be easily shown that the structure is equivalent to a factor-of-R decimator with a length-RN running sum decimation filter. Further flexibility in the design of a CIC decimator is obtained by including K feedback paths before and K feedforward paths after the down-sampler, as indicated in Figure 15.67. The corresponding transfer function is then given by

$$H(z) = \left(\frac{1 - z^{-RN}}{1 - z^{-1}} \right)^K. \tag{15.139}$$

The parameters N and K can be adjusted for a given down-sampling factor R to yield the desired out-of-band attenuation.

---

[17]The integrator overload caused by the input adder overflow can be easily handled with binary arithmetic.

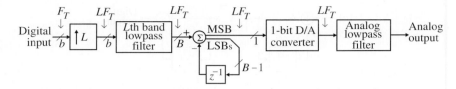

Figure 15.68: Block diagram representation of an oversampling sigma-delta D/A converter.

In some applications, it may be preferable to use a multistage decimation process in which all but the last stage employs the CIC decimators in various forms, followed by an FIR lowpass filter providing a much sharper cutoff before the final down-sampling. For example, in digital telephone applications, the following IIR transfer function provides a very good frequency response and can be used to design a factor-of-4 decimator [Can92]:

$$H(z) = (1 - \tfrac{3}{2}z^{-1} + z^{-2})(1 - \tfrac{11}{8}z^{-1} + \tfrac{5}{8}z^{-2})(1 - \tfrac{5}{4}z^{-1} + z^{-2})$$

$$\times (1 - \tfrac{101}{64}z^{-1} + \tfrac{7}{8}z^{-2}) \cdot \frac{(1 + z^{-1})(1 + z^{-2})(1 - z^{-5})}{(1 - z^{-1})}. \tag{15.140}$$

Further details on first- and higher-order sigma-delta converters can be found in Candy and Temes [Can92].

15.12   Oversampling D/A Converter

As indicated earlier in Section 4.1, the digital-to-analog conversion process consists of two steps: the conversion of input digital samples into a staircase continuous-time waveform by means of a D/A converter with a zero-order hold at its output, followed by an analog lowpass reconstruction filter. If the sampling rate F_T of the input digital signal is the same as the Nyquist rate, the analog lowpass reconstruction filter must have a very sharp cutoff in its frequency response, satisfying the requirements of Eq. (4.78). As in the case of the anti-aliasing filter, this involves the design of a very high-order analog reconstruction filter requiring high-precision analog circuit components. To get around the above problem, here also an oversampling approach is often used, in which case a wide transition band can be tolerated in the frequency response of the reconstruction filter allowing its implementation using low-precision analog circuit components, while, requiring a more complex digital interpolation filter at the front end.

Further improvement in the performance of an oversampling D/A converter is obtained by employing a digital sigma-delta 1-bit quantizer at the output of the digital interpolator, as indicated in Figure 4.50 and repeated in Figure 15.68 for convenience [Can86], [Lar93]. The quantizer extracts the MSB from its input and subtracts the remaining LSBs, the quantization noise, from its input. The MSB output is then fed into a 1-bit D/A converter and passed through an analog lowpass reconstruction filter to remove all frequency components beyond the signal band of interest. Since the signal band occupies a very small portion of the baseband of the high-sample-rate signal, the reconstruction filter in this case can have a very wide transition band, permitting its realization with a low-order filter that, for example, can be implemented using a Bessel filter to provide an approximately linear phase in the signal band.[18]

The spectrum of the quantized 1-bit output of the digital sigma-delta quantizer is nearly the same as that of its input. Moreover, it also shapes the quantization noise spectrum by moving the noise power out of the signal band of interest. To verify this result analytically, consider the sigma-delta quantizer shown

---

[18]See Section 4.4.5.

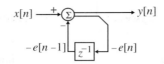

Figure 15.69: The sigma-delta quantizer.

separately in Figure 15.69. It follows from this figure that the input–output relation of the quantizer is given by

$$y[n] - e[n] = x[n] - e[n-1],$$

or, equivalently, by

$$y[n] = x[n] + e[n] - e[n-1], \tag{15.141}$$

where $y[n]$ is the MSB of the nth sample of the adder output, and $e[n]$ is the nth sample of the quantization noise composed of all bits except the MSB. From Eq. (15.141), it can be seen that the transfer function of the quantizer with no quantization noise is simply unity, and the noise transfer function is given by $G(z) = 1 - z^{-1}$, which is the same as that for the first-order sigma-delta modulator employed in the oversampling A/D converter discussed in the previous section.

Examples 15.14 and 15.15 illustrate the operation of a sigma-delta D/A converter for a discrete-time sinusoidal input sequence.

EXAMPLE 15.14   Illustration of the Oversampling D/A Conversion

Let the input to the D/A converter be a sinusoidal sequence of frequency 100 Hz operating at a sampling rate F_T of 1 kHz. Figure 15.70(a) shows the digital input sequence and the analog output generated by a D/A converter operating at F_T from this input. Figure 15.70(b) depicts the interpolated sinusoidal sequence operating at a higher sampling rate of 5 kHz, obtained by passing the low-sampling-rate sinusoidal signal through a factor-of-5 digital interpolator and the corresponding analog output generated by a D/A converter operating at $5F_T$ rate. If we compare the two D/A converter outputs, we can see that the staircase waveform of the oversampling D/A converter output is much smoother with smaller jumps than that of the lower-rate D/A converter output. Thus, the oversampling D/A converter output has considerably smaller high-frequency components in contrast to the lower-rate D/A converter. This fact can be easily verified by examining their spectra.

The high-frequency components in the baseband outside the signal band of interest can be removed by passing the D/A converter output through an analog lowpass filter, which also eliminates any leftover replicas of the baseband not completely removed by the zero-order hold in the D/A converter. Since the signal band of interest occupies a small portion of the baseband, the replicas of the signal band immediately outside the baseband are widely separated from the signal band inside the baseband. Hence, the lowpass filter can be designed with a very wide transition band. Moreover, due to reduced high-frequency components in the D/A converter output caused by oversampling, the stopband attenuation also does not have to be very large. On the other hand, the replicas of the signal band in the spectrum of the output of the low-rate D/A converter are closely spaced, and the high-frequency components are relatively large in amplitudes. In this case, the lowpass filter must have a sharp cutoff with much larger stopband attenuation to effectively remove the undesired components in the D/A converter output.

Figure 15.71 shows the filtered outputs of the conventional lower-rate and oversampled D/A converters when the same lowpass filter with a wide transition band is used in both cases. As can be seen from this figure, the analog output in the case of the low-rate D/A converter still contains some high-frequency components, while that in the case of the oversampled D/A converter is very close to a perfect sinusoidal signal. A much better output response is obtained in the case of a conventional D/A converter if a sharp cutoff lowpass filter is employed, as indicated in Figure 15.72.

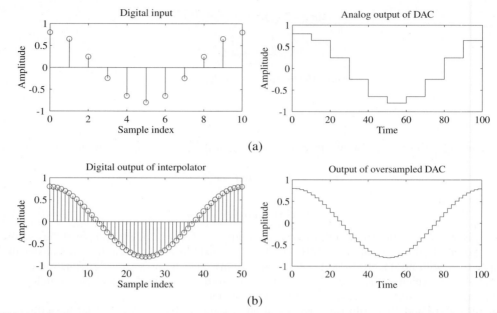

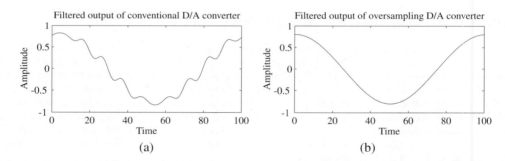

Figure 15.70: Input and output signals of (a) lower-rate D/A converter and (b) oversampling D/A converter.

Figure 15.71: Lowpass filtered output signals of (a) conventional D/A converter and (b) oversampling D/A converter.

Program 15_13.m

EXAMPLE 15.15   Illustration of the Operation of the Sigma-Delta D/A Converter Using MATLAB

In this example, we verify using MATLAB the operation of the sigma-delta D/A converter for a sinusoidal input sequence of frequency 100 Hz operating at a sampling rate F_T of 5 kHz. The signal is clearly oversampled since the sampling rate is much higher than the Nyquist rate of 200 Hz. Program 15_13 first generates the input digital signal, then generates a two-valued digital signal by quantizing the output of the sigma-delta quantizer, and finally, develops the output of the D/A converter by lowpass filtering the quantized output. As in the case of the sigma-delta converter of Example 15.14, the filtering operation here has also been performed in the DFT domain due to the short length of the input sequence [Sch91].

Figure 15.73 shows the digital input signal, the quantized digital output of the sigma-delta quantizer, and the filtered output of the D/A converter generated by this program. As can be seen from these plots, the lowpass filtered output is nearly a scaled replica of the desired sinusoidal analog signal.

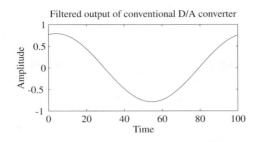

Figure 15.72: Filtered output signals of the conventional D/A converter employing a sharp cutoff lowpass filter.

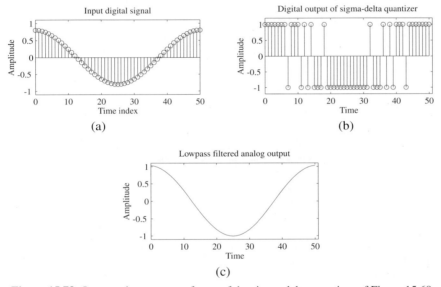

Figure 15.73: Input and output waveforms of the sigma-delta quantizer of Figure 15.69.

One of the most common applications of the oversampling sigma-delta D/A converter is in the compact disk (CD) player. Figure 15.74 shows the block diagram of the basic components in the signal processing part of a CD player, where typically a factor-of-4 oversampling D/A converter is employed for each audio channel [Goe82]. Here, the 44.1-kHz input digital audio signal is interpolated first by a factor of 4 to the 176.4-kHz rate and then converted into an analog audio signal.

15.13 Sparse Antenna Array Design

Linear-phased antenna arrays are used in radar, sonar, ultrasound imaging, and seismic signal processing. Sparse arrays with certain elements removed are economical and, as a result, are of practical interest. There is a mathematical similarity between the far-field radiation pattern for a linear antenna array of equally spaced elements and the frequency response of an FIR filter. This similarity can be exploited to design sparse arrays with specific beam patterns. In this section, we point out this similarity and outline a few simple designs of sparse arrays. We restrict our attention here on the design of sparse arrays for ultrasound

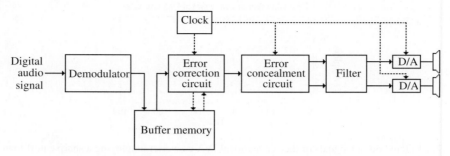

Figure 15.74: Signal processing part of a CD player.

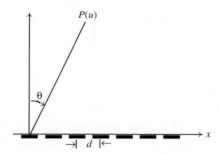

Figure 15.75: Uniform linear antenna array.

scanners.

Consider a linear array of N isotropic, equispaced elements with inter-element spacing d and located at $x_n = n \cdot d$ for $0 \leq n \leq N - 1$, as shown in Figure 15.75. The far-field radiation pattern at an angle θ away from the broadside (i.e., the normal to the array), is given by

$$P(u) = \sum_{n=0}^{N-1} w[n] e^{j[2\pi(u/\lambda)d]n}, \tag{15.142}$$

where $w[n]$ is the complex excitation or weight of the nth element, λ is the wavelength, and $u = \sin\theta$. The function $P(u)$ thus can be considered as the discrete-time Fourier transform of $w[n]$, with the frequency variable given by $2\pi(u/\lambda)d$. The array element weighting $w[n]$ as a function of the element position is called the *aperture function*. For a uniformly excited array, $w[n] = $ a constant, and the grating lobes in the radiation pattern are avoided if $d \leq \lambda/2$. Typically, $d = \lambda/2$, in which case the range of u is between $-\pi$ and π. From Eq. (15.142), it can be seen that the expression for $P(u)$ is identical to the frequency response of an FIR filter of length N. An often used element weight is $w[n] = 1$ whose radiation pattern is this same as the frequency response of a running-sum or boxcar FIR filter.

Sparse arrays with fewer elements are obtained by removing some of the elements, which increases the interelement spacing between some consecutive pairs of elements to more than $\lambda/2$. This usually results in an increase of sidelobe levels and can possibly cause the appearance of grating lobes in the radiation pattern. However, these unwanted lobes can be reduced significantly by selecting array element locations appropriately. In the case of ultrasound scanners, a two-way radiation pattern is generated by a transmit array and a receive array. The design of such arrays is simplified by treating the problem as the design of an "effective aperture function" $w_{eff}[n]$, which is given by the convolution of the transmit aperture function $w_T[n]$ and the receive aperture function $w_T[n]$ [Loc96]:

$$w_{eff}[n] = w_T[n] \circledast w_R[n]. \tag{15.143}$$

If the number of elements (including missing elements) in the the transmit and receive arrays are, respectively, L and M, then the number of elements N in a single array with an effective aperture function $w_{eff}[n]$ is $L + M - 1$. The design problem is thus to determine $w_T[n]$ and $w_T[n]$ for a desired $w_{eff}[n]$.

15.13.1 The Polynomial Factorization Approach

In the z-domain, Eq. (15.143) is equivalent to

$$P_{eff}(z) = P_T(z) P_R(z), \tag{15.144}$$

where

$$P_{eff}(z) = \sum_{n=0}^{N-1} w_{eff}[n]z^{-n}, \qquad P_T(z) = \sum_{n=0}^{L-1} w_T[n]z^{-n}, \qquad P_R(z) = \sum_{n=0}^{M-1} w_R[n]z^{-n}. \tag{15.145}$$

As a result, the sparse antenna array design problem can be formulated as the factorization of the polynomial $P_{eff}(z)$ into factors $P_T(z)$ and $P_R(z)$ with missing coefficients. We first consider the design of a uniform array for which $w_{eff}[n] = 1$. To this end, we can make use of the following factorization of $P_{eff}(z)$ for values of N that are powers-of-2 [Mit2004a]:

$$P_{eff}(z) = (1 + z^{-1})(1 + z^{-2}) \cdots (1 + z^{-2^{K-1}}), \tag{15.146}$$

where $N = 2^K$. A proof of Eq. (15.146) is left as an exercise (Problem 13.24).

15.13.2 Uniform Effective Aperture Function

We now illustrate the application of the above factorization approach to sparse array design for the case $N = 16$; that is, $K = 4$. From Eq. (15.146) we then have

$$P_{eff}(z) = (1 + z^{-1})(1 + z^{-2})(1 + z^{-4})(1 + z^{-8}).$$

Three possible choices for $P_T(z)$ and $P_R(z)$ are as follows:

$$\begin{aligned}
\text{Design \#1}: \quad & P_T(z) = 1, \\
& P_R(z) = (1 + z^{-1})(1 + z^{-2})(1 + z^{-4})(1 + z^{-8}) \\
& \qquad = 1 + z^{-1} + z^{-2} + z^{-3} + z^{-4} + z^{-5} + z^{-6} + z^{-7} \\
& \qquad + z^{-8} + z^{-9} + z^{-10} + z^{-11} + z^{-12} + z^{-13} + z^{-14} + z^{-15}, \\
\text{Design \#2}: \quad & P_T(z) = 1 + z^{-1}, \\
& P_R(z) = (1 + z^{-2})(1 + z^{-4})(1 + z^{-8}) \\
& \qquad = 1 + z^{-2} + z^{-4} + z^{-6} + z^{-8} + z^{-10} + z^{-12} + z^{-14}, \\
\text{Design \#3}: \quad & P_T(z) = (1 + z^{-1})(1 + z^{-8}) = 1 + z^{-1} + z^{-8} + z^{-9}, \\
& P_R(z) = (1 + z^{-2})(1 + z^{-4}) = 1 + z^{-2} + z^{-4} + z^{-6}.
\end{aligned}$$

Additional choices for $P_T(z)$ and $P_R(z)$ can be found elsewhere [Mit2004a].

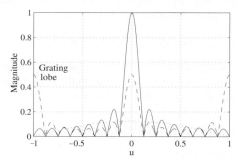

Figure 15.76: Radiation patterns of transmit array (dotted line), receive array (dashed line), and two-way radiation pattern (solid line). The radiation patterns have been scaled by a factor of 16 to make the value of the two-way radiation pattern at $u = 0$ unity.

Design #1 consists of a single-element transmit array and a 16-element nonsparse receive array and thus requires a total of 17 elements. The remaining designs given above result in sparse transmit and/or receive arrays. For example, the transmit and receive aperture functions for Design #2 are given by [Loc96]

$$w_T[n] = \{1 \quad 1\}, \qquad w_R[n] = \{1 \quad 0 \quad 1 \quad 0 \quad 1 \quad 0 \quad 1 \quad 0 \quad 1 \quad 0 \quad 1 \quad 0 \quad 1 \quad 0 \quad 1\},$$

where 0 in $w_R[n]$ indicates the absence of an element and requires a total of 10 elements. Figure 15.76 shows the radiation patterns of the individual arrays and the two-way radiation pattern of the composite array. Note that the grating lobes in the radiation pattern of the receive array are being suppressed by the radiation pattern of the transmit array.

Most economic sparse array design with eight elements is obtained with the Design #3, requiring a total of eight elements. For example, the transmit and receive aperture functions for Design #3 are given by [Loc96]:

$$w_T[n] = \{1 \quad 1 \quad 0 \quad 0 \quad 0 \quad 0 \quad 0 \quad 0 \quad 1 \quad 1\}, \qquad w_R[n] = \{1 \quad 0 \quad 1 \quad 0 \quad 1 \quad 0 \quad 1\}.$$

15.13.3 Linearly Tapered Effective Aperture Function

The shape of the effective aperture function can be made smoother to reduce the grating lobes by controlling the shape of the transmit and receive aperture functions. For the design of a sparse array pair with a linearly tapered effective aperture function $P_{eff}(z)$, one can choose[Mit2004b]

$$P_{eff}(z) = P_1(z) P_2(z), \tag{15.147}$$

where

$$P_1(z) = \frac{1}{R} \sum_{n=0}^{R-1} z^{-n}, \qquad P_2(z) = \sum_{n=0}^{S-1} z^{-n}. \tag{15.148}$$

The number of elements in the effective aperture function is then $N = R + S - 1$. The number of apodized elements in the beginning and at the end of the effective aperture function is $(R - 1)$ each. The values of the apodized elements are $\frac{1}{R}, \frac{2}{R}, \ldots, \frac{R-1}{R}$. Moreover, the parameter S must satisfy the condition $S > R - 1$. For a sparse antenna pair design, the value of either R or S or both must be power-of-2.

We consider the design of a linearly tapered array for $R = 3$ and $S = 8$, which results in an effective aperture function given by

$$w_{eff}[n] = \{\tfrac{1}{3} \quad \tfrac{2}{3} \quad 1 \quad 1 \quad 1 \quad 1 \quad 1 \quad 1 \quad \tfrac{2}{3} \quad \tfrac{1}{3}\}.$$

A possible design for the transmit and receive arrays is given by

$$w_T[n] = \{1 \quad 1 \quad 0 \quad 0 \quad 1 \quad 1\},$$
$$w_R[n] = \{\tfrac{1}{3} \quad \tfrac{1}{3} \quad \tfrac{2}{3} \quad \tfrac{1}{3} \quad \tfrac{1}{3}\}.$$

The corresponding scaled radiation patterns are shown in Figure 15.77(a)

15.13.4 Staircase Effective Aperture Function

Sparse antenna array pairs with a staircase effective aperture function also exhibit reduced grating lobes. For designing such arrays, there are two possible forms of the factor $P_1(z)$ in Eq. (15.147) [Mit2004b]. One form is for an even number of steps in the effective aperture function, and the other form is for an odd number of steps. We consider here the first form for which

$$P_1(z) = \tfrac{1}{2\ell+1}[1 + z^{-k_1}(1 + z^{-k_2}(1 + \ldots + z^{-k_\ell}(1 + \ldots + z^{-k_2}(1 + z^{-k_1})\ldots)))]. \qquad (15.149)$$

The number R of elements (including zero-valued ones) in $P_1(z)$ is given by $R = 2\sum_{i=1}^{\ell} k_i + 1$. Moreover, for a staircase effective aperture function, the number S of elements in $P_2(z)$ of Eq. (15.147) must satisfy the condition $S > 2\sum_{i=1}^{\ell} k_i$. The number of apodized elements in the beginning and at the end of the effective aperture function is $2\sum_{i=1}^{\ell} k_i$ each. The values of the apodized elements are $\tfrac{1}{2\ell+1}, \tfrac{2}{2\ell+1}, \ldots, \tfrac{2\ell}{2\ell+1}$. For a sparse antenna pair design, the value of S must be a power-of-2.

For example, consider the design of an array with $k_1 = 1$, $k_2 = 2$, and $S = 8$. Here

$$P_1(z) = \tfrac{1}{5}[1 + z^{-1}(1 + z^{-2}(1 + z^{-2}(1 + z^{-1})))] = \tfrac{1}{5}[1 + z^{-1} + z^{-3} + z^{-5} + z^{-6}],$$
$$P_2(z) = 1 + z^{-1} + z^{-2} + z^{-3} + z^{-4} + z^{-5} + z^{-6} + z^{-7}.$$

The effective aperture function is then of the form

$$w_{eff}[n] = \{0.2 \quad 0.4 \quad 0.4 \quad 0.6 \quad 0.6 \quad 0.8 \quad 1 \quad 1 \quad 0.8 \quad 0.6 \quad 0.6 \quad 0.4 \quad 0.4 \quad 0.2\}.$$

One possible choice for the transmit and the receive aperture functions is given by

$$w_T[n] = \{\tfrac{1}{5} \quad \tfrac{1}{5} \quad \tfrac{1}{5} \quad \tfrac{2}{5} \quad 0 \quad \tfrac{2}{5} \quad \tfrac{1}{5} \quad \tfrac{1}{5} \quad \tfrac{1}{5}\},$$
$$w_R[n] = \{1 \quad 1 \quad 0 \quad 0 \quad 1 \quad 1\}.$$

The corresponding scaled radiation patterns are shown in Figure 15.77(b).

15.14 Summary

This chapter has touched upon a variety of practical applications of digital signal processing. The first application discussed is in the efficient and robust detection of dual-tone multifrequency (DTMF) tones employed in dial-pulse telephone signaling, ATM machines, voice mains, and so on. The next application treated here is in the spectral analysis of stationary signals and is the basis of most commercial spectrum analyzers. For the spectral analysis of nonstationary signals, such as speech and radar signals, the DFTs of

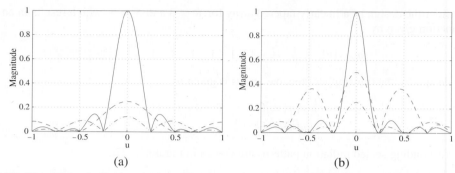

Figure 15.77: Illustration of effective aperture smoothing by shaping transmit and receive aperture functions. The radiation patterns have been scaled to make the value of the two-way radiation pattern at $u = 0$ unity.

small windowed segments of these signals are computed, and a three-dimensional display of the resulting spectrums, called spectrogram, is employed. This type of signal analysis is more popularly called the short-term Fourier transform, or the time-dependent Fourier transform, which is covered next. This is followed by a brief discussion of spectral analysis of random signals. Here both nonparametric and parametric spectral analysis methods are reviewed.

A number of applications of digital signal processing for spectral shaping digital audio and for the generation of special audio effects are outlined. These are followed by a discussion on the use of digital filtering for preemphasis of audio signals.

The discrete-time analytic signal has a zero-valued spectrum for all negative frequencies and, as a result, is a complex signal. Such a signal can be generated from a real signal by passing the latter through a discrete-time Hilbert transformer. Several methods for designing the Hilbert transformer are described. One application of the analytic signal considered in this chapter is in the design of a digital single-sideband communication system.

Several applications discussed here involve multirate digital signal processing. The first two applications are concerned with image compression and the design of transmultiplexers for interconnecting frequency-division multiplex (FDM) and time-division multiplex (TDM) communication systems. A method for the efficient digital data transmission based on multirate techniques is discussed next. This method makes use of the discrete Fourier transform (DFT) computation and, as a result, can be implemented using fast Fourier transform (FFT) algorithms. The next two applications considered in this chapter are the oversampling sigma-delta analog-to-digital (A/D) and digital-to-analog (D/A) converter design. Such converters are being increasingly employed in many systems because of their improved signal-to-noise ratio performances.

Finally, the chapter shows the similarity between the far-field pattern of a linear antenna array of equally spaced elements and the frequency response of an FIR filter. This similarity is then made use of in the design of sparse antenna arrays.

There are numerous other applications, with many requiring a knowledge of other subjects, and as a result, they are beyond the scope of this book.

A Discrete-Time Random Signals

The random discrete-time signal is a random process and is thus a sequence of random variables and consists of a typically infinite collection or ensemble of discrete-time sequences. We review in this appendix the important statistical properties of the random variable and the random process.

A.1 Statistical Properties of a Random Variable

The statistical properties of a random variable depend on its probability distribution function or, equivalently, on its probability density function, which are defined next. The probability that the random variable X takes a value in a specified range from $-\infty$ to α is given by its *probability distribution function*

$$P_X(\alpha) = \text{Probability}[X \le \alpha]. \tag{1.1}$$

The *probability density function* of X is defined by

$$p_X(\alpha) = \frac{\partial P_X(\alpha)}{\partial \alpha}, \tag{1.2}$$

if X can assume a continuous range of values. From Eq. (1.2). the probability distribution function is therefore given by

$$P_X(\alpha) = \int_{-\infty}^{\alpha} p_X(u)\, du. \tag{1.3}$$

Three more commonly used statistical properties characterizing a random variable are the *mean* or *expected value* m_X, the *mean-square value* $E(X^2)$, and the *variance* σ_X^2 as defined below:

$$m_X = E(X) = \int_{-\infty}^{\infty} \alpha\, p_X(\alpha)\, d\alpha, \tag{A.4a}$$

$$E(X^2) = \int_{-\infty}^{\infty} \alpha^2 p_X(\alpha)\, d\alpha, \tag{A.4b}$$

$$\sigma_X^2 = E\left([X - m_X][X - m_X]^*\right)$$
$$= \int_{-\infty}^{\infty} (\alpha - m_X)(\alpha - m_X)^* p_X(\alpha)\, d\alpha = E(X^2) - |m_X|^2, \tag{A.4c}$$

where $E(\cdot)$ denotes the expectation operator and where * denotes complex conjugation. These three properties provide adequate information about a random variable in most practical cases. The square root of the variance, σ_X, is called the *standard deviation* of the random variable X.

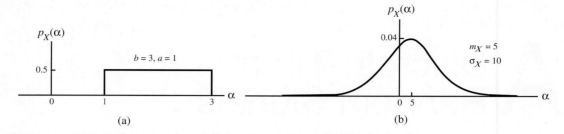

Figure A.1: (a) Uniform and (b) Gaussian probability density functions.

Two probability density functions, commonly encountered in digital signal processing applications, are the *uniform density function* defined by

$$p_X(\alpha) = \begin{cases} \frac{1}{b-a}, & \text{for } a \leq \alpha \leq b, \\ 0, & \text{otherwise,} \end{cases} \tag{A.5}$$

and the *Gaussian density function*, also called the *normal density function*, defined by

$$p_X(\alpha) = \frac{1}{\sigma_X \sqrt{2\pi}} e^{-(\alpha - m_X)^2 / 2\sigma_X^2}, \tag{A.6}$$

where the parameters m_X and σ_X are, respectively, the mean value and the standard deviation of the real random variable X and lie in the range $-\infty < m_X < \infty$ and $\sigma_X \geq 0$. These density functions are plotted in Figure A.1.

It can be shown that the mean and the variance of a uniformly distributed real random variable X defined by Eq. (A.5) are given by

$$m_X = \frac{1}{b-a} \int_a^b \alpha \, d\alpha = \frac{b+a}{2}, \tag{A.7a}$$

$$E(X^2) = \frac{1}{b-a} \int_a^b \alpha^2 \, d\alpha = \frac{b^2 + a^2 + ab}{3}, \tag{A.7b}$$

$$\sigma_X^2 = \frac{(b-a)^2}{12}. \tag{A.7c}$$

In the case of two random variables X and Y, their joint statistical properties, as well as their individual statistical properties are of practical interest. The probability that X takes a value in a specified range from $-\infty$ to α and that Y takes a value in a specified range from $-\infty$ to β is given by their *joint probability distribution function*

$$P_{XY}(\alpha, \beta) = \text{Probability } [X \leq \alpha, Y \leq \beta] \tag{A.8}$$

or, equivalently, by their *joint probability density function*

$$p_{XY}(\alpha, \beta) = \frac{\partial^2 P_{XY}(\alpha, \beta)}{\partial \alpha \, \partial \beta}. \tag{A.9}$$

The joint probability distribution function is thus given by

$$P_{XY}(\alpha, \beta) = \int_{-\infty}^{\alpha} \int_{-\infty}^{\beta} p_{XY}(u, v) \, du \, dv. \tag{A.10}$$

The joint statistical properties of two random variables X and Y are described by their cross-correlation and cross-covariance, as defined by

$$\phi_{XY} = E\left(XY^*\right) = \int_{-\infty}^{\infty} \int_{-\infty}^{\infty} \alpha\beta p_{X,Y}(\alpha, \beta^*)\, d\alpha\, d\beta, \tag{A.11a}$$

$$\gamma_{XY} = E\left([X - m_X][Y - m_Y]^*\right)$$

$$= \int_{-\infty}^{\infty} \int_{-\infty}^{\infty} (\alpha - m_X)(\beta - m_Y)^* p_{X,Y}(\alpha, \beta)\, d\alpha d\beta$$

$$= \phi_{XY} - m_X m_Y^*, \tag{A.11b}$$

where m_X and m_Y are, respectively, the mean of the random variables X and Y. The two random variables X and Y are said to be *linearly independent or uncorrelated* if

$$E(XY) = E(X)E(Y) \tag{A.12a}$$

and *statistically independent* if

$$P_{X,Y}(\alpha, \beta) = P_X(\alpha)P_Y(\beta). \tag{A.12b}$$

It can be shown that if the random variables X and Y are statistically independent, then they are also linearly independent. However, if X and Y are linearly independent, they may not be statistically independent.

A.2 Statistical Properties of a Random Signal

As mentioned earlier, the random discrete-time signal consists of a typically infinite collection or ensemble of discrete-time sequences. The statistical properties of the random signal $\{X[n]\}$ at time index n are given by the statistical properties of the random variable $X[n]$. Thus, the *mean* or *expected value* of $\{X[n]\}$ at time index n is given by

$$m_{X[n]} = E(X[n]) = \int_{-\infty}^{\infty} \alpha p_{X[n]}(\alpha; n)\, d\alpha. \tag{A.13}$$

The *mean-square value* of $\{X[n]\}$ at time index n is given by

$$E\left(X[n]^2\right) = \int_{-\infty}^{\infty} \alpha^2 p_{X[n]}(\alpha; n)\, d\alpha. \tag{A.14}$$

The *variance* $\sigma_{X[n]}^2$ of $\{X[n]\}$ at time index n is defined by

$$\sigma_{X[n]}^2 = E\left(\{X[n] - m_{X[n]}\}^2\right) = E\left(X[n]^2\right) - \left(m_{X[n]}\right)^2. \tag{A.15}$$

In general, the mean, mean-square value, and variance of a random discrete-time signal are functions of the time index n, and can be considered as sequences.

Often, the statistical relation of the samples of a random discrete-time signal at two different time indices m and n is of interest. One such relation is the *autocorrelation* defined by

$$\phi_{XX}[m, n] = E\left(X[m]X^*[n]\right). \tag{A.16}$$

Another relation is the *autocovariance* of $\{X[n]\}$, defined by

$$\gamma_{XX}[m, n] = E\left((X[m] - m_{X[m]})(X[n] - m_{X[n]})^*\right)$$

$$= \phi_{XX}[m, n] - m_{X[m]}(m_{X[n]}^*). \tag{A.17}$$

As can be seen from Eqs. (A.16) and (A.17), both the autocorrelation and the autocovariance are functions of two time indices m and n and can be considered as two-dimensional sequences.

The correlation between two different random discrete-time signals $\{X[n]\}$ and $\{Y[n]\}$ is described by the *cross-correlation function*

$$\phi_{XY}[m, n] = E\left(X[m]Y^*[n]\right) = \int_{-\infty}^{\infty} \int_{-\infty}^{\infty} \alpha \beta^* p_{X[m],Y[n]}(\alpha, m, \beta, n)\, d\alpha\, d\beta \tag{A.18}$$

and the *cross-covariance function*

$$\begin{aligned}
\gamma_{XY}[m, n] &= E\left((X[m] - m_{X[m]})(Y[n] - m_{Y[n]})^*\right) \\
&= \phi_{XY}[m, n] - m_{X[m]}(m_{Y[n]}^*),
\end{aligned} \tag{A.19}$$

where $p_{X[m],Y[n]}(m, \alpha, n, \beta)$ is the joint probability density function of $X[n]$ and $Y[n]$. Both the cross-correlation and the cross-covariance functions can also be considered as two-dimensional sequences. The two random discrete-time signals $\{X[n]\}$ and $\{Y[n]\}$ are uncorrelated if $\gamma_{XY}[m, n] = 0$ for all values of the time indices m and n.

A.3 Wide-Sense Stationary Random Signal

In general, the statistical properties of a random discrete-time signal $\{X[n]\}$, such as the mean and variance of the random variable $X[n]$, and the autocorrelation and the autocovariance functions are time-varying functions. The class of random signals often encountered in digital signal processing applications are the so-called *wide-sense stationary* (WSS) random processes for which some of the key statistical properties are either independent of time or of the time origin. More specifically, for a wide-sense stationary random process $\{X[n]\}$, the mean $E(X[n])$ has the same constant value m_X for all values of the time index n, and the autocorrelation and the autocovariance functions depend only on the difference of the time indices m and n; that is,

$$m_X = E(X[n]), \text{ for all } n, \tag{A.20a}$$

$$\phi_{XX}[\ell] = \phi_{XX}[n + \ell, n] = E(X[n + \ell]X^*[n]), \text{ for all } n \text{ and } \ell, \tag{A.20b}$$

$$\begin{aligned}
\gamma_{XX}[\ell] = \gamma_{XX}[n + \ell, n] &= E\left((X[n + \ell] - m_X)(X[n] - m_X)^*\right) \\
&= \phi_{XX}[\ell] - |m_X|^2, \text{ for all } n \text{ and } \ell.
\end{aligned} \tag{A.20c}$$

Note that in the case of a WSS random process, the autocorrelation and the autocovariance functions are one-dimensional sequences.

The mean-square value of a WSS random process $\{X[n]\}$ is given by

$$E\left(|X[n]|^2\right) = \phi_{XX}[0], \tag{A.21}$$

and the variance is given by

$$\sigma_X^2 = \gamma_{XX}[0] = \phi_{XX}[0] - |m_X|^2. \tag{A.22}$$

The cross-correlation and cross-covariance functions between two WSS random processes $\{X[n]\}$ and $\{Y[n]\}$ are given by

$$\phi_{XY}[\ell] = E\left(X[n + \ell]Y^*[n]\right), \tag{A.23a}$$

$$\begin{aligned}
\gamma_{XY}[\ell] &= E\left((X[n + \ell] - m_X)(Y[n] - m_Y)^*\right) \\
&= \phi_{XY}[\ell] - m_X(m_Y^*).
\end{aligned} \tag{A.23b}$$

The symmetry properties satisfied by the autocorrelation, autocovariance, cross-correlation, and cross-covariance functions are

$$\phi_{XX}[-\ell] = \phi_{XX}^*[\ell], \tag{A.24a}$$

$$\gamma_{XX}[-\ell] = \gamma_{XX}^*[\ell], \tag{A.24b}$$

$$\phi_{XY}[-\ell] = \phi_{YX}^*[\ell], \tag{A.24c}$$

$$\gamma_{XY}[-\ell] = \gamma_{YX}^*[\ell]. \tag{A.24d}$$

From the above symmetry properties, it can be seen that sequences $\phi_{XX}[\ell]$, $\gamma_{XX}[\ell]$, $\phi_{XY}[\ell]$, and $\gamma_{XY}[\ell]$ are always two-sided sequences.

Some additional useful properties concerning these functions are

$$\phi_{XX}[0]\phi_{YY}[0] \geq |\phi_{XY}[\ell]|^2, \tag{A.25a}$$

$$\gamma_{XX}[0]\gamma_{YY}[0] \geq |\gamma_{XY}[\ell]|^2, \tag{A.25b}$$

$$\phi_{XX}[0] \geq |\phi_{XX}[\ell]|, \tag{A.25c}$$

$$\gamma_{XX}[0] \geq |\gamma_{XX}[\ell]|. \tag{A.25d}$$

A consequence of the above properties is that the autocorrelation and autocovariance functions of a WSS random process assume their maximum values at $\ell = 0$. In addition, it can be shown that, for a WSS signal with nonzero mean, that is, $m_{X[n]} \neq 0$, and with no periodic components,

$$\lim_{|\ell| \to \infty} \phi_{XX}[\ell] = \left| m_{X[n]} \right|^2. \tag{A.26}$$

If $X[n]$ has a periodic component, then $\phi_{XX}[\ell]$ will contain the same periodic component as illustrated in Example 2.45.

A.4 Concept of Power in a Random Signal

The average power of a deterministic sequence $x[n]$ was defined earlier and is given by Eq. (2.31). To compute the power associated with a random signal $\{X[n]\}$, we use instead the following definition:

$$\mathcal{P}_X = E\left(\lim_{N \to \infty} \frac{1}{2N+1} \sum_{n=-N}^{N} |X[n]|^2 \right). \tag{A.27}$$

In most practical cases, the expectation and summation operators in Eq. (A.27) can be interchanged, resulting in a more simple expression given by

$$\mathcal{P}_X = \lim_{N \to \infty} \frac{1}{2N+1} \sum_{n=-N}^{N} E\left(|X[n]|^2 \right). \tag{A.28}$$

In addition, if the random signal has a constant mean-square value for all values of n, as in the case of a WSS signal, then Eq. (A.28) reduces to

$$\mathcal{P}_X = E\left(|X[n]|^2 \right). \tag{A.29}$$

From Eqs. (A.21) and (A.22), it follows that for a WSS signal, the average power is given by

$$\mathcal{P}_X = \phi_{XX}[0] = \sigma_X^2 + |m_X|^2. \tag{A.30}$$

A.5   Ergodic Signal

In many practical situations, the random signal of interest cannot be described in terms of a simple analytical expression to permit computations of its statistical properties, which invariably involves the evaluation of definite integrals or summations. Often a finite portion of a single realization of the random signal is available, from which some estimation of the statistical properties of the ensemble must be made. Such an approach can lead to meaningful results if the ergodicity condition is satisfied. More precisely, a stationary random signal is defined to be an *ergodic signal* if all its statistical properties can be estimated from a single realization of sufficiently large finite length. It should be noted that an important class of signals encountered in practice, namely, those with finite memory (i.e., those that satisfy the so-called mixing conditions) are ergodic.

For an ergodic signal, time averages equal ensemble averages derived via the expectation operator in the limit as the length of the realization goes to infinity. For example, for a real ergodic signal, we can compute the mean value, variance, and autocovariance as

$$m_X = \lim_{M \to \infty} \frac{1}{2M+1} \sum_{n=-M}^{M} x[n], \tag{A.31a}$$

$$\sigma_X^2 = \lim_{M \to \infty} \frac{1}{2M+1} \sum_{n=-M}^{M} (x[n] - m_X)^2, \tag{A.31b}$$

$$\gamma_{XX}[\ell] = \lim_{M \to \infty} \frac{1}{2M+1} \sum_{n=-M}^{M} (x[n] - m_X)(x[n+\ell] - m_X). \tag{A.31c}$$

The limiting operation required to compute the ensemble averages by means of time averages is still not practical in most situations and therefore, replaced with a finite sum to provide an estimate of the desired statistical properties. For example, approximations to Eqs. (A.31a) through (A.31c) that are often used are

$$\hat{m}_X = \frac{1}{M+1} \sum_{n=0}^{M} x[n], \tag{A.32a}$$

$$\hat{\sigma}_X^2 = \frac{1}{M+1} \sum_{n=0}^{M} (x[n] - m_X)^2, \tag{A.32b}$$

$$\hat{\gamma}_{XX}[\ell] = \frac{1}{M+1} \sum_{n=0}^{M} (x[n] - m_X)(x[n+\ell] - m_X). \tag{A.32c}$$

A.6   Transform-Domain Representations of Random Signals

The notion of random discrete-time signals was introduced in Section A.2, along with their statistical characterizations in the time domain. These infinite-length signals have infinite energy and do not have transform-domain characterizations like the deterministic signals. However, the autocorrelation and the autocovariance sequences of stationary random signals, defined by Eqs. (A.16) and (A.17), are of finite energy, and, in most practical cases, their transform-domain representations do exist. We discuss these representations in this section.

A.6.1   Discrete-Time Fourier Transform Representation

The discrete-time Fourier transform of the autocorrelation sequence $\phi_{XX}[\ell]$ of a WSS sequence $X[n]$, defined in Eq. (A.20b), is given by

$$\Phi_{XX}(e^{j\omega}) = \sum_{\ell=-\infty}^{\infty} \phi_{XX}[\ell]e^{-j\omega\ell}, \qquad |\omega| < \pi, \tag{A.33}$$

and is usually referred to as the *power density spectrum* or, simply, the *power spectrum*[1] of $X[n]$. It is denoted by $\mathcal{P}_{XX}(\omega)$. The above relation between the autocorrelation sequence and the power spectrum is more commonly known as the *Wiener–Khintchine theorem*. A sufficient condition for the existence of the power spectrum $\mathcal{P}_{XX}(\omega)$ is that the autocorrelation sequence $\phi_{XX}[\ell]$ be absolutely summable. Likewise, the discrete-time Fourier transform of the autocovariance sequence $\gamma_{XX}[\ell]$ of a WSS sequence $x[n]$, defined in Eq. (A.20c), is given by

$$\Gamma_{XX}(e^{j\omega}) = \sum_{\ell=-\infty}^{\infty} \gamma_{XX}[\ell]e^{-j\omega\ell}, \qquad |\omega| < \pi. \tag{A.34}$$

A sufficient condition for the existence of $\Gamma_{XX}(e^{j\omega})$ is that the autocovariance sequence $\gamma_{XX}[\ell]$ be absolutely summable. Applying the inverse discrete-time Fourier transform to Eq. (A.33) and using the notation $\mathcal{P}_{XX}(\omega) = \Phi_{XX}(e^{j\omega})$, we arrive at

$$\phi_{XX}[\ell] = \frac{1}{2\pi} \int_{-\pi}^{\pi} \mathcal{P}_{XX}(\omega)e^{j\omega\ell}\,d\omega. \tag{A.35}$$

It follows from Eqs. (A.35) and (A.21) that

$$E\left(|X[n]|^2\right) = \phi_{XX}[0] = \frac{1}{2\pi} \int_{-\pi}^{\pi} \mathcal{P}_{XX}(\omega)\,d\omega. \tag{A.36}$$

Thus, $\phi_{XX}[0]$ represents the *average power* in the random signal $X[n]$. Similarly, the inverse transform of Eq. (A.34) yields

$$\gamma_{XX}[\ell] = \frac{1}{2\pi} \int_{-\pi}^{\pi} \Gamma_{XX}(e^{j\omega})e^{j\omega\ell}\,d\omega. \tag{A.37}$$

From Eqs. (A.37) and (A.22), we get

$$\sigma_X^2 = \gamma_{XX}[0] = \frac{1}{2\pi} \int_{-\pi}^{\pi} \Gamma_{XX}(e^{j\omega})\,d\omega$$
$$= \frac{1}{2\pi} \int_{-\pi}^{\pi} \mathcal{P}_{XX}(\omega)\,d\omega - |m_X|^2. \tag{A.38}$$

Applying the discrete-time Fourier transform to both sides of Eq. (A.24a), we can show that the power spectrum $\mathcal{P}_{XX}(\omega)$ of a WSS random discrete-time signal $\{X[n]\}$ is a real-valued function of ω. In addition, if $\{X[n]\}$ is a real random signal, $\mathcal{P}_{XX}(\omega)$ is an even function of ω; that is, $\mathcal{P}_{XX}(\omega) = \mathcal{P}_{XX}(-\omega)$.

Likewise, the discrete-time Fourier transform of the cross-correlation sequence $\phi_{XY}[\ell]$ of two jointly stationary random signals $\{X[n]\}$ and $\{Y[n]\}$, given by

$$\Phi_{XY}(e^{j\omega}) = \sum_{\ell=-\infty}^{\infty} \phi_{XY}[\ell]e^{-j\omega\ell}, \qquad |\omega| < \pi, \tag{A.39}$$

---

[1]Also called the *power spectral density*.

is usually referred to as the *cross-power spectral density* or *cross-power spectrum*. It is denoted by $\mathcal{P}_{XY}(\omega)$, and, in general, it is a complex function of ω. A sufficient condition for the existence of $\mathcal{P}_{XY}(\omega)$ is that the cross-correlation sequence $\phi_{XY}[\ell]$ be absolutely summable. Applying the discrete-time Fourier transform to both sides of Eq. (A.24c), we can show that $\mathcal{P}_{XY}(\omega) = \mathcal{P}_{YX}^*(\omega)$. Similarly, we can define the discrete-time Fourier transform of the cross-covariance sequence $\gamma_{XY}[\ell]$ as

$$\Gamma_{XY}(e^{j\omega}) = \sum_{\ell=-\infty}^{\infty} \gamma_{XY}[\ell]e^{-j\omega\ell}, \qquad |\omega| < \pi. \tag{A.40}$$

A sufficient condition for the existence of $\Gamma_{XY}(e^{j\omega})$ is that the cross-covariance sequence $\gamma_{XY}[\ell]$ be absolutely summable.

The relation between the discrete-time Fourier transforms of the autocorrelation sequence and the autocovariance sequence can be derived from Eq. (A.20c) and is given by

$$\Gamma_{XX}(e^{j\omega}) = \mathcal{P}_{XX}(\omega) - 2\pi \, |m_X|^2 \, \delta(\omega), \qquad |\omega| < \pi, \tag{A.41}$$

where we have used the notation $\mathcal{P}_{XX}(\omega) = \Phi_{XX}(e^{j\omega})$. Likewise, the relation between the discrete-time Fourier transforms of the cross-correlation sequence and the cross-covariance sequence follows from Eq. (A.23b) and is given by

$$\Gamma_{XY}(e^{j\omega}) = \mathcal{P}_{XY}(\omega) - 2\pi m_X m_Y^* \delta(\omega), \qquad |\omega| < \pi, \tag{A.42}$$

where we have used the notation $\mathcal{P}_{XY}(\omega) = \Phi_{XY}(e^{j\omega})$.

A.6.2   *z*-Transform Representation

As can be seen from Eqs. (A.41) and (A.42), the Fourier transforms of the sequences $\gamma_{XX}[\ell]$ and $\gamma_{XY}[\ell]$ contain impulse functions. As a result, their *z*-transforms do not exist in general. However, for zero-mean stationary random signals, the *z*-transform of the autocorrelation sequence, $\Phi_{XX}(z)$, and that of the cross-correlation sequence, $\Phi_{XY}(z)$, may exist under certain conditions. Since the autocorrelation and cross-correlation sequences are two-sided sequences, their region of convergence must be an annular region of the form

$$R_1 < |z| < \frac{1}{R_1}. \tag{A.43}$$

We can generalize some of the results of the previous section if the *z*-transforms exist. For example, from the symmetry properties of the power spectrum $\mathcal{P}_{XX}(\omega)$ and cross-power spectrum $\mathcal{P}_{XY}(\omega)$, it follows that $\Phi_{XX}(z) = \Phi_{XX}^*(1/z^*)$ and $\Phi_{XY}(z) = \Phi_{YX}^*(1/z^*)$. It also follows from Eqs. (A.38) and (A.41) that

$$\sigma_X^2 = \frac{1}{2\pi j} \oint_C \Phi_{XX}(z)z^{-1} \, dz, \tag{A.44}$$

where C is a closed counterclockwise contour in the ROC of $\Phi_{XX}(z)$.

A.7   White Noise

A random process $\{X[n]\}$ for which any pair of samples, $X[m]$ and $X[n]$, are uncorrelated, with $m \neq n$; that is, $E(X[m]X[n]) = E(X[m])E(X[n])$, is called a *white random process*. For a WSS white random process, the autocorrelation sequence is given by

$$\phi_{XX}[\ell] = \sigma_X^2\delta[\ell] + m_X^2, \tag{A.45}$$

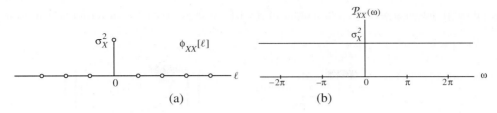

Figure A.2: (a) Autocorrelation sequence and (b) power spectral density of a white noise.

and the corresponding power spectrum is given by

$$\mathcal{P}_{XX}(\omega) = \sigma_X^2 + 2\pi m_X^2 \delta(\omega), \qquad |\omega| < \pi. \tag{A.46}$$

A zero-mean white WSS random process has an autocorrelation sequence $\phi_{XX}[\ell]$ that is an impulse sequence of area σ_x^2 and a power spectrum $\mathcal{P}_{XX}(\omega)$ that is of constant value σ_x^2 for all values of ω, as indicated in Figure A.2. Such a random process is more commonly called *white noise* and plays an important role in digital signal processing.

A.8 Discrete-Time Processing of Random Signals

As we shall observe later in the text, there are occasions when we need to study the effect of processing a random discrete-time signal by a linear time-invariant discrete-time system. More precisely, we need to determine the statistical properties of the output signal $\{y[n]\}$ generated by a stable LTI system with an impulse response $\{h[n]\}$ when its input $x[n]$ is a particular realization of a wide-sense stationary (WSS) random process $\{X[n]\}$. For simplicity, we assume the input sequence and the impulse response to be real-valued. Now, the output of an LTI system is given by the linear convolution of the input and the impulse response of the system; that is,

$$y[n] = \sum_{k=-\infty}^{\infty} h[k]x[n-k]. \tag{A.47}$$

Since the function of a random variable is also a random variable, it follows from Eq. (A.47) that the output $y[n]$ is also a sample sequence of an output random process $\{Y[n]\}$.

A.8.1 Statistical Properties of the Output Signal

Since the input $x[n]$ is a sample sequence of a stationary random process, its mean m_x is a constant independent of the time index n.[2] The mean $E(y[n])$ of the output random process $y[n]$ is then given by

$$m_y = E(y[n]) = E\left(\sum_{k=-\infty}^{\infty} h[k]x[n-k]\right)$$

$$= \sum_{k=-\infty}^{\infty} h[k]E(x[n-k]) = m_x \sum_{k=-\infty}^{\infty} h[k] = m_x H(e^{j0}), \tag{A.48}$$

which is a constant independent of the time index n.

---

[2]For convenience, we drop the notational difference between the random process $\{X[n]\}$ and its particular realization $\{x[n]\}$.

The autocorrelation function of the output of the LTI discrete-time system for a real-valued input is given by

$$\phi_{yy}[n + \ell, n] = E(y[n + \ell]y[n])$$

$$= E\left(\left\{\sum_{i=-\infty}^{\infty} h[i]x[n + \ell - i]\right\}\left\{\sum_{k=-\infty}^{\infty} h[k]x[n - k]\right\}\right)$$

$$= \sum_{i=-\infty}^{\infty} h[i] \sum_{k=-\infty}^{\infty} h[k]E(x[n + \ell - i]x[n - k])$$

$$= \sum_{i=-\infty}^{\infty} h[i] \sum_{k=-\infty}^{\infty} h[k]\phi_{xx}[n + \ell - i, n - k]. \tag{A.49}$$

Since the input is a sample sequence of a WSS random process, its autocorrelation sequence depends only on the difference $\ell + k - i$ of the time indices $n + \ell - i$ and $n - k$; that is,

$$\phi_{xx}[n + \ell - i, n - k] = \phi_{xx}[\ell + k - i]. \tag{A.50}$$

Substituting the above in Eq. (A.49), we arrive at

$$\phi_{yy}[n + \ell, n] = \sum_{i=-\infty}^{\infty} h[i] \sum_{k=-\infty}^{\infty} h[k]\phi_{xx}[\ell + k - i] = \phi_{yy}[\ell], \tag{A.51}$$

indicating that the output autocorrelation sequence depends on the difference ℓ of the time indices $n + \ell$ and n. As a result of Eqs. (A.48) and (A.51), it follows that the output $y[n]$ is also a sample sequence of a WSS random process.

Substituting $m = i - k$ in Eq. (A.51), we arrive at

$$\phi_{yy}[\ell] = \sum_{m=-\infty}^{\infty} \phi_{xx}[\ell - m] \sum_{k=-\infty}^{\infty} h[k]h[m + k]$$

$$= \sum_{m=-\infty}^{\infty} \phi_{xx}[\ell - m]r_{hh}[m], \tag{A.52}$$

where

$$r_{hh}[m] = \sum_{k=-\infty}^{\infty} h[k]h[m + k] = h[m] \circledast h[-m] \tag{A.53}$$

is called the *aperiodic autocorrelation sequence* of the impulse response sequence $\{h[n]\}$. It should be noted that $r_{hh}[m]$ is the autocorrelation of a finite-energy deterministic sequence and is not the same as the autocorrelation of an infinite-energy WSS random signal.

The cross-correlation function between the output and the input sequences of the LTI system for a real-valued input is given by

$$\phi_{yx}[n + \ell, n] = E(y[n + \ell]x[n])$$

$$= E\left(\sum_{i=-\infty}^{\infty} h[i]x[n + \ell - i]x[n]\right)$$

$$= \sum_{i=-\infty}^{\infty} h[i] E\left(x[n+\ell-i]\,x[n]\right)$$

$$= \sum_{i=-\infty}^{\infty} h[i]\phi_{xx}[\ell-i] = \phi_{yx}[\ell]. \tag{A.54}$$

The result of Eq. (A.54) indicates that the cross-correlation sequence depends on ℓ, the difference of the time indices $n+\ell$ and n.

A.8.2 z-Transform Domain Representation

We now consider the z-transform representation of Eq. (A.51). As indicated in Section A.6.2, the z-transform of $\phi_{xx}[\ell]$ may exist if the input random signal is of zero mean. From Eq. (A.48), for a zero-mean random input, the output of an LTI system is also a zero-mean random signal. In this case, we obtain from Eq. (A.51), by taking the z-transform of both sides,

$$\Phi_{yy}(z) = \Psi(z)\Phi_{xx}(z), \tag{A.55}$$

where $\Phi_{xx}(z)$, $\Phi_{yy}(z)$, and $\Psi(z)$ are, respectively, the z-transforms of $\phi_{xx}[\ell]$, $\phi_{yy}[\ell]$, and $\psi[r]$. But, from Eq. (A.53), $\Psi(z) = H(z)H(z^{-1})$, which when substituted in Eq. (A.55) yields

$$\Phi_{yy}(z) = H(z)H(z^{-1})\Phi_{xx}(z). \tag{A.56}$$

On the unit circle, Eq. (A.56) reduces to

$$\Phi_{yy}(e^{j\omega}) = |H(e^{j\omega})|^2 \Phi_{xx}(e^{j\omega}). \tag{A.57}$$

Using the notations $\mathcal{P}_{xx}(\omega)$ and $\mathcal{P}_{yy}(\omega)$ to denote the input and output power spectral densities $\Phi_{xx}(e^{j\omega})$ and $\Phi_{yy}(e^{j\omega})$, respectively, we can rewrite Eq. (A.57) as

$$\mathcal{P}_{yy}(\omega) = |H(e^{j\omega})|^2 \mathcal{P}_{xx}(\omega). \tag{A.58}$$

Now, from Eq. (A.21), for a zero-mean WSS process $y[n]$, the total average power is given by the mean-square value $E(y^2[n]) = \phi_{yy}[0]$. But, $\phi_{yy}[\ell]$ is given by the inverse Fourier transform of $\Phi_{yy}(e^{j\omega})$:

$$\phi_{yy}[\ell] = \frac{1}{2\pi}\int_{-\pi}^{\pi} \Phi_{yy}(e^{j\omega})e^{j\omega\ell}\,d\omega. \tag{A.59}$$

Therefore, from Eqs. (A.57) and (A.58), the total average power in the output signal $y[n]$ is given by

$$E(y^2[n]) = \phi_{yy}[0]$$

$$= \frac{1}{2\pi}\int_{-\pi}^{\pi} \Phi_{yy}(e^{j\omega})\,d\omega$$

$$= \frac{1}{2\pi}\int_{-\pi}^{\pi} |H(e^{j\omega})|^2 \mathcal{P}_{xx}(\omega)\,d\omega. \tag{A.60}$$

For a real-valued WSS random signal $x[n]$, the autocorrelation sequence $\phi_{xx}[\ell]$ is an even sequence, and hence, $\Phi_{xx}(e^{j\omega})$ is an even function of ω. Assume the LTI system $h[n]$ to be an ideal filter with a square magnitude response

$$|H(e^{j\omega})|^2 = \begin{cases} 1, & \omega_{c1} \le |\omega| \le \omega_{c2}, \\ 0, & 0 \le |\omega| < \omega_{c1}, \omega_{c2} < |\omega| < \pi. \end{cases} \tag{A.61}$$

In this case, Eq. (A.60) reduces to

$$\phi_{yy}[0] = \frac{1}{\pi} \int_{-\omega_{c2}}^{-\omega_{c1}} \mathcal{P}_{xx}(\omega) \, d\omega + \frac{1}{\pi} \int_{\omega_{c1}}^{\omega_{c2}} \mathcal{P}_{xx}(\omega) \, d\omega$$

$$= \frac{2}{\pi} \int_{\omega_{c1}}^{\omega_{c2}} \mathcal{P}_{xx}(\omega) \, d\omega. \tag{A.62}$$

Since the total average output power $\phi_{yy}[0]$ is always nonnegative independent of the bandwidth of the linear filter, it follows from Eq. (A.62) that

$$\mathcal{P}_{xx}(\omega) \geq 0, \tag{A.63}$$

proving that the power spectral density function of a real WSS random signal is also nonnegative, in addition to being a real and even function.

Likewise, from the z-transforms of both sides of Eq. (A.54), we obtain

$$\Phi_{yx}(z) = H(z) \, \Phi_{xx}(z), \tag{A.64}$$

where $\Phi_{yx}(z)$ is the z-transform of $\phi[\ell]$. On the unit circle, Eq. (A.64) reduces to

$$\Phi_{yx}(e^{j\omega}) = H(e^{j\omega}) \, \Phi_{xx}(e^{j\omega}). \tag{A.65}$$

The function $\Phi_{yx}(e^{j\omega})$ is the *cross-spectral density* or *cross-power spectrum,* denoted by $\mathcal{P}_{yx}(\omega)$. Note that if $x[n]$ is a WSS white noise sequence, its power spectrum is a constant K at all frequencies. In this case, Eq. (A.65) reduces to

$$\Phi_{yx}(e^{j\omega}) = K \cdot H(e^{j\omega}).$$

Variance of the Output Signal

Now we develop the expression for the variance of the output random signal when the input to the LTI system is a real-valued white random process. From Eq. (A.38), we get

$$\sigma_y^2 = \gamma_{yy}[0] = \frac{1}{2\pi} \int_{-\pi}^{\pi} \mathcal{P}_{yy}(\omega) \, d\omega - m_y^2. \tag{A.66}$$

Substituting Eq. (A.58) in Eq. (A.66) and making use of Eq. (A.46), we arrive at

$$\sigma_y^2 = \frac{\sigma_x^2}{2\pi} \int_{-\pi}^{\pi} |H(e^{j\omega})|^2 \, d\omega, \tag{A.67}$$

which can be alternatively written as

$$\sigma_y^2 = \frac{\sigma_x^2}{2\pi j} \oint_C H(z) H(z^{-1}) z^{-1} \, dz, \tag{A.68}$$

where C is a counterclockwise closed contour in the ROC of $H(z)H(z^{-1})$.

Using Parseval's relation in Eq. (A.67), we can also express the output variance as

$$\sigma_y^2 = \sigma_x^2 \sum_{n=-\infty}^{\infty} |h[n]|^2. \tag{A.69}$$

Bibliography

[Abr72]   M. Abramowitz and I.A. Stegun, editors. *Handbook of Mathematical Functions*. Dover Publications, New York NY, 1972.

[Abr97]   E. Abreu, S.K. Mitra, and R. Marchesani. Nonminimum phase channel equalization using noncausal filters. *IEEE Trans. on Acoustics, Speech, and Signal Processing*, 45:1–13, January 1997.

[Aca83]   A. Acampora. Wideband picture detail restoration in a digital NTSC comb-filter system. *RCA Engineer*, 28(5):44–47, September/October 1983.

[Ada83]   J.W. Adams and A.N. Willson, Jr. A new approach to FIR digital filters with fewer multipliers and reduced sensitivity. *IEEE Trans. on Circuits and Systems*, CAS-30:277–283, May 1983.

[Ada91]   J.W. Adams. FIR digital filters with least-squares stopbands subject to peak-gain constraints. *IEEE Trans. on Circuits and Systems*, 39:376–388, April 1991.

[Aga75]   R.C. Agarwal and C.S. Burrus. New recursive digital filter structures having very low sensitivity and low roundoff noise. *IEEE Trans. on Circuits and Systems*, CAS-22:921–927, December 1975.

[Ahm74]   N. Ahmed, T. Natarajan, and K.R. Rao. Discrete cosine transform. *IEEE Trans. on Computers*, C-23:90-93, January 1974.

[Aka92]   A.N. Akansu and R.A. Haddad. *Multiresolution Signal Decomposition*. Academic Press, New York NY, 1992.

[Ala93]   M.A. Al-Alaoui. Novel digital integrator and differentiator. *Electronic Letters*, 29:376–378, 18 February 1993.

[All77]   J.B. Allen and L.R. Rabiner. A unified approach to short-term Fourier analysis and synthesis. *Proc. IEEE*, 65:1558–1564, November 1977.

[All80]   H.G. Alles. Music synthesis using real time digital techniques. *Proc. IEEE*, 68:436–449, April 1980.

[Ans85]   R. Ansari. An Extension of the discrete Fourier transform. *IEEE Trans. on Circuits & Systems*, CAS32:618–619, June 1985.

[Ans91]   R. Ansari, C. Guillemot, and J.F. Kaiser. Wavelet construction using Lagrange halfband filters. *IEEE Trans. on Circuits & Systems*, 38:1116–1118, September 1991.

[Ans93]   R. Ansari and B. Liu. Multirate signal processing. In S.K. Mitra and J.F. Kaiser, editors, *Handbook for Digital Signal Processing*, chapter 14, pages 981–1084. Wiley-Interscience, New York NY, 1993.

[Ant93]   A. Antoniou. *Digital Filters: Analysis, Design, and Applications*. McGraw-Hill, New York NY, 2nd edition, 1993.

[Bag98]   S. Bagchi and S.K. Mitra. *Nonuniform Discrete Fourier Transform and Its Signal Processing Applications*, Kluwer Academic Publishers, Norwell MA, 1998.

[Bar48]   M.S. Bartlett. Smoothing periodograms from the time series with continuous spectra. *Nature (London)*, 161:686-687, 1948.

[Bel76]   M. Bellanger, G. Bonnerot, and M. Coudreuse. Digital filtering by polyphase network: Application to sample-rate alteration and filter banks. *IEEE Trans. Acoustics, Speech, and Signal Processing*, ASSP-24:109–114, April 1976.

[Bel81]   M. Bellanger. On computational complexity in digital filters. *Proc. The European Conference on Circuit Theory & Design*, The Haugue, The Netherlands, pp. 58–63, August 1981.

[Bel2000]   M. Bellanger. *Digital Processing of Signals:Theory and Practice*. Wiley, New York NY, Third Edition, 2000.

[Ben48]   W.R. Bennett. Spectra of quantized signals. *Bell System Technical Journal*, 27:446–472, 1948.

[Ber76]   P.A. Bernhardt, D.A. Antoniadis, and A.D. Da Rossa. Lunar perturbations in columnar electron content and their interpretations in terms of dynamo electrostatic fields. *Journal of Geophysics Research*, 43:5957–5963, December 1976.

[Bin90]   J.A.C. Bingham. Multicarrier modulation for data transmission: An idea whose time has come. *IEEE Communications Magazine*, pages 5–14, May 1990.

[Bjo98]   G. Bjontegaard. Response to call for proposals for H.26L. ITU-T SG16 Doc. Q15-F-11, October 1998.

[Bla65]   R.B. Blackman. *Linear Data Smoothing and Prediction in Theory and Practice*. Addison-Wesley, Reading MA, 1965.

[Ble78]   B. Blesser and J.M. Kates. Digital processing in audio signals. In A.V. Oppenheim, editor, *Applications of Digital Signal Processing*, chapter 2. Prentice Hall, Englewood Cliffs NJ, 1978.

[Boi81]   R. Boite and H. Leich. A new procedure for the design of high-order minimum-phase FIR digital or CCD filters. *Signal Processing*, 3:101–108, 1981.

[Bol93]   B.A. Bolt. *Earthquakes*. W.H. Freeman, New York NY, 1993.

[Bon76]   G. Bongiovanni, P. Corsini, and G. Forsini. One-dimensional and two-dimensional generalized discrete Fourier transform. *IEEE Trans. on Acoustics, Speech and Signal Processing*, ASSP-24:97–99, February 1976.

[Boo51]   A.D. Booth. A signed binary multiplication technique. *Quart. J. Mech. Appl. Math.*, 4(Part 2):236–240, 1951.

[Box70]   G.E.P. Box and G.M. Jenkins. *Time Series Analysis: Forecasting and Control*. Holden-Day, San Francisco CA, 1970.

[Bra83]   R.N. Bracewell. The discrete Hartley transform. *Journal of the Optical Society of America*, 73:1832–1835, December 1983.

[Bur73]   R.S. Burrington. *Handbook of Mathematical Tables and Formulas*. McGraw-Hill, New York NY, 5th edition, 1973.

[Bur72]   C.S. Burrus. Block realization of digital filters. *IEEE Trans. on Audio and Electroacoustics*, AU-20:230–235, October 1972.

[Bur77]   C.S. Burrus. Index mappings for multidimensional formulation of the DFT and convolution. *IEEE Trans. on Acoustics, Speech, and Signal Processing*, ASSP-25:239–242, June 1977.

[Bur81]   C.S. Burrus and P.W. Eschenbacher. An in-place, in-order prime factor FFT algorithm. *IEEE Trans. on Acoustics, Speech, and Signal Processing*, ASSP-29:806–817, April 1981.

[Bur92]   C.S. Burrus, A.W. Soewito, and R.A. Gopinath. Least squared error FIR filter design with transition bands. *IEEE Trans. on Signal Processing*, 40:1327–1340, 1992.

[Büt77]   M. Büttner. Elimination of limit cycles in digital filters with very low increase in the quantization noise. *IEEE Trans. on Circuits and Systems*, CAS-24:300–304, 1977.

[Cad87]   J.A. Cadzow. *Foundations of Digital Signal Processing and Data Analysis*. Macmillan, New York NY, 1987.

[Cag56]   P. Cagan. The monetary dynamics of hyperinflation. In M. Friedman, editor,*Studies in the Quantity Theory of Money*, University of Chicago Press, Chicago, Illinois, 1956.

[Can86]   J.C. Candy and A-N. Huynh. Double interpolation for digital-to-analog conversion. *IEEE Trans. on Communications*, COM-34:77–81, January 1986.

[Can92]   J.C. Candy and G.C. Temes. Oversampling methods for A/D and D/A conversion. In J.C. Candy and G.C. Temes, editors, *Oversampling Delta-Sigma Data Converters*, pages 1–25. IEEE Press, New York NY, 1992.

[Cav2000]   T.J. Cavicchi. *Digital Signal Processing*. Wiley, New York NY, 2000.

[Cha73]   D.S.K. Chan and L.R. Rabiner. Analysis of quantization errors in the direct form for finite impulse response digital filters. *IEEE Trans. on Audio and Electroacoustics*, AU-21:354–366, August 1973.

[Cha81]   T-L. Chang and S.A. White. An error cancellation digital-filter structure and its distributed-arithmetic implementation. *IEEE Trans. on Circuits and Systems*, CAS-28:339–342, April 1981.

[Cha80]   C. Charalambous and A. Antoniou. Equalisation of recursive digital filters. *IEE Proc.*, Part G, 127:219-225, October 1980.

[Che2001]   C-T. Chen. *Digital Signal Processing:Spectral Computation and Filter Design*. Oxford University Press, New York NY, 2001.

[Chr66]   E. Christian and E. Eisenmann. *Filter Design Tables and Graphs*. Wiley, New York NY, 1966.

[Chu85]   P.L. Chu. Quadrature mirror filter design for an arbitrary number of equal bandwidth channels. *IEEE Trans. on Acoustics, Speech, and Signal Processing*, ASSP-33:203–218, February 1985.

[Chu90]   R.V. Churchill and J.W. Brown.   *Introduction to Complex Variables and Applications*. McGraw-Hill, New York NY, 5th edition, 1990.

[Chu95]   J. Chun and N.K. Bose. Fast evaluation of an integral occuring in digital filtering applications. *IEEE Trans. on Signal Processing*, 43:1982–1986, August 1995.

[Cio91]   J.M. Cioffi. *A multicarrier primer*. ANSI T1E1.4 Committee Contribution, Boca Raton FL, November 1991.

[Cio93]   J.M. Cioffi and Y-S Byun. Adaptive filtering. In S.K. Mitra and J.F. Kaiser, editors, *Handbook for Digital Signal Processing*, chapter 15, pages 1085–1142. Wiley Interscience, New York NY, 1993.

[Cla73]   T.A.C.M. Classen, W.F.G. Mecklenbräuker, and J.B.H. Peek. Some remarks on the classifications of limit cycles in digital filters. *Philips Research Reports*, 28:297–305, August 1973.

[Coh86]   A. Cohen. *Biomedical Signal Processing*, volume II. CRC Press, Boca Raton FL, 1986.

[Coh22]   A. Cohn. Über die anzahl der wurzeln einer algebraischen gleichung in einem kreise. *Math.* 2, 14:110–148, August 1922.

[Con69]   A.C. Constantinides. Digital notch filters. *Electronics Letters*, 5:198–199, 1 May 1969.

[Con70]   A.C. Constantinides. Spectral transformations for digital filters. *Proc. IEE*, 117:1585–1590, August 1970.

[Coo65]   J.W. Cooley and J.W. Tukey. An algorithm for the machine calculation of complex Fourier series. *Math. Computation*, 19:297–301, 1965.

[Cou83]   L.W. Couch II. *Digital and Analog Communication Systems*. Macmillan, New York NY, 1983.

[Cox83]   R.V. Cox and J.M. Tribolet. Analog voice privacy systems using TFSP scrambling: Full duplex and half duplex. *Bell System Technical Journal*, 62:47–61, January 1983.

[Cro75]   R.E. Crochiere and A.V. Oppenheim. Analysis of linear digital networks. *Proc. IEEE*, 62:581–595, April 1975.

[Cro76a] A. Croisier, D. Esteban, and C. Galand. Perfect channel splitting by use of interpolation/decimation/tree decomposition techniques. In *Proc. International Symposium on Information Science and Systems*, Patras, Greece, 1976.

[Cro76b] R.E. Crochiere and L.R. Rabiner. On the properties of frequency transformations for variable cutoff linear phase digital filters. *IEEE Trans. on Circuits and Systems*, CAS-23:684–686, 1976.

[Cro83]   R.E. Crochiere and L.R. Rabiner. *Multirate Digital Signal Processing*. Prentice Hall, Englewood Cliffs NJ, 1983.

[Cuc91]   S. Cucchi, F. Desinan, G. Parladori, and G. Sicuranza. DSP implementation of arbitrary sampling frequency conversion for high quality sound application. In *Proc. IEEE International Conference on Acoustics, Speech, and Signal Processing*, pages 3609–3612, Toronto Canada, May 1991.

[Dar76]   G. Daryanani. *Principles of Active Network Synthesis and Design*. Wiley, New York NY, 1976.

[Dau88]   I. Daubechies. Orthonormal bases of compactly supported wavelets. *Comm. Pure Appl. Math.*, 41:909–996, 1988.

[DeF88]   D.J. DeFatta, J.G. Lucas, and W.S. Hodgkiss. *Digital Signal Processing: A System Design Approach*. Wiley, New York NY, 1988.

[Del93]   J.R. Deller, Jor., J.G. Proakis, and J.H.L. Hansen. *Discrete-Time Processing of Speech Signals*. Macmillan, New York NY, 1993.

[Dep80]   E. Deprettere and P. DeWilde. Orthogonal cascade realization of a real multiport digital filter. *International Journal on Circuit Theory Appl*, 8:245–277, 1980.

[Doo99]   S. R. Dooley, R. W. Stewart, and T. S. Durrani. Fast on-line B-spline interpolation. *Electronics Letters*, 35:1130-1131, 8 July 1999.

[Dre90]   J.O. Drewery. Digital filtering of television signals. In C.P. Sandbank, editor, *Digital Television*, chapter 5. Wiley, New York NY, 1990.

[Dug80]   J.P. Dugré, A.A.L. Beex, and L.L. Scharf. Generating covariance sequences and the calculation of quantization and roundoff error variances in digital filters. *IEEE Trans. on Acoustics, Speech, and Signal Processing*, ASSP-28:102–104, 1980.

[Dug82]   J.P. Dugré and E.I. Jury. A note on the evaluation of complex integrals using filtering interpretations. *IEEE Trans. on Acoustics, Speech, and Signal Processing*, ASSP-30:804–807, 1982.

[Duh86]   P. Duhamel. Implementation of "split-radix" FFT algorithms for complex, real, and real-symmetric data. *IEEE Trans. on Acoustics, Speech, and Signal Processing*, ASSP-34:285–295, April 1986.

[Dur59]   J. Durbin. Efficient estimation of parameters in moving average model. *Biometrika*, 46:306–316, 1959.

[Dut80]   D.L. Duttweiler. Bell's echo-killer chip. *IEEE Spectrum*, 17:34–37, October 1980.

[Dut83]   S.C. Dutta Roy. Comments on "On the construction of a digital transfer function from its real part on the unit circle." *Proceedings of the IEEE (Letters)*, 71:1009–1010, August 1983.

[Ear76]   J. Eargle. *Sound Recording*. Van Nostrand Reinhold, New York NY, 1976.

[Ear86]   J.M. Eargle. *Handbook of Recording Engineering*. Van Nostrand Reinhold, New York NY, 1986.

[Ebe69]   P.M. Ebert, J.E. Mazo, and M.G. Taylor. Overflow oscillations in digital filters. *Bell System Technical Journal*, 48:2999–3020, November 1969.

[Egg84]   L.D.J. Eggermont and P.J. Berkhout. Digital audio circuits: Computer simulations and listening tests. *Philips Technical Review*, 41(3):99–103, 1983/84.

[Est77]   D. Esteban and C. Galand. Application of quadrature mirror filters to split-band voice coding schemes. In *Proc. IEEE International Conference on Acoustics, Speech, and Signal Processing*, pages 191–195, May 1977.

[Far88]   C.W. Farrow. A continuously variable digital delay element. In *Proc. IEEE International Symposium on Circuits and Systems*, Helsinki, Finland, pages 2641–2645, June 1988.

[Fet71]   A. Fettweis. Digital filter structures related to classical filter networks. *Archiv für Elektrotechnik und Übertragungstechnik*, 25:79–81, 1971.

[Fet72]   A. Fettweis. A simple design method of maximally flat delay digital filters. *IEEE Trans. on Audio and Electroacoustics*, AU-20:112–114, June 1972.

[Fet86]   A. Fettweis. Wave digital filters: Theory and practice. *Proc. IEEE*, 74:270–316, February 1986.

[Fli94]   N.J. Fliege. *Multirate Digital Signal Processing*. Wiley, New York NY, 1994.

[Fot2001]   I. Fotinopoulos, T. Stathaki, and A. Constantinides. A method for FIR filter design from joint amplitude and group delay characteristics. In *Proc. 35th Asilomar Conference on Signals, Systems, and Computers*, pages 621–625, November 2001.

[Fre78]   S.L. Freeny, J.F. Kaiser, and H.S. McDonald. Some applications of digital signal processing in telecommunications. In A.V. Oppenheim, editor, *Applications of Digital Signal Processing*, chapter 1, pages 1–28. Prentice Hall, Englewood Cliffs NJ, 1978.

[Fre94]   M.E. Frerking. *Digital Signal Processing in Communication Systems*. Van Nostrand Reinhold, New York NY, 1994.

[Gaz85]   L. Gazsi. Explicit formulas for lattice wave digital filters. *IEEE Trans. on Circuits and Systems*, CAS-32:68–88, January 1985.

[Goe58]   G. Goertzel. An algorithm for evaluation of finite trigonometric series. *American Mathematical Monthly*, 65:34–35, January 1958.

[Goe82]   D. Goedhart, R.J. Van de Plassche, and E.F. Stikvoort. Digital-to-analog conversion in playing a compact disc. *Philips Technical Review*, 40(6):174–179, 1982.

[Gol68]   B. Gold and K.L. Jordan. A note on digital filter synthesis. *Proc. IEEE*, 56:1717–1718, October 1968.

[Gol69a]   B. Gold and K.L. Jordan. A direct search procedure for designing finite duration impulse response filters. *IEEE Trans. Audio and Electroacoustics*, AU-17:33–36, March 1969.

[Gol69b]   B. Gold and C.M. Radar. *Digital Processing of Signals*. McGraw-Hill, New York, NY, 1969.

[Gon2002]   R.C. Gonzalez and P. Wintz. *Digital Image Processing*, Second Edition. Prentice-Hall, Upper Saddle River NJ, 2002.

[Goo77]   D.J. Goodman and M.J. Carey. Nine digital filters for decimation and interpolation. *IEEE Trans. on Acoustics, Speech, and Signal Processing*, ASSP-25:121–126, April 1977.

[Gra73]   A.H. Gray Jr. and J.D. Markel. Digital lattice and ladder filter synthesis. *IEEE Trans. on Audio and Electroacoustics*, AU-21:491–500, December 1973.

[Ham89]   R.W. Hamming. *Digital Filters*. Prentice Hall, Englewood Cliffs NJ, 3rd edition, 1989.

[Haa10]   A. Haar. Zur theorie der orthogonalen funktionen-systeme. *Math. Annalen.*, 69:331-371, 1910.

[Hay70]   S.S. Haykin and R. Carnegie. New method of synthetising linear digital filters based on convolution integral. *IEE Proc.*, 117:1063–1072, June 1970.

[Hee82]   J.P.J. Heemskerk and K.A.S. Immink. Compact disc: System aspects and modulation. *Philips Technical Review*, 40(6):157–165, 1982.

[Hel68]   H.D. Helms. Nonrecursive digital filters: Design methods for achieving specifications on frequency response. *IEEE Trans. on Audio and Electroacoustics*, AU-16:336–342, September 1968.

[Hen83]   D. Henrot and C.T. Mullis. A modular and orthogonal digital filter structure for parallel processing. In *Proc. IEEE International Conference Acoustics, Speech, and Signal Processing*, pages 623–626, 1983.

[Her70]   O. Herrmann and H.W. Schüssler. Design of nonrecursive digital filters with minimum phase. *Electronics Letters*, 6:329–330, 1970.

[Her71]   O. Herrmann. On the approximation problem in nonrecursive digital filter design. *IEEE Trans. Circuit Theory*, CT-18:411–413, 1971.

[Her73]   O. Herrmann, L.R. Rabiner, and D.S.K. Chan. Practical design rules for optimum finite impulse response lowpass digital filters. *Bell System Tech. J.*, 52:769–799, 1973.

[Hig84]   W.E. Higgins and D.C. Munson Jr. Optimal and suboptimal error spectrum shaping for cascade form digital filters. *IEEE Trans. on Circuits and Systems*, CAS-31:429–437, May 1984.

[Hir73]   K. Hirano, T. Saito, S. Nishimura, and S.K. Mitra. Time-sharing realization of Butterworth digital filters. In *Monograph of the Circuits and Systems Group*. Institution of Electronic and Communication Engineers (Japan), No. CST 73–59, December 1973. (In Japanese).

[Hir74]   K. Hirano, S. Nishimura, and S.K. Mitra. Design of digital notch filters. *IEEE Trans. on Circuits and Systems*, CAS-21:540–546, July 1974.

[Hog81]   E.B. Hogenauer. An economical class of digital filters for decimation and interpolation. *IEEE Trans. Acoustics, Speech, and Signal Processing*, ASSP-29:155–162, April 1981.

[Hsi87]   C-C. Hsiao. Polyphase filter matrix for rational sampling rate conversions. In *Proc. IEEE International Conference on Acoustics, Speech, and Signal Processing*, pages 2173–2176, Dallas TX, April 1987.

[Hub89]   D.M. Huber and R.A. Runstein. *Modern Recording Techniques*. Howard W Sams, Indianapolis IN, 3rd edition, 1989.

[Hwa79]   K. Hwang. *Computer Arithmetic: Principles, Architecture and Designs*. Wiley, New York NY, 1979.

[IEEE85]   Institute of Electrical and Electronic Engineers. IEEE Standard for Binary Floating-Point Arithmetic, 1985.

[Ife2001]   E.C. Ifeachor and B.W. Jervis. *Digital Signal Processing: A Practical Approach*, Second Edition. Prentice-Hall, Harlow, England, 2001.

[ITU84]   International Telecommunication Union. *CCITT Red Book*, volume VI. Fascicle VI.1, October 1984.

[Jac69]   L.B. Jackson. An analysis of limit cycles due to multiplicative rounding in recursive digital filters. In *Proc. 7th Allerton Conference on Circuit and System Theory*, pages 69–78, Monticello IL, 1969.

[Jac70a]   L.B. Jackson. On the interaction of roundoff noise and dynamic range in digital filters. *Bell System Technical Journal*, 49:159–184, February 1970.

[Jac70b]   L.B. Jackson. Roundoff-noise analysis for fixed-point digital filters realized in cascade or parallel form. *IEEE Trans. on Audio and Electroacoustics*, AU-18:107–122, June 1970.

[Jac96]   L.B. Jackson. *Digital Filters and Signal Processing*. Kluwer, Boston MA, 3rd edition, 1996.

[Jac2003]   E. Jacobsen and R. Lyons. The sliding DFT. *IEEE Signal Processing Magazine*, 20:74–80, March 2003.

[Jaf83]   D.A. Jaffe and J.O. Smith. Extensions of the Karplus-Strong plucked-ptring algorithm. *Computer Music Journal*, 9:26–23, 1983.

[Jai89]   A.K. Jain. *Fundamentals of Digital Image Processing*. Prentice Hall, Englewood Cliffs NJ, 1989.

[Jar88]   P. Jarske, Y. Neuvo, and S.K. Mitra. A simple approach to the design of FIR digital filters with variable characteristics. *Signal Processing*, 14:313–326, 1988.

[Jay84]   N.S. Jayant and P. Knoll. *Digital Coding of Waveforms*. Prentice Hall, Englewood Cliffs NJ, 1984.

[Jen91]   Y-C. Jenq. Digital convolution algorithm for pipelining multiprocessor system. *IEEE Trans. on Computers*, C-30:966–973, December 1991.

[Jia97]   Z. Jiang and A.N. Willson, Jr. Efficient digital filtering architectures using pipelining/interleaving. *IEEE Trans. on Circuits and Systems*, Part II. 44:110–119, February 1997.

[Joh80]   J.D. Johnston. A filter family designed for use in quadrature mirror filter banks. In *Proc. IEEE International Conference on Acoustics, Speech, and Signal Processing*, pages 291–294, April 1980.

[Joh89]   J.R. Johnson. *Introduction to Digital Signal Processing*. Prentice Hall, Englewood Cliffs NJ, 1989.

[Jos99]   Y.V. Joshi and S.C. Dutta Roy. Design of IIR multiple notch filters based on all-pass filters. *IEEE Trans. on Circuits and Systems*, Part II. 46:134–138, February 1999.

[Jur73]   E.I. Jury. *Theory and Application of the z-Transform Method*. Robert E. Krieger, Huntington, NY, 1973.

[Kah98]   M. Kahrs and K. Brandenburg, editors. *Application of Digital Signal Processing to Audio and Acoustics*. Kluwer, Boston MA, 1998.

[Kai66]   J.F. Kaiser. Digital filters. In F. Kuo and J.F. Kaiser, editors, *System Analysis by Digital Computers*, chapter 7. Wiley, New York NY, 1966.

[Kai74]   J.F. Kaiser. Nonrecursive digital filter design using the I_0-sinh window function. In *Proc. 1974 IEEE International Symposium on Circuits and Systems*, pages 20–23, San Francisco CA, April 1974.

[Kai77]   J.F. Kaiser and R.W. Hamming. Sharpening the response of a symmetric nonrecursive filter by multiple use of the same filter. *IEEE Trans. on Acoustics, Speech, and Signal Processing*, ASSP-25:415–422, October 1977.

[Kai80]   J.F. Kaiser. On a simple algorithm to calculate the 'energy' of a signal. *Proc. IEEE International Conference on Acoustics, Speech, and Signal Processing*, pages 381–384, Albuquerque NM, April 1980.

[Kam86]   J.J. Van der Kam. A digital "decimating" filter for analog-to-digital conversion of hi-fi audio signals. *Philips Technical Review*, 42:230–238, 1986.

[Kan71]   E.P.F. Kan and J.K. Aggarwal. Error analysis in digital filters employing floating-point arithmetic. *IEEE Trans. on Circuit Theory*, CT-18:678–686, November 1971.

[Kar83]   K. Karplus and A. Strong. Digital synthesis of plucked-string and drum timbres. *Computer Music Journal*, 7:43–55, Summer 1983.

[Kau98]   J. Kauraniemi, T.I. Laakso, I. Hartimo, and S.J. Ovaska. Delta operator realizations of direct-form IIR filters. *IEEE Trans. on Circuits and Systems-II: Analog and Digital Signal Processing*, 45:41–52, January 1998.

[Kel62]   J.L. Kelly Jr. and C. Lochbaum. Speech synthesis. In *Proc. Stockholm Speech Communication Seminar*, Stockholm, Sweden, September 1962. Royal Institute of Technology.

[Kin72]   N. Kingsbury. Second-order recursive digital filter element for poles near the unit circle and the real-z axis. *Electronics Letters*, 8: 155-156, March 1972.

[Kin2001] N. Kingsbury. *Image Coding Course Notes*. Department of Engineering, University of Cambridge, Cambridge, U.K., July 13, 2001.

[Knu69]   D.E. Knuth. *The Art of Computer Programming: Volume 2 – Seminumerical Algorithms*. Addison-Wesley, Reading MA, 2nd edition, 1969.

[Ko97]   N. Ko, D.J. Shpak, and A. Antoniou. Design of recursive delay equalizers using constrained optimization. *Proc. IEEE Pacific Rim Conference on Communications, Computers and Signal Processing*, Victoria, B.C., Canada, pp. 173-177, August 1997.

[Koi91]   R.D. Koilpillai, T.Q. Nguyen, and P.P. Vaidyanathan. Some results in the theory of crosstalk-free transmultiplexers. *IEEE Trans. on Signal Processing*, 39:2174–2183, October 1991.

[Kol77]   D.P. Kolba and T.W. Parks. A prime factor FFT algorithm using high speed convolution. *IEEE Trans. on Acoustics, Speech, and Signal Processing*, ASSP-25:281–294, August 1977.

[Kor93]   I. Koren. *Computer Arithmetic Algorithms*. Prentice Hall, Englewood Cliffs NJ, 1993.

[Kum93]   R. Kumaresan. Spectral analysis. In S.K. Mitra and J.F. Kaiser, editors, *Handbook for Digital Signal Processing*, chapter 16, pages 1143–1242. Wiley-Interscience, New York NY, 1993.

[Laa96]   T.I. Laakso, V. Välimäki, M. Karjalainen, and U.K. Laine. Splitting the unit delay. *IEEE Signal Processing Magazine*, 13:30–60, January 1996.

[Lag82]   R. Lagadec, D. Pelloni, and D. Weiss. A 2-channel, 16-bit digital sampling frequency converter for professional digital audio. In *Proc. IEEE International Conference on Acoustics, Speech, and Signal Processing*, pages 93–96, April 1982.

[Lar93]   L.E. Larson and G.C. Temes. Signal conditioning and interface circuits. In S.K. Mitra and J.F. Kaiser, editors, *Handbook for Digital Signal Processing*, chapter 10, pages 677–720. Wiley-Interscience, New York NY, 1993.

[Lar99]   J. Laroche. A modified lattice structure with pleasant scaling properties. *IEEE Trans. on Signal Processing*, 47:3423–3425, December 1999.

[Lat98]   B.P. Lathi. *Signals Processing and Linear Systems*. Berkley-Cambridge, Carmichael CA, 1998.

[Law78]   V.B. Lawrence and K.V. Mina. Control of limit cycle oscillations in second-order recursive digital filters using constrained random quantization. *IEEE Trans. on Acoustics, Speech, and Signal Processing*, ASSP-26:127–134, April 1978.

[LeG88]   D. LeGall and A. Tabatabai. Subband coding of images using symmetric short kernel filters and arithmetic coding techniques. In *Proc. IEEE International Conference on Acoustics, Speech and Signal Processing*, pages 761–764, 1988.

[Ler83]   E.L. Lerner. Electronically synthesized music. *IEEE Spectrum*, 17:46–51, June 1983.

[Lev47]   N. Levinson. The Wiener RMS criterion in filter design and prediction. *J. Math. Phys.*, 25:261–278, 1947.

[Lim90]   J.S. Lim. *Two-Dimensional Signal and Image Processing*. Prentice Hall, Englewood Cliffs NJ, 1990.

[Lim86]   Y.C. Lim. Frequency-response masking approach for the synthesis of sharp linear phase digital filters. *IEEE Trans. on Circuits and Systems*, CAS-33:357-364, April 1986.

[Liu69]   B. Liu and T. Kaneko. Error analysis of digital filters realized in floating-point arithmetic. *Proc. IEEE*, 57:1735–1747, October 1969.

[Loc96]   G.R. Lockwood, P-C. Li, M. O'Donnell, and F.S. Foster. Optimizing the radiation pattern of sparse periodic linear arrays. *IEEE Trans. on Ultrasonics Ferroelectrics, and Frequency Control*, 43:7-14, January 1996.

[Lon73]   J.L. Long and T.N. Trick. An absolute bound on limit cycles due to roundoff errors in digital filters. *IEEE Trans. on Audio and Electroacoustics*, AU-21:27–30, February 1973.

[Lut91]   A. Luthra and G. Rajan. Sampling rate conversion of video signals. *SMPTE J.*, pages 869–879, November 1991.

[Lüt91]   H. Lütkepohl. *Introduction to Multiple Time Series Analysis*. Springer-Verlag, New York NY 1991.

[Mak75]   J. Makhoul. Linear prediction: A tutorial review. *Proc. IEEE*, 62:561–580, April 1975.

[Mak2001]   A. Makur and S.K. Mitra. Warped discrete Fourier transform: Theory and applications. *IEEE Trans. on Circuits and Systems I: Fundamental Theory and Applications*, 48:1086–1093, September 2001.

[Mal2002]   H. Malvar, A. Hallapuro, M. Karczezwicz, and L. Kerofsky. Low-complexity transform and quantization with 16-bit arithmetic for H.26L. In *Proc. IEEE International Conference on Image Processing*, pages II-489–II-4924, 2002.

[Mar87]   S.L. Marple, Jr. *Digital Spectral Analysis with Applications*. Prentice Hall, Englewood Cliffs NJ, 1987

[Mar92]   A. Mar, editor. *Digital Signal Processing Applications Using the ADSP-2100 Family*. Prentice Hall, Englewood Cliffs NJ, 1992.

[Mar94]   S.A. Martucci. Symmetric convolution and the discrete sine and cosine transforms. *IEEE Trans. on Signal Processing*, 42:1038–1051, May 1994.

[Mes84]   D.G. Messerschmitt. Echo cancellation in speech and data transmission. *IEEE J. on Selected Areas in Communications*, SAC-2:283–297, March 1984.

[Mia82]   G.A. Mian and A.P. Naider. A fast procedure to design equiripple minimum-phase FIR filters. *IEEE Trans. on Circuits and Systems*, CAS-29:327–331, May 1982.

[Mid86]   R.H. Middleton and G.C. Goodwin. Improved finite word length characteristics in digital control using delta operators. *IEEE Trans. on Automatic control*, AC-31:1015-1021, November 1986.

[Mik92]   N. Mikami, M. Kobayashi, and Y. Tokoyama. A new DSP-oriented algorithm for calculation of the square root using a nonlinear digital filter. *IEEE Trans. on Signal Processing*, 40:1663–1669, July 1992.

[Mil78]   W.L. Mills, C.T. Mullis, and R.A. Roberts. Digital filter realizations without overflow oscillations. *IEEE Trans. on Acoustics, Speech, and Signal Processing*, ASSP-26:334–338, August 1978.

[Min85]   F. Mintzer. Filters for distortion-free two-band multirate filter banks. *IEEE Trans. on Acoustics, Speech, and Signal Processing*, ASSP-33:626–630, June 1985.

[Mit73a]   S.K. Mitra and R.J. Sherwood. Digital ladder networks. *IEEE Trans. on Audio and Electroacoustics*, AU-21:30–36, February 1973.

[Mit73b]   S.K. Mitra. On reciprocal digital two-pairs. *Proc. IEEE (Letters)*, 61:1647–1648, November 1973.

[Mit74a]   S.K. Mitra and K. Hirano. Digital all-pass networks. *IEEE Trans. on Circuits and Systems*, CAS-21:688–700, 1974.

[Mit74b]   S.K. Mitra, K. Hirano, and H. Sakaguchi. A simple method of computing the input quantization and the multiplication round off errors in digital filters. *IEEE Trans. on Acoustics, Speech, and Signal Processing*, ASSP-22:326–329, October 1974.

[Mit74c]   S.K. Mitra and R.J. Sherwood. Estimation of pole-zero displacements of a digital filter due to coefficient quantization. *IEEE Trans. on Circuits and Systems*. CAS-21:116–124, January 1974.

[Mit75]   S.K. Mitra, K. Hirano, and K. Furuno. Digital sine-cosine generator. In *Proc. Second Florence International Conference on Digital Signal Processing*, pages 142–149, Florence, Italy, September 1975.

[Mit77a]   S.K. Mitra and C.S. Burrus. A simple efficient method for the analysis of structures of digital and analog systems. *Archiv für Elektrotechnik und Übertrangungstechnik*, 31:33–36, 1977.

[Mit77b]   S.K. Mitra, P.S. Kamat, and D.C. Huey. Cascaded lattice realization of digital filters. *International Journal on Circuit Theory and Applications*, 5:3–11, 1977.

[Mit77c]   S.K. Mitra, K. Mondal, and J. Szczupak. An alternate parallel realization of digital transfer functions. *Proc. IEEE (Letters)*, 65:577–578, April 1977.

[Mit80]   S.K. Mitra. *An Introduction to Digital and Analog Integrated Circuits, and Applications*. Harper and Row, New York NY, 1980.

[Mit87]   S.K. Mitra, K. Hirano, and K. Mensa-Ababio. Theory and applications of all-digital N-path filters. *IEEE Trans. on Circuits and Systems*. CAS-34:1045–1052, September 1987.

[Mit90a]   S.K. Mitra, K. Hirano, S. Nishimura, and K. Sugahara. Design of digital bandpass/bandstop digital filters with tunable characteristics. *Frequenz*, 44:117–121, March/April 1990.

[Mit90b]   S.K. Mitra, Y. Neuvo, and H. Roivainen. Design and implementation of recursive digital filters with variable characteristics. *International Journal on Circuit Theory and Applications*, 18:107–119, 1990.

[Mit93]   S.K. Mitra, A. Mahalanobis, and T. Saramäki. A generalized structural subband decomposition of FIR filters and its application in efficient FIR filter design and implementation. *IEEE Trans. on Circuits and Systems II: Analog and Digital Signal Processing*, 40:363–374, June 1993.

[Mit98]   S.K. Mitra and H. Babic. Partial-fraction expansion of rational z-transforms. *Electronics Letters*, 34:1726, 3 September 1998.

[Mit2004a] S.K. Mitra, M.K. Tchobanou, and G. Jovanovic-Dolecek. A simple approach to the design of one-dimensional sparse antenna arrays. In *Proc. IEEE International Symposium on Circuits & Systems*, May 2004, pages III-541–III-544, Vancouver, B.C., Canada.

[Mit2004b] S.K. Mitra, G. Jovanovic-Dolecek, and M.K. Tchobanou. On the design of one-dimensional sparse arrays with apodized end elements. In *Proc. 12th European Signal Processing Conference*, pages 2239-2242, Vienna, Austria, September 2004.

[Moo77]   J.A. Moorer. Signal processing aspects of computer music: A survey. *Proc. IEEE*, 65:1108–1137, August 1977.

[Moo79]   J.A. Moorer. About this reverberation business. *Computer Music Journal*, 3(2):13–28, 1979.

[Mun81]   D.C. Munson, Jr., and B. Liu. Narrowband recursive filters with error spectrum shaping. *IEEE Trans. on Circuits and Systems*, CAS-28:160–163, February 1981.

[Naw88]   S.H. Nawab and T.F. Quatieri. Short-time Fourier transform. In J.S. Lim and A.V. Oppenheim, editors, *Advanced Topics in Signal Processing*, chapter 6. Prentice Hall, Englewood Cliffs NJ, 1988.

[Neu84a]   Y. Neuvo and S.K. Mitra. Complementary IIR digital filters. In *Proc. IEEE International Symposium on Circuits and Systems*, pages 234–237, Montreal, Canada, May 1984.

[Neu84b]   Y. Neuvo, C-Y. Dong, and S.K. Mitra. Interpolated finite impulse response filters. *IEEE Trans. on Acoustics, Speech, and Signal Processing*, ASSP-32:563–570, June 1984.

[Oet75]   G. Oetken, T.W. Parks, and H.W. Schüssler. New results in the design of digital interpolators. *IEEE Trans. on Acoustics, Speech, and Signal Processing*, ASSP-23:301–309, June 1975.

[Opp75]   A.V. Oppenheim and R.W. Schafer. *Digital Signal Processing*. Prentice Hall Englewood Cliffs NJ, 1975.

[Opp76]   A.V. Oppenheim, W.F.G. Mecklenbräuker, and R.M. Mersereau. Variable cutoff linear phase digital filters. *IEEE Trans. on Circuits and Systems*, CAS-23:199–203, 1976.

[Opp78a]   A.V. Oppenheim, editor. *Applications of Digital Signal Processing*. Prentice Hall, Englewood Cliffs NJ, 1978.

[Opp78b]   A.V. Oppenheim. Digital processing of speech. In A.V. Oppenheim, editor, *Applications of Digital Signal Processing*, chapter 2. Prentice Hall, Englewood Cliffs NJ, 1978

[Opp83]   A.V. Oppenheim and A.S. Willsky. *Signals and Systems*. Prentice Hall, Englewood Cliffs NJ, 1983.

[Opp89]   A.V. Oppenheim and R.W. Schafer. *Discrete-Time Signal Processing*. Prentice Hall, Englewood Cliffs NJ, 1989.

[Ora2002]   S. Oraintara, Y-J Chen, and T.Q. Nguyen. Integer fast Fourier transform. *IEEE Trans. on Signal Processing*, 50:607–618, March 2002.

[Orc2003]   H.J. Orchard and A.N. Willson, Jr. On the computation of a minimum-phase spectral factor. *IEEE Trans. on Circuits and Systems*, 50:365–375, March 2003.

[Orf96]   S.J. Orfanidis. *Introduction to Signal Processing*. Prentice Hall, Englewood Cliffs NJ, 1996.

[Orm61]   J.F.A. Ormsby. Design of numerical filters with applications to missile data processing. *Journal of ACM*, 8:440-466, July 1961.

[Pap62]   A. Papoulis. *The Fourier Integral and Its Applications*. McGraw-Hill, New York NY, 1962.

[Par60]   E. Parzen. *Modern Probability Theory and Its Applications*. Wiley, New York NY, 1960.

[Par72]   T.W. Parks and J.H. McClellan. Chebyshev approximation for nonrecursive digital filters with linear phase. *IEEE Trans. on Circuit Theory*, CT-19:189–194, 1972.

[Par87]   T.W. Parks and C.S. Burrus. *Digital Filter Design*. Wiley, New York NY, 1987.

[Pat80]   R.K. Patney and S.C. Dutta Roy. A different look at round-off noise in digital filters. *IEEE Trans. on Circuits and Systems*, CAS-27:59–62, January 1980.

[Pei98]   S-C. Pei and C-C. Tseng. A comb filter design using fractional-sample delay. *IEEE Trans. on Circuit and Systems*, 45:649–653, June 1998.

[Pel80]   A. Peled and A. Ruiz. Frequency domain data transmission using reduced computational complexity algorithms. In *Proc. IEEE International Conference on Acoustics, Speech and Signal Processing*, pages 964–967, Denver CO, April 1980.

[Pie96]   J.W. Pierre. A novel method for calculating the convolution sum of two finite length sequences. *IEEE Trans. on Education*, 39:77–80, February 1996.

[Por97]   B. Porat. *A Course in Digital Signal Processing*. Wiley, New York NY, 1997.

[Pou87]   K. Poulton, J.J. Corcoran, and T. Hornak. A 1-GHz 6-bit ADC system. *IEEE J. of Solid-State Circuits*, SC-22:962–970, December 1987.

[Pri80]   D.H. Pritchard. A CCD comb filter for color TV receiver picture enhancement. *RCA Review*, 41:3–28, 1980.

[Pro96]   J.G. Proakis and D.G. Manolakis. *Digital Signal Processing: Principles, Algorithms and Applications*. Prentice Hall, Upper Saddle River NJ, 3rd edition, 1996.

[Rab69]   L.R. Rabiner, R.W. Schafer, and C.M. Rader. The chirp-z transform algorithm. *IEEE Trans. on Audio and Electroacoustics*, AU-17:86–92, June 1969.

[Rab73]   L.R. Rabiner. Approximate design relationships for low-pass FIR digital filters. *IEEE Trans. on Audio and Electroacoustics*, AU-21:456–460, 1973.

[Rab74a]   L.R. Rabiner and R.W. Schafer. On the behavior of minimax relative error FIR digital differentiators. *Bell System Technical Journal*, 53:333-362, February 1974.

[Rab74b]   L.R. Rabiner and R.W. Schafer. On the behavior of minimax relative error FIR digital Hilbert transformers. *Bell System Technical Journal*, 53:363-394, February 1974.

[Rab74c]   L.R. Rabiner, J.F. Kaiser, and R.W. Schafer. Some considerations in the design of multiband finite-impulse-response digital filters. *IEEE Trans. on Acoust., Speech, Signal Processing*, 22:462–472, December 1974.

[Rab75a]   L.R. Rabiner and B. Gold. *Theory and Application of Digital Signal Processing*. Prentice Hall, Englewood Cliffs NJ, 1975.

[Rab75b]   L.R. Rabiner and M.R. Sambur. An algorithm for determining the endpoints of isolated utterances. *Bell System Tech. J.*, 54:297-315, February 1975.

[Rab78]   L.R. Rabiner and R.W. Schafer. *Digital Processing of Speech Signals*. Prentice Hall, Englewood Cliffs NJ, 1978.

[Rab2001]   R. Rabenstein and L. Trautmann. Digital sound synthesis by physical modelling. In *Proc. 2nd International Symp. on Image and Signal Processing and Analysis*, pages 12-23, Pula, Croatia, June 2001.

[Ram84]   T. Ramstad. Digital methods for conversion between arbitrary sampling frequencies. *IEEE Trans. Acoustics, Speech and Signal Processing*, ASSP-32:577–591, June 1984.

[Ram89]   V. Ramachandran. Determination of discrete transfer function from its real (or imaginary) part on the unit circle. *IEEE Trans. Acoustics, Speech and Signal Processing*, ASSP-37:440–442, March 1989.

[Rao84]   S.K. Rao and T. Kailath. Orthogonal digital filters for VLSI implementation. *IEEE Trans. on Circuits and Systems*, CAS-31:933–945, 1984.

[Reg87a]   P.A. Regalia, S.K. Mitra, and J. Fadavi-Ardekani. Implementation of real coefficient digital filters using complex arithmetic. *IEEE Trans. on Circuits and Systems*, CAS-34:345–353, April 1987.

[Reg87b]   P.A. Regalia and S.K. Mitra. Tunable digital frequency response equalization filters. *IEEE Trans. Acoustics, Speech, and Signal Processing*, ASSP-35:118–120, January 1987.

[Reg87c]   P.A. Regalia and S.K. Mitra. A class of magnitude complementary loudspeaker crossovers. *IEEE Trans. on Acoustics, Speech, and Signal Processing*, ASSP-35:1509–1515, November 1987.

[Reg88]   P.A. Regalia, S.K. Mitra, and P.P. Vaidyanathan. The digital all-pass filter: A versatile signal processing building block. *Proc. IEEE*, 76:19–37, January 1988.

[Reg93]   P.A. Regalia. Special filter designs. In S.K. Mitra and J.F. Kaiser, editors, *Handbook for Digital Signal Processing*, chapter 13, pages 967–980. Wiley-Interscience, New York NY, 1993.

[Ren87]   M. Renfors and T. Saramäki. Recursive *n*-th band digital filters, Parts I and II. *IEEE Trans. on Circuits and Systems*, CAS-34:24–51, January 1987.

[Rob80]   E.A. Robinson and S. Treitel. *Geophysical Signal Analysis*. Prentice Hall, Englewood Cliffs NJ, 1980.

[Rob82]   E.A. Robinson. A historical perspective of spectrum estimation. *Proc. IEEE*, 70:885-907, 1982.

[Ros71]   A.E. Rosenberg. Effect of glottal pulse shape on the quality of natural vowels. *J. Acoust. Soc. Am.*, 49:583–590, February 1971.

[Ros75]   J.P. Rossi. Digital television image enhancement. *SMPTE J.*, 84:545–551, July 1975.

[Rot83]   J.H. Rothweiler. Polyphase quadrature filters, a new subband coding technique. In *Proc. IEEE International Conference on Acoustics, Speech and Signal Processing*, pages 1980–1983, Boston MA, April 1983.

[San67]   I.W. Sandberg. Floating-point-roundoff accumulation in digital filter realization. *Bell System Technical Journal*, 46:1775–1791, October 1967.

[Sar88]   T. Saramäki, Y. Neuvo, and S.K. Mitra. Design of computationally efficient interpolated FIR filters. *IEEE Trans. on Circuits & Systems*, CAS-35:70–88 1988.

[Sar93]   T. Saramäki. Finite impulse response filter design. In S.K. Mitra and J.F. Kaiser, editors, *Handbook for Digital Signal Processing*, chapter 4, pages 155–278. Wiley-Interscience, New York NY, 1993.

[Sch62]   M.R. Schroeder. Natural sounding artificial reverberation. *Journal of the Audio Engineering Society*, 10:219–223, 1962.

[Sch18]   I. Schur. Über potenzreihen, die im innern des einheitskreises beschränkt sind. *Journal für Mathematik,* 147:205–232, 1917.

[Sch72]   H.W. Schüssler. On structures for nonrecursive digital filters. *Archiv für Electrotechnik und Überstragunstechnik*, 26:255–258, June 1972.

[Sch75]   M. Schwartz and L. Shaw. *Signal Processing: Discrete Spectral Analysis, Detection, and Estimation*. McGraw-Hill, New York NY, 1975.

[Sch91]   R. Schreier. *Noise-Shaped Coding*. PhD thesis, University of Toronto, Toronto Canada, 1991.

[Sha81]   A.F. Shackil. Microprocessors and the MD. *IEEE Spectrum*, 18:45–49, April 1981.

[She95]   K. Shenoi. *Digital Signal Processing in Telecommunication*. Prentice Hall, Englewood Cliffs NJ, 1995.

[Shp90]   D. J. Shpak and A. Antoniou. A generalized Remez method for the design of FIR digital filters. *IEEE Trans. on Circuits & Systems*, 37:161-174, February 1990.

[Sko62]   M.I. Skolnik. *Introduction to Radar Systems*. McGraw-Hill, New York NY, 1962.

[Skw65]   J.K. Skwirzynski. *Design Theory and Data for Electrical Filters*. Van Nostrand Reinhold, New York NY, 1965.

[Smi91]   J.O. Smith III. Viewpoints on the history of digital synthesis. In *Proc. International Computer Music Conference*, pages 1–10, Montreal, Que., Canada, October 1991.

[Smi84]   M.J.T. Smith and T.P. Barnwell III. A procedure for designing exact reconstruction filter banks for tree-structured subband coders. In *Proc. IEEE Conf. on Acoustics, Speech, and Signal Processing*, pages 27.1.1–27.1.4, San Diego, CA, March 1984.

[Ste93]   K. Steiglitz. Mathematical foundations of signal processing. In S.K. Mitra and J.F. Kaiser, editors, *Handbook for Digital Signal Processing*, chapter 2, pages 57–99. Wiley-Interscience, New York NY, 1993.

[Ste96]   K. Steiglitz. *A Digital Signal Processing Primer*. Addison Wesley, Menlo Park CA, 1996.

[Sto66]   T.G. Stockham Jr. High speed convolution and correlation. *1966 Spring Joint Computer Conference, AFIPS Proc*, 28:229–233, 1966.

[Sto94]   G. Stoyanov and H. Clausert. A comprative study of first-order digital allpass filter sections. *Frequenz*, 48:221–226, September–October 1994.

[Str96]   G. Strang and T. Nguyen. Wavelets and Filter Banks. Wellesley-Cambridge Press, Wellesley MA, 1996.

[Swe96]   W. Sweldens. The lifting scheme: A custom-design construction of biorthogonal wavelets. *Applied and Computational Harmonic Analysis*, 3:186-200, 1996.

[Szc75a] J. Szczupak and S.K. Mitra. Digital filter realization using successive multiplier-extraction approach. *IEEE Trans. Acoustics, Speech, and Signal Processing*, ASSP-23:235–239, April 1975.

[Szc75b]   J. Szczupak and S.K. Mitra. Detection, location, and removal of delay-free loops in digital filter configurations. *IEEE Trans. Acoustics, Speech, and Signal Processing*, ASSP-23:558–562, December 1975.

[Szc88]   J. Szczupak, S.K. Mitra, and J. Fadavi-Ardekani. A computer-based synthesis method of structurally LBR digital allpass networks. *IEEE Trans. on Circuits and Systems*, CAS-35:755–760, 1988.

[Tem73]   G.C. Temes and S.K. Mitra, editors. *Modern Filter Theory and Design*. Wiley, New York NY, 1973.

[Tem77]   G.C. Temes and J.W. LaPatra. *Introduction to Circuit Synthesis and Design*. McGraw-Hill, New York NY, 1977.

[Tha98]   M. T. Tham. *Dealing with measurement noise.* http://lorien.ncl.ac.uk/ming/filter/filter.htm.

[Thô77]   T. Thông and B. Liu. Error spectrum shaping in narrow-band recursive digital filters. *IEEE Trans. on Acoustics, Speech, and Signal Processing*, ASSP-25:200–203, 1977.

[Thu2000]   S. Thurnhofer. Two-dimensional Teager filters. In S.K. Mitra and G. Sicuranza, editors, *Nonlinear Image Processing*, chapter 6. Academic Press, New York NY, 2000.

[Tom81]   W.J. Tompkins and J.G. Webster, editors. *Design of Microcomputer-Based Medical Instrumentation*. Prentice Hall, Englewood Cliffs NJ, 1981.

[Tri77]   J.M. Tribolet. A new phase unwrapping algorithm *IEEE Trans. Acoustics, Speech, and Signal Processing*, ASSP-25:170–177, April 1977.

[Tri79]   J.M. Tribolet. *Seismic Applications of Homomorphic Signal Processing*. Prentice Hall, Englewood Cliffs NJ, 1979.

[Tuk74]   J. W. Tukey. Nonlinear (nonsuperposable) methods for smoothing data. *Cong. Rec. EASCON*, 73, 1974.

[Ünv75]   Z. Ünver and K. Abdullah. A tighter practical bound on quantization errors in second-order digital filters with complex conjugate poles. *IEEE Trans. on Circuits and Systems*, CAS-22:632–633, July 1975.

[Urk58]   H. Urkowitz. An extension to the theory of airborne moving target indicators. *IRE Trans*, ANE-5:210–214, December 1958.

[Vai84]   P.P. Vaidyanathan and S.K. Mitra. Low passband sensitivity digital filters: A generalized viewpoint and synthesis procedures. *Proc. IEEE*, 72:404–423, 1984.

[Vai85a]   P.P. Vaidyanathan. The doubly terminated lossless digital two-pair in digital filtering. *IEEE Trans. on Circuits and Systems*, CAS-32:197–200, 1985.

[Vai85b]   P.P. Vaidyanathan and S.K. Mitra. Very low-sensitivity FIR filter implementation using "structural passivity" concept. *IEEE Trans. on Circuits and Systems*, CAS-32:360–364, April 1985.

[Vai85c]   P.P. Vaidyanathan. On power-complementary FIR filters. *IEEE Trans. on Circuits and Systems*, CAS-32:1308–1310, December 1985.

[Vai86a]   P.P. Vaidyanathan, S.K. Mitra, and Y. Neuvo. A new approach to the realization of low-sensitivity IIR digital filters. *IEEE Trans. on Acoustics, Speech, and Signal Processing*, ASSP-34:350–361, April 1986.

[Vai86b]   P.P. Vaidyanathan. Passive cascaded-lattice structures for low-sensitivity FIR filter design, with application to filter banks. *IEEE Trans. on Circuits and Systems*, CAS-33:1045–1064, November 1986.

[Vai87a]   P.P. Vaidyanathan and T.Q. Nguyen. Eigenfilters: A new approach to least-squares FIR filter design and applications including Nyquist filters. *IEEE Trans. on Circuits and Systems*, CAS-34:11–23, January 1987.

[Vai87b]   P.P. Vaidyanathan and T.Q. Nguyen. A 'TRICK' for the design of FIR half-band filters. *IEEE Trans. on Circuits and Systems*, CAS-34:297–300, March 1987.

[Vai87c]   P.P. Vaidyanathan. Low-noise and low-sensitivity digital filters. In D.F. Elliot, editor, *Handbook of Digital Signal Processing*, chapter 5. Academic Press, New York NY, 1987.

[Vai87d]   P.P. Vaidyanathan. Quadrature mirror filter banks, M-band extensions and perfect-reconstruction techniques. *IEEE ASSP Magazine*, 4:4–20, 1987.

[Vai87e]   P.P. Vaidyanathan and S.K. Mitra. A unified structural interpretation and tutorial review of stability test procedures for linear systems. *Proc. IEEE*, 75:478–497, April 1987.

[Vai87f]   P.P. Vaidyanathan, P.A. Regalia, and S.K. Mitra. Design of doubly-complementary IIR digital filters using a single complex allpass filter, with multirate applications. *IEEE Trans. on Circuits and Systems*, CAS-34:378–389, April 1987.

[Vai88a]   P.P. Vaidyanathan and P-Q. Hoang. Lattice structures for optimal design and robust implementation of two-channel perfect-reconstruction QMF banks. *IEEE Trans. on Acoustics, Speech, and Signal Processing*, ASSP-36:81–94, January 1988.

[Vai88b]   P.P. Vaidyanathan and S.K. Mitra. Polyphase networks, block digital filtering, LPTV systems, and alias-free QMF banks: A unified approach based on pseudocirculants. *IEEE Trans. on Acoustics, Speech, and Signal Processing*, ASSP-36:381–391, March 1988.

[Vai89]   P.P. Vaidyanathan, T.Q. Nguyen, Z. Doğanata, and T. Saramäki. Improved technique for design of perfect reconstruction FIR QMF filter banks with lossless polyphase matrices. *IEEE Trans. on Acoustics, Speech, and Signal Processing*, ASSP-37:1042–1056, July 1989.

[Vai90]   P.P. Vaidyanathan. Multirate digital filters, filter banks, polyphase networks, and applications: A tutorial. *Proc. IEEE*, 78:56–93, January 1990.

[Vai93]   P.P. Vaidyanathan. *Multirate Systems and Filter Banks*. Prentice Hall, Englewood Cliffs NJ, 1993.

[Vai2001]   P.P. Vaidyanathan and B. Vrcelj. Biorthogonal partners and applications. *IEEE Trans. on Acoustics, Speech, and Signal Processing*, 49:1013–1027, May 2001.

[Vet88]   M. Vetterli. Running FIR and IIR filtering using multirate filter banks. *IEEE Trans. on Acoustics, Speech and Signal Processing*, ASSP-36:730–738, May 1988.

[Vet89]   M. Vetterli and D. LeGall. Perfect reconstruction FIR filter banks: Some properties and factorization. *IEEE Trans. on Acoustics, Speech and Signal Processing*, 37:1057–1071, July 1989.

[Vet95]   M. Vetterli and J. Kovacevic. *Wavelets and Subband Coding*. Prentice Hall, Englewood Cliffs NJ, 1995.

[Vla69]   J. Vlach. *Computerized Approximation and Synthesis of Linear Networks*. Wiley, New York NY, 1969.

[Vla83]   J. Vlach and K. Singhal. *Computer Methods for Circuit Analysis and Design*. Van Nostrand Reinhold, New York NY, 1983.

[Wei69a]   C.J. Weinstein and A.V. Oppenheim. A comparison of roundoff noise in fixed point and floating point digital filter realizations. *Proc. IEEE*, 57:1181–1183, June 1969.

[Wei69b]   C.J. Weinstein. Roundoff noise in floating point fast Fourier transform computation. *IEEE Trans. on Audio and Electroacoustics*, AU-17:209–215, September 1969.

[Wel67]   P.D. Welch. The use of fast Fourier transform for the estimation of power spectra: A method based on time averaging over short, modified periodograms. *IEEE Trans. on Audio and Electroacoustics*, AU-15:70–73, 1967.

[Wel69]   P.D. Welch. A fixed-point fast Fourier transform error analysis. *IEEE Trans. on Audio and Electroacoustics*, AU-17:151–157, June 1969.

[Whi58]   W.D. White. Synthesis of comb filters. *Proc. National Electronics Conference*, pages 279–285, 1958.

[Whi71]   S.A. White. New method of synthesizing linear digital filters based on convolution integral. *IEE Proc. (Corr.)*, 118:348, February 1971.

[Wid56]   B. Widrow. A study of rough amplitude quantization by means of Nyquist sampling theory. *IRE Trans. on Circuit Theory*, CT-3:266–276, December 1956.

[Wid61]   B. Widrow. Statistical analysis of amplitude-quantized sampled-data systems. *AIEE Trans. (Appl. Industry)*, 81:555–568, January 1961.

[Wor89]   J.M. Worham. *Sound Recording Handbook*. Howard W. Sams, Indianapolis IN, 1989.

[Yan82]   G-T. Yan and S.K. Mitra. Modified coupled form digital-filter structures. *Proc. IEEE (Letters)*, 70:762–763, July 1982.

[Yel96]   D. Yellin and E. Weinstein. Multichannel signal separation: Methods and analysis. *IEEE Trans. on Signal Processing*, ASSP-44:106–118, January 1996.

[Yu90]   T-H. Yu, S.K. Mitra, and H. Babic. Design of linear-phase FIR notch filters. *Sadhana*, 15:133–155, November 1990.

[Zöl95]   U. Zölzer and T. Boltze. Parametric digital filter structures. *Proc. 99th Audio Engineering Society Convention*, New York NY, Preprint No. 4099, October 1995.

[Zöl97]   U. Zölzer. *Digital Audio Signal Processing*. Wiley, New York NY, 1997.

[Zve67]   A.I. Zverev. *Handbook of Filter Synthesis*. Wiley, New York NY, 1967.

Index